ICHEP 2000
Volume I

Proceedings of the 30th International Conference
on
High Energy Physics

ICHEP 2000
Volume I

Proceedings of the 30th International Conference
on
High Energy Physics

Osaka, Japan   27 July – 2 August 2000

Editors

C. S. Lim
Kobe University, Japan

Taku Yamanaka
Osaka University, Japan

World Scientific
Singapore • New Jersey • London • Hong Kong

Published by

World Scientific Publishing Co. Pte. Ltd.

P O Box 128, Farrer Road, Singapore 912805

USA office: Suite 1B, 1060 Main Street, River Edge, NJ 07661

UK office: 57 Shelton Street, Covent Garden, London WC2H 9HE

Library of Congress Cataloging-in-Publication Data
Proceedings of the 30th International Conference on High Energy Physics, Osaka, Japan,
27 July–2 August 2000 / editors C.S. Lim, Taku Yamanaka.
        p. cm.
    ISBN: 9810249756 (Vol. 1) : alk. paper
           9810249764 (Vol. 2) : alk. paper
           9810245335 (set) : alk. paper
     1. Particles (Nuclear physics)--Congresses. I. Lim, C. S. II. Yamanaka, Taku,
   1957– . III. International Conference on High Energy Physics (30th : 2000 : Osaka,
Japan)

QC793.3.H5 I56 2000
539.7'2--dc21   2001017781

British Library Cataloguing-in-Publication Data
A catalogue record for this book is available from the British Library.

Printed in Singapore by World Scientific Printers

FOREWORD

The XXXth International Conference on High Energy Physics (ICHEP2000) was held in Osaka from July 27 to August 2, 2000, under the aegis of the International Union of Pure and Applied Physics(IUPAP), C11 Commission. This is the second time the conference was held in Japan after Tokyo in 1978. In all, there were 950 participants from 36 countries together with 260 accompanying persons. The conference took place at International House of Osaka and employed the traditional format with 6 parallel sessions running through the first three days, followed by three days of plenary talks. Total of 1,025 abstracts were submitted. In all, 341 parallel talks and 27 plenary talks were given, most of which are reproduced in this Proceedings. Contributed papers were registered either at Los Alamos preprint archive or at big collaborations' official home page. They were made accessible via World Wide Web first to session conveners and plenary speakers and once the session started to all the participants. We also provided video recordings of all the plenary talks for world wide audience. They are still available at http://ichep2000.hep.sci.osaka-u.ac.jp/.

While announcement of scheduled closing of LEP which had produced a wealth of remarkable precision data on the standard model signified the end of an era, new results from many fronts were presented with a glimpse beyond the standard model. First results from BaBar and Belle of newly constructed B-factories suggested a rich harvest on CP violation in B sector. A long baseline experiment K2K gave indication of neutrino oscillation thus confirming the atmospheric neutrino data. Data from RHIC which are expected to shed lights on QCD phase transition began to appear. Theorists talked about a new tower of gauge hierarchy based on the idea of extra dimension. We also heard that a consistent and converging picture of big bang cosmology is being formulated with non-zero cosmological constant.

A public lecture was given by Professor Jerome Friedman of Massachusetts Institute of Technology with a title "Are we really made of quarks? " in the midst of the conference when all the participants took a holiday on excursion to Kyoto, Nara and other places. It was carried out as the Nishina Memorial Lecture with Japanese translation and captured a local audience of more than 600 including high school students. The lecture was given under joint sponsorship of ICHEP2000 Organizing Committee, Nishina Memorial Foundation, Osaka City and International House of Osaka.

In addition, the social program included evenings of traditional Japanese arts, Bunraku puppet show, Yokobue (Japanese flute) and Guitar, visit to Osaka Castle and a buffet style banquet at Hotel New Otani. Volunteer based programs for accompanying persons included classes to introduce flower arrangement and tea ceremony, strolls in the downtown area and visit to Nara which also involved Koto(Japanese harp) and Noh (traditional dance) play by Nara Women's University students.

The conference was held under joint auspices of the Science Council of Japan and the Physical Society of Japan. It was also sponsored by High Energy Accelerator Research Organization(KEK) and Osaka University. It would not have been made possible without generous contributions from Grant-in-aid of Mombusho, Inoue Foundation, Osaka Convention Bureau and many private industries. The conference was organized and operated by hard working people from Osaka University, Kobe University, Kyoto University, Osaka City University, Nara Women's University, Kinki University and KEK.

Yorikiyo Nagashima, Chairman
ICHEP2000 Organizing Committee
October 27, 2000

Conference Organization

International Advisory Committee

A. Astbury	(TRIUMF)	B. Barish	(Caltech)
R.J. Cashmore	(CERN)	E. Fernandez	(Barcelona)
V.G. Kadyshevsky	(JINR)	J.S. Kang	(Korea)
T. Kirk	(BNL)	J. Lefrancois	(LAL)
J. Peoples	(Fermilab)	B. Richter	(SLAC)
D.H. Saxon	(Glasgow)	H. Sugawara	(KEK)
S.N. Tovey	(Melbourne)	A. Wagner	(DESY)
A.K. Wroblewski	(Warsaw)	Z. Zheng	(IHEP)

Program Committee

T. Maskawa   (YITP, Chairman)

J.A. Appel	(Fermilab)	R.J. Cashmore	(CERN)
M. Danilov	(ITEP)	J. Dorfan	(SLAC)
T. Eguchi	(Tokyo)	K. Higashijima	(Osaka)
T. Inami	(Chuo)	S. Iwata	(KEK)
S. Komamiya	(Tokyo)	T. Kugo	(Kyoto)
S. Kurokawa	(KEK)	C.S. Lim	(Kobe)
K. Nakamura	(KEK)	M. Ninomiya	(YITP)
R. Peccei	(UCLA)	N. Sakai	(TIT)
N. Sasao	(Kyoto)	F. Takasaki	(KEK)
H. Takeda	(Kobe)	M. Yoshimura	(Tohoku)

Local Organizing Committee

Y. Nagashima	(Osaka, Chairman)	Y. Kimura	(KEK)
C.S. Lim	(Kobe)	A. Maki	(KEK)
S. Noguchi	(Nara Women's U.)	T.K. Ohska	(KEK)
T. Okusawa	(Osaka City)	E. Takasugi	(Osaka)
S. Yamada	(KEK)	T. Yamanaka	(Osaka)

Conference Sponsors

IUPAP
Science Council of Japan
The Physical Society of Japan
Osaka University
KEK

CONTENTS

VOLUME I

PLENARY SESSIONS

PARALLEL SESSIONS

PA-03 Hard Interactions (high Q^2: DIS, jet, perturbative QCD)

VOLUME II

PA-07 Heavy Flavor Physics (charm, bottom, top and τ)

PA-10 Beyond the Standard Model (theory)

PA-11 Search for New Particles and New Phenomena

PA-12 New Detectors and Techniques

PA-15 Field Theory

PA-16 Superstring Theory

Editors would like to thank Dr. Kiyotomo Kawagoe for helping us prepare the proceedings.

Plenary Session 1

New Results from e^+e^- B-factories

Chair: Alan Astbury (TRIUMF)
Scientific Secretary: Hiroyasu Tajima

FIRST *CP* VIOLATION RESULTS FROM *BABAR*

DAVID G. HITLIN

(for the *BABAR* Collaboration)
California Institute of Technology
Pasadena, CA 91125 USA
E-mail: hitlin@hep.caltech.edu

We present a preliminary measurement of time-dependent *CP*-violating asymmetries in $B^0 \to J/\psi K_S^0$ and $B^0 \to \psi(2S)K_S^0$ decays recorded by the *BABAR* detector at the PEP-II asymmetric B Factory at SLAC. The data sample consists of $9.0\,\mathrm{fb}^{-1}$ collected at the $\Upsilon(4S)$ resonance and $0.8\,\mathrm{fb}^{-1}$ off-resonance. One of the neutral B mesons, produced in pairs at the $\Upsilon(4S)$, is fully reconstructed. The flavor of the other neutral B meson is tagged at the time of its decay, mainly with the charge of identified leptons and kaons. The time difference between the decays is determined by measuring the distance between the decay vertices. Wrong-tag probabilities and the time resolution function are measured with samples of fully-reconstructed semileptonic and hadronic neutral B final states. The value of the asymmetry amplitude, $\sin 2\beta$, is determined from a maximum likelihood fit to the time distribution of 120 tagged $B^0 \to J/\psi K_S^0$ and $B^0 \to \psi(2S)K_S^0$ candidates: $\sin 2\beta = 0.12 \pm 0.37(\mathrm{stat}) \pm 0.09(\mathrm{syst})$.

1 Introduction

The *BABAR* detector at the PEP-II asymmetric B Factory at SLAC has been taking data since the end of May, 1999. As of the date of this conference a data sample of $\sim 9.0\,\mathrm{fb}^{-1}$ has been collected at the $\Upsilon(4S)$ resonance, with an additional $0.8\,\mathrm{fb}^{-1}$ taken off-resonance. A variety of preliminary B physics results using this data sample have been submitted to this conference and reported in parallel sessions. This paper will discuss in some detail *BABAR*'s first measurements of *CP*-violating asymmetries in $B^0 \to J/\psi K_s^0$ and $B^0 \to \psi(2S)K_s^0$ decays.

2 Motivation and Overview

The *CP*-violating phase of the three-generation Cabbibo-Kobayashi-Maskawa (CKM) quark mixing matrix can provide an elegant explanation of the well-established *CP*-violating effects seen in K_L^0 decay[1]. However, studies of *CP* violation in neutral kaon decays and the resulting experimental constraints on the parameters of the CKM matrix[2] do not, in fact, yet provide a test of whether the CKM phase describes *CP* violation[3].

The unitarity of the three generation CKM matrix can be expressed in geometric form as six triangles of equal area in the complex plane. A nonzero area[4] directly implies the existence of a *CP*-violating CKM phase. The most experimentally accessible of the unitarity relations, involving the two smallest elements of the CKM matrix, V_{ub} and V_{td}, has come to be known as the (B) Unitarity Triangle. Because the lengths of the sides of this Unitarity Triangle are of the same order, the angles can be large, leading to potentially large *CP*-violating asymmetries from phases between CKM matrix elements.

The *CP*-violating asymmetry in $b \to c\bar{c}s$ decays of the B^0 meson such as $B^0/\bar{B}^0 \to J/\psi K_s^0$ (or $B^0/\bar{B}^0 \to \psi(2S)K_s^0$) is caused by the interference between mixed and unmixed decay amplitudes. A state initially prepared as a B^0 ($\bar{B}^0$) can decay directly to $J/\psi K_s^0$ or can oscillate into a $\bar{B}^0$ (B^0) and then decay to $J/\psi K_s^0$. With little theoretical uncertainty, the phase difference between these amplitudes is equal to twice the angle $\beta = \arg\left[-V_{cd}V_{cb}^*/V_{td}V_{tb}^*\right]$ of the Unitarity Triangle. The *CP*-violating asymmetry can thus provide a crucial test of the Standard Model. The interference between the two amplitudes, and hence the CP asymmetry, is maximal when the mixing probability is at its highest, *i.e.*, when the lifetime t is approximately 2.2 B^0 proper lifetimes.

In e^+e^- storage rings operating at the $\Upsilon(4S)$ resonance a $B^0\bar{B}^0$ pair produced in $\Upsilon(4S)$ decay evolves in a coherent P-wave until one of the B mesons decays. If one of the B mesons (B_{tag}) can be ascertained to decay to a state of known flavor at a certain time t_{tag}, the other B is *at that time* known to be of the opposite flavor. For this measurement, the other B (B_{CP}) is fully reconstructed in a CP eigenstate ($J/\psi K_S^0$ or $\psi(2S)K_S^0$). By measuring the proper time interval $\Delta t = t_{CP} - t_{tag}$ from the B_{tag} decay time to the decay of the B_{CP}, it is possible to determine the time evolution of the initially pure B^0 or $\bar{B}^0$ state. The time-dependent rate of decay of the B_{CP} final state is given by

$$f_\pm(\Delta t; \Gamma, \Delta m_d, \mathcal{D}\sin 2\beta) = \frac{1}{4}\,\Gamma\,e^{-\Gamma|\Delta t|}$$
$$\times\,[1 \pm \mathcal{D}\sin 2\beta \times \sin\Delta m_d\Delta t]\,, \quad (1)$$

where the $+$ or $-$ sign indicates whether the B_{tag} is tagged as a B^0 or a $\bar{B}^0$, respectively. The dilution factor $\mathcal{D}$ is given by $\mathcal{D} = 1 - 2w$, where w is the mistag fraction, *i.e.*, the probability that the flavor of the tagging B is identified incorrectly. A term proportional to $\cos\Delta m_d\,\Delta t$ would arise from the interference between two decay mechanisms with different weak phases. In the Standard Model, the dominant diagrams (tree and penguin) for the decay modes we consider have no relative weak phase, so no such term is expected.

To account for the finite resolution of the detector, the time-dependent distributions $f_\pm$ for B^0 and $\bar{B}^0$ tagged events (Eq. 1) must be convoluted with a time resolution function $\mathcal{R}(\Delta t; \hat{a})$:

$$\mathcal{F}_\pm(\Delta t; \Gamma, \Delta m_d, \mathcal{D}\sin 2\beta, \hat{a}) =$$
$$f_\pm(\Delta t; \Gamma, \Delta m_d, \mathcal{D}\sin 2\beta) \otimes \mathcal{R}(\Delta t; \hat{a}), \quad (2)$$

where $\hat{a}$ represents the set of parameters that describe the resolution function.

In practice, events are separated into different tagging categories, each of which has a different mean dilution $\mathcal{D}_i$, determined individually for each category.

It is possible to construct a CP-violating observable

$$\mathcal{A}_{CP}(\Delta t) = \frac{\mathcal{F}_+(\Delta t) - \mathcal{F}_-(\Delta t)}{\mathcal{F}_+(\Delta t) + \mathcal{F}_-(\Delta t)}\,, \quad (3)$$

which is proportional to $\sin 2\beta$:

$$\mathcal{A}_{CP}(\Delta t) \sim \mathcal{D}\sin 2\beta \times \sin\Delta m_d\,\Delta t\,. \quad (4)$$

Since no time-integrated CP asymmetry effect is expected, an analysis of the time-dependent asymmetry is necessary. At an asymmetric-energy B Factory, the proper decay-time difference Δt is, to an excellent approximation, proportional to the distance Δz between the two B^0-decay vertices along the axis of the boost, $\Delta t \approx \Delta z/c\,\langle\beta\gamma\rangle$. At PEP-II the average boost of B mesons, $\langle\beta\gamma\rangle$, is 0.56. The distance Δz is 250 μm per B^0 lifetime, while the typical Δz resolution for the *BABAR* detector is about 110 μm.

Since the amplitude of the time-dependent CP-violating asymmetry in Eq. 4 involves the product of $\mathcal{D}$ and $\sin 2\beta$, one needs to determine the dilution factors $\mathcal{D}_i$ (or equivalently the mistag fractions w_i) in order to extract the value of $\sin 2\beta$. The mistag fractions can be extracted from the data by studying the time-dependent rate of $B^0\bar{B}^0$ oscillations in events in which one of the neutral B mesons is fully reconstructed in a self-tagging mode and the other B (the B_{tag}) is flavor-tagged using the standard CP analysis flavor-tagging algorithm. In the limit of perfect determination of the flavor of the fully-reconstructed neutral B, the dilution in the mixed and unmixed amplitudes arises solely from the B_{tag} side, allowing the values of the mistag fractions w_i to be determined.

The value of sin2β is extracted by maximizing the likelihood function

$$\ln \mathcal{L}_{CP} =$$
$$\sum_i \left[\sum_{B^0\text{tag}} \ln \mathcal{F}_+(\Delta t; \Gamma, \Delta m_d, \hat{a}, \mathcal{D}_i \sin 2\beta) \right.$$
$$\left. + \sum_{\bar{B}^0\text{tag}} \ln \mathcal{F}_-(\Delta t; \Gamma, \Delta m_d, \hat{a}, \mathcal{D}_i \sin 2\beta) \right],$$
$$(5)$$

where the outer summation is over tagging categories i.

2.1 Overview of the analysis

The measurement of the CP-violating asymmetry has five main components :

- Selection of the signal $B^0/\bar{B}^0 \to J/\psi\, K_S^0$ and $B^0/\bar{B}^0 \to \psi(2S)K_S^0$ events, as described in detail in Ref. 5.

- Measurement of the distance Δz between the two B^0 decay vertices along the $\Upsilon(4S)$ boost axis, as described in detail in Refs. 6 and 7.

- Determination of the flavor of the B_{tag}, as described in detail in Ref. 6.

- Measurement of the dilution factors $\mathcal{D}_i$ from the data for the different tagging categories, as described in detail in Ref. 6.

- Extraction of the amplitude of the CP asymmetry and the value of sin2β with an unbinned maximum likelihood fit.

Whenever possible, we determine time and mass resolutions, efficiencies and mistag fractions from the data.

3 Sample selection

For this analysis we use a sample of $9.8\,\text{fb}^{-1}$ of data recorded by the BABAR detector[8] between January 2000 and the beginning of July 2000, of which $0.8\,\text{fb}^{-1}$ was recorded 40 MeV below the $\Upsilon(4S)$ resonance (off-resonance data).

A brief description of the BABAR detector and the definition of many general analysis procedures can be found in Ref. 8. Charged particles are detected and their momenta measured by a combination of a central drift chamber (DCH) filled with a helium-based gas and a five-layer, doubled-sided silicon vertex tracker (SVT), in a 1.5 T solenoidal field produced by a superconducting magnet. The charged particle momentum resolution is approximately $(\delta p_T/p_T)^2 = (0.0015\, p_T)^2 + (0.005)^2$, where p_T is measured in GeV/c. The SVT, with typical $10\,\mu$m single-hit resolution, provides vertex information in both the transverse plane and in the z direction. Vertex resolution is typically $50\,\mu$m in z for a fully reconstructed B meson, depending on the decay mode, and of order 100 to $150\,\mu$m for a generic B decay. Leptons and hadrons are identified with measurements from all the BABAR components, including the energy loss dE/dx from a truncated mean of up to 40 samples in the DCH and at least 8 samples in the SVT. Electrons and photons are identified in the barrel and the forward regions by the CsI electromagnetic calorimeter (EMC). Muons are identified in the instrumented flux return (IFR). In the central polar region the Cherenkov ring imaging detector (DIRC) provides K-π separation with a significance of at least three standard deviations over the full momentum range for B decay products above $250\,\text{MeV}/c$.

3.1 Particle identification

An electron candidate must be matched to an electromagnetic cluster of at least three crystals in the CsI calorimeter. The ratio of the cluster energy to the track momentum, E/p, must be between 0.88 and 1.3. The lateral moment of the cluster must be between 0.1 and 0.6, and the Zernike moment of order $(4,2)^a$ must be smaller than 0.1. In addition the electron candidate track in the drift chamber must have a dE/dx measurement consistent with that of an electron and, if measured, the Cherenkov angle in the DIRC must be consistent with that of an ultra-relativistic particle.

Muon identification relies principally on the measured number of interaction lengths, N_λ, penetrated by the candidate in the IFR iron, which must have a minimum value of 2.2 and, at higher momenta, must be larger than $N_\lambda^{exp} - 1$, where N_λ^{exp} is the expected number of interaction lengths for a muon. The number of IFR layers with a "hit" must be larger than two. To reject hadronic showers, we impose criteria on the number of IFR strips with a hit as a function of the penetration length, and on the distance between the strips with hits and the extrapolated track. In the forward region, which suffers from accelerator-related background, extra hit-continuity criteria are applied. In addition, if the muon candidate is in the angular region covered by the EMC, the energy deposited by the candidate in the calorimeter must be larger than 50 MeV and smaller than 400 MeV. (The expected energy deposited by a minimum ionizing particle is about 180 MeV.)

Particles are identified as kaons if the ratio of the combined kaon likelihood to the combined pion likelihood is greater than 15. The combined likelihoods are the product of the individual likelihoods in the SVT, DCH and DIRC subsystems. In the SVT and DCH tracking detectors, the likelihoods are based on the measured dE/dx truncated mean compared to the expected mean for the K and π hypotheses, with an assumed Gaussian distribution. The dE/dx resolution is estimated on a track-by-track basis, based on the direction and momentum of the track and the number of energy deposition samples. For the DIRC, the likelihood is computed by combining the likelihood of the measured Cherenkov angle compared to the expected Cherenkov angle for a given hypothesis, with the Poisson probability of the number of observed Cherenkov photons, given the number of expected photons for the same hypothesis. DIRC information is not required for particles with momentum less than $0.7\,\mathrm{GeV}/c$, where the DCH dE/dx alone provides good K/π discrimination.

3.2 Data samples

We define three event classes[b]:

- the *CP* sample, containing B^0 candidates reconstructed in the *CP* eigenstates $J/\psi K_S^0$ or $\psi(2S)K_S^0$. The charmonium mesons J/ψ and $\psi(2S)$ are reconstructed through their decays to e^+e^- and $\mu^+\mu^-$. The $\psi(2S)$ is also reconstructed through its decay to $J/\psi\pi^+\pi^-$. The K_S^0 is reconstructed through its decay to $\pi^+\pi^-$ and $\pi^0\pi^0$. The selection criteria for the *CP* sample are described in the next section.

- the fully reconstructed B^0 samples, containing B^0 candidates in either semileptonic or hadronic flavor eigenstates. The sample of semileptonic decays contains candidates in the $B^0 \to D^{*-}\ell^+\nu_\ell$ mode ($\ell^+ = e^+$ or μ^+); the sample of hadronic neutral decays contains B^0 candidates in the $D^{(*)-}\pi^+$, $D^{(*)-}\rho^+$ and $D^{(*)-}a_1^+$ modes; the sample of hadronic charged decays contains B^+ candidates in the $\overline{D}^0\pi^+$, and $\overline{D}^{*0}\pi^+$ (with $\overline{D}^{*0} \to \pi^0\overline{D}^0$) modes. The selection criteria for these samples are described in Refs. 6 and 7. We reconstruct $\approx 7500\ B^0 \to D^{*-}\ell^+\nu_\ell$ candidates, ≈ 2500 candidates in hadronic B^0 final states, and ≈ 2300 candidates in hadronic B^+ final states.

---

[a]The lateral and Zernike moments are cluster shape variables introduced in Ref. 8.

[b]Throughout this paper, conjugates of flavor-eigenstate modes are implied.

- the charmonium control samples, containing fully reconstructed neutral or charged B candidates in two-body decay modes with a J/ψ in the final state, such as $B^+ \rightarrow J/\psi K^+$ or $B^0 \rightarrow J/\psi (K^{*0} \rightarrow K^+\pi^-)$. The selection criteria for these samples are described in Ref. 5. We reconstruct 570 $B^+ \rightarrow J/\psi K^+$ candidates and 237 $B^0 \rightarrow J/\psi (K^{*0} \rightarrow K^+\pi^-)$ candidates.

Signal event yields and purities for the individual samples are summarized in Table 1.

3.3   The CP sample

We select events with a minimum of four reconstructed charged tracks in the region defined by $0.41 < \theta_{lab} < 2.41$. Events are required to have a reconstructed vertex within 0.5 cm of the average position of the interaction point in the plane transverse to the beamline, and a total energy greater than 5 GeV in the fiducial regions for charged tracks and neutral clusters. To reduce continuum background, we require the second-order normalized Fox-Wolfram moment[9] ($R_2 = H_2/H_0$) of the event to be less than 0.5.

The selection criteria for the $J/\psi K_s^0$ and $\psi(2S)K_s^0$ events are optimized by maximizing the ratio $\mathcal{S}/\sqrt{\mathcal{S}+\mathcal{B}}$, where $\mathcal{S}$ (the number of signal events that pass the selection) is determined from signal Monte Carlo events, and $\mathcal{B}$ (the number of background events that pass the selection) is estimated from a luminosity-weighted average of continuum data events and nonsignal $B\bar{B}$ Monte Carlo events.

For the J/ψ or $\psi(2S) \rightarrow e^+e^-$ candidates, at least one of the decay products is required to be positively identified as an electron or, if outside the acceptance of the calorimeter, to be consistent with an electron according to the drift chamber dE/dx information. If both tracks are within the calorimeter acceptance and have a value of E/p larger than 0.5, an algorithm for the re-

covery of Bremsstrahlung photons[5] is used.

For the J/ψ or $\psi(2S) \rightarrow \mu^+\mu^-$ candidates, at least one of the decay products is required to be positively identified as a muon and the other, if within the acceptance of the calorimeter, is required to be consistent with a minimum ionizing particle.

We select J/ψ candidates with an invariant mass greater than 2.95 GeV/c^2 and 3.06 GeV/c^2 for the e^+e^- and $\mu^+\mu^-$ modes, respectively, and smaller than 3.14 GeV/c^2 in both cases. The $\psi(2S)$ candidates in leptonic modes must have a mass within 50 MeV/c^2 of the $\psi(2S)$ mass. The lower bound is relaxed to 250 MeV/c^2 for the e^+e^- mode.

For the $\psi(2S) \rightarrow J/\psi \pi^+\pi^-$ mode, mass-constrained J/ψ candidates are combined with pairs of oppositely charged tracks considered as pions, and $\psi(2S)$ candidates with mass between 3.0 GeV/c^2 and 4.1 GeV/c^2 are retained. The mass difference between the $\psi(2S)$ candidate and the J/ψ candidate is required to be within 15 MeV/c^2 of the known mass difference.

K_s^0 candidates reconstructed in the $\pi^+\pi^-$ mode are required to have an invariant mass, computed at the vertex of the two tracks, between 486 MeV/c^2 and 510 MeV/c^2 for the $J/\psi K_s^0$ selection, and between 491 MeV/c^2 and 505 MeV/c^2 for the $\psi(2S)K_s^0$ selection.

For the $J/\psi K_s^0$ mode, we also consider the decay of the K_s^0 into $\pi^0\pi^0$. Pairs of π^0 candidates, with total energy above 800 MeV and invariant mass, measured at the primary vertex, between 300 and 700 MeV/c^2, are considered as K_s^0 candidates. For each candidate, we determine the most probable K_s^0 decay point along the path defined by the K_s^0 momentum vector and the primary vertex of the event. The decay-point probability is the product of the χ^2 probabilities for each photon pair constrained to the π^0 mass. We require the distance from the decay point to the primary vertex to be between -10 cm and $+40$ cm and the K_s^0 mass measured at this point to be between 470 and 536 MeV/c^2.

B_{CP} candidates are formed by combining mass-constrained J/ψ or $\psi(2S)$ candidates with mass-constrained K_S^0 candidates. The cosine of the angle between the K_S^0 three-momentum vector and the vector that links the J/ψ and K_S^0 vertices must be positive. The cosine of the helicity angle of the J/ψ in the B rest frame must be less than 0.8 for the e^+e^- mode and 0.9 for the $\mu^+\mu^-$ mode.

For the $\psi(2S)K_S^0$ candidates, the helicity angle of the $\psi(2S)$ must be smaller than 0.9 for both leptonic modes. The K_S^0 flight length with respect to the $\psi(2S)$ vertex is required to be greater than 1 mm. In the $\psi(2S) \to J/\psi\pi^+\pi^-$ mode, the absolute value of the cosine of the angle between the B_{CP} candidate three-momentum vector and the thrust vector of the rest of the event, in the center-of-mass frame, must be less than 0.9.

B_{CP} candidates are identified with a pair of nearly uncorrelated kinematic variables: the difference ΔE between the energy of the B_{CP} candidate and the beam energy in the center-of-mass frame, and the beam-energy substituted mass[8] m_{SE}. The signal region is defined by 5.270 GeV/c^2 < m_{SE} < 5.290 GeV/c^2 and an approximately three-standard-deviation cut on ΔE (typically $|\Delta E| < 35$ MeV).

Distributions of ΔE and m_{SE} are shown in Fig. 1, 2 and 3 for the CP samples and in Fig. 4 and 5 for the charmonium control samples. Signal event yields and purities, determined from a fit to the m_{SE} distributions after selection on ΔE, are summarized in Table 1.

The CP sample used in this analysis is composed of 168 candidates: 121 in the $J/\psi K_S^0$ ($K_S^0 \to \pi^+\pi^-$) channel, 19 in the $J/\psi K_S^0$ ($K_S^0 \to \pi^0\pi^0$) channel and 28 in the $\psi(2S)K_S^0$ ($K_S^0 \to \pi^+\pi^-$) channel.

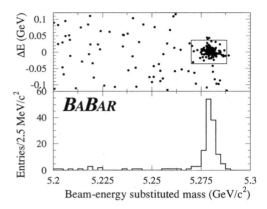

Figure 1. $J/\psi K_S^0$ ($K_S^0 \to \pi^+\pi^-$) signal.

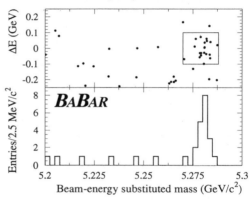

Figure 2. $J/\psi K_S^0$ ($K_S^0 \to \pi^0\pi^0$) signal.

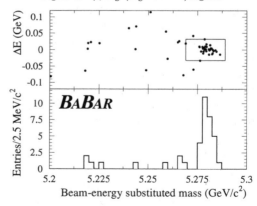

Figure 3. $\psi(2S)K_S^0$ ($K_S^0 \to \pi^+\pi^-$) signal.

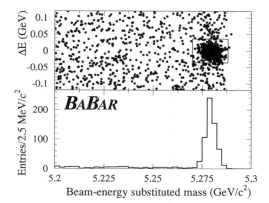

Figure 4. $J/\psi K^+$ signal.

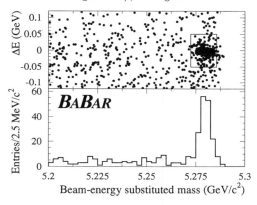

Figure 5. $J/\psi K^{*0}$ ($K^{*0} \to K^+\pi^-$) signal.

4 Time resolution function

The resolution of the Δt measurement is dominated by the z resolution of the tagging vertex. The tagging vertex is determined as follows. The three-momentum of the tagging B and its associated error matrix are derived from the fully reconstructed B_{CP} candidate three momentum, decay vertex and error matrix, and from the knowledge of the average position of the interaction point and the $\Upsilon(4S)$ four-momentum. This derived B_{tag} three-momentum is fit to a common vertex with the remaining tracks in the event (excluding those from B_{CP}). In order to reduce the bias due to long-lived particles,

all reconstructed V^0 candidates are used as input to the fit in place of their daughters. Any track whose contribution to the χ^2 is greater than 6 is removed from the fit. This procedure is iterated until there are no tracks contributing more than 6 to the χ^2 or until all tracks are removed. Events are rejected if the fit does not converge for either the B_{CP} or B_{tag} vertex. We also reject events with large Δz ($|\Delta z| > 3\,\mathrm{mm}$) or a large error on Δz ($\sigma_{\Delta z} > 400\,\mu\mathrm{m}$).

The time resolution function is described accurately by the sum of two Gaussian distributions, which has five independent parameters:

$$\mathcal{R}(\Delta t; \hat{a}) =$$
$$\sum_{i=1}^{2} \frac{f_i}{\sigma_i\sqrt{2\pi}} \exp\left(-(\Delta t - \delta_i)^2/2\sigma_i^2\right). \quad (6)$$

A fit to the time resolution function in Monte Carlo simulated events indicates that most of the events ($f_1 = 1 - f_2 = 70\%$) are in the core Gaussian, which has a width $\sigma_1 \approx 0.6\,\mathrm{ps}$. The wide Gaussian has a width $\sigma_2 \approx 1.8\,\mathrm{ps}$. Tracks from forward-going charm decays included in the reconstruction of the B_{tag} vertex introduce a small bias, $\delta_1 \approx -0.2\,\mathrm{ps}$, for the core Gaussian.

A small fraction of events have very large values of Δz, mostly due to vertex reconstruction problems. This is accounted for in the parametrization of the time resolution function by a very wide unbiased Gaussian with fixed width of $8\,\mathrm{ps}$. The fraction of events populating this component of the resolution function, f_w, is estimated from Monte Carlo simulation as $\sim 1\%$.

In likelihood fits, we use the error $\sigma_{\Delta t}$ on Δt that is calculated from the fits to the two B vertices for each individual event. However, we introduce two scale factors $\mathcal{S}_1$ and $\mathcal{S}_2$ for the width of the narrow and the wide Gaussian distributions ($\sigma_1 = \mathcal{S}_1 \times \sigma_{\Delta t}$ and $\sigma_2 = \mathcal{S}_2 \times \sigma_{\Delta t}$) to account for the fact that the uncertainty on Δt is underestimated due to effects such as the inclusion of particles

Table 1. Event yields for the different samples used in this analysis, from the fit to m_{SE} distributions after selection on ΔE. The purity is quoted for $m_{SE} > 5.270\,\mathrm{MeV}/c^2$ (except for $D^{*-}\ell^+\nu$).

Sample	Final state	Yield	Purity (%)
CP	$J/\psi\,K_S^0\ (K_S^0 \to \pi^+\pi^-)$	124 ± 12	96
	$J/\psi\,K_S^0\ (K_S^0 \to \pi^0\pi^0)$	18 ± 4	91
	$\psi(2S)K_S^0$	27 ± 6	93
Hadronic	$D^{*-}\pi^+$	622 ± 27	90
(neutral)	$D^{*-}\rho^+$	419 ± 25	84
	$D^{*-}a_1^+$	239 ± 19	79
	$D^-\pi^+$	630 ± 26	90
	$D^-\rho^+$	315 ± 20	84
	$D^-a_1^+$	225 ± 20	74
	total	2438 ± 57	85
Hadronic	$\overline{D}^0\pi^+$	1755 ± 47	88
(charged)	$\overline{D}^*\pi^+$	543 ± 27	89
	total	2293 ± 54	88
Semileptonic	$D^{*-}\ell^+\nu$	7517 ± 104	84
Control	$J/\psi\,K^+$	597 ± 25	98
	$\psi(2S)K^+$	92 ± 10	93
	$J/\psi\,K^{*0}\ (K^{*0} \to K^+\pi^-)$	251 ± 16	95

from D decays and possible underestimation of the amount of material traversed by the particles. The scale factor $\mathcal{S}_1$ and the bias δ_1 of the narrow Gaussian are free parameters in the fit. The scale factor $\mathcal{S}_2$ and the fraction of events in the wide Gaussian, f_2, are fixed to the values estimated from Monte Carlo simulation by a fit to the pull distribution ($\mathcal{S}_2 = 2.1$ and $f_2 = 0.25$). The bias of the wide Gaussian, δ_2, is fixed at $0\,\mathrm{ps}$. The remaining set of three parameters:

$$\hat{a} = \{\, \mathcal{S}_1,\, \delta_1,\, f_w \,\} \qquad (7)$$

are determined from the observed vertex distribution in data.

Because the time resolution is dominated by the precision of the B_{tag} vertex position, we find no significant differences in the Monte Carlo simulation of the resolution function parameters for the various fully reconstructed decay modes, validating our approach of determining the resolution function parameters $\hat{a}$ with the relatively high-statistics fully-reconstructed B^0 data samples, and fixing these parameters in the likelihood fit for the determination of $\sin 2\beta$ with the low-statistics CP sample. The differences in the resolution function parameters in the different tagging categories are also small.

Table 2 presents the values of the parameters obtained from a fit to the hadronic B^0 sample. These values are used in the final fit for $\sin 2\beta$.

5 B flavor tagging

Each event with a CP candidate is assigned a B^0 or $\overline{B}^0$ tag if the rest of the event (*i.e.*, with the daughter tracks of the B_{CP} removed) satisfies the criteria for one of several tagging categories. The figure of merit for each tagging category is the effective tagging efficiency $Q_i = \varepsilon_i\,(1 - 2w_i)^2$, where ε_i is the fraction of events assigned to category i and

Table 2. Parameters of the resolution function determined from the sample of events with fully-reconstructed hadronic B candidates.

Parameter		Value	
δ_1	(ps)	-0.20 ± 0.06	from fit
$\mathcal{S}_1$		1.33 ± 0.14	from fit
f_w	(%)	1.6 ± 0.6	from fit
f_1	(%)	75	fixed
δ_2	(ps)	0	fixed
$\mathcal{S}_2$		2.1	fixed

w_i is the probability of misclassifying the tag as a B^0 or $\overline{B}^0$ for this category. w_i is called the mistag fraction. The statistical error on $\sin 2\beta$ is proportional to $1/\sqrt{Q}$, where $Q = \sum_i Q_i$.

Three tagging categories rely on the presence of a fast lepton and/or one or more charged kaons in the event. Two categories, called neural network categories, are based upon the output value of a neural network algorithm applied to events that have not already been assigned to lepton or kaon tagging categories.

In the following, the tag refers to the B_{tag} candidate. In other words, a B^0 tag indicates that the B_{CP} candidate was in a $\overline{B}^0$ state at $\Delta t = 0$; a $\overline{B}^0$ tag indicates that the B_{CP} candidate was in a B^0 state.

5.1 Lepton and kaon tagging categories

The three lepton and kaon categories are called Electron, Muon and Kaon. This tagging technique relies on the correlation between the charge of a primary lepton from a semileptonic decay or the charge of a kaon, and the flavor of the decaying b quark. A requirement on the center-of-mass momentum of the lepton reduces contamination from low-momentum opposite-sign leptons coming from charm semileptonic decays. No similar kinematic quantities can be used to discriminate against contamination from opposite-sign kaons. Therefore, for kaons the optimization of Q relies principally on the balance between kaon identification efficiency and the purity of the kaon sample.

The first two categories, Electron and Muon, require the presence of at least one identified lepton (electron or muon) with a center-of-mass momentum greater than $1.1 \, \mathrm{GeV}/c$. The momentum cut rejects the bulk of wrong-sign leptons from charm semileptonic decays. The value is chosen to maximize the effective tagging efficiency Q. The tag is B^0 for a positively-charged lepton, $\overline{B}^0$ for a negatively-charged lepton.

If the event is not assigned to either the Electron or the Muon tagging categories, the event is assigned to the Kaon tagging category if the sum of the charges of all identified kaons in the event, ΣQ_K, is different from zero. The tag is B^0 if ΣQ_K is positive, $\overline{B}^0$ otherwise.

If both lepton and kaon tags are present and provide inconsistent flavor tags, the event is rejected from the lepton and kaon tagging categories.

5.2 Neural network categories

The use of a second tagging algorithm is motivated by the potential flavor-tagging power carried by non-identified leptons and kaons, correlations between leptons and kaons, multiple kaons, softer leptons from charm semileptonic decays, soft pions from D^* decays and more generally by the momentum spectrum of charged particles from B meson decays. One way to exploit the information contained in a set of correlated quantities is to use multivariate methods such as neural networks.

We define five different neural networks, called feature nets, each with a specific goal. Four of the five feature nets are track-based : the L and LS feature nets are sensitive to the presence of primary and cascade leptons, respectively, the K feature net to that of charged kaons and the SoftPi feature net to that of soft pions from D^* decays. In addition, the Q feature net exploits the charge of the fastest particles in the event.

The variables used as input to the neural network tagger are the highest values of the L, LS and SoftPi feature net outputs multiplied by the charge, the highest and the second highest value of the K feature net output multiplied by the charge, and the output of the Q feature net.

The output of the neural network tagger, x_{NT}, can be mapped onto the interval $[-1, 1]$. The tag is B^0 if x_{NT} is negative, $\overline{B}^0$ otherwise. Events with $|x_{NT}| > 0.5$ are classified in the NT1 tagging category and events with $0.2 < |x_{NT}| < 0.5$ in the NT2 tagging category. Events with $|x_{NT}| < 0.2$ have very little tagging power and are excluded from the sample used in the analysis.

6 Measurement of mistag fractions

The mistag fractions are measured directly in events in which one B^0 candidate, called the B_{rec}, is fully reconstructed in a flavor eigenstate mode. The flavor-tagging algorithms described in the previous section are applied to the rest of the event, which constitutes the potential B_{tag}.

Considering the $B^0\overline{B}^0$ system as a whole, one can classify the tagged events as *mixed* or *unmixed* depending on whether the B_{tag} is tagged with the same flavor as the B_{rec} or with the opposite flavor. Neglecting the effect of possible background contributions, and assuming the B_{rec} is properly tagged, one can express the measured time-integrated fraction of mixed events χ as a function of the precisely-measured $B^0\overline{B}^0$ mixing probability χ_d :

$$\chi = \chi_d + (1 - 2\chi_d)\,w \qquad (8)$$

where $\chi_d = \frac{1}{2}\,x_d^2/(1+x_d^2)$, with $x_d = \Delta m_d/\Gamma$. Thus one can deduce an experimental value of the mistag fraction w from the data.

A time-dependent analysis of the fraction of mixed events is even more sensitive to the

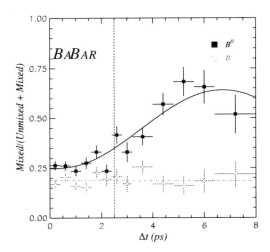

Figure 6. Fraction of mixed events $m/(u + m)$ as a function of $|\Delta t|$ (ps) for data events in the hadronic sample, for neutral B mesons (full squares) and charged B mesons (open circles). All tagging categories are included. This rate is a constant as a function of Δt for charged B mesons, but develops a mixing oscillation for neutral B mesons. The rate of mixed events extrapolated to $\Delta t = 0$ is governed by the mistag fraction w. The dot-dashed line at $t_{cut} = 2.5\,\text{ps}$ indicates the bin boundary of the time-integrated single-bin method.

mistag fraction. The mixing probability is smallest for small values of $\Delta t = t_{rec} - t_{tag}$ so that the apparent rate of mixed events near $\Delta t=0$ is governed by the mistag probability (see Fig. 6). A time-dependent analysis can also help discriminate against backgrounds with different time dependence.

By analogy with Eq. 2, we can express the density functions for unmixed $(+)$ and mixed $(-)$ events as

$$\mathcal{H}_\pm(\Delta t;\,\Gamma,\,\Delta m_d,\,\mathcal{D},\,\hat{a}\,) =$$
$$h_\pm(\Delta t;\,\Gamma,\,\Delta m_d,\,\mathcal{D})\otimes\mathcal{R}(\Delta t;\,\hat{a}), \quad (9)$$

where

$$h_\pm(\Delta t;\,\Gamma,\,\Delta m_d,\,\mathcal{D}) =$$
$$\frac{1}{4}\,\Gamma\,e^{-\Gamma|\Delta t|}\,[\,1 \pm \mathcal{D}\times\cos\Delta m_d\,\Delta t\,].(10)$$

These functions are used to build the log-likelihood function for the mixing analysis:

$$\ln \mathcal{L}_M =$$
$$\sum_i \left[\sum_{\text{unmixed}} \ln \mathcal{H}_+(t; \Gamma, \Delta m_d, \hat{a}, \mathcal{D}_i) \right.$$
$$\left. + \sum_{\text{mixed}} \ln \mathcal{H}_-(t; \Gamma, \Delta m_d, \hat{a}, \mathcal{D}_i) \right],$$

$$(11)$$

which is maximized to extract the estimates of the mistag fractions $w_i = \frac{1}{2}(1 - \mathcal{D}_i)$.

The extraction of the mistag probabilities for each tagging category is complicated by the possible presence of mode-dependent backgrounds. We deal with these by adding specific terms in the likelihood functions describing the different types of backgrounds (zero lifetime, non-zero lifetime without mixing, non-zero lifetime with mixing). described in Ref. 6.

A simple time-integrated single bin method is used as a check of the time-dependent analysis for the determination of dilutions from the fully reconstructed B^0 sample. The mistag fractions are deduced from the number of unmixed events, u, and the number of mixed events, m, in a single optimized Δt interval, $|\Delta t| < t_{cut}$. The bin boundary t_{cut}, chosen to minimize the statistical uncertainty on the measurement, is equal to 2.5 ps, *i.e.*, 1.6 B^0 lifetimes. (t_{cut} is indicated by a dot-dashed line in Fig. 6.) The resulting mistag fractions based on this method are in good agreement with the mistag fractions obtained with the maximum-likelihood fit[6].

6.1 Tagging efficiencies and mistag fractions

The mistag fractions and the tagging efficiencies are summarized in Table 3. We find a tagging efficiency of $(76.7 \pm 0.5)\%$ (statistical error only). The lepton categories have the lowest mistag fractions, but also have low efficiency. The Kaon category, despite having a larger mistag fraction (19.7%), has a

higher effective tagging efficiency; one-third of events are assigned to this category. Altogether, lepton and kaon categories have an effective tagging efficiency $Q \sim 20.8\%$. Most of the separation into B^0 and $\overline{B}^0$ in the NT1 and NT2 tagging categories derives from the SoftPi and Q feature nets. Simulation studies indicate that roughly 40% of the effective tagging efficiency occurs in events that contain a soft π aligned with the B_{tag} thrust axis, 25% from events which have a track with $p^* > 1.1$ GeV/c, 10% from events which contain multiple leptons or kaons with opposite charges and are thus not previously used in tagging, and the remaining 25% from a mixture of the various feature nets. The neural network categories increase the effective tagging efficiency by $\sim 7\%$ to an overall $Q = (27.9 \pm 0.5)\%$ (statistical error only).

Of the 168 CP candidates, 120 are tagged: 70 as B^0 and 50 as $\overline{B}^0$. The number of tagged events per category is given in Table 4.

7 Systematic uncertainties and cross checks

Systematic errors arise from uncertainties in input parameters to the maximum likelihood fit, incomplete knowledge of the time resolution function, uncertainties in the mistag fractions, and possible limitations in the analysis procedure. We fix the B^0 lifetime to the nominal PDG[10] central value $\tau_{B^0} = 1.548$ ps and the value of Δm_d to the nominal PDG value $\Delta m_d = 0.472\,\hbar\,\text{ps}^{-1}$. The errors on $\sin 2\beta$ due to uncertainties in τ_{B^0} and Δm_d are 0.002 and 0.015, respectively. The remaining systematic uncertainties are discussed in the following sections.

7.1 Uncertainties in the resolution function

The time resolution is measured with the high-statistics sample of fully-reconstructed B^0 events. The time resolution for the CP

Table 3. Mistag fractions measured from a maximum-likelihood fit to the time distribution for the fully-reconstructed B^0 sample. The `Electron` and `Muon` categories are grouped into one `Lepton` category. The uncertainties on ε and Q are statistical only.

Tagging Category	ε (%)	w (%)	Q (%)
Lepton	11.2 ± 0.5	$9.6 \pm 1.7 \pm 1.3$	7.3 ± 0.7
Kaon	36.7 ± 0.9	$19.7 \pm 1.3 \pm 1.1$	13.5 ± 1.2
NT1	11.7 ± 0.5	$16.7 \pm 2.2 \pm 2.0$	5.2 ± 0.7
NT2	16.6 ± 0.6	$33.1 \pm 2.1 \pm 2.1$	1.9 ± 0.5
all	76.7 ± 0.5		27.9 ± 1.6

Table 4. Categories of tagged events in the CP sample.

Tagging Category	$J/\psi K_S^0$ $(K_S^0 \to \pi^+\pi^-)$ B^0	$\overline{B}^0$	all	$J/\psi K_S^0$ $(K_S^0 \to \pi^0\pi^0)$ B^0	$\overline{B}^0$	all	$\psi(2S)K_S^0$ $(K_S^0 \to \pi^+\pi^-)$ B^0	$\overline{B}^0$	all	CP sample (tagged) B^0	$\overline{B}^0$	all
Electron	1	3	4	1	0	1	1	2	3	3	5	8
Muon	1	3	4	0	0	0	2	0	2	3	3	6
Kaon	29	18	47	2	2	4	5	7	12	36	27	63
NT1	9	2	11	1	0	1	2	0	2	12	2	14
NT2	10	9	19	3	3	6	3	1	4	16	13	29
Total	50	35	85	7	5	12	13	10	23	70	50	**120**

sample should be very similar, especially to that measured for the hadronic sample. We verify that the resolution function extracted in the hadronic sample is consistent with the one extracted in the semileptonic sample. We assign as a systematic error the variation in $\sin 2\beta$ obtained by changing the resolution parameters by one statistical standard deviation. The corresponding error on $\sin 2\beta$ is 0.019.

We use a full Monte Carlo simulation to verify that the Bremsstrahlung recovery procedure in the $J/\psi \to e^+e^-$ mode does not introduce any systematic bias in the Δt measurement, nor does it affect the vertex resolution and pull distributions.

In order to check the impact of imperfect knowledge of the bias in Δt on the measurement, we allow the bias of the second Gaussian to increase to 0.5 ps. The resulting

change in $\sin 2\beta$ of 0.047 is assigned as a systematic error. The sensitivity to the bias is due to the different number of events tagged as B^0 and $\overline{B}^0$.

7.2 Uncertainties in flavor tagging

The mistag fractions are measured with uncertainties that are either correlated or uncorrelated between tagging categories. We study the effect of uncorrelated errors (including statistical errors) on the asymmetry by varying the mistag fractions individually for each category, using the full covariance matrix. For correlated errors, we vary the mistag fractions for all categories simultaneously.

The main common source of systematic uncertainties in the measurement of mistag fractions is the presence of backgrounds, which are more significant in the semileptonic sample than in the hadronic sample. The largest background is due to random combinations of particles and can be studied with mass sidebands. Additional backgrounds arise in the semileptonic sample from misidentified leptons, from leptons incorrectly associated with a true D^* from B decays, and from charm events containing a D^* and a lepton. The details of the procedure for accounting for the backgrounds and the uncertainties on the background levels, and the estimates of resulting systematic errors on the mistag fractions are given in Ref. 6. We estimate the systematic error on $\sin 2\beta$ due to the uncertainties in the measurement of the mistag fractions to be 0.053, for our CP sample.

In the likelihood function, we use the same mistag fractions for the B^0 and $\bar{B}^0$ samples. However, differences are expected due to effects such as the different cross sections for K^+ and K^- hadronic interactions. For equal numbers of tagged B^0 and $\bar{B}^0$ events, the impact on $\sin 2\beta$ of a difference in mistag fraction, $\delta w = w_{B^0} - w_{\bar{B}^0}$, is insignificant. From studies of charged and neutral B samples, we find that the mistag differences are ≤ 0.02 for the NT1 category, ≤ 0.04 for the Kaon category, and negligible for the lepton categories. However, for the NT2 category, there is a significant difference between the B^0 and $\bar{B}^0$ mistag fractions, $\delta w = 0.16$, which is not predicted by our simulation. Although this would lead to a negligible systematic shift in $\sin 2\beta$, we cover the possibility of different mistag fractions in the CP sample and the fully-reconstructed sample used to measure the mistag fractions by assigning as a systematic uncertainty the shift in $\sin 2\beta$ resulting from using the measured mistag fraction for the NT2 category from the sample of $J/\psi K^{*0}$ events only. The resulting conservative systematic uncertainty on $\sin 2\beta$ is 0.050.

For a small sample of events, there can be a significant difference in the number of B^0 and $\bar{B}^0$ events, $\Delta N = N_{B^0} - N_{\bar{B}^0}$.

For a single tagging category, the fractional change in $\sin 2\beta$ from such a difference is $\Delta \sin 2\beta / \sin 2\beta \approx \delta w \Delta N / N$. In the CP sample, $\Delta N / N$ is significant only in the Kaon and NT1 categories (see Table 4). Taking into account their relative weight in the overall result, we assign a fractional systematic error of 0.005 on $\sin 2\beta$.

The systematic uncertainties on the mistag fractions due to the uncertainties on τ_{B^0} and Δm_d are negligible.

7.3 Uncertainties due to backgrounds

The fraction of background events in the CP sample ($J/\psi K_s^0$ and $\psi(2S)K_s^0$) is estimated to be $(5 \pm 3)\%$. The portion of this background that occurs at small values of Δt (e.g., contributions from u, d and s continuum events) does not contribute substantially to the determination of the asymmetry. We estimate that this reduces the effective background to 3%. We correct for the background by increasing the apparent asymmetry by a factor of 1.03. In addition, we assign a fractional systematic uncertainty of 3% on the asymmetry, to cover both the uncertainty in the size of the background and the possibility that the background might have some CP-violating component.

8 Extracting $\sin 2\beta$

8.1 Blind analysis

We have adopted a blind analysis for the extraction of $\sin 2\beta$ in order to eliminate possible experimenter's bias. We use a technique that hides not only the result of the unbinned maximum likelihood fit, but also the visual CP asymmetry in the Δt distribution. The error on the asymmetry is not hidden.

The amplitude of the asymmetry $\mathcal{A}_{CP}(\Delta t)$ from the fit is hidden from the experimenter by arbitrarily flipping its sign and adding an arbitrary offset. The sign flip hides whether a change in the analysis increases or

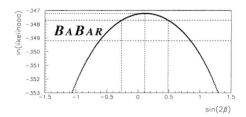

Figure 7. Variation of the log likelihood as a function of $\sin 2\beta$. The two horizontal dashed lines indicate changes in the log likelihood corresponding to one and two statistical standard deviations.

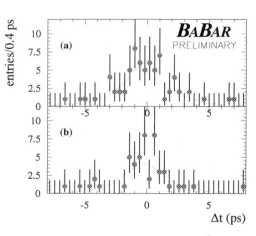

Figure 8. Distribution of Δt for (a) the B^0 tagged events and (b) the $\overline{B}^0$ tagged events in the CP sample. The error bars plotted for each data point assume Poisson statistics. The curves correspond to the result of the unbinned maximum-likelihood fit and are each normalized to the observed number of tagged B^0 or $\overline{B}^0$ events.

decreases the resulting asymmetry. However, the magnitude of the change is not hidden.

The visual CP asymmetry in the Δt distribution is hidden by multiplying Δt by the sign of the tag and adding an arbitrary offset.

With these techniques, systematic studies can be performed while keeping the numerical value of $\sin 2\beta$ hidden. In particular, we can check that the hidden Δt distributions are consistent for B^0 and $\overline{B}^0$ tagged events. The same is true for all the other checks concerning tagging, vertex resolution and the correlations between them. For instance, fit results in the different tagging categories can be compared to each other, since each fit is hidden in the same way. The analysis procedure for extracting $\sin 2\beta$ was frozen, and the data sample fixed, prior to unblinding.

8.2 Cross checks of the fitting procedure

We submitted our maximum-likelihood fitting procedure to an extensive series of simulation tests. The tests were carried out with two different implementations of the fitting algorithm to check for software errors. The validation studies were done on two types of simulated event samples.

- "Toy" Monte Carlo simulation tests. In these samples, the detector response is not simulated. Monte Carlo techniques are used with parametrized resolution

functions and tagging probabilities. We validated the fitting procedure on large samples of simulated CP events, for various numbers of tagging categories, values of mistag fractions and values of $\sin 2\beta$. We also simulated a large number of 100-event experiments, with the purpose of investigating statistical issues with small samples, including values of $\sin 2\beta$ near unphysical regions. We checked that the fitter performs well in the presence of backgrounds for the extraction of the mistag fractions. We exercised the combined CP and mixing fits, and found that although combined fits perform well, they do not significantly improve the statistical sensitivity of the result.

- Full Monte Carlo simulation tests. We studied samples of $J/\psi K_s^0$, $J/\psi K^+$, $D^*\pi$ and $D^*\ell\nu$ events produced with the BABAR GEANT3 detector simulation and reconstructed with the BABAR reconstruction program. $J/\psi K_s^0$ events were generated with various values of

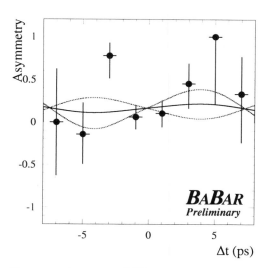

Table 5. Result of fitting for CP asymmetries in the entire CP sample and in various subsamples.

sample	$\sin2\beta$
CP sample	**0.12±0.37**
$J/\psi K_s^0$ $(K_s^0 \to \pi^+\pi^-)$	-0.10 ± 0.42
other CP events	0.87 ± 0.81
Lepton	1.6 ± 1.0
Kaon	0.14 ± 0.47
NT1	-0.59 ± 0.87
NT2	-0.96 ± 1.30

Figure 9. The raw B^0-$\overline{B}^0$ asymmetry ($N_{B^0} - N_{\overline{B}^0})/(N_{B^0} + N_{\overline{B}^0})$, with binomial errors, is shown as a function of Δt. The time-dependent asymmetry is represented by a solid curve for our central value of $\sin2\beta$, and by two dotted curves for the values at plus and minus one statistical standard deviation from the central value. The curves are not centered at $(0,0)$ in part because the probability density functions are normalized separately for B^0 and $\overline{B}^0$ events, and our CP sample contains an unequal number of B^0 and $\overline{B}^0$ tagged events (70 B^0 versus 50 $\overline{B}^0$). The χ^2 between the binned asymmetry and the result of the maximum-likelihood fit is 9.2 for 7 degrees of freedom.

$\sin2\beta$. We extracted the "apparent CP-asymmetry" for the charged B's and found it to be consistent with zero. We studied the difference in tagging efficiencies and in mistag fractions between the charged and neutral B samples. We also tested the procedure for extracting the mistag fractions from hadronic and semileptonic samples of fully simulated events ($D^*\pi$ and $D^*\ell\nu$).

9 Results

The maximum-likelihood fit for $\sin2\beta$, using the full tagged sample of $B^0/\overline{B}^0 \to J/\psi K_s^0$ and $B^0/\overline{B}^0 \to \psi(2S)K_s^0$ events, gives:

$$\sin2\beta = 0.12 \pm 0.37(\text{stat}) \pm 0.09(\text{syst})\,.$$

For this result, the B^0 lifetime and Δm_d are fixed to the current best values[10], and Δt resolution parameters and the mistag rates are fixed to the values obtained from data as summarized in Tables 2 and 3. The log likelihood is shown as a function of $\sin2\beta$ in Fig. 7, the Δt distributions for B^0 and $\overline{B}^0$ tags in Fig. 8, and the raw asymmetry as a function of Δt in Fig. 9. The results of the fit for each type of CP sample and for each tagging category are given in Table 5. The contributions to the systematic uncertainty are summarized in Table 6.

We estimate the probability of obtaining the observed value of the statistical uncertainty, 0.37, on our measurement of $\sin2\beta$ by generating a large number of toy Monte Carlo experiments with the same number of tagged CP events, and distributed in the same tagging categories, as in the CP sample in the data. We find that the errors are distributed around 0.32 with a standard deviation of 0.03, and that the probability of obtaining a value of the statistical error larger than the one we observe is 5%. Based on a large number of full Monte Carlo simulated experiments with the same number of events as our data sample, we estimate that the probability of finding a lower value of the likelihood than our observed value is 20%.

Table 6. Summary of systematic uncertainties. We compute the fractional systematic errors using the actual value of our asymmetry increased by one statistical standard deviation, that is $0.12 + 0.37 = 0.49$. The different contributions to the systematic error are added in quadrature.

Source of uncertainty	Uncertainty on $\sin 2\beta$
Uncertainty on τ_B^0	0.002
Uncertainty on Δm_d	0.015
Uncertainty on Δz resolution for CP sample	0.019
Uncertainty on time-resolution bias for CP sample	0.047
Uncertainty on measurement of mistag fractions	0.053
Different mistag fractions for CP and non-CP samples	0.050
Different mistag fractions for B^0 and $\overline{B}^0$	0.005
Background in CP sample	0.015
Total systematic error	**0.091**

10 Validating analyses

To validate the analysis we use the charmonium control sample, composed of $B^+ \rightarrow J/\psi K^+$ events and events with self-tagged $J/\psi K^{*0}$ ($K^{*0} \rightarrow K^+\pi^-$) neutral B's. We also use the event samples with fully-reconstructed candidates in charged or neutral hadronic modes. These samples should exhibit no time-dependent asymmetry. In order to investigate this experimentally, we define an "apparent CP asymmetry", analogous to $\sin 2\beta$ in Eq. 3, which we extract from the data using an identical maximum-likelihood procedure.

The events in the control samples are flavor eigenstates and not CP eigenstates. They are used for testing the fitting procedure with the same tagging algorithm as for the CP sample and, in the case of the B^+ modes, with self-tagging based on their charge. We also perform the fits for B^0 and $\overline{B}^0$ (or B^+ and B^-) events separately to study possible flavor-dependent systematic effects. For the charged B modes, we use mistag fractions measured from the sample of hadronic charged B decays.

In all fits, including the fits to charged samples, we fix the lifetime τ_{B^0} and the oscillation frequency Δm_d to the PDG values[10].

Table 7. Results of fitting for apparent CP asymmetries in various charged or neutral flavor-eigenstate B samples.

Sample	Apparent CP-asymmetry
Hadronic $B^\pm$ decays	0.03 ± 0.07
Hadronic B^0 decays	-0.01 ± 0.08
$J/\psi K^+$	0.13 ± 0.14
$J/\psi K^{*0}$, $\quad K^{*0} \rightarrow K^+\pi^-$	0.49 ± 0.26

The results of a series of validation checks on the control samples are summarized in Table 7.

The two high-statistics samples and the $J/\psi K^+$ sample give an apparent CP asymmetry consistent with zero. The 1.9 σ asymmetry in the $J/\psi K^{*0}$ is interpreted as a statistical fluctuation.

Other *BABAR* time-dependent analyses presented at this Conference demonstrate the validity of the novel technique developed for use at an asymmetric B Factory. The measurement of the B^0-$\overline{B}^0$ oscillation frequency described in Ref. 6 uses the same time resolution function and tagging algorithm as the CP analysis. Fitting for Δm_d in the maximum-likelihood fit for the fully-reconstructed hadronic and semileptonic neutral B decays, we measure

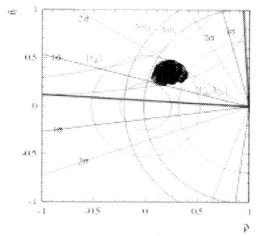

Figure 10. Present constraints on the position of the apex of the Unitarity Triangle in the $(\bar{\rho}, \bar{\eta})$ plane. The fitting procedure is described in Ref[2]. Our result $\sin2\beta = 0.12 \pm 0.37$(stat) is represented by cross-hatched regions corresponding to one and two statistical standard deviations.

$$\Delta m_d =$$
$$0.512 \pm 0.017(\text{stat}) \pm 0.022(\text{syst})\,\hbar\,\text{ps}^{-1}\,,$$

which is consistent with the world average[10] $\Delta m_d = 0.472 \pm 0.017\,\hbar\,\text{ps}^{-1}$. The B^0 lifetime measurement described in Ref. 7 uses the same inclusive vertex reconstruction technique as the CP analysis. We measure

$$\tau_{B^0} = 1.506 \pm 0.052(\text{stat}) \pm 0.029(\text{syst})\,\text{ps}\,,$$

also consistent with the world average[10] $\tau_{B^0} = 1.548 \pm 0.032\,\text{ps}$.

11 Conclusions and prospects

We have presented BABAR's first measurement of the CP-violating asymmetry parameter $\sin2\beta$ in the B meson system:

$$\sin2\beta = 0.12 \pm 0.37(\text{stat}) \pm 0.09(\text{syst})\,.$$

Our measurement is consistent with the world average[c] $\sin2\beta = 0.9 \pm 0.4$, and is cur-

rently limited by the size of the CP sample. We expect to more than double the present data sample in the near future.

Figure 10 shows the Unitarity Triangle in the $(\bar{\rho}, \bar{\eta})$ plane, with BABAR's measured central value of $\sin2\beta$ shown as two straight lines; there is a two-fold ambiguity in deriving a value of β from a measurement of $\sin2\beta$. Both choices are shown with cross-hatched regions corresponding to one and two times the one-standard-deviation experimental uncertainty. The ellipses correspond to the regions allowed by all other measurements that constrain the Unitarity Triangle. Rather than make the common, albeit unfounded, assumption that our lack of knowledge of theoretical quantities, or differences between theoretical models, can be parametrized (typically as a Gaussian or flat distribution), we have chosen to display the ellipses corresponding to measurement errors at a variety of representative choices[d] of theoretical parameters. The fitting procedure is described in Ref. 2.

While the current experimental uncertainty on $\sin2\beta$ is large, the next few years will bring substantial improvements in precision, as well as measurements for other final states in which CP-violating asymmetries are proportional to $\sin2\beta$, and measurements for modes in which the asymmetry is proportional to $\sin2\alpha$.

12 Acknowledgments

We wish to thank our PEP-II colleagues for their superb accomplishment in achieving excellent peak luminosity and remark-

[c]Based on the OPAL result[11] $\sin2\beta = 3.2^{+1.8}_{-2.0} \pm 0.5$ and the CDF result[12] $\sin2\beta = 0.79^{+0.41}_{-0.44}$. See also ALEPH's preliminary result[13] $\sin2\beta = 0.93^{+0.64\ +0.36}_{-0.88\ -0.24}$.

[d]We use the following set of measurements: $|V_{cb}| = 0.0402 \pm 0.017$, $|V_{ub}/V_{cb}| = \langle|V_{ub}/V_{cb}|\rangle \pm 0.0079$, $\Delta m_d = 0.472 \pm 0.017\,\hbar\,\text{ps}^{-1}$ and $|\epsilon_K| = (2.271 \pm 0.017) \times 10^{-3}$, and for Δm_s the set of amplitudes corresponding to a 95%CL limit of $14.6\,\hbar\,\text{ps}^{-1}$. We scan the model-dependent parameters $\langle|V_{ub}/V_{cb}|\rangle$, B_K, $f_{B_d}\sqrt{B_{B_d}}$ and ξ_s, in the range $[0.070, 0.100]$,$[0.720, 0.980]$, $[185, 255]$ MeV and $[1.07, 1.21]$, respectively.

able efficiency, enabling us to accumulate a $\Upsilon(4S)$large data sample of excellent quality in a remarkably short time. *BABAR* has received support from the Natural Sciences and Engineering Research Council (Canada), The Institute of High Energy Physics (China), Commissariat à l'Energie Atomique and Institut National de Physique Nucléaire et de Physique des Particules (France), Bundesministerium für Bildung and Forshung (Germany), Instituto Nazionale di Fisica Nucleare (Italy), The Research Council of Norway, The Ministry of Science and Technology of the Russian Federation, The Particle Physics and Astronomy Research Council (United Kingdom) and the US Department of Energy and National Science Foundation.

References

1. J. H. Christenson *et al.*, Phys. Rev. Lett. **13**, 138 (1964);
   NA31 Collaboration, G. D. Barr *et al.*, Phys. Lett. **317**, 233 (1993);
   E731 Collaboration, L. K. Gibbons *et al.*, Phys. Rev. Lett. **70**, 1203 (1993).

2. See, for instance, "Overall determinations of the CKM matrix", in P. H. Harrison and H. R. Quinn, eds., "The *BABAR* physics book", SLAC-R-504 (1998) section 14 and references therein.

3. For an introduction to *CP* violation, see, for instance, "A *CP* violation primer", in P. H. Harrison and H. R. Quinn, eds., "The *BABAR* physics book", SLAC-R-504 (1998) section 1 and references therein.

4. C. Jarlskog, in *CP Violation*, C. Jarlskog ed., World Scientific, Singapore (1988).

5. *BABAR* Collaboration, B. Aubert *et al.*, "Exclusive *B* decays to charmonium final states", *BABAR*-CONF-00/05, submitted to the XXXth International Conference on High Energy Physics, Osaka, Japan.

6. *BABAR* Collaboration, B. Aubert *et al.*, "A measurement of the $B^0\overline{B}^0$ oscillation frequency and determination of flavor-tagging efficiency using semileptonic and hadronic B decays", *BABAR*-CONF-00/08, submitted to the XXXth International Conference on High Energy Physics, Osaka, Japan.

7. *BABAR* Collaboration, B. Aubert *et al.*, "A measurement of the charged and neutral B meson lifetimes using fully reconstructed decays", *BABAR*-CONF-00/07, submitted to the XXXth International Conference on High Energy Physics, Osaka, Japan.

8. *BABAR* Collaboration, B. Aubert *et al.*, "The first year of the *BABAR* experiment at PEP-II", *BABAR*-CONF-00/17, submitted to the XXXth International Conference on High Energy Physics, Osaka, Japan.

9. G. C. Fox and S. Wolfram, Phys. Rev. Lett. **41**, 1581 (1978).

10. Particle Data Group, D. E. Groom *et al.*, Eur. Phys. Jour. C **15**, 1 (2000).

11. OPAL Collaboration, K. Ackerstaff *et al.*, Eur. Phys. Jour. C **5**, 379 (1998).

12. CDF Collaboration, T. Affolder *et al.*, Phys. Rev. **D61**, 072005 (2000).

13. ALEPH Collaboration, ALEPH 99-099 CONF 99-54 (1999).

A MEASUREMENT OF CP VIOLATION IN B^0 MESON DECAYS AT BELLE

HIROAKI AIHARA

Department of Physics, University of Tokyo,
Tokyo 113-0033, Japan
E-mail:aihara@phys.s.u-tokyo.ac.jp
Representing the Belle Collaboration

We present a preliminary measurement of the Standard Model CP violation parameter $\sin 2\phi_1$ at the KEKB asymmetric e^+e^- collider using a data sample of 6.2 fb^{-1} recorded at the $\Upsilon(4S)$ resonance with the Belle detector. One of the neutral B mesons was fully reconstructed through its decay to a CP eigenstate: $J/\psi K_S$, $\psi(2S)K_S$, $\chi_{c1}K_S$, $J/\psi K_L$ or $J/\psi\pi^0$. The flavor of B was identified mainly by the charge of a high momentum lepton or $K^\pm$. The time interval between two B decays was determined from the measurement of the distance between the decay vertices. A maximum likelihood fitting method was used to extract $\sin 2\phi_1$ from the asymmetry in the time interval distribution. We obtained a preliminary result of $\sin 2\phi_1 = 0.45^{+0.43}_{-0.44}(\text{stat})^{+0.07}_{-0.09}(\text{syst})$.

1 Introduction

The Belle experiment at the KEKB asymmetric energy e^+e^- B-meson factory recently completed a successful first year of operation. The KEKB luminosity, which was about 10^{31} cm^{-2}s^{-1} at the time of startup in June 1999, reached a level of 2×10^{33} cm^{-2}s^{-1} by the end of July 2000. Belle collected a total integrated luminosity of 6.8 fb^{-1}, 6.2 fb^{-1} on $\Upsilon(4S)$ resonance and 0.6 fb^{-1} off resonance. Based on this sample we present a preliminary measurement of the Standard Model CP violation parameter $\sin 2\phi_1$ using $B^0_d \rightarrow J/\psi K_S, \psi(2S)K_S, \chi_{c1}K_S, J/\psi\pi^0$ and $J/\psi K_L$ decays[a]. The Standard Model predicts CP violation through a mechanism of Spontaneous Symmetry Breaking of the electroweak symmetry, which results in the Cabibbo-Kobayashi-Maskawa (CKM) quark mixing matrix. In the system involving the b quark, CP violation is expected to have a large effect. The interference between the direct $B^0_d \rightarrow f_{CP}$ decay amplitude and the mixing-induced $B^0_d \rightarrow \overline{B}^0_d \rightarrow f_{CP}$ decay amplitude (where f_{CP} is a CP eigenstate to which both B^0_d and $\overline{B}^0_d$ can decay) gives rise

to an asymmetry in the time-dependent decay rate:

$$A(t) \equiv \frac{dN/dt(\overline{B}^0_{t=0}\rightarrow f_{CP}) - dN/dt(B^0_{t=0}\rightarrow f_{CP})}{dN/dt(\overline{B}^0_{t=0}\rightarrow f_{CP}) + dN/dt(B^0_{t=0}\rightarrow f_{CP})}$$
$$= -\eta_f \sin 2\phi_1 \sin \Delta m_d t, \quad (1)$$

where t is the proper time, $dN/dt(\overline{B}^0_{t=0})$ $(B^0_{t=0}) \rightarrow f_{CP})$ is the decay rate for $\overline{B}^0 (B^0)$ produced at $t = 0$ to decay to f_{CP} at time t, η_f is the CP-eigenvalue of f_{CP}, -1 for $J/\psi K_S, \psi(2S)K_S$ and $\chi_{c1}K_S$; and $+1$ for $J/\psi\pi^0$ and $J/\psi K_L$, Δm_d is the mass difference between two B^0 mass eigenstates, and ϕ_1 is one of three angles[1] of the CKM Unitarity Triangle, defined as $\phi_1 \equiv \pi - \arg(\frac{-V^*_{tb}V_{td}}{-V^*_{cb}V_{cd}})$.

In $\Upsilon(4S)$ decays, B^0 and $\overline{B}^0$ are pair-produced and remain in a coherent p-state until one of them decays. When one of the B^0 mesons decays to a final state f_1 at time t_1, this projects the remaining B^0 meson onto an orthogonal state at that time, which then propagates in time and decays to f_2 at time t_2. CP violation can be measured if one of the B mesons decays to a tagging state, f_{tag}, a final state unique to B^0 or $\overline{B}^0$, at time t_{tag} and the other decays to f_{CP} at time t_{CP}. The time-dependent asymmetry, $A(\Delta t)$, which is obtained by replacing time t in (1) with the proper time interval $\Delta t \equiv t_{CP} - t_{tag}$, can be

---

[a]Throughout this paper, when a mode is quoted the inclusion of a charge conjugate mode is implied unless otherwise stated.

observed in $\Upsilon(4S)$ decays. Because a $B^0\overline{B}^0$ pair is produced nearly at rest in the $\Upsilon(4S)$ center of mass system (CMS), Δt can be determined from the distance between the f_{CP} and f_{tag} decay vertices, z_{CP} and z_{tag}, in the boost direction (z) as $\Delta t \sim \Delta z/\beta\gamma c$, where $\beta\gamma$ is a Lorentz boost factor of the $\Upsilon(4S)$ and is equal to 0.425 at KEKB.

This asymmetry becomes diluted by experimental factors including the background in the reconstructed f_{CP} states, the wrong tag fraction (ω), that is the probability that the flavor of the tagging B is incorrectly identified, and the dilution due to the resolution of the decay vertex determination (d_{res}):

$$A_{observed} = \{\frac{1}{1+B/S}(1-2\omega)d_{res}\}A = DA.$$
(2)

Here, B/S is the background[b] to signal ratio and $D(<1)$ is called the dilution factor. The statistical error of $\sin 2\phi_1$ is inversely proportional to D: $\delta \sin 2\phi_1 = \frac{1}{\sqrt{S+B}}\frac{1}{D}$.

2   The Belle Detector

Belle is a large-solid-angle magnetic spectrometer [2]. Charged particle tracking is provided by a three-layer, double-sided silicon vertex detector (SVD) and a small-cell cylindrical drift chamber (CDC). The CDC consists of 50 layers of anode wires, of which 18 layers are inclined at small angles. Both SVD and CDC are immersed in a 1.5 T solenoidal field. The charged particle acceptance is $17° < \theta < 150°$ (where θ is the polar angle in the laboratory frame with respect to the beam axis) corresponding to $\sim 92\%$ of the full CMS solid angle. The impact parameter resolutions are $\sigma_{r\phi}^2 = (21)^2 + (\frac{69}{p\beta \sin^{3/2}\theta})^2$ μm in the plane perpendicular to the beam axis, and $\sigma_z^2 = (39)^2 + (\frac{51}{p\beta \sin^{5/2}\theta})^2$ μm in the direction along the beam axis, where p is the momentum measured in GeV/c and β is the velocity measured in units of c. The momen-

tum resolution is $(\sigma_{p_t}/p_t)^2 = (0.0019p_t)^2 + (0.0034)^2$, where p_t is the transverse momentum. Charged hadron identification is provided by dE/dx measurements in the CDC, aerogel Cherenkov counters (ACC) and a barrel of 128 time-of-flight scintillation counters (TOF). The dE/dx measurements have a resolution for hadron tracks of $\sigma(dE/dx) = 6.9\%$ and are useful for π/K separation for $p < 0.8$ GeV/c. The TOF system has a time resolution of 95 ps (rms) and provides π/K separation for $p < 1.5$ GeV/c. The indices of refraction of the ACC elements vary from 1.01 to 1.03 with polar angle, providing π/K separation for 1.5 GeV/$c < p < 3.5$ GeV/c. Particle identification probabilities are calculated from the combined response of the three detectors. The efficiency for $K^\pm$ is $\sim 80\%$ with a π fake rate of $\sim 10\%$ up to 3.5 GeV/c.

An array of 8736 CsI(Tl) crystals provides electromagnetic calorimetry and covers the same solid angle as the charged particle tracking system. The energy resolution for photons, estimated from beam tests, is $(\sigma_E/E)^2 = (0.013)^2 + (0.0007/E)^2 + (0.008/E^{1/4})^2$, where E is measured in GeV. Neutral pions are detected via their decay to $\gamma\gamma$. The π^0 mass resolution varies slowly with energy, averaging $\sigma_{m_{\pi^0}} = 4.9$ MeV/c^2. For a $\pm 3\sigma$ mass selection requirement, the overall detection efficiency including geometric acceptance for π^0s from $B\overline{B}$ events is 40\%. Electron identification is based on a combination of the CDC dE/dx information, the response of the ACC, and the position, shape and energy deposit of its associated CsI shower. The electron identification efficiency is above 90\% for $p > 1.0$ GeV/c while a π fake rate is below 0.5\%.

The magnetic field is returned via an instrumented iron yoke consisting of alternating layers of resistive plate counters and 4.7 cm thick iron plates. A total of 65.8 cm thick iron plates plus the CsI calorimeter corresponds to 4.7 nuclear interaction length at normal incidence. This system, called the

---

[b]Background in (2) s assumed to have no asymmetry.

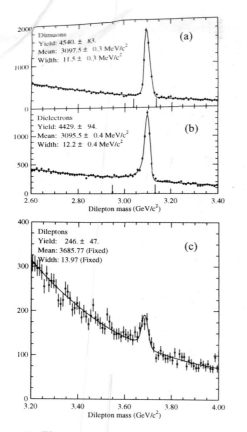

Figure 1. The invariant mass distributions for (a) $J/\psi \rightarrow \mu^+\mu^-$, (b) $J/\psi \rightarrow e^+e^-$ and (c) $\psi(2S) \rightarrow \mu^+\mu^-, e^+e^-$.

KLM, detects muons and K_L mesons in the region of $20° < \theta < 155°$. The overall muon identification efficiency is above 90% for $p > 1$ GeV/c tracks detected in the CDC while a π fake rate is below 2%. The K_L mesons are identified by the presence of KLM hits originating from hadronic interaction between K_L and the CsI and/or iron. The K_L direction is determined from the energy-weighted center of gravity of CsI hits if CsI hits associated with the KLM hits are observed, otherwise from the average position of the KLM hits. The angular resolution of the K_L direction is estimated to be $\sim 1.5°$ and $\sim 3°$ with and without associated CsI hits.

3 Selection of B^0 Decays to CP Eigenstates

We reconstructed B^0 decays to the following CP eigenstates[3]: $J/\psi K_S$, $\psi(2S)K_S$, $\chi_{C1}K_S$ for $CP - 1$ states and $J/\psi\pi^0$, $J/\psi K_L$ for $CP + 1$ states. The $B^0\overline{B}^0$ hadronic events were selected requiring (i) at least three tracks with a minimum p_t of 0.1 GeV/c originating within 2.0 cm and 4.0 cm of the run-by-run average interaction point (IP) in the plane perpendicular to the beam axis (xy plane) and along the beam axis (z axis), respectively, (ii) at least two neutral energy clusters with a minimum energy of 0.1 GeV in the barrel CsI calorimeter, (iii) a sum of all CsI cluster energies within 10 % and 80 % of the CMS energy, (iv) the total visible (charged and neutral) energy greater than 20 % of the CMS energy, (v) $|\sum p_z^{\text{CMS}}|$ less than 50 % of the CMS energy/c, where p_z^{CMS} is the z component of the momentum calculated in the CMS frame, (vi) a reconstructed event vertex within 1.5 cm and 3.5 cm of the IP in the xy plane and along the z axis, respectively. In addition, we required an event topology cut, $H_2/H_0 \leq 0.5$, where H_2 and H_0 are the second and zeroth Fox-Wolfram moments, to reject continuum background.

The J/ψ and $\psi(2S)$ mesons were reconstructed through their decays to $\mu^+\mu^-$ and e^+e^-. Dimuon candidates are oppositely charged track pairs where at least one track was positively identified as a muon by the KLM system and the other was either positively identified as a muon or had a CsI energy deposit consistent with a minimum ionizing particle. Similarly, dielectron candidates are oppositely charged track pairs where at least one track was well identified as an electron and the other track satisfied at least the dE/dx or the CsI E/p electron identification requirements. Dielectron candidates were corrected for final state radiation or bremsstrahlung in the inner parts of the detector by including the four-momentum of every photon, detected within 0.05 radian of the original electron direction, in the e^+e^- invariant mass calculation. The invariant mass distributions for $J/\psi \to \mu^+\mu^-$, $J/\psi \to e^+e^-$ and $\psi(2S) \to \mu^+\mu^-, e^+e^-$ are shown in Fig. 1(a), (b) and (c), respectively. The $\psi(2S)$ was also reconstructed through its decay to $J/\psi\pi^+\pi^-$. In addition, the χ_{c1} was reconstructed through its decay to $J/\psi\gamma$. Figures 2 (a) and (b) show the mass difference distributions of $M_{\ell^+\ell^-\pi^+\pi^-} - M_{\ell^+\ell^-}$ and $M_{\ell^+\ell^-\gamma} - M_{\ell^+\ell^-}$.

The $K_S \to \pi^+\pi^-$ candidates were selected by requiring oppositely charged track pairs to have an invariant mass between 482 MeV/c^2 and 514 MeV/c^2, corresponding to $\pm 3\sigma$ region of the $\pi^+\pi^-$ distribution around the K_S mass peak. The $K_S \to \pi^0\pi^0$ candidates were also included for the $J/\psi K_S$ mode. They were selected among photons with a minimum energy of 50 MeV and 200 MeV in the barrel and endcap regions, respectively, by requiring, assuming K_S decayed at the IP, (i) a minimum π^0 momentum of 100 MeV/c, (ii) 118 MeV/c^2 < $M_{\gamma\gamma}$ < 150 MeV/c^2, and (iii) 300 MeV/c^2 < $M_{\pi^0\pi^0}$ < 1000 MeV/c^2. For each candidate, we determined the most probable K_S decay point by minimizing the sum of the χ^2 for each photon pair constrained to the π^0 mass, varying the K_S decay point along the K_S flight direction defined by the sum of four photon momenta and the IP. We then recalculated the invariant masses of the two photons and two π^0s. We required the recalculated K_S mass be between 470 MeV/c^2 and 520 MeV/c^2. For the $J/\psi\pi^0$ mode, the π^0 candidates were selected from photons with a minimum energy of 100 MeV.

To fully reconstruct B mesons we calculated the beam-constrained mass $M_{beam} \equiv \sqrt{E_{beam}^2 - p_B^2}$, and the energy difference $\Delta E \equiv E_B - E_{beam}$, where E_{beam} is the CMS beam energy and p_B and E_B are the B candidate momentum and energy calculated in the CMS. Figure 3

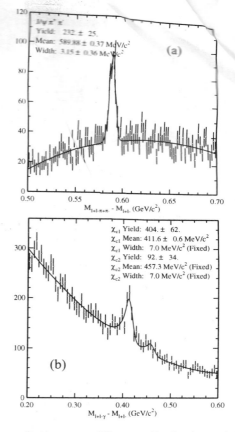

Figure 2. the mass difference distributions of (a) $M_{\ell^+\ell^-\pi^+\pi^-} - M_{\ell^+\ell^-}$ and (b) $M_{\ell^+\ell^-\gamma} - M_{\ell^+\ell^-}$.

shows the M_{beam} distribution of sum of $B \to J/\psi K_S(\pi^+\pi^-)$, $J/\psi K_S(\pi^0\pi^0)$, $\psi(2S)K_S(\pi^+\pi^-)$, $\chi_{c1}K_S(\pi^+\pi^-)$, and $J/\psi\pi^0$, after imposing 3.5σ cut on $|\Delta E|$ (40 MeV for modes with $K_S \to \pi^+\pi^-$ and 100 MeV for modes containing π^0). The signal region of B meson was defined as $|M_{beam}- < M_{beam} > | < 0.01$ GeV/c^2, where $< M_{beam} >$ is the mean value of observed M_{beam}. Table 1 lists the number of signal candidates (N_{ev}) and the backgrounds (N_{bkgd}) estimated based on the extrapolation of events in the non-signal region of ΔE vs M_{beam} distributions to the signal region, and the full simulation results.

The $B^0 \to J/\psi K_L$ candidates were se-

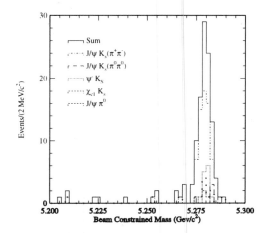

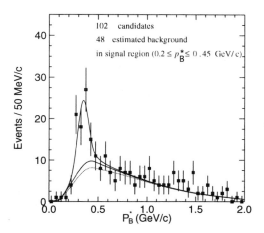

Figure 3. The beam-constrained mass distribution of sum of $B \to J/\psi K_S(\pi^+\pi^-)$, $J/\psi K_S(\pi^0\pi^0)$, $\psi(2S)K_S(\pi^+\pi^-)$, $\chi_{c1} K_S(\pi^+\pi^-)$, and $J/\psi\pi^0$ events.

Figure 4. The p_B^* distribution with the fit result. Upper solid line is a sum of the signal and background. The total background (lower solid line) is divided to those coming from $J/\psi K^{*0}(K_L\pi^0)$ and $J\psi$ + non-resonant $K_L\pi^0$ (above the dotted line) and those coming from all other sources (below the dotted line).

lected requiring the J/ψ momentum and the K_L direction consistent with the two-body kinematics. Selecting events with the J/ψ CMS momentum within 1.42 GeV/c and 2.0 GeV/c, we calculated the B momentum in the CMS, p_B^*. The p_B^* for signal must be equal to ~ 0.34 GeV/c. Figure 4 shows the p_B^* distribution. Also shown are expected distributions of signal and background derived based on the full simulation. The background was found to be dominated by $B \to J/\psi X$ events including $B \to J/\psi K^{*0}(K_L\pi^0)$ and $B \to J/\psi$ + non-resonant $K_L\pi^0$, which are a mixture of $CP+1$ and $CP-1$ eigenstates. We obtained 102 $J/\psi K_L$ candidates in the signal region defined as $0.2 \leq p_B^* \leq 0.45$ GeV/c. By fitting the data with the expected shapes, we

found a total of 48 background events in the signal region, of which 8 events were from $J/\psi K^{*0}(K_L\pi^0) + J/\psi$ non-resonant $K_L\pi^0$.

4 Flavor Tagging

Each event with a CP eigenstate was tested if the rest of the event, defined as the tag side (f_{tag}), contains a signature specific to B^0 or $\overline{B}^0$. Our tagging methods are based on the correlation between the flavor of the decaying B mesons and the charge of a prompt leptons in $b \to c\nu\ell$ decays, the charge of Kaon originating from $b \to c \to s$ decays, or the charge of π from $B \to D^*(\to \pi D)\ell\nu$ decays. A $B^0(\overline{B}^0)$ flavor for f_{tag} indicates that f_{CP} was $\overline{B}^0(B^0)$ at $\Delta t = 0$. We tested events by applying the following four tagging methods in a descending order: if the event failed the method (1), it was tested with the method (2) and so on.

1. A high momentum lepton: If f_{tag} contains a lepton ($\ell^\pm = e^\pm$ or $\mu^\pm$) with the CMS momentum $p^* \geq 1.1$ GeV/c, we assign $f_{tag} = B^0(\overline{B}^0)$ for $\ell^+(\ell^-)$.

Table 1. Summary of reconstructed CP eigenstates

Mode	N_{ev}	N_{bkgd}
$J/\psi K_S(\pi^+\pi^-)$	70	3.4
$J/\psi K_S(\pi^0\pi^0)$	4	0.3
$\psi(2S)K_S(\pi^+\pi^-)$	5	0.2
$\psi(2S)(J/\psi\pi^+\pi^-)K_S(\pi^+\pi^-)$	8	0.6
$\chi_{c1}(\gamma J/\psi)K_S(\pi^+\pi^-)$	5	0.75
$J\psi\pi^0$	10	1
Total	102	6.25

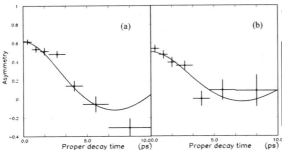

Table 2. Tagging efficiency (ϵ) and wrong tag fraction (ω)

Method	ϵ	ω
High $p^*\,\ell$	0.0142 ± 0.021	0.071 ± 0.045
$K^{\pm}$	0.279 ± 0.042	0.199 ± 0.070
Med. $p^*\,\ell$	0.029 ± 0.015	0.29 ± 0.15
Soft π	0.070 ± 0.035	0.34 ± 0.15

Figure 5. The $B^0 - \overline{B}^0$ mixing amplitude as a function of the proper-time interval Δt of two neutral B mesons obtained using (a) $B^0 \to D^*\ell\nu$ decays and (b) $B^0 \to D\ell\nu$ decays. Also shown are the results of the fit (solid lines). We obtained $\Delta m_d = 0.488 \pm 0.026$ pb^{-1}.

2. **A charged Kaon:** If f_{tag} contains no high momentum $\ell^{\pm}$, the sum of the charges of all identified Kaons, Q_K, in f_{tag} is investigated. We assign $f_{tag} = B^0(\overline{B}^0)$ if $Q_K > 0$ ($Q_K < 0$). If $Q_K = 0$, the event fails this method.

3. **A medium momentum lepton:** If f_{tag} contains a lepton with $0.6 \le p^*_\ell < 1.1$ GeV/c, we calculate the CMS missing momentum (p^*_{miss}) as an approximation of the ν CMS momentum. If $p^*_\ell + p^*_{miss} \ge 2.0$ GeV/c, we assume f_{tag} is from $b \to c\nu\ell$ decay and assign its flavor based on the charge of ℓ as in the method (1).

4. **A soft pion:** If f_{tag} contains a low momentum ($p^* < 200$ MeV/c) charged track consistent with π of $D^* \to D\pi$ decay, we assign $f_{tag} = B^0(\overline{B}^0)$ for π^- (π^+).

The efficiency (ϵ) and the wrong tag fraction (ω) were obtained using a sample of exclusively reconstructed $B^0 \to D^{*-}(D^-)\ell^+\nu$ decays, which are self-tagging, and the full simulation. Due to the $B^0 - \overline{B}^0$ mixing the probability of finding the opposite flavor (OF) neutral B meson pair and that of the same

flavor (SF) neutral meson pair are given as

$$P_{OF}(\Delta t) \propto 1 + (1 - 2\omega)\cos(\Delta m_d \Delta t)$$
$$P_{SF}(\Delta t) \propto 1 - (1 - 2\omega)\cos(\Delta m_d \Delta t). \quad (3)$$

Therefore, the ω can be determined from a measurement of the $B^0 - \overline{B}^0$ oscillation amplitude:

$$A_{mix} \equiv \frac{P_{OF} - P_{SF}}{P_{OF} + P_{SF}} = (1 - 2\omega)\cos(\Delta m_d \Delta t). \quad (4)$$

The flavor of one of the two neutral B mesons was identified using $B^0 \to D^{*-}\ell^+\nu$ where D^{*-} decays to $\overline{D}^0\pi^-$ followed by $\overline{D}^0$ decays to either $K^+\pi^-$, $K^+\pi^-\pi^0$, or $K^+\pi^+\pi^-\pi^-$. The mode $B^0 \to D^-\ell^+\nu$ where D^- decays to $K^+\pi^-\pi^-$ was also used. We then identified the flavor of another B meson by applying tagging methods described above. The vertex position of $D^*\ell\nu$ was determined by requiring the ℓ track and the reconstructed D momentum vector form a common vertex. The vertex position of the tagging B meson was determined using the method described in the next section. The proper-time interval Δt of two neutral mesons was derived from the distance between two decay vertices. We obtained ω and Δm_d by fitting the Δt distributions of the OF and SF events with the expected functions including the Δt resolution and the background. We found $\Delta m_d = 0.488 \pm 0.026$ ps^{-1}, which is in good agreement with the world average[5]. Figure 5 shows the measured A_{mix} distributions together with the fitted functions.

Table 2 summarizes the tagging efficiency and the wrong tag fractions. The total tagging efficiency was measured to be 0.52, and

Table 3. Summary of tagged events

f_{CP}	N_{ev}
$J/\psi K_S(\pi^+\pi^-)$	40
$J/\psi K_S(\pi^0\pi^0)$	4
$\psi(2S)K_S(\pi^+\pi^-)$	2
$\psi(2S)(J/\psi\pi^+\pi^-)K_S(\pi^+\pi^-)$	3
$\chi_{c1}(\gamma J/\psi)K_S(\pi^+\pi^-)$	3
$J\psi\pi^0$	4
$J/\psi K_L$	42
Total	98

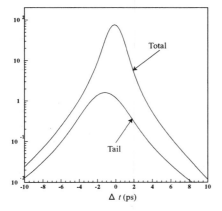

Figure 6. The average shape of the event-by-event resolution function obtained by summing over 300 $B^\pm \rightarrow J/\psi K^\pm$ events.

5 Proper-time Interval Reconstruction

The f_{CP} vertex was determined by the two lepton tracks from J/ψ. We required that at least one of two tracks must have SVD hits and that the vertex be consistent with the IP profile. The f_{tag} vertex was formed using remaining tracks in the event. In order to reduce a bias due to long-lived particles, the track was excluded from the fit if it formed the invariant mass consistent with K_S with any other charged track, if the track had large tracking error in z direction ($\sigma_z > 0.5$ mm), or if the minimum distance between the track and ·the reconstructed f_{CP} vertex was too large, $\delta z > 1.8$ mm or $\delta r > 0.5$ mm (in the $r\phi$ plane). In addition, the track with the worst χ^2 was removed from the fit if the reduced χ^2 of the vertex fit was more than 20. This procedure was iterated until the reduced χ^2 became no more than 20. The expected vertex resolutions were ~ 40 μm and ~ 85 μm for f_{CP} and f_{tag}, respectively.

The most probable Δt was estimated as $\Delta z/\beta\gamma c$. The resolution of Δt, $R(\Delta t)$, was parameterized by a sum of two Gaussians, a main Gaussian arising from the intrinsic SVD resolution and the f_{tag} vertex smearing due to the finite lifetime of the secondary charmed meson, and a tail Gaussian arising from a few poorly measured tracks:

$$R(\Delta t) = \frac{f_{main}}{\sigma\sqrt{2\pi}} \exp(-\frac{(\Delta t-\mu)^2}{2\sigma^2}) + \frac{f_{tail}}{\sigma_{tail}\sqrt{2\pi}} \exp(-\frac{(\Delta t-\mu_{tail})^2}{2\sigma_{tail}^2}). \quad (5)$$

The mean values (μ, μ_{tail}) and widths (σ, σ_{tail}) of those Gaussians were calculated from the event-by-event f_{CP} and f_{tag} vertex errors and by taking into account the error due to the approximation of $\Delta t \sim \Delta z/\beta\gamma c$. The Gaussian parameters and $f_{tail}(= 1 - f_{main})$ were determined based on the full simulation studies and a multi-parameter fit to $B \rightarrow D^*\ell\nu$ data. In addition, by measuring the lifetime of the $D^0 \rightarrow K^-\pi^+$ decays using only z coordinate information, we studied the intrinsic z vertex resolution. In order to show the average shape of the $R(\Delta t)$, Fig. 6 was drawn by summing event-by-event $R(\Delta t)$ functions over 300 $B^\pm \rightarrow J/\psi K^\pm$ (real) events. The $R(\Delta t)$ was dominated by the

the total effective tagging efficiency (ϵ_{eff}), defined as a sum of $\epsilon(1-2\omega)^2$ over all tagging methods, was found to be 0.22. The statistical error of $\sin 2\phi_1$ is proportional to $1/\sqrt{\epsilon_{eff}}$. Table 3 lists the number of tagged events for each f_{tag}. We obtained 98 tagged events in total, of which 14 events were tagged by a high momentum e, 12 by a high momentum μ, 48 by $K^\pm$, 3 by a medium momentum e, 3 by a medium momentum μ, and 18 by a soft π.

main Gaussian ($f_{main} = 0.96 \pm 0.04$) and we found $< \sigma > \sim 1.11$ ps, $< \sigma_{tail} > \sim 2.24$ ps, $< \mu > = -0.19$ ps, where $<>$ indicates the average over all events. (The μ_{tail} was fixed to -1.25 ps based on the simulation studies.) The non-zero negative mean values of the Gaussians reflect the bias in f_{tag} vertex position due to the secondary charmed mesons.

Based on the above mentioned vertex reconstruction and the resolution function, we measured lifetimes of neutral and charged B mesons[4]. Figure 7 shows Δt distributions of some of measured decay modes with the results of the lifetime fit. Table 4 summarizes the results. The obtained values for different decay modes are consistent with each other and are in good agreement with the world averages[5], $\tau_{B^0} = 1.548 \pm 0.032$ ps, and $\tau_{B\pm} = 1.653 \pm 0.028$ ps. This verifies the validity of our Δz measurement and $R(\Delta t)$.

6   Extraction of $\sin 2\phi_1$

An unbinned maximum likelihood method was used to extract the best value for $\sin 2\phi_1$. The probability density function expected for the signal distribution with a CP eigenvalue of η_f is given by :

$$Sig(\Delta t, \eta_f, q) = \frac{1}{\tau_{B^0}} \exp(-|\Delta t|/\tau_{B^0})$$
$$\times \{1 - q(1 - 2\omega)\eta_f \sin 2\phi_1 \sin(\Delta m_d \Delta t)\},$$
$$(6)$$

where $q = 1(-1)$ if $f_{tag} = B^0(\overline{B}^0)$. The ω depends on the method of the flavor tagging as given in Table 2. The values of τ_{B^0} and Δm_d were fixed to the world averages[5], 1.548 ± 0.032 ps and 0.472 ± 0.017 ps^{-1}, respectively. By investigating the events in the background dominated regions (the side bands in ΔE vs M_{beam} distributions), we found the probability density function for background events for all f_{CP} events except for $B^0 \to J/\psi K_L$ events was consistent with $Bkg(\Delta t) = \frac{1}{2\tau_{bkg}} \exp(-|\Delta t|/\tau_{bkg})$, where $\tau_{bkg} = 0.73 \pm 0.12$ ps. The likelihood

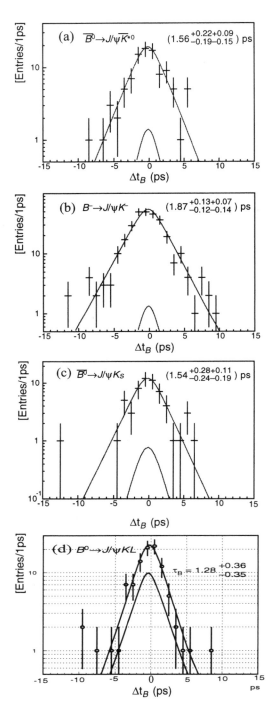

Figure 7. Δt distributions and the results of the lifetime fit for (a) $\overline{B}^0 \to J/\psi \overline{K}^{*0}$, (b) $B^- \to J/\psi K^-$, (c) $\overline{B}^0 \to J/\psi K_S$ and (d) $\overline{B}^0 \to J/\psi K_L$. The lower solid curve represents the background distribution.

Table 4. Summary of B meson lifetime measurements

Decay mode	lifetime (ps)
$\overline{B}^0 \to D^{*+}\ell^-\nu$	$1.50 \pm 0.06^{+0.06}_{-0.04}$
$\overline{B}^0 \to D^{*+}\pi^-$	$1.55^{+0.18+0.10}_{-0.17-0.07}$
$\overline{B}^0 \to D^+\pi^-$	$1.41^{+0.13}_{-0.12} \pm 0.07$
$\overline{B}^0 \to J/\psi \overline{K}^{*0}$	$1.56^{+0.22+0.09}_{-0.19-0.15}$
Combined	$1.50 \pm 0.05 \pm 0.07$
$\overline{B}^0 \to J/\psi K_S$	$1.54^{+0.28+0.11}_{-0.24-0.19}$
$\overline{B}^0 \to J/\psi K_L$	$1.28^{+0.36}_{-0.35}$
$B^- \to D^{*0}\ell^-\overline{\nu}$	$1.54 \pm 0.10^{+0.14}_{-0.07}$
$B^- \to D^0\pi^-$	$1.73 \pm 0.10 \pm 0.09$
$B^- \to J/\psi K^-$	$1.87^{+0.13+0.07}_{-0.12-0.14}$
Combined	$1.70 \pm 0.06^{+0.11}_{-0.10}$

of an event, i, was then calculated as:

$$\rho_i = p_{sig} \int_{-\infty}^{+\infty} Sig(s, \eta_f, q) R(\Delta t - s)ds$$
$$+(1 - p_{sig}) \int_{-\infty}^{+\infty} Bkg(s) R(\Delta t - s)ds, \quad (7)$$

where p_{sig} is the probability for the event being a signal, and $R(\Delta t)$ is the resolution function described in the previous section. Summing over all signal events the log-likelihood $-\sum_i \ln \rho_i$ was calculated. By scanning $\sin 2\phi_1$ to minimize this log-likelihood function, the most probable $\sin 2\phi_1$ value was found.

To test possible bias in the analysis we applied the same analysis program including tagging, vertexing and log-likelihood fitting to control data samples with null intrinsic asymmetry, $B^0 \to J/\psi K^{*0}(K^{*0} \to K^+\pi^-)$, $B^- \to J/\psi K^-$, $B^- \to D^0\pi^-$ and $B^0 \to D^{*-}\ell^+\nu$ events. Figure 8 shows the Δt distributions of $B^0 \to J/\psi K^{*0}(K^{*0} \to K^+\pi^-)$, $B^- \to J/\psi K^-$, and $B^- \to D^0\pi^-$ events. The results of the fit for apparent CP asymmetry are given in Table 5 and are all consistent with null asymmetry. Results of the fit to the tagged f_{CP} events are summarized in Table 6. To display the fitted results, $dN/d\Delta t$ distribution for $q = +1$ events and $dN/d(-\Delta t)$ distribution for $q = -1$ events

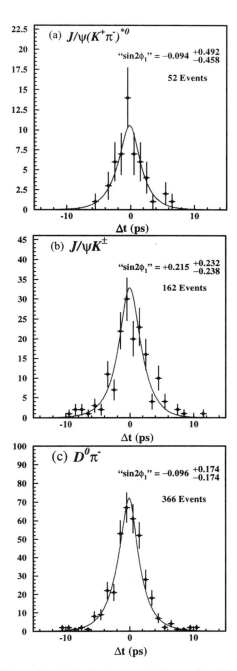

Figure 8. Δt distributions and the results of the CP fit for (a) $B^0 \to J/\psi K^{*0}(K^{*0} \to K^+\pi^-)$, (b) $B^- \to J/\psi K^-$, and (c) $B^- \to D^0\pi^-$ events.

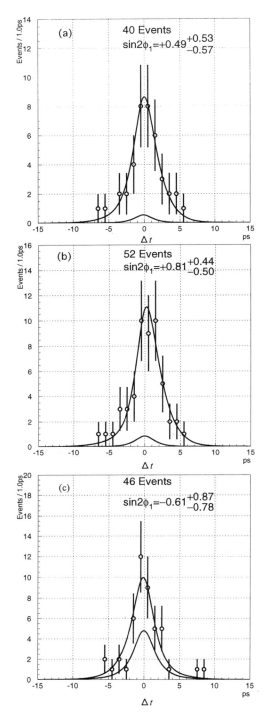

Figure 9. $dN/d\Delta t|_{q=+1} + dN/d(-\Delta t)|_{q=-1}$ distributions and the results of the CP fit for (a) $B^0 \rightarrow J/\psi K_S(\pi^+\pi^-)$ only, (b) all $CP-1$ f_{CP} modes combined, and (c) all $CP+1$ f_{CP} modes combined. A lower solid line in each figure indicates background contribution.

Table 5. Results of CP fit to control data.

Decay mode	Apparent $\sin 2\phi_1$
$\overline{B}^0 \rightarrow J/\psi(K^+\pi^-)^{*0}$	$-0.094^{+0.492}_{-0.458}$
$B^- \rightarrow J/\psi K^-$	$+0.215^{+0.232}_{-0.238}$
$B^- \rightarrow D^0\pi^-$	-0.096 ± 0.174
$B^0 \rightarrow D^{*-}\ell^+\nu$	$+0.09 \pm 0.18$

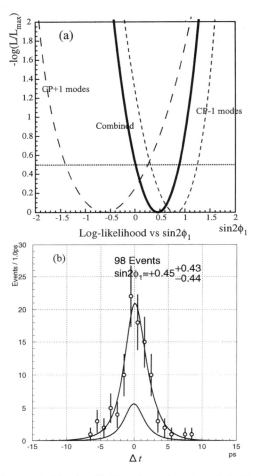

Figure 10. The log-likelihood as a function of $\sin 2\phi_1$ (a) and a sum of $dN/d(+\Delta t)|_{q=+1} + dN/d(-\Delta t)|_{q=-1}$ for $\eta_f = -1$ and $dN/d(-\Delta t)|_{q=+1} + dN/d(+\Delta t)|_{q=-1}$ for $\eta_f = +1$ (b) for combined $CP-1$ and $CP+1$ events.

Table 6. Results of CP fit to tagged f_{CP} events.

Decay mode	$\sin 2\phi_1$
$J/\psi K_S(\pi^+\pi^-)$ only	$+0.49^{+0.53}_{-0.57}$
All $CP-1$ modes	$+0.81^{+0.44}_{-0.50}$
All $CP+1$ modes	$-0.61^{+0.87}_{-0.78}$
All combined	$+0.45^{+0.43}_{-0.44}$

were added:

$$dN/d\Delta t|_{q=+1} + dN/d(-\Delta t)|_{q=-1}$$
$$\propto \exp(-|\Delta t|/\tau_{B^0}) \qquad (8)$$
$$\times\{1 - (1-2\omega)\eta_f \sin 2\phi_1 \sin(\Delta m_d \Delta t)\}$$

Figures 9 (a) and (b) show the results for only $f_{CP} = J/\psi K_S(\pi^+\pi^-)$ events and for all $CP-1$ events combined. Figure 9 (c) shows the result for $CP+1$ events, that is for $J/\psi K_L$ and $J/\psi \pi^0$ combined. In fitting to $f_{CP} = J/\psi K_L$ events, the background due to $J/\psi K^*(K_L\pi^0)$ + non-resonant $J/\psi K_L\pi^0$, which amounts to $\sim 17\%$ of the total background, was assumed as a mixture of $CP-1$ (73%) and $CP+1$ (27%) states, based on the $B \to J/\psi K_S\pi^0$ analysis[6]. A fit to 52 events in the $J\psi K_L$ sideband (i.e. $1.0 < p_B^* < 2.0$ GeV/c region), where the non-CP $J/\psi X$ events dominate, gave the result of $\sin 2\phi_1 = +0.02^{+0.48}_{-0.49}$, consistent with null asymmetry. Finally we performed the simultaneous fit to $CP-1$ and $CP+1$ events to extract the best $\sin 2\phi_1$ value. Figure 10 (a) shows the log-likelihood as a function of $\sin 2\phi_1$ for a total of 98 $CP-1$ and $CP+1$ combined events together with the results for $CP-1$ and $CP+1$ separately. We found $\sin 2\phi_1 = +0.45^{+0.43}_{-0.44}$. To display the results of the fit, $dN/d(+\Delta t)|_{q=+1} + dN/d(-\Delta t)|_{q=-1}$ for $\eta_f = -1$ and $dN/d(-\Delta t)|_{q=+1} + dN/d(+\Delta t)|_{q=-1}$ for $\eta_f = +1$ were added so that the distribution becomes approximately proportional to $1 + (1-2\omega)\sin 2\phi_1 \sin(\Delta m_d \Delta t)$. Figure 10 (b) shows its distribution.

We generated 1000 toy Monte Carlo experiments with the same number of tagged CP events, having the same composition of the tags and the same resolutions as in the CP data sample, for an input value of $\sin 2\phi_1 = 0.45$. Figures 11 (a) and (b) show the distributions of the central $\sin 2\phi_1$ value and the statistical errors, $+$ side and $-$ side separately. We found that the probability of obtaining a value of the statistical error greater than the observed value was $\sim 5\%$.

Table 7 lists systematic errors. The largest error was due to uncertainty in the wrong tag fraction estimation. It was studied by varying ω individually for each tagging method. The error due to uncertainty in Δt resolutions for both signal and background was studied by varying parameters in $R(\Delta t)$. Also included are the error due to uncertainties in estimation of the background fraction, and in the world average τ_B and m_d values. The imperfect knowledge of the event-by-event IP profile could cause systematic error in $\sin 2\phi_1$ through the vertex reconstruction. It was studied by repeating the entire fitting procedure by varying the IP envelope by $\pm 1\sigma$ in all 3 dimensions. The total systematic error of $\sin 2\phi_1$ was found to be $+0.07 - 0.09$.

7 Conclusion

Based on a 6.2 fb^{-1} data sample collected on $\Upsilon(4S)$ during the first year of the KEKB operation, we presented a preliminary measurement of $\sin 2\phi_1$ using 98 flavor-tagged events consisting of 40 $J/\psi K_S(\pi^+\pi^-)$, 4 $J/\psi K_S(\pi^0\pi^0)$, 5 $\psi(2S)K_S(\pi^+\pi^-)$, 3 $\chi_{C1}K_S(\pi^+\pi^-)$, 4 $J/\psi \pi^0$ and 42 $J/\psi K_L$ events. We found

$$\sin 2\phi_1 = 0.45^{+0.43}_{-0.44}(\text{stat})^{+0.07}_{-0.09}(\text{syst}).$$

Figure 12 shows the region covered by this measurement covers within $\pm 1\sigma$ ($\sin 2\phi_1 = 0.45^{+0.44}_{-0.45}$) in the complex plane showing the Unitarity Triangle $V_{ud}V_{ub}^* + V_{cd}V_{cb}^* + V_{td}V_{tb}^* = 0$, together with the constraints derived from other measurements[5]. While the current statistical uncertainty does not allow anything conclusive, this preliminary result is consis-

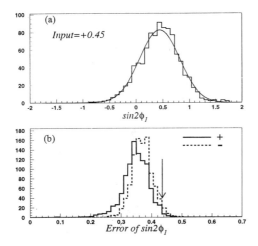

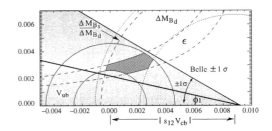

Figure 12. The area that this measurement covers by $\sin 2\phi_1 = 0.45^{+0.44}_{-0.45}$ in the complex plane showing the Unitarity Triangle. This was made by modifying Figure 11.2 of the reference[5].

Figure 11. Results of 1000 toy Monte Carlo experiments generated for $\sin 2\phi_1 = 0.45$, distributions of the fitted $\sin 2\phi_1$ (a) and statistical errors (b). In (b) distributions for errors on + side and − side are shown separately. Each experiment contains 98 events with the same tag composition as the real data.

Table 7. List of systematic errors of $\sin 2\phi_1$

Source	$\sigma+$	$\sigma-$
Wrong tag	0.050	−0.066
$R(\Delta t)$	0.026	−0.025
Background shape	0.029	−0.042
Background fraction	0.029	−0.032
τ_{B^0}, Δm_d	0.005	−0.006
IP profile	0.004	−0.000
Total	+0.07	−0.09

tent with the Standard Model prediction.

Acknowledgments

We are grateful to the conference organizers for their hospitality. It is a pleasure to thank the KEKB group and the KEK computing research center. We acknowledge support from the Ministry of Education, Science, Sports and Culture of Japan and the Japan Society for the Promotion of Science; the Australian Research Council and the Australian Department of Industry, Science and Resources; the Department of Science and Technology of India; the BK21 program of the Ministry of Education of Korea and the Basic Science program of the Korea Science and Engineering Foundation; the Polish State Committee for Scientific Research; the Ministry of Science and Technology of Russian Federation; the National Science Council and the Ministry of Education of Taiwan; the Japan-Taiwan Cooperative Program of the Interchange Association; and the U.S. Department of Energy.

References

1. H. Quinn and A. I. Sanda, Eur.Phys.J. C **15**, 625 (2000).
2. Belle Collaboration, "The Belle Detector," to be submitted to Nucl.Instr. and Methods.
3. S. Schrenk (Belle Collaboration), in this Proceedings.
4. H. Tajima (Belle Collaboration), *ibid.*
5. Particle Data Group, D.E. Groom *et al.*, Eur.Phys.J. C **15**, 1(2000).
6. Belle Collaboration, Contributed paper (#285) to the XXXth International Conference on High Energy Physics, July 27 - August 2, 2000, Osaka, Japan. http://bsunsrv1.kek.jp/conferences /ichep2000.html

NEW RESULTS FROM CLEO

D. CINABRO

Wayne State University, Department of Physics and Astronomy, Detroit, MI 48201, USA
E-mail: cinabro@physics.wayne.edu

The latest results from the CLEO collaboration are summarized. An update on the status of the upgraded CLEO III detector is also included.

1 Introduction

The $\Upsilon(4S)$ has proved to be a rich mine of physics results. The study of rare B decays has been a constant challenge to beyond standard model extensions. Semileptonic B decays have provided crucial information on the CKM matrix elements V_{cb} and V_{ub}. Hadronic B decays are also a challenge and provide insight into QCD. With the beginning of the programs at the asymmetric B factories at SLAC and KEK the new avenue of CP-violation studies has opened up. There is also a wealth of physics under the resonance especially in charm mesons, charm baryons, and taus. The energy range is also ideally suited to two photon production studies, and of course the study of the properties of the Υ and other $b\bar{b}$ resonances themselves. CLEO entered into its twentieth year of data taking at the end of 1999. I summarize here the latest, that is since the 99-00 winter conference season, results from CLEO. Even limiting myself to this time slice leaves no room for tau physics and many results can be only referenced.

I will first discuss our new results in rare B decays where we have an unambiguous observation of the gluonic penguin mode, $B \to \phi K$, new limits on B decays to $\pi^0\pi^0$, a pair of charged leptons, $\tau\nu$, and $K\nu\nu$. Turning to semileptonic B decays I present our preliminary results on V_{cb} in $B \to D^*\ell\nu$ decay, and a new measure of B mixing parameters combining a lepton tag with a partial reconstruction hadronic tag. In hadronic B decays we have observed $B \to D^{(*)}4\pi$ and studied the resonance substructure of the 4π system. We have many new results on $B \to$ Charmonium including observations of $B \to \eta_c K$ and limits on χ_{c0} in an attempt to understand our anomalous results on $B \to \eta$. Also we test charmonium production models by observing $B \to \chi_{c1}$ and limiting χ_{c2}. We have measurements of $B \to D_s^{(*)}D^*$ and evidence for $D_s^{(*)}D^{**}$. In the $D_s D^*$ decay and $D^*4\pi$ we measure the polarization of the D^* to test a prediction of the factorization ansatz. We have observed the first exclusive B decays to nucleons in $D^{*-}p\bar{p}\pi^+$ and $D^{*-}p\bar{n}$. In charm physics we have observed wrong sign $D^0 \to K\pi\pi^0$ and are working hard on more $D\bar{D}$ mixing and Doubly Cabibbo Suppressed Decay (DCSD) studies. In charm baryons we have many new results including first observations of the Σ_c^{*+} and the pair Ξ_{c1}^+ and Ξ_{c1}^0. These complete the set of L = 0 charm baryons, the bulk of which were first observed by CLEO. In resonance physics we have a measure of the rate for the $\Upsilon(4S)$ to decay to charged and neutral B meson pairs. We have measurements of η_c parameters based on its two photon production to investigate a puzzle in PQCD. Finally CLEO went through a major upgrade to the third major version of our detector. This began taking physics data in July of 2000, and I will briefly review the status of CLEO III.

The results discussed below are based on the CLEO II data set taken with the CLEO II detector from 1989 to 1999. The detector is described in detail elsewhere. [1] About two thirds of the data were taken from 1995 to 1999 in the CLEO II.V configuration which replaced the innermost straw-tube detector of CLEO II with a high precision silicon vertex detector.[2] Most of the analyses discussed use the entire CLEO II data set with exceptions for systematic error limited studies and analyses that depend on the precision vertex measuring capabilities only available in CLEO II.V. The total data set has an integrated luminosity of roughly 14/fb with two thirds taken at a center of mass energy of about 10.58 GeV on the peak of $\Upsilon(4S)$ resonance, corresponding to roughly ten million $B\bar{B}$ events, and one third taken at an energy 60 MeV below the $\Upsilon(4S)$ peak and well below the $B\bar{B}$ threshold.

2   Rare B Decays

2.1 $B \to \phi K$

The decay $b \to sg$ produced by the gluonic penguin can be uniquely tagged when the gluon splits into an $s\bar{s}$ pair as no other b decay can produce this final state. The mode $B \to \phi K$ is one such tag of the gluonic penguin and its rate is a vital piece to the rare B decay puzzle. We search for the signal in both the charged and neutral modes pairing a reconstructed $\phi \to KK$ candidate with a charged track that has a specific ionization (dE/dx) and time-of-flight consistent with a kaon or a reconstructed $K_s^0 \to \pi\pi$ candidate. We extract the yield of signal events by performing an unbinned, maximum likelihood fit to the six variables shown in Figure 1 where the PDF for the signal is taken from simulation and the background is taken from off resonance data. The yield, significance, efficiency for this procedure and the preliminary branching fractions for the two modes are displayed in Table 1. Including systematics, we interpret the neutral mode as an upper limit of 1.2×10^{-5} at the 90% C.L., measure $(6.4^{+2.5+0.5}_{-2.1-2.0}) \times 10^{-6}$ for the charged mode, and $(6.2^{+2.0+0.7}_{-1.8-1.7}) \times 10^{-6}$ for the average. The average has a significance of 5.56σ. The systematics are dominated by the fit procedure.

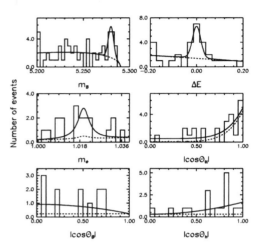

Figure 1. Projections of the ϕK^- data on the six variables used in the maximum likelihood fit. The solid lines show the total fit while the dashed lines show the contribution of the background.

Table 1. Results of likelihood fit. Note that the efficiencies do not include any branching fractions

	$B^- \to \phi K^-$	$B^0 \to \phi K_s^0$
Signal Yield	$15.8^{+6.1}_{-5.1}$	$4.3^{+2.0}_{-2.1}$
Significance (σ)	4.72	2.94
Efficiency (%)	49	31
B ($\times 10^{-6}$)	$6.4^{+2.5}_{-1.8}$	$5.9^{+4.0}_{-2.9}$

Table 2. Limits on the indicated branching fractions from the CLEO search for $B^0 \to \ell\ell$ compared with the predicted rates from the Standard Model.

Mode	90% C.L.U.L.	Prediction[8]
e^+e^-	8.3×10^{-7}	1.9×10^{-15}
$e^\pm \mu^\mp$	15×10^{-7}	0
$\mu^+\mu^-$	6.1×10^{-7}	8.0×10^{-11}

All the results are preliminary. They agree with theoretical expectations for $B \to \phi X_s{}^3$ if the K fraction of X_s is 6-10%.

2.2 $B \to \pi^0\pi^0$

Continuing our program of search for all the $B \to \pi\pi$, $K\pi$, and KK modes[4] to try to gain information on the angles of the standard unitarity triangle,[5] we have a new preliminary results on the all neutral mode $B^0 \to \pi^0\pi^0$. No signal is observed, it is expected to be much smaller than our previously observed $B^0 \to \pi^+\pi^-$ signal as it is color suppressed, and we set a 90% C.L. upper limit of 5.6×10^{-6} on the branching fraction. It is interesting to note that in this analysis we have to account for possible feed down from $B \to \pi\rho$ modes which we have recently observed.[6]

2.3 $B \to \ell\ell$

Higgs doublet and SUSY extensions to the Standard Model and leptoquark models predict a large enhancement in the rate of neutral B decays to two charged leptons. The leptoquark models can even allow the lepton flavor violating mode $B^0 \to e^\pm\mu^\mp$ to occur. We have searched for these modes and see no evidence for them.[7] The results are summarized in Table 2.

2.4 $B \to \tau\nu$ and $B \to K\nu\nu$

The charged B can decay via annihilation to a W of its two internal quarks to a lepton and neutrino. The observation and measurement of these decays are among the most

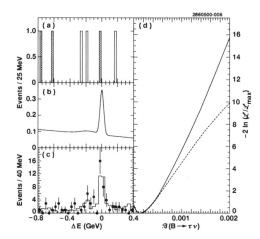

Figure 2. The results of the $B \to \tau\nu$ search. (a) The six events that pass all selections. In the three shaded events the single track is not consistent with a lepton and these are also candidates for the $K\nu\nu$ search. (b) The fitting shape used. (c) A comparison between data and simulation for events opposite tagging B's, but with two extra charged tracks. (d) The log likelihood used to calculate the upper limit. The solid curve considers statistical errors only; the dashed includes systematic effects.

important rare B decays because they provide a unique constraint on the unitarity triangle when combined with the B mixing measurements[9] and provide the cleanest way to measure the decay constant, f_B. At CLEO we have searched for the decay $B \to \tau\nu$, the largest mode because of helicity suppression, by looking for a single track from the τ decay opposite an inclusively reconstructed charged B decay to D or D^* plus up to four pions only one of which is allowed to be a π^0.[10] We look for eight decay modes of the D. We calibrate this tagging efficiency by running the analysis on a $B \to D^*\ell\nu$ sample and measure the tagging efficiency with a relative accuracy of 24% where this error is dominated by the statistics of the check sample. The result is shown in Figure 2 in terms of the difference between the energy of the tagging B and the beam energy. Also shown is the expected distribution from simulated signal events, and the result of a likelihood fit for the most probable branching fraction. We obtain an upper limit of 8.4×10^{-4} for the branching fraction at 90% C.L. Standard model expectations are in the mid-10^{-5} range.

This search can be easily modified to look for $B \to K^{\pm}\nu\nu$ which can be mediated by an electroweak penguin. Single tracks that are consistent with leptons are excluded, as shown in Figure 2, and we obtain an upper limit on the branching fraction of 2.4×10^{-4} at 90% C.L.

3   Semileptonic B Decays

3.1 $|V_{cb}|$ in $B \to D^*\ell\nu$

The measurement of V_{cb} is vital to our understanding of the unitarity triangle as it sets the scale of the entire triangle. The favorite technique is to consider the decay $B \to D^*\ell\nu$ in the context of the Heavy Quark Effective Theory (HQET). The prediction is that at the kinematic end point where the D^* is at rest with respect to the decaying B, q^2 is maximal and $w \equiv v_B \cdot v_{D^*}$ is minimal at 1, the rate for the decay is proportional to $(|V_{cb}|F(1))^2$. Here $F(w)$ is the universal form factor of HQET. Thus the strategy of this analysis is measure $d\Gamma/dw$ for $B \to D^*\ell\nu$, extrapolate to the end point, and appeal to theory to calculate the proportionality to measure $|V_{cb}|$. In CLEO this analysis critically depends on the tracking efficiency for the low momentum pion from the D^* decay, is systematically limited mainly by our ability to measure this, and thus only uses the first third of the CLEO II data set, the data taken in the CLEO II configuration. It is preliminary and is documented more completely elsewhere.[11]

The $d\Gamma/dw$ distribution is measured by fitting the distribution of $\cos\theta_{B-D^*\ell}$ which can be computed from observed D^*-ℓ pairs and the known beam energy and B mass assuming the only missing particle from the B decay is a massless neutrino. Backgrounds are determined from the data from non-$B\bar{B}$

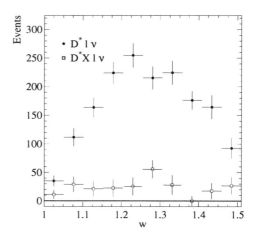

Figure 3. The $D^*\ell\nu$ and $D^*X\ell\nu$ yields in each w bin.

events using off resonance data and combinatorics from D^* sidebands. We use the simulation to model correlated backgrounds, where both the D^* and ℓ are from the same B decay, uncorrelated backgrounds where they are from different B decays, and we ignore the small contribution from fakes. Combinatorics are the largest background source, about 6%, continuum and uncorrelated are about 4% each, and correlated background is about 0.5%. The result is shown in Figure 3. There remains a background of $D^*X\ell\nu$ with contributions from B semileptonic decay to the so called D^{**} resonances and to non-resonant D^* plus at least one pion. These are modeled with the simulation and generously varied to test their systematic effect.

The partial width is given by[12]

$$\frac{d\Gamma}{dw} = \frac{G_F^2|V_{cb}|^2}{48\pi^3}G(w)F(w)^2 \qquad (1)$$

where $G(w)$ is a known kinematic function. The universal HQET form factor $F(w)$ depends on two form factor ratios $R_1(w)$ and $R_2(w)$, and a normalization $h_{A_1}(w)$. These can be constrained with dispersion relations.[13] The dependence can then be reduced to a "slope," ρ^2, and $R_1(1)$, and $R_2(1)$. We use our measured values for $R_1(1)$ and $R_2(1)$.[14] To extract the intercept we start

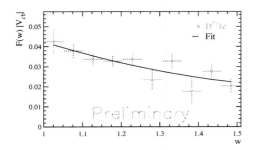

Figure 4. The result of the fit to the corrected $d\Gamma/dw$ distribution.

with Figure 3 subtract the remaining background, correct for efficiency and resolution, and fit to Equation 1 with the constraints discussed above. The result is shown in Figure 4.

Systematic errors are dominated by the uncertainty in the slow pion detection efficiency as a function of momentum. The combinatoric background and the lepton ID efficiency are also important effects. For the slope the uncertainties on $R_1(1)$ and $R_2(1)$ are the largest effect and the branching fraction has significant uncertainty from the measurements of the $D^* \to D\pi$ and the lepton ID efficiency. We find

$$|V_{cb}|h_{A_1}(1) = 0.0424 \pm 0.0018 \pm 0.0019, \quad (2)$$

$$\rho^2 = 1.67 \pm 0.11 \pm 0.22, \quad (3)$$

$$\mathcal{B}(\bar{B}^0 \to D^{*+}\ell^-\bar{\nu}) = (5.66 \pm 0.29 \pm 0.33)\%, \quad (4)$$

with a correlation coefficient of 0.90 between $|V_{cb}|h_{A_1}(1)$ and ρ^2. This slope is defined differently than our previous analysis which assumed a linear dependence. If we use a linear fit we obtain consistent results. Using $h_{A_1}(1) = 0.913 \pm 0.042$[15] this gives

$$|V_{cb}| = 0.0464 \pm 0.0020 \pm 0.0021 \pm 0.0021 \quad (5)$$

where the last error is due to the uncertainty on $h_{A_1}(1)$. This result is somewhat higher than our previous result, but this is partly due to the different assumption made about

the shape of $d\Gamma/dw$. If the same assumption is made we obtain consistent results.

3.2 B^0 Mixing

The measurement of B_d-mixing is an important cross check and provides a valuable input to the extraction CP-violation parameters in the b system. We have used a new technique at the $\Upsilon(4S)$[16] which combines the traditional lepton tag with a partial reconstruction of $\bar{B}^0 \to D^{*+}\pi^-$ or ρ^-. This partial reconstruction technique only observes the fast π^- or ρ^- from the B decay and the slow pion from the D^* decay. This increases the statistics over the dilepton method and reduces the dilution due to charged B contamination. Also the partial reconstruction is very clean. The complete analysis has statistics of about 2000 doubly tagged events with a dilution of 13% and only 3% mistagging. This leads to the best single measure for the probability of B_d mixing of $\chi_d = 0.198 \pm 0.013 \pm 0.014$. Note that this measurement has very different sources of systematic errors that the lifetime based measurements from LEP and the asymmetric B factories. We can also do the analysis comparing $B^0 B^0$ events versus $\bar{B}^0 \bar{B}^0$ and obtain limits on $|\Re(\epsilon_B)| < 0.034$, the B system analog of the ϵ parameter in the K system, and combined with LEP measures of Δm_d and lifetime $\Delta\Gamma_d/2\Gamma_d = |y_d| < 0.41$, both at the 95% C.L. This is the first nontrivial limit on y_d.

4 Hadronic B Decays

4.1 $B \to D^{(*)}4\pi$

Despite the large progress in the understanding of hadronic B decays, only a small fraction of the hadronic branching ratio has been measured. The majority of the measured modes are low multiplicity, and thus we are motivated to search for higher multiplicity modes. We investigate $B \to D^{(*)}4\pi$[17] as shown in Figure 5 and see a clear signal.

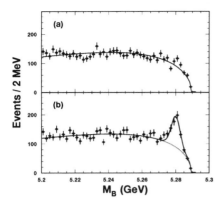

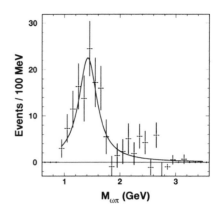

Figure 5. The beam constrained mass for $B \rightarrow D^{*+}\pi^+\pi^+\pi^-\pi^-\pi^0$ with $D^0 \rightarrow K\pi$. The top distribution is for a ΔE sideband, and the bottom is for ΔE consistent with zero.

Figure 6. The invariant mass spectrum of the $\omega\pi^-$ for the final state $D^{*+}\pi^+\pi^-\pi^-\pi^0$ combining all D decay modes. This is the spectrum determined from fitting the yield in the beam constrained mass distribution and displaying a fit to Breit-Wigner function.

Table 3. Measured branching fractions

Mode	$\mathcal{B}(\%)$
$\bar{B}^0 \rightarrow D^{*+}\pi^+\pi^-\pi^-\pi^0$	$1.72 \pm 0.14 \pm 0.24$
$\bar{B}^0 \rightarrow D^{*+}\omega\pi^-$	$0.29 \pm 0.03 \pm 0.04$
$\bar{B}^0 \rightarrow D^+\omega\pi^-$	$0.28 \pm 0.05 \pm 0.03$
$B^- \rightarrow D^{*0}\pi^+\pi^-\pi^-\pi^0$	$1.80 \pm 0.24 \pm 0.25$
$B^- \rightarrow D^{*0}\omega\pi^-$	$0.45 \pm 0.10 \pm 0.07$
$B^- \rightarrow D^0\omega\pi^-$	$0.41 \pm 0.07 \pm 0.04$

The decays $B \rightarrow D^{(*)}(2-3)\pi$ are dominated by resonant decays to the ρ and the a_0. This motivates a search for substructure in the 4π system. We do see a clear signal of an ω in the $2\pi\pi^0$ system. Combining the ω with the remaining charged π we obtain Figure 6. We clearly see a resonance and determine it to have a mass of $1418 \pm 26 \pm 19$ MeV and width of $388 \pm 41 \pm 32$ MeV. By studying the angular distribution of the decays to this resonance we find that it has $J^P = 1^-$. We identify this resonance as the ρ'. This ρ' mass and width measurement are the most accurate and have little model dependence.

Table 3 summarizes our observations in this mode. The ρ' saturates the $\omega\pi$ final states. In the $D^*\rho'$ modes we have measured

the longitudinal polarization of the D^* to be $63 \pm 9\%$. This can be compared to a prediction of the factorization model that this polarization should be the same as in the semileptonic decay $B \rightarrow D^*\ell\nu$ at q^2 equal to the mass of the ρ'. The comparison of this measurement with this prediction of factorization is shown in Figure 7. The $D^*\rho$ measure is from a previous CLEO result[18] and the D^*D_s result is discussed below.

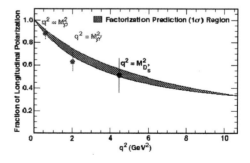

Figure 7. The fraction of D^* longitudinal polarization for three hadronic B decays. The measurements are compared with the prediction of the factorization model.

Table 4. Measured branching fractions or upper limits. The third error is due to η_c branching fractions.

Mode	$\mathcal{B}$ or 90% C.L.U.L($\times 10^{-3}$)
$B^+ \to \eta_c K^+$	$0.69 \pm 0.24 \pm 0.08 \pm 0.20$
$B^0 \to \eta_c K^0$	$1.09 \pm 0.49 \pm 0.12 \pm 0.31$
$B^+ \to \chi_{c0} K^+$	< 0.48
$B^0 \to \chi_{c0} K^0$	< 0.50

Table 5. Measured branching fractions. Direct means that feed down from $B \to \psi(2S)$ has been subtracted

Mode	$\mathcal{B}(\times 10^{-3})$
$B^0 \to \chi_{c1} X)$	$4.14 \pm 0.31 \pm 0.40$
$B^0 \to \chi_{c1}[\text{direct}]X)$	$3.83 \pm 0.31 \pm 0.40$

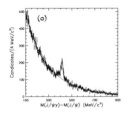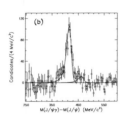

Figure 8. The $M(J/\psi\gamma) - M(J/\psi)$ distribution. The plot on the right subtracts off the background fit displayed in the plot on the left.

4.2 $B \to$ Charmonium

We have many new results on B decays to charmonium. Motivated by our observation of surprisingly large B decay rates to $\eta' X$[19] we search for B decays to $\eta_c K$ and with trivial modifications to $\chi_{c0} K$.[20] If as has been suggested that the large $\eta' X$ rate is due to intrinsic charm content of the η' then we expect an enhancement in the $\eta_c K$ rate. Our observations are summarized in Table 4. The rates for $\eta_c K$ are similar to those for J/ψ indicating no unexpected enhancement. Invoking factorization we can measure $f_{\eta_c} = 335 \pm 75$ MeV also agreeing with expectation. It seems that the charm content of the η' is not the explanation for the anomalous $B \to \eta' X$ decay rates.

The measurement of a large rate for high-P_T charmonium at the Tevatron is a challenge to theoretical models of charmonium production.[21] An especially clean test is provided by measuring the χ_{c2}-to-χ_{c1} production ratio in B decays.[22] Figure 8 shows our basic result. We do not see significant χ_{c2} production, and our results are sum-

marized in Table 5. We also limit $\mathcal{B}(B^0 \to \chi_{c2}[\text{direct}]X)/\mathcal{B}(B^0 \to \chi_{c1}[\text{direct}]X) < 0.44$ at the 95% C.L. The results are preliminary and do not support the prediction of the color-octet model that the χ_{c2}-to-χ_{c1} production ratio should be larger than 0.5.

We have new measurement of the exclusive two body B decays to charmonium that are being used in CP violation measurements at the asymmetric B factories: $\mathcal{B}(B \to J/\psi K^0) = (9.5 \pm 0.8 \pm 0.6) \times 10^{-4}$; $\mathcal{B}(B \to J/\psi \pi^0) = (2.5 \pm 1.0 \pm 0.2) \times 10^{-5}$; and $\mathcal{B}(B \to \chi_{c1} K^0) = (3.9 \pm 1.6 \pm 0.4) \times 10^{-4}$.[23] We have searches for direct CP violation in charged B decays to charmonium: $\mathcal{A}_{CP}(B^\pm \to J/\psi K^\pm) = (+1.8 \pm 4.3 \pm 0.4)$ and $\mathcal{A}_{CP}(B^\pm \to J/\psi(2S) K^\pm) = (+2.0 \pm 9.1 \pm 1.0)$.[24] We also have precision measurements of the B meson masses in $B \to \psi^{(\prime)} K$: $m(B^0) = 5279.1 \pm 0.7 \pm 0.3$ MeV and $m(B^+) = 5279.1 \pm 0.4 \pm 0.4$ MeV.[25]

4.3 $B \to D_s^{(*)} D^*$

There is a substantial discrepancy between the inclusive and sum of the exclusive B decays to D_s. We have used a new technique where a fully reconstructed $D_s \to \phi\pi$ is combined with a partially reconstructed D^*, whose soft pion is the only observed decay product, to measure exclusive B decays to D_s.[26] This partial reconstruction technique results in much higher statistics than previous analysis, and much more accurate measurements can be made. We observe evidence for $B \to D_s^{(*)} D^{**0}$ decays. The results are summarized in Table 6. We also measure the longitudinal polarization of the $D_s^* D^{*-}$ production to be $(51 \pm 14 \pm 4)\%$. This is com-

Table 6. Measured branching fractions. The third error is due to the $D_s \to \phi\pi$ branching fraction.

Mode	$\mathcal{B}(\%)$
$B^0 \to D_s D^{*-}$	$1.01 \pm 0.18 \pm 0.10 \pm 0.28$
$B^0 \to D_s^* D^{*-}$	$1.82 \pm 0.37 \pm 0.24 \pm 0.46$
$B^+ \to D_s^{(*)} D^{**0}$	$2.73 \pm 0.78 \pm 0.48 \pm 0.68$

Table 7. Measured branching fractions.

Mode	$\mathcal{B}(\times 10^{-4})$
$B^0 \to D^{*-} p\bar{p}\pi^+$	$6.6 \pm 1.4 \pm 1.0$
$B^0 \to D^{*-} p\bar{n}$	$14.5 \pm 3.2 \pm 2.7$

pared with the prediction of factorization in Figure 7, and agrees well.

4.4 $B \to$ Nucleons

A unique feature of the B meson is that the large mass of the b quark allows weak decays to baryon-anti-baryon pairs. Measurements of the inclusive rate of $B \to \Lambda_c X$ lead to estimates that the $B \to \bar{\Lambda}_c N X$, where N is a proton or neutron, rate accounts for only half of the baryon production observed in B decays. We are thus motivated to search for $B \to DN\bar{N}X$.[27] We do observe such decays in the modes and with rates given in Table 7, and the signals are displayed in Figure 9. Note that the $B^0 \to D^{*-} p\bar{n}$ mode is observed via the annihilation of the $\bar{n}$ in our calorimeter. We are not sensitive to the charge conjugated mode with a n. These measurements account for a substantial fraction of the non-Λ_c B to baryon decay rate.

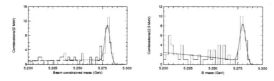

Figure 9. The beam constrained mass for the $B \to DN\bar{N}X$ signals On the left is $D^{*-} p\bar{p}\pi^+$ and on the right is $D^{*-} p\bar{n}$.

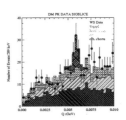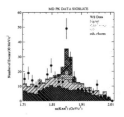

Figure 10. The wrong sign $D \to K\pi\pi^0$ signal. On the right is the Q distribution and on the left is the $K\pi\pi^0$ invariant mass.

5 Charm Physics

5.1 D^0 Mixing and Doubly Cabibbo Suppressed Decays

One of the most exciting areas in the last year has been new probes with a factor of three more sensitivity for D-mixing and Doubly Cabibbo Suppressed Decays (DCSD).[28] We are working hard at CLEO to make use of our clean event environment to search in many D decay modes to improve on our results. All of these analyses are only done with the superior vertex resolution of the CLEO II.V detector.

The latest work is preliminary. We do have a significant wrong sign signal in $D \to K\pi\pi^0$ decay, shown in Figure 10, of $39 \pm 10 \pm 7$ events compared to a right sign yield of over 9000. We are working hard to turn this yield into a rate, but the Dalitz structure of the wrong sign decay may be different than the right sign decay leading to a difference in efficiency for mixed and DCSD wrong sign events. The resonance substructure for $D \to K\pi\pi^0$ is very rich with three dominant modes clearly visible in the Dalitz plot for the right sign signal shown in Figure 11, clear signs of interference, and many other smaller amplitudes that can contribute.

We are also working on the lifetime analyses of the $K\pi\pi^0$ mode; the CP even eigenstates KK and $\pi\pi$; CP odd eigenstates $K_s^0\phi$, $K_s^0\rho^0$, and $K_s^0\omega$; and in the the semileptonic decays $K\ell\nu$ and $K^*\ell\nu$. One of the first

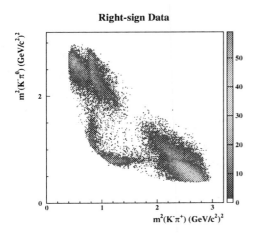

Right-sign Data

Figure 11. The Dalitz plot for right sign $D \to K\pi\pi^0$ decay in the CLEO data.

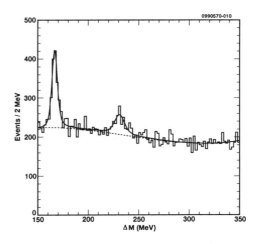

Figure 12. The distribution of $M(\Lambda_c\pi^0) - M(\Lambda_c)$ showing the narrow Σ_c^+ and wide Σ_c^{*+}.

steps on this road is a preliminary measure of the CP asymmetries $\mathcal{A}_{CP}(D^0 \to KK) = (+0.04 \pm 2.18 \pm 0.84)\%$ and $\mathcal{A}_{CP}(D^0 \to \pi\pi) = (+1.94 \pm 3.22 \pm 0.84)\%$.

5.2 Charm Baryons

There are a great many results on charm baryons, and I can only cover the very highest points, and even these only briefly. We have a first observation of the Σ_c^{*+} and a new measure of Σ_c^+ mass both in $\Lambda_c\pi^0$ decay.[29] The data is shown in Figure 12 showing the clear signals.

We make the first observation of the $J^P = 1/2^-$ pair Ξ_{c1}^+ and Ξ_{c1}^0.[30] Note that this result combined with the observation of the Σ_c^{*+} discussed above means that all of the $L = 0$ single charm baryons have been observed. The majority of these have had their first observation by CLEO.

In a preliminary result we see two states in decays to $\Lambda_c\pi^+\pi^-$; a wide state decaying to Σ_c and Σ_c^*, possibly an orbitally excited Σ_{c1}, and a narrow state decaying to $\Sigma_c\pi$ and non-resonant, possibly an excited baryon with $L = 1$ between the light quarks.[31] The data is shown in Figure 13. Our inter-

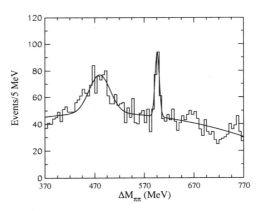

Figure 13. The distribution of $M(\Lambda_c\pi^+\pi^-) - M(\Lambda_c)$ showing the two new states.

pretation of these two states is guided by the observed mass and width rather than any observation of the angular distribution of their decays.

CLEO has observed the Ω_c and made a precision measure of its mass,[32] and we have a new measurement of the Λ_c branching fraction into $pK\pi$.[33]

6 Other Physics Results

6.1 $\Upsilon(4S)$ Charged and Neutral Decay Fraction

From the relative rates for $B^0 \to J/\psi K^{(*)0}$ and $B^+ \to J/\psi K^{(*)+}$ we are able to measure the relative branching fraction of the $\Upsilon(4S)$ to charged and neutral B mesons.[34] We find

$$\frac{f_{+-}}{f_{00}} \equiv \frac{\Gamma(\Upsilon(4S) \to B^+ B^-)}{\Gamma(\Upsilon(4S) \to B^0 \bar{B}^0)} \qquad (6)$$
$$= 1.04 \pm 0.07 \pm 0.04$$

and assuming $f_{+-} + f_{00} = 1$, that is the $\Upsilon(4S)$ only decays to $B\bar{B}$, $f_{+-} = 0.49 \pm 0.02 \pm 0.01$ and $f_{00} = 0.51 \pm 0.02 \pm 0.014$.

6.2 η_c in 2γ

We have observed 300 events of the two photon production of the η_c.[35] We measure $m(\eta_c) = (2980.4 \pm 2.3 \pm 0.6)$ MeV, $\Gamma(\eta_c) = (27.0 \pm 5.8 \pm 1.4)$ MeV, and the two photon partial width $\Gamma_{2\gamma}(\eta_c) = (7.6 \pm 0.8 \pm 0.4 \pm 2.3)$ keV, where the last error is from the branching fraction of the η_c into $K_s^0 K^\mp \pi^\pm$. This last observation agrees much better with the prediction of Peturbative QCD based on the $e^+ e^-$ partial width of the J/ψ.

7 CLEO III Status

The CLEO II.V detector ceased data taking in February of 1999. It has been upgraded to the CLEO III detector. The upgrade consists of a new four layer double sided silicon drift detector, a new 47 layer drift chamber, much stronger particle ID with the addition of a new barrel RICH, refurbished CsI calorimeter and muon system, and a new data acquisition and trigger system that are designed to handle delivered luminosity of up to $5 \times 10^{33}/\text{cm}^2$–sec. The upgraded detector was completed in April of 2000 and started taking physics data in July of 2000.

The new detector is performing well with a tracking system already working as well as the CLEO II configuration, and a much improved resolution on photons in the calorimeter endcaps due to the reduction of support material in the upgrade from the old to the new drift chamber. The RICH is also performing very well with a preliminary efficiency of around 90%, and an intrinsic resolution on the Cherenkov angle of 2-4 mrad.

8 Conclusion

There is a vast array of physics results to be had from the $\Upsilon(4S)$. Highlights from CLEO in the last six months are an unambiguous observation of the gluonic penguin, the best single measure of $|V_{cb}|$, and the first observation of a high multiplicity hadronic B decay mode. There are many other results and with the beginning of CLEO III data taking and first results from the asymmetric B factories we can all look forward to much more exciting physics from the $\Upsilon(4S)$ in the future.

References

1. Y. Kubota, *et al.* (CLEO Collaboration), *Nuclear Instruments and Methods* A **820**, 66 (1990).

2. T.S. Hill, *Nuclear Instruments and Methods in Physics Research* A **418**, 32 (1998).

3. N.G. Deshpande and Xiao-Gang He, *Physics Letters* B **336**, (1994) 471.

4. D. Cronin-Hennessy *et al.* (CLEO Collaboration) *Physical Review Letters* **85**, (2000) 515.

5. W.-S.Hou, J.G.Smith, and F.Wurthwein, hep-ex/9910014, and references therein.

6. C.P. Jessop *et al.* (CLEO Collaboration), hep-ex/9910014.

7. T. Bergfeld *et al.* (CLEO Collaboration), hep-ex/0007042.

8. A.Ali, C.Greub, and T.Mannel in ECFA-93-151.

9. T.Draper, *Nuclear Physics Proceedings*

Supplement **73** (1999) 43.

10. T.E.Browder *et al.* (CLEO Collaboration), hep-ex/0007057.

11. J.P.Alexander *et al.* (CLEO Collaboration), hep-ex/0007052.

12. J.D.Richman and P.R.Burchat, *Reviews of Modern Physics* **67**, (1995) 483.

13. I. Caprini, L. Lellouch, and M. Neubert, *Nuclear Physics* B **530**, (1998) 153.

14. J. Dubosq, *et al.* (CLEO Collaboration), *Physical Review Letters* **76**, (1996) 3898.

15. *BaBar Physics Book*, P.F. Harrison and H. R. Quinn ed. (1998) SLAC-R-504.

16. B.H. Behrens *et al.* (CLEO Collaboration), hep-ex/0005013.

17. M. Artuso *et al.* (CLEO Collaboration), hep-ex/0006018.

18. G. Bonivicini *et al.* (CLEO Collaboration), CLEO CONF 98-23.

19. S.J. Richichi *et al.* (CLEO Collaboration), *Physical Review Letters* **85**, (2000) 520.

20. K.W. Edwards *et al.* (CLEO Collaboration), hep-ex/0007012.

21. F. Abe *et al.* (CDF Collaboration), *Physical Review Letters* **79**, (1997) 572; S. Abachi *et al.* (D0 Collaboration), *Physics Letters* B **370**, (1996) 239; G.A. Schuler CERN-TH-7170-94.

22. G. Brandenburg *et al.* (CLEO Collaboration), hep-ex/0007046.

23. P. Avery *et al.* (CLEO Collaboration), *Physical Review* D **62**, (2000) 051101.

24. G. Bonvicini *et al.* (CLEO Collaboration), *Physical Review Letters* **84**, (2000) 5940.

25. S.E. Csorna *et al.* (CLEO Collaboration), *Physical Review* D **61**, (2000) 111101.

26. S.Ahmed *et al.* (CLEO Collaboration), hep-ex/0008015.

27. S.Anderson *et al.* (CLEO Collaboration), hep-ex/0009011.

28. R. Godang *et al.* (CLEO Collaboration), *Physical Review Letters* **84**, (2000) 5038.

29. R. Ammar *et al.* (CLEO Collaboration), hep-ex/0007041.

30. P. Avery *et al.* (CLEO Collaboration), hep-ex/0007050.

31. P. Avery *et al.* (CLEO Collaboration), hep-ex/0007049.

32. S. Ahmed *et al.* (CLEO Collaboration), hep-ex/0007047.

33. D. Jaffe *et al.* (CLEO Collaboration), hep-ex/0004001.

34. J. P. Alexander *et al.* (CLEO Collaboration), hep-ex/0006002.

35. G. Brandenburg *et al.* (CLEO Collaboration), hep-ex/0006026.

Plenary Session 2

CP Violation and Rare Decays

Chair: Vladimir G. Kadyshevsky (JINR)
Scientific Secretary: Kazunori Hanagaki

RECENT EXPERIMENTAL RESULTS ON CP VIOLATING AND RARE K AND μ DECAYS

P. DEBU

DSM/DAPNIA, CEA Saclay, F-91191 Gif-Yvette cedex, France
E-mail: pdebu@cea.fr

Results on rare K and μ decays obtained since 1998 are summarized. Forthcoming projects are mentioned. Emphasis is put on CP violation.

1 Introduction

In the past two years, an impressive amount of experimental results in the K and μ sectors have been obtained. They cover direct CP violation, CP violating and related rare decays, searches for T violation and for Lepton Flavor Violation. In addition, several very challenging projects are in preparation, and those will also be mentioned in this report.

2 Direct CP violation

In the Standard Model (SM) with three quarks and leptons families, CP violation arises through one unique complex parameter in the Cabbibo Kobayashi Maskawa (CKM) quark mixing matrix [1]. It allows to accomodate CP violation in the $K^0 \overline{K^0}$ mixing, and naturally induces direct CP violation in $K \to 2\pi$ decays. However, the $\Delta I = 1/2$ rule damps the effect, since CP violation in the $\Delta S = 1$ transitions arises from a phase difference between $\Delta = 1/2$ and $\Delta I = 3/2$ decay amplitudes. Direct CP violation in $K \to 2\pi$ decays is parametrized by ϵ'. The theoretical estimate of ϵ' basically needs three ingredients :

- The value of the CKM matrix elements combination $\text{Im}(V_{td}V_{ts}^*)$, which is constrained by the measurements of ϵ, the charmless B decays branching ratio, the $b \to c$ branching ratio, and the mass difference of neutral B mesons. The resulting relative uncertainty on ϵ' is of order 15 % .

- The calculation of the short distance part of the dominant so-called penguin diagrams responsible for direct CP violation has been made at the Next to Leading Order in pertubative QCD. Uncertainties on the t quark mass and on the strength of the QCD coupling constant lead again to a $\sim$ 15 % uncertainty on ϵ'.

- The uncertainties coming from the Long Distance contributions are fully dominating, the effect being enhanced by a partial cancellation between the electroweak and the gluonic penguin diagrams. Many theoretical groups present results based on the Wilson expansion of the $K \to 2\pi$ amplitude :

$$< \pi\pi|\mathcal{L}|K^0 >= \sum_{i=1}^{10} C_i(\mu)Q_i(\mu)$$

where the Q_i's represent the contributions from 10 effective 4 quark operators, and the C_i's are the Wilson coefficients, μ being the scale at which the C_i's are evaluated. Different methods are used to estimate the Q_i's. In principle, calculation on the lattice should be the most satisfying one, but, at present, one of the major contributions, the gluonic penguin operator Q_6, cannot be extracted. Values for ϵ'/ϵ range from small negative values up to .003 or even higher [2].

Even though the magnitude of ϵ' is difficult to predict, direct CP violation is naturally present in the standard model. Given the fundamental nature of this symmetry breakdown, experimentalists have been trying to establish the existence of direct CP violation for more than two decades. A first evidence was reported in 1988 by the NA31 collaboration at CERN, which published in 1993 its final result [3] : $\epsilon'/\epsilon = (23.0 \pm 6.5) \times 10^{-4}$. Just before, the E731 experiment at FNAL had reported [4] : $\epsilon'/\epsilon = (7.4 \pm 5.9) \times 10^{-4}$.

The NA48 experiment at CERN, E832 at FNAL, and KLOE at Frascati were launched to resolve this ambiguous situation. The basic principles of NA48 and E832 rely on the direct measurement of the double ratio R :

$$R = \frac{\Gamma(K_L \to \pi^0\pi^0)/\Gamma(K_S \to \pi^0\pi^0)}{\Gamma(K_L \to \pi^+\pi^-)/\Gamma(K_S \to \pi^+\pi^-)}$$
$$= 1 - 6 \; \mathrm{Re}(\epsilon'/\epsilon)$$

The use of simultaneous "K_L" and "K_S" beams reduces systematic uncertainties from K fluxes, detector inefficiencies, acquisition dead time, losses of events due to accidental activity, calibration drifts.

High performance data acquisition systems, of order 100 Mbyte/s, allow the recording of several 10^7 $K \to 2\pi$ decays and many more 3 body decays for monitoring and calibration purposes.

Most of the background from K_L decays is identified with a high resolution spectrometer and a precision electromagnetic calorimeter in both experiments.

The comparison of charged and neutral decay rates at a precision below 10^{-3} requires a precise matching of the energy scales between both decay modes. This is the most challenging constraint. It lead to the construction of high resolution, linear, fine grain calorimeters. The absolute energy scale for $\pi^0\pi^0$ decays is fixed by the reconstruction of the leading edge of the vertex distribution of $K_S \to \pi^0\pi^0$ decays, which is precisely defined by the position of a γ converter followed by a scintillating counter which defines the beginning of the fiducial decay region.

Because of their very different lifetimes, the vertex distribution of K_L and K_S decays are not similar. To account for the resulting detector acceptance bias, the E832 collaboration uses a very detailed Montecarlo simulation of their setup. The quality of the simulation is checked by producing pseudo experimental Ke3 and $K_L \to 3\pi^0$ decays and comparing the energy and vertex distributions to those of corresponding very large data samples. NA48 uses a different technique : they weight K_L decay events so that their longitudinal vertex distribution is similar to that of K_S decays, reducing to less than 3×10^{-3} the acceptance correction, due to a simple geometrical effect induced by the relative positions and divergences of the K_L and K_S beams.

In 1999, both groups published a result with their first recorded data :

E832 [5] : $\epsilon'/\epsilon = (28.0 \pm 4.1) \times 10^{-4}$

NA48 [6] : $\epsilon'/\epsilon = (18.5 \pm 7.3) \times 10^{-4}$

establishing direct CP violation. The significance of these results is however impaired by the marginal agreement of the various measurements. The world average reads $(21.2 \pm 2.8) \times 10^{-4}$ with a χ^2 of 8.4 for three degrees of freedom. This corresponds to a 4 % confidence level.

In february 2000, NA48 announced a new preliminary result [7] with data taken in 1998 : $\epsilon'/\epsilon = (12.2 \pm 4.9) \times 10^{-4}$, leading to the combined NA48 $\epsilon'/\epsilon = (14.0 \pm 4.3) \times 10^{-4}$ and the new world average of $\epsilon'/\epsilon = (19.2 \pm 2.5) \times 10^{-4}$ and $\chi^2/dof = 10.4/3$ (1.5 % CL). Figure 1 shows the spread of those recent measurements.

E832 did not present a new value, but have measured the charged ratio $\Gamma(K_L \to \pi^+\pi^-)/\Gamma(K_S \to \pi^+\pi^-)$ to be consistent with their published measurement. They have improved the detector simulation, and this reduces significantly the systematic uncertainty in the acceptance correction in the charged mode, where a slight discrepancy between data and Montecarlo had been found in the vertex distribution of $K_L \to \pi^+\pi^-$ events.

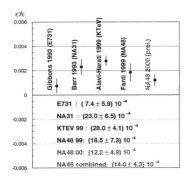

Figure 1. Recent ϵ'/ϵ results.

From data to be analysed, E832 and NA48 will be able to significantly improve the precision on ϵ'/ϵ. One can hope that this will clarify the present situation. The KLOE experiment at Frascati has started taking data in 1999. $\Phi \to K_L K_S$ decays at rest are used to measure ϵ'/ϵ and most of the other phenomenological parameters of the K^0 $\overline{K^0}$ system. Given the present luminosity of DAΦNE of about 10^{31} cm^{2}s^{-1}, their objective is to collect the equivalent of 1 pb^{-1} in a year and reach 10^{-3} precision on ϵ'/ϵ.

One can notice that the observed direct CP violation in neutral K to 2 π decays is somewhat large. Non Standard Model contributions might be present, and it is important to try to search for such new effects. The NA48 collaboration proposes [8] to search for a difference in the Dalitz plot slope parameters of $K^+ \to \pi^+\pi^+\pi^-$ and $K^- \to \pi^-\pi^-\pi^+$ decays with simultaneous K^+ and K^- beams. Many other measurements can be made with these beams, in particular the study of Ke4 decays.

3 CP violation and related K decays

3.1 $K \to \pi^+\pi^- e^+ e^-$

Like the $K \to \pi\pi\gamma$ decay, $K \to \pi^+\pi^- e^+ e^-$ can proceed through the direct emission of a virtual γ or the radiative emission by one pion

in the final state of the $K \to \pi^+\pi^-$ decay.

For K_L, the interference of those amplitudes leads to a CP violating polarization of the γ^* and to a T-odd term in the differential decay rate :

$$d\Gamma/d\Phi = \Gamma_1 \cos^2\Phi + \Gamma_2 \sin^2\Phi + \mathbf{\Gamma_3 \sin 2\Phi}$$

where Φ in the angle between the $\pi\pi$ and ee decay planes. The asymmetry parameter $\mathcal{A}$ is defined by :

$$\mathcal{A} = \frac{N(\sin 2\Phi > 0) - N(\sin 2\Phi < 0)}{N(\sin 2\Phi > 0) + N(\sin 2\Phi < 0)}$$

It has been measured by the KTEV collaboration [9] : $\mathcal{A} = (13.6 \pm 2.5 \pm 1.2)\%$, in full agreement with the expectation of 14.4 % from CP violation in $K^0 \overline{K^0}$ mixing [10].

For K_S, the decay is fully dominated by the CP conserving radiative 2 π decay, and no significant asymmetry is expected. The first observation of the $K_S \to \pi^+\pi^- e^+ e^-$ decay has been made by NA48 [11] : BR($K_S \to \pi^+\pi^- e^+ e^-$) = $(5.1 \pm .9 \pm .3) \times 10^{-5}$ (prelim.). Figure 2 shows the signal seen in the 1998 and 1999 data. The observed asymmetry is consistent with 0 within a few % .

3.2 $K_L \to \pi^0 e^+ e^-$

Direct CP violation can be searched for in $K_L \to \pi^0 e^+ e^-$ decays. However 3 contributions have to be disentangled. The CP conserving one from the $\pi^0 \gamma\gamma$ intermediate state, the

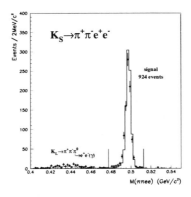

Figure 2. The observation of $K_S \to \pi^+\pi^- e^+ e^-$ by NA48.

indirect CP violating one from the K_1 component of the K_L state, and the direct CP violating part.

The study of $K_L \to \pi^0 \gamma\gamma$ decay is used to estimate the first part. KTEV has published [12] : $BR(K_L \to \pi^0 \gamma\gamma) = (1.68 \pm .07 \pm .08) \times 10^{-6}$, and $a_V = -.72 \pm .05 \pm .06$, where a_V is the effective vector coupling not accounted for in Chiral Perturbation Theory (χP_T) [13]. The new NA48 preliminary result is [14] :

$$BR(K_L \to \pi^0 \gamma\gamma) = (1.51 \pm .05 \pm .20) \times 10^{-6}.$$

The indirect CP violating contribution is estimated by using the measured $K^+ \to \pi^+ e^+ e^-$ BR and using isospin invariance, but this procedure has large theoretical uncertainties. An improved limit on, or a measurement of, $BR(K_S \to \pi^0 e^+ e^-)$, would fix this part. NA48 plans to search for this decay with a 6×10^{-10} SES per year in a dedicated run with a high intensity K_S beam. Such a run would allow in addition a measurement of η^{000} with $\mathcal{O}(10^{-2})$ precision [15].

During a 2 days test run in 99, NA48 already performed a search for $K_S \to \pi^0 e^+ e^-$. They reported [16] a 90 % CL limit on the BR of 1.6×10^{-7}, a factor of 10 improvement over the previous limit. The measurement of $BR(K_L \to \pi^0 e^+ e^-)$ is limited by the background from $K_L \to e^+ e^- \gamma\gamma$. This decay is well measured by KTEV ($BR(K_L \to e^+ e^- \gamma\gamma$, $E_\gamma^* > 5$ MeV$) = (6.31 \pm .14 \pm .43) \times 10^{-7}$ (prelim.) [17]) and NA48 ($BR(K_L \to e^+ e^- \gamma\gamma$, $E_\gamma^* > 5$ MeV$) = (6.32 \pm .31 \pm .20 \pm .29(normalization)) \times 10^{-7}$ (prelim.) [18]), in good agreement with the Standard Model prediction. The important variables to discriminate $\pi^0 e^+ e^-$ and $e^+ e^- \gamma\gamma$ decays are the angle between the γ's in the π^0 rest frame and the smallest angle between one γ and one electron, which should be smaller for the $e^+ e^- \gamma\gamma$ decay. In the 1997 data, KTEV finds 2 events for 1.1 expected background, leading to the 90 % CL limit [19] :

$$BR(K_L \to \pi^0 e^+ e^-) < 5.1 \times 10^{-10}.$$

Plots in figure 3 show $m_{\gamma\gamma}$ versus

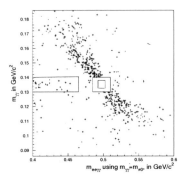

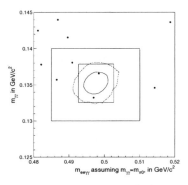

Figure 3. KTEV search for $K_L \to \pi^0 e^+ e^-$ (see text).

$m_{ee\gamma\gamma}$ after all selection criteria except the phase space fiducial cut to suppress the $K_L \to e^+ e^- \gamma\gamma$ background. Events in the central boxes are only plotted in the zoom around the signal region in the bottom plot.

3.3 $K_L \to \pi^0 \mu^+ \mu^-$

This decay is similar to the previous one, with an available phase space about a third, but better experimental acceptance and different backgrounds (π decay or punch through, smaller $\mu^+ \mu^- \gamma\gamma$ BR).

KTEV finds 4 $K_L \to \mu^+ \mu^- \gamma\gamma$ events in the 1997 data, with .1 expected background, and gets [20] : $BR(K_L \to \mu^+ \mu^- \gamma\gamma) = (10.4 \, ^{+7.5}_{-4.5} \pm 1) \times 10^{-9}$

This is the first measurement of this branching ratio.

They found 2 candidates for $K_L \to \pi^0 \mu^+ \mu^-$ with $.87 \pm .15$ expected background, and published [21] :

$BR(K_L \rightarrow \pi^0 \mu^+ \mu^-) < 3.8 \times 10^{-10}$ (90 % CL)

3.4 $K \rightarrow \gamma^{(*)} \gamma^{(*)}$

These decays are used to test various theoretical models and to estimate the Long Distance contributions to the K_L to ee and $\mu\mu$ decays. The short distance contribution to these two FCNC decays is proportional to $A^4(1-\rho)$ (A and ρ are two parameters of the CKM matrix in the Wolfenstein parametrization [22]). An improved knowledge of the long distance contributions would allow a better interpretation of the BNL-AGS-E871 results [23] :
$BR(K_L \rightarrow \mu\mu) = (7.18 \pm .17) \times 10^{-9}$
($\sim$ 6200 events)
$BR(K_L \rightarrow ee) = (8.7 ^{+5.7}_{-4.1}) \times 10^{-12}$
(4 signal events for .17 $\pm$ .10 expected background)
This is the smallest branching ratio ever measured.

3.5 $K \rightarrow \gamma^* \gamma^*$

Three leptonic decays proceed through the $\gamma^* \gamma^*$ intermediate state : $K_L \rightarrow 4e$, $K_L \rightarrow e^+ e^- \mu^+ \mu^-$, $K_L \rightarrow 4\mu$. χP_T predicts $BR(K_L \rightarrow 4e) = 3.9 \times 10^{-8}$. Both KTEV and NA48 have consistent (preliminary) results :
KTEV [24] : $(3.77 \pm .18 \pm .27) \times 10^{-8}$
(436 events)
NA48 [25] : $(3.67 \pm .32 \pm .23 \pm .08) \times 10^{-8}$

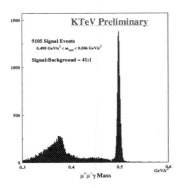

Figure 4. KTEV : invariant mass of $K_L \rightarrow \mu^+ \mu^- \gamma$ candidates.

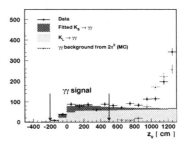

Figure 5. NA48 : longitudinal vertex distribution of $K \rightarrow \gamma\gamma$ candidates.

(132 events)

The last uncertainty for NA48 comes from the branching ratio of the normalization channel ($K_L \rightarrow \pi^+ \pi^- \pi^0_D$).

From KTEV, $BR(K_L \rightarrow e^+ e^- \mu^+ \mu^-) = (2.50 \pm .41 \pm .15) \times 10^{-9}$ (prelim.) [26] with 38 events and .2 background from $\mu^+ \mu^- \gamma$ (converted), $\pi^+ \pi^- \pi^0_D$, $\pi^+ \pi^- e^+ e^-$. NA48 found 21 candidate events, but did not give a branching ratio.

3.6 $K_L \rightarrow \gamma\gamma^* \rightarrow \mu^+ \mu^- \gamma$, $e^+ e^- \gamma$

The $K_L \rightarrow \gamma\gamma^*$ vertex is parametrized in various models. In the BMS model [27], the amplitude is the sum of a pseudoscalar pseudoscalar transition and a vector vector transition, with a relative strength parametrized by α_{K^*}. In χP_T, one parameter α is introduced to account for a symmetry breaking effect at a scale Λ. In the phenomenological description of D'Ambrosio et al. [28], two parameters α and β are introduced to describe the $K_L \rightarrow \gamma^* \gamma^*$ form factor. The discussion of these models is outside the scope of this report. The experimental data available are the $K_L \rightarrow \mu^+ \mu^- \gamma$ and the $K_L \rightarrow e^+ e^- \gamma$ branching ratio and the invariant mass distribution of the lepton pair. KTEV has presented [29] :
$BR(K_L \rightarrow \mu^+ \mu^- \gamma) = (3.70 \pm .04 \pm .07) \times 10^{-7}$
with more than 9000 events (see figure 4). With this value and the dimuon mass distribution, they get [29] $\alpha_{K^*} = -.163 ^{+.026}_{-.027}$, in rough agreement with previous measure-

52

ments, although somewhat lower than those from $K_L \to \mu^+ \mu^- \gamma$. This might indicate that the BMS model is not adequate to describe the observed spectra.

3.7 $K_S \to \gamma\gamma$

One of the most important test of χP_T is the measurement of the $K_S \to \gamma\gamma$ branching ratio. Indeed, this decay is completely dominated by long distance contributions. The χP_T prediction is $(2.4 \pm .2) \times 10^{-6}$. The previous measurement was made by NA31 [30] : $(2.4 \pm .9) \times 10^{-6}$. NA48 improved this result (see figure 5) to $(2.4 \pm .4 \pm .2) \times 10^{-6}$ (prelim.) [31] using data from the short 99 test run in a high intensity K_S beam.

3.8 Summary

A large amount of new data on K_L and K_S decays have appeared in the past two years. They are summarized in table 1. Limits on new physics phenomena are pushed and χP_T is confirmed as an excellent model.

In the near future, KTEV and NA48 will keep on providing improved results, since both experiments have on tape data corresponding to about twice the sensitivity for most channels. On a longer term, KLOE and the proposed NA48 extensions (if accepted) will provide interesting measurements on K_S and charged K decays.

4 $K \to \pi\nu\bar\nu$

When the $K_L \to \pi^0\nu\bar\nu$ decay appeared in the discussions many years ago, it looked like a "Graal" to experimentalists.

This is a CP violating decay, the interest of which lies in the fact that theoretical predictions are very clean. Indeed, the relevant hadronic matrix elements can be extracted by using the measured $K^+ \to \pi^0 e^+ \nu$ branching ratio. This leads to a few % uncertainty on $BR(K_L \to \pi^0\nu\bar\nu)$, which reads [32] :
$$3 \; 10^{-11} \; (\eta/.39)^2 \; (m_t/170\text{GeV})^{2.3} \; (|V_{cb}|/.04)^4$$

A measurement of this BR would allow a clean determination of the CP violating Wolfenstein parameter η of the CKM matrix. The present 90 % CL limit is $5.9 \; 10^{-7}$ [33], many orders of magnitude above the expectation.

The KEK-PS-E391a experiment is in preparation with the goal to achieve 10^{-10} single event sensitivity (SES). It is meant as a test experiment before launching a more ambitious program at the planned 50 GeV PS at KEK. The KOPIO project at BNL has been recommended. Their objective is to detect about 50 events for a BR of 3×10^{-11}. The KAMI collaboration is preparing a proposal with the same kind of sensitivity. The techniques are very different : full reconstruction of low energy K decays for KOPIO, measurement of the missing transverse momentum of high energy K's from a carefully designed pencil beam and with a hermetic apparatus for KAMI. The efficiency to detect low energy γ's is one of the critical parameters.

The $K^+ \to \pi^+\nu\bar\nu$ decay has somewhat larger theoretical uncertainties than the previous one, but they remain below 10 % [32]. The BNL E787 experiment has detected one event with negligible background in their 1995 data. They have found no other candidate in their 96 and 97 samples, leading to the updated measurement [34] :

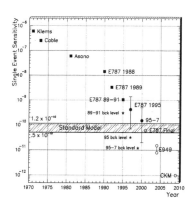

Figure 6. History of $K_L \to \pi^+\nu\bar\nu$ search.

Table 1. Rare K_L and K_S decays results summary.

K_L decay mode	BR or 90 % CL	Experiment	PDG98 [39]
$\pi^+\pi^-e^+e^-$	$(3.63 \pm .18) \times 10^{-7}$	FNAL E799	$< 4.6 \times 10^{-7}$
$\pi^0\gamma\gamma$	$(1.68 \pm .11) \times 10^{-6}$	FNAL E799	$(1.70 \pm .28) \times 10^{-6}$
	$(1.51 \pm .21) \times 10^{-6}$	CERN NA48	
$e^+e^-\gamma\gamma$	$(6.31 \pm .44) \times 10^{-7}$	FNAL E799	$(6.5 \pm 1.2) \times 10^{-7}$
	$(6.32 \pm .47) \times 10^{-7}$	CERN NA48	
$\pi^0e^+e^-$	$< 5.1 \times 10^{-10}$	FNAL E799	$< 4.3 \times 10^{-9}$
$\mu^+\mu^-\gamma\gamma$	$(10.4^{+7.5}_{-4.5} \pm .1) \times 10^{-9}$	FNAL E799	$-$
$\pi^0\mu^+\mu^-$	$< 3.8 \times 10^{-10}$	FNAL E799	$< 5.1 \times 10^{-9}$
$\mu^+\mu^-\gamma$	$(3.70 \pm .08) \times 10^{-7}$	FNAL E799	$(3.25 \pm .28) \times 10^{-7}$
$e^+e^-\gamma$	$(1.06 \pm .05) \times 10^{-5}$	CERN NA48	$(.91 \pm .05) \times 10^{-5}$
$e^+e^-e^+e^-$	$(3.77 \pm .32) \times 10^{-8}$	FNAL E799	$(4.1 \pm .8) \times 10^{-8}$
	$(3.67 \pm .40) \times 10^{-8}$	CERN NA48	
$e^+e^-\mu^+\mu^-$	$(2.50 \pm .44) \times 10^{-9}$	FNAL E799	$(2.9^{+6.7}_{-2.4}) \times 10^{-9}$
$\mu^+\mu^-$	$(7.18 \pm .17) \times 10^{-9}$	BNL E871	$(7.2 \pm .5) \times 10^{-9}$
e^+e^-	$(8.7^{+5.7}_{-4.1}) \times 10^{-12}$	BNL E871	$< 4.1 \times 10^{-11}$
K_S decay mode			
$\pi^+\pi^-e^+e^-$	$(5.1 \pm .9) \times 10^{-5}$	CERN NA48	$-$
$\pi^0e^+e^-$	$< 1.6 \times 10^{-7}$	CERN NA48	$< 1.1 \times 10^{-6}$
$\gamma\gamma$	$(2.6 \pm .4) \times 10^{-6}$	CERN NA48	$(2.4 \pm .9) \times 10^{-6}$

$$BR(K^+ \to \pi^+\nu\bar{\nu}) = (1.5^{+3.4}_{-1.2}) \times 10^{-10}$$

This measurement gives a constraint in the $(\bar{\rho}, \bar{\eta})$ plane. If approved, two projects should bring more information in the future : BNL-E949, which aims at 10^{-10} SES, and CKM at FNAL, with 10^{-12} SES objective. Figure 6 shows the historical progress of this search.

To conclude, the $K \to \pi\nu\bar{\nu}$ decays can provide information complementary to the study of B decays, the value of which is at the same level because of the cleanliness of the interpretation of the measurements. It is to be hoped that the experimental effort will match the scientific interest.

5 Non Standard Model T violation

It is generally believed that CP violation seen in the K system and interpreted as the presence of an irreducible complex parameter in the CKM matrix is not sufficient to account for the observed matter-antimatter asymmetry in the universe. This is a strong motivation to search for violation of the discrete symmetries in phenomena where it is not expected.

5.1 T violation in $K_{\mu3}$ decays

The transverse polarization P_T of the μ from the $K^+ \to \pi^0\mu^+\nu_\mu$ decay, defined as the component of the μ polarization normal to the π μ plane, is expected to be essentially 0 in the SM, final state interactions being negligible.

With their 1997 data, KEK-PS-E246 has recently published [35]:
$$P_T = .0042 \pm .0049 \pm .0009$$

The precision is already 30 % better than the previous measurement. They should present soon another measurement with their 1998 data, and a new run will start end of year 2000. The final statistical uncertainty should be more than a factor of 2 smaller.

The present beam intensity is 3×10^5 K$^+$ delivered during a .6 s spill every 3 s. They detect K decays at rest, and the μ transverse polarization is measured by detecting the direction of the e$^+$ from the μ decay. The apparatus has a 12-rotational symmetry. The analysis technique allows a strong cancellation of most systematic uncertainties. They measure the following asymmetry :

$$A_T \equiv \left(\frac{(N_{cw}/N_{ccw})_{fwd}}{(N_{cw}/N_{ccw})_{bwd}} - 1 \right)/4$$

where :

$N_{(ccw)cw}$ is the number of electrons emitted in the (counter)clockwise direction

$(N_{cw}/N_{ccw})_{fwd(bwd)}$ is the ratio of these numbers for forward (backward) emitted π^0's.

Some extensions of the SM predict a value of P$_T$ only one order of magnitude below the attained precision. This motivates the BNL-E923 experiment, under construction, which will use an intense 2 GeV K$^+$ beam (2×10^7 K$^+$ every 3.6 s).

5.2 $\mu^+ \rightarrow e^+ \nu_e \bar{\nu}_\mu$

Muons can be produced copiously and hence allow precision tests of the SM and the search for new physics. Two experiments at the Paul Scherrer Institute (R-94-10 and R-97-06) are dedicated to the measurement of the polarization of the e$^+$ from the $\mu^+ \rightarrow e^+ \nu_e \bar{\nu}_\mu$ decay.

The R-94-10 experiment (μP$_T$) measures the transverse polarization of the e$^+$ from stopped polarized muons : a non zero component orthogonal to the direction of the μ polarization (P$_{T2}$) would be a non ambiguous sign of T violation.

The μ spin is rotated at a frequency $\omega/2\pi$, the e$^+$ are annihilated in a magnetized foil and the γ's are detected in a BGO crystal calorimeter. The distribution of the $\gamma\gamma$ plane angle ψ with respect to the plane defined by the muon polarization direction and the e$^+$ momentum direction depends on the e$^+$ polarization. An oscillatory dependance of the spectrum at the same frequency for a fixed value of ψ is the signal which is looked for. From a first test run, they have measured $< $ P$_{T2}$ $>= .009 \pm .022$, a precision equivalent to the previous measurement.

R-97-06 aims at measuring the longitudinal polarization of the e$^+$ and will look for non SM Michel parameters.

Another experiment at TRIUMF, E614, is in preparation. The objective is to record 10^9 μ decays and measure all the Michel parameters with precisions three to ten times better than the present ones. More details on the physics of muon decays and on the experimental status can be found in the recent review of Y. Kuno and Y. Okada [36].

6   Lepton Flavor Violation

In the SM with massless neutrinos, Lepton Flavor Violation (LFV) is forbidden. It is allowed in the simplest extension with massive neutrinos, but, even with maximal mixing, it would not be measurable for masses of order the eV/c^2. LFV is thus a good place to search for physics beyond the SM at scales well above the TeV.

6.1   LFV in K decays

One should first recall here the limit set by BNL-AGS-E871 [37] :
BR(K$_L \rightarrow \mu$ e) $< 4.7 \times 10^{-12}$ (90 % CL).

FNAL-E799 just obtained two new limits (90 % CL) :
BR(K$_L \rightarrow \pi^0 \mu$ e) $< 3.1 \times 10^{-9}$
BR(K$_L \rightarrow e^\pm$ e$^\pm$ $\mu^\mp$ $\mu^\mp$) $< 1.36 \times 10^{-10}$.

BNL-AGS-E865 has many new results from their 1996 data. This experiment is dedicated to the search for LFV in K$^+$ $\rightarrow \pi^+ \mu^+ e^-$ decay. A spectrometer with 16 planes of proportional chambers provides the measurement of all charged particle momenta. Particle identification, which is a critical issue in this kind of experiment, relies on Cerenkov counters, a 15 r.l. deep "Shash-lik" calorimeter, and a 24 planes proportional tube - iron plate range stack. The beam intensity was 10^{13} protons on target, producing 10^8 K^+ per 1.6 s pulse. This corresponds to 2.2×10^{10} K $\rightarrow 3$ π decays seen in the detector. A likelihood analysis is used to evaluate the probability of various hypotheses for selected events, using experimental distributions of the relevant variables to produce the probability density functions. The background comes from 3 π decays (the normalizing channel), Dalitz decays and accidentals. It is expected to be 2.6 ± 1.0. Three events survive all cuts. This translates into BR($K^+ \rightarrow \pi^+ \mu^+ e^-$) $< 3.9 \times 10^{-11}$ (90 % CL). When combined with the 1995 data and the older E777 limit, this leads to : BR($K^+ \rightarrow \pi^+ \mu^+ e^-$) $< 2.8 \times 10^{-11}$ (90 % CL).

The E865 experimental setup allows the search for many other decays with LFV ($K^+ \rightarrow \pi^+ e^+ \mu^-$, $\pi^0 \rightarrow e^+ \mu^-$), total lepton number violation ($\pi\mu\mu$ and πee decays), and effects of Majorana ν's in the second generation ($\mu^+ \mu^+ \pi^-$ decay). Table 2 summarizes these results [38].

6.2 LFV in μ decays

The best limit on the most simple $\mu \rightarrow e\gamma$ decay has been obtained by the LANL MEGA collaboration [40] : BR $< 1.2 \times 10^{-11}$ (90 % CL). The total flux of stopped muons was 1.2×10^{14}. The limiting factor of the experiment was the instantaneous muon intensity of 2.5×10^8 s^{-1} with 6% duty factor.

A new proposal at PSI (R99-05) has been approved, aiming at a 10^{-14} SES. Data taking should start in 2003. The main improvements are :

- A 100 % duty factor continuous beam, leading to same instantaneous beam intensity as in MEGA for a factor 16 increase in the total number of available muons.

- A liquid xenon calorimeter.

- A constant bending radius spectrometer : a superconducting coil is arranged to produce a graded magnetic field in which the e^+ from the $\mu \rightarrow e\gamma$ decay follows a trajectory with a radius independant of the emission angle. This allows to only install drift chambers in the outer part of the solenoïd volume, reducing by several orders of magnitude the rate of Michel positrons in the detector (see figure 7).

Muons also allow to search for LFV in looking at $\mu^- N \rightarrow e^- N$ conversion of a muonic atom. The BNL MECO experiment aims at a 10^{-16} sensitivity in the process $\mu^- + Al \rightarrow e^- + Al$. The present best limit is BR($\mu^- + Ti \rightarrow e^- + Ti$) $< 6.1 \times 10^{-13}$ (90 % CL). If the $e\gamma$ form factors relevant to the $\mu \rightarrow e\gamma$ process are also dominant in the muonic atoms conversion, the relation BR($\mu^- N \rightarrow e^- N$) $= (B(A,Z)/400) \times$ BR($\mu^- \rightarrow e^- \gamma$) can be established, where B(A,Z) is a correction factor of order unity (1.2 for Al, 1.8 for Ti). It is convenient to compare the sensitivity of various experiments. No new result has been presented on these processes.

Perspectives for muon physics are good since several R&D programs dedicated to the development of very high intensity proton sources have been launched worldwide. They can be used to produce intense muon beams, with a large number of applications, including a neutrino factory or a muon collider in the far future.

7 Final remarks

Direct CP violation is now established by many precise measurements, although with a large spread of results. Those have stimulated a lot of theoretical efforts, but lattice QCD, the cleanest method, in principle, cannot yet provide a solid prediction for ϵ'/ϵ. KTEV and NA48 have still a lot of data to analyze, which hopefully will clarify soon the

Table 2. Recent E865 results : BR's or 90 % CL limits.

Decay mode	E865	PDG98 [39]
$K^+ \to \pi^+ ee$	$(2.94 \pm .14) \times 10^{-7}$	$(2.7 \pm .2) \times 10^{-7}$
$K^+ \to \pi^+ \mu\mu$	$(9.2 \pm .6) \times 10^{-8}$	$(5 \pm 1) \times 10^{-8}$
$K^+ \to \pi^+ \mu^+ e^-$	$< 2.8 \times 10^{-11}$	$< 2 \times 10^{-10}$
$\pi^0 \to \mu^+ e^-$	$< 3.8 \times 10^{-10}$	$< 1.6 \times 10^{-8}$
$K^+ \to \mu^+ \mu^+ \pi^-$	$< 3 \times 10^{-9}$	$< 1.5 \times 10^{-4}$
$K^+ \to e^+ e^+ \pi^-$	$< 6.3 \times 10^{-10}$	$< 1 \times 10^{-8}$
$K^+ \to \pi^+ e^+ \mu^-$	$< 5.1 \times 10^{-10}$	$< 7 \times 10^{-9}$
$K^+ \to \mu^+ e^+ \pi^-$	$< 4.9 \times 10^{-10}$	$< 7 \times 10^{-9}$

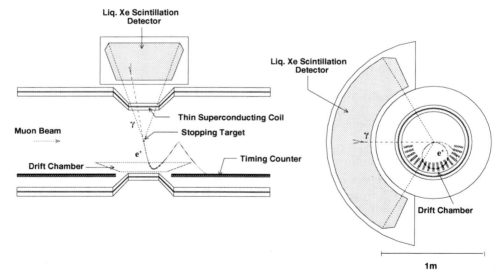

Figure 7. PSI R99-05 experimental setup.

experimental situation.

On the rare K_L decays side, most branching ratios and limits have been updated (KTEV, NA48, BNL E871), bringing no surprise but rather confirming models predictions. NA48 has started a K_S program, which still needs approval to be continued.

KLOE has started taking data and should provide complementary measurements in a near future.

Many projects are in preparation to pursue the $K \to \pi\nu\bar{\nu}$ search beyond the BNL E787 and KTEV sensitivities : BNL E926 (KOPIO), BNL E949, FNAL CKM, FNAL KAMI, KEK E391a. They should provide clean independant constraints on the CKM matrix and complement measurements from the B sector.

The search for CP or T violation outside the Standard Model is being pushed with K^+ (CERN NA48 proposal, KEK E246, BNL E923 project) and μ^+ decays (PSI μP_T, PSI μP_L, TRIUMF E614).

Many new limits on lepton flavor violation have been given (BNL E865, BNL E871, FNAL KTEV, LANL MEGA) and several projects are being launched (PSI R99-05, BNL MECO). They should reach sensitivities allowing, for example, the test of some supersymmetric extensions of the Standard Model.

Progresses in light flavor physics are steady. How much time will the Standard Model still resist ?

References

1. N. Cabbibo, *Phys. Rev. Lett.* **10**, 531 (1963)
   M. Kobayashi and T. Maskawa, *Progr. Theor. Phys.* **49**, 652 (1973)
2. M. Ciuchini et al., *TUM-HEP-376-00, RM3-TH-00-10* (2000)
   S. Bosch et al., *Nucl. Phys.* **B565**, 3 (2000)
   S. Bertolini et al., *hep-ph/0002234* (2000)

T. Hambye et al., *Nucl. Phys.* **B564**, 391 (2000)
3. G.D. Barr et al., *Phys. Lett.* **B317**, 233 (1993)
4. L.K. Gibbons et al., *Phys. Rev. Lett.* **70**, 1203 (1993)
5. A. Alavi-Harati et al., *Phys. Rev. Lett.* **83**, 22 (1999)
6. V. Fanti et al., *Phys. Lett.* **B465**, 335 (1999)
7. A. Ceccucci, CERN seminar, February 29th 2000
8. R. Batley et al., **CERN/SPSC 2000-003**
9. A. Alavi-Harati et al., *Phys. Rev. Lett.* **84**, 408 (2000)
10. L.M. Sehgal and M. Wanninger, *Phys.Rev.* **D46**, 1035 (1992), *ibid.* **D46**, 5209(E) (1992)
11. V. Kekelidze, *these proceedings*
12. A. Alavi-Harati et al., *Phys. Rev. Lett.* **83**, 917 (1999)
13. P. Heiliger and L.M. Sehgal, *Phys. Rev.* **D47**, 4920 (1993)
14. V. Kekelidze, *these proceedings*
15. R. Batley et al., **CERN/SPSC 2000-002**
16. V. Kekelidze, *these proceedings*
17. T. Yamanaka, *XXXIVth Rencontres de Moriond*, March 13-20, 1999
18. V. Kekelidze, *these proceedings*
19. A. Alavi-Harati et al., **Fermilab-Pub-00/225-E** (2000), submitted to Phys. Rev. Lett.
20. A. Alavi-Harati et al., **Fermilab-Pub-00/023-E** (2000), submitted to Phys. Rev. Lett.
21. A. Alavi-Harati et al., *Phys. Rev. Lett.* **84**, 5279 (2000)
22. L. Wolfenstein, *Phys. Rev. Lett.* **51**, 1945 (1983)
23. D. Ambrose et al., *Phys. Rev. Lett* **81**, 4309 (1998)
    D. Ambrose et al., *Phys. Rev. Lett* **84**, 1389 (2000)
24. Yau W. Wah, *these proceedings*

25. V. Kekelidze, *these proceedings*

26. Yau W. Wah, *these proceedings*

27. L. Bergström, E. Massò, P. Singer, *Phys. Lett.* **B131**, 229 (1983)
    L. Bergström, E. Massò, P. Singer, *Phys. Lett.* **B249**, 141 (1990)

28. D'Ambrosio, G. Isidori, J. Portoles, *Phys. Lett* **B423**, 385 (1998)

29. Yau W. Wah, *these proceedings*

30. G.D. Barr et al., *Phys. Lett.* **B351**, 579 (1995)

31. V. Kekelidze, *these proceedings*

32. A.J. Buras, R. Fleisher, **TUM-HEP-275-97** (1997)

33. A. Alavi-Harati et al., *Phys. Rev.* **D61** 072006 (2000)

34. S. Adler et al., **TRI-PP-00-04**, *to be published in Phys. Rev. Lett.*

35. M. Abe et al., *Phys. Rev. Lett.* **83**, 4253 (1999)

36. Y. Kuno and Y. Okada, **KEK-TH-639**, *submitted to Rev. Mod. Phys.*

37. D. Ambrose et al., *Phys. Rev. Lett.* **81**, 5734 (1998)

38. M. Zeller, *private communication*

39. C. Caso et al., *Eur. Phys. J.* **C3**, 1 (1998)

40. M.L. Brooks et al., *Phys. Rev. Lett.* **83**, 1521 (1999)

Plenary Session 3

Heavy Flavor Physics
(charm, bottom, top and τ)

Chair: Vladimir G. Kadyshevsky (JINR)
Scientific Secretary: Mikihiko Nakao

HEAVY FLAVOUR PHYSICS

A. GOLUTVIN

ITEP, B. Cheremushkinskaya 25
Moscow 07661, Russia
E-mail: golutvin@iris1.itep.ru

This paper reviews the status of our knowledge of the (CKM) matrix elements V_{cb}, V_{ub}, V_{td} and V_{ts}. The recent results on the charm and bottom lifetimes and searches for the $D^0\bar{D}^0$ mixing are also included.

1 Introduction

The experimental highlight of the heavy flavour physics this year is the new results of the *BaBar* and *BELLE* experiments. At this conference we celebrate their successful startup. In particular, the measurements of the sin 2β $C\mathcal{P}$-violating parameter have been presented in great details[1,2].

The *beauty* part of this review will focus mainly on the present knowledge of the *Cabibbo − Kobayashi − Maskawa (CKM)* matrix elements V_{cb}, V_{ub}, V_{td} and V_{ts}. Precise evaluation of these elements is crucial to provide redundant tests of the validity of the Standard Model (SM).

The *charm* part is not intended to review exhaustively the status of charm physics but limited to a few highlights presented at this conference.

Due to limited space the results on the τ and *top*-physics are ignored. The reviews of these parts of heavy flavour physics can be found in [3,4]. The recent measurements were presented in the parallel sessions and contributed papers. Of note is the first direct evidence for the tau neutrino [5] that completes the table of the fundamental fermions.

2 Charm Physics

After many years of experimental studies charm physics remains a very exciting field, in particular because of strong motivation to search for new phenomena in the processes dominated by loop contributions. The sensitivity is greatly increased since the SM loop diagrams vanish in the charm sector. Additionally charm physics is an important framework for beauty physics in our understanding of the SM. Some measurements in *b*-physics are still limited by systematic errors associated with an imperfect knowledge of the branching ratios of the charmed particles.

This brief review will focus only on two experimental aspects of charm physics:

- Measurements of the charmed particle lifetimes which help to understand the contribution of non-spectator effects involved in charm decays.

- Searches for $D^0\bar{D}^0$ mixing.

2.1 Lifetimes of charm particles

Since the 1998 *PDG* review[6] new measurements of the charm lifetimes had been made by $E791$[7], $CLEO$[8], $SELEX$[9], $FOCUS$[10] and $BELLE$[11]. Figure 1 shows averaged lifetime ratios[a] with respect to the lifetime of the D^0 meson. The results of the $FOCUS$ experiment, which achieved by far the best statistical accuracy, are shown separately since the analysis of the systematic uncertainties of this experiment is still underway.

The ratio $\tau(D^+)/\tau(D^0) = 2.54 \pm 0.03$ can serve as a tool to estimate the *Pauli*-type interference, which could extend the life-

---

[a]The *FOCUS* mesurements were not included into the averages.

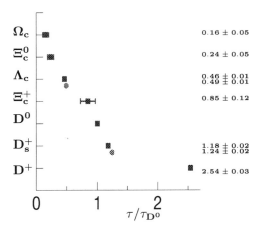

Figure 1. Lifetimes of charm particles normalised to the D^0 lifetime. Resulst of *FOCUS* experiment are shown by circles.

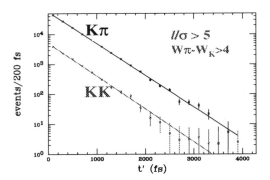

Figure 2. Decay time distributions for $D^0 \rightarrow K^- K^+$ and $D^0 \rightarrow K^- \pi^+$ decays measured by *FOCUS*.

time of the D^+. This explanation is further validated by the relative enhancement of the Double Cabibbo Suppressed Decays (DCSD) of D^+, which are not suppressed by the *Pauli* principle. Another handle of such effects is provided by the relatively small lifetimes of charmed baryons, in particular by the Ω_c with two identical strange quarks in the initial state. Besides that, the W exchange for baryons in neither color nor helicity suppressed.

The ratio $\tau(D_s^+)/\tau(D^0) = 1.18 \pm 0.02$ estimates a contribution from the weak annihilation diagram which is not Cabibbo suppressed for the case of charm and anti-strange quarks in D_s^+ decays. The present value of the D_s^+ lifetime is about 20% higher than that of the D^0 lifetime. The *FOCUS* measurement, 1.24 ± 0.02 of comparable to world average precision gives an even larger difference.

2.2 $D^0 \bar{D}^0$ mixing

In the Standard model the $D^0 \bar{D}^0$ mixing is dominated by the loop diagrams with an internal strange quark and therefore highly suppressed. The short distance and other

known contributions to the mixing parameter $r_m = (x^2 + y^2)/2$, where $x = \frac{\Delta M}{\Gamma}$ and $y = \frac{\Delta \Gamma}{2\Gamma}$, are expected to be very small, much below the experimental sensitivity. Therefore any observation of a sizeable $D^0 \bar{D}^0$ mixing would indicate a contribution from new physics.

At this conference the *FOCUS*[12] experiment reported the results of their measurement of the lifetimes of $\mathcal{CP}$ eigenstates in the decay modes $D^0 \rightarrow K^+ K^-$, which is $\mathcal{CP}$ even with $\Gamma = \Gamma_2$, and $D^0 \rightarrow K^- \pi^+$, which is a $\mathcal{CP}$ mixed state with $\Gamma = (\Gamma_1 + \Gamma_2)/2$. Then the y parameter is determined as:

$$y = \frac{\Gamma(D^0 \rightarrow K^- K^+)}{\Gamma(D^0 \rightarrow K^- \pi^+)} - 1.$$

The lifetime difference should be seen as a difference in slopes of the decay times shown in figure 2. The y parameter was found to be more than two σ away from zero, $y = (3.42 \pm 1.39 \pm 0.74)\%$. The confidence in this result could be further increased by measuring the lifetimes of other $\mathcal{CP}$ mixed states, $D^0 \rightarrow K\pi\pi, K3\pi$.

E791[7] and *BELLE*[11] also reported the measurements of y using the same method. They found respectively $(0.8 \pm 2.9 \pm 1.0)\%$ and $(0.8 \pm^{3.8}_{3.5} \pm^{1.1}_{2.1})\%$. Both results are consistent with zero.

CLEO[13] recently performed a study of the $D^0 \bar{D}^0$ mixing by searching for wrong sign

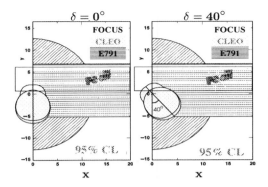

Figure 3. Summary of results on $D^0\bar{D}^0$ mixing.

signals, namely for produced $D^0(\bar{D}^0)$ that decays as $\bar{D}^0(D^0)$. The flavour of the produced D was tagged by the charge of the daughter pion in $D^{*+} \to D^0\pi^+$, and then D^0 decays into $K^+\pi^-$. The use of the silicon vertex detector and helium, as the drift chamber gas, substanially improved the detection efficiency of $D^{*+} \to D^0\pi^+$ and $D^0 \to K^+\pi^-$ compared to the earlier $CLEO$ work.

The wrong sign final state can also result from $DCSD$, $D^0 \to K^+\pi^-$. However the time dependences of mixed and $DCSD$ events are quite different. The fraction of mixed events evolves with a time proportionally to $t^2\exp(-t/\tau)$, while $DCSD$ events have an ordinary exponential time dependence. The different time dependence was used to separate contributions from the mixed and $DCSD$ events. Additionally the $DCSD$ might have a strong phase shift δ with respect to the favoured decay leading to rotation of the effective x and y parameters to $x' = x\cos\delta + y\sin\delta$ and $y' = y\cos\delta - x\sin\delta$.

The fit to $CLEO$ data resulted in the allowed region, shown in figure 3, and the mixing parameters:

$$y' = (-2.5 \pm^{1.4}_{1.6} \pm 0.3)\%$$

$$x' = (0.0 \pm 1.5 \pm 0.2)\%.$$

There is almost no overlap between the $CLEO$ and $FOCUS$ results for $\delta = 0°$. For larger values of δ (the case of $\delta = 40°$ is also presented in figure 3) the agreement becomes better. The $FOCUS$ hint for a $D^0\bar{D}^0$ mixing signal is very intriguing but clearly requires verification.

3 Beauty Physics

The *beauty* physics today is a substantial part of the scientific program of all major laboratories. A tremendous amount of experimental data, presented at this conference, includes:

- Determination of the basic parameters of the b-hadrons, for example the precision measurements of the B^0 and B^+ masses[1,14] and lifetimes.

- Observation of new decay channels.

- Determination of the CKM parameters in a variety of processes.

- Detailed studies of CP-violation.

3.1 Lifetimes of beauty hadrons

Measurements of beauty hadron lifetimes are reaching an impressive precision. Figure 4 shows the current data, as reported by the *B Lifetime Working Group*[16]. Not included are the first measurements of the B^+ and B^0 lifetimes by $BaBar$[17,18] and $BELLE$[11] presented at this conference and listed in table 1. The $\Upsilon(4S)$ data agree well with results obtained at higher energies.

The ratios of $B_{d,s}$ lifetimes are close to unity as predicted. Of note is presently significant observation that B^+ lives longer than B^0, $\tau(B^+)/\tau(B^0) = 1.07 \pm 0.02$, which is in accordance with expectations from the *Pauli*-type interference. Keeping in mind the rapid growth of the $BaBar$ and $BELLE$ data samples, one would soon expect substantial improvement in the accuracy of this ratio. Finally there are no new measurements of b-baryon lifetimes and no theoretical consensus on their predictions[15]. So, a relatively

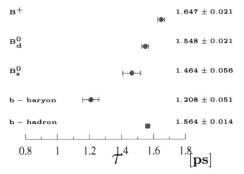

Figure 4. Lifetimes of beauty hadrons.

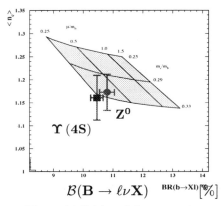

Figure 5. The multiplicities of charm quarks per b-decay vs the inclusive semileptonic branching ratios as measured at the $\Upsilon(4S)$ and Z^0.

Table 1. First measurements of B lifetimes at the $\Upsilon(4S)$

τ_{B^0} [ps]	τ_{B^+} [ps]
dileptons, BaBar	
$1.506 \pm 0.052 \pm 0.029$	$1.602 \pm 0.049 \pm 0.035$
dileptons, BELLE	
$1.50 \pm 0.05 \pm^{0.07}_{0.06}$	$1.70 \pm 0.06 \pm^{0.10}_{0.06}$
$D^*\pi$ events, BaBar	
$1.55 \pm 0.05 \pm 0.07$	
$D^*l\nu$ events, BaBar	
$1.62 \pm 0.02 \pm 0.09$	

short lifetime of b-baryons remains a point of controversy.

3.2 Inclusive semileptonic decays

The inclusive semileptonic branching fractions have been measured at the $\Upsilon(4S)$ and Z^0.

At the $\Upsilon(4S)$ the shape of the lepton spectrum, as well as the branching fraction, has been determined using a lepton-tagged technique, pioneered by $ARGUS$[22]. The charge and angular correlations of two leptons allow a separation of the contributions from the primary and secondary leptons. While statistical uncertainties are larger than in the single lepton analysis, the model de-

pendence is much smaller. The average of the $ARGUS$ and $CLEO$ results[6] is

$$\mathcal{B}^{\Upsilon(4S)}_{B \to \ell\nu X} = (10.45 \pm 0.21)\%.$$

A number of new measurements of $\mathcal{B}(B \to \ell\nu X)$ have been performed at the Z^0 using the tagging with the second lepton, b-vertex and jet charge or neural nets in order to separate the primary, secondary leptons and backgrounds. The combined $\mathcal{B}(B \to \ell\nu X)$ at Z^0 has been found[20] to be:

$$\mathcal{B}^{Z^0}_{b \to \ell\nu X} = (10.564 \pm 0.106 \pm 0.134 \pm 0.115_{mod.})\%.$$

Assuming that

$$\mathcal{B}^{Z^0}_{B \to \ell\nu X} = \mathcal{B}^{Z^0}_{b \to \ell\nu X} \times \frac{\tau_B}{\tau_b},$$

one obtains $\mathcal{B}^{Z^0}_{B \to \ell\nu X} = (10.79 \pm 0.25)\%$, which is somewhat smaller then the previous Z^0 result[6].

The measurements of the multiplicity of charm quarks per b-decay (n_c) have been reviewed in[19]. As seen from figure 5, the measurements of n_c and $\mathcal{B}_{B \to \ell\nu X}$ at the Z^0 now agree well with those at $\Upsilon(4S)$ and the theoretical predictions[21], shown as a shaded area in figure 5.

A combination of the measurements at the $\Upsilon(4S)$ and Z^0 leads to

$$\langle \mathcal{B}_{B \to \ell\nu X} \rangle = (10.59 \pm 0.16)\%. \qquad (1)$$

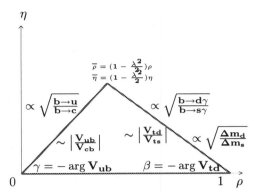

Figure 6. Unitarity triangle in the $(\rho - \eta)$ plane.

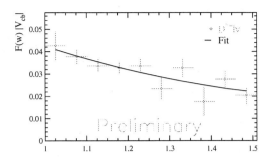

Figure 7. The CLEO data for the decay $B \to D^* l\nu$ used to extract $|V_{cb}|$. The data points represent the product $|V_{cb}|F(\omega)$.

3.3 Unitarity triangles

The mixing of quarks is described in the Standard Model by the unitary CKM matrix. The matrix is determined by four fundamental parameters (λ, A, ρ and η in the Wolfenstein expansion) which have to be determined experimentally. The unitarity of the matrix imposes certain relations on its elements, which can be presented geometrically as six unitarity triangles. Present experimental precision is adequate to use only one non-degenerated triangle shown in figure 6.

The length of the sides and the magnitude of the angles are directly related to specific measurements in the b-sector. The side opposite to the angle β is proportional to the ratio of $|V_{ub}| / |V_{cb}|$. The side opposite to the angle γ is proportional to the ratio $|V_{td}| / |V_{ts}|$. The angles can be derived from the measurements of $\mathcal{CP}$-violating asymmetries in B decays. It is important to perform precision measurements of the two angles and two sides in order to check if they are really consistent with a triangle.

3.4 Determination of $|V_{cb}|$

Apart from the Cabibbo angle, $|V_{cb}|$ is the most precisely measured element of the CKM matrix. $|V_{cb}|$ is measured using both inclusive and exclusive semileptonic decays.

For the inclusive method, the required experimental inputs are the inclusive semilep-

tonic branching ratio from equation (1), $\mathcal{B}_{B \to \ell\nu_\ell X}$ and the average B lifetime, τ_B. Using the receipt of the LEP working group and the values of $\mathcal{B}_{B \to \ell\nu_\ell X}$ and τ_B mentioned in sections 3.1 and 3.2 one arrives at $|V_{cb}| = (40.66 \pm 0.36) \times 10^{-3} \times (1 \pm 0.015_{pert.} \pm 0.010_{m_b} \pm 0.012_{1/m_Q^3})$.

The measurement of the $B \to D^* \ell\nu_\ell$ decay is the best way to determine $|V_{cb}|$ using the exclusive approach. $CLEO$[23] reported at this conference the update of their previous analysis of this decay channel. The differential decay rate of $B \to D^*\ell\nu_\ell$, with respect to the boost $\omega = (m_B^2 + m_D^2 - q^2)/(2m_B m_D)$ of D^* in the B rest frame, is given by:

$$\frac{d\Gamma}{d\omega} = \frac{G_F^2}{48\pi^3} k(\omega)|V_{cb}^2|\mathcal{F}_{D^*}^2(\omega), \qquad (2)$$

where $k(\omega)$ is a phase space function and $\mathcal{F}_{D^*}(\omega)$ is the universal form factor. The HQET limit at zero recoil, corresponding to $\omega = 1$, allows a calculation of $\mathcal{F}_{D^*}(1)$ up to corrections of the order of $1/m_Q^2$[25]. Experimentally, measuring the differential rate at $\omega = 1$ is difficult because at that point the D^{*+} is moving slowly, leading to a very slow decay pion for experiments operating at the $\Upsilon(4S)$. Understanding the efficiency for this slow pion as a function of its momentum is the main source of systematic error. The $CLEO$ data are used to extract $|V_{cb}|$ and the results are shown in fig.7.

The new $CLEO$ measurement $\mathcal{F}_{D^*}(1) \times$

Table 2. The $|V_{ub}|$ values obtained by the inclusive method.

	$ISGWII$	ACM
$|V_{ub}| \times 10^3$	3.22 ± 0.24	3.30 ± 0.24

$|V_{cb}| \times 10^3 = 42.2 \pm 1.8 \pm 1.9$ is about 20% higher than their previous value and only marginally consistent with the LEP^{24} average, $\mathcal{F}_{D^*}(1) \times |V_{cb}| \times 10^3 = 34.5 \pm 0.7 \pm 1.5$. The accuracy of both measurements is limited by relatively large systematic uncertainties, which have completely different origin at the $\Upsilon(4S)$ and Z^0 energies. The averaging of the extracted $|V_{cb}|$ values leads to $|V_{cb}| = (41.8 \pm 1.6 \pm 2.1_{\mathcal{F}_{D^*}(1)}) \times 10^{-3}$.

3.5 Determination of $|V_{ub}|$

$|V_{ub}|$ is an order of magnitude smaller than $|V_{cb}|$ and correspondingly more difficult to measure. Three methods have been probed experimentally to determine $|V_{ub}|$.

The inclusive approach utilises the isolation of a $B \to X_u \ell \nu$ signal in the endpoint region above the kinematical boundary for $B \to X_c \ell \nu$ decays. However a large background from charm decays remains in this region. Using this method $ARGUS^{26}$ and $CLEO^{27}$ observed semileptonic b $\to$ u decays for the first time. Later $CLEO\ II$ measured the lepton yield in the range 2.3 - 2.6 GeV/c with improved precision[28]. The extraction of $|V_{ub}|$ is highly model dependent, primarily due to uncertainty in relating the measured endpoint branching fraction to the CKM element. The $CLEO\ II$ results obtained for $ISGW\ II$ and ACM models are shown in table 2.

Recently $ALEPH$, $DELPHI$ and $L3$ exploited[29] a new approach based on an idea to extend the available phase space by using information on the measured hadronic mass, X_u. It is possible at LEP to reconstruct inclusively the hadronic system in semileptonic b decays, since the two b quarks

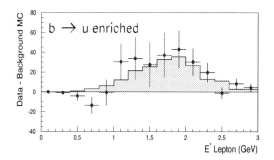

Figure 8. Lepton energy spectrum in the B hadron rest frame measured in the $b \to u$ enriched sample. The background has been predicted using Monte Carlo and subtracted.

are boosted and their decay products are well separated. In addition to cutting on $M_X < M_{X_c}$, other discriminating variables have been constructed to improve the isolation of the b $\to$ u signal in the presence of dominant b $\to$ c background. Even after suppression, understanding of this background remains to be a crucial factor for the reliability of the method. The value of $|V_{ub}|$ is extracted by a binned likelihood fit to the yield and shape of the lepton spectrum. Theoretical uncertainties can be controlled at the 10% level provided that the cut on the hadronic mass M_X is kept above 1.5 GeV/c^2.

The lepton energy spectrum in the B hadron rest frame for $B \to X_u \ell \bar{\nu}$ decays, as measured by $DELPHI$, is shown in figure 8. The resulting LEP average value[29] is $|V_{ub}| = (40.9 \pm 6.1 \pm 1.9) \times 10^{-4}$.

Exclusive charmless semileptonic decays provide an alternative method to determine $|V_{ub}|$. Unfortunately $HQET$ can not be effectively applied for the description of the heavy-to-light quark transition; poor knowledge of the form factors implies a large model dependence.

CLEO performed a new measurement[30] of the decay $B \to \rho \ell \nu$ using high momentum leptons paired with π, ρ and ω candidates. The $\rho \ell \nu$ branching fraction was extracted from the simultaneous fit which contained all five possible decay modes and the background. The signal shapes and relative normalizations for the fit were calculated using various form factor models and isospin relations, respectively. Combining the fit result with an earlier statistically independent analysis $CLEO$ obtained: $\mathcal{B}(B^0 \to \rho^- l^+ \nu) = (2.57 \pm 0.29 \pm^{0.33}_{0.46} \pm 0.41) \times 10^{-4}$ and $|V_{ub}| = (3.25 \pm 0.14 \pm^{0.21}_{0.29} \pm 0.55) \times 10^{-3}$, where the last and largest error is due to model dependence. The $CLEO$ analysis includes a first measurement of the q^2 distribution for $B \to \rho \ell \nu$ decay. Further improvement of statistical accuracy is very important to tune models for precision determination of $|V_{ub}|$.

3.6 Rare charmless B decays

Rare charmless B decays involve contributions from both tree $b \to uW$ and penguin $b \to dg, sg$ amplitudes. The branching fractions themselves and resulting $\mathcal{CP}$-violating asymmetries carry information on the weak and strong phases. Therefore studying several decay modes should, in principle, allow constrains on the phases. A complication to this naive picture arises from possible contributions from electroweak penguins and, mainly, long distance rescattering effects.

The decay $B \to \phi K$ is dominated by the gluonic penguin diagram and, thus, provides information on the strength of the $b \to sg$ transition when the gluon splits into an $s\bar{s}$ pair. Both $BELLE$[33] and $CLEO$[31] reported the evidence for $B^+ \to \phi K^+$ decay. The $BELLE$ result $\mathcal{B}(B \to \phi K) = (17.2 \pm^{6.7}_{5.4} \pm 1.8) \times 10^{-6}$ is substantially higher than the updated $CLEO$ result $\mathcal{B}(B \to \phi K) = (6.4 \pm^{2.5}_{2.1} \pm^{0.5}_{2.0}) \times 10^{-6}$ and only marginally consistent with the previously reported $CLEO$[35] upper limit $\mathcal{B}(B \to \phi K) < 5.9 \times 10^{-6}$ (90%CL). No significant signal has been observed for the $B^0 \to \phi K^0$ decay which has substantially lower detection efficiency.

Many of the rare charmless B decays discovered by $CLEO$[31], have been also measured by $BaBar$[32]. The updated $CLEO$ and new $BaBar$ results are summarized in table 3.

$BaBar$ observe significant signal for $B^+ \to \eta' K^+$ confirming the earlier $CLEO$ observation of an unexpectedly large $B \to \eta' K$ rate (see figure 9). One of the possible explanations could be an intrinsic charm content of the η' meson, resulting, as a consequence, in an enhancement in the $B \to \eta_c K$ rate. However the recent $CLEO$[34] measurement shows no enhancement with respect to $B \to J/\psi K$. So, there is no quantitative explanation for the large $B \to \eta' X$ rate. Another mysterious observation is the fact that apart from $\eta' K$ and ηK^*, no other decay modes involving η' or η were found.

The branching fractions of B decays into $K\pi$ and $\pi\pi$ final states, as measured by $CLEO$[31], $BaBar$[32] and $BELLE$[33], are listed in table 4. Figure 10 shows corresponding beam constrained mass distributions for the $K^+\pi^-$ and $\pi^+\pi^-$ decay modes.

The results of all three experiments agree well for the sum of $B^0 \to K^+\pi^-$ and $B^0 \to \pi^+\pi^-$ branching fractions while the sharing between the $K^+\pi^-$ and $\pi^+\pi^-$ modes is somewhat different. The $CLEO$ and $BELLE$ data point to significant enhancement for the $B^0 \to K^+\pi^-$ decay rate while

Table 3. $\mathcal{B}(B \to hh) \times 10^6$ or upper limit at 90% confidence.

	CLEO	BaBar
$\rho^0 \pi^\pm$	$10.4 \pm^{3.4}_{3.3} \pm 2.1$	$24 \pm 8 \pm 3$
$\rho^\pm \pi^\mp$	$27.6 \pm^{8.4}_{7.4} \pm 4.2$	$24 \pm 13 \pm^{6}_{5}$
$\omega \pi^+$	$11.3 \pm^{3.3}_{2.9} \pm 1.4$	< 24
$\eta' K^\pm$	$80 \pm^{10}_{9} \pm 7$	$62 \pm 18 \pm 8$
$\eta' K^0$	$89 \pm^{18}_{16} \pm 9$	< 112
$\eta K^{*\pm}$	$26.4 \pm^{9.6}_{8.2} \pm 3.3$	—
ηK^{*0}	$13.8 \pm^{5.5}_{4.6} \pm 1.6$	—

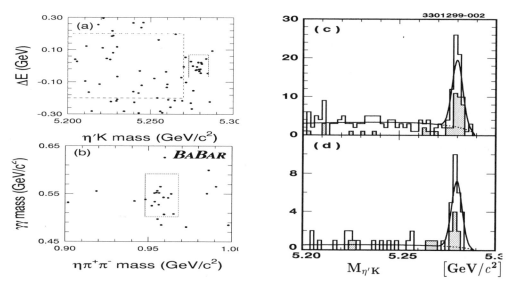

Figure 9. The *BaBar* measurement of the B $\to \eta'$K$^\pm$ decay (a), followed by the $\eta' \to \eta\pi^+\pi^-$(b). The *CLEO* measurement of B $\to \eta'$K$^\pm$ (c) and B $\to \eta'$K^0 (d) modes.

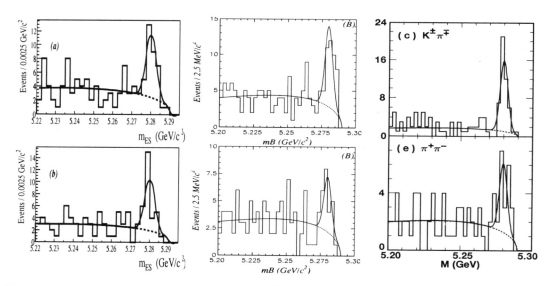

Figure 10. The decays of *B* mesons to K$^\pm\pi^\mp$ and $\pi^\pm\pi^\mp$ as observed by experiments at the $\Upsilon(4S)$. Left column: The *BaBar* measurement of B $\to \pi^+\pi^-$ (top) and B $\to$ K$^\mp\pi^\mp$(bottom). Middle column: The *BELLE* measurement of K$^\pm\pi^\mp$ (top) and $\pi^\pm\pi^\mp$ (bottom). Right column: The *CLEO* measurement of K$^\pm\pi^\mp$ (top) and $\pi^\pm\pi^\mp$ (bottom).

Table 4. $\mathcal{B}(B \to K\pi\,\pi\pi) \times 10^6$ or upper limit at 90% confidence

	CLEO	BELLE	BaBar	Average
$\pi^\pm\pi^\mp$	$4.3 \pm^{1.6}_{1.4} \pm 0.5$	$6.3 \pm^{3.9}_{3.5} \pm 1.6$	$9.3 \pm^{2.6}_{2.3} \pm^{1.2}_{1.4}$	5.5 ± 1.3
$\pi^\pm\pi^0$	< 12.7	< 11		
$\pi^0\pi^0$	< 5.7			
$K^\pm\pi^\mp$	$17.2 \pm^{2.5}_{2.4} \pm 1.2$	$17.4 \pm^{5.1}_{4.6} \pm 3.4$	$12.5 \pm^{3.0}_{2.6} \pm^{1.3}_{1.7}$	15.5 ± 2.0
$K^0\pi^\pm$	$18.2 \pm^{4.6}_{4.0} \pm 1.6$	$16.6 \pm^{9.8}_{7.8} \pm^{2.2}_{2.4}$		17.8 ± 4.1
$K^\pm\pi^0$	$11.6 \pm^{3.0}_{2.7} \pm^{1.4}_{1.3}$	$18.8 \pm^{5.5}_{4.9} \pm 2.3$		13.0 ± 2.7
$K^0\pi^0$	$14.6 \pm^{5.9}_{5.1} \pm^{2.4}_{3.3}$	$21.0 \pm^{9.3}_{7.8} \pm^{2.5}_{2.3}$		16.8 ± 5.1

the *BaBar* measurement shows rather balanced $K^+\pi^-$ and $\pi^+\pi^-$ yields. It is important to verify with higher statistics how large is the contribution of the penguin amplitude in B decays to two pseudoscalar mesons. A possibility to measure the angle $(\beta + \gamma)$ of the unitarity triangle depends crucially on the $\mathcal{B}(B \to \pi^+\pi^-)/\mathcal{B}(B \to K^+\pi^-)$ ratio.

The branching ratios of the decays listed in tables 3 and 4 can be used to constrain the angle γ. Present data favour large values of γ and large final state interactions.

3.7 Determination of $|V_{td}|$

The most effective way to determine $|V_{td}|$ is a precise measurement of the $B^0\bar{B}^0$ mixing. Within the Standard Model, the box diagrams with a top quark in the internal loop dominate the contribution to the mass difference:

$$\Delta m_q = \frac{G_F^2}{6\pi^2} m_t^2 m_{B_q} \eta^{\text{QCD}}$$

$$\times f_B^2 B_B \mathcal{F}\left(\frac{m_t^2}{m_W^2}\right) |V_{tb}^* V_{tq}|^2 \quad (3)$$

This expression holds true for both $|V_{td}|$ (q = d) and $|V_{ts}|$ (q = s). With the top quark mass relatively well known[3], $m_t = 174.3 \pm 3.4 \pm 4.0$ GeV/c^2, the largest uncertainties in the evaluation of Δm_q come from an imperfect knowledge of the f_B and B_B factors. A part of the common uncertainties is cancelled out in the ratio $\frac{\Delta m_d}{\Delta m_s}$, which today provides the best tool to obtain the ratio $\left|\frac{V_{td}}{V_{ts}}\right|$.

Experimentally two techniques were used to measure $B^0\bar{B}^0$ oscillations. $B_d^0\bar{B}_d^0$ mixing was first discovered by *ARGUS*[37] and confirmed by *CLEO*[38] using the time-integrated analysis at the $\Upsilon(4S)$. Later the *LEP* experiments, *CDF* and *SLD* performed the time-dependent mixing measurement[16,39]. The large boost of the *B* mesons and use of the silicon vertex detectors made possible an improved precision of the Δm_d measurement. Using a variety of techniques the combined *ALEPH, CDF, DELPHI, L3, OPAL* and *SLD* result[29] is $\Delta m_d = (0.486 \pm 0.015)$ ps^{-1}.

At this conference new results on Δm_d have been reported using both a time integrated analysis by *CLEO*[40] and first measurements of the time evolution of $B_d^0\bar{B}_d^0$ mixing at the $\Upsilon(4S)$ by *BaBar*[17,18] and *BELLE*[41].

In order to increase statistical power and reduce the dilution due to charged *B* contamination, *CLEO* applied a partial reconstruction of $B^0 \to D^{*+}\pi^-(\rho^-)$. The flavour of the second B is tagged with a lepton. The selected data sample consists of about 2000 events with a dilution of 13% and only 3% mistagging. The resulting value for the mix-

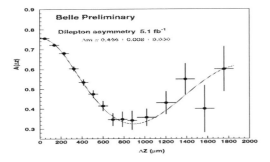

Figure 11. The dilepton asymmetry as a function of Δz as measured by *BELLE*.

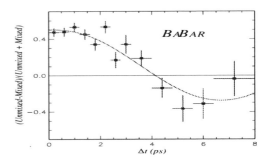

Figure 13. The asymmetry as a function of Δt for reconstructed events as measured by *Babar*.

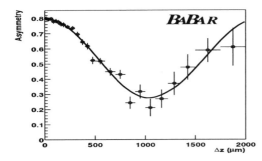

Figure 12. The dilepton asymmetry as a function of Δz as measured by *BaBar*.

ing parameter $\chi_d = 0.198 \pm 0.013 \pm 0.014$ is the most precise one obtained in a single experiment. Assuming that $|y| \ll x$ (defined in section 2.2), this value of χ_d transforms to $\Delta m_d = 0.523 \pm 0.029 \pm 0.031$ ps^{-1}. Comparison of the charge states of the like sign events also allows to restrict $\mathcal{CP}$ violation in B_d^0 state mixing by $|\text{Re}(\epsilon_B)| < 3.4\%$ at 95% CL.

The time evolution of the $B_d^0 \bar{B}_d^0$ mixing probability is given by:

$$\text{Prob}(B_d^0 \to \bar{B}_d^0) \approx e^{-|\Delta t|/\tau_{B_d^0}}$$

$$\times (1 - \cos \Delta m_d \Delta t) \quad (4)$$

where $\Delta t \approx \Delta z / \beta\gamma c$ is the proper time difference between the two B decays.

In order to measure Δm_d, *BELLE* selected events with the same sign (SS) and opposite sign (OS) dileptons[41]. These samples include dileptons resulting from

both $B_d^0 \bar{B}_d^0$ and $B^+ B^-$ pairs; their relative composition depends on the ratios of the $\Upsilon(4S)$ branching fractions to neutral- and charged-B pairs, $f_0/f_\pm$, and semileptonic branching fractions of B_d^0 and $B^\pm$, $b_0/b_\pm$. The ratios were fixed in the fit to their measured values: $f_\pm/f_0 = 1.07$, and $b_\pm/b_0 = \tau_{B^\pm}/\tau_{B_d^0} = 1.04$. The value of $\tau_{B_d^0}$ was fixed to the world average value[6]. The fit yielded $\Delta m_d = (0.456 \pm 0.008)$ ps^{-1}.

Figure 11 shows the asymmetry, $\mathcal{A} = \frac{N_{OS} - N_{SS}}{N_{OS} + N_{SS}}$, together with a result of the fit. Taking into account the systematic error, the *BELLE* result is $\Delta m_d = (0.456 \pm 0.008 \pm 0.030)$ ps^{-1}, where the systematic uncertainty arises mainly due to uncertainty in $\tau_{B^\pm}/\tau_{B_d^0}$ (± 0.022 ps^{-1}) and in $f_0/f_\pm$ (± 0.012 ps^{-1}).

The corresponding distribution of $\mathcal{A}$, as measured by *BaBar* using an inclusive dilepton sample[17], is shown in figure 12.

In contrast to the *BELLE* analysis, the ratio $\frac{b_\pm^2 f_\pm}{b_0^2 f_0}$ was left as a free parameter in the *BaBar* fit. Such a fit procedure leads to a redistribution between statistical and systematic uncertainties. Despite comparable numbers of selected dileptons, the statistical error of the *BaBar* measurement $\Delta m_d = (0.507 \pm 0.015 \pm 0.022)$ ps^{-1} is larger than that of *BELLE*.

BaBar also performed a measurement of the time-dependent $B^0 \bar{B}^0$ oscillation frequency using a sample of B^0 mesons re-

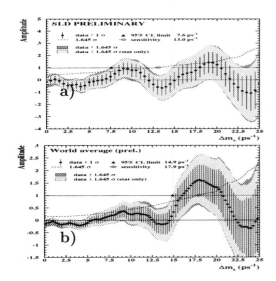

Figure 14. The amplitude A (see text) as a function of Δm_s obtained in (a) the SLD analysis and (b) in combination of the SLD, CDF and LEP data

Table 5. Summary of results on Δm_s

	Excluded Δm_s $[ps^{-1}]$ (95% CL)	Sensitivity $[ps^{-1}]$
ALEPH	< 10.1	11.5
CDF	< 5.8	
DELPHI	< 7.3	10.6
OPAL	< 5.2	6.0
SLD	< 7.6 [11.8, 14.8]	13.0
Combined	< 14.9	17.9

constructed in hadronic decay channels[18] and in the semileptonic decay mode[17] $B^0 \to D^{*-}\ell^+\nu$. Four different flavour tags were used for this analysis. Two of them relied on the presence of a prompt lepton or charged kaon in the event while two others were based on neural network algorithms. The *BaBar* results for the hadronic and semileptonic samples are respectively $\Delta m_d = (0.516 \pm 0.031 \pm 0.018)$ ps^{-1} and $\Delta m_d = (0.508 \pm 0.020 \pm 0.022)$ ps^{-1}.

The present level of precision in Δm_d measurements is so high that a combination of results obtained in different experiments requires a detailed accounting of the correlated systematic uncertainties inherent in the analysis assumptions.

3.8 Determination of $|V_{ts}|$

Analogous to $|V_{td}|$, the magnitude of $|V_{ts}|$ can be determined from the measurement of Δm_s using equation (3). The strength of the $B_s^0\bar{B}_s^0$ oscillations has not yet been measured although the experimental sensitivity has substantially extended to

higher values of Δm_s over past few years. The sensitivity is defined as the largest value of Δm_s that would have been excluded if the amplitude A in the expression $\text{Prob}(B_s^0 - \bar{B}_s^0) \approx 1 - A\cos(\Delta m_s t)$ were set to zero at all values of Δm_s. The excluded intervals of Δm_s as well as the experimental sensitivity are given in table 5 for various experiments[16].

The amplitude A obtained in the most sensitive SLD analysis[39] is plotted in figure 14a as a function of Δm_s.

The observation of $B_s^0\bar{B}_s^0$ oscillations should manifest itself as a peak with amplitude of 1 for a given value of Δm_s. When combining the SLD with LEP and CDF data, a tantalizing 2.5 σ hint for the oscillation signal appears at $\Delta m_s \approx 17.5$ ps^{-1}, very close to the sensitivity edge. The data from the *Tevatron Run II* and $HERA - B$ are extremely important to extend the interval where Δm_s can be reliably measured.

3.9 Electromagnetic penguin decays and $|V_{td}|/|V_{ts}|$

The ratio of the branching ratios $\mathcal{B}(B \to \rho/\omega\gamma)/\mathcal{B}(B \to K^*\gamma)$ is related to $|V_{td}|/|V_{ts}|$ via:

$$\frac{\mathcal{B}(B \to \rho\gamma)}{\mathcal{B}(B \to K^*\gamma)} = \xi \times \left|\frac{V_{td}}{V_{ts}}\right|^2, \qquad (5)$$

Table 6. $\mathcal{B}(B \to K^*\gamma) \times 10^5$.

$B \to K^{*0}\gamma$	
CLEO	$4.55 \pm^{0.72}_{0.68} \pm 0.34$
BELLE	$4.94 \pm 0.93 \pm^{0.55}_{0.52}$
BaBar	$5.4 \pm 0.8 \pm 0.5$
$B \to K^{*+}\gamma$	
CLEO	$3.76 \pm^{0.89}_{0.83} \pm 0.28$
BELLE	$2.87 \pm 1.20 \pm^{0.55}_{0.40}$
$B \to K_2^*(1430)\gamma$	
CLEO	$1.66 \pm^{0.59}_{0.53} \pm 0.13$

where predictions for ξ range from 0.58 to 0.81[45]. There however remains a question-mark whether long-distance effects are significant. The signal of $B \to K^*\gamma$, first observed by $CLEO$[42], has been confirmed by both $BaBar$[43] and $BELLE$[44] using the excellent performance of the electromagnetic calorimeters and large available data sets. The corresponding distibutions of the beam constrained mass of the $K^*\gamma$ system, shown in figure 15, exibit signals on top of a low level background.

The branching ratios measured by $CLEO$, $BaBar$ and $BELLE$ are summarized in table 6.

The $B \to \rho\gamma$ search was performed by $CLEO$[42] and $BELLE$[44]. Taking advantage of good π/K separation over the full kinematically allowed momentum range, $BELLE$ applied essentially no cuts to suppress feeddown from the dominant $B \to K^*\gamma$ decay. Only 3 and 0 events were found in the $\rho^+\gamma$ and $\rho^0\gamma$ signal windows, respectively (with the expected $K^*\gamma$ background of only 0.25 and 0.08 events, respectively) leading to an upper limit $\mathcal{B}(B \to \rho\gamma)/\mathcal{B}(B \to K^*\gamma) < 0.28$.

An absence of π/K separation at high momenta decreases the $CLEO$ sensitivity. Despite having more than 2 times larger data sample the $CLEO$ upper

limit, $\mathcal{B}(B \to \rho/\omega\gamma)/\mathcal{B}(B \to K^*\gamma) < 0.32$, is slightly less restrictive than that of $BELLE$.

Using equation (5), the $BELLE$ upper limit and the prediction $\xi = 0.58$ one obtains:

$$|\frac{V_{td}}{V_{ts}}| < 0.69.$$

3.10   Determination of $\sin 2\beta$

The new $BaBar$[1] and $BELLE$[2] measurements of $sin(2\beta)$ are summarized in table 7 together with earlier data[46]. The combination of these results leads to:

$$\langle\sin(2\beta)\rangle = 0.47 \pm^{0.22}_{0.23}.$$

Table 7. *Measurements of $sin(2\beta)$*.

	$sin(2\beta)$
OPAL	$3.2 \pm^{1.8}_{2.0} \pm 0.5$
CDF	$0.79 \pm^{0.41}_{0.44}$
ALEPH	$0.86 \pm^{0.82}_{1.05} \pm 0.20$
BaBar	$0.12 \pm 0.37 \pm 0.09$
BELLE	$0.45 \pm^{0.43}_{0.44} \pm^{0.07}_{0.09}$

4   Conclusions

Much progress has been made in heavy flavour physics this year. In particular, the first $BaBar$ and $BELLE$ contributions to *beauty-*, as well as to *charm-* and τ-, physics have been received. The successful startup of both experiments demonstrates their potential for providing even better results in the near future. $DONUT$ observed the last fundamental fermion, ν_τ.

The recent precision measurements of the lifetimes of charm hadrons should further sharpen theoretical understanding of the charm particle decay mechanism. The experimental precision of the D^0, D_s and Λ_c lifetimes is now dominated by the $FOCUS$ measurements.

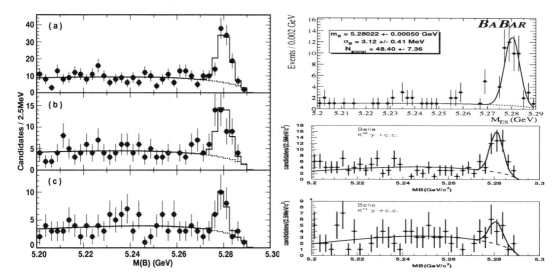

Figure 15. The measurements of B → K*γ at Υ(4S). Left column: The *CLEO* measurement of B → K*⁰γ (top), B → K*⁺γ (middle), B → K₂*(1430)γ (bottom). Right column: The *BaBar* measurement of B → K*⁰γ (top) and the *BELLE* measurement of B → K*⁰γ (middle), B → K*⁺γ (bottom).

One of the most exciting results in *charm* physics is a 2.2 σ hint for a D⁰D̄⁰ oscillation signal reported by *FOCUS*. However the new *CLEO* analysis of comparable sensitivity shows no evidence for the oscillation signal. Direct verification of the *FOCUS* result should be done either by measuring the lifetimes of *CP*-eigenstates or using semileptonic D decays. Such measurements can be performed by *BaBar*, *BELLE* and *CLEO*.

Beauty physics is entering a new phase of precision tests of the SM. The data presented at this conference provide information on *all* four elements of the unitarity triangle: two sides and two angles.

The precision of the $|V_{ub}/V_{cb}|$ side, of about 20%, is completely determined by theoretical uncertainties in the calculation of $|V_{ub}|$. The most perspective way to improve precision here is by measurement of the exclusive semileptonic decay modes provided that an adequate model to describe heavy-to-light quark transitions will be developed.

A measurement of $|V_{ts}|$ is missing to construct the $|V_{td}/V_{ts}|$ side. Δm_d is measured to a few percent precision. The analyses of *LEP*, *CDF* and particularly *SLD* achieved impressive sensitivity on Δm_s, which is at the edge of a real observation. In addition, $|V_{td}/V_{ts}|$ can be also determined from the $\frac{B \to \rho\gamma}{B \to K^*\gamma}$ ratio. The data from the *Tevatron Run II*, *HERA − B* and from the e^+e^- B-factories are extremely important to provide reliable determination of $|V_{td}/V_{ts}|$.

The results on rare charmless B decays reported by *CLEO*, *BaBar* and *BELLE* show that the measurement of decays with branching fractions at the level of 10^{-6} is now experimentally possible. A number of attempts has been made to constrain the angle γ using these data. However the consensus on the proper theoretical procedure remains to be found.

BaBar and *BELLE* reported their first measurements of the angle β. The precision of both measurements is strongly dominated by statistical uncertainty. Present experience with the operation of the *PEP − II* and *KEK B* factories provides a confidence that the precision of sin 2β will be significantly improved in the near future.

Finally there are excellent prospects to

see deviations from SM predictions, if any, at the time of the next ICHEP.

Acknowledgments

I am gratefull to I.Belyaev for his help in preparation of this report. I also thank S.Barsuk, M.Danilov, N.Harnew and T.Nakada for their assistance in various ways and interesting discussions.

References

1. D.Hitlin, *Talk at this Conference.*
2. H.Aihara, *Talk at this Conference.*
3. S. Willenbrock, hep-ph/0008189.
4. A.Pich, hep-ph/9912294.
5. M. Nakamura, *Talt at this Conference.*
6. C.Caso *et al.*, *Eur.Phus.J.* **C 3** (1998).
7. E.M.Aitala *et al.*, *Phys.Lett.* **B 445** 449 (1998);
   E.M.Aitala *et al.*, *Phys.Rev.Lett.* **83** 32 (1999).
8. G. Bonvicini *et al.*, *Phys.Rev.Lett.*, **82** 4586 (1999).
9. J.S.Russ, *Talk at this Conference.*
10. H.W.K.Cheung, FERMILAB-CONF-99-344, Jul 1999.
11. H.Tajima, *Talk at this conference.*
12. J.M.Link *et al.*, *Phys.Lett.* **B 485** 62 (2000).
13. H.N.Nelson, *Talk at this Conference.*
14. S.Csorna *et al.*, *Phys.Rev.* **D 61** 111101 (2000).
15. I.Bigi, hep-ph/0001003;
    B.Melic, *Talk at this Conference*; C.-S.Huang, C.Liu, S.-L.Zhu, hep/ph-9906300.
16. A.Stocchi, *Talk at this Conference.*
17. C.Yeche, *Talk at this Conference*
18. F.Martinez-Vidal, *Talk at this Conference*
19. G.Barker, *Talk at this Conference.*
20. S.Blyth, *Talk at this Conference.*
21. M.Neubert, hep/ph-9702375.
22. H.Albrecht *et al.*, *Phys.Lett.* **B 318** 397 (1993).
23. K.Ecklund, *Talk at this Conference.*
24. E.Barberio, *Talk at this Conference.*
25. M.Shifman and M.Voloshin, *Yad.Fiz.* **47** 801 (1988).
26. H.Albrecht *et al.*, *Phys.Lett.* **B 234** 409 (1990).
27. R.Fulton *et al.*, *Phys.Rev.Lett.* **64** 16 (1990).
28. J.Bartelt *et al.*, *Phys.Rev.Lett.* **71** 4111 (1993).
29. D.Abbaneo *et al.*, SCALC-PUB-8492, CERN-EP-2000-0.96
30. B.BEhrens *et.al.* *Phys.Rev.* **D 61** 052001 (2000).
31. R.Stroynowski, *Talk at this Conference.*
32. T.Champion, *Talk at this Conference.*
33. P.Chang, *Talk at this Conference.*
34. S.Stone, *Talk at this Conference.*
35. M.Bishai et al., CLEO-CONF 99-13, (1999).
36. I.Bigi, *Talk at this Conference*
37. H.Albrecht *et al.*, *Phys.Lett.* **B 192** 245 (1987).
38. M. Artuzo *et al.*, *Phys.Rev.Lett.* **62** 2233 (1989).
39. T.Usher, *Talk at this Conference*
40. B.H. Behrens *et al.*, *Phys.Lett.* **B 490** 36 (2000).
41. J.-I. Suzuki, *Talk at this Conference*
42. T.Coan, *Talk at this Conference*
43. C.Jessop, *Talk at this Conference*
44. M. Nakao, *Talk at this COnference*
45. A.Ali, V.Braun and H.Simma, *Z.Phys.* **C 63** 437 (1994);
    A.Narison, *Phys.Lett.* **B 327** 354 (1994);
    J.Soares, *Z.Phys.* **C 63** 437 (1994)
46. K. Ackerstaff *et al.*, *Eur.Phys.J.* **C 5** 379 (1998);
    F.Abe *at al.*, *Phys.Rev.Lett.* **81** 5513 (1998);
    R.Barate *at al.*, hep-ex/0009058.

Plenary Session 4

Flavor Dynamics
(theory)

Chair: Tao Huang (Chinese Academy of Sciences)
Scientific Secretary: Naoyuki Haba

FLAVOUR DYNAMICS – CENTRAL MYSTERIES OF THE STANDARD MODEL

I.I. BIGI

Physics Dept., Univ. of Notre Dame du Lac, Notre Dame, IN 46556, U.S.A.
E-mail: bigi@undhep.hep.nd.edu

After pointing out the amazing success of the CKM description in accommodating the phenomenology of flavour changing neutral currents I review the present status of theoretical technologies for extracting CKM parameters from data and their anticipated refinements. I sketch novel directions, namely attempts to deal with quark-hadron duality in a (semi)quantitative way and to develop a QCD description of two-body modes of B mesons. After commenting on predictions for ϵ'/ϵ and CP asymmetries in B decays I address indirect probes for New Physics in D^0 oscillations and CP violation, in $K_{\mu3}$ decay and electric dipole moments. I describe in which way searching for New Physics in B decays provides an exciting adventure with novel adventures not encountered before.

Flavour dynamics involve central mysteries of the Standard Model (SM): Why is there a family structure relating quarks and leptons? Why is there more than one family, why three, is three a fundamental parameter? What is the origin of the observed pattern in the quark masses and the CKM parameters?

There are two different strategies for obtaining answers to these questions:
(**A**) One argues that one has already enough data and therefore can turn one's energy towards the last fundamental challenge, namely to bring gravity into the quantum world; flavour dynamics with its family structure will then emerge as a 'side product'.
(**B**) Suspecting that nature has a few more surprises up her sleeves one commits oneself to elicit more answers from her.

My talk is geared towards strategy (**B**) and its necessary theoretical tools. I will merely list experimental numbers; details can be found in Golutvin's talk [1].

1 New Landmarks and Challenges

Since ICHEP98 new landmarks have been reached:

- *Direct* CP violation has been established experimentally – a first rank discovery irrespective of theoretical interpretations.

- We are on the brink of observing CP violation in B decays.

- We are reaching fertile ground for finding New Physics in D^0 oscillations and CP violation.

On the theory side we are learning lessons of humility, increasing the sophistication of our theoretical technologies, and pushing back new frontiers.

The new challenges for theory are to regain theoretical control over ϵ'/ϵ; to develop reliable quantitative predictions for CP asymmetries in B decays and to refine them into precise ones; to establish theoretical control over D^0 oscillations and CP violation and finally, to develop comprehensive strategies to not only establish the intervention of New Physics, but also identify its salient features.

My talk will address extracting numerical values of CKM parameters; I will discuss possible limitations to quark-hadron duality and refer to the lifetimes of charm and beauty hadrons as validation studies before describing new attempts to describe exclusive nonleptonic B decays; I will sketch the difficulties inherent in predicting ϵ'/ϵ before addressing CP violation in B decays; I will comment on future searches for New Physics based on CKM trigonometry and the nature of theoretical uncertainties before describing 'exotic' searches for transverse polarization of muons in $K_{\mu 3}$ decays, electric dipole moments and CP violation in charm transitions.

2 The Charged Current Dynamics of Quarks

2.1 The 'unreasonable' Success of the CKM Description

The observation of the 'long' B lifetime together with the dominance of $b \to c$ over $b \to u$ revealed a hierarchical structure in the KM matrix that is expressed in the Wolfenstein representation in powers of $\lambda = \mathrm{tg}\theta_C$. We often see plots of the CKM unitarity triangle where the constraints coming from various observables appear as broad bands [2]. While the latter is often bemoaned, it obscures a more fundamental point: the fact that these constraints can be represented in such plots at all is quite amazing! Let me illustrate that by an analogy first: plotting the daily locations of the about 1000 high energy physicists attending this meeting on a city map of Osaka produces fairly broad bands. On a map of Japan instead those bands shrink to lines and their intersection to a point – for which there is of course a very good reason. Likewise one should look at the bigger picture of flavour dynamics. The quark box *without* GIM subtraction yields a value for Δm_K exceeding the experimental number by more than a factor of thousand; it is the GIM mechanism that brings it down to within a factor of two or so of experiment. The GIM subtracted quark box for ΔM_B coincides with the data again within a factor of two. Yet if the beauty lifetime were of order 10^{-14} sec while $m_t \sim 180$ GeV it would exceed it by an order of magnitude; on the other hand it would undershoot by an order of magnitude if $m_t \sim 40$ GeV were used with $\tau(B) \sim 10^{-12}$ sec; i.e., the observed value can be accommodated because a tiny value of $|V(td)V(ts)|$ is offset by a large m_t.

This amazing success is repeated with ϵ.

Over the last 25 years it could always be accommodated (apart from some very short periods of grumbling mostly off the record) whether the *correct* set $[m_t = 180$ GeV with $|V(td)| \sim \lambda^3$, $|V(ts)| \sim \lambda^2]$ or the *wrong* one $[m_t = 40$ GeV with $|V(td)| \sim \lambda^2$, $|V(ts)| = \lambda]$ were used. Yet both $m_t = 180$ GeV with $|V(td)| = \lambda^2$, $|V(ts)| = \lambda$ as well as $m_t = 40$ GeV with $|V(td)| = \lambda^3$, $|V(ts)| = \lambda^2$ would have lead to a clear inconsistency!

Thus the phenomenological success of the CKM description has to be seen as highly nontrivial or 'unreasonable'. This cannot have come about by accident – there must be a good reason.

2.2 Extracting CKM Parameters

A crucial element in extracting CKM parameters defined for the quark degrees of freedom from data involving hadrons is the quality of our theoretical technologies to deal with the strong forces. For strange mesons with $m_s < \Lambda_{QCD}$ one invokes chiral perturbation theory, for beauty hadrons with $m_b \gg \Lambda_{QCD}$ the heavy quark expansion (HQE) which might be extended to charm hadrons in a semiquantitative fashion $(m_c > \Lambda_{QCD})$ [a]. Lattice QCD on the other hand deals with the nonperturbative dynamics of all quark flavours.

Both HQE and lattice QCD (to be discussed in Kenway's talk [3]) represent mature technologies with both operating in Euklidean space that are complementary to each other. There has already been fruitful feedback between the two on the conceptual as well as numerical level; this interaction is about to intensify. While quark models no longer represent state-of-the-art, they still serve useful purposes in the diagnostics of our results if employed properly.

The main tool for numerical results so far have been the HQE. The last few years

---

[a] The situation is qualitatively different for top states: with $\Gamma_t \sim \mathcal{O}(\Lambda_{QCD})$ top quarks decay before they can hadronize and they are therefore controlled by perturbative QCD [4].

have seen a conceptual convergence among its practitioners: most of them accept the argument that HQE allow to describe in principle nonleptonic as well as semileptonic beauty decays as long as an operator product expansion can be relied upon. At the same time one expects the numerical accuracy to decrease when going from $B \to l\nu q\bar{q}'$ to $B \to c\bar{u}d\bar{q}'$ and on to $B \to c\bar{c}s\bar{q}'$ for fundamental as well practical reasons (the latter meaning that the energy release is lowest for $b \to c\bar{c}s$.). Considerable progress has been achieved also in the numerical value of basic quantities the most important one being the beauty quark mass. Last year three groups extending earlier work by Voloshin [5] have presented new determinations, which – when expressed in terms of the socalled 'kinetic' mass – read as follows:

$$m_b^{\text{kin}}(1\,\text{GeV}) = 4.56 \pm 0.06\,\text{GeV} \; [6],$$

$$4.57 \pm 0.04\,\text{GeV} \; [7], \; 4.59 \pm 0.06\,\text{GeV} \; [8] \quad (1)$$

The error estimates of 1 - 1.5 % might be overly optimistic, but not foolish. Since all three analyses use basically the same input from the $\Upsilon(4S)$ region, they could suffer from a common systematic uncertainty, though. This can be checked by analysing the shape of the lepton spectrum in $B \to l\nu X$. More concretely one forms two moments both of the lepton and of the hadron energies [9]; each set yields $\bar{\Lambda}$ and μ_π^2, where $\bar{\Lambda} \to M_B - m_b$ as $m_b \to \infty$ and $\mu_\pi^2 \equiv \langle B|\bar{b}(iD)^2 b|B\rangle/2M_B$. Comparing those two sets of values with each other and with the m_b values listed above represents a crucial self-consistency check. An early CLEO analysis appeared to yield inconsistent values. It is being redone now, and I eagerly await their results; yet I do that with considerable confidence, in particular since a recent lattice study [10] has yielded numbers that are in agreement with those inferred from the SV sum rules [11].

Two methods exist with excellent theoretical credentials for determining $V(cb)$:

(i) Extrapolating the rate of $B \to l\nu D^*$ to zero recoil one extracts $V(cb)F_{D^*}(0)$. The form factor $F_{D^*}(0)$ has nice features: it is normalized to unity in the infinite mass limit and the leading nonperturbative correction is of order $1/m_Q^2$. Unfortunately it is m_c that sets the scale here rather than m_b, and that is one of the challenges in evaluating it. Three estimates provide representative numbers:

$$F_{D^*}(0) = 0.89 \pm 0.08 \; [13],$$

$$0.913 \pm 0.042 \; [14], \; 0.935 \pm 0.03 \; [15] \quad (2)$$

I will use here

$$F_{D^*}(0) = 0.90 \pm 0.05 \quad (3)$$

as a convenient *reference* point. CLEO has presented a new analysis that yields a considerably larger number than before [16]:

$$|V(cb)F_{D^*}(0)| = (42.4 \pm 1.8|_{stat} \pm 1.9|_{syst}) \times 10^{-3} \quad (4)$$

The updated LEP number on the other hand has hardly changed [17]:

$$|V(cb)F_{D^*}(0)| = (34.9 \pm 0.7|_{stat} \pm 1.6|_{syst}) \times 10^{-3} \quad (5)$$

There is now about a 20% difference between the two central values, which means that 'stuff happens'. With Eq.(3) one gets:

$$|V(cb)|_{excl,CLEO} =$$

$$= (47.1 \pm 2.0|_{stat} \pm 2.1|_{syst} \pm 2.1_{th}) \times 10^{-3} \quad (6)$$

$$|V(cb)|_{excl,LEP} =$$

$$= (38.8 \pm 0.8|_{stat} \pm 1.8|_{syst} \pm 1.7_{th}) \times 10^{-3} \quad (7)$$

I view a theoretical error of 5% as on the optimistic side, and I am skeptical about being able to reduce it below this level.

(ii) The *inclusive* semileptonic width of B mesons can be calculated in the HQE: $\Gamma_{SL}(B) \propto m_b^5 \cdot (1 + \mathcal{O}(1/m_b^2) + \mathcal{O}(\alpha_S))$. Again there is no correction $\sim \mathcal{O}(1/m_Q)$. The advantage over the previous case is that the expansion parameter is effectively the inverse energy release $\sim (m_b - m_c)^{-1}$ rather than the larger $1/m_c$; the challenge is provided by the fact that the leading term depends on the fifth power of the b *quark* mass. It was only the great conceptual and technical progress in HQE that made this method competitive.

LEP has updated its analysis and finds:

$$|V(cb)|_{incl} = (40.76 \pm 0.41|_{stat} \pm 2.0_{th}) \times 10^{-3}$$
$$(8)$$

The theoretical error has been evaluated in a careful way [12]; I am optimistic that it can be cut in half in the foreseeable future; but even then it would appear to represent the limiting factor. Yet it is mandatory to check the small overall experimental error. CLEO has amassed a huge amount of data on tape; I am most eager to see their findings.

The first *direct* evidence for $V(ub) \neq 0$ came from the endpoint spectrum in *inclusive* semileptonic B decays. Such studies yielded $|V(ub)|_{end} = (3.2 \pm 0.8) \times 10^{-3}$ with a heavy reliance on theoretical models which makes both the central value and the error estimate suspect. Yet with huge new data sets becoming available, this avenue should be re-visited due to the following two observations:

- The AC^2M^2 model constitutes a good implementation of QCD, in particular for $b \rightarrow u$ transitions [18]. The main caveat is that one should not determine the two model parameters p_F and m_{sp} from the $b \rightarrow c$ spectrum and then apply it blindly to $b \rightarrow u$ decays. With sufficient statistics one can fit it directly to the $b \rightarrow u$ spectrum even over the very limited kinematical regime where it can be cleanly separated from $b \rightarrow c$.

- A few years ago it has been suggested [43] to extract the required shape function for $b \rightarrow u$ from the measured photon spectrum in $B \rightarrow \gamma X$. This might become a feasible procedure with future data. Some more theoretical work is needed, though, a point I will return to.

¿From the exclusive channels $B \rightarrow l\nu\pi$ and $B \rightarrow l\nu\rho$ one has inferred

$$|V(ub)|_{excl} = (3.25 \pm 0.14|_{stat.} \pm 0.27|_{syst.}$$

$$\pm 0.55|_{th}) \times 10^{-3} \qquad (9)$$

There is a very strong model dependance, and it is quite unclear to me whether the theoretical uncertainty has been evaluated in a reliable fashion by comparing the findings from various quark models and QCD sum rules. One hopes that lattice QCD will provide the next step forward.

LEP groups have made heroic efforts to extract the width $\Gamma(H_b \rightarrow l\nu X_{no\ charm})$. Their findings read as follows [1]:

$$|V(ub)|_{\Gamma_{SL}} = (4.04 \pm 0.44|_{stat} \pm 0.46|_{b \rightarrow c, syst}$$

$$\pm 0.25|_{b \rightarrow u, syst} \pm 0.02|_{\tau_b} \pm 0.19|_{HQE}) \times 10^{-3}$$
$$(10)$$

The theoretical uncertainties in this fully integrated width are under good control [19]; however it is an experimental tour de force with the uncertainty in the modelling for $b \rightarrow c$ the central one.

The main drawback in using the charged lepton energy as a kinematical discriminator is its low efficiency: about 90% of the $b \rightarrow u$ events are buried under the huge $b \rightarrow c$ background. The hadronic recoil mass spectrum $\frac{d}{dM_X}\Gamma(B \rightarrow l\nu X)$ provides a much more efficient filter with only about 10% of $b \rightarrow u$ being swamped by $b \rightarrow c$ as first suggested within a parton model description [20]. Using HQE methodology it has been shown that the theoretical description can be based more directly on QCD [21,22]. Furthermore the fraction of $b \rightarrow u$ events below $M_X \sim 1.6$ GeV appears to be fairly stable. The predicted M_X spectrum can be compared with data – if one 'smears' the latter over energy intervals $\sim \Lambda$. Refinements of these ideas are under active theoretical study [23].

(i) $|V(td)|$ can be inferred from B_s oscillations [b]

$$\frac{x_d}{x_s} \simeq \frac{|V(td)|^2}{|V(ts)|^2}\frac{Bf^2(B_d)}{Bf^2(B_s)}, \qquad (11)$$

---

[b]The 3-family unitarity constraint $|V(ts)| \simeq |V(cb)|$ is assumed throughout this talk unless stated otherwise.

although even the relative size of B_d and B_s oscillations could be affected significantly by New Physics. (ii) Another approach is to compare *exclusive* radiative decays $B \to \gamma\rho/\omega$ vs. $B \to \gamma K^*$. Yet one has to keep in mind here that long distance physics could affect $B \to \gamma\rho$ much more than $B \to \gamma K^*$. (iii) The cleanest way theoretically is provided by the width for $K^+ \to \pi^+\nu\bar{\nu}$. With the hadronic matrix element inferred from $\Gamma(K^+ \to \pi^0 l^+\nu)$ the contributions from intermediate charm quarks provide the irreducible theoretical uncertainty estimated to be around several percent. With the present loose bounds on $|V(td)|$ one expects [24]

$$\mathrm{BR}(K^+ \to \pi^+\nu\bar{\nu}) = (0.82 \pm 0.32) \cdot 10^{-10} \tag{12}$$

One candidate has been observed by E787 at BNL corresponding to

$$\mathrm{BR}(K^+ \to \pi^+\nu\bar{\nu}) = (1.5^{+3.4}_{-1.2}) \cdot 10^{-10} \tag{13}$$

The single event sensitivity is supposed to go down to $0.7 \cdot 10^{-10}$; the successor experiment E949 hopes for a sensitivity of $\sim 10^{-11}$.

In summary:

- There are two ways for extracting $|V(cb)|$ from semileptonic B decays where the *theoretical* uncertainty has been reduced to about 5% with a further reduction appearing feasible. This theoretical confidence cannot be put to the test yet due to a divergence in the available data.

- PDG2K quotes a $\sim 40\%$ error on $V(ub)$. The situation will improve qualitatively as well as quantitatively: reducing uncertainties down to the 10% level seems feasible, and in the long run one can dream to go even beyond that!

- Observing B_s oscillations and $B \to \gamma\rho/\omega$ would elevate our knowledge of $|V(td)|$ to a new level: in particular the former should yield a value with an error not exceeding 10 %, although it could be affected very significantly by New Physics;

an intriguing long term prospect is provided by $K^+ \to \pi^+\nu\bar{\nu}$.

2.3 Quark-Hadron Duality – a New Frontier

When extracting the value of CKM parameters with few percent errors only, one has to be concerned about several sources of systematic uncertainties. A fundamental one is the assumption of *quark-hadron duality (QHDu)* that enters the theoretical reasoning. When calculating a rate on the quark-gluon level QHDu is invoked to equate the result with what one should get for the corresponding process expressed in hadronic quantities. QHDu *cannot* be exact: it is an approximation the quality of which is process-dependant – it should work better for semileptonic than nonleptonic transitions – and increases with the amount of averaging or 'smearing' over hadronic channels. There is a lot of folklore that leads to several useful concepts – but no theory. That is not surprising: for QHDu can be addressed in a quantitative fashion only *after* nonperturbative effects have been brought under control, and that has happened only relatively recently in beauty decays.

Developing such a theory for QHDu thus represents a new frontier requiring the use of new tools. Considerable insight exists into the physical origins of QHDu violations: (i) They are caused by the exact location of hadronic thresholds that are notoriously hard to evaluate. Such effects are implemented through 'oscillating terms'; i.e., the fact that innocuous, since suppressed contributions $\exp(-m_Q/\Lambda)$ in Euclidean space turn into dangerous while unsuppressed $\sin(m_Q/\Lambda)$ terms in Minkowski space. (ii) There is bound to be some sensitivity to 'distant cuts' [11]. (iii) The validity of the $1/m_c$ expansion arising in the description of $B \to l\nu D^*$ is far from guaranteed.

The OPE *per se* is insensitive to QHDu

violations (although it provides some indirect qualitative insights). One can probe QHDu in exactly solvable model field theories among which the 't Hooft model – QCD in 1+1 dimensions with $N_C \to \infty$ – has gained significant consideration. It had been suggested [25], based on a numerical analysis, that nonleptonic transitions exhibit significant or even large QHDu violations; yet analytical studies revealed such violations to be tiny only [26], even in spectra [27].

A more convincing probe for QHDu violations would be to determine the same basic quantity in different ways. I have already listed one example, namely to extract m_b from $\Upsilon(4S)$ spectroscopy as well as the leptonic and hadronic moments in B decays. One very telling implementation of such a program would be to determine CKM parameters in B_s decays and compare the results with the findings in $B_{u,d}$ decays. For practical reasons one would probably be limited to compare the leptonic and hadronic moments in semileptonic B_s decays and to infer $|V(bc)|$ from $\Gamma_{SL}(B_s)$ and $B_s \to l\nu D_s^*$. Comparing B_s with B results is much more revealing than comparing B_d with B_u decays. For a likely source of QHDu violations in $b \to c$ is provided by the presence of a 'near-by' resonance with appropriate quantum numbers. If B_d decays are affected, so will be those of B_u and by the same amount, but not B_s. Likewise a resonance near the B_s could affect its transitions, but not those of $B_{u,d}$. Nature had to be truly malicious to place one resonance next to $B_{u,d}$ and a second one next to the B_s. Barring that a comparison of the values of $|V(cb)|$ obtained from B and B_s decays would allow us to gauge QHDu violations in those transitions. The situation is more complex for $b \to u$, though, as already alluded to. An isoscalar or isovector resonance would affect B_u and B_d modes differently.

2.4 Lifetimes as Validation Studies

Among the many several important lessons to be derived from the lifetimes of charm and beauty hadrons I will emphasize just one aspect: with QHDu violations expected to be larger in nonleptonic than semileptonic decays, one can view studies of lifetimes as validation studies. The new measurements reported on D^0, D^+, D_s and Λ_c are in line with previous mesasurements and do not change the overall picture [28]: (i) The D^0-D^+ lifetime difference is given mainly by Pauli intererference yielding a ratio of $\sim 2 \cdot (f_D/200\,\mathrm{MeV})^2$. (ii) Weak annihilation should contribute in *mesons* on the 10 - 20 % level. (iii) The ratio $\tau(D_s)/\tau(D^0)$ is fully consistent with such a semiquantitative picture. (iv) What is missing for a full evaluation are more accurate Ξ_c lifetimes: measurements of $\tau(\Xi_c^{0,+})$ with 10 - 15 % accuracy are needed for this purpose.

The situation of beauty lifetimes has changed in one respect: the world average for the B^+-B_d lifetime ratio now shows a significant excess over unity in agreement with a prediction using factorization [28]:

$$\frac{\tau(B^-)}{\tau(B_d)} = 1.07 \pm 0.02 \ \mathrm{exp.} \ \ vs.$$

$$1 + 0.05 \cdot ft \left(\frac{f_B}{200\,\mathrm{MeV}}\right)^2 \ \ \mathrm{theor.} \quad (14)$$

The discrepancy for $\tau(\Lambda_b)$ has remained [28]:

$$\frac{\tau(\Lambda_b)}{\tau(B_d)} = 0.794 \pm 0.053 \ \mathrm{exp.} \ vs. \ 0.88 - 1.0 \ \mathrm{theor.} \quad (15)$$

While this could signal a significant limitation to QHDu, I like to reserve my judgement till CDF and D0 measure $\tau(\Lambda_b)$ & $\tau(\Xi_b^{0,-})$ in the next run.

The most striking success has been the apparently correct prediction of the B_c lifetime: $\tau(B_c) \sim 0.5$ psec [29] vs. the CDF findings 0.46 ± 0.17 psec with $\tau(B_c)/\tau(B_d) \sim 1/3$: the absence of a $1/m_Q$ correction is essential here. The B_s lifetime deserves further dedicated scrutiny since theoretically one expects with confidence $\bar{\tau}(B_s)/\tau(B_d) = 1 \pm \mathcal{O}(0.01)$ vs. the experimental value of 0.945 ± 0.039.

2.5 Exclusive Nonleptonic B Decays – another New Frontier

In describing nonleptonic two-body modes $B \to M_1 M_2$ valuable guidance has been provided by symmetry considerations based on $SU(2)$ and to a lesser degree $SU(3)$. Phenomenological models have played an important role; often they involve factorization as a central assumption. Such models still play an important role in widening our horizon when used with common sense [30]. Yet the bar has been raised for them by the emergence of a new theoretical framework for dealing with these decays. The essential pre-condition for this framwork is the large energy release, and it invokes concepts like 'colour transparency'; while those have been around for a while [31], only now they are put into a comprehensive framework. Two groups have presented results [32,33]. The common feature in their approaches is that the decay amplitude is described by a kernel containing the 'hard' interaction given by a perturbatively evaluated Hamiltonion folded with form factors, decay constants and ligh-cone distributions into which the long distance effects are lumped; this *factorization* is symbolically denoted by

$$\langle M_1 M_2 | H | B \rangle = f_{B \to M_1} f_{M_2} T^H * \Phi_{M_2} + ... \tag{16}$$

The two groups differ in their dealings with the soft part: BBNS regularize the divergent IR integrals they encounter at the price of introducing low energy parameters. KLS on the other hand invoke Sudakov form factors to shield them against IR singularities. It is not surprising that the two groups arrive at different conclusions: while BBNS infer final state interactions to be mostly small in $B \to \pi\pi, K\pi$ with weak annihilation being suppressed, KLS argue for weak annihilation to be important with final state interactions *not* always small.

The trend of these results have certainly the ring of truth for me: e.g., while factorization represents the leading effect in most cases (including $B \to D\pi$), it is not of universal quality. One should also note that the *non*-factorizable contributions move the predictions for branching ratios towards the data – a feature one could not count on *a priori*. It is not clear to me yet whether the two approaches are complementary or irreconcilable. Secondly one should view these predictions as preliminary: a clear disagreement with future data should be taken as an opportunity for learning rather than for discarding the whole approach. This is connected with a third point: there are corrections of order Λ/m_b which are beyond our computational powers. Since Λ might be as large as 0.5 - 1 GeV, they could be sizeable.

2.6 Radiative B Decays

The transition $B \to \gamma X$ has been the first correctly predicted penguin footprint. The CLEO number is still the most accurate one, but the BELLE result is not far behind

$$\mathrm{BR}(B \to \gamma X_{no\ charm}) = (3.15 \pm 0.35 \pm 0.32$$
$$\pm 0.26) \cdot 10^{-4} \quad \mathrm{CLEO} \tag{17}$$

$$\mathrm{BR}(B \to \gamma X_{no\ charm}) = (3.34 \pm 0.5 \pm 0.35$$
$$\pm 0.28) \cdot 10^{-4} \quad \mathrm{BELLE} \tag{18}$$

The SM prediction as summarized in an illuminating talk by Misiak reads [34]

$$\mathrm{BR}(B \to \gamma X_{no\ charm})|_{\mathrm{SM}} = (3.29 \pm 0.33) \cdot 10^{-4} \tag{19}$$

While the central value and the uncertainty have hardly changed over the last four years, an impressive theoretical machinery has been developed resulting in many new calculations – with the result that new contributions largely cancel. Careful analysis of the photon spectrum is under way, which is necessary to determine the branching ratio even more precisely and to determine the shape function needed to extract $|V(ub)|$ from the lepton endpoint spectrum [43].

The results and caveats for $B \to l^+l^- X$ have been updated. One should note that New Physics in general impacts $B \to \gamma X$ and $B \to l^+l^- X$ quite differently.

3 CP Violation in $\Delta S, \Delta B \neq 0$

The quantity ϵ'/ϵ describes the difference in CP violation between $K_L \to \pi^+\pi^-$ and $K_L \to \pi^0\pi^0$:

$$\text{Re}\frac{\epsilon'}{\epsilon} = \frac{1}{6} ft \left[\frac{|\eta_{+-}|^2}{|\eta_{00}|^2} - 1 \right] \qquad (20)$$

Within the KM ansatz direct CP violation has to exist, yet it is suppressed by the $\Delta I = 1/2$ rule and the large top mass: $0 < \epsilon'/\epsilon \ll 1/20$. A guesstimate suggests $\epsilon'/\epsilon \sim \mathcal{O}(10^{-3})$ [36]. The effective CP odd $\Delta S = 1$ Lagrangian has been calculated with high accuracy on the quark level [35]; eight operators emerge. Evaluating their hadronic matrix elements with the available techniques one finds four positive and four negative contributions of roughly comparable size giving rise to large cancellations and thus enhanced uncertainties with central values typically below 10^{-3}. While such studies found sizeable $\Delta I = 1/2$ enhancements they fell well short of the observed size; various rationalizations were given for this failure, and overcoming it was left as a homework assignment for lattice QCD. However there were dissenting voices arguing for a more phenomenological approach where reproducing the $\Delta I = 1/2$ rule is imposed as a goal. Not surprisingly this required the enhancement of some operators more than others thus reducing the aforementioned cancellations and increasing the prediction for ϵ'/ϵ [37]. The first KTeV data gave considerable respectability to this approach and lead to re-evaluations of other studies leading to somewhat larger predictions, as discussed at this conference [38].

This illustrates that theoretical uncertainties are very hard to estimate reliably, although in fairness two things should be pointed out: (i) Due to the large number on contributions with different signs one is facing an unusually complex situation. (ii) While there is no doubt that $\epsilon' \neq 0$ holds, its exact size is still uncertain:

$$\text{Re}\left(\frac{\epsilon'}{\epsilon'}\right) = (2.80 \pm 0.41) \cdot 10^{-3} \text{ KTeV}$$

$$= (1.40 \pm 0.43) \cdot 10^{-3} \text{ NA48(21)}$$

some of the earlier theoretical expectations might experience some vindication still. In any case we are eagerly awaiting the new results from KTeV.

Our interpretation of the data is thus still in limbo: it might represent another striking success for the KM scheme with the $\Delta I = 1/2$ rule explained in one fell swoop – or it might be dominated by New Physics. I am not very confident that analytical methods can decide this issue, although some interesting new angles have been put forward on the $\Delta S = 1/2$ rule [39]. One has to hope for lattice QCD to come through, yet it has to go beyond the quenched approximation, which will require more time.

Although CP violation implies T violation due to the CPT theorem, I consider it highly significant that more direct evidence has been obtained through the 'Kabir test': CPLEAR has found [40]

$$A_T \equiv \frac{\Gamma(K^0 \to \bar{K}^0) - \Gamma(\bar{K}^0 \to K^0)}{\Gamma(K^0 \to \bar{K}^0) + \Gamma(\bar{K}^0 \to K^0)} =$$

$$= (6.6 \pm 1.3 \pm 1.0) \cdot 10^{-3} \qquad (22)$$

versus the value $(6.54 \pm 0.24) \cdot 10^{-3}$ inferred from $K_L \to \pi^+\pi^-$. Of course, some assumptions still have to be made, namely that *semileptonic* K decays obey CPT or that the Bell-Steinberger relation is satisfied with *known* decay channels only. Avoiding both assumptions one can write down an admittedly contrived scheme where the CPLEAR data are reproduced *without* T violation; the price one pays is a large CPT asymmetry $\sim \mathcal{O}(10^{-3})$ in $K^\pm \to \pi^\pm\pi^0$ [42].

KTeV and NA48 have analyzed the rare decay $K_L \to \pi^+\pi^- e^+ e^-$ and found a large T-*odd* correlation between the $\pi^+\pi^-$ and e^+e^- planes in full agreement with predictions [41]. Let me add just two comments here: (i) This agreement cannot be seen as a success for the KM ansatz. Any scheme reproducing η_{+-} will do the same. (ii) The argument that strong final state interactions (which are needed to generate a T odd correlation above 1% with T invariant dynamics) cannot affect the relative orientation of the e^+e^- and $\pi^+\pi^-$ planes fails on the quantum level [42].

One often hears that observing a CP asymmetry in $B \to \psi K_S$ is no big deal since it is confidently expected – unless it falls outside the predicted range – and likewise in $B \to \pi^+\pi^-$ since it cannot be interpreted cleanly due to Penguin 'pollution' and the value of its asymmetry is hardly constrained. Such sentiments, however, miss the paradigmatic character of such observations: (a) An asymmetry in $B \to \psi K_S$ would be the first one observed outside K_L decays, it would have to be big to be established in the near future and it would establish the KM ansatz as a major agent. (b) Likewise an asymmetry in $B \to \pi^+\pi^-$ again would have to be big, and it would probably reveal *direct* CP violation to be big as well in beauty decays.

These CP asymmetries are described in terms of the angles of the usual unitarity triangle. An ecumenical message in PDG2000 endorses two different notations, namely

$$\phi_1 \equiv \beta = \pi - \arg\left(\frac{V_{tb}^* V_{td}}{V_{cb}^* V_{cd}}\right) \quad (23)$$

$$\phi_2 \equiv \alpha = \arg\left(\frac{V_{tb}^* V_{td}}{-V_{ub}^* V_{ud}}\right) \quad (24)$$

$$\phi_3 \equiv \gamma = \arg\left(\frac{V_{ub}^* V_{ud}}{-V_{cb}^* V_{cd}}\right). \quad (25)$$

From CP insensitive rates one can deduce the sides of this triangle and from CP asymmetries the angles: e.g., from $\epsilon/\Delta m(B_d)$ one can infer $\sin 2\phi_1$. A whole new industry has sprung up for doing these fits. Typical examples are (I will discuss caveats below):

$$\sin 2\phi_1 = 0.716 \pm 0.070 \;[2] , \; 0.7 \pm 0.1 \;[44] \quad (26)$$
$$\sin 2\phi_2 = -0.26 \pm 0.28 \;[2] , \; -0.25 \pm 0.6 \;[44] \quad (27)$$

The first results from the asymmetric B factories leave us in limbo:

$$\sin 2\phi_1 = 0.45^{+0.43+0.07}_{-0.44-0.09} \;\; \text{BELLE} \quad (28)$$
$$\sin 2\beta = 0.12 \pm 0.37 \pm 0.09 \;\; \text{BaBar} \quad (29)$$

Nevertheless one can raise the question what we would learn from a 'Michelson-Morley outcome', if, say, $|\sin 2\phi_1| < 0.1$ were established? Firstly, we would know that the KM ansatz would be ruled out as a major player in $K_L \to \pi\pi$ – there would be no plausible deniability! Secondly, one would have to raise the basic question why the CKM phase is so suppressed, unless there is a finely tuned cancellation between KM and New Physics forces in $B \to \psi K_S$; this would shift then the CP asymmetry in $B \to \pi\pi$, $\pi\rho$.

4 Probing for New Physics

$\Delta S = 1, 2$ dynamics have provided several examples of revealing the intervention of features that represented New Physics *at that time*; it thus has been instrumental in the evolution of the SM. This happened through the observation of 'qualitative' discrepencies; i.e., rates that were expected to vanish did not, or rates were found to be smaller than expected by several orders of magnitude. Such an indirect search for New Physics can be characterised as a 'King Kong' scenario: one might be unlikely to encounter King Kong; yet once it happens there can be no doubt that one has come across someting out of the ordinary. Such a situation can be realized for charm and $K_{\mu3}$ decays and EDMs.

4.1 D^0 Oscillations & CP Violation

It is often stated that D^0 oscillations are slow and CP asymmetries tiny within the SM and that therefore their analysis provides us with zero-background searches for New Physics.

Oscillations are described by the normalized mass and width differences: $x_D \equiv \frac{\Delta M_D}{\Gamma_D}$, $y_D \equiv \frac{\Delta\Gamma}{2\Gamma_D}$. A conservative SM estimate yields x_D, $y_D \sim \mathcal{O}(0.01)$. Stronger bounds have appeared in the literature, namely that the OPE contributions are completely insignificant and that long distance contributions *beyond* the OPE provide the dominant effects yielding x_D^{SM}, $y_D^{SM} \sim \mathcal{O}(10^{-4} - 10^{-3})$. A recent detailed analysis [45] revealed that a proper OPE treatment reproduces also such long distance contributions with

$$x_D^{SM}|_{OPE}, \, y_D^{SM}|_{OPE} \sim \mathcal{O}(10^{-3}) \qquad (30)$$

and that $\Delta\Gamma$, which is generated from on-shell contributions, is – in contrast to Δm_D – insensitive to New Physics while on the other hand more susceptible to violations of QHDu.

Four experiments have reported new data on y_D [1]:

$$y_D = (0.8 \pm 2.9 \pm 1.0)\% \quad \text{E791} \qquad (31)$$
$$= (1.0^{+3.8+1.1}_{-3.5-2.1})\% \quad \text{BELLE} \qquad (32)$$
$$= (3.42 \pm 1.39 \pm 0.74)\% \quad \text{FOCUS} \quad (33)$$
$$y'_D = (-2.5^{+1.4}_{-1.6} \pm 0.3)\% \quad \text{CLEO} \qquad (34)$$

E 791 and FOCUS compare the lifetimes for two different channels, whereas CLEO fits a general lifetime evolution to $D^0(t) \to K^+\pi^-$; its y'_D depends on the strong rescattering phase between $D^0 \to K^-\pi^+$ and $D^0 \to K^+\pi^-$ and thus could differ substantially from y_D [46]. The FOCUS data contain a suggestion that the lifetime difference in the $D^0 - \bar{D}^0$ complex might be as large as $\mathcal{O}(1\%)$. If y_D indeed were ~ 0.01, two scenarios could arise for the mass difference. If $x_D \leq$ few $\times 10^{-3}$ were found, one would infer that the $1/m_c$ expansion yields a correct semiquantitative result while blaming the large value for y_D on a sizeable and not totally surprising violation of QHDu. If on the other hand $x_D \sim 0.01$ would emerge, we would face a theoretical conundrum: an interpretation ascribing this to New Physics would hardly be convincing since $x_D \sim y_D$. A more sober interpretation would be to blame it on

QHDu violation or on the $1/m_c$ expansion being numerically unreliable. Observing D^0 oscillations then would not constitute a 'King Kong' scenario.

Searching for *direct* CP violation in Cabibbo suppressed D decays as a sign for New Physics would also represent a very complex challenge: within the KM description one expects to find some asymmetries of order 0.1 %; yet it would be hard to conclusively rule out some more or less accidental enhancement due to a resonance etc. raising an asymmetry to the 1% level.

The only clean environment is provided by CP violation involving D^0 oscillations, like in $D^0(t) \to K^+K^-$ and/or $D^0(t) \to K^+\pi^-$. For the asymmetry would depend on the product $\sin(\Delta m_D t) \cdot \text{Im}[T(\bar{D} \to f)/T(D \to \bar{f})]$: with both factors being $\sim \mathcal{O}(10^{-3})$ in the SM one predicts a practically zero effect.

4.2 $P_\perp(\mu)$ in $K^+ \to \mu^+\pi^0\nu$

The muon polarization transverse to the decay plane in $K^+ \to \mu^+\pi^0\nu$ represents a T-odd correlation $P_\perp(\mu) = \langle \vec{s}(\mu) \cdot (\vec{p}(\mu) \times \vec{p}(\pi))/|\vec{p}(\mu) \times \vec{p}(\pi)| \rangle$, which in this case could not be faked realistically by final-state interactions and would reveal genuine T violation. With $P_\perp(\mu) \sim 10^{-6}$ in the SM, it would also reveal New Physics that has to involve chirality breaking weak couplings: $P_\perp(\mu) \propto \text{Im}\xi$, where $\xi \equiv f_-/f_+$ with $f_-[f_+]$ denoting the chirality violating [conserving] decay amplitude. There are 'ancient' data yielding

$$\text{Im}\xi = -0.01 \pm 0.019 \qquad (35)$$

A new preliminary result from the ongoing experiment was reported here:

$$\text{Im}\xi = -0.013 \pm 0.016 \pm 0.003 \,. \qquad (36)$$

4.3 EDM's

Electric dipole moments d of non-degenerate systems represent direct evidence for T violation. The present bounds read:

$$d_{neutron} < 9.7 \cdot 10^{-26} \; ecm \qquad (37)$$

$$d_{electron} = (-0.3 \pm 0.8) \cdot 10^{-26} \ ecm \quad (38)$$

With the KM scheme predicting unobservably tiny effects (with the only exception being the 'strong CP' problem), and many New Physics scenarios yielding $d_{neutron}$, $d_{electron}$ $\geq 10^{-27}$ ecm, this is truly a promising zero background search for New Physics!

4.4 KM Trigonometry

There certainly could be 'qualitative' discrepancies in the CP asymmetries for B decays. The cleanest case is given by the CP asymmetry in $B_s(t) \to \psi\eta$ or $B_s(t) \to \psi\phi$, which is Cabibbo suppressed [47] and thus below 4% due to three-family unitarity.

Yet otherwise the situation in $\Delta B = 1, 2$ is more complex meaning it provides more opportunites, yet also more challenges. For one will be looking for *quantitative* discrepancies between predictions and the data that can*not* amount to orders of magnitude.

With three families there are six unitarity triangles containing three types of angles:

1. Angles of order unity like $\phi_{1,2,3}$; they differ from each other in order λ^2.

2. Angles that themselves are of order λ^2; the most accessible representative is an angle in the bs triangle denoted by χ:

$$\chi = \phi_1^{bs} = \pi + \left(\frac{V_{cs}^* V_{cb}}{V_{ts}^* V_{tb}} \right) \simeq \lambda^2 \eta \quad (39)$$

which controls the aforementioned asymmetry in $B_s(t) \to \psi\phi, \ \psi\eta$ [47].

3. Angles $\sim \mathcal{O}(\lambda^4)$, the least unaccessible one being in the cu triangle often referred to as χ' [48]

$$\chi' = \phi_3^{cu} = \left(\frac{-V_{ud}^* V_{cd}}{V_{us}^* V_{cs}} \right) \simeq -\lambda^4 A^2 \eta \quad (40)$$

controling CP asymmetries in D decays.

A comprehensive program will have to undertake three steps:

- measure the large angles $\phi_{1,2,3}$ (and their 'cousins') and check their correlations with the sides of the triangle;

- check whether the small [tiny] angle χ [χ'] is indeed small [tiny];

- attempt to measure the $\mathcal{O}(\lambda^2)$ differences between $\phi_{1,2,3}$ and their cousins.

All of these represent searches for New Physics with in particular the last item probing features of such New Physics beyond its 'mere' existence.

With many of the SM effects being large, one is looking for deviations from expectations that are mostly of order unity. A typical scenario would be that an asymmetry of, say, 40 % is expected, yet 80% is observed; how confident could we be in claiming New Physics? What about 40% vs. 60% or even 50%? The situation is thus qualitatively different from K decays where *original* expectations and data differed by orders of magnitude! Therefore we have to be very conscious of three scourges: (i) Systematic experimental uncertainties; (ii) experiments could be wrong – an issue addressed by the 'combiner' program [44]; (iii) theoretical uncertainties!

4.5 On Theoretical Uncertainties

While considerable experience exists concerning the *quantitative* aspects of experimental shortcomings, this is not so with respect to theoretical uncertainties. My understanding behind quoting the latter is the following: "I would be very surprised if the true value would fall outside the stated range." Such a statement is obviously hard to quantify.

An extensive literature on how to evaluate them has em erged over the last two years in particular (see, e.g., [2,44]). It seems to me that the passion of the debate has overshadowed the fact that a lot of learning has happened. For example it is increasingly understood that any value within a stated range has to be viewed as equally likely. While concerns are legitimate that some actors might be overly aggressive in stating constraints on the KM triangle, it would be unfair to characterize them as silly. I also view it as counterproductive to bless one approach while anathematizing all others 'ex cathedra'. I believe many different paths should be pursued since "good decisions come from experience that often is learnt from bad decisions".

Our most powerful weapon for controlling theoretical uncertainties will again be *over*determining basic quantities by extracting their values from more than one independant measurement. The situation is more favourable in B than in K decays since there are fewer free parameters *relative* to the number of available decay modes. Once the investment has been made to collect the huge number of decays required to obtain a sufficient number of the transitions of primary interest, then we have also a host of other channels that can provide us with information about hadronization effects etc. Finally one should clearly distinguish the goal one has in mind: does one want to state the most likely expectation – or does one want to infer the presence of New Physics from a discrepancy between expectations and data? The latter goal is of course much more ambitious where for once being conservative is a virtue!

4.6 *Looking into the Crystal Ball*

I expect various large CP asymmetries to be found in B decays – including direct CP violation – over the next 15 years that agree with the KM expectations to first order, yet exhibit smallish, though definite deviations thus revealing the intervention of New Physics. However it is conceivable that the whole future beauty phenomenology can be accommodated in the CKM ansatz. Would that mean our efforts will have been wasted?

My answer is an emphatic no! The pattern in the Yukawa couplings often referred to as 'textures' is presumably determined by very high scale dynamics. They provide the seeds for the quark mass matrix arising when Higgs fields develop vacuum expectation values at lower scales. The quark mass matrix yields the quark masses and the CKM parameters. My conjecture is that such textures follow a simple pattern yielding 'special' CKM parameters. From the observed values of CKM quantities one can thus infer information on the dynamics at very high scales.

Yet what is a manifestly simple pattern at very high energies will look quite different at the electroweak scales that can be probed: renormalization will tend to wash out striking features. This again calls for *precise* extractions of these fundamental parameters.

5 Conclusions & Outlook

We have reached an exciting and even decisive phase in flavour dynamics.

- Since the phenomenological success of the CKM description is a priori quite surprising, it must contain a deep, albeit hidden message.

- New (sub)paradigms have been established or are about to be established: direct CP violation has been found, intriguing hints for the first CP asymmetry outside K_L decays have emerged and the CKM predictions for CP violation in B decays are about to be tested. These represent *high sensitivity* probes of dynamics and contain many possible portals to New Physics.

- Basic quantities have become known with good accuracy and the promise for even more: the beauty quark mass is known to within about 1.5 % – the most precise quark mass; the top mass is known to within 3% [10 %] due to direct observation [radiative corrections]; $|V_{cb}|$ has been extracted with about 5 % or so accuracy with a reduction down to ~ 2 %

appearing feasible; the error on $|V_{ub}|$ of presently about 40 % should be reduced to the 10% level with 5% not appearing to be impossible in the long run; for $|V_{td}|$ with its present uncertainty $\sim$ 60 % a reduction down to 10 % again might not be impossible. Thus B physics will develop into a *high precision* probe for New Pghysics as well.

- These developments have been made possible by *practical* theoretical technologies having been greatly improved: there has been increasing sophistication in treating semileptonic and radiative B decays; a new frontier has emerged in treating exclusive nonleptonic B decays with intriguing classification schemes truly based on QCD that might allow us to calculate these transitions in the real world.

- Theoretical uncertainties are mostly of a systematic nature. They can reliably be evaluated only through overconstraints. Prior to that they should be considered *preliminary*; in that context I would like to ask the community to accord us theorists the same professional courtesies that is granted to experimental analyses.

- To make good use of such developments we need experimental programs that allow *precise* measurements in a *comprehensive* way rather than just one or two precise ones. It will be an exciting adventure to find out how far such a program can be pushed. In this context I applaud the managements of CERN and FNAL for their wisdom in approving LHC-b and BTeV.

- There are other areas that might well contain portals to New Physics: dedicated searches for CP violation in charm decays, EDMs and transverse muon polarization in $K_{\mu 3}$ decays are an absolute

must since any improvement in experimental sensitivity might reveal an effect. This is even more so in light of recent efforts to explain baryogenesis as being driven by leptogenesis in the Universe.

- There is mounting evidence for neutrino oscillations implying lepton flavour eigenstates to differ from lepton mass eigenstates; the saw-see mechanism provides an attractive framework for explaining the smallness of neutrino masses. There are intriguing connections between the atmospheric neutrino anomaly and $\tau \to \mu\gamma$ and between the solar neutrino anomaly and $\mu \to e\gamma$ in the context of SUSY GUTs [49].

In future meeting there will be detailed discussions of the lepton analogue to the Cabibbo-Kobayashi-Maskawa matrix, the Maki-Nakagawa-Sakata [50] matrix indicating that leptons after all are 'exactly like quarks – only different!'.

Acknowledgements

The organizers deserve thanks for their smooth organization of this conference in the lively city of Osaka. I have benefitted from discussions with A. Golutvin, M. Beneke and A. Sanda. This work has been supported by the NSF under the grant PHY 96-05080.

References

1. A. Golutvin, these Proc.
2. see, for example: A. Stocchi, these Proc.
3. R. Kenway, these Proc.
4. I. Bigi et al., *Phys.Lett.* **B181**(1986)157.
5. M. Voloshin, *Int. J. Mod. Phys.* **A10** (1995) 2865.
6. K. Melnikov, A. Yelkhovsky, *Phys. Rev.* **D 59** (1999) 114009.
7. A. Hoang, *Phys.Rev.***D61**(2000)034005.
8. M. Beneke, A. Signer, *Phys. Lett.* **B 471** (1999) 233.

9. A. Falk, M. Luke, M. Savage, *Phys. Rev.* **D 53** (1996) 2491.

10. A.Kronfeld, J.Simone, hep-ph/0006345.

11. I. Bigi, M. Shifman, N. Uraltsev, A. Vainshtein, *Phys. Rev.* **D52** (1995) 196.

12. I. Bigi, M. Shifman, N. Uraltsev, *Annu. Rev. Nucl. Part. Sci.* **47**(1997)591.

13. I. Bigi, hep-ph/9907270.

14. The BABAR Physics Book, P. Harrison & H. Quinn (eds.), SLAC-R-504.

15. J. Simone et al., *Nucl. Phys. Proc. Supp.* **83** (2000) 334.

16. K. Ecklund, these Proc.

17. E. Barberio, these Proc.

18. I. Bigi, M. Shifman, N. Uraltsev, A. Vainshtein, *Phys. Lett.* **B328** (1994) 431.

19. N. Uraltsev, *Int. J. Mod. Phys.* **A14** (1999) 4641.

20. V. Barger, C.S. Kim, G. Phillips, *Phys. Lett.* **B 251** (1990) 629.

21. R. Dikeman, N. Uraltsev,*Nucl. Phys.* **B509** (1998)378; I. Bigi, R. Dikeman, N. Uraltsev,*Eur. Phys. J.***C4** (1998)453.

22. A. Falk, Z. Ligeti, M. Wise,*Phys. Lett.* **B 406** (1997) 255.

23. C. Bauer, Z. Ligeti, M. Luke, hep-ph/0002161.

24. G. Buchalla, A. Buras, *Nucl. Phys.* **B 548** (1999) 309.

25. B. Grinstein, R. Lebed,*Phys.Rev.* **D57** (1998) 1366; **D59** (1999) 054022.

26. I. Bigi, M. Shifman, N. Uraltsev, A. Vainshtein, *Phys. Rev.* **D 59** (1999) 054011; I. Bigi, N. Uraltsev, *Phys. Rev.* **D 60** (1999) 114034; I. Bigi, N. Uraltsev, *Phys. Lett.* **B 457** (1999) 163.

27. R. Lebed, N. Uraltsev, hep-ph/0006346.

28. G. Bellini, I. Bigi, P. Dornan, *Phys. Rep* **289** (1997) 1.

29. I. Bigi, *Nucl. Instr. & Meth. in Phys. Res.* **A 351** (1994) 240; *Phys. Lett.***B 371** (1996) 105; M. Beneke, G. Buchalla, *Phys. Rev.* **D 53**(1996) 4991.

30. R. Fleischer, H.-Y. Cheng, G. Hou, these Proc.

31. See, e.g., B. Ward, *Phys. Rev.* **D 51** (1995) 6253.

32. M. Beneke, G. Buchalla, M. Neubert, C. Sachrajda, hep-ph/0007256.

33. Y.-Y. Keum, H.-n. Li, A. Sanda, hep-ph/0004173.

34. M. Misiak, these Proc.

35. A. Buras, P. Gambino, M. Gorbahn, S. Jager, L. Silvestrini, hep-ph/0007313.

36. M. Fabbrichesi, these Proc.

37. S. Bertolini, J. Eeg, M. Fabbrichesi, *Rev. Mod. Phys.* **72** (2000) 65.

38. H.-Y. Cheng, G. Golowich, J. Prades ,G. Valencia,these Proc.

39. J. Prades, *Nucl. Phys. B (Proc. Suppl.)* **86** (2000) 294.

40. A. Apostolakis et al., *Phys. Lett.* **B444** (1998)43; *Phys. Lett.* **B456** (1999)297.

41. L. Sehgal, M. Wanninger, *Phys. Rev.* **D 46**(1992) 1035; 5209 (E).

42. I. Bigi, A. Sanda,*Phys. Lett.* **B466** (1999) 33.

43. M. Neubert, *Phys. Rev.* **D 49**(1994) 4623; I.I. Bigi, M.Shifman, N. Uraltsev, A. Vainshtein, *Int. J. Mod. Phys.* **A 9** (1994) 2467.

44. A. Hoecker, H. Lacker, S. Laplace, F. Le Diberder, preliminary results obtained using a non-Bayesian approach, paper in preparation.

45. I. Bigi, N. Uraltsev, hep-ph/0005089.

46. S. Bergmann, Y. Grossman, Z. Ligeti, Y. Nir, A. Petrov, *Phys. Lett.* **B 486** (2000) 418.

47. I. Bigi, A.I. Sanda, *Nucl. Phys.* **B 193** (1981) 85.

48. I.Bigi,in: Proc. of the Tau-Charm Factory Workshop,1989, SLAC-Report-343.

49. S. Baek, T. Goto, Y. Okada, K.-I. Okumura, hep-ph/0002141.

50. Z. Maki, M. Nakagawa, S. Sakata, *Prog. Theor. Phys.* **30** (1963) 727.

Plenary Session 5

Lattice Field Theory

Chair: Tao Huang (Chinese Academy of Sciences)
Scientific Secretary: Tetsuya Onogi

LATTICE FIELD THEORY

RICHARD KENWAY

Department of Physics and Astronomy, The University of Edinburgh, Edinburgh EH9 3JZ, Scotland

E-mail: r.d.kenway@ed.ac.uk

This review concentrates on progress in lattice QCD during the last two years and, particularly, its impact on phenomenology. The two main technical developments have been successful implementations of lattice actions with exact chiral symmetry, and results from simulations with two light dynamical flavours which provide quantitative estimates of quenching effects for some quantities. Results are presented for the hadron spectrum, quark masses, heavy-quark decays and structure functions. Theoretical progress is encouraging renewed attempts to compute non-leptonic kaon decays. Although computing power continues to be a limitation, projects are underway to build multi-teraflops machines over the next three years, which will be around ten times more cost-effective than those of today.

1 Introduction

1.1 Overview

Progress in lattice QCD tends to be incremental. Through improved formulations and increased computer power, we have been steadily gaining control over all the systematic approximations inherent in numerical simulations. Rarely, progress is revolutionary. Happily, we are in the midst of such a major leap forward, through recent demonstrations that lattice formulations, which preserve exact chiral symmetry, work in practice. Combined with increasingly cost-effective computing technology, a period of accelerated progress, impacting directly on phenomenologically important QCD calculations, can be foreseen over the next five years.

In this review, I describe recent progress demonstrating that chirally symmetric formulations are feasible and results for phenomenologically relevant quantities. I omit results for QCD thermodynamics and, with one exception, for non-QCD theories. Although there has been a lot of work on the challenging problem of the confinement mechanism, the understanding achieved so far is partial and the picture is still too confusing to do it justice in a review such as this. For results in these areas, other exploratory phenomenological applications and most of the technical details, I refer you to the latest in the annual series of lattice conference proceedings[1].

1.2 Objectives of Lattice Field Theory

The primary objective of lattice field theory is to determine the parameters of the Standard Model and, thereby, to seek signals of new physics. Due to confinement, the quark sector is not directly accessible by experiment and numerical simulations of QCD are needed to provide the missing link. In principle, lattice QCD offers model-independent computations of hadronic masses and matrix elements. Ultimately, it should test QCD as the theory of strong interactions and provide an understanding of confinement. The name of the game is the control of systematics, particularly to quantify dynamical quark effects and reliably simulate at, or extrapolate to the physical values of the light and heavy quark masses.

The second objective is to determine the phase structure of hadronic matter. Both the location and the order of the line of phase transitions, separating the confined, hadronic phase from the deconfined, quark-gluon plasma phase, in the temperature (T), chemical potential (μ) plane, are sensitive to the flavour content. Since T_c is close to the strange quark mass, it is particularly important to simulate the strange quark accurately, and this awaits simulations with more realistic dynamical light flavours. Also, the $\mu \neq 0$ plane is not accessible to simulations with current algorithms, because the action is complex and Monte Carlo importance sampling fails. Significant progress has been made in QCD thermodynamics at $\mu = 0$, and new approaches using anisotropic lattices should yield spectroscopic results to help better understand leptonic decays of vector mesons, strangeness production and J/ψ suppression.

Finally, looking beyond QCD, we aim to develop simulations into a general purpose non-perturbative tool. Recent progress in formulating lattice chiral symmetry has reawakened hopes of being able to simulate chiral and SUSY theories.

2 Theoretical Progress

2.1 Lattice Chiral Symmetry

Major progress has been achieved over the past eight years following the rediscovery of the Ginsparg-Wilson (GW) relation[2]. This states that if the lattice Dirac operator, D, is chosen to satisfy

$$\gamma_5 D + D\gamma_5 = aD\gamma_5 D, \qquad (1)$$

where a is the lattice spacing, then the theory possesses an exact chiral symmetry. Such a formulation has the great virtue that it separates the chiral and continuum limits. It also forbids $O(a)$ terms, so that the resulting fermion actions are improved. The GW relation languished for many years, because there was no practical implementation. The breakthrough was the discovery of three constructions: the overlap[3], domain wall[4] and perfect action[5]. As a consequence, lattice simulations with exact chiral symmetry are now a reality (see Kikukawa's talk[6] and other reviews[7]).

For vector theories like QCD, this means that we can maintain a global chiral symmetry at non-zero lattice spacing, so that simulations should be able to approach the physical u and d quark masses in a controlled fashion. More importantly, the mixing of operators of different chiralities, which has plagued kaon mixing and decay calculations, is avoided. Formally, abelian and non-abelian chiral gauge theories have been constructed on the lattice[8], so the Standard Model can now be defined non-perturbatively. Somewhat more speculatively, for SUSY theories without scalar fields, lattice chiral symmetry forbids relevant SUSY-violating terms in the action, and so offers the prospect of lattice simulations without fine tuning[9].

2.2 Overlap Quarks

In the overlap formulation, the lattice Dirac operator has the form

$$D_{\text{ov}}(\mu) = \left(\frac{1+\mu}{2}\right) + \left(\frac{1-\mu}{2}\right) \gamma_5 \text{sgn}(H_{\text{W}}),$$
$$(2)$$

where μ is the bare quark mass, and H_{W} is the (hermitian) Wilson-Dirac matrix with a large negative mass. $D_{\text{ov}}(0)$ obeys the GW relation (Eq. (1)). Numerically, the challenge is to compute accurately the sign of the large sparse matrix, H_{W}, eg using a rational approximation[10]:

$$\text{sgn}(H_{\text{W}}) \approx \sum_{s=1}^{N} \frac{1}{c_s^2 + \frac{b_s^2}{H_{\text{W}}}}. \qquad (3)$$

This approximation breaks down for small eigenvalues, so it is necessary to project out the lowest eigenvectors and treat their signs exactly.

The resulting implementation has been tested for quenched QCD[11]. The results in Figure 1 show that, as expected, there is no additive quark mass renormalisation, ie

$$m_q^{\overline{\text{MS}}} a = Z_m^{-1} \mu \left[1 + O(a^2)\right]. \qquad (4)$$

The quenched light hadron spectrum has also been computed[11], albeit in a small volume,

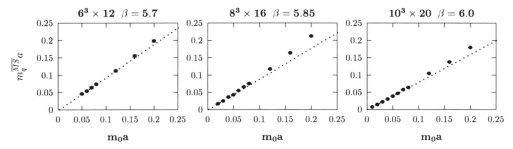

Figure 1. The renormalised quark mass, obtained from the axial Ward identity, versus bare quark mass ($\mu = m_0 a$) in quenched QCD, computed using the overlap formulation at three different lattice spacings[11].

and evidence is found that the mass ratios roughly scale at fixed quark mass, showing that it is now possible to do phenomenologically interesting calculations in these new chirally symmetric formulations.

Unfortunately, even for quenched QCD, the computational cost of simulating overlap quarks is comparable to that of dynamical simulations[12], due to the extra work involved in computing the sign function (Eq. (3)). Little is known yet about the cost of simulating dynamical quarks this way.

The approximation of the sign function in Eq. (3) may be replaced by local interations amongst a set of $2N$ auxiliary fields:

$$\bar{\psi}\, \text{sgn}(H_{\text{W}})\, \psi \approx \bar{\psi} \sum_{s=1}^{N} \frac{1}{c_s^2 + \frac{b_s^2}{H_{\text{W}}}} \psi$$
$$\rightarrow \sum_s \left[(\bar{\psi}_s \chi_s + \bar{\chi}_s \psi_s) + \bar{\chi}_s (c_s^2 H_{\text{W}}) \chi_s \right.$$
$$\left. + b_s (\bar{\chi}_s \phi_s + \bar{\phi}_s \chi_s) - \bar{\phi}_s H_{\text{W}} \phi_s \right]. \quad (5)$$

This shows that the overlap formulation may be thought of as five-dimensional, in which the fifth dimension is like flavour[13].

2.3 Domain Wall Quarks

Here the fermions live in five dimensions and are coupled to a mass defect located on a four-dimensional hyperplane. On a finite lattice, with L_s sites in the fifth dimension, the Dirac operator may be written

$$D_{\text{DW}}(\mu) = \left(\frac{1+\mu}{2} \right) + \left(\frac{1-\mu}{2} \right)$$

$$\times \gamma_5 \tanh \left(-\frac{L_s}{2} \log T \right). \quad (6)$$

Here T is the transfer matrix in the fifth dimension and μ is again the bare mass. The strong similarity with the overlap, Eq. (2), is evident. In fact, the two formulations are identical in the limits of $L_s \rightarrow \infty$ and zero lattice spacing in the fifth direction.

For finite L_s, the chiral modes of opposite chirality are trapped on the four-dimensional domain walls at each end of the fifth dimension. Chiral symmetry is broken, but the breaking is exponentially suppressed by the size of the fifth dimension. Several groups have tested this in quenched QCD[14,15]. It is found that the pion mass does not always vanish with quark mass, in the limit $L_s \rightarrow \infty$, due to near unit eigenvalues of T, which allow unsuppressed interactions between the LH and RH fermions. This is a strong-coupling effect, which goes away for weak enough coupling, or using a renormalisation-group improved action[15]. The problem may be controlled numerically by projecting out the low eigenvectors and taking their contribution with infinite L_s[16].

Thus, there is now a good understanding of how to achieve a close approximation to exact chiral symmetry in lattice QCD simulations. Although numerically relatively expensive, it is early days and more efficient algorithms may yet be found. Already, the extra degree of control given by exact symmetry probably outweighs the cost for matrix

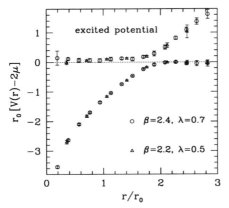

Figure 2. String breaking in the four-dimensional SU(2) Higgs model[17]. The results for the potential, $V(r)$, versus separation, r, at two lattice spacings are in excellent agreement.

element calculations, and we can anticipate rapid progress over the next few years.

3   String Breaking

For the remainder of this review, I will describe results obtained using traditional formulations of lattice QCD, focusing on the effects of dynamical quarks in order to try to quantify quenching errors for as many quantities as possible.

Perhaps, the first effect of dynamical quarks we might hope to see is string breaking. This flattening of the potential between two static charges, as they are separated, has been observed as level crossing in the confinement phase of the SU(2) Higgs model[17]. At a particular separation, the string state becomes degenerate with the state of two static-light "mesons", as shown in Figure 2.

Demonstrating string breaking in QCD at zero temperature is proving to be much more challenging than expected. As yet there is no completely convincing signal. This is because there is poor overlap between the string states used to compute the static quark potential and the broken-string state, comprising two static-light mesons. Including the latter is computationally very costly, because

it requires quark propagators at all sites.

However, the mixing matrix element has been computed for two dynamical flavours and found to be non-zero[18]. Using only string states, there are hints of a flattening potential, just as the signal becomes swamped by noise[19], but a recent high-statistics calculation provides pretty conclusive evidence that such attempts are doomed[20] and the two-meson state must be included (as was done for the Higgs model[17]). The situation is frustrating, but it is only a matter of time before sufficient computing resources are brought to bear.

4   Hadron Spectrum

The important result that the quenched light hadron spectrum disagrees with experiment was finally established by the CP-PACS Collaboration[21] in 1998 and announced at ICHEP98. This had proved difficult, requiring high statistics, because the deviation is less than 10%. This small deviation is good news for phenomenological applications of lattice QCD, which still rely heavily on the quenched approximation. The main symptom of quenching is that it is not possible consistently to define the strange quark mass – the two spectra obtained from using the K and the ϕ meson to determine the strange quark mass disagree.

Since 1998, the focus has been on simulations with two degenerate dynamical flavours, which are identified with the u and d quarks. The strange and higher-mass quarks are still treated in the quenched approximation. At this conference, CP-PACS reported that the resulting strange meson spectrum is much closer to experiment[22], as shown in Figure 3.

The glueball spectrum has only been computed in quenched QCD and the results[23] reported at ICHEP98 remain state of the art. This calculation was hard because of strong scaling violations. Better scaling has

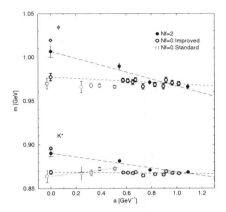

Figure 3. Comparison of continuum extrapolations of the K^* and ϕ meson masses in two-flavour (filled symbols) and quenched (open symbols) QCD[22]. The K mass was used to fix the strange quark mass.

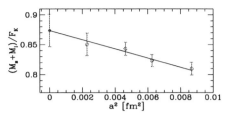

Figure 4. Continuum extrapolation of the RG-invariant mass, $M_s + M_l$, where $M_l = (M_u + M_d)/2$, in units of the K decay constant for quenched QCD[29].

been reported recently using the fixed-point action[24]. Mixing with quark states should be very important. A first attempt to compute the mixing of the lightest scalar glueball with the lightest scalar quarkonium states has concluded that the $f_0(1710)$ is 74% glueball, whereas the $f_0(1500)$ is 98% quarkonium[25].

Now that two-flavour simulations are possible, it is interesting to try to compute the flavour-singlet meson masses. A first attempt has produced a result that the η, η' mixing angle is around $45°$ in the $\{(\bar{u}u + \bar{d}d)/\sqrt{2}, \bar{s}s\}$ basis, but, despite sophisticated variance reduction techniques, at least ten-times better statistics is needed[26].

5 Quark Masses

Quark masses are encoded within hadron masses. They are scale and renormalisation-scheme dependent. To be useful for phenomenology, it is necessary to compute the scale evolution of the quark mass, from the lattice scale at which it is determined, to a suitably high scale where it can be matched to a perturbative scheme.

The usual approach is to define an intermediate scheme, such as the Schrödinger Functional (SF), in which the scale depen-

dence can be computed non-perturbatively (see Heitger's talk[27]). Once this scale dependence is known to a high enough energy, it can be continued to infinite energy, using the perturbative renormalisation group, to define the ratio of the lattice mass to the renormalisation-group (RG) invariant mass, M[28]. Thus the lattice quark mass fixes M and, from it, perturbation theory may be used to determine the quark mass in any chosen scheme.

Figure 4 shows the sum of the RG-invariant average u and d, and s quark masses, computed in this way for quenched QCD (with the K mass as input)[29]. The resulting estimate for the strange quark mass is

$$m_s^{\overline{\text{MS}}}(2\text{ GeV}) = 97(4)\text{ MeV} \quad (N_f = 0). \quad (7)$$

At this conference, CP-PACS announced results for light quark masses in two-flavour QCD[22], updating earlier results[30]. They use an improved action, with a fixed lattice size of $(2.5\text{ fm})^3$, and extrapolate downwards in the sea-quark mass, from masses corresponding to pseudoscalar-to-vector meson mass ratios above 0.6, to the average u and d quark mass. Currently, non-perturbative matching is not available for two-flavour QCD, so CP-PACS uses mean-field improved 1-loop matching. The results are shown in Figure 5, for three different definitions of quark mass at non-zero lattice spacing, and compared with quenched QCD. The different definitions permit consis-

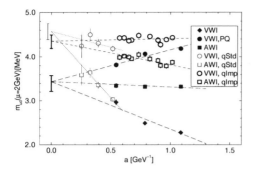

Figure 5. Continuum extrapolation of the average u and d quark masses for $N_f = 2$ (filled symbols) and $N_f = 0$ (open symbols) QCD[22,30].

tent continuum extrapolations and the final results are from combined fits with a single limit. They find a big effect from the inclusion of dynamical quarks. The average u and d quark mass is reduced by roughly 25%:

$$m_{ud}^{\overline{\mathrm{MS}}}(2\mathrm{GeV}) = 3.44^{+0.14}_{-0.22}\mathrm{MeV}\,(N_f = 2) \quad (8)$$

$$m_{ud}^{\overline{\mathrm{MS}}}(2\mathrm{GeV}) = 4.36^{+0.14}_{-0.17}\mathrm{MeV}\,(N_f = 0). \quad (9)$$

Treating the strange quark in the quenched approximation, CP-PACS finds that the 20% inconsistency in the strange quark mass in quenched QCD disappears with dynamical u and d quarks (within 10% errors), as can be seen in Figure 6. The continuum estimates are

$$m_s^{\overline{\mathrm{MS}}}(2\mathrm{GeV}) = 88^{+4}_{-6}\mathrm{MeV} \ (K \text{ mass}) \quad (10)$$

$$m_s^{\overline{\mathrm{MS}}}(2\mathrm{GeV}) = 90^{+5}_{-11}\mathrm{MeV} \ (\phi \text{ mass}). \quad (11)$$

Again, the mass is reduced substantially compared to quenched QCD. Such a low strange quark mass suggests a large value of ϵ'/ϵ, and raises the interesting question how much lower the strange quark mass would be if it were treated dynamically. The result $m_s/m_{ud} = 26(2)$[22] agrees with the chiral perturbation theory estimate of 24.4(1.5).

The b quark mass, in this world in which only the u and d quarks are dynamical, obtained from the B_s binding energy at leading order in $1/m_b$ and using NNLO perturbative matching, is[31]

$$m_b^{\overline{\mathrm{MS}}}(m_b^{\overline{\mathrm{MS}}}) = 4.26(9) \text{ GeV.} \quad (12)$$

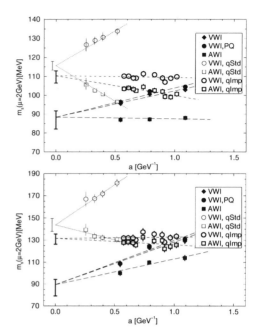

Figure 6. Continuum extrapolation of the strange quark mass for $N_f = 2$ (filled symbols) and $N_f = 0$ (open symbols) QCD, determined from the K mass (upper figure) and from the ϕ mass (lower figure)[22,30].

6 Heavy-Quark Decays

The calculation of hadronic matrix elements associated with the weak decays of b quarks is the most successful phenomenological application of lattice QCD (see Kronfeld's talk[32]).

6.1 Leptonic Decays and Mixing

The top-quark CKM matrix elements and the neutral B_q mass difference are related through the hadronic matrix element $f_{B_q}\sqrt{\hat{B}_{B_q}}$:

$$\Delta m_q = \frac{G_F^2}{6\pi^2} M_W^2 \eta_B S_0(m_t^2/M_W^2) M_{B_q}$$
$$\times |V_{tq}V_{tb}^*|^2 f_{B_q}^2 \hat{B}_{B_q}. \quad (13)$$

Traditionally, f_B and $\hat{B}_B$ have been computed separately in lattice QCD. Quenched estimates for f_B have stabilised in recent years, but suffer a relatively large irreducible scale uncertainty due to the quenched

Table 1. Summary of lattice results for B decay constants, presented by Hashimoto at Lattice 99[33].

	$N_f = 2$	$N_f = 0$
f_B (MeV)	210(30)	170(20)
f_{B_s} (MeV)	245(30)	195(20)
f_{B_s}/f_B	1.16(4)	1.15(4)

approximation. Including two dynamical flavours increases f_B by around 20%, although statistical errors currently overwhelm systematic effects. Table 1 shows the best estimates of the B decay constants, presented at Lattice 99. The only direct comparison with experiment is for f_{D_s}, and here the lattice estimates,

$$f_{D_s} = 241(30) \text{ MeV } (N_f = 0)^{34} \quad (14)$$

$$f_{D_s} = 275(20) \text{ MeV } (N_f = 2)^{22}, \quad (15)$$

are consistent with experiment (eg, ALEPH's result of 285(45) MeV[35]), although the errors are too large to expose systematic effects.

The combination $f_B \sqrt{\hat{B}_B}$ may be computed directly in lattice QCD and, to the extent that systematic errors in f_B and $\hat{B}_B$ are correlated, this may be more reliable than separate determinations of f_B and $\hat{B}_B$. A recent non-perturbatively-renormalised result in quenched QCD is[36]

$$f_B \sqrt{\hat{B}_B} = 206(29) \text{ MeV}. \quad (16)$$

Systematics should also cancel in ratios, so that quenched results such as[36]

$$\frac{f_B}{f_{D_s}} = 0.74(5) \quad (17)$$

$$\frac{f_{B_s}\sqrt{\hat{B}_{B_s}}}{f_B\sqrt{\hat{B}_B}} = 1.16(7) \quad (18)$$

are probably the most reliable.

6.2 Lifetimes

Lifetime calculations are at an exploratory stage. Two groups have computed the B_s lifetime difference, $\Delta\Gamma_{B_s}/\Gamma_{B_s}$, obtaining $0.047(15)(16)^{37}$ and $0.107(26)(14)(17)^{38}$, to

be compared with the experimental upper bound of 0.31. Although they use quite different lattice techniques, the matrix element calculations are consistent, and the discrepancy in the final results is due to one using the quenched, and the other using the unquenched value of f_{B_s}.

The Λ_b lifetime is a puzzle, because the experimental measurement for the ratio $\tau(\Lambda_b)/\tau(B_0) = 0.79(5)$ is significantly different from one, whereas, to leading order in the heavy-quark mass, all b hadrons have the same lifetime. A preliminary lattice calculation, however, does indicate that spectator effects are significant at the 6–10% level[39].

6.3 Exclusive Semileptonic Decays

The main purpose of computing exclusive semileptonic B decay form factors is to extract model-independent estimates of $|V_{ub}|$ and $|V_{cb}|$ from experiment. All the results I present were computed in quenched QCD, although dynamical-quark effects are now beginning to be explored[40]. They are expected to be around 10%.

During the past two years, the most progress has been in calculating the form factors for $B \to \pi\ell\nu$, defined by

$$\langle \pi(p')|V^\mu|B(p)\rangle = \frac{M_B^2 - M_\pi^2}{q^2}q^\mu f_0(q^2)$$

$$+ \left(p^\mu + p'^\mu - \frac{M_B^2 - M_\pi^2}{q^2}q^\mu\right)f_+(q^2)\,(19)$$

$$q = p - p', \quad V^\mu = \bar{b}\gamma^\mu u. \quad (20)$$

Here, lattice QCD fixes the normalisation of the form factors, unlike heavy-quark effective theory (HQET). However, today's lattice spacings are too large to represent a high-momentum pion accurately. So the kinematic range is restricted to near zero recoil, and model-independent results are only possible for the differential decay rate[41,42], as shown in Figure 7. Differential rates should be measured experimentally soon, at which point, direct comparison with the lattice re-

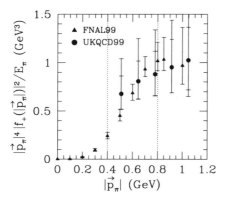

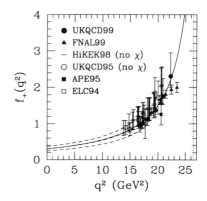

Figure 7. Results for the $B \to \pi \ell \nu$ differential decay rate, compiled by Lellouch[43], from FNAL[41] and UKQCD[42]. Vertical lines indicate the momentum range, 0.4 GeV$\leq |\vec{p}_\pi| \leq 0.8$ GeV, within which the lattice artefacts are minimised and comparison with experiment should be most reliable[41].

Figure 8. Results from various lattice groups for the $B \to \pi \ell \nu$ form factor f_+, compiled by Lellouch[43]. The solid line is a fit to the UKQCD data and the dashed lines are light-cone sum rule results.

sults will provide a model-independent estimate for $|V_{ub}|$.

Model-dependent extrapolation is needed to obtain the full kinematic range. A dipole fit to the UKQCD results for $f_+(q^2)$ is shown in Figure 8 (a simultaneous pole fit to $f_0(q^2)$ imposes the constraint $f_0(0) = f_+(0)$). This fit gives a total decay rate

$$\Gamma/|V_{ub}|^2 = 9^{+3+2}_{-2-2} \text{ ps}^{-1}. \qquad (21)$$

There is good agreement with the results from other groups and with light-cone sum rules at low q^2. The form factor f_0 provides an important consistency check on the lattice results through the soft-pion theorem:

$$f_0(q^2_{\text{max}}) = f_B/f_\pi \qquad (22)$$

in the limit of zero pion mass. Whether this is satisfied by current simulations is somewhat controversial[44,40], but this will have to be resolved if we are to have full confidence in the lattice results.

There have been no new results for $B \to \rho \ell \nu$ and $B \to K^* \gamma$ and the present status is described in Lellouch's review[43].

The form factors for the heavy-to-heavy decays, $B \to D^{(*)} \ell \nu$, are better suited to lattice calculations, because the recoil is smaller and present-day lattices can cover the full kinematic range. However, HQET is able to determine the normalisation at zero recoil in the heavy-quark symmetry limit. So, lattice QCD is left with the tougher task of quantifying the deviations from the symmetry limit at physical quark masses, needed to extract $|V_{cb}|$ from experiment, which requires few percent accuracy. The FNAL group has devised a technique for determining these power corrections at zero recoil, from ratios of matrix elements, in which many statistical and systematic errors cancel[32,45], obtaining, eg,

$$F_{D^*}(1) = 0.935(22)(^8_{11})(8)(20). \qquad (23)$$

From the recoil dependence of the form factors, it is possible to extract the Isgur-Wise function, $\xi(\omega)$. UKQCD has done this in quenched QCD[46], using the $B \to D \ell \nu$ form factor. The resulting estimate of $\xi(\omega)$, shown in Figure 9, is independent of the heavy-quark masses, for masses around that of the charm quark, and is insensitive to the lattice spacings used. Thus, it is demonstrably the Isgur-Wise function for quenched QCD.

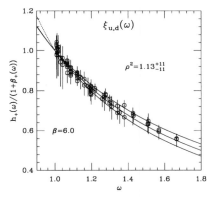

Figure 9. The Isgur-Wise function for quenched QCD[46], as a function of the recoil ω. ρ is the slope parameter at $\omega = 1$.

7 Kaon Physics

7.1 Mixing

The B-parameter for neutral kaon mixing,

$$
B_K = \frac{3}{8} \frac{\langle \overline{K}^0 | \bar{s}\gamma_\mu(1-\gamma_5)d\,\bar{s}\gamma_\mu(1-\gamma_5)d | K^0 \rangle}{\langle \overline{K}^0 | \bar{s}\gamma_\mu\gamma_5 d | 0\rangle \langle 0 | \bar{s}\gamma_\mu\gamma_5 d | K^0 \rangle},
\tag{24}
$$

is probably the best determined weak matrix element in quenched QCD. The most reliable result comes from staggered quarks, because the mixing due to chiral symmetry breaking is non-trivial for Wilson quarks. The error in the continuum result[47],

$$
B_K(2\text{ GeV}) = 0.628(42), \tag{25}
$$

is mostly from perturbative matching, and should be reduced when non-perturbative renormalisation becomes available. Dynamical quark effects raise B_K by around 5% at fixed lattice spacing, but it is not known yet how this affects the continuum result.

7.2 Non-Leptonic Decays

Although there are no new results, there is renewed optimism for lattice calculations of $K \to \pi\pi$ decays (see Testa's talk[48]), because we now have more sophisticated techniques, which may afford control over the severe cancellations between the matrix ele-

ments concerned. The new chirally symmetric lattice formulations should avoid the mixing between operators of different chiralities, and the large measured value for ϵ'/ϵ reassures us that a signal should exist.

The fundamental problem for lattice QCD is that, according to the Maiani-Testa no-go theorem, there is no general method for dealing with multi-hadron final states in Euclidean space. The traditional approach around this is to use chiral perturbation theory to relate those matrix elements which can be computed, such as $K \to$ vacuum, π, or $\pi\pi$ at unphysical momenta, to the desired physical matrix elements[49]. A new proposal is to tune the lattice volume so that one of the (discrete) energy levels of the two pions equals the K mass, and then relate the transition matrix element to the decay rate in infinite volume[50].

The impending flood of experimental data for non-leptonic B decays presents a formidable challenge to lattice QCD. Chiral perturbation theory no longer helps. Perhaps, the $B \to \pi\pi$ factorisation[51], proved for $m_b \gg \Lambda_{\text{QCD}}$, can be exploited in some way?

8 Structure Functions

Lattice QCD can provide the normalisation for parton densities. Ultimately, this should test QCD and the validity of perturbation theory. It enables us to disentangle power corrections and, where experimental information is scarce, such as for the gluon distribution for $x > 0.4$, lattice QCD can help phenomenology. Dynamical quark effects are presumably crucial. Although results so far are for quenched QCD, this will soon change (see Jansen's talk[52]).

The traditional approach uses the operator product expansion to relate moments of structure functions to hadronic matrix elements of local operators:

$$
\mathcal{M}_n(q^2) = \int_0^1 dx\, x^{n-2} F_2(x, q^2)
$$

$$= C_n^{(2)}(q^2/\mu^2, g(\mu)) A_n^{(2)}(\mu)$$
$$+ O(1/q^2). \tag{26}$$

The Wilson coefficients, $C_n^{(2)}$, are determined in perturbation theory and the hadronic matrix elements, $A_n^{(2)}$ are determined on the lattice. Renormalisation is the major source of systematic error, since the product $C(\mu)A(\mu)$ must be independent of the scale μ. This is achieved using a non-perturbative intermediate scheme, in the same way as for quark masses:

$$A(\mu) = Z_{\mathrm{INT}}^{\overline{\mathrm{MS}}}(\mu) Z_{\mathrm{latt}}^{\mathrm{INT}}(\mu a) A^{\mathrm{latt}}(a). \tag{27}$$

The INT=SF scheme[53] uses a step scaling function to relate the matrix element, renormalised at the lattice scale, to a high scale where perturbation theory can be used to determine the RG-invariant matrix element in the limit $\mu \to \infty$. This, in turn, may be related via perturbation theory to $\overline{\mathrm{MS}}$. At this conference, Jansen[52] reported that the average momentum of partons in the pion in quenched QCD, computed in this way, is

$$\langle x \rangle (2.4\ \mathrm{GeV}) = 0.30(3), \tag{28}$$

to be compared with the experimental result of 0.23(2), and confirming early lattice results that the quenched estimate is larger than experiment.

A quite different method involves computing the current-current matrix element, $\langle h|J_\mu J_\nu|h\rangle$, which appears in the cross-section, directly on the lattice[54]. The Wilson coefficients are determined non-perturbatively from matrix elements between quark states, by inverting

$$\langle p|J_\mu(q)J_\nu(-q)|p\rangle = \sum_{m,n} C_{\mu\nu,\mu_1\ldots\mu_n}^{(m)}(a,q)$$
$$\times \langle p|O_{\mu_1\ldots\mu_n}^{(m)}(a)|p\rangle, \tag{29}$$

thereby avoiding mixing and renormalon ambiguities. Using 62 operators and 70 momenta to extract the C's, and reconstructing $\langle N|J_\mu J_\nu|N\rangle$ from them and nucleon matrix elements, QCDSF obtained the lowest non-trivial moment of the unpolarised structure

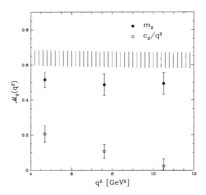

Figure 10. The leading contribution, m_2, and the power correction, c_2/q^2, to the lowest non-trivial moment of the unpolarised nucleon structure function in quenched QCD[54]. The hatched strip is the result without the inclusion of higher-twist effects (the width of the strip indicates the error).

function[54], shown in Figure 10. This indicates large power corrections and strong mixing between twist-2 and twist-4 operators.

9 Machines and Prospects

Progress in lattice QCD, and particularly its application to phenomenology, continues to be critically dependent on increasing computer power. Three machines dominate lattice QCD today. Historically, the first was CP-PACS's 300 Gflops (sustained) Hitachi SR2201. This has been operating since 1996 and cost approximately \$70/Mflops. The second was the QCDSP, custom built using 32-bit digital signal processors, which has been sustaining 120 Gflops and 180 Gflops at Columbia and Brookhaven, respectively, since 1998. Its cost was around \$10/Mflops. This year, APE's latest fully-customised 32-bit machine, called APEmille, began operation at Pisa, Rome and Zeuthen, sustaining around 70 Gflops in the largest configuration so far (this will double by the end of 2000). Its cost is \$5/Mflops. These machines show an encouraging trend towards greater cost-effectiveness.

In December 1999, an ECFA Working

Panel concluded[55] that "the future research programme using lattice simulations is a very rich one, investigating problems of central importance to the development of our understanding of particle physics". It also concluded that "to remain competitive, the community will require access to a number of 10 Tflops machines by 2003" and "it is unlikely to be able to procure a 10 Tflops machine commercially at a reasonable price by 2003".

Two new projects are targeting 10 Tflops 64-bit machines, with a price/performance of $1/Mflops, by 2003. The QCDOC project, involving Columbia and UKQCD will employ PowerPC nodes in a 4-dimensional mesh interconnect. The apeNEXT project, involving INFN and DESY, will continue the APE architecture of custom nodes in a 3-dimensional mesh. Two US projects, Cornell-Fermilab-MILC and JLAB-MIT, are exploring Alpha and Pentium clusters using commodity (Myrinet) interconnect, in the hope that these commodity components can be made to scale to many thousands of processors and that the intense market competition will drive the price very low. These developments, together with the highly parallel algorithms employed for QCD, suggest there will be no obstacle to multi-teraflops machines for QCD except money!

However, we still do not understand the scaling of our algorithms well enough to predict how much computer power will be needed. Our best estimates are that to achieve comparable precision to quenched QCD, in simulations with two dynamical flavours with masses around 15 MeV, will require between 15 and 150 Tflops years[56]. However, we know nothing about simulations with light enough quarks for $\rho \to \pi\pi$! We still have a great deal to learn.

In conclusion, the range of phenomenological applications of lattice QCD continues to expand. Key developments have been improved actions (reported at ICHEP98) and non-perturbative renormalisation, both of which have considerably increased our confidence in matrix element calculations. Lattice QCD continues to drive the development of cost-effective high-performance computing technology and there is no technological limit in sight. The primary objective will be to extend the range of quark masses which may be simulated reliably, and it is hard to see how we will get away with less than 100 Tflops machines. Finally, in the discovery of lattice chiral symmetry, we are witnessing the "Second Lattice Field Theory Revolution", and this will vastly increase the reach of *ab initio* computer simulations.

References

1. Lattice '99, eds M. Campostrini et al., *Nucl. Phys. B Proc. Suppl.* **83** (2000).

2. P.H. Ginsparg and K.G. Wilson, *Phys. Rev.* D **25**, 2649 (1982).

3. R. Narayanan and H. Neuberger, *Phys. Lett.* B **302**, 62 (1993); *Nucl. Phys.* B **443**, 305 (1995); H. Neuberger, *Phys. Lett.* B **427**, 353 (1998).

4. D.B. Kaplan, *Phys. Lett.* B **288**, 342 (1992); Y. Shamir, *Nucl. Phys.* B **406**, 90 (1993); Y. Shamir and V. Furman, *Nucl. Phys.* B **439**, 54 (1994).

5. P. Hasenfratz and F. Niedermayer, *Nucl. Phys.* B **414**, 785 (1994); P. Hasenfratz, *Nucl. Phys. B Proc. Suppl.* **63**, 53 (1998).

6. Y. Kikukawa, this conference.

7. F. Niedermayer, *Nucl. Phys. B Proc. Suppl.* **73**, 105 (1999) hep-lat/9810026.

8. M. Lüscher, *Phys. Lett.* B **428**, 342 (1998); *Nucl. Phys.* B **549**, 295 (1999); *Nucl. Phys.* B **568**, 162 (2000) hep-lat/9904009; *Nucl. Phys. B Proc. Suppl.* **83**, 34 (2000) hep-lat/9909150; H. Neuberger, *Nucl. Phys. B Proc. Suppl.* **83**, 67 (2000) hep-lat/9909042.

9. D.B. Kaplan and M. Schmaltz, hep-lat/0002030.

10. R.G. Edwards et al., hep-lat/0001013.

11. S.J. Dong et al., hep-lat/0006004.

12. P. Hernández and K. Jansen, hep-lat/0001008.

13. R. Narayanan and H. Neuberger, hep-lat/0005004.

14. L. Wu, *Nucl. Phys. B Proc. Suppl.* **83**, 224 (2000) hep-lat/9909117.

15. CP-PACS: Ali Khan et al., hep-lat/0007014.

16. R.G. Edwards and U.M. Heller, hep-lat/0005002.

17. F. Knechtli, *Nucl. Phys. B Proc. Suppl.* **83**, 673 (2000) hep-lat/9909164.

18. UKQCD: P. Pennanen and C. Michael, hep-lat/0001015.

19. MILC: S. Tamhankar, *Nucl. Phys. B Proc. Suppl.* **83**, 212 (2000) hep-lat/9909118.

20. SESAM-TχL: B. Bolder et al., hep-lat/0005018.

21. CP-PACS: S. Aoki et al., *Phys. Rev. Lett.* **84**, 238 (2000) hep-lat/9904012.

22. K. Kanaya, this conference.

23. C. Morningstar and M. Peardon, *Phys. Rev.* D **60**, 034509 (1999).

24. F. Niedermayer et al., hep-lat/0007007.

25. W. Lee and D. Weingarten, *Phys. Rev.* D **61**, 014015 (2000) hep-lat/9910008.

26. UKQCD: C. McNeile and C. Michael, hep-lat/0006020.

27. J. Heitger, this conference.

28. Alpha: S. Capitani et al., *Nucl. Phys.* B **544**, 669 (1999).

29. Alpha and UKQCD: J. Garden et al., *Nucl. Phys.* B **571**, 237 (2000) hep-lat/9906013; *Nucl. Phys. B Proc. Suppl.* **83**, 168 (2000) hep-lat/9909098.

30. CP-PACS: Ali Khan et al., hep-lat/0004010.

31. V. Giménez et al., hep-lat/0002007.

32. A.S. Kronfeld, this conference.

33. S. Hashimoto, *Nucl. Phys. B Proc. Suppl.* **83**, 3 (2000) hep-lat/9909136.

34. UKQCD: K.C. Bowler et al., hep-lat/0007020.

35. A. Golutvin, this conference; F. Parodi et al., *Nuovo Cim.* A **112**, 833 (1999) hep-ex/9903063.

36. D. Becirevic et al., hep-lat/0002025.

37. D. Becirevic et al., hep-ph/0006135.

38. S. Hashimoto et al., hep-lat/0004022.

39. UKQCD: M. Di Pierro et al., hep-lat/9906031.

40. MILC: C. Bernard et al., *Nucl. Phys. B Proc. Suppl.* **83**, 274 (2000) hep-lat/9909076.

41. S. Ryan et al., *Nucl. Phys. B Proc. Suppl.* **83**, 328 (2000) hep-lat/9910010.

42. UKQCD: K.C. Bowler et al., *Phys. Lett.* B **486**, 111 (2000) hep-lat/9911011.

43. L. Lellouch, hep-ph/9912353.

44. UKQCD: C.M. Maynard, *Nucl. Phys. B Proc. Suppl.* **83**, 322 (2000) hep-lat/9909100; JLQCD: S. Aoki et al., *Nucl. Phys. B Proc. Suppl.* **83**, 325 (2000) hep-lat/9911036.

45. S. Hashimoto et al., *Phys. Rev.* D **61**, 014502 (2000) hep-ph/9906376; J.N. Simone et al., *Nucl. Phys. B Proc. Suppl.* **83**, 334 (2000) hep-lat/9910026.

46. UKQCD: G. Douglas, *Nucl. Phys. B Proc. Suppl.* **83**, 280 (2000) hep-lat/9909126.

47. Y. Kuramashi, *Nucl. Phys. B Proc. Suppl.* **83**, 24 (2000) hep-lat/9910032.

48. M. Testa, this conference.

49. M. Golterman and E. Pallante, hep-lat/0006029.

50. L. Lellouch and M. Lüscher, hep-lat/0003023.

51. M. Beneke et al., hep-ph/0006124.

52. K. Jansen, this conference.

53. M. Guagnelli et al., *Phys. Lett.* B **457**, 153 (1999) hep-lat/9901016; *Phys. Lett.* B **459**, 594 (1999) hep-lat/9903012.

54. QCDSF: S. Capitani et al., *Nucl. Phys. B Proc. Suppl.* **79**, 173 (1999) hep-ph/9906320.

55. F. Jegerlehner et al., ECFA Report, ECFA/99/200.

56. C.T. Sachrajda, *Nucl. Phys. B Proc. Suppl.* **83**, 93 (2000), hep-lat/9911016.

Plenary Session 6

Precision Tests of the Standard Model

Chair:	Vera G. Luth (SLAC) and
	Walter Hoogland (Amsterdam)
Scientific Secretaries:	Toru Iijima and
	Masaharu Aoki

PRECISION TESTS OF THE ELECTROWEAK GAUGE THEORY

A. GURTU

Tata Institute of Fundamental Research, Homi Bhabha Road, Bombay 400 005, India
E-mail: gurtu@tifr.res.in

The current status of electroweak measurements and related theoretical developments is presented, mainly reflecting updates and new results obtained during the past one year. Fits to all the electroweak data within the framework of the standard model are given demonstrating the internal consistency of the standard model and predictions for the Higgs mass. The overall conclusion is that the standard model describes the precision data well and the predicted mass of the Higgs continues to be low. Finally the short and long term outlook in this area is presented.

1 Introduction

The past decade has produced precision electroweak data at high energy from e^+e^- and $p\bar{p}$ colliders and from neutrino interactions: these consist of results from the LEP collaborations, ALEPH, DELPHI, L3 and OPAL, the TEVATRON collaborations, CDF and DZERO, the SLD collaboration at SLC, BES-II at BEPC and NuTeV at FNAL. Theoretical developments have kept pace with this experimental march towards precision enabling one to make meaningful precision tests of the standard model.

This talk will mainly report and review new results submitted to this conference, details of which appear in the parallel sessions. Most of these have come from LEP, SLC and BEPC. Brief mention of recent important theoretical advances in this area will be made and results from fits within the standard model framework will be presented. Owing to the restricted length of talks in the proceedings, all the material presented at the conference[1] is not covered here. For further details one may also consult the presentations made at the conference in the relevant parallel sessions[2], the papers contributed to the conference by the LEP collaborations[3,4,5,6], by the SLD collaboration[7] and the combinations made by the LEP Electroweak Working Group[8].

2 Results from LEP-I and SLC

2.1 LEP-I: Z lineshape, lepton F-B asymmetries

Three LEP collaborations ALEPH[9](A), DELPHI[10](D), L3[11](L) have final results and the fourth, OPAL (O), has a near-final analysis[12]. The full LEP data sample has 17 million hadronic and 1.7 million leptonic events collected between 1990-95. Basic experimental data consists of hadronic and leptonic cross sections and leptonic forward-backward (F-B) asymmetry measurements at various energies around the Z-mass. Figure 1 illustrates the measurement of hadronic cross section as a function of the e^+e^- c.m. energy.

With such large statistics available, systematics dominate the errors on extracted Z parameters. Every effort has been made to reduce the systematic errors:
– LEP experiments upgraded the detectors, particularly for luminosity and tracking
– to accurately determine the LEP beam energy the LEP machine/energy groups[13] developed a model incorporating environmental effects, e.g., earth tides, leakage currents due to passing trains etc, in addition to the machine related parameters
– notable contributions were made by theorists in refining SM calculations so that theory dependent systematic errors in the fit are minimised: ZFITTER[14], TOPAZ0[15], ALIBABA[16], BHLUMI[17] packages due to

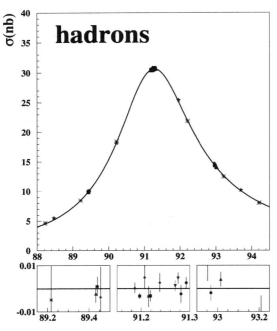

Figure 1. Hadronic lineshape: ALEPH

Table 1. LEP average Z parameters

M_Z	91187.6	$\pm$	2.1 MeV
Γ_Z	2495.2	$\pm$	2.3 MeV
σ_{had}°	41.541	$\pm$	0.037 nb
R_e	20.804	$\pm$	0.050
R_μ	20.785	$\pm$	0.033
R_τ	20.764	$\pm$	0.045
$A_{FB}^{\circ,e}$	0.0145	$\pm$	0.0025
$A_{FB}^{\circ,\mu}$	0.0169	$\pm$	0.0013
$A_{FB}^{\circ,e}$	0.0188	$\pm$	0.0017
Under lepton universality:			
R_ℓ	20.767	$\pm$	0.025
$A_{FB}^{\circ,\ell}$	0.01714	$\pm$	0.00095

Bardin et al, Montagna et al, Beenaker et al, Jadach et al and Ward et al, respectively, and important contributions from Degrassi, Kniehl, Kuhn, Berends, Passarino, Was and many others. Thus theory dependent errors are very small: 0.3 MeV on M_Z, 0.2 MeV on Γ_Z, 0.02% on the hadronic pole cross section, σ_h°, and 0.004 on $R_\ell(=\Gamma_{hadrons}/\Gamma_\ell)$.
– procedures for combining LEP data including common errors have been developed by the LEP ElectroWeak Working Group[18].

With the definitions: $\sigma_{had}^\circ = \frac{12\pi}{M_Z^2}\frac{\Gamma_e\Gamma_{had}}{\Gamma_Z^2}$, $A_{FB}^{\circ,\ell} = \frac{3}{4}A_eA_\ell$, $A_\ell = \frac{2\bar{g}_V^\ell\bar{g}_A^\ell}{(\bar{g}_V^\ell)^2+(\bar{g}_A^\ell)^2}$, the fit values of the Z parameters are as in table 1.

The Z partial decay widths, Γ_e, Γ_μ, Γ_τ, Γ_ℓ and Γ_{had} are derived to be 83.92±.12, 83.99±.18, 84.08±.22, 83.984±.086, and 1744.4±2.0 MeV respectively and the invisible width, Γ_{inv}= 499.0±1.5 MeV. This leads to the number of light neutrino species, N_ν = 2.984 ± .008, and an upper limit on the invisible width due to decay into unknown particles $\Gamma_{inv}^X < 2.0$ MeV at 95% C.L.

2.2 Tau polarization at LEP-I

τ-polarization as a function of polar angle, θ, of the τ^- w.r.t. the e^- beam, is expressed as

$$P_\tau(\cos\theta) = -\frac{A_\tau(1+\cos^2\theta)+2A_e\cos\theta}{1+\cos^2\theta+2A_\tau A_e\cos\theta}$$

Almost uncorrelated measurements of A_τ and A_e are obtained from a fit to $P_\tau(\cos\theta)$; averaged over $\cos\theta$ it yields A_τ. LEP average values are: $A_\tau = 0.1439 \pm$ .0042, $A_e = 0.1498 \pm$ .0048 whose average is $A_\ell = 0.1465 \pm$ .0032. Combining with $A_\ell = 0.1512 \pm$ .0042 derived from LEP averaged $A_{FB}^{\circ,\ell}$, the overall LEP average $\mathbf{A_\ell = 0.1481 \pm .0026}$.

2.3 Asymmetry measurements at SLD

550,000 $e^+e^- \rightarrow$ hadrons events with polarised e^-, yield Left-Right Asymmetry: $A_{LR} \equiv A_e = 0.1514 \pm 0.0022$. Measurement of Left-Right F-B Asymmetry in $e^+e^- \rightarrow \ell^+\ell^-$ events yields: $A_e = 0.1544 \pm 0.0060$, $A_\mu = 0.142 \pm 0.015$ and $A_\tau = 0.136 \pm 0.015$. Combining with A_e from A_{LR} measurement, the overall SLD average for $\mathbf{A_\ell = 0.1513 \pm 0.0021}$.

2.4 Lepton universality in g_A, g_V

The lepton asymmetry and partial width values obtained above can be used to derive the

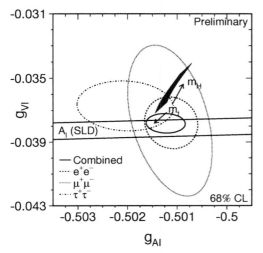

Figure 2. LEP, SLD g_A vs g_V

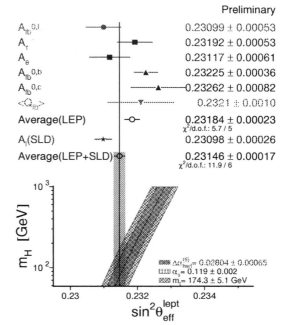

Figure 3. Determinations of $\sin^2\theta_{\text{eff}}^{\text{lept}}$

axial and vector couplings of the Z to the 3 species of charged leptons. Contours of g_A vs g_V are shown in figure 2. Contours for the 3 leptons overlap indicating lepton universality of Z couplings. The contour assuming lepton universality is also shown.

2.5 Z couplings to b, c quarks

Both LEP and SLD have determined the Z decay fractions into b and c quarks as ratios R_b, R_c, with $R_q = \frac{\Gamma(Z \to q\bar{q})}{\Gamma(Z \to \text{hadrons})}$. At LEP the F-B aysmmetries $A_{FB}^{o,b}$ and $A_{FB}^{o,c}$ have also been measured whereas SLD has directly determined A_b and A_c asymmetry parameters taking advantage of the polarised electron beam of SLC. The average values of R_b, R_c, $A_{FB}^{o,b}$, $A_{FB}^{o,c}$, A_b, A_c are: $0.21653 \pm .00069$, $0.1709 \pm .0034$, $0.0990 \pm .0020$, $0.0689 \pm .0035$, $0.922 \pm .023$, $0.631 \pm .026$ respectively. Except for $A_{FB}^{o,b}$, which is 2.4σ away from its SM value, all others agree very well with SM expectations.

2.6 $\sin^2\theta_{\text{eff}}^{\text{lept}}$ determinations

All the asymmetry measurements mentioned above can be converted into a common parameter, $\sin^2\theta_{\text{eff}}^{\text{lept}}$, the effective electroweak mixing angle at the Z mass:

$$\sin^2\theta_{\text{eff}}^{\text{lept}} = \tfrac{1}{4}\left(1 - \frac{\bar{g}_V^\ell}{\bar{g}_A^\ell}\right)$$

Figure 3 depicts the various determinations of this quantity along with its dependence on the mass of the Higgs within the SM.

2.7 A_s measurement at SLD

A beautiful new result from SLD[7] is the asymmetry measurement of the s quark tagged via high momentum $K^\pm, K_S^0$ and absence of B, D mesons. Polar angle distributions of the tagged events with positive and negative electron polarization yield $A_s = 0.895^{+.066}_{-.063}$, yielding $A_b/A_s = 1.02 \pm 0.10$, compatible with down-type quark universality.

3 Results from LEP-II

LEP operated at c.m. energies between 130-140 GeV during end of 1995, and above the W-pair threshold (161 GeV) from 1996 onward. By now a total of $\sim$475 pb^{-1}/experiment has been collected at energies between 161 and 208 GeV.

3.1 Theoretical advances for LEP-II

To match the superb performance of the machines (LEP,SLC,Tevatron,...), and excellent high resolution detectors and experimental analyses, one needs matching theoretical predictions in order to test the SM meaningfully. A large amount of work has been reported at the LEP2MC workshop[19] at CERN.
Some highlights are:

– RacoonWW[20], YFSWW3[21] packages which calculate $\sigma(W^+W^-)$ to 0.4% by using the Double-Pole Approximation - valid much above the W-pair production threshold

– WTO[22], WPHACT[23], grc4f[24] packages for σ(1-W) to 4-5% accuracy using the Fermion Loop Scheme

– YFSZZ[25], ZZTO[19] packages for $\sigma(ZZ)$ to 2% accuracy

– GENTLE v2.10[26] now corrects for overestimated Coulomb correction and agrees well with RacoonWW; there is still ~0.75% ISR related uncertainty

– for 2-fermion processes ZFITTER[14], KKMC[27] give better than 0.2% accuracy in σ(tot), of hadrons and leptons and 0.2-0.4% on A_{FB}

– KKMC is the first MC generator for LEP, LC, μ-colliders, τ, b factories

There is now a good match between experimental and theoretical accuracies.

3.2 2-fermion processes at LEP-II

The presence of initial state radiation (ISR) results in "return to the Z" events at c.m. energies far above the Z mass. Figure 4 shows the distribution of events as a function of $\sqrt{s'/s}$, where $\sqrt{s'}$ is the mass of outgoing lepton pair or Z/γ^* propagator. The Z peak is clearly seen. Events are called inclusive for $s'/s > 0.01$ and exclusive, or high energy, for $s'/s > 0.7\text{-}0.8$.

Cross sections & asymmetries are determined for hadronic & leptonic events & heavy flavours (b,c). Figure 5 shows OPAL

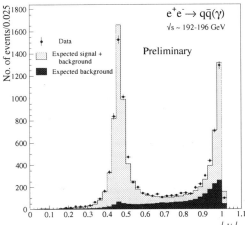

DELPHI

Figure 4. $q\bar{q}\gamma$ at high energies

results for lepton asymmetry measurements from LEP-I to LEP-II energies (note the difference in shape for the e^+e^- final state owing to additional t-channel and s-t interference). Comparison with SM provides limits on new physics: Contact Interactions, Z', $\tilde{\nu}$ exchange, extra dimensions,... These topics are covered under search for new physics.

3.3 W-pair production, BR's

Using SM branchings $B(W \to q\bar{q'}) = 67.6\%$, & $B(W \to \ell\,\bar{\nu}) = 10.8\%$ per lepton flavour:

– $B(W^+W^- \to q\bar{q'}q\bar{q'}) = 45.6\%$ (4 jet high multiplicity, balanced events)

– $B(W^+W^- \to q\bar{q'}\ell\,\bar{\nu}) = 14.6\%$ (2 jets, 1 lepton - hadronic events with 1 energetic lepton or narrow τ jet) for each lepton flavour

– $B(W^+W^- \to \ell\,\bar{\nu}\ell'\,\nu') = 10.6\%$ (2 leptons, low multiplicity non-hadronic events with acoplanar isolated leptons) summed over lepton flavours.

The LEP averaged W^+W^- cross section as a function of c.m. energy is shown in figure 6. The data agrees very well with the latest theoretical models.

LEP average measurements of the hadronic and leptonic BR's of the W are as follows

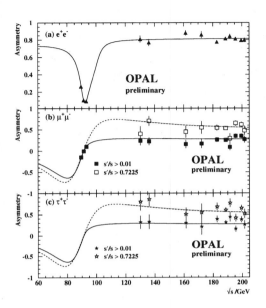

Figure 5. Lepton pair asymmetry measurements

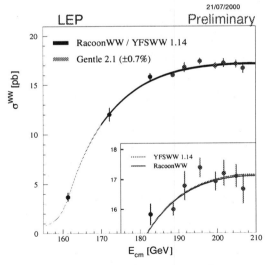

Figure 6. W^+W^- cross section

B(W→hadrons) = 67.78±0.32 %,
B(W→ $e\nu$) = 10.62±0.20 %,
B(W→ $\mu\nu$) = 10.60±0.18 %,
B(W→ $\tau\nu$) = 11.07±0.25 %,
B(W→ $\ell\nu$) = 10.74±0.10 %.

Using the average B(W→hadrons) the CKM matrix element $\mid V_{cs} \mid$ can be determined indirectly:
$B_{had}/(1 - B_{had}) = \sum \mid V_{ij}^2 \mid (1 + \frac{\alpha_s}{\pi})$,
leading to $\mid V_{cs} \mid = 0.989 \pm 0.016$, where the other matrix elements are taken from Groom et al[28]. $\mid V_{cs} \mid$ can also be directly determined by identifying charm in W jets. A new measurement from OPAL yields a value of $0.91 \pm 0.07 \pm 0.11$, and the LEP average of such direct measurements is 0.95±0.08.

Tevatron measurements of B(W → $e\nu$) from D0 and CDF are 10.50±0.30% and 10.39±0.35% respectively giving an average of 10.43±0.25% (this conference) which agrees very well with the LEP result.

3.4 WWV Triple gauge couplings

There are 7 complex couplings for each of the vertices γWW and ZWW: g_1^V, g_4^V, g_5^V, κ_V, λ_V, $\tilde{\kappa}_V$, $\tilde{\lambda}_V$ ($V = Z$ or γ). Three of these conserve C- & P-, three violate CP and one violates C-, P- but conserves CP (g_5^V). Initial studies determined only the C- and P- conserving TGC's g_1^Z, κ_γ, λ_γ and fitted one TGC at a time setting others to their SM values (1,1,0 respectively). With increased statistics and higher c.m. energies C-, P- and CP- violating couplings are also determined and simultaneous 2- and 3- parameter fits are being carried out. The sensitive observables are the cross sections, energy and angular distributions. Figure 7 depicts the processes at LEP which are used to determine TGC's. Figure 8 exemplifies the use of the photon energy distribution in the single-photon final state to determine the WWγ coupling.

The single-W cross section at various LEP energies is another good observable for the determination of the WWγ coupling. In

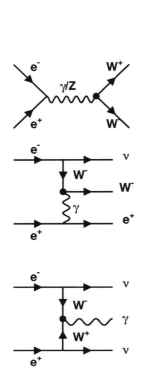

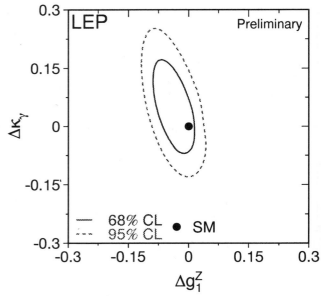

Figure 9. LEP: Δg_1^Z vs $\Delta\kappa_\gamma$

Figure 7. Final states for TGC studies at LEP

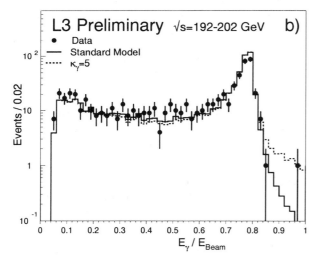

Figure 8. E_γ/E_{beam} in the $\nu\nu\gamma$ final state

the W-pair final state the various angular distributions are the most sensitive observables. The LEP averages for the three C- and P- conserving couplings obtained from 1-parameter fits are:

$\Delta\kappa_\gamma = -0.002^{+.067}_{-.065}$, $\Delta g_1^Z = -0.025 \pm .026$, $\lambda_\gamma = -0.036^{+.028}_{-.027}$.

Example of a 2-parameter fit is shown in figure 9.

Details on the limits on TGC's from various LEP experiments are given in one of the EW parallel session talks at this conference[2]. From Tevatron there is no submission to this conference. D0 published results combine $W\gamma; WW \rightarrow$ dilepton; $WW/WZ \rightarrow e\nu jj, \mu\nu jj; WZ \rightarrow$ trilepton data and obtain: $\Delta\kappa_\gamma = -0.08 \pm 0.34$, $\lambda_\gamma = 0.00^{+0.10}_{-0.09}$, and $-0.37 < \Delta g_1^Z < 0.57$, fixing $\lambda_Z = \Delta\kappa_Z = 0$ & assuming SM values for WWγ couplings.

3.5 LEP-II: ZZV Triple gauge couplings

The ZZV TGC's are studied via ZZ and Zγ production. In the SM the ZZV couplings practically do not exist and ZZ and Zγ production occurs predominantly via t-channel

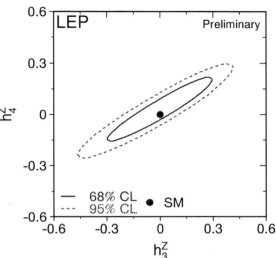

Figure 12. LEP: ZγV coupling contours

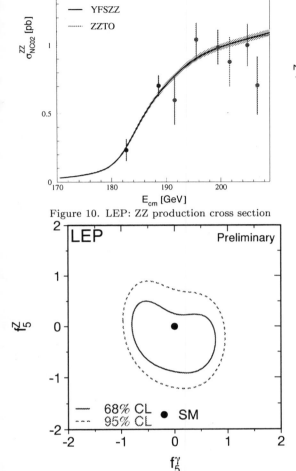

Figure 10. LEP: ZZ production cross section

Figure 11. LEP: ZZV coupling contours

For ZγV (Zγ production) one uses the general formalism of Mery et al[30] and Gounaris et al[31]. This incorporates an energy scale of new physics (Λ):

$$\frac{\sqrt{\alpha}h_i^V}{m_Z^2} \equiv \frac{1}{\Lambda_{iV}^2}, i = 1,3$$

$$\frac{\sqrt{\alpha}h_i^V}{m_Z^4} \equiv \frac{1}{\Lambda_{iV}^4}, i = 2,4$$

i = 1,2 are CP violating and i = 3,4 are CP conserving. Requirement of unitarity leads to $h_i^V \to 0$ as $s \to \infty$. Example of LEP combined contours of h_3^Z vs h_4^Z are shown in figure 12.

3.6 LEP-II: W/Z Quartic couplings

The SM predictions for WWWW, WWZZ, WWZγ, WWγγ couplings are small at LEP, but expected to become important at a TeV linear Collider. At LEP such possible couplings, a_0, a_c, a_n, are expressed in terms of six dimensional operators:

$$L_6^0 = -\frac{e^2}{16\Lambda^2}a_0 F^{\mu\nu}F_{\mu\nu}\vec{W}^\alpha \cdot \vec{W}_\alpha,$$
$$L_6^c = -\frac{e^2}{16\Lambda^2}a_c F^{\mu\alpha}F_{\mu\beta}\vec{W}^\beta \cdot \vec{W}_\alpha,$$
$$L_6^n = -\frac{e^2}{16\Lambda^2}a_n \epsilon_{ijk}W_{\mu\alpha}^{(i)}W_\nu^{(j)}W^{(k)\alpha}F^{\mu\nu}$$

where F, W are photon and W fields, L_6^0 & L_6^c

exchange of e. Anomalous ZZV (s-channel type) couplings, if they exist, increase the cross sections of ZV production and modify the γ polar angle, specially at large angles.

For ZZV couplings the formalism of Hagiwara et al[29] is used which expresses these couplings in terms of two parameters, f_4^V, which is CP-violating and f_5^V, which is CP-conserving. The ZZ cross section at LEP-II is shown in figure 10.

Contours of f_i^Z vs f_i^γ are obtained for combined LEP data. Example of f_5^Z vs f_5^γ is shown in figure 11.

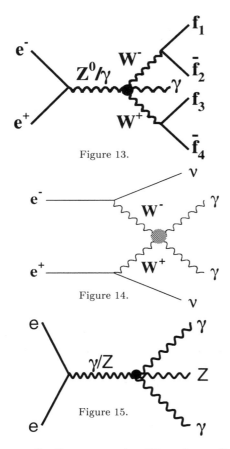

Figure 13.

Figure 14.

Figure 15.

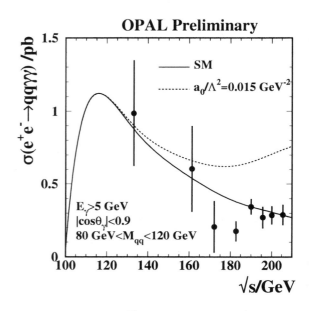

Figure 16.

conserve C-, P- separately, L_6^n violates CP and Λ is a scale for new physics.

The processes studied in search of these quartic couplings are illustrated in figures 13-15, i.e., one looks for the $WW\gamma$, $2\text{-}\gamma(\nu\nu)$ and $Z\gamma\gamma$ final states. The cross section for $Z\gamma\gamma$ with $Z\to q\bar{q}$ is shown in figure 16 for OPAL who used it to set limits on a_0/Λ^2. The LEP combined 95% C.L. limits, in GeV^{-2}, using 1-parameter fits utilizing all channels are as follows: $-0.0049 < a_0/\Lambda^2 < 0.0056$, $-0.0054 < a_c/\Lambda^2 < 0.0098$ and $-0.45 < a_n/\Lambda^2 < 0.41$.

3.7 LEP-II: W mass & width

Precision determination of M_W and Γ_W is a major goal of LEP II, firstly to check of internal consistency of the SM and, secondly, to improve the prediction for the Higgs mass. At LEP the c.m. energy is precisely known and the final states are clean thus enabling the reconstruction of W's using kinematic fits. The reconstructed W invariant mass distribution can then be fitted to obtain the best value of M_W and Γ_W. The four final states $qqqq$, $qqe\nu$, $qq\mu\nu$ and $qq\tau\nu$ are used. An example of a reconstructed mass distribution from Delphi is shown in figure 17.

In individual LEP experiments the systematic error is now comparable to the statistical error ($\simeq 50$ MeV). Thus, when one combines LEP results it is the common systematics which now dominates. Systematic errors on M_W may be classified as (1) uncorrelated between channels & experiments, (2) correlated between channels within an experiment, e.g., detector calibration, simulation, etc, and (3) correlated between experiments, e.g., those due to (a) Bose-Einstein, Color Reconnection effects in the $qqqq$ channel, (b) Initial State Radiation, fragmentation effects and (c) LEP energy error, its correlation between years.

Of major concern are the correlated errors between experiments. Of these the error on M_W due to LEP energy uncertainty

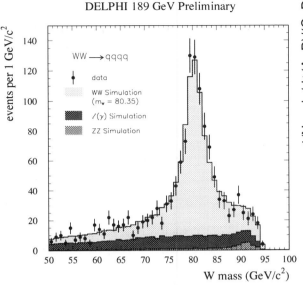

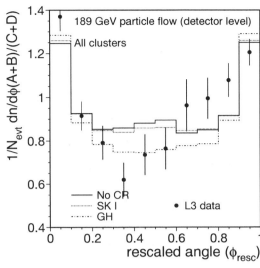

Figure 18. L3 Color Reconnection study

Figure 17. qq invariant mass in WW→qqqq

is ~17 MeV currently and it is hoped it will reduce to ~12 MeV eventually. Error due to ISR is small; due to fragmentation uncertainties is ~25 MeV. Presently the major uncertainty on M_W comes due to unknown Bose-Einstein (BE) correlation and color reconnection (CR) effects in the qqqq channel: ~25 and 50 MeV respectively. As qqqq forms ~50% of all events used in W-mass determination, the overall error on M_W due to these is ~30 MeV.

Color reconnection is the exchange of gluons between quarks from decays of different W's. BE correlation is the enhanced production of identical bosons at small 4-momentum difference. Both these may cause distortions in the qq invariant mass distribution in the qqqq channel leading to a possible shift in fitted the M_W. If one could measure these effects and correct for them then one would be left with a residual error on the correction applied.

For color reconnection L3 and Aleph have investigated the inter-jet particle and energy flow within the framework of the Sjostrand-Khoze[32] (SK) and Gustafson-Hakkinen[33] (GK) models. The effect of CR is to migrate particles from the angular regions (A,B) between quarks from the *same* W to angular regions between quarks from *different W's* (C,D). Figure 18 shows the L3 data on the ratio of particles in regions (A+B) to (C+D) as a function of the "rescaled" angle, ϕ_{resc}, which shows a significant deficit at central values of the angle as expected from the SK-I and GH models with CR. In the SK-I model the color reconnection probablity obtained is ~40%. Aleph obtain compatible results with ≤45% CR probability at 1σ level. This corresponds to a 40 MeV shift in M_W.

BE correlation is a well known effect seen already at LEP-I. Aleph, L3 and Opal see BE effects within the same W→qq, but *inter-W* effects have not been seen yet (Aleph, L3 *disfavour* such effects). In a new result presented to this conference, Delphi[4] indicate a ~ 2σ evidence for inter-W Bose-Einstein correlations in the qqqq state. The 2-particle correlation function $R(p_1, p_2) = \frac{P(p_1, p_2)}{P_0(p_1, p_2)}$, where P is the 2-particle probability density, p_i the 4-mom, and P_0 is a reference 2-particle density like P, but without BE correlation is

W-Boson Mass [GeV]

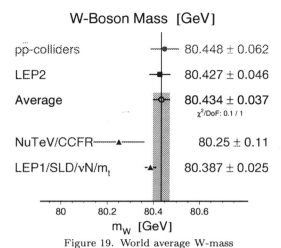

Figure 19. World average W-mass

parametrized as $R(Q) = K.(1 + \lambda e^{-r^2 Q^2})$, with $Q^2 = -(p_1 + p_2)^2 = M_{\pi\pi}^2 - 4m_\pi^2$, r = radius of a spherically symmetric pion source and $\lambda =$ strength of correlation.

Delphi measured $R(Q)$ in hadronic Z decays, WW→qqℓν and qqqq decays. Comparison sample: event mixing, taking (qq)'s from 2 qqℓν events to fake qqqq with no inter-W BEC's. Correlation strength difference between 4q data and (a) mixed events: $\Delta\lambda = 0.062 \pm .025 \pm .021$, (b) qqℓν : $\Delta\lambda = 0.077 \pm .026 \pm .020$.

On both CR and BE fronts more work is needed but the prospects of measuring these effects to correct M_W now look brighter than before.

Results on M_W, Γ_W LEP average value of $M_W(qqqq) = 80.432\pm0.073$ GeV and $M_W(qq\ell\nu) = 80.427\pm0.051$ GeV, with difference $M_W(qqqq) - M_W(qq\ell\nu) = 5\pm51$ MeV, indicating that CR and BE effects can't be too large. The overall average LEP M_W = 80.427±0.046 GeV and $\Gamma_W = 2.12\pm0.11$ GeV. This value of Γ_W may be compared with CDF direct measurement of 2.055 ± 0.125 GeV and CDF+D0 "Extracted" value of 2.171 ± .052 GeV. The world average of W-mass is shown in figure 19.

4 Global Fits to Electroweak data

Fits are made to all electroweak data within the SM framework. The fitted parameters are M_{top}, M_Z, α_s, $1/\alpha^{(5)}(M_Z^2)$, and M_{Higgs} and the ZFITTER and TOPAZ0 packages are used.

In addition to the results given in sections 2 and 3.7, the additional data used is $\sin^2\theta_{eff}^{lept} = 0.2321\pm0.0010$ from $< Q_{FB} >$ at LEP[18], $\sin^2\theta_w = 0.2255\pm0.0021$ from NuTeV[34], $M_{top} = 174.3\pm5.1$ GeV[35], and $\Delta\alpha_{had}^{(5)}(M_Z^2) = 0.02804\pm0.00065$[36].

The first fit using only LEP data yields $M_{top} = 179^{+13}_{-10}$ and $M_{Higgs} = 135^{+262}_{-83}$ GeV. Fitted M_{top} value agrees well with the direct measurement. The second fit uses all data except directly measured M_W & M_{top} and yields $M_{top} = 169^{+10}_{-8}$ & $M_W = 80.376\pm0.034$ GeV, both values showing agreement with the direct measurements. The third fit excludes only the direct M_W measurement and yields $M_W = 80.386\pm0.025$ GeV. The quality of fits is acceptable with χ^2/DOF ranging between 1.4 and 1.7. These fits thus demonstrate the compatibility and internal consistency of the SM with the existing precision electroweak measurements. Finally the fit to all data yields $M_{top} = 174.3^{+4.4}_{-4.1}$, $M_{Higgs} = 60^{+52}_{-29}$, $M_W = 80.402\pm0.021$ GeV and $\alpha_s = 0.1183\pm0.0027$, the χ^2/DOF being 1.4. The 95% C.L. upper limit on M_{Higgs} is 165 GeV, as seen from figure 20.

5 Outlook, Summary

LEP-I and SLD results are (almost) final; LEP-II will increase statistics at the highest energies (>200 GeV) leading to more sensitive tests of couplings; and it is hoped that the CR & BE systematics on M_W will be better controlled & the final LEP error on M_W, ΔM_W, will be $\simeq$35 MeV.

In the SM fits sector, new data on R (=hadrons/$\mu^+\mu^-$) from BES-II (BEPC) in the energy range 2-5 GeV was presented

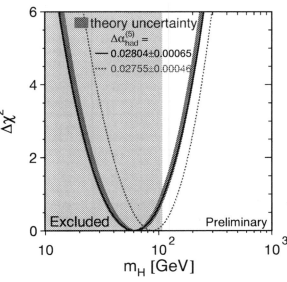

Figure 20. χ^2 variation with M_{Higgs}

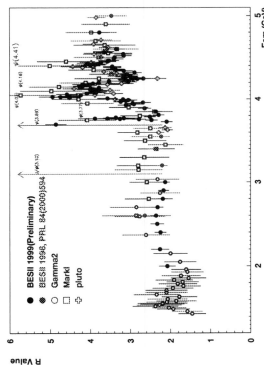

Figure 21. R($=$hadrons$/\mu^+\mu^-$) vs c.m. energy

at this conference[2]. This is shown in figure 21. This input is critically used in determining $1/\alpha^{(5)}(M_Z^2)$. Including the new BES-II data yields the value $\Delta\alpha_{had}^{(5)}(M_Z^2) = 0.02755\pm0.00046$ after a *very preliminary analysis*. Used in the SM fit one obtains $M_{Higgs}= 88^{+60}_{-37}$ GeV and a 95% C.L. upper limit of 203 GeV^2 (instead of 165 GeV). One hopes to finalize such an analysis soon.

To summarize, modulo ν oscillations, the SM is still in excellent shape and the Higgs is still predicted to be light ($\leq$200 GeV). From the upcoming Tevatron Run-II one expects reduced errors on M_W and possible Higgs discovery (if hints at LEP are not borne out). Eventually we await the LHC with its agenda of physics at the TeV scale.

References

1. Slides presented at the conference: http://ichep2000.hep.sci.osaka-u.ac.jp/scan/0731/pl/gurtu/index.html, /afs/cern.ch/user/g/gurtu/public/osaka-talk.ps.gz
2. Osaka conference parallel sessions slides: http://ichep2000.hep.sci.osaka-u.ac.jp/Program.html
   Session PA-03 Hard Interactions
   Session PA-05 Tests of the Electroweak Gauge Theory,
   Session PA-07 Heavy flavour physics
3. ALEPH contributions to the conference: http://alephwww.cern.ch/ALPUB/ oldconf/osaka00/osaka.html
4. DELPHI contributions to the conference: http://delphiwww.cern.ch/ pubxx/ delwww/www/delsec/conferences/osaka/
5. L3 contributions to the conference: http://l3www.cern.ch/conferences/ Osaka2000/index.html
6. OPAL contributions to the conference: http://www1.cern.ch/Opal/pubs/ ichep2000/abstract/abstr.html
7. SLD contributions to the conference:

http://www-sld.slac.stanford.edu/ sldwww/physics/conf00/index.html

8. The LEP Electroweak Working Group: http://lepewwg.web.cern.ch/LEPEWWG/

9. ALEPH Collab.: R. Barate et al, *Euro. Phys. Jour.* **C14**, 1 (1999).

10. DELPHI Collab.: P. Abreu et al, *Euro. Phys. Jour.* **C16**, 371 (2000).

11. L3 Collab.: M. Acciarri et al, *Euro. Phys. Jour.* **C16**, 1 (2000).

12. OPAL Collab.: Contribution to this conference, Abstract no. 204, OPAL Physics Note PN442, July 20, 2000.

13. The LEP Energy Working Group: R. Assmann et al., *Euro. Phys. Jour.* C **6**, 187 (1999).

14. D. Bardin et al, *Z. Phys.* C **44**, 493 (1989); *Comp. Phys. Comm.* **59**, 303 (1990); ZFITTER v.6.21: DESY 99-070 (1999) [hep-ph/9908433] to appear in *Comp. Phys. Comm.*

15. G. Montagna et al, *Comp. Phys. Comm.* **117** 278 (1999).

16. W. Beenaker et al, *Nucl. Phys.* **B 349** 323 (1991).

17. S. Jadach et al, *Phys. Rev.* **D** 40 3582 (1989); later versions: S. Jadach et al, *Phys. Lett.* **B** 353 362 (1995), B.F.L. Ward et al, *Phys. Lett.* **B** 450 262 (1999).

18. The LEP Collaborations ALEPH, DELPHI, L3, OPAL, and the LEP Electroweak Working Group and the SLD Heavy Flavour and Electroweak Groups, *CERN-EP-2000-016*, January 21, 2000.

19. M. Gruenewald and G. Passarino, et al, [hep-ph/0005309].

20. A. Denner et al, [hep-ph/0007245]; *Nucl. Phys. Proc. Suppl.* **89** 100 (2000); [hep-ph/0006307]; [hep-ph/9912447]; [hep-ph/0007245]; *J. Phys.* G 26 593 (2000)

21. M. Skrzypek & Z. Was, *Comp. Phys. Comm.* **125**, 8 (2000); M. Skrzypek et al, *Proc. IVth Intern. Symp. on Radiative Corrections, Barcelona, Spain*, Ed. J. Sola, World Sci. p. 316 (1999); and earlier references.

22. G. Passarino, *Comp. Phys. Comm.* **97**, 261 (1996); [hep9810416]; *Nucl. Phys.* B 574 451 (2000); *Nucl. Phys.* **B** 578 451 (2000).

23. E. Accomando & A. Ballestrero, *Comp. Phys. Comm.* **99**, 270 (1997)

24. T. Ishikawa et al, *GRACE* manual, KEK report 92-19 (1993)

25. S. Jadach, W. Placzek & B.F.L. Ward, *Phys. Rev.* **D** 56 6369 (1997)

26. D. Bardin et al, *Comp. Phys. Comm.* **104**, 161 (1997)

27. S. Jadach, B.F.L. Ward & Z. Was, *Nucl. Phys. Proc. Suppl.* **89** 106 (2000); *Comp. Phys. Comm.* **130**, 260 (2000)

28. PDG Collab.: D.E. Groom et al, *Euro. Phys. Jour.* **C15**, 1 (2000).

29. K. Hagiwara et al, *Nucl. Phys.* **B** 282 253 (1987).

30. P. Mery, M. Perrotet & F.M. Renard, *Z. Phys.* **C** 38 579 (1988)

31. G.J. Gounaris, J. Layssac & F.M. Reynard, *Phys. Rev.* **D** 61 073013 (2000)

32. T. Sjostrand & V. Khoze, *Z. Phys.* **C** 62 281 (1994); V. Khoze & T. Sjostrand, *Euro. Phys. Jour.* **C** 6 271 (1999).

33. G. Gustafson & J. Hakkinen, *Z. Phys.* **C** 64 659 (1994).

34. NuTeV Collab.: K. McFarland, presentation at XXXIIIth Rencontres de Moriond, Les Arcs, France, 15-21 March 1998, hep-ex/9806013. Result is a combination of NuTeV and CCFR results.

35. R. Partridge, *Heavy quark production and decay (t and b onia)*, talk presented at ICHEP 98, Vancouver, Canada, 23-29 July 1998.

36. S. Eidelman & F. Jegerlehner *Z. Phys.* **C** 67 585 (1995), M. Steinhauser, *Phys. Lett.* **B** 429 158 (1998).

NEW RESULTS FROM THE MUON $G-2$ EXPERIMENT

R. M. CAREY FOR THE MUON G-2 COLLABORATION

Boston University Physics Department, Boston, MA 02215 USA
E-mail: carey@bu.edu

The E821 collaboration has submitted for publication a new 5 ppm measurement of the muon anomalous magnetic moment. The result is consistent with previous work at CERN and BNL and with the standard model. A 0.7 ppm (statistical) result is expected from data collected in 1999 and 2000. Steady progress in the measurement of $R(s)$, which limits the theoretical prediction, was also reported by groups from CMD2, SND and BES.

1 Introduction

The gyrmomagnetic ratio, g, relates the spin and magnetic moment of an elementary particle,

$$\vec{\mu} = g\frac{e}{mc}\vec{s}. \tag{1}$$

In the Dirac theory of a spin 1/2, pointlike particle, g is 2. Since for real spin 1/2 particles, g is never exactly 2, we define the anomalous moment, a as

$$a \equiv \frac{g-2}{2} \tag{2}$$

Studies of anomalous moments have provided important insights into the nature of elementary particles. For example, the fact that baryon g factors were notably different from 2, indicated that baryons were not point-like but possessed some sort of substructure. The electron, by contrast, has never shown any signs of compositeness. Nonetheless, it has a small but finite anomalous magnetic moment which can be attributed to the cloud of virtual particles which always surround it, principally electrons, photons and positrons, which change, in some sense, the distribution of charge and angular momentum. Fifty years of ever more refined experiments and ever more sophisticated QED calculations have made the electron anomaly perhaps the best understood quantity in all of physics[1].

The muon anomaly is of the same order as that of the electron but not so well known. Because of its much larger mass, the muon's anomaly is far more sensitive to the effects of heavier particles and provides a better test for new physics. Three famous experiments at CERN in the 1960s and 70s measured the muon anomaly to 7 ppm, establishing the existence of contributions from virtual hadrons.

2 Theoretical Overview

The goal of the new experiment[2] at BNL is to improve on the CERN result by a factor of 20, to measure the muon anomaly at the level of 0.35 ppm. On the theoretical side, the dominant QED contribution is well established - the success of the closely related electron anomaly calculation gives us confidence in the result. The lowest order QED term is shown in figure 1. First calculated by Schwinger for the electron, its contribution to both the electron and muon anomalies is $\frac{\alpha}{2\pi}$.[3] Taking the value of α from the electron $(g-2)$,[1] the total QED contribution is $a_\mu(\text{QED}) = 116\,584\,705.7(1.8)(0.5) \times 10^{-11}$.[4]

The dominant first-order electroweak diagrams are shown in figure 2 and have been known for some time.[5,6,7,8] More recent second-order calculations have demonstrated that the total of the first and second-order weak contributions is 20 percent smaller than the first-order result alone.[9,10,11] The full second-order result is $a_\mu(\text{weak}) = 151(4) \times 10^{-11}$ which is 1.30 ± 0.03 ppm of a_μ.

The least certain part of the theory concerns the hadronic corrections, two of which are pictured in figure 3. After a stormy history in the 1980s and 90s, the hadronic light-by-light scattering diagram, at right, is now thought to be under control[12,13]. The largest contribution (about 60 ppm) and the largest uncertainty comes from the hadronic vacuum polarization diagram, at left, which cannot be calculated from first principles. Instead, measurements of R

$$R(s) = \frac{\sigma_{tot}(e^+e^- \to \text{Hadrons})}{\sigma_{tot}(e^+e^- \to \mu^+\mu^-)} \tag{3}$$

are used as input to the dispersion relation

$$a_\mu(\text{Had}; 1) = \left(\frac{\alpha m_\mu}{3\pi}\right)^2 \int_{4m_\pi^2}^{\infty} \frac{ds}{s^2} K(s)R(s) \tag{4}$$

where the factor $\frac{K(s)}{s^2}$ is largest at small s. Using published data from $e^+e^- \to$ hadrons *only*, the total error on this integral is about 0.8 ppm,[14] with some uncertainty arising from the techniques used to combine results from the many experiments which contribute data. The integral can be divided up into three kinematic regions: Region 1 goes from the two pion threshold to 1.4 GeV, Region 2 from 1.4 to 2.0 GeV and Region 3 from 2.0 GeV to 3.1 GeV. (One must also consider the contributions from the J/ψ and Υ resonance regions but they are relatively small and

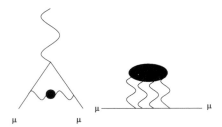

Figure 3. Hadronic vacuum polarization and hadronic light-by-light scattering diagrams

Figure 1. First order QED term

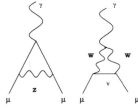

Figure 2. Dominant first-oder electroweak contributions to muon anomaly

very well known.) The uncertainties in these three regions (to be added in quadrature) are roughly 0.7 ppm, 0.25 ppm and 0.3 ppm.

Significant progress in regions 1 and 3 was reported at this conference. In region 3, as part of an energy sweep from 2 to 5 GeV, the BES collaboration measured $R(s)$ at nine energy points between 2.0 and 3.1 GeV, with errors between 5 and 10 percent. Their results[15], shown in figure 4, are consistent with but noticeably lower than those from previous measurements and have a much smaller uncertainty. No one has yet calculated the effect of the BES results on the total error in region 3, but it will probably be smaller by at least a factor of two.

Good news came also from Novosibirsk where the CMD2 and SND detectors reported preliminary results on energy region 1[16]. The integral and error in Region 1 are dominated by the two pion cross-section, or pion form-factor. The most recent results from Novosibirsk are shown in figure 5. Preliminary results and anticipated errors from newly analyzed data are shown in table 1. The improvement on the error of the two pion cross section reflects a complete reevaluation of the radiative corrections. Still, the remaining error is largely systematic.

	Channel	Contribution (ppm)	Cross Section Error (%)	a_μ error
1	$\pi^+\pi^-$	43.19	0.6	0.26
2	$\pi^+\pi^-\pi^0$	3.88	1.5	0.06
3	K^+K^-	1.81	5.2	0.09
4	$K_L K_S$	1.12	1.9	0.02
5	$\pi^+\pi^-\pi^0\pi^0$	0.77	7	0.05
6	$\pi^+\pi^-\pi^+\pi^-$	0.53	7	0.04
7	$\pi^0\gamma, \eta\gamma$	0.31	6	0.02
	Total	51.61		0.29

Table 1. Expected contributions and errors: hadronic vacuum polarization integral

No results pertaining to region 2 were reported at the conference. However, it was announced that the

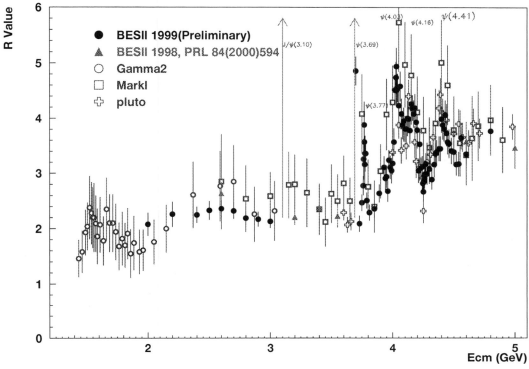

Figure 4. Results of BES R scan (preliminary)

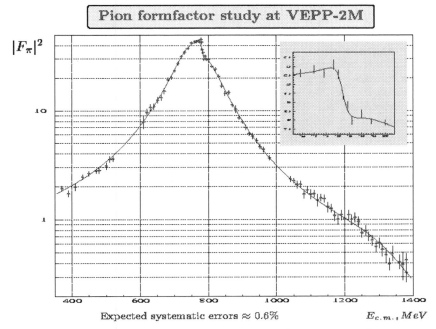

Figure 5. Two pion cross section in the ρ region (CMD2 preliminary)

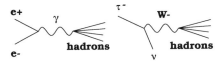

Figure 6. Schematic illustration of CVC

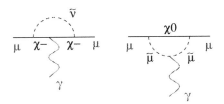

Figure 7. Supersymmetric contributions to the muon anomalous moment

Budker Institute at Novosibirsk is now officially committed to the VEPP-2000 project, which will raise the center-of-mass energy to 2 GeV, at a luminosity approaching 10^{32}/cm²sec and will provide for the detector upgrades required by the increase in luminosity. This will allow high precision measurements of R in kinematic regions not currently being investigated, where the error is still significant, and will minimize the problems associated with combining results from different experiments. First data from VEPP-2000 can be expected in the next two to three years.

Using the conserved vector current hypothesis, the cross sections for $e^+e^- \to 2n$ pions can be related to the spectral function for the associated tau decay, as shown suggestively in figure 6. Although this isovector piece of the cross section is not the whole story (see table 1), the tau decay data forms so large a part of the total and is measured with such precision, that it can reduce significantly the total error in region 1. Using the tau decay data as well as that from e^+e^-, the total contribution is $a_\mu(\text{had}; v.p.) = 6924(62) \times 10^{-11}$ which is 59.39 ± 0.53 ppm of a_μ[17]. No new developments were reported at the conference but we can anticipate a complete reanalysis of the tau decay data from LEP sometime this coming winter.[18] It is worth mentioning that discrepancies between the two pion results from e^+e^- and tau decay, reported last year at the lepton moments conference in Heidelburg, have not grown more significant with the new data from CMD2. Large discrepancies in the 4 pion data do remain but they are not significant in interpreting the $g - 2$ result. We earnestly hope that all the parties involved will work together to arrive at an answer which has a minimal error and is both correct and utterly unassailable.

With the theory firmly established and a significantly improved measurement of the muon anomaly in hand, one can search for evidence of new physics.[19,20] $g-2$ provides competitive bounds on the possibility of anomalous couplings, muon substructure, and leptoquarks to name but a few. One kind of theory to which the anomalous magnetic moment is particularly sensitive is a supersymmetric model with large $\tan\beta$, which can arise as the low-energy limit of a grand unified theory at higher energy[21]. (See figure 7). In such theories, contributions to a_μ can be

expressed as

$$a_\mu(\text{SUSY}) \approx \frac{\alpha}{8\pi \sin^2\theta_W} \frac{m_\mu^2}{\tilde{m}^2} \tan\beta$$
$$\approx 140 \times 10^{-11} \left(\frac{100 \text{ GeV}}{\tilde{m}}\right)^2 \tan\beta \tag{5}$$

The functional form $m_\mu^2/\tilde{m}^2$ where $\tilde{m}$ sets the energy scale of the new physics is a universal feature of changes to a_μ and is required of any theory which generates lepton masses by chiral symmetry breaking. It indicates the tremendous advantage in energy reach which measurements of the muon anomaly enjoy over those of the electron. At the same time, it also implies the inherent disadvantage of anomaly measurements. The energy reach of the experiments scale only with the square root of the sensitivity.

3 Experimental Technique

3.1 Precession dynamics: the connection between a_μ and ω_a

The dynamics of spin precession are straightforward. A charged particle moving in a uniform magnetic field executes cyclotron motion with the orbital cyclotron frequency $\omega_c = (eB)/(m\gamma)$. The spin precession frequency in a magnetic field, ω_s, including both the Larmor and Thomas terms, is given by $\omega_s = \frac{geB}{2m} + (1 - \gamma)\frac{eB}{m\gamma}$. The spin vector of a charged particle moving in a uniform magnetic field will precess, relative to the momentum vector, with a frequency ω_a, which is given by the difference between the orbital cyclotron frequency ω_c and the spin precession frequency ω_s,

$$\omega_a = \omega_s - \omega_c = \frac{e}{m}a_\mu B.$$

ω_a is directly proportional to the anomalous moment and is independent of the particle's momentum.

3.2 Measuring ω_a, the anomalous precession frequency

Because the muon's lifetime is relatively long, and because muons are produced fully-polarized along their direction of motion in pion decay at rest, it is possible to store large numbers of them in a storage ring and measure the anomalous precession frequency. In the three body decay $\mu^+ \to e^+ \bar{\nu}_\mu \nu_e$, the highest energy positrons in the muon rest frame are preferentially emitted parallel to the muon spin direction. Furthermore, the highest energy positrons in the muon rest frame, when emitted parallel to the muon momentum, are Lorentz boosted to become the highest energy positrons in the lab frame. Therefore, the number of high energy positrons is a maximum when the muon spin is parallel to the momentum, and a minimum when it is antiparallel. We measure the frequency of oscillation in the number of high energy decay positrons, which is the same as the muon precession frequency.

The muon decay positrons, generally having lower momenta than the stored muons, are swept by the B-field into 24 scintillating fiber-lead electron calorimeters, which are spaced evenly around the interior of the ring. Each calorimeter is read out by four phototubes. Timing and pulse height information are recorded without deadtime. Some detector stations are equipped with position-sensitive scintillators or wire chambers to give information on the spatial distribution of decay positrons entering the calorimeters.

The precession frequency is extracted from the "time of arrival" histogram of the highest energy positrons, where it modulates the exponential decay $N(t) = N_0 e^{-t/\gamma\tau}(1 + A\cos(\omega_a t + \phi))$. A, the product of the decay asymmetry and the stored muon polarization, is about 0.35 for positrons above 1.8 GeV (the maximum positron energy is 3.1 GeV). Ultimately, $g-2$ plans to collect $\sim 2 \times 10^{10}$ decay positrons spread over more than ten, time-dilated muon lifetimes, or 640 μs.

3.3 Kicking the beam onto a stored orbit

A beam of muons cannot be injected onto a stored orbit by static elements alone. The situation in our experiment is shown in figure 8. The injected muon beam arrives through an inflector magnet[22] at the edge of the storage region and encounters an extremely uniform dipole field. If left alone, the beam describes the solid circular orbit shown in the figure and slams back into the inflector. To put the beam on a circular orbit centered on the storage ring's center, we must give it a 10 mrad outward magnetic kick on its first turn. In the CERN III experiment and in the first year of our own experiment, we injected charged pions into the ring and used the momentum kick from the decay to store a small number of muons. The new technique of muon injection produces an order of magnitude more stored muons per injection with a factor 50 less injection-related background. Also, us-

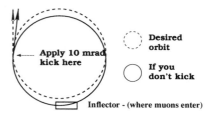

Figure 8. Kicking the beam onto a stored orbit

ing the Faraday effect, we have determined that the residual field from eddy currents produced by the kick is negligible.

3.4 Beam storage and the magic gamma

The B field in the equation above is the average field seen by the ensemble of muons in the storage ring. Vertical focusing must be provided to keep the muon beam stored, which can be accomplished with magnetic multipoles, or with an electrostatic quadrupole field. However, if magnetic multipoles are used, it is difficult to determine the average B field felt by the muons to the accuracy needed for a precision measurement of a_μ. We use a uniform B-field with electrostatic focusing instead. In a region in which both magnetic and electric fields are present, the relativistic formula for the precession is given by

$$\vec{\omega}_a = \frac{d\Theta_R}{dt} = -\frac{e}{m}\left[a_\mu \vec{B} - \left(a_\mu - \frac{1}{\gamma^2 - 1}\right)\vec{\beta} \times \vec{E}\right] (6)$$

where Θ_R is the angle between the muon spin direction in its rest frame and the muon velocity direction in the laboratory frame. The other quantities refer to the laboratory frame. If the muon beam has the magic value of $\gamma = 29.3$, then the coefficient of the $\vec{\beta} \times \vec{E}$ term is zero, and the motional magnetic field produced by $\vec{E}$ does not cause spin precession. The corresponding magic radius is 711.2 cm. Thus the magnetic field determines the spin precession frequency while the electrostatic quadrupoles alone provide vertical focusing. The magnetic field can be a pure dipole and the field felt by the muons can be determined very accurately. There is a small second-order correction (see the fast rotation analysis below) from the electric field because the cancellation occurs only for the central momentum of the storage ring. The technique of using a uniform B-field with electrostatic focusing was first used in the third CERN $(g-2)$ experiment [23].

Each set of electrostatic focusing quadrupoles[24] consists of 4 aluminum electrodes placed on the top, bottom, inside and outside of the storage volume as shown in figure 9. There are four sets in all, placed with four-fold symmetry around the storage ring, each

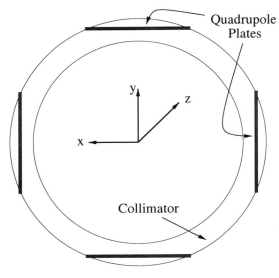

Figure 9. Beam view of focusing quadrupoles

in vacuum, on a movable trolley. The strength of the dipole field is described in terms of the proton precession frequency, w_p. During data-taking periods the field is monitored with 360 fixed probes which sit on the outside of the storage region.

The uniformity of the magnetic field has improved markedly over the four runs. Figure 11 shows the azimuthally averaged dipole field for two of the four data runs. In 1998, the spread of the magnetic field was 4 ppm, more uniform by an order of magnitude, than the CERN III field.[23] By the 2000 run, intensive shimming had made the field uniform to better than 2 ppm. For the 1997 run, the geometrical uniformity was an important uncertainty. In the 1998 and 1999 runs, which were separated by less than 4 months, our knowledge of the magnetic field was limited at the 0.2 ppm level by the calibration procedure itself. An even larger factor in 1998 was uncertainties in tracking the magnetic field with the fixed NMR probes between trolley runs. While the feedback circuitry ensures an average dipole field measured by the fixed probes remains very constant in time, the weighting assigned to each probe (a function of where it sits and how well-behaved it is) was somewhat arbitrary and different but reasonable choices of weights lead to a 0.3 ppm systematic error. The procedure for tracking the field between trolley runs was improved for 1999 and more trolley runs were taken. The corresponding error was cut to 0.15 ppm. The total error from the magnetic field analysis was 0.5 ppm in 1998 and estimated at 0.4 ppm in 1999.

covering about 39 degrees of the circumference. The flattop voltges are $\pm$ 24 kV. Some of the pairs of plates (top-bottom,inside-outside) are ramped to full voltage in an asymmetric manner so that the periphery of the beam is scraped against circular or half-circular collimators. In this way, the rate of lost muons during our measurement period ($t > 30$ μs) is minimized.

4   Experimental Status

4.1   Magnetic field

A cross section of the g-2 storage ring magnet is shown in figure 10.[25] It is a continuous C magnet, about 8 meters in radius and 2 meters high. The 1.45 tesla dipole field is excited by two superconducting coils sitting in the storage ring midplane. Its main goal is to provide a highly uniform and stable dipole field in the storage volume. To that end, the pole pieces are fabricated of very soft, high-quality steel, machined flat and placed so that their faces are perfectly aligned. Passive shimming elements include movable wedges, pole bumps on the pole tips, and thin pieces of steel affixed to the pole faces in regions where the pole pieces meet. Active elements include coils wound around 15 degree sections of the pole tips which are used to maintain a constant dipole field as well as current sheets on the pole faces themselves, which are used to remove unwanted higher moments in the field.

The field is mapped periodically with a set of 17 NMR probes which ride through the storage region,

5   Analysis of the anomalous precession frequency

As mentioned above, the anomalous precession frequency is determined by the time of arrival of high energy decay positrons in our scintillating fiber calorimeters[26], (See figure 12.) The scalloped shape of the vacuum chamber helps preserve the energy resolution, which is about 10 percent at 1 GeV. The rate of high energy decay positrons incident on the calorimeters is largest when the muon spin points tangentially to the orbit. The signals from the calorimeter PMTs are read out with 400 MHz waveform digitizers, and the energies and times-of-arrival are calculated from the waveform record. The energy spectrum of all reconstructed pulses, shown in figure 13, is peaked at small values with a high energy endpoint at 3.1 GeV. The asymmetry, the correlation of the direction of the muon spin vector with the decay positron's laboratory energy, is largest for high energy positrons. A time spectrum for part of the 1999 data set, about 750 million decay positrons in all, is shown in figure

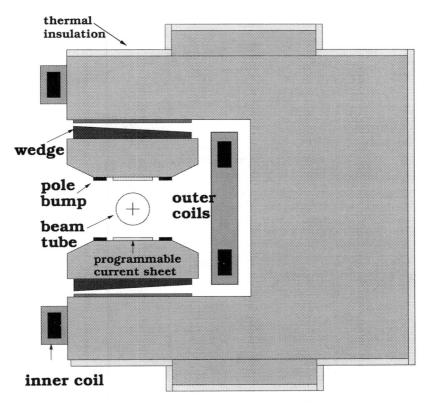

Figure 10. g-2 storage ring magnet in cross section

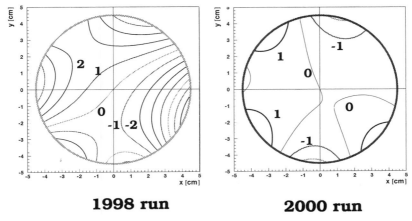

Figure 11. Variation in azimuthally averaged magnetic field over aperture, 1998 and 2000, 1 ppm contours

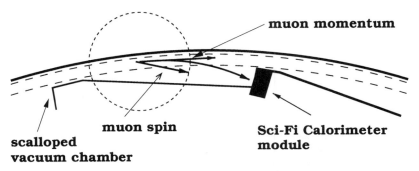

Figure 12. Muon decay in g-2 vacuum chamber

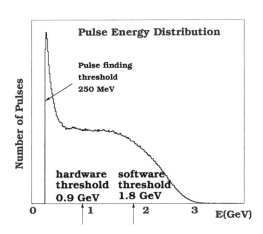

Figure 13. Energy spectrum for decay positrons

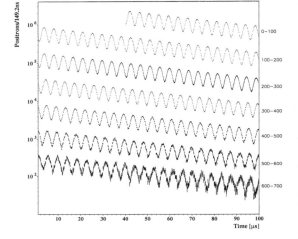

Figure 14. Time spectrum for 750 million high energy decay positrons - 1999 data

14.

5.1 Anomalous Precession Analysis

In contrast to the magnetic field, the ω_a analysis has been statistically limited. However, a number of systematic issues have been addressed, the most important of which are listed in table 5.1. Some are related to the dynamics of the stored muon beam and others to the performance of the electron calorimeters and their associated PMTs, as well as to the procedures used to reconstruct the electron times and energies from the waveform record.

Excellent timing stability is required of the electronics. In particular, the average pickoff time must be stable to 20 ps over the first 200 μs of data-taking. For closely related reasons, the gain must be stable to 0.2 percent over the same period. Using our laser calibration system, and for gain stability, the positron energy data itself, we have demonstrated that both

goals have been met.

At high rates, which come early in the data-taking cycle, pulse pileup is also a problem. Smaller pulses which follow close on the heels of a larger trigger pulse can be missed. See figures 15 and 16. Two small pulses which lie directly on top of one another can masquerade as a single high energy pulse. Because pileup occurs more often at early times than later on, it can produce a systematic error in ω_a, just like a shift in gain. In our 1998 analysis, pileup was the largest single systematic error. Since then, careful study of our pulse and precession fitting procedures has made it possible to model and correct for the effect of pileup on ω_a.

A knowledge of beam dynamics is also critical to the ω_a analysis. Because not all muons have a momentum corresponding to the magic gamma, a small correction must be applied to the measured precession frequency to account for the effect of the elec-

Systematic Effect	ϵ (ppm)
1. Magnetic Field B	0.5
2. Timing Shifts	<0.1
3. Backgrounds	<0.5
4. Gain Changes	<0.1
5. Coherent Betatron Oscillations	0.2
6. Radial E Field, Pitch Correction	0.3
7. Fitting Start Time	0.3
8. Binning Effects	0.2
9. Pileup	<0.6
10. Lost Muons	<0.1
Total Systematic Error	1.0

Table 2. Systematic Errors for 1998 run

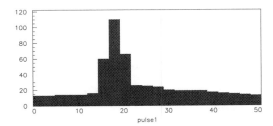

Figure 15. A typical pulse, x-axis is ns

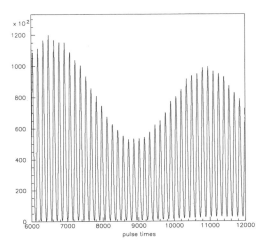

Figure 17. Clear fast rotation structure about 6 microseconds after injection. x-axis is ns

tric field. That correction, 0.5 ppm in the 1999 run, is proportional to the rms momentum spread of the stored muon beam, which we can measure by examining the rate at which the beam debunches at early times. There is also a small "pitch" correction [27] which arises because the muons execute vertical betatron motion, rather than traveling precisely in the storage ring midplane, perpindicular to the magnetic field.

Figure 17 shows the high energy positron time spectrum early in the data-taking cycle. The interval

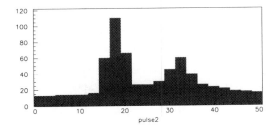

Figure 16. Pileup of positron pulses in waveform record

between the peaks is the beam's 150 ns "fast-rotation" time around the storage ring. Figure 18 shows the time spectrum about 30 μs later, by which time the debunching of the beam has noticeably washed out the fast rotation structure. Figure 19 shows the inferred equilibrium radius distribution which peaks at the radius corresponding to the magic gamma, 711.2 cm, but is not symmetric about it, having something of a high side shoulder.

Another important aspect of the beam's motion did not become fully clear until we had examined data from the scintillating fiber harps. The harps consist of seven 0.5 mm scintillating fibers strung either radially or horizontally, and are used to measure the beam profile. The measurement is destructive and is not peformed during standard data runs but still gives a very reliable idea of the beam profile for many tens of microseconds.

Data from the harps is shown in figure 20. The two data sets, with beam scraping on and off, have different characteristic frequencies but send the same message: the beam undergoes a collective radial betatron motion, here of amplitude 0.4 cm. These coherent betatron oscillations are most prominent at early times but persist for more than 100 microseconds. The oscillations arise because the kick we give to the incoming beam is not quite adequate. Instead of arriving on a stored circular orbit centered on the middle of the storage ring, the beam describes a series of roughly circular orbits with a slowly moving center, rather like the motion of a hula-hoop around a

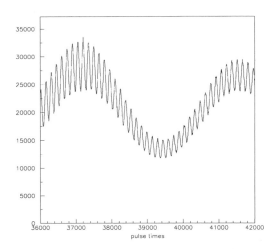

Figure 18. Fast rotation structure about 36 microseconds after injection

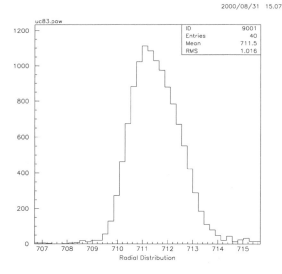

Figure 19. Distribution of equilibrium radii (cm)

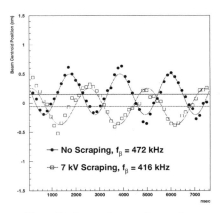

Figure 20. Average beam radius vs. time

person's waist. A small radial acceptance bias transforms the oscillatory motion into a small modulation of the basic $g - 2$ precession spectrum. The effect is small, but must be accounted for in our 1999 analysis.

6   Results

In the honorable tradition of precision experiments, the magnetic field and ω_a analyses were performed double blind. All values discussed in group meetings were reported with hidden offsets, so that there would be no pressure to arrive at a particular final answer for ω_a or the average magnetic field, ω_p. In fact, three independent (and consistent) analyses were made of ω_a and two independent (and also consistent) analyses were made of the average magnetic field. The anomalous magnetic moment is obtained from the frequency ratio $R = \omega_a/\omega_p$ by

$$a_{\mu^+} = \frac{R}{\lambda - R} = 116\ 591\ 91(59) \times 10^{-10} \qquad (7)$$

where $\lambda = \mu_\mu/\mu_p = 3.183\ 345\ 39(10)$[29]. Our 1998 result, newly accepted for publication by Physical Review D, [30] is consistent with our 1997 result[31] and both results (on μ^+ and μ^-) from CERN III[28]. The 5 ppm error is overwhelmingly statistical. Assuming CPT symmetry we can combine the four measurements to obtain a new world average of

$$a_{\mu^+} = 116\ 592\ 05(46) \times 10^{-10} \quad (\pm 4\text{ppm}) \qquad (8)$$

The standard model value is $a_\mu(SM) = 116\ 591\ 63(8) \times 10^{-10}$ (± 0.7ppm) where the error is dominated by the uncertainty in the contribution arising from hadronic vacuum polarization[17]. The weighted average of the experimental results agrees with the standard model where

$$a_\mu(\text{Expt.}) - a_\mu(\text{Theory}) = (42 \pm 47) \times 10^{-10} \qquad (9)$$

The analysis of our data from 1999 is in its final stages. We hope to remove the offsets and combine the two results early this fall. In addition, a preliminary production pass for our 2000 data run has been completed and we expect that the full production will be finished by the end of November 2000. Although the number of decay positrons in our data sets has grown from 20 million in 1997 to almost 10 billion in 2000 and the systematic error has remained neglible, the time required to complete each analysis has not grown significantly larger.

Our goal of 0.30 ppm statistical error will require that we collect an additional 6 billion decay positrons (E>1.8 GeV) in our 2001 run, which is scheduled to start in February. This will be the first time that we run in parallel with the RHIC collider and while no particular problems are anticipated, how well we will do remains an open question.

7 Conclusions

The $g - 2$ experiment continues to make steady progress toward its goal of measuring the muon anomaly to 0.35 ppm. Our 1998 result, with an error of 5 ppm, is more precise than any other previous measurement but it will be superseded in the fall of 2000 by our 1999 result. Analysis of the 2000 data has begun and more data is expected in the winter of 2001. At the same time, thanks to the efforts of physicists from BES, CMD2, SND, LEP and CLEO, the theoretical calculation of the hadronic vacuum polarization becomes ever more precise. We are confident that their continued hard work will make possible an unambiguous interpretation of our final result.

Acknowledgments

I would like to thank my collaborators for their many years of hard work on this very challenging experiment, as well as the management and staff of Brookhaven National Laboratory, for their tireless and cheerful support.

References

1. T. Kinoshita, *Rep. Prog. Phys.* **59**, 1459 (1996).
2. The $(g - 2)$ Collaboration consists of: R.M. Carey, W. Earle, E. Efstathiadis, E.S. Hazen, F. Krienen, I. Logashenko, J.P. Miller, J. Paley, O. Rind, B.L. Roberts, L.R. Sulak, A. Trofimov (Boston University), H.N. Brown, G. Bunce, G.T. Danby, R. Larsen, Y.Y. Lee, W. Meng, J. Mi, W.M. Morse, C. Özben, C. Pai, R. Prigl, R. Sanders, Y. Semertzidis, D. Warburton (Brookhaven National Laboratory) Y. Orlov (Cornell University) D. Winn (Fairfield University), A. Grossmann, K. Jungmann, G. zu Putlitz (University of Heidelberg) P.T. Debevec, W. Deninger, F. Gray, D.W. Hertzog, C. Onderwater, C. Polly, S. Sedykh, D. Urner (University of Illinois) U. Haeberlen (Max Planck Institiute für Med. Forschung, Heidelberg) A. Yamamoto (KEK) P. Cushman, L. Duong, S. Giron, J. Kindem, I. Kronkvist, R. McNabb, C. Timmermans, D. Zimmerman (University of Minnesota) V.P. Druzhinin, G.V. Fedotovich, B.I. Khazin, N.M. Ryskulov, Yu.M. Shatunov, E. Solodov (Budker Institute of Nuclear Physics, Novosibirsk), M. Iwasaki, M. Kawamura (Tokyo Institute of Technology) H. Deng, S.K. Dhawan, F.J.M. Farley, M. Grosse-Perdekamp, V.W. Hughes, D. Kawall, J. Pretz, S.I. Redin, A. Steinmetz (Yale University)
3. J. Schwinger, *Phys. Rev.* **73**, 416 (1948) and *Phys. Rev.* **75**, 898 (1949).
4. V.W. Hughes and T. Kinoshita, *Rev. Mod. Phys.* **71**, S133 (1999).
5. W.A. Bardeen, R. Gastmans and B Lautrup, *Nucl. Phys.* B **46**, 319 (1972).
6. R. Jackiw and S. Weinberg, *Phys. Rev.* D **5**, 157 (1972).
7. I. Bars and M. Yoshimura,*Phys. Rev.* D **6**, 374 (1972).
8. J. Calmet, S. Narison, M. Perrottet and E. De Rafael, *Rev. Mod. Phys.* **49**, 21 (1977).
9. A. Czarnecki, B. Krause and W.J. Marciano, *Phys. Rev.* D **52**, R2619 (1995).
10. A. Czarnecki, B. Krause and W.J. Marciano, *Phys. Rev. Lett.* **76**, 3267 (1996).
11. G. Degrassi and G.F. Giudice, *Phys. Rev.* D **58**, 53007 (1998).
12. M. Hayakawa and T. Kinoshita, *Phys. Rev.* D **57**, 465 (1998) and ref. therein.
13. J. Bijnens, E. Pallante and J. Prades, *Nucl. Phys.* B **474**, 379 (1996) and ref. therein.
14. Simon Eidelman, private communication.
15. Presented by Zheng-guo Zhao at this conference.
16. Presented by Alexander Bondar, at this conference.
17. M. Davier and A. Höcker, *Phys. Lett.* B **435**, 427 (1998) and ref. therin.
18. M. Davier, private communication.
19. An overview of non-standard model physics is given by T. Kinoshita and W.J. Marciano in *Quantum Electrodynamics* (Directions in High Energy Physics, Vol. 7), ed. T. Kinoshita, (World Scientific, Singapore, 1990), p. 419.
20. F. Renard, et al., *Phys. Lett.* B **409**, 398 (1997).
21. A. Czarnecki and W. Marciano, Nucl. Phys. B

(Proc. Suppl.) **76**, 245 (1999).

22. F. Krienen, D. Loomba and W. Meng, *Nucl. Instrum. Methods* A **283**, 5 (1989).

23. J. Bailey, et. al, *Nucl. Phys.* B **150**, 1 (1979).

24. Y. Semertzidis, et al., "The Brookhaven Muon g-2 Storage Ring High Voltage Quadruples", to be submitted to Nuclear Instruments and Methods in Physics Research.

25. Gordon T. Danby, et. al., in preparation, and G.T. Danby and J.W. Jackson, Proceedings of the 1987 IEEE Particle Accelerator Conference, ed. E.R. Lindstron and L.S. Taylor, (IEEE, 1987) p. 1517.

26. S. Sedykh et al, "Electromagnetic calorimeters for the BNL muon (g-2) experiment", Nuclear Instruments and Methods in Phsyics Research, in press.

27. F.J.M. Farley and E. Picasso in *Quantum Electrodynamics*, ed. T. Kinoshita, (World Scientific, Singapore, 1990), p. 479.

28. Particle Data Group, *Eur. Phys. J.* C **3**, 1 (1998).

29. W. Liu, et al., *Phys. Rev. Lett.* **82**, 711 (1999).

30. H. Brown et al., to be published November 1, 2000, in Phys. Rev. D.

31. R.M. Carey et al., *Phys. Rev. Lett.* **82**, 1632 (1999).

32. Bernd Krause, *Phys. Lett.* B **390**, 392 (1997).

Plenary Session 7

Search for New Particles and New Phenomena

Chair: Walter Hoogland (Amsterdam)

Scientific Secretaries: Satoshi Mihara and
Takeshi Komatsubara

SEARCHES FOR NEW PARTICLES AND NEW PHYSICS: RESULTS FROM E^+E^- COLLIDERS

P. IGO-KEMENES

*Physikalisches Institut, Universität Heidelberg, Heidelberg, Germany
and CERN, Geneva, Switzerland*

Most of the recent results on searches for new particles and new phenomena in e^+e^- collisions are provided by the LEP experiments. Indirect searches are attentive to traces of new physics produced by virtual effects in Standard Model processes. Direct searches are driven by extensions of the Standard Model, of which Supersymmetry is the most prominent today; however, other models are also investigated, and model independent results are also presented. The status of Higgs boson searches, within and beyond the Standard Model, is reviewed.

1 Introduction

Approaching the shutdown of the LEP e^+e^- collider at CERN, the four collaborations ALEPH, DELPHI, L3 and OPAL have considerably broadened the scope of their searches for new particles and phenomena. Theoretical models are considered which have not received due attention so far. Seeking more completeness, the experiments also include in their investigations regions of parameter space which are difficult to access experimentally.

Beyond the Standard Model (SM), supersymmetric (SUSY) models embedded in a grand unification theory (GUT) are attracting most of the interest; however, models of compositeness, with excited fermions, leptoquarks or technicolor, are also addressed.

In the SUSY domain, various messenger interactions are investigated by which the symmetry breaking is transmitted from the hidden to the visible sector. The mediating interaction could be gravity (MSUGRA) or the ordinary gauge interactions (GMSB).

In the past, most of the searches assumed R-parity conservation, but now detailed studies assuming R-parity violation (RPV) are also available.

The data accumulated at LEP correspond to more than 2 fb^{-1} of luminosity at centre-of-mass energies (E_{cm}) exceeding 180 GeV. In the year 2000, the four experiments have collected together about 500 pb^{-1} of data, at energies between 200 and 210 GeV. Most of the results presented below include these new data. The four LEP collaborations have developed procedures to combine their data for a better statistical significance. Wherever available, the combined LEP results will be quoted.

At the conference there was no claim of a statistically significant signal for new physics. Thus, the results presented below represent bounds on particle masses, topological cross sections, or limits on model parameters. Unless explicitly mentioned, these are quoted at the 95% confidence level.

2 Indirect limits on new physics

Indications for new phenomena could be obtained indirectly, by the detection of deviations from SM predictions, in fermionic, photonic or bosonic final states. Fits to cross-sections and charge asymmetries as a function of E_{cm}, or to angular distributions, could reveal virtual effects due to new particles or interactions. These indirect searches typically exceed the kinematic reach of direct searches. In global terms, the data are rather well described by the SM [1], leaving little room for new phenomena.

(a) The process $e^+e^- \rightarrow f\bar{f}$ is sensitive to four-fermion contact interactions, which pro-

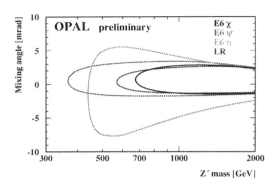

Figure 1. Bounds on the mass and Z^0-Z' mixing predicted by various models, from OPAL.

vides a generic description of new phenomena. It is characterized by a mass scale Λ, coupling g, and a certain Lorentz structure. Fixing the coupling ($g^2 = 4\pi$), lower bounds are obtained for Λ. Of these, the most stringent ones are obtained in the cases where the Lorentz structure is *vector-vector* ($\Lambda >11$-18 TeV) or *axial-axial vector* ($\Lambda >9$-15 TeV).

(b) Deviations from QED can be sought by analysing the rate and distributions of the process e$^+$e$^- \to \gamma\gamma$, which is an almost pure QED process. Possible deviations are expressed as a generic ultra-violet cutoff parameter, for which lower bounds in the range 330-330 GeV are quoted by the LEP experiments. The same studies also exclude the *t-channel* exchange of an excited electron with mass less than 300-320 GeV, assuming a unit form factor.

(c) Additional Z^0 bosons, Z', predicted by GUT, left-right symmetric or sequential extensions of the SM, would mix with the standard Z^0 boson and affect the distributions of e.g. the process e$^+$e$^- \to$f$\bar{\text{f}}$. The Z^0 lineshape measurement have already excluded large Z^0-Z' mixing. High-energy LEP2 data place new bounds on the mass and mixing of such new vector bosons. For illustration, see Fig. 1.

(d) Theories with gravitational interaction at a low energy scale [2] are widely discussed [3]. One of their virtues is to avoid the "scale-hierarchy problem" of the SM without introducing SUSY or compositeness. The models assume δ extra space dimensions of range R and a fundamental scale M close to the electroweak scale ($\approx M_W$). These parameters are linked to M_{Planck} (which is not fundamental any more but becomes merely an effective scale in 3-dimensional space) by the relation

$$M_{Planck} = R^\delta M^{\delta+2}.$$

Models with $\delta = 1$ have already been excluded by $\sim 1/r^2$ tests of the gravitational force. $\delta \geq 2$ is tested at LEP either by searching for the graviton G in the reactions e$^+$e$^- \to \gamma G$ and e$^+$e$^- \to$(Z$^0 \to$f$\bar{\text{f}}$)G, or *via* virtual effects on e$^+$e$^- \to$f$\bar{\text{f}}$, $\gamma\gamma$, W^+W^- and Z^{0}Z^0 rates and distributions. Lower bounds on M of 1.0-1.3 TeV, 0.7-0.8 TeV and $\approx$0.5 TeV are obtained for $\delta = 2$, 4 and 6, respectively. Thus, current tests in e$^+$e$^-$ collisions do not favour low-scale gravity with its fundamental scale close to the electroweak scale.

3 Direct searches

Searches for SUSY and Higgs bosons are discussed in separate sections. Here we review new data on heavy and excited leptons, leptoquarks [4], technicolor [5] and single-production of top quarks [6].

(a) The possibility of having neutrinos with non-zero mass generates renewed interest for heavy leptons with masses around 100 GeV which can be generated e.g. by the *see-saw* mechanism. A rather complete update on heavy leptons is presented by L3 [7] addressing sequential, vector and mirror leptons, Dirac and Majorana neutrinos. For such objects, lower mass bounds of 80-100 GeV are obtained. The strongest bounds are obtained for *vector* leptons with muon and electron flavour.

The production of an isosinglet neutrino

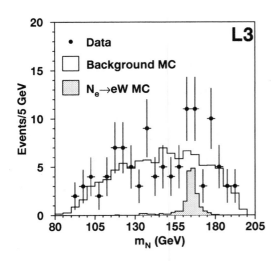

Figure 2. Distribution of the reconstructed mass of the isosinglet heavy neutrino N_e, obtained by L3.

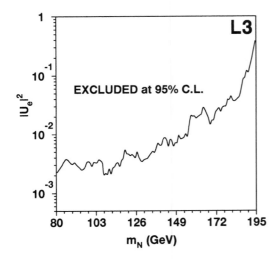

Figure 3. Upper bound on the mixing amplitude U_e, obtained by L3.

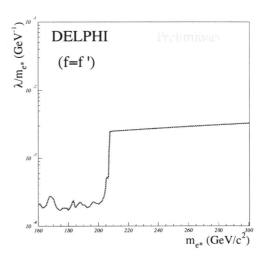

Figure 4. Upper bound on the ratio λ/m_{e*} combining direct and indirect limits, by DELPHI.

N_e, predicted by GUT, superstring and left-right symmetric models, is also addressed by L3. The process $e^+e^- \to N_e + \nu_e$ occurs by mixing (amplitude U_e) with the associated isodoublet neutrino. Previous limits on the mixing amplitude, from LEP1 data and low-energy measurements, were at about 0.1. In the first fermion generation, the cross-section is enhanced by the *t-channel* contribution. As for the decay, L3 chooses the channel $N_e \to e^\pm + W^\mp$ ($W \to q\bar{q}$) where the mass is fully reconstructed, see Fig. 2. Previous limits on $|U_e|^2$ are improved by more than an order of magnitude, see Fig. 3.

(b) Excited leptons (or quarks) would be a clear signal for substructure. The current emphasis is on single-production, $e^+e^- \to \ell^*\ell$, with a mass reach close to E_{cm}. The new interaction is characterized by a compositeness scale Λ, by charged- and neutral-current couplings g and g', and weights f and f' which control the rate of radiative and weak decays $\ell^*(\nu^*) \to \ell(\nu)\gamma$, $\nu(\ell)W$. Upper bounds for $\lambda/m_{e*} = f/\Lambda\sqrt{2}$ are obtained, see Fig. 4. In the figure, direct and indirect (see Section 2) limits have been combined.

(c) Leptoquarks are coloured particles

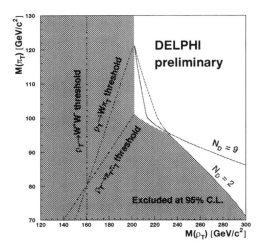

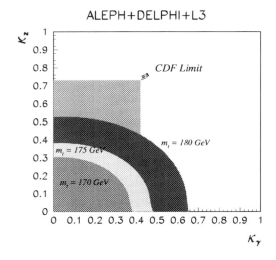

Figure 5. Exclusions in the (ρ_T, π_T) mass plane obtained by DELPHI. N_D is the number of TC-quark doublets in the model.

Figure 6. Exclusions in the plain of couplings $(\kappa_\gamma, \kappa_Z)$, for various top quark masses, obtained by combining LEP data, compared to those from CDF.

carrying both baryon and lepton number. Such particles appear in GUT, composite models and technicolor. They may be scalar or vector particles, with electric charges from 1/3 to 5/3. New results have been submitted both on pair- and single-production. If one assumes electroweak coupling strength, $\lambda = \sqrt{4\pi\alpha_{em}} \approx 0.3$, lower bounds on the mass in a range 150-185 GeV are obtained.

(d) Technicolor (TC) models provide an alternative way to the Higgs mechanism to produce massive vector bosons by assuming a new, very strong interaction ($\Lambda_{TC} >> \Lambda_{QCD}$). "Standard" technicolor with QCD-like "running" of Λ_{TC} conflicts with LEP1 data, but more elaborate models with slower running ("walking technicolor") provide still viable solutions. DELPHI [8] and L3 [9] have searched for techni-particles, π_T, ρ_T, ω_T, and obtained exclusions in the (π_T, ρ_T) mass plane, see Fig. 5.

(e) Single-production of the top quark by the processes $e^+e^- \to t\bar{c}$ and $t\bar{u}$ became kinematically possible with LEP2 energies in the vicinity of 200 GeV. In the SM, these FCNC processes have very low cross-sections ($\approx 10^{-9}$ fb); their detection would be a sign

of anomalously large tcZ or $tc\gamma$ couplings (κ_Z, κ_γ). The searches at LEP have been combined [6], giving rise to upper bounds in the plane of couplings, see Fig. 6, which improve slightly previous limits from CDF.

4   Searches for Supersymmetry

One of the attractive features of SUSY is to provide a possible solution of the scale hierarchy problem. SUSY attributes to each fermion/boson of the SM a bosonic/fermionic "sparticle" partner. Quantum corrections from particles and sparticles appear with opposite sign and cancel each other, providing stability to the electroweak energy scale. SUSY, embedded in a GUT scheme, postulates unification of masses and couplings at Λ_{GUT}. In minimal SUSY (MSSM), the number of parameters is reduced by GUT unification from 124 to five.

The link between the unified parameters and the masses and couplings at the electroweak energy scale is given by the renormalisation group equations (RGE). These predict heavy squarks and gluinos, and light ($\mathcal{O}(100)$ GeV) sleptons, gauginos and higgsinos. The gauginos and higgsinos mix to produce the physical neutralino ($\tilde{\chi}^0$) and chargino ($\tilde{\chi}^\pm$) states. The mass spectrum at the electroweak scale is strongly influenced, further, by mixing of left- and right-handed fields and by Yukawa couplings, especially in the third sfermion generation. The phenomenology is even more complex for sfermion masses comparable to the chargino and neutralino masses, in which case the decay pattern may change substantially for small variations of the fundamental parameters.

SUSY appears to be softly broken in nature. "Softness" is required to avoid overcompensation of the quantum corrections. It implies the existence of a *messenger sector* with a weak messenger interaction mediating between the *hidden sector* where the symmetry is broken and the *visible sector*. In minimal supergravity (MSUGRA) the messenger role is attributed to the eminently weak gravitational interaction, but mediation by the usual gauge interactions (GMSB) at loop level may be an attractive alternative, giving rise to rather different phenomenologies at the electroweak scale. In MSUGRA the gravitino ($\tilde{G}$) is heavy and has no influence. The phenomenology depends on the following fundamental parameters defined at the GUT scale: common scalar and fermion masses m_0 and $m_{1/2}$, one trilinear coupling A_0, the ratio $\tan\beta = v_2/v_1$ of the vacuum expectation values of the two Higgs fields (the field with vev v_2 (v_1) couples to *up* (*down*) fermions), and the sign of the Higgs mass term μ. The lightest SUSY particle (LSP) is likely to be the $\tilde{\chi}^0$ or perhaps a slepton ($\tilde{\ell}$). In GMSB, $\tilde{G}$ is very light ($M_{\tilde{G}} \sim \mathcal{O}(\text{eV})$) and almost certainly the LSP.

R-parity, $P_R = (-1)^{3B+L+2s}$, combines baryon number, lepton number and spin. Generally, it is assumed to be conserved, predicting sparticle production in pairs only, and a stable LSP. (In MSUGRA, the $\tilde{\chi}^0$ LSP is a good cold dark matter candidate). Recently more attention is devoted to *R*-parity violation (RPV), giving rise to an unstable, thus detectable, LSP.

4.1 MSUGRA searches

Currently the LEP experiments put more emphasis on investigating difficult "corners" of the phase space which have been neglected so far. A small mass difference $\Delta m = m - m_{LSP}$ (< 10 GeV) gives rise to barely detectable events mixed with a high background from two-photon processes. Special analyses have been developed using e.g. an energetic initial-state radiation photon as a supplementary tag. Often a small Δm leads to delayed decays for which the standard search procedures are not sensitive and new approaches had to be worked out. More attention is also devoted to the case of small m_0 which predicts light sleptons and a complex sparticle decay pattern. All these detailed investigations lead to a more global picture which is fairly independent of a precise choice of model parameters.

(a) For charginos and neutralinos, the most general mass bounds are obtained [10] by scanning the parameter space and combining the searches for charginos, neutralinos and sleptons. This is illustrated for example in Fig. 7. As a general lower bound for the mass of the neutralino LSP, the LEP experiments quote values in the range 36-38 GeV. Limits of about 40 GeV are obtained if constraints on $\tan\beta$ from MSSM Higgs searches (see next section) are also used. From the same combination of searches, lower bounds in the range 90-95 GeV are obtained for the chargino mass, see e.g. Fig. 8.

(b) In the case of sfermions, the LEP data have been combined. The searches for stop and sbottom postulate a heavy gluino, excluding strong decay processes and leading to the weak decays $\tilde{t} \to c\tilde{\chi}^0$, $b\tilde{\chi}^\pm$ and $\tilde{b} \to b\tilde{\chi}^0$. The mass limits depend on the coupling of the squark to the Z^0 boson. The most

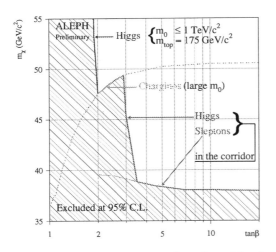

Figure 7. Derivation of a general lower bound for the mass of the neutralino LSP, by ALEPH

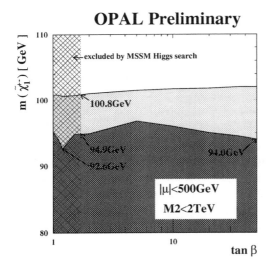

Figure 8. Derivation of a general lower bound for the mass of the chargino, by OPAL.

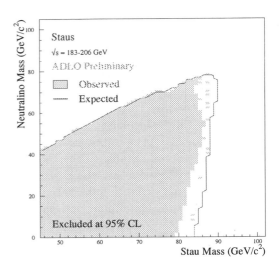

Figure 9. Excluded domain in the stau-neutralino mass plane obtained by combining the LEP data.

general bounds, $m_{\tilde{t}} > 95$ GeV, $m_{\tilde{t}} > 89$ GeV and $m_{\tilde{b}} > 85$ GeV, correspond to the cases of complete decoupling.

DELPHI has reported [11] a search for $\tilde{t} \rightarrow c\tilde{g}$ in the context of a light gluino and exclude large domains of the stop- gluino-mass plane.

Sleptons produced via $e^+e^- \rightarrow \tilde{\ell}^+\tilde{\ell}^-$ would give rise to a clear signature with a pair of acoplanar charged leptons plus missing energy. The LEP-combined lower bounds of the slepton masses are $m_{\tilde{e}} > 98$ GeV, $m_{\tilde{\mu}} > 94$ GeV and $m_{\tilde{\tau}} > 79$ GeV (see Fig. 9). A slight excess in earlier data, suggesting stau production, has unfortunately not been confirmed by the year 2000 data.

4.2 GMSB searches

The light gravitino being the LSP, the GMSB phenomenology is determined by the nature of the NLSP, which could be the $\tilde{\chi}^0$ or perhaps the $\tilde{\tau}_R$. The final state should thus contain isolated photons or $\tau^{\pm}$ leptons from the NLSP decays $\tilde{\chi}^0 \rightarrow \gamma\tilde{G}$ or $\tilde{\tau}_R \rightarrow \tau^{\pm}\tilde{G}$. This supplementary signature provides good visibility for GMSB processes. The NLSP lifetime is essentially arbitrary in GMSB. Delayed decays are characterized by non-

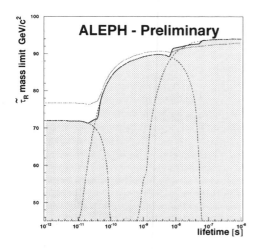

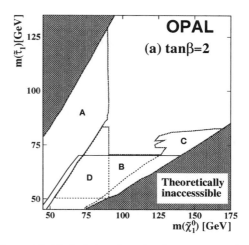

Figure 10. Lower bound for the stau NLSP mass in GMSB, as a function of its lifetime, by ALEPH.

Figure 11. Excluded regions in the stau-neutralino NLSP mass plane, for tan β=2, obtained by OPAL.

pointing photons or tracks, or tracks with anomalous ionization. The current NLSP mass limits [12], independent of the lifetime, are $m_{\tilde{\chi}^0} > 95$ GeV, $m_{\tilde{\tau}} > 75$ GeV (see e.g. Fig. 10).

Scans over the messenger sector parameters [13] are performed by ALEPH and OPAL to obtain lower bounds for sparticle masses, in particular for the NLSP ($\tilde{\chi}^0$ or $\tilde{\tau}$), see e.g. Fig. 11. These scans also determine a lower bound for the messenger scale, $\Lambda > 30$ (10) TeV, if the number of SU(5) multiplets in the messenger sector are fixed to $N_5 = 1$ (5). The corresponding lower bounds on the gravitino mass are of 0.2 (0.02) eV.

4.3 Searches for RPV-SUSY

R-parity violation predicts 45 new e^+e^- processes with baryon- and/or lepton-number violation, and characteristic couplings λ, λ' and λ'' [14]. Proton lifetime and low-energy data impose severe constraints on these couplings but e^+e^- data can test even smaller values. In the associated production and subsequent decay of SUSY particles, the small RPV couplings would affect mainly the decay of the LSP. A sfermion LSP would decay *directly* to ordinary fermions ($\tilde{f} \to ff'$). If $\tilde{\chi}^0$ is the LSP, the RPV decay of the sfermion is *indirect*: the decay to the LSP, $\tilde{f} \to f\tilde{\chi}^0$ (R-parity conserving), is followed by the RPV decay of the LSP, $\tilde{\chi}^0 \to ff'f''$.

Since the LSP becomes visible, the general SUSY signature of missing energy is lost; instead, RPV would manifest itself through an increase in the number of leptons and quark jets (up to 10) due to the LSP decay. At LEP, most of these decay topologies have been investigated. However, the RPV searches are by no means complete: they assume that only one of the couplings is nonzero at a time; that the decays occur typically within 1 cm from the e^+e^- interaction point, limiting the sensitivity to RPV couplings larger than $10^{-4}/10^{-7}$ for *indirect/direct* sfermion decays; they do not include regions of small Δm. Within these constraints, the currently available mass limits [14] for sfermions and gauginos are similar to those obtained in R-parity conserving scenarios.

R-parity violation also predicts the lepton-number violating *s-channel* process $e^+e^- \to \tilde{\nu}$. Resonances in the distributions

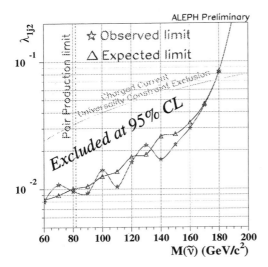

Figure 12. Upper bound for the RPV coupling λ from searches for single sneutrino production, by ALEPH.

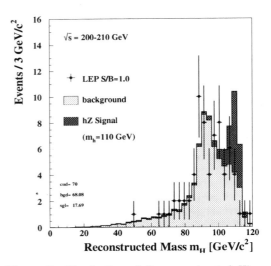

Figure 13. Distribution of the reconstructed Higgs mass for the data of the four LEP experiments pooled, and for a selection which produces a signal-to-background ratio of one in the range above 100 GeV.

of $e^+e^- \rightarrow \tilde{\nu} \rightarrow e^+e^-$, $\mu^+\mu^-$ and $\tau^+\tau^-$ have been searched at LEP [14]. ALEPH has also reported a searches for the decays $\tilde{\nu} \rightarrow \tilde{\chi}^0\nu$, $\tilde{\chi}^\pm\mu^\mp$ [15]. The bound on the relevant λ coupling is more restrictive than those from low-energy constraints, see Fig. 12.

R-parity may also be broken spontaneously in the third generation via the bilinear Higgs term ϵLH in the Lagrangian [16]. The resulting massless Majoron J has been searched by DELPHI in the process $\tilde{\chi}^\pm \rightarrow \tau^\pm J$ [17], and large domains of $m_{\tilde{\chi}^\pm}$ have been excluded.

5 Searches for Higgs bosons

The LEP collaborations presented updated results including their most recent data. These have been combined [18] using a statistical procedure worked out by the LEP-wide Higgs working group. The method consists of comparing the data to the *signal* + *background* and the *background only* hypotheses. A modified frequentist approach is used where the ratio of the corresponding likelihoods is the test statistic. The signal and background confidence levels CL_s and CL_b

for the two hypotheses are used as indicators for a possible signal and to establish lower bounds for the Higgs masses. No significant excess of events has been claimed at this conference that would suggest a Higgs boson signal.

5.1 The SM Higgs boson

The main process is $e^+e^- \rightarrow Z^0H^0$ with the Higgs boson decaying into $b\bar{b}$. The "four-jet", "missing energy" and "leptonic" final states, given by the decay of the associated Z^0 boson ($Z^0 \rightarrow q\bar{q}$, $\bar{\nu}\nu$, $\ell^+\ell^-$), encompass more than 90% of the cross-section [19]. The LEP experiments quote the following lower bounds for the mass: 110.8 GeV (ALEPH), 109.0 GeV (DELPHI), 107.6 GeV (L3), 109.5 GeV (OPAL).

The distribution of the reconstructed Higgs mass (Fig. 13), the behaviour of the test statistic [here $-2ln(Q)$] (Fig. 14) and of CL_b (Fig. 15) show full compatibility with the SM background prediction, up to the edge of sensitivity, where the signal+background and the background hypotheses cannot be

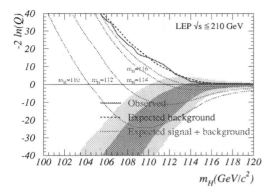

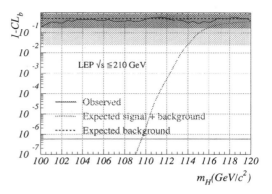

Figure 14. Behaviour of the test statistic (here $-2lnQ$) as a function of the "test mass". The full curve (data) follows closely the SM background expectation (dashed) and deviates from the predictions including a Higgs boson of various hypothetical masses (dotted lines). The shaded bands show the median expected signal behaviour at the test mass, with its $\pm 1\sigma$ and $\pm 2\sigma$ bands.

Figure 15. The background confidence level $1 - CL_b$ as a function of the "test mass". The full line (data) follows closely the SM background expectation (dashed line, $\pm 1\sigma$ and $\pm 2\sigma$ bands) and deviates from the signal plus background expectation (dotted line).

distinguished any more. The combined LEP limit for the SM Higgs boson mass is 113.3 GeV. It is intriguing to compare this lower bound to the value $m_H = 62^{+53}_{-39}$ GeV preferred by the electroweak precision measurements [1].

Note: At the time of writing the situation is radically different. Based on new data, ALEPH has reported an excess of Higgs-like events [20] at a mass of 113-115 GeV. The currently available data of the four experiments do not allow to confirm or exclude the signal hypothesis without ambiguity. In combination, the LEP data "show" a 2.6σ excess [21]. The LEP shutdown has been postponed to allow more data to be accumulated.

5.2 Supersymmetric Higgs bosons

In the MSSM, the two Higgs field doublets give rise to five physical Higgs bosons, h^0, H^0, A^0 and $H^\pm$. The lightest CP-even boson h^0 (CP conservation is assumed) is predicted to have its mass m_h below 135 GeV. The processes $e^+e^- \to h^0 Z^0$ and $e^+e^- \to h^0 A^0$ with complementary cross-sections are addressed by the searches [22]. The negative re-

sults are interpreted in a constrained MSSM-GUT framework which assumes unification of masses and couplings at the GUT scale. Representative scans of the parameter space [23] have been devised; in particular, the "$m_h - max$" scan maximizes the possible values of m_h, tending to producing conservative bounds on Higgs boson masses and MSSM parameters.

The exclusions shown in Fig's. 16 and 17 are obtained for the $m_h - max$ scan by combining the LEP data. The lower mass bounds for h^0 and A^0 are of 90.5 GeV. Fixing the top mass to 175 GeV, values of $\tan\beta$ of 0.5-2.3 are excluded (0.7-1.9 is excluded for $m_t = 180$ GeV).

5.3 Charged Higgs bosons

The searches at LEP for $e^+e^- \to H^+H^-$ are conducted [24] in the framework of models with two Higgs field doublets (2HDM) where the mass m_{H+} is not restricted (in the MSSM it is predicted to be larger than M_W). It is assumed, further, that the decays $H^+ \to c\bar{s}$ and $\tau^+\nu_\tau$ are exhaustive. Their relative branching ratio being free, the results are presented as exclusions in the

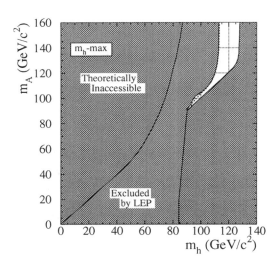

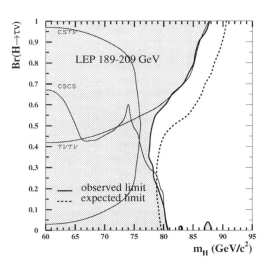

Figure 16. Exclusion obtained by combining the LEP data, in the MSSM framework and the $m_h - max$ benchmark scenario, projected onto the (m_h, m_a) plane.

Figure 18. Exclusions for charged Higgs bosons of the 2HDM, in the $(m_{H^+}, B(\tau^+ \nu_t au))$ plane, obtained by pooling the LEP data.

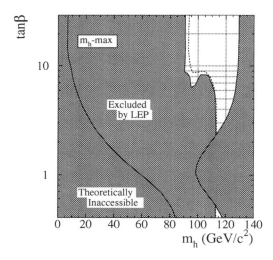

Figure 17. Exclusion obtained by combining the LEP data, in the MSSM framework and the $m_h - max$ benchmark scenario, projected onto the $(m_h, \tan\beta)$ plane.

$[m_{H^+}, B(\tau^+ \nu_\tau)]$ plane. The combined LEP results are shown in Fig. 18. Background from $e^+e^- \to W^+W^-$ limits the search sensitivity to $m_{H^+} \approx 80$ GeV.

5.4 Higgs search "exotica"

A wide range of searches for "exotic" Higgs boson beyond the SM and the MSSM have been reported [24].

Higgs bosons may decay into "invisible" (weakly interacting neutral) particles such as the $\tilde{\chi}^0$ LSP. In the process $e^+e^- \to Z^0(h^0 \to \tilde{\chi}^0 \tilde{\chi}^0)$ the Higgs mass is obtained as the mass recoiling against the reconstructed Z^0 boson. The LEP data exclude invisibly decaying Higgs bosons of mass less than 107.6 GeV if produced with SM rate.

In some 2HDM variants, the Higgs-fermion couplings can be "tuned" to zero. The channel $h^0 \to \gamma\gamma$, which occurs in the SM at loop level, can thus be enhanced. For such *fermiophobic* Higgs bosons, the LEP data provide a lower mass bound of 106.4 GeV. DELPHI and L3 discuss photonic events in the broader context [25] and exclude large values of anomalous Higgs-gauge couplings, see

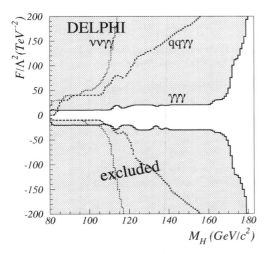

Figure 19. Bounds on anomalous Higgs-gauge couplings from the DELPHI analysis of photonic events.

Fig. 19.

L3 has searched for the bosonic decay process $e^+e^- \rightarrow Z^0 H^0 \rightarrow \nu\bar{\nu} W W^*$ [26], producing upper bounds for the cross-section, see Fig. 20.

Other Higgs "exotica" include the OPAL searches for the decay $H^+ \rightarrow W^+ A^0$ relevant to 2HDM [27], and for doubly charged Higgs bosons [28] predicted e.g. by left-right symmetric models (the $H^{++} \rightarrow \tau^+\tau^+$ channel is used).

Special searches have been developed by each of the LEP experiments where b-flavour is not required in the decay of the neutral Higgs bosons. Results from such flavour independent searches, as well as on model-independent bounds on cross-sections for specific decay topologies, are of great importance in preparing the future after LEP. The OPAL search [29] for the process $e^+e^- \rightarrow Z^0 S^0$ (S^0 is a generic scalar particle) is an example of decay-mode independence. The upper bound on the cross-section, see Fig. 21, extends over the mass range of 500 MeV to 100 GeV and is valid for any combination of hadronic, leptonic, photonic or invisible decays of the generic scalar S^0, which can be unstable or stable.

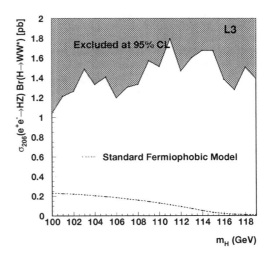

Figure 20. Upper bounds on the topological cross-section for the process $e^+e^- \rightarrow H^0 Z^0 \rightarrow \nu\bar{\nu} W W^*$, obtained by L3.

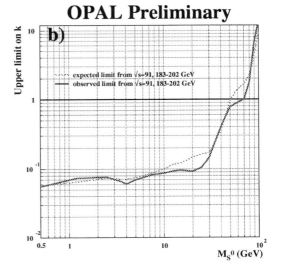

Figure 21. Upper limit on the cross-section ($k = \sigma/\sigma_{HZ}^{SM}$) for the production of a generic scalar S^0, valid for arbitrary decays, obtained by OPAL.

6 Summary

After a decade of efficient exploitation, the LEP e^+e^- collider is approaching its final shutdown. With no compelling evidence for Higgs bosons so far, the scalar sector remains virgin territory. Supersymmetry has not been seen either, although several scenarios of soft symmetry breaking, with R-parity conservation and violation, have been considered.

In the absence of a signal for new physics, the rich harvest of relevant bounds on masses, topological cross-sections and model parameters produced by LEP constitutes a legacy of inestimable value to the next generation of accelerators and experiments.

At the time of writing, LEP is still running at its highest energies. Current data may suggest that in the Higgs sector the last word has not yet been pronounced!

The author wishes to congratulate the organizers of this successful conference. Many thanks to my colleagues and friends from the LEP experiments and the LEP-wide working groups for providing early insight into their submissions. I am grateful to Satoshi Mihara for his kind assistance during the conference.

References

1. A. Gurtu, PL-06a, these proceedings.
2. N. Arkani-Hamed, S. Dimopoulos and G. Dvali, Phys. Lett. **B429** (1998) 263.
3. G. Landsberg, PA-11, these proceedings.
4. M.A. Falagan, PA-11, these proceedings.
5. W. Lohmann, PA-11, these proceedings.
6. V. Obraztsov, PA-11, these proceedings.
7. L3 Note 2495 (2000).
8. DELPHI 2000-074 CONF 373.
9. L3 Note 2428 (1999).
10. K. Nagai, PA-11, these proceedings.
11. DELPHI 2000-086 CONF 385.
12. V. Hedberg, PA-11, these proceedings.
13. ALEPH Coll., CERN-EP/99-171; OPAL Coll., CERN-EP-2000/078.
14. I. Fleck, PA-11, these proceedings.
15. ALEPH 2000-027 CONF 2000-022.
16. J.W.F. Valle, PA-11, these proceedings.
17. DELPHI 2000-021 CONF 342.
18. LEP Higgs Working Group, http://lephiggs.web.cern.ch/LEPHIGGS/pap *Note for ICHEP'2000...*
19. Shan Jin, PA-11, these proceedings.
20. D. Schlatter, ALEPH Status Report, LEPC, Sept. 5, 2000.
21. Ch. Tully, Status of LEP-wide Higgs searches, LEPC, Sept. 5, 2000, http://lephiggs.web.cern.ch/LEPHIGGS/tal
22. I.M. Fisk, PA-11, these proceedings.
23. M. Carena, S. Heinemeyer, C.E.M. Wagner and G. Weiglein, CERN-TH/99-374, hep-ph/9912223.
24. A.P.P. Kiiskinen, PA-11, these proceedings.
25. DELPHI 2000-082 CONF 381, L3 Note 2430 (2000), L3 Note 2558 (2000).
26. L3 Note 2592 (2000)
27. OPAL Coll., CERN-EP/2000-092.
28. OPAL Physics Note PN-445 (2000), PN-446 (2000).
29. OPAL Physics Note PN-449 (2000).

SEARCHING FOR NEW PARTICLES AND PHENOMENA AT HADRON COLLIDERS

JOHN S. CONWAY

Dept. of Physics and Astronomy, Rutgers University, Piscataway, NJ, USA

The high-energy frontier in the search for new particles is at the hadron colliders, including the Tevatron, HERA, and, in the future, the LHC. Though no definitive signals for new particles have yet been observed in the experiments, past and present hints in the data of new physics have stimulated a great deal of experimental and theoretical work. This paper reviews some recent highlights in the search for new phenomena at hadron colliders, including the various hints and anomalies, the search for the Higgs, supersymmetric particles, and the effects of possible large extra dimensions.

1 The Accelerators and the Experiments

Hadron colliders offer a unique hunting ground in the search for new particles and phenomena. Since it is technically possible to reach a much higher center of mass energy with hadron machines compared with electron-positron machines, phenomena at much higher mass scales can be explored. However, the parton structure of hadrons leads to a steeply falling spectrum for the effective luminosity as a function of the center of mass energy of parton-parton collisions. Lack of precise knowledge of this c.m. energy, as well as the debris from the "underlying event," (the collision of the remnants of the protons and antiprotons), makes efficient selection of new particle signals difficult. Nevertheless, with higher energy reach hadron machines play a complementary role to e^+e^- colliders such as LEP.

In the 1990's two high energy accelerators, the Tevatron and HERA, began producing copious data samples in which to search for new phenomena. The Tevatron at Fermilab collided protons and antiprotons at c.m. energies of 1.8 TeV, and the two large general purpose detectors, CDF and DØ , recorded about 100 pb^{-1} integrated luminosity. This sample led to the discovery of the top quark[1] in 1995, a very precise determination of the W boson mass[2], and numerous publications on searches for new particles.

The HERA accelerator at DESY collides protons with up to 920 GeV energy with electrons or positrons at 27.6 GeV. The two large experiments at HERA, called ZEUS and H1, have recorded samples corresponding to a total of 120 pb^{-1} integrated luminosity. The most recent sample recorded in 1999 and 2000, and still accumulating, is with positrons on protons, and the sample on which data are reported here corresponds to 50 pb^{-1}.

In 2005 the LHC will begin operating at CERN, colliding protons with protons at $\sqrt{s} = 14$ TeV. This large leap in c.m. energy and instantaneous luminosity will allow the two very large detectors, ATLAS and CMS, to study new phenomena up to the TeV scale, and address definitively the question of the origin of electroweak symmetry breaking.

2 The NuTeV Dimuon Anomaly

It is important to realize that as physicists we are not necessarily sure where new phenomena will first reveal themselves. A recent example is the observation of anomalous dimuon events by the NuTeV experiment at Fermilab, a large calorimeter/muon spectrometer designed to study neutrino charged-current interactions at extremely high rates.

To look for possible new long-lived neutral particles such as neutral heavy leptons or neutralinos, the experimenters built a helium-filled decay region upstream of the

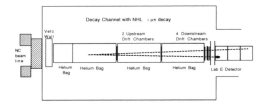

Figure 1. Schematic diagram of NuTeV decay channel.

Table 1. Result of NuTeV search for two-charged-track vertex events in the decay channel.

channel	observed	expected
$\mu\mu$	3	0.04±0.01
$e\mu$	0	0.14±0.02
$\pi\mu$	0	0.13±0.02

detector, instrumented with large single-wire drift chambers, shown schematically in Fig. 1. A long-lived neutral particle accompanying the neutrinos could decay in this volume, leaving a vertex with two final state charged particles. The calorimeter and spectrometer downstream provide identification and energy measurements for electrons, muons and pions with very good efficiency and purity.

Table 1 shows the result of a blind analysis of the data sample. Three dimuon events are observed in the fiducial volume (which excludes the chamber region) but no electron-muon or pion-muon events. The expected number of such events is 0.04±0.01; the probability for seeing 3 or more is on the order of 10^{-5}, though it is important to note that such an excess could have appeared in any of the three bins with probability around 10^{-3}.

The properties of the events are similar in that they all have one very large energy muon and one low-energy muon, and large missing transverse momentum. These features would tend to disfavor their interpretation as the decay of long-lived neutral heavy leptons. That they are neutrino-induced background events is ruled out by the fact that relaxing the fiducial cut, and allowing into the sample over 20 times more material for neutrinos to interact in, results in a small number of events consistent with the expectation.

3   Signatures Versus Models

The open-minded physicist must clearly admit that she knows not where new physics will first appear, and thus a trend has developed in the field towards signature-based rather than model-based searches for new particles and phenomena.

The advantages of a definite model, such as compositeness or supersymmetry, lie in that one has a clear *a priori* hypothesis for what one is looking for. This allows one to perform statistical tests of the significance of a new signal, if one is observed, without bias, and with a well-defined model for the detector acceptance for the new particle. The disadvantage of relying upon models to tell what to look for is that there is no empirical reason to believe any model over any other, and the space of possible models is enormous.

What the collider experiments can do is to trigger on and reconstruct various "physics objects": electrons, muons (together referred to colloquially as leptons), photons, hadronic decay products of taus, and jets, including the tagging of heavy flavor jets with reasonably high efficiency. Also, the experiments can all typically measure missing transverse energy (which more properly should be called missing transverse momentum) which is often an ingredient of signals for new physics.

The experiments must, however, demand that two, three, or more of these objects be present in an event and have large energy (typically over 20 GeV) to avoid backgrounds and retain good efficiency. If one considers that there are 50 combinations of five of these objects (photons, leptons, tau decays, jets, or b-quark jets), and asks how many of these possible signatures have been studied, say, at the Tevatron, it is interesting to note that only about half have been explored! Many of these signatures are relevant for several possible new physics processes, and thus searching the signatures themselves without reference to models, simply comparing the observed events in each channel with the standard model expectation, may be the most sure way not to miss new phenomena.

4 Any Hints in the Data?

Given the signatures which can be studied, one can ask whether there are in fact any hints of new physics, any unexplained excess events, in the existing data samples. The field has experienced a number of these in the past decades, and with few exceptions (the top quark, B_c meson, etc.) such excesses have proven to be mere statistical fluctuations of Standard Model background processes. This is quite healthy: very often the observation of an anomaly in the data stimulates theoretical activity which might otherwise be dormant or as yet unexplored.

The Tevatron and HERA experiments have in fact observed various interesting anomalies in their data, and we highlight a few of them here.

4.1 The CDF $e\slashed{p}\gamma\gamma$ Event

The observation of an event with four high-energy electromagnetic clusters and large missing transverse energy recorded by CDF was first reported in 1995.[3] Though now well-known in the community, this is by far the most difficult-to-explain event in the entire CDF Run 1 data set. The forward electromagnetic cluster, originally thought to be a positron, is actually more likely to be, for example, a tau decay product such as $\tau^+ \to \rho^+\nu_\tau, \rho^+ \to \pi^+\pi^0$. The central electron and two photons are near-ideal, and there is very little underlying event activity. An event display of the calorimeter energy deposits appears in Fig. 2

The main Standard Model process which could account for such an event is $WW\gamma\gamma$

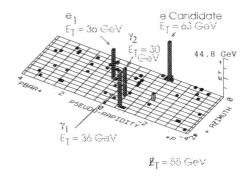

Figure 2. Event display showing calorimeter energy deposits in the CDF $e\slashed{p}\gamma\gamma$ event from Run 1.

production. The expected rate, however, is very small, on the order of 10^{-6} such events expected in 100 pb^{-1} integrated luminosity. There is no reason to believe that any detector failure modes could account for the event. The DØ collaboration observes no such events.

This event generated numerous studies within CDF for related events, and tests of specific theoretical models generated to explain the candidate; indeed the observation of the event gave life to several minor theoretical industries, including gauge-mediated supersymmetry. All subsequent studies, however, have failed to turn up more evidence of a consistent pattern of anomaly in either CDF or DØ data. The experiments await the order-of-magnitude-larger samples expected in Run 2 to search for more related events.

4.2 The HERA High-p_T Lepton Excess

In their 1994-1997 data sample both H1 and ZEUS observed events with a high-p_T lepton (e or μ), one or more jets, and missing transverse momentum. Such events are expected in the Standard Model (SM) in direct W production processes such as that shown in Fig. 3. Beyond the Standard Model such events could be a signal of R–parity violat-

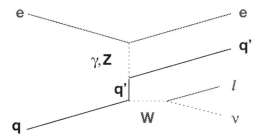

Figure 3. Diagram of direct W production in high energy $e-p$ collisions at HERA.

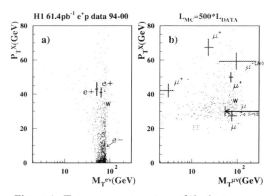

Figure 4. Transverse momentum of the lepton versus transverse mass for selected H1 and ZEUS high-p_T lepton events, from the full 1994-2000 data set.

Table 2. Number of observed (expected) high-p_T lepton ($p_T > 25$ GeV) events in H1 and ZEUS, full 1994-2000 data sample.

	H1	ZEUS
e	3 (0.8±0.2)	1 (0.78)
μ	6 (0.9±0.3)	0 (0.82)
total	9 (1.8±0.4)	1 (1.6)

ing supersymmetry, or anomalous single-top production.

Remarkably, the H1 experiment recorded five events with a high-p_T muon and one with an electron. Three of the muon events seemed to lie outside the expected kinematic region for Standard Model W production. The ZEUS experiment, by contrast, observed three high-p_T positron events, in good agreement with the expectation.

Both experiments now report results based on 82 pb^{-1} from data collected in 1994-2000 (1999 for ZEUS).[4] In the original analyses, the ZEUS selection had a higher SM background expectation than H1, and so to facilitate direct comparison both experiments made changes to the analysis. Table 2 shows the result for the case where the leptons have at least 25 GeV/c p_T.

As the table shows, the H1 selection yields an excess of events with high-p_T muons, whereas the ZEUS results are in good agreement with the predictions. The situation cries out for further study and especially more data, which, fortunately, is still being collected.

4.3 The HERA High-Q^2 Excess

In their 1994-1997 sample, the HERA experiments also observed an excess of neutral-current events with very high Q^2. For example, Fig. 5 shows the observed Q^2 spectrum from ZEUS. There are 49 events with invariant mass of the e-jet system above 210 GeV/c^2. Such events could signal various new physics processes, for example s-channel leptoquark production or R-parity-violating SUSY. The observations stimulated a great deal of activity both on the experimental and theoretical side.

Recent data, however, fail to confirm any significant excess. The H1 data from e^+p data shown in Fig. 5 reveal no high-Q^2 excess; these data can be used to set the limits on leptoquarks shown in Fig. 7, and other new physics processes.

5 Search for the Higgs

The origin of the mechanism of electroweak symmetry breaking remains one of the great outstanding questions in high energy physics. In the Standard Model the Higgs mechanism accounts for this, and predicts the existence of a single neutral scalar Higgs boson. In extensions to the Standard Model such as supersymmetry the Higgs sector is more complicated, predicting two neutral scalar

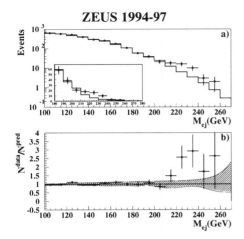

Figure 5. Observed Q^2 spectrum in the ZEUS 1994-1997 sample.

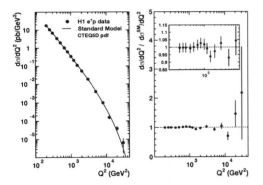

Figure 6. Observed Q^2 spectrum and data to simulation ratio in the H1 high-Q^2 study.

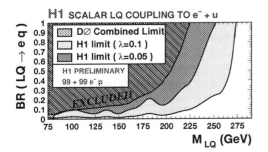

Figure 7. Limits on scalar and vector leptoquarks from the H1 high-Q^2 study.

Higgses, one pseudoscalar, and two charged scalar Higgses. The Tevatron and LHC both have sensitivity to the Higgs, both in the Standard Model and beyond.

5.1 Standard Model Higgs

If the Higgs is a single scalar, then combining the world's electroweak measurements implies that its mass is 62^{+53}_{-30} GeV/c^2, with a 95% CL upper limit at 170 GeV/c^2. However, very preliminary new measurements of α_{QED} from BES indicate that the central value is actually larger, and the upper limit moves up to 210 GeV/c^2 at 95% CL.[5] Direct searches for the Higgs at LEP2 have excluded a Standard Model Higgs with mass less than 113.4 GeV/c^2 with 95% confidence.[6]

At the Tevatron the Higgs is produced dominantly by gluon-gluon fusion, but since it decays mainly to b-quark pairs the background from ordinary QCD dijet processes is overwhelming. Then next highest rate is for the associated production of a Higgs with a vector boson (W or Z) leading to several distinct final states which depend on the decay modes of the vector bosons. In Run 1 CDF and DØ have both studied the final state $\ell\nu b\bar{b}$, and CDF has in addition studied the $\nu\bar{\nu}b\bar{b}$, $\ell^+\ell^- b\bar{b}$ and $q\bar{q}b\bar{b}$ channels. No signal is observed, and Fig. 8 shows the limits on the cross section times branching ratio obtained for the individual channels and all channels combined. The limit is far above the prediction for Standard Model Higgs, and thus no mass limit can be obtained.

The Tevatron Run 2 SUSY/Higgs Workshop studied the question of the Higgs reach in Run 2 at the Tevatron.[7] With an order of magnitude higher instantaneous luminosity and upgraded detectors, by combining all channels and both experiments' data it is possible to discover or exclude a Standard Model Higgs, using the same set of channels as in Run 1. In addition, for larger Higgs masses it is possible to use the WW decay modes and extend the reach.

Fig. 9 shows the integrated luminosity required to exclude or discover a Standard Model Higgs as a function of Higgs mass. With 10 fb^{-1} integrated luminosity, which should take on the order of five years, a 95% CL exclusion can be set up to about 190 GeV/c^2 Higgs mass. To discover a Higgs with 5σ significance requires a great deal more integrated luminosity, however; with 15 fb^{-1} the discovery reach is about 120 GeV/c^2. Clearly, this is a difficult search, and though optimistic assumptions have been made, new ideas and techniques may, as in the past, result in significant improvements to the reach projected so far.

The LHC experiments, with very high instantaneous luminosity and very high energy, can observe a Standard Model Higgs in the mass range below about 200 GeV/c^2 predicted by the electroweak combination with about 30 fb^{-1}. Fig. 10 shows the statistical significance attainable by combining the various search channels. The most sensitive channels in the lower mass range are $H \to ZZ \to \gamma\gamma$ and $H \to \gamma\gamma$.

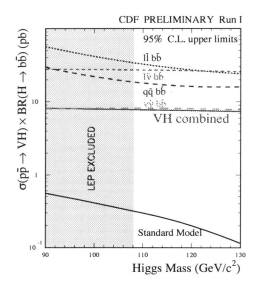

Figure 8. CDF Run 1 Standard Model Higgs combined cross section times branching ratio limits.

5.2 Higgs in Supersymmetry

In the minimal supersymmetric extension to the Standard Model (MSSM), the lighter of the two neutral scalar Higgs mentioned above can behave essentially the same way as the SM Higgs. Thus it is possible to directly apply the results of the search for the SM Higgs to the MSSM case, and express the limits obtained in terms of the parameters of the model, typically $m(A)$ (the mass of the pseudoscalar Higgs) and $\tan\beta$ (the ratio of the vacuum expectation values of the two Higgs doublets in the model. LEP2 has already excluded a large range of parameter space; if no observation is made the Tevatron can exclude or discover the MSSM Higgs over more of the parameter space, as depicted in Figs. 11 and 12.[7]

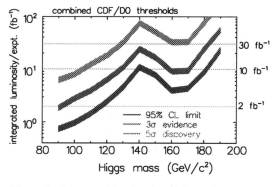

Figure 9. Integrated luminosity (delivered per experiment) required in the Tevatron Run 2 to exclude or discover a Standard Model Higgs.

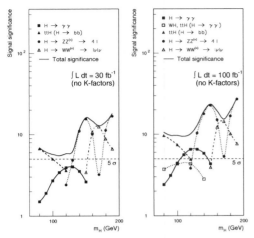

Figure 10. ATLAS significance for Standard model Higgs search after 30 fb^{-1} and 100 fb^{-1}.

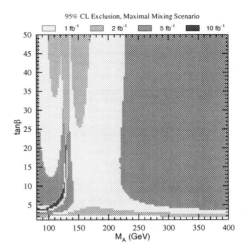

Figure 11. Projected Tevatron limits on MSSM Higgs using the SM Higgs search, for the case of maximal stop mixing. This prediction assumes that both experiments' data are combined, for all SM Higgs search channels.

The LHC experiments will be able to discover a SUSY Higgs over the full parameter space with about 30 fb^{-1}.

One advantage hadron colliders have over e^+e^- colliders is in the case of the search for the MSSM Higgses when $\tan\beta$ is very large. Since the coupling of the pseudoscalar Higgs (A) to $b\bar{b}$, for example, is proportional to $\tan\beta$, the cross section for the production of $Ab\bar{b}$ is enhanced above the SM rate by the factor $\tan^2\beta$, which can be very large. The rates for $hb\bar{b}$ and $Hb\bar{b}$ are also enhanced by a similar factor, and all lead to a very distinct four-b-jet final state.

CDF has searched for this process in Run 1 data; no signal is observed and the resulting limits are shown in Fig. 13. This search is statistics-limited, and in Run 2 a large region of parameter space can be excluded, extending down to $\tan\beta = 15$ for low $m(A)$ given 15 fb^{-1}.[7]

6 Supersymmetry Searches

Evidence for the existence of supersymmetric partners to the quarks, leptons, and gauge bosons has been sought at every major particle collider for the past two decades, yet none has yet been found. Whereas e^+e^- machines

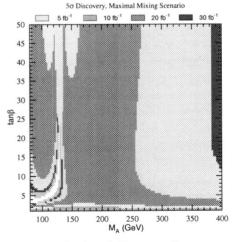

Figure 12. Projected Tevatron discovery reach for MSSM Higgs using the SM Higgs search, for the case of maximal stop mixing. This prediction assumes that both experiments' data are combined, for all SM Higgs search channels.

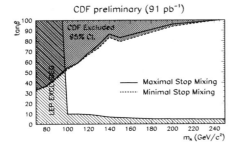

Figure 13. CDF limits on MSSM Higgs using $b\bar{b}b\bar{b}$ final state.

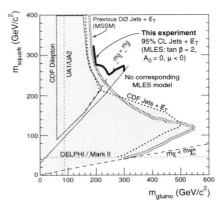

Figure 15. DØ and CDF limits on squarks and gluinos.

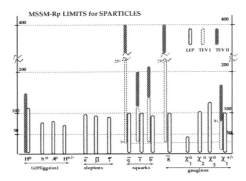

Figure 14. Approximate range of expected discovery reach for various supersymmetric particles at LEP2 and the Tevatron.

have well-defined c.m. energies and a clean environment, hadron colliders can reach to higher energies and produce strongly interacting sparticles copiously but suffer inefficiencies when the final state products have low energy. This leads to a situation where the two machines are sensitive to complementary regions of parameter space. Fig. 14 shows schematically the range of sensitivity to various supersymmetric particles for LEP2 and the Tevatron.

The Tevatron excels in the search for squark and gluino pair production. The results from DØ are the most stringent over most of the parameter space; these appear in Fig. 15. The limits come mainly from the search for jets plus missing E_T, the classic signature for this process.

To see another specific and striking example of the complementarity of LEP2 and the Tevatron, consider the production of stop and sbottom quark pairs. If these are heavier than the neutralino and other decay modes are kinematically accessible, the events bear the distinct signature of two heavy-flavor jets recoiling against missing energy. Fig. 17 shows the limit obtained in Run 1 on this process by CDF for $\tilde{b}\tilde{b}$, and Fig. 16 shows the limit for $\tilde{t}\tilde{t}$ production. Also shown in the plots are the regions of parameter space excluded by ALEPH. These plots illustrate the higher mass reach obtainable at the Tevatron, but the fact that the limit does not extend to squark masses near that of the neutralino is due to the kinematic cuts on the jet E_T necessary to trigger on the signal and distinguish it from background. In Run 2, though, the upper edge of the sensitive region will move to higher masses, near $200\,\text{GeV}/c^2$ with large integrated luminosity.

7 Large Extra Dimensions

An interesting new prospect which has arisen in the past two years is that if, as predicted by string theory, the world has more than the normal three space and one time dimension, the extra dimensions could be large,

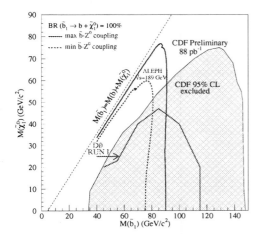

Figure 16. CDF limits on stop decaying to bottom plus neutralino.

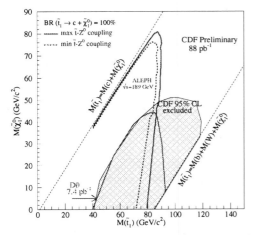

Figure 17. CDF limits on stop decaying to charm plus neutralino.

with radii on the order of millimeters. Surprisingly such a situation is not ruled out by astronomical or other observations and leads to measurable effects at lepton and hadron colliders.[8]

Gravity is weak in the usual 3+1 spacetime dimensions (called "the wall"), and strong in the extra dimensions ("the bulk"). In addition the Standard Model fields (electromagnetic, weak and strong) are confined to the wall. The extra dimensions can be characterized by their radius, which can be related to the effective Planck scale in the extra dimensions, usually denoted M_s or M_H in the literature. The presence of such extra dimensions with effective Planck scale in the TeV range in fact solves the hierarchy problem by removing it. Another parameter, usually called λ, and of scale $\pm\mathcal{O}(1)$, is calculable only with full knowledge of the full quantum gravity theory.

Various direct and indirect effects could signal the presence of such large extra dimensions. Massive gravitons could appear as a narrowly spaced set of Kaluza-Klein excitations in the (closed) extra dimensions and interact with the Standard Model fields. Thus, in $f\bar{f}$ collisions, for example, an s-channel graviton could be exchanged, modifying the usual rates and angular distributions of the final state fermion pairs. Or single gravitons could be emitted, leading to signals with large missing energy. Typical effects such as the modified angular distribution for e^+e^- pair production depend on λ/M_s^4.

The most stringent lower limits to date come from the DØ experiment, which looked for deviations from the SM rate and angular distribution of photon and e^+e^- pairs.[9] They set limits on the effective Planck scale of between 1.0 and 1.4 TeV depending on model parameters and the number of extra dimensions.

In a variant of the model by Randall and Sundrum[10] there are two "walls" connected by the "bulk" extra dimensions. One wall has

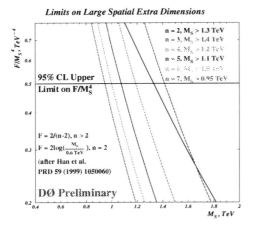

Figure 18. DØ limits on large extra dimensions from the search for anomalous high-mass photon and e^+e^- pairs.

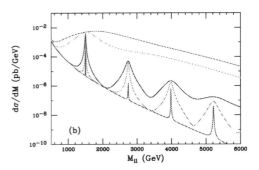

Figure 20. Large resonances in fermion pair production at the LHC in the presence of large extra dimensions, in the Randall-Sundrum model.

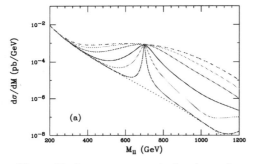

Figure 19. Large resonance in fermion pair production at the Tevatron in the presence of large extra dimensions, in the Randall-Sundrum model.

8 Summary

Though no new signals for any particles or phenomena outside the Standard Model have yet been observed, the hunt is on at hadron colliders in the last decade and the next. The large data sets collected at the Tevatron and HERA have been scoured, albeit incompletely, for signs of new physics, and the main hints of any signal, none overwhelmingly statistically significant, are the NuTeV dimuon anomaly, the HERA high-p_T lepton excess, and the CDF $ee\gamma\gamma$ event. Such observations have been key in stimulating theoretical activity and further experimental investigations, and provide strong motivations to continue the hunt, both within the predictions of definite models and in the realm of the possible. With more data, the questions raised will be answered definitively.

The experiments at the Tevatron and HERA have also searched for a wealth of other new particles and phenomena predicted by specific models. The hadron machines have the best sensitivity for supersymmetric particles which couple strongly, the best sensitivity to leptoquark production, R-parity violating supersymmetric processes, and many other phenomena and regions of parameter space complementary to those covered at LEP2.

The Tevatron and HERA experiments

our normal three space dimensiona, and the other is hidden.

Among the experimental consequences of this model is the prediction that, at the Tevatron and the LHC, large resonances in fermion pair prodiction will be observable, as shown in Figs. 19 and 20. Indeed, the LHC will be able to see the effects of any large extra dimensions with minimal integrated luminosity.

will enjoy the enhancements of major upgrades to the accelerators and detectors, and will press on with the work of determining which, if any, of the present hints in the data might signal new physics. The new capabilities will allow the experiments to extend the reach for new phenomena to higher mass scales. The Tevatron in particular has a five- to six-year window to search for the Higgs, which likely lies at a mass within its reach. Ultimately, however, in the new decade the LHC will begin the real work of exploring up to the TeV scale for what lies beyond the Standard Model, and will elucidate the true mechanism of electroweak symmetry breaking.

Acknowledgments

I would like to thank the organizers of the conference for providing a wonderfully organized and pleasant venue, and a stimulating program of physics events as well as a unique cultural experience. I would also like to thank my scientific secretary, Takeshi Komatsubara, for his help in preparing my talk. This work is supported by a grant from the National Science Foundation.

References

1. F. Abe, *et al.* (CDF Collaboration), Phys. Rev. Lett. 74 (1995) 2626. S. Abachi, *et al.* (DØ Collaboration), Phys. Rev. Letters 74 (1995) 2632.

2. F. Abe, *et al.* (CDF Collaboration), Phys. Rev. D52:4784-4827, 1995; S. Abachi, *et al.* (DØ Collaboration), Phys. Rev. Letters 77, 3309 (1996).

3. F. Abe *et al.* (CDF Collaboration), Phys. Rev. D59:032001, 1999.

4. A. Mehta, these proceedings.

5. B. Pietrzyk, these proceedings.

6. K. Hoffman, these proceedings.

7. Report of the Higgs Working Group of the Tevatron Run 2 SUSY/Higgs Workshop, in preparation.

8. J.L. Hewett, Phys. Rev. Lett. 82 (1999) 4765; G. Giudice, R. Ratazzi, and J. Wells, Nucl. Phys. B544 (1999) 3; T. Han, J.D. Lykken, and R.-J. Zhang, Phys. Rev. D59 (1999), 105006; E.A. Mirabelli, M. Perelstein, and M.E. Peskin, Phys. Rev. Lett. 82 (1999) 2236.

9. B. Abbott, *et al.* (DØ Collaboration), hep-ex/0008065, submitted to Phys. Rev. Lett.

10. L. Randall and R. Sundrun, Phys. Rev. Lett. 83 (1999), 3370.

Plenary Session 8

Neutrino Physics

Chair:	J. S. Kang (Korea)
Scientific Secretaries:	Takanobu Ishii and
	Toshiyuki Toshito

ACCELERATOR AND REACTOR NEUTRINO EXPERIMENTS

KOICHIRO NISHIKAWA

Kyoto university, Department of Physics, Kitashirakawa,Sakyo-ku
Kyoto, Japan
E-mail: nishikaw@scphys.kyoto-u.ac.jp

In this article results of neutrino oscillation from accelerator and reactor experiments are reviewed. The remaining questions and future experiments, which will address these questions, are described.

In this conference, the detection of five charged current ν_τ events in nuclear emulsion was reported by DONUT group at Fermilab the beam dump experiment [1]. Now the existence of the final member of fermions in the Standard Model is confirmed.

The Standard Model has been tested with exceptional precision. However it does not address the question of generations and their mixing. This article reviews results and future experiments presented at Parallel session of the International Conference on High Energy Physics (ICHEP2000). Section 1 provides an introduction to neutrino oscillation. Section 2 reviews results of neutrino oscillation searches for various Δm^2 regions in various oscillation modes. Section 3 describes questions, which could be answered in the near future. The determination of sign of Δm^2 and CP violation are described in Section 4.

1 Neutrino oscillation

The possibility of neutrino oscillation was pointed out by Pontecorvo[2] for ν and $\bar{\nu}$ oscillation in an analogy with $K^0 - \overline{K^0}$ oscillation. The possibility of oscillation between different flavors was first pointed out by Maki-Nakagawa-Sakata[3] in 1961. The existence of neutrino oscillation proves that neutrinos are massive and the lepton flavors are not conserved quantum numbers. The oscillation assume that the three known neutrino flavor states, ν_e, ν_μ, and ν_τ are not mass eigen-states but they are quantum mechanical superposi-

tion of three mass eigen-states, ν_1, ν_2, and ν_3 with mass eigen-values m_1, m_2, and m_3, respectively:

$$\begin{pmatrix} \nu_e \\ \nu_\mu \\ \nu_\tau \end{pmatrix} = \begin{bmatrix} U_{\alpha i} \end{bmatrix} \begin{pmatrix} \nu_1 \\ \nu_2 \\ \nu_3 \end{pmatrix}$$

where $\alpha = e, \mu, \tau$ (flavor index), i=1,2,3 are the index of the mass eigen-states, and U is a unitary 3x3 matrix that can be defined by a product of three rotaion matrix with three angles (θ_{12}, θ_{23}, and θ_{31}) and complex phase (δ) as in Cabibbo-Kobayashi-Maskawa matrix [4]. Neutrinos are produced as a flavor eigen-state and the each component of mass eigen-state gets different phase, $m_i^2 L/2E_\nu$ after traveling a distance L. The detection of neutrino by the charged current interactions projects this new state back to a flavor eigen-state. The probability of oscillation ($P(\nu_\alpha \rightarrow \nu_\beta)$) is given by the formula,

$$P(\nu_\alpha \rightarrow \nu_\beta) = \delta_{\alpha\beta}$$
$$- 4 \sum_{i>j} Re(U_{\alpha i}^* U_{\beta i} U_{\alpha j} U_{\beta j}^*) sin^2 \left[\frac{1.27\Delta m_{ij}^2 L}{E_\nu} \right]$$
$$\pm 2 \sum_{i>j} Im(U_{\alpha i}^* U_{\beta i} U_{\alpha j} U_{\beta j}^*) sin \left[\frac{2.54\Delta m_{ij}^2 L}{E_\nu} \right]$$

where $\Delta m_{ij}^2 = m_i^2 - m_j^2$ in eV^2, L is the distance in km, and E_ν is the neutrino energy in GeV. The $\pm$ sign in the third term is the CP violation effect, - for neutrinos and + for anti-nautinos. The neutrino oscillation is the most promising way to search for very small neutrino mass by choosing a large L/E_ν and to study the relation of flavors and generations in lepton sector. Also neutrino oscillation experiments are the only way to search for possible new neutrino species (light sterile neutrinos ν_s) which does not interacts even with weak interaction. If there are only three kind of neutrinos and if the two Δm^2 are those suggested by solar and atmospheric neutrino experiments, the two Δm^2 are $\Delta m_{12}^2 \equiv \Delta m_{sol}^2 = 10^{-5} \sim 10^{-10} eV^2$ and $\Delta m_{23}^2 \simeq \Delta m_{31}^2 \equiv \Delta m_{atm}^2 = 5 \times 10^{-3} \sim 10^{-3} eV^2$. For an oscillation measurement with L/E of the order of 100, the effect of Δm_{12}^2 term can be neglected. The oscillation probabilities can be expressed by three mixing angles:

$$P\,(\nu_\mu \to \nu_e) = sin^2 2\theta_{13} s_{23}^2 S$$
$$P\,(\nu_\mu \to \nu_\tau) = sin^2 2\theta_{23} c_{13}^4 S$$
$$P\,(\nu_e \to \nu_\tau) = sin^2 2\theta 13 c_{23}^2 S$$
$$P\,(\nu_e \to \nu_e) = 1 - sin^2 2\theta_{13} S$$
$$S = sin^2 \frac{\Delta m_{23}^2}{4E}, c_{ij} = cos\theta_{ij}, s_{ij} = sin\theta_{ij}$$

Solar neutrino experiments are measuring essentially θ_{12} in this scheme. Any inconsistent results from the above formalism point to the existence of sterile neutrinos. The atmospheric neutrino experiments [5] (near maximal mixing for $\nu_\mu \to \nu_\tau$ and non-observation of $\nu_\mu \to \nu_e$) suggest the following MNS matrixes for three generation mixing.

$$\begin{pmatrix} c_{12} & s_{12} & 0 \\ -\frac{s_{12}}{\sqrt{2}} & \frac{c_{12}}{\sqrt{2}} & \frac{1}{\sqrt{2}} \\ \frac{s_{12}}{\sqrt{2}} & -\frac{c_{12}}{\sqrt{2}} & \frac{1}{\sqrt{2}} \end{pmatrix}$$

The lepton mixing is vastly different from quark mixing. Neutrino physics has still a large room for surprises. The theoretical conjectures such as extra dimension, neutrino decays [6] have not been rejected on the experimental basis. The tasks of accelerator and reactor neutrino oscillation experiments are:

- to establish clear signatures of neutrino oscillation by either observing oscillation pattern in disappearance or by observing appearance,

- to determine number of neutrinos,

- to determine neutrino mass hierarchy,

- to determine mixing angles between flavor and mass eigen-states. This will impose a strong constraint on the lepton-quark unification scheme.

2 Results from neutrino oscillation experiments

2.1 Reactor experiments : searches for $\nu_e \to \nu_x$ at Δm_{atm}^2

Nuclear reactors are source of $\overline{\nu_e}$ produced by β decay of fission fragments. The flux and spectrum of $\overline{\nu_e}$ can be predicted from the power of the reactor and the history of the operation of the reactor. The flux can be predicted with an uncertainty of 2.7%. The neutrino energy peaks at about 2.5 MeV and extend to about 6 MeV. Recently two long baseline (with baseline ~ 1 km) reactor neutrino experiments have finished data taking. The CHOOZ experiment[7] published their final results and the Palo Verde experiment reported the preliminary results at NEUTRINO2000 [8]. Both experiments detect neutrinos through the inverse β decay followed by delayed γ emission from neutron capture by Gd nucleus:

$$\overline{\nu_e} + p \to e^+ + n$$
$$n + Gd \to Gd + \gamma(\sim 8 MeV)$$

The energy-integrated ratio of the measured and the expected event rates are found to be:

$$CHOOZ : 1.010 \pm 0.028 \pm 0.027$$
$$Palo\,Verde : 1.04 \pm 0.03 \pm 0.08$$

where the first error is statistical and the second one is systematic. Figur 1 shows the 90% confidence level excluded region of $\overline{\nu_e} \to \overline{\nu_x}$ oscillation parameters. Thus $\overline{\nu_e}$ does not oscillate to other neutrinos with more than 10% probability at $\Delta m^2 \geq 10^{-3} eV^2$. The reactor experiments have shown that ν_e disap-

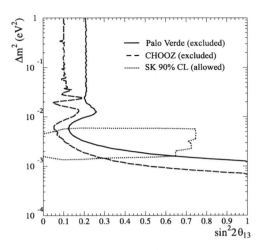

Figure 1. Excluded region of $\overline{\nu_e} \to \overline{\nu_x}$ oscillation parameters by the CHOOZ and Palo Verde result. Also shown is the allowed region (to the left of the contour) by a three generation atmospheric neutrino analysis from Super-Kamiokande.

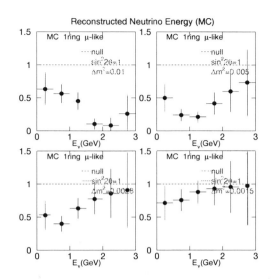

Figure 2. The effect of oscillation on neutrino spectrum. The expected ratios of with and without oscillation for various Δm^2s.

pearance in the solar neutrino and the atmospheric neutrino deficit should have different Δm^2 values and that $sin^2 2\theta_{13} \leq 0.1$ in the three neutrino scheme.

2.2 K2K : ν_μ disappearance and $\nu_\mu \to \nu_e$ search at Δm^2_{atm}

The K2K experiment is the first accelerator-based long base-line neutrino oscillation experiment that can explore Δm^2 below $10^{-2}eV^2$ with ν_μ beam to draw a definite conclusion on the neutrino oscillation. The oscillation modes to be searched for are $\nu_\mu \to \nu_x$ (ν_μ disappearance, where ν_x can be ν_τ or ν_s) and $\nu_\mu \to \nu_e$ (ν_e appearance). The experiment consists of a fast-extracted 12 GeV proton beam (about 5.5×10^{12} protons/pulse, every 2.2 seconds, with 1μsecond spill), a double horn magnets system, a 200 m decay pipe, beam monitors, and near detectors (at 300 m from the production target). The K2K uses Super-Kamiokande (SK) as the far detector. The baseline is 250km and the mean neutrino energy is about 1.4 GeV. The ν_μ disappearance will induce ν_μ spectrum distortion at SK. The expected effects as a function of neutrino energy is shown in Figure 2 for various Δm^2. Data has been accumulated from June 1999 to June 2000 for effective 100 days. The total accumulated number of protons on target is 2.3×10^{19}.

The K2K collaboration reported their first results about neutrino events rate at SK over the distance of 250 km. The critical points for the experiment to be successful are:

- the stability and the monitoring of the beam direction,

- how to predict the neutrino flux at SK from the measurements at the near site (far/near ratio)

- the method of identifying the beam associated neutrino events.

The requirement for the precision of the direction can be estimated from the expected neutrino distribution at SK. Figure 3 shows the neutrino profile at SK. Due to rather wide π decay angles at low energies, the precision of 1 mrad.(250m at 250km) is more than enough to make this source of error negligible. The experimental apparatus consists

Table 1. Summary of the neutrino detectors used in K2K

Detectors	Fiducial(tons)	Purpose
1 kton	50	Study neutrino interactions in water Cherenkov detector and directly compare with events in Super-Kamiokande. Also good sensitivity for low energy neutrinos.
SFT	6	The SFT is a stack of the water containers (5.6cm of water and 4mm aluminum tube) sandwiched with sheets of staggered scintillation fiber, read out by image intensifier tubes and CCD chain to measure neutrino flux with good vertex resolution.
LG	60	Measure ν_e contamination in the beam.
MUC	450	Monitor neutrino beam profile and rate with high statistics.
SK	22500	The far detector at 250km away from the neutrino source.

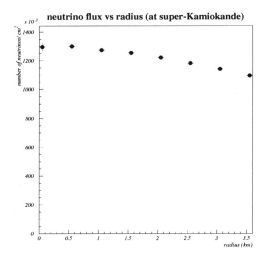

Figure 3. The expected neutrino beam profile as a function of distance from beam center at 250km.

of beam monitors and near detector system. They were place according to the GPS survey. The precision of the survey corresponds to about 10^{-5} radian. The direction and the stability of the beam are monitored by the profile of high energy muons due to $\pi \rightarrow \mu\nu_\mu$ decay in the decay volume and by the vertex distribution of neutrino interactions. The muon profile at 200 m from the target, which is measured by a segmented ionization chamber and a set of silicon pads, gives pulse-by-pulse measurements of the direction and the beam stability. The center of the distribution is stable within 20 cm (1 mrad).and has been stable within 1 mrad during the run, despite the fact that the horn system magnifies the fluctuation of extraction by about a factor of fifty

Figure 4 shows the near detector system, placed at 300m from the target. The detector consists of a 1 kton water Cherenkov detector and a fine grain detector system. Water Cherenkov detectors (SK and 1kton detector) have high detection efficiencies for low energy neutrino interactions and have excellent particle identification capability (separate e-

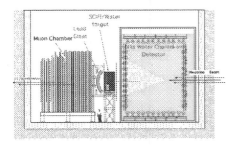

Figure 4. Near detector system at 300m from the production target.

μ). Most of the systematic errors (neutrino cross sections, spectrum shape) cancel after taking the ratio of number of event in the 1kton detector and SK. The fine grain detector consists of water-scintillating fiber tracking detector(SFT), lead glass counters(LG), and muon range detector(MUC).

Table 1 summarizes the near detector system and SK. The MUC data is the most suitable to monitor neutrino profile and rate stability due to its large mass. The center of the neutrino beam can be determined to about 10cm accuracy on a daily basis. Figure 5 and Figure 6 show the vertical and the horizontal profile of the neutrino beam, respectively. Both the center and the width of the neutrino beam are well reproduced by the Monte-Carlo calculation. The stability of the beam center is shown in Figure 7.

The ratio of the neutrino flux at near and at far site was predicted by the following principle. Once momentum and angular distributions of pions are known, the neutrino energy can be calculated at any distance. The neutrino energy is given by

$$E_\nu = \frac{0.48 \times E_\pi}{1 + \theta_\pi^2 \times \gamma_\pi^2}$$

where E_π, θ_π, and γ_π are energy, angle, and gamma factor of pion, respectively. Since maximum proton momentum is 13 GeV/c, pions whose momentum are more than 2 GeV/c can be separated by gas Cherenkov technique. A gas Cherenkov detector with a spherical mirror with a phototube array has

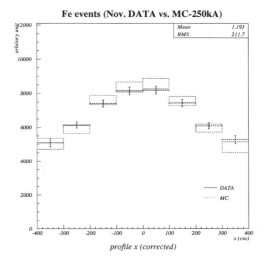

Figure 5. Horizontal profile of the neutrino beam measured in MUC. The boxes are the results of a beam simulation.

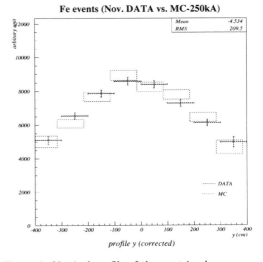

Figure 6. Vertical profile of the neutrino beam, measured in MUC. The boxes are the results of a beam simulation.

Table 3. The number of observed events and expected events.

	Obs.	Expected no oscillation	$\Delta m^2(\times 10^{-3}eV^2), sin^2 2\theta = 1$		
			3	5	7
22.5kton Fiducial	27	40.3 $^{+4.7}_{-4.6}$	26. 6 $^{+3.4}_{-3.3}$	17.8 $^{+2.3}_{-2.2}$	14.9 $^{+1.9}_{-1.9}$
1-ring	15	24.3 ± 3.6	14.4 ± 2.3	9.4 ± 1.5	8.6 ± 1.4
(μ-like)	14	21.9 ± 3.5	12.4 ± 2.1	7.5 ± 1.3	6.8 ± 1.2
(e-like)	1	2.4 ± 0.5	2.1 ± 0.4	1.9 ± 0.4	1.8 ± 0.4
multi-ring	12	16.0 ± 2.7	12.2 ± 2.1	8.4 ± 1.5	6.3 ± 1.1

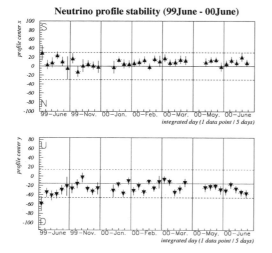

Figure 7. Stability of the neutrino beam direction. The dotted lines coresppond to 1 mrad.

been used. The pion angular divergence (relative to the beam axis) and the pion momentum distribution were measured with various indexes of refraction. Figure 8 show a comparison of the neutrino spectrum prediction by pion monitor data and the one based on a proton interaction model (Sanford-Wang model with Cho's parameters [9]). Figure 9 show the prediction of the ratio of flux at near and at SK detectors as a function of neutrino energy. The expected number of events at SK without neutrino oscillation is estimated by the Monte Carlo calculation. The beam Monte Carlo, whose reliability is confirmed by pion monitor measurements above 1 GeV, to obtain spectrum and far/near ratio. A neutrino event simulator was used for the calculation of detection efficiencies and the neutrino cross-sections for given neutrino energy bin. Note that the neutrino cross-sections and the detection efficiencies are almost canceled in 1kton and SK detectors. Table 2 lists the expected number of events in SK based on the 1kton, SFT, and MUC detectors. The events in SK are selected by requiring no entering particle from out-side and about 30MeV electron-equivalent energy deposit in the inner part of SK. A clock referenced to the timing signal from GPS (global

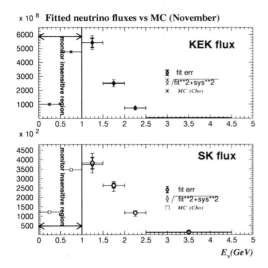

Figure 8. Prediction of neutrino flux at near and at Super-Kamiokande, based on pion monitor data. The pion monitor is not sensitive below 2 GeV pions (1 GeV neutrinos). Above 1 GeV the prediction by the pion monitor and that by Monte Carlo agree each other.

Table 2. The expected number of events without neutrino oscillation

Detectors	SK expected	error
1 kton	40.3	+4.7-4.6
MUC	41.4	+6.2-6.4
SFT	40.0	+5.2 -5.5

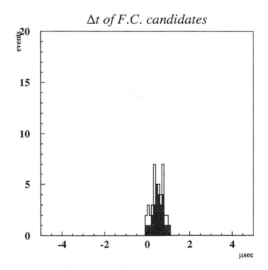

Figure 10. The time difference between the beam extraction time and the observed time in Super-Kamiokande. The hatched histogram is for fiducial contained events. The figure clearly shows that there are no backgrounds.

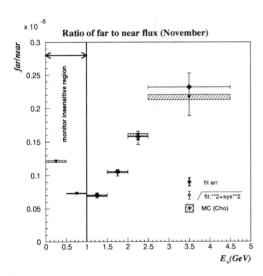

Figure 9. Prediction of the ratio of neutrino flux at near and at Super-Kamiokande, based on pion monitor data and by Monte Carlo calculation. The pion monitor is not sensitive below 2 GeV pions (1 GeV neutrinos).

positioning system) was used for identification of the events.

Figure 10 show the difference of expected arrival time of neutrinos and the event time. The figure clearly shows that there are no backgrounds, such as atmospheric neutrino events. The estimated background due to atmospheric neutrinos for this entire run was estimated to be 10^{-3}. The results are summarized in Table 3. The observed number of events is 2σ away from what would be expected if there is no oscillation. K2K has established the principles of a long baseline neutrino oscillation experiment, especially the neutrino flux and the direction can be controlled and monitored pulse-by-pulse, daily, and several month periods and GPS (Global Positioning System) can be used to identify neutrino events that are associate with accelerator source with a timing resolution of less than 100nsec.

The collaboration now proceeds to the spectrum analysis.

2.3 CHORUS and NOMAD : Investigation at high Δm^2

Two experiments search for $\nu_\mu \rightarrow \nu_\tau$ oscillation in the $\Delta m^2 > 10eV^2$ region, have reported their final results. Both experiments used the wide-band neutrino beam from the CERN 450 GeV proton synchrotron. The average neutrino energy is about 40 GeV and the distance of the detector from the production target is 820m. The ν_τ is identified by detecting τ^- produced by ν_τ charged current interaction with the sensitivity about 10^{-5} of the total neutrino events. The conventional ν_τ source in the beam is expected to be about 5×10^{-6} relative to the ν_μ. Thus the observation of τ could only result from $\nu_\mu \rightarrow \nu_\tau$ oscillation. The two experiments, CHORUS [10] and NOMAD [11], employ two different approaches to identify τ^-.

CHORUS experiment aims at detecting the decay of the short-lived τ^- in nuclear emulsion which provides a space resolution of about $1\mu m$. While the average τ^- decay length is 1mm. Thus this experiment can identify τ^- decay in the emulsion as a kink in the track. The 770 kg of emulsion target is followed by a magnetic spectrometer, electromagnetic and hadronic calorimeter, and muon spectrometer allowing momentum reconstruction and particle identification. Automatic emulsion scanning system have been developed to handle the large amount of data ($\sim$ 6 million).

NOMAD experiment aims at detecting τ^- using kinematical signature associated with τ decay. A precise measurement of secondary particle momentum is required. The detector

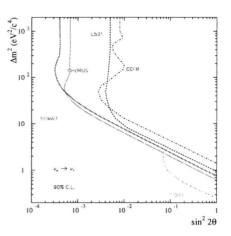

Figure 11. Excluded region of $\nu_\mu \rightarrow \nu_\tau$ oscillation parameters by the NOMAD and by CHORUS experiments. The CHORUS result uses statistical procedure by Junk, while NOMAD uses that of Feldman and Cousins.

is a fine-grained electric detector composed of a large aperture dipole magnet ($3m \times 3m \times 7m$ with 0.7T magnetic field). Inside the magnet, 2.7 tons of drift chambers used act as both target and tracking device, followed by transition radiation detectors. An electromagnetic, hadronic calorimeter and muon chambers are placed after the magnet. The separation of the τ signal from backgrounds uses ratios of likelihood functions of kinematics variables.

Both experiments reported negative results for $sin^2 2\theta >\sim 4 \times 10^{-4}$ and $\Delta m^2 > 0.5eV^2$, as shown in Figure 11. In interpreting the results, a care must be taken that the CHORUS results is based on Junk [12] statistical procedure, while NOMAD took Feldman-Cousins [13] method. The NOMAD experiment also get a exclusion region for $\nu_\mu \rightarrow \nu_e$ oscillation, which is shown in Figure 12. The possibility that the massive neutrinos are dark matter is becoming less likely, unless the mixing is very small.

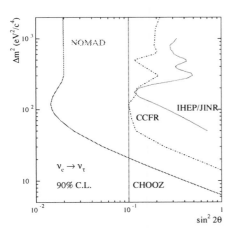

Figure 12. Excluded region of $\nu_\mu \to \nu_e$ oscillation parameters by the NOMAD result.

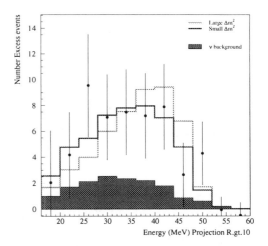

Figure 13. The e^+ energy distribution of the 70 events observed by LSND. The expected distributions from backgrounds and from $\nu_{mu} \to \nu_e$ oscillation for two different Δm^2 values are also shown.

2.4 LSND and KARMEN : Investigation at medium Δm^2

The LSND experiment [15] and the KARMEN experiment [14] have completed their experiments. The LSND re-analyzed their data. The new analysis involves fitting all of the conventional neutrino processes with theoretical predictions. The KARMEN increased their statistics with new data that has been taken in 2000.

These experiment use neutrinos produced by the following decay processes in low energy proton beam:

$\pi^+ \to \mu^+ \nu_\mu$ (in flight or at rest)

$\mu^+ \to e^+ \nu_e \overline{\nu_\mu}$ (at rest)

$\pi^- \to \mu^- \overline{\nu_\mu}$ (in fligt)

$\mu^- \to \nu_\mu e^- \overline{\nu_e}$ (at rest)

The $\overline{\nu_e}$ flux is of the order of 4×10^{-4} with respect to ν_μ because π^-s at rest are immediately captured by nuclei, and most stopping μ^- in high-Z materials undergo the capture process $\mu^- p \to \nu_\mu n$. The oscillation signal consists of the detection of positron and the delayed γ signal results from nuclear capture of thermal neutron.

$$n + p \to d + \gamma(2.2MeV)(LSND)$$
$$n + Gd \to Gd + \gamma(8MeV)(KARMEN)$$

In KARMEN experiment, the time correlation between e^+ signal and the beam pulse is also required. Figure 13 shows the e^+ energy distribution by LSND, together with the distributions expected from backgrounds and from $\nu_\mu \to \nu_e$ oscillations for two different Δm^2 values. Figure 14 shows the time and energy distribution observed in KARMEN experiment. The distributions are consistent with the expected backgrounds. The possible differences in experimental conditions are:

- spill structure
  LSND : $\sim 500\mu sec$, 8.3msec apart, KARMEN : 100nsec, 320nsec apart with 50Hz

- distance (LSND:29m, KARMEN : 17m)

- scintillator (LSND : mineral oil, KARMEN : Gd loaded scintillator)

The LSND result gives evidence for an excess of ν_e events , while KARMEN result does not support this. The allowed region of oscillation parameters describing LSND result is shown in Figure 15, together with the region excluded by KARMEN result and by Bugey reactor experiment [16]. A narrow band with

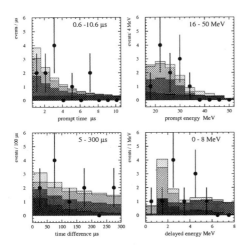

Figure 14. The time and energy distribution of e^+ observed by KARMEN. The distribution is consistent with the expected distributions from backgrounds

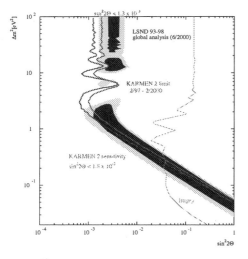

Figure 15. Allowed region of $\nu_{mu} \rightarrow \nu_e$ oscillation parameters by the LSND result. Also shown are the excluded regions by the KARMEN result and Bugey experiment.

Δm^2 between 0.2 and $2eV^2$ and $sin^2 2\theta$ between 0.002 and 0.04 still remain as a possible region for $\overline{\nu_\mu} \rightarrow \overline{\nu_e}$ oscillation.

3 Experiments in near future

There are five experiments that are being constructed or will be constructed. The questions, which will be addressed in the near future are the number of neutrinos, the clear evidence of oscillation by τ appearance, the precision measurement of spectrum distortion in ν_μ disappearance, the discrimination of $\nu_\mu \rightarrow \nu_\tau$ and $\nu_\mu \rightarrow \nu_s$ in Δm^2_{atm} region, the precision measurements of mixing angles, and the test of large angle solution of the solar neutrino deficit by long baseline reactor neutrino experiment.

3.1 Mini-BooNE

The main purpose of the Mini-BooNE is to either confirm or reject LSND results for $\nu_\mu \rightarrow \nu_e$ or $\overline{\nu_\mu} \rightarrow \overline{\nu_e}$ oscillation. The Mini-BooNE detector is being installed at a distance of 500 m from the production target. The detector is a 100 ton mineral oil based Cherenkov detector. The neutrino beam is produced by 8 GeV protons from Fermilab Booster Synchrotron with horn focusing system. The neutrinos distribute from 0.3 to 2 GeV with a small contamination ($\sim$0.3%) of ν_e. A $\nu_\mu \rightarrow \nu_e$ oscillation with the parameters required to explain the LSND results would produce an excess of about 1500 ν_e CC events, while $5 \times 10^5 \nu_\mu$ CC events and about $1700 \nu_e$ CC events due to ν_e contamination in the original beam are expected. The dominant backgrounds will be due to π^0 production by ν_μ neutral current events. A rejection of π^0 is required to be more than 95% level. The sensitivity is shown in Figure 16. The experiment is scheduled to start data taking in 2002. If a signal is observed, the oscillation parameter will be determined by installing second detector at a different distance.

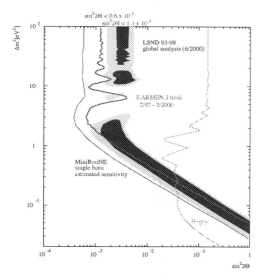

Figure 16. The expected sensitivity of $\nu_\mu \to \nu_e$ oscillation parameters by the MiniBooNE experiment.

3.2 CERN-Gran Sasso

In CERN-Gran Sasso project [17], use the neutrino beam produced by the CERN 450 GeV SPS. The baseline is 732 km. The main aim of this project is to search for ν_τ appearance at Δm^2_{atm}. The beam line is designed to optimize the τ production rate. The rate of ν_μ charged current interactions is 3200/year for a 1000 ton detector. The expected yearly rates of τ^- is estimated to be 25, if Δm^2 is $3.5 \times 10^{-3} eV^2$. The production rate varies as $(\Delta m^2)^2$. Two experiments are planned to be installed. ICANOE is a new 1200 ton detector based on a liquid Argon time projection chamber with a spatial resolution of about 1 mm in three dimensional reconstruction capabilities. The τ^- are identified by using kinematical criteria. The required rejection of 10^2 is sufficient for $\tau^- \to e^-$ channel. The efficiency for this channel is expected to be about 50%. The background is estimated to be 0.25 events. Another experiment for a τ appearance search is OPERA. This experiment is based on a detector with nuclear emulsion interleaved with 1 mm thick lead plates. The total mass will be about

1000 tons. The τ^- are identified by kink in space. The τ^- detection efficiency is estimated to vary 0.29 at small Δm^2 and 0.33 at $\Delta m^2 \sim 10^{-2}eV^2$. For a run of one year (4.5×10^{19} protons on target) the backgrounds is expected to be 0.4 events, mostly from charm production and decay in events where the primary μ was not identified.

3.3 NuMI/Minos

The NuMI project [18] uses neutrinos from the Fermilab Main Injector, a 120 GeV proton synchrotron. The intensity of proton beam will be 5×10^{13} with a repetition rate of 1/1.9 sec. The expected number of protons on target is 3.6×10^{20}/year. The beam will be sent to Soudan mine at a distance of 730 km. The MINOS experiment will use two detectors (the near detector located at Fermilab and far detector). The far detector has a total mass of 5400 tons with a fiducial mass of 3300 ton. It consists of 2.54 cm thick magnetized iron interleaved with 4 cm wide, 8 m long scintillator strip with wave shifting fiber read-out, which provides both calorimetric and tracking information. Three different neutrino beams have been designed as shown in Figure 17. The oscillation minimum will occur at neutrino energy of about 2 GeV for $\Delta m^2 \sim 3 \times 10^{-3}$ at 730 km distance. NC/CC ratio can discriminate $\nu_\mu \to \nu_\tau$ and $\nu_\mu \to \nu_s$. A 5% measurement of $sin^2 2\theta_{23}$ in ν_μ disappearance and a search of $\nu_\mu \to \nu_e$ down to a % level are expected.

3.4 JHF

The JHF (Japan Hadron Facility) is proposal to have a 50 GeV high intensity proton synchrotron at Tokai, 295km from Super-Kmiokande. The intensity of proton beam will be 3.2×10^{14} /pulse with a repetition rate of 0.3 Hz. The far detector will be Super-Kamiokande for the first phase of the experiment. With the very high-intense beam, two types of neutrino beams are planned

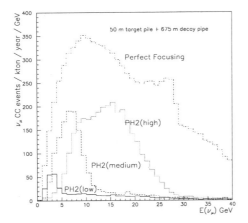

Figure 17. At Fermilab, three options of neutrino beams have been designed.

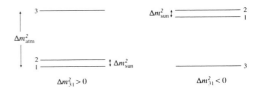

Figure 18. Two possibilities of mass ordering of three mass eigen-state

to build: a wide band beam and a narrow band beam. To be most sensitive to Δm^2 of $2 \sim 5 \times 10^{-3} eV^2$, a neutrino beam design will be optimized to about 1 GeV which about 50 times more intense than K2K. The primary physics goals will be

(1) the observation of clear oscillation pattern above $2 \times 10^{-3} eV^2$ with high statistics,
(2) the measurement of Δm^2 with the ν_μ disappearance with an accuracy of $2 \times 10^{-4} eV^2$,
(3) the measurement of $cos^4\theta_{13} sin^2\theta_{23}$ with an accuracy of 1%
(4) Search for $\nu_\mu \rightarrow \nu_e$ down to about 0.003,
(5) The measurement of neutral π^0 events to provide stringent limit on the existence of sterile neutrinos converted from ν_μ.
(6) If θ_{13} is large enough to be detected in the first stage, with much more massive detector of the order of 1 Mton, it may be possible to search for CP violation in lepton sector.

3.5 Kamland

Another important issue in neutrino oscillation is the still ambiguous solar neutrino deficit. Kamland [19] is a long baseline reactor neutrino experiment with $L/E \sim 5 \times 10^4$. The detector is a 1000 ton liquid scintilla-

tor. The total reactor power within the distance of 200 km is 127 GW. With full reactor power, the event rate is expected to be $\sim$ 2/day with a signal noise ratio of 10. The experiment can sensitive to $\Delta m^2 > 7 \times 10^{-6} eV^2$ and $sin^2 2\theta > 0.1$, a region which includes the large mixing angle solution of the solar neutrino deficit. The experiment will start in 2001.

4 Future

4.1 Mass and generation

The sign of Δm^2_{13} cannot determined by neutrino oscillation in vacuum. In three generation mixing scheme, ν_1 and ν_2 are associate with solar neutrinos, and ν_3 is supposed to be responsible to atmospheric neutrino observations. The possible ordering is shown in Figure 18/ While neutrinos propagate in matter, the set of mass eigen-states are different from those in vacuum. Ignoring the small mass splitting Δm^2_{12}, the mixing angle θ^m_{13} and oscillation length L^m can be written as

$$sin^2 2\theta^m_{13} = \frac{sin^2 2\theta_{13}}{(\frac{2EV}{m_3^2 - m_1^2} - cos2\theta_{13})^2 + sin^2 2\theta_{13}}$$

$$L^m = L^v \frac{1}{\sqrt{(\frac{2EV}{m_3^2 - m_1^2} - cos2\theta_{13})^2 + sin^2 2\theta_{13}}}$$

$V = \pm\sqrt{2} G_F N_e$, where $\pm$ is for neutrino and anti-neutrino, L^v is the oscillation length in vacuum ($L^v = \frac{4\pi E}{\Delta m^2}$), G_F is the Fermi coupling constant, and N_e is the electron density in matter. The sign of Δm^2 can be determined from $sin^2 2\theta_{13}^m$ and $sin^2 2\theta_{13}$. For small θ_{13}, the optimal sensitivity will be achieved at $2EV = \Delta m_{13}^2 cos2\theta_{13}$. The optimal neutrino energy is

$$E_{opt} = 15 GeV \frac{\Delta m_{13}^2}{3.5 \times 10^{-3} eV^2} \frac{2.8g/cm^3}{\rho}$$

The oscillation length increases by a factor $\frac{1}{sin2\theta}$ at this setting. Thus this measurement requires high-energy neutrino beam and very long-base line. Neutrino factories would be an ideal tool for the above studies. But some studies show conventional beam with a large detector can access to this effect even before neutrino factories are build[20].

4.2 CP violation

CP violation in lepton sector may be related with baryon-anti-baryon asymmetry in the universe. CP violation can occur only in case more than three neutrinos are involved. Also from unitarity the oscillation probabilities for disappearance measurements do not contain CP violation term. The CP asymmetry in ν_e appearance can be expressed in the three generation case as :

$$A_{CP} = \frac{(\nu_\mu \to \nu_e) - (\overline{\nu_\mu} - \overline{\nu_e})}{(\nu_\mu \to \nu_e) + (\overline{\nu_\mu} - \overline{\nu_e})}$$

$$\sim \frac{4sin2\theta_{12}sin\delta}{sin\theta_{13}} sin\frac{\Delta m_{12}^2 L}{4E}$$

The asymmetry is proportional to L/E and depends on the θ_{12}, Δm_{12}^2, and θ_{13} beside CP phase (δ). For the CP violation to be measurable in neutrino oscillation, the solar neutrino solution must be so-called large angle solution (Δm_{sol}^2 and θ_{12} must be large) and $\nu_\mu \to \nu_e$ must be measureable with Δm_{atm}^2 region. The future developments depend heavily on the results in coming decade.

5 Conclusion

The results on atmospheric neutrinos revolutionize neutrino physics. However, we are far from having a coherent picture of neutrino oscillation. It is clear that progress in understanding what is going on in the neutrino sector can only come from further data. Fortunately, new data be forthcoming in all the possible Δm^2 regions in the coming decade. The future developments depend heavily on the results in coming decade.

Acknowledgements

I thank Guido Drexlin, Manfred Lindner, Masahiro Komatsu, Adam Para, Ken Peach, Chiara Roda, Myungkee Sung, and Yifang Wan for the help provided in preparing this article.

References

1. M.Nakamura, these proceedings
2. B.Pontecorvo, Zh.Eksp.Teor.Fiz. 33,549 (1957)
3. Z.Maki, M.Nakagawa, S.Sakata, Prog. Theor. Phys. 28,870 (1962)
4. M.Kobayashi, T.Maskawa, Prog. Theor. Phys. 49,652 (1973)
5. E.Kerns, these proceedings
6. L.J.Hall, these proceedings
7. M.Apollino et. al. Phys. Lett. B466,415 (1999)
8. G.Gratta, Proceedings of the Neutrino2000
9. Y.Cho et. al. Phys. Rev. D4,7 (1971)
10. M.Komatsu, these proceedings
11. C.Rhoda, these proceedings
12. T.Junk, Nucl.Inst.Meth., A343,435 (1999)
13. G.J.Feldman and R.D.Cousins, Phys.Rev. D57,3873(1998)
14. C.Oehler, these proceedings
15. R.Imlay, these proceedings
16. B.Achtar et.al., Nucl.Phys. B434,503 (1995)
17. A.Bettini, this proceedings
18. A.Para, these proceedings
19. J.Shirai, these proceedings
20. M.Lindner these proceedings

RESULTS FROM ATMOSPHERIC AND SOLAR NEUTRINOS EXPERIMENTS

EDWARD T. KEARNS

Physics Department, Boston University
590 Commonwealth Avenue, Boston MA 02215, USA
E-mail: kearns@hep.bu.edu

This paper is a summary of non-accelerator neutrino physics, concentrating on the latest experimental results using atmospheric and solar neutrinos to study neutrino oscillation. Neutrino oscillation is well-established in atmospheric neutrinos and current efforts aim to better measure and understand the phenomenon. Solar neutrinos continue to present an unsolved puzzle, with the latest data from Super-Kamiokande exhibiting a large flux deficit, but no significant day-night flux difference or spectral distortion. For both atmospheric and solar neutrinos, the data prefers oscillation between active flavors $(\nu_e, \nu_\mu, \nu_\tau)$ and disfavors a 2-flavor oscillation to ν-sterile as the primary effect.

1 Introduction

This talk summarizes recent neutrino experiments performed without a man-made accelerator, i.e. using natural sources of neutrinos such as cosmic ray showers and nuclear reactions in the sun. These experiments, generally performed deep underground to escape the large flux of cosmic rays on the earth's surface, have provided a rich supply of new physics and unexplained phenomena. The most common application is to study neutrino mass via neutrino oscillation. Neutrino oscillation studies are facilitated by long flight distances which, until recently, were not available to accelerator or reactor experiments. In fact, both atmospheric and solar neutrinos provide experimental evidence that lead us to believe that neutrino oscillation is taking place and that neutrinos have mass. The majority of this paper will review this evidence, and summarize the latest results that are revealing and measuring the detailed nature of neutrino oscillations.

To do justice to the exciting positive results from atmospheric and solar neutrinos, I can only briefly summarize other non-accelerator neutrino experiments which have yielded only negative results and limits. No distant astrophysical source has been observed since the detection of neutrinos from SN1987A by Kamiokande and IMB. No point source of high energy neutrinos has been observed, and there is no evidence for neutralino annihilation in the sun, earth, or galactic center. The most significant non-observation, as far as neutrino mass goes, involves neutrinoless double beta decay. This experiment generally consists of several kilograms of one of a few special isotopes such as ^{76}Ge or ^{100}Mo. The signature is a peak in the energy of the two beta-decay electrons corresponding to virtual annihilation of neutrinos in the transition diagram. This is possible only if the neutrinos have mass and are Majorana particles. To date, no signal has been observed and the limit on the effective Majorana neutrino mass is < 0.2 eV[1]. Note that this is directly sensitive to the mass, not the mass splitting, as in the case of neutrino oscillation. As we will soon see, a possible neutrino mass scale is given by $\sqrt{\Delta m^2_{atm.}} = 0.05$ eV, which is a factor of 40 away, providing an interesting target for future experiments.

The theory of neutrino oscillation is well developed[2], based on the unitary transformation between two eigenbases, in this case the mass basis and flavor basis that describe the neutrino fields. Even though there are three known neutrinos, and we will shortly consider a fourth sterile neutrino, we can often describe the experimental result in terms of a two-flavor oscillation, where the transformation matrix is a 2×2 rotation matrix by a mixing angle θ. Then, the oscillation probability from neutrino flavor a to b is given by:

$$P_{ab} = \sin^2 2\theta \sin^2 \frac{1.27 \Delta m^2 L}{E_\nu}. \quad (1)$$

This reduction from the general form, based on a 3×3 matrix, is possible for a couple of different reasons. The Δm^2 splitting between one state in question and a third state might be so small that their mixing does not play a role, given the experimental baseline, L. Or the mixing angle to the third state might be so small that the relative probability of mixing is not observable. Finally, even if 3-flavor mixing is operating, if the experiment measures only the disappearance of a neutrino flavor, the data may be interpreted as though due to two-flavor mixing, albeit improperly. Nonetheless, until the entire mixing matrix is explored, it is practical to interpret and compare results based on the above formula.

2 Atmospheric Neutrinos

Large underground detectors originally built to search for proton decay are also exposed to a flux of neutrinos with energies of a few GeV created by cosmic ray showers in the atmosphere. A byproduct of the background studies necessary for a proton decay search has been the recognition of an anomalous relative rate of ν_μ and ν_e interactions, including a flight length dependent distortion suggestive of neutrino oscillation.

Two normalizing principles are used to understand these data. First, cosmic ray showers consist mostly of pions, which decay to $\mu + \nu_\mu$, and the μ decays to $e + \nu_\mu + \nu_e$, resulting in a flux ratio $(\nu_\mu/\nu_e) \sim 2$. This ratio grows to larger values at high energy, as the time-dilated muon may strike the ground before decaying (removing ν_e from the beam), but at all relevant energies the ratio is predicted to about 5%[3]. The second principle is that primary cosmic rays arrive isotropically.

Although low energy primaries, less than a few GeV, are affected by the geomagnetic field, above a few GeV the infall is isotropic. The details of hadronic shower development break the isotropy, where horizontal showers contain pions that spend more time in rarefied atmosphere, and which are more likely to decay. But the geometry of the situation dictates that at equal zenith and nadir angles the flux of high energy neutrinos must be nearly identical. Again, this is predicted with an accuracy of a few percent. These simple foundations are automatically considered by the detailed calculations[4] of neutrino flux that are used for comparison with the experimental data.

2.1 Evidence for Neutrino Oscillation

The Super-Kamiokande experiment obtained the first convincing evidence of atmospheric neutrino oscillation[5], using a 22.5 kton (fiducial mass) water Cherenkov detector. They currently currently report[6] a 70 kton-year (1144 day) sample of 11672 events with parent neutrino energies ranging from 100 MeV to 10 TeV. To cancel systematic uncertainties, the flux ratio of ν_μ to ν_e is compared to a detailed Monte Carlo prediction by forming a double-ratio, R. For the sub-GeV sample of events ($E_\nu < 1$ GeV),

$$R = \frac{(N_\mu/N_e)_{data}}{(N_\mu/N_e)_{m.c.}} = 0.65 \pm 0.02 \pm 0.05. \quad (2)$$

The up-down asymmetry is expected to be zero for the multi-GeV muon sample ($E_\nu > 1.3$ GeV), but is measured to be:

$$A = \frac{N_{up} - N_{down}}{N_{up} + N_{down}} = -0.30 \pm 0.03 \pm 0.01, \quad (3)$$

whereas the asymmetry is consistent with zero for the multi-GeV electron sample and the multi-GeV muon Monte Carlo. These two independent indications are significant by 6.5 and 9.5 standard deviations, respectively.

Independent comfirmation is available by measuring the flux of upward-going muons from neutrino interactions in the rock below and around the detector. One figure of merit is the flux ratio of stopping to through-going muons. The stopping upward-going muons come from ν_μ with mean neutrino energy of ~ 10 GeV, whereas through-going muons arise from $\langle E_\nu \rangle \sim 100$ GeV. By studying this ratio, one cancels uncertainty in the flux as well as for neutrino interaction and muon propagation in rock. The measured ratio is:

$$R(s/t) = \frac{\Phi_{stop}}{\Phi_{through}} = 0.242 \pm 0.017^{+0.013}_{-0.011},$$

(4)

which can be compared to a predicted ratio of 0.37 ± 0.05. The data also shows evidence for neutrino oscillations in the distortion of the zenith angle shape[7].

Further confirmation is found by experiments using quite a different technology. The Soudan 2 experiment uses a honeycomb lattice of iron to form a tracking calorimeter that shows good detail in track trajectory and dE/dx. Based on a 5.1 kton-year exposure and 326 contained interactions, they report[8] a background subtracted double ratio R equal to $0.68 \pm 0.11 \pm 0.06$, in good agreement with the Super-K result. To study neutrino oscillation they select a high resolution sample, including low energy quasi-elastic muons if the recoil proton is visible, otherwise restricted to single tracks above 600 MeV and higher energy multi-track events. For this sample they achieve 20° to 30° pointing resolution and thus a $\log(L/E)$ resolution of ~ 0.5. This data sample is binned in zenith, as shown by crosses in Fig. 1. In this preliminary analysis, it is found that the best fit is around $\Delta m^2 = 8 \times 10^{-3}$ eV^2 with nearly maximal mixing.

The MACRO detector operates as a large area time-of-flight detector with streamer tube tracking and tanks of liquid scintillator providing fast timing. MACRO uses the sign of the time-of-flight (up versus down) to isolate neutrino induced upward-going muons from the million times greater flux of

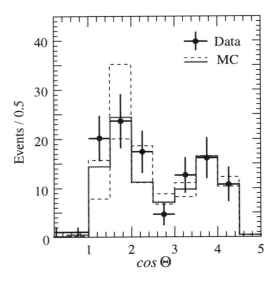

Figure 1. The high resolution sample of atmospheric muons measured by Soudan 2. The data (crosses) are compared to oscillated Monte Carlo predictions with $\sin^2 2\theta = 1$ and $\Delta m^2 = 10^{-3}$ eV^2 (upper dashed line), 7×10^{-3} eV^2 (solid line), and 10^{-1} eV^2 (lower dashed line)

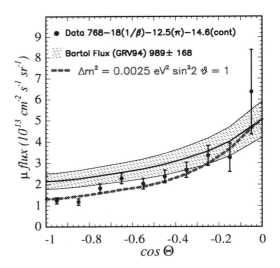

Figure 2. The zenith angle distribution of upward-going muons in MACRO. The shaded region is the prediction with no oscillation (showing a 17% uncertainty in normalization) and the dashed line shows the prediction for $\sin^2 2\theta = 1$ and $\Delta m^2 = 2.5^{-3}$ eV^2.

downward-going muons. As shown in Fig. 2, they observe a zenith angle dependent distortion in the rate of upward-going muons consistent with maximal $\nu_\mu \leftrightarrow \nu_\tau$ oscillation[9]. A sample of low energy neutrino interactions in the absorbing planes of the detector has also recently been studied[10]. The low energy sample consists of two topologies. First are internal upward-going muons (IU) that start in the absorber volume (the lower half of the detector) and are timed by passing through two planes of scintillator in the upper half of the detector. Second are upward-going stopping muons (US), but these only pass through a single lower plane of scintillator, so they are mixed with downward-going muons induced by neutrino interactions in the absorber (ID). However all events are neutrino induced and can be compared to a detailed Monte Carlo. After background subtraction, both classes of events are separately found to be consistent with neutrino oscillation, and a combined figure of $IU/(US + ID) = 0.59 \pm 0.07$ is measured, compared to a no-oscillation prediction of 0.75 ± 0.06.

2.2 Measuring the oscillation parameters

To estimate the mixing parameters of neutrino oscillation, the Super-K group takes all of the single-ring ν_e and ν_μ events and forms numerous bins in log(energy) and zenith angle (effectively path length L). Additional ν_μ zenith bins are included with partially contained events, stopping upward-going muons, and through-going upward-going muons. In total, 155 bins are formed, too many to easily display. Figure 3 shows 95 bins in 10 zenith angle distributions, where some of the 155 bins used in the fit have been combined. This figure clearly illustrates the significant disagreement as a function of zenith angle between the measured counting rate of muon events and the Monte Carlo prediction. However the data fits the prediction if two-flavor

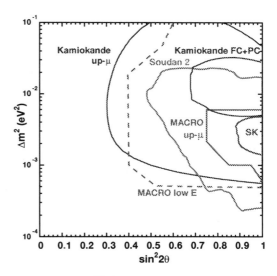

Figure 4. The 90% C.L. contours for several experiments reporting positive evidence of $\nu_\mu \leftrightarrow \nu_\tau$ oscillation.

$\nu_\mu - \nu_\tau$ oscillation is applied to the neutrino flux with $\sin^2 2\theta = 1$ and $\Delta m^2 = 3 \times 10^{-3}$ eV^2.

The confidence intervals for $\sin^2 2\theta$ and Δm^2 for these experiments are overlayed in Fig. 4. The results from the earlier and smaller Kamiokande experiment are included; although the FC+PC result[11] does not graphically overlap the Super-Kamiokande best fit region, as a 90% C.L. contour it is not a statistically significant separation. No systematic cause for this separation is known, other than the *prima facie* flatness of the low energy (sub-GeV) zenith angle distribution of $R(\mu/e)$, which was also observed to be flat in IMB[12] data.

2.3 Comparing $\nu_\mu \leftrightarrow \nu_\tau$ and $\nu_\mu \leftrightarrow \nu_s$

The zenith distribution of high energy events in Super-K disfavors $\nu_\mu \leftrightarrow \nu_e$ oscillation, but fits $\nu_\mu \leftrightarrow \nu_\tau$ because the high threshold (~ 3.5 GeV) for tau production and the complicated final states of ν_τ interactions are manifested as a disappearance of ν_μ events. Although nearly maximal, two-flavor, $\nu_\mu \leftrightarrow \nu_\tau$ oscillation provide a consistent picture for

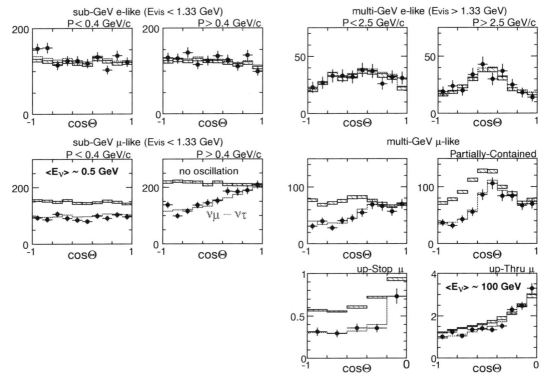

Figure 3. The zenith angle rate of atmospheric neutrino events in 70 kton-years of Super-Kamiokande data. The rectangular boxes show the prediction by a detailed Monte Carlo, with the height of the box equal to the statistical uncertainty in the number of generated events. The solid line represents the best fit to $\nu_\mu \leftrightarrow \nu_\tau$ oscillation.

current experimental measurements of atmospheric neutrinos, there is an alternative hypothesis that must be considered. Other models of ν_μ disappearance, such as neutrino decay or exotic oscillation (e.g. $L \times E$) do not fit the data, with some exceptions[13]. However, another two-flavor fit that provides reasonable results is ν_μ mixing with a sterile neutrino (ν_s). An additional motivation for this scenario is related to the LSND experiment, which reports[14] evidence of $\overline{\nu}_\mu \leftrightarrow \overline{\nu}_e$ mixing with Δm^2 around 1 eV^2. With only three neutrino mass eigenstates, it is impossible to reconcile the sum of Δm^2_{LSND}, $\Delta m^2_{atm} \sim 10^{-3}$ eV^2, and $\Delta m^2_{solar} \sim 10^{-4}$ to 10^{-10} eV^2. However, this can be resolved if solar or atmospheric oscillations involved a fourth neutrino state, necessarily sterile to

satisfy the LEP bounds on 3 neutrinos that couple to the Z^0.

Fortunately, it is possible to distinguish $\nu_\mu \leftrightarrow \nu_s$ oscillations from $\nu_\mu \leftrightarrow \nu_\tau$. One way is to look for the absence of neutral current (NC) interactions, which will disappear along with charged current (CC) interactions for $\nu_\mu \leftrightarrow \nu_s$, but will not be depleted for $\nu_\mu \leftrightarrow \nu_\tau$. Another way is to look for a suppression of oscillation effects at high energy due to the interference of ν_μ and ν_s forward scattering as the neutrino state passes through matter. This matter effect for sterile neutrinos[15] is analogous to the MSW matter effect[16] that may play a role in solar neutrino oscillations. Figure 5 shows the oscillation probability as a function of zenith angle for several different neutrino energies, fixing the

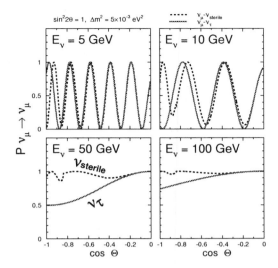

Figure 5. The oscillation probability of $\nu_\mu \leftrightarrow \nu_\tau$ compared to $\nu_\mu \leftrightarrow \nu_s$. The x-axis shows the cosine of the zenith angle, with upward-going neutrinos at -1. Matter effects suppress the oscillation probability for ν_s at high energy.

mixing parameters at $\Delta m^2 = 5 \times 10^{-3}$ eV^2 and $\sin^2 2\theta = 1$. Below about 10 GeV, the oscillation probability for the two cases (ν_τ and ν_s) are similar; above 10 GeV, the oscillation probability for $\nu_\mu \leftrightarrow \nu_s$ is strongly suppressed.

Super-K has performed an analysis where they isolate three sub-samples that behave differently for $\nu_\mu \leftrightarrow \nu_\tau$ oscillations and $\nu_\mu \leftrightarrow \nu_s$ oscillations. One sample is based on multiring events, with cuts that enhance the fraction of neutral current interactions and reduce the fraction of charged current ν_μ. Here, the up-down flux is expected to be nearly symmetric for $\nu_\mu \leftrightarrow \nu_\tau$, with neither the CC-$\nu_e$ nor the NC affected by ν_μ oscillation. The measured ratio, $N_{up}/N_{down} = 0.96 \pm 0.07_{stat}$, whereas the expectation for ν_s oscillations at the best-fit value is about $0.8 \pm 0.02_{sys}$. There is another sample rich in neutral current that has been discussed previously: single π^0 events identified by the invariant mass peak found for two showering Cherenkov rings from the decay to two gam-

mas. This sample is not used yet, because (a) the neutrino directionality for events where the two rings are distinct is insufficient to study the up/down ratio, and (b) the production cross section is poorly known. The latter uncertainty, of about 20%, should be reduced after detailed measurements of 1 GeV neutrino interactions using the K2K near detector.

The other two samples consist of high energy neutrino interactions, susceptible to matter effects. One is the through-going upward-going muon sample which should show essentially no signs of zenith distortion, due to matter effect suppression (see the bottom two panes of Fig. 5), for $\nu_\mu \leftrightarrow \nu_s$ mixing at the preferred value of $\Delta m^2 = 3 \times 10^{-3}$ eV^2. Another high energy sample was formed from the partially contained events; by requiring visible energy greater than 5 GeV, the mean E_ν is raised to ~ 20 GeV. Again, this sample should show up-down symmetry due to matter effect suppression of $\nu_\mu \leftrightarrow \nu_s$, but does not. This analysis is documented in a recent publication[17], where the individual data samples are statistically combined, and which concludes that the region of parameters favored by the low energy events (a few times 10^{-3} eV^2 for either ν_τ or ν_s) is excluded at the 99% C.L. by the data from the isolated samples of NC and high energy neutrinos.

3 Solar neutrinos

Solar neutrinos provide a measurable indication of the nuclear reactions that take place in the sun. The reaction $p + p \rightarrow {}^2\text{H} + e^+ + \nu_e$ is responsible for 98% of the energy production of the sun, and therefore flux of pp neutrinos is the most reliably estimated. The reaction ${}^7\text{Be} + e^- \rightarrow {}^7\text{Li} + \nu_e$ produces dichromatic ${}^7\text{Be}$ neutrinos of energy 383 keV or (predominantly) 861 keV. A small fraction of the time (0.1%), ${}^7\text{Be}$ captures a proton, instead of an electron, to form ${}^8\text{B}$. The beta decay of ${}^8B \rightarrow {}^8\text{Be} + e^+ + \nu_e$ produces a spectrum of ${}^8\text{B}$

neutrinos that extends to 15 MeV. The top frame in Fig. 6 shows the spectrum of solar neutrinos for the experimentally important reactions[18]. The arrows at the top of the figure indicate the threshold for the three types of experiments that have been performed so far.

3.1 Measurements of the total flux

The simplest figure of merit for solar neutrino experiments is the total flux, measured in SNU (disintegration/s/10^{36} atoms) for radiochemical experiments and cm^{-2} s^{-1} for electron scattering experiments. The measured flux can be directly compared to the standard solar model (SSM) flux prediction[19], by integrating the cross section weighted contribution from each solar reaction above the experimental threshold. The latest such results are reported in Table 1. The most notable feature is the small ratio with respect to the SSM found by all experiments. In fact, the low flux ratio is especially problematic for the gallium experiments, since 70±1% of the flux is predicted to come from the *pp* reaction, so (a) the measured ratio of 58% is inconsistent with that contribution alone and (b) the inconsistency grows when one accounts for the *measured* flux of neutrinos from ^{7}Be and ^{8}B. In fact, the flux measured for ^{8}B is also incompatible with the flux measured for ^{7}Be, the progenitor nucleus for ^{8}B production in the sun. Finally, there is a noticeable energy dependence, where the very low ratio for the chlorine experiment suggests greater suppression around 1 MeV. These many inconsistencies comprise the solar neutrino puzzle[25].

Neutrino oscillations provide a plausible explanation for the solar neutrino puzzle. If the solar ν_e converts to ν_μ, ν_τ, or ν_s, it cannot initiate the charged current radiochemical reactions on gallium or chlorine. Similarly, the neutrino electron scattering reaction in water is eliminated for ν_s, and is significantly reduced for ν_μ and ν_τ (retaining 17% of the

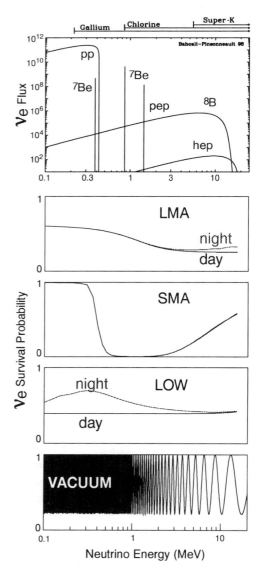

Figure 6. Top panel: the SSM flux prediction as a function of energy. Next three panels: the survival probability calculated using the transition amplitude for ν_e conversion in the sun due to matter (MSW) effects. Last panel: the survival probability valid for Δm^2 below the MSW region, based on equation 1.

Table 1. *Integrated solar neutrino flux measurements compared to theory.*

Experiment	Technique	Measurement(Prediction)	$\frac{\text{Data}}{\text{Theory(BP98)}}$
Gallex+GNO[20] SAGE[21]	Gallium, $E_\nu > 230$ keV	$74.1 \pm 5.4^{+4.0}_{-4.2}$ SNU $75.4^{+7.0}_{-6.8}\ ^{+3.5}_{-3.0}$ SNU (SSM $= 129^{+9}_{-7}$ SNU)	$57\% \pm 7\%$ $58\% \pm 6\%$
Homestake[22]	Chlorine, $E_\nu > 810$ keV	$2.56 \pm 0.16 \pm 0.16$ SNU (SSM $= 7.7^{+1.3}_{-1.1}$ SNU)	$33\% \pm 9\%$
Kamiokande[23] Super-K[24]	water, $E_e > 6.5 - 7$ MeV water, $E_e > 5.5$ MeV	$2.80 \pm 0.19 \pm 0.33 \ 10^6 \text{cm}^{-1}\text{s}^{-1}$ $2.40 \pm 0.03^{+0.08}_{-0.07} \ 10^6 \text{cm}^{-1}\text{s}^{-1}$ (SSM $= 5.15^{+0.20}_{-0.16} 10^6 \text{cm}^{-1}\text{s}^{-1}$)	$54\% \pm 14\%$ $47\% \pm 2\%$

cross section from the neutral current contribution). The calculation of the ν_e survival probability requires the inclusion of matter effects[16], which actually provide an elegant mechanism for achieving the required energy dependent suppression due to resonant conversion in the dense matter of the sun. When a global analysis[26] of the flux data from all three types of experiments is performed, four regions of neutrino mixing parameter space are identified: large mixing angle (LMA), small mixing angle (SMA), low Δm^2 (LOW), and vacuum oscillations (VAC). The survival probability for typical mixing parameters in each of these three regions is shown in the lower panes of Fig. 6.

3.2 The search for smoking gun evidence for neutrino oscillations

The second generation of solar neutrino experiments might be defined as those capable of finding so-called "smoking gun" evidence of neutrino oscillation, i.e. some distortion that does not rely on the calculation of the integrated flux. The possible smoking guns are: (a) matter-effect regeneration in the earth, causing a different flux during the day and night, (b) energy spectrum distortion from the well-predicted shape due to nuclear physics, (c) a seasonal variation in flux differing from the $\pm 3.5\%$ expected from the eccentricity of the earth's orbit, and (d) evidence of neutrino flavor dependence by com-

paring charged current and neutral current reactions. The last method is the unique goal of the SNO heavy-water Cherenkov experiment, which has just begun running and has shown a promising glimpse of their data[27]. The first three methods have been studied by Super-Kamiokande, although the seasonal variation does not yet have much statistical sensitivity.

Super-Kamiokande reports[24] an overall day-night asymmetry of:

$$\frac{\Phi_{day} - \Phi_{night}}{\Phi_{day+night}} = -0.034 \pm 0.022^{+0.13}_{-0.14}, \quad (5)$$

which is is only significant by 1.3 standard deviations. The data has also been divided into 10 angular bins with respect to the vertical, to check the nadir bin for isolated enhancement due to the dense iron core of the earth, as expected for certain solutions in the SMA region. However, no such enhancement is seen.

The energy spectrum is measured by separating the data into 19 energy bins of 0.5 MeV width. A solar peak is found in the $\cos \Theta_{sun}$ distribution for each bin and the flux for that bin is calculated much as the integrated flux is calculated– by measuring the rate of events pointing towards the sun over an uncorrelated background due to radioactive decays in the water (see Fig. 8). The result is shown in Fig. 7, plotted as the bin-by-bin ratio of the measured flux to that predicted by the SSM. The data from 5.5 to

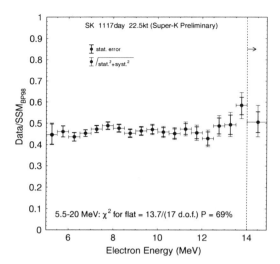

Figure 7. The Super-Kamiokande energy spectrum. Each bin is determined by fitting the $\cos\Theta_{sun}$ distribution, subtracting background, and comparing to the SSM prediction based on the ^{8}B spectrum and differential cross section for $\nu - e$ scattering.

20 MeV is consistent with flat with a χ^2 of 13.7 for 17 degrees of freedom. A preliminary data point from 5.0 to 5.5 MeV is not used in this, or later, analysis, but supports the apparently energy independent suppression of the data.

Observant readers may compare this result to a preliminary energy distribution based on 825 days of Super-K data[28] or a published result based on 504 days of data[29]. The new spectrum is statistically and systematically compatible with the earlier results, although earlier results had larger χ^2 with respect to flat. In particular, the energy bins above 13 MeV were somewhat higher with respect to the average data/SSM level. Besides the obvious increase in statistics, what has changed in the analysis? First, new selection criteria have been developed, particularly to extract the solar signal from the larger background rate found at lower electron energies. Previously, these tighter cuts were only applied to "super-low-energy" data below 6.5 MeV; now the tight cuts are applied over the entire energy range. This results in an improved signal to background ratio. Figure 8 shows the daily event rate in the $\cos\Theta_{sun}$ distribution for the new and old analysis, where the decrease in the background level is evident. The Monte Carlo simulation has been refined and retuned; the overall shift in the absolute energy scale was only 0.27%. From studies where the old and new analysis is applied to the older data set, it seems that some of the change is due to statistics (the new data) and some of the change is due to the new analysis. In any case, let us take this preliminary result at face value, and in the next section we will see what it has to say about neutrino oscillation.

There is one other physical effect to consider when interpreting the high energy bins in the neutrino spectrum. The SSM predicts a small flux for the reaction ^{3}He $+ p \rightarrow{}^4$He $+ e^+ + \nu_e$ which has an endpoint at $E_\nu = 18.8$ MeV, just beyond the 15 MeV ^{8}B endpoint. However, the predicted hep neutrino flux is very uncertain, due to delicate cancellations and other model dependencies[30]; due to these difficulties, no uncertainty is estimated[25,19]. The Super-K spectrum has been studied allowing the hep flux to be a free contribution, which results in an estimated hep flux of $5.4\pm4.6 \times$SSM. Analysis of the data above 18 MeV sets a limit of $\Phi_{hep} < 13.2\times$SSM. It is interesting (but probably coincidental) that a recent recalculation[30] of the hep flux is about five times larger than the BP98 value.

3.3 Oscillation interpretation of solar neutrino results

A two-flavor oscillation analysis for $\nu_e \leftrightarrow \nu_\mu$ (or ν_τ) is performed with 18 bins in energy for daytime exposure and 18 bins in energy for nighttime exposure. The lowest energy bin considered is from 5.5 to 6 MeV. A fit is done in which only the shape is studied, removing one degree of freedom for normalization. This defines an excluded region in neutrino mixing parameter space, as shown in Fig. 9a

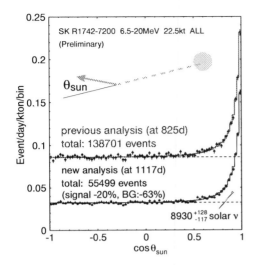

Figure 8. The daily rate of low energy neutrino candidates versus the cosine of the angle with the sun (see inset). The lower curve represents the current analysis tools used to measure the energy spectrum.

within which a sizeable day-night difference or energy distortion is expected but not observed. In comparison with the global flux solutions, the 95% C.L. contour from this flux independent exclusion covers the entire SMA and VAC regions, leaving all of the LOW region and the upper half of the LMA solution.

Another fit is performed that includes one additional piece of information: the SSM ^8B flux prediction, weighted by the 19% uncertainty[19], is used as an additional constraining term in the χ^2. This analysis finds a more restrictive region along the $\theta = \pi$ axis, where the energy spectrum is flat. The result favors large mixing angle for two large regions of Δm^2, where no significant day-night difference is present. The best fit location is near $\Delta m^2 = 10^{-7} eV^2$ (the LOW region) with a $\chi^2 = 31.5/34DOF$, however this is indistinguishable from the local minimum of $31.6/34DOF$ in the LMA region. The contours described are also shown in Fig. 9a.

The same analysis is repeated for the case of $\nu_e \leftrightarrow \nu_s$, where different results are expected due to the absence of a neutral cur-

rent contribution to the electron scattering. In fact, the analysis of the gallium, chlorine, and electron scattering flux shows an incompatibility with any large mixing solution, except for a small region of vacuum oscillation. This is caused by a tension between the suppression to 47% measured by Super-K and to 33% measured by Homestake, each of which are now strictly charged current ν_e reactions. This can be achieved with the complete suppression of the ^7Be flux as in the SMA scenario, but at 95% C.L. it cannot be achieved with any LMA parameters, which tend to have energy independent suppression from 1-15 MeV. As shown in Fig. 9b, the flat spectrum measurement is incompatible with the SMA region, and in fact no two-flavor sterile solution allowed by the global flux analysis (red or dark region) is compatible with the flux independent test on spectrum shape and day-night effect (hollow region with thick border). Combining the two arguments, two-flavor $\nu_e \leftrightarrow \nu_s$ is disfavored as an explanation for the solar neutrino puzzle. The qualification is that it is possible to find acceptable solutions by considering four flavor mixing[31] that includes partial contribution from ν_s.

4 Conclusions

The natural source of neutrinos provided by high energy cosmic rays and nuclear reactions in the sun has afforded us positive indications that neutrinos of one flavor mix with another as they travel through matter or across large distances. The case for atmospheric neutrinos is bolstered by multiple signatures seen in multiple experiments. The data can be explained by two-flavor $\nu_\mu \leftrightarrow \nu_\tau$ mixing, but the alternative hypothesis of two-flavor $\nu_\mu \leftrightarrow \nu_s$ mixing is unsupported. Although the picture is compelling and nearly complete, there are two missing signatures: appearance of CC tau interactions and a measurable recovery in the oscillation probability. In addition, there is yet to be any evidence

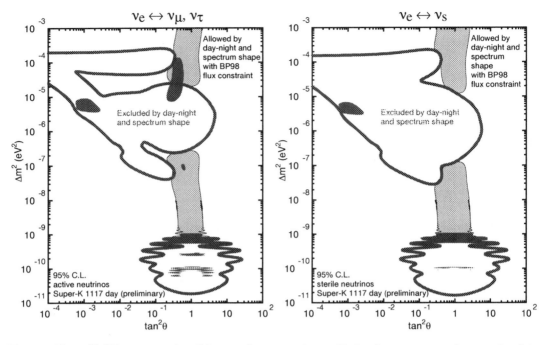

Figure 9. The 95% C.L. contours describing two-flavor neutrino oscillation fits to current solar neutrino data. The solid red (dark) regions correspond to the SMA, LMA, LOW, and VAC solutions based on the combined flux measurements of gallium, chlorine and water experiments. The large hollow regions are excluded by the flatness and lack of day-night asymmetry seen in Super-K data. The solid cyan (light) region is allowed by the Super-K data including a term for the ^8B flux constraint. The left figure (a) considers ν_e oscillation with ν_μ or ν_τ, the right figure (b) considers ν_e oscillation with a sterile neutrino, ν_s.

for a small mixture of $\nu_\mu \leftrightarrow \nu_e$ mixing, as one expects from a general 3×3 matrix. So there is still major experimental work to do, much of it now best accomplished using man-made neutrinos from accelerators or reactors.

The case for solar neutrino oscillation is supported by multiple experiments, but only one clear signature: a large deficit from the predicted flux. The relative rates between three very different experiments suggest an energy dependent suppression, however that is not supported by the new (preliminary) energy spectrum reported by Super-Kamiokande. This is statistically acceptable in an oscillation scenario that has large mixing between ν_e and ν_μ or ν_τ. However, it restricts the opportunity to easily see some "smoking gun" signature; the next best opportunity should come with the ratio of

neutral current to charged current events to be measured by SNO. Again, there is no direct evidence that supports ν_s as playing the major role in solar ν_e mixing.

In summary, the main positive evidence for neutrino oscillation remains with atmospheric neutrinos, where a consistent picture of nearly maximal, essentially two-flavor, $\nu_\mu \leftrightarrow \nu_\tau$ oscillation is preferred. Neutrino oscillation remains an attractive solution to the solar neutrino puzzle, however no single experiment has yet shown any sort of flux independent evidence. Nevertheless, the possible mixing parameters are becoming quite constrained by the absence of smoking gun evidence. In particular, there is no positive indication to suggest that two-flavor mixing with sterile neutrinos plays a role in either the atmospheric or solar neutrino measurements.

Acknowledgments

The author thanks the organizers of the ICHEP conference, particularly the support staff of the computing facility. He is also grateful to many participants in the neutrino session, as well as his own collaborators on MACRO and Super-Kamiokande, for figures and helpful discussions. The author acknowledges support by the U.S. Department of Energy.

References

1. L. Baudis, it et al., *Phys. Rev. Lett.* **83**, 41 (1999).
2. B. Kayser, these proceedings, and references therein.
3. M. Honda, *Nucl. Phys. Proc. Suppl.* **77**, 140 (1999), hep-ph/9811504.
4. M. Honda *et al.*, *Phys. Rev. D* **52**, 4985 (1995); V. Agrawal *et al.*, *Phys. Rev. D* **53**, 1314 (1996).
5. Y. Fukuda *et al.*, *Phys. Rev. Lett.* **81**, 1562 (1998).
6. T. Toshito, these proceedings.
7. Y. Fukuda *et al.*, *Phys. Rev. Lett.* **82**, 2644 (1999); Y. Fukuda *et al.*, *Phys. Lett. B* **467**, 185 (1999).
8. G. Pearce, these proceedings; see also A. Mann *et al*, Proceedings of Neutrino 2000, hep-ex/0007031.
9. F. Ronga, these proceedings.
10. M. Ambrosio *et al.*, *Phys. Lett. B* **478**, 5 (2000).
11. Y. Fukuda *et al.*, *Phys. Lett. B* **355**, 237 (1994).
12. R. Becker-Szendy *et al.*, *Phys. Rev. D* **46**, 3720 (1992).
13. G.L. Fogli, *et al.*, Proceedings of Neutrino 2000, hep-ph/0009269.
14. C. Athanassopoulos, *et al*, *Phys. Rev. Lett.* **81**, 1774 (1998).
15. E. Akhmedov, P. Lipari, and M. Lusignolli, *Phys. Lett. B* **300**, 128 (1993); P. Lipari and M. Lusignolli, *Phys. Rev.*

D **58**, 73005 (1998); Q.Y. Liu and A. Yu. Smirnov, *Nucl. Phys. B* **524**, 505 (1998); Q.Y. Liu, S.P. Mikheyev, and A.Yu. Smirnov, *Phys. Lett. B* **440**, 319 (1998).
16. S.P. Mikheyev and A.Yu. Smirnov, *Sov. Jour. Nucl. Phys.* **42**, 913 (1985); L. Wolfenstein, *Phys. Rev. D* **17**, 2369 (1978).
17. S. Fukuda *et al.*, submitted to PRL, hep-ex/0009001.
18. Considerable resources are available at: http://www.sns.ias.edu/ jnb/.
19. J. Bahcall, S. Basu, and M. Pinsonneault, *Phys. Lett. B* **433**, 1 (1998); J. Bahcall and M. Pinsonnealt, *Rev. Mod. Phys.* **67**, 781 (1995).
20. C. Catadorri, these proceedings. See also: M. Altmann *et al.*, *Phys. Lett. B* **490**, 16 (2000).
21. N. Gavrin, Proceedings of Neutrino 2000; see also: J.N. Abdurashitov *et al.*, *Phys. Atom. Nucl.* **63**, 943 (2000).
22. B.T. Cleveland *et al.*, *Astrophys. J.* **496**, 505 (1998).
23. Y. Fukuda *et al.*, *Phys. Rev. Lett.* **77**, 1683 (1996).
24. Y. Takeuchi, these proceedings.
25. For a review, see: J.N. Bahcall, P.I. Krastev, and A.Yu. Smirnov, *Phys. Rev. D* **58**, 096016 (1998).
26. C. Gonzalez Garcia, these proceedings.
27. J. Klein, these proceedings.
28. Y. Suzuki, Proceedings of 19th Lepton-Photon, Stanford, (1999), eConf C990809:201-228,2000
29. Y. Fukuda *et al.*, *Phys. Rev. Lett.* **82** 2430, (1999).
30. L.E. Marcucci *et al.*, *Phys. Rev. Lett.* **84** 5959 (2000); L.E. Marcucci, nucl-th/0009066.
31. O. Yasuda, these proceedings; M.C. Gonzalez-Garcia, these proceedings; C. Giunti, M.C. Gonzalez-Garcia, and C. Pena-Garay, *Phys. Rev D* **62**, 013005 (2000);

Plenary Session 9

Beyond the Standard Model

Chair: J. S. Kang (Korea)
Scientific Secretary: Motoi B. Tachibana

BEYOND THE STANDARD MODEL

LAWRENCE J. HALL

Physics Department and LBNL, University of California, Berkeley, CA 94720, USA
ljhall@lbl.gov

Theories with large extra dimensions are reviewed and compared to theories with weak-scale supersymmetry, including experimental signatures, gauge coupling unification, neutrino masses, proton decay and grand unification, the hierarchy problem, flavor and electroweak symmetry breaking. In each of these cases, the physics of the energy desert is replaced with the physics of the spatial bulk.

1 Two Towers

What are the big questions raised by the Standard Model? Why are the gauge couplings, quark and lepton masses and mixings what they are? These are the *numerical* questions. There are also the *structural* questions: why is the gauge group what it is, and why do the quarks and leptons have such apparently *ad hoc* quantum numbers for these gauge interactions? Why are there three generations? Why is the electroweak symmetry broken at the TeV scale, V, to electromagnetism? Of all these questions, the origin of the scale V, and why it is so much smaller than the gravitational scale, M_P, is perhaps the most critical – the answer to this question provides the basic framework within which many of the other big questions are to be addressed.

All mass scales of the standard model, and therefore essentially all of high energy physics, are derived directly or indirectly from the scale of electroweak symmetry breaking (EWSB), $V \approx 1000$ GeV [a]. On the other hand, dimensional analysis of the Newtonian coupling

$$G_N = \frac{1}{M_P^2} \qquad (1)$$

leads to an enormous gravitational mass scale, $M_P \approx 10^{19}$ GeV. The conventional paradigm for understanding this hierarchy of masses is to extend the Standard Model and

not to change gravity. At the scale M_P, which is taken to be fundamental, a new perturbative dimensionless coupling is introduced. Under renormalization this coupling grows strong logarithmically at lower energies, just like the QCD coupling. Eventually, it becomes strong at some new scale which it thereby generates as a dimensional transmutation. The weak scale, V, is related to this newly generated scale. How directly depends on whether we are talking about technicolor or supersymmetry, and how supersymmetry breaking is being transmitted to the standard model particles. But in all cases, the small ratio V/M_P is related to some new coupling constant becoming strong. It is the logarithmic evolution of this coupling which makes the hierarchy so large – I will call the interval between M_P and V the *energy desert*. At any accessible energy, E, the graviton remains very weakly coupled, with interactions proportional to E/M_P. Gravity is therefore irrelevant for high energy physics experiments – for example, the branching ratio for $K \to (\pi + \text{graviton})$ is proportional to $(m_K/M_P)^2$ and is of order 10^{-38}.

During the 1990s, our field has apparently "voted" in favor of an energy desert that is supersymmetric – there were about 35 talks at the parallel sessions on one or another aspect of supersymmetry, both theoretical and experimental. Over the last two decades theorists have constructed a *weak-scale supersymmetry tower*, which seems to incorporate our collective wisdom on the likely grand

---

[a] Directly for the W and electron masses, but indirectly for the QCD scale, which can be viewed to some degree as being triggered by EWSB.

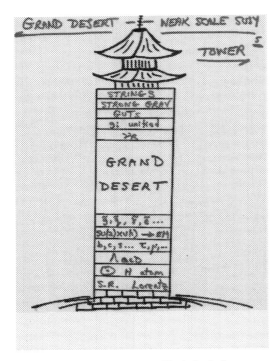

Figure 1. The Grand Desert – Weak-Scale Supersymmetry Tower.

which some feel is already a cause for concern.

The main message of this talk is that a new idea has allowed a solution to the smallness of V/M_P *without the need for an energy desert*. In early 1998, I would have betted heavily against a completely new alternative to supersymmetry appearing before the end of the 90s, let alone an alternative to the energy desert. Yet, over the last two years the *Tower of Large Extra Dimensions* has been constructed, and it is the explosion of creativity about this idea which is the theme of my talk.

The possibility of extra compact spatial dimensions is an old one, pioneered by Kaluza and Klein in the 20s, with a resurgence in the early 80s in the context of supergravity, and followed by superstrings shortly thereafter. So what is new? The proposal of Arkani-Hamed, Dimopoulos and Dvali (ADD)[2] has two alarmingly simple elements

- The new dimensions are large, with radius R between a mm and a fm.

- The standard model particles are confined to a single point in the extra dimensions. Thus quarks and leptons and their gauge interactions live on a 3d subspace, which I will call "our brane".

Given this, you may guess that these extra dimensions are irrelevant to us. Wrong: gravity is changed as it can propagate in these extra dimensions (in the "bulk"), changing the inverse square law of gravity to inverse $(2+n)$ on distance scales less than R. As shown in Figure 2, the gravitational coupling unifies with the gauge couplings at the fundamental scale, M_F, which can be taken in the TeV domain.

The gauge hierarchy is an illusion – the Planck scale, M_P, is a non-physical scale derived from an invalid extrapolation. From Figure 2, we find the physically meaningful expression for the Newton coupling to be

$$G_N = \frac{1}{M_F^2} \frac{1}{(M_F R)^n} \tag{2}$$

picture for our field, as illustrated in Figure 1. The picture includes radiative heavy top quark dynamics for EWSB, the economical Yukawa couplings for transmitting EWSB to the quarks and leptons without running afoul of precision electroweak data, dark matter, see-saw neutrino masses and ultimate unification into superstring theory. Grand unification is an option for understand the quark and lepton quantum numbers, but whether it is present or not, the crowning achievement of this remarkable edifice is a successful correlation between the weak mixing angle and the QCD coupling at the percent level of accuracy[1].

For theorist and experimentalist alike, the tower is not a closed subject – the question of how supersymmetry is broken and mediated to give the superpartner spectrum is an open issue, which I will survey in section 2.2, and there is the small issue of why we haven't discovered any superpartners yet,

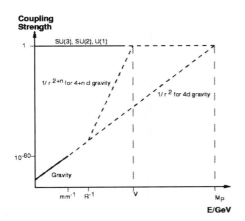

Coupling Strength

SU(3), SU(2), U(1)

1/r^{2+n} for 4+n d gravity

1/r^2 for 4d gravity

Gravity

mm^{-1}   R^{-1}   V   M_P

E/GeV

Figure 2. Gauge coupling unification. The solid lines represent measured values of the dimensionless couplings as a function of energy. Dashed lines show extrapolations for 4 and $n+4$ dimensions.

for n extra dimensions of radius R. The faster power law running of the gravitational force at distances smaller than R, leads to the factor of $(M_F R)^n$ in (2), which reduces M_F below M_P of (1), providing the volume of the extra dimensions, R^n, is large in fundamental units. For the fundamental scale to be a TeV, R must be chosen to be of order (mm, 10 nm, 10 fm) for $n = 2, 4, 6$. It is remarkable that such large dimensions are not excluded[3], and I return in section 8 to the reformulated hierarchy problem of why $R \gg 1$ in fundamental units.

The physical simplicity of this idea demands that it be explored, and this talk will attempt to provide a critical comparison of these two towers: the supersymmetric energy desert vs. large extra dimensions. I begin with a brief summary of direct experimental tests in section 2, and then examine the large extra dimension idea from the viewpoint of all the areas where the supersymmetric tower has been so successful: gauge coupling unification, neutrino masses, proton stability and grand unification, flavor, EWSB and the theoretical origin of the small number V/M_P, in sections 3 – 8.

Important ideas often emerge after others have been heading in similar directions, certainly large extra dimensions has many precursor ideas. The idea that we live on a 3d subspace in a larger space is old[4], as is the idea that the extra dimensions may be very much larger than the Planck scale, for example the weak scale[5]. The Horava-Witten idea, that extra dimensions somewhat larger than the inverse unification scale could reduce the fundamental string scale from Planck to unification scales[6], is close in spirit. Lykken speculated about strings at the TeV scale without a concrete proposal to accomplish it[7], while Dvali and Shifman studied how gauge interactions could be localized on subspaces[8].

There have been many exciting developments over the last year or two that I will not cover. Chief amongst these are the Randall-

Sundrum extra dimensional schemes[10], cosmology of theories in extra dimensions, and recent work on the cosmological constant problem.

2 Direct Experimental Probes

2.1 Sub-mm Gravity

Searching for deviations from the inverse square law of gravity has always been an active experimental field; now it has the bonus that it might overthrow the predominant paradigm of particle physics of the last quarter century. The current experimental situation is that order unity deviations from gravity are excluded on distance scales above a millimeter, but on shorter scales the limits become rapidly weaker. At 10^{-2} mm there could be a new force 10^7 times stronger than gravity. It is interesting that the millimeter scale is also chosen theoretically: from (2), for the case of two extra dimensions and a fundamental scale in the TeV domain, the size of these dimensions is of order a millimeter. This has provided motivation for several new experiments[9].

However, the possibility that gravity changes from inverse square to $1/r^4$ at any scale above a micron seems improbable from astrophysical and cosmological arguments. For the case of two extra dimensions of equal size, supernovae cool too rapidly by the emission of gravitons into the extra dimensions unless $R < 7 \times 10^{-4}$ mm[11]. Such gravitons would also have been produced in the early universe and would decay to photons leading to too large a diffuse gamma ray background unless $R < 5 \times 10^{-5}$ mm[12]. While this cosmological bound is more stringent than the astrophysical one, it is more easily evaded.

In view of this, what positive signal might we expect from the current experiments? One possibility is that the extra dimensions are not all equal in size: perhaps one is just beneath a mm and we will discover a transition from $1/r^2$ to $1/r^3$. Although we have no special theoretical argument for the mm scale in this case, superstring theory tells us that 6 extra dimensions are to be expected, and if the fundamental scale is at a TeV there are about 15 orders of magnitude in size for these 6 dimensions to populate, so it may not be improbable for one of them to show up in the first 1 or 2 orders of magnitude improvement of the limit.

There *are* interesting arguments for the deviation of gravity in the sub-mm domain – when scalars or vectors propagate in the bulk. For example, a gauge boson associated with baryon number would make short distance gravity repulsive[3].

2.2 Collider Searches: Supersymmetry

Perhaps the most important feature of the supersymmetric tower is that there is a firm link between the overall scale of the superpartner masses and the mass of the W and Z. On the other hand, the precise value of the overall scale is not predicted, and the details of the spectrum is model dependent.

The firm link arises because EWSB is understood as a radiative correction to the Higgs mass induced by heavy top quark dynamics[13]. The crucial diagram is a top squark loop generating a negative Higgs mass squared. By supersymmetry, the strength of the coupling in this diagram is known – it is the same as the top quark Yukawa coupling – and hence the only unknown is the scale of the top squark mass. The Tevatron and LHC will either confirm or severely corner the supersymmetric tower over the coming decade.

This goes a long way to explaining the current interest in theories of weak scale supersymmetry. We heard talks on searches for supersymmetry at LEP[14], at other colliders[15], a host of related theoretical issues[16], and direct detection of supersymmetric dark matter[17].

In the early days of supersymmetric model building, it was realized that the best strategy was to insert supersymmetry breaking into the theory by hand by a set of operators which gave superpartner masses without reintroducing quadratic divergences for the Higgs boson mass[18]. For the last two decades, theorists have been pondering the origin of these operators – in the supersymmetry tower the big question is no longer EWSB, but rather supersymmetry breaking and its mediation. Somewhere there is a sector of the theory which spontaneously breaks supersymmetry – presumably dynamically via new strong interactions – many such theories are now known. The interesting part is how this supersymmetry breaking is communicated – or mediated – to the squarks, sleptons, Higgsinos and gauginos. One of the earliest ideas was that the mediation occurs via the standard model gauge interactions[19], but this was rapidly displaced by gravity mediation in supergravity theories[20]. In fact gravity mediation dominated thinking in the field until about 1995, when gauge mediation reappeared[21], and in the last few years several new mediation mechanisms have been proposed. This is a very healthy situation – there are no really solid arguments to prefer one mediation mechanism over another. This means that we are left with many options for the detailed pattern of the superpartner spectrum. Sample comparison spectra[22] show just how different the schemes are. For example, the sleptons tend to be heavier in gravity mediation than in other schemes. Experimentalists must be alert to the variety of possibilities for how these superpartners may appear, and it is experiment which will decide the mediation question, at which point we will gain a great insight into the direction chosen by nature.

Four mediation mechanisms are shown in Table 1. Anomaly mediation[23] is a Planck scale effect expected in all theories. In the simplest such theories the slepton mass-squareds are negative, so extensions are necessary to prevent spontaneous breaking of

Table 1. Mediation of Supersymmetry Breaking. Gravity mediation requires a flavor symmetry to give squark degeneracy to overcome FCNC and CP violation difficulties, but is more easily able to account for the lightness of the Higgs boson.

Scheme	Ref.	$\widetilde{q}$ degen	Light Higgs?
Gravity	20	–	+
Gauge	19 21	+	–
Anomaly	23	+	–
Gaugino	24	+	–

electromagnetism. In gaugino mediation[24], the standard model gauge interactions also propagate in an extra dimension, but in this case the extra dimension is extremely small, so that we still have an energy desert. Supersymmetry is broken on a brane which is spatially separated in the bulk from the standard model brane, where the quarks and leptons are located. The gauginos propagate in this bulk, so they are the first standard model superpartners to feel the effect of supersymmetry breaking, and they communicate the breaking to the squarks and sleptons. The picture is both simple and predictive. There are other extra-dimensional ideas: supersymmetry breaking may be shone to us from another brane[25], or supersymmetry may be broken by the radius modulus[26].

Much of the motivation for gauge, anomaly and gaugino mediation is usually quoted as the perceived difficulties that gravity mediation has with flavor-changing and CP violating phenomena. If squarks and sleptons of a given charge are not degenerate, there are one loop diagrams involving superpartners that typically contribute to rare processes, such as $K\bar{K}$ mixing, at too high a rate. However, I do not view this as a problem for gravity mediation: the required degree of degeneracy is easy to obtain from the same flavor symmetries which are responsible for yielding the observed hierarchy in fermion masses. Indeed, interesting theories of flavor can be constructed where the superpartners contribute to flavor and CP violating phenomena at an interesting level.

A question for supersymmetry has always been why there is a light Higgs and why it is different from the sneutrino. Gravity mediation does have the advantage that it is relatively straightforward to understand why there is a light Higgs[27]; whereas a certain amount of effort is required in gauge, anomaly and gaugino mediation schemes as indicated in the Table. Nevertheless, FCNC and light Higgs arguments are relatively weak – only experiment will settle the mediation question.

2.3 KK Modes

Suppose that a particle is in an eigenstate of momentum in a compact extra dimension of radius R. Unwrapping the circle, we have quantum mechanics in a box of size $2\pi R$ with periodic boundary conditions. The nth eigenstate has momentum n/R. What would such particles look like from the viewpoint of observers fixed in our 4d spacetime subspace? How would we interpret the quantity that corresponds to momentum in the extra dimension? Since $E^2 = p_4^2 + p_5^2 + m^2$ the momentum p_5^2 appears to us as an extra contribution to the m^2 of the particle. Thus all the different p_5 momentum eigenstates appear to us as particles with different masses. From the 4d viewpoint, the quantum number n labels the members of a tower of different particles having squared mass $m^2 + (n/R)^2$. This is the famous KK tower. Notice that the spacing is governed by the size of the extra dimensions. For ADD it is only the graviton that propagates in the extra dimension, so only the graviton has a KK tower, and, since R is large, the members of the tower are very closely spaced in mass.

We will also be interested in extra dimensions with much smaller radii, $r \approx TeV$, in which standard model particles can propa-

gate. The particles which we have discovered are those with $p_5 = 0$ – the zero modes. Since they have masses much less than $1/R$ the masses of the particles in the KK tower will be n/R. When spacetime is enlarged it is not surprising that the known particles acquire partners – we have become used to that with antimatter and superpartners. The distinctive feature of extra compact spatial dimensions is that each particle acquires a whole tower of partners.

2.4 Collider Searches: Extra Dimensions

The collider signals for weak scale supersymmetry involve superpartners, with coupling strengths governed by the standard model couplings. The collider signal for extra dimensions involve the KK towers, with coupling strengths also governed by the standard model couplings. Over the last two years, there has been an explosion in studies of KK mode collider signals. Excellent reviews of this were given in the parallel sessions[29], where the relevant references can be found, so here I will simply make a few remarks.

For the ADD scheme of large extra dimensions, the signal is the production of graviton KK modes. While each KK mode has a tiny coupling proportional to $1/M_P$, there are huge numbers of these modes, since the mass spacing of the tower is $1/R$, of order 10^{-3}eV to a GeV, for $n = 2$ to 6. The net effect is to give a production cross-section in sub-processes of energy E of order $\sigma \approx (1/M_F^2)(E/M_F)^n$ for n large dimensions. The relevant sub-processes give ($\gamma+$ KK mode) at e^+e^- colliders, and (jet + KK mode) at hadron colliders. Since the KK modes are gravitational coupled with strength $1/M_P$, they escape from the detector leaving a characteristic missing energy spectrum, steeper for larger n. Current limits on the fundamental scale, M_F, range from 0.5 TeV (n=6) to 1 TeV ($n = 2$) from searches at LEP, to 1 TeV (n=6) to 1.5 TeV ($n = 2$) from Tevatron searches. At LHC these can be improved to about 4 TeV (n=2) to 7 TeV (n=6) with a luminosity of 10 fb^{-1}, and are little improved by a larger luminosity of 100 fb^{-1} or a 1 TeV linear e^+e^- collider[28].

Does this mean that LHC will have the power to essentially exclude large extra dimensions as well as weak scale supersymmetry? That is the big question – and in my opinion it has not been adequately addressed. For weak-scale supersymmetry, with a perturbative energy desert, EWSB occurs via heavy top quark radiative dynamics which provides naturalness constraints on at least some of the superpartners. For large extra dimensions it is not clear what EWSB mechanism is to be used – some possibilities will be reviewed in section 7. Are there schemes where we have solid arguments for $M_F < 4-7$ TeV? Are there others which can evade this LHC reach?[b]

In the next section, I argue that, to preserve an approximate gauge coupling unification with large extra dimensions, the standard model gauge particles should propagate in some other extra dimensions of inverse size, $1/r$, about an order of magnitude less than M_F, i.e. on the order of a TeV. In this case, there are collider signals for W, Z and γ KK modes. Although they are too heavy to have been directly produced, their virtual effects contribute to all the usual precision electroweak observables, and so the current limits on $1/r$ are strong. For 1 extra dimension, the current limits are $1/r > 3$ to 4 TeV, with some dependence on whether the Higgs propates in the bulk[31]. At future colliders, sub-process collisions can be affected by the virtual KK tower exchange. This appears to be most powerful at lep-

ton colliders: limits on $1/r$ of 13,23,31 TeV could be reached with 500 fb^{-1} of data at an e^+e^- collider with $\sqrt{s} = 500, 1000, 1500$ GeV, from studies of cross sections and polarization asymmetries[31].

The present 3 TeV limit on $1/r$ suggests that power-law gauge coupling constant unification within the ADD scheme is only possible if M_F is above 10 – 30 TeV, which is above the reach of LHC for KK gravitons, and requires some explanation of the mini-desert from M_Z to M_F. However, there are other ideas for gauge coupling unification which may not need such a large fundamental scale[32]. Also, if the fundamental scale is below 10 TeV, it could give rise to higher dimensional operators in the low energy theory which reduce the precision electroweak limit on $1/r$.

3 Gauge Coupling Unification

The weak-scale supersymmetry tower leads to a successful relation between the standard model gauge couplings[1]. Despite such diverse differences in strength at the GeV scale, or even at the 100 GeV scale, when evolved to extremely high energies the coupling strengths are found to become accurately equal. Inputting the strong and electromagnetic couplings, this can be given as a prediction for the weak mixing angle

$$\sin^2 \theta_{susy} = 0.233 \pm 0.002, \qquad (3)$$

to be compared with the experimental value

$$\sin^2 \theta_{exp} = 0.2312 \pm 0.0002. \qquad (4)$$

After 25 years of theoretical studies beyond the standard model, this is the only successful prediction of any of the 18 free parameters of the standard model to the 1% level of accuracy[c].

---

[b]There are also signals from virtual KK graviton exchange, but these have larger theoretical uncertainties[29]. Presumably M_F is the scale of string theory, in which case one can also hope to study the effects of string modes at colliders[30].

[c]When phrased in terms of the strong coupling the prediction appears to only be accurate to 10%. However, the prediction should really be viewed as a correlation between the strong coupling and the weak

With a fundamental scale in the TeV domain, there is little room for the couplings to evolve, so that this prediction is apparently lost. In fact, Dienes, Dudas and Gherghetta have shown that if the gauge interactions propagate in extra dimensions a rapid power law running of the coupling constants is possible, so that unification can be achieved in a small energy interval[33]. Hence we can imagine the following picture. There are large extra dimensions in which only gravity propagates, bringing the fundamental scale down to, say, 10 TeV. Above this scale we have superstrings propagating in a 10d spacetime. What does the theory look like below 10 TeV? Perhaps it is not a 4d quantum field theory of strong and electroweak gauge bosons! Perhaps the standard model gauge particles propagate in some extra dimensions of size $r \approx 1/\text{TeV}$. In this case, the gluon, W, Z and γ could carry momentum in the extra dimensions – ie from our 4d perspective there would be KK towers of the standard model gauge particles. When we calculate the renormalization of the gauge couplings, we must include radiative corrections from diagrams containing these towers of states, and this causes a rapid power law running of the gauge couplings: $g \propto E^p$ rather than the more familiar logarithmic running of 4d: $g \propto \ln E$. Indeed, numerical studies show that the strength of the gauge couplings can become comparable at an energy of only $20/r$, with one such extra dimension, and at even lower energies with several extra dimensions.

While the picture of gauge coupling unification can be preserved with a fundamental scale as low as 10 TeV, the precision of the prediction is lost. The simplest scheme [33], with the Higgs and gauge particles in the bulk, gives the central prediction

$$\sin^2 \theta_{powerlaw} = 0.243, \qquad (5)$$

which is about as far from the data as the

non-supersymmetric grand unification result. In addition, the size of the error bar is not under control. Power law running implies that the evolution is dominated by the highest energies – but at the highest energies the theory cannot be trusted since it is only the low energy effective theory of some more fundamental theory such as superstrings. We must conclude that while the unification picture can be maintained, the prediction for the weak mixing angle is much less significant than in the case of weak-scale supersymmetry.

You may be tempted to read no further – but that would be a mistake, since this is the only significant strike against large extra dimensions, and it is clearly not fatal. While it prevents large extra dimensions toppling weak-scale supersymmetry from its pedestal, the possibility of a fundamental scale, and 4d quantum gravity, at a TeV deserves a more thorough exploration, both theoretically and experimentally. There are already other ideas about how gauge couplings can unify[32].

4 Neutrino Masses

Neutrino masses have become one of the most exciting areas of physics beyond the Standard Model. Over the last two years, theoretical effort has concentrated on ideas for why $\theta_{\mu\tau}$ is of order unity while V_{cb} is small. The announcements we have heard here on solar neutrinos suggest that this will now be extended to mixing between the first two generations. There are now several highly predictive theories of flavor, some involving grand unification; it is conceivable that measurements of neutrino mixing over the coming decade could give sufficient information for one of these theories to be selected as a leading candidate.

Viewing the standard model as an effective theory below some mass scale, M, we expect the dimension 5 interaction $LLHH/M$ to induce neutrino masses at the level $m_\nu \approx V^2/M$. If there is a large energy desert, then

---

mixing angle, and in this plane, using any sensible measure, the correlation is good to about 1%.

$M \gg V$ and we expect neutrinos to be very much lighter than the charged fermions. As long as neutrinos were exactly massless there was an alternative possibility: a low value of M with no energy desert, with the full theory above M possessing exact lepton number conservation. The discovery of small, non-zero neutrino masses seems to suggest that there is an energy desert, which would strike at the heart of the large extra dimensions scheme.

In fact, the bulk has its own see-saw mechanism: as the volume of the bulk gets large so the neutrino Yukawa coupling becomes small. Consider an extra dimensional scheme with right-handed neutrinos. The standard model gauge interactions are taken to propagate only in our three space dimensions, and this prevents the quarks and charged leptons from escaping out into the bulk. However, the distinctive feature of the right-handed neutrinos is that they do not interact with the standard model gauge bosons, so they are free to propagate also in the bulk. How large a Yukawa coupling is expected between the bulk right-handed neutrino and the lepton and Higgs doublets? Since the wavefunction of the right-handed neutrino is spread over the entire volume of the bulk, it has a size $1/\sqrt{\text{vol. of bulk}}$. The volume of the bulk is also responsible for reducing the fundamental scale to a value below the Planck mass, so this volume suppression leads to a Yukawa coupling of order M_F/M_P. Thus the conventional formula for the Majorana mass of the neutrino, $(V/M)V$ is replaced by one for the Dirac mass of the neutrino: $(M_F/M_P)V$.

The weakness of gravity can be due to an energy desert, $M \gg V$, or to a large bulk $M_F \ll M_P$; in either case neutrinos are made much lighter than the weak scale by the same thing that makes gravity much weaker than the gauge interactions. This is just one example of a very general phenomenon: physics which is conventionally done at very large energies in the energy desert can be replaced by physics occuring at very large distances in the bulk. One might have guessed that we need an energy desert for neutrino masses, baryogenesis, inflation, theories of flavor, etc, but one finds that the bulk is an alternative arena for this physics.

I find it exciting that the bulk physics is so different from the desert physics, with different predictions. In the neutrino case we see this clearly as the light neutrinos are Dirac not Majorana. In fact, one can ask whether neutrino physics might provide a valuable probe of the whole extra-dimensional scenario. Could there be new ways to understand atmospheric and solar neutrino oscillations, which differ from the well-known conventional possibilities with three light Majorana neutrinos[35]? Supernova cooling by emission of KK modes of the right-handed neutrino are quite stringent, and do not allow large mixings to the KK tower, so the atmospheric oscillation is unchanged. However, it may just allow small angle solar mixing to the tower, giving a new distinctive possibility.

5 Proton Stability and Grand Unification

An energy desert is apparently an obvious requirement for understanding why new physics, especially gravitational physics, does not give rapid proton decay, and for an understanding of gauge quantum numbers given by grand unification.

Many ideas for physics beyond the standard model destabilize the proton, and therefore lead to theories with some special ingredient designed to restore proton stability. It is perhaps humbling that the two most obvious cases of this are weak-scale supersymmetry and large extra dimensions! In the supersymmetric case there are renormalizable interactions between quarks, leptons and their superpartners which induce proton decay. For theories with strong gravity in the

TeV domain, black holes are expected to violate anomalous global symmetries such as baryon number: there is nothing to prevent a black hole from swallowing a proton and radiating a positron and pion. In both cases the relevant scale for proton decay is a TeV – the proton would have a weak decay rate!

For supersymmetry the standard ingredient for proton stability is R parity, which forbids the dangerous operators. For large extra dimensions, it is similarly possible to stabilize the proton by introducing an extra symmetry, for example a discrete gauge symmetry[36] or baryon number gauged in the bulk[3]. In all of these cases, at a deeper level one would like to gain an insight into the origin of what seems to be a specially designed symmetry. Without such deeper understandings, supersymmetry and large extra dimensions appear to be on comparably bad footings with regard to proton stability. For supersymmetry, imposing R symmetry has the bonus of interesting experimental signatures – superpartners are always pair produced at colliders, and the lightest is stable and is a good cold dark matter candidate. Similarly, in the extra-dimensional case, gauged baryon number in the bulk leads to a long range force on distance scales $(M_F^2/M_P)^{-1} \approx$ mm, nomatter how many large extra dimensions there are. This provides another impetus for pursuing sub-mm tests of gravity.

However, large extra dimensions offer the prospect of proton stability without the need for imposing some special symmetry. Matter is necessarily localized at some point in the bulk, and it is possible that the quarks and leptons are localized at different places! Such a spatial separation can easily lead to an exponential suppression of proton decay, and has the virtue that matter localization is already an inherent feature of the large extra dimension scheme[37]. It could be that fermion mass hierarchies are due to smaller spatial separations amongst the quarks and amongst the leptons – localization offers an exciting

alternative to flavor symmetries. Hierarchical patterns of flavor symmetry breaking in the energy desert is replaced by a hierarchy of spatial locations in the bulk. The physics is very different – and hence so are the predictions for rare phenomena. It is surprising that this aspect of the bulk has not received more attention.

Quark-lepton separation does not conflict with an underlying grand unified symmetry – the separation may be a consequence of the unified symmetry breaking. SO(10) theories at the TeV scale have been constructed which generate quark-lepton separation, a t–b mass hierarchy and small neutrino masses[38]. There is no doublet-triplet splitting problem (maybe the triplet Higgs will be discovered before the the doublet Higgs!) and the extra gauge bosons induce neutral D meson mixing rather than proton decay. The extraordinary insight into quark and lepton quantum numbers provided by grand unified theories is equally possible in supersymmetric and large extra dimensional theories.

6 Flavor

If the fundamental scale is in the TeV domain, there are two immediate concerns about flavor and CP violation. There is very little room in energy above the top mass to incorporate a theory for the hierarchy of fermion masses, such as a Froggatt-Nielsen theory based on flavor symmetries. Secondly, such theories, and local higher-dimensional operators, would be expected to give disastrously large contributions to flavor and CP violating processes. There are two possible viewpoints. One is that our inability to discover rare processes beyond the standard model places significant constraints on large extra dimensional theories and suggests that a low fundamental energy scale is not the preferred solution to the hierarchy problem[39]; the other is that the physics of the bulk opens up the first really new possibilities for understanding flavor in many years[40,41], and provides mechanisms for protecting us from rare processes.

One new idea is that flavor is broken strongly on branes that are located far from us across the bulk. We observe small breaking of flavor because there is an exponential suppression arising from the propagation (or shining) of flavor breaking across the bulk[40,41]. Another inherently extra-dimensional idea is orthogonal to the idea of flavor symmetry. The quarks and leptons of different generations are taken to be located at different points in the bulk. Small fermion masses and mixings arise from the overlap of wavefunctions which are peaked at different locations. Both types of theory could be criticized as not being sufficiently constrained: as well as the usual freedom to choose multiplets and interactions, there is now the additional questions of the geometrical configuration of the branes and the bulk. However, bulk dynamics may favor certain symmetrical structures. It is certainly possible that elegant, predictive theories are just waiting to be discovered.

7 Electroweak Symmetry Breaking

Large extra dimensions allows new options for electroweak symmetry breaking, however, no new option has appeared which successfully competes with radiative EWSB in supersymmetry.

This should not be taken as a strike against large extra dimensions, since it can incorporate weak-scale supersymmetry with radiative EWSB. For example, the fundamental scale could be 3 TeV, the hidden sector may break supersymmetry at 1 TeV, so that the superpartner masses are of order 300 GeV. However, if weak-scale supersymmetry is incorporated, it is not so easy to argue for a fundamental scale low enough for KK modes of the graviton to be accessible at the LHC — why not push the fundamental scale beyond 10 TeV? It would be extremely exciting to find a new scheme for EWSB, without any fine tuning, which demands a low fundamental scale.

There are several cases where a low fundamental scale makes a real difference to EWSB:

7.1 Top quark radiative EWSB in the standard model

Suppose the Higgs boson of the standard model is massless at tree level. The quadratically divergent top quark loop leads to a radiative correction $\delta m_H^2 \approx -\Lambda^2/\pi^2$. It is just conceivable that the cutoff, Λ, is given by physics at a low fundamental scale such that the resulting Higgs is perturbatively coupled. The Higgs will have a mass which violates the precision electroweak bound for the standard model. But this is not quite the standard model – it is not very unlikely for operators suppressed by powers of $1/M_F$ to remove this bound on the light Higgs[42]. While the standard model could indeed be the correct low energy effective theory, without fine tuning and without supersymmetry, this picture fails to give an adequate theory of EWSB – the quadratic divergence indicates that the physics of EWSB is at the fundamental scale.

7.2 Technicolor, with shining of flavor breaking

Any theory of technicolor has three main challenges. The strong dynamics for the technicolor condensate must not violate the precision electroweak data; extended technicolor, which transmits EWSB to the quarks and leptons, must not lead to too large FCNC effects; and a sufficiently large top quark mass must be generated by the techniconden-sate. Extra dimensions could certainly solve the FCNC problems of extended technicolor. The primordial flavor breaking could occur on a brane which is distant from us, and could be shone to our brane, which contains the technicolor interaction[40].

7.3 Topcolor from KK exchange

The strong dynamics of 4 fermion operators can trigger EWSB[43]. The necessary fine tuning is reduced as this dynamics occurs at lower and lower scales – but a big question is what generates the 4 fermion operators. With extra dimensions, there is an extremely elegant and economical possibility: these operators can be generated by the exchange of the KK tower of the ordinary gluons[44]. Furthermore, in the 1 gluon exchange approximation, the most tightly bound scalar resulting from the exchange of the entire KK tower of the standard model gauge interactions has the quantum numbers of the Higgs boson[45]. While such a calculation cannot be justified, the scheme does have a controlled prediction for both Higgs and top masses. The Higgs boson mass is predicted to lie in the range 165 – 230 GeV.

7.4 Scherk-Schwarz supersymmetry breaking

The Scherk-Schwarz method of supersymmetry breaking is inherently extra dimensional[46]. In a 5th dimension of size r, the fermionic modes can have free boundary conditions, while the bosonic modes can have fixed boundary conditions, and supersymmetry is broken by the differnt profiles of the wavefunctions in the bulk. With quarks and leptons propagating in this 5th dimension, the exciting thing is that this breaking *directly* becomes a mediation mechanism also. Usually the primordial breaking has to be somewhat isolated from the standard model fields, hence the need for a mediation mechanism. In Scherk-Schwarz breaking, no such isolation is necessary. Furthermore, since the superpartner masses are given by $1/r$, these dimensions must have a size of order an inverse TeV. Such theories can be remarkably predictive; for example,

simple ones are already excluded by the top mass[47] and by the recent Higgs mass limit announced at this conference from LEP[48]. Certainly there are other interesting theories of this type. I think they merit much further study, as they really should be thought of as a new, inherently extra-dimensional method of EWSB. The method still relies on radiative corrections involving the large top Yukawa coupling. But instead of a single squark in the loop, there are now whole towers of diagrams involving the KK towers. This means that there is no longer the usual logarithmic running of the Higgs boson mass-squared, rather one finds a finite one loop result proportional to $1/r^2$.

Should Scherk-Schwarz be included as a mediation mechanism in Table 1? I think not: the standard model gauge couplings are propagating in the TeV-sized bulk, so that they get strong at about 10 TeV. This is not the perturbative supersymmetric desert; this idea finds a more natural home in theories with large extra dimensions and a low fundamental scale.

8 Origin of Gauge Hierarchy

An initial reaction to the idea of large extra dimensions is: this is just a rephrasing of the hierarchy problem, not a solution. It does not explain why the scale $1/R \ll M_F$. I have two comments: any rephrasing of the hierarchy problem is very interesting; this rephrasing particularly so, as it leads to all sorts of new physics ideas. Secondly, it is possible to understand why R is exponentially large in fundamental units. One example involves a scalar field in the bulk which has certain brane source terms. For a 2d bulk, the potential for this scalar field depends on $\ln R$, leading to an exponentially large determination of R[49]. The solutions of the desert hierarchy problem involve quantum mechanics: the logarithmic evolution of some radiatively corrected coupling induces dimensional

Figure 3. The Weak Scale-Supersymmetry Tower under attack?

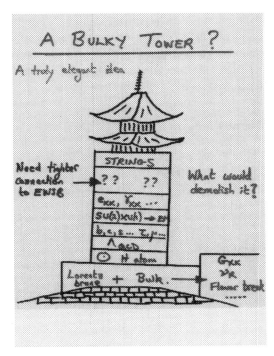

Figure 4. The Large Extra Dimension Tower.

transmutaion. However, in the above example the exponential generation of the scale R is a purely classical result.

Another origin for exponentially large radii is provided by a cosmic string in a 2d bulk[50].

9   Summary

The last two years has seen an explosion of activity in studying the consequences of theories with extra spatial dimensions. In this talk I have focussed on theories where the size of the compact dimensions is large enough that the fundamental scale of gravity is reduced to near the TeV domain. The immediate objections to such a view, involving proton decay, FCNC and CP violation, gauge quantum numbers and neutrino masses prove to be unfounded – the bulk provides a new arena for understanding this physics. On the other hand, while a picture of coupling constant unification can be preserved, the significance of the prediction of the weak mixing angle is greatly reduced. The supersymetry tower has come under attack, from both theory and experiment, but for now it is held firmly in place by the weak mixing angle, as shown in Figure 3.

The most natural understanding of electroweak symmetry breaking is still provided by heavy top radiative corrections, although in the context of large extra dimensions this may occur via the inherently extra dimensional Scherk-Schwarz mechanism. Extra dimensions suggest exciting new options for electroweak symmetry breaking, but none of these have been fully implemented in a natural way. While large extra dimensions has a variety of characteristic collider signals, until further progress is made on pinning down electroweak symmetry breaking, none of these appears guaranteed to be accessible to currently funded accelerators. Large extra dimensions have justly led to much creative excitement – further progress may hinge on exploring new connections with electroweak symmetry breaking. The bulky tower, shown in Figure 4, has made a remarkable emergence, but is not yet complete.

Acknowledgments

Thanks are due the organisers for a stimilating and well organised meeting, and my collaborators for numerous insights and interactions.

References

1. H. Georgi, H. Quinn and S. Weinberg, *Phys. Rev. Lett.* **33** 451 (1974); S. Dimopoulos, S. Raby and F. Wilczek, *Phys. Rev.* **D24** 1681 (1981); S. Dimopoulos and H. Georgi, *Nucl. Phys.* **B193** 150 (1981); L. Ibanez and G. Ross, *Phys. Lett.* **105B** 439 (1981).
2. N. Arkani-Hamed, S. Dimopoulos and G. Dvali, *Phys.Lett.* **B429** 263 (1998), hep-ph/9803315.
3. N. Arkani-Hamed, S. Dimopoulos and G. Dvali, *Phys. Rev.* **D59** 086004 (1998), hep-ph/9807344.
4. K. Akama, Lecture notes in Physics, Nara (1982) 267, hep-th/0001113; V. Rubakov and M. Shaposhnikov, *Phys. Lett.* **B125** 136 (1983).
5. I. Antoniadis, *Phys. Lett.* **B246** 377 (1990).
6. P. Horava and E. Witten *Nucl.Phys.* **B460** (1996) 506, hep-th/9510209; *Nucl.Phys.* **B475** (1996) 94, hep-th/9603142.
7. J. Lykken, *Phys. Rev.* **D54** 3693 (1996).
8. G. Dvali and M. Shifman, *Phys.Lett.* **B396** (1997) 64, errat.**B407** (1997) 452, hep-th/9612128.
9. J. Long, A Kapitulnik, Talks given at *Beyond 4d* ICTP, Trieste, July 2000.
10. L. Randall and R. Sundrum, *Phys. Rev. Lett.* **83** 3370 (1999), hep-ph/9905221; *Phys.Rev.Lett.* **83** (1999) 4690, hep-th/9906064.
11. S. Cullen and M. Perelstein, *Phys. Rev. Lett.* **83** 268 (1999) hep-ph/9903422; C. Hanhart, D.R. Phillips, S. Reddy, M.J. Savage, nucl-th/000716.
12. L.J. Hall and D. Smith, *Phys.Rev.* **D60** (1999) 085008, hep-ph/9904267.
13. L. Ibanez and G. Ross, *Phys. Lett.* **B110** 215 (1982); L. Alvarez-Gaume, M. Claudson and M. Wise, *Nucl. Phys.* **B207** 96 (1982); K. Inoue, A. Kakuto, H. Komatsu and S. Takeshita, *Prog. Theor. Phys.* **67** 1889 (1982); **68** 927 (1982).
14. M. Antonelli, I. Fisk, I. Fleck, V. Hedberg, K. Nagai, these proceedings.
15. T. LeCompte, S. Rajagopalan, A. Rozanov, L. Stanco, these proceedings.
16. M. Aoki, E. Berger, G. Cho, W. de Boer, O. Eboli, S. Kanemura, P. Ko, O. Kong, Y. Okada, W. Porod, T. Takeuchi, X. Tata, Y. Yamada, these proceedings.
17. R. Arnowitt, M. Drees, M. Nojiri, K. Olive, these proceedings.
18. S. Dimopoulos and H. Georgi, *Nucl. Phys.* B (1981).
19. M. Dine, W. Fischler, M. Srednicki, *Nucl. Phys.* **B189** 575 (1981); S. Dimopoulos and S. Raby *Nucl. Phys.* **B192** 353 (1981); L. Alvarez, M. Claudson and M. Wise *Nucl. Phys.* **B207** 96 (1982).
20. A.H. Chamseddine, R. Arnowitt and P. Nath, *Phys. Rev. Lett.* **49** 970 (1982); R. Barbieri, S. Ferrara and C.A. Savoy, *Phys. Lett.* **119B** 343 (1982); L.J. Hall, J. Lykken and S. Weinberg, *Phys. Rev.* **D27** 2359 (1983).
21. M. Dine and A. Nelson *Phys. Rev.* **D48** 1277 (1993), hep-ph/9303230; M. Dine, A. Nelson and Y. Shirman *Phys. Rev.* **D51** 1362 (1995), hep-ph/9408384.
22. M. Peskin, hep-ph/0002041.
23. L. Randall and R. Sundrum, *Nucl.Phys.* **B557** (1999) 79, hep-th/9810155; G.

Giudice, M. Luty, H. Murayama and R. Rattazzi, *JHEP* 9812 (1998) 027, hep-ph/9810442.

24. D.E. Kaplan, G.D. Kribs, M. Schmaltz, hep-ph/9911293; Z. Chacko, M.A. Luty, A.E. Nelson and E. Ponton,*JHEP* 0001:003 (2000), hep-ph/9911323.

25. N. Arkani-Hamed, L. Hall, D. Smith and N. Weiner, hep-ph/9911421.

26. T. Kobayashi and K. Yoshioka, hep-ph/0008069; Z. Chacko and M. Luty, hep-ph/0008103.

27. G. Giudice and A. Masiero, *Phys. Lett.* **206B** 480 (1988).

28. G. Giudice, R. Rattazzi and J. Wells, *Nucl.Phys.* **B544** (1999) 3, hep-ph/9811291.

29. J. Hewett, G. Landsberg, these proceedings.

30. S. Cullen, M. Perelstein and M. Peskin, hep-ph/0001166.

31. T. Rizzo and J. Wells, *Phys.Rev.***D61** (2000) 016007, hep-ph/9906234.

32. N. Arkani-Hamed, S. Dimopoulos and J. March-Russell, hep-th/9908146.

33. K.R. Dienes, E. Dudas and T. Gherghetta, *Phys. Lett.* **B436** 55 (1998); *Nucl. Phys.* **B537** 47 (1999).

34. N. Arkani-Hamed, S. Dimopoulos, G. Dvali and J. March-Russell, hep-ph/9811448; K. Dienes, E. Dudas and T. Gherghetta, *Nucl.Phys.***B557** (1999) 25, hep-ph/9811428.

35. G. Dvali and A. Smirnov, *Nucl.Phys.* **B563** (1999) 63-81, hep-ph/9904211; R. Barbieri, P. Creminelli and A. Strumia hep-ph/0002199; A. Lukas, P. Ramond, A. Romanino and G. Ross, hep-ph/0008049.

36. I. Antoniadis, N. Arkani-Hamed, S. Dimopoulos and G. Dvali, *Phys.Lett.* **B436** 263 (1998), hep-ph/9804398.

37. N. Arkani-Hamed and M. Schmaltz, Phys.Rev. D61 (2000) 033005, hep-ph/9903417; E. A. Mirabelli and M. Schmaltz, *Phys.Rev.* **D61** (2000)

113011, hep-ph/9912265.

38. N. Arkani-Hamed, L.J. Hall and M. Schmaltz, in preparation.

39. T. Banks, A. Nelson, and M. Dine, *JHEP* 9906 (1999) 014, hep-th/9903019.

40. N. Arkani-Hamed and S. Dimopoulos, hep-ph/9811353.

41. N. Arkani-Hamed, L. Hall, D. Smith and N. Weiner, *Phys.Rev.* **D61** (2000) 116003, hep-ph/9909326.

42. L.J. Hall and C. Kolda, *Phys. Lett.* **B459** 213 (1999), hep-ph/9904236; R. Barbieri and A. Strumia, *Phys. Lett.* **B462** 144 (1999), hep-ph/9905281.

43. Y. Nambu, Talk at Kazimierz Conference, EFI88-39 (1988); W. Bardeen, C. Hill and M. Lindner, *Phys. Rev.* **D41** 1647 (1990).

44. B. A. Dobrescu, *Phys.Lett.* **B461** (1999) 99, hep-ph/9812349; H.-C. Cheng, B.A. Dobrescu and C.T. Hill, hep-ph/9912343.

45. N. Arkani-Hamed, H. Cheng, B. A. Dobrescu and L. J. Hall, hep-ph/0006238.

46. J. Scherk and J. Schwarz, *Phys. Lett.* **B82** 60 (1979).

47. I. Antoniadis, C. Munoz and M. Quiros, *Nucl.Phys.* **B397** (1993) 515, hep-ph/9211309.

48. A. Pomarol and M. Quiros, *Phys. Lett.* **B438** 255 (1998), hep-ph/9806263; A. Delgado, A. Pomarol and M. Quiros, hep-ph/9812489.

49. N. Arkani-Hamed, L. Hall, D. Smith and N. Weiner, hep-ph/9912453.

50. A. Cohen and D. B. Kaplan, *Phys.Lett.* **B470** (1999) 52, hep-th/9910132.

Plenary Session 10

Recent Progress in Field Theory

Chair: Luciano Miani (CERN)
Scientific Secretary: Muneto Nitta

THE UTILITY OF QUANTUM FIELD THEORY

MICHAEL DINE

Santa Cruz Institute for Particle Physics, Santa Cruz CA 95064 USA
E-mail: dine@scipp.ucsc.edu

This talk surveys a broad range of applications of quantum field theory, as well as some recent developments. The stress is on the notion of effective field theories. Topics include implications of neutrino mass and a possible small value of $\sin(2\beta)$, supersymmetric extensions of the standard model, the use of field theory to understand fundamental issues in string theory (the problem of multiple ground states and the question: does string theory predict low energy supersymmetry), and the use of string theory to solve problems in field theory. Also considered are a new type of field theory, and indications from black hole physics and the cosmological constant problem that effective field theories may not completely describe theories of gravity.

1 Introduction: Why a Talk on Quantum Field Theory?

Before the advent of string theory, theorists tended to distinguish themselves as "phenomenologists" and "quantum field theorists." Phenomenologists were rather lowly sorts, who dealt with questions having to do with experiment; the quantum field theorists dealt with "deep" questions such as anomalies and solitons. Presumably this is why it is traditional at this meeting to have sessions on quantum field theory and a summary talk. Today, the dividing line is more between string theory and phenomenology. It is a rare theorist who describes him or herself as a quantum field theorist. There is, however, some logic to having these sessions and this talk. First, quantum field theory is an essential tool both to those interested in phenomenological questions and those interested in the difficult questions of string theory and quantum gravity. Second, despite its current, orphaned status, it remains a subject of great fascination in itself, and great strides continue to be made in its development.

This reasoning gives the speaker license to speak about anything and everything, from phenomenology to string theory, and that is what I will do. Below is an outline of the topics to be covered:

- Quantum Field Theories as Effective Theories

- Quantum Theory and Experiment: The Standard Model

- Quantum Theory and Experiment: Beyond the Standard Model

  A. Neutrinos

  B. What if $\sin(2\beta) \approx 0$ (more precisely, what are the implications of a small CP asymmetry in $B \to \psi K_s$?)

- Applications of Quantum Field Theory to Fundamental Problems in Physics: String (M) Theory

- Applications of String Theory to Problems in Quantum Field Theory

- New Ideas in Quantum Field Theory: Large Dimensions, Non-Commutative Field Theory

- Limitations of Quantum Field Theory: The Cosmological Constant Problem.

- Beyond Quantum Field Theory: Holography

2 Quantum Field Theory as Effective Theory

In the past 30 years, quantum theory has emerged triumphant as the framework in which to describe nature in the very small. At the same time, we have come to view quantum field theories as *effective* theories, valid up to some energy scale. This idea is familiar in the Fermi theory, a theory which breaks down at scales of order 10's of GeV, where it is supplanted by a larger theory.

The development of this viewpoint may account for the current lowly status of these theories, but this is somewhat unfair, because effective field theory accounts for virtually everything we currently understand about nature. The standard model follows from simply postulating a gauge symmetry and particle content. Renormalizability is not a principle to be enforced, but rather the effects of non-renormalizable operators are suppressed by some scale (compare 250 GeV in the Fermi theory), Λ, where some new particles, interactions, or other phenomena must appear. The effects of physics at scales above Λ can be absorbed into the parameters of the effective lagrangian of the low energy field theory.

These ideas have a utility which extends beyond applications to the standard model. They give us a framework in which to understand the limitations of the model, and also the physics which may lie beyond. They also provide powerful tools to understand basic questions which we confront in theoretical physics. By thinking about the low energy behavior of theories which, at a macroscopic level, are very complicated, we have been able, during the last several years, to:

- Make significant progress in understanding field theories themselves. It has proven possible to make exact statements about a variety of field theories, even strongly interacting ones, particularly in cases where the theories are supersymmetric. There are quantum field theories where we can study phenomena such as confinement and electric-magnetic duality in *controlled* approximations.

- Make exact statements about strongly interacting limits of string theory, even though we don't have a non-perturbative setup in which to describe the microscopic theory.

3 The Standard Model as an Effective Field Theory

Physics as we know it has $SU(3) \times SU(2) \times U(1)$ symmetry. There are three generations of quarks and leptons. The unsuppressed (i.e. renormalizable) terms are exactly the gauge and Yukawa interactions of the standard model In addition, there are a variety of possible suppressed – non-renormalizable – interactions. The most interesting of these – those which we have the best chance of observing – are those which violate cherished symmetry principles, e.g. baryon number, such as:

$$\mathcal{L}_{\not{B}} = \frac{1}{\Lambda^2} QQQL \tag{1}$$

or lepton number:

$$\mathcal{L}_{\not{L}} = \frac{1}{\Lambda} \phi L \phi L \tag{2}$$

In each case, the question is: what is Λ? If we are lucky, Λ is not so large that we can't observe the corresponding phenomenon. We have some theoretical guesses: Λ might be the Planck scale of the scale of grand unification. It might be some scale associated with supersymmetry breaking (10^{11} GeV? 10^3 GeV?)

The operator of eqn. 2 gives rise to neutrino masses. The increasing evidence for neutrino masses suggests that Λ is not too small. We might expect, for example, that this operator is suppressed by quark or lepton masses, just as are typical Yukawa couplings. For example, we might guess they are suppressed by y^2, $y \sim m_\tau/v$, so

$$m_\nu = \frac{m_\tau^2}{\Lambda} \tag{3}$$

In this simple-minded view, neutrino masses suggest that there is a new scale in nature, perhaps at $10^{11} GeV$. Of course, without a theory, y can vary over a huge range, and so can Λ. Still, neutrino masses are our first glimpse at a new scale of physics – real physics beyond the standard model.

4 Standard Model and CP Violation?

Shortly before coming to this conference, one of my experimental colleagues asked me what I would think if $\sin(2\beta) \ll 1$. He seemed to believe that I would be very troubled by such a finding, since it is not compatible with the standard model. Instead, my reaction was one of great excitement. First, such an observation would have a natural description in the language of effective field theory. Second, it would definitely mean that there is new physics, at a not too distant energy scale. Third, various people, myself included, have long proposed that this is quite a reasonable – even likely – possibility, if nature exhibits low energy supersymmetry[1,2,3]. Let me explain each of these points.

If $\sin(2\beta)$ is small, then so is the CKM phase. In the limit of vanishing CKM phase, there is no CP violation in the standard model, so there must be some new physics. (In this situation, of course, the measurement of the CP asymmetry in $B \to \psi K_s$ is not necessarily a measurement of $\sin(2\beta)$, since there are contributions to this process beyond those of the standard model.) In the $K - \bar{K}$ and $B - \bar{B}$ systems, it must be possible to describe the effects of this new physics by operators in the low energy effective lagrangian. An operator such as

$$\mathcal{L}_{\Delta s=2} = \frac{e^{i\phi}}{\Lambda^2} s\bar{d}s\bar{d} + \text{h.c.} \qquad (4)$$

for example, could be responsible for ϵ, where Λ, as always, represents the scale of the new physics.

Now suppose that nature is supersymmetric, with supersymmetry broken at a scale $\Lambda = M_{susy} < TeV$. In supersymmetry, there are many new sources of CP violation. CP-violating, $\Delta s = 2$ operators are generated in the low energy theory, for example, by exchanges of gluinos and squarks. The possibility that phases in the squark and gluino mass matrices might be the origin of the observed CP violation has been explored for some time[2], and was discussed by Ko at this meeting[4].

Not only is it *possible* that this is the case, but one might even argue that it is *likely*. The line of argument goes as follows. If nature is supersymmetric with supersymmetry broken near the weak scale, one typically obtains too large a value for the neutron electric dipole moment, unless CP-violating phases are small, of order 10^{-2}. CP violating phases can naturally be small if CP is a good symmetry of the microscopic theory, spontaneously broken at some lower scale. In this case, the KM angle is small. This picture finds support in string theory, where CP is a gauge symmetry, which must be spontaneously broken[5,6].

The final ingredient in this picture comes from the physics of flavor conservation/violation. In supersymmetric theories, the absence of flavor-changing neutral currents is not automatic; some additional structure must be imposed. In many of the proposals for this additional structure, $K - \bar{K}$ mixing is nearly saturated by supersymmetric contributions, so the supersymmetric contribution to ϵ is automatically of the correct order if the CP-violating phases are of order 10^{-2} (ϵ' can also be accomodated).

So the answer to my colleague is: few things could be as exciting as the discovery that $\sin(2\beta)$ doesn't agree with the Standard Model prediction, and there is at least one well-motivated framework which predicts that $\sin(2\beta)$ should be quite small.

5 The Hierarchy Problem

Thinking about the Standard Model as an effective field theory with a cutoff Λ leads immediately to a puzzle connected with the Higgs field. Dimensional analysis implies

$$m_H^2 = c\Lambda^2, \qquad (5)$$

and this is obtained from the Feynman graphs of the effective theory. The fact that $m_H < TeV$ (and is most likely much lighter) suggests that new physics should not be too far away.

Previous guesses about this physics were:

- New Strong Interactions – Technicolor. Here $\Lambda \sim 1\text{TeV}$.

- A New Symmetry – Supersymmetry. Here, we expect $\Lambda \sim M_z - 1\text{TeV}$.

Of course, one of the likely possibilities has always been: something we have not guessed. In the past two years, there has been extensive exploration of two new possibilities:

- Large extra dimensions. This was the subject of Lawrence Hall's excellent review at this meeting. The basic idea here is that m_H *is* the fundamental scale; the Planck mass is large because some or all of the extra dimensions are large[7,8].

- Warped dimensions: here the idea is that the hierarchy of scales results from an exponential dependence of the metric on the distance in an additional dimension[10]. We will discuss this possibility further shortly.

6 Supersymmetry and Our Understanding of Field Theory

Field theories such as real QCD are complex, and it is difficult to extract even qualitative information. In the last few years, however, it has been possible to solve, at least in part, many non-trivial field theories with supersymmetry. Supersymmetry turns out to give a great deal of mathematical control. This has allowed an attack on basic issues in quantum field theory such as duality and

confinement, and on the basic problem of supersymmetry phenomenology: understanding the origin of supersymmetry breaking.

In supersymmetric theories, the coupling constants are often *complex numbers*; the reason one can learn so much about these theories, is that many physical quantities are analytic functions of these numbers.

Supersymmetric gauge theories provide an example of this phenomenon[11]. The imaginary part of the coupling constant is the θ parameter. The low energy effective coupling is an analytic function of the "bare coupling,"

$$\tau = \frac{8\pi^2}{g^2} + i\theta \qquad g_{eff}^{-2} = f(\tau). \qquad (6)$$

This, by itself, does not allow one to say much. But the the low energy theory is often symmetric under:

$$\tau \to \tau + 2\pi i \qquad (7)$$

(corresponding to the 2π periodicity of the θ parameter). This highly restricts f;

$$f = \tau + \sum_{n>0} a_n e^{-n\tau} \qquad (8)$$

This equation, for example, says there are no corrections to the coupling constant in perturbation theory (the subtle meaning of this statement was made clear in [11], where it was shown how one can compute a β-function to all orders in perturbation theory). Further considerations in some cases determine f completely.

This type of analysis has given control many seemingly impossible problems in supersymmetric field theories. These include:

- Exact solutions of theories with N=2 supersymmetry (Seiberg and Witten[12]). In these theories many quantities can be computed exactly. In the strongly coupled region, for example, one can study confinement as the result of monopole condensation. Further progress in this area was reported at the meeting, including exquisite tests of these ideas (Fucito[13], Khoze[14]) and determination of patterns of symmetry breaking in particular examples (by Murayama[15] and Yasue[16]).

- Theories with N=1 Supersymmetry might provide the solution to the hierarchy problem. Not only do they give a way of understanding why there are not big corrections to the Higgs mass $\Lambda \ll M_p$, but they can naturally produce very large hierarchies. Further progress on such theories was reported at this conference (Kazakov[17], Nitta[18], Tachibana[19]).

7 Applied Duality

Apart from addressing fundamental questions in field theory, we can try to use these ideas to understand, e.g., how supersymmetry might be realized in nature. This is an area which has been developing for some time, but the past two years have seen some interesting new ideas:

- Models with dynamical supersymmetry breaking have been put forward in which the partners of the first two generations of fermions are composite and quite massive, while the partners of the third generation are light[20]. These models, first, implement both the ideas of dynamical supersymmetry breaking and quark and lepton compositeness. They are readily compatible with bounds coming from direct searches as well as processes such as $b \to s + \gamma$. They are simpler and less contrived than earlier proposals.

- Models in which nature is approximately conformally invariant over a range of energies can address not only supersymmetry breaking, but also provide models of flavor[21]. Yukawa hierarchies arise because fields in different generations possess different anomalous dimensions. Many problems of flavor physics are readily understood in this context.

8 String Theory as a Tool for the Investigation of Field Theory

Over the past several years, it has proven fruitful to consider certain problems in quantum field theory from the perspective of string theory. This is illustrated by simple configurations of D-branes. In Type II string theory, a configuration of N parallel $D3$ branes describes a theory with gauge group $SU(N)$ and $N = 4$ supersymmetry. More interesting models, with less supersymmetry, can be constructed along these lines. Problems which are very difficult from a field theory perspective take on quite a different character (e.g. geometric) in the string picture.

Recent new developments have been based on the "AdS-CFT" correspondence. This correspondence asserts that conformally invariant QFT is equivalent to string theory in AdS space[22]. This can be used to provide insight into a variety of field theories with conformal invariance, but we would like to understand real QCD, which is certainly not conformally invariant. It is necessary to perturb the system in some way.

Early approaches to this problem involved, for example, finite temperature in five dimensions (in the high temperature limit, the system becomes essentially four dimensional)[22]. These methods were of limited power. Recently, Polchinski and Strassler have exhibited cases where one can perturb the conformal theory by adding non-conformally invariant operators, and where the physics on the supergravity side is completely under control (non-singular spaces)[23]. These are theories where confinement, flux tubes, glueballs, and other interesting phenomena can be thoroughly studied in the gravity dual.

9   A New Type Of Field Theory

In the past year, much attention has been focused on a new type of field theory, known as "non-commutative field theory" (NCFT). These theories arise in some cases as the low energy limits of string theories, and seem to incorporate some of the non-locality of string theory[24]. They exhibit bizarre connections between the infrared and the ultraviolet. These features are interesting in themselves, and might be relevant to understanding difficult problems such as the cosmological problem and issues in black hole physics.

The basic feature of these theories is that space coordinates do not commute:

$$[x, y] = i\theta. \tag{9}$$

This sort of relation arises in string theory in the presence of a background magnetic field. NCFT's can't be local. They exhibit peculiar connections between the infrared and ultraviolet – which have come to be called the *infrared-ultraviolet* connection. For example, typical Feynman graphs behave as

$$\int \frac{d^4 k}{(2\pi)^4} \frac{1}{k^2} e^{i\theta p_1 k_2} \tag{10}$$

For $\theta = 0$, this diagram would be highly divergent in the ultraviolet, but for $\theta \neq 0$, it behaves as $\frac{\theta}{p^2}$. In other words, an ultraviolet divergence gets replaced by a divergence as $p^2 \to 0$.

It is fair to say that the significance of these theories is only beginning to be understood. Could there be real phenomena which might be described by such theories? Might they give some insight into the cosmological constant problem? Could these structures have relevance to other areas of physics? Time will tell.

10   Field Theory as a Tool for Understanding String Theory

The pictures which have been described above can be viewed from a different perspective: One can hope to use one's understanding of field theory in order to understand difficult questions in string (M) theory. (See the talk of Paul Townsend at this meeting.)

These ideas have a long history. The easiest way to prove the finiteness of string theory is to study the effective field theory. Indeed, even though there is much that we do not understand about the fundamental structure of the theory, many questions can be addressed by considering the low energy field theory limit.

Here are just a few of the areas in which field theory has proven useful to understanding outstanding problems in string theory:

- Much of the understanding of duality in string theory has been obtained from the study of the low energy effective field theory.

- String theory has a host of possible vacuum states which are uncovered in various approximations. These are characterized by the number of dimensions (2-11), the amount of supersymmetry ($N = 0, \ldots 4$), the number of generations, as well as sets of continuous parameters ("moduli"). The hope is that some dynamical effects pick out one vacuum or another. From considerations of the low energy effective field theory, however, we know that all of the vacua with some supersymmetry in $d \geq 5$ or with $N > 1$ supersymmetry in $d = 4$ are true, stable vacua of string theory, *exactly*.

- We can make many exact statements about more promising vacua which, in some approximation, have $N = 1$ supersymmetry. We can often compute the ground state energy as a function of the moduli reliably using effective field theory. We can sometimes argue that couplings unify even if the theory is strongly coupled.

11 Unconventional Approaches to Outstanding Problems

11.1 Large Dimensions

In the past two years, two new approaches have been put forth to the hierarchy problem. While the underlying justification for both is string or M theory, both are firmly based on pictures developed by considering the low energy field theory.

The premise of each of these proposals is that the fundamental scale of physics might be close to the weak scale. This obviates the need for supersymmetry as a solution to the hierarchy problem, and, indeed, in both of these approaches, low energy supersymmetry (at least as it is conventionally discussed) is not a likely outcome.

Lawrence Hall has discussed the large dimension possibility at some length at this meeting. The basic idea is that the fundamental scale of the theory is of order a TeV. The Planck scale, in this view, is large because some set of extra dimensions are large. I will not review this proposal in detail here. However, I would like to mention two sets of ideas about supersymmetry breaking which have emerged from thinking about large, but not extremely large, extra dimensions. These start from the idea of two separated walls, with the standard model on one wall, supersymmetry-breaking on another. The first of these is known as Anomaly Mediation[25]. Precursors of this idea arose from four-dimensional, field theoretic reasoning[26]. In this picture, one finds an approximate degeneracy between squarks, necessary to understand the suppression of flavor violating processes. In the simplest version, some sleptons are tachyonic, however, and it is necessary to consider rather complicated models. The second is known as gaugino mediation[27]. Again, this idea has field theoretic precursors[28], but finds a firmer motivation in the large dimension picture. Here, the idea is that certain gauge multiplets propagate in the bulk, and are natural candidates to mediate supersymmetry breaking. Again, this is a way to obtain a spectrum with a suitable degree of degeneracy and other distinct predictions for the low energy soft breakings.

11.2 Warped extra dimensions: The Randall-Sundrum Model(s)

The second of these new proposals to understand the hierarchy problem is known as the Randall Sundrum model. Actually, there are several versions of this model. The simplest to describe is set in five dimensions, with two walls. With the walls as sources of stress-energy, if one tunes parameters, Einstein's equations admit a solution:

$$ds^2 = e^{-2kr_o|y|}dx^\mu dx_\mu + r_o^2 dy^2 \quad (11)$$

Four dimensions are flat, but the fifth, described by y, is curved, or "warped." The standard model sits on the wall at $y = 1$; the wall at $y = 0$ is referred to as the "Planck Brane."

In the effective theory in four dimensions, Newton's constant is given by:

$$G_N = \frac{k}{M^3} \frac{1}{1 - e^{-kr_o\pi}} \quad (12)$$

while the typical scales on our brane are of order

$$m_H^2 = M^2 e^{-2kr_o}. \quad (13)$$

So the hierarchy is due to the warping of space, and it is large because it is the exponential of a rather modest number (compare technicolor, susy approaches). (New solutions of this type were reported at this meeting by Ichinose[29].)

What fixes the separation of the walls which determines the exponential? Goldberger and Wise have shown that it can arise from plausible scalar field dynamics in the low energy theory[30].

There are a number of versions of these ideas currently being explored. These include the possibility that the extra dimensions are in infinite, with gravity localized on a brane[31], or that, viewed from far enough away, the extra dimension is simply flat[32]. Surprisingly, these ideas are not easily ruled out, and if correct, these lead to distinctive phenomenologies (reviewed by Hewett at this meeting[33]), with some features in common with the large dimension picture.

It should be noted that this structure, unlike the large dimensions structure, has not been derived from string theory, though there is much effort along these lines.

11.3 An Effective Field Theory Critique

The large dimension and warped dimension ideas are exciting, and are plausible alternatives to supersymmetry as solutions to the hierarchy problem. Experiment might produce a smoking gun for one of them.

On the theoretical side, there are many questions which must be settled. All of these are problems of the effective field theory:

- Proton decay, $\mu \to e + \gamma$, etc. One can certainly imagine various ways of suppressing proton decay. Refs. [34] provide several proposals, and if these are operative, they provide more than adequate suppression. One can debate whether these are more or less plausible than R-parity, for example, in supersymmetric models.

- Flavor changing neutral currents.

- High precision electroweak experiments.

In each case, one expects operators to appear in the effective low energy theory which contribute at a dangerous level, unless the fundamental scale is sufficiently large. Precision electroweak experiments provided the most model-independent limits on the fundamental scale in the TeV range. This is perhaps troubling for hierarchies, and is reminiscent of some of the problems of techni-

color. Scenarios have been proposed to suppress other effects; these are typically tied in to ideas of how the KM matrix, with its various peculiar features, is generated. One possibility is that there is a large flavor symmetry, with symmetry breaking occurring on branes located far from the brane on which the standard model sits[35]. While the original models of this sort were rather elaborate, more elegant models were proposed in [36].

11.4 Solutions to the Hierarchy Problem: A Scorecard

It is interesting to compare the various solutions which have been proposed for the hierarchy problem, and to compare with the minimal standard model. One can score them according to:

- Do they solve the hierarchy problem?

- Do explicit models exist?

- Do they explain unification of couplings in a robust, generic way?

- Can they explain the absence of flavor changing processes in a simple way?

- Do they explain the absence of proton decay in a simple way?

- Do they lead naturally to a dark matter candidate?

I will let you do the scoring yourself (I offered my own at the conference) but I think it is clear that the standard model and supersymmetry score the highest in any such ranking. Still, nature will ultimately decide. It is hard to imagine anything which would be more exciting than the experimental discovery of extra dimensions; I'll let you choose where to place your bets.

12 Is Low Energy Supersymmetry A Prediction of String Theory?

It is often said that low energy supersymmetry is a prediction of string theory, and indeed string theory rather naturally produces this sort of structure. But the large and warped dimension ideas are plausible alternatives and need not exhibit the states (squarks, sleptons, neutralinos) expected there.

¿From studies of low energy field theory limit of strings, however, there *is* some evidence that non-supersymmetric states have problems. Fabinger and Horava[37] have shown that many non supersymmetric states of string theory undergo catastrophic decay. This instability is closely related to an instability of the simplest Kaluza-Klein theory, discussed some years ago by Witten[38]. If this problem is generic, low energy supersymmetry is a prediction of string theory.

Are there other problems with non-supersymmetric theories? Perhaps related to the instability discussed above, non-supersymmetric string theories rather typically have tachyons somewhere in their classical moduli spaces. One might imagine that this is not so serious; perhaps there is simply a nearby vacuum. But a little thought shows that the problem is deeper. Even if the potential has a minimum as a function of the tachyon field, the energy associated with this minimum is of order $V_o = -\frac{1}{g^2}$. Since g^2 is dynamical in M theory, the system can attain arbitrarily low energy by moving to small enough coupling.

All of this sounds serious, but with the current state of our knowledge, it is hardly a proof that non-supersymmetric string states don't make sense. We simply don't understand string theory well enough to decide whether it might be possible that the universe sits in a state far from one of the tachyonic states, or that the lifetime of the universe for the catastrophic vacuum decay of Fabinger and Horava is much greater than the age of the universe. It would be interesting to exhibit some sort of disease of the non-supersymmetric vacua, such as an anomaly, which would decisively indicate such an inconsistency. I have spoken about some possible candidates for such anomalies elsewhere, and am currently engaged in a search for examples.

13 Limitations of Effective Field Theory?

13.1 Two Problems for Effective Field Theory

There is growing evidence that the ideas of effective field theory do not apply to gravity. This evidence arises from the study of Black Holes and the problem of the Cosmological Constant.

One of the most exciting recent developments in physics is the observation of what appears to be a non-vanishing cosmological constant, λ. This is a quantity one would think one could compute from particle physics. However, the same sort of dimensional analysis we used before suggests that

$$\lambda = a\Lambda^4 \tag{14}$$

So even if Λ is as small as 100 GeV, we obtain an estimate 55 orders of magnitude larger than the reported observation! (Alternatively, if λ were this large, our horizon would be about 10 cm!)

In field theory, even if, for some reason, there is no cosmological constant at the classical level, one expects a large value for λ quantum mechanically. This is simply because (for weak coupling) one can think of a quantum field theory, as a collection of harmonic oscillators, one for each particle type and momentum $\vec{k}$. The vacuum energy, which is just the cosmological constant, then gets a contribution from the zero point fluctuations of each oscillator:

$$E_o = \lambda = \sum \int^\Lambda \frac{d^3k}{(2\pi)^3} \frac{1}{2}(-1)^F \sqrt{k^2 + m^2} \tag{15}$$

$$\propto \Lambda^4.$$

$((-1)^F$ is $+1$ for fermions, -1 for bosons; it arises because, in the case of fermions, rather than considering the zero point energy, one must compute the energy of the filled fermi sea). Supersymmetry might act as some sort of cutoff. If susy were exact, the bosonic and fermionic contributions to this expression would cancel. However, from our failure to observe any supersymmetric particles to date, we know that we can safely take the cutoff to be as large as 100 GeV. So the low energy contribution to the cosmological constant is at least 56 orders of magnitude too large! At our present level of understanding, we must somehow imagine that this is miraculously cancelled (to a part in 10^{56}!) by high energy contributions.

Many attempts to solve this problem have failed. It seems likely that this represents a breakdown of our ideas about effective field theory.

There is a good deal of evidence that the usual rules of quantum mechanics break down near black holes. The problem is connected with Hawking radiation. Hawking showed many years ago that black holes evaporate. If one imagines a black hole created in a pure state, than in the far future, one has a thermal system. One might imagine that this is no different than, say, burning a piece of coal: the original information in the quantum system must be encoded in subtle correlations among the outgoing photons. This, however, turns out to violate our usual notions of *locality*. String theory seems to possess some degree of non-locality, and there is growing evidence that string theory provides a consistent quantum mechanical framework in which to understand black holes.

13.2 *The Holographic Principle*

From considerations of black hole physics, 't Hooft and Susskind have suggested that in a theory with gravity, there are not as many degrees of freedom in a volume V as we might expect; they argue that a consistent theory of gravity must be *holographic* – the number of degrees of freedom is proportional to the surface area of V[39,40].

This holographic principle is in many ways mysterious, but it can sometimes be seen to hold in string theory. At low energies, for many purposes, string theory is well described by an effective field theory, but perhaps not for everything?

If these ideas are correct, some important questions in nature cannot be answered by the methods of effective field theory. This might be crucial to understanding not only black holes but also the cosmological constant problem, since it means that there are far less degrees of freedom than in eqn. 15. By itself, however, the holographic principle does not answer the question of the cosmological constant. Apart from conceptual issues, there is a numerical one. Even if the cosmological constant is suppressed from some naive estimate, say 10^8GeV^4, by a factor of the current horizon in Planck units, we still miss the observed value by more than 10 orders of magnitude!

14 New Ideas About the Cosmological Constant

Can field theory in four dimensions resolve the cosmological constant problem?

Through the years, a number of ideas have been put forward:

- Perhaps the dynamics of a light particle cancels the cosmological constant, in a manner reminiscent of the axion solution of the strong CP problem? There have been many attempts along these lines, but Weinberg has proven a no go theorems which shows that this cannot occur, at least in conventional field theories[41].

- Interesting gravitational dynamics such as Euclidean wormholes[42] have been proposed, but are not completely satisfactory.

In the last few years, a number of new ideas have been put forward.

- Kachru and Silverstein[43], motivated by the AdS/CFT correspondence, have exhibited a number of string models without supersymmetry in which the cosmological constant cancels at low orders of perturbation theory, and have argued that this cancellation may be general. The known examples, however, have Fermi-Bose degeneracy and it is not clear whether or not this is an essential part of the proposal.

- As noted above, the notion of holography is likely to have some implications for the cosmological constant problem, but it is not clear, at present, precisely what those implications might be. Moreover, while the present horizon is very large, it is not large enough:

$$d \approx 10^{10} \text{ light years} \qquad (16)$$

so $d \times (100\text{GeV}) \sim 10^{36}$ (we need about 10^{55}). Still, there have been some interesting suggestions about how this might work[44].

- Warped geometries: there have been a number of suggestions that the gravitational equations in this framework permit solutions with vanishing four dimensional cosmological constant, going back to[45], and more recently [46]. But troubling singularities appear, and it is not yet clear whether these solutions make sense. It has been argued that these singularities are at best just a rephrasing of the fine-tuning problems[47]. This problem is under intense investigation.

15 More Embarassing Proposals

15.1 What are we trying to explain?

If recent observations of a cosmological constant are correct, then the value of the constant is just such that the cosmological constant is becoming important in the present epoch of cosmic history,

$$\Omega_\lambda \approx 0.7 \, \Omega_{crit} \qquad (17)$$

In thinking about λ, there is a piece of numerology about the cosmological constant which is often invoked:

$$\lambda \approx \frac{(\text{TeV})^8}{M^4}. \qquad (18)$$

Here M is the reduced Planck mass, $M \approx 10^{18}$.

So if we had a theory in which this relation held, we would have Ω_λ in the right ballpark. But while the order of magnitude is correct, we are confronted today with a very close coincidence. If we change TeV to 2.7 TeV, for example, in this formula

$$\Omega_\Lambda \approx 10^3 \Omega_{crit} \qquad (19)$$

So even if we had a theory in which 19 held, we would still be confronted with a significant puzzle.

15.2 The Anthropic Principle Rears Its Ugly Head

These remarks are suggestive of an anthropic explanation of the cosmological constant. I believe that the only scientifically defensible form of anthropic explanation is what Weinberg calls the "Weak Anthropic Principle." Suppose a theory has many metastable (or stable) ground states. The universe in its history may sample all of these states. Only some may develop in a way which can allow for even rudimentary forms of life; most might collapse, for example, long before structure can form.

Even within this framework, as we will see, we are treading on dangerous ground. As Weinberg has remarked: A physicist talking about the anthropic principle runs the same risk as a cleric talking about pornography: no matter how much you say you're against it, some people will think you're a little too interested.

How would we apply this sort of weak anthropic explanation to the cosmological constant? For this to make sense, the underlying theory must have lots and lots (10^{120} = zillions and zillions – to borrow a phrase from Carl Sagan) of reasonably stable ground states. We live in one with a small cosmological constant because that's the only place intelligent beings can evolve.

Weinberg originally argued that this sort of weak anthropic explanation was not good enough to explain the cosmological constant; this could only explain why $\Omega_\Lambda < 10^2 - 10^3 \Omega_{crit}$[41]. Garriga and Vilenkin have argued that a more refined argument gives the right order of magnitude[48]. So we have to face the possibility that this *might* provide an explanation of the observed facts.

Whether you like this sort of explanation or not, we need to ask: do we know of any theories with these properties? The short answer is no, but recently Bousso and Polchinski, and Donoghue have pointed out a way in which such a vast set of metastable states might arise in string theory[49,50]. The analysis is based on considerations of effective field theory, and in particular of certain gauge fields with three indices, $A_{\mu\nu\rho}^{[i]}$, whose flux is quantized (compare monopoles):

$$F_{\mu\nu\rho\sigma}^{[i]} = q^{[i]} n^{[i]} \epsilon_{\mu\nu\rho\sigma}. \qquad (20)$$

The vacuum energy takes on values:

$$E = \sum_i^N n^{[i]\,2} q^{[i]} - \lambda_o \qquad (21)$$

where λ_o represents the other contributions to the cosmological constant. The number of states grows rapidly with N; Bousso and Polchinski argue that if $N \sim 120$, for example, there may a sufficient number of states.

Whether these states actually exist as (meta)stable states is an open question, but within our current understanding, we must acknowledge that it is conceivable. When one delves further into this type of picture[51], one finds that in some versions, everything in this suggestion becomes anthropic. In others, it is only the cosmological constant. Determining whether such a vast set of states truly exists is a problem which cannot be settled in effective field theory.

15.3 A Much Milder Use of the Anthropic Principle?

Whether or not string theory has the vast set of metastable states required for the application of the anthropic principle, it is certain that it contains a large number of ground states, only a small fraction of which – if any – resemble the real world. We have already noted that string theory definitely contains states with more than four dimensions and more than four supersymmetries. A milder application of the anthropic principle might be to understand how nature selects among these possible vacua. It could be that in most of them, one can not develop even the most minimal structures one would imagine are necessary to sustain life, and in fact that many of them would be subject to gravitational collapse. We are a long way from being able to answer this question completely, but partial (positive) answers can already be given, using methods of effective field theory[51].

16 Conclusions

Field theory continues to enjoy an extraordinary level of utility. It gives the standard model, and suggests possible extensions and new phenomena. It gives a way of organizing our questions about new experimental discoveries, and suggests possible explanations. It suggests broad ranges of new phenomena. It is a crucial tool in our study of candidates for a fundamental theory.

Yet field theory also has limitations. We probably need to go beyond quantum field theory if we are to understand:

- The problems of Black Holes

- The Cosmological Constant Problem

- The principles which determine the ground states of M theory, and what selects among them.

Acknowledgements:

This work supported in part by the U.S. Department of Energy. M.D. wishes to thank Nima Arkani-Hamed, Jon Bagger, Yossi Nir and Scott Thomas for discussions and comments on the manuscript.

References

1. This topic is nicely reviewed in Y. Nir, "Lessons from Recent Measurements of CP Violation", hep-ph/008226. This article includes extensive references beyond those mentioned below.

2. A. Pomrarol, Phys. Rev. **D47** (1993) 273, hep-ph/9208205; M. Dine, R. Leigh and A. Kagan, Phys. Rev. **D48** (1993) 4269, hep-ph/9304299; K.S. Babu and S.M. Barr, Phys. Rev. Lett. **72** (1994) 2831, hep-ph/9309249; S.A. Abel and J.M. Frere, Phys. Rev. **D55** (1997) 1623, hep-ph/9608251; G. Eyal and Y. Nir, Nucl. Phys. **B528** (1998) 21, hep-ph/9801411; G. Eyal, A. Masiero, Y. Nir and L. Silvestrini, JHEP **11** (1999) 032, hep-ph/9908296; K.S. Babu, B. Dutta and R. N. Mohapatra, Phys. Rev. **D61** (2000) 091701, hep-ph/9905464

3. A. Masiero and H. Murayama, Phys. Rev. Lett. **83** (1999) 907, hep-ph/9906374; S. Baek, J.-H. Jang, P. Ko and J.H. Park, hep-ph/9907572.

4. See the talk by Ko, these proceedings, and also in [3].

5. A. Strominger and E. Witten, "New Manifolds for Superstring Compactification," Commun.Math.Phys. **101** (1985) 341.

6. M. Dine, R.G. Leigh and D.A. MacIntire, "Of CP and Other Gauge Symmetries in String Theory," Phys.Rev.Lett.

69 (1992) 2030, hep-th/9205011; K. Choi, D.B. Kaplan and A.E. Nelson, "Is CP a Gauge Symmetry", Nucl.Phys. **B391** (1993) 515, hep-ph/9205202.

7. P. Horava and E. Witten, "Heterotic and Type I String Dynamics From Eleven Dimensions", *Nucl. Phys.* **B460**, (1996) 506, hep-th/9510209.

8. N. Arkani-Hamed, S. Dimopoulos and G. Dvali, "The Hierarchy Problem and New Dimensions at a Millimeter," Phys. Rev. Lett. **B429** (1998) 263, hep-ph/9803315; . Antoniadis, N. Arkani-Hamed, S. Dimopoulos and G. Dvali, "New dimensions at a millimeter to a Fermi and superstrings at a TeV," Phys. Lett. **B436**, 257 (1998) hep-ph/9804398. Some precursors of these ideas can be found in [9].

9. Some precursors of these ideas can be found in V.A. Rubakov and M.E. Shaposhnikov, "Do We Live Inside A Domain Wall?," Phys. Lett. **125B**, 136 (1983); G. Dvali and M. Shifman, "Dynamical compactification as a mechanism of spontaneous supersymmetry breaking," Nucl. Phys. **B504**, 127 (1996) hep-th/9611213; G. Dvali and M. Shifman, "Domain walls in strongly coupled theories," Phys. Lett. **B396**, 64 (1997) hep-th/9612128.

10. L. Randall and R. Sundrum, "A Large Mass Hierarchy From a Small Extra Dimension", Phys.Rev.Lett. **83** (1999) 3370, hep-ph/9905221.

11. M.A. Shifman and A.I. Vainshtein, Nucl. Phys. **B277** (1986) 456; Nucl. Phys. **B359** (1991) 571.

12. N. Seiberg and E. Witten, "Electric-Magnetic Duality, Monopole Condensation and Confinement in N=2 Supersymmetric Yang-Mills Theory," Nucl. Phys. **B426** (1994) 19, hep-th/947087; N. Seiberg and E. Witten, "Monopoles, Duality and Chiral Symmetry Breaking in N=2 Supersymmetric QCD," **B431**

(1994) 484, hep-th/9408099.

13. D. Bellisai, F. Fucito, A. Tanzini and G. Travlini, "Multi-Instantons, Supersymmetry and Topological Field Theories", Phys.Lett. **B480** (2000) 365, hep-th/0002110.

14. N. Dorey, T.J. Hollowood, V.V. Khoze, M. Mattis and S. Vandoren, "Multi-Instanton Calculus and the AdS/CFT Correspondence in $N = 4$ Superconformal Field Theory", Nucl.Phys. **B552** (1999) 88, hep-th/9901128.

15. G. Carlino, K. Konishi and H. Murayama, "Dynamics of Supersymmetric $SU(N_c)$ and $USP(2N_c)$ Gauge Theories", JHEP **0002** (2000) 004, hep-th/0001036.

16. Y. Honda and M. Yasue, "New Electric Description of Supersymmetry Quantum Chromodynamics," Phys. Lett. **B466** (1999) 244, hep-th/9909193

17. D.I. Kazakov and V.N. Velizhanin, "Massive Ghosts in Softly Broken SUSY Gauge Theories", Phys.Lett. **B485** (2000) 393, hep-ph/0005185.

18. K. Higashijima and M. Nitta, "Supersymmetric Nonlinear Sigma Models as Gauge Theories", hep-th/9911139.

19. M. Sakamoto, M. Tachibana and K. Takenaga, "A New Mechanism of Spontaneous SUSY Breaking", hep-th/9912229.

20. A.G. Cohen, D.B. Kaplan and A.E. Nelson, "The More Minimal Supersymmetric Standard Model", Phys.Lett. **B388** (1996) 588, hep-ph/9607394; M.A. Luty, J. Terning and A.K. Grant, "Electroweak Symmetry Breaking by Strong Supersymmetric Dynamics at the TeV Scale", hep-ph/0006224.

21. A.E. Nelson and M.J. Strassler, "Suppressing Flavor Anarchy", hep-ph/0006251.

22. O. Aharony, S.S. Gubser, J. Maldacena, H. Ooguri and Y. Oz, "Large N Field Theories, String Theory and Gravity", Phys.Rept. **323** (2000) 183, hep-th/9905111.

23. J. Polchinski and M.J. Strassler, "The String Dual of a Confining Four-Dimensional Gauge Theory", hep-th/0003136.

24. A. Connes, M.R. Douglas and A. Schwarz, "Noncommutative Geometry and Matrix Theory: Compaticficaction on Tori", JHEP **02** (1998) 003, hep-th/9711162; M.R. douglas and C. Hull, "D-Branes and the Noncummative Torus", JHEP **02** (1998) 008, hep-th/9711165; M. Li, "Strings from IIB Matrices", Phys. Lett. **B499** (1997) 149, hep-th/9612222; N. Seiberg and E. Witten, "String Theory and Non-Commutative Geometry, JHEP **09** (1999) 032, hep-th/9908142; S. Minwalla, M. Van Raamsdonk and N. Seiberg, "Noncommutative Pertrubative Dynamics," hep-th/9912072.

25. L. Randall and R. Sundrum, "Out of This World Supersymmetry Breaking", Nucl.Phys. **B557** (1999) 79, hep-th/9810155; G.F. Giudice, M.A. Luty, H. Murayama and R Rattazzi, "Gaugino Mass Without Singlets", JHEP **9812** (1998) 027, hep-ph/9810442.

26. M. Dine and D. MacIntire, "Supersymmetry, Naturalness and Dynamical Supersymmetry Breaking", Dine and MacIntire Phys.Rev. **D46** (1992) 2594, hep-ph/9205227.

27. D.E. Kaplan, G.D. Kirbs and M. Schmaltz, "Supersymmetry Breaking Through Transparent Extra Dimensions", Phys.Rev. **D62** (2000) 035010, hep-ph/9911293; Z. Chacko, M.A. Luty, A.E. Nelson and E. Ponton, "Gaugino Mediated Supersymmetry Breaking", JHEP **0001** (2000) 003, hep-ph/9911323; M. Schmaltz and W. Skiba, "Minimal Gauge Mediation", hep-ph/0001172.

28. M. Dine, A. Kagan and S. Samuel,

"Naturalness in Supersymmetry, or Raising the Supersymmetry Breaking Scale, Phys.Lett.B243:250-256,1990 (such spectra also arise in "No Scale Supergravity Models", J. Ellis, C. Kounnas and D.V. Nanopoulos, Nucl. Phys. **B247** (1984) 373).

29. S. Ichinose, "An Exact Solution of the Randall-Sundrum model and the Mass Hierarchy Problem," hep-th/0003275.

30. W.D. Goldberger and M.B. Wise, "Modulus Stabilization with Bulk Fields", Phys.Rev.Lett. **83** (1999) 4922, hep-ph/9907447.

31. L. Randall and R. Sundrum, "An Alternative To Compactification", Phys.Rev.Lett. **83** (1999) 4690, hep-th/9906064.

32. R. Gregory, V.A. Rubakov and S.M. Sibiryakov, Opening Up Extra Dimensions at Ultra Large Scales", ("quasilocalization."– Gregoriev, Phys.Rev.Lett. **84** (2000) 5928, hep-th/0002072.

33. H. Davoudiasl, J.L. Hewett and T.G. Rizzo, "Experimental Probes of Localized Gravity: On and Off the Wall," hep-ph/0006041.

34. N. Arkani-Hamed and M. Schmaltz, "Hierarchies Without Symmetries from Extra Dimensions, Phys.Rev. **D61** (2000) 033005, hep-ph/9903417; N. Arkani-Hamed, S. Dimopoulos and G. Dvali, "Phenomenology, Astrophysics and Cosmology of Theories with Submillimeter Dimensions and TeV Scale Quantum Gravity, Phys.Rev. **D59** (1999) 086004, hep-ph/9807344.

35. See, for example, N. Arkani-Hamed, S. Dimopoulos, "New Origin for Approximate Symmetries from Distant Breaking in Extra Dimensions," hep-ph/9811353.

36. N. Arkani-Hamed, L. Hall, D. Smith and N. Weiner, "Flavor at the TeV Scale with Extra Dimensions," Phys.Rev. **D61** (2000) 116003, hep-ph/9909326.

37. M. Fabinger and P. Horava, "Casimir Effect Between World Branes in Heterotic M Theory", Nucl.Phys. **B580** (2000) 243, hep-th/0002073.

38. E. Witten, "Instability of the Kaluza-Klein Vacuum," Nucl.Phys.B195:481,1982.

39. G. 't Hooft, "Dimensional Reduction in Quantum Gravity", gr-qc/9310026.

40. L. Susskind, "The World as a Hologram", hep-th/9409089.

41. S. Weinberg, "The Cosmological Constant Problem", Rev.Mod.Phys. (1989) 61.

42. S. Coleman, "Why is There Nothing Rather Than Something: A Theory of the Cosmological Constant", Coleman, Banks wormhole references. Nucl.Phys. **B310** (1988) 643; T. Banks, "Prolegomena to a Theory of Bifurcating Universes: A Non-local Solution to the Cosmological Constant Problem of λ Goes Back to the Future", Nucl.Phys. **B309** (1988) 493.

43. S. Kachru, J. Kumar and E. Silverstein, "Vacuum Energy Cancellation in a Non-Supersymmetric String," Phys.Rev.**D59** (1999)106004, hep-th/9807076; S. Kachru and E. Silverstein, "On Vanishing Two Loop Cosmological Constants in Non-Supersymmetric Strings", JHEP **9901** (1999) 004, hep-th/9810129.

44. A.G. Cohen, D.B. Kaplan and A. E. Nelson, "Effective Field Theory, Black Holes and the Cosmological Constant", Published in Phys.Rev.Lett.82:4971-4974,1999, hep-th/9803132.

45. V.A. Rubakov and M.E. Shaposhnikov, "Do we Live Inside a Domain Wall", Phys.Lett. **B125** (1983) 136.

46. N. Arkani-Hamed, S. Dimopoulos, N. Kaloper and R. Sundrum, "A Small Cosmological Constant from a Large Extra Dimension", Phys.Lett. **B480** (2000) 193, hep-th/0001197; S. Kachru, M. Schulz and E. Silverstein, "Self-Tuning

Flat Domain Walls in 5-D Gravity and String Theory", Phys.Rev.**D62** (2000) 045021, hep-th/0001206; S.H. Henry Tye and I. Wasserman, "A Brane World Solution to the Cosmological Constant Problem," hep-th/0006068; E.I. Guendelman, these proceedings, and "Scale Invariance, Inflation and the Present Vacuum Energy of the Universe," gr-qc/0004011.

47. See, for example, [23], Forst, Lalak, Lavignac, Nilles S. Forste, Z. Lalak, S. Lavignac, H.P. Nilles, "The Cosmological Constant Problem From a Brane World Perspective", e-Print Archive: hep-th/0006139.

48. J. Garriga and A. Vilenkin, "On Likely Values of the Cosmological Constant", Phys. Rev. **D61** (2000) 083502," astro-ph/9908115; J. Garriga and A. Vilenkin, "The Cosmological Constant and the Time of its Dominance", Phys.Rev. **D61** (2000) 023503, astro-ph/9906210; S. Weinberg, "The Cosmological Constant Problems," astro-ph/0005265.

49. R. Bousso and J. Polchinski, "Quantization of Four Form Fluxes and Dynamical Neutralization of the Cosmological Constant", JHEP **0006** (2000) 006, hep-th/0004134.

50. J.F. Donoghue, "Random Values of the Cosmological Constant," hep-ph/0006088.

51. T. Banks, M. Dine and L. Motl, "On Anthropic Solutions of the Cosmological Constant Problem," hep-th/0007206

Plenary Session 11

Hard Interaction Processes
(high Q^2: DIS, jets)

Chair: M. Turala (Institute of Nuclear Physics)
Scientific Secretary: Kunihiro Nagano

HARD INTERACTIONS AT HIGH Q^2: DIS, JETS

R. NANIA

INFN Sezione di Bologna , via Irnerio 46 , 40226 Bolgona , ITALY
E-mail: nania@bo.infn.it

Recent results on hard interactions at Tevatron, LEP and HERA are reviewed. The new measurements of proton structure functions, jets and event shapes allow more stringent tests of QCD.

1 Introduction

At this conference 128 abstracts were submitted to Parallel Session 03 (PA03) where the latest results on hard interactions were presented and discussed.

A complete summary of the 30 presentations is certainly not possible and I will only underline the main results, referring to the specific talks for the details.

In the following the reference to a submitted abstract will be indicated by brackets {}, while reference to a specific talk will be done in the same way but with the name of the speaker.

2 Proton Structure Functions

Figure 1 shows the kinematic quantities relevant to deep inelastic processes (DIS). At this conference new results have been presented both from fixed target experiments and from the electron/positron-proton collider HERA. The impressive kinematic region now covered by the data is summarized in figure 2. Equations 1 and 2 give the cross-section formulae for Neutral Current (NC) and Charged Current (CC) interactions in the case of electron/positron proton scattering:

$$\frac{d\sigma_{NC}^{e^{\pm}p}}{dxdQ^2} = \frac{2\pi\alpha^2}{xQ^4}[Y_+ F_2 \mp Y_- xF_3 - y^2 F_L] \quad (1)$$

with $Y_{\pm} = 1 \pm (1-y)^2$ and $F_i = F_i(x, Q^2)$

$$\frac{d\sigma_{CC}(e^{\pm}p)}{dxdQ^2} = \frac{G_F^2}{2x\pi}[\frac{M_W^2}{Q^2 + M_W^2}]^2 \Phi_{CC}^{\pm} \quad (2)$$

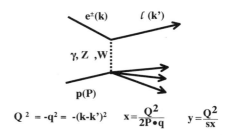

Figure 1. Definition of the kinematic variables of DIS. s is the center-of-mass energy.

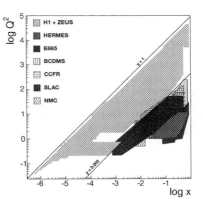

Figure 2. Q^2-x plane with the different regions covered by the experiments.

with $\Phi_{CC}^+(x, Q^2) = \bar{u} + \bar{c} + (1-y)^2(d+s)$ and $\Phi_{CC}^-(x, Q^2) = u + c + (1-y)^2(\bar{d} + \bar{s})$

In the case of NC processes the structure functions F_2 , xF_3 and F_L are sensitive to the sum and difference of the quark and antiquark densities and to the gluon density in the proton respectively. In the following I will review the present knowledge of the quantities entering equations 1 and 2.

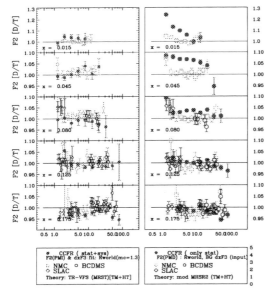

Figure 3. Ratio of F_2 data/theory for different x regions as a function of Q^2 as measured by the CCFR, BCDMS, NMC and SLAC. Right before CCFR reanalysis; left after reanalysis.

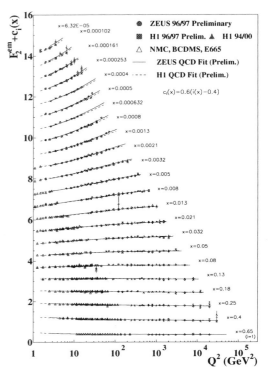

Figure 4. F_2 measurement (H1 and ZEUS) as a function of Q^2 for different intervals of x.

2.1 Fixed target measurements

The CCFR experiment presented a reanalysis of their F_2^ν measurements {Yang}. As shown in figure 3, there was a considerable disagreement between the CCFR measurements and the electron or muon scattering data at low x values. Thanks to a new model independent extraction of the structure functions, the discrepancy is corrected and this allows a more coherent use of the different data sets in the extraction of the parton density functions (PDF). The CCFR collaboration also presented a first measurement of $\Delta x F_3^\nu = x F_3^\nu - x F_3^{\bar{\nu}}$ that, at their center-of-mass energy, is essentially sensitive to the strange content of the proton.

First results on F_2^ν were presented at this conference by the CHORUS collaboration studying neutrino-iron interactions {Oldeman}. Although the results are quite preliminary, it is noticeable that the experiment has on tape a statistic comparable to CCFR and the data extend to lower Q^2.

2.2 HERA results on structure functions

The ZEUS and H1 experiments working at HERA presented at this conference many new results on structure functions. The F_2 data {1048,944} have reached a systematic error of about 3% in most of the bins, improving considerably the precision with respect to the previous results and extending the data to almost 30000 GeV^2 [a]. Figure 4 reports the new measurements as a function of Q^2 for different x intervals and very clearly shows the effect of scaling violations.

QCD relates this effect to the gluon density in the proton and the value of the strong coupling constant α_s. Up to this conference these two quantities were determined separately by fixing one and determining the other from QCD fits to F_2 data. The H1 collaboration has presented a first attempt to simultaneously extract these two quanti-

---

[a]See {Schleper} for a discussion of F_2 at very low Q^2.

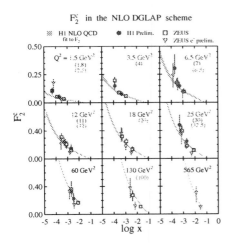

F$_2^c$ in the NLO DGLAP scheme

Figure 5. H1 and ZEUS F_2^{charm} as a function of x for different Q^2 regions.

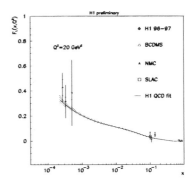

Figure 6. H1: F_L as a function of x at $Q^2 = 20$ GeV.

ties from a QCD fit of the F_2 measurements {944}. To improve the convergence of the fit, data at high x from the BCDMS collaboration are also used. The measured value of the strong coupling constant is $\alpha_s(M_Z) = 0.1150 \pm 0.0017(st) \pm 0.0012(sy)$ with an additional 0.005 error from the scale uncertainty.

This determination of the gluon distribution in the proton can be cross-checked by independent experimental measurements. This is performed at HERA measuring F_2^{charm}, the charm contribution to F_2, and F_L.

The production of heavy-flavors proceeds in DIS mainly via boson-gluon-fusion processes. Using the gluon density obtained from scaling violations, QCD predicts an F_2^{charm} that can be compared to the measurements. Experimentally charm can be tagged via D^* {983,858,855} or with electrons from its semileptonic decay {853}. The result of this comparison is shown in figure 5. The figure also shows the QCD predictions from F_2 fits and a good agreement is observed, with the possible small exception of the lower x and lower Q^2 region (see {Schleper}).

Another method to check the gluon density is the measurement of F_L which contributes at high y (see equation 1). This measurement would be optimally performed by fixing Q^2 and x bins and varying the center-of-mass energy in order to vary y. At HERA this was not possible up to now, but the H1 collaboration {944} has measured F_L determining F_2 and $\partial F_2/\partial \ln Q^2$ with a QCD fit in the low y region and then extrapolating the result into the high y region where the comparison to the data allows the extraction of F_L. Figure 6 shows the final result at one particular Q^2 value together with low energies data. Within the present experimental errors, the QCD prediction agrees with the measurements.

2.3 Parton density uncertainties

An appropriate determination of the errors associated to the parton densities is extremely important when comparing the QCD predictions with data. An example is the study of jets production in hadron-hadron and lepton-hadron processes. At this conference the CTEQ group has presented its first attempt to quantify the error associated to the fit procedures using two different approaches based on Hessian and Lagrange-multipliers respectively {Tung}.

The correlated systematic experimental errors also contribute to the uncertainties of the parton densities, as shown, for example, in figure 7 [1], where the d/u ratio is plotted as a function of x. This ratio is the result of a QCD fit to ZEUS, NMC and CCFR data with the inclusion of the correlated system-

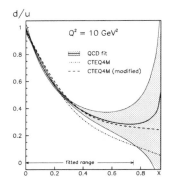

Figure 7. The d/u ratio as a function of x as obtained by a QCD fit to F_2 data. The uncertainties include the correlated experimental errors.

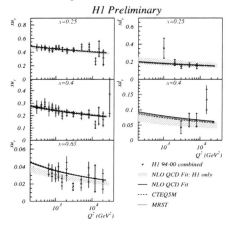

Figure 8. H1: the xu (right) and xd (left) distribution as a function of Q^2 for different x intervals.

atic errors. Beyond $x=0.3$ the error in the ratio increases dramatically, finally blowing up toward $x=1$ (see also [2]). New measurements in the high x region are clearly needed to improve our knowledge of this ratio. With the increased luminosity available, H1 {975} has recently analysed CC events (see equation 2) to extract the first u and d parton densities at x values up to 0.65 (figure 8).

2.4 Very high Q^2 results at HERA

Figures 9 and 10 show the latest update from HERA experiments on the NC and CC cross-sections for electron and positron proton scattering {1049,1050,971,975}. The unification of electroweak forces at very high

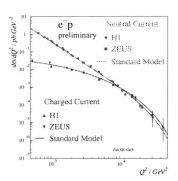

Figure 9. H1 and ZEUS NC and CC cross-section for e^-p interactions as a function of Q^2.

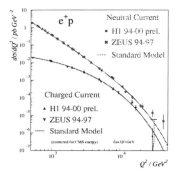

Figure 10. H1 and ZEUS NC and CC cross-section for e^+p interactions as a function of Q^2.

Q^2 is beautifully shown in the figures. Also shown are the Standard Model predictions, that very well match the data up to $Q^2 = 30000$ GeV^2. For both collaborations, the new and statistically independent data sample does not confirm the excess observed in NC interactions with the 1994-1996 data [3].

The availability of HERA data with both electron and positron beams allows the observation of the γZ-interference term at high Q^2 {1049,971}. This leads to the first measurement of xF_3 at HERA (figure 11): the error is still quite large but encouraging in view of the coming upgrade of HERA.

2.5 Conclusions on Structure Functions measurements

A reanalysis of the CCFR data show now a better agreement with other lepton-nucleon data in the low Q^2, low x region. More pre-

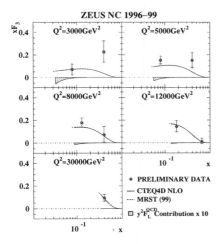

Figure 11. ZEUS: the xF_3 as a function of x for different Q^2 intervals.

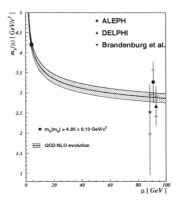

Figure 12. m_b as measured at LEP and lower energies compared to the NLO QCD predictions of a running mass.

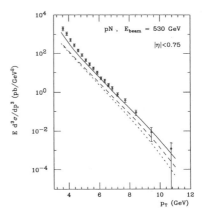

Figure 13. Prompt photon production data from E706 compared to NLO QCD calculations: the dotted line is the full NLO, the dashed line includes threshold resummation and the solid one threshold and recoil resummation.

cise results on F_2 are available from HERA experiments, extending the measurements to Q^2 up to 30000 GeV². New measurements of F_2^{charm} , F_L , xF_3 agree with the predictions of the Standard Model in a very wide Q^2 interval. First direct extractions of quarks densities at HERA have been presented. A first attempt to simultaneously determine the gluon density and α_s from QCD fits to F_2 data has been performed. An improved evaluation of the errors associated to the parton densities determined with the QCD fits is presently underway.

3 General QCD tests and event shapes

In this section I will review few general tests of QCD not involving studies of structure functions or jets. Here follow three particularly interesting examples.

The new measurements of the b mass from ALEPH and DELPHI {142,1034} confirm, within the large errors, the Next-to-Leading Order (NLO) QCD prediction for a running mass (see figure 12).

A new D0 measurement first establishes QCD contributions to the W angular decay distribution {462}.

Instantons are non perturbative fluctua-

tions of non-abelian gauge theories and hence represent a basic feature of QCD. The H1 collaboration presented first interesting results on Instanton searches in DIS {972}, which make use of the available theoretical predictions for Instanton production in DIS [5].

3.1 Prompt photon production

Prompt photon production was originally thought to be one of the best processes to study the gluon distribution in the proton at high x, because of the missing hadronization uncertainties in the final state[b]. However, NLO QCD predictions fail to describe the data, especially at the lower fractional transverse momentum of the photon $x_T = \frac{2p_T^\gamma}{\sqrt{s}}$ measured {Magill,Ferbel}. In this region it is conceivable that simple NLO calculations are not enough. On the other hand not all data sets are consistent with each other and an underestimate of the errors given by the experiments is possible. As a consequence, recently the CTEQ group has no longer considered prompt photon data in their fits to extract parton densities. Similar problems are observed in π^0 production, although the data look more consistent to each other.

However two results indicate that there could be missing contributions in the QCD calculations. In {561}, from an analysis of the results from many experiments at different center-of-mass energies, it is shown that the ratio γ/π^0 is in better agreement with the NLO QCD predictions, especially at higher $\sqrt{s}$. Moreover, in the past years, the prompt photon problem has been phenomenologically tackled introducing with success an *ad hoc* 'intrinsic' parton transverse distribution $k_\perp$ with an average value increasing from fixed target experiments to collider energies from 1.5 to 3 GeV. Also at HERA the introduction of an average $k_\perp$ improves the LO MC description of the data {910}.

The result of a recent attempt to include threshold-resummed calculations and corrections for the recoil from soft radiation {130} is shown in figure 13. The method is not yet fully developed, but the improvement goes certainly in the right direction.

3.2 Event shapes with power corrections.

Some years ago a new method was suggested[6] in order to describe the event shape variables through QCD calculations that sum perturbative and non perturbative contributions.

[b]See also [4] and {Berger} for alternative measurements via lepton pairs

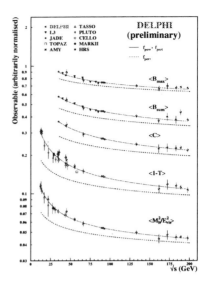

Figure 14. DELPHI: average event-shape variables versus the center-of-mass energy. The predictions of a power correction calculation are also shown.

Defining as $\mathcal{S}$ a generic event shape variable (Thrust, Jet Broadening....), its average value can be written as:

$$\langle \mathcal{S} \rangle = \langle \mathcal{S}_{pert} \rangle + \langle \mathcal{S}_{npert} \rangle \quad (3)$$

where $\mathcal{S}_{pert}$ is calculable via NLO QCD and $\mathcal{S}_{npert}$ is defined as $\mathcal{S}_{npert} = c \cdot \mathcal{P}$. The coefficients c are calculable and depend on the particular variable and $\mathcal{P}$ is a calculable function which depends on a universal parameter α_0, the mean effective value of α_s in the non-perturbative region. In principle, fitting the data with equation 3, it should be possible to determine simultaneously α_s and α_0.

New analyses have been performed using this method at LEP {638,164,626,630} where data at energies up to 208 GeV have been studied. An example of these fits is shown in figure 14. Although the fits look reasonable, the α_s and α_0 values extracted have a considerable spread, as shown in figure 15 that summarizes the different LEP experiments results {Banerjee}.

Similar spreads in the fitted parameters are obtained from HERA experiments

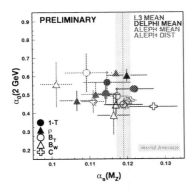

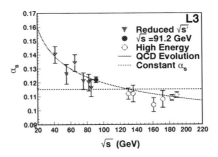

Figure 15. Scatter plot of the measured α_s and α_0 from the different LEP experiments and different event-shape variables.

Figure 17. L3: running of α_s as measured from QCD fits to event shape distributions.

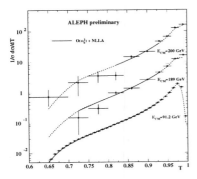

Figure 16. ALEPH: thrust distributions at different energies compared to the QCD NLO+NLLA calculations.

{1006,909}.

Although interesting for its simplicity, clearly this approach is not yet mature for a reliable extraction of QCD parameters and will require further work.

3.3 QCD fits of event shapes distributions.

The measured event shape distributions, corrected for hadronization via MC, can be compared to NLO+NLLA (Next-to-Leading-Log-Approximation) calculations to extract α_s {113,164,626,630,638}. Here three uncertainties arise: i) the definition of the matching procedure between NLO and NLLA results, ii) the contribution from higher order corrections and iii) the definition, for each variable,

Table 1. $\alpha_s(M_Z)$ from the four LEP experiments using NLO+NLLA fits to event shape distributions.

ALEPH	0.1195±0.0035
DELPHI	0.1227±0.0034(st+sy)±0.0052(sc)
L3	0.1215±0.0012(ex)±0.0061(th)
OPAL	0.1173±0.0053

of the range where to perform the fits.

The range of the fits varies from experiment to experiment due to the necessity to exclude particular regions where the QCD calculations are not yet able to describe the data or the hadronization corrections are large. The only exception is the L3 collaboration which instead claims that a fit using all the available ranges of the measured variables is still valid. An agreement on the experimental procedure must be reached by the four collaborations.

An example of the quality of the fits is reported in figure 16 that shows the thrust distribution for three LEP energies. QCD NLO+NLLA calculations are also shown and agree well with the data. From measurements at different energies it is possible to measure the running of α_s and an example is given in figure 17. The agreement of the theory is good from 30 GeV to the highest LEP energies[c]. Table 1 show the latest $\alpha_s(M_Z)$ presented from the four LEP collaborations at this conference.

---

[c] Here the energies lower than the Z peak are obtained using events with initial photon radiation

3.4 Conclusions on the general QCD tests and event shapes

New theoretical efforts on prompt photon production open new possibilities for a more satisfactory QCD description of this process, but more work is still needed.

Basic QCD tests like the running b-mass and the W angular decay distribution have been performed with success at LEP and Tevatron. First searches on Instanton production are in progress at HERA.

Event shapes analyses using QCD power corrections have been performed both at LEP and HERA, but the spread in the results is still quite large and the predictive power has to be improved.

Event shape analyses using NLO+NLLA QCD calculations are in better shape, although not all the experimental and theoretical aspects are completely under control.

4 Jet physics

Studies of processes with jets in the final states allow important tests for QCD predictions. However in order to make meaningful comparisons between data and theory it is important to take into consideration the uncertainties, both experimental and theoretical, which enter in these studies.

As a general example let us consider jets production at Tevatron. Here the total jet cross-section is given by the following formula:

$$\sigma^{jet} = \sum_{i,j,a,b} \int f_i(x_1) f_j(x_2) \sigma_{ij} \mathcal{P}_a^J \mathcal{P}_b^J dx_1 dx_2 \tag{4}$$

Equation 4 needs three main inputs: the parton densities functions in the proton ($f_{i,j}$), the matrix elements to define the cross-section between two partons (σ_{ij}) and the hadronization step that produces the final jet from the final state parton ($\mathcal{P}_{a,b}^J$). In section 2.3 it was shown that a complete evaluation of the errors in the PDFs may bring substan-

tial variations in the predictions and, moreover, the scale at which the PDF should be evaluated is not always clear. Also the evaluation of the matrix elements requires the definition of the scale to be used, which in principle may be different from the previous one. In addition, the importance of orders beyond NLO depends on the process studied and a careful comparison with the experiments is required[d]. The effect of the hadronization of a parton into a jet represents a further uncertainty when comparing data with theoretical predictions.

4.1 Studies on jets structure

I will mention here only two of the many interesting studies on final states presented at this conference.

Results on particle production and fragmentation functions of light and heavy quarks have been presented by SLD and LEP {Gary,DeAngelis,Burrows}. Particularly important are the new measurements of the b-quark fragmentation function {173,690} from ALEPH and SLD, because this is a fundamental ingredient of all heavy-flavor cross-section predictions.

The structure of gluon and quark jets has been studied in photoproduction at HERA {1066} and at LEP {116,640}. In particular it is known that the measured ratio of the mean multiplicity of gluon and quark jets $R = n_g/n_q$ is lower than the QCD asymptotic value $R = 9/4$, indicating the need for higher order and/or non perturbative contributions in the QCD calculations. Recent new analytic NNNLO [7] calculations reduce the QCD predictions and approach the experimental value of ≈ 1.74. Alternatively, it is suggested to describe the data adding non perturbative contributions [8].

---

[d]At this conference the completion of the NLO 3-jets calculation has been announced {Kilgore} and the progresses in NNLO have been reviewed in {Bern}.

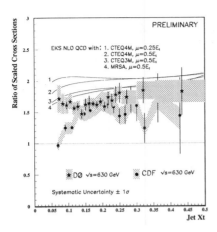

Figure 18. D0 and CDF: ratio of inclusive jet cross-sections at 630 and 1800 GeV as a function of the fractional jet transverse energy. NLO QCD predictions are also displayed.

4.2 Results from Colliders

New results on forward jet production have been presented by the D0 collaboration {465}. The measurements extend now to pseudorapidities of 3, where a greater sensitivity to the different PDFs is expected.

Last year the D0 and CDF collaborations have presented measurements of the inclusive jet ratio $\mathcal{R}$ at 630 and 1800 GeV. The measured value of $\mathcal{R}$ is lower than QCD expectations (figure 18), almost independent of $x_T = 2E_T^{jet}/\sqrt{s}$. The discrepancy between the two experiment at lower x_T values is not completely understood yet, but in this region the experimental correction procedure is more delicate. The variation of the QCD predictions with the scale has been studied and may improve the data-theory comparison. The new information is a preliminary analysis from D0 {466}. The comparison of the data with the NLO QCD predictions is performed evaluating the χ^2 with respect to several PDFs. The analysis makes use of the full covariance matrix that includes the correlated systematic uncertainties . The results, presented in table 2, show reasonable agreement between data and QCD predictions for

Table 2. D0: χ^2 from NLO QCD fits to the inclusive jet ratio 630/1800 .

PDF	χ^2 (20 deg)	Prob(%)
CTEQ3M	20.5	42.5
CTEQ4M	22.4	31.9
CTEQ4HJ	21.0	40.0
MRST	22.0	33.0

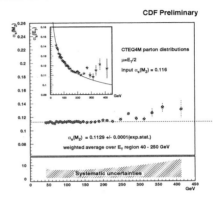

Figure 19. CDF: running of α_s as measured using inclusive jets data. The discrepancy at higher energies depends on the PDF used.

a wide range of PDFs. Similar conclusions were obtained in an analysis of the inclusive jets production [9]. The normalization discrepancy of figure 18 has still to be resolved.

The CDF collaboration has presented a measurement of the running α_s using the inclusive jet measurements and comparing data to NLO calculations {1080}. The result is presented in figure 19 and beautifully shows the running of the strong coupling constant from 50 to almost 400 GeV. The collaboration quotes $\alpha_s(M_Z) = 0.1129 \pm 0.0001(st) \pm^{0.0078}_{0.0089} (sy)$. In addition a theoretical error comparable with the systematic one must be added.

In ep interactions at HERA the new data at high and low Q^2 compare differently with the QCD predictions .

At high Q^2 the experimental uncertainties are small and the data show a very good agreement with respect to NLO QCD predictions (figure 20). It is then possible to extract α_s in a wide range of Q^2: figure 21

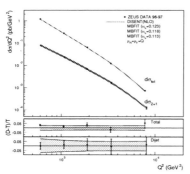

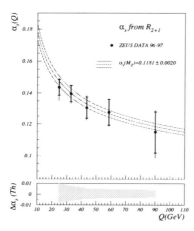

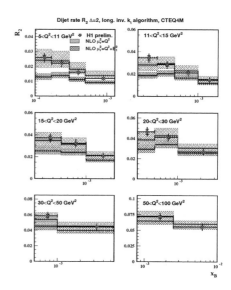

ZEUS 96-97 PRELIMINARY

Figure 20. ZEUS high Q^2 DIS : dijet and total cross-sections measured as a function of Q^2. The lower part shows the comparison to the theory.

ZEUS 96-97 PRELIMINARY

Figure 21. ZEUS: running of α_s measured from dijet cross-sections at high Q^2 DIS.

shows the running of α_s and the nice agreement with the QCD expectations. These results lead to the measurement of $\alpha_s(M_Z)$: $0.1221 \pm 0.0034(ex) \pm^{0.0054}_{0.0059} (th) \pm^{0.0033}_{0.0016} (pdf)$ (H1)[10] and $0.1166 \pm 0.0019(st) \pm 0.0038(ex) \pm 0.0050(th)$ (ZEUS){891}. Note that the errors include the uncertainties from the PDFs. The measurements show a precision comparable to that obtained at LEP.

Even in this case, as it was for the structure functions, the measurement of α_s assumes a gluon distribution. The H1 collaboration {1000} has attempted a simultaneous measurement of the two quantities, this time fitting dijet, inclusive jets and F_2 data. The

Figure 22. H1: dijet cross-sections as a function of x for different low Q^2 regions. The NLO QCD predictions are also shown with two choices for the scale.

results are consistent with the expectations, although the errors are still large.

At low Q^2 in DIS the situation is not as good as for high Q^2 {997,888}. The precision of the data is still quite good but the uncertainties related to the choice of the scale are high. As an example figure 22 shows the dijet cross-section as a function of x for different Q^2 values. At low Q^2 the NLO QCD predictions based on a Q^2 or a $Q^2 + E_T^2$ scale differ substantially, indicating that NLO calculations seem not to be enough to describe this kind of processes.

Jets in γp and $\gamma\gamma$ interactions are studied at HERA and LEP and the availability of NLO QCD predictions allows tests of the theory and of the structure functions of the photon. In photoproduction {1066,977} two processes contribute to dijet production: resolved photon interactions where a parton in the photon interacts with a parton in the proton and only a fraction of the photon energy is involved in the scattering ($x_\gamma \leq 1$), and direct photon interactions where the full energy of the photon enters in the interaction ($x_\gamma = 1$). Looking at the inclusive dijet cross-section measurements as a function of

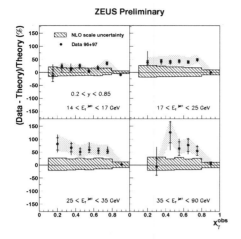

Figure 23. ZEUS photoproduction dijets: comparison (Data-Theory)/Theory as a function of x_γ for different intervals of jet transverse momentum.

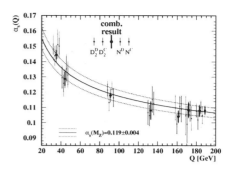

Figure 24. OPAL and JADE: running of α_s from jets measurements.

x_γ (figure 23), thanks to the precision now reached in the measurements, a discrepancy with NLO QCD calculations is observed in the lower x_γ region. This is true even at high transverse energies of the jets. The high x_γ region is instead well described. Different PDFs of the photon may reduce the discrepancy, but still the data lie above the calculations. The origin of this problems is not yet clear, although the fact that the direct contribution is well reproduced by QCD would point toward not correct PDFs of the photon.

The study of jet production in $\gamma\gamma$ also helps to investigate the photon structure. Here there are three possible contributions to dijet production: direct-direct, direct-resolved and resolved-resolved photon interactions. The OPAL collaboration presented a preliminary comparison of dijet cross-sections to NLO and the result, as for HERA, indicates problems with the resolved part {255}. In fact, in the range $7 \leq E_T^{jet} \leq 11$ GeV and for $x_\gamma \geq 0.75$ NLO predicts a cross-section of $87.0\,pb$ while data give $77.4\pm2.6\,pb$. For $x_\gamma \leq 0.75$ NLO predicts $32.6\,pb$ against $71.5 \pm 2.2\,pb$ in the data.

New results on jet production in e^+e^- have been presented by the LEP collaborations at energies up to 204 GeV

{114,626,630,638,164}. These data are compared to QCD calculations, including NLO+NLLA with different matching procedures, to extract α_s. In this context the new result from the OPAL and JADE collaborations {114} is particularly interesting and spans energies between 30 and 189 GeV. Jet variables have been measured with different jet algorithms and compared to the QCD predictions. Then, for each energy, α_s is measured, taking into account restricted intervals of the variables where the agreement is better. The results are plotted in figure 24. The observed running agrees well with the QCD predictions and the final value extracted is $\alpha_s(M_Z) = 0.1187\pm^{0.0034}_{0.0019}$. It is worth noting that this important result has been made possible thanks to the skill of people recovering the old Jade data and corresponding MC programs: an example underlying the importance to keep our data available also after the experiments have ended.

Other two α_s measurements have been performed at LEP using jets. OPAL {251} has attempted a nice simultaneous fit of the color factors (C_A and C_F) and α_s, although the errors are still quite large. ALEPH {162} makes use of 4-jet rates to extract the strong coupling constant, the final measurement yielding $\alpha_s(M_Z) = 0.1172 \pm 0.0009(st + sy + had) \pm 0.0040(th)$.

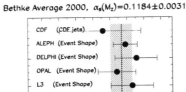

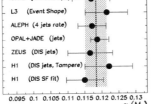

Figure 25. Summary of the α_s measurements presented at the Parallel Session 03.

4.3 Conclusions on jet physics

Jets measurements in DIS at high Q^2 and at LEP are well described by perturbative QCD and give reliable estimations of the strong coupling constant and of its running with the energy.

Tevatron jets cover now also the forward regions and the running of α_s has been measured over a huge range of energies. However the experimental errors still limit the precision of the measurements.

NLO predictions for jets at low Q^2 in DIS suffer from substantial scale uncertainties when compared to data: NLO calculations in this kinematic region are not enough.

Photoproduction results on dijet from HERA have reached a good experimental precision and demand improved calculations and/or possible revisions of the photon PDFs. LEP data on dijet production in $\gamma\gamma$ interactions show the same trend, although the precision of the measurement is lower.

5 Summary

Figure 25 summarizes the results on α_s presented in the parallel session on hard interactions. It includes data from structure function, jets and event shape analyses. The agreement between the measurements is good and the values compare well with the latest world average [11].

In general data and theory agree in most of the analyses performed. However, in some cases, like jets at HERA, the precision of the data demand improved calculations and/or revisions of PDFs. On the other hand, other measurements, like very high Q^2 DIS and jets at Tevatron, call for improvements in precision. In this context the new data from the updated HERA machine and from Run II at Tevatron will certainly allow us to further test the Standard Model in more detail.

6 Acknowledgements

I would like to thank the organizers for giving me the opportunity to present this review, meet new friends and discuss with them in the charming (and very warm) atmosphere of Osaka. I also thank the many friends who helped me in preparing this summary and in particular G. Bruni.

References

1. M. Botje, *Eur. Phys. J.* C **14**,285(2000).
2. S. Kuhlman et al, hep-ph/9912283
3. ZEUS collab., *Zeit. Phys* C **74**, 207 (1997). H1 collab., *Zeit. Phys.* C **74**, 191 (1997).
4. P. Auranche et al., *Phys. Let.* B **209**, 375 (1988).
5. A. Ringwald and F. Schrempp,*Phys. Let.* B **438**,217 (1998); *ibid* B **459**,249 (1999); hep-ph/9909338 and 9911516.
6. Y.L.Dokshitzer and B.R.Webber, *Phys. Let.* B **352**, 229 (1995).
7. I.M.Dremin and J.W.Gary, hep-ph/0004215
8. P. Eden, G. Gustafson and V. Khoze, *JHEP.* **09**, 015 (1998).
9. D0 Collab., *Phys. Rev. Let.* **82**, 2451 (1999).
10. H1 Collab., contributed paper 157, EPS-HEP99, Tampere
11. S. Bethke, hep-ex/0004021.

Plenary Session 12

Soft Interaction Processes
(low Q^2: diffraction, two-photon and spin structure)

Chair:　M. Turala (Institute of Nuclear Physics)

Scientific Secretary:　Yuji Yamazaki

SOFT HADRONIC INTERACTIONS

PETER SCHLEPER

DESY, Notkestr. 85, 22607 Hamburg, Germany
E-mail: Peter.Schleper@desy.de

Recent developments in soft hadronic interactions are reviewed. Emphasis it put on measurements of the proton structure at low x, photon structure, diffraction and exclusive processes such as vector-meson production and their interpretation in approaches to QCD dynamics like BFKL or CCFM.

1 Introduction

Quantum Chromodynamics is the generally accepted field theoretical prescription of strong interactions and is successfully applied to processes where a hard scale is present, given by either a highly virtual particle, a large transverse momentum or a large mass of the exchanged particles[1]. In such processes the strong coupling constant is small enough to allow for perturbative calculations and QCD is predictive. In soft processes, where such a hard scale is not present, α_s becomes large and perturbative techniques are not applicable. This prohibits predictions of such fundamental quantities as the size and mass of the proton, the total cross-section for hadron-hadron scattering or cross-sections for elastic scattering of hadrons. All these questions are closely connected to confinement and still belong to the least well understood properties of strong interactions.

In the past many phenomenological models have been developed to describe soft interactions. More recently, driven by new data obtained in the transition region between soft and hard processes from the HERA, LEP and Tevatron experiments, the theoretical interpretation within perturbative QCD (pQCD) has made considerable progress. This is also the focus of this review[a], and hardly any reference is given to phenomenological prescriptions of soft physics.

---

[a]Invited Plenary Talk given at ICHEP 2000, Osaka.

Deep Inelastic Scattering

To introduce the concepts it is useful to start with the example of deep inelastic scattering (DIS) of a lepton on a proton (Fig. 1). Here

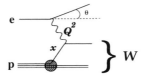

Figure 1. Feynman diagram for deep inelastic scattering.

Q^2 denotes the virtuality of the exchanged photon γ^*, the Bjorken scaling variable x corresponds in lowest order to the momentum fraction of the struck quark in the proton and W is the total centre of mass energy in the γ^*-proton system. For $Q^2 \gg \Lambda_{QCD}^2$ the high γ^* virtuality provides the hard scale and the structure of the proton is resolved into partons, i.e. the cross-section is proportional to the structure function $F_2(x, Q^2)$ which measures the quark momentum distribution in the proton. Fig. 2 shows the data on F_2 as obtained by fixed target experiments and at HERA[2], which now cover a Q^2 range from several 10^4 GeV^2 down to $\simeq \Lambda_{QCD}^2$, and include momentum fractions x as low as 10^{-6}. The data nicely demonstrate the point-like nature of quarks in the region of approximate scaling at $x \approx 0.1$, modified by negative scaling violations at higher x which are attributed to the quark splitting $q \to qg$ and, at low x, by positive scaling violation due to the prevailing gluon splitting $g \to gg, q\bar{q}$.

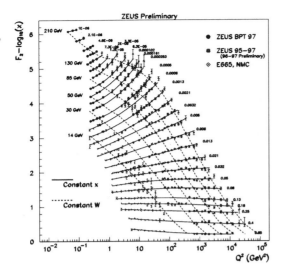

Figure 2. Proton structure function F_2 as a function of Q^2 with lines of constant x and W from [2].

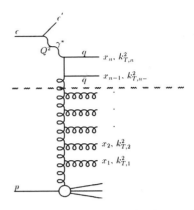

Figure 3. The parton ladder in a hard scattering process. The dashed line denotes the factorisation into a hard matrix element and the parton density.

The sharp increase of F_2 with decreasing x at Q^2 values of a few GeV^2, the main result of the first HERA data [5], is of great importance as this slope is directly related to the gluon density in the proton. The proton at low x is thus a system of very high gluon density and thereby a unique environment for the understanding of QCD dynamics. Since $W^2 = Q^2(1 - x)/x$, the extrapolation towards low x at fixed Q^2 corresponds to the high energy (W) limit of QCD, which is interesting in itself but also has impact for e.g. cosmic ray experiments, heavy ion collisions and Higgs production at the LHC via $gg \to H$ [9].

QCD Dynamics at Low x

To better understand the present excitement about low x dynamics it is worth recalling the assumption of *factorisation* of hadronic cross-sections into a matrix element of the hard process (which is calculable in pQCD) and parton distribution functions which represent all other soft parts of the diagrams (Fig. 3). In the standard Altarelli-Parisi (DGLAP[11]) evolution the phase-space for parton emission is approximated by summing those contributions where the transverse momenta $k_{T,i}^2$ along the ladder strictly increase from the proton to the hard scattering, and therefore is applicable only at larger Q^2. To this approximation the process factorises into calculable coefficient functions and supposedly universal parton distribution functions which obey the DGLAP evolution equations.

It is apparent that other momentum configurations will contribute both at low Q^2 and at low x. At low Q^2 (i.e. Q^2 not much larger than Λ_{QCD}^2) at any point along the ladder $k_{T,i}^2$ might exceed Q^2 which destroys the required strong k_T ordering. Even at higher Q^2 but low x_n, the large possible differences between the longitudinal momenta x_i imply sufficient phase-space also for large transverse momenta k_T somewhere along the ladder, again affecting strong k_T ordering. The transition region from high to low Q^2 and the low x limit thereby elucidate other approaches to pQCD and hence to QCD dynamics.

If the strong k_T ordering criterion is relaxed the approximations become more complex. For the BFKL[13] and CCFM[14] evolution equations ordering in x or in the emission angle of partons along the ladder is assumed, respectively. The resulting "unintegrated" parton densities depend directly on the transverse momenta k_T as shown in Tab. 1. The

Table 1. QCD evolution equations based on different ordering schemes, the dependence of the parton density functions (pdf), the terms summed and the kinematic range of application.

	DGLAP	BFKL	CCFM
order	k_T	x	angle
pdf	$f(x,Q^2)$	$f(x,k_T^2)$	$f(x,k_T^2,Q^2)$
$\sum$	$\ln Q^2$	$\ln 1/x$	$\ln Q^2 +$
			$\ln 1/x$
valid	high Q^2	$k_T^2 \approx Q^2$	low x
	high x	low x	

BFKL approximation is expected to be valid only at low x, since it does not contain the DGLAP terms. It is not known yet how low Q^2 or x have to be to yield sizeable BFKL type contributions. The CCFM equation would in principle enable a smooth extrapolation between the DGLAP and the BFKL regime as it contains both parts. Up to now it is only applicable at low x since the quark splitting terms are not known yet.

Vector-Meson Production

An intriguing view of the interplay between soft and hard physics is derived from *elastic photoproduction* $(Q^2 \approx 0)$ of vector-mesons at HERA[6], $\gamma p \to V p$. For vector-mesons consisting of light quarks (ρ, ω) the energy dependence of the cross-section $\sigma_{\gamma p \to V p}$ is very weak and similar to the total photon-proton cross-section (Fig. 4). For the J/Ψ, however, the energy dependence is significantly stronger, indicating that the charm quark mass provides a hard scale. The size of the cross-section is hence not of geometrical nature but can be associated with the partonic content of the proton. In this case the process is assumed to be dominated by the exchange of a pair of gluons which together form a colour singlet to yield an elastically scattered proton (Fig. 5). The increase of the cross-section towards large W reflects the increase

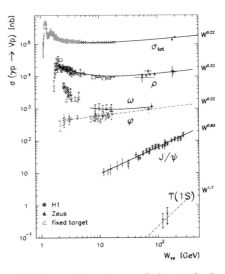

Figure 4. Energy dependence of the total photon-proton cross-section σ_{tot} in comparison to the cross-section for elastic photoproduction of vector-mesons from HERA [6].

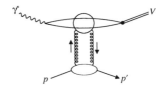

Figure 5. Feynman diagram for the production of vector-mesons V on an elastically scattered proton via 2-gluon exchange.

of the gluon density (squared) towards low x.

It is evident that in the extreme high energy limit this behaviour must change since the J/Ψ contribution should never exceed the total cross-section. New dynamics must therefore dampen the J/Ψ cross-section in the high W (low x) limit.

In QCD calculations[15] the cross-section is explained as a three step process: the splitting of the photon into a $q\bar{q}$ dipole, the interaction of this dipole with the gluon pair and finally the formation of the vector-meson. At low Q^2 and low x the three steps take place on very different time scales suggesting that the cross-section factorises into the probabilities for each of the individual steps.

This factorisation allows the same dipole cross-section to be applied also in other processes at low x such as inclusive DIS, jet production or diffraction[7,8].

2 The Proton at Low x

The proton structure at low x is of interest not only as a new domain in QCD, where fundamental insight into the dynamics within and beyond the perturbative regime is still to be gained. It is also of relevance for the program at the Tevatron and the LHC[9]. For the small Higgs mass expected due to the indirect and direct measurements at LEP [10] the dominant production process $gg \to H$ (with $M_H = x_1 x_2 s_{LHC}$) at LHC energies implies gluon momenta x_i in the range $1 > x > 10^{-4}$, or $0.1 > x > 10^{-3}$ if the angular acceptance for the Higgs decay products is restricted to rapidities $|y| < 2$. Sensitive tests of the Higgs sector therefore crucially depend on the gluon density at low x and the reliability of the the oretical extrapolation from low to high Q^2 [9]. At and below $x = 10^{-3}$ the only process giving access to the gluon density in the proton is DIS at HERA.

HERA Data and the Gluon Density

Both HERA collaborations H1 and ZEUS have released new preliminary data[2,3] on their structure function measurements in the low x and low Q^2 region. The H1 data[3] shown in Fig. 6 have now in a large range a statistical precision of $\approx 1\%$ and systematic errors of about 3% due to new instrumentation such as silicon tracking and high granularity calorimetry. This represents important progress in comparison to existing published data, and the precision is probably close to the final results attainable by the experiments in parts of the phase-space. To determine simultaneously α_s and the gluon density at low x, the H1 collaboration has subjected this data to an elab-

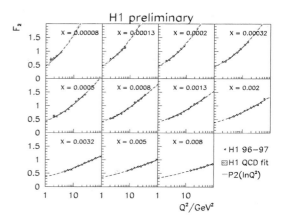

Figure 6. The proton structure function F_2 at low x as a function of Q^2 from H1 [3,4].

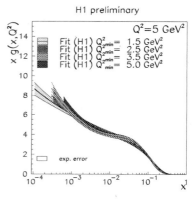

Figure 7. The gluon density obtained by H1 [4] in a DGLAP fit for different minimal Q^2_{min} values of the input data.

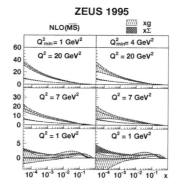

Figure 8. The gluon density (xg) and quark singlet density ($x\Sigma$) at medium and low Q^2 from ZEUS[17].

orate NLO QCD fit taking into account all experimental and theoretical systematic uncertainties [4]. A value of $\alpha_s(M_Z^2) = 0.1150 \pm 0.0017(stat.) \pm 0.0011(model) \pm 0.005(scale)$ was obtained. The largest uncertainty, arising from the choice of the renormalisation and factorisation scale, is expected to be reduced significantly once NNLO calculations become available[12]. The gluon density is constrained to $\approx 3\%$ for $10^{-3} < x < 0.1$ at $Q^2 = 20\,\mathrm{GeV}^2$. The fit describes the data well[4] down to $Q^2 \gtrsim 1\,\mathrm{GeV}^2$, but below $Q^2 = 5\,\mathrm{GeV}^2$ the resulting gluon distribution becomes sensitive to the inclusion of data at smaller and smaller Q^2 (Fig. 7). This more precise result confirms the previous finding [16,17] that the role of the gluon and quark density change when going to low values of Q^2 (Fig. 8). At low x the gluon density dominates the proton structure for $Q^2 \gg 1\,\mathrm{GeV}^2$, but tends to vanish at $Q^2 \approx 1\,\mathrm{GeV}^2$. The parton density functions appear to be flexible enough for the NLO DGLAP fit to accommodate the inclusive F_2 data down to $Q^2 \approx 1\,\mathrm{GeV}^2$, and also describe the longitudinal structure function F_L [4]. Nevertheless this "valence"-like behaviour of the gluon density is likely to signal the limit of applicability of the perturbative series in DGLAP. It might imply that higher twist effects or new dynamics such as described by the BFKL equation become sizeable. More exclusive data would be highly welcome in this low x region [19].

The Low x and Low Q^2 Limit

Since the CMS energy at HERA is limited to $\sqrt{s} = 320\,\mathrm{GeV}$ kinematics imply that values of $x < 10^{-4}$ are accessible only at *very* low $Q^2 < xs$. Both H1 and ZEUS have equipped the region close to the beam axis with small calorimeters and silicon trackers which are able to measure down to $x \approx 10^{-6}$ for $Q^2 \gtrsim \Lambda_{QCD}^2$ (Fig. 9). New data from ZEUS [2] complement previous measurements

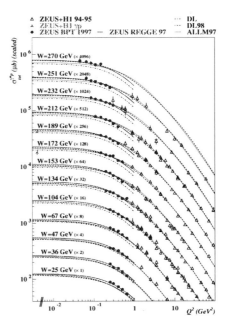

Figure 9. The total cross-section for γ^*-proton scattering as a function of Q^2 for constant values of W from HERA[2]. Also shown are two measurements for $Q^2 \approx 0$, from [20].

on $\sigma_{\gamma^* p} \sim F_2/Q^2$ and allow interpolation between the steep fall of the cross-section at high Q^2 and the photoproduction region at $Q^2 \approx 0$, where the cross-section must become independent of Q^2 because of conservation of the electromagnetic current. In other terms, the photon in the limit $Q^2 \to 0$ fluctuates into a hadronic object of similar size as the proton, such that the "dipole"-proton cross-section becomes independent of the exact value of Q^2 [b]. The precise form of $\sigma_{\gamma^* p}$ has been subject to considerable discussions[18,8] about the onset of a possible recombination (or saturation) of the gluon density in this region of low x and high gluon density. Fig. 10 shows slopes $(\partial F_2/\partial \log Q^2)_{\mathrm{fixed}\,x}$, which to first approximation are directly proportional

---

[b]Unfortunately this constant behaviour of the dipole cross-section in the limit $Q^2 \to 0$ is sometimes referred to as "saturation", which is not the same as saturation of parton densities in the proton at higher Q^2 due to recombination effects.

to the gluon density. For all Q^2, and es-

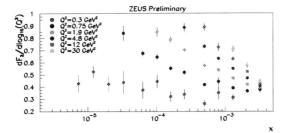

Figure 10. The derivative $(\partial F_2/\partial \log Q^2)_{\text{fixed } x}$, as a function of x in bins of Q^2 based on data from fixed target experiments and from ZEUS [2].

pecially $Q^2 \gtrsim 1$ GeV^2 where the picture of a proton resolved into partons is applicable, this slope rises linearly towards low x. No deviation from this behaviour is visible in the energy range accessible at HERA[c]. In summary the inclusive HERA data do not provide evidence for saturation effects of parton densities for Q^2 above a few GeV^2. At smaller Q^2 where the photon itself develops hadronic structure, low x effects are difficult (if not impossible) to disentangle from low Q^2 effects.

3 The Photon as a Hadronic Object

A quasi-real photon of very small virtuality ≈ 0 not only couples to other particles directly as a gauge boson or as a $q\bar{q}$ dipole of small transverse size (the point-like component which is calculable in pQCD) but can also fluctuate into a hadron-like object of large transverse size. The hadronic structure of the photon can be measured in the processes[21,22]:

- $e^+e^- \to e^+e^- \gamma^*\gamma \to e^+e^- X$ where the virtual γ^* resolves the structure of the real γ (Fig. 11). This process is mainly

[c]Note that the data shown here is identical to that in a much debated figure where the same slope is shown in bins of W. Due to the kinematic relation $W^2 = Q^2(1-x)/x$ this figure however shows a peak at Q^2 values of a few GeV^2, which should not be interpreted as saturation of parton densities.

sensitive to the quark densities in the γ. The gluon density in the photon is only accessible indirectly via the scale dependence which requires very precise data.

- $ep \to e\gamma p \to e + jets + X$ at HERA where jets with large transverse energy are required to resolve the γ structure. Since coloured partons of the proton enter the hard interaction, this process is sensitive directly to both the quark and gluon density of the γ.

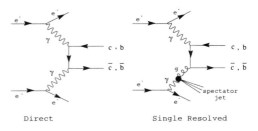

Figure 11. Example for a direct (left) and single resolved (right) $\gamma\gamma \to$ hadrons process in e^+e^- interactions, shown here for heavy quarks.

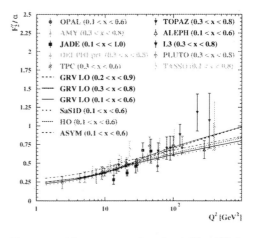

Figure 12. The structure function $F_2^\gamma(x, Q^2)$ for real photons from e^+e^- scattering[21,23] as a function of the scale Q^2 at which the γ is probed. Note that Q^2 here denotes the virtuality of the γ^* which probes the quasi-real photon of virtuality ≈ 0.

The LEP experiments have recently made substantial progress in the understanding of

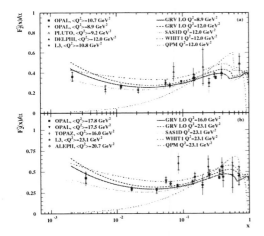

Figure 13. The photon structure function F_2^γ from e^+e^- scattering[24] at low and high Q^2 as a function of x. The QPM line which vanishes at low x corresponds to the quark-parton model approximation to the point-like component.

both the simulation of the hadronic final state and the detector response close to the beam direction. Improved unfolding methods have led to much more precise measurements and also to better consistency between the experiments. The LEP II data now are superior to all previous data from e.g. the PETRA experiments, and in addition give access to the photon structure at much larger scales and at lower x. Fig. 12 and 13 show that the photon structure function F_2^γ is now known with a precision of $\approx 10\%$. The basic expectations for the behaviour of F_2^γ are:

- F_2^γ is dominated at high x by the point-like part.

- F_2^γ rises with Q^2, in contrast to the proton case, for all x_γ due to the point-like contribution.

- At low x the hadron-like component is expected to dominate and the photon becomes similar to the proton, i.e. F_2^γ rises strongly towards low x and the photon is dominated by gluons.

The first two points are clearly borne out in the data shown in Fig. 12 and 13 when comparing with the expectation for the point-like

component. The hadron-like component is seen at low x as the data clearly exceed the point-like part, although the expected rise of F_2^γ at very low x is not significant in the accessible x range. Note that existing parameterisations of the γ structure like GRV(LO) describe the data well for $Q^2 > 5$ GeV^2. Heavy flavour production data from LEP[25] are shown in Fig. 14. For charm produc-

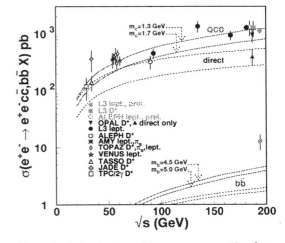

Figure 14. The charm and bottom cross-section from $\gamma^*\gamma$ scattering at LEP[25].

tion the data agree well with QCD expectations based on the same parameterisations. A first measurement of the charm structure function $F_{2,c}^\gamma$ has been obtained by OPAL[26] (Fig. 15). The bottom cross-section as mea-

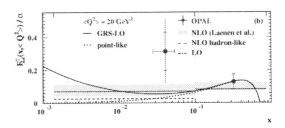

Figure 15. The charm structure function $F_{2,c}^\gamma$ from OPAL[26].

sured by L3 [25] however is larger than expected (Fig. 14), a very interesting observation as the same trend is observed also by the Tevatron experiments and in photoproduction at HERA. These observations require the theoretical description of heavy flavour production processes in hadronic collisions to be reconsidered.

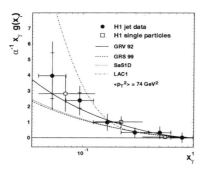

Figure 16. The gluon density in the photon as measured by H1[28].

Further access especially to the gluon component in the γ is obtained from jet production at HERA. In an analysis tailored towards the low x region, the H1 experiment has used data at relatively low transverse jet energies $E_T > 6$ GeV which was corrected for the substantial effects of secondary interactions between the photon remnant and the proton remnant in resolved photon processes. Subtracting the direct and quark induced parts based on expectation from e^+e^- data, the gluon distribution in the γ is extracted (Fig. 16). Albeit only in LO this is the only experimental evidence for a rise of parton densities in the photon at low x.

A complementary analysis by ZEUS[29] uses jets at much larger E_T, which kinematically excludes the low x region. Here the experimental and theoretical uncertainties are smaller however. In the range $0.3 < x < 0.8$ the ZEUS data are consistent with the NLO calculations for $14 < E_T < 17$ GeV, but exceed the calculations for larger E_T (Fig. 17). A similar effect was seen in a first jet measurement in e^+e^- scattering by OPAL [30].

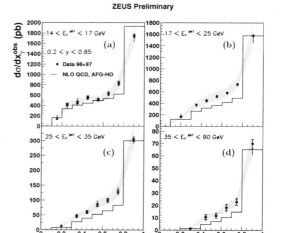

Figure 17. Jet cross-section in γp interactions from ZEUS[29].

Note that this poses a question on the overall consistency of the photon structure data. The LEP F_2^γ data, which are sensitive mainly to quarks up to scales of $\lesssim 800$ GeV^2, as well as the H1 data on gluons and quarks at low x, agree with e.g the GRV parameterisation. The ZEUS jet data, also sensitive to quarks and gluons but at high x and high E_T where the point-like component should dominate and little freedom due to the hadron-like component is expected, indicate an increased scale dependence of the parton densities, an effect which is difficult to understand. Improved precision of the data in an extended x and Q^2 range calls for a new effort in the understanding of the photon structure in NLO QCD. Whether parton densities can be derived which consistently describe all data sets within errors remains to be seen.

4 QCD Dynamics at Low x

In spite of the fact that the inclusive proton structure function F_2 is compatible with the DGLAP evolution equations even at the lowest accessible x values (for $Q^2 \gtrsim 1$ GeV^2), it is still expected that $\ln 1/x$ terms

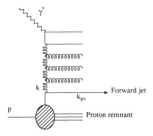

Figure 18. Parton ladder with a hard "forward" jet close to the proton direction.

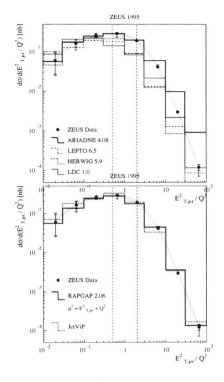

Figure 19. Cross-section for forward jet production from ZEUS[34] in comparison to calculations based on: upper figure: DGLAP (LEPTO and HERWIG) and colour-dipole model (ARIADNE); lower figure: resolved virtual photons in LO (RAPGAP) and NLO (Jetvip).

must become sizeable in comparison with the $\ln Q^2/\Lambda^2_{QCD}$ terms if only x is small enough. Considerable effort is therefore expended at HERA, as well as at LEP and Tevatron, into the investigation of more exclusive processes [32,31]. The advocated[33] test case at HERA is the production of "forward jets" (Fig. 18), i.e. a jet close in rapidity to the proton. In such events, when the jet transverse energy E_T is comparable to the photon virtuality at the other end of the parton ladder, k_T ordered radiation should be suppressed and parton emission might dominate which is not ordered in k_T.

Fig. 19 shows the cross-section for forward jets as a function of E_T^2/Q^2 from ZEUS[34]. None of the DGLAP based calculations are able to explain the data everywhere. Calculations based on structure functions for *virtual photons* have been available since several years, where for the case $E_T^2 \gg Q^2$ the γ^* is assumed to be resolved by the E_T of the jets. Starting from the highest E_T somewhere along the ladder, two parton cascades are then evolved towards the photon and the proton. As this corresponds to $non - k_T$ ordering, it may be viewed as an approximation to new QCD dynamics such as those predicted by the BFKL or CCFM equations. Calculations based on resolved virtual photons are indeed able to explain the data both in the DGLAP regime at $E_T \ll Q^2$ or $E_T \gg Q^2$ and in the BFKL regime at $E_T \approx Q^2$. Apart from these data only weak

evidence for BFKL type dynamics exists up to now [31,32].

Similar to the forward jet case, deep inelastic production of charm is a two-scale process and thus a test-case for effects beyond k_T ordering. Fig. 20 shows the charm production cross-section in comparison to DGLAP and CCFM based calculations. Note that the calculation makes use of unintegrated parton densities $f(x, k_T^2, Q^2)$ which are obtained from a fit to the H1 F_2 data[35]. While deviations from the data are present in both approaches, the CCFM calculation does better especially in the forward direction at large pseudorapidities η close to the proton. This test of CCFM dynamics looks promising also in details of the final state.

H1 preliminary

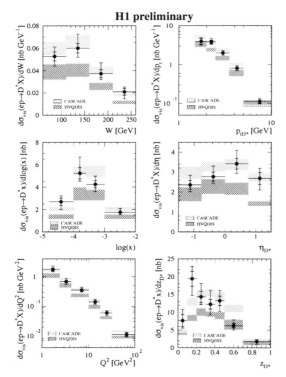

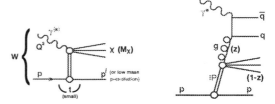

Figure 21. Feynman diagrams for diffractive scattering in $\gamma^* p$ collisions at HERA; inclusive scattering (left) and di-jet production (right).

Figure 20. Cross-section for D^* mesons from H1[36] as a function of W, x and Q^2 together with the transverse momentum P_{T,D^*}, pseudorapidity η_{D^*} and momentum fraction z of the D^*. HVQDIS and CASCADE are NLO DGLAP and CCFM based calculations, respectively.

5   Diffraction

Processes in which a proton is scattered elastically are a challenge to pQCD calculations as they must proceed via a colour singlet exchange, in contrast to the standard approximation of single quark or gluon exchange. These processes are of fundamental interest as, in the end, they address the nature of colour confinement in QCD[38].

From *soft* hadronic processes it is known that beyond the expected exchange of photons and mesons, an additional component must be present, which can not be associated to any known particle[d]. This colour singlet

exchange has generally been assumed to be dominated by gluons, however its precise nature remained unclear.

The interest in diffraction was renewed when *hard* diffractive processes were observed in $p\bar{p}$ collisions[39] and ep collisions[40], which provide the resolution to resolve the partonic structure of the colour singlet exchange.

Hard Diffraction at HERA

Fig. 21 shows the diagram for deep inelastic, inclusive diffractive scattering. The corresponding cross-section depends on four kinematic quantities: x, Q^2, the momentum transfer squared t at the proton vertex and the momentum fraction $x_{I\!P}$ of the colour singlet relative to the proton.

Important theoretical progress was recently achieved by the proof of QCD hard scattering factorisation[41] for diffractive ep scattering[e], which states that the cross-section factorises into hard partonic cross-sections and universal diffractive parton distributions $f_i^D(x, Q^2, x_{I\!P}, t)$. The latter should evolve (at fixed $x_{I\!P}$ and t) according to the DGLAP evolution equations. This proof puts diffraction on a solid basis for a treatment in pQCD and experimental verification is highly desirable.

For inclusive diffractive scattering the HERA experiments usually integrate over t

---

[d]In the framework of Regge theory[37] this exchange was labelled "pomeron" ($I\!P$).

[e]Note that factorisation is expected to break down in $p\bar{p}$ collisions due to secondary interactions between the two protons.

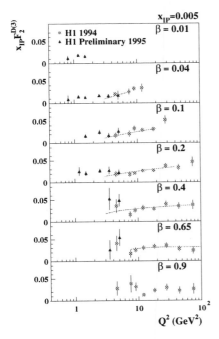

Figure 22. The diffractive structure function $F_2^{D(3)}$ from H1[42,43].

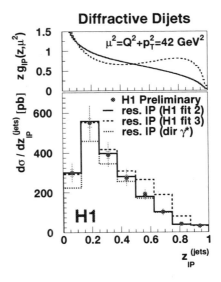

Figure 23. Diffractive di-jet cross-section from H1[44] (bottom) in comparison to predictions based on gluon densities (top) extracted by two QCD fits to the H1 $F_2^{D(3)}$ data. Here $z_{I\!P}$ corresponds to β for inclusive diffractive scattering.

and present their result as a structure function $F_2^{D(3)}(\beta, Q^2, x_{I\!P})$ (Fig. 22), where $\beta = x/x_{I\!P}$ corresponds to the momentum fraction of the struck parton in the colour singlet exchange. The data[42] show positive scaling violations up to large β, indicating the dominance of gluons in the diffractive exchange.

Assuming in addition Regge factorisation[37], $F_2^{D(3)}$ can be written as a flux factor for the colour singlet $f_{(x_{I\!P}, t)}$ times a structure function for the exchange $F_2^D(\beta, Q^2)$. QCD fits of the scaling violation in $F_2^D(\beta, Q^2)$ yielded gluon densities[42] as shown in Fig. 23 (top).

A direct measure of the gluon distribution can be obtained from di-jet production in diffraction (Fig. 21). The new data from H1[44] (Fig. 23) are compatible with both Regge and QCD factorisation and nicely confirm the previous $F_2^{D(3)}$ analysis which assumed Regge factorisation. A new, QCD based calculation of this diffractive cross-

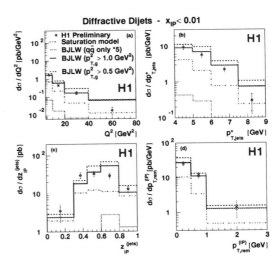

Figure 24. Diffractive di-jet cross-section from H1[44] in comparison to calculations (BJLW) based on the exchange of 2 gluons which interact dominantly with a $q\bar{q}g$ fluctuation of the γ.

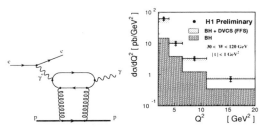

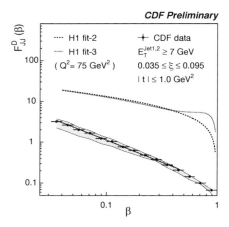

Figure 25. Feynman diagram and cross-section for Deeply Virtual Compton Scattering (DVCS) from H1[46]. The process interferes with the Bethe-Heitler (BH) process $ep \to e\gamma p$ (where the γ is emitted from the electron line) for which the contribution is shown separately.

Figure 26. The diffractive structure function extracted from jet data by CDF[48] in comparison to an expectation based on the H1 diffractive parton densities.

section, which assumes dominant exchange of two gluons interacting with the $q\bar{q}(g)$ system emitted by the virtual photon (c.f. Fig. 21), leads to a reasonable description of the data at small $x_{I\!P}$ (Fig. 24). Again unintegrated gluon densities are employed here. Similar calculations have also become available for the diffractive production of vector-mesons (Fig. 4) and for Deeply Virtual Compton Scattering $\gamma^*p \to \gamma p$ (DVCS) (Fig. 25). It is remarkable that perturbative calculations are now able, with only a few free parameters, to describe a number of hard diffractive processes in ep scattering. It would be of high interest to complement these data with measurements at high t where the colour-singlet itself might be calculable in pQCD.

Hard Diffraction at the Tevatron

Both D0 and CDF have investigated processes where jets with large E_T are employed to study the partonic structure of the diffractive exchange. The most striking observation is that the overall rate of diffractive processes is much smaller (by factors 5 to 20[47]) in comparison to the findings at HERA, where they contribute as much as 10% to the DIS cross-section.

Fig. 26 shows the effective structure function from jets F_{jj}^D from CDF for events where the elastically scattered proton was tagged

at very small scattering angles. The data are far below a calculation which is based on the H1 parton densities extracted from $F_2^{D(3)}$ (Fig. 23). QCD factorisation obviously is badly broken for diffractive $p\bar{p}$ collisions. The same conclusion is obtained using only Tevatron data from the ratios of double- to single and single to non-diffractive cross-sections, which are significantly different (Fig. 27). It is noted that the QCD fac-

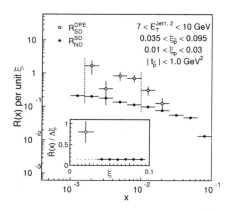

Figure 27. Ratios of double to single diffractive cross-sections R_{DPE}^{SD} and single to non-diffractive cross-sections R_{SD}^{ND} from CDF[49].

torisation proof for ep scattering does not apply to $p\bar{p}$ collisions. The large factorisation breaking observed when extrapolating from HERA results to Tevatron processes reflects that a point-like virtual photon (or a $q\bar{q}$ fluctuation of small transverse size) is able to pass through the proton without destroying it, whereas two large proton remnants will destroy each other in most cases. While this prohibits at present an interpretation of the $p\bar{p}$ data in terms of universal diffractive parton densities, the mechanism of factorisation breaking is in itself of interest and a challenge for the understanding of diffraction in hadron-hadron collisions[50].

6 Conclusion

Among the most active fields of QCD are the regions where perturbative approaches are difficult, namely at large distances close to the confinement limit, at high energies or low x, in regions of large parton densities and where multi-parton exchange becomes crucial. The entire field is driven by the availability of very precise data which are needed for detailed tests of QCD.

At low $x < 10^{-3}$ the structure function data from HERA constrain the gluon density with a precision of better than 3% at high Q^2, which paves the way for significant tests of the Higgs sector at the Tevatron/LHC. While conventional DGLAP evolution describes the inclusive data down to $Q^2 \gtrsim 1$ GeV, more exclusive measurements of the hadronic final state indicate the need for calculations beyond strong k_T ordering. Evidence for BFKL effects from HERA, LEP or Tevatron are still weak, in spite of the fact that theoretical uncertainties seem to be better controlled[31]. The first CCFM calculations based on parton densities unintegrated in k_T look promising when compared with HERA data on charm production.

The LEP and HERA experiments have provided data on the photon structure with much improved precision. As existing photon parton densities seem not to be sufficient to describe all measurements a new effort in the understanding of the photon structure in NLO QCD is required.

In diffraction the comparison between hard inclusive and exclusive processes at HERA has led to a consistent picture of the structure of diffractive colour singlet exchange which is dominated by gluons. A challenge here is the application of the HERA results to hadron collisions, where QCD factorisation in diffraction is shown to be broken.

Acknowledgements

It is a pleasure to thank the organizers, Y.Yamazaki and the technical staff in Osaka for their help, J. Dainton, E. Elsen, B. Foster, M. Klein, P. Newman and R. Nisius for useful comments to the manuscript and many others for valuable discussions.

References

1. R. Nania, these proceedings.

2. A. Pellegrino, these proceedings; also [3].

3. M. Klein, proceedings Lepton-Photon 1999, SLAC.

4. F. Zomer, these proceedings.

5. H1 Coll., *Nucl.Phys.* **B407** (1993) 515; ZEUS Coll., *Phys.Lett.* **B316** (1993) 412.

6. B. Mellado, these proceedings.

7. B.L.Ioffe, *Phys.Lett.* **30B** (1969) 123.

8. K.Golec-Biernat, M.Wüsthoff, *Phys.Rev.* **D59** (1999) 014017; M.McDermott, hep-ph/9912547, and references therein.

9. S. Catani et. al., hep-ph/0005025, hep-ph/0005114.

10. A.Gurtu, these proceedings; P. Igo-Kemenes, these proceedings, and talk at the LEPC Nov.3rd, 2000.

11. Y. Dokshitzer, *Sov. Phys. JETP* **46** (1977) 641; V. Gribov, L. Lipatov, *Sov. J. Nucl. Phys.* **15** (1972) 438 and 675; G. Altarelli, G. Parisi, *Nucl.Phys.*

B126 (1977) 298.

12. W.L. van Neerven, A. Vogt, hep-ph/0006154; A.D. Martin et. al., hep-ph/0007099.

13. E.Kuraev, L.Lipatov, V.Fadin, *Sov. Phys. JETP* **45** (1977) 199; Y. Balittski, L. Lipatov, *Sov. J. Nucl. Phys* **28** (1978) 822; L. Lipatov, *Sov. Phys. JETP* **63** (1986) 904.

14. M.Ciafaloni, *Nucl.Phys.* **B296** (1988) 49; S. Catani, F. Fiorani, G. Marchesini, *Phys.Lett.* **B234** (1990) 339, *Nucl.Phys.* **B336** (1990) 18; G. Marchesini, *Nucl.Phys.* **B445** (1995) 45.

15. M. Ryskin et. al., *Z. Phys.* **C76** (1997) 231; L. Frankfurt et. al., *Phys.Rev.* **D57** (1998) 512.

16. H1 Coll., contr. paper ICHEP 1997, Jerusalem.

17. ZEUS Coll., *Eur.Phys.J.* **C7** (1999) 609.

18. B. Foster, hep-ex/0008069, and references therein.

19. U. Maor, these proceedings.

20. S. Levonian, these proceedings.

21. R. Nisius, *Phys.Rept.* **332** (2000) 165.

22. M. Erdmann, Springer Tracts in Modern Physics 138 (1996), DESY-96-090.

23. R. Nisius, private communication.

24. OPAL Coll., , CERN-EP-2000-82.

25. A. Böhrer, these proceedings.

26. OPAL Coll., *Eur.Phys.J.* **C16** (2000) 579.

27. H1 Coll., *Phys.Lett.* **B467** (1999) 156.

28. H1 Coll., *Phys.Lett.* **B483** (2000) 36.

29. ZEUS Coll., *Eur.Phys.J.* **C11** (1999) 35.

30. Th. Wengler, these proceedings.

31. L.H. Orr, these proceedings.

32. H1 Coll., *Phys.Lett.* **B462** (1999) 440; D0 Coll., B.Pope, these proceedings; L3 Coll., M.Wadhwa, these proceedings.

33. A. Mueller, *Nucl.Phys. (Proc.Suppl.)* **18C** (1991) 125, *J.Phys.* **G17** (1991) 1443; J. Kwiecinski, *Eur.Phys.J.* **C9** (1999) 611.

34. ZEUS Coll., *Phys.Lett.* **B474** (2000) 223.

35. H. Jung, DIS workshop 1999, hep-ph/9905554; S.Baranow, H.Jung, N. Zotov, hep-ph/9910210.

36. E. Tzamariudaki, these proceedings.

37. P.D.B. Collins, *An Introduction to Regge Theory and High-Energy Physics*, Cambridge 1977.

38. J.D.Bjorken, hep-ph/0008048; J.Bartels, H.Kowalski, hep-ph/0010345.

39. UA8 Coll., *Phys.Lett.* **B211** (1988) 239; *Phys.Lett.* **B297** (1992) 417; *Phys.Lett.* **B421** (1998) 395.

40. ZEUS Coll., *Phys.Lett.* **B315** (1993) 481; H1 Coll., *Nucl.Phys.* **B429** (1994) 477.

41. J. Collins, *Phys.Rev.* **D57** (1998) 3051, erratum.

42. H1 Coll., *Z. Phys.* **C76** (1997) 613.

43. H1 Coll., contr. paper to ICHEP 1998, Vancouver.

44. H1 Coll., F.P. Schilling, proceedings DIS 2000, Liverpool.

45. ZEUS Coll., contr. paper to EPS99, Tampere.

46. H1 Coll., R. Stamen, proceedings DIS 2000, Liverpool; L. Favart, these proceedings.

47. A. Sznadjer, these proceedings.

48. CDF Coll., *Phys.Rev.Lett.* **84** (2000) 5043.

49. CDF Coll., *Phys.Rev.Lett.* **85** (2000) 4215.

50. F. Hautmann, D.E. Soper, *Phys.Rev.* **D63** (2000) 011501; V. Khoze, A. Martin, M. Ryskin, hep-ph/0007083; and references therein.

Plenary Session 14

Future Accelerators

Chair:	Michel Davier (LAL-Orsay)
Scientific Secretary:	Satoru Yamashita

FUTURE ACCELERATORS

M. TIGNER

Newman Laboratory, Cornell University, Ithaca, NY 14853
E-mail: mt52@cornell.edu

The pace of advance of the energy frontier has slowed, requiring special attention. R&D and plans for several new initiatives give hope that the situation can be remedied. In addition, some new ways of involving the experimental particle physics community in accelerator conception, planning and R&D may be instrumental in advancing the field.

1 Introduction

The pace of advance of the energy frontier has lagged in recent years as shown by the Livingston chart of Fig. 1. We do have some cause for hope, however. There is much activity in the field of accelerators for particle physics. Much of this activity involves upgrades of existing facilities. We will mention in this survey only accelerators which are considered to be leading candidates for future frontier accelerators and which are not now in construction but rather in the R&D and planning stages. An additional sign of hope is that the broader experimental particle physics community is becoming active in the conception, planning and R&D phases of accelerators. A particularly dramatic example may be found by scanning the author list of a recent "feasibility" study for a neutrino factory based on a muon storage ring[1]. Last, but far from least in this category, is the growth of inter and intra regional collaborations in accelerator concept development and early R&D activity. In this regard, and interesting new initiative, known as the Global Accelerator Network[2], is receiving considerable attention. Using technical means for fast data transfer it is envisioned that it may be possible for collaborators in major frontier accelerator creation to share in operational responsibility from their home sites. A major challenge will be making the needed social adjustments, large among them being conveyance of the needed pride of ownership to the governments involved.

The new accelerator activity encompasses proton machines, electron-positron colliders, muon colliders and neutrino factories. In the following we mention the most visible of these and outline the status of each[3].

2 Muon Based Machines

2.1 Muon Collider

The original motivation for a muon based machine was to make a lepton collider free, to the greatest possible extent, of the radiation that plagues high energy e+e- machines[4]. In such an accelerator a multi-megawatt proton beam consisting of two, or a very few, nanosecond proton bunches impinge on a liquid metal or rotating band target to produce pions. These are collected in a very high field collecting magnet, drift and decay to muons, the muons being phase rotated, bunched and ionization cooled in subsequent beam line apparatus followed by acceleration in a series of superconducting recirculating linacs and, finally, injection into a storage ring where they circulate for up to 1000 turns, colliding in a detector and producing interactions. Technical studies of this future possibility continue.

With to the recent excitement in neutrino physics from SuperK et al, the focus has shifted. It was soon recognized that the first part of the collider, with much more modest ionization

254

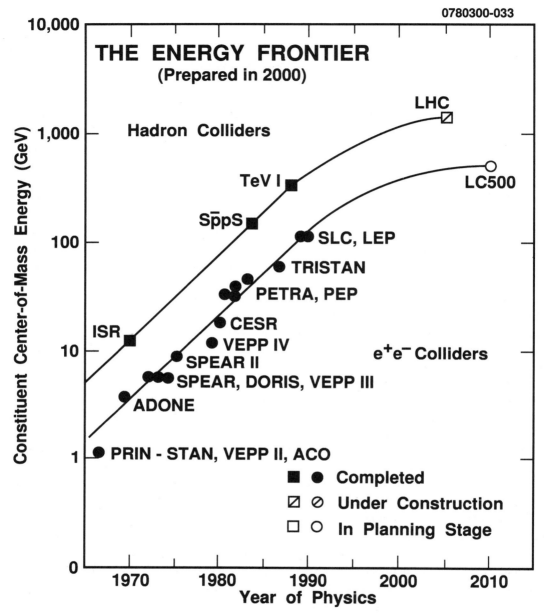

Fig. 1 Livingston chart showing facility capabilities vs time

cooling, could provide an unprecedentedly intense beam of neutrinos i.e. $10^{19} - 10^{21}$ per year launched toward detectors, by taking the muon beam out of the chain at 20 to 50 GeV and inserting it into a storage ring with very long straight sections, the straights being aimed at detectors perhaps as much as 7000 km distant through the earth. Interest in pursuing such a neutrino factory is high throughout the world community of particle physics. The major R&D issues are: measurement of the pion production cross-sections relevant to the target materials (special carbon rods or heavy metal liquids or rotating bands) at proton driver energies (2 – 24 GeV) and relevant pion energies (50 to100's of MeV) and, likewise, of precision measurements of the muon scattering and energy loss in the 100 MeV region; the high current, short bunch, proton driver; the targetry for the driver beam; ionization cooling; very high gradient acceleration with normal and super-conducting cavities at low frequency; beam instrumentation suitable for the very diffuse muon beams that will be present at the front end of these devices.

2.2 R&D in Japan

In Japan[5], plans are being made for using the 50 GeV proton synchrotron of the JHF as the proton driver. Alternatives for the subsequent collection and acceleration are a scheme similar to that being pursued in Europe and the US, to be described briefly below, as well as a scheme based on the use of a chain of FFAG accelerators without need for cooling between the pion production target and the muon decay storage ring. Already a 1 MeV FFAG model proton accelerator has been commissioned as part of the R& D program to assess this approach.

2.3 R&D in Europe

In Europe, efforts are centered at CERN and RAL[6], Here, very active work on the important cross-section measurements is now under way with the HARP experiment at CERN and the MUSCAT measurement spearheaded by RAL at TRIUMF. Data from these experiments will be available within the year. Other R&D work focuses on targetry, use of a superconducting linac as proton driver followed by rings for accumulation and phase rotation with very low frequency, normal conducting cavities for rf capture and cooling. Recirculating linacs are envisioned for the subsequent acceleration of the muons prior to their introduction into a "bow-tie" form storage ring which permits both neutrino beams to be downward directed. Design and some engineering work on these systems is underway and study of high gradient, low frequency cavity behavior in high radiation fields is being carried out.

2.4 R&D in the US

In the US a feasibility study for an entry level neutrino factory[7] has recently been completed under the aegis of FNAL: with the collaboration of the Neutrino Factory and Muon Collider Collaboration (MC) and other individuals and laboratories. A feasibility study for a "high end" neutrino factory is now being launched under the aegis of BNL with MC collaboration. A high intensity targetry experiment has been approved at BNL and is under construction with a focus on liquid metal. A complementary study featuring a high tech carbon target with complete infrastructure for dealing with the radiation challenges is also under way as is a program of R&D on ionization cooling theory, simulation and component development and diagnostic instrumentation. The overall aim is to devise a satisfactory demonstration of ionization cooling on the needed scale with appropriate boundary conditions. Other US efforts include, among other things, high field solenoids, normal and superconducting and materials for these magnets that can withstand high radiation environments. Acceleration systems are also an important part of this program. For phase rotation, induction linac modules with superconducting solenoid focusing are under study as are high gradient, 200 MHz, normal conducting cavities. It is hoped that sufficient support will be forth coming that a zeroeth order design (ZDR) can be accomplished in about three years to be followed by a full conceptual design on the scale of about five years.

3 e⁺e⁻ Linear Colliders

3.1 Four Distinct approaches

Currently four distinct approaches to the linear collider are receiving great attention. CM energies from a few hundred GeV to a few TeV are being discussed with the emphasis on ~500 GeV.

Major R&D issues for the linear colliders are: particle sources; damping rings; emittance preservation; high gradients; collimation and stability at the IP. For the X band machines klystrons and modulators are of prime importance. For the 30 GHz machine, power generation is also of prime importance.

3.2 R&D in Japan

Programs of development for linear collider apparatus at C band (~6 GHz) and X band (~11 GHz) are well underway. In addition, an accelerator test facility (ATF) houses a very low emittance damping ring driven by a 1.5 GeV S band (~3 GHz) linac.

Regarding the C band work[9] it is reported that the klystron, modulator and accelerating structure are now well developed but that the pulse compressor needs high power testing. The X band work[10] continues intensely with substantial progress in klystron, modulator, accelerating structure and pulse compressor areas. A 75 MW, 1.5 μs klystron has been demonstrated and a solid state switch, replacing the thyratron in the modulator, has been developed. Damped structures have been built and power tested. Pulse compressors have been tested as well. Further developments in gradient capability are expected. Modulator, rf pulse compression and accelerating structures are joint projects with SLAC. The klystrons are being developed in parallel efforts.

Recently the ATF[11] has demonstrated an emittance in the range needed by most of the linear collider candidates, some two orders of magnitude less than needed by the SLC. While this achievement was with a single, relatively weak, bunch whereas ultimately multiple high charge bunches are needed, this does represent a signal achievement.

3.3 R&D in Europe

In Europe, two major approaches are currently being pursued: TESLA using superconducting cavities at 1.3 GHz and CLIC utilizing 30 GHz normal conducting cavities.

TESLA[12], being developed by a consortium centered at DESY, includes an x-ray FEL as part of its features. The TESLA test facility (TTF) is now in operation at >200 MeV and is under upgrade to 1 GeV. While the specifications for the TTF included 15 MV/m as the target gradient, the third cryomodule, eight nine cell cavities in one cryogenic housing, is very close to meeting the ultimate TESLA goal of 25 MV/m. In a recent experiment the SASE operation of an FEL at ~100 nm has been demonstrated in the TTF at ~ 200 MeV. A first

multi-beam klystron delivering very close to final design parameters has been delivered and tested. A full proposal for a 500 GeV CM TESLA collider is expected in the spring of 2001.

CLIC[13], the Compact Linear Collider is being developed at CERN. Its special features are the very high frequency, 30 GHz and high gradient planned, 150 MV/m. A novel method for generating the needed power at 30 GHz is being employed. A "low" energy, ~ 1 GeV, high current beam carrying the full energy needed to power the main linacs is accelerated in a ~ 1 GHz, low frequency linac. The very short bunches in this beam are regrouped into trains of closely spaced bunches by means of an rf deflector followed by two rings of different circumference for verniering. These trains are then propagated in a direction opposite to the direction of acceleration in the main linacs and peeled off at the appropriate point and injected into 30 GHz decelerating structures coupled to the main linac by waveguides where the kinetic energy of the drive beam is converted into electromagnetic energy which subsequently drives the main linac beam. The feasibility of two beam acceleration has been demonstrated in the CLIC Test Facility (CTF) in which a beam formed in an S band linac accelerated a beam in a 30 GHz linac structure. Now under construction is an advanced version of the CTF in which the formation of the closely spaced bunch trains can be demonstrated. Completion of this demonstration is expected in 2004/5. Meanwhile, prototyping of the important components proceeds together with system design work for the full accelerator complex.

3.4 R&D in the US

The X band NLC[14] work is centered at SLAC and, as mentioned in the JLC paragraph, several aspects are being done jointly with KEK. Envisioned is a 500 GeV CM machine with attention given to extending it later into the TeV regime.

In the SLAC parallel effort on klystrons, a 75 MW, 3 μs klystron has been demonstrated. This wide pulse could result in considerable cost saving through decreasing the number of klystrons needed. Special features of the NLC

R&D program are the ASSET facility allowing direct measurement of the wakefields induced by short beam bunches, the FFTB which allows investigation of the technical problems involved in the final focus of a linear collider and the NLCTA which allows investigation of the full scale properties of accelerating structures and strategies for beam loading compensation. Most particularly, the design being developed benefits from the great reservoir of experience with the SLC.

4 VLHC

4.1 Definition

The collaborative group involved in R&D and concept development for this proton – proton collider have defined it to be a collider at 100 TeV CM with capability for about 500 pb^{-1} per day[15].

4.2 Concept Development and R&D

While the accelerator physics challenges have been examined in some depth by the interested parties, the bulk of the R&D focuses on magnet technology that has resulted in several very innovative designs[16]. A "low field" concept using a 1.8 T field with a resulting circumference of 646 km shapes the field with iron, combined function, poles, the flux being driven through both gaps by a superconducting transmission line. In this warm iron design the 75 kA transmission line is housed in a tubular cryostat. The line is a helical cable of Rutherford cables. Several "high field" concepts have been suggested with fields ranging above 12 T. A cosineθ design has been put forward and studied as have several versions of the "common coil" design which uses flat pancake coils so that Nb3Sn material can be used more easily. A 12.2 T model has been demonstrated. Photos and diagrams of the various magnets are to be found in the references.

References

1 http://www.fnal.gov/projects/muon_collider/nu/study/report/machine_report/00_fermilab_study_title+author_rev8.PDF
2 http://www.ccrn.ch link to CERN Courier, June 2000, link to "accelerators to span the globe"
3 http://ichep2000.hep.sci.osaka-u.ac.jp/ follow link to "program" then to Friday parallel sessions then to PA13, or, Wednesday plenary.
4 http://www.cap.bnl.gov/mumu/ under Status Report on muon collider R/D
5 http://www-jhf.kek.jp
6 http://muonstoragerings.cern.ch/Welcome.html
7 op cit (4)
8 http://www.cap.bnl.gov/mumu/mu_home_page.html
9 http://c-band.kek.jp
10 http://lcdev.kek.jp
11 http://lcdev.kek.jp/ATF
12 http://tesla.desy.de
13 http://cern.web.cern.ch/CERN/Divisions/PS/CLIC/Welcome.html
14 http://www-project.slac.stanford.edu/lc/nlc.html; http://lcdev.kek.jp/ISG
15 http://VLHC.org
16 http://VLHC.org link to VLHC Mini Symposium, etc.

Plenary Session 15

High Energy Heavy Ion Collisions

Chair: Michel Davier (LAL-Orsay)
Scientific Secretary: Jun-Ichi Suzuki

HIGH ENERGY HEAVY ION COLLISIONS: THE PHYSICS OF SUPER-DENSE MATTER

BARBARA V. JACAK

Department of Physics and Astronomy, SUNY Stony Brook, Stony Brook, NY, 11794, USA
E-mail: jacak@nuclear.physics.sunysb.edu

I review experimental results from ultrarelativistic heavy ion collisions. Signals of new physics and observables reflecting the underlying collision dynamics are presented, and the evidence for new physics discussed. Measurements of higher energy collisions at RHIC are described, and I give some of the very first results.

1 Introduction

High energy heavy ion collisions aim to recreate the conditions which existed a few microseconds following the big bang, and determine the properties of this super-dense matter. The density of produced hadrons is very high; at energy densities of 2-3 GeV/fm^3, the inter-hadron distance is smaller than the size of the hadrons themselves. Interactions among hadrons under such conditions are unlikely to be the same as in the familiar dilute hadron gas. QCD predicts that at sufficiently high energy density and temperature, the vacuum "melts" into numerous $q\bar{q}$ pairs.

Such matter is expected to leave the realm where quarks and gluons are confined in colorless hadrons, and form, instead, a quark-gluon plasma. The experiments explore two fundamental puzzles of QCD, namely the confinement of quarks and gluons into hadrons, and the breaking of chiral symmetry which produces mass of the constituent quarks. We aim to study experimentally the nature of deconfined matter, investigate the confinement phase transition, and determine its temperature. The chiral transition is expected to occur under similar conditions. Use of the heaviest ions maximizes the volume and lifetime of matter at high energy density, enhancing signals of new physics. Understanding the background from high energy hadronic collisions, as well as the underlying dynamics and nuclear structure is

accomplished via p+p and p+nucleus collisions in the same detectors.

Solutions of QCD on the lattice have been used to estimate the energy density required for deconfinement.[1] In calculations with three massless quark flavors, a rapid change in the energy density occurs at a critical temperature of 170 ± 10 MeV. The energy density at which the system is fully in the new phase is approximately 3 GeV/fm^3. With 2 massless and one strange quark, the critical energy density is 15% lower. Studies have shown that the mass of the $< q\bar{q} >$ condensate falls to zero, signifying restoration of chiral symmetry, at about the same temperature.

1.1 Early stage and evolution of the collision

Experimental access to information about the high energy density phase is complicated by the subsequent expansion, cooling and rehadronization of the matter. Theoretically, however, one may consider several separate stages of a heavy ion collision. Interpenetration and initial nucleon-nucleon collisions are complete in less than 1 fm/c. This is accompanied by multiple parton collisions leading, probably, to local thermal equilibration. The hot, dense matter expands longitudinally and transversely, cooling until the quarks rehadronize. The hadrons continue to interact among themselves until the system is sufficiently dilute that their mean free path exceeds the size of the collision zone. At this point, hadronic interactions cease and the system "freezes out".

Elementary nucleon-nucleon collisions have long been studied, and a wealth of data on $p - p$ and $\bar{p} - p$ collisions are in the literature. Quantitative understanding of the initial parton production in heavy ion collisions requires starting with the nucleon quark and gluon structure functions, which are now rather well known from deep inelastic e-p and from p-nucleus experiments.[2] A steep rise of F_2, the quark structure function, was discovered[2] toward low x for $Q^2 \geq 2$ GeV^2. This rise is understood to indicate the dominance of gluons, and implies very large numbers of gluon-gluon interactions when two nuclei collide at high energy. H1 at HERA has unfolded the gluon distribution from their data,[3] and finds a steep rise at small x. Following this observation, we may expect significant enhancement of gluon fusion processes, such as charm production, for example, in heavy ion collisions. The x and Q^2 regions of interest at RHIC are $x \geq 0.01$ and $Q^2 \approx$ 10-20 GeV^2.

It has been observed, however, that structure functions of quarks in nuclei differ from those of nucleons. There is a depletion of quarks in the small-x region, known as "nuclear shadowing"; this effect is expected in gluon distributions also. Shadowing is usually attributed to parton fusion preceding the hard scattering which probes the parton distribution. As the overcrowding at small x is larger in nuclei than in individual nucleons, saturation should be more evident for heavy nuclei, causing shadowing to strengthen with nucleon number. This is indeed the case, and measurements show a 20% modification is heavy nuclei at $x = 0.03$. For Au nuclei, the shadowing in this x region should be a 30 % effect. Shadowing may reduce the gluon momentum requiring corresponding enhancement in the large x region if the momen-

tum fraction of gluons is to be conserved. Such "anti-shadowing" has been predicted by Eskola and co-workers, using a DGLAP evolution.[4] The total number of charged particles produced in a heavy ion collision is sensitive to the magnitude of shadowing and anti-shadowing effects, and can be used to constrain the evolution calculations.

Even with nuclear shadowing, the density of partons after the initial hard nuclear collision is truly enormous, leading us to expect a large amount of multiple parton scattering. Such multiple scattering is already visible in proton-nucleus collisions as the Cronin effect, which hardens the pion p_T spectra above 1.5 GeV/c. The higher parton density in nucleus-nucleus collisions should drive the system toward thermal equilibrium by thermalizing mini-jets and increasing the multiplicity of soft particles. Indeed, parton cascade descriptions of the collision dynamics predict that equilibration among the partonic degrees of freedom happens within 0.3-1 fm/c in collisions at $\sqrt{s} = 200$ GeV/A.[5]

The dense medium should affect fast quarks traversing it, and in fact a medium-induced energy loss of partons is expected. As first predicted by Gyulassy and coworkers,[6] and Baier, Dokshitzer, Mueller, Peigne and Schiff,[7] the energy loss of a fast quark increases with the density of the medium, due to an accumulation of the transverse momentum transferred. The energy loss dE/dx may exceed 1 GeV/fm, and BDMPS calculated that it could reach $3/times(L/10fm)$ GeV/fm at $T = 250$ MeV, where L is the path length through the dense medium. Experimental measurements of this energy loss will thus reflect the density of the medium early in the collision.

The quark gluon plasma expands and cools, whereupon the system hadronizes. The deconfined and mixed phases are expected to last approximately 3 fm/c, after which the system becomes a dense, interacting hadron gas. Expansion continues, and the system finally becomes sufficiently dilute that the hadrons cease to interact approximately 10 fm/c after the start of the collision.[5]

Of course, these values depend strongly upon the assumptions in the models, and the boundaries between phases are not sharp in either time or space. A major experimental challenge is to determine the timescales, along with the duration of hadron emission following freezeout. The expansion velocity is accessible via interferomety; scaling longitudinal expansion ($\beta \approx 1$) along with radial expansion at approximately half the longitudinal velocity have been observed.[8]

1.2   Predicted signals of quark gluon plasma

A number of key predictions for quark gluon plasma signatures were made prior to experiments at CERN and Brookhaven. Color screening by a quark gluon plasma was predicted to suppress bound $c\bar{c}$ pairs, resulting in decreased J/ψ, ψ' and χ_c production.[9] Observation of this effect is subject to understanding final state interactions of the charmed mesons with nucleons and co-moving hadrons, which break up the bound state.

Rafelski and Mueller predicted in 1982 that production of strange hadrons should be enhanced by formation of quark gluon plasma.[10] The rate of gluon-gluon collisions rises in a hot gluon gas, thereby increasing the cross section of gluon fusion processes and production of strange and charmed quarks. An important hadronic background to this measurement is associated production of strange particles in the dense hadron gas, primarily via $\pi + N \rightarrow K + \Lambda$. Strangeness exchange also complicates the picture.

Thermal electromagnetic radiation reflects the initial temperature of the system, via quark-antiquark annihilation to virtual photons which decay to lepton pairs, and via quark-gluon Compton scattering. The

rate, proportional to T^4, should be dominated by the initial temperature, T_{init}; the shape of the spectrum will reflect this maximum temperature. Measurements are difficult because of the large lepton and photon backgrounds from hadron decays, hadronic bremsstrahlung, D meson decays and Drell-Yan pairs.

Possible observable effects of chiral symmetry restoration include modification of meson masses[11] (visible through their leptonic decays) and formation of disoriented chiral condensate domains. Such a condensate should result in modified ratios of charged to neutral pions and enhanced production of soft pions with $p_T \leq 100$ MeV/c.[12]

The predicted large energy loss of quarks traversing a very dense medium would result in "quenching" of jets,[6,7] which can be observed experimentally via the spectrum of hadrons at high p_T. Since such hadrons are dominantly the leading particles in jet fragmentation, their spectrum reflects the spectrum of quarks exiting the medium. This observable reflects the density of the medium, rather than its confinement properties, but experimental evidence for existence of a superdense medium would be most exciting.

1.3   Experimental observables

The charged particle multiplicities in high energy heavy ion collisions are enormous. At SPS energy of $\sqrt{s} = 18\text{-}20$ GeV/A, there are more than 1000 hadrons produced, while at RHIC the number is closer to 10000.

Experimentally accessible observables fall into two classes. The first characterize the system formed and ascertain that the conditions warrant a search for new physics. These observables are primarily hadronic and reflect the system late in the collision. Detailed analysis of hadrons also yields dynamical information about the collision, allowing one to extrapolate the hadronic final state back to the hottest, densest time when quark-gluon plasma may have existed.[13]

The second class of observables comprises signals of new physics. Lepton pairs and photons (*i.e.* virtual and real photons) decouple from the system early, and emerge undisturbed by the surrounding hadronic matter. Consequently, their distributions are dominated by the early time in the collision, and the rate reflects the initial temperature. Strangeness production can be detected via K, Λ and other hadrons containing strange quarks; multistrange anti-baryons are particularly promising indicators of strangeness enhancement, as it is difficult to affect their production via hadronic means.[14] In a high temperature plasma with many gluons, the gluon fusion reaction $g + g \rightarrow c\bar{c}$ should be important.[15,16,5] Measurement of semileptonic, or perhaps even fully reconstructed, decays of charmed mesons would indicate whether charm production reflects enhanced gluon fusion.

2 Results for $\sqrt{s} \leq 20$ GeV/nucleon

2.1 Energy Density

Before searching for evidence of deconfinement, we must determine whether appropriate values of energy density are in fact reached. Estimating the energy density 1 fm/c into the collision from measured quantities requires some assumptions. However, this can be done from measured production of energy transverse to the beam direction, $E_T = \Sigma E \sin \theta$. E_T reflects the randomization of the incoming longitudinal energy of the beam. For collisions undergoing scaling longitudinal expansion, the energy density may be estimated via $\epsilon = \mathrm{d}E_T/\mathrm{d}\eta \times 1/\mathrm{volume}$. The volume is given by the cross sectional area of the nucleus involved, and a length defined by the formation of particles, τ, generally taken to be 1 fm, though this is likely $\sqrt{s}$ dependent, becoming smaller for high energy collisions.

Selecting as central collisions, the few percent of the total cross section producing the largest multiplicity, one may estimate the relevant nuclear area to be $\approx 90\%$ of the total. Using the E_T value of 450 GeV at this point, as measured in Pb + Pb collisions by the NA49 collaboration [17], and $R = 1.2A^{1/3}(0.9)$, with $V = (\pi R^2)\tau$, the energy density, ϵ, is found to be ≈ 3.2 GeV/fm^3. This is sufficiently high compared to the predicted transition point, to encourage searching for signals of new physics.

2.2 Color screening

As charmed quarks are produced in the initial hard collisions, they traverse the dense matter and therefore probe its properties. The screening length is directly related to the temperature and energy density, so $c\bar{c}$ bound states of different radius will be screened under different conditions. The J/ψ, with radius of 0.29 fm and binding energy of approximately 650 MeV should be more stable than the ψ' with binding energy of 60 MeV and nearly twice the radius.

Suppression of J/ψ production has been observed by NA50.[18] In light systems, the suppression is consistent with expectations from initial and final state effects on production and binding of the $c\bar{c}$ pairs. However, in Pb + Pb collisions, the suppression is 25% greater than that expected from conventional processes. The anomalous suppression sets in for semi-peripheral Au + Au collisions, and increases in strength with the volume of the excited system.

The observed suppression has been compared with hadronic models as a function of collision centrality (determined by measurement of transverse energy).[19] The measured J/ψ to Drell-Yan ratio decreases in more central collisions, and by $E_T \approx 100$ GeV has fallen well below models which assume charmonium states are absorbed by interactions with comoving hadrons.[19] Figure 1 shows the

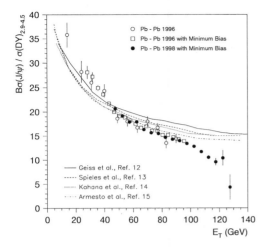

Figure 1. Comparison of NA50 measurement of the ratio of J/ψ and Drell-Yan cross section as a function of E_T (i.e. centrality) in Pb + Pb collisions with several conventional descriptions of J/ψ suppression [19].

measured ratio of J/ψ to Drell-Yan in 158 GeV/A Pb + Pb collisions, as a function of E_T; high E_T corresponds to central Pb + Pb collisions.[19] The points are data, while the curves show J/ψ production in models which assume that the charmonium states are absorbed by interactions with comoving hadrons.[20] Discussion continues within the community regarding discontinuities and thresholds, and a possible second drop of the J/ψ production at $E_T \geq 110$ GeV. Nevertheless, the data very clearly deviate from the hadronic models for central collisions.

2.3 Strangeness enhancement

Strangeness production in a heavy ion collision may not be subject to the well-known strangeness suppression observed in elementary nucleon collisions. This can be easily understood from simple energy level considerations. If the energy levels of u and d quarks are empty, it is energetically favorable to produce these light quarks, since the s quark levels have an energy gap of twice the strange quark mass. In nucleus-nucleus collisions, however, the dense matter causes the lowest u and d quark levels to be filled, resulting in a relative enhancement of strange quark number over p-p collisions.

Enhancement of several species of strange hadrons has been observed. Kaon and Λ enhancement may be expected in a dense hadron gas,[21] as a result of associated production in hadron multiple scattering, but the excess production of $\overline{\Lambda}$, Ξ and Ω observed by WA97 is not easily explained without new physics.[22,23] It is particularly remarkable that the enhancement compared to p-nucleus collisions of multiply strange antibaryons increases with the number of strange quarks. In order to achieve such production rates via hadronic equilibration, a dense hadron gas would need to live for about 100 fm/c. Given the measured expansion velocities, such a lifetime is ruled out.

2.4 Thermal radiation

Measurable rates of thermal dileptons and thermal photons were predicted by Shuryak and others.[24] The rate is proportional to the fourth power of the temperature, and is thus dominated by the initial (highest) temperature of the system. The distribution of the radiation should have an exponential shape

$$\propto e^{-M/T},$$

and the rate should depend on the square of the particle multplicity. Consequently, detection of thermal radiation depends strongly on T_{init} achieved in the collisions. Complicating both measurement and interpretation, and boosting the photon emission rate, is partonic bremsstrahlung, which also contributes.[25]

It is important to note that a hot hadron gas will radiate thermal photons and dileptons as well. Consequently, thermal radiation does not indicate the phase of the matter, but reflects its highest temperature. Experimentally, the goal is to measure the presence of real photons or dileptons, beyond those from hadronic decays, and from the yield and distributions extract T_{init}. This value is then to be compared to the expected transition temperature.

WA98 observed direct photons beyond contributions from hadronic decays in central Pb + Pb collisions at 158 GeV/nucleon at $p_T \gtrsim 2$ GeV/c.[26] A decomposition of the excess photons does not clearly show whether the p_T distribution differs from photons in p-p and p-nucleus collisions, but the enhanced yield appears consistent with $T_{init} > T_C$.[26,27]

The spectrum of lepton pairs below the ρ meson mass is of considerable interest. Though in p+Be and p+Au collisions the invariant mass distribution of electron-positron pairs below 1 GeV is well described by hadronic decays, a clear excess is observed in S + Au and Pb + Pb collisions by CERES.[28] It is difficult to reproduce the observed distribution without allowing the mass of the ρ to change or invoking a tremendous amount of collision broadening, which opens new phase space and effectively lowers the ρ mass.[29] Such observations may indicate partial chiral symmetry restoration in the dense matter created in collisions at the SPS.

It may be that the low mass dileptons arise from thermal radiation.[30] Excess dileptons in the intermediate mass range between 1 and 3 GeV have also been observed. The intermediate mass lepton pair cross section can be explained if thermal radiation is added to Drell-Yan and charm decay sources.[31] Both sets of data imply initial temperatures in the range 170-200 MeV. Both calculations make use of parton/hadron duality in the dense system to predict the thermal radiation from simple $q\bar{q}$ annihilation rates, integrating over the time evolution of the collision. Such explanations do not prove that the observed dileptons are thermal in origin, but indicate that the spectra are consistent with an initial temperature near or above the predicted phase transition temperature.

2.5 Evolution of the hadronic phase

The high density of produced particles should create high pressure in the collision, leading to rapid expansion. The expansion velocity can be extracted by combining measurements of single particle m_T distributions with two particle correlations;[13] requiring a simultaneous fit of both distributions to disentangle flow from thermal motion.

The inverse slopes of single particle spectra are given by

$$T \approx T_{freezeout} + \frac{1}{2}m_0 < v_T >^2$$

where $T_{freezeout}$ is the temperature at which the hadronic system decouples (i.e. hadron collisions cease) and $< v_T >$ is the average radial expansion velocity.

The two particle correlations measure the size of the region of hadron homogeneity (i.e. full information transport) at freezeout. Position-momentum correlations from expansion case this to be smaller than the entire hadron gas volume. Large statistics are needed for 3-dimensional analysis of the correlation functions, binned in m_T of the particles. The results of such analyses follow approximately

$$R_T^2 = \frac{R^2}{1 + \xi \frac{m_T}{T_{freezeout}} < v_T^2 >}$$

The freezeout temperature is approximately 100 MeV and the average radial expansion velocity 0.5 c. The data indicate that the system expands by a factor of 3 radially while undergoing a scaling expansion longitudinally. Back-extrapolation from the freezeout conditions, combined with the measured transverse energy yields energy densities of 2-3 GeV/fm^3.

If the radial expansion velocity, v_T, indicates pressure created in the collision, v_T should increase with the number of produced particles. Burward-Hoy[32] performed a global study of v_T with system size by analyzing single particle spectra below $m_T = 1$ GeV/c; correlation functions are only available for

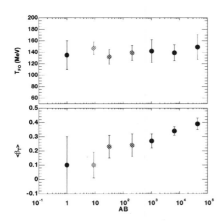

Figure 2. Dependence of freezeout temperature and expansion velocity extracted from π, K, and p spectra below $m_T = 1$ GeV on colliding system size [32].

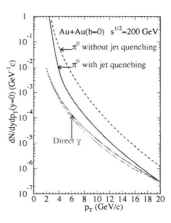

Figure 3. p_T distribution for π^0 with and without parton energy loss as compared to direct photons in central Au + Au at $\sqrt{s} = 200$ GeV/nucleon. $dE/dx = 1$ GeV/fm was used [33].

a small subset of projectile-target combinations. The radial expansion velocity indeed increases with system size, as can be seen in Figure 2, which shows the extracted freezeout temperature and v_T as a function of the number of possible nucleon-nucleon collisions ($A_{projectile} \times A_{target}$). The apparent freezeout temperature is 140 MeV, approximately independent of system size. Radial expansion in the large colliding systems boosts the particles, thus soft physics processes reach larger p_T than in elementary collisions. Consequently, observations of hard scattering will require higher p_T. Burward-Hoy extrapolated the soft spectrum using the T_{FO} and v_T parameters and found that for Pb + Pb at CERN, hard scattering is only a partial contribution to the spectrum below $p_T \approx 5$ GeV/c.[32]

2.6 Quark energy loss

Figure 3 shows predictions by X.N. Wang of the effect of quark energy loss on the single particle p_T spectrum.[33] At sufficiently high p_T, where the spectrum is dominated by leading particles from jet fragmentation, energy loss or jet quenching will decrease the yield

of particles by lowering the energy of the fragmenting jet. Comparing the solid and dashed π^0 curves indicates that the difference could be easily measurable already at $p_T = 4$ GeV/c. The lower pair of curves illustrates the small difference expected in the direct photon p_T distribution, indicating that another effect of jet quenching will be to increase γ/π^0.

At CERN, WA98 measured the π^0 spectrum to nearly 4 GeV/c and did not observe any evidence of jet quenching.[34] However, as discussed above, the soft physics likely still contributes significantly at this p_T, masking energy loss effects. As the cross section for hard processes will be considerably larger at RHIC, the spectra should be measurable to considerably higher p_T. Jet quenching will thus be a very important observable at RHIC.

3 Prospects for RHIC

The Relativistic Heavy Ion Collider (RHIC) at Brookhaven National Laboratory began operation in summer 2000. RHIC collided Au beams at $\sqrt{s} = 130$ GeV per nucleon, which will be increased to 200 GeV per nucleon in the next run. RHIC will also collide smaller nuclei, protons on nuclei, and two polarized proton beams at $\sqrt{s}$ up to 500 GeV. The design luminosity is $2 \times 10^{26}/\text{cm}^2/\text{sec}$ for Au + Au, $10^{31}/\text{cm}^2/\text{sec}$ for p + p and $10^{29}/\text{cm}^2/\text{sec}$ for p + A. Luminosity achieved during the first run reached 10% of design value.

Higher energy and long running time at RHIC will allow in-depth investigation of the currently tantalizing observables. With the factor ten increase in center of mass energy, every collision should be well above the phase transition threshold. The initial temperature can be expected to significantly exceed estimates of T_C. Furthermore, hard processes which provide probes of the early medium have considerably higher cross sections. Consequently, experiments will be able to measure J/ψ and other hard processes to higher p_T with better statistical significance than before. Charged and neutral pion spectra at $p_T \geq 5 GeV$ to look for evidence of jet quenching will be accessible, and the γ/π^0 ratio will reach the p_T range where direct photon yields are calculable.

3.1 Experiments at RHIC

To cover the full range of experimental observables, RHIC has a suite of four experiments. There are two large and two small experiments, each optimized differently. Together, they form a comprehensive program to fully characterize the heavy ion collisions and search for all the predicted signatures of deconfinement and chiral symmetry restoration.

Each experiment is outfitted with two zero-degree calorimeters of identical design. These calorimeters measure neutral particles produced at zero degrees, allowing a common method of selecting events according to centrality. An event sample with interesting behavior observed by one experiment can therefore be checked by the other experiments. Many of the hadronic observables are mea-

sured by two or more of the experiments, so a complete picture of the collisions at RHIC will be investigated.

The two small experiments, PHOBOS and BRAHMS, focus on difficult-to-measure regions of rapidity and p_T. PHOBOS is optimized to measure and identify hadrons at very low p_T and fits on a (large) table top. The low p_T capability provides good sensitivity to formation of disoriented chiral condensates. In addition, PHOBOS has a full coverage multiplicity measurement, allowing analysis of fluctuations and selection of events with unusually numerous particles. Particle tracking is done primarily with highly granular silicon detectors, allowing very short flight paths and minimizing decay of the low p_T hadrons. Identification is accomplished by time-of-flight measurements.

BRAHMS maps particle production over a wide range of rapidities with good p_T coverage. BRAHMS has two movable, small acceptance spectrometers to sample the particle distributions; a typical event has only a few particles in each spectrometer. Tracking is provided by modest size time projection chambers and drift chambers. Particle identification is performed via time-of-flight measured by scintillator hodoscopes and via gas Cherenkov threshold counters.

The two large experiments each measure many of the predicted QGP signatures, along with observables to map the hadronic phase. STAR has maximum acceptance for hadrons, allowing event-by-event analyses of the final state and reconstruction of multi-strange hadron decays. PHENIX is optimized for photon and lepton detection and has high rate capability and selective triggers to collect statistics on rare processes.

STAR consists of a large acceptance time projection chamber, covering full azimuth over two units of rapidity centered around mid-rapidity (90 degrees in the laboratory). The TPC sits in a solenoidal magnetic field. In the second and third years of RHIC running, a silicon vertex tracker and electromagnetic calorimeters will be added to improve the efficiency of finding secondary vertices and to allow measurement of jets. In the first year, STAR had a partial acceptance ring-imaging Cherenkov counter to identify a subset of the particles and trigger on high p_T hadrons. STAR events include dense information on each of many charged particle tracks and are consequently very large; approximately one event per second is written.

PHENIX has multiple subsystems to track, identify, and trigger on leptons, photons, and hadrons. At midrapidity, there is an axial field magnet with two detector sectors, each covering 90 degrees in azimuth. Drift, pad, and time expansion chambers provide tracking, scintillator hodoscope time-of-flight detectors for hadron identification, a large ring-imaging Cherenkov counter identifies electrons, and a highly granular electromagnetic calorimeter is used for electron and photon identification and triggers. Charged particle multiplicity and fluctuations are measured with silicon detectors. Forward and backward, PHENIX has two cone-shaped magnets outfitted with cathode strip detectors for tracking and Iarocci tubes interleaved with steel plates for muon identification. The pole tips of the central magnet absorb approximately 90% of the hadrons. PHENIX began running with the central arms and silicon detectors, with muon measurements commencing in 2001.

3.2 First Results

Figure 4 shows a central Au + Au collision at $\sqrt{s} = 130$ GeV/nucleon, recorded in the STAR TPC.[35] The large number of tracks illustrates the challenges for the experiments. All experiments reconstruct tracks with good efficiency. STAR has demonstrated successful particle identification via dE/dx in these collisions. PHENIX, with excellent granularity ($\Delta\eta = \Delta\phi = 0.01$) and resolu-

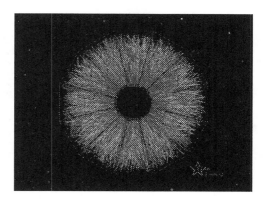

Figure 4. Display of a central Au + Au collision at $\sqrt{s} = 130$ GeV/nucleon in the STAR time projection chamber [35].

tion ($\approx 8\%/\sqrt{E}$) calorimetry, reconstructs π^0 and transverse energy distributions from such events with high particle multiplicity.

The PHOBOS Collaboration has measured the charged particle rapidity density at midrapidity for the 6% most central Au + Au collisions. They find $dN/d\eta = 555 \pm 12(\text{stat.}) \pm 35(\text{syst.})$ at $\sqrt{s} = 130$ GeV/nucleon.[36] The importance of this first measurement can be appreciated by looking at the variation in predicted particle multiplicity for $\sqrt{s} = 200$ GeV/nucleon in the literature[37] and for $\sqrt{s} = 130$ GeV/nucleon in Figure 5.[37] The range of predictions is almost a factor of 2! Three important factors control the total number of charged particles produced at midrapidity: parton multiple scattering (which increases the multiplicity), nuclear shadowing (the effect is very sensitive to the x and Q^2 dependence), and energy loss in the dense medium (energy loss tends to increase the number of soft particles at the expense of p_T in the tail of the distribution).

The PHOBOS result shows excellent agreement with the HIJING model of Wang using quark energy loss dE/dx = 1 GeV/fm, gluon dE/dx = 0.5 GeV/fm and nuclear shadowing taken from lower $\sqrt{s}$ measurements.[33] The anti-shadowing prescription of Eskola and coworkers clearly

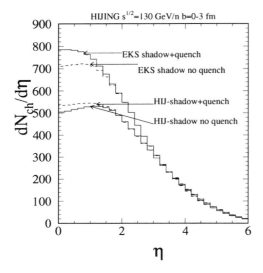

Figure 5. Charged particle multiplicity distribution predicted by HIJING for central Au + Au collisions at $\sqrt{s} = 130$ GeV/nucleon. Different curves correspond to different assumptions of nuclear shadowing and parton energy loss [37].

overpredicts the multiplicity. Of course, it is difficult to crisply separate three components with a single data point. The p_T spectrum of hadrons will constrain the parton energy loss in these collisions; analyses are currently underway. The shadowing can be determined directly by measurement of hadron yields at $p_T = 2 - 6 GeV/c$ in proton-nucleus and proton-proton collisions.[38] However, at this point we may tentatively conclude that (unless Nature has conspired to provide some exact cancellations) that nuclear shadowing appears to saturate and no anti-shadowing occurs.

4 Conclusions

I have shown that experiments produce dense interacting matter in the laboratory and that we can extract physics from the very complex interactions between heavy ions. One may ask whether the quark gluon plasma has been observed in collisions near $\sqrt{s} = 20$ at CERN, and the answer must needs be "prob-

ably". Several predicted signatures have been independently measured which defy currently available conventional explanations. Correlated onset has not been demonstrated, however. The lack of a coherent theoretical description and the incompleteness of appropriate dynamic theories make unambiguous conclusions difficult. Still missing is experimental determination of the energy threshold for deconfinement, and characterization of the properties of the quark gluon plasma state.

The experimental program in the coming years has its work clearly cut out: We must determine T_{init} from electromagnetic radiation, measure the jet quenching and learn to untangle the soft from the hard physics. Observation of multiple signatures at the same condition will be crucial, and a measurement of the hadron formation transition would be most helpful. RHIC has begun operation, and will contribute greatly via an experimental program with common event selection to constrain theory via a suite of observables.

Acknowledgments

I would like to thank Axel Drees, Xin-Nian Wang, Thomas Ullrich and Sam Aronson for valuable discussions and figures for the talk. This work was supportedby the U.S. Department of Energy under grant number DE-FG02-96ER40988.

References

1. F. Karsch, E. Laermann and A. Peikert, *Phys. Lett.* B **478**, 447 (2000).

2. M. Klein in *Proceedings of Lepton-Photon Symposium*, Stanford, Calif. August 1999, and references therein.

3. C. Adloff, et al. (H1 Collaboration), *Nucl. Phys.* B **545**, 21 (1999).

4. K.J. Eskola, V.J. Kolhinen and P.V. Ruuskanen, *Nucl. Phys.* B **535**, 351 (1998); K.J. Eskola, V.J. Kolhinen and C.A. Sal-

gado, *Eur. Phys. J.* C **9**, 61 (1999).

5. K. Geiger, *Phys. Rev.* D **48**, 4129 (1993).

6. M. Gyulassy and M. Pluemer, *Phys. Lett.* B**243**, 432 (1990); X.N. Wang and M. Gyulassy, *Phys. Rev. Lett.* **68**, 1480 (1992).

7. R. Baier, Yu.L. Dokshitzer, A.H. Mueller, S. Peigne and D. Schiff, *Nucl. Phys.* B **483**, 291 (1997); *Nucl. Phys.* B **484**, 265 (1997).

8. H. Appelshäuer, et al. (NA49 Collaboration), *Eur. Phys. J* C **2**, 661 (1998).

9. T. Matsui and H. Satz, *Phys. Lett.* B **178**, 416 (1986).

10. J. Rafelski and B. Mueller, *Phys. Rev. Lett.* **48**, 1066 (1982).

11. G.E. Brown and Mannque Rho, *Phys. Lett.* B **237**, 3 (1990).

12. K. Rajagopal and F. Wilczek, *Nucl. Phys.* B**399**, 395 (1993); *Nucl. Phys.* B**404**, 577 (1993).

13. U. Heinz and B.V. Jacak, *Ann. Rev. of Nucl. Part. Sci.* **49**, 529 1999.

14. H.C. Eggers and J. Rafelski, *Int. J. Mod. Phys.* A **6**, 1067 (1991).

15. B. Mueller and X.N. Wang, *Phys. Rev. Lett.* **68**, 2437 (1992).

16. B. Kaempfer, et al., *J. Phys.* G**23**, 2001 (1997).

17. H. Appelshaueser, et al., *Phys. Rev. Lett.* **75**, 3814 (1995).

18. M.C. Abreu, et al. (NA50 Collaboration), *Phys. Lett.* B **410**, 327 and 337 (1997).

19. M.C. Abreu, et al. (NA50 Collaboration), *Phys. Lett.* B **477**, 28 (2000).

20. J. Geiss, et al. *Phys. Lett.* B **447**, 31 (1999); C. Spieles, et al., *Phys. Rev.* C **60**, 054901 (1999); D. E. Kahana and S. H. Kahana, *Prog. Part. Nucl. Phys.* **42**, 269 (1999) and nucl-th/9908063; N. Armesto, A. Capella and E. G. Ferreiro, *Phys. Rev.* C **59**, 395 (1999).

21. H. Sorge, et al., *Phys. Lett.* B **271**, 37 (1991).

22. S. Abatzis, et al. (WA85 Collaboration), *Phys. Lett.* B **270**, 123 (1992).

23. E. Andersen, et al. (WA97 Collaboration), *Phys. Lett.* B **433**, 209 (1998).

24. E. Shuryak, *Phys. Lett.* B **78**, 150 (1978).

25. P. Aurenche, D. Gelis, H. Zavaket and R. Kobes, *Phys. Rev.* D *58*, 085003 (1998).

26. M. M. Aggarwal, et al. (WA98 Collaboration), nucl-ex/0006008 and nucl-ex/0006007, (2000).

27. D. Srivastava and B. Sinha, nucl-th/0006008 (2000).

28. G. Agakichiev et al. (CERES Collaboration), *Phys. Rev. Lett.* **75**, 1272 (1995); *Phys. Lett.* B **422**, 405 (1998).

29. G.E. Brown, G.Q. Li, R. Rapp, M. Rho and J. Wambach, *Acta Phys. Polon.* B **29**, 2309 (1998).

30. B. Kaempfer, K. Gallmeister and O.P. Pavlenko, hep-ph/0001242.

31. R. Rapp and E. Shuryak, *Phys. Lett.* B **473**, 13 (2000).

32. J. Burward-Hoy and B. Jacak, to be published.

33. X.N. Wang, *Phys. Rev.* C **58**, 2321 (1998).

34. M.M. Aggarwal, et al. (WA98 Collaboration), *Phys. Rev. Lett.* **81**, 4087 (1998).

35. http://www.star.bnl.gov, and T. Ullrich, private communication.

36. B. Back, et al. (PHOBOS Collaboration), *Phys. Rev. Lett.* **85**, 3100 (2000).

37. X.N. Wang, *Nucl. Phys.* A **661**, 210c (1999); X.N. Wang and Miklos Gyulassy, nucl-th/0008014 (2000); X.N. Wang, private communication.

38. X.N. Wang, *Phys. Rev.* C **61**, 064910 (2000).

Plenary Session 16

Experimental Particle Astrophysics

Chair: Enzo Iarocci (INFN)
Scientific Secretaries: Takuya Hasegawa and
Yasuo Takeuchi

HIGH ENERGY PARTICLES FROM THE UNIVERSE

B. DEGRANGE

LPNHE Ecole polytechnique and IN2P3/CNRS, 91128 Palaiseau cedex, France
E-mail: degrange@poly.in2p3.fr

The old problem of the origin of cosmic rays, whose energies extend up to 10 Joules per particle, has recently been renewed due to experiments detecting and identifying a small sample of very meaningful particles: gamma-rays, neutrinos and ultra-high-energy cosmic rays (which provide directional information from sources) and also antimatter particles. During the last ten years, gamma-ray astronomy has revealed several sources of cosmic rays, leading to new insights into emission mechanisms with the help of data at all wavelengths from radiowaves to TeV gamma-rays. Neutrino astronomy is still in the exploratory phase, but large new detectors are under construction. Similarly, ambitious projects are now underway which will significantly improve our knowledge of the ultra-high-energy region. New experiments in space, using magnetic spectrometers, will provide more accurate information on antiprotons and possible antinuclei as well as on the chemical and isotopic composition of ordinary cosmic rays. This report summarizes the recent progress of this new field of astrophysics, in which the community of particle physicists has been strongly involved through instrumental developments.

1 Introduction

As is well known, charged cosmic rays detected at the Earth provide no information on their initial directions except perhaps at the highest energies; this is due to their diffusion in the interstellar medium of the Galaxy threaded by irregular magnetic fields (of the order of 10^{-10} T). The energy spectrum (figure 1) is affected by propagation effects; the interpretations of the observed breaks — the "knee" at $\sim 5 \times 10^{15}$ eV and the "ankle" at $\sim 5 \times 10^{19}$ eV — are still controversial. The chemical composition of cosmic rays provides indirect clues on their origin but direct measurements in satellite- or balloon-borne experiments are available only up to 100 TeV; an important aspect of this field is the search for the rare particles of antimatter, for obvious cosmological reasons. In order to address the origin problem the importance of neutral and stable particles, namely gamma-rays and neutrinos, was realized early, but, due to experimental problems, significant progress in these fields has only been made over the last ten years. On the other hand, charged cosmic rays with energies greater than 10^{19} eV have a Larmor radius larger than the dimension of the Galaxy and should keep some memory of

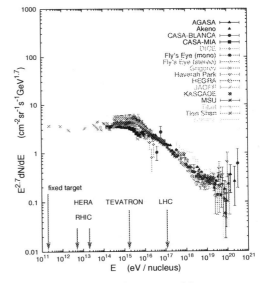

Figure 1. Cosmic-ray intensity $\times E^{2.7}$ as a function of primary energy (from T.K. Gaisser).

their initial direction; a strong effort is now underway to increase the presently limited statistics of ultra-high-energy (UHE) events.

The present report will thus focus on the detection of this minority of meaningful cosmic rays: antiparticles, gamma-rays, neutrino and UHE particles.

Table 1. Most recent results on antihelium.

BESS	AMS-01
1993-1998	1998 flight
Altit. 37 km	Altit. 400 km
Balloon	Space shuttle
$0.30 \text{ m}^2 \times \text{sr}$	$0.30 \text{ m}^2 \times \text{sr}$
$\bar{H}e/He$	
$< 0.74 \times 10^{-6}$	$< 1.1 \times 10^{-6}$

2 Antimatter in cosmic rays

Experiments searching for antimatter in cosmic rays generally measure the particle rigidity (through a magnetic spectrometer), velocity and electric charge (through energy loss and time of flight) and use Cherenkov counters for additional discrimination. Until recently, these experiments were balloon-borne, due to the important payload. The first magnetic spectrometer in space was successfully operated in 1998 by the AMS collaboration on a single flight on the space shuttle "Discovery" (AMS-01)[1].

In the absence of primary antiparticles, antiprotons should all be produced by cosmic-ray interactions in the interstellar medium. The latest antiproton spectrum presented at the time of this conference was obtained by the BESS collaboration[2] on the basis of about 400 $\bar{p}$'s in the interval 0.2 - 3 GeV; the spectrum is compatible with a secondary origin but one cannot exclude a significant amount of "exotic" antiprotons (e.g. from the annihilation of weakly-interacting massive particles[3] (WIMP's) or from the evaporation of primordial black holes). Another important aspect is the search for antinuclei. Table 1 shows that upper bounds on the antihelium-to-helium ratio in cosmic rays found by the BESS and the AMS-01[1] collaborations are at the level of 10^{-6} (Table 1).

New satellite experiments with magnetic spectrometers, larger detection areas and better particle identification capabilities are scheduled in the following years: PAMELA (Russian-Italian mission)[4] to be launched in 2002 and AMS-02, the final version of the AMS experiment[1], to be operated on the International Space Station in 2003. This last experiment should collect 10^6 $\bar{p}$'s within 3 years and push the bound on the antihelium-to-helium ratio down to 10^{-8}; in addition, it will provide direct studies of cosmic-ray composition and spectra with a high accuracy up to about 1 TeV.

3 Cosmic-ray sources revealed by gamma-ray astronomy

An important breakthrough in gamma-ray astronomy was brought about by the Compton Gamma-Ray Observatory (CGRO) satellite which operated between 1991 and 2000. The high-energy instrument (100 MeV - 30 GeV) on board CGRO, EGRET, produced a catalog of about 300 point-like sources[5], more than 60 of which are extragalactic and belong to a very particular class of radio-loud active galactic nuclei (AGN), the "blazars". These objects are characterized by relativistic plasma jets[6] ejected from the central region and directed close to the line of sight. Gamma-rays emitted from the jet pointing towards the Earth are thus boosted; the Lorentz factor of the bulk motion must be of the order of 10 in order to reconcile the high γ-ray luminosity and the high variability (and thus the compact nature of the emitting region); otherwise, γ-rays would produce $e^{\pm}$ pairs before escaping. The ejected plasma is a natural site of particle acceleration through the Fermi mechanisms; due to the additional Lorentz boost provided by the bulk motion, particles can reach very high energies. Another recent breakthrough concerns gamma-ray bursts[7], which, up to now, have been mainly studied in the low-energy range. Multiwavelength observations (X-ray satellites, optical telescopes) following the alerts given by the BATSE detector on board CGRO, have been able in favourable cases to identify the host-galaxies and to measure their redshifts. Large redshifts (thus "cosmological" distances) were found for several bursts. Although the triggering phenomenon is still a matter of controversy, the burst, as it is observed, is well described as a fireball expanding with a relativistic bulk motion; the same "opacity argument" used for blazar jets leads to a still higher Lorentz factor of the order of 100; it is also likely that the emission from gamma-ray bursts is beamed in the direction of observation. This feature makes gamma-ray bursts, as well as blazars, possible candidates for accelerating cosmic rays at the highest energies.

As far as the more common galactic cosmic rays are concerned, the favoured candidates are supernova remnants (SNR's), the acceleration region being either the blast wave produced in the interstellar medium or, when an active pulsar is present in the remnant, the shock wave produced by the pulsar wind in the ejected matter. This view is supported by the fact that 10% of the mechanical energy from supernovae explosions matches the power needed to maintain the cosmic-ray flux in the Galaxy; on the other hand, this mechanism is only effective up to about 100 TeV[8]. The gamma-ray emission of a few SNR's has been observed by EGRET at GeV energies, but studies at TeV energies are highly desirable to check that SNR's are actually the main sources of galactic cosmic rays. Pulsars are also interesting accelerator candidates due to the high electric fields which may exist in some regions of their magnetospheres, but their γ-ray emission seems to suffer a cut-off above 10 GeV. Finally, no X-ray binaries were detected by EGRET; however, some of them, the so-called "micro-quasars"[9] exhibit relativistic jets (albeit on a parsec scale) and are considered as interesting targets for experiments.

The γ-ray emission can be due to electromagnetic processes (e.g. synchrotron or inverse-Compton radiations) or to hadronic interactions of protons or ions with the surrounding matter or radiation and subsequent π^0 decays. In the latter case, neutrinos are also produced and their detection will allow us to discriminate between the different emission mechanisms. Neutrinos are particularly interesting in the case of extragalactic sources since γ-rays suffer absorption by pair-producing on the background radiation fields. The optical and infrared background radiations already affect extragalactic observations in the TeV range, and the microwave background radiation precludes extragalactic γ-ray astronomy in the 100-1000 TeV region, which should be the preferred energy domain of the forthcoming neutrino astronomy. The microwave background radiation also affects the propagation of UHE protons or nuclei due to pion photoproduction (Δ resonance). Above 10^{20} eV, the absorption length for protons is reduced to about 50 Mpc, and UHE hadrons produced further away should be degraded to lower energies. If sources are mostly located beyond 50 Mpc, the so-called Greisen-Zatsepin-Kuzmin (GZK) cut-off[10] should be observed in the spectrum at about 10^{20} eV.

4 Gamma-ray astronomy : present and future

EGRET data are limited to about 30 GeV by the lack of statistics. Large detection areas are required to compensate for very weak γ-ray fluxes at higher energies. Since the end of the 1980's, gamma-ray astronomy has been extended to the TeV region by ground-based experiments[11] with typical flux sensitivities of the order of $10^{-8}\gamma$ m^{-2} s^{-1} at 1 TeV, four orders of magnitude below that of EGRET at 1 GeV. These experiments are based on the detection of

Cherenkov light emitted by γ-ray-initiated atmospheric showers, the detection area being given by the size of the Cherenkov light pool on the ground ($\sim 3 \times 10^4$ m^2). The main difficulty, namely selecting the rare γ-ray events among the common hadronic cosmic-ray showers, was overcome at the end of the 1980's by the "imaging" technique, pioneered by the group of the Whipple Observatory[12]. The selection of γ-ray events is based on the shape, light profile and direction of the shower image obtained in the focal plane of the telescope. Present Imaging Atmospheric Cherenkov Telescopes (IACT's) combine an efficient rejection of the hadronic background and good angular resolution ($\sim 0.2°$). Their energy thresholds (set by the fluctuations of the night-sky background) have been progressively lowered, presently reaching about 150-200 GeV in the best cases, which is still higher than the EGRET upper energy range; as will be shown below, lower thresholds are now obtained with new Cherenkov techniques. The drawbacks of atmospheric Cherenkov telescopes are their low duty cycle (moonless clear nights) and their limited fields of view (3° to 5°). However, extensive air shower (EAS) arrays which do not suffer from these constraints have still higher thresholds and cannot discriminate hadrons easily. The Tibet array[13] was the first EAS array to obtain a 5 σ signal on the Crab nebula with 502 days of observation, whereas typical IACT's only need about one hour to get a signal of the same significance.

Improvements in the imaging technique were brought about by several groups: fine grain imaging cameras were implemented by the CAT[14], Whipple[15] and CANGAROO[19] groups; moreover, stereoscopic observations pioneered by the HEGRA[16] experiment with 5 telescopes (also used by the Seven Telescope Array[17]) improve the hadronic rejection by almost two orders of magnitude resulting in a much better flux sensitivity. Most of these detectors are located in the northern hemi-

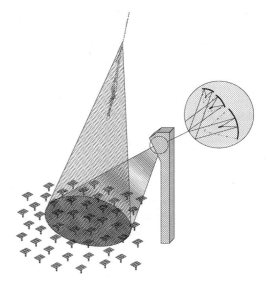

Figure 2. Principle of the use of a solar plant as a sampling array.

sphere and only two of them (both in Australia) explore the southern sky: Narrabri[18] and CANGAROO[19]. In order to reach energy thresholds below 100 GeV and to match the EGRET upper energy range, very large reflector areas are needed as well as fast timing electronics in order to minimize the effects of the night-sky background. Several experiments are now using the heliostats of former solar plants as collectors to sample the Cherenkov light front in both time and amplitude. Secondary optics located at the top of the tower of the solar farm disentangles the light beams coming from different heliostats, the light from each mirror being viewed by one phototube (figure 2); the signals are then corrected in time according to the different light pathlengths. At low energies, the light front is close to a sphere, the centre of which corresponds to the shower maximum; the light distribution on the ground, given by the amplitudes of the different signals, yields the position of the shower core; the two preceding points allow the reconstruction of the shower direction. At the time of this Conference, one of these sampling arrays, CELESTE[20], operating on the same site as the CAT imaging telescope in the French Pyrenees, actually achieved a 30 GeV threshold at the trigger level (50 GeV after event analysis) with 40 heliostats of 54 m^2 each. This result opens the previously unexplored energy region between 30 GeV and 300 GeV which turns out to be crucial for the study of several sources (e.g. pulsars and AGN's with significant redshifts). A similar experiment in the U.S., STACEE[21], presently working with a threshold of about 100 GeV, should eventually reach 30 to 50 GeV. Another solar plant in the U.S., SOLAR II[22], will soon be operated for γ-ray astronomy.

Only four γ-ray sources have been confirmed above 300 GeV: two plerions (i.e. filled-centre SNR's), the Crab nebula and PSR 1706-44 and two blazars, Markarian 421 (Mkn 421) and Markarian 501 (Mkn 501). A plerion is a supernova remnant in which the wind of an active pulsar produces a shock within the nebula, a site of particle acceleration; the radio map shows a filled structure due to the intense synchrotron radiation of high-energy electrons in the nebula. The Crab nebula, a typical plerion, is the source of a steady γ-ray signal, quite distinct from the pulsed signal originating from the magnetosphere of the Crab pulsar; the latter, observed by EGRET, is not found above 300 GeV. On the other hand, the steady signal from the nebula extends up to the highest observed γ-ray energies (> 20 TeV). The corresponding energy flux per decade of energy E $d\Phi/d(\ln E) = E^2 \, d\Phi/dE$ is shown in figure 3 as a function of the γ-ray energy E. Figure 3 includes the new CELESTE measurement[20] at 50 GeV as well as the results from IACT's at higher energies. In this range, a new precision measurement was obtained from the HEGRA stereoscopic system[23]. The wide-band spectrum in figure 3 shows a two-bump structure: the low-energy bump (below $\sim$ 20 MeV) is naturally attributed to synchrotron radiation of

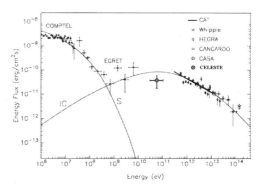

Figure 3. Distribution of the energy flux per decade of energy from the Crab nebula (E^2 dΦ/dE) as a function of γ-ray energy E.

very-high-energy electrons ($\sim 10^{15}$ eV) and the second one to γ-rays produced by inverse Compton scattering of the preceding electrons on synchrotron photons. This "synchrotron self-Compton" model (SSC) is in good agreement with the data from satellites and ground-based Cherenkov experiments. The Crab γ-ray emission at high energy is thus explained by purely electromagnetic processes. In the more common shell-like supernova remnants, the shock wave is produced by the expansion of the hot gas in the interstellar medium. Since these objects are supposed to be the main accelerators of galactic protons or ions, one expects π^0's (thus γ-rays) to be produced through nuclear interactions with the interstellar medium or radiation fields. Until now, only two positive observations of shell-type SNR's have been reported, both needing confirmation: SN 1006 by the CANGAROO group[24] and Cassiopeia A by the HEGRA experiment[11], the latter with a 4.9 σ signal on the basis of 232 hours of observation. The preceding observations could still be explained by electromagnetic processes and do not provide a proof of the hadronic origin of γ-rays. The case for SNR's as the main cosmic ray accelerators clearly requires more sensitive detectors.

The two observed TeV blazars are remarkable in many respects. Firstly, both have small redshifts (z = 0.031 for Mkn 421 and z = 0.034 for Mkn 501); in fact, it is expected that γ-rays from sources with redshifts greater than 0.1 suffer significant absorption on their way to the Earth, due to pair production on the infrared or optical background radiation. Gamma-ray astronomy could actually probe this poorly known background which is of strong cosmological interest. Secondly, similar to EGRET blazars, Mkn 421 and Mkn 501 exhibit a two-bump structure in the distribution of the energy flux per decade of energy $E^2\, d\Phi/dE$ as a function of energy. However, for EGRET blazars ("red" blazars), the first bump peaks in the infrared or optical range, whereas the second one peaks in the GeV region. In TeV blazars ("blue" blazars) on the other hand, peak energies are shifted to the UV or X-ray range for the first bump and to ~ 0.1 to 0.5 TeV for the second one. Recently, it was noticed[25] that blazars could be considered as members of a continuous sequence from red blazars (which give rise to higher intensities) to extreme blue blazars (whose emission extends to much higher energies, albeit with a lower intensity). The low-energy bump is clearly due to synchrotron radiation whereas the origin of the γ-ray bump is still controversial. Above 300 GeV, a significant curvature is found in the energy flux distribution of Mkn 501 by all experiments (Whipple[28], HEGRA[29] and CAT[30]; see figure 4 from HEGRA). Figure 6 from the CAT experiment shows that, in the most intense flare of 1997, the peak energy of the γ-ray bump was around 500 GeV. In all observations of Mkn 421, on the contrary, spectra were found to be well described by power laws, indicating that the peak energy is located below 200 GeV. TeV blazars are also found to be strongly variable sources (with doubling times ranging from 15 mn to 1 day). Figure 5 shows the γ-ray rate as a function of time for the shortest flare of Mkn 421 ob-

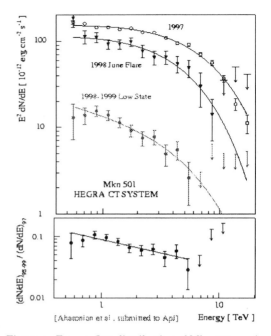

Figure 4. Energy flux distribution of Mkn 501 as observed by the HEGRA experiment at different periods.

served by the Whipple group[26] in 1996; this flare allowed a limit to be set on "in vacuo" dispersion[27], a possible quantum gravity effect. Evidence for spectral variability has been reported on Mkn 501 by CAT on time scales of a few days[30] and by HEGRA[11] on much longer time scales (figure 4). On Mkn 501, a strong correlation is observed between X-ray and γ-ray emissions[31]. Simultaneous spectra measured on two different days in April 1997 by the X-ray satellite Beppo-SAX and the CAT telescope[30] are shown in figure 6; the two-bump structure is clearly visible, with the synchrotron peak in the 100 keV region; during these intense flares, the source was below the EGRET sensitivity in the GeV range, which, for this object, corresponds to the dip in the spectrum. This remarkable correlation between the two bumps suggests a "synchrotron-self-Compton" scenario (as in the Crab nebula) in which X-rays and γ-rays originate from a single population of electrons. An alterna-

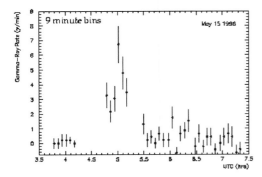

Figure 5. Gamma-ray rate (γ min^{-1}) from Mkn 421 on May 15 1996, as a function of time (from the Whipple Observatory).

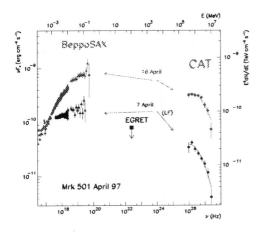

Figure 6. Energy flux distribution of Mkn 501 from X-rays to high energy γ-rays on April 16 and April 7 1997.

tive view interpretes the γ-ray bump as due to synchrotron radiation from UHE proton cascades accelerated with the electrons in the blazar jet.

The next major space mission foreseen for the high-energy domain is NASA's GLAST project[32], which should be launched in 2005. With a detection area seven times larger than EGRET and a very large field of view, and using the most advanced techniques from particle physics, this experiment should be two orders of magnitude more sensitive in flux and should extend space obser-

vations up to about 300 GeV. Before 2005, the Italian γ-ray satellite AGILE[33] (to be launched in 2002) will have a sensitivity comparable to EGRET; a similar performance is expected from the AMS-02 detector on the International Space Station which can be adapted to γ-ray astronomy[34]. On the ground, a new large mirror (17 m diameter) IACT project, MAGIC[35], should reach a threshold of 60 GeV (30 GeV in a further step), thus competing with solar farms. By 2003, several stereoscopic systems of large IACT's will be in place, which will permit a gain in sensitivity by a factor of 10 over current detectors. The american VERITAS project[36] foresees installing seven telescopes in Arizona. In the southern hemisphere, the German and French groups will build a set of four telescopes (up to 16 at a later date) in Namibia for the HESS project[37]; the Japanese CANGAROO group has already commissioned the first instrument of a similar four-telescope system[38] in Australia. New or upgraded extensive air shower arrays with thresholds of the order of 3 TeV will also contribute to γ-ray astronomy: among these the Tibet array[13] (4300 m a.s.l.), soon to be completed by the ARGO array of resistive plate chambers[39], and the MILAGRO experiment[40] in Los Alamos (2300 m a.s.l.), using a 80 m × 60 m water pool, 8 m deep, equipped with two layers of phototubes. These arrays with high duty cycles and large fields of view are well adapted to the detection of transient emission.

5 The birth of neutrino astronomy

Neutrino astronomy presents us with considerable experimental challenges due to the very small neutrino cross sections (10^{-35} cm^2 to 10^{-33} cm^2 from TeV to PeV energies) together with the low fluxes expected at TeV energies. Experiments actually detect high-energy muons produced by muon-neutrinos. This extends the effective target much be-

Table 2. Towards a km² neutrino detector. D is the site depth.

Effective area A $\leq 10^3$ m²			
MACRO[41] since 1989	LNGS Italy	Liquid scintillator + streamer tubes	$E_\mu > 1$ GeV 1100 ↑ μ's
BAIKAL[42] NT36 to NT200 1993-1998	Lake Baikal D = 1.1 km	Cherenkov 192 O.M.'s in 8 strings	$E_\mu > 10$ GeV
Effective area A $\sim 10^4$ m²			
AMANDA[43] B4 → B10 1996-1998	Ice South Pole D = 1.5 to 2 km	Cherenkov 418 O.M.'s in 13 strings	Upgraded to AMANDA II
NESTOR[44]	Mediterranean Greece D = 3.8 km	Cherenkov 168 O.M.'s in 1 tower	Under development
Effective area A $\sim 10^5$ m²			
AMANDA II[43] 2000 → ...	Ice South Pole D = 1.5 to 2 km	Cherenkov 681 O.M.'s	Taking data
ANTARES[45] Starting 2003	Mediterranean France D = 2.4 km	Cherenkov 1000 O.M.'s in 13 strings	Construction phase 1st string in 2001

yond the detector itself but results in a loss of angular resolution in the neutrino direction below 10 TeV. Fluxes expected from TeV γ-ray astronomy require an effective detection area of the order of 1 km²; therefore, the detecting medium must be sea water or polar ice. It is necessary to restrict to upward-going muons in order to remove penetrating muons directly produced by cosmic-ray showers in the atmosphere above the detector; the Cherenkov technique in water or ice is well adapted to this selection. Neutrino telescopes are nevertheless subject to the diffuse background of atmospheric neutrinos which is a strong limitation at low energies. The exper-

imental setup generally consists of strings of "optical modules" (O.M.'s) deployed in deep sea or polar ice. These modules are equipped with photodetectors recording timing and amplitude information from Cherenkov light. The depth must be chosen as a compromise between technical difficulties and the necessity to minimize the rate of downward-going muons. The latter are $\sim 10^5$ times more frequent than upward-going muons at depths ranging from 1 to 2 km and experiments must provide equally large rejection factors. This was succesfully achieved by the NT96 experiment in lake Baikal[42] (with 12 upward-going muons in 70 days) and by the AMANDA B10

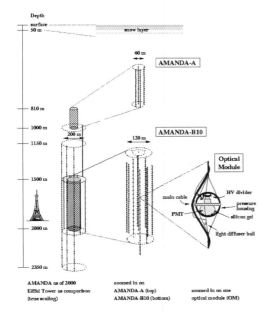

Figure 7. The AMANDA detectors.

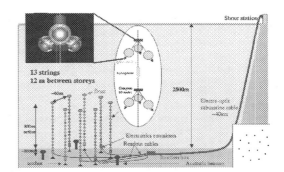

Figure 8. The ANTARES detector.

experiment[43] in South Pole ice (with $\sim$ 200 upward-going muons from 1997 data). Ice and sea water provide complementary advantages and drawbacks: the attenuation length is larger in ice ($\sim$ 100 m) and there is no background light, but the scattering length is shorter (24 m) which broadens time signals thus affecting the angular resolution. In sea water, on the other hand, the attenuation length is $\sim$ 60 m and, due to the large scattering length (in excess of 100 m), an angular resolution of $\sim$ 0.4° can be achieved. Table 2 shows the progression towards larger and larger neutrino detectors, on the basis of a crude estimate of the effective detection area which depends upon energy and zenith angle. The AMANDA II experiment at the South Pole has already reached a size of $\sim$ 0.1 km^2; the next step of this collaboration should be the Ice Cube project of 1 km^2 detector (2002-2008). The ANTARES project in the Mediterranean, close to Toulon, now in the construction phase, should achieve a detection area of 0.1 km^2 in 2003; this experiment should open the way to a further

1 km^2 detector not necessarily on the same site. A 0.1 km^2 neutrino detector has already an interesting discovery potential. If gamma-ray bursts were actually accelerators of UHE cosmic rays, they should also be very likely neutrino sources and a few bright bursts per year could be detected[46] in AMANDA II or ANTARES. Since bursts are well localized in direction and in time by satellites, the expected background of atmospheric neutrinos is small in such a search. The diffuse flux of active galactic nuclei is also an important area in which neutrino experiments complement γ-ray astronomy. First, they are sensitive to the central regions of AGN's which are opaque to γ-rays. In addition, the signal is expected to dominate the background of atmospheric neutrinos above $\sim$ 100 TeV, an energy range in which extragalactic γ-rays are absorbed by the cosmic microwave radiation. Finally, neutrino telescopes provide a way to search for non-baryonic dark matter if WIMPs annihilate in the Sun or in the Earth[3].

6 The ultra-high energy domain

Above 10^{19} eV, cosmic rays are no longer confined in the Galaxy and their arrival directions are expected to be meaningful. Unfortunately, in this energy range, fluxes are at the level of one event per km^2 per year. If one extrapolates the power-law dependence of the spectrum to arbitrary high energies, one would expect one event per km^2 per century above 10^{20} eV. Experiments in this field must provide very large "apertures" (i.e. detection area × solid angle) in excess of 1000 km^2× sr. Two complementary detection techniques are used[10]. In extensive air shower (EAS) arrays, charged particles ($e^{\pm}$ and $\mu^{\pm}$) are sampled by a large number of detectors (e.g. scintillators or water Cherenkov detectors) over at least 100 km^2 on the ground; the energy measurement is obtained indirectly from the number of electrons and muons as a function of the distance to the shower core. An alternative method consists of detecting the fluorescence light produced by the shower in the atmosphere; this light, isotropically emitted, is detected by an array of telescopes equipped with a cluster of phototubes in the focal plane. Each telescope monitors a sector of the sky; these sectors are similar to the facets of a fly's eye, hence the name "Fly's Eye" given to the technique. Fluorescence detectors provide a rather direct energy measurement but require moonless clear nights, hence a low duty cycle. The most recent results come from an EAS array (AGASA[47]) and a fluorescence detector ("Hi-Res" fly's eye[48]). The AGASA array has been operated in Japan since 1990 and consists of 111 scintillators (2.2 m^2 each) deployed over 100 km^2; the energy is measured with 30% accuracy and the angular resolution is 3° at 10^{19} eV. The high resolution fly's eye (Utah, USA) has been working since 1997; it achieves an energy resolution of 10% and an angular accuracy of 0.3°. This detector should be completed this year by a second one allowing for stereoscopic measurements. Both AGASA and "Hi-Res" fly's eye experiments have reached similar sensitivities above 10^{20} eV, namely 1000 km^2× sr × year; both have collected 7 events with an energy in excess of 10^{20} eV, thus beyond the GZK cutoff (figure 9). Unfortunately, the statistics are too scarce to derive conclu-

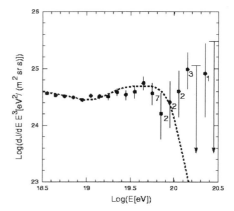

Figure 9. The AGASA UHE spectrum × E^3. The dotted curve shows the spectrum expected from extragalactic sources uniformly distributed in the Universe (GZK cutoff).

sions from the angular distribution at these energies[49]; moreover, the southern sky has not been been observed by recent experiments on UHE cosmic rays. The "Pierre Auger Observatory" project, presently in the construction phase, intends to bring a breakthrough in this field. It should combine a wide EAS array of 1600 water Cherenkov detectors deployed over 3000 km^2 and a fluorescence detector consisting of 30 electronic telescopes with 13000 pixels[50]. The use of both techniques allows cross calibration, thus a much better control of systematics; an angular resolution of ∼ 0.3° is expected; the energy resolution in hybrid mode should be 10% to 20%. The experiment will provide about 3000 events per year above 10^{19} eV and about 30 events per year above 10^{20} eV. The long-term goal is to have two identical experiments, one in each hemisphere. The southern experiment is presently under construction at El Nihuil (Argentina); the northern one should be in Utah (USA). An "engineering array" including 40 water Cherenkov detectors and one fluorescence telescope deployed over 40 km^2 is scheduled for mid 2001. The construction phase should last until 2004 but data from a partial setup are expected in 2002.

7 Conclusion

High-energy astrophysics is clearly developing rapidly. In all the fields discussed in the present report, new generation experiments are expected to provide much better flux sensitivities. The new astronomies (based on γ-rays, neutrinos and UHE cosmic rays) should complement each other in the study of many objects (e.g. AGN's, gamma-ray bursts). Simultaneous observations in γ-rays and at other wavelengths have already proved very powerful; similar studies with different "messengers" will probably open the way to understanding the origin of cosmic rays.

Acknowledgments

I would like to thank the conference organizers for inviting me to act as a rapporteur and for their warm hospitality. In the preparation of this talk, I am indebted to Prs. T. K. Gaisser, F. Halzen and T. Kifune and Drs. D. Cowen, J. J. Hernandez, M. Kleifges, A. Kohnle and M. Nozaki for very valuable discussions and/or exchange of data.

References

1. V. Choutko, parallel session PA 09, this Conference.
   J. Alcaraz et al., *Phys. Lett.* B **461**, 387 (1999).
2. T. Maeno et al. ,Proc. 33th COSPAR Scientific Assembly, Warsaw, 16-23 July 2000, to be published.
   S. Orito et al., Phys. Rev. Lett. **84**,1078 (2000).
3. N. Smith, rapporteur's talk, this Conference.
4. S. Piccardi et al., Proc. 26th ICRC (Salt Lake City 1999), **5**, 96 (1999).
5. R. C. Hartman et al., Ap.J. Sup. **123**, 79 (1999).
6. M. Sikora, M. Begelman and M. Rees, Ap.J. **421**, 153 (1994).
7. P. Mészáros, Nucl. Phys. B (Proc. Suppl.) **80**, 63 (2000).
8. P.-O. Lagage and C. Cesarsky, Astron. Astrophys. **118**, 223 and **125**, 249 (1983).
9. I.F. Mirabel and L.F. Rodriguez, Nucl. Phys. B (Proc. Suppl.) **80**, 143 (2000).
10. M. Nagano and A. A. Watson, Rev. Mod. Phys. **72**, 689 (2000).
11. A. Kohnle, parallel session PA-09, this Conference.
12. T. C. Weekes et al., Ap. J. **342**, 379 (1989).
13. M. Amenomori et al., Ap. J. **525**, L93 (2000).
14. A. Barrau et al., Nucl. Instr. and Meth. **A 416**, 278 (1998).
15. J.P. Finley et al., Proc. GeV-TeV Astrophysics Workshop, Snowbird, Utah, AIP Conference Proceedings **515**, 301 (1999).
16. A. Daum et al., Astropart. Phys. **8**, 1 (1997).
17. S. Aiso et al., Proc. 25th ICRC (Durban, South Africa) **4**, 181 (1997).
18. P. Armstrong et al., Experimental Astron., **9**, 51 (1999).
19. H. Kubo et al., Proc. GeV-TeV Astrophysics Workshop, Snowbird, Utah, AIP Conference Proceedings **515**, 313 (1999).
20. M. de Naurois et al., Proc. TeV Gamma-Ray Symposium (Heidelberg, July 2000), to be published.
21. R. Ong, Proc. GeV-TeV Astrophysics Workshop, Snowbird, Utah, AIP Conference Proceedings **515**, 401 (1999).
    D. Williams et al., Proc. TeV Gamma-Ray Symposium (Heidelberg, July 2000), to be published.
22. J. A. Zweerink et al., Proc. GeV-TeV Astrophysics Workshop, Snowbird, Utah, AIP Conference Proceedings **515**, 426 (1999).
23. F. Aharonian et al., preprint astro-ph/0003182 (2000).
24. T. Tanimori et al., Ap.J. Lett. **497**, L25 (1998).

25. R. M. Sambruna, Ap.J. **487**, 536 (1997). R. M. Sambruna, Proc. GeV-TeV Astrophysics Workshop, Snowbird, Utah, AIP Conference Proceedings **515**, 19 (1999).

26. J. A. Gaidos et al., Nature **383**, 319 (1996).

27. S. D. Biller et al., Phys. Rev. Lett. **83**, 2108 (1999).

28. F. Krennrich et al., Ap.J. **511**, 149 (1999).

29. F. A. Aharonian et al., Astron. Astrophys. **349**, 11 (1999).

30. A. Djannati-Ataï et al., Astron. Astrophys. **350**, 17 (1999).

31. H. Krawczynski et al., Astron. Astrophys. **353**, 97 (2000).

32. D. A. Kniffen, D. L. Bertsch and N. Gehrels, Proc. GeV-TeV Astrophysics Workshop, Snowbird, Utah, AIP Conference Proceedings **515**, 492 (1999).

33. S. Mereghetti et al., Proc. GeV-TeV Astrophysics Workshop, Snowbird, Utah, AIP Conference Proceedings **515**, 467 (1999).

34. R. Battiston, Proc. GeV-TeV Astrophysics Workshop, Snowbird, Utah, AIP Conference Proceedings **515**, 474 (1999).

35. E. Lorenz, Proc. GeV-TeV Astrophysics Workshop, Snowbird, Utah, AIP Conference Proceedings **515**, 510 (1999).

36. F. Krennrich et al., Proc. GeV-TeV Astrophysics Workshop, Snowbird, Utah, AIP Conference Proceedings **515**, 515 (1999).

37. W. Hofmann, Proc. GeV-TeV Astrophysics Workshop, Snowbird, Utah, AIP Conference Proceedings **515**, 500 (1999).

38. M. Mori et al., Proc. GeV-TeV Astrophysics Workshop, Snowbird, Utah, AIP Conference Proceedings **515**, 485 (1999).

39. D. Martello, Proc. TeV Gamma-Ray Symposium (Heidelberg, July 2000), to be published.

40. G. B. Yodh, Proc. GeV-TeV Astrophysics Workshop, Snowbird, Utah, AIP Conference Proceedings **515**, 453 (1999).

41. M. Ambrosio et al., preprint astro-ph 0002492 (2000), submitted to Ap.J.

42. V. A. Balkanov et al., Astropart. Phys. **12**, 75 (1999).

43. D. Cowen, parallel session PA 08, this Conference.

44. B. Monteleoni, Proc. 17th International Conference on Neutrino Physics and Astrophysics (Helsinki, Finland, 1996), World Scientific 1997.

45. J. J. Hernandez, parallel session PA 08, this Conference.

46. F. Halzen and D.W. Hooper, Ap.J. Lett. **527**, L93 (1999).

47. M. Takeda et al., Phys. Rev. Lett. **81**, 1163 (1998).

48. T. Abu-Zayyad et al., Proc. 26th ICRC (Salt Lake City 1999), **5**, 349 (1999).

49. M. Takeda et al., Ap.J. **522**, 225 (1999).

50. M. Kleifges, parallel session PA-09, this Conference.

LOW ENERGY PARTICLES FROM THE UNIVERSE

N.J.T. SMITH

*Particle Physics Department, Rutherford Appleton Laboratory,
Chilton, Didcot, Oxfordshire, U.K. OX11 0QX
E-mail: n.j.t.smith@rl.ac.uk*

The current status of searches for low energy particles from the Universe is briefly reviewed. The scope of this review is limited to the invisible hypothetical particles that may constitute the Galactic dark matter and focuses on the search for axions and direct and indirect searches for WIMPs. The justification for these solutions to the dark matter problem is outlined, with an overview of the detection techniques used. Current search results and the prospects for future generation detectors are reviewed.

1 Introduction.

All viable cosmological models require a significant contribution from non baryonic cold dark matter[1]. Within Galactic haloes it is expected that the dark matter is predominantly non-baryonic and cold, with a composition that mimics that in the Universe as a whole[2,3].

Recent results from CMBR surveys[4,5], when constrained by large scale structure observations[6] and high redshift type Ia supernovae observations[7,] indicate the overall matter density in the Universe is $\Omega_m \approx 0.3$[8]. The high precision CMB measurements from BOOMERanG and MAXIMA also indicate, from the position of the first acoustic peak, that the Universe is flat, i.e. $\Omega_m + \Omega_\lambda \approx 1$. The contribution of baryonic matter to Ω_m has been constrained by quasar absorption studies[9] and BBN models[10] to $\Omega_b h^2 \approx 0.02$, clearly a small fraction of that required.

Within galaxies the halo density is given[11] by $\Omega_m h \approx 0.03.(R_h/100pc)$, where R_h is the dark halo radius. The 200kpc halo observed in some galaxies[12] would exceed that available from baryonic material alone. Direct searches for baryonic dark matter utilising gravitational microlensing[13,14] searches indicate the most likely fraction to be 20%, although the stellar population this infers is problematic[15].

There is thus clear evidence that within both the Galaxy and the Universe there is a significant contribution from non-baryonic, cold dark matter. The three main types of non-baryonic cold dark matter postulated are a massive neutrino with mass $\approx 10eV$, axions with mass $\approx 10^6 eV$ or WIMPS with mass $\approx 100GeV$. The evidence for massive neutrinos is outside the scope of this article and is reviewed elsewhere in these proceedings[16]. This article will review the current experimental status of searches for the axion and WIMP.

2 The search for axions.

The non-baryonic axion arises as the massive Goldstone boson associated with the breaking of the Peccei-Quinn U(1) symmetry proposed to solve the strong CP problem. The non-Abelian nature of quantum chromodynamics should introduce T, P and CP violating effects in the strong interaction, such as an electric dipole moment of the neutron. The vanishingly small edm of the neutron[17] indicates that these effects are suppressed. One mechanism for this suppression proposed by Peccei and Quinn[18,19] is the addition of a global U(1) symmetry which cancels the CP-violating terms in the QCD Lagrangian. The massive Goldstone boson associated with the breaking of this global symmetry is the axion[20,21]. The coupling of the axion to charged leptons is model dependent. Those models where the tree level lepton coupling strength is similar to the quark coupling are known as DFSV axions[22,23] those where no tree level lepton coupling occurs are known as hadronic axions, or KSVZ axions[24,25]. The mass of the axion is given by[26]

$$m_a \approx 6\,\mu eV \left(\frac{10^{21}eV}{f_{PQ}} \right)$$

where f_{PQ} is the scale at which symmetry breaking occurs. Constraints on the axion mass and coupling strengths have been placed by direct reactor and accelerator searches[27], through the influence of axions on energy dissipation in stellar evolution and cosmological requirements to prevent overclosure of the Universe[28]. Figure 1 shows the limits on the axion mass from these arguments, delineating the favoured mass range of interest for direct searches between 10^{-6}eV and 10^{-2}eV.

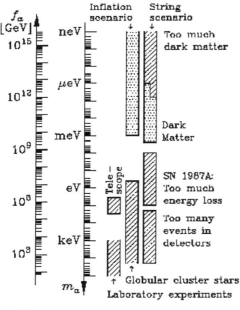

Exclusion Range

Plausible Dark-Matter Range

Figure 1[28]. Allowed mass region of axions from cosmological and astronomical limits

The detection of Galactic halo dark matter axions is performed through the Primakov conversion of the axion to a microwave photon in an intense magnetic field[29]. In this type of experiments the B-field is maintained in a high-Q resonance cavity held at ≈K temperatures to minimise thermal noise. The cavity is tuneable by a movable rod allowing the resonance frequencies to be scanned whilst the cavity output power is measured to search for the resonance signature of axion conversion. The effective temperature (thermal + electronic) determines the S/N, hence limiting the observable coupling strength for a given scan rate.

Following proof of principle experiments[30,31] two second generation detectors have been constructed to probe the cosmologically significant axion mass and coupling strength space. The US Large Scale Axion Search[32] utilises HEMT amplifiers with a $T_{eff} \approx 1.5$K to scan a high purity annealed copper $0.2m^3$ cavity containing a 7.6T field. The power spectrum of the extracted signal is determined in two frequency resolutions, 125Hz and 0.02Hz. This allows a search to be made for both the Maxwellian energy spectrum expected for virialised halo axions and also the discrete high energy lines due to axions infalling into the Galactic potential well. Figure 2 gives a schematic of this detector.

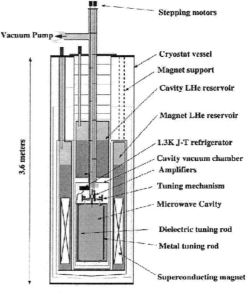

Figure 2[27]. The US Large Scale Axion Search

The ability to re-scan frequency ranges, and to remove the B-field, allows potential signals to be verified. The results from a search over a 700-800MHz frequency range which yielded no candidates on verification are shown in figure 3, which also indicates the expected strengths for DFSV and KSVZ axions. It can be seen that this cavity experiment is now reaching the sensitivity of interest for KSVZ axions. The recent development[32] of high frequency capable in-line DC SQUID amplifiers with a $T_{eff} \approx 50mK$, the quantum noise limit, and multiple cavity arrays promises to extend the sensitivity of the search into the DSFZ region over the favoured axion mass region.

The second generation CARRACK II detector developed by the University of Kyoto[33] uses a smaller $5.10^{-3}m^3$ 7T strength Primakov conversion cavity housed in a dilution refridgerator maintained at 15mK. The conversion photons are coupled from the conversion cavity into a detection cavity, which is field free and tuned to the same frequency. These photons are then detected through the use of Rydberg atoms. A beam of rubidium atoms is excited to a Rydberg state by two-step laser excitation and are then fed through the detection cavity. Here they absorb the axion conversion photon and are excited to an upper state, the transition frequency being defined by Stark electrodes. Selective field ionisation is used to extract the promoted electron from the excited atoms, which are then accelerated and detected. The aim of the CARRACK II experiment is to cover the 2-30μeV mass range down to DFSV coupling strengths[34].

3 The direct search for WIMPs.

The Weakly Interacting Massive Particle (WIMP) is a generic classification for a group of non-baryonic relic particles, which includes heavy Dirac neutrinos. The most favoured candidates in this classification arise from the supersymmetric extension to the Standard Model of particle physics, in which a global symmetry exists between fermions and bosons[35]. This extension was independently postulated to solve the mass hierarchy problem, in which higher order corrections to the Higgs scalar mass precludes the Higgs mechanism occurring at independent differing mass scales. The introduction of a higher global symmetry between fermions and bosons allows these corrections to be re-normalised.

Figure 3[32]. Current axion mass and coupling limits, with KSVZ, DSFV expectations.

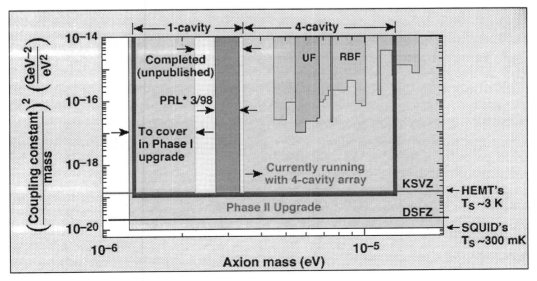

In most supersymmetric models, such as the minimal extension (MSSM), conservation of the multiplicative quantum number R-parity results in the lightest supersymmetric particle being stable. In the MSSM the LSP is the lightest neutralino, a superposition of the super-partners of the Higgs and gauge bosons – the higgsinos and gauginos. The large mass of the top quark implies the neutralino is predominantly gaugino, mainly bino[36]. Accelerator constraints[37,38] from direct neutralino, chargino and Higgs searches at LEP II and the Tevatron and naturalness arguments limit the neutralino mass range to 50-300GeV, with a neutralino-proton cross section $\sigma_{\chi\text{-}p} > 10^{-10}$pb.

Direct searches for halo WIMPs are undertaken by studying the low energy elastic nuclear recoils produced in various detector targets[39]. Early limits on neutralino cross sections were afforded by low background Ge double-beta decay experiments[40]. Recent detector programmes have focussed on techniques to identify WIMP recoils through discrimination, and rejection, of background electron recoils due to gamma and beta interactions or through the effects of the Earth's motion through the dark matter halo. The latter effect will give rise to a small, few %, modulation in the WIMP recoil spectrum due to the annual modulation of the Earth's velocity in the Galactic frame. Through measurement of the recoil direction of the target nucleii a correlation to the motion of the Earth in the Galactic frame would give a definitive signature, and may be of greater importance if the dark matter halo distribution is non-Maxwellian as has been postulated[41,42].

Nuclear recoil discrimination may be undertaken by comparison of the ionisation, scintillation or phonon signal from differing species interacting in the target mass. Examples are discrimination in NaI is due to differences in the time constant of the scintillation light, as used by the DAMA[43] and UKDMC[44] collaborations, and discrimination

due to differences in thermal and ionisation yield in Ge and Si detectors, as used by CDMS[45] and Edelweiss[46].

The DAMA collaboration has operated $\approx$100kg of NaI target at the Gran Sasso laboratory, searching for an annual modulation signal[47]. Four years of operation have yielded 60 tonne.days of data which, after only photomultiplier noise rejection, contains a low energy annually modulated component up to 6keV in the total observed counting rate. Figure 4 illustrates the 2-6keV bin residuals in the counting rate on a measured background rate of 2 dru (events/kg/day). The interpretation of this modulating component as a WIMP interaction would indicate a neutralino mass of 52±8 GeV, with a $\sigma_{\chi\text{-}p}$=7x10^{-6}pb, as illustrated in figure 5. Models have been proposed[48] which would make this signature consist with relic neutralinos, although there has been considerable debate within the community[49,50] about the validity of ascribing the observed annual modulation to a WIMP signal rather than some, as yet unidentified, systematic.

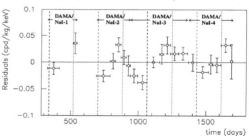

Figure 4[47]. Residuals in 2-6keV energy bin for the DAMA four year annual modulation search.

The UKDMC Collaboration have identified[44], through pulse shape discrimination, a population of events with fast time constants in low background, shielded, NaI detectors, figure 6. The ratio of the anomalous events' time constant to the gamma calibration is $\approx$0.65, with the neutron:gamma ratio $\approx$0.75. These events have been observed in a series of detectors, and have also been observed in a crystal operated by the Saclay group, a twin of the NaI crystals within the

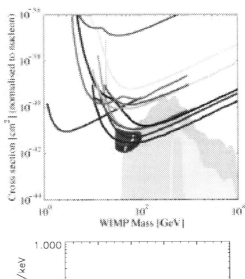

Figure 5. The DAMA allowed cross section and mass, with additional experiment limits. Plot from http://cdms.berkeley.edu/limitplots.

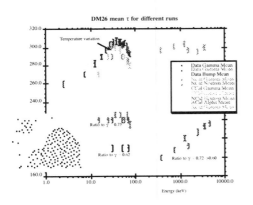

Figure 6. The fast scintillation time constant events seen in UKDMC NaI crystals compared to a gamma and neutron calibrations.

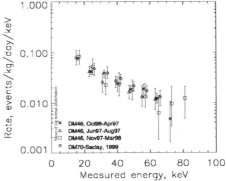

Figure 7[39]. The energy spectra of fast time constant events from two crystals (UKDMC 5kg – DM46, Sacaly 9.7kg – DM70)

DAMA array[51]. Figure 7 shows the event rates of these anomalous events in two crystals, including the Saclay one. The origin of these events has been postulated through extensive species irradiation tests[52,53] to be due to a 0.1 micron layer of alpha emitter. Outgoing alphas from such a layer would yield an E^{-2} energy spectrum, as observed[54]. A mechanism for the implantation of these alpha emitters due to the decay of atmospheric radon contamination[54],

although producing a correct layer depth, would not yield the observed rate from realistic values of the radon exposure. For a continuously replenished radon environment the additional short-lived daughters could contribute and would give the observed rate[55]. This latter mechanism would be highly correlated to radon concentration.

The CDMS Collaboration operates Ge and Si thermal/ionisation detectors utilising the greater ionisation yield in electron recoils to provide species discrimination. Results from 1.6 kg.days of Si and 10.6 kg.days of Ge data have been presented[45] from a 16 mwe deep site at Stanford University. A stack of four detectors is operated within a modified dilution refrigerator at 20mK, the Ge detectors using

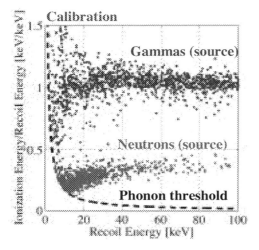

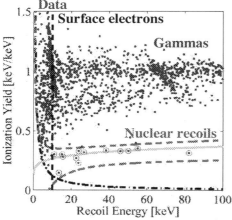

Figure 8[45]. Discrimination potential of CDMS Ge detectors showing ionisation to thermal species dependence. The punch through neutron events are seen as nuclear recoils.

NTD Ge thermistors and electrode readout for thermal and charge signals. A 99.99% efficient muon veto surrounds the detector chamber. Figure 8 illustrates the discrimination potential of these detectors, comparing gamma and neutron source data with background observations. Analysis of the Ge data indicate there are 17 nuclear recoil candidates, of which 4 are multiple scatters through the four detector stack, the Si data yielding another 4 nuclear recoils. The interpretation of the multiple scattered events is that they are due to high energy neutrons, which punch through the muon veto. This is supported by Monte Carlo simulations of the muon induced neutron flux. The single scatter events, if interpreted solely as WIMP recoils would be compatible with the DAMA signal. However the presence of the multiple scatters indicates that many, if not all, of these events are also due to punch through neutrons. The Monte Carlo assessment of the neutron interactions within the detector stack, normalised to the observed 17 Ge events, would indicate 1.3 multiple scatters and 2.7 Si events[45]. Figure 5 shows the upper limit on the neutralino interaction cross section, based on subtraction of the nuclear recoil events, based on the simulations of the muon induced high energy neutron flux. On the basis of this analysis the CDMS and DAMA results are in conflict at the 84% level.

The experimental techniques utilised in the search for Galactic halo WIMPs outlined above are all reaching sensitivities of $\sigma_{\chi\text{-}p} \approx 10^{-6}$pb as outlined in figure 5. These experiments are all observing some artefact within the data, either intrinsic to the technique or external. The DAMA collaboration are to continue operation of their NaI array, whilst the UKDMC have shown that by careful surface treatment of the NaI crystals the alpha induced events may be removed, allowing nuclear recoil events to be statistically extracted in a 50kg array[55]. The CDMS collaboration are to relocate their experiment to the Soudan underground site from mid-2001, which will reduce the cosmic ray flux by 5.10^5, allowing a significant induced neutron background reduction.

Additional techniques are under development worldwide which will reach sensitivities of $\sigma_{\chi\text{-}p} \approx 10^{-8}$pb and beyond. Liquid xenon scintillation detectors have been constructed by the UKDMC and DAMA groups[56,57] which have improved discrimination over NaI due to the dE/dx dependence on the energy distributed between the ionisation and excitation channels[58]. The UKDMC detector is

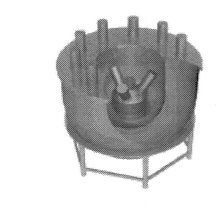

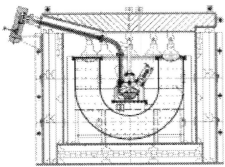

Figure 9. Schematic of the UKDMC 4kg liquid xenon single phase detector, illustrating the target volume, active Compton veto and passive shielding.

a 4kg fiducial volume chamber, viewed by self shielding turrets of liquid xenon, surrounded by an active Compton and passive gamma shield as illustrated in figure 9. This detector has been deployed at the Boulby mine during the Autumn of 2000.

Two phase xenon detectors, in which the ionisation yield is measured through secondary scintillation[59] or electro-luminescence in the gaseous phase are under development by the UKDMC in collaboration with UCLA, ITEP, Torino, Columbia, Coimbra, LLNL and Padova. These detectors are scheduled for deployment during 2001.

The CRESST collaboration[60] had initially developed a conventional cryogenic bolometer target based on Al_2O_3 for operation in the Gran Sasso laboratory. They have recently been developing recoil discrimination detectors based on $CaWO_4$ in which the thermal and scintillation signals are measured giving discrimination down to 10keV. A target mass of 10kg is proposed to be operational by 2002.

The EDELWEISS collaboration have developed, as CDMS, a low temperature Ge ionisation-thermal detector based on NTD Ge thermistors[46]. Three 320g modules have been operational in the Fréjus underground laboratory, although contamination by implantation of surface alpha emitters has been observed.

A directional detector based on recoils in a low pressure TPC is under development by the UKDMC, Temple and Occidental groups[61]. This detector will utilise a CS_2 target which produces −ve ions that, when drifted to a readout plane, will maintain the track profile as diffusion is limited. This precludes the need for a magnetic field to surround the detector, as in conventional TPCs, which aids deployment underground. A $1m^3$ device will be installed at the Boulby facility during 2001 to assess the background, predictions and initial tests indicating zero background capability.

4 The indirect search for WIMPs.

In addition to the direct detection of WIMPs through the elastic nuclear recoil within a target material, there is also the possibility of indirect detection through neutralino annihilation within the Galactic halo, or within the centre of the Sun or Earth where they have been gravitationally attracted and their density enhanced.

$$\chi\chi \to WW, ZZ, \gamma\gamma, gg,... \to e^+, p, \gamma, \nu,..$$

An initial signature used to search for neutralino annihilation was the low energy anti-proton flux[62]. However cosmic ray induced antiprotons have been found to populate this low energy region[63], making extraction of any neutralino induced signal difficult. As with other secondary decay products the resultant spectrum is featureless, although the distribution of events will follow the halo and not the galactic disk, which will allow some discrimination against the cosmic ray background.

The annihilation into gammas and Z, gamma through loop processes are found to be form factor free, which allows this spectral line feature to be searched in the 10^{11}eV gamma ray spectrum[64]. The flux due to these annihilation processes will be within the sensitivity of the new generation air Cherenkov telescopes and space-bourne detectors such as GLAST.

The annihilation of neutralinos within the centre of the Sun or Earth will give, as secondary decay products, high energy neutrinos which will be liberated isotropically from the centre of the gravitationally attracting body. The neutrinos will maintain the direction of production, allowing the annihilation signature to be identified against the general background of the cosmic ray induced neutrinos in the Earth's atmosphere. The detection principle used to search for the annihilation neutrinos is to study the neutrino induced muon flux from the direction of the centre of the Earth and Sun. These muons may be detected through the Cerenkov radiation liberated in a target material such as ice or water. Several telescopes are being built to exploit this technique, including the AMANDA[65], ANTARES[66], and NT-200[67] experiment at lake Baikal, with energy thresholds of order tens of GeV. Neutrino telescopes with lower energy thresholds at 1GeV, such as Baksan, MACRO and Super-Kamiokande are more sensitive to the lower mass neutralino than the Cerenkov telescopes. Figure 10 shows the current limits on the muon flux from the Sun for the AMANDA[68], MACRO[69], Baksan and SuperK[70] detectors, figure 11 showing the limits from the Earth. SUSY models with varying neutralino spin independent cross sections[71] are superimposed illustrating that the indirect searches are beginning to constrain the neutralino models.

Should the neutralino velocity distribution be different from the simple Maxwellian profile usually assumed there are enhancement mechanisms which would affect the indirect detection rate[72]. Neutralinos scattered from an interaction with the Sun may have their orbits perturbed by the large outer planets, giving a population with velocities close to the orbital velocity of the Earth. This then allows for efficient gravitational capture of neutralinos with masses below 150GeV, enhancing the indirect signature.

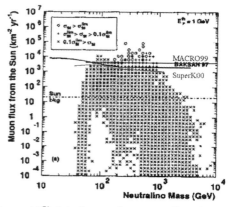

Figure 10[71]. Limits on the muon flux from the centre of the Sun, with SUSY neutralino annihilation rates superimposed.

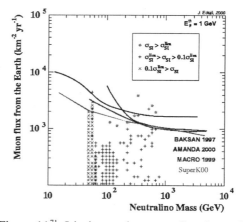

Figure 11[71]. Limits on the muon flux from the centre of the Earth, with SUSY neutralino annihilation rates superimposed.

5 Conclusions.

The search for the hypothetical low energy particles that may contribute to the galactic dark matter is coming of age. The direct searches for axions are now reaching the required sensitivity in certain mass ranges to exclude the KSVZ axion, with improvements in conversion photon detection promising to improve sensitivities to that required to exclude the DSFZ axion over the cosmologically significant mass region. The direct WIMP searches through elastic nuclear recoil scattering now have sensitivities of $\sigma_{\chi-p} \approx 10^{-6}$ pb with three experiments observing artefacts within their data. Improvements to these detector systems and the development of improved discrimination technologies will see the origin of these artefacts resolved and sensitivities pushed to $\sigma_{\chi-p} \approx 10^{-9}$ pb within a few years. The indirect searches for neutrinos and gamma rays from WIMP annihilation in the centre of the Sun and Earth are also beginning to constrain the supersymmetric models which describe the neutralino.

References

1. M. Fukugita. These proceedings
2. B. Moore *et al* astro-ph/9903164 (1999)
3. S. Ghinga *et al* Ap. J. in press astro-ph/09910166 (2000)
4. De Bernardis *et al* Nature **404** 995 (2000)
5. S. Hanany *et al* Ap. J. Lett. *In press* astro-ph/0005123 (2000)
6. W. Saunders *et al*. Astro-ph/0001117 (2000)
7. S. Perlmutter *et al*. Ap. J. **517** 565 (1999)
8. M. Tegmark *et al* hep-ph/0008145 (2000)
9. S. Burles and D. Tytler Ap. J. **499** 699 (1998)
10. S. Burles *et al*. Ap. J. **507** 732 (1999)
11. B. Carr Proc. Microlensing 2000 ASP Conf. Series. (2000)
12. D. Zaritsky *et al* Ap. J. **405** 464 (1993)
13. C.Alcock astro-ph/0001272 (2000)
14. J.F. Glicenstein These proceedings
15. K. Freese *et al* astro-ph/0007444 (2000)
16. E. Kearns. These proceedings.
17. P.G. Harris *et al.*, Nucl. Inst. & Meth. **A440** 479 (2000).
18. R. Peccei, H. Quinn, Phys. Rev. Lett. **38** 1440 (1977).
19. J.E. Kim, Phys. Rep. **150** 1 (1987).
20. S. Weinberg. Phys. Rev, Lett **40** 223 (1978).
21. F. Wilczek. Phys. Rev, Lett **40** 279 (1978).
22. A.R. Zhitnitsky. Sov. J. Nucl. Phys. **31** 260 (1980).
23. M Dine *at el* Phys. Lett. B **156** 199 (1981).
24. J. Kim. Phys. Rev, Lett **4** 103 (1979).
25. M.A. Shifman *et al*. Nucl. Phys. B **166** 493 (1980).
26. E.W. Kolb and M. S. Turner. Phys. Rev. Lett. **67** 5 (1991).
27. L.J. Rosenberg and K.A. van Bibber. Phys. Rep. **325** 1 (2000).
28. G.G. Raffelt. Stars as laboratories for fundamental physics. Uni. of Chigaco Press, Chicago (1996).
29. P. Sikivie. Phys. Rev. Lett. **51** 1415 (1983).
30. W. Wuensch *et al*. Phys. Rev. D. **40** 3153 (1989).
31. C. Hagmann *et al* Phys. Rev. D **42** 1297 (1990).
32. K.A.van Bibber. These proceedings.
33. S. Matsuki and K. Yamamoto. Phys. Lett. B **263** 523 (1991)
34. M. Tada *et al*. Nucl. Phys. B (Proc. Suppl.) **72** 164 (1999)
35. G. Jungmann *et al*. Phys. Rep. **267** 195 (1996)
36. L. Roszkowski Phy. Lett. B **262** 59 (1991)
37. J. Ellis *et al*. Phys. Rev. D **62** 075010 (2000)
38. M. Drees, K.A. Olive, R Arnowitt. These proceedings.

39. N.J.C. Spooner. These proceedings.
40. M. Beck *et al* Phys. Lett. B **336** 141 (1994)
41. P. Sikivie *et al*. Phys. Rev. D **56** 1863 (1997)
42. C. Copi *et al*. Phys. Lett. B **461** 43
43. R. Bernabei *et al*. Phys. Lett B **389** 757 (1996)
44. P.F. Smith *etal*. Phys. Lett. B **379** 299 (1996)
45. R. Abusaidi *etal*. Phys. Rev. Lett **84** 25 (2000)
46. P. Di Stefano *et al*, Astropart. Phys. **14** 329 (2001)
47. R. Bernabei *et al* Phys. Lett. B **480** 23 (2000)
48. A. Bottino *et al*. Astropart. Phys. **10** 203 (1999)
49. G. Gerbier *et al* Proc. DM2000 Marina del Rey (2000)
50. R. Gaitskell, Proc. Inner Space/Outer Space, Fermilab (2000)
51. N.J.C. Spooner *et al*. Proc. DM2000 Marina del Rey (2000)
52. V.A. Kudryavtsev *et al*. Phys. Lett. B **452** 167 (1999)
53. N.J.T. Smith *et al*. Phys. Lett B **467** 132 (1999)
54. N.J.T. Smith *et al* Phys. Lett. B **485** 9 (2000)
55. N.J.C. Spooner *et al* Phys. Lett. B **473** 330 (2000)
56. R. Bernabei *et al*. Phys. Lett. B **436** 379 (1998)
57. T.J. Sumner *et al* 26th ICRC **2 516** (1999)
58. G.J. Davies *et al* Phys. Lett. B **320** 395 (1994)
59. H. Wang *et al*. Proc. DM2000 Marina del Rey (2000)
60. J. Jochum *et al*. Proc. DM2000 Marina del Rey (2000)
61. C.J. Martoff *et al*. Nucl. Instr. & Meth. A **440** 335 (2000)
62. G. Jungmann *et al* Phys. Rev. D **49** 2316 (1994)
63. L. Bergström *et al* Atrophys. J. **526** 215 (1999)
64. L. Bergström *et al* Astropart. Phys. **9** 137 (1998)
65. F. Halzen *et al* Nucl. Phys. B Proc. Supp. **38** 472 (1995)
66. F. Blanc *et al* astro-ph/9707136
67. L.A. Belolaptikov *et al* Nucl. Phys. Proc. Supp. **35** 290 (1994)
68. D. F. Cowan *et al* These proceedings.
69. T. Montaruli *et al* astro-ph/9905021
70. A. Okada *et al* astro-ph/0007003
71. L. Bergström *et al* Phys. Rev. D **58** 103519 (1998)
72. T. Damour and L.M. Krauss. Phys. Rev. D **59** 063509 (1999)

Plenary Session 17

Cosmology and Particle Physics

Chair: Peter I.P. Kalmus (London)
Scientific Secretary: Toru Goto

COSMOLOGY AND PARTICLE PHYSICS

MASATAKA FUKUGITA

Institute for Cosmic Ray Research, University of Tokyo, Kashiwa 2778582, Japan
Institute for Advanced Study, Princeton, NJ 08540, USA

The state of our understanding of cosmology is reviewed from an astrophysical cosmologist point of view with a particular emphasis given to recent observations and their impact. Discussion is then presented on the implications for particle physics.

1 Overview of astrophysical cosmology

By 1970 astrophysicists were already reasonably confident that the universe began as a fireball and the hot universe then cooled as it expanded. The 2.7K radiation and the 25% helium abundance are among the strongest fossil evidence supporting this scenario. This is beautifully described in Weinberg's book *The First Three Minutes*[1] published in 1977. It is remarkable that the work over the last three decades has not found anything which would invalidate this view, but only strengthened the evidence for it. Furthermore, the subsequent research has made it possible to delineate the story beyond the first three minutes up to the present, which was missing in Weinberg's book, namely the story that is dominated by the formation of cosmic structure. Attempts to understand cosmic structure formation have greatly enriched cosmological tests both for structure formation itself and for the evolution of the universe as a background to the structure. Successful results of a number of key tests lead us to conclude that we are approaching understanding of the evolution of the universe and cosmic structure.

Cosmology today is based on three paradigms: (i) the hot Big Bang and the subsequent Friedmann-Lemaître expanding universe, (ii) the universe today being dominated by cold dark matter (CDM)[2], and (iii) the presence of inflation in some early period[3]. (ii) is still hypothetical and (iii) is even more so. Yet, we cannot construct a reasonable model of the universe without the aid of these two concepts. The most important implication of inflation is the generation of density fluctuations over superhorizon scales, the presence of which is firmly established by the observations of the cosmic microwave background (CMB) by the COBE satellite[4].

The formation of structure basically reads as follows. At some early epoch density fluctuations are generated adiabatically. The most promising idea ascribes the origin to thermal fluctuations of the Hawking radiation in the de Sitter phase of inflation, and these fluctuations are frozen into classical fluctuations in the inflation era[5]. The observed fluctuations are close to Gaussian noise with their spectrum usually represented as

$$P(k) = \langle |\delta_k|^2 \rangle \propto k^n , \qquad (1)$$

where n is close to unity. This noise is amplified by self gravity in an expanding universe[6]. The fluctuations grow only when the universe is matter dominated and the scale considered (i.e., $2\pi/k$) is within the horizon. Therefore, the perturbations are modified by a scale dependent factor as they grow

$$P(k, z) = D(z)k^n T(k) , \qquad (2)$$

where $D(z)$ is the growth factor as a function of redshift, and the transfer function $T(k) \sim 1$ for small k and $\sim k^{-4}$ for large k. The transition takes place at $k \simeq k_{eq} \simeq 2\pi/ct_{eq}$, where

$$ct_{eq} = 6.5(\Omega h)^{-1}h^{-1}\mathrm{Mpc} \qquad (3)$$

is the horizon scale at matter - radiation equality[7]. $T(k)$ is called the transfer function. Hence, the universe acts as a low-pass gravitational amplifier of cosmic noise. The amplitude of the fluctuations that enter the horizon is nearly constant $(n \simeq 1)$[8] and is of the order of 10^{-5}. The small-scale fluctuations become non-linear at $z \simeq 10 - 20$, and the first objects form from high peaks of rare Gaussian fluctuations. As time passes, lower peaks and larger scale fluctuations enter the non-linear regime, and eventually form gravitationally bound systems which decouple from the expansion of the universe. We call this state 'collapsed'. At the present epoch $(z = 0)$ objects larger than $\sim 8h^{-1}\mathrm{Mpc}$ are still in the linear regime.

The fluctuations and their evolution are described by a single function of the power spectrum $P(k)$ scaled to today, with the normalisation represented by rms mass fluctuations within spheres of radius of $8h^{-1}\mathrm{Mpc}$:

$$\sigma_8 = \langle (\delta M/M)^2 \rangle^{1/2}|_{R=8h^{-1}\mathrm{Mpc}} \qquad (4)$$

$$= \int_0^\infty 4\pi k^2 dk |\delta_k|^2 W(kR) , \qquad (5)$$

where $W(kR)$ is called the window function. The fluctuations (adiabatic fluctuations) before recombination epoch $(z \sim 1000)$ are imprinted on the CMB, and they are conveniently represented by multipoles of the temperature field as,

$$\langle (\Delta T/T)^2 \rangle = \sum_\ell \frac{2\ell + 1}{4\pi} C_\ell . \qquad (6)$$

For illustration the power spectrum is shown in Figure 1.

The empirical match of the power spectra estimated from large scale galaxy clustering and from CMB (COBE), which generically differ by a factor of 10^5, has brought us confidence that we are working with the correct theory[10,11]. Here the CDM hypothesis plays a crucial role. Without CDM this matching is

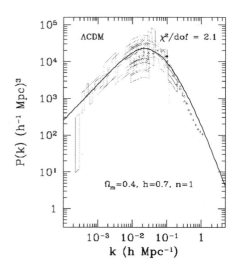

Figure 1. Power spectrum derived from large scale clustering of galaxies (shown with data points) and CMB temperature fluctuations (shown with boxes). The curve is the model power spectrum of a flat CDM universe with a cosmological constant. The figure is taken from Gawiser and Silk [9].

impossible, or more precisely we do not know any alternatives yet.

As fluctuations grow, they enter a non-linear regime. This phase was first extensively studied by the use of N body simulations. The statistical results of simulations are very well described with an approach called the Press-Schechter formalism[12], in which Gaussian fluctuations, when they exceed some threshold[13,7], follow nonlinear evolution as described by a spherical collapse model[14]. This allows us to treat statistical aspects of non-linear growth in an analytic way. Figure 2 shows the mass fraction of collapsed objects with mass $> M$ as a function of redshift z.

Whether the collapsed object forms a brightly shining single entity (galaxy) or an assembly of galaxies depends on the cooling time scale (t_{cool}) compared to the dynamical scale, $t_{\mathrm{dyn}} \sim (G\rho)^{-1/2}$ [15]. For $t_{\mathrm{cool}} < t_{\mathrm{dyn}}$, the object cools and shrinks by dissipation and stars form, shining as a (proto)galaxy. Otherwise, the object remains a virialised

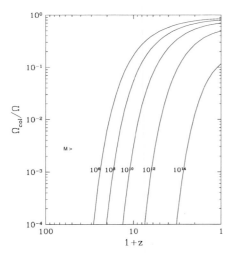

Figure 2. Fraction of gravitationally collapsed objects with mass greater than a given value M (in solar mass units) as a function of redshift $1 + z$. The calculation uses the Press-Schechter formalism and assumes the CDM model with parameters $\Omega = 0.3$, $\lambda = 0.7$ and $\sigma_8 = 0.9$.

cloud, and is observed as a group or a cluster of galaxies. In the latter case only gravity works efficiently, so that the system is sufficiently simple to serve as a test for gravitational clustering theory. Galaxy formation is very complicated due not only to the action of the cooling process, which eventually leads to star formation, but also to feedback effects, such as UV radiation and supernova winds from stars. We expect that the first galaxies form at around $z \sim 10$. The period between $z \sim 1000$ and the epoch of first galaxy formation constitutes a dark age of cosmological history. Observationally, the highest redshift securely measured is $z = 5.8$ for a quasar[16]. Even higher redshift galaxies have been reported, though the redshift measurement is not as secure as for quasars. How galaxies formed and evolved is the most important arena for astrophysical cosmologists today, both theoretically and observationally. I omit to discuss this subject in this talk, however, since it does not seem to give us insights into particle physics; it is entirely a world of astrophysics.

Before concluding this section I would emphasise that crucial cosmological tests can be made by the convergence of cosmological parameters, notably the Hubble constant, H_0, the cosmic matter density ρ in units of the Einstein-de Sitter (EdS) closure value $\rho_{\rm crit}$, Ω, and the cosmological constant or vacuum energy density in units of $\rho_{\rm crit}$, $\lambda = \Lambda/3H_0^2 = \rho_V/\rho_{\rm crit}$; $\Omega + \lambda = 1$ defines a flat (zero curvature) universe. H_0 is often represented by $h = H_0/100$ km s^{-1}Mpc^{-1}.

2 Hubble constant

The Hubble constant, which has dimension of inverse time, sets the scale of the size and age of the Universe. Recent efforts have almost solved the long-standing discrepancy concerning the extragalactic distance scale; at the same time, however, significant uncertainties are newly revealed in the distance scale within the Milky Way and to the Large Magellanic Cloud (LMC), the mile stone to the extragalactic distance.

The global value of H_0 was uncertain by a factor of two for several decades. The discovery of a few new distance indicators around 1990 has made possible an estimation of the systematic error for each indicator by cross-correlating the resulting distances (For a review of the methods, see [17]). This greatly enhanced the reliability of Hubble constant determinations. The error shrunk, notwithstanding there was still a dichotomous discrepancy between $H_0 = 80$ and 50 depending on the method used[18]. This meant that there still existed systematic effects that were not understood. The next major advance was brought with the refurbishment mission of HST, which enabled one to resolve Cepheids in galaxies as distant as 20 Mpc (HST Key Project[19]). This secured the distance to the Virgo cluster and tightened the calibrations of the extragalactic distance indicators, and resulted in $H_0 = (70 - 75) \pm 10$, 10% lower than the 'high value'[20]. Another important contribution was the discovery that the maximum brightness of type Ia supernovae (SNeIa) is not an absolute standard candle, but correlates with the decline rate of brightness[21], along with the direct calibration of the maximum brightness of several SNeIa with HST Cepheid observations[22]. The resulting H_0 was 64 ± 3, appreciably higher than 50. These results nearly resolved the long-standing controversy.

Extragalactic distance ladders are calibrated with the Cepheid period-luminosity relation. The majority of observations are provided by the HST-KP, whereas those of SNIa host galaxies are given by Saha, Sandage and collaborators. It was found by the HST-KP group that the Saha-Sandage distances that calibrate the SNIa brightness are all systematically longer by 5-10%[23] for different reasons for different galaxies[24]. The result of HST-KP is confirmed by [25]. This makes the Hubble constant from SNeIa 69 ± 4.

Another distance indicator that allows an accurate estimation is surface brightness fluctuations. The current best result based on 300 galaxies is $H_0 = 77 \pm 7$, or 74 ± 4 with a model of the peculiar velocity flow from galaxy density distributions[26]. Taking SN and SBF results we may conclude $H_0 = 71 \pm 7$ (2σ)[27], in agreement with the 2000 summary of the HST-KP group[28]. HST-KP group slightly updated H_0 in their later report[29]: $H_0 = 74 \pm 7$. Further reduction of the error needs accurate understanding of interstellar extinction corrections and metallicity effects, which is by no means easy.

All the methods mentioned above use distance ladders and take the distance to the Large Magellanic Cloud (LMC) to be 50 kpc (distance modulus $m - M = 18.5$) as the zero point. Before 1997 few doubts were cast on this. With the exception of RR Lyraes, the distances have converged to $m - M = 18.5 \pm 0.1$, i.e., within 5% error, and the RR Lyr discrepancy was blamed on its larger calibration error. The work over the last three years, notably that by the Hipparcos astrometric satellite (ESA 1997), revealed that the distance to LMC is not as secure as has been thought. The current estimate of the LMC distance varies from 43 to 53 kpc. This means that the Hubble constant is uncertain by a factor $0.95 - 1.15$[27]. Leaving this uncertainty I conclude the Hubble constant to be

$$H_0 = 71 \pm 7 \times \begin{matrix} 1.15 \\ 0.95 \end{matrix} \ \mathrm{km \ s^{-1} Mpc^{-1}} \ . \quad (7)$$

Cosmic age

The estimate of the age of the universe uses the position of turn off from main sequence tracks in the HR diagram of globular clusters. The age thus estimated also turned out to be more uncertain than had been thought. The major elements of uncertainties are the zero point of RR Lyr (20%) and the interpretation as to the formation of globular clusters, whether their heavy element abundances indicate the formation epoch or their formation was coeval independent of the heavy element abundance (20%). The first uncertainty is related to that of the LMC distance: the calibration giving a long distance to LMC gives a shorter age of clusters. The minimum of the estimated age is 12 ± 1 Gyr[30] and the maximum is 18 ± 2 Gyr, where $\pm$ reflects errors other than are discussed here; see [27].

3 Ω and Λ

The mass density parameter Ω, as measured in units of the critical density, controls the cosmic structure formation. From the cosmic structure formation point of view the role of the cosmological constant λ is subdominant: it partially compensates the slow speed of structure formation in a low density universe.

Whether Ω and λ inferred from the geometry of the universe or dynamics agree with those with the aid of structure formation models serves as an important cosmological test not only for the validity of the Friedmann universe but also for the model of cosmic structure formation.

Determinations of Ω and λ which can be carried out without resorting to specific structure formation models are: ,

(1) $H_0 - t_0$ matching using $t_0 = H_0^{-1} f(\Omega, \lambda)$, which gives $\Omega < 0.8 - 0.9$. This at least excludes the Einstein-de Sitter ($\Omega = 1$) universe.

(2) Luminosity density and the average mass to light ratio of galaxies, $\Omega = \mathcal{L} \langle M/L \rangle / \rho_{\text{crit}}$. This gives $\Omega = 0.2 - 0.5$. A slightly larger value compared to those in the literature is due to a correction for unclustered components[31].

(3)* Cluster baryon fraction, as inferred from X-ray emissivity[32,33] or the Zeldovich-Sunyaev effect[34]. This should match with the global value Ω_b/Ω, where Ω_b is the baryon density inferred from primordial nucleosynthesis.

(4) Peculiar velocity - overdensity relation, $\nabla \cdot v_p = -H_0 \Omega^{0.6} \delta$, a direct derivative from gravitational instability theory[35]. The result of this test is still grossly controversial; the estimate varies from $\Omega = 0.2$ to 1.

(5) Type Ia supernova Hubble diagram, which measures the luminosity distance, $d_L = d_L(z; \Omega, \lambda)$. The result is summarised as $\Omega = 0.8\lambda - 0.4 \pm 0.4$ [36,37].

(6) Gravitational lensing frequency. The image of distant quasars occasionally splits into two or more images due to foreground galaxy's gravitational potential. The frequency is sensitive to λ, while it is nearly independent of Ω. The current limit[31] is $\lambda < 0.8$.

Determinations that depend on specific structure formation models are:

(7) Shape parameter of the transfer function. The break of the transfer function depends on the shape parameter $\Gamma \simeq \Omega h$, and this is estimated from large scale galaxy clustering, as $\Gamma + (n-1)/2 = 0.15 - 0.3$ [38,39]. This means $\Omega \simeq 0.35$.

(8)* Matching of the cluster abundance with the COBE normalisation. The cluster abundance at $z \approx 0$ requires the rms mass fluctuation to satisfy[40] $\sigma_8 \approx 0.5 \Omega^{-0.5}$. σ_8 is also accurately determined by the large-scale fluctuation power imprinted in the CMB with the aid of eq.(2). The result depends on the power n, but notwithstanding $\Omega > 0.5$ cannot be reconciled with observations unless n is largely deviated from unity[11].

(9)* Multipoles of CMB temperature fields: the position of the acoustic peak is roughly proportional to $\ell_1 \approx 220[(1 - \lambda)/\Omega]^{1/2}$. A compilation of C_ℓ measurements favours $\Omega + \lambda \approx 1$ [41,42].

(10) Evolution of cluster abundance[43]. The evolution of rich cluster to $z \approx 0.5 - 0.8$ is more sensitive to σ_8, and one can separately determine σ_8 and Ω by studying the evolution of the cluster abundance. The results, however, are at present dichotomous: $\Omega = 0.2 - 0.45$ [44] and ≈ 1 [45].

The items with asterisks will be revisited in the next section. Among the ten tests, (5), (6) and (10) are particularly important tests for the cosmological constant. We have omitted well-known 'classical tests' such as the number count of galaxies, angular-diameter redshift relation, and the redshift distribution of galaxies, since these tests depends on the galaxy evolution, the understanding of which is still far from complete.

The conclusion we can draw from this list is a gross convergence to $\Omega = 0.2 - 0.5$ and indications for a finite λ (SNIa Hubble diagram and CMB multipoles). We shall see in the next section that these conclusions are corroborated by the new data of CMB observations, as seen in Figure 6 below.

4   Impacts of the new CMB experiments

The hot news of this year is the data release of two high resolution CMB anisotropy experiments using balloon flights. One is BOOMERanG[46] flight in Antarctica observing the southern sky and the other is MAXIMA[47], a North American flight exploring the northern sky. The two experiments have the beam sizes of the order of 10′, significantly smaller than that of COBE, and explore CMB multipoles for $\ell = 26 - 625$ and $36 - 785$, respectively. These data cover both first and second peak regions, and MAXIMA marginally reaches the third peak region. The one sigma error is about 10−20% for each data point, and normalisation errors are 10% (BOOMERanG) and 4% (MAXIMA). BOOMERanG gives more restrictive information for low ℓ.

BOOMERanG presents a map of CMB sky at 90, 150, 240 GHz (at which CMB is supposed to dominate) and 400 GHz (at which dust emission dominates). The maps at the first three frequencies show patterns in a remarkable agreement, verifying that the fluctuations are indeed intrinsic to CMB. On the other hand, the map at 400 GHz shows a completely different pattern, correlated well with interstellar dust emission known from infrared observations[48]. The 400 GHz map agrees with a map obtained from the difference of the two maps at 240 and 150 GHz.

The multipoles extracted from two observations, which observe the opposite sides of sky with respect to the Galactic plane, show an excellent agreement with each other (except for one point at $\ell \approx 140$) once we shift the data within the normalisation errors (see Figure 3). These data are also quite consistent with the earlier experiments if the average is taken over the data with large errors and large scatters. The two experiments have brought important improvement in the accuracy, which allows us to derive solid conclusions on the cosmological parameters from CMB.

A number of extensive analyses followed the data release, and the conclusions, while they are expressed in different ways, agree with each other [49−54]. Most authors use a

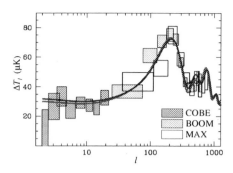

Figure 3. CMB multipoles $\Delta T_\ell = \sqrt{\ell(\ell + 1)C_\ell/2\pi}$ from BOOMERanG and MAXIMA experiments (the normalisations are shifted within one sigma calibration error), together with the COBE 4 year data. The curves show the prediction of the CDM structure formation model. The thick solid curve represents the model that satisfies the joint constraint: $\Omega = 0.35$, $\lambda = 0.65$, $h = 0.75$, $\Omega_b h^2 = 0.023$ and $n = 0.95$. The grey curve is the model that is a good fit to CMB alone: $\Omega = 0.3$, $\lambda = 0.7$, $h = 0.9$, $\Omega_b h^2 = 0.03$ and $n = 1$. Note a high baryon abundance for the latter. Figure is taken from Hu et al.[53].

general likelihood analysis in multiparameter (typically 7 parameter) space imposing varieties of prior conditions, while Hu et al. [53] developed a parametric approach to make correlations among parameters more visible.

The major conclusions we can derive from these CMB data alone are:

(1) The position of the first peak is securely measured to be $\ell = 206 \pm 6$. This means that the universe is close to flat. See Figure 4. See also (9) of section 3.

(2) The spectral index is close to unity: $n = 1 \pm 0.08$.

(3) The height of the second peak is significantly smaller than was expected from the standard set of cosmological parameters, pointing to a baryon abundance that is higher than is inferred from primordial nucleosynthesis[55]. The best fit requires $\Omega_b h^2 \simeq 0.03$ compared with $\Omega_b h^2 < 0.023$ (95% confidence) from nucleosynthesis. The CMB data are consistent with the upper limit from nucleosynthesis only at a 2 sigma level with a red tilt of the perturbation spectrum:

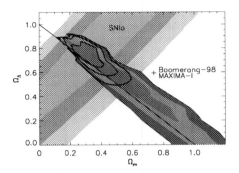

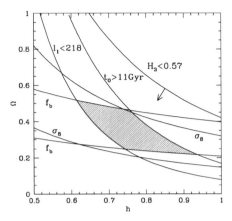

Figure 4. Constraints on the $\Omega - \lambda$ plane derived from CMB multipoles and a type Ia supernova Hubble diagram, taken from Jaffe et al.[54]. The labels are: $\Omega_m \equiv \Omega$ and $\Omega_\Lambda \equiv \lambda$. The three levels of shading mean 1, 2 and 3 sigma. The contours show the joint constraint. The straight line indicates flat universes.

Figure 5. Constraints on the $h-\Omega$ plane derived from the new CMB experiments. ℓ_1 stands for the position of the first peak, H_3 is the ratio of the heights of the third peak to the second, the curves labelled by 'σ_8' are obtained by the match of CMB data with the cluster abundance, and those with 'f_b' are the constraint from the CMB and the cluster baryon fraction. Cosmic age $t_0 > 11Gyr$ is also plotted. The curves are taken from [53].

$$0.85 < n < 0.98. \qquad (8)$$

The 2 σ lower limit derived from CMB is $\Omega_b h^2 > 0.019$.

(1) is one of the most straightforward constraints derived from CMB, and based only on geometry and acoustic physics. In the flat universe the derived constraint agrees with $t_0 < 13.5$ Gyr.

(1) and (2) are taken to be a strong support for standard cosmology based on the CDM dominance of matter and adiabatic density perturbations. They also support the idea of inflation as the origin of density fluctuations; fluctuations from pure defects (cosmic strings, textures) are excluded. On the other hand, (3) indicates marginal consistency with the baryon abundance in our current standard understanding; inconsistency would become acute if the accuracy of the data increases with the central values fixed.

The CMB data alone are not sufficient to uniquely determine the cosmological parameters. When they are supplemented with the information on large scale structure (either (7) or (8)), we are led to:

$$\Omega = 0.4\pm0.2, \quad H_0 = 75\pm15, \quad \lambda = 1-\Omega^{+0.2}_{-0.1}. \qquad (9)$$

Figure 5 presents the constraints derived from the CMB either directly or indirectly with the aid of external constraints (Ω_b from nucleosynthesis, cluster abundance, and cluster baryon fraction) in the $h - \Omega$ plane. The allowed parameter region agrees with what are discussed in section 3, as shown in Figure 6, where it is shown together with the constraints independent of CMB. The significance is that the cosmological parameters derived via the structure formation model agree with each other but also with model-independent analysis, corroborating our understanding of cosmology and structure formation. This argument will be elaborated (or falsified) upon the completion of the currently on-going CMB experiments, DASI, CBI and MAP.

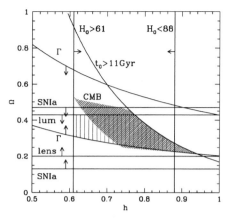

Figure 6. Summary of constraints shown in the $h - \Omega$ plane. Constraints shown by lines correspond to (1), (2), (5)-(7) of §3 and the range of the Hubble constant in §2. The allowed region derived from CMB (corresponding to (3), (8) and (9) of §3 and cosmic age) is shown by thick shading, while those derived independent of CMB are indicated by light shading.

5 Matter content of the universe

5.1 Baryons

The total baryon abundance, as represented by the baryon to photon ratio $\eta = n_b/n_\gamma$, is inferred from nucleosynthesis of light elements d, ^{4}He, and ^{7}Li. With $T = 2.728$K, we have $\Omega_b h^2 = 0.00367(\eta/10^{-10})$. A recent review of Olive et al.[56] gives two solutions $0.004 < \Omega_b h^2 < 0.010$ and $0.015 < \Omega_b h^2 < 0.023$ as 2 sigma allowed ranges. The major change over the last five years is the new input from deuterium abundance measured for Lyman α absorbing clouds (Lyman limit systems) at high redshift, and a higher He abundance reported by Izotov & Thuan[57]. Deuterium lines are observed for five Lyman limit systems; three of them gives low deuterium abundance, while the other two (including the one observed at the first time[58]) give high abundance. Assuming the lower abundance to represent the true value, Tytler takes an average of the three and concludes D/H$= 3.4 \pm 0.25 \times 10^{-5}$, which turns into $0.019 < \Omega_b h^2 < 0.021$ [59]. The primordial He abundance of Izotov & Thuan is

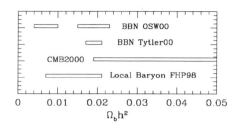

Figure 7. Baryon abundances inferred from CMB (discussed above), nucleosynthesis (BBN)[56,59],and accounting the local baryon distribution[60].

$Y_p = 0.245 \pm 0.002$, which is compared with the traditional value 0.235 ± 0.003. The difference primarily arises from the use of different calculation of the helium recombination rates and different corrections for collisional excitation than were used in the past. So the difference is of systematic nature rather than due to errors in the observations. It seems that more thorough studies of systematic errors are needed for the extraction of primordial elemental abundances.

The relative height of even to odd harmonic peaks of CMB multipoles is sensitive to the baryon abundance, and the upper limit from BOOMERanG and MAXIMA clearly rules out the low baryon option, but also it is marginally consistent with the high baryon option (see Figure 7).

We note that only 10% of baryons are frozen in stars which are visible in optical observations $\Omega_{star} = 0.004 \pm 0.002$ at $h = 0.7$; baryons in the hot gas component which is visible through X-ray emission is a similar amount. It is inferred that the rest is present around the galaxies in the form of warm gas that is not easily detectable. It seems that the high baryon option is barely consistent with the amount which is obtained by adding all budget list for baryons[60].

5.2 Dark matter

The presence of 'non-baryonic' dark matter is compelling. Among others the most important evidence is (i) the mismatch of $\Omega \approx 0.3$

with Ω_b from nucleosynthesis by one order of magnitude, and (ii) the matching of fluctuations in CMB at $z = 1000$ with those inferred from large scale structure at $z \approx 0$. If dark matter is baryonic and couples to photons, this agreement of (ii) is completely lost: yet we do not know any theories that give a correct matching between CMB and large scale structure without the aid of the CDM dominance.

A promising candidate of the dark matter is weak interacting massive particles as a relic of the hot universe, as discussed widely by particle physicists (see talks by Drees, Olive and Arnowitt[61]). If these particles were in thermal equilibrium the dark matter density would be $\Omega \sim 3 \times 10^{-27} \langle \sigma_{ann} v_{rel} \rangle^{-1}$, where average is taken over thermal distributions at the epoch of the decoupling of dark matter, which is about $T \sim 0.05 m_{dark}$. The important fact is that the desired amount of dark matter is obtained with physics of typical weak interaction scale: $\langle \sigma_{ann} v_{rel} \rangle^{-1} \sim G_F^2 T^2 \sim 3 \times 10^{-26}$ cm^{3}s^{-1} for $m_{dark} \sim 100$ GeV. This makes the lowest supersymmetric particle a natural candidate (see [62] for a review). The current most promising candidate is the neutralino that is a mixture of the photino, zino and Higgsino (tg$\beta > 3$, $M_\chi > 50$ GeV)[63,64].

Massive neutrino.

We are now convinced that neutrinos are massive. The mass density corresponding to $m_\nu \simeq 0.05$ eV is $\Omega_\nu \simeq 0.001$. This is the lower limit and the mass density could be larger if neutrino oscillation experiments are observing the difference of two or more degenerate neutrino masses. From the view point of cosmology, neutrinos can no longer be a candidate for the dominant component of dark matter. The universe dominated by neutrinos does not give correct structure formation, due to free streaming in the early universe that smooths out small scale fluctuations. It has been discussed within the EdS universe that a small admixture ($\sim 20\%$)

of light neutrino component would enhance relatively the large scale power required by observation[65]. In a low matter density universe, however, sufficient large scale power is expected without massive neutrinos, and addition of massive neutrinos only disturbs the CMB cluster abundance matching[66,67,68]. Figure 8 shows the effect of massive neutrinos on the power spectrum. The effect on small scale is apparent even if the neutrino mass is as small as 1 eV or less. Accepting the conventional baryon abundance upper limit, the CMB-cluster abundance matching leads to

$$\sum_i m_{\nu_i} < 4 eV \qquad (10)$$

at 95 % C.L.[53]. A stronger limit is derived if mass density is smaller, say $\Omega < 0.4$ [68].

Strongly interacting dark matter?

The possibility is recently discussed that dark matter might be strongly interacting. The motivation is that N body simulations with CDM models predict halo profiles more singular at the centre of the core and more small-scale objects than are observed[69]. This problem would be solved ^{a} if dark matter is strongly interacting[70] (see [71], however) or it undergoes annihilation[72]. Either scenario requires the cross section to be of the order of strong interaction. While this problem would offer another arena for particle physics, it seems much more surprising if particles with such properties exist in nature. I would like to ascribe it to our incomplete understanding of astrophysics at small scales which is certainly more complicated than large-scale dynamics.

5.3 MACHO

MACHO's (massive astrophysical compact halo objects) are a possible form of cold dark matter. These objects are collisionless, do not couple to photons, and have no cutoff scale for perturbations; so they would satisfy

---

^{a}Warm dark matter is another possibility[73].

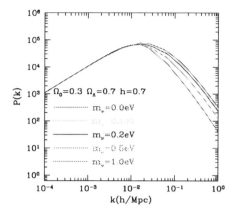

Figure 8. Effect of massive neutrinos on the power spectrum. The three species of neutrinos are assumed to have equal mass, i.e., $\sum_i m_{\nu_i}$ is three times the value indicated by labels. The curve on the top is with $m_\nu = 0$.

requirements for structure formation. The most likely candidate is aborted stars or stellar remnants. This is nearly the only possibility for baryonic dark matter, although the nucleosynthesis constraint on total baryonic matter still have to be evaded.

The novel feature of this form of dark matter is that it would cause gravitational lensing when it passes through the line of sight to background stars. The deflection angle is too small to be observed, but it causes a large magnification of the flux if it crosses the line of sight with a small impact parameter[74]. The duration of 'flash' is $\approx 70(M/M_\odot)^{1/2}$ days for background stars in LMC. Possible sources of confusion are variable or flare stars that would mimic the effect. This confusion, however, can be avoided by observing events with more than one wavelength passband, since the stellar variation is associated with the variation of temperature and the variations in different colour bands are not identical, in contrast to the gravitational lensing which is purely geometrical.

Observations have been made mainly by two groups. The MACHO collaboration conducted the observation for 5.7 years and found $13 - 17$ events towards LMC which are

compared with 70 if the halo consists entirely of MACHO[75]. The EROS group found 3 events compared to 27 expected[76]. It is not clear whether detected 'MACHO's are a part of the halo dark matter or not. From the time duration of the events these objects should have mass between 0.1 to $1M_\odot$, the mass typical of ordinary stars. The MACHO collaboration concludes the fraction f of MACHO to the total halo matter to be about 0.2, while EROS group only quotes $f < 0.4$ (at 95% C.L.) as an upper limit. The definitive and important conclusions are that (a) objects with a mass range of $10^{-7} - 10^{-2}M_\odot$ cannot be a candidate of dark matter, and (b) MACHO, if any, accounts only for a minor fraction of dark matter. We add a remark that the lensing cannot be associated with main sequence stars. The number is also too large if they are to be ascribed to white dwarfs (see [77] in this connection).

6 Theory of cosmological constant

We have now almost compelling evidence for a non-vanishing cosmological constant. The traditional problem is why the cosmological constant is so small, but now we are faced with anther problem why it is non-vanishing and it is close to matter density. Traditional attempts to understand the first problem are summarised by Weinberg[78]. Many further attempts have recently been made, but the situation seems still far from the solution. Here I quote three categories of the attempts.

(1) Time varying cosmological constant (Quintessence).

The cosmological constant is decreasing with time, as realised, for example, by a scalar field (quintessence field) slowly evolving down a potential. In this view, we can start with a large cosmological constant. The problem is why vacuum energy density and matter density are approximately equal. Peebles & Ratra[79] have considered the potential of the form,

$$V \sim M^{4+\alpha}\phi^{-\alpha} . \qquad (11)$$

This model was studied more recently by Zlatev, Wang & Steinhardt[80] who showed that the ϕ field rolls down this potential works as an attractor-like solution to the equation of motion, in the sense that the field and its derivative approach a common evolutionary track for a wide range of initial conditions ('tracker solution'). With an appropriate choice of M we get $\rho_V \sim \rho_m$. The weak points are that one has to tune M to give $\rho_V \sim \rho_m$, and an addition of a constant term spoils the desired behaviour. Armendariz-Picon et al.[81] further developed a model so that negative pressure automatically becomes effective after the epoch of matter-radiation equality. This model needs a modification of the kinetic term (k-essence).

(2) Use of exact symmetry.

With exact supersymmetry the cosmological constant vanishes. In supersymmetry theories in 4 space-time dimension, however, the disparity of fermion masses and their boson partners means breaking of supersymmetry, which inevitably results in a positive cosmological constant of the order of the supersymmetry breaking scale. This constant may be cancelled by some counter term in supergravity theory with an extreme fine-tuning, but this does not solve the problems posed above.

Witten observed in 3 dimensional theory that the disparity arises between fermion and boson masses when matter interacts with gravity, while supersymmetry is maintained[82]. Namely, $Q_{SUSY}|0\rangle = 0$ and $[Q_{SUSY}, H] = 0$, so that the cosmological constant vanishes. When this model is embedded into supergravity theory having a dilaton field, the compactified dimension is stretched to the fourth dimension in the strong coupling limit of the dilaton coupling, making the theory full four dimensional.

Once we have zero cosmological constant we must consider a mechanism to generate a small vacuum energy. An idea is to consider ultra mini-chaotic inflation. If the potential for a scalar field is very flat with the mass of the order of Hubble constant, the initial state at $\phi \approx M_{pl}$ still remains at the same value due to the Hubble viscosity, giving a very small vacuum energy. Such a minuscule mass may be generated by electroweak instanton effects. This proposal in [83] is justified by an explicit calculation within a supergravity model[84], which yields $m \sim G_F^{5/4} m_q^{5/2} M_{SUSY}^{3/2} M_{pl}^{-1/2} \exp(-4\pi^2/g_2^2)$ (G_F is the Fermi constant, m_q the quark mass, M_{SUSY} is the SUSY breaking scale, and g_2 is the SU(2) gauge coupling constant). This gives about the correct order of magnitude for the cosmological constant.

(3) Anthropic principle.

If our universe is one member of an ensemble, and if the vacuum density varies among the different members of the ensemble, the value observed by us is conditioned by the necessity that the observed value should be suitable for the evolution of our Galaxy and of intelligent life. This argument is called the anthropic principle, as explicitly stated by Carter[85]. (More discussion will be made in section 8.) An application of the anthropic principle to the vacuum energy is first discussed by Weinberg[86].

The argument is that galaxies would not have formed if vacuum energy were larger than some critical value, since vacuum energy, providing a repulsive force, hinders evolved perturbations from collapsing into galaxies. The condition is roughly expressed as $\rho_V < \rho_m$ at $1 + z \sim 4 - 5$, where galaxies formed. This translates to $\rho_V < 100\rho_m$ today, a dramatic narrowing of the allowed range for the cosmological constant.

Some authors further elaborated the argument by adopting the hypothesis called 'principle of mediocrity' which says that we should expect to find ourselves in a big bang that is typical of those in which intelligent life is possible[87]. A working assumption to calculate the probability of civilisation is that it is proportional to the number of baryons frozen into a galaxy, and the *a priori* probability of a universe having ρ_V, $P(\rho_V) \simeq$ constant. Calculations are made for galaxy formation according to the formalism described in section 1. The result depends on further input assumptions, but it generally is not too far from the value with which we actually live[88]. For instance Martel et al. obtained the probability of finding ourselves in a universe with $\lambda < 0.7$ to be 5-12%. Whether $P(\rho_V) \simeq$ constant is realised is investigated within scalar field theory in [89].

7   Theory of inflation

Cosmological inflation gives a universe a number of desirable features. The most important among them is the generation of density fluctuations over superhorizon scales. Quantum particle creations in the de Sitter phase generate thermal fluctuations corresponding to the effective temperature $T = \hbar H/2\pi$ (H being the expansion rate) and they are frozen into classical fluctuations when the scale considered goes outside the horizon in the inflation era. Inflation is the only known mechanism that can generate empirically viable fluctuations.

Theory of inflation assumes the presence of one or more scalar fields, called inflatons, which obey the field equation,

$$\ddot{\phi} + 3H\dot{\phi} + \frac{\partial V}{\partial \phi} = 0 . \qquad (12)$$

Inflation takes place if the second term that works as viscosity is large enough so that roll down of the state is sufficiently slow: This slow roll regime is realised when

$$\epsilon = \frac{M_{pl}^2}{16\pi}(V'/V)^2 \ll 1, \quad \eta = \frac{M_{pl}^2}{8\pi^2}(V''/V) \ll 1. \qquad (13)$$

There are several points to be satisfied to make the model observationally viable:

(1) $\Omega + \lambda = 1$

(2) $N_{e-\text{fold}} \geq 50$, where N is the logarithmic ratio of the scale factors before and after inflation.

(3) $V^{3/2}/M_{pl}^3 V' = 4 \times 10^{-6}$ as required by the COBE observation, $Q = 2 \times 10^{-5}$. This is a crucial condition to have successful structure formation and the presence of ourselves.

(4) The spectral index $n \approx 1$. The CMB observations indicate $n = 1 \pm 0.15$ at 95% C.L. If we require the consistency with primordial nucleosynthesis, only red tilt is allowed as shown in eq.(8). The theory of inflation predicts $n = 1 - 6\epsilon + 2\eta$ which is close to, but *not* quite unity. Simple (one field) models of inflation predict $n < 1$ (red tilt), whereas $n > 1$ requires a more complicated class of models, such as hybrid inflation.

(5) Tensor modes. Inflation may generate tensor perturbations, which contribute to the CMB fluctuation for small ℓ. Excessive tensor modes cause the CMB harmonics increasing too rapidly towards a small ℓ. A reasonable guess for the limit from the current data is $r = T/S < 0.5$, but detail statistical analyses are yet to be carried out. Inflation that takes place significantly below the gravity scale does not generate tensor perturbations[90].

It would be instructive to impose our constraints on models of inflation. For example, chaotic inflation with the potential $V = m^2 \phi^2$ (mass term only) predicts $n = 0.96$ just consistent with the upper limit of eq.(8). The value of the ϕ field at the epoch that the physical scale goes out of the horizon is $\phi_{\text{phys}} = 2.8 M_{pl}$ from (2). Inflaton mass $m = 2 \times 10^{13}$ GeV from (3). The model predicts $T/S = 0.12$, which is consistent with the observation.

There is a generic prediction of slow-roll inflation, $T/S \approx -6n_T$ with n_T the spectral index of the tensor mode. Unfortunately, this n_T is the quantity most difficult to measure. The relation

$$T/S \approx 6(1 - n) \qquad (14)$$

often quoted in the literature holds only for specific classes of models. It is argued that most models of inflation predicts the relation either close to eq.(14) or $T/S \approx 0$ [91].

Many hundreds of inflation models have been considered by now[92]. I do not intend here to discuss model building, but it seems that there are no satisfactory models. I only briefly mention the outline of models which seem more generic and why they are not satisfactory; see [92] for details.

(1) ϕ^α potential. This is the prototype for chaotic inflation[93]. The slow roll condition and e-fold require that ϕ_{phys} be larger than a few times M_{pl} and α be reasonably small. A red tilt $n < 1$, and appreciable tensor perturbations that nearly satisfy eq.(14) are predicted. We must deal with super-Planck scale physics, which is beyond the understanding of particle physics today. The real problem, however, is that there is no principle to forbid higher order terms of the form $\phi^n/M_{\mathrm{pl}}^{n-4}$, which spoils the slow roll condition for $\phi \gtrsim M_{pl}$. An attempt to forbid such terms by introducing symmetry is presented at this Conference[94], but it lacks a particle physics motivation.

(2) $V_0[1 - (\frac{\phi}{\mu})^p]$ type potential. This is typical of 'new inflation'[95]. Inflation starts with $\phi \simeq 0$, and ends with $\phi_{\mathrm{end}} \ll M_{pl}$. A red tilt is predicted, while tensor perturbations are very small due to a low energy scale involved. The difficulty is that one needs fine tuning for the initial condition. Furthermore, in most models of this type the inflaton field is not in thermal equilibrium, so that symmetry restoration at high temperature does not work. There is also a fundamental problem as to why universe has not collapsed long before the onset of this inflation.

(3) Hybrid inflation with two fields. The model is $V(\phi, \sigma) = \lambda(\sigma^2 - M^2)^2 + m^2\phi^2/2 + g\phi^2\sigma^2/2$. This is a model which combines chaotic and new inflation features[96]. For ϕ greater than some critical value ϕ_c, $\sigma = 0$ is the minimum and the model behaves as the chaotic type; ϕ remains large for a long time. At the moment when ϕ becomes smaller than ϕ_c, symmetry breaking occurs and rapid rolling of the field σ takes place. One nice feature with this model is that it can be embedded into SUSY or supergravity models, and the energy scale of the phase transition appears to agree with the unification scale[97]. The problems have been pointed out in more recent studies, however, that the model needs spatial homogeneity in the superhorizon scale in the preinflation era[98] and that the choice of the initial condition needs fine tuning to keep the ϕ filed in the desired valley[99]. This model predicts blue tilt $n > 1$ in the tree level, but the tilt can be blue or red after loop corrections, depending on input parameters[92].

The general problem with inflation is a lack of satisfactory models motivated from particle physics. For example, such an idea that simply combines inflation with supergravity theory is liable to fail because the Kähler potential is too curved with the exponential dependence of the field. Most models discussed in the literature are constructed without regarding low energy physics; so the models are those just to do it for its own sake alone. For the view pint of astrophysical applications, the discovery that inflation does not exclude open universes[100] greatly diminished its predictive power. The observation of the tilt and the strength of the tensor mode will offer an important test for the model, though the current data are not yet accurate enough for this purpose. Astrophysicists may not feel comfortable, however, unless particle physics would explain why $V^{3/2}/V'$ takes a specific value as referred to in (3). A misprediction by an order of magnitude leads to a disaster for us (see below).

One philosophically interesting consequence arises from the fact that inflation never ends (eternal inflation) whichever inflation one considers[101]. This would result in different patches of the universe expanding differently; inflation leads to great inhomogeneity at superhorizon scales. Many universes are born at arbitrary instants in many different patches; after all we are living in just one of them and observe this 'small' patch as an 'entire universe'. This 'multiverse' picture would give a base to the speculation that many physical parameters may vary in different universes. In this picture the Big Bang is no longer given any special position.

8 Anthropic principle: use or misuse?

There are many constants that appear to be so tuned that they are just appropriate for the evolution of intelligent life. We would wonder whether what we expect to observe is restricted by the condition necessary for our presence as observers[85]. We have discussed that the vacuum energy is one such example. This is also true with the matter density. If the lightest neutrino would have mass larger than 10 keV, $\Omega_\nu \sim 100$ and the age of the universe would be too short for intelligent life to have developed. If, on the other hand, $\Omega < 0.01$, say, the galaxies would not have formed. In fact, $\Omega_{CDM} \sim \Omega_b \sim \Omega_\nu \sim \lambda \sim O(1)$ up to only three orders of magnitude is an intriguing coincidence.

Another cosmologically important parameter is the strength of initial density fluctuations $Q \sim O(10^{-5})$. If this were larger by one order of magnitude, galaxies would be dominated by vast black holes; no stars or solar system could survive. If it were smaller by one order of magnitude, cooling does not efficiently work, and galaxies would not have formed[102]. From a view point of particle theory this is an obscure quantity $\sim V^{3/2}/V'$ in terms of the inflaton potential. Why this quantity takes a specific value which makes *us* habitable is puzzling.

There is a similar tricky coincidence (or providence) also in particle physics parameters, which is crucial to the evolution of in-

telligent life. The central issue is the parameters that affect element synthesis in the early universe and in stars. A small change of quark mass and/or the QCD strength stabilises or destabilises neutron, proton, deuterium, di-proton or di-neutron. Furthermore, the production of elements heavier than carbon just depends on the luck of the existence of a resonance in the ^{12}C system, which makes the bottleneck nuclear reaction $3\alpha \rightarrow {}^{12}C$ possible. A similar situation also exists with ^{16}O. Agrawal et al.[103] focused on the aspect that weak interaction scale is close to QCD scale rather than the Planck scale. Hogan[104] argued for the arrangement of mass difference among m_u, m_d and m_e, and a ± 1 MeV change of $m_d - m_u$ would disturb the existence of complex elements. He radically claims that the correct unification scheme should *not* allow calculation of $(m_d - m_u)/m_{proton}$ from first principles. Rees[102] formulated the requirement in a way that fractional binding energy of helium, $\epsilon = BE(^4He)/M(^4He)$, be tuned between 0.006 and 0.008.

It is clear that our existence hinges on delicate tuning of many parameters irrespective of whether it is a result of the anthropic principle or not. I refer the reader to Rees' book[102] for more arguments. Of course, the view on the anthropic principle is wildly divided. Hawking considers that why we are living in 3+1 dimension but not in 2+1 or 2+2 and why low energy theory is $SU(3) \times SU(2) \times U(1)$ etc. are all results of the anthropic principle[105]. Physicists usually hope that all parameters are derived up to only one from fundamental principles, and the anthropic argument appears for them to be equivalent to giving up this effort. For instance, Witten[82] states that "I want to ultimately understand that, with all the particle physics one day worked out, life is possible in the universe because π is between 3.14159 and 3.1416. To me, understanding this would be the real anthropic principle ..." It is disappointing if the anthropic principle is the solution to many problems, but such a possibility is not excluded.

9 Baryogenesis

Before concluding this talk let me mention briefly baryogenesis, which is in principle in the interface between cosmology and particle physics. The real contact between the two disciplines, however, is a subject in the future: astrophysical cosmologists argue about an error of 10% for the baryon abundance, whereas particle physicists struggle to understand the order of magnitude.

Four major scenarios proposed so far and their state-of-the-art are:

(1) Grand unification. This prototype baryogenesis idea does not receive much support. Unless baryogenesis takes place with $B - L$ violated, the baryon excess is erased above the electroweak scale under the effect of Kuzmin-Rubakov-Shaposhnikov's (KRS) sphalerons. With the presence of inflation, whether the reheating temperature is sufficiently high to produce coloured Higgs is also a non-trivial problem. In SUSY GUT the reheat temperature T_R needs to be $> m_H^c \sim 10^{17}$ GeV, which is the lower limit on m_H^c to avoid fast proton decay. In supergravity theory the reheat temperature cannot be sufficiently high ($T_R < 10^9$ GeV) to avoid copious gravitino production.

(2) Electroweak baryogenesis with the KRS effect. The necessary condition is that the electroweak phase transition is of first order. Within the standard model this requires the Higgs mass to be lower than 70GeV, which is already much lower than the current experimental limit. The possibility is not yet excluded in supersymmetric extension. The electroweak transition can be strong if the stop mass is lower than the top quark mass[106]. A possible large relative phase between the vacuum expectation values of two Higgs doublets may bring CP violation large enough to give the observed magnitude of baryon number.

(3) Leptogenesis from heavy Majorana neutrino decay and the KRS mechanism. This mechanism works in varieties of unified models with massive Majorana neutrinos. For a recent review see [107]. Another mechanism is proposed for leptogenesis via neutrino oscillation[108].

(4) Affleck-Dine baryo/leptogenesis[109]. When the flat direction of the SUSY potential is lifted by higher-dimensional effective operator, coherent production of slepton and squark fields that carry baryon and lepton number takes place in the reheat phase of inflation. The oscillation starts and ends earlier than was thought in the original paper due to thermal plasma effect[110,111,112], leading to some suppressions of the baryon or lepton number production. Notwithstanding, this is still a viable scenario, although proper treatment of leptogenesis requires the lightest neutrino mass to be smaller than 10^{-8} eV for $T_R < 10^9$ GeV [112], the limit being stronger than was obtained in [110].

10 Conclusion

Over the last few years our understanding of cosmic structure formation based on the CDM dominance and statistical description has significantly tightened. The new CMB experiments reported this year further corroborated the validity of the model. Concerning the world model of the universe we may conclude that (1) open universes are excluded, (2) the EdS universe is excluded, and (3) a non-zero cosmological constant is present. These conclusions seem to be compelling in so far as we keep the current cosmic structure formation model. Note that the CDM model is the only model known today that successfully describes widely different observations. The cosmological parameters are converging to $H_0 = 62-83$ km s^{-1}Mpc^{-1}, $\Omega = 0.25 - 0.48$ and $\lambda = 0.75 - 0.52$.

For astrophysical cosmology an interesting problem is the baryon abundance. The CMB experiments indicate the optimum value of the baryon abundance higher than is inferred from nucleosynthesis by 50%, although the two are still consistent at a 95% confidence level. It is also interesting to notice that the dominant fraction of energy density of the universe is 'invisible' (see Table 1). The vacuum energy and CDM mass occupies over 95%. Even 3/4 of baryons are invisible. That visible with optical and X-ray is less than 1% of the total energy density.

I emphasise that the standard model of the universe involves three basic ingredients which are poorly understood in particle physics: (i) the presence of small vacuum energy ($\rho_V \simeq 3$ (meV)^4), (ii) the presence of cold dark matter, and (iii) the presence of scalar fields that cause inflation. We cannot have successful cosmology without these three substances.

Although the subjects I have discussed here serve as an interface between cosmology and particle physics, the particle physics part is still poorly understood for most aspects. More successful among others are speculation of candidate dark matter, and to some extent baryogenesis. At least we have a number of successful models which are related with low energy phenomenology; yet we cannot choose among these models. The attempt to understand inflation is much poorer: most models are constructed without regarding low energy phenomenology or even unified theories. Furthermore, theorists seem to assume too easily *ad hoc* mechanisms that are not internally motivated in order to solve 'difficulties' the own model create. Most attempts look like 'particle-physics-independent' models.

Acknowledgements

I am grateful to W. Hu, A. Jaffe, M. Kawasaki, N. Sugiyama, T. Yanagida for discussions. and to D. Jackson, M. Kawasaki, T. Yanagida and Ed. Turner for their comments on the manuscript. This work is supported in part by the Raymond and Beverly Sackler Fellowship in Princeton and Grant-in-Aid of the Ministry of Education of Japan.

TABLE 1
COSMIC ENERGY DENSITY BUDGET

entity		fraction		observation
vacuum		70%		invisible
CDM		26%		invisible
baryon		4%		
	warm gas		3%	invisible
	stars		**0.5%**	optical
	hot gas		**0.5%**	X-rays
neutrino		>0.1%		invisible

Note:–bold face means observable components

References

1. Weinberg, S. 1977, *The First Three Minutes* (André Deutsch, London)
2. Peebles, P. J. E. 1982, ApJ, 263, L1
3. Guth, A. H. 1981, Phys. Rev. D23, 347
4. Smoot, G. F. et al. 1992, ApJ 396, L1
5. Hawking, S. W. 1982, Phys. Lett. 115B, 295; Starobinsky, A. A. 1982, Phys. Lett. 117B, 175; Guth, A. H. & Pi, S.-Y. 1982, Phys. Rev. Lett. 49, 1110
6. Lifshitz, E. 1946, J. Phys. U.S.S.R. 10, 116
7. Bardeen, J. M., Bond, J. R., Kaiser, N. & Szalay, A. S. 1986, ApJ 304, 15
8. Harrison, E. R. 1970, Phys. Rev. D1, 2726; Zeldovich, Ya. B. 1972, MNRAS, 160, 1p
9. Gawiser, E. & Silk, J. 2000, astro-ph/0002044
10. Wright, E. L. et al. 1992, ApJ, 396, L13
11. Efstathiou, G., Bond, J. R. & White, S. D. M. 1992, MNRAS, 258, 1p
12. Press, W. H. & Schechter, P. L. 1974, ApJ, 187, 425

13. Kaiser, N. 1984, ApJ, 284, L9

14. Gunn, J. E. & Gott, J. R. 1972, ApJ 176, 1

15. Rees, M. J. & Ostriker, J. P. 1977, MN-RAS 179, 541; Silk, J. 1977, ApJ, 211, 638; Blumenthal, G. R., Faber, S. M., Primack, J. R. & Rees, M. J. 1984, Nature, 311, 517

16. Fan, X. et al. 2000, AJ 120, 1167

17. Jacoby, G. H. et al. 1992, PASP, 104, 599

18. Fukugita, M., Hogan, C. J. & Peebles, P. J. E. 1993, Nature, 366, 309

19. Freedman, W. L. et al. 1994, Nature, 371, 757

20. Freedman, W. L., Madore, B. F. & Kennicutt, R. C. 1997, in *The Extragalactic Distance Scale*, eds. Livio, M., Donahue, M. & Panagia, N. (Cambridge University Press, Cambridge), p.171

21. Pskovskiĭ, Yu. P. 1984, Astron. Zh., 61, 1125 (Sov. Astron. 28, 658); Phillips, M. M. 1993, ApJ, 413, L105

22. Saha, A. et al. 1999, 522, 802

23. Gibson, B. K. et al. 2000, ApJ, 529, 723

24. Freedman, W. L. 2000, personal communication

25. Willick, J. A. & Batra, P. 2000, astro-ph/0005112

26. Blakeslee, J. P. et al. 1999, ApJ, 527, L73

27. Fukugita, M. & Hogan, C. J. 2000, in *Review of Particle Properties*, ed. D. E. Groom et al., Eur. Phys. J. C15, 136; Fukugita, M. 2000, in *Structure Formation in the Universe*, Proc. of the NATO ASI, Cambridge 1999, astro-ph/0005069

28. Mould, J. R. et al. 2000, ApJ, 529, 786

29. Freedman, W. L. 2000, in *New Cosmological Data and the Value of the Fundamental Parameters*, Proceedings of the IAU Symposium 201 (to be published)

30. Gratton, R. G. et al. 1997, ApJ, 491, 749; Reid, I. N. 1997, AJ, 114, 161; Chaboyer, B., Demarque, P. Kernan, P. J. & Krauss, L. M. 1998, ApJ, 494, 96

31. Fukugita, M. 2000, in *New Cosmological Data and the Value of the Fundamental Parameters*, Proceedings of the IAU Symposium 201 (to be published)

32. White, S. D. M., Navarro, J. F., Evrard, A. E. & Frenk, C. S. 1993, Nature 366, 429

33. White, D. A. & Fabian, A. C. 1995, MN-RAS, 273, 72; Arnaud, M. & Evrard, A. E. 1999, MNRAS, 305, 631

34. Myers, S. T. et al. 1997, ApJ, 485, 1; Grego, L. et al. 1999, presented at the AAS meeting (194.5807G)

35. Peebles, P. J. E. 1980, *The Large Scale Structure of the Universe* (Princeton University Press, Princeton)

36. Riess, A. G. et al. 1998, AJ, 116, 1009

37. Perlmutter, S. et al. 1999, ApJ, 517, 565

38. Efstathiou, G., Sutherland, W. J. & Maddox, S. J. 1990, Nature, 348, 705

39. Peacock, J. A. & Dodds, S. J. 1994, MN-RAS, 267, 1020; Bond, J. R. & Jaffe, A. H. 1999, Phil. Trans. Roy. Soc. London, 357, 57

40. White, S. D. M., Efstathiou, G. & Frenk, C. S. 1993, MNRAS, 262, 1023; Eke, V. R., Cole, S., & Frenk, C. S. 1996, MN-RAS, 282, 263; Pen, U.-L. 1998, ApJ, 498, 60

41. Efstathiou, G. et al. 1999, MNRAS, 303, 47

42. Lineweaver, C. H. 1998, ApJ, 505, L69

43. Oukbir, J. & Blanchard, A. 1992, A& A, 262, L21

44. Blanchard, A. & Bartlett, J. G. 1998, A&A, 332, L49; Reichart, D. E. et al. 1999, ApJ, 518, 521

45. Bahcall, N. A. & Fan, X. 1998, ApJ, 504,1; Eke, V. R., Cole, S., Frenk, C. S. & Henry, J. P. 1998, MNRAS, 298, 1145

46. de Bernardis, P. et al. 2000, Nature, 404, 955

47. Hanany, S. et al. 2000, astro-ph/0005123

48. Schlegel, D. J., Finkbeiner, D. P. & Davis, M. 1998, ApJ 500, 525

49. Lange, A. E. et al. 2000, astro-ph/0005004

50. Balbi, A. et al. 2000, astro-ph/0005124

51. Tegmark, M. & Zaldarriaga, M. 2000, astro-ph/0004393

52. Bridle, S. L. et al. 2000, astro-ph/0006170

53. Hu, W., Fukugita, M., Zaldarriaga, M. & Tegmark, M. 2000, astro-ph/0006436

54. Jaffe, A. H. et al. 2000, astro-ph/0007333

55. Hu, W. et al. 2000, Nature 404, 939

56. Olive, K. A., Steigman, G. & Walker, T. P. 2000, Phys. Rep. 333-334, 389-407

57. Izotov, Y. I. & Thuan, T. X. 1998, ApJ, 500, 188

58. Songaila, A., Cowie, L. L., Hogan, C. J., & Rugers, M. 1994, Nature 368, 599

59. Tytler, D., O'Meara, J. M., Suzuki, N. & Lubin, D. 2000, astro-ph/0001318

60. Fukugita, M., Hogan, C. J. & Peebles, P. J. E. 1998, ApJ, 503, 518

61. Olive, K. A. 2000, these proceedings; Drees, M. 2000, these proceedings Arnowitt, R. 2000, these proceedings

62. Jungman, G., Kamionkowski, M. & Griest, K. 1996, Phys. Rep. 267, 195

63. Ellis, J., Falk, T., Olive, K. A. & Schmitt, M. 1997, Phys. Lett. B413, 355

64. Drees, M. et al. 2000, astro-ph/0007202

65. Jing, Y. P., Mo. H. J., Börner, G. & Fang, L. Z. 1993, ApJ, 411, 450; Klypin, A., Holtzman, J., Primack, J. & Regős, E. 1993, ApJ. 416, 1

66. Hu, W., Eisenstein, D. J., Tegmark, M. 1998, Phys. Rev. Lett. 80, 5255

67. Valdarnini, R., Kahniashvili, T. & Novosyadlyj, B. 1998, A& Ap. 336, 11

68. Fukugita, M. Liu, G.-C. & Sugiyama, N. 2000, Phys. Rev. Lett. 84, 1082

69. Navarro, J. F. & Steinmetz, M. 2000, ApJ, 528, 607; Moore, B. et al. 1999, ApJ, 524, L19; McGaugh, S. S. & de Block, W. J. G. 1998, ApJ, 499, 41; Ghigna, S. et al. 1999, astro-ph/9910166

70. Spergel, D. N. & Steinhardt, P. J. 2000, Phys. Rev. Lett. 84, 3760; Goodman, J. 2000, New Astron. 5, 103

71. Moore, B. et al. 2000, ApJ, 535, L21; Yoshida, N., Springel, V., White, S. D. M. & Tormen, G. 2000, astro-ph/0002362

72. Kaplinghat, M., Knox, L. & Turner, M. S. 2000, Phys. Rev. Lett. 85, 3335

73. Sommer-Larsen, J. & Dolgov, A. 1999, astro-ph/9912166; Hannestad, S. B. & Scherrer, R. B. 2000, Phys. Rev. D62, 043522

74. Paczyński, B. 1986, ApJ 304, 1

75. Alcock, C. et al. 2000, astro-ph/0001272; see also astro-ph/0003392

76. Lasserre, T. et al. 2000, astro-ph/0002253 (A& A in press); Glicenstejn, J. F. 2000, these proceedings

77. Ibata, R. et al. 2000, ApJ, 532, L41

78. Weinberg, S. 1989, Rev. Mod. Phys. **61**, 1

79. Peebles, P. J. E. & Ratra, B. 1988, ApJ, 325, L17; Ratra, B. & Peebles, P. J. E. 1988, Phys. Rev. D37, 3406

80. Zlatev, I., Wang, L. & Steinhardt, P. J. 1999, Phys. Rev. Lett. 82, 896; Steinhardt, P. J., Wang, L. & Zlatev, I. 1999, Phys. Rev. D59, 123504

81. Armendariz-Picon, C., Mukhanov, V. & Steinhardt, P. J. 2000, astro-ph/0004134; ibid 0006373

82. Witten, E. 2000, astro-ph/0002297; see also Witten, E. 1995, Int. J. Mod. Phys. A, 10, 1247

83. Fukugita, M. & Yanagida, T. 1994, Kyoto preprint YITP-K-1098 (unpublished); idem 1996, in *Cosmological Constant and the Evolution of the Universe*, ed. K. Sato et al. (Universal Academy Press, Tokyo) p.127

84. Nomura, Y., Watari, T. & Yanagida, T. 2000, Phys. Rev. D61, 105007

85. Carter, B. 1974 in *Confrontation of Cosmological Theories with Observational Data*, Proceeding of IAU Symposium 63,

ed. M. S. Longair, p.291; Carr, B. J. & Rees, M. J. 1979, Nature, 278, 605

86. Weinberg, S. 1987, Phys. Rev. Lett., 59, 2607

87. Vilenkin, A. 1995, Phys. Rev. Lett. 74, 846

88. Efstathiou, G. 1995, MNRAS, 274, L73; Martel, H., Shapiro, P. & Weinberg, S., 1998, ApJ, 492, 29; Garriga, J., Livio, M. & Vilenkin, A., 1999, astro-ph/9906210; Weinberg, S. 2000, astro-ph/0005265

89. Garriga, J. & Vilenkin, A. 2000, Phys. Rev. D61, 083502; Weinberg, S. 2000, Phys. Rev. D61, 103505

90. Lyth, D. H. 1997, Phys. Rev. Lett. 78, 1861

91. Hoffman, M. B. & Turner, M. S. 2000, astro-ph/0006321

92. Lyth, D. H. & Riotto, A. 1999, Phys. Rep. 314, 1; Liddle, A. R. & Lyth, D. H. 2000, *Cosmological Inflation and Large-Scale Structure* (Cambridge Uniersity Press, Cambridge); see also Linde, A. D. 1990, *Particle Physics and Inflationary Cosmology* (Harwood, Chur)

93. Linde, A. D. 1983, Phys. Lett. 129B, 177

94. Kawasaki, M. 2000, these proceedings; Kawasaki, M., Yamaguchi, M. & Yanagida, T. 2000, Phys. Rev. Lett. 85, 3572

95. Albrecht, A. & Steinhardt, P. J. 1982, Phys. Rev. Lett. 48, 1220; Linde, A. D. 1982, Phys. Lett. 108B, 389

96. Linde, A. 1991, Phys. Lett. B259, 38

97. Copeland, E. J. et al. 1994, Phys. Rev. D49, 6410; Stewart, E. D. 1995, Phys. Rev. D51, 6847; Linde, A. & Riotto, A. 1997, Phys. Rev. D56, R1841

98. Vachaspati, T. & Trodden, M. 2000, Phys. Rev. D61, 023502

99. Tetradis, N. 1998, Phys. Rev. D57, 5997; Mendes, L. E. & Liddle, A. R. 2000, Phys. Rev. D62, 103511

100. Sasaki, M., Tanaka, T., Yamamoto, K. & Yokoyama, J. 1993, Phys. Lett. B317, 510; Linde, A. D. 1995, Phys. Lett. B351, 99; see also Coleman, S. & De Luccia, F. 1980, Phys. Rev. D21, 3305; Gott, J. R. 1982, Nature 295, 304 (1982).

101. Linde, A. D. 1986, Phys. Lett. B175, 395

102. Rees, M. 1999, *Just Six Numbers*, (Basic Books, New York)

103. Agrawal, V., Barr, S. M., Donoghue, J. F. & Seckel, D. 1998, Phys. Rev. Lett. 80, 1822

104. Hogan, C. J. 1999, astro-ph/9909295

105. Hawking, S. W. 2000, in *Structure Formation in the Universe*, Proc. of the NATO ASI, Cambridge 1999

106. Carena, M., Quirós, M. & Wagner, C. E. M. 1998, Nucl. Phys. B524, 3

107. Buchmüller, W. & Plümacher, M. 2000, hep-ph/0007176

108. Akhmedov, E. Kh., Rubakov, V. A. & Smirnov, A. Yu. 1998, Phys. Rev. Lett. 81, 1359

109. Affleck, I. & Dine, M. 1985, Nucl. Phys. B249, 361

110. Dine, M., Randall, L. & Thomas, S. 1996, Nucl. Phys. B458, 291

111. Allahverdi, R., Campbell, B. A. & Ellis, J. 2000, hep-ph/0001122

112. Asaka, T., Fujii, M., Hamaguchi, K. & Yanagida, T. 2000, hep-ph/0008041

Plenary Session 18

Reports on
IUPAP and ICFA

Chair: Enzo Iarocci (INFN)

IUPAP
THE INTERNATIONAL UNION OF PURE AND APPLIED PHYSICS

PETER I. P. KALMUS

Department of Physics, Queen Mary, University of London, London E1 4NS, UK
Chair : C11 Commission on Particles and Fields

The aims of IUPAP are the following:

- To stimulate and promote international co-operation

- To sponsor suitable international meetings

- To foster free circulation of scientists

- To reach international agreement on physical symbols and units

- To encourage research and education

The first three of these are of particular relevance to this conference.

The President of IUPAP is Burton Richter, and he has already said a few words about IUPAP, so I will concentrate on C11. There are about 20 Commissions and a few Associate Commissions covering the various areas of physics. They are labelled C1 ..... C20. C11 is the Commission on Particles and Fields.

The membership of C11 is intended to cover a wide geographical spread of the regions in which particle physics research is carried out. Members serve for 3 years, but can be re-elected for a further term, particularly if they then become Secretary or Chair. The Commission also suggests candidates for the future, but the actual election is carried out once every 3 years at the IUPAP General Assembly. Commissions can also request Associate Members from other commissions which have overlapping interests. C11 has Associate Members from C4 (Cosmic Rays), C12 (Nuclear Physics) and C19 (Astrophysics).

The present membership is

P. I. P. Kalmus	UK	Chair
H.-U. Klein	Germany	Vice-Chair
S. Yamada	Japan	Secretary
M. Davier	France	
W. Hoogland	Netherlands	
T. Huang	China	
E. Iarocci	Italy	
G. Mikenberg	Israel	
V. Luth	USA	
A. Santaro	Brazil	
A. S. Sissakian	Russia	
M. Zeller	USA	

Associate Members	
T. K. Gaisser	C 4
P. Kienle	C12
B. Sadoulet	C19

The Commission meets once per year, at the ICHEP or Lepton-Photon Symposium in alternate years. The rest of the business is carried out by email, phone, post or fax.

One of the main items of the Commission is to decide the location of the ICHEP and LP conferences. The Commission invites and receives bids from interested groups of physicists, and tries to ensure that a reasonable rotation of regions is achieved. The Commission also checks that the general IUPAP rules are observed, which include ensuring that physicists are not denied attendance at the conference on grounds of nationality.

C11 also sponsors ICFA - the International Committee for Future Accelerators, which will be described in the next talk, and co-sponsors (with other commissions) PANAGIC - the Particle and Nuclear Astrophysics and Gravitation International Committee.

A list of past and future ICHEP and LP

conferences sponsored by C11 is given below.

1990	Singapore	ICHEP
1991	Geneva	LP
1992	Dallas	ICHEP
1993	Cornell	LP
1994	Glasgow	ICHEP
1995	Beijing	LP
1996	Warsaw	ICHEP
1997	Hamburg	LP
1998	Vancouver	ICHEP
1999	Stanford	LP
2000	Osaka	ICHEP
2001	Rome	LP
2002	Amsterdam	ICHEP
2003	Fermilab	LP

Comments and queries about C11 can be sent to

| Sakue Yamada | (Secretary) | sakue.yamada@kek.jp |
| Peter Kalmus | (Chair) | p.i.p.kalmus@qmw.ac.uk |

There is a Web page
http://ccwww.kek.jp/iupap.c11/

REPORT ON ICFA

H. SUGAWARA

KEK, High Energy Accelerator Research Organization
Tsukuba, Ibaraki

A. ICFA decided to form two task forces to study the feasibility of "global accelerator network" or "global laboratory": one from social and the other from technical point of view.

1. ICFA Task Force for Global Laboratory (social)

task

(a) How to maintain active interest and participation in project in all member institutes.
(b) Identify areas that need to be developed.
(c) Work out mechanisms for co-operation and decision taking.

Members

A.Astbury (TRIUMF,coordinator)
J.Colas (ATRAS/CMS)
L.Evans (CERN)
Y.Kimura (KEK)
E.Paterson (SLAC)
M.Shochet (CDF)
D.Trines (DESY)

We obtained an interim report from the task force 1.(social issues) which contains the following contents.

Interim report

(a)

construction stage	heavy involvement of all the participating laboratories
operation stage	Just participating in the remote operation does not guarantee the raison d'etre of participating laboratories Ex) Rutherford Lab TRIUMF

(b) to be considered later
(c) necessity of a host nation
$\rightarrow$ necessity of a host laboratory
to avoid "SSC mistake"

host nation : pay $1/3 \sim 1/2$ of the costs
partners : in kind contribution
$\times$ LHC

No. of partners ≤ 10
host laboratory   technical contribution
civil engineering
power,   water
legal issues
environmental problems

◯ Experience from LEP/ LHC/ TEVATRON/ detectors
Two principles
(1) consensus
(2) peer pressure
$\Rightarrow$ cost problems and delays

"not suitable for the machine building"
o hierarchical line management
o minimum number of funding sources

2. The task force 2.(technical issues) gave us the interim report which can be summarized in the following way.

Members

E.Balakin (INP)
P.Czarapta (FNAL)
D.Harthill (Cornell)
S.Myers (CERN)
N.Phinney (SLAC)
M.Serio (INFN)
N.Toge (KEK)
F.Willeke (DESY,coordinator)
C.Zhang (IHEP)

(Introduction)

(1) Next machine requires a new mode of international and inter-laboratory collaboration.

(2) An obvious consequence of multi-laboratory project is that the accelerator will be located a considerable distance from the contributing institutions.

(3) Once the implications of remote operation are understood, the location of any new project may become much less important than it is today.

Goals

A) a model of remote operation
B) examination of the experience available
C) existing examples
D) technical developments required
E) communication to lab directors

(A) example of e^+e^- - collider

(1) general structure

hardware components : injectors, damping
                      ring main linacs,
                      beam delivery final
                      focusing

responsibilities of collaborators
    (1) control system infrastructure
    (2) all BPM electronics
    (3) all power supplies of particular type
    (4) all rf controls(or all of a particular frequency)

(2) Control management

    responsible for the coordination of design and construction, operation and maintenance management of common infrastructure.

    ← control board

(3) design board

(4) machine operation

control center
   ∘ need not to be at a fixed geographical location
   ∘ virtual center

conditions
   ∘ each collaborator will have a fully functioning control room capable of operating the entire accelerator
   ∘ "permanent video conference"

mechanism

1. machine operation board

2. maintenance

3. skilled maintenance

4. trouble shooting and intervention on components

5. communications

6. radio active safety

7. exceptional situation

B. ICFA discussed the issue of Global Science Forum which can be summarized in the following.

GLOBAL SCIENCE FORUM

(Formerly MEGASCIENCE F.)
(Members from OECD Countries)
(Reports to Governments)
April 2000 Workshop to consider forming Working Group on HEP.
ICFA expressed concern that there be input from HEP community. All activities truly Global, not just OECD countries.
ICFA representative at all activities.

Recommendation to GSF that a
"consultative Group" on HEP be set up
GSF approved

Report to GSF, and all participating
Governments, no later than June 2002.

C. ICFA discussed the communication
among regional studies on the future of high
energy physics.

STUDIES — FUTURE HEP PROJECTS

Japan	Update JLC Milestones
	— End 00
US	Update 98Gilmann Report
	— 00
	Form New HEPAP Panel
	— Early 01
	Report — Late 01
	Snowmass Study
	— Summer 01
Europe	ECFA/DESY Study
	— Sept 00
	TDR — March 01
	ECFA Special Panel on Long Term
	Perspectives of Particle Physics in Europe (L.FOA – Chair) June 01

> ICFA suggests that these regional study groups collaborate to reach a global consensus.

Plenary Session 19

Summary

Chair: Peter I.P. Kalmus (London)
Scientific Secretaries: Makoto Sakamoto and
Masahiro Kuze

SUMMARY

H. SUGAWARA

KEK, High Energy Accelerator Research Organization
Tsukuba, Ibaraki

1. Introduction

2. Gauge Interaction

3. Higgs Interaction

4. Cosmology

5. Theoretical Issues

6. Future Projects

7. Social Issues

1 Introduction

It is often said that the 20th century was a century of physics and the 21st century will be a century of biology - a statement which physicists can hardly accept without any reservation. While it can easily be dismissed when biologists use it as a catchline to justify their budget request, it must be taken seriously when it means that biology finally is becoming accessible to physics: the 21st century may be the century for biology to be understood in physical terms just as chemistry was in the 20th century. This does not, however, mean that living systems can be understood within the existing physics framework. In fact the generic definition of a complex system of which the living system[1] is a typical example, implies that the straight forward application of neither statistical physics nor modern chaos or fractal idea is adequate.

On practical side the size of data sets which biologists must handle is getting to be comparable to that which high energy physicists have been handling. The present status of biology may be compared to that of the transition period when high energy physics shifted from cosmic ray observations to accelerator based experiments. The complete data set in biology includes the complete determination of the genes and of the 3-dimensional structures and the functions of all the proteins which they produce. Even the technical methods are becoming closer to physics. For example, in the determination of protein structure, the utilization of various beams as X-rays, neutrons or muons is becoming essential, thus making an accelerator an indispensable tool for biology.

The foregoing discussion is presented to point out the importance of understanding on the larger scientific environment surrounding our high energy physics community. Among the major laboratories, DESY is ahead in this respect thanks to the brilliant insight of late Bjorn Wiik who initiated the TESLA project. By adopting the superconducting technology they are able to combine a next generation high energy physics project with one of structural biology. Other regions are also not far behind in their own x-ray FEL projects[2].

It is true that the ability of higher energy physicists to share the technology they developed with the structural biologists stems from the fact that the physicists have a long tradition of technological strength in achieving their physics goals. But this does not mean that high energy physicists are leading in every aspect of technology. The utilization of the internet, for example, for the purpose of sharing data with other researchers is far more advanced in biology, space science, ocean science etc[3]. A mechanism to liberate the data from the possession of each experimental group must be seriously considered thus changing the research style of high energy physics in a way which is more appropriate for the 21st century.

Can high energy physics exist forever? At least one can say that science in general was initiated and has developed in a completely detached manner from eschatology. History tells us that eschatological ideas occasionally leaped over scientific thought, making the latter usually less fertile. I hope that the superstring theory will stay uncontaminated by eschatologists' ideas since it has enormous potential to be fertile in the 21st century, as E.Witten claims. It may become common knowledge for all the high energy physicists a few generations ahead without which even a detector design would be impossible. In the mean time, we must proceed step by step towards the complete understanding of the standard model and also of what could be slightly beyond it.

The development of high energy physics in the last two years was not as dramatic as in some other periods. But we observe many projects which either have started construction or set forth to data taking. While the Tevatron, LEP, HERA, CESR, BEPC and VEPP-2M are vigorously taking data primarily to understand the gauge interaction, RHIC, $DA\Phi NE$ and the Nuclotron in Dubna are now engaged in their initial operations. We can expect a lot of new results from them at the next High Energy Physics Conference. LEP is about to end its very productive career leaving CERN able to concentrate on LHC project. In neutrino physics K2K is vigorously taking data with some preliminary results already being available. The SNO detector has also begun to provide us some results which are complementary to the SK data. Two long base line experiments, Minos and Opera, are either under construction or under serious consideration. In B physics, two B-factories, PEPII and KEKB, performed a very successful commissioning and provided us with some preliminary data. In the field of cosmology,

Boomerang and Maxima-I made observations on the anisotropy of CMBR and determined the cosmological constant and the spectral index parameter, impact of which can be enormous to the foundations of particle physics.

2 Gauge Interactions

Gauge interactions are better understood than Higgs interactions. Still the understanding is more or less confined to the individual "low energy" electroweak or QCD interaction and the evidence for unification is far from concrete. A LEP–SLC result (Fig.1) is the only available indication of the unification at the scale of 10^{16}GeV within the framework of MSSM although this does not exclude the possibility of a premature unification at the TeV scale as is predicted by the theory of large extra dimensions[4]. The existence of the extra dimensions will be investigated in future experiments to study the running property of the Higgs coupling.

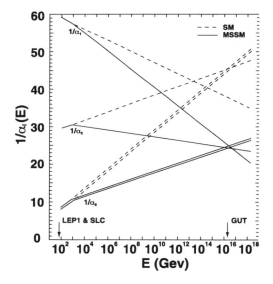

Figure 1. Running gauge couplings in the SM and the MSSM with SUSY particles ($\geq 1TeV$)(prepared by G.C.Gho of Ochanomizu University)

The LEP project contributed enormously to our understanding of the electroweak interaction, and it also provided some evidence for possible supersymmetry, thus indicating a relatively low lying Higgs particle.

The termination of the LEP operation is scheduled to be October 1, 2000 [a] and we are all thankful to those who have been involved in this outstanding project, particularly to the technical staff who contributed to:

(1) The technology of superconducting cavities with niobium film on a copper substrate, which made heat transfer a lot more efficient.

(2) The invention of non-evaporable getter pump which was adopted by later projects such as KEKB.

Some theorists are so convinced of the LEP result on the supersymmetric unification that they have come up with several possible mechanisms for broken supersymmetry which leads to the MSSM at lower energies (Table 1). Typically we have gravity mediated models and gauge mediated models. New additions are based on the proposals[5] that our 4-dimensional space-time is a brane world in some space-time with much higher dimension [b]. There is not much study in the case of intersecting branes although the broken supersymmetry seems to be most natural in this case[5]. Neither gravity mediated nor the gauge mediated models give us the understanding of why the cosmological constant is so small although these models use this fact in an essential way. Notice also that they give different candidates for the LSP, which might be investigated in a future facility such as a "neutralino factory" as proposed by K.Maki and S.Orito[6] some time ago.

Let us turn now to QCD. The short-to-medium distance QCD is being studied in the LEP, HERA and TEVATRON machines presently. The issues include the quark or gluon distribution functions, photon structure function, spin structure function, NC and CC cross sections, etc.. So far, the agreement between theory and experiment is quite remarkable. There was a report of the measurement of xF_3 for the first time in this conference where xF_3 is defined by,

$$\frac{d^2\sigma_{NC}(e^{\pm}P)}{dxdQ^2} =$$

$$\frac{2\pi\alpha^2}{xQ^4}[Y_-F_2(x,Q^2)\mp Y_-xF_3(x,Q^2)-y^2F_L(x,Q^2)],$$

where $Y_{\pm} = 1 \pm (1-y)^2$ and $y = sin^2(\theta^*/2)$ with θ^* the CM scattering angle. x is the usual Bjorken parameter: $x = Q^2/(2p \cdot q)$.

Statistical errors will be substantially reduced when the HERA luminosity upgrade project[7] is realized in the future, aiming at an integrated luminosity of

$$\int_{2001}^{2006} Ldt \sim 1(fb)^{-1}.$$

Long distance QCD is being studied in such machines as the SPS, TEVATRON, BEPC, RHIC, $DA\Phi NE$, and the Nuclotron. The last three are new-comers which promise to be productive in the next several years to deepen our understanding of long distance QCD. RHIC is currently operated with the energy of 65+65 GeV/n with 5% luminosity of the design value, but the lifetime has already achieved the design value of 10 hours. The suppression of J/ψ production due to the screening of the confining potential, already observed in SPS[8], should be investigated here also to see if it really implies the existence of a quark-gluon plasma. Some preliminary physics results have been reported in this conference[8].

$DA\Phi NE$ certainly needs some improvement in the machine parameters and it is on its way. Some new results from the KLOE detector have been reported[9]. The Nuclotron[10] in Dubna has begun to produce heavy ion beams, although the energy/n is much lower than that of RHIC.

---

[a]This is at the time of Osaka Conference. The situation may change depending on the last minute physics output.

[b]Brane world or intersecting branes may not need supersymmetry to understand the hierarchy of mass scale. They are shown here for comparison.

	Gravity mediated model	Wyle anomaly	Gauge mediated model	brane world	intersecting branes
Cosmological constant	×	×	×	×	
FCNC	m (gravitino mass) > several to 10 TeV m_f becomes large or universal scalar mass	○	○ (masses are generation independent) $m_f \sim m_g$ (gaugino mass)		
μ problem	× (but Giudice etal)		×		
μ · B connection	○		△ { D-N : × / MMM : renormalization / DTM : ?		
Gravitino mass	large (FCNC)		small (1 KeV ~ 5 GeV)		
connection to string theory	○		where are the messengers? ($E_8 \times E_8$)	?	
Moduli problem	○ if m > 10 TeV		?		
$SU_3 \times SU_2 \times U_1$ breaking	renomalization		3 - loop		
LSP	neutralino	neutralino	gravitino		

Table 1. Comparison of models of supersymmetry breaking

The issues which should be addressed in long distance QCD studies include hadron spectroscopy, determination of weak parameters such as ε'/ε or B→Kπ or K$\pi\pi$ amplitudes, QCD phase transitions, screening of the QCD potential etc..

On the theoretical side, the standard calculational technique is lattice QCD. Its present status and the future possibilities are summarized in Table 2. The current machine speed is of the order of 1 Tflops and in the near future it will be increased to 10–100 Tflops. The cost will be almost comparable to or even higher than some major high energy experiments. All the issues of long distance QCD can be addressed with this technology.

An important contribution of theorists to facilitate the analysis of experimental data comes from a group of researchers doing automatic computation of Feynman diagrams (Table 3). As one can see from the table, there is already a program to compute up to 1-loop level for some electro-weak processes.

The next topic is the precision measurement of the gauge interaction. Muon g-2 is being measured in Brookhaven as has been reported in this conference[11]. 1PPM precision which can be attained in the year 2000 data will give us a useful information on the standard model. A new way[12] of measuring the fine structure constant in addition to the methods of quantum hall effect[13], ac Josephson effect[14], atom interferometry[15] or electron g-2[16] has been proposed. It is based on the relation

$$\alpha^2 = \frac{2R_\infty}{e} \frac{m_n}{m_p} \frac{m_p}{m_e} \frac{h}{m_n}, \left(R_\infty = \frac{\alpha^2 m_e c}{2h}\right)$$

The de Broglie wavelength of the neutron together with a velocity measurement can give the value of $h/m_n (h/m_n = \lambda v)$. α in this case is given by

$$\alpha^{-1} = 137.0360119(51)(37ppb).$$

The best value for α is still from the electron g-2 and is given by

$$\alpha^{-1} = 137.03599958(52)(3.8ppb).$$

The precise measurement of the low energy pion form factor in VEPP-2M[17] and the measurement of R at 2 GeV in BEPC[18] were reported in this conference. Both give an important contribution to the muon g-2 and the latter is reported also to change the value of predicted Higgs mass when combined with the $sin^2\theta_w$ measurement of SLC and LEP experiments[19].

The measurement of 1s-2s or the hyperfine splitting of antihydrogen is an excellent way to test the CPT theorem. CERN experiments (ASACUSA, ATHENA, ATRAP)[20]

Group	Present machine	Future plan
QCDSP	dedicated: 1TFlops	6TFlops, till 2002
CP-PACS	dedicated: 0.6TFlops	Plan of 100TFlops, proposed
JLQCD	Hitac SR8000 : 1.2 TFlops	next upgrade scheduled in 2006
APE	dedicated: 0.1 - 1 TFlops	combined proposal
UKQCD	CRAY T3E/152	based on ApeNEXT
German groups	Quadrics	10TFlops, till 2003
Fermilab	dedicated: ACPMAPS	combined proposal
MILC	commercial machines	

Table 2. LATTICE QCD: present and near future

1. Automatic computation systems working in the world

   ALPHA(Italy), CompHEP(Russia), FDC(China), FeynArts/FeynCalc series (Germany),

   GEFICOM(Germany/Russia), GRACE(Japan), MadGraph(USA), NIKHEF setup (Holland)...

2. Examples of Achievements

   - 4-fermion generators (76 processes) for LEP-2 experiments (ALPHA, CompHEP, GRACE).

   - $e^-e^+ \rightarrow$ 6-fermion (ALPHA, GRACE), $e^-e^+ \rightarrow \chi_1^0\chi_1^0 \bar{q}_1 q_2 q_3 \bar{q}_4$ (GRACE), $\gamma\gamma \rightarrow$ 4-fermion (CompHEP).

   - $ep \rightarrow el^+\Gamma X$(GRACE).

   - $p\bar{p} \rightarrow Wb\bar{b}j$ (CompHEP), $pp \rightarrow W^+W^-b\bar{b}j$ (MadGraph), $gg, q\bar{q} \rightarrow 8g$(ALPHA).

   - 1-loop calculation for $e^-e^+ \rightarrow W^-W^+$, $\gamma\gamma \rightarrow W^+W^-$, $W^+W^- \rightarrow W^+W^-$ (FeynArts/FeynCalc, GRACE), $e^-e^+ \rightarrow W^+\mu^-\bar{\nu}_\mu$(GRACE).

   - Hadronic Higgs decay in $O(\alpha_s^3)$, $O(\alpha\alpha_s)$ corrections to $Z \rightarrow b\bar{b}$, etc. (GEFICOM).

   - 4-loop β-function ($\sim$50,000 diagrams) (NIKHEF setup).

Table 3. Current Status of Automatic Computation

have started and we can anticipate much from these experiments on this issue.

3 Higgs Interaction

Unlike the gauge interaction, the Higgs interaction does not seem to be well understood even in the low energy region. Theorists hope that it will be unified with the gauge interactions at the energy where we start seeing extra-dimensions. In fact the Higgs bosons are supposed to be the extra-dimensional components of the gauge bosons. It is extremely important therefore, to see in the future experiment if the running Higgs coupling will imply this unification with the gauge coupling. At the present stage of development, we still do not know at what mass range the Higgs particle exists, although there is an indirect constraint from LEP and SLC experiments combined with the top-quark mass measurement at TEVATRON[19].

The standard model states that the current quark masses and the lepton masses come from the interaction of these particles with the Higgs boson, perhaps except for the right handed neutrino masses, if they exist. It is, therefore, very important to measure the neutrino masses to understand the Higgs-lepton Yukawa couplings and the physics beyond the standard model through the right handed neutrino masses.

Other indirect information on the Higgs quark, Yukawa couplings comes from the study of the CKM matrix especially from the study of CP-violation for its phase.

3.1 Neutrino Oscillations

In the frame of diagonal lepton matrix the neutrino mixing matrix[21] can be written as,

$$U_{MNS} =$$
$$\begin{pmatrix} C_{13}C_{12} & C_{13}S_{12} & S_{13}e^{-i\delta} \\ -C_{23}S_{12}-S_{13}S_{23}C_{12}e^{i\delta} & C_{23}C_{12}-S_{13}S_{23}S_{12}e^{i\delta} & C_{13}S_{23} \\ -S_{23}S_{12}-S_{13}C_{23}C_{12}e^{i\delta} & -S_{23}C_{12}-S_{13}C_{23}S_{12}e^{i\delta} & C_{13}C_{23} \end{pmatrix}$$

$\nu_e \rightarrow \nu_e$
$P = 1 - sin^2 2\theta_{13} sin^2 \Delta_{32} - cos^4 \theta_{13} sin^2 2\theta_{12} sin^2 \Delta_{21}$
$\sim 1 - sin^2 2\theta_{13} sin^2 \Delta_{32}$
for Chooz, Kamland, PaloVerde
$\sim 1 - cos^4 \theta_{13} sin^2 2\theta_{12} sin^2 \Delta_{21}$
for solar neutrino (neglecting matter modification)

$\nu_\mu \rightarrow \nu_\mu$
$P = 1 - cos^2 \theta_{13} sin^2 2\theta_{23} sin^2 \Delta_{32}$
for atmospheric disappearance, K2K

$\nu_\mu \rightarrow \nu_\tau$
$P = sin^2 2\theta_{23} cos^4 \theta_{13} sin^2 \Delta_{32}$ for Minos, Opera

$\nu_\mu \rightarrow \nu_e$
$P = sin^2 2\theta_{13} sin^2 \theta_{23} sin^2 \Delta_{32}$ for Minos, Opera, JHF

CP violation
$P = (\nu_\alpha \rightarrow \nu_\beta) - P(\overline{\nu}_\alpha \rightarrow \overline{\nu}_\beta)$
$\cong 1/2 sin2\theta_{13} sin2\Delta_{21} sin\delta$
For bimaximal case(neglecting matter effect)

Table 4. Examples of the probability of important processes in the neutrino oscillation.

We know from the solar neutrino experiments and from the atmospheric experiments that

$$\Delta_{sun} = |\Delta_{21}| \ll |\Delta_{31}| \sim |\Delta_{32}| = \Delta_{atm},$$

with $\Delta_{ij} = (m_i^2 - m_j^2)L/4E$. The transition probability for $(\nu_\alpha \rightarrow \nu_\alpha)$ can be written as

$$P_\alpha = 1 - 4 \sum_{k>j} P_{\alpha j} P_{\alpha k} \sin^2 \Delta_{jk},$$

and a similar formula for $\nu_\alpha \rightarrow \nu_\beta$ ($\alpha \neq \beta$) where $P_{\alpha i} = |U_{\alpha i}|^2$.

Important examples are listed in Table4. From the second and the third boxes of this table and from Fig.2, we know that the atmospheric and the long base line experiments give an information on $(\sin^2 2\theta_{23})$ and Δ_{32}. The value of $(\sin^2 2\theta_{23})$ seems to be close to 1 and $|\Delta_{32}|^2 \simeq 3 \times 10^{-3} eV$. In Fig.2 the K2K line is the sensitivity line and the measured value will be available soon. Future long baseline experiments like Minos, Opera, ICARUS or JHF can measure the value of $(\sin^2 2\theta_{13})$ which is already known to be small by combining the reactor data and the SK data (Fig.3).

Solar neutrino observation can give an information on $(\sin^2 2\theta_{12})$ and (Δ_{21}) as shown in the first box of Table 4. The Fig.4 shows that the MSW large mixing angle solution

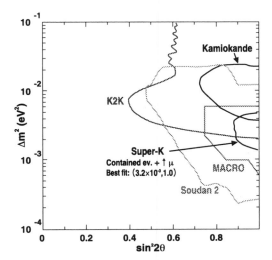

Figure 2. $\nu_\mu \to \nu_\tau @90\% CL$. Curves show allowed regions for atmospheric neutrino experiments and sensitivity region for K2K (10^{20} pot)

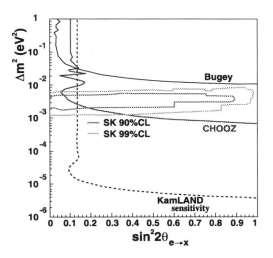

Figure 3. Bounds on $\sin^2 2\theta_{13}$. The curves show excluded regions (90% CL) for reactor experiments and allowed regions for Super-Kamiokande

is the most favored choice due to the recent day-night effect observation in Super Kamiokande. We are, therefore, in a situation where both ($\sin^2 2\theta_{12}$) and ($\sin^2 2\theta_{23}$) are close to 1, which is somewhat amusing and gives some hint to the model building.

Tantalizing SNO experiment is now underway. Observation of the neutral current process

$$\nu + D \to \nu + P + n,$$

will start after installing the chloride scintillator in the next year. Charged current events

$$\nu + D \to e^- + P + P$$

in addition to the SuperK $\nu + e \to \nu + e$ elastic events have already been observed and clearly indicate the usefulness of the heavy water target[22].

Models for the finite neutrino masses can be classified into several categories:
(1) SU(2)$_L \times$ SU(2)$_R \times$U(1)
(2) SO(10)
(3) Supersymmetric models with or without R symmetry
(4) extra-dimensional models.

Let me choose two examples of mechanisms which lead to small neutrino masses.
1) SO(10)model by C.H.Albrecht & S.M. Barr (P.L.B461(1999)218):

$$U_{\alpha J} = (U_L^\dagger U_\nu)_{\alpha J}$$
$$M_\nu = -N^T M_R^{-1} N \qquad (see-saw).$$

Interplay between U_L and U_ν can lead to bimaximal mixing angle.
2) Model based on the Randall– Sundrum metric by Y.Grassman & M.Nuebert, (P.L.B474(2000)361).

$$ds^2 = e^{-2kr_c|\Phi|}\eta_{\mu\nu}dX^\mu dX^\nu - r_c^2 d\Phi^2$$

$$S = \int d^4X d\Phi \sqrt{G}(E_a^A i\bar{\psi}\gamma^a \partial_A \psi - m\bar{\psi}\psi)$$

$$+ \int d^4X \sqrt{-g_{vis}}\{g\bar{\psi}_L(X)H\psi_R(X,\Phi=\pi)\},$$

here $\psi = \Sigma\psi_n f_n$, $f_0 \propto \left(\frac{\langle H\rangle}{m_{pl}}\right)^{\frac{\nu-1}{2}}$, $\nu = \frac{m}{k}$,

$$g_{vis} = e^{-kr_c|\Phi|}G$$

$$m = \begin{pmatrix} gvf_0 & gvf_1 & \cdots & gvf_n \\ m_1 & \cdot & & gvf_n \\ & & \cdot & \\ & & & m_n \end{pmatrix}$$

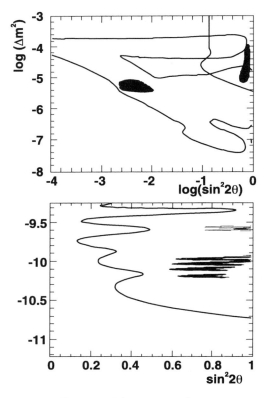

Figure 4. Solar neutrino data

As one can see from these models there are typically two mechanisms to obtain small neutrino masses. The first is the standard see-saw mechanism[23] which leads to the small Majorana neutrino mass. The second model depends on the two miracles: one is the smallness of the warp factor $e^{-2kr_c|\Phi|}$ and the other is that of the ground state wave function f_0 on the brane. This case leads to a Dirac neutrino mass.

Experimentally the Majorana mass limit comes from the observation of the neutrino-less double beta decay[24]:

$$|\Sigma_i|U_{ei}|^2 m_i| \leq 0.2eV.$$

There are also two sources for the absolute mass limit.

1) Cosmological:

$$m_\nu < 5.5eV$$

$$m_\nu < 2.4(\Omega_m/1.7 - 1)eV \text{ for } 0.2 < \Omega_m < 0.5$$

R.A.C.Croft etal, P.R.L83(1999)1092

$$m_\nu < 0.6eV, \Omega_m = 0.3, H_0 \leq 80kmsec^{-1}M_{pc}^{-1}$$

M.Fukugita etal, P.R.L84(1999)1082

2) Tritium:

$$m_\nu < 2.5eV, 2.8eV \quad \text{Mainz, I.N.R}$$

We have seen growing enthusiasm to construct a neutrino factory in the last several months in connection with a muon collider. The question is do we really need it? From the physics point of view it is necessary to determine all the elements of U_{MNS} which includes the CP phase δ. Except for δ all the other parameters may be determined using the existing methods of solar, atmospheric, reactor and the accelerator neutrino from π decay. The justification of the neutrino factory comes from the measurement of CP phase and this is a very important parameter. It is, therefore desirable to continue the R/D effort especially to reduce the cost of a neutrino factory.

3.2 CP Violation in K and B Decays

The Higgs couplings in the standard model is supposed to give

$$V_{CKM} = \begin{pmatrix} 1 - \lambda^2/2 & \lambda & A\lambda^3(\rho - i\eta) \\ -\lambda & 1 - \lambda^2/2 & A\lambda^2 \\ A\lambda^3(1 - \rho - i\eta) & -A\lambda^2 & 1 \end{pmatrix}$$

There was a report on the new value for ε'/ε from the CERN NA48 experiment but KTEV confirms the old value. The present status can be summarized in Fig.5.

$$Re(\varepsilon'/\varepsilon) = \begin{cases} (28.0 \pm 4.1) & \times 10^{-4} & \textbf{KTEV} \\ (12.2 \pm 4.9) & \times 10^{-4} & \textbf{NA48} \\ (14.0 \pm 4.3) & \times 10^{-4} \end{cases}$$

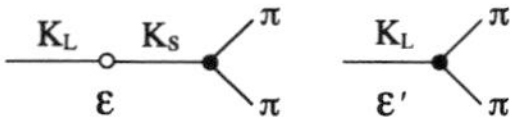

Figure 5. Measurement of $Re(\varepsilon'/\varepsilon)$

Final States	No.evt (bgd) / No.tagged evt (bgd)	
	BABAR	Belle
∫ L	14.8 fb⁻¹ total to 7/28 9.1 fb⁻¹ used for CP	6.2 fb⁻¹
$B \to J/\psi K_S$	$K_S \to \pi^+\pi^-$ 124±12/85 $\pi^0\pi^0$ 18±4/12	74 / 44
$B \to J/\psi K_L$	82±14±9 not used for CP	102 (47.6) / 42 (20)
$B \to \psi' K_S$	27±6 / 23	13 / 5
$B \to \chi_{c1} K_S$	44±9 not used for CP	5 / 3
$B \to J/\psi \pi^0$	/	10 / 4
$B \to J/\psi K_S \pi^0$	/	/
$B \to J/\psi K_S \rho^0$	/	/
Sin (2Φ_1)	0.12±0.37±0.09	0.80 $^{+0.44}_{-0.50}$ (without $J/\psi K_L$) 0.45 $^{+0.43}_{-0.44}$ ($J/\psi K_L$;combined)

CDF : sin2Φ_1 = 0.79 $^{+0.41}_{-0.44}$

Table 5. New Results from B-factories

	PEP-II	**KEKB**
Energy e^+/e^- (GeV)	3.1/9.0	3.5/8.0
Peak luminosity (10^{33}cm^{-2}s^{-1})	2.28	2.04
Current e^+/e^- (A)	1.25/0.75	0.47/0.42
Number of bunches	606	1146
Beta function at IP β_x^*/β_y^* (cm)	50/1.25	70/0.7
Beam sizes at IP σ_x^*/σ_y^* (μm)	170/7.0	112(e^+), 145(e^-)/1.7
Beam-beam tuneshift $e^+ \xi_x/\xi_y$ $e^- \xi_x/\xi_y$	0.06/0.04 0.04/0.02	0.036/0.037 0.029/0.023
Max int. luminosity /day(1/pb)	151	90
Max int. luminosity /week(1/pb)	890	505
Int. luminosity by July 25(1/fb)	16.0	6.9
Number of days for physics run	~ 300 since 5/99	~ 200 since 6/99

Table 6. Performances of PEP-II and KEKB

The penguin calculation is underway for ε'/ε by various groups using lattice QCD(Table 2). Both experiments and theories must eventually come to an agreement but it seems that the process is very tedious and rather subtle.

Let us turn to the issue of CP violation in B decays. There have been various efforts at the CESR, LEP and TEVATRON machines to study CP violation in B decays. For example, the CLEO collaboration recently studied whether there is an obvious deviation from the standard model in the parameter:

$$A_{cp} = \frac{P(B^- \to J/\psi K^-) - P(B^+ \to J/\psi K^+)}{P(B^- \to J/\psi K^-) + P(B^+ \to J/\psi K^+)}$$

which is predicted to be of the order of λ^2 in the standard model. They found it to be $[+1.8\pm4.3(stat)\pm0.4(syst)]\%$ which is perfectly consistent with the standard model.

Following theoretical suggestions[25] on one hand and the idea for an asymmetric collider[26] on the other, the two B-factories have been constructed to extensively study the CP-violation scheme of Kobayashi and Maskawa[27]. The clean angle measurement of the unitary triangle has been started in both B-factories PEPII at SLAC and KEKB at KEK. The preliminary results from BABAR and BELLE are summarized in Table 5. Notice that the integrated luminosity is already $14.8 fb^{-1}$ for BABAR and $6.2 fb^{-1}$ for BELLE. Even with this huge luminosity the value of $sin(\phi_1)$, or $sin(2\beta)$ still suffers from large errors. The machine status of PEPII and KEKB is compared in Table 6. Both machines are aiming at going beyond the $10^{34}cm^{-2}sec^{-1}$ luminosity from the current value of $2 \times 10^{33}cm^{-2}sec^{-1}$. Such new devices as the superconducting-crab-cavities may be needed to reach the 10^{34} level. It is also essential to understand and to be able to respond to various types of instabilities. When they go beyond the 10^{34} level, the study of very rare non penguin contaminated processes may become realistic.

4 Cosmology

Cosmology is an important subject by itself but it is also an essential source of information on some of the crucial particle physics parameters. Here I would like to briefly sum-

	Leptogenesis-sphaleron (Fukugita - Yanagida)	Affleck-Dine	Sphaleron (Kuzmin, Rubakov Shaposhnikov)
ΔB	$\Delta L \to \Delta B$ by $\Delta(B-L)=0$ sphaleron	flat potential $\to <q_i> \neq 0$	Sphaleron
$\not{CP}$	lepton secton	gaugino mass matrix	hadron sector
non -thermal	Out-of-equilibrium decay of ν_R	Out-of-equilibrium decay of " "	
role of supersymmetry	$\times$	$\bigcirc$	

Table 7. Models of Baryogenesis

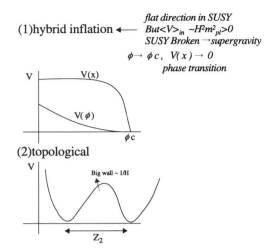

Figure 6. Models of inflation scenario

marize the recent developments in 1) baryo-genesis, 2) CMBR measurements and 3) Dark matter searches.

4.1 Baryogenesis

There is no observational development since this subject is concerned with a rather well established single parameter.

Table 7 summarizes the well known pro-posals to explain the baryogenesis. This table is to show how each model tries to respond to the three Sakharov conditions. The role of supersymmetry in each model is also pointed out. The study of CP violation in the lep-ton sector can be important in distinguishing these models. This gives a strong support to the construction of the neutrino factory.

4.2 Measurement of Anisotropy of CMBR

Both the Boomerang collaboration and the Maxima-1 collaboration seem to give a non-zero value for the cosmological constant[28]. This probably is the first rather convincing evidence for the non-vanishing cosmological constant although there was some indication in the past, for example, from the measure-ment of the typeI supernova. The content of both Boomerang and Maxima-1 is so rich

that it gives information on the baryonic or dark matter content of the universe and also on the value of the spectral index which may distinguish various models of inflation (Fig.6). In Fig.6 the hybrid inflation makes essential use of the fact that the supersym-metric models can have certain flat directions in a potential and it should also be a su-pergravity model so that the flat direction corresponds to $V \neq 0$. A phase transition pulls this down to V(x)=0 at certain value of ϕ. Although the hybrid inflation model appeals most favorably to particle theorists, this is not the only possibility. The topolog-ical inflation model utilizes a discrete sym-metry which divides the universe into many domains with a thick network of domain walls which in turn provide a large enough vacuum energy for the universe to inflate. Either way the slow roll condition gives rise to a value of spectral index close to 1 which seems to be confirmed by Boomerang and by Maxima-1.

4.3 Dark Matter and Axion Searches

The WIMP searches are continuing.
Dama result was first reported in the last High Energy Conference in Vancou-ver. It is unfortunate that its result has al-

ready been almost excluded by the CDMS collaboration[29].

Various types of axion searches are also going on: microwave, x-ray, laser, nuclear transition, beam dump, reactor, telescope searches, etc., as is discussed in the report by L.J.Rosenberg and K.A.van Bibber[30]. There is a recent theoretical study[31] on f as large as $10^{16} GeV$, rather than $10^{12} GeV$, in which case the detection may be beyond the current technology.

5   Theoretical Issues

There was a report on the idea of extra dimensions in a parallel session at the Vancouver conference. On the other hand one of the plenary talks is totally devoted to this idea in this conference[32]. In fact we have seen a huge number of theoretical papers on this subject in the last year and a half. A rather different, but not entirely disconnected idea on the brane world hierarchy possibility, was suggested by Randall and Sundrum[33] and this work also received much attention.

The latter model is along the line of conventional string theory and should be understood within that scheme. The hierarchical mass scales are understood in terms of the renormalization flow of the gauge coupling in the standard string theory[34]. If one can understand the "warp factor" of Randall-Sundrum model in this way, we are more or less certain of the relevance of this model within the scheme of string theory. Another aspect is the similarity of this model with the proposal of Horava and Witten[35] on the $E_8 \times E_8$ heterotic string models within the M-theory framework although the latter theory is an 11-dimensional theory and so requires some reasonable compactification of the extra dimensions (Fig.7).

The observation of anisotropy of CMBR implies that the cosmological constant may be non-zero. How can this be reconciled or can be computed within our theoretical

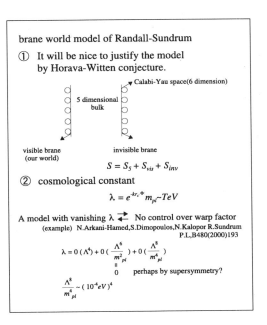

Figure 7. Brane world issues

framework? The currently popular Randall-Sundrum brane world model depends crucially on the interplay of cosmological constants in our world, in the invisible brane and in the bulk. It is amazing that there is already a proposal (Fig.7) to get a vanishing cosmological constant in this scheme although it is likely that the proposed model may suffer from the existence of a naked singularity and a possible loss of control of the warp factor. This model is interesting in another way because the proof of the vanishing constant is valid only in the zeroth order in the expansion:

$$\lambda = 0(\Lambda^4) + 0(\frac{\Lambda^6}{m_{pl}^2}) + 0(\frac{\Lambda^8}{m_{pl}^4}) + \cdots$$

where Λ is the mass scale of the order of TeV. If one can prove the vanishing of the $0(\frac{\Lambda^6}{m_{pl}^2})$ term in the expansion assuming supersymmetry for example, we end up with

$$\lambda \propto \frac{\Lambda^8}{m_{pl}^4} \sim (10^{-4} eV)^4,$$

which is the right order of magnitude. There are, of course, many other ideas as is discussed by M.Dine[36] in one of the plenary sessions of this conference.

6 Future Projects

The importance of the neutrino factory has already been pointed out in this talk. The interest is global and certainly needs some coordination if one takes into account its cost and the difficulty of the technology. To obtain a sufficiently strong neutrino beam, the proton source must be powerful enough $(4 \sim 5MW)$ and the beam width must be large to make the cooling and phase rotation easier. All these require extensive R&D (See "A Feasibility Study of a Neutrino Source Based on a Muon Storage Ring" edited by N.Holtkamp and D.Finley, FermiLab). We should keep in mind that even the necessity of cooling is debatable, as is touched upon in the talk by M.Tigner[37]. Brute force acceleration by a fixed field alternating gradient synchrotron with a large momentum acceptance may work, although this itself requires a lot of R&D.

The next subject is the LHC. It is already under construction and a lot of studies are being done on its physics potentiality. As is summarized very briefly in Fig.8, the study of Higgs interaction may not be as extensive as in the case of linear collider especially when one is restricted to the gluon fusion processes. It was pointed out recently by D.Zeppenfeld (Fig.8) that WW fusion processes may add extra information. Study of supersymmetry may be rather dramatic. For example, there is a possibility of substantial deviation from the non-supersymmetric value in the cross section for the four-jet processes as a function of "missing energy" which is defined to be:

$$M_{eff} = P_{T,1} + P_{T,2} + P_{T,3} + P_{T,4} + E_T,$$

as is shown in Fig.8.

◎Higgs decay branching ratio

- $gg \to H$ $H \to \gamma\gamma$ $M_H < 150 GeV$
- $ \to ZZ$ $M_H > 130 GeV$
- $ \to WW$ $M_H > 130 GeV$
- $WW \to H$ $H \to \gamma\gamma$ $M_H > 150 GeV$
- $ \to ZZ$ $M_H < 140 GeV$
- $ \to WW$ $M_H > 120 GeV$

D.ZEPPENFELD
hep-ph 0005151

◎Superparticle

Missing energy signature

$m_0 = m_{1/2} = 400 GeV, A_0 = 0, \tan\beta = 2.0, sign\mu = +$

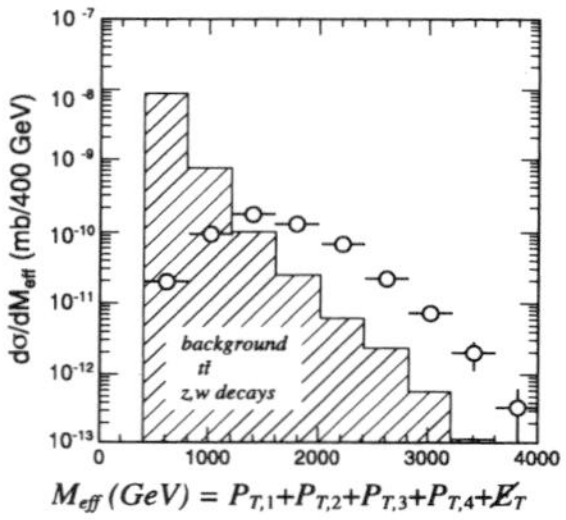

I.Hinchliffe,
F.E.Paige,
M.D.Shapiro,
J.Soderqvist, and
W.Yao

P.R.D55 (97) 5520

$M_{eff} (GeV) = P_{T,1} + P_{T,2} + P_{T,3} + P_{T,4} + E_T$

Figure 8. LHC physics

It is gratifying that all the major technological difficulties are now under control and the magnets and cryogenic systems are either expecting the delivery soon or being ordered. The target date of the machine commissioning still remains July 2005 although we are informed of some delay in the civil engineering schedule. World wide effort is being made to handle an enormous amount of data either on-line or off-line. It is hoped that the Atlas and the CMS collaborations provide a good model for the research style in future large high energy physics projects.

The last item to be discussed is the linear collider which is supposed to supplement a hadron machine like the LHC. A supersymmetric Higgs can be extensively studied with the linear collider which is energetic enough to produce the Higgs particles. Fig.9 exemplifies the claim and as one can see from this figure we may be able to pinpoint the mass of the CP-odd Higgs particle M_A although we do not find it directly. The sound approach seems to be to construct a linear

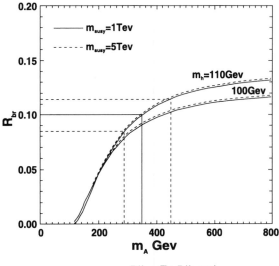

Figure 9. $R_{br} = \frac{B(h \to c\bar{c}) + B(h \to gg)}{B(h \to b\bar{b})}$

	C	X	Tesla
W	2856×2 $\sim 6GHz$	2856×4 $\sim 11GHz$	$1.3GHz$
V^2/P	$\sqrt{\omega}$		$\omega / (A\omega^2 + B)$
$\Delta E(\sigma \Delta E \simeq \varepsilon_L)$	large		small
pulse length	$\sim \mu s$ (high peak power)		$10^3 \mu s$
bunches/pulse	10^2		10^3
bunch spacing	$\sim ns$		$10^2 ns$
bunch length	$< 10^2 \mu m$		$> 10^2 \mu m$
$\varepsilon_{x,y}$	$10, 10^2 (\times 10^{-6} \, mrad)$		$\lesssim (10, 10^2) \times 10^{-6} \, mrad$
accelerating field	50MV/m		25MV/m
emittance growth (wake power)	ω^3 large		ω^3 small
cost construction			higher(?)
cost operation	beam power		beam helium

<u>damping systems</u>
⚪Tesla

General layout of the "dogbone"-shaped damping ring

⚪C or X   2GeV ring

Table 8. Comparison of LC projects

collider which is capable of producing the Higgs, with as little investment as possible for extendability. It is, of course, desirable to choose a technology which has the maximum extendability, as long as that technology is feasible.

So far there are basically two technologies proposed to reach the energy range of the Higgs ($\approx 500 GeV$). A very unique two beam linac is being considered for much higher energy. Table 8 compares the two linear collider technologies: warm and cold. In the warm case, the higher frequency is desirable to avoid surface absorption, but the cold case has some optimum value and 1.3 GHz was chosen for TESLA. The lower frequency has some advantage as is seen in the table. There is a three order of magnitude difference in the pulse length which makes the TESLA power source much easier. In fact the klystron development is one of the major issues either in C or X band. 50 to 75MW class klystron of a few micro-second pulse length are now being developed at SLAC and at KEK. TESLA also has the advantage in making the energy spread smaller because of the large bunch length. This makes it more suitable for FEL applications. Another advantage for TESLA comes from the wake power which is proportional to ω^3. Especially tight requirements apply to X band, and tolerances on the structure are less than $10\mu m$. The advantage of a warm machine comes from its ability to reach higher accelerating fields compared to the cold machine. C or X can reach more than 50MV/m but the current TESLA design in 25MV/m. TESLA also must depend on a huge damping and injection system to respond to 10^3 bunches/pulse, with 10^2 ns bunch spacing. It is probably premature however to ask experts to compare these technologies base on the current R&D status.

7   Social Issues

When it became clear that the LHC project needed some contributions from non-member countries and the physicists from non-member countries resolutely expressed their

342

willingness to participate in the project, European community still wanted to keep CERN as it is, i.e.the European Center rather than the global center. This regional or national sentiment still exists globally.

Under this circumstance we are roughly left with two choices in building the next generation of large machine in our field to avoid unnecessary competition. One is to take the regional sentiment seriously and to try to share several big projects within our field or even including big projects of other fields — basket method. The other possibility is to go to the global laboratory idea by rendering the advantage of ownership of the backyard as small as possible. Both technical and social aspects of this possibility are being investigated in the task forces set up in ICFA. It is undoubtedly the duty of high energy physicists to maintain the culture of high energy physics in all regions and in all interested nations. To achieve this goal within the framework of a global laboratory is not at all easy. The interim report of a task force on social issue clearly identifies this problem.

A big project certainly involves the process of decision making by government officials and politicians, but it is my hope that the scientists can take a leadership roll in this process. It requires a good deal of imagination and the readiness for compromise with the idealistic goals hidden deep in our minds. This must be an important part of a lesson we all learned out of the demise of the SSC (Fig.10). It is crucial for our future not to repeat the SSC-LHC saga. An old soldier like myself who kept fighting for so many years in this field and is soon to reach the end of his career is greatly worried that the nightmare (Fig.11) may become a reality sometime in the 21st century.

Acknowledgment

It is my pleasure to thank all the people who helped me prepare this talk by providing information, by preparing materials, by discussing the content or by checking English. They include Asacusa group, B.Jacak, Babar group, BELLE group, A.E.Bondar, G.C.Cho, M.Dine, K.Hagiwara, K.Higashijima, D.Hitlin, K.Hubner, Y.Kimura, T.Kinoshita, M.Kobayashi, T.Kondo, S.Kurokawa, M.Kuze, K2K group, J.Learned, I.Maiani, K.Nakamura, T.K.Ohska, Y.Okada, S.Olsen, S.Ozaki, S.Pakvasa, B.Richter, M.Sakamoto, T.Shintake, B.Sieman, F.Takasaki, N.Toge, S.Yamada, A.Wagner.

I also want to express my sincere gratitude to the organizing committee for providing me a chance to review the entire field of high energy physics.

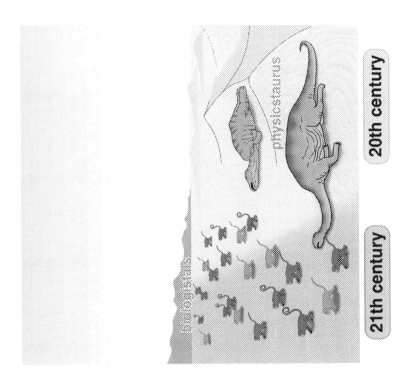

Figure 11: *nightmare?*

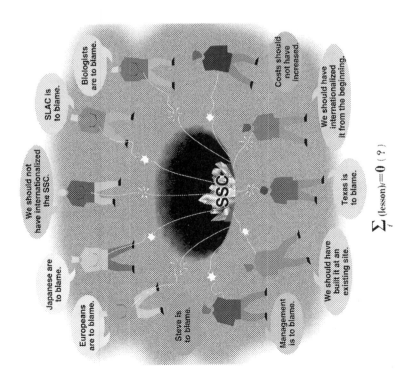

$$\sum_i (lesson)_i = 0 \ (?)$$

Figure 10: *Did we learn a lesson from the SSC?*

References

1. M.Gell-Mann and S.Lloyd, Conplexity 2, 44 (1996)
2. Linac Coherent Light Source(LCLS) Design Report, SLAC-R-521(1998) ;Several unpublished notes in KEK
3. The panel discussion chaired by B.Richter at the 1998 ICFA Seminar at Fermilab
4. N.Arkani-Hamed, S.Dimopoulos, and G.Dvali, phys. Lett. B 429, 263 (1998)
   L.Randall and R.Sundrum, Phys.Rev. Letters, 83, 3370 (1999)
   For the early unification see,
   K.R.Dienes, E.Dudas, T.Gherghetta, Phys. Letters, B436, 55 (1998); Nucl. Phys, B537, 47 (1999)
   D.Ghilencea and G.G.Ross, phys.letters B442, 165 (1998)
5. Within the framework of superstring theory the intersecting branes are discussed extensively. See, J.G.Polchinski, String Theory, Vol1 and 2 (Cambrige Monographs on Mathematical Physics), (1998). From phenomenological side we have N.Arkani-Hamed, S.Dimopoulos, G.Dvali and N.Kaloper, Phys. Rev. Letters 84,586 (1999)
6. K.Maki and S.Orito, Phys.Rev, D57, 554 (1998)
7. See, for example, K.R.Long, Presentation at the XIVth International Workshop on High Energy Physics and Quantum Field Theory, Moscow, June 1999
8. B.Jacak, these Proceedings
9. A.Passeri, these Proceedings; P.Gauzzi, these Proceedings
10. A.I.Malakhov, these Proceedings
11. R.Carey, these Proceedings
12. Kruegen etal, Metro logia 36, 147(1999)
13. Teffery etal, Metro logia 35, 83(1998)
14. Mohr and Taylor, RMP 72, 351 (2000)
15. Young, thesis (1997)
16. Hughs and Kinoshita, RMP71, S133 (1999)
17. A.E.Bondar, these Proceedings
18. Z.Zhao, these Proceedings
19. A.Gurtu, these Proceedings
20. http://asacusa.web.cern/ASACUSA/, http://athena.web.cern/athena/, http://hussle.harvard.edu/natrap/
21. Z.Maki, M.Nakagawa and S.Sakata, Prog. Theor. Phys. 28, 870(1962)
22. J.R.Klein, these Proceedings
23. T.Yanagida, Proc. Workshop on Unified Theories and Baryon number in the Universe, KEK, Tsukuba, Japan, 1979, edS. O.Sawada and A.Sugamoto; M.Gell-Mann, P.Slansky and P.Ramond, Supergravity, North - Holland, 1979, edS. P.van Nieuwenliuizen and D.Freedman
24. Heidelberg -Moscow Collaboration, Phys. Rev. Lettes 83, 41(1999)
25. A.Carter and A.O.Sanda, Phys. Rev. Letters 45, 952 (1980)
26. P.Oddone, Proceedings of the UCLA workshop on linear collider B-factory conceptual design,ed. D.H.stork,World Scient January (1987)
27. M.Kobayashi and T.Maskawa, Prog. Theor. Phys. 49, 652 (1973)
28. M.Fukugita, these Procceddings
29. R.Abusaidi etal, Phys.Rev.Letters 84, 5699 (2000)
30. L.J.Rosenberg and K.A.van Bibber, Phys. Reports 325, 1 (2000)
31. M.Hashimoto, K.-I,Izawa, M.Yamaguchi and T.Yanagida Phys. Letters B437, 44 (1998)
32. L.Hall, these Proceedings
33. L.Randall and R.Sundrum, Phys. Rev. Letters 83, 3370 (1999)
34. E.Verfinde and H.Verlinde, hep-th/9912018
35. P.Horava and E.Witten. Nucl. Physics B460, 506(1996)
36. M.Dine, these Proceedings
37. M.Tigner, these Proceedings

Parallel Session 1

Hadron Spectroscopy
(light and heavy quarks, glueballs, exotics)

Conveners: Alexandre M. Zaitsev (IHEP) and
Zhengguo Zhao (Chinese Academy of Sciences)

THE MIXING OF THE $f_0(1370)$, $f_0(1500)$ AND $f_0(1710)$ AND THE SEARCH FOR THE SCALAR GLUEBALL

A. KIRK

School of Physics and Astronomy, Birmingham University, U.K.
E-mail: ak@hep.ph.bham.ac.uk

For the first time a complete data set of the two-body decays of the $f_0(1370)$, $f_0(1500)$ and $f_0(1710)$ into all pseudoscalar mesons is available. The implications of these data for the flavour content for these three f_0 states is studied. We find that they are in accord with the hypothesis that the scalar glueball of lattice QCD mixes with the $q\bar{q}$ nonet that also exists in its immediate vicinity. We show that this solution also is compatible with the relative production strengths of the $f_0(1370)$, $f_0(1500)$ and $f_0(1710)$ in pp central production, $p\bar{p}$ annihilations and J/ψ radiative decays.

1 Introduction

It is now generally accepted that glueballs will mix strongly with nearby $q\bar{q}$ states with the same J^{PC} and that this will lead to three isoscalar states of the same J^{PC} in a similar mass region. In general these mixings will negate the naive folklore that glueball decays would be "flavour blind ".

Lattice gauge theory calculations (in the quenched approximation) predict that the lightest glueball has $J^{PC} = 0^{++}$ and that its mass is in the $1.45 - 1.75$ GeV region. This means that the three states in the glueball mass range are the $f_0(1370)$, $f_0(1500)$ and the $f_0(1710)$.

2 Data from the WA102 experiment

Recently the WA102 collaboration has published [1], for the first time in a single experiment, a complete data set for the decay branching ratios of the $f_0(1370)$, $f_0(1500)$ and $f_0(1710)$ to all pseudoscalar meson pairs (see fig. 1).

A coupled channel fit to this data yields sheet II pole positions of $M(f_0(1370)) = (1310 \pm 19 \pm 10) - i (136 \pm 20 \pm 15)$ MeV $M(f_0(1500)) = (1508 \pm 8 \pm 8) - i (54 \pm 7 \pm 6)$ MeV and $M(f_0(1710)) = (1712 \pm 10 \pm 11) - i (62 \pm 8 \pm 9)$ MeV.

The relative decay rates $\pi\pi : K\overline{K} : \eta\eta :$ $\eta\eta' : 4\pi$ are for the $f_0(1370)$: $1 : 0.46 \pm 0.19 : 0.16 \pm 0.07 : 0.0 : 34.0^{+22}_{-9}$ for the $f_0(1500)$: $1 : 0.33 \pm 0.07 : 0.18 \pm 0.03 : 0.096 \pm 0.026 : 1.36 \pm 0.15$ and for the $f_0(1710)$: $1 : 5.0 \pm 0.7 : 2.4 \pm 0.6 :< 0.18 (90 \% \ CL) :< 5.4 (90 \% \ CL)$

These data will be used as input to a fit to investigate the glueball-quarkonia content of the $f_0(1370)$, $f_0(1500)$ and $f_0(1710)$.

3 The fit

In the $|G\rangle = |gg\rangle$, $|S\rangle = |s\bar{s}\rangle$, $|N\rangle = |u\bar{u} + d\bar{d}\rangle/\sqrt{2}$ basis, the three physical states can be read as

$$\begin{pmatrix} |f_0(1710)\rangle \\ |f_0(1500)\rangle \\ |f_0(1370)\rangle \end{pmatrix} = \begin{pmatrix} x_1 \ y_1 \ z_1 \\ x_2 \ y_2 \ z_2 \\ x_3 \ y_3 \ z_3 \end{pmatrix} \begin{pmatrix} |G\rangle \\ |S\rangle \\ |N\rangle \end{pmatrix},$$

(1)

where the parameters x_i, y_i and z_i are related to the partial widths of the observed states [2] as given in table 1.

We then perform a χ^2 fit based on the measured branching ratios. The details of the fit are given in ref. [2]. For this presentation the parameter r_3, used in ref. [2], has been set to zero. As input we use the masses of the $f_0(1500)$ and $f_0(1710)$. In this way seven parameters, M_G, M_N, M_S, M_3, f, r_2 and ϕ are determined from the fit. The mass of the $f_0(1370)$ is not well established so we have left it as a free pa-

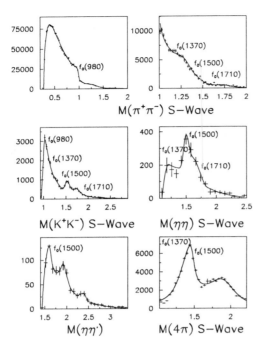

Figure 1. The observation of the $f_0(1370)$, $f_0(1500)$ and $f_0(1710)$ in the WA102 experiment.

Table 1. The theoretical reduced partial widths.

$\gamma^2(f_i \to \eta\eta')$	$2[2\alpha\beta(z_i - \sqrt{2}y_i)]^2$
$\gamma^2(f_i \to \eta\eta)$	$[2\alpha^2 z_i + 2\sqrt{2}\beta^2 y_i + r_2 x_i]^2$
$\gamma^2(f_i \to \pi\pi)$	$3[z_i + r_2 x_i]^2$
$\gamma^2(f_i \to K\bar{K})$	$4[\frac{1}{2}(z_i + \sqrt{2}y_i) + r_2 x_i]^2$

rameter (M_3). The χ^2/NDF of the fit is 13.9/7 and the largest contribution comes from the $\frac{f_0(1710)\to\eta\eta}{f_0(1710)\to K\bar{K}}$ branching ratio which contributes 6.1 to the χ^2.

The mass parameters determined from the fit are $M_G = 1446 \pm 16$ MeV, $M_S = 1664 \pm 9$ MeV, $M_N = 1374 \pm 28$ MeV and $M_3 = 1248 \pm 31$ MeV. The output

masses for M_N and M_S are consistent with the $K^*(1430)$ being in the nonet and with the glueball mass being at the lower end of the quenched lattice range. The mass found for the $f_0(1370)$ (1256 ± 31 MeV) is at the lower end of the measured range for this state. The pseudoscalar mixing angle is found to be $\phi = -25 \pm 4$ degrees consistent with other determinations.

The physical states $|f_0(1710)\rangle$, $|f_0(1500)\rangle$ and $|f_0(1370)\rangle$ are found to be

$$|f_0(1710)\rangle = 0.42|G\rangle + 0.89|S\rangle + 0.17|N\rangle,$$

$$|f_0(1500)\rangle = -0.61|G\rangle + 0.37|S\rangle - 0.69|N\rangle,$$

$$|f_0(1370)\rangle = 0.65|G\rangle - 0.15|S\rangle - 0.73|N\rangle.$$

Other authors have claimed that $M_G > M_S > M_N$ [3]. This scenario is disfavoured as, if in the fit we require $M_G > M_S > M_N$, the χ^2 increases to 57. In any event, we are cautious about such claims [3] as they are likely to be significantly distorted by the presence of a higher, nearby, excited $n\bar{n}$ state (N^*) such that $M_{N^*} > M_G > M_S$: the philosophy of dominant mixing with the nearest neighbours would then lead again to the "singlet - octet - singlet" scenario that we have found above.

4 Predictions for production mechanisms

Our preferred solution has implications for the production of these states in $\gamma\gamma$ collisions, $p\bar{p}$ annihilations, in central pp collisions and in radiative J/ψ decays. These are interesting in that they are consequences of the output and were not used as constraints.

4.1 $\gamma\gamma$ production

The most sensitive probe of flavours and phases is in $\gamma\gamma$ couplings. In the spirit of ref. [4], ignoring mass-dependent effects, the above imply $\Gamma(f_1(1710) \to \gamma\gamma)$: $\Gamma(f_1(1500) \to \gamma\gamma)$: $\Gamma(f_1(1370) \to \gamma\gamma) = 3.8\pm0.9 : 6.8\pm0.8 : 16.6\pm0.9$. The $\gamma\gamma$ width of $f_0(1500)$ exceeding that of $f_0(1710)$ arises because the glueball is nearer to the N than the S. This shows how these $\gamma\gamma$ couplings have the potential to pin down the input pattern.

4.2 $p\bar{p}$ production

The production of the f_0 states in $p\bar{p} \to \pi + f_0$ is expected to be dominantly through the $n\bar{n}$ components of the f_0 state, possibly through gg, but not prominently through the $s\bar{s}$ components. The above mixing pattern implies that $\sigma(p\bar{p} \to \pi + f_0(1710)) < \sigma(p\bar{p} \to \pi + f_0(1370)) \sim \sigma(p\bar{p} \to \pi + f_0(1500))$ Experimentally [5] the relative production rates are, $p\bar{p} \to \pi + f_0(1370) : \pi + f_0(1500)) \sim 1 : 1$. and there is no evidence for the production of the $f_0(1710)$. This would be natural if the production were via the $n\bar{n}$ component. The actual magnitudes would however be model dependent; at this stage we merely note the

consistency of the data with the results of the mixing analysis above.

4.3 Central production

For central production, the cross sections of well established quarkonia in WA102 suggest that the production of $s\bar{s}$ is strongly suppressed relative to $n\bar{n}$. The relative cross sections for the three states of interest here are pp $\to$ pp $+ (f_0(1710) : f_0(1500) : f_0(1370)) \sim 0.14 : 1.7 : 1$. This would be natural if the production were via the $n\bar{n}$ and gg components.

In addition, the WA102 collaboration has studied the production of these states as a function of the azimuthal angle ϕ, which is defined as the angle between the p_T vectors of the two outgoing protons. An important qualitative characteristic of these data is that the $f_0(1710)$ and $f_0(1500)$ peak as $\phi \to 0$ whereas the $f_0(1370)$ is more peaked as $\phi \to 180$ [6]. If the gg and $n\bar{n}$ components are produced coherently as $\phi \to 0$ but out of phase as $\phi \to 180$, then this pattern of ϕ dependence and relative production rates would follow; however, the relative coherence of gg and $n\bar{n}$ requires a dynamical explanation. We do not have such an explanation and open this for debate.

4.4 Radiative J/ψ decays

In J/ψ radiative decays, the absolute rates depend sensitively on the phases and relative strengths of the G relative to the $q\bar{q}$ component, as well as the relative phase of $n\bar{n}$ and $s\bar{s}$ within the latter. As discussed in ref. [2], based on the mixings found, we expect that the rate for $f_0(1370)$ will be smallest and that the rate of $J/\psi \to \gamma f_0(1500)$ rate will be comparable to $J/\psi \to \gamma f_0(1710)$.

In ref. [7], the branching ratio of BR($J/\psi \to \gamma f_0$)($f_0 \to \pi\pi + K\bar{K}$) for the $f_0(1500)$ and $f_0(1710)$ is presented. These can be used to show that [2]: $J/\psi \to f_0(1500) : J/\psi \to f_0(1710) = 1.0 : 1.1 \pm 0.4$ which is consistent with the prediction above based on our mixed state solution.

4.5 $\pi^- p$ and $K^- p$ production

In these mixed state solutions, both the $f_0(1500)$ and $f_0(1710)$ have $n\bar{n}$ and $s\bar{s}$ contributions and so it would be expected that both would be produced in $\pi^- p$ and $K^- p$ interactions. The $f_0(1500)$ has clearly been observed in $\pi^- p$ interactions and there is also evidence for the production of the $f_0(1500)$ in $K^- p \to K_S^0 K_S^0 \Lambda$.

There is evidence for the $f_0(1710)$ in the reaction $\pi^- p \to K_S^0 K_S^0 n$, originally called the $S^{*\prime}(1720)$. One of the longstanding problems of the $f_0(1710)$ is that in spite of its dominant $K\bar{K}$ decay mode it was not observed in $K^- p$ experiments. In ref. [8] it was demonstrated that if the $f_0(1710)$ had $J = 0$, as it has now been found to have, then the contribution in $\pi^- p$ and $K^- p$ are compatible. One word of caution should be given here: the analysis in ref. [8] was performed with a $f_0(1400)$ not a $f_0(1500)$ as we today know to be the case. As a further test of our solution, it would be nice to see the analysis of ref. [8] repeated with the mass and width of the $f_0(1500)$ and the decay parameters of the $f_0(1710)$ determined by the WA102 experiment.

5 Summary

In summary, based on the hypothesis that the scalar glueball mixes with the nearby $q\bar{q}$ nonet states, we have determined the flavour content of the $f_0(1370), f_0(1500)$ and $f_0(1710)$ by studying their decays into all pseudoscalar meson pairs. The solution we have found is also compatible with the relative production strengths of the $f_0(1370), f_0(1500)$ and $f_0(1710)$ in pp central production, $p\bar{p}$ annihilations and J/ψ radiative decays.

References

1. D. Barberis *et al.*, Phys. Lett. **B479** (2000) 59.
2. F.E. Close and A. Kirk, Phys. Lett. **B483** (2000) 345.
3. D. Weingarten, Nucl. Phys. Proc. Suppl. **53** (1997) 232; **63** (1998) 194; **73** (1999) 249;
   M. Strohmeier-Presicek, T. Gutsche, R. Vinh Mau, Amand Faessler, Phys. Rev. **D60** (1999) 054010;
   L. Burakovsky, P.R. Page, Phys. Rev. **D59** (1999) 014022.
4. F.E. Close, G. Farrar and Z.P. Li, Phys. Rev. **D55** (1997) 5749.
5. U. Thoma, Proceedings of Hadron 99, Beijing, China 1999.
6. D. Barberis *et al.*, Phys. Lett. **B467** (1999) 165.
7. W. Dunwoodie, Proceedings of Hadron 97, AIP Conf. Series 432 (1997) 753.
8. S. Lindenbaum and R.S. Longacre, Phys. Lett. **B274** (1992) 492.

STUDY OF MESON RESONANCES IN DIFFRACTIVE-LIKE REACTIONS

D.AMELIN, E.BERDNIKOV, A.FENYUK, YU.GOUZ, V.DOROFEEV, R.DZHELYADIN, A.KARYUKHIN, I.KACHAEV, YU.KHOKHLOV, A.KONOPLYANNIKOV, V.KONSTANTINOV, S.KOPIKOV, V.KOSTYUKHIN, I.KOSTYUHINA, V.MATVEEV, V.NIKOLAENKO, A.OSTANKOV, D.RYABCHIKOV, B.POLYAKOV, O.SOLOVIYANOV, A.A.SOLODKOV, A.V.SOLODKOV, E.STARCHENKO, E.VLASSOV, A.ZAITSEV

Institute for High Energy Physics,
142280, Protvino, Moscow Region, Russia
E-mail: zaitsev@mx.ihep.su

The recent results of VES-collaboration on the resonances produced in the diffractive-like reactions $\pi^- Be \to X Be$ at $P_\pi = 37$ GeV/c are presented. The exotic wave with $J^{PC} = 1^{-+}$ is studied in $\pi^+\pi^-\pi^-$, $b_1\pi$, $\eta'\pi$, $\eta\pi$ and $\eta\pi^+\pi^-\pi^-$ systems. A structure at $M \approx 1.6$ GeV is observed in the first three channels. This signal could be regarded as a strong candidate for the hybrid state.

Introduction.

The quantum numbers of $q\bar{q}$ states could be expressed by two equations:

$$P = (-1)^{L+1} \qquad C = (-1)^{L+S}$$

States with quantum numbers forbidden by these equations are known as exotics: $J^{PC}_{exotics} = 1^{-+}, 0^{+-}, 2^{+-}, 0^{--}...$ These states could be built from more than two quarks or from $q\bar{q}$ and gluons. These last states are called hybrids. In phenomenological models the quantum numbers of the lightest hybrids are $J^{PC} = 1^{-+}$. The lightest isovector state with these quantum numbers could decay to $\pi\eta$, $\pi\eta'$, $\pi\rho$, πb_1, πf_1, $a_0\rho$, $a_1\eta$, $\pi\rho'$.

The diffractive reactions could be especially effective for hybrid meson production as these mesons have some gluon-like components in addition to quarks and therefore could have a strong coupling with pomeron. The experimental data to study of the hybrid mesons in diffractive reactions were collected by VES facility - a large aperture magnetic spectrometer running with the negatively charged particles beam with the momentum of 37 GeV/c. The exclusive reactions $\pi^- Be \to X Be$ (where X means the mentioned above decay channels) were selected and treated with the partial wave analysis (PWA).

The wave $J^{PC} = 1^{-+}$ in the $\eta\pi$ channel.

The results of the VES-collaboration on this channel was published in [1]. The dominant wave $J^{PC} = 2^{++}$ is peaking at the mass of $a_2(1320)$ meson[1]. There is a clear signal in the exotic wave $J^{PC} = 1^{-+}$ centered at $M \approx 1.4$ GeV (Fig. 1a). Later this signal was observed by E852 and interpreted as a resonance [2]. As different experiments lead to different conclusions on this resonance existence and on the resonance parameters the $J^{PC} = 1^{-+}$ wave in the $\eta\pi$ channel deserves more study.

The wave $J^{PC} = 1^{-+}$ in the $\eta'\pi$ channel.

The wave 1^{-+} is the dominant one [1] in diffractive production of $\eta'\pi^-$ system at $P_\pi = 37$ GeV/c. It has the maximum at $M \approx 1.6$ GeV (Fig. 1b). Due to the smallness of other waves and poor knowledge of their parameters at $M \approx 1.6$ GeV the phase motion of the 1^{-+} wave is not known and the resonance nature of the signal could not be established from the data on this channel. Assuming that this signal is Breit-Wigner resonance the parameters were found: $M = 1.57 \pm 0.03$ GeV, $\Gamma = 0.55 \pm 0.08$ GeV.

The most striking feature of this wave is that its intensity at $M \approx 1.6$ GeV in the $\eta'\pi$ channel is higher than that in the $\eta\pi$ channel.

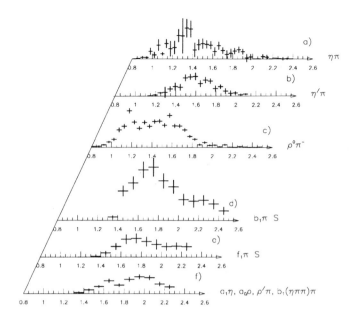

Figure 1. The intensity of exotic wave $J^{PC} = 1^{-+}$.

The PWA of the reaction $\pi^- A \to \pi^+ \pi^- \pi^- A$. Our previous results on the analysis of the $\pi^- Be \to \pi^+ \pi^- \pi^- Be$ reaction were published in [3]. The modified version of the Illinois PWA program [5] has been used for the analysis. Up to 45 waves with total angular momentum from $J = 0$ to $J = 4$ were included in the PWA model. The isobars $\rho(770)$, $f_2(1270)$, $\rho_3(1690)$ have been described by relativistic Breit-Wigner functions with standard parameters. The S-wave in the channel $\pi^+ \pi^-$ has been parameterized by two different states, namely a "narrow" resonance with the $f_0(980)$ parameters and a "broad" wave which is the AMP M-solution [6] with $f_0(980)$ pole removed. Well known resonances $a_1(1260)$, $a_2(1320)$, $\pi_2(1670)$, $\pi(1300)$, $\pi(1800)$, $a_4(2050)$ are clearly seen in the PWA results. The results point also to the existence of new resonances: $a_1(1750)$, $\pi_2(2100)$, $a_3(1850)$. The $J^{PC} = 1^{-+} \rho^0 \pi^-$ wave was introduced in

the PWA with three different states: $M\eta = 0-$, $1-$ and $1+$. The dominant component is $M\eta = 1+$. It has a nonzero spin projection on Gottfrid-Jakson z-axis and is better seen at high t' (Fig. 1c). The intensity of this wave is very small and does not exceed 2% of the sum of all waves at any $M_{3\pi}$. At $M \approx 1.6$ GeV a broad shoulder is seen. Its intensity at $M \approx 1.6$ GeV is only 3% of the intensity of the $a_2(1320)$ signal at its maximum. The phase of 1^{-+} signal is rising at $M \approx 1.6$ GeV. The shape of the 1^{-+} signal and its phase rising at $M \approx 1.6$ GeV piont to the possible presence of the resonance in this region. Due to its smallness and broadness this signal taken alone can not have unambiguous resonance interpretation. With our PWA model we do not see a narrow signal at $M \approx 1.6$ GeV in the 1^{-+} wave reported by the E852 [7]. Moreover in our PWA the intensity of the signal at $M \approx 1.6$ GeV in the unnatural parity exchange sector ($M\eta = 0-, 1-$) is compatible with zero as one could expect.

The wave $J^{PC} = 1^{-+}$ in the $b_1\pi$ channel.
The results of the VES collaboration on the reaction $\pi^- Be \rightarrow \omega\pi^-\pi^0 Be$ were published in [8]. The signal $J^{PC} = 1^{-+}$ is seen in the $b_1\pi$ channel at high t' (Fig. 1d). The phase of this signal is rising near $M \approx 1.6$ GeV. The mass-dependent fit of the ρ-matrix elements correponding to the waves $J^{PC} = 1^{-+}$ $b_1\pi$ and $J^{PC} = 2^{++}$ $\omega\rho$ was carried out. The $b_1\pi$ amplitude was saturated by a Breit-Wigner resonance and a coherent background. The $2^{++}\omega\rho^-$ amplitude was described by the $a_2(1320)$-meson and a broad background. The results of the fit point out to the resonance nature of the $b_1\pi$ signal. The range of the parameters variation is large due to the freedom of the 2^{++} wave model.

The wave $J^{PC} = 1^{-+}$ in the $f_1\pi$ channel.
The results on the reaction $\pi^- Be \rightarrow \eta\pi^+\pi^-\pi^- Be$ were reported by the VES collaboration [4]. The signal in the wave $J^{PC} = 1^{-+}$ is clearly seen in the channel $f_1\pi$ (Fig. 1e). There are also some signals in other channels: $a_1\eta$, $a_0\rho$, $\rho'\pi, \pi b_1 \rightarrow(\eta\pi\pi)$. The sum of intensities in these channels is plotted at Fig. 1f. The most intensive signal $f_1\pi$ showes a bump near $M \approx 1.7$ GeV but it does not demonstrate rising phase at this mass.

Momentum dependence.
The results on the reactions $\pi^- Be \rightarrow \eta'\pi^- Be$ and $\pi^- Be \rightarrow b_1\pi^- Be$ at two beam momenta $P_\pi = 28$ GeV and $P_\pi = 37$ GeV show that the ratio of the crosssections $X(1^{-+})/a_2(1320)$ is rising with beam momentum. This observation points to the pomeron exchange dominance in the production of the exotic $J^{PC} = 1^{-+}$ wave.

Parameters of the exotic resonance.
Clear bumps seen at $M \approx 1.6$ GeV in two channels $\eta'\pi$ [1] and $b_1(1235)\pi$ [8] as well as shape and phase of 1^{-+}-wave in $\rho^0\pi^-$ -

channel point to the possible existence of exotic resonance at $M \approx 1.6$ GeV. To estimate the parameters of this resonance the intensities of $\eta'\pi$, $b_1\pi$ and $\rho\pi$ waves were fitted with incoherent sum of a single Breit-Wigner resonance and backgrounds in each channel. The fit results in the following parameters: $M = 1.56 \pm 0.06$ GeV, $\Gamma = 0.34 \pm 0.05$ GeV.

Despite large uncertainty in the estimations of branching fractions for the 1^{-+} state the enhanced decay to $\eta'\pi$ looks very peculiar [9]. On the basis of this feature we consider the hybrid interpretation of the observed state as preferable.

Acknowledgments
This work was supported in part by INTAS-RFBR grant no. 97-0232 and RFBR grant no. 98-02-16392.

References

1. G.Beladidze et al, VES-collaboration, Phys. Lett. B313 (1993), 276
2. Tompson at al. E852 collaboration, Phys.Rev.Lett. V79 (1997), 1630
3. D.V. Amelin et al., Phys. Lett. **B356** (1995) 595.
4. D.Ryabchikov et al, VES-collaboration, HADRON-95, Proceedings,Manchester, 1995
   A.Zaitsev et al, VES-collaboration, ICHEP96, Proceedings, Warsawa, 1996
5. J.D. Hansen et al., Nucl. Phys. **B81** (1974) 403.
6. Au, Morgan, Pennington, Phys. Rev. **D35** (1987) 1633.
7. G.S. Adams et al., Phys. Rev. Lett. **V81, N 26** (1998) 5760
8. D. Amelin et al., VES-collaboration, Phys. of At. Nucl. 62 (1999) 445
9. F.E. Close and H.J. Lipkin, Phys. Lett. B, V196, (1987), 245
   J.-M. Frére and S.Titard, Phys. Let. B214 (1988) 463

PARTIAL WAVE STUDY OF $\pi^- P \to K^+ K^- \pi^0 N$ AT 18 GEV/C

J. HU

Rensselaer Polytechnic Institute, Troy, NY 12180, USA

for the **E852** *Collaboration*

We have measured the reaction $\pi^- p \to K^+ K^- \pi^0 n$ at 18.0 GeV/c. A partial wave analysis of the $K^+ K^- \pi^0$ system shows evidence for three pseudoscalar resonances, $\eta(1295)$, $\eta(1415)$, and $\eta(1485)$, as well as two axial vectors, $f_1(1285)$, and $f_1(1420)$. Their observed masses, widths and decay properties are reported. We see no signal for $C(1480)$, a state previously reported in this channel.

1 Introduction

There is a long history of controversy surrounding the $D(1285)$ and E/ι [1]. Results from previous experiments [2,3,4,5] indicate that there are at least six isoscalar states coupling to the $K\overline{K}\pi$ decay channel in the region from 1280 to 1530 MeV. But it is hard to interpret them in the framework of the quark model [6]. Thus, we need more experimental information to help clarify the situation. Moreover, the $C(1480)$ [7], a state with a great theoretical interest [8], also awaits to be confirmed.

This paper reports the results of a partial wave analysis of the $(K^+ K^- \pi^0)$ system from the reaction $\pi^- p \to K^+ K^- \pi^0 n$ at 18 GeV/c. The data were taken by the Experiment 852 (E852) Collaboration during the 1997 running period of the Alternating Gradient Synchrotron (AGS) at Brookhaven National Laboratory (BNL).

2 Partial Wave Analysis

After data selection a total of 19,576 $K^+ K^- \pi^0$ events survive for PWA analysis. The formalism used in this analysis is based on the isobar models [9]. C-parity eigenstates are fitted to the data.

The PWA study in this report is mainly focused on the D and E region, *i.e.* $1.2 < M(K^+ K^- \pi^0) < 1.55$ GeV. We choose to include K^*, a_0 and ϕ isobars in a set of $J < 3$ partial waves, plus an incoherent

Table 1. Resonant parameters and decay modes of the observed states

Resonance	M, Γ (MeV)	Decay Mode
$f_1(1285)$	1288 ± 4 48 ± 10	$a_0 \pi^0$
$\eta(1295)$	1301 ± 9 59 ± 23	$a_0 \pi^0$
$\eta(1415)$	1415 ± 4 43 ± 10	$a_0 \pi^0, K^* \overline{K}$
$f_1(1420)$	1428 ± 4 39 ± 9	$K^* \overline{K}$
$\eta(1485)$	1485 ± 5 102 ± 24	$K^* \overline{K}$

isotropic background amplitude. Because of the $K^* \overline{K}$ threshold we divided the fit region into two parts. The low mass fit (1.2–1.4 GeV) doesn't include the $K^* \overline{K}$ waves. A fit was performed using a bin width of 20 MeV with $|t| > 0.1$ GeV^2. The main features of the PWA fit together with the mass-dependent fits are shown in Figs 1 and 2. Three separate coupled mass-dependent fits were made to the intensities of two waves and their phase difference. The resonance parameters of the $\eta(1415)$ and the $f_1(1420)$ obtained from the fit in Fig. 1 b,d,f were fixed when using the two Breit-Wigner structures to fit the $0^{-+} K^* \overline{K}$ wave to extract the $\eta(1485)$ in the fit in Fig. 2 a-c. The results of the mass-dependent fits are shown in Table 1. As a test, we also did a single-pole fit over this wave. The fitted width comes out

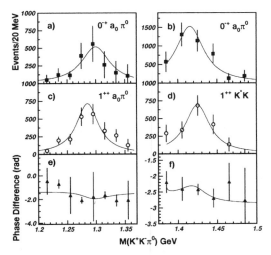

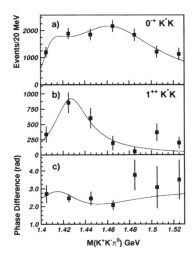

Figure 1. Coupled mass-dependent Breit-Wigner fits. (a) $0^{-+}[a_0(980)]S0^+$ and (c) $1^{++}[a_0(980)]P0^+$ wave intensity; (b) $0^{-+}[a_0(980)]S0^+$ and (d) $1^{++}[K^*(892)]S0^+$ wave intensity; (e) phase difference between (a) and (c); (f) phase difference between (b) and (d).

Figure 2. A coupled mass-dependent Breit-Wigner fit of $0^{-+}[K^*(892)]P0^+$ (a) and $1^{++}[K^*(892)]S0^+$ (b) waves; (c) phase difference between (a) and (b).

to be ~ 300 MeV and the quality of the fit becomes worse ($\chi^2/d.o.f$ changed from 1.07 to 1.38).

Contrary to some other analyses [4], we did not include the $(K\pi)_S\overline{K}$ waves in our final fit. We found that the PWA results changed very little when these waves were substituted for the $a_0\pi^0$ waves. If both sets were included in the fit a large interference term resulted, producing severe fluctuations in the bin to bin amplitude values.

As to the $C(1480)$, both the side-band study and PWA analysis were applied. There is no indication of its existence. We will report this in more detail in our future paper.

3 Summary

We have performed a partial wave analysis of the reaction $\pi^- p \to K^+ K^- \pi^0 n$. All expected well-known states ($f_1(1285)$, $\eta(1415)$ and $f_1(1420)$) are observed. In addition, we confirm the $\eta(1295)$ and a broad $\eta(1485)$ and also clearly see their phase motions. No evidence for the $C(1480)$ is observed.

References

1. Particle Data Group, *The European Phys. Journal* **3**, 1 (1998).

2. M.G. Rath *et al.*, *Phys. Rev. Lett.* **61**, 802 (1988). A. Birman *et al.*, *Phys. Rev. Lett.* **61**, 1557 (1988).

3. G. Adams *et. al.*, *Phys. Rev.* **D62**, 012003 (2000). J.H. Lee *et. al.*, *Phys. Lett.* **B323**, 227 (1994).

4. A. Bertin *et al.*, *Phys. Lett.* **B400**, 226 (1997); *Phys. Lett.* **B361**, 187 (1995).

5. J. Bai *et al.*, *Phys. Lett.* **B440**, 217 (1998); Z. Bai *et al.*, *Phys. Rev. Lett.* **65**, 2507 (1990). J.-E. Augustin *et al.*, *Phys. Rev.* **D46**, 1951 (1992).

6. S. Godfrey and J. Napolitano, *Rev. Mod. Phys.* **71**, 1411 (1999).

7. S.I. Bityukov *et. al.*, *Phys. Lett.* **B188**, 383 (1987).

8. F.E. Close and H.J. Lipkin, *Phys. Lett.* **B196**, 245 (1987).

9. J.P. Cummings and D.P. Weygand, "The New BNL Partial Wave Analysis Programs", unpublished(1997); S.U. Chung, "Analysis of $K\overline{K}\pi$ systems", BNL-QGS-98-901, unpublished(1998).

RECENT RESULTS OF CHARMONIUM PHYSICS FROM BES

YONGSHENG ZHU

Institute of High Energy Physics, P.O.Box 918, Beijing 100039, China
E-mail: zhuys3@hpws3.ihep.ac.cn

Based on $\psi(2S)$ and J/ψ events collected at BEPC/BES, we report the preliminary results of the branching fractions (or upper limits) for $\psi(2S)$ radiative decays, hadronic decays, baryon pairs final states, and the mass and fullwidth of the η_c meson. Recent progress are also presented, including the $\psi(2S)$ resonance scan, and the inclusive gamma spectrum of the J/ψ radiative decay.

1 Introduction

The results presented in this report are based on 7.8 million J/ψ and 3.96 million $\psi(2S)$[a] events collected by BESI[1] detector. The upgraded version, BESII [2], improves its vertex chamber spacial resolution from 220 μ to 90 μ, time resolution of TOF from 375 ps to 180 ps, and DAQ readout time per event from 20 ms to 10 ms, therefore greatly enhanced its PID capability and data collection efficiency. With BESII detector, we made a R scan in 2-5 GeV range [3], $\psi(2S)$ fine scan, and collect 22 million J/ψ data sample.

2 $\psi(2S)$ Hadronic Decays

Branching fractions of 12 hadronic decays are measured for the first time. The preliminary results of their branching fractions are listed in Table 1.

One expects that the J/ψ and $\psi(2S)$ decays into light hadrons via ggg, or γ^*, or γgg. In either case, the partial width for the decay is proportional to $\mid \psi(0) \mid^2$, where $\psi(0)$ is the wave function at the origin in the non-relativistic quark model for $c\bar{c}$. Thus, it is reasonable to expect[4] on the basis of perturbative QCD that, for any hadronic final state h

---

[a]The produced number of $\psi(2S)$ events is calculated by BES measured number of $\psi(2S) \to \pi^+\pi^- J/\psi$ events of $(1.227 \pm 0.003 \pm 0.017) \times 10^6$ [6] and its branching fractions of 0.310 ± 0.028 from PDG2000.

Table 1. BES preliminary $B(\psi(2S) \to h)$ and 15% rule test (limit at C.L.90%)

h	$B_h(10^{-5})$	$Q_h(\%)$
$\omega K^+ K^-$	12.5 ± 5.6	16.9 ± 9.4
$\omega p\bar{p}$	6.4 ± 2.6	5.0 ± 2.2
$\phi \pi^+ \pi^-$	16.8 ± 3.2	21.0 ± 5.1
$\phi K^+ K^-$	5.8 ± 2.2	7.0 ± 2.9
$\phi p\bar{p}$	0.82 ± 0.52	18.1 ± 12.8
ϕf_0	6.3 ± 1.8	19.6 ± 7.8
$K^* K^- \pi^+ + c.c.$	60.4 ± 9.0	?
$K^* \overline{K}^* + c.c.$	3.92 ± 1.03	13.6 ± 4.9
$K^* \overline{K_2}^* + c.c.$	7.98 ± 5.28	1.20 ± 0.93
$\pi^0 \pi^+ \pi^- p\bar{p}$	34.9 ± 6.4	15.2 ± 6.6
$\eta \pi^+ \pi^- p\bar{p}$	24.7 ± 9.6	?
$\eta p\bar{p}$	$< 18.$	< 8.6

$$Q_h \equiv \frac{B(\psi(2S) \to h)}{B(J/\psi \to h)} \simeq \frac{B(\psi(2S) \to e^+e^-)}{B(J/\psi \to e^+e^-)}$$
$$= (14.8 \pm 2.2)\%.$$

where the leptonic branching fractions are taken from PDG2000[5]. This relation is sometimes refered to as PQCD 15% rule. The Q_h values for these 12 channels are also shown in Table 1. There are no severe suppressions seen comparing to the PQCD prediction, except for $K^* \overline{K_2}^* + c.c.$ channel.

3 $\psi(2S)$ Baryon Pair Decays

BES measured 8 channels branching fractions(5 for the first time), listed in Table 2,together with corresponding Q_h's, which agree

Table 2. Branching fractions of $\psi(2S) \to B\overline{B}$ (limit at C.L.90%)

Decay	B (10^{-4})	Q_h (%)	$\mid M_i \mid^2_{\psi(2S)}$ (10^{-4})
$p\bar{p}$	$2.16 \pm .39$	10.1 ± 1.9	$1.60 \pm .29$
$\Lambda\overline{\Lambda}$	$1.81 \pm .34$	13.4 ± 2.9	$1.45 \pm .27$
$\Sigma^0\overline{\Sigma}^0$	$1.2 \pm .6$	9.4 ± 4.6	$1.0 \pm .5$
$\Xi^-\overline{\Xi}^+$	$0.94 \pm .31$	10.4 ± 4.1	$0.86 \pm .28$
$\Delta^{++}\overline{\Delta}^{--}$	$1.28 \pm .35$	11.6 ± 4.5	$1.10 \pm .30$
$\Sigma^{*-}\overline{\Sigma}^{*+}$	$1.1 \pm .4$	11 ± 4	$1.1 \pm .4$
$\Xi^{*0}\overline{\Xi}^{*0}$	$< .81$		< 0.93
$\Omega^-\overline{\Omega}^+$	$< .73$		< 1.11

Table 3. Preliminary branching fractions from $\psi(2S)$ radiative decay (limit at C.L.90%)

Process	$B(10^{-4})$
$\psi(2S) \to \gamma f_2(1270)$	2.27 ± 0.43
$\to \gamma f_J(1710) \to \gamma\pi\pi$	$.336 \pm .165$
$\to \gamma f_J(1710) \to \gamma K^+ K^-$	0.55 ± 0.21
$\to \gamma f_J(1710) \to \gamma K^0_S K^0_S$	0.21 ± 0.15
$\chi_{c0} \to \pi^0\pi^0$	26.8 ± 6.5
$\chi_{c2} \to \pi^0\pi^0$	8.8 ± 5.6
$\chi_{c0} \to \eta\eta$	19.4 ± 10.0
$\chi_{c2} \to \eta\eta$	< 12.2

with PQCD 15% expectations.

In the context of flavor SU(3), a pure $c\bar{c}$ state is a flavor singlet. In the limit of SU(3) flavor symmetry,the phase-space-corrected reduced branching ratio $\mid M_i \mid^2$

$$\mid M_i \mid^2 = \frac{B(\psi(2S) \to B_i\overline{B_i})}{\pi p^*/\sqrt{s}}$$

should be the same for every baryon B_i in the same multiplet (p^* is the momentum of the baryon in the $\psi(2S)$ rest frame). This relation works reasonably well for $J/\psi \to B\overline{B}$, but has not been tested for the $\psi(2S)$.

The $\mid M_i \mid^2$ for $\psi(2S) \to B\overline{B}$ have been calculated based on BES data and also listed in Table 2. The results show a trend to smaller values for the higher masses, only marginally consistent to flavor-SU(3) symmetry prediction, similar to J/ψ case.

4 $\psi(2S)$ **Radiative Decays**

By studying the radiative decays of $\psi(2S) \to \gamma(\pi\pi, K\overline{K}, \eta\eta)$, BES has measured 8 branching fractions for the first time, and their preliminary results are listed in Table 3. The branching fractions for $\psi(2S) \to \gamma f_2(1270)$ and $\psi(2S) \to \gamma f_J(1710) \to \gamma K\overline{K}$ agree with PQCD 15% prediction. The ratio of $\chi_{c0} \to \eta\eta$ to $\chi_{c0} \to \pi^0\pi^0$ is equal to 0.73 ± 0.39, consistent with the theoretical prediction of 0.95 based on the flavor SU(3) symmetry.

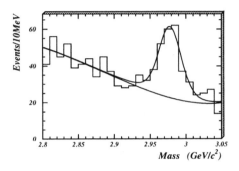

Figure 1. Combined 4 prong invariant mass in η_c region for $J/\psi \to \gamma\eta_c$, $\eta_c \to \pi^+\pi^-\pi^+\pi^-, \pi^+\pi^- K^+ K^-, \phi\phi, K^\pm K^0_S \pi^\mp$, and $\psi(2S) \to \gamma\eta_c, \eta_c \to \pi^+\pi^-\pi^+\pi^-, \pi^+\pi^- K^+ K^-, K^\pm K^0_S \pi^\mp, K^+ K^- K^+ K^-$

5 η_c **Mass and Full Width**

With a sample of 7.8 million J/ψ decays collected by BESI detector, the process $J/\psi \to \gamma\eta_c$ is observed for five different η_c decay channels: $K^+ K^-\pi^+\pi^-, \pi^+\pi^-\pi^+\pi^-, K^0_S K^\pm\pi^\mp$ (with $K^0_S \to \pi^+\pi^-$), $\phi\phi$(with $\phi \to K^+ K^-$) and $K^+ K^-\pi^0$. We determine the mass of η_c to be $2976.6 \pm 2.9 \pm 1.3$ MeV. Combining this study with a previously reported results from $\psi(2S) \to \gamma\eta_c$ [7] leads to $M_{\eta_c} = 2976.3 \pm 2.3 \pm 1.2$ MeV and $\Gamma_{\eta_c} = 11.0 \pm 8.1 \pm 4.1$ MeV. Fig. 1 shows the combined four-charged-track invariant mass distribution in the η_c region from $J/\psi, \psi(2S) \to \gamma\eta_c$ decays.

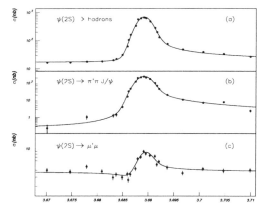

Figure 2. The cross sections for $e^+e^- \to$ hadrons, $\pi^+\pi^-J/\psi$, $\mu^+\mu^-$ in the vicinity of $\psi(2S)$ resonance. The solid curves represent the results of the fit to the data

6 $\psi(2S)$ Decay Widths

BESII has made a scan in the vicinity of $\psi(2S)$ resonance $E_{cm} \sim (3.67, 3.71)$ GeV. The events of $e^+e^- \to$ hadrons, $\pi^+\pi^-J/\psi$, $\mu^+\mu^-$, e^+e^- are collected at 24 energy points with total integrated luminorsity of $\sim 790nb^{-1}$. The data set of $e^+e^- \to e^+e^-$ is used to determine the integrated luminorsity for each energy point, while the other three data sets are fitted simultaneously to obtain the partial widths of $\psi(2S)$ to corresponding channels as shown in Fig. 2. The final fit results and the systematic errors are still under study.

7 Inclusive γ Spectrum of J/ψ Radiative Decays

We identify photons which convert into e^+e^- pairs inside BESII detector near the beam pipe . The momentum of produced e^+e^- pair is measured by BESII main drift chamber inside a 0.4 Tesla solenoidal magnetic field with the resolution of $\sigma_p/p = 0.0178\sqrt{1+p^2(GeV)}$. By this method, the energy resolution of converted photons can reach $\sim 1.5\%\sqrt{E_\gamma}$ at $E_\gamma = 0.75$ GeV. With newly collected 22 million J/ψ data sample,

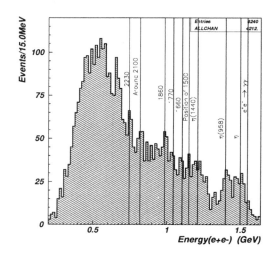

Figure 3. The distribution of $E_{e^+e^-}$ from 22 million J/ψ data sample. The vertical lines indicate the energy of converted photons corresponding to the X particles in $J/\psi \to \gamma X$ radiative decays

the measured inclusive γ spectrum of J/ψ radiative decays is shown in Fig.3, clear signals near $E_\gamma = 1.21, 1.40, 1.50, 1.548$ GeV are found corresponding to $\eta(1440), \eta', \eta$ and $e^+e^- \to \gamma\gamma$, respectively. A narrow signal appears at $E_\gamma = 0.745$ GeV, which suggests a narrow resonance may exist near 2.230 GeV.

References

1. J.Z. Bai et al., BES Collab., *Nucl. Instr. Meth.* A **344**, 319 (1994).

2. J.Z. Bai et al., BES Collab., *The BES Upgrade*, submitted to *Nucl. Instr. Meth.* A.

3. Z.G. Zhao, *Measurement of R at 2-5 GeV*, This Proceedings.

4. S.J. Brodsky and M.K. karliner, *Phys. Rev. Lett.* **78**, 468 (1997); Yu-Qi Chen and Eric Braaten, *Phys. Rev. Lett.* **80**, 5060 (1998); and references therein.

5. D.E.Groom et al., Particle Data Group, *Eur. Phys. J.* C **15**, 1 (2000).

6. J.Z.Bai et al., BES Collab., *Phys. Rev.* D **58**, 092006 (1999).

7. J.Z.Bai et al., BES Collab., *Phys. Rev.* D **60**, 072001 (1999).

A DETERMINATION OF THE S-WAVE PION-PION SCATTERING LENGTHS FROM THE MEASUREMENT OF THE LIFETIME OF PIONIUM

A. LANARO

CERN, CH-1211 Geneva 23, Switzerland
E-mail: armando.lanaro@cern.ch

on behalf of
"The DIRAC Collaboration"

B. ADEVA[O], L. AFANASEV[L], M. BENAYOUN[D], Z. BERKA[B], V. BREKHOVSKIKH[N],
G. CARAGHEORGHEOPOL[M], T. CECHAK[B], M. CHIBA[J], S. CONSTANTINI[P], S. CONSTANTINESCU[M],
A. DOUDAREV[L], D. DREOSSI[F], D. DRIJARD[A], M. FERRO-LUZZI[A], T. GALLAS TORREIRA[A,O],
J. GERNDT[B], R. GIACOMICH[F], P. GIANOTTI[E], F. GOMEZ[O], A. GORIN[N], O. GORTCHAKOV[L],
C. GUARALDO[E], M. HANSROUL[A], R. HOSEK[B], M. ILIESCU[E,M], M. JABITSKI[L], N. KALININA[R],
V. KARPOUKHINE[L], J. KLUSON[B], M. KOBAYASHI[G], P. KOKKAS[P], V. KOMAROV[L], A. KOULIKOV[L],
A. KOUPTSOV[L], V. KROUGLOV[L], L. KROUGLOVA[L], K.-I. KURODA[K], A. LAMBERTO[F], A. LANARO[A],
V. LAPSHIN[N], R. LEDNICKY[C], P. LERUSTE[D], P. LEVI SANDRI[E], A. LOPEZ AGUERA[O], V. LUCHERINI[E],
T. MAKI[I], I. MANUILOV[N], L. MONTANET[A], J.-L. NARJOUX[D], L. NEMENOV[A,L], M. NIKITIN[L], T. NUNEZ
PARDO[O], K. OKADA[H], V. OLCHEVSKII[L], A. PAZOS[O], M. PENTIA[M], A. PENZO[F], J.-M. PERREAU[A],
C. PETRASCU[E,M], M. PLO[O], T. PONTA[M], D. POP[M], G.F. RAPPAZZO[F], A. RIAZANTSEV[N],
J.M. RODRIGUEZ[O], A. RODRIGUEZ FERNANDEZ[O], V. RYKALIN[N], C. SANTAMARINA[O], J. SABORIDO[O],
J. SCHACHER[Q], C. SCHUETZ[P], A. SIDOROV[N], J. SMOLIK[C], F. TAKEUTCHI[H], A. TARASOV[L],
L. TAUSCHER[P], M.J. TOBAR[O], S. TROUSOV[R], P. VAZQUEZ[O], S. VLACHOS[P], V. YAZKOV[R],
Y. YOSHIMURA[G], P. ZRELOV[L]

[A] *CERN, Geneva, Switzerland;* [B] *Czech Technical University, Prague, Czech Republic;* [C] *Institute of Physics ASCR, Prague, Czech Republic;* [D] *LPNHE des Universites Paris VI/VII, IN2P3-CNRS, France;* [E] *INFN - Laboratori Nazionali di Frascati, Frascati, Italy;* [F] *Trieste University and INFN-Trieste, Italy;* [G] *KEK, Tsukuba, Japan;* [H] *Kyoto Sangyou University, Japan;* [I] *UOEH-Kyushu, Japan;* [J] *Tokyo Metropolitan University, Japan;* [K] *Waseda University, Japan;* [L] *JINR Dubna, Russia;* [M] *National Institute for Physics and Nuclear Engineering IFIN-HH, Bucharest, Romania;* [N] *IHEP Protvino, Russia;* [O] *Santiago de Compostela University, Spain;* [P] *Basel University, Switzerland;* [Q] *Bern University, Switzerland;* [R] *Skobeltsyn Institute for Nuclear Physics of Moscow State Univeristy*

The aim of the DIRAC Experiment at CERN is to provide an accurate determination of the S-wave $\pi\pi$ scattering lengths from the measurement of the lifetime of $\pi^+\pi^-$ atoms. The precision of the measurement will submit low-energy predictions from Chiral Perturbation Theory to a stringent test. We illustrate the performance of the experimental apparatus and present some preliminary results from the analysis of a 1999 data sample.

1 Introduction

Atomic states of two strongly interacting particles ($\pi\pi, \pi K, \pi N, KN, ...$) are a source of model independent data for testing QCD predictions at low energy.

Pionium ($A_{\pi^+\pi^-}$) is a weakly-bound Coulomb system. The involved dynamical scales ($m_{\pi^\pm}$, p_B: the Bohr momentum ~ 0.5 MeV, E_b: the binding energy ~ 1.9 keV) are small compared to the typical hadronic scale (~ 1 GeV). Hence, an approach based on non-relativistic effective lagrangians is used, in the context of Chiral Perturbation Theory (χPT), to formulate theoretical predictions.

The pionium lifetime is dominated by the charge-exchange process to two neutral pions. Its decay width is proportional to the square of the atom wave function at zero pion separation and to the square of $\Delta = a_0 - a_2$, the difference between the isoscalar and isotensor

S-wave $\pi\pi$ scattering lengths. The predicted value for the $A_{\pi\pi}$ lifetime in the ground state, according to χPT at leading and next-to-leading order in isospin breaking, is of the order of 3×10^{-15} s [1] [2] [3].

The goal of the DIRAC Experiment [4] is to measure the $A_{\pi\pi}$ lifetime with 10% accuracy in order to determine $|a_0 - a_2|$ with 5% precision in a model independent way [5]. The determination of the S-wave $\pi\pi$ scattering lengths will yield the size of the two-flavour quark condensate [6]

2 Experimental method

There exists a finite probability for two oppositely charged pions to produce a bound state when there is a sizeable overlap in space between their wave functions. If both pions come closer than the atom Bohr radius (~ 387 fm) this probability is enhanced. Relativistic $A_{\pi\pi}$ produced in high-energy hadron-nucleus interaction will move in the target material before their decay. The electromagnetic interaction with the atoms of the medium might, however, lead to the pionium excitation or breakup (ionization). From the knowledge of the atomic interaction cross sections, for a given target material and thickness, one can calculate the breakup probability (P_{br}) for arbitrary values of the atom momentum and lifetime. In Fig. 1 the dependency of P_{br} on the atom lifetime is shown for an atom momentum of 4.7 GeV/c.

A comparison between the measured value of P_{br}, from the ratio of detected " atomic pairs" and produced $A_{\pi^+\pi^-}$, and the calculated dependence of P_{br} on τ, will allow to determine the atom lifetime, and hence $\Delta = a_0 - a_2$.

"Atomic pairs" produced from the breakup of $A_{\pi^+\pi^-}$ have very distinct features. They are characterised by a small relative momentum in their c.m. system ($Q < 3$ MeV/c), a small opening angle ($\theta \approx 0.35$ mrad) and nearly equal energies ($E_+ \approx E_-$).

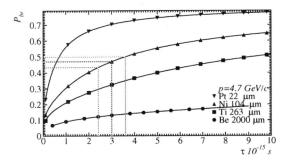

Figure 1. Probability of pionium breakup in the target as a function of its lifetime.

3 The DIRAC apparatus

The DIRAC setup is located at CERN, at the ZT8 beam area of the PS East Hall. It became operational at the end of 1998 and has been collecting data since the summer of 1999, using the 24 GeV/c proton beam extracted from the PS accelerator.

DIRAC is a fixed-target experiment and its apparatus is designed to detect charged pion pairs with high efficiency. It consists of a straight detector section between the target station and the analyzing magnet (2.3 Tm bending power), and of a double arm spectrometer downstream the magnet. The upstream straight section of the apparatus is instrumented with high resolution microstrip gas chambers and scintillating fibre detectors, followed by a scintillating ionization hodoscope. Along each downstream arm are located: a set of drift chambers, two orthogonal planes of scintillation hodoscopes, a gas threshold Cherenkov counter, a preshower detector and a muon detector consisting of an iron absorber followed by a scintillator wall.

A multi-level trigger system has been designed to select atomic pairs from the overwhelming background of uncorrelated $\pi^+\pi^-$ pairs and to reduce the recorded data flow to the level manageable by the DAQ system. The Cherenkov, preshower and muon detectors ensure rejection of e^+e^-, $\mu^+\mu^-$ pairs at

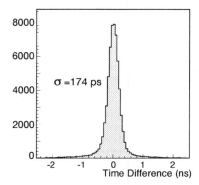

Figure 2. Relative time difference of e^+e^- pairs in vertical hodoscope.

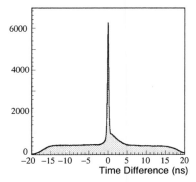

Figure 3. Relative time difference of hadron pairs in vertical hodoscope.

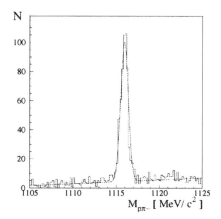

Figure 4. Invariant mass spectrum of $p\pi^-$ pairs.

trigger level. Pairs of $p\pi^-$ are identified by off-line analysis of time-of-flight measurements.

4 Experimental results

The number of ionised $A_{\pi^+\pi^-}$ is measured directly from the excess of $\pi^+\pi^-$ pairs with very low relative momentum ($Q < 3$ MeV/c) compared to the expected number of "free" pairs (pairs generated from continuum). The Q-distribution of $\pi^+\pi^-$ pairs is sensitive to the setup momentum resolution, the resolution on the pair opening angle and to the multiple scattering in the target, membranes and detectors.

An investigation of the performances of the setup has been done from the analysis of

data collected in 1999 with a platinum target.

Fig. 2 shows the accuracy in the time measurement by displaying the time difference between correlated e^+e^- pairs in the vertical hodoscope arms. A similar distribution, but for hadron pairs, is shown in Fig. 3. The peak of time correlated particles is due to $\pi^+\pi^-$ and admixture of $p\pi^-$, $K^+\pi^-$ pairs (shoulder on the right side of central peak). The flat background provides a sample of accidental coincidences needed for the analysis.

Calibration using particle decay was performed by reconstructing the effective mass of $p\pi^-$ pairs. Fig. 4 shows a clear Λ signal with a standard deviation $\sigma_\Lambda = 0.43$ MeV/c^2 which yields the best precision compared to all previous determinations [7].

The distribution of the longitudinal component Q_L (the projection of Q along the total momentum of the pair) is shown in Fig. 5 for correlated (including some accidental) $\pi^+\pi^-$ pairs (Fig. 5 a), and for accidental pairs (Fig. 5 b). To reject contamination from unresolved protons a cut on the momentum of the positive particle is applied ($p_+ < 5$ GeV/c). In the region $|Q_L| < 10$ MeV/c there is a noticeable enhancement of correlated $\pi^+\pi^-$ pairs due to Coulomb attraction in the final state (Fig. 5 a).

It is preferable to introduce a correla-

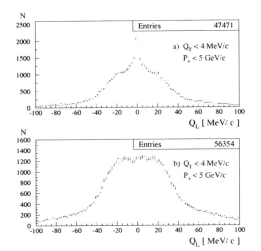

Figure 5. Q_L distribution for a) correlated (including some accidental) and b) accidental $\pi^+\pi^-$ pairs.

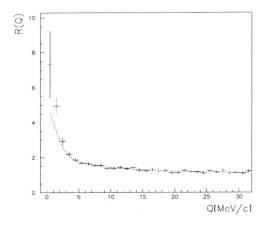

Figure 6. The experimental correlation function (points with error bars) and the free pairs correlation function (dashed curve) obtained from the shape of Q-distribution of free pairs with $Q > 4$ MeV/c.

tion function $R(Q)$ defined as the ratio between the Q-distributions of free and accidental pairs. Its dependence on Q is shown in Fig. 6. The dashed curve represents the correlation function for free pairs which best fits the experimental $R(Q)$ distribution in the region where atomic pairs are absent ($Q > 4$ MeV/c). For small values of Q free pairs are mostly "Coulomb" pairs as shown in Fig. 6 by the Coulomb enhancement at very low Q. In this low-Q region, where atomic pairs are expected, the experimental distribution shows a clear excess of $\pi^+\pi^-$ pairs compared to the fitted function (dashed curve).

Conclusion

The DIRAC Experiment is at present collecting statistics to achieve the final goal of measuring the lifetime of pionium with 10% accuracy. The quality of the data is excellent and a preliminary analysis of data collected last year has already shown an excess of $\pi^+\pi^-$ pairs coming from the breakup of $A_{\pi^+\pi^-}$ in a platinum target. At the present rate of $\sim$ 3-4 detected atomic pairs per million triggers we expect to fulfill our experimental programme within the year 2002.

References

1. A. Gall et al., *Phys. Lett.* B **462**, 335 (1999).
2. J. Gasser et al., *hep-ph/9910438*, (1999).
3. G. Colangelo et al., *hep-ph/0007112*, (2000).
4. B. Adeva et al., Proposal to the SP-SLC, *CERN/SPSLC 95-1, SPSLC/P 284*, (1994).
5. L.L. Nemenov, *Sov. J. Nucl. Phys.* **41**, 629 (1985).
6. N.H. Fuchs et al., *Phys. Lett.* B **269**, 183 (1991); *Phys. Rev.* D **47**, 3814 (1993).
7. E. Hartouni et al., *Phys. Rev. Lett.* **72**, 1322 (1994)

ON THE η' GLUONIC ADMIXTURE

E. KOU

Department of Physics, Ochanomizu University, Tokyo 112-0012, Japan
Physics Department., Nagoya University, Nagoya 464-8602, Japan
E-mail: kou@eken.phys.nagoya-u.ac.jp

The η' which is mostly $SU(3)_F$ singlet state can contain a pure gluon component, gluonium. We examine this possibility by analysing all available experimental data. It is pointed out that the η' gluonic component may be as large as 26%. We also show that the amplitude for $J/\psi \to \eta'\gamma$ decay obtains a notable contribution from gluonium.

1 Introduction

1.1 Notation

It is convenient to write the η and η' states as follows [1] [2]

$$|\eta\rangle = X_\eta |\frac{u\bar{u} + d\bar{d}}{\sqrt{2}}\rangle + Y_\eta |s\bar{s}\rangle$$

$$|\eta'\rangle = X_{\eta'} |\frac{u\bar{u} + d\bar{d}}{\sqrt{2}}\rangle + Y_{\eta'} |s\bar{s}\rangle + Z_{\eta'} |gluonium\rangle,$$

where $X_{\eta(\eta')}$, $Y_{\eta(\eta')}$ and $Z_{\eta'}$ parameterize the $(u\bar{u} + d\bar{d})$, $s\bar{s}$, and gluonic component of η' (η'), respectively. These parameters are related to the pseudoscalar mixing angles θ_p as $X_\eta = \cos\alpha_p, Y_\eta = -\sin\alpha_p$, $X_{\eta'} = \cos\phi\sin\alpha_p$, $Y_{\eta'} = \cos\phi\cos\alpha_p$ and $Z_{\eta'} = \sin\phi$, where $\alpha_p = \theta_p - \theta_I + \frac{\pi}{2}$ with the ideal mixing angle $\theta_I = \tan^{-1}\frac{1}{\sqrt{2}}$. The angle ϕ is introduced to parameterize the gluonic component. We have the following conditions from the normalization:

$$X_\eta^2 + Y_\eta^2 = 1 \qquad (1)$$
$$X_{\eta'}^2 + Y_{\eta'}^2 + Z_{\eta'}^2 = 1 \qquad (2)$$

2 Result and conclusion

2.1 Determination of $X_{\eta^{(')}}$, $Y_{\eta^{(')}}$ and $Z_{\eta'}$

Using the vector meson dominance model and the two mixing angle definition of the decay constants for η and η' [3], the radiative decay rates of the light mesons can be written in terms of $X_{\eta(\eta')}$, $Y_{\eta(\eta')}$ and $Z_{\eta'}$ (see the original paper[1] for details). The result

are shown in Figure 1 and Figure 2. The experimental data used for the computation is listed in Table 1. The values I to VII are averages of PDG98 [4] and VIII is the latest data from CMD-2 collaboration [5].

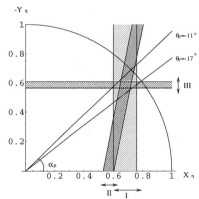

Figure 1: Determination of X_η and Y_η. The experimental bounds for $\omega \to \eta\gamma$ (I), $\eta \to \gamma\gamma$ (II) and $\phi \to \eta\gamma$ (III) are depicted.

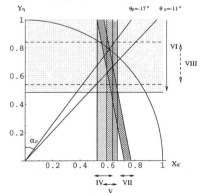

Figure 2: Determination of $X_{\eta'}$, $Y_{\eta'}$ and $Z_{\eta'}$. The experimental bounds for $\eta' \to \omega\gamma$ (IV), $\eta' \to \rho\gamma$ (V), $\phi \to \eta'\gamma$ (VI, VIII) and $\eta' \to \gamma\gamma$ (VII) are depicted.

The constraint on the pseudoscalar mixing angle obtained in this analysis is $-17° < \theta_p < -11°$. The maximum 26 % of the gluonic component in η' is observed at $\theta_p = -11°$.

I	$Br(\omega \to \eta\gamma) = (6.5 \pm 1.0) \times 10^{-4}$
II	$\Gamma(\eta \to \gamma\gamma) = (0.46 \pm 0.04)\text{KeV}$
III	$Br(\phi \to \eta\gamma) = (1.26 \pm 0.06) \times 10^{-2}$
IV	$Br(\eta' \to \omega\gamma) = (3.0 \pm 0.30) \times 10^{-2}$
V	$Br(\eta' \to \rho\gamma) = (3.0 \pm 0.13) \times 10^{-1}$
VI	$Br(\phi \to \eta'\gamma) = (1.2 ^{+0.7}_{-0.5}) \times 10^{-4}$
VII	$\Gamma(\eta' \to \gamma\gamma) = (0.20 \pm 0.016)\text{MeV}$
VIII	$Br(\phi \to \eta'\gamma) = (8.2 ^{+2.1}_{-1.9} \pm 1.1) \times 10^{-5}$

Table 1: The experimental data used for the computation.

2.2 The radiative J/ψ decays

Now we analyse the radiative J/ψ decays into η and η' and see the influence of the allowed amount of gluonic admixture in Section 2.1 on the amplitudes. The ratio of the two

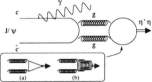

Figure 3: The radiative $J/\psi \to \eta^{(\prime)}\gamma$ decays.

decay rates $R_{J/\psi}$ can be written as

$$R_{J/\psi} = \frac{\Gamma(J/\psi \to \eta\gamma)}{\Gamma(J/\psi \to \eta'\gamma)}$$
$$= \left(\frac{1 - m_\eta^2/m_{J/\psi}^2}{1 - m_{\eta'}^2/m_{J/\psi}^2}\right)^3 |\frac{\sqrt{2}\xi X_\eta - \zeta(-Y_\eta)}{(\sqrt{2}\xi X_{\eta'} + \zeta Y_{\eta'}) + g'_r Z_{\eta'}}|^2$$

where ξ, ζ and g'_r are f_π/f_x, f_π/f_y, and the coupling of two gluons to gluonium, respectively. The terms $\sqrt{2}\xi X_{\eta^{(\prime)}}$ and $\zeta Y_{\eta^{(\prime)}}$ represent the contributions from the intermediate state Figure 3(a) and $g'_r Z_{\eta'}$ from Figure 3(b). We define the ratio between the amplitudes for the process Figure 3(b) and Figure 3(a) by r:

$$r = \frac{g'_r Z_{\eta'}}{(\sqrt{2}\xi X_{\eta'} + \zeta Y_{\eta'})} \qquad (3)$$

First, we examine the case of $r = 0$. The result is shown in Figure 4. We observe that for $g'_r Z_{\eta'} = 0$, the θ_p angle is determined in a region $\theta_p = -13° \pm 1.0°$, while in Figure 2, $Z_{\eta'} = 0$ is allowed only when θ_p is in a narrow region around $-17°$. This disagreement indicates that $Z_{\eta'} = 0$ should be excluded.

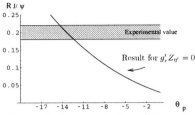

Figure 4: The result for $g'_r Z_{\eta'} = 0$.

Now let us examine the case of $g'_r Z_{\eta'} \neq 0$ for three sets of parameters allowed in Section 2.1, $\theta_p = -17°$ with 6% of the glue content, $\theta_p = -14°$ with 17% and $\theta_p = -11°$ with 26%. The result is shown in Figure 5. In a case when we choose $\theta_p = -17°$ with 6% of gluonium in η', we have observed that the 20% of the amplitude of $J/\psi \to \eta'\gamma$ comes from gluonium.

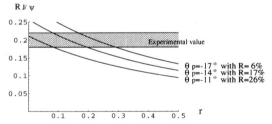

Figure 5: The result for $g'_r Z_{\eta'} \neq 0$.

Acknowledgments

The author gratefully thanks Professor A. I. Sanda for suggesting the presented project. This work is supported by the Japanese Society for the Promotion of Science.

References

1. E. Kou, *hep-ph/9908214*. The analysis here is based on this paper.

2. J.L. Rosner, *Phys. Rev.* D **27**, 1101 (1983).

3. H. Leutwyler, *hep-ph/9709408*, T. Feldmann, P. Kroll and B. Stech, *Phys. Rev.* D **58** 114006 (1998) and *Phys. Lett.* B **449** 339 (1999).

4. Review of Particle Physics, *Eur. Phys. J.* C **3** 1 (1998).

5. CMD-2 Collabration, R.R. Akhmetshin, *Phys. Lett.* B **473** 337 (2000).

SEARCH FOR THE GLUEBALL CANDIDATES $F_0(1500)$ AND $F_J(1710)$ IN $\gamma\gamma$ COLLISIONS

SHAN JIN

University of Wisconsin-Madison, Madison, WI 53706, USA
Corresponding adress: CERN / EP division, CH-1211 Geneva 23, Switzerland
E-mail: Shan.Jin@cern.ch

Data taken with the ALEPH detector at LEP1 have been used to search for $\gamma\gamma$ production of the glueball candidates $f_0(1500)$ and $f_J(1710)$ via their decay to $\pi^+\pi^-$. No signal is observed and upper limits to the product of $\gamma\gamma$ width and $\pi^+\pi^-$ branching ratio of the $f_0(1500)$ and the $f_J(1710)$ have been measured to be $\Gamma(\gamma\gamma \to f_0(1500)) \cdot BR(f_0(1500) \to \pi^+\pi^-) < 0.31 keV$ and $\Gamma(\gamma\gamma \to f_J(1710)) \cdot BR(f_J(1710) \to \pi^+\pi^-) < 0.55 keV$ at 95% confidence level.

For the detail content of this talk, please see *Physics letters B 472 (2000) 189-199.*

RESONANCE PRODUCTION IN TWO-PHOTON COLLISIONS AT LEP WITH THE L3 DETECTOR

M.N. KIENZLE-FOCACCI

*Département de Physique Nucléaire et Corpusculaire, 24 quai Ernest-Ansermet,
CH-1211 GENEVE 4, Switzerland
E-mail: maria.kienzle@cern.ch*

A spin-parity-helicity analysis shows the dominance of the $\gamma\gamma \to \rho^0\rho^0$ state at threshold, in the wave $(J^P, J_z) = (2^+, \pm 2)$ and a significant signal in the 0^+ wave. The negative parity states and the $(2^+, 0)$ wave are negligible. For high two-photon center-of-mass energies a strong production of $f_2(1270)$ is also observed. The study of the reaction $e^+e^- \to e^+e^- K_S^0 K^\pm \pi^\mp$ shows evidence of the formation of $\eta(1440)$, $f_1(1285)$ and $f_1(1420)$ while only the $f_1(1285)$ is formed in the $e^+e^- \to e^+e^-\eta\pi^+\pi^-$ interaction. The measured value of the $\eta(1440)$ two-photon width indicates a strong gluon content.

1 Exclusive two-photon reactions

The LEP e^+e^- collider is a copious source of two-photon interactions. Using the L3 detector we have studied the exclusive channels[1,2] $e^+e^- \to e^+e^-\pi^+\pi^-\pi^+\pi^-$, $e^+e^- \to e^+e^- K_S^0 K^\pm \pi^\mp$ and $e^+e^- \to e^+e^-\eta\pi^+\pi^-$ with an integrated luminosity of $\simeq 450$ pb^{-1} at beam energies from 160 GeV to 202 GeV.

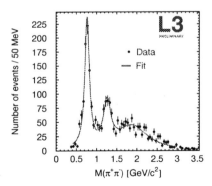

Figure 1. The $\pi^+\pi^-$ mass spectrum for $W_{\gamma\gamma} > 3$ GeV, after subtraction of the equal sign $\pi^\pm\pi^\pm$ combinations.

1.1 $e^+e^- \to e^+e^-\pi^+\pi^-\pi^+\pi^-$

To select the events, four charged particles, originating from the interaction vertex, with a net charge zero, are required. The two electrons, scattered at small angles, are undetected. To ensure an exclusive interaction, the total transverse momentum of the observed particles must be small: $P_T^2 = (\sum \vec{p_t})^2 \leq 0.02$ GeV^2. The 4π mass spectrum of the selected events ($\simeq 56000$) is dominated by a broad enancecement at the $\rho^0\rho^0$ threshold. For two-photon center-of-mass energies, $W_{\gamma\gamma}$, greater than 3 GeV, we observe not only the ρ^0, but also a clear production of $f_2(1270)$ (fig.1). The presence of the f_2 and of the higher $\pi^+\pi^-$ masses is not expected by the Vector Dominance Model of $\gamma\gamma$ interactions. A spin-parity-helicity (J^P, J_z) analysis is performed for $W_{\gamma\gamma} < 3$ GeV. Following the model first proposed by TASSO[3], we consider an isotropic production of $\pi^+\pi^-\pi^+\pi^-$ and $\rho^0\pi^+\pi^-$ and only the orbital angular momenta 0 and 1 for the $\rho^0\rho^0$ system. The allowed $\rho^0\rho^0$ states in quasi-real two-photon interactions are therefore limited to $0^+, 0^-, (2^+, \pm 2), (2^+, 0), (2^-, 0)$. All final states are assumed to be produced incoherently. The result of a maximum likelihood fit, performed in 50 MeV mass bins, shows that the contributions of the negative parity states and of the $(2^+, 0)$ wave are negligible.

The cross section of the dominant $(2^+, \pm 2)$ wave and of the other non negligible contributions are shown in fig.2. Due to the high statistics, we may be able in future to find evidence of resonance formation in this mass region by studying interference terms and the complementary channel $e^+e^- \rightarrow e^+e^-\rho^+\rho^-$.

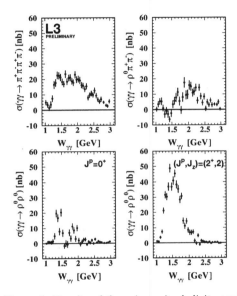

Figure 2. Results of the spin-parity-helicity analysis.

1.2 $e^+e^- \rightarrow e^+e^- K_S^0 K^\pm \pi^\mp$

To select the events four charged particles and no photon are required. Two tracks must form a secondary vertex, separated from the interaction vertex by at least 3 mm, the lifetime and the mass of this system must be compatible with a K_S^0. The dE/dx measurement of the two tracks from the primary vertex must be compatible with the $K^\pm\pi^\mp$ hypothesis. The $K_S^0 K^\pm \pi^\mp$ mass spectrum, subdivided in four P_T^2 intervals, is shown in fig.3. Monte Carlo studies prove that $P_T^2 \simeq Q^2$, where Q^2 is the maximum virtuality of the two photons. In all spectra a peak at ~ 1450 MeV is present. In the higher Q^2 interval, $1 \le Q^2 \le 7$ GeV^2, the $f_1(1285)$ is also

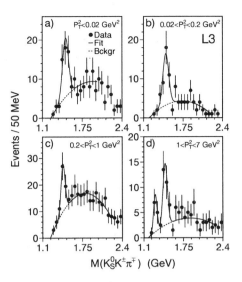

Figure 3. The $K_S^0 K^\pm \pi^\mp$ mass spectrum in four P_T^2 intervals.

present. The P_T^2 dependence of this peak (fig.4) cannot be explained by a pure 0^- or 1^+ state. The data require the presence of both the $\eta(1440)$ and the $f_1(1420)$ resonance. From the events with $P_T^2 < 0.02$ GeV^2, we

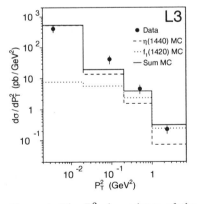

Figure 4. The P_T^2 dependence of the data requires the presence of $\eta(1440)$ and $f_1(1420)$.

measure the $\eta(1440)$ parameters:
$M = 1481 \pm 12$ MeV, $\quad \Gamma = 80 \pm 16$ MeV, $\Gamma_{\gamma\gamma} \cdot BR(\eta(1440) \rightarrow K\bar{K}\pi) = 212 \pm 55$ eV. Inspection of the $K\pi$ and $K_S^0 K$ spectra show a dominance of the $K^*\bar{K}$ decay. These ob-

servations point to the identification of this resonance with the η_H [4].

1.3 $e^+e^- \rightarrow e^+e^-\eta\pi^+\pi^-$

The events are selected by requiring two charged particles and two photons. Both photons must deposit more than 0.1 GeV in the electromagnetic calorimeter and their effective mass must be in the range $0.47 - 0.62$ GeV. A kinematical fit, with the η mass constraint, is then applied. The $\eta\pi^+\pi^-$ mass spectrum of 6400 selected events (fig.5a) shows the presence of the η' and of the $f_1(1285)$. By selecting quasi-real photons, with a $P_T^2 < 0.02$ GeV^2 cut, no signal is observed for the $\eta(1440)$ nor for the $\eta(1295)$ (fig.5b). The upper limits of their two-photon width, at 95% CL are:

$\Gamma_{\gamma\gamma} \cdot BR(\eta(1440) \rightarrow \eta\pi\pi) < 95$ eV

$\Gamma_{\gamma\gamma} \cdot BR(\eta(1295) \rightarrow \eta\pi\pi) < 66$ eV .

This observation corroborates the interpretation of the resonance observed in $K\bar{K}\pi$ as the η_H state.

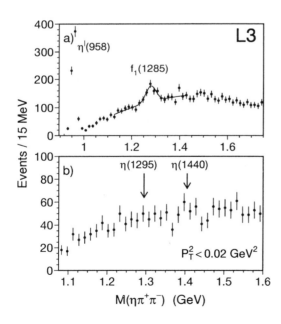

Figure 5. The $\eta\pi^+\pi^-$ mass spectrum. (a) All selected events. (b) Events with $P_T^2 < 0.02$ GeV^2.

2 Tests of gluonium

We have observed, for the first time, the $\eta(1440)$ meson in two-photon formation. The value of its two-photon width is compatible with a recent calculation[5] of the first radial excitation of the pseudoscaler nonet. However, by comparing this measurement to the J/ψ radiative decay[4], we find a stickiness[6] value of 79±26 and a gluiness[7] value of 41±14. Both values are higher than expected for a normal $q\bar{q}$ state, thus indicating a strong gluon contribution.

Acknowledgments

All my thanks go to Oleg Fedin and Igor Vodopianov who performed the analyses presented here.

References

1. L3 Coll. *"Analysis of the $\pi^+\pi^-\pi^+\pi^-$ final state in $\gamma\gamma$ collisions at LEP"*, L3 Note 2571 submitted to this Conference.

2. L3 Coll. *"Formation of light resonances in $K_S^0 K^\pm\pi^\mp$ and $\eta\pi^+\pi^-$ final states in $\gamma\gamma$ collisions at LEP"*, L3 Note 2564 submitted to this Conference.

3. M. Althoff *et al.*, *Z. Phys.*C **16**,13(1982).

4. Particle Data Group, D. E. Groom *et al.*, *Eur. Phys. J* C **15**, 1 (2000).

5. A.V. Anisovich *et al.*, *Eur. Phys. J* A **6**, 247 (1999).

6. M. Chanowitz, in *Proceedings of the VI International Workshop on $\gamma\gamma$ Collisions* (World Scientific,1985).

7. F.E.Close *et al.*, *Phys. Rev.D* **55**, 5749 (1997).

TWO PHOTON COLLISIONS AT LEP: $K_S^0 K_S^0$ FINAL STATE, GLUEBALL SEARCHES AND $\Lambda\bar{\Lambda}$ PRODUCTION

A. J. BARCZYK

ON BEHALF OF THE L3 COLLABORATION

PSI Villigen, CH-5232 Villigen, and ETH Zürich, CH-8093 Zürich, Switzerland
Current address: CERN, CH-1211 Genève, Switzerland
E-mail: Artur.J.Barczyk@cern.ch

The exclusive $K_S^0 K_S^0$ final state produced by two photon interactions is studied by the L3 experiment at LEP. The mass spectrum of the $K_S^0 K_S^0$ system is dominated by resonance production: the $f_2 - a_2$ interference, the $f_2'(1525)$ tensor meson and a clear enhancement in the region around 1750 MeV. The spin and helicity of the two last states are determined. No signal is found at 2230 MeV.

Also presented is the first measurement at LEP of the $\Lambda\bar{\Lambda}$ production in $\gamma\gamma$ interactions.

1 Introduction

Two-photon interactions $e^+e^- \rightarrow e^+e^-\gamma\gamma \rightarrow e^+e^-R$ are particularly useful to study the properties of hadron resonances (R). In untagged events, where the outgoing $e^\pm$ escape undetected along the beam pipe, the two interacting photons are quasi-real ($Q^2 \approx 0$). The produced resonance R must therefore be a neutral unflavoured meson with $C = +1$ and $J \neq 1$. The cross section is proportional to the two-photon partial width of the resonance: $\sigma(e^+e^- \rightarrow e^+e^-R) = \mathcal{K} \times \Gamma_{\gamma\gamma}(R)$. In contrast to $q\bar{q}$ resonances, glueballs have a very small two-photon width $\Gamma_{\gamma\gamma}(G)$, and are not expected to be observed in $\gamma\gamma$ interactions. The analysis of $e^+e^- \rightarrow e^+e^-K_S^0 K_S^0$ process[1] is presented in Sec. 2.

The study of $\Lambda\bar{\Lambda}$ production[2] at LEP is presented in Sec. 3. Baryon production in two-photon interactions has been studied by the CLEO collaboration and compared to theoretical predictions. While CLEO's analysis of the $\gamma\gamma \rightarrow p\bar{p}$ data[3] favours the quark-diquark model[6], the measured $\gamma\gamma \rightarrow \Lambda\bar{\Lambda}$ cross section[4] is higher than predicted by the model. A comparison of our data with CLEO's results, as well as with new theoretical predictions[5] is performed.

Both analyses presented here are based on the integrated luminosity of 588pb^{-1} col-

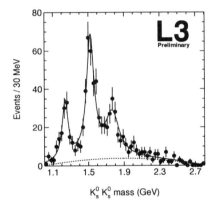

Figure 1. The $K_S^0 K_S^0$ mass spectrum.

lected at centre-of-mass energies in the range from 91 GeV to 202 GeV.

2 $K_S^0 K_S^0$ final state

In the analysis of the reaction $e^+e^- \rightarrow e^+e^-K_S^0 K_S^0$, only the decay $K_S^0 \rightarrow \pi^+\pi^-$ is considered. The selection of exclusive $K_S^0 K_S^0$ events is based on the requirement of four charged tracks with a total charge equal zero. The presence of two secondary vertices and a low transverse momentum imbalance $|\Sigma \vec{p}_T|^2 < 0.1$ GeV are required. The selected data sample consists of 801 events. The resulting $K_S^0 K_S^0$ invariant mass spectrum

Table 1. Results of the maximum likelihood fit to the $K_S^0 K_S^0$ mass spectrum. Mass and width are given in MeV.

	$f_2 - a_2$	$f_2'(1525)$	1750 MeV
Mass	1239 ± 6	1523 ± 6	1767 ± 14
Width	78 ± 19	100 ± 15	187 ± 60
events	123 ± 22	331 ± 37	220 ± 55

Table 2. Confidence level for different spin/helicity hypotheses.

	$J = 0$	$J = 2$ $\lambda = 0$	$J = 2$ $\lambda = 2$
$f_2'(1525)$	10^{-6}	10^{-6}	99.9%
R(1750)	8%	10^{-6}	33%

is shown in Fig. 1.

Four regions of interest can be identified in the spectrum: the $f_2(1270) - a_2^0(1320)$ destructive interference, the dominating peak of the $f'(1525)$, a new resonance at 1750 MeV, and the region around 2230 MeV, where no excess is observed.

An unbinned maximum likelihood fit using three Breit-Wigner functions plus a second order polynomial for the background is performed on the spectrum. The result of the fit is superimposed on the data points in Fig. 1, while Tab. 1 gives the numerical results.

2.1 The $f_2'(1525)$ tensor meson and the 1750 MeV mass region.

In order to determine the spin and helicity state of the resonance, the angular distribution of the K_S^0's in the two-photon centre-of-mass frame is compared to the expected distributions for spin zero, spin two helicity zero and spin two helicity two cases. The confidence levels for these three hypotheses are shown in Tab. 2 for the $f_2'(1525)$ tensor meson resonance and the fit in the region at 1750 MeV. The mass intervals chosen are $1400 - 1640$ MeV and $1640 - 2000$ MeV for the two peaks, respectively.

While in the $f_2'(1525)$ region the spin two helicity two state is clearly dominant, in the region around 1750 MeV also a small content of spin zero can be present. A mixed state fit in this region, considering both the $J = 0$ and $J = 2, \lambda = 2$ waves leads to a confidence level of 84% for a $J = 0$ contribution of $24 \pm 16\%$.

This value is in good agreement with the BES results[7].

The two-photon partial width times branching ratio for the $f_2'(1525)$ resonance is found to be

$$\Gamma_{\gamma\gamma}(f_2') \times \mathrm{Br}(K\bar{K}) = 0.076 \pm 0.006 \pm 0.011 \, \mathrm{keV}$$

2.2 The 2230 MeV mass region

The $\xi(2230)$ resonance observed by the MARK III and BES collaborations[8] in the J/Ψ radiative decay is a potential glueball candidate. Since no signal is observed in this mass region, an upper limit on the two-photon width is obtained: $\Gamma_{\gamma\gamma}(\xi) \times \mathrm{Br}(\xi \to K_S^0 K_S^0) < 1.4$ eV at 95% C.L, under the hypothesis of a pure spin-two helicity-two state. In addition, the stickiness parameter[a] is computed using this value and the results of MARK III and BES collaborations. A lower limit is obtained to be $S_{\xi(2230)} > 73$ at 95% C.L. This value is much larger than the values measured for all the well established $q\bar{q}$ states, and supports the interpretation of the $\xi(2230)$ as the tensor glueball.

3 $\Lambda\bar{\Lambda}$ formation

To analyse the reaction $e^+e^- \to \Lambda\bar{\Lambda}$, only the decays $\Lambda \to p\pi^-$ and $\bar{\Lambda} \to \bar{p}\pi^+$ are considered. The main requirements in the event selection are: four charged tracks with a net

---

[a]Stickiness S_R is defined as the squared ratio of the coupling of a resonance to two gluons relative to two photons. It is expected to be small for $q\bar{q}$ and large for gluonium states.

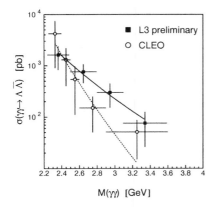

Figure 2. Measured $\sigma(\gamma\gamma \rightarrow (\Lambda/\Sigma^0)(\bar{\Lambda}/\bar{\Sigma}^0))$.

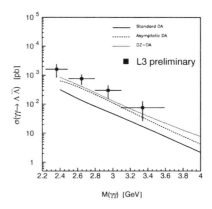

Figure 3. $\sigma(\gamma\gamma \rightarrow (\Lambda/\Sigma^0)(\bar{\Lambda}/\bar{\Sigma}^0))$ compared to calculations of C. Berger et al.

charge of zero, two secondary vertices and the calorimeter energy associated with the $\bar{p}$ to be above $E_{em} > 0.35$ GeV, due to the anti-proton annihilation in the electromagnetic calorimeter. No Σ^0 identification is performed, thus the cross section measurement refers to the channel $\gamma\gamma \rightarrow (\Lambda/\Sigma^0)(\bar{\Lambda}/\bar{\Sigma}^0)$.

In the data sample, 44 events have been selected. The cross section is measured to be $\sigma = 4.13 \pm 0.65 \pm 0.69$ pb. Deconvoluting the two-photon luminosity function and extrapolating the measurement to the interaction of real photons, we extract the cross section shown in Fig. 2. Superimposed is the result of a fit to the data of the form $\sigma \propto M^{-n}$. A comparison of our data with CLEO's measurement[4] shows a good agreement within experimental errors, the mass dependence, however, is slightly different for the two experiments. The mentioned fit to our data gives a value of $n = 8.7 \pm 3.8$, while a similar fit to CLEO gives $n = 16.1 \pm 5.8$.

Comparing our data to the calculations performed recently by C. Berger et al.[5], we observe a good description of the mass dependence, c.f. Fig. 3. In their quark-diquark model they consider three different distribution amplitudes for the octet baryons. Our data lie above the predictions, as expected due to the Σ^0 contamination in the selection.

Acknowledgments

The main people behind the presented analyses were S. Braccini ($K_S^0 K_S^0$) and B. Echenard ($\Lambda\bar{\Lambda}$). I would like to thank them for their support in preparing this presentation.

References

1. L3 Collab., "$K_S^0 K_S^0$ Final State in Two-Photon Collisions and Glueball Searches", L3 Note 2557 submitted to this conference.
2. L3 Collab., "$\Lambda\bar{\Lambda}$ Formation in Two-Photon Collisions at LEP", L3 Note 2566 submitted to this conference.
3. CLEO Collab., *Phys. Rev.* **D 50** (1994) 5484.
4. CLEO Collab., *Phys. Rev.* **D 56** (1997) 2485.
5. C. Berger, et al., *Fizika* **B 8** (1999) 371.
6. M. Anselmino et al., *Int. J. Mod. Phys.* **A 4** (1989) 5213.
7. BES Coll., *Phys. Rev. Lett.* **77** (1996) 3959.
8. MARK III Collab., *Phys. Rev. Lett.* **56** (1986) 107.
   BES Coll., *Phys. Rev. Lett.* **76** (1996) 3502.

MEASUREMENT OF K^+K^- PRODUCTION IN TWO-PHOTON COLLISIONS AT BELLE

S. UEHARA

for the Belle Collaboration

KEK-IPNS, Institute of Particle and Nuclear Studies,
High Energy Accelerator Research Organization, 1-1 Oho, Tsukuba 305-0801 Japan
E-mail: uehara@post.kek.jp

K^+K^- production in two-photon collisions has been studied with the Belle detector at KEKB. We have obtained the first high statistics data sample in the invariant mass range above 1.6 GeV. We report preliminary results of the cross section for $\gamma\gamma \to K^+K^-$ in the c.m. energy range between 1.36 and 2.30 GeV. In addition, we also present preliminary results for the $\gamma\gamma \to K_S^0 K_S^0$ process.

In $K\bar{K}$ production in two-photon collisions, contributions from the three ground-state tensor mesons, $f_2(1270)$, $a_2(1320)$ and $f_2'(1525)$, have been observed, and no established meson state is found above the $f_2'(1525)$ mass[1,2].

Studies in the energy region above 1.6 GeV are very important for exploring the meson states whose existence or properties have not yet been established. The L3 experiment has reported a resonance-like peak near 1.75 GeV in the $K_S^0 K_S^0$ channel[2]. Furthermore, some glueball candidates around 2 GeV have been observed in the decays of J/ψ into the $\gamma K\bar{K}$ state. In the K^+K^- channel, no structures have been identified above 1.6 GeV, due to the limited statistics of previous experiments.

We have measured K^+K^- production in the $\gamma\gamma$ c.m. energy range between 1.36 and 2.30 GeV. In addition, $K_S^0 K_S^0$ production in a similar range has also been measured. All the results are preliminary. Details of the measurements are reported in contributed papers for the conference[3].

The experiment was carried out using the Belle detector at the KEKB asymmetric e^+e^- collider. The c.m. energy of the beams was set around 10.58 GeV.

The tracking of charged particles is performed using the central drift chamber(CDC) and the silicon vertex detector. For kaon identification, we use mainly the TOF counters which can separate kaons from pions up to 1.2 GeV/c with a time resolution of 100 ps as well as information from the Aerogel Cherenkov counters(ACC), CsI electromagnetic calorimeters, and dE/dx information from the CDC.

We used data samples corresponding to an integrated luminosity of 3.1 fb^{-1} for this analysis.

We select events with only two charged particles, and require each track to have a transverse momentum $p_t > 0.4$ GeV/c and a polar angle $-0.34 < \cos\theta_{\text{lab}} < +0.82$. We impose a strict p_t-balance cut on the final state, $|\sum \mathbf{p}_t^*|^2 < 0.01$ GeV^2, in order to select events induced by quasi-real two photons.

K^+K^- events are selected using an E/p cut to reject electron-pair backgrounds, and information from TOF and ACC (in the form of likelihood functions) to separate kaons from pions, muons and protons.

We show the cross section for the $\gamma\gamma \to K^+K^-$ process derived from the invariant mass distribution for selected K^+K^- events in Fig. 1. The $\gamma\gamma$ c.m. angular range for the kaons is restricted to $|\cos\theta^*| < 0.6$.

The present results have a systematic error of 20% mainly coming from an uncer-

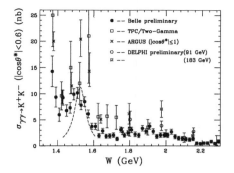

Figure 1. The cross section for $\gamma\gamma \rightarrow K^+K^-$. The results from previous experiments are from Ref.s[1,4]. The dashed curve is the expected contribution from the $f_2'(1525)$ (see the text).

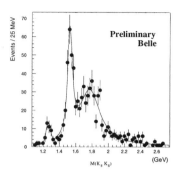

Figure 2. Invariant mass distribution for the $K_S^0 K_S^0$ events. The curve is a fit described in the text.

tainty of particle-ID efficiency. We have confirmed that the background contamination is not large compared with the systematic error.

In the cross section, we find a $f_2'(1525)$ peak, as shown in Fig. 1. For comparison, the dashed curve shows the expected contribution of $f_2'(1525)$ calculated by the $\Gamma_{\gamma\gamma}(f_2')BR(f_2' \rightarrow K\bar{K})$ value determined in the $K_S^0 K_S^0$ channel by L3[2].

We find a broad bump in the 1.7 - 2.1 GeV region: this is the first time that such a structure has been seen. With a fit of a single Breit-Wigner curve to the structure, we obtain parameters $M = 1.88 \pm 0.02$ GeV/c^2, $\Gamma = 0.47 \pm 0.08$ GeV and $\Gamma_{\gamma\gamma}BR(K^+K^-) = 104 \pm 13 \pm 21$ eV ($0.84 \pm 0.11 \pm 0.17$ keV) for the $(J, \lambda) = (2, 2)$ ($J = 0$) case.

We find no significant enhancement around 2.23 GeV. We obtain a 95%CL upper limit, $\Gamma_{\gamma\gamma}(f_J(2220))BR(f_J(2220) \rightarrow K\bar{K}) < 3.2$ eV, under the assumption of $(J, \lambda) = (2, 2)$.

We have also measured the $K_S^0 K_S^0$ final state. We identify each K_S^0 by its $\pi^+\pi^-$ decay, and apply cuts on invariant mass and decay length. The p_t-balance of the K_S^0 pair is required to be $|\sum \mathbf{p}_t^*|^2 < 0.1$ GeV^2. Data corresponding to $\int \mathcal{L}dt = 5$ fb^{-1} are used for this analysis.

Fig. 2 shows the $K_S^0 K_S^0$ invariant mass distribution in the final sample. We fit the distribution with the sum of a Gaussian (representing the $f_2(1270)$-$a_2(1320)$ interference), two Breit-Wigners, and a constant background. We find masses and widths of two peak structures, $(M, \Gamma) = (1526 \pm 4, 52 \pm 11)$ MeV and $(1771 \pm 13, 264 \pm 29)$ MeV for the two Breit-Wigners. The former corresponds to the $f_2'(1525)$, and the latter to the structure reported by the L3 experiment[2].

We have measured K^+K^- production in two-photon collisions. We find a broad structure in the 1.7 - 2.1 GeV region. An upper limit for the $f_J(2220)$ is obtained. The $K_S^0 K_S^0$ final state has also been measured. We confirm a peak structure near 1.77 GeV.

References

1. ARGUS Collab., H.Albrecht et al., *Z. Phys.* **C48** 183 (1990); TPC/Two-Gamma Collab., H.Aihara et al., *Phys. Rev. Lett.* **57** 404 (1986).

2. L3 Collab., M.Acciarri et al., *Phys. Lett.* **B363** 118 (1995); L3 Collab., S.Saremi, *Nucl. Phys.* **B82** 344 (2000).

3. Belle Collab., A.Abashian et al., KEK Preprint 2000-86 (2000); Belle Collab., A.Abashian et al., KEK Preprint 2000-87 (2000).

4. DELPHI Collab., K.Grzelak, *Nucl. Phys.* **B82** 316 (2000).

KLOE FIRST RESULTS ON HADRONIC PHYSICS

THE KLOE COLLABORATION[a]

presented by P. GAUZZI

Dipartimento di Fisica Università "La Sapienza" and INFN Sezione di Roma,
P.le A.Moro 2, 00185 - Rome (Italy)
E-mail: Paolo.Gauzzi@roma1.infn.it

The KLOE experiment has collected 2.4 pb^{-1} of integrated luminosity during the 1999 data-taking. This data sample has been used to study the radiative decays of the $\phi(1020)$ meson: $\phi \to \pi^0\pi^0\gamma$, $\pi^+\pi^-\gamma$, $\eta\pi^0\gamma$ and $\phi \to \eta\gamma$, $\eta'\gamma$. The preliminary results are reported in this paper. The analysis of the Dalitz plot of the decay $\phi \to \pi^+\pi^-\pi^0$ is also presented.

1 Introduction

The KLOE experiment [a,1] has started taking data at the DAΦNE ϕ-factory[2] in April 1999. An integrated luminosity of 2.4 pb^{-1} has been collected during the 1999 run, that correspond to about 7 millions of ϕ produced. Such statistics has been used to perform some studies of the ϕ radiative decays, providing also a good check of the detector performances. The decays $\phi \to \pi^0\pi^0\gamma$, $\pi^+\pi^-\gamma$, $\eta\pi^0\gamma$ have been exploited to study the scalar mesons $f_0(980)$ and $a_0(980)$. With $\phi \to \eta\gamma$, $\eta'\gamma$ a measurement of the $\eta-\eta'$ mixing angle has been performed. The Dalitz plot of the decay $\phi \to \pi^+\pi^-\pi^0$ has been analysed, looking for a possible contribution of the direct decay, and to obtain a new measurement of the ρ line-shape parameters.

A more detailed description of the analyses presented in this paper can be found in ref.[3].

2 Radiative decays

2.1 The scalar sector: $\phi \to f_0\gamma$, $a_0\gamma$

The nature of the scalar mesons $f_0(980)$ and $a_0(980)$ is not well established, alternative hypotheses with respect to the standard $q\bar{q}$ structure have been proposed: a $q\bar{q}q\bar{q}$ state or a $K\bar{K}$ bound state. According to the theory

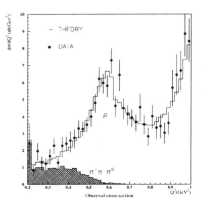

Figure 1. $\phi \to \pi^+\pi^-\gamma$ cross-section as a funcion of $Q^2 = M_{\pi\pi}^2$, compared to the theoretical one for pure QED contribution.

the branching ratios of $\phi \to f_0\gamma$ and $\phi \to a_0\gamma$ are sensitive to the structure of the particles, their measurement can help in understanding the nature of $f_0(980)$ and $a_0(980)$.

Three processes are expected to contribute to the $\phi \to \pi^+\pi^-\gamma$ decay: *(i)* the $f_0\gamma$ intermediate state, with $f_0 \to \pi^+\pi^-$, *(ii)* the Initial State Radiation in which the photon is emitted by an incoming electron or positron, and *(iii)* the Final State Radiation in which the photon is emitted by a pion (interference with $\phi \to f_0\gamma$ is expected).

$\pi^+\pi^-\gamma$ events have been selected from a partial sample of 1.8 pb^{-1} with an overall

---

[a]The author list is in e-print hep-ex/0006036

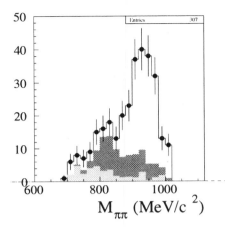

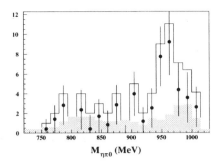

Figure 3. Invariant mass of the $\eta\pi^0$ system, histogram: data; shadowed histogram: expected background (MC); points: data-background difference.

Figure 2. Invariant mass of the $\pi^0\pi^0$ system for $\phi \to f_0\gamma$ selected events. Points: data; histograms: backgrounds (mainly from $e^+e^- \to \omega\pi^0$).

efficiency of 50%, evaluated by Montecarlo (MC). The invariant mass spectrum of the two pions (fig.1) has been fitted to the theoretical prediction, excluding the f_0 region, $M_{\pi\pi}^2 > 0.84$ GeV^2. By extrapolating to the signal region, no evidence of the f_0 has been found, thus the upper limit Br($\phi \to f_0\gamma \to \pi^+\pi^-\gamma$)< $1.6\cdot10^{-4}$ at 90% C.L. can be set.

The 5 photon events have been selected to study both $\phi \to f_0\gamma$ (with $f_0 \to \pi^0\pi^0$) and $\phi \to a_0\gamma$ (with $a_0 \to \eta\pi^0$ and $\eta \to \gamma\gamma$) decays. The main background for the first decay is the non resonant process $e^+e^- \to \omega\pi^0$ with $\omega \to \pi^0\gamma$. The two processes can be separated by using a kinematic fit. In fig.2 is shown the invariant mass spectra of the two pions for $\phi \to f_0\gamma$ selected events. 307 ± 18 $f_0\gamma$ events have been selected from the 1.8 pb^{-1} sample, with 40% efficiency and with an estimated background of 112 ± 11 (from MC), corresponding to Br($\phi \to f_0\gamma \to \pi^0\pi^0\gamma$)=$(0.81\pm0.09_{stat}\pm0.06_{syst})\cdot10^{-4}$.

With different topological cuts 529 ± 23 $\omega\pi^0$ events have been selected (ε=35%) with a background of 93 ± 10 events. The cross section at the ϕ peak can be extracted: $\sigma(e^+e^- \to \omega\pi^0)$=$0.67\pm0.04_{stat}\pm0.05_{syst}$ nb.

Concerning the $\phi \to \eta\pi^0\gamma$ decay, the whole 2.4 pb^{-1} sample has been analysed,

and 74 ± 9 events have been selected with an efficiency of 23%. The expected contamination is 21 ± 6 events. In the $\eta\pi$ invariant mass spectrum, shown in fig.3, is clearly visible the peak of the $a_0(980)$. Using the value Br($\eta \to \gamma\gamma$)=39.2% from PDG, one obtains Br($\phi \to \eta\pi^0\gamma$)=$(0.77\pm0.15_{stat}\pm0.10_{syst})\cdot10^{-4}$.

2.2 The pseudoscalar sector: $\phi \to \eta\gamma,\ \eta'\gamma$

The reason for studying the radiative ϕ decays into η and η' is twofold since the measurement of Br($\phi \to \eta'\gamma$) can help in defining the gluonic content of η' while from the ratio R=Br($\phi \to \eta'\gamma$)/Br($\phi \to \eta\gamma$) the value of the $\eta - \eta'$ mixing angle ϑ_p can be extracted[4].

Charged ($\phi \to \eta\gamma \to \pi^+\pi^-\pi^0\gamma$ and $\phi \to \eta'\gamma \to \eta\pi^+\pi^-\gamma$ with $\eta \to \gamma\gamma$) and neutral ($\phi \to \eta\gamma \to \pi^0\pi^0\pi^0\gamma$ and $\phi \to \eta'\gamma \to \eta\pi^0\pi^0\gamma$ with $\eta \to \gamma\gamma$) final states have been exploited.

In the charged final state, 21 ± 4.6 $\eta'\gamma$ events have been selected (see fig.4), with an overall efficiency of 19%, togheter with about 6700 $\eta\gamma$ events (ε=32%). From these numbers a value of R=$(7.1\pm1.6_{stat}\pm0.3_{syst})\cdot10^{-3}$ has been obtained. From the known Br($\phi \to \eta\gamma$), one can obtain also Br($\phi \to \eta'\gamma$)=$(8.9\pm2.0_{stat}\pm0.6_{syst})\cdot10^{-5}$.

Concerning the 7 photon final state, the decay chain $\phi \to \eta'\gamma \to \eta\pi^0\pi^0\gamma$ has been detected for the first time: $6^{+3.3}_{-2.2}$ events have

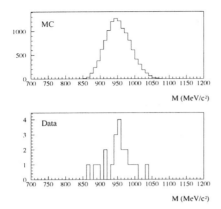

Figure 4. Invariant mass of the two charged tracks and the two most energetic photons for the 21 events selected as $\phi \to \eta'\gamma$, compared to MC.

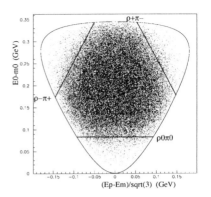

Figure 5. Dalitz plot of $\phi \to \pi^+\pi^-\pi^0$.

Table 1. Results of the fit to the Dalitz plot.

Parameter	Fit result	PDG value
M_{ρ^0} (MeV)	776 ± 1.0	776.0 ± 0.9
ΔM (MeV)	-0.5 ± 0.7	0.1 ± 0.9
Γ_ρ (MeV)	145.6 ± 2.2	150.9 ± 2.0
$A_{direct}/A_{\rho\pi}$	0.10 ± 0.01	-0.15 ± 0.11
phase$_{direct}$/phase$_{\rho\pi}$	$(114\pm12)^o$	

been selected with an estimated efficiency of 13%. From these events and from the about 11000 $\phi \to \eta\gamma \to \pi^0\pi^0\pi^0\gamma$ selected (ε=41%), a value R=$(6.9^{+3.3}_{-2.2}\ _{stat}\pm0.9_{syst})\cdot10^{-3}$ can be derived. Combining the two R measurements one obtains for the mixing angle :
ϑ_p=$(-18.9^{+3.6}_{-2.8}\ _{stat}\pm0.6_{syst})^o$.

3 $\quad \phi \to \pi^+\pi^-\pi^0$

About 15% of the ϕ decay into $\pi^+\pi^-\pi^0$. Three different processes are expected to contribute to this final state: (i) $\phi \to \rho\pi$ in all the three possible charge states, (ii) the direct decay $\phi \to \pi^+\pi^-\pi^0$ and (iii) $e^+e^- \to \omega\pi^0$ with $\omega \to \pi^+\pi^-$. The Dalitz plot (fig.5) has been filled with 330000 events selected in a 2.1 pb^{-1} partial sample and then fitted to the function (X and Y are the two Dalitz plot variables, and $\vec{p}^{\ +}$, $\vec{p}^{\ -}$ are the momenta of π^+ and π^-):

$$f(X,Y) =$$
$$= |\vec{p}^{\ +} \times \vec{p}^{\ -}|^2 \cdot |\mathbf{A}_{\rho\pi} + \mathbf{A}_{\mathbf{direct}} + \mathbf{A}_{\omega\pi}|^2$$

The free parameters of the fit are: the ρ^0 mass, the mass difference ΔM between the charged and neutral ρ, the ρ width, a modulus and a phase for the direct term $\mathbf{A}_{\mathbf{direct}}$, and a modulus and a phase for $\mathbf{A}_{\omega\pi}$ term, while the ω mass and width have been fixed

to the PDG values. The results of the fit are shown in tab.1 where the parameters are compared to the PDG values: the ρ line-shape parameters are in agreement with the PDG ones, the mass difference is compatible with zero, and there is an indication of a sizeable direct term (10% of the $\rho\pi$ term) with a relative phase close to 90°.

References

1. S. Bertolucci, *A Status Report of KLOE at DAΦNE*, hep-ex/0002030 (2000).

2. S. Guiducci et al., *DAΦNE Operating Experience*, Proceedings of PAC99, N.Y., March 1999.

3. The KLOE Collaboration, Contributed paper No.220 to this Conference, hep-ex/0006036 (2000).

4. A. Bramon et al., *Eur.Phys.J.* C **7**, 271 (1999).

RESULTS ON CHARMED MESON SPECTROSCOPY FROM FOCUS

FRANCO L. FABBRI

Laboratori Nazionali di Frascati via E.Fermi, 40 - Frascati (Rome) - 00044 Italy
ON BEHALF OF THE FOCUS(E831) COLLABORATION

We report the preliminary measurement by the FOCUS Collaboration (E831 at Fermilab) of masses and widths of the L=1 charm mesons: a D_2^{*0} state of mass (width) $2463.5 \pm 1.5 \pm 1.5 (30.5 \pm 1.9 \pm 3.8)$ MeV/c^2 decaying to $D^+\pi^-$, and a D_2^{*+} state of mass (width) $2468.2 \pm 1.5 \pm 1.4 (28.6 \pm 1.3 \pm 3.8)$ MeV/c^2 decaying to $D^0\pi^+$. The fit of the invariant mass distribution requires an additional term to account for a broad structure over background.

In this paper we present preliminary results from the FOCUS experiment [a] (E831 at Fermilab) on the spectroscopy of bound states of a charm quark and a lighter quark with orbital angular momentum $L = 1$, called $D_2^*(c\bar{u}, c\bar{d})$. With the high-statistics,

---

[a]Coauthors: J.Link, V.S. Paolone, M. Reyes, P.M. Yager (**UC DAVIS**); J.C. Anjos, I. Bediaga, C. Göbel, J. Magnin, J.M. de Miranda, I.M. Pepe, A.C. dos Reis, F. Simão (**CPBF, Rio de Janeiro**); S. Carrillo, E. Casimiro, H. Mendez, A.Sánchez-Hernández,, C. Uribe, F. Vasquez (**CINVESTAV, México City**); L. Cinquini, J.P. Cumalat, J.E. Ramirez, B. O'Reilly, E.W. Vaandering (**CU Boulder**); J.N. Butler, H.W.K. Cheung, I. Gaines, P.H. Garbincius, L.A. Garren, E. Gottschalk, S.A. Gourlay, P.H. Kasper, A.E. Kreymer, R. Kutschke (**Fermilab**); S. Bianco, F.L. Fabbri, S. Sarwar, A. Zallo (**INFN Frascati**); C. Cawlfield, D.Y. Kim, K.S. Park, A. Rahimi, J. Wiss (**UI Champaign**); R. Gardner (**Indiana**); Y.S. Chung, J.S. Kang, B.R. Ko, J.W. Kwak, K.B. Lee, S.S. Myung, H. Park (**Korea University, Seoul**); G. Alimonti, M. Boschini, D. Brambilla, B. Caccianiga, A. Calandrino, P. D'Angelo, M. DiCorato, P. Dini, M. Giammarchi, P. Inzani, F. Leveraro, S. Malvezzi, D. Menasce, M. Mezzadri, L. Milazzo, L. Moroni, D. Pedrini, F. Prelz, M. Rovere, A. Sala, S. Sala (**INFN and Milano**); T.F. Davenport III (**UNC Asheville**); V. Arena, G. Boca, G. Bonomi, G. Gianini, G. Liguori, M. Merlo, D. Pantea, S.P. Ratti, C. Riccardi, P. Torre, L. Viola, P. Vitulo (**INFN and Pavia**); H. Hernandez, A.M. Lopez, L. Mendez, A. Mirles, E. Montiel, D. Olaya, J. Quinones, C. Rivera, Y. Zhang (**Mayaguez, Puerto Rico**); N. Copty, M. Purohit, J.R. Wilson (**USC Columbia**); K. Cho, T. Handler (**UT Knoxville**); D. Engh, W.E. Johns, M. Hosack, M.S. Nehring, M. Sales, P.D. Sheldon, K. Stenson, M.S. Webster (**Vanderbilt**); M. Sheaff (**Wisconsin, Madison**); Y. Kwon (**Yonsei University, Korea**)

high-mass resolution experiments attaining maturity, emphasys has been shifted from the ground state (0^- and 1^-) $c\bar{q}$ mesons and ($1/2^+$ and $3/2^+$) cqq baryons to the orbitally- and, only very recently, radially-excited states[b]. A consistent theoretical framework for the spectrum of heavy-light mesons is given by the ideas of Heavy Quark Symmetry (HQS), later generalized by Heavy Quark Effective Theory in the QCD framework. The basic idea (mediated from the JJ coupling in atomic physics) is that in the limit of infinite heavy quark mass: a) the much heavier quark does not contribute to the orbital degrees of freedom, which are completely defined by the light quark(s) only; and b) properties are independent of heavy quark flavor. Heavy Quark Symmetry provides explicit predictions on the spectrum of excited charmed states[1].

In the limit of infinite heavy quark mass, the spin of the heavy quark $\mathbf{S_Q}$ decouples from the light quark degrees of freedom (spin $\mathbf{s_q}$ and orbital $\mathbf{L}$), with $\mathbf{S_Q}$ and $\mathbf{j_q} \equiv \mathbf{s_q} + \mathbf{L}$ the conserved quantum numbers. Predicted excited states are formed by combining $\mathbf{S_Q}$ and $\mathbf{j_q}$. For $L = 1$ we have $j_q = 1/2$ and $j_q = 3/2$ which, combined with S_Q, provide prediction for two $j_q = 1/2$ (J=0,1) states, and two $j_q = 3/2$ (J=1,2) states. These four states are named respectively D_0^*, $D_1(j_q = 1/2)$, $D_1(j_q = 3/2)$ and D_2^*. Finally,

---

[b]In the past, these excited states were called generically and improperly D^{**}.

parity and angular momentum conservation favor the $(j_q = 1/2)$ states to decay to the ground states mainly via S-wave transitions (broad width), while $(j_q = 3/2)$ states would decay via D-wave (narrow width). While the narrow states are well established, the evidence for the broad states (both in the c-quark and in the b-quark sector) is much less stringent [2].

The data for this paper were collected in the Wideband photoproduction experiment FOCUS during the Fermilab 1996–1997 fixed-target run. FOCUS is a considerably upgraded version of a previous experiment, E687 [3]. In FOCUS, a forward multi-particle spectrometer is used to measure the interactions of high energy photons on a segmented BeO target. We obtained a sample of over 1 million fully reconstructed charm particles in the three major decay modes: $D^0 \rightarrow K^-\pi^+$, $K^-\pi^+\pi^-\pi^+$ and $D^+ \rightarrow K^-\pi^+\pi^+$.

The FOCUS detector is a large aperture, fixed-target spectrometer with excellent vertexing, particle identification, and reconstruction capabilities for photons and π^0's. A photon beam is derived from the bremsstrahlung of secondary electrons and positrons with an ≈ 300 GeV endpoint energy produced from the 800 GeV/c Tevatron proton beam. The charged particles which emerge from the target are tracked by two systems of silicon microvertex detectors. The upstream system, consisting of 4 planes (two views in 2 stations), is interleaved with the experimental target, while the other system lies downstream of the target and consists of twelve planes of microstrips arranged in three views. These detectors provide high resolution separation of primary (production) and secondary (decay) vertices with an average proper time resolution of ≈ 30 fs for 2-track vertices. The momentum of a charged particle is determined by measuring its deflections in two analysis magnets of opposite polarity with five stations of multiwire proportional chambers. Three multicell threshold

Čerenkov counters are used to discriminate between pions, kaons, and protons.

The decays $D^0 \rightarrow K^-\pi^+$, $D^+ \rightarrow K^-\pi^+\pi^+$, $D^0 \rightarrow K^-\pi^+\pi^-\pi^+$, $D^{*+} \rightarrow D^0\pi^+$ were selected (Fig.1). To ensure clean charm samples, candidate events were selected with a large ℓ/σ_ℓ, being ℓ the separation of the primary and weak decay vertices, and σ_ℓ its uncertainty; the primary multiplicity was required to be greater than 1, and the primary to be located within one of the interaction targets; the kaon and pion candidates to be consistent with the kaon and pion hypotesis, based on the Čerenkov identification system; the weak decay vertex to be outside of the interaction targets ($\sigma_{OoM} > 0$) for the $D^0 \rightarrow K^-\pi^+\pi^-\pi^+$ mode only; and $|\cos\theta_K| < 0.7$ for the $D^0 \rightarrow K^-\pi^+$ mode only, where the decay angle θ_K is defined as the angle which, in the D rest frame, the kaon momentum forms with the D momentum in the lab. The results in this paper have been shown to be insensitive to the detailed choice of selection parameters.

The principal preliminary result in this paper relates to a study of the $D^+\pi^-$ and $D^0\pi^+$ mass spectra. The D^+ or D^0 candidates were combined with the pion tracks in the primary vertex to form L=1 D-meson candidates. Events with charm candidates coming from D^* decays were rejected by applying a $\pm 3\sigma$ cut around the $D^* - D$ mass difference. Figure 2 shows the distribution in the invariant mass difference

$$\Delta M_0 \equiv M(D^+\pi^-) - M(D^+) + M_{PDG}(D^+).$$

The plot shows a pronounced peak, consistent with being due to a D_2^{*0} of mass $M \approx 2460$ MeV/c^2. Because of the narrow width, this state has traditionally been identified as the $J = 2^+$ state. The additional enhancement at $M \approx 2300$ MeV/c^2 is consistent, as verified from Monte Carlo simulations, with arising from the feed-down of the states D_1^0 and D_2^{*0} decaying to $D^{*+}\pi^-$, with the D^{*+} subsequently decaying to D^+ and undetected

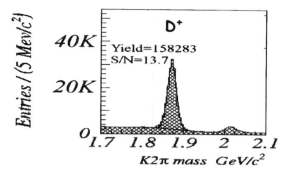

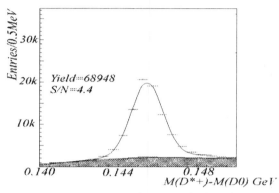

Figure 1. Mass plots for $D^+ \to K^-\pi^+\pi^+$ (top); and $D^{*+} \to D^0\pi^+$ (bottom).

Table 1. Preliminary measurements of masses and widths for narrow structures in $D^+\pi^-$ and $D^0\pi^+$ invariant mass spectra.

	Mass MeV/c^2	Width MeV/c^2
D_2^{*0}	$2463.5 \pm 1.5 \pm 1.5$	$30.5 \pm 1.9 \pm 3.8$
PDG2000	2458.9 ± 2.0	23 ± 5
D_2^{*+}	$2468.2 \pm 1.5 \pm 1.4$	$28.6 \pm 1.3 \pm 3.8$
PDG2000	2459 ± 4	25^{+8}_{-7}

neutral pion. The D_2^{*0} signal was fitted with a relativistic D-wave Breit-Wigner function, convoluted with a gaussian resolution function ($\sigma = 7$MeV). The background was fitted with the sum of an exponential, and two gaussians for the feed-downs described above, whose peaks and widths were fixed at the Monte Carlo values. The slope of the exponential was fixed to the value determined by a fit to the wrong-side events mass distribution, which is very well described by a single-slope exponential in the entire fitting interval $2250 - 3000 \,\text{MeV}/c^2$. For this fit we get a $\chi^2/\text{dof} = 2$, and a $\Gamma = 55 \pm 3 \,\text{MeV}/c^2$ D_2^{*0} width statistically non compatible with the PDG2000 world average of $\Gamma = 23 \pm 5 \,\text{MeV}/c^2$. We then add an S-wave relativistic Breit-Wigner function to the fit, which improves the fit quality $\chi^2/\text{dof} = 0.9$, and provides a width $\Gamma = 30 \pm 2 \,\text{MeV}/c^2$ compat-

ible to the PDG2000 value.

The mass difference

$$\Delta M_+ \equiv M(D^0\pi^+) - M(D^0) + M_{PDG}(D^0)$$

spectrum (Fig. 2 shows structures similar to those in the ΔM_0 spectrum. The prominent peak is consistent with being due to a D_2^{*+} of mass $M \approx 2460 \,\text{MeV}/c^2$. The additional enhancement at $M \approx 2300 \,\text{MeV}/c^2$ is consistent, as verified from Monte Carlo simulations, with arising from the feed-down of the states D_1^0 and D_2^{*0} decaying to $D^{*0}\pi^+$, with the D^{*0} subsequently decaying to D^0 and undetected neutral pion. The fitting procedure for the ΔM_+ spectrum follows the same guidelines as the ΔM_0. Several systematics checks have been performed to verify the stability of our measurements of masses and widths, such as fit variants varying the selection cuts over an extended range, and the stability of our mass measurements when performed on statistically independent subsamples (ℓ/σ_ℓ greater and less than 30, particle vs. antiparticle, momentum of the pion from the D_2^* decay greater and less than $18\text{GeV}/c$, momentum of the D meson greater and less than $70\text{GeV}/c$). Table 1 summarizes the preliminary results on the measurements of masses and widths from the study of $D^0\pi^+$ and $D^+\pi^-$ final states.

In conclusion, FOCUS has collected a large sample of D_2^{*0} and D_2^{*+} L=1 mesons out of the total sample of about 10^6 D meson states, and $0.5\,10^4$ D^* meson states. The study of the $D\pi$ mass spectrum provides new preliminary values of the masses and widths

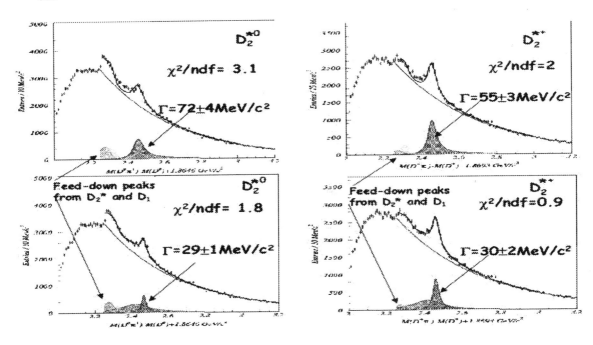

Figure 2. The $D^+\pi^-$ ($D^0\pi^+$) mass spectra is shown on the left (right). The invariant mass variable is defined as $\Delta M_0 \equiv M(D^+\pi^-) - M(D^+) + M_{PDG}(D^+)$ and $\Delta M_+ \equiv M(D^0\pi^+) - M(D^0) + M_{PDG}(D^0)$, respectively.

for the D_2^* meson (Tab. 1). The $D\pi$ mass spectrum (once subtracted the background, the D_2^* signal, and the expected feed-downs) shows an excess of events centered around 2420 MeV/c^2 and about 185 MeV/c^2 wide. A broad ($\sim 100 - 200$ MeV/c^2) state (the D_0^*) is predicted by HQS at about 2350 MeV/c^2. The observed excess could be reminiscent of this state, or of a feed-down from another broad state such as the $D_1(j_q = 1/2)$, possibly interfering. Work is in progress to verify such hypothesis.

We wish to acknowledge the assistance of the staffs of Fermi National Accelerator Laboratory, the INFN of Italy, and the physics departments of the collaborating institutions. This research was supported in part by the U. S. National Science Foundation, the U. S. Department of Energy, the Italian Istituto Nazionale di Fisica Nucleare and Ministero dell'Università e della Ricerca Scientifica e Tecnologica, the Brazilian Conselho Nacional de Desenvolvimento Científico e Tecnológico, CONACyT-México, the Korean Ministry of Education, and the Korean Science and Engineering Foundation.

References

1. S. Godfrey and N. Isgur, Phys. Rev. **D32**, 189 (1985); S. Godfrey and R. Kokoski, Phys. Rev. **D43**, 1679 (1991); N. Isgur and M.B. Wise, Phys. Rev. Lett. **66**, 1130 (1991); E.J. Eichten, C.T. Hill and C. Quigg, Phys. Rev. Lett. **71**, 4116 (1993); J. Bartelt and S. Shukla, Annu. Rev. Nucl. Part. Sci 1995 **45** 133-61.

2. F. L. Fabbri, *In *Frascati 1999, Hadron spectroscopy* 627-639*; S. Bianco, hep-ex/9911034.

3. P. L. Frabetti *et al.* (E687 Coll.), Nucl. Instrum. Meth. **A320**, 519 (1992).

CHARMED BARYONS PHOTOPRODUCED IN FOCUS AT FERMILAB

SERGIO P. RATTI*

Dip. di Fisica Nucleare and Sezione I.N.F.N. I27100 Pavia (Italy)
E-mail: ratti@pv.infn.it

FOCUS collected over 7×10^7 triggers and more than 10^6 fully reconstructed charm particles in a photoproduction experiment at Fermilab. The experimental setup is an upgraded version of a multiparticle spectrometer used in the previous experiment E687[1].

Data on charmed meson spectroscopy have been presented by F.L. Fabbri in this Section. Here data on photoproduction of charmed baryons are presented.

1 Σ_c: Mass Splitting and Widths

The selection of the $\Sigma_c \to \Lambda_c^+ \pi^\pm$ decays is performed on a sample of over $25,000$ fully reconstructed $\Lambda_c^+ \to pK^-\pi^+$ baryons selected through a "candidate driven" algorithm[1]. In this analysis we exploit a sample of $(12,410 \pm 180)$ Σ_c events requiring that the Λ_c^+ candidates have a momentum $P > 40 GeV$ and a proper decay time less than 10 times the Λ_c^+ lifetime. The Σ_c baryons are reconstructed combining Λ_c^+ candidates with a mass within 2.1σ of the Λ_c^+ average mass, with a charged pion. As usual, to minimize systematic effects, we compute the invariant mass differences ΔM: $\Delta M^{o,++} = M(pK^-\pi^+\pi^\pm) -$

Table 1. Mass difference $\Delta M(\Sigma_c) - m(\Lambda_c^+)$.

State	Exp	$\Delta(M)$
	E791[2]	$167.76 \pm 0.29 \pm 0.15$
ΔM^o	CLEO II[3]	$168.20 \pm 0.30 \pm 0.20$
	FOCUS	$167.38 \pm 0.21 \pm 0.13$
	PDG[4]	167.31 ± 0.21
	E791[2]	$167.38 \pm 0.29 \pm 0.15$
ΔM^{++}	CLEO II[3]	$167.10 \pm 0.30 \pm 0.20$
	FOCUS	$167.35 \pm 0.19 \pm 0.12$
	PDG[4]	167.87 ± 0.20
$\Delta M =$	E791[2]	$+0.38 \pm 0.40 \pm 0.15$
$M(\Sigma_c^{++})$	CLEO II[3]	$+1.10 \pm 0.40 \pm 0.10$
$-$	FOCUS	$-0.03 \pm 0.28 \pm 0.11$
$M(\Sigma_c^o)$	PDG[4]	0.66 ± 0.28

$M(pK^-\pi^+)$. The mass difference distributions $\Delta M^o = M(\Sigma_c^o) - M(\Lambda_c^+)$ (fig. 1a) and $\Delta M^{++} = M(\Sigma_c^{++}) - M(\Lambda_c^+)$ (fig. 1b) are shown in fig.s 1. The results of the fits (also shown in fig.s 1) are compared to the results of other experiments in Table 1. We find no evidence of isospin splitting for the Σ_c baryons; our measurements do not support the conclusion that the doubly charged Σ_c is more massive than its neutral counterpart. The Σ_c isospin multiplet appears to be fully degenerate. This excludes a number of recent theoretical predictions[5]. On the same data Focus has preliminary estimates of the mass widths for the Σ_c baryons. By selecting two somewhat larger samples, a signal of 425 ± 55 Σ_c^o (fig. 1c) and a signal of 540 ± 59 Σ_c^{++} (fig. 1d) were obtained on which we

*The co-authors are: J.M.Link, M.Reyes, P.M.Yager (**UC DAVIS**); J.Anjos, I.Bediaga, C.Gobel, J.Magnin, I.M.Pepe, A.C.Reis, A.Sánchez-Hernández, F.R.A.Simão (**CPBF, Rio de Janeiro**); S.Carrillo, E.Casimiro, H.Mendez, M.Sheaff, C.Uribe, F.Vasquez (**CINVESTAV, México City**); L.Cinquini, J.P.Cumalat, J.E.Ramirez, B.O'Reilly, E.W.Vaandering (**CU Boulder**); J.N.Butler, H.W.K.Cheung, I.Gaines, P.H.Garbincius, L.A.Garren, S.A.Gourlay, P.H.Kasper, A.E.Kreymer, R.Kuschke (**Fermilab**); S.Bianco, F.L.Fabbri, S.Sarwar, A.Zallo (**INFN Frascati**); C.Cawlfield, E.Gottschalk, K.S.Park, A.Rahimi, J.Wiss (**UI Champaign**); R.Gardner (**IU Bloomington**); Y.S.Chung, J.S.Kang, B.R.Ko, J.W.Kwak, K.B.Lee, S.S.Myung, H.Park (**Korea University, Seoul**); G.Alimonti, M.Boschini, B.Caccianiga, A.Calandrino, P.D'Angelo, M.DiCorato, P.Dini, M.Giammarchi, P.Inzani, F.Leveraro, S.Malvezzi, D.Menasce, M.Mezzadri, L.Milazzo, L.Moroni, D.Pedrini, F.Prelz, A.Sala, S.Sala (**INFN and Milano**); T.F.Davenport III (**UNC Asheville**); V.Arena, G.Boca, G.Bonomi, G.Gianini, G.Liguori, M.Merlo, D.Pantea, C. Riccardi, L.Viola, P.Vitulo (**INFN and Pavia**); A.M.Lopez, L.Mendez, A.Mirles, E.Montiel, D.Olaya, C.Rivera, W.Johns, Y.Zhang (**Mayaguez, Puerto Rico**); N.Copty, M.V.Purohit, J.R.Wilson (**USC Columbia**); K.Cho, T.Handler (**UT Knoxville**); D.Engh, M.Hosack, M.Nehring, M.Sales, P.D.Sheldon, M.Webster (**Vanderbilt**); K.Stenson (**Wisconsin, Madison**); Y.Kwon (**Yonsei University, Korea**).

382

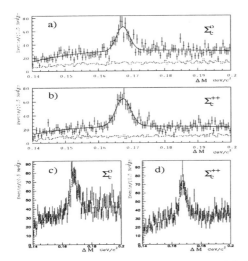

Figure 1. Mass difference distributions: a- Σ_c^o; b- Σ_c^{++}. The lower histograms in a,b, are from data in sidebands away from the Λ_c^+ signal; b,d- Σ_c^{++}. In c and d, larger samples for width measurements. The full lines are fit of a smooth background plus a Breit-Wigner.

fit the mass widths. The preliminary values of the fit are: $\Gamma(\Sigma_c^o) = 2.68 \pm 0.79$ MeV and $\Gamma(\Sigma_c^{++}) = 2.63 \pm 0.77$ MeV respectively for the two charged states. The experimental resolution is $\sigma \approx 1.5$ MeV. These values are well consistent with the recent results provided by the CLEO II collaboration.

2 Σ_c and Λ_c^+ Excited States

Expanding the ΔM distributions we find preliminary evidence of 593 ± 146 Σ^{*o} events above background, at a mass difference $\Delta M^o = 232.7 \pm 1.2$ MeV and evidence of 1284 ± 225 Σ^{*++} events above background, at a mass difference $\Delta M^{++} = 234.2 \pm 1.5$ MeV, well compatible with the CLEO data reported in the 1998 P.D.G. Book [4], i.e.: $\Delta M^o = 232.6 \pm 1.0 \pm 0.8$ MeV (504 events) and $\Delta M^{++} = 234.5 \pm 1.1 \pm 0.8$ MeV (677 events). The E687 investigation of the $\Lambda_c^{*+} \to \Lambda_c^+ \pi^- \pi^+$ has been extended to the improved statistics. The new preliminary signals consist of 100 ± 20 $\Lambda_c^{*+}(2593)$ events at $\Delta M = 308.1 \pm 0.7$ MeV and 371 ± 32 $\Lambda_c^{*+}(2593)$ events at $\Delta M = 314.6 \pm 0.3$. The

final analyses are still in progress.

3 Some Λ_c^+ Decays

A number of decay channels of the Λ_c^+ baryons are also under investigation. The statistics in FOCUS is about an order of magnitude better than in E687 and exceeds by a factor 2 the statistics of competing experiments.

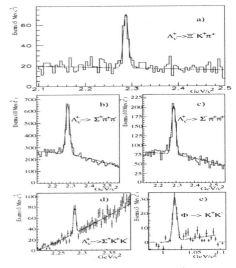

Figure 2. Mass distributions for Λ_c^+ search: a- $\Lambda_c^+ \to \Xi^- K^+ \pi^+$; b- $\Lambda_c^+ \to \Sigma^+ \pi^- \pi^+$; c- $\Lambda_c^+ \to \Sigma^- 2\pi^+$; d- $\Lambda_c^+ \to \Sigma^+ K^+ K^-$; e- the $\phi(K^+ K^-)$ signal for the events in the Λ_c^+ peak.

Fig. 2a shows our improved signal of 138 ± 14 events (signal/noise ratio ≈ 3) for the channel $\Lambda_c^+ \to \Xi^- K^+ \pi^+$, which more than double the evidence for this decay reported in 1998 P.D.G. Book[4]. Fig.s 2b,c show our improved signals for the decay channels $\Lambda_c^+ \to \Sigma^+ \pi^- \pi^+$ (1418 ± 63 events, fig. 2b) and $\Lambda_c^+ \to \Sigma^- 2\pi^+$ (728 ± 35 events, fig. 2c). The data of fig.s 2b,c might provide material for a Dalitz plot search for 2-body decays. From the data of fig.s 2d,e, we derive a very preliminar estimate of $R = N(\Sigma^+ \phi^o)/N(\Sigma^+ K^+ K^-) \approx 0.56 \pm 0.11$ that seems to be hardly compatible with a dominance of the quasi two-body decay channel. The measurement of the relative branching ratios for these channels are in progress.

4 Some Ξ_c^+, Ξ_c^o decays

We investigated also several decay channels of the strange-charmed baryons Ξ_c. Fig. 3 shows our final signal of 200 ± 39 event evidence for the Cabibbo suppressed decay $\Xi_c^+ \to pk^-\pi^+$. In order to extract the tiny signal from background (signal/noise ratio of 0.4 ± 0.1 we applied tight selection criteria for the vertex detachment ($l > 8.0\sigma(l)$, where l is the decay measured decay length) and for the proton and K Cerenkov identification. In this way the Λ_c^+ signal consists of about 7,000 events compared to over 25,000 in the initial fully reconstructed sample.

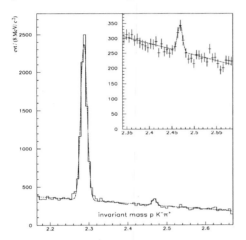

Figure 3. $M(pK^-\pi^+)$ distribution. Insert is the enlargement of the region around the Ξ_c^+ mass.

A preliminary value of the ratio $R = \Gamma(\Xi_c^+ \to pK^-\pi^+)/\Gamma(\Xi_c^+ \to \Xi^-2\pi^+) = 0.13 \pm 0.03 \pm 0.02$. is to be compared to SELEX[6] $R = 0.20 \pm 0.04 \pm 0.02$.

Ξ_c signals have been detected in different Cabibbo allowed channels. In fig. 4 collects four decay channels of the Ξ_c^+ including the **first observation** of 25 ± 7 events of the channel $\Xi_c^+ \to \Omega^-K^+\pi^+$ (fig. 4d). The distributions of fig.s 4b,c ($\Xi_c^+ \to \Lambda^oK^-2\pi^+$ and $\Xi_c^+ \to \Sigma^+K^-\pi^+$ respectively) are still on partial statistics. All masses, with uncertainties of about $1.5MeV$ are consistent with $M = 2469MeV$, with σ's 6-10 MeV.

Finally, the signals for the decay modes

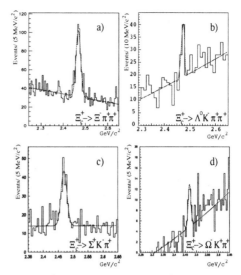

Figure 4. Ξ_c^+ masses: a- $M(\Xi^-2\pi^+)$, 320 ± 29 ev.; b- $M(\Lambda^oK^-2\pi^+)$, 31 ± 12 ev.; c- $M(\Sigma^+K^-\pi^+)$, 147 ± 15 ev.; d- $M(\Omega^-K^+\pi^+)$, 25 ± 7 ev.

$\Xi^-\pi^+$ (fig. 5a) and Ω^-K^+ (fig. 5b) of the Ξ_c^o baryon are collected in fig.s 5. The masses,

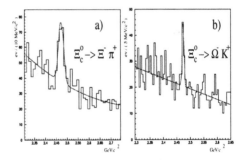

Figure 5. Ξ_c^0 mass distributions: a- $M(\Xi^-\pi^+)$, 117 ± 20 ev.; b- $M(\Omega^-K^+)$, 421 ± 11 ev.

measured with an error of about 2 MeV are very close to 2470 MeV.

References

1. E687 Coll., N.I.M. **A320**, 519 (1992);
2. E791 Coll., Phys.Lett **379B**, 292 (1996);
3. Cleo II Coll., Phys.Rev.Lett. **71**, 3259 (1993);
4. P.D.Group, Eur.Phys.J. **C3**, 1 (1998);
5. Focus Coll., Phys.Lett. **B488**, 218 (2000);
6. Selex Coll., Phys.Rev.Lett. **84**, 1857 (2000).

LIGHT MESON AND BARYON SPECTROSCOPY FROM CHARM DECAYS IN FERMILAB E791

M. V. PUROHIT

FOR THE FERMILAB E791 COLLABORATION

Department of Physics & Astronomy,

University of South Carolina, Columbia, SC 29208, USA

e-mail: purohit@sc.edu

We present results from Fermilab experiment E791. We extracted the fractions of resonant components in the $\Lambda_c^+ \to pK^-\pi^+$ decays, and found a significant polarization of the Λ_c^+ using a fully 5-dimensional resonant analysis. We also did resonant analyses of D^+ and D_s^+ decays into $\pi^+\pi^-\pi^+$. We observed an insignificant asymmetry in the Breit Wigner describing the $f_0(980)$ and found good evidence for a light and broad scalar resonance in the D^+ decays.

1 Experiment E791 at Fermilab

Fermilab E791 is a fixed-target pion-production charm collaboration involving 17 institutions. Over 20 billion events were collected during 1991-92. The experiment used a segmented target with five thin foils (1 Pt, 4 C) with about 1.5 cm center-to-center separation. With this target configuration and with 23 planes of silicon detectors we were able to suppress the large backgrounds due to random track combinations and secondary interactions. Complementing this vertex detector is a complete 2-magnet spectrometer with 35 planes of drift chambers and with Cherenkov detectors and calorimeters for particle identification and energy measurement.[1]

2 $\Lambda_c^+ \to pK^-\pi^+$ decays

This analysis[2] is unique in at least two respects: this is the first coherent amplitude analysis of a charm baryon decay and is the first fully 5-dimensional resonant analysis of any decay.[3] The most obvious components of the $\Lambda_c^+ \to pK^-\pi^+$ decays (and charge conjugate decays, which are implied throughout this paper) include the nonresonant $pK^-\pi^+$ decay, and the $p\overline{K}^{*0}(890)$ and $\Lambda(1520)\pi^+$ two-body decays. These three decays can be described by spectator and W-exchange

amplitudes. However, in lowest order, the $\Delta^{++}(1232)K^-$ decay can occur only via the exchange amplitude. Thus, a significant presence of $\Delta^{++}(1232)K^-$ decays would indicate that exchange amplitudes are important. Unlike charm meson decays, helicity and form-factor suppression are not expected to inhibit exchange amplitudes for charm baryons.

The Λ_c^+ and its decay proton carry spin, and the Λ_c^+ may be polarized upon production. Therefore an analysis of these decays requires five kinematic variables for a complete description. As a by-product of the analysis, the production polarization of the Λ_c^+, $\mathbf{P}_{\Lambda_c}$, is also measured.

Fitting to 886 ± 43 events, we find the fractions of resonant components listed in Table 1, indicating that exchange amplitudes (signaled by the $\Delta^{++}(1232)K^-$ decays) contribute significantly to charm baryon decays.

Table 1. The decay fractions for $\Lambda_c^+ \to pK^-\pi^+$ with statistical and systematic errors from the final fit.

Mode	Fit Fraction (%)
$p\overline{K}^{*0}(890)$	$19.5\pm2.6\pm1.8$
$\Delta^{++}(1232)K^-$	$18.0\pm2.9\pm2.9$
$\Lambda(1520)\pi^+$	$7.7\pm1.8\pm1.1$
Nonresonant	$54.8\pm5.5\pm3.5$

The projections of the data and of the fit

on the traditional m^2 variables as well as on the angular variables introduced by the spin-dependent 5-dimensional analysis show that we get had a good fit (the χ^2/DF is 1.06). The only discrepancy between data and the fit is in the low m^2_{pK} region. This may be due to the tail of a $\Lambda(1405)$ decaying to pK^- (see Ref. [4]). We also searched for other resonances and found that the data weakly favor a $\frac{1}{2}^-$ resonance in pK with mass 1556 ± 19 MeV/c^2 and width 279 ± 74 MeV/c^2. However, no such resonance is known to exist.

In Figure 1 we show the result of measuring the polarization as a function of p_T^2. The average polarization P for all our data is consistent with zero but there is a clear fall in P as a function of p_T^2.

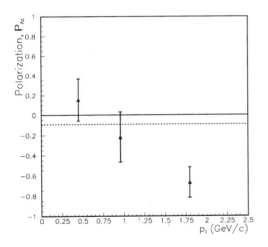

Figure 1. The polarization of the Λ_c as a function of the Λ_c's transverse momentum.

3 Dalitz analyses of D^+, D_s^+ decays to $\pi^+\pi^-\pi^+$

In Figure 2 below, the $\pi^+\pi^-\pi^+$ mass plot has 1172 ± 61 events above background in the D^+ signal region and 848 ± 44 events above background in the D_s^+ signal region.

A fit to the D_s^+ decays[5] to $\pi^+\pi^-\pi^+$ followed the WA76 parameterization for the $f_0(980)$ resonance[6] and yields the parameters

$m[f_0(980)] = 977 \pm 3 \pm 2$ MeV, $g_K = 0.02 \pm 0.04 \pm 0.03$ and $g_\pi = 0.09 \pm 0.01 \pm 0.01$, in disagreement with results from WA76. The fit fractions are shown in Table 2. Note that there is no significant fraction of decays into purely non-strange modes, indicating that the annihilation amplitude is small.

The fit to the $D^+ \to \pi^+\pi^-\pi^+$ decays had poor quality until we tried to include a scalar resonance. We found that including such a resonance improves the fit quality from a χ^2/DF of 254/162 (poor) to an acceptable χ^2/DF of 138/162. When the mass and width of this scalar (the σ) are allowed to float, we obtain $m_\sigma = 478^{+24}_{-23}\pm17$ MeV/c^2 and $\Gamma_\sigma = 324^{+42}_{-40}\pm21$ MeV/c^2. As additional tests, we fit for a new resonance with a real scalar amplitude, or as a vector or a tensor particle. In all three cases, the fit was poorer than the fit with a "normal" scalar resonance. We conclude that we have good evidence for this scalar resonance and a good physical environment for measuring its mass and width.

Table 2. The decay fractions for D^+, D_s^+ decays to $\pi^+\pi^-\pi^+$

Mode	D_s^+ Fit Fraction (%)	D^+ Fit Fraction (%)
Non-resonant	0.5±1.4±1.7	7.8±6.0±2.7
$f_0(980)\pi^+$	56.5±4.3±4.7	6.2±1.3±0.4
$\rho^0(770)\pi^+$	5.8±2.3±3.7	33.6±3.2±2.2
$f_2(1270)\pi^+$	19.7±3.3±0.6	19.4±2.5±0.4
$f_0(1370)\pi^+$	32.4±7.7±1.9	2.3±1.5±0.8
$\rho^0(1450)\pi^+$	4.4±2.1±0.2	0.7±0.7±0.3
$\sigma\pi^+$		46.3±9.0±2.1

4 Acknowledgements

I would like to thank members of my collaboration (Fermilab E791). This work was supported by a grant from the U.S. Department of Energy.

386

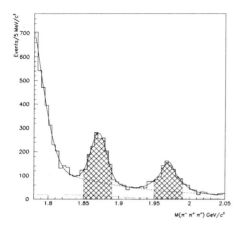

Figure 2. The three pion mass showing the D^+ and D_s^+ peaks used for the Dalitz analysis

References

1. E791 Collaboration, E. M. Aitala et al., EPJdirect C4 (1999) 1. S. Amato et al., Nucl. Instrum. Meth. A324 (1993) 535.
2. E. M. Aitala et al., Phys. Lett. B471 (2000) 449.
3. Particle Data Group, D.E. Groom et al., "The Review of Particle Physics", Eur. Phys. J. C15 (2000) 1 and references therein.
4. R. Dalitz, *ibid* p. 676.
5. E. M. Aitala et al., hep-ex/0007027 and hep-ex/0007028.
6. T. A. Armstrong et al., Z. Phys. C51 (1991) 351.

FIRST OBSERVATION OF THE Σ_c^{*+} CHARMED BARYON, AND NEW MEASUREMENTS OF THE Σ_c^0, Σ_c^+, Σ_c^{++}, AND Ω_c^0 CHARMED BARYONS

ANDREAS WARBURTON
(REPRESENTING THE CLEO COLLABORATION)

Laboratory of Nuclear Studies, Cornell University, Ithaca, New York 14853, USA
E-mail: andreas.warburton@cornell.edu

Using $\sim$13.7 fb^{-1} of e^+e^- collision data recorded at near the $\Upsilon(4S)$ resonance by the CLEO II and CLEO II.V detector configurations on the Cornell Electron Storage Ring, we present the world's most precise measurements of the Σ_c^0, Σ_c^+, and Σ_c^{++} charmed-baryon masses as well as the first measurements of the intrinsic widths of the Σ_c^0 and Σ_c^{++} baryons. We also report on the first observation and mass measurement of the Σ_c^{*+} charmed baryon, $M(\Sigma_c^{*+}) - M(\Lambda_c^+) = (231.0 \pm 1.1[\text{stat}] \pm 2.0[\text{syst}])$ MeV/c^2, and the first CLEO observation of the Ω_c^0 baryon, for which we measure a mass $M(\Omega_c^0) = (2694.6 \pm 2.6[\text{stat}] \pm 2.4[\text{syst}])$ MeV/c^2 from a sample of $(40.4 \pm 9.0[\text{stat}])$ candidate events. All new results are preliminary.

1 Introduction

The charmed baryons consist of a c quark and a light diquark with a specific $\mathrm{J^P}$ spin-parity configuration and can be organized in terms of isospin and strangeness using the theory of SU(4) multiplets. Several new observations and improved measurements in charmed-baryon spectroscopy have been made in the past decade[a]. We report here on new CLEO measurements of the Σ_c and Ω_c^0 states [1,2].

2 The Experiment

The results described herein come from studies of e^+e^- collisions conducted at the Cornell Electron Storage Ring (CESR) operating near the $\sim$10.58 GeV/c^2 $\Upsilon(4S)$ bottomonium resonance. The 4π general purpose CLEO II detector configuration [3], comprising a cylindrical drift chamber system for charged-track detection inside a 1.4 T solenoidal magnetic field and a CsI electromagnetic calorimeter for π^0 detection, was employed to take a data sample with time-integrated luminosity $\int \mathcal{L}dt \simeq 4.7$ fb^{-1}. An additional 9.0 fb^{-1} of data were taken with the CLEO II.V detector configuration [4], which had improved charged-track measurement capabilities.

For the searches and measurements described in this paper, the analysis approaches have sought to optimize signal efficiency and background suppression. Identification of p, K^+, and π^+ candidates[b] was achieved through the use of specific ionization dE/dx in the drift chamber and, when available, time-of-flight information. Hyperons, in the modes $\Xi^- \to \Lambda \pi^-$, $\Xi^0 \to \Lambda \pi^0$, $\Omega^- \to \Lambda K^-$, $\Sigma^+ \to p\pi^0$, and $\Lambda \to p\pi^-$, were reconstructed by detecting their decay points separated from the primary event vertex.

Charmed baryons at CESR are either produced from the secondary decays of B mesons or directly from e^+e^- annihilation to $c\bar{c}$ jets. Combinatorial backgrounds, which are highest for low-momentum charmed baryons, are reduced by requiring candidates to exceed an optimum scaled momentum x_p, defined as $x_p \equiv |\vec{p}|/p_{\max}$, where $\vec{p}$ is the momentum of the charmed baryon candidate, $p_{\max} \equiv \sqrt{E_{\text{beam}}^2 - M^2}$, E_{beam} is the beam energy, and M is the reconstructed candidate's mass. We use decay-mode dependent scaled-momentum criteria[c] of $x_p > 0.5$ or $x_p > 0.6$, based on the optimization results.

---

[a] Note, however, that none of the $\mathrm{J^P}$ quantum numbers of the charmed baryons has yet been measured.

[b] Charge conjugate modes are implicit throughout.

[c] Charmed baryons with threshold-produced B-meson parents are kinematically limited to $x_p < 0.4$.

388

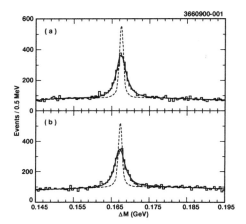

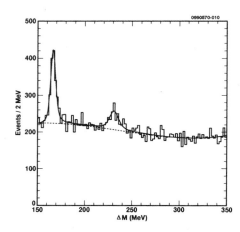

Figure 1. The $\Delta M \equiv M(\Lambda_c^+ \pi^\pm) - M(\Lambda_c^+)$ mass difference, where in (a) the Λ_c and π candidates have like charges (Σ_c^{++} candidates) and in (b) they have opposite charges (Σ_c^0 candidates). The solid curve shows the results of the fit described in the text. The dashed curve represents the resolution function of the detector response.

Figure 2. The $\Delta M \equiv M(\Lambda_c^+ \pi^0) - M(\Lambda_c^+)$ mass-difference distribution. The solid line represents a fit to a third-order polynomial background shape and two p-wave Breit-Wigner functions smeared by Gaussian resolution functions for the two signal shapes. The dashed curve indicates the background function.

3 The Σ_c and Σ_c^* Baryon Isotriplets

The Σ_c baryons [5] consist of one charm and two light (u or d) valence quarks in an $I = 1$ isospin configuration. The Σ_c states in this study decay strongly into final states with a Λ_c^+ baryon and either a π^+ or π^0 transition meson. We reconstruct $\sim$58 000 Λ_c^+ signal candidates using 15 different decay modes [6].

3.1 Σ_c Final States involving $\Lambda_c^+ \pi^\pm$

Candidates in the Λ_c^+ sample described in Sec. 3 were combined with $\pi^\pm$ charged-track candidates and the mass difference, $\Delta M \equiv M(\Lambda_c^+ \pi^\pm) - M(\Lambda_c^+)$, was calculated for those combinations satisfying an $x_p > 0.5$ criterion. Each of the resulting mass-difference distributions, depicted in Fig. 1, indicated an unambiguous signal of $\sim$2000 Σ_c^{++} and Σ_c^0 candidates, respectively. The two distributions were fit to a sum of a polynomial background function with threshold suppression and a p-wave Breit-Wigner line shape convolved with a double-Gaussian detector resolution function.

The fitted masses and widths are summarized in Tab. 1. Scaling from the strange-baryon widths, Rosner [9] has predicted $\Gamma(\Sigma_c) \simeq 1.32$ MeV/c^2, somewhat narrower than our measured values. Other authors [10] have employed measurements of the Σ_c^* widths to derive $\Gamma(\Sigma_c)$ predictions consistent with our values listed in Tab. 1.

3.2 $\Sigma_c^{(*)}$ Final States involving $\Lambda_c^+ \pi^0$

In a manner similar to that described in Sec. 3.1 but with an $x_p > 0.6$ requirement [1], Λ_c^+ candidates were combined with transition π^0 candidates, which were required to have momenta greater than 150 MeV/c and masses consistent with and kinematically fit to the known π^0 mass [11], to form the mass difference $\Delta M \equiv M(\Lambda_c^+ \pi^0) - M(\Lambda_c^+)$ shown in Fig. 2. Note the higher background level.

The lower-mass peak[d] in Fig. 2, containing $(661^{+63}_{-60}[\text{stat}])$ events, is due to the Σ_c^+ baryon. The second peak in Fig. 2,

---

[d]Note that, for the purpose of yield determination, the fit function described in Fig. 2 was replaced with a third-order Chebyshev polynomial background function and two Gaussian signal line shapes.

Table 1. A summary of the CLEO mass and width measurements, in MeV/c^2, of the $J^P = \frac{1}{2}^+$ and $J^P = \frac{3}{2}^+$ Σ_c charmed baryons. The first uncertainties are statistical and the second are systematic. Values marked with a † were reported in Ref. [7]. We note that the isospin mass splittings are consistent with theoretical expectation [8].

J^P	Observable	$\Sigma_c^{(*)0}$	$\Sigma_c^{(*)+}$	$\Sigma_c^{(*)++}$
$\frac{1}{2}^+$	$M(\Sigma_c) - M(\Lambda_c^+)$	$167.2 \pm 0.1 \pm 0.2$	$166.4 \pm 0.2 \pm 0.3$	$167.4 \pm 0.1 \pm 0.2$
	$\Gamma(\Sigma_c)$	$2.4 \pm 0.2 \pm 0.4$	< 4.6 (90% CL)	$2.5 \pm 0.2 \pm 0.4$
$\frac{3}{2}^+$	$M(\Sigma_c^*) - M(\Lambda_c^+)$	$232.6 \pm 1.0 \pm 0.8^\dagger$	$231.0 \pm 1.1 \pm 2.0$	$234.5 \pm 1.1 \pm 0.8^\dagger$
	$\Gamma(\Sigma_c^*)$	$13.0\,^{+3.7}_{-3.0} \pm 4.0^\dagger$	< 17 (90% CL)	$17.9\,^{+3.8}_{-3.2} \pm 4.0^\dagger$

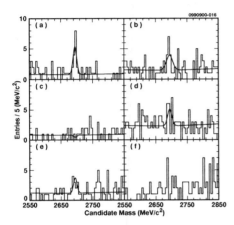

Figure 3. The invariant mass distribution and simultaneous fit to the five Ω_c^0 search modes: (a) $\Omega^- \pi^+$, (b) $\Omega^- \pi^+ \pi^0$, (c) $\Omega^- \pi^+ \pi^+ \pi^-$, (d) $\Xi^0 K^- \pi^+$, and (e) $\Xi^- K^- \pi^+ \pi^+$. The final state (f) $\Sigma^+ K^- K^- \pi^+$ was not included in the fit. The signal region was fitted with a fixed-width Gaussian while the background was fitted to a second-order polynomial.

sistency of the claimed masses, however, is marginal [2,11].

4.1 First Ω_c^0 Observation by CLEO

Based on patterns observed in other charmed baryon decays and considerations of reconstruction efficiency and combinatorial background, we searched [2] for Ω_c^0 candidates in the five weak decay modes $\Omega^- \pi^+$, $\Omega^- \pi^+ \pi^0$, $\Omega^- \pi^+ \pi^+ \pi^-$, $\Xi^0 K^- \pi^+$, and $\Xi^- K^- \pi^+ \pi^+$. We separately investigated a sixth channel, $\Sigma^+ K^- K^- \pi^+$, because E687 [2,11] showed a significant signal in this mode. The invariant mass distributions of the six modes are shown in Fig. 3, the sum of the five search modes is depicted in Fig. 4, and the fitted event yields are listed in Tab. 2. We observe $(40.4 \pm 9.0[\text{stat}])$ Ω_c^0 candidates.

4.2 Measurement of the Ω_c^0 Mass

The mass of our Ω_c^0 candidates we measure by performing an unbinned maximum-likelihood fit using the sum of a single Gaussian signal and a second-order polynomial background. We find the Ω_c^0 mass to be $M(\Omega_c^0) = (2694.6 \pm 2.6[\text{stat}] \pm 2.4[\text{syst}])$ MeV/c^2, where the systematic uncertainty is dominated by our sensitivity to the fitting method employed.

5 Conclusion

The CLEO collaboration has made new measurements of the Σ_c^0, Σ_c^+, and Σ_c^{++} baryon

with $(327^{+78}_{-73}[\text{stat}])$ signal candidates, we ascribe to the first observation of Σ_c^{*+} baryons. The fitted masses and limits on the intrinsic widths of the Σ_c^+ and Σ_c^{*+} states are listed in Tab. 1.

4 The Ω_c^0 Baryon Isosinglet

The Ω_c^0 baryon is a $J^P = \frac{1}{2}^+$ ground state with valence quarks $(c\{ss\})$, where $\{ss\}$ denotes a symmetry in the wave function under the exchange of the light-quark spins. Several experimental groups have reported the observation of an Ω_c^0 state; the mutual con-

390

Table 2. The Ω_c^0 charmed baryon search results in the six decay modes. The fitted yields were computed with a mode-dependent x_p criterion, whereas the relative branching fraction ($\mathcal{B}$) and cross-section-branching-fraction product ($\sigma \cdot \mathcal{B}$) were determined with a uniform $x_p > 0.5$ requirement. The first uncertainties are statistical and the second are systematic.

Ω_c^0 Search Channel	Fitted Yield	Relative $\mathcal{B}$	$\sigma \cdot \mathcal{B}$ [fb]
$\Omega^- \pi^+$	13.3 ± 4.1	1.0	$11.3 \pm 3.9 \pm 2.3$
$\Omega^- \pi^+ \pi^0$	11.8 ± 4.9	$4.2 \pm 2.2 \pm 1.2$	$47.6 \pm 18.0 \pm 2.8$
$\Xi^0 K^- \pi^+$	9.2 ± 4.9	$4.0 \pm 2.5 \pm 0.6$	$45.1 \pm 23.2 \pm 4.1$
$\Xi^- K^- \pi^+ \pi^+$	7.0 ± 3.7	$1.6 \pm 1.1 \pm 0.4$	$18.2 \pm 10.6 \pm 3.8$
$\Omega^- \pi^+ \pi^+ \pi^-$	-0.9 ± 1.4	< 0.56 (90% CL)	< 5.1 (90% CL)
Above 5 modes combined	40.4 ± 9.0	–	–
$\Sigma^+ K^- K^- \pi^+$	2.8 ± 4.1	< 4.8 (90% CL)	< 53.8 (90% CL)

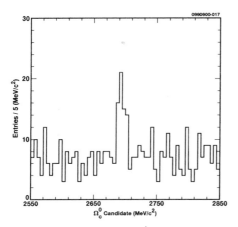

Figure 4. The invariant mass distribution for the sum of the $\Omega^- \pi^+$, $\Omega^- \pi^+ \pi^0$, $\Omega^- \pi^+ \pi^+ \pi^-$, $\Xi^0 K^- \pi^+$, and $\Xi^- K^- \pi^+ \pi^+$ search modes.

masses as well as preliminary measurements of the Σ_c^0 and Σ_c^{++} intrinsic widths. We report the first observation of the Σ_c^{*+} baryon and determine its mass difference to be $M(\Sigma_c^{*+}) - M(\Lambda_c^+) = (231.0 \pm 1.1[\text{stat}] \pm 2.0[\text{syst}])$ MeV/c^2. We also observe, for the first time, a significant signal for the Ω_c^0 baryon and measure its mass to be $M(\Omega_c^0) = (2694.6 \pm 2.6[\text{stat}] \pm 2.4[\text{syst}])$ MeV/c^2.

Acknowledgments

My colleagues in the CLEO collaboration, the staff at CESR, and their funding sources made these results possible. I thank Basit Athar and John Yelton for useful discussions.

References

1. R. Ammar *et al.* (CLEO), CLNS-00/1681, [hep-ex/0007041], submitted to *Phys. Rev. Lett.*

2. S. Ahmed *et al.* (CLEO), CLEO-CONF-00-4, [hep-ex/0007047v2], and citations therein.

3. Y. Kubota *et al.* (CLEO), *Nucl. Instrum. Methods Phys. Res.* **A320**, 66 (1992).

4. T. S. Hill, *Nucl. Instrum. Methods Phys. Res.* **A418**, 32 (1998).

5. The symbols Σ_c and Σ_c^* refer to the $\Sigma_c(2455)$ and $\Sigma_c(2520)$ baryons, respectively.

6. P. Avery *et al.* (CLEO), *Phys. Rev. D* **43**, 3599 (1991); *idem*, *Phys. Rev. Lett.* **71**, 2391 (1993); *idem*, *Phys. Lett.* **B325**, 257 (1994); M. S. Alam *et al.*, *Phys. Rev. D* **57**, 4467 (1998).

7. G. Brandenburg *et al.* (CLEO), *Phys. Rev. Lett.* **78**, 2304 (1997).

8. J. Franklin, *Phys. Rev. D* **59**, 117502 (1999).

9. J. L. Rosner, *Phys. Rev. D* **52**, 6461 (1995).

10. D. Pirjol and T.-M. Yan, *Phys. Rev. D* **56**, 5483 (1997); S. Tawfiq *et al.*, *Phys. Rev. D* **58**, 054010 (1998); M. A. Ivanov *et al.*, *Phys. Lett. B* **442**, 435 (1998).

11. D. E. Groom *et al.* (PDG), *Eur. Phys. J. C* **15**, 1 (2000).

RECENT RESULTS FROM CLEO: EXCITED CHARMED BARYONS

R.D. EHRLICH

Wilson Synchrotron Laboratory, Cornell University, Ithaca, NY 14853, USA
E-mail: rde4@cornell.edu

Using data recorded by the CLEO detector at CESR, we report preliminary evidence for previously unknown charmed baryons. In the charm-strange sector, we find two states, one decaying into $\Xi_c^{0\prime}\pi^+$ with the subsequent decay $\Xi_c^{0\prime} \to \Xi_c^0\gamma$, and its isospin partner decaying into $\Xi_c^{+\prime}\pi^-$ followed by $\Xi_c^{+\prime} \to \Xi_c^+\gamma$. We measure the following mass differences for the two states: $M(\Xi_c^0\gamma\pi^+) - M(\Xi_c^0) = 319.3 \pm 1.7 \pm 2.5$ MeV, and $M(\Xi_c^+\gamma\pi^-) - M(\Xi_c^+) = 323.0 \pm 1.9 \pm 2.5$ MeV. We interpret these states as the $J^P = \frac{1}{2}^-$ Ξ_{c1} particles, the charmed-strange analogs of the $\Lambda_{c1}^+(2593)$. We have also investigated the spectrum of charmed baryons more massive than the Λ_{c1}, which may decay into $\Lambda_c^+\pi^-\pi^+$, finding evidence for two new states. One is wide and has an invariant mass about 480 MeV above the Λ_c^+; the other is comparatively narrow and has an invariant mass of $596 \pm 1 \pm 1$ MeV above the Λ_c^+.

1 Introduction

Neither the rich phenomonology of charmed baryons, nor the analysis techniques employed can be described fully here, for lack of space. Pedagogical and experimental details can be found in the ICHEP2000 conference submissions.[1].

Depending on the symmetry properties of the wave function under interchange of the two light-quark spins (or flavors), baryons consisting of a charmed quark and two light (up or down) quarks, are denoted Λ_c (antisymmetric, isospin zero) and Σ_c (symmetric), respectively; those which have a single strange quark are correspondingly termed Ξ_c and Ξ_c'. All Σ_c, Λ_c, and Ξ_c ground (s-wave) states have been identified; knowledge of orbitally excited states in the sequence is much more limited. Two states decaying into $\Lambda_c^+\pi^+\pi^-$ have been identified with the $J^P = \frac{1}{2}^-, \frac{3}{2}^-$ Λ_{c1}^+ particles[2], where the numerical subscript denotes one unit of orbital angular momentum. In the Ξ_c sector, CLEO has found candidates for both isospin states of the $J^P = \frac{3}{2}^-$ p-wave doublet, decaying to $\Xi^*\pi$.[3]

The data presented here were taken by the CLEO II and CLEO II.V detector configurations[4] operating at the Cornell Electron Storage Ring (CESR). The sample used in this analysis corresponds to an integrated luminosity of 13.7 fb^{-1} of data taken on the $\Upsilon(4S)$ resonance and in the nearby continuum. Of this data, 4.7 fb^{-1} was taken with the CLEO II detector, and the remainder with the CLEO II.V configuration.

2 $\Xi_{c1}^{+,0}, J^p = \frac{1}{2}^-$ search

We search for these states in their decay to $\Xi_c'\pi^-$ [a]. We use a large sample of reconstructed Ξ_c^+ and Ξ_c^0 particles, decaying to Λ, Ξ^-, Ω^- and Ξ^0 hyperons as well as K's, π's and protons. We combine those Ξ_c candidates within two standard deviations (σ) of the known mass of the Ξ_c^+ or Ξ_c^0) with a photon, calculating the mass differences $\Delta M = M(\Xi_c^0\gamma) - M(\Xi_c^0)$ and $M(\Xi_c^+\gamma) - M(\Xi_c^+)$. Those combinations within 8 MeV ($\approx 2\sigma$) of the measured mass differences for the Ξ_c' particles are retained for further analysis.

We combine these Ξ_c' candidates with a correctly charged pion in the event, and plot $M(\Xi_c'\pi) - M(\Xi_c)$ for both isospin states. Figure 1(a) shows $M(\Xi_c^{0\prime}\pi^+) - M(\Xi_c^0)$, and Figure 1(b) shows $M(\Xi_c^{+\prime}\pi^-) - M(\Xi_c^+)$, each with a requirement of $x_p > 0.7$ on the final combination. In both figures there is a

---

[a] Charge conjugate modes are implicit throughout.

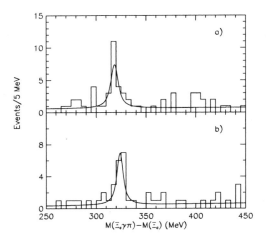

Figure 1. $M(\Xi_c^0\gamma\pi^+)$ $-$ $M(\Xi_c^0)$ (upper) and $M(\Xi_c^+\gamma\pi^-) - M(\Xi_c^+)$ (lower) each showing distinct peaks. Note that the Ξ_c' mass has been fixed in these plots, so that we could equivalently show $M(\Xi_c'\pi) - M(\Xi_c)$ as this would just be a translation of the x-axis.

peak at around 320 MeV, indicative of the decay of a Ξ_{c1}^+ or a Ξ_{c1}^0. We have fit the plots to Breit-Wigner functions convoluted with a Gaussian resolution function, together with a polynomial background. The results of this fit are: $M(\Xi_c^{0\prime}\pi^+) - M(\Xi_c^0) = 319.3 \pm 1.7$ MeV, and $\Gamma = 7.7^{+5.0}_{-3.5}$ MeV, and for Figure 1a, and $M(\Xi_c^{+\prime}\pi^-) - M(\Xi_c^+) = 323.0 \pm 1.9$ MeV, and $\Gamma = 5.8^{+4.3}_{-2.8}$ MeV for Figure 1b. The results for Γ are limited in their accuracy by the low statistics, but indicate that it is very likely that these states have an intrinsic width of several MeV. The resolution of the detector for these decays is known from a Monte Carlo simulation program to be 1.2 MeV in the CLEO II.V data, and 1.4 MeV in the CLEO II data. If we release the cuts on $M(\Xi_c\gamma) - M(\Xi_c)$, select combinations within 8 MeV of our final signal peaks, and plot $M(\Xi_c\gamma) - M(\Xi_c)$ (not shown), we find that our data is consistent with all our Ξ_{c1} decays proceeding via an intermediate Ξ_c'.

The systematic errors in the mass differences with respect to the ground state are dominated by the uncertainty in the Ξ_c' masses, which we estimate to be 3.0 MeV. Quoting the mass difference with respect to the Ξ_c' states, we find $M(\Xi_c^{0\prime}\pi^+) - M(\Xi_c^{0\prime}) = 212.3 \pm 1.7 \pm 1.0$ MeV and $M(\Xi_c^{+\prime}\pi^-) - M(\Xi_c^{+\prime}) = 215.2 \pm 1.9 \pm 1.0$ MeV.

Although we do not measure the spin or parity of these states, the observed decay modes, masses, and widths are all consistent with the new states being the $J^P = \frac{1}{2}^-$ Ξ_{c1}^+ and Ξ_{c1}^0 states, the charmed-strange analogues of the $\Lambda_{c1}^+(2593)$.

3 Structures in $\Lambda_c\pi^+\pi^-$

As noted in the introduction, our knowledge of orbitally-excited Λ_c and Σ_c states is limited to the two $J^P = \frac{1}{2}^-, \frac{3}{2}^-$ Λ_{c1}^+ particles, where the numerical subscript denotes one unit of light quark angular momentum. Here we search for additional states that decay into a Λ_c^+ and two oppositely charged pions.

To obtain adequate statistics, we reconstructed the Λ_c^+ baryons using 15 different decay modes, where the analysis used was optimized for high efficiency and low background. The Λ_c^+ candidates were then combined with two oppositely-charged π candidates in the event. The known hardness of the momentum spectrum of charmed baryons led us to place a cut of $x_p > 0.7$ on the combination to mimimize combinatoric backgrounds. Figure 2 shows the mass difference spectrum, $\Delta M_{\pi\pi} = M(\Lambda_c^+\pi^+\pi^-) - M(\Lambda_c^+)$ in the region above the well-known Λ_{c1} resonances. The upper plot in Figure 1 cannot be reasonably fit to a second order polynomial; however, a fit to the sum of a second order polynomial and two Gaussian signals has a χ^2 of 59, for 71 degrees of freedom. The lower-mass signal has a yield of 997^{+141}_{-129}, $\Delta M_{\pi\pi} = 480.1 \pm 2.4$ MeV, and a width of $\sigma = 20.9 \pm 2.6$ MeV. The upper signal has a yield of 350^{+57}_{-55}, $\Delta M_{\pi\pi} = 595.8 \pm 0.8$ MeV and $\sigma = 4.2 \pm 0.7$ MeV. All these uncertainties are statistical, coming from the fit. The mass

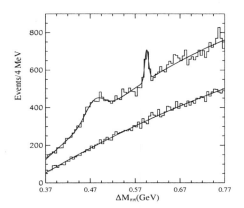

Figure 2. The upper histogram shows $\Delta M_{\pi\pi} = M(\Lambda_c^+ \pi^+ \pi^-) - M(\Lambda_c^+)$ above the Λ_{c1} range; the fit is to a quadratic background shape plus two Gaussian signal functions. The lower histogram shows the same distribution for scaled Λ_c^+ sidebands.

resolutions in these regions are ≈ 2.0 and ≈ 2.8 MeV, respectively, based on our Monte Carlo simulation. The lower peak clearly has a width greater than the experimental resolution. If we fit it to a Breit-Wigner function, we obtain a width, Γ, of ≈ 50 MeV, though it may actually be the sum of more than one wide peak. If we fit the upper peak to a Breit-Wigner convoluted with a double-Gaussian detector resolution function, we obtain a width of $\Gamma = 4 \pm 2 \pm 2$ MeV where the errors are statistical and systematic, respectively. This experimental width is not significantly different from zero, hence we derive an upper limit: $\Gamma < 8$ MeV at 90% confidence level.

The upper state decays to $\Sigma_c \pi$ approximately $\frac{1}{3}$ of the time, but does not decay to $\Sigma^* \pi$, while the lower, broad state decays comparably to $\Sigma_c \pi$, $\Sigma^* \pi$, and non-resonant $\Lambda_c \pi \pi$. Noting the narrow width of the upper state, we are led to consider p-wave states with no allowed two-body decays to less massive charmed baryons, in the heavy-quark effective theory (HQET) limit.[5] There is one such state, the $\Lambda_{c0}^{+\prime}$. With one unit of an-

gular momentum *between* the light quarks, $J^P = \frac{1}{2}^-$, and $J_{light}^P = 0^-$, it is a candidate for the upper peak. Conservation of J_{light}^P, required by HQET, forbids the decay of $\Lambda_{c0}^{+\prime}$ to $\Sigma_c \pi$. However, its sister state, the $\Lambda_{c1}^{+\prime}$, has the same overall quantum numbers (but with $J_{light}^P = 1^-$), and may mix slightly with the first state, thus permitting some decay via S-wave $\Sigma_c \pi$.

Identification of the lower, wider, state is also open to interpretation. One possibility is that it consists of a pair of Σ_{c1}^+ particles, with overall $J^P = \frac{1}{2}^-$ and $J^P = \frac{3}{2}^-$. These particles might be expected to be split in mass by around 30 MeV, and should have preferred decay modes of $\Sigma_c \pi$ and $\Sigma_c^* \pi$, respectively. Their decay widths have been predicted to be about 100 MeV.[6] Of course, other interpretations of our data exist, including the possible decay of radially-excited charmed baryons.

4 Acknowledgments

We thank the CESR staff for its dedication, the U.S. National Science Foundation for its support, and the ICHEP2000 organizers for their hospitality.

References

1. CLEO Collaboration, hep-ex 00070049, hep-ex 0007050,(2000)
2. H. Albrecht *et al.*, Phys. Lett. **B**317 227 (1993), P. Frabetti *et al.*, Phys. Rev. Lett **72** 961 (1994), K. Edwards *et al.*, Phys. Rev. Lett **74** 3331 (1995), H. Albrecht *et al.*, Phys. Lett. **B** 402 207 (1997).
3. J. Alexander et al., PRL **83** 3390 (1999).
4. Y. Kubota et al., Nucl. Instr. and Meth. **A320**, 66., T. Hill et al., Nucl. Instr. and Meth. **A418**, 32 (1998).
5. N. Isgur and M. Wise. PRL **66** 1130 (1991).
6. D. Pirjol and T-M. Yan, Phys. Rev. D56, 5483 (1997).

RARE Ξ^0 DECAYS: NEW RESULTS FROM NA48

R. WANKE

(FOR THE NA48 COLLABORATION)

Institut für Physik, Universität Mainz, D-55099 Mainz, Germany

E-mail: Rainer.Wanke@uni-mainz.de

The NA48 Collaboration has investigated rare decays of the Ξ^0 hyperon. Using data collected during a special high intensity run at the end of 1999, we determined the following preliminary branching fractions of the weak radiative decays to $\Lambda\gamma$ and $\Sigma^0\gamma$: $\text{Br}(\Xi^0 \to \Lambda\gamma) = (1.9 \pm 0.1 \pm 0.2) \times 10^{-3}$ and $\text{Br}(\Xi^0 \to \Sigma^0\gamma) = (3.7 \pm 0.5) \times 10^{-3}$. In addition, we report on the observation of the Ξ^0 beta decay and a search for the $\Delta S = 2$ transition $\Xi^0 \to p\pi^-$.

1 Introduction

Weak radiative Ξ^0 decays can be used to study the mechanisms of SU(3) symmetry breaking. However, past investigations have not led to a conclusive picture. On the theoretical side, a variety of models exist with predictions varying by an order of magnitude.[1,2,3,4,5] Experimentally, the situation is unclear, too: Branching ratio measurements agree for $\Xi^0 \to \Sigma^0\gamma$,[6,7] but do not for $\Xi^0 \to \Lambda\gamma$.[6,8]

The interest in $\Xi^0 \to \Sigma^+ e^- \bar{\nu}_e$, the analogue to neutron beta decay, stems from its potential to study flavor symmetry violation.

Last but not least, the theoretical prediction for the $\Delta S = 2$ transition $\Xi^0 \to p\pi^-$ is of order 10^{-17}.[9] An observation would therefore be an unambiguous sign for new physics.

2 Experimental Set-Up

2.1 Beam-Line Set-Up

The data used for the analyses described here were taken with the NA48 detector at CERN during a two-day special data taking period in September 1999. The NA48 experiment has been designed to measure the parameter $\mathcal{R}(\epsilon'/\epsilon)$ of direct CP violation in the decay of neutral kaons to two pions. In the usual data taking mode, a 450 GeV/c proton beam from the CERN-SPS is directed to two different targets. The first ("K_L") target, around

120 m upstream of the decay volume, is exposed to the full proton beam intensity and produces a pure K_L beam. A small fraction of the protons which pass the K_L target is redirected to a second ("K_S") target — about 6 m upstream of the fiducial volume — to generate a neutral beam consisting mainly of K_S mesons and neutral hyperons. However, during the special period used for the investigations decribed here, this beamline was altered: The K_L target was removed and a proton beam with about 200 times the usual intensity was directed to the K_S target.

2.2 The NA48 Detector

The detector components most relevant to measure Ξ^0 decays are the magnetic spectrometer and the liquid krypton calorimeter.

The spectrometer consists of four drift chambers, each comprised of eight planes of sense wires in 4 different views. Between the second and the third drift chamber a dipole magnet applies a kick of 265 MeV/c to the horizontal transverse momenta of charged particles. The momentum resolution is $\Delta p/p = 0.5\% \oplus 0.009\% \times p[\text{GeV/c}]$.

The quasi-homogeneous liquid krypton electromagnetic calorimeter consists of an electrode structure of copper/beryllium ribbons in longitudinal direction, forming 13212 read-out cells with 2×2 cm^2 cross section each. The energy resolution is $\Delta E/E = 3.2\%/\sqrt{E[\text{GeV}]} \oplus 0.5\% \oplus 100\,\text{MeV}/E$.

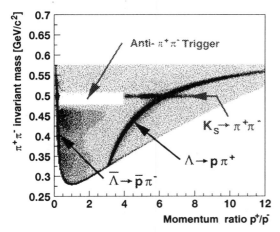

Figure 1. Invariant 2-track mass under the assumption of two pions versus the momentum ratio between the positive and the negative track. The $K_S \to \pi^+\pi^-$ anti-trigger is clearly visible as well as the 'banana-shape' $\Lambda \to p\pi^-$ and $\bar{\Lambda} \to \bar{p}\pi^+$ candidates. The flatly distributed background is mostly due to $K_L \to \pi^+\pi^-\pi^0$ decays.

2.3 Trigger Conditions

For the K_S and hyperon data taking, the trigger conditions were changed with respect to normal ϵ'/ϵ running. To suppress $K_S \to \pi^+\pi^-$, a special two-track trigger was applied, requiring the invariant mass of the two charged tracks (under $\pi^\pm$ assumption) to be inconsistent with the nominal K^0 mass *and* the ratio p^+/p^- between the momenta of the positive and the negative track to be greater than 4 or smaller than 0.25. The effect of these trigger conditions is shown in Fig. 1.

In total, 41 million 2-track events were collected, including 17 million $\Lambda \to p\pi^-$ and about 2 million $\bar{\Lambda} \to \bar{p}\pi^+$ decays.

3 Event Selection and Normalization

The following essential selection criteria were applied: The charged tracks have to be less than 4.5 ns apart in time and are required to form a vertex within 40 m of the target. Under a $p\pi^-$ assumption, the invariant mass of the track pair has to be within 2.5 MeV/c² of the nominal Λ or Ξ^0 mass, respectively.

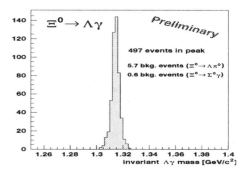

Figure 2. Invariant $\Lambda\gamma$ mass for $\Xi^0 \to \Lambda\gamma$ candidates.

Photon candidates in the calorimeter are required to be within 3 ns in-time with the tracks, to have an energy greater than 3 GeV (15 GeV for $\Xi^0 \to \Lambda\gamma$), and to be well separated from the track extrapolations and from the calorimeter edges. Candidate π^0's have to have an invariant $\gamma\gamma$ mass within 10 MeV/c² of the nominal π^0 mass. In addition, all $K_L \to \pi^+\pi^-\pi^0$ candidates are rejected.

The abundant decay $\Xi^0 \to \Lambda\pi^0$ is used for normalization. With two charged tracks plus calorimeter activity it has a topology similar to those of the investigated decays. As the dominant Ξ^0 decay, it is virtually free of any backgrounds. Applying the selection criteria described above, we find about 115,000 accepted candidates. With an acceptance of 4.7%, derived from Monte Carlo simulation, this number corresponds to a total of about 2.45 million Ξ^0 decays in the fiducial volume.

4 Results

4.1 $\Xi^0 \to \Lambda\gamma$

We observe 497 candidates for the decay $\Xi^0 \to \Lambda\gamma$ (Fig. 2). The background estimate is 5.7 events from $\Xi^0 \to \Lambda\pi^0$ and 0.6 events from $\Xi^0 \to \Sigma^0\gamma$ decays with missing photons in the signal region. The detection efficiency is 10.5%. From this, we obtain a preliminary branching ratio of

$$\mathrm{Br}(\Xi^0 \to \Lambda\gamma) = (1.9 \pm 0.1 \pm 0.2) \times 10^{-3}.$$

396

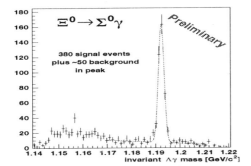

Figure 3. Invariant $\Lambda\gamma$ mass for $\Xi^0 \to \Sigma^0\gamma$ candidates. The Ξ^0 mass was used as constraint. The line is the fit to signal and background.

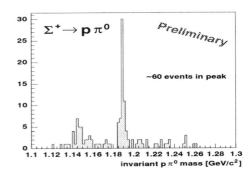

Figure 4. Invariant $p\pi^0$ mass for $\Xi^0 \to \Sigma^+ e^- \bar{\nu}_e$ candidates.

The systematic error is dominated by uncertainties in the Ξ^0 polarization and the $\Xi^0 \to \Lambda\gamma$ decay asymmetry. The result is in agreement with a previous NA48 measurement on a different data set.[6] However, it is inconsistent with the Fermilab measurement.[8]

4.2 $\Xi^0 \to \Sigma^0\gamma$

The Σ^0 directly decays to $\Lambda\gamma$. The $\Xi^0 \to \Sigma^0\gamma$ decay therefore has a signature similar to $\Xi^0 \to \Lambda\pi^0$. Our Σ^0 signal, shown in Fig. 3, contains roughly 380 events on top of a background of about 50 events. The background is consistent with arising exclusively from $\Xi^0 \to \Lambda\pi^0$ decays. The detection efficiency has been determined to be 4.1%. We obtain

$$\mathrm{Br}(\Xi^0 \to \Sigma^0\gamma) = (3.7 \pm 0.5) \times 10^{-3}$$

(with statistical and systematic error combined), in agreement with a previous Fermilab measurement.[7]

4.3 $\Xi^0 \to \Sigma^+ e^- \bar{\nu}_e$

The Σ^+ candidates are selected via their decay to $p\pi^0$. The background stems from $\Xi^0 \to \Lambda\pi^0$ with subsequent $\Lambda \to p\pi^-$ or $\Lambda \to p e^- \bar{\nu}_e$ decays. Fig. 4 shows the Σ^+ signal for events with an additional e^-. We find roughly 60 signal candidates and expect to have a future measurement of the branching fraction with an accuracy of $\sim 15\%$.

4.4 $\Xi^0 \to p\pi^-$

For $\Xi^0 \to p\pi^-$ we expect backgrounds only at small decay lengths and therefore require the two-track vertex to have a distance greater than 10 m from the target. We find no event in the signal mass region and expect to obtain an upper limit of $\mathcal{O}(10^{-6})$, which is about an order of magnitude below the existing limit.[10]

References

1. F.J. Gilman and A.B. Wise, *Phys. Rev.* D **19** (1979) 976.

2. K.G.Rauh, *Z. Phys.* C **10** (1981) 81.

3. A.N. Kamal and R.C. Verma, *Phys. Rev.* D **26** (1982) 190.

4. R.C. Verma and A. Sharma, *Phys. Rev.* D **38** (1988) 1443.

5. P. Zenczykowski, *Phys. Rev.* D **40** (1989) 2290.

6. V. Fanti et al., *Eur. Phys. Jour.* C **12** (2000) 69.

7. S. Teige et al., *Phys. Rev. Lett.* **63** (1989) 2717.

8. C. James et al., *Phys. Rev. Lett.* **64** (1990) 843.

9. X.-G. He and G. Valencia, *Phys. Lett.* B **409** (1997) 469, Erratum ibid. B **418** (1998) 443.

10. C. Geweniger et al., *Phys. Lett.* B **57** (1975) 193.

MEASUREMENT OF FORM FACTOR OF THE NEUTRAL CASCADE HYPERON USING FINAL STATE POLARIZATION

ROLAND WINSTON (FOR THE KTEV COLLABORATION)

Department of Physics and Enrico Fermi Institute, The University of Chicago, Chicago, IL 60637, USA

E-mail: r-winston@uchicago.edu

The KTeV collaboration has made a first observation of neutral cascade beta decay ($\Xi^0 \to \Sigma^+ e^- \overline{\nu}$). We present measurements of the branching ratio and form factors for this semi leptonic process. The results agree well with the Cabibbo $SU(3)_f$ theory and do not favor attempts to include symmetry breaking effects.

1 Introduction

Since the bare quarks are not accessible for carrying out direct measurements, the Hyperon Octet affords the best available laboratory for studying the weak interactions between the light quarks, u,d,s and the leptons. Up to now, neutral cascade beta decay was the last missing entry of the obviously accessible processes. This decay is of particular interest because it is the $SU(3)_f$ analog of the well-studied neutron decay. This is therefore an excellent case for probing $SU(3)_f$ symmetry breaking effects. The KTeV collaboration has now observed this decay and reports results on both branching and form factors.

2 Experimental method

The KTeV apparatus is designed to study CP violation in neutral kaon decay as well as rare kaon decays. In addition, there is a sufficient flux of neutral cascade Hyperon to also study semi-leptonic and radiative decays. To study the hyperons, we instrumented the beam region in order to trigger on the high momentum forward proton characteristic of all decay channels of interest.

Our events of interest are neutral cascades that decay to Σ^+, electron, antineutrino followed by the subsequent decay of $\Sigma^+ \to p\pi^0$. The decay $\Sigma^+ \to p\pi^0$ has a de-

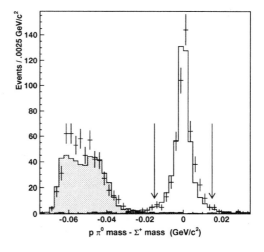

Figure 1. The mass peak, after all selection criteria have been applied. The background is well modelled by Monte Carlo.

cay asymmetry of 98% making this the equivalent of an essentially fully polarized beta decay measurement. Only the semileptonic decay produces a Σ^+ which thereby serves as a unique signature.

3 Branching Ratio

The figure shows our two-body and beta decay event samples. The back ground is about 2% and well simulated as resulting from neutral kaon decays. Our result is: B.R. = [2.54 ± 0.11 (stat) ± 0.16 (sys)] x 10^{-4} (A. Affolder

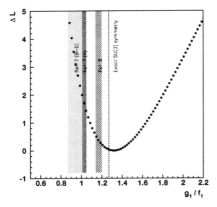

Figure 2. Maximum likelihood fit to g_1/f_1. The shaded bands indicate the theoretical predictions found in the references, the vertical line is the exact $SU(3)_f$ value.

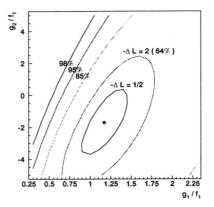

Figure 3. Maximum likelyhood fit to both g1 and g2 form factors

et. al. Phys. Rev. Lett. 82, 3751(1999) which agrees well with the $SU(3)_f$ value of $[2.61 \pm 0.11] \times 10^{-4}$.

4 Form Factors

The observables in our beta decays in addition to branching ratios and energy spectra are several angular correlations: electron neutrino, electron proton and neutrino proton. To interpret these in terms of the form factors of the hadron current, f_1, g_1 g_2 requires expressing the angular distribution in terms of the polarization of the final state baryon. Accordingly we have derived highly transparent expressions which are accurate to second order in q/M of the decay, that is to a few percent. (S. Bright et. al. Phys Rev D 60,117505(1999)) To demonstrate our ability to measure angular correlations we analyzed the proton lambda spin correlation for the 2-body mode. Our result $\alpha_{\Xi^0}\alpha_\Lambda = -.288 \pm 0.007(stat) \pm 0.016(sys)$ agrees well

with the accepted PDG value of -0.264 +/- 0.013. Our beta decay events are remarkably low in background (only 2 %) and the background is well simulated from K_L decays. We study our beta decay correlations in the Σ^+ rest frame. Our electron correlation is 1.0 (as large as possible). We analyze the form factors under two assumptions. First, making the conventional choice $g_2 = 0$ we obtain the fit shown in Figure 2. However, since we have correlations with polarization available, we may also fit for the presence of a (second-class) term g_2. The result shown in Figure 3 is consistent with $g_2 = 0$.

5 Discussion

Our results are in better agreement with the Cabibbo $SU(3)_f$ value (same as neutron decay) than with any of the various models that attempt to include $SU(3)_f$ breaking effects. The beta decays of the octet of hyperons agree remarkably well with the original Cabibbo $SU(3)_f$ predictions.

PREDICTIONS FOR THE DECAYS OF RADIALLY-EXCITED BARYONS

CARL E. CARLSON

Nuclear and Particle Theory Group, Physics Department, College of William and Mary,
Williamsburg, VA 23187-8795, USA
E-mail: carlson@physics.wm.edu

We consider decays of the lowest-lying radially excited baryons. Assuming a single-quark decay approximation, and negligible configuration mixing, we make model-independent predictions for the partial decay widths to final states with a single meson. Masses of unobserved states are predicted using an old mass formula rederived using large-N_c QCD. The momentum dependence of the one-body decay amplitude is determined phenomenologically by fitting to observed decays. Comparison of these predictions to experiment may shed light on whether the Roper resonance can be interpreted as a three-quark state.

1 Introduction

I will report results of some studies of excited baryon masses and decays[1,2], concentrating mainly on the radially-excited baryon multiplet that includes the Roper resonance[3].

Of course, the fundamental QCD degrees of freedom are quarks and gluons, but we must deal with observed states that are baryons and mesons. Our response is to use effective field theory. Here one first writes down all operators that are consistent with all known symmetries, and then use some method—in our case large N_C—to provide a size estimate for each operator. We discard small operators, keep as many of the large operators as possible and use them to calculate masses or decay amplitudes.

To illustrate how effective field theory and large N_C are used, the next section outlines a modern derivation of the Gürsey-Radicati[4] mass formula. The result is in itself useful for estimating masses of undiscovered radially excited baryons. After that, we show how we make predictions for decay widths of radially excited baryons, without assumptions about spatial wave functions.

2 Mass formula

We look at radially excited baryons where the spatial state, and so also the spin-flavor state, is totally symmetric. There are 56 totally symmetric 3-quark states that one can make from $u_\uparrow$, $u_\downarrow$, $d_\uparrow$, $d_\downarrow$, $s_\uparrow$, and $s_\downarrow$, where the arrows indicate the spin projection. The ground states form the **56**, and the radially-excited states form the **56′**. The states are the N, Λ, Σ, Ξ, Δ, Σ^*, Ξ^*, and Ω.

The mass operators for these states are built from the spin $S^i = \sum_\alpha \sigma_\alpha^i/2$ (the sum is over the quarks α), the flavor operators $T^a = \sum_\alpha \tau_\alpha^a/2$ (where the τ^a are a set of 3×3 matrices), and the SU(6) operators

$$G^{ia} = \sum_\alpha \frac{1}{2}\sigma_\alpha^i \cdot \frac{1}{2}\tau_\alpha^a . \tag{1}$$

Terms in mass operators must be rotation symmetric, and flavor symmetric to leading order. Not all terms should be included. For example, in symmetric states matrix elements of T^2 and G^2 are linearly dependent on those of S^2 and the unit operator[5].

Flavor symmetry is not exact. The mass of the strange quark allows non-flavor symmetric terms in the effective mass operator, visible as unsummed flavor indices $a = 8$ below. The effective mass operator is

$$H_{eff} = a_1 1 + \frac{a_2}{N_C}S^2 + \epsilon a_3 T^8 + \frac{\epsilon}{N_C}a_4 S^i G^{i8}$$

$$+ \frac{\epsilon}{N_C^2}a_5 S^2 T^8 + \frac{\epsilon^2}{N_C}a_6 T^8 T^8$$

$$+ \frac{\epsilon^2}{N_C^2}a_7 T^8 S^i G^{i8} + \frac{\epsilon^3}{N_C^2}T^8 T^8 T^8 . \tag{2}$$

There is an ϵ for each violation of flavor symmetry, where $\epsilon \approx 1/3$. Also, a term that is a product of two or three operators comes from an interaction that has at least one or two gluon exchanges, and the strong coupling falls with number of colors as $g^2 \sim 1/N_C$. (A crucial theorem is that no perturbation theory diagrams fall slower in $1/N_C$ than the lowest order ones[5].)

Keeping the first four terms, taking the matrix elements, and reorganizing leads to

$$M = A + BN_s + C[I(I+1) - \frac{1}{4}N_s^2] + DS(S+1) \tag{3}$$

where N_s is the number of strange quarks. This is the Gürsey-Radicati[4] mass formula. We use it to predict masses of 4 undiscovered members of the **56′**, given that 4 are known.

3 The Decays 56′ → 56 + meson

Four of the 8 states in the **56′** are undiscovered or unconfirmed, and existing measurements have large uncertainty. However, we anticipate new results soon from the CLAS detector at CEBAF. One member of the **56′** is the Roper or N(1440), whose composition has been debated. Might it be a qqqG state[6], a non-resonant cross section enhancement[7], or just a 3-quark radial excitation[8,9,10]? Our predictions depend upon the last possibility.

We assume that only single quark operators are needed. Two quark operators were studied for decays of orbitally-excited states[2], and found unnecessary. There is only one single quark operator here, so

$$H_{eff} \propto G^{ia}k^i\pi^a \;, \tag{4}$$

where k^i is the meson 3-momentum and π^a is a meson field operator.

One gets for the decay widths,

$$\Gamma = \frac{M_f}{6\pi M_i}k^3 f(k)^2 \sum |\langle B_f|G_{ja}|B_i\rangle|^2, \tag{5}$$

where $f(k)$ parameterizes the momentum dependence of the amplitude. For the 7 measured decays it is well fit by $f = (2.8 \pm 0.2)/k$.

With this in hand, we can predict the widths for 22 decays. The detailed results are in[3].

To summarize, we have shown how large N_C ideas provide a modern derivation of the old Gürsey-Radicati mass formula, and have predicted decay widths of the **56′**. The success of our predictions would bolster the view of the Roper as a 3-quark state.

Acknowledgments

I thank the conference organizers for their excellent work; Chris Carone, José Goity, and Rich Lebed for pleasant times collaborating; and the National Science Foundation for support under Grant No. PHY-9900657.

References

1. C. E. Carlson, C. D. Carone, J. L. Goity and R. F. Lebed, Phys. Lett. B **438**, 327 (1998); Phys. Rev. D **59**, 114008 (1999).
2. C. E. Carlson and C. D. Carone, Phys. Rev. D **58**, 053005 (1998); Phys. Lett. B **441**, 363 (1998).
3. C. E. Carlson and C. D. Carone, Phys. Lett. B **484**, 260 (2000).
4. F. Gürsey and L. Radicati, Phys. Rev. Letters **13**, 173 (1964).
5. E. Jenkins and R. F. Lebed, Phys. Rev. D **52** (1995) 282; A. Manohar, hep-ph/9802419.
6. Zhen-ping Li, V. Burkert, and Zhu-jun Li, Phys. Rev. D **46**, 70 (1992); C. E. Carlson and N. C. Mukhopadhyay, Phys. Rev. Lett. **67**, 3745 (1991).
7. O. Krehl, C. Hanhart, S. Krewald, and J. Speth, nucl-th/9911080.
8. N. Isgur and G. Karl, Phys. Rev. D **19**, 2653 (1979).
9. L. Ya. Glozman and D. O. Riska, Phys. Rept. **268**, 263 (1996).
10. S. Sasaki, *et al.*, Talk at NSTAR2000, Newport News, VA, February 2000, hep-ph/0004252.

DETERMINING THE QUANTUM NUMBERS OF EXCITED HEAVY MESONS

GAD EILAM

Physics Department, Technion–Israel Institute of Technology, 32000 Haifa, Israel
E-mail: eilam@physics.technion.ac.il

We discuss the decays $X^* \to X e^+ e^-$ ("Dalitz decays") of excited heavy mesons into their ground states and an electron–positron pair. We argue that the measurement of the invariant mass spectrum of the lepton pair gives clear indication on the quantum numbers of the excited meson and thus provides an experimental test of the quark model predictions. All the branching ratios ($\mathcal{B}$) are $\mathcal{O}(0.5)\% \times \mathcal{B}(X^* \to X\gamma)$. Furthermore, the relevance of Dalitz decays of heavy mesons for B and D precision experiments is pointed out .

In this talk I present a somewhat shortened version of our work[1] on the importance of the reactions $X^* \to X e^+ e^-$, where $X = D, D_s, B, B_s$. The outline of the talk is as follows: First I discuss our motivation for studying these reactions, then present the assumptions and results and finally I conclude and summarize.

Motivation

We were led by the following reasons to study the problem at hand:

1. All J^Ps of X^*s are assumed by PDG[2] to be 1^-. It would be nice to confirm experimentally these basic predictions of the quark model.

2. The quantum numbers of even heavier mesons depend on the quantum numbers of the X^*s they decay into.

3. In PDG the central values of the branching ratios for each of the three D^*s, sum up to exactly 100 %. A closer look reveals that this was one of the assumptions in the experimental analyses! As long as the error on the central values of the branching ratios is $\mathcal{O}$ (a few %), it is justified to neglect a branching ratio which is 0.5% of the corresponding branching ratio with real photon.

4. In view of the importance of accurate data for the B, and since Bs decay mostly to D^*s, we should know the identity and the size of the leading decays to better than 1%.

Define:

$$R \equiv \frac{\Gamma(X^* \to X e^+ e^-)}{\Gamma(X^* \to X\gamma)}. \tag{1}$$

More than 40 years ago[3] R was found to be a good measure of the quantum numbers of baryons. Results were close to ours: $R \approx 0.5 \times 10^{-2}$. More recently, R for $X = D$, was calculated[4] to be about 5 time smaller than ours.

Model and Results

Let us parametrize the matrix element as follows $\mathcal{M}(X^* \to X\gamma) = g_{X^* X\gamma} \cdot \mathcal{F}(q^2)$. Then assume that $\mathcal{F}$ is independent of q^2. This is justified by: $m_{X^*} - m_X < 150$ MeV $<< m_\rho$. We then calculate the matrix element squared, summed over final and averaged over initial states, for $J^P = 1^-$ & 2^+ and obtain as mentioned above, R values which are around 5×10^{-3} for all cases. We also obtain $R(J^P = 1^-) > R(J^P = 2^+)$ everywhere, and the difference between these two values of R is about 5%. In addition, we get that R for Bs is approximately 1.4 times larger than R for D mesons.

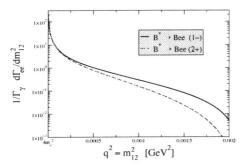

Figure 1. m_{12}^2–distribution of the electron–positron pairs in the Dalitz decay $B^* \to Be^-e^+$ with q the invariant mass of the pairs. Clearly, the quantum numbers affect the tail of the distribution.

A typical example for the B system is $X^* = B^{*0}$ (for which $m_{X^*} - m_X = 45.78$ MeV), where we obtained that $R(1^-) = 4.69 \times 10^{-3}$ while $R(2^+) = 4.38 \times 10^{-3}$; note that in this case we only know[2] that $\mathcal{B}(X^* \to X\gamma)$ is dominant. Out of the 5 e^+e^- modes relevant to the present study only a single case has been observed in experiment, namely[5] $R(X = B^0) = (4.7 \pm 1.1 \pm 0.9) \times 10^{-3}$, in excellent agreement with the value of $R(1^-)$ above.

For the D system we look for example at the D_s complex ($m_{X^*} - m_X = 143.8$ MeV), where we found that $R(1^-) = 6.45 \times 10^{-3}$ and $R(2^+) = 6.14 \times 10^{-3}$. This is an interesting case since[2] $\mathcal{B}(D_s^* \to D_s\gamma)$ attains a relatively high value of 0.942 ± 0.025, and therefore we get a large e^+e^- rate.

It is important to note that though total rates for Dalitz pair production may serve as good indicator for quantum numbers, it is better to use $q^2 = m_{e^+e^-}^2$ distributions. Let us take, for example, the case $X = B^0$. In Fig. 1 we plot the ratio between the differential rate for $B^{*0} \to B^0 e^+e^-$ and the rate for $B^{*0} \to B^0\gamma$. We see a marked difference between the predictions for $J^P(B^*) = 1^-$ and for 2^+.

Summary and Conclusions

The ratio between the widths $X^* \to Xe^+e^-$ and $X^* \to X\gamma$ for heavy flavor mesons is – based on a QED like calculation – approximately equal to 0.5%. This ratio has a role in determining the quantum numbers of X^* for $X = D$, B and may become essential with the advent of new precision B and D experiments.

Acknowledgments

This research has been supported in part by the Israel Science Foundation founded by the Israel Acacademy of Sciences and Humanities and by the US–Israel Binational Science Foundation. I would like to thank Frank Krauss for a fruitful collaboration.

References

1. G. Eilam and F. Krauss, *Phys. Lett.* B **482**, 374 (2000).
2. D.E. Groom *et al*, *Europ. Phys. J.* C **15**, 1 (2000).
3. G. Feinberg, *Phys. Rev.* **109**, 1019 (1958).
4. T.M. Aliev *et al.*, *Zeit. f. Phys.* C **64**, 683 (1994).
5. G. Eigen, hep-ex/9901007.

Parallel Session 2

Soft Interaction Processes
(including diffraction, two-photon, small x)

Conveners: Alexei B. Kaidalov (ITEP) and
David J. Miller (College London)

PHOTON STRUCTURE

STEFAN SÖLDNER-REMBOLD

CERN, CH-1211 Geneva 23, Switzerland
E-mail: stefan.soldner-rembold@cern.ch

The LEP experiments measure the QED and QCD structure of the photon in deep-inelastic electron-photon scattering. The status of these measurements is discussed in this short review.

1 Kinematics

At LEP the virtuality of the "probing" photon is $Q^2 = -q^2$ (the negative squared four-momentum of the photon) and the virtuality of the "probed" photon is $P^2 = -p^2 \approx 0$. The deep-inelastic scattering cross-section is written as

$$\frac{\mathrm{d}^2\sigma_{e\gamma \to e+\text{hadrons}}}{\mathrm{d}x\mathrm{d}Q^2} = \frac{2\pi\alpha^2}{x\,Q^4} \quad (1)$$

$$\left[\left(1 + (1-y)^2\right) F_2^\gamma(x, Q^2) - y^2 F_L^\gamma(x, Q^2)\right],$$

where α is the fine structure constant, x and y are are the usual dimensionless variables of deep-inelastic scattering and $W^2 = (q + p)^2$ is the squared invariant mass of the hadronic final state. The scaling variable x is given by

$$x = \frac{Q^2}{Q^2 + W^2 + P^2}. \quad (2)$$

The term proportional to $F_L^\gamma(x, Q^2)$ is small and is therefore usually neglected. In leading order the structure function $F_2^\gamma(x, Q^2)$ can be identified with the sum over the parton densities of the photon weighted by the square of the parton's charge.

2 QED Structure Functions

The QED structure function F_2^γ has been measured in the process $e^+e^- \to e^+e^-\mu^+\mu^-$. In addition, the measurement of the distribution of the azimuthal angle χ between the electron scattering plane and the plane containing the muon pair in the $\gamma^*\gamma$ centre-of-mass system gives access to the structure functions F_A^γ and F_B^γ. They are related to the transverse-longitudinal (A) and

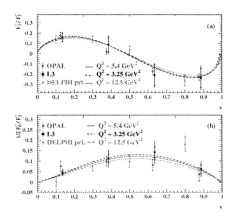

Figure 1. The measured ratios F_A^γ/F_2^γ and $1/2 \cdot F_B^\gamma/F_2^\gamma$ compared to the QED prediction [1].

the transverse-transverse (B) interference in the interaction of the transverse real photon with the virtual photon. The LEP measurements [2,3,4] are shown in Fig. 1. Both structure functions are found to be significantly different from zero and the ratios are well described by QED [1].

3 Hadronic Structure Functions

The measurement of hadronic structure functions is considerably more difficult due to the necessity to reconstruct x from the hadronic final state in the detector (Eq. 2). Significant progress has been made recently in reducing the systematic errors due to unfolding and hadronisation uncertainties. ALEPH, L3 and OPAL have compared their combined data to the PHOJET and HERWIG generators [5].

An unbiased tune using informations from HERA has improved HERWIG significantly.

Furthermore new methods for regularised unfolding like the maximum entropy method or the singular value decomposition method have been used. ALEPH and OPAL have introduced two-dimensional unfolding and improved treatment of hadronic energy in the forward region. L3 is applying energy-momentum conservation using kinematic information from both hadrons and the electrons. The uncertainty on the measurements

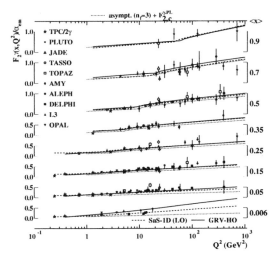

Figure 3. The Q^2 dependence of the hadronic structure function F_2^γ in bins of x compared to the GRV-HO, SaS-1D and the asymptotic prediction.

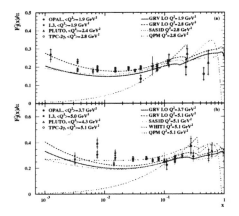

Figure 2. The measured hadronic structure function F_2^γ compared to the GRV-LO [8], SaS-1D [9] and the WHIT1 [10] parametrisations and to the QPM model. In the case of L3 the values obtained using PHOJET and TWOGAM for unfolding are shown separately.

shown in Fig. 2 are therefore considerably reduced [6,7].

The hadron-like component dominates at low x and there could be a first indication of the low x rise of the photon structure function expected from QCD evolution.

The Q^2 dependence of the structure function F_2^γ in bins of x is shown in Fig. 3 for all currently available measurements. The data are compared to the GRV-HO and the SaS-1D parametrisation, and to the sum of the asymptotic prediction [11] for 3 light flavours and the point-like part of the charm structure function taken from GRV. Positive scaling violation of the photon structure function

is observed in all x ranges - different from the proton - due to the regular QCD evolution at low x and due to the inhomogeneous term ($\gamma \to q\bar{q}$) at larger x. As expected, the asymptotic prediction fails to describe the data at low x, where the non-perturbative hadron-like contribution dominates, whereas all models give a reasonable description of the medium to high x, Q^2 region.

4 Charm Structure Function

In Fig. 3 the charm threshold is clearly visible. Above the kinematic threshold for charm production, the c and u contribution to the point-like part of the photon structure function are of similar size. OPAL has measured the charm structure function of the photon for the first time using D^* decays [12]. The region $x > 0.1$ - which is dominated by the point-like component - is in good agreement with a NLO calculation [13]. In the region $x < 0.1$ the measurement suggests the existence of a hadron-like component with currently large errors. These uncertainties are expected to be significantly reduced in the future due to higher statistics and better MC

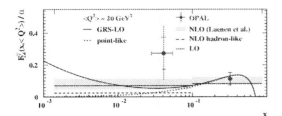

Figure 4. Charm structure function $F_{2,c}^\gamma$ of the photon as function of x for $\langle Q^2 \rangle = 20$ GeV^2.

modelling of charm production.

5 Virtual Photon Structure

In addition to the structure function of (quasi-)real photons, i.e. $P^2 \approx 0$, the effective structure function of virtual photons can be measured if $Q^2 >> P^2 >> \Lambda_{QCD}^2$. This was first done by PLUTO [14]. For real pho-

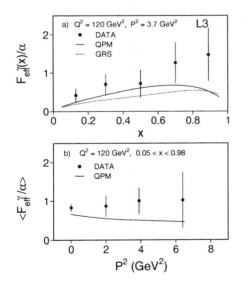

Figure 5. Effective structure function of the virtual photon as function of x and P^2.

tons only the cross-sections σ_{LT} and σ_{TT} contribute, where the indices refer to the longitudinal and transverse helicity states of the probe and target photon, respectively, i.e. $F_2^\gamma \simeq \sigma_{LT} + \sigma_{TT}$. For $P^2 >> 0$ other helicity states have to be taken into account,

leading to the definition of the effective structure function $F_{eff}^\gamma \simeq \sigma_{LT} + \sigma_{TT} + \sigma_{TL} + \sigma_{LL}$ (interference terms are neglected). This effective structure function measured by L3 [15] is shown in Fig. 5. We expect the hadron-like part of the parton densities at low x to decrease with increasing virtuality of the photon. In Fig. 5b the QPM approximation of the point-like part therefore fails to describe the data point at $P^2 = 0$. The shape of the P^2 dependence is consistent with the QPM ansatz but the errors are still large. Much more precise data is to be expected from LEP on virtual photon structure in the next years.

References

1. R. Nisius, Phys. Rept. 332 (2000) 165.
2. DELPHI note 2000-135, abstract 644 submitted to this conference.
3. L3 Coll., Phys. Lett. B438 (1998) 363.
4. OPAL Coll., Eur. Phys. J. C11 (1999) 409.
5. ALEPH, L3 and OPAL Coll., CERN-EP-2000-109.
6. L3 Coll., Phys. Lett. B447 (1999) 147; Phys. Lett. B436 (1998) 403.
7. OPAL Coll., hep-ex/0007018.
8. M. Glück et al., Phys. Rev. D46 (1992) 1973; Phys. Rev. D45 (1992) 3986.
9. G.A. Schuler et al., Z. Phys. C68 (1995) 607.
10. K. Hagiwara et al., Phys. Rev. D51 (1995) 3197.
11. E. Witten, Nucl. Phys. B120 (1977) 189; in the parametrisation of L.E. Gordon et al., Z. Phys. C56 (1992) 307.
12. OPAL Coll., hep-ex/9911030, see also A. Böhrer, hep-ph/0009121, these proceedings.
13. E. Laenen et al., Phys. Rev. D49 (1994) 5753; E. Laenen, S. Riemersma, Phys. Lett. B376 (1996) 169.
14. PLUTO Coll., C. Berger et al., Phys. Lett. B142 (1984) 119.
15. L3 Coll., Phys. Lett. B483 (2000) 373.

DIJET PHOTOPRODUCTION

J. M. BUTTERWORTH

Department of Physics & Astronomy, University College London, Gower St., WC1E 6BT, London, UK

E-mail: J.Butterworth@ucl.ac.uk

REPRESENTING THE H1 AND ZEUS COLLABORATIONS

Recent jet photoproduction measurements are presented, including measurements of jet substructure which are sensitive to parton type. The implications for real and virtual photon structure are discussed.

1 Introduction

The last several years of HERA running have provided a wealth of data on jet production in photon-proton scattering. The data have implications for many areas of physics including photon structure, jet substructure [1], heavy flavour production [2], diffraction [3,4] and colour coherence effects [5]. These data have recently been joined by jet data from photon-photon collisions at LEP [6]. Here, the focus is on the first area - photon structure - although jet substructure is used as a tool to unravel the nature of the partons.

The leading order processes for jet, particle and heavy quark production can be divided into two classes - those where the photon enters the hard scatter as a point-like particle (direct) and those where it acts as a source of partons (resolved). Separation of the two becomes a matter of convention at higher orders (and therefore in real life), but the distinction remains useful and can be implemented by measuring suitable variables such as x_γ^{OBS}, the fraction the photon's energy manifest in the dijet system [7].

2 Dijet Cross Sections

The transverse energy of the jet (E_T^{Jet}) determines the scale at which the structure is probed, and thus the applicability of perturbative QCD. Kinematically it also determines the minimum x values which can be probed. Thus low E_T^{Jet} events are predominantly low

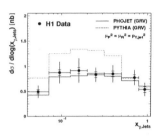

Figure 1. The dijet photoproduction cross section for $E_T^{\mathrm{Jet}} > 4$ GeV.

x_γ^{OBS} events and are sensitive to the gluon distribution in the photon. However, in this region there is also a strong sensitivity to non- or semi-perturbative effects such as the "underlying event", hadronisation and the transverse momentum cut-off ($\hat{p}_T^{\mathrm{min}}$) used to regularise the partonic cross section.

These data shown in figure 1 will strongly constrain models of jet production, but it is difficult to disentangle the effect of the photon pdf and comparisons to parton-level QCD calculations are dangerous. Attempts have been made to reduce the dependence of non-perturbative effects.

2.1 LO Parton Distribution

H1 [8] subtract an "underlying event" contribution, then demand $E_T^{\mathrm{Jet}} > 6$ GeV. Since the underlying event contribution can be large, this means that the true minimum (hadron level) E_T^{Jet} is close to 7.5 or 8 GeV. The correction is taken from the ambient energy density in photoproduction events, some

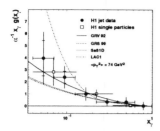

Figure 2. Extracted leading order gluon distribution in the photon.

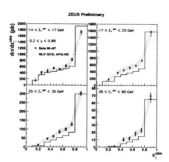

Figure 3. Dijet cross sections.

of which will come from perturbative QCD radiation. Thus points altered in this manner can only be compared to LO QCD without parton showers. H1 then assume that all subprocesses have the same angular distribution (this is reasonable for resolved processes since most involve gluon exchange) and extrapolate to the full phase space to extract an effective parton distribution. Assuming the light quark distribution is given by e^+e^- data, it is then subtracted, and the remaining contribution is assumed to be the gluon distribution in the photon at LO. The result is shown in figure 2 and exhibits the rise at low-x expected in (for example) the GRV pdf. It should be noted that the model dependency is large and difficult to quantify.

2.2 High E_T^{Jet} cross sections

Going to still higher E_T^{Jet} reduces the sensitivity to low-x, but also reduces the effect of non-perturbative physics and thus the model dependence in a comparison to perturbative QCD. ZEUS have measured cross sections for $E_T^{\text{Jet}} > 14$ GeV. The x_γ^{OBS} distribution is shown in figure 3, for four different E_T^{Jet} regions.

In this region underlying event and hadronisation effects are less than 15% according to a range of Monte Carlo models. An excess is seen at low x_γ^{OBS} when compared to NLO QCD [9]. This effect is also seen in cross sections differential in η^{jet}, and is not decreasing with increased E_T^{Jet}. To investigate the source of this discrepancy, ZEUS

have also studied the dijet angular distributions of resolved and direct events separately at these E_T^{Jet} values. They find that the shape of this distribution is well described by NLO QCD. This implies that the excess events have the same dynamics as the rest and maybe that the source of the discrepancy is the photon parton distribution.

3 Dynamics and Jet Structure

In general, as observed in e^+e^- experiments, gluon-initiated jets are broader than quark-initiated jets. This fact can be used to study the subprocess composition in photoproduction. The jet shape (e.g. the fraction of E_T^{Jet} within 0.3 units of the axis) or subjet multiplicity have both been used by ZEUS to define "thick" and "thin" jet samples. The dominant subprocesses in direct ($\gamma g \to q\bar{q}$) and resolved ($q(\bar{q})g \to q(\bar{q})g$) photoproduction imply that selecting thin jets leads to quark exchange processes dominating, with and angular distribution $\approx (1 - |\cos\theta^*|)^{-1}$. Conversely selecting thick jets means that gluon exchange processes dominate, with an angular distribution $\approx (1 - |\cos\theta^*|)^{-2}$. The angular distributions in these two cases are shown in figure 4. A steeper rise is seen for thick jets than for thin jets, consistent with the expectation.

4 From Real to Virtual Photons

With respect to direct photon processes, it is expected that the perturbative part of re-

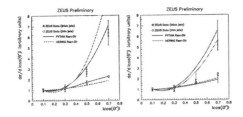

Figure 4. Dijet angular distribution for fat and thin jets, as defined by a cut of jet shape (left) and subjet multiplicity (right).

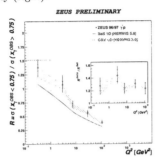

Figure 5. Ratio of low to high x_γ^{OBS} events as a function of photon virtuality.

solved component will fall $\approx \ln(E_T^{\mathrm{Jet}\,2}/\mathrm{P}^2)$ whilst non-perturbative ("vector meson") part should fall $\approx m_v^2/(m_v^2 + \mathrm{P}^2)$, where P^2 is the photon virtuality. This expectation is implemented in SaS pdfs [10]. The x_γ^{OBS} distribution fairly well modelled at all except the lowest photon virtualities.

The data confirm previous data [11] in showing the need for a 'resolved' component at high photon virtuality, and that this component is gradually suppressed as virtuality increases. There is a discrepancy w.r.t. SaS in the ratio, but this discrepancy is not changing rapidly with photon virtuality (see inset to figure).

5 Summary and Prospects

The theoretical and experimental uncertainties are now getting small at high E_T^{Jet}, and the dynamics and subprocess composition are well understood. There are more resolved events than expected in NLO QCD, which could be an indication of a need for new

photon pdfs. New NLO QCD fits including HERA and LEP data for the photon pdfs are needed. At low E_T^{Jet}, accurate measurements exist with sensitivity in principle to the gluon distribution in the photon. The interpretation of these results in terms of parton distribution is however limited by the current understanding of the underlying event and low $\hat{p}_T^{\mathrm{min}}$ physics. As in the high E_T^{Jet} region there is a need to improve models and to compare data consistently across experiments. At non-zero virtualities, data continue to be produced with improved accuracy. They qualitatively confirm QCD based expectations for evolution from photoproduction to DIS and we are developing a consistent picture of these interactions as a function of photon virtuality.

References

1. ZEUS Collab., Eur. Phys. J. C2 (1998) 1, 61-75

2. H1 Collab., Phys. Lett. B467 (1999) 156-164, ZEUS Collab., Eur. Phys. J C6 (1999) 67-83.

3. ZEUS Collab., Phys. Lett. B 369 (1996) 55-68.

4. ZEUS Collab., Phys. Lett. B 356 (1995) 129-146.

5. ZEUS Collab., Phys. Lett. B 443 (1998) 394-408.

6. Thorsten Wengler, these proceedings.

7. ZEUS Collab., Phys. Lett. B 348 (1995) 665-680.

8. H1 Collab., Phys. Lett. B483 (2000) 36-48.

9. S.Frixione and G.Ridolfi, Nucl. Phys. B507(97) 315.

10. G. Schuler and T Sjöstrand, Phys. Lett. B 376 (1996) 193.

11. H1 Collab., Eur. Phys. J. C13 (2000) 397-414, Phys. Lett. B415 (1997) 418-434. ZEUS Collab., Phys. Lett. B479 (2000) 37-52.

THE HADRONIC PICTURE OF THE PHOTON

THORSTEN WENGLER

CERN, EP division, CH-1211 Geneve 23, Switzerland
email: Thorsten.Wengler@cern.ch

The hadronic interactions of the photon are studied in terms of new measurements of the total hadronic cross section and di-jet production in photon-photon collisions at LEP2.

1 Total hadronic $\gamma\gamma$ cross section

The reaction of $e^+e^- \rightarrow e^+e^- \gamma^*\gamma^* \rightarrow e^+e^- + hadrons$ is analysed for quasi-real photons using data collected by L3 and OPAL at LEP2. From this measurement the total hadronic cross section $\sigma_{\gamma\gamma}(W)$ is extracted using a luminosity function for the photon flux and extrapolating to $Q^2=0$ GeV^2. The L3 data [1] are shown in Figure 1, representing an integrated luminosity of 392.6 pb^{-1} taken at $\sqrt{s}$ from 189 to 202 GeV. The total hadronic cross section, $\sigma_{\gamma\gamma}$, is shown as a function of the invariant mass W of the photon-photon system. Also shown as the solid line is the result of a fit using the parameterisation proposed by Donnachie and Landshoff [2]: $\sigma_{\text{tot}} = As^\epsilon + Bs^{-\eta}$. The second term represents the Reggeon exchange dominant at low W, while at high W the Pomeron exchange driven by the exponent ϵ prevails. If photons behave predominantly like hadrons the values of $\epsilon = 0.095\pm 0.002$ and $\eta = 0.34\pm 0.02$ obtained from a universal fit to hadron-hadron cross sections [3] should be valid for $\sigma_{\gamma\gamma}(W)$ as well. Indeed this is not the case for the L3 data, as demonstrated by the dashed line in Figure 1. Instead a fit with A, B, and ϵ as free parameters yields $\epsilon = 0.250\pm0.016$, which is more than a factor of two higher than the universal value, a rise significantly steeper than expected from the universal fit of hadron-hadron cross sections. The L3 data is compared to an OPAL measurement [4] using an integrated luminosity of 74.3 pb^{-1} taken at $\sqrt{s}$ from 161 to 183 GeV. Both measurements are consistent

inside their respective uncertainties. A similar fit as described above to the OPAL data however yields $\epsilon = 0.101\pm 0.025$, in agreement with the universal fit. The steeper rise observed by L3 is mainly determined by their data points at highest and lowest W.

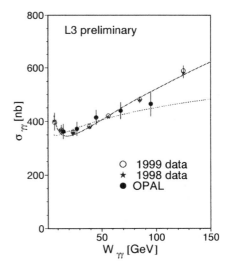

Figure 1. The photon-photon total hadronic cross section, $\sigma_{\gamma\gamma}(W)$. The solid line is a fit to the L3 data, the dashed line uses the parameters from a universal fit to hadron-hadron cross sections (see text). Here the L3 points do not include the dominant systematic error due to MC model dependencies. With this error included the total uncertainties on the OPAL and L3 measurements are of comparable size.

2 Di-jet Production

The production of di-jets in the collisions of two quasi-real photons has been measured by ALEPH and OPAL. Jet production is calculable in perturbative QCD. The measurements can therefore be used to study the per-

412

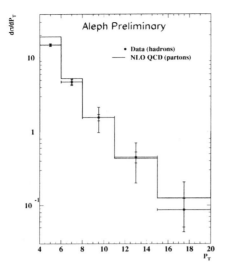

Figure 2. Differential di-jet cross section as measured by ALEPH using 59.2 pb^{-1} taken at $\sqrt{s}$ =183 GeV.

formance of NLO predictions for these processes. ALEPH [5] has analysed 59.2 pb^{-1} taken at $\sqrt{s}$ =183 GeV using the Durham $k_\perp$ clustering algorithm [6] with Y_{cut} =0.018, chosen to maximise the rate of di-jet events. Di-jet events are required to have at least two jets with $|\eta_{\text{jet}}| < 1.4$, and a minimum jet transverse momentum of 4 (3) GeV for the first (second) jet. The differential di-jet cross section as a function of the highest p_T jet in the event is shown in Figure 2. The data is compared to an NLO calculation of Klasen et al.[7], performed at the level of QCD partons. The agreement of the calculation with the data is good, except at the lowest p_T^{jet}.

OPAL [8] has analysed 384 pb^{-1} taken at $\sqrt{s}$ from 189 GeV to 202 GeV. Jets are found using the inclusive $k_\perp$ [6] clustering algorithm. Jets entering the analysis have a pseudorapidity $|\eta_{\text{jet}}| < 2.0$. The two jets with the highest E_T^{jet} in each event are taken. An average transverse energy of the two leading jets in the event of $\bar{E}_T^{\text{jet}} > 5$ GeV is required. The additional condition $|E_{T,1}^{\text{jet}} - E_{T,2}^{\text{jet}}|/(E_{T,1}^{\text{jet}} + E_{T,2}^{\text{jet}}) < 1/4$ keeps low E_T^{jet} jets from entering the distributions and ensures asymmetric E_T^{jet} thresholds for the two jets. The differ-

ential cross section as a function of $\bar{E}_T^{\text{jet}}$ (not shown) is reasonably well described by both the LO MC models PHOJET and PYTHIA and the NLO calculation [7]. Di-jet events have the particular advantage that the fraction of the photon momentum, x_γ, entering the hard scattering can be estimated from the di-jet system. The observable quantity $x_\gamma^\pm \equiv (\sum_{\text{jets}=1,2}(E \pm p_z))/(\sum_{\text{hadrons}}(E \pm p_z))$ is defined for this purpose. OPAL has for the first time measured differential cross sections as a function of x_γ, where the data has been fully unfolded for detector effects. Here x_γ refers to both x_γ^+ and x_γ^- entering the distribution. The region of small x_γ is expected to be dominated by resolved photon interactions, and is hence particularly sensitive to the gluon density in the photon. The OPAL data is shown in Figure 3 in three regions of $\bar{E}_T^{\text{jet}}$. The data is compared to three predictions of the PHOJET generator using the GRV LO, SaS1D, and LAC1 parton distribution functions respectively. The sensitivity of this observable to the different parton distributions is clearly visible. The already disfavoured LAC1 set is shown to demonstrate the effect of a high gluon density and clearly overshoots the data especially at low $\bar{E}_T^{\text{jet}}$. GRV and SaS1D on the other hand seem to underestimated the gluon density. A preliminary study of the underlying event us-

	data [pb]	NLO [pb]
5 GeV $< \bar{E}_T^{\text{jet}} < 7$ GeV		
$x_\gamma > 0.75$	111.0±3.8	206.5
$x_\gamma < 0.75$	205.5±4.8	84.4
7 GeV $< \bar{E}_T^{\text{jet}} < 11$ GeV		
$x_\gamma > 0.75$	77.4±2.6	87.0
$x_\gamma < 0.75$	71.5±2.2	32.6
11 GeV $< \bar{E}_T^{\text{jet}} < 25$ GeV		
$x_\gamma > 0.75$	32.0±2.5	27.4
$x_\gamma < 0.75$	15.8±1.7	7.5

Table 1. OPAL di-jet cross sections compared to NLO predictions in regions of x_γ and $\bar{E}_T^{\text{jet}}$.

OPAL Preliminary

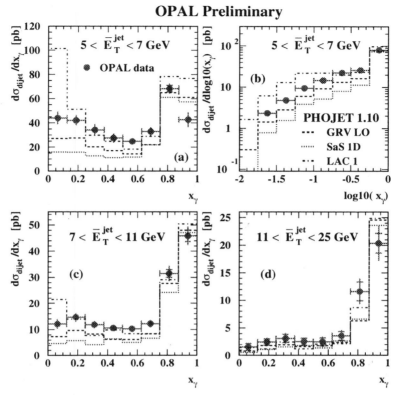

Figure 3. OPAL differential di-jet cross section as a function of x_γ in several regions of $\bar{E}_T^{jet}$.

ing PYTHIA with and without multiple interactions show a visible effect only at the lowest values of $\bar{E}_T^{jet}$ and x_γ. Table 1 compares the OPAL measurement with an NLO calculation [7] for small and large x_γ in the three regions of $\bar{E}_T^{jet}$. It should be stressed that hadronisation corrections are not taken into account in this calculation. The NLO calculation predicts much too large a cross section for $x_\gamma > 0.75$ and small $\bar{E}_T^{jet}$, while being too low by about a factor of two for $x_\gamma < 0.75$. With increasing $\bar{E}_T^{jet}$ the discrepancy for $x_\gamma > 0.75$ largely disappears, while for $x_\gamma < 0.75$ the NLO prediction remains too low by a factor of two for the highest $x_\gamma > 0.75$ considered. This also suggests that the parton density functions used in the NLO calculation (GRV) underestimate the gluon density in the photon.

References

1. L3 Coll., L3 note 2570, Abstract 622 submitted to this conference.

2. A. Donnachie and P.V. Landshoff, Phys. Lett. B296 (1992) 227.

3. PDG, Eur. Phys. J. C3 (1998) 1.

4. Opal Coll., G. Abbiendi et al., Eur. Phys. J. C14 (2000) 199.

5. ALEPH Coll., ALEPH 2000-052, Abstract 265 submitted to this conference.

6. S. Catani, Yu.L. Dokshitzer, M.H. Seymour and B.R. Webber, Nucl. Phys. B406 (1993) 187; S.D. Ellis, D.E. Soper, Phys. Rev. D48 (1993) 3160

7. M. Klasen, T. Kleinwort and G. Kramer, Eur. Phys. J. Direct C1 (1998) 1; M. Klasen, private communication.

8. OPAL Coll., OPAL PN443, Abstract 583 submitted to this conference.

DOUBLE TAG EVENTS IN TWO-PHOTON COLLISIONS AT LEP

M. WADHWA

University of Basel, Klingelbergstrasse 82,
CH-4056 Basel, Switzerland
E-mail: Maneesh.Wadhwa@cern.ch

Double tag events in two photon collisions are studied using the L3 detector at the LEP center of mass energies $\sqrt{s} \simeq 189 - 202$ GeV. The cross-section of $\gamma^*\gamma^*$ collisions is measured at an average photon virtuality $\langle Q^2 \rangle = 15$ GeV^2. The results are in agreement with Monte Carlo predictions based on perturbative QCD, while the Quark Parton Model alone is insufficient to describe the data. The measurements are compared to the LO and the NLO BFKL calculations.

1 Introduction

In this paper we present new results on double-tag two-photon events $e^+e^- \rightarrow e^+e^-\ hadrons$. The data, collected at centre-of-mass energies $\sqrt{s} \simeq 189 - 202$ GeV, correspond to an integrated luminosity of 401 pb^{-1}. Both scattered electrons [a] are detected in the small angle electromagnetic calorimeters. The virtuality of the two photons, Q_1^2 and Q_2^2, is in the range of 4 GeV$^2 < Q_{1,2}^2 < 40$ GeV^2.

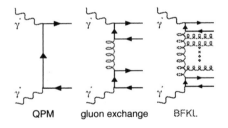

Figure 1. Diagrams for the a) QPM, b) one-gluon exchange and c) BFKL Pomeron processes in a $\gamma^*\gamma^*$ interaction.

The centre-of-mass energy of the two virtual photons, $\sqrt{\hat{s}} = W_{\gamma\gamma}$, ranges from 5 GeV to 90 GeV. The cross-section measurement of the two virtual photons is considered as "golden" process to test the BFKL dynamics [2]. For this scheme the $\gamma^*\gamma^*$ interaction can be seen as the interaction of two $q\bar{q}$ pairs scattering off each other via multiple gluon

---

[a] Electron stands for electron or positron throughout this paper.

exchange. (Fig. 1c). In the leading order approximation (LO), the cross-section in the saddle point approximation for the collision of two virtual photons is [3,4]:

$$\sigma_{\gamma^*\gamma^*} = \frac{\sigma_0}{Q_1 Q_2 Y}\left(\frac{s}{s_0}\right)^{\alpha_P - 1} \qquad (1)$$

Here

$$\sigma_0 = \text{const}$$
$$s_0 = \frac{K Q_1 Q_2}{y_1 y_2} \quad , \quad Y = \ln\left(s/s_0\right) \qquad (2)$$
$$y_i = 1 - (E_i/E_b)\cos^2(\theta_i/2)$$

where E_b is the beam energy, E_i and θ_i are the energy and polar angle of the scattered electrons and α_P is the "hard Pomeron" intercept; K is a scale factor which accounts for uncertainity in the BFKL energy scale s_0. The centre-of-mass energy of the two-photon system is related to the e^+e^- centre-of-mass energy s by $\hat{s} = W_{\gamma\gamma}^2 \approx sy_1 y_2$. In leading order $(\alpha_P - 1) = (4\ln 2)N_c\alpha_s/\pi$, where N_c is the number of colours. Using $N_c = 3$ and $\alpha_s = 0.2$, $(\alpha_P - 1) \simeq 0.53$. The born cross-section of one gluon exchange (see Fig. 1b) is independent of $W_{\gamma\gamma}$. Recently, effort has been devoted to improve the exact leading order calculation [2] by studying the effect of charm mass and the contribution of longitudinal photon polarization states [5]. Still these effects are not sufficient to describe our previous measurement [6]. One needs next to leading order corrections(NLO). It turns out that the NLO corrections [7] to the intercept

"$\alpha_P - 1$" are negative for $\alpha_s > 0.16$. Different techniques [8,9,10,11,12,13,14] have been proposed to improve the NLO calculations in a suitable renormalization scheme thus giving values of $(\alpha_P - 1)$ in the range $0.17-0.33$.

2 Double-tag cross-section

After selection cuts described in ref [17], we have selected 336 candidate events. The estimated background is 56 events, mainly due to $e^+e^- \to e^+e^-\tau^+\tau^-$ and misidentified single-tag events. The contamination from annihilation processes and lepton channels in two photon collisions is negligible. The preliminary cross-section is measured in the kinematic region limited by:

- $E_{1,2} > 30$ GeV, 30 mrad $< \theta_{tag} <$ 66 mrad and $2 \leq Y \leq 7$

The data is then corrected for efficiency and acceptance with two Monte-Carlo models; PHOJET [15] and Vermaseren(QPM) [16] respectively. The differential cross-sections $d\sigma(e^+e^- \to e^+e^- + \text{hadrons})/dY$ are measured in four ΔY intervals. As one can be seen in Table 1 and in Fig. 2, none of the models are sufficient to describe the data. The value of the cross-section at $5 < Y < 7$ exceeds the Monte Carlo prediction by about 3.5 standard deviations.

Table 1. The differential cross-section, $d\sigma(e^+e^- \to e^+e^- + \text{hadrons})/dY$ in picobarn measured in the kinematic region defined in the text, at $\sqrt{s} \simeq 189 - 202$ GeV. The predictions of the PHOJET and the QPM Monte Carlo models are also listed. The first error is statistical and the second is systematic.

ΔY	DATA $d\sigma/dY$	PHOJET $d\sigma/dY$	QPM $d\sigma/dY$
$2.0 - 2.5$	$0.50 \pm 0.07 \pm 0.03$	0.40	0.32
$2.5 - 3.5$	$0.30 \pm 0.03 \pm 0.02$	0.29	0.17
$3.5 - 5.0$	$0.15 \pm 0.02 \pm 0.01$	0.14	0.05
$5.0 - 7.0$	$0.08 \pm 0.02 \pm 0.01$	0.03	0.006

From the measurement of the $e^+e^- \to e^+e^- + \text{hadrons}$ cross-section, σ_{ee}, we extract

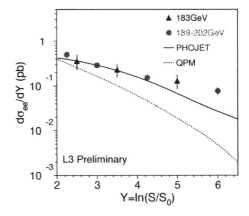

Figure 2. The cross-section of $e^+e^- \to e^+e^-\,hadrons$ as a function of Y in the kinematical region defined in the text at $\sqrt{s} \simeq 189 - 202$ GeV compared to our previous results $\sqrt{s} \simeq 183$ GeV. In the figure the predictions of PHOJET (continuos line) and of the QPM (dashed line) are indicated.

the two-photon cross-section, $\sigma_{\gamma^*\gamma^*}$, by using only the transverse photon luminosity function, $\sigma_{ee} = L_{TT} \cdot \sigma_{\gamma^*\gamma^*}$. In Fig. 3 we show $\sigma_{\gamma^*\gamma^*}$, after subtraction of the QPM contribution as a function of Y. Using an average value of Q^2, $\langle Q^2 \rangle = 15$ GeV^2 at $\sqrt{s} \simeq 189-202$ GeV, we calculate the one-gluon exchange contribution with the asymptotic formula. The expectations are below the data. The leading order expectations of the BFKL model,, shown as a dotted line in Fig. 3, are too high. By leaving α_P as a free parameter and $K = 1$, a fit to the data, taking into account the statistical, yields:

$$\alpha_P - 1 = 0.36 \pm 0.02, \quad \chi^2/d.o.f = 0.98/3$$

with $\chi^2/d.o.f = 0.98/3$ and if the energy scale factor K is a free parameter and $(\alpha_P - 1)=0.53$, a fit to data yields:

$$K = 6.4 \pm 1.0, \quad \chi^2/d.o.f = 1.34/3$$

These results are shown in Fig. 3 as a soild and dashed lines respectively. The value of $(\alpha_P - 1)$, smaller than expected from the LO BFKL calculation at the saddle point approximation, and the scale factor K much

416

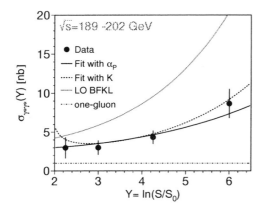

Figure 3. Two-photon cross-sections, $\sigma_{\gamma^*\gamma^*}$, after the subtraction of the QPM contribution at $\sqrt{s} \simeq 189 - 202$ GeV ($\langle Q^2 \rangle = 15$ GeV^2). The data are compared to the predictions of the LO BFKL calculation at saddle point approximation(eq.1)(dotted line) with K=1 and $(\alpha_P - 1) = 0.53$ and the solid line is the fit to the data of the LO BFKL (eq.1) with K=1 and the coefficient $(\alpha_P - 1)$ as a free parameter. The dashed line is the fit with $(\alpha_P - 1) = 0.53$ and the scale factor K as a free parameter.

larger than unity indicate that higher order corrections are important. NLO calculations are in progress [14,9,13] which agree better with the experimental results.

Acknowledgements

I would like to thank C. H. Lin of his collaboration. This work is supported by the Swiss National Science Foundation.

References

1. PLUTO Coll., C. Berger *et al.*, Phys. Lett. **B 142** (1984) 119; TPC/2γ Coll., D. Bintinger *et al.*, Phys. Rev. Lett. **54** (1985) 763; MD-1 Coll., S.E. Baru *et al.*, Z. Phys. **C 53** (1992) 219; TOPAZ Coll., R. Enomoto *et al.*, Phys. Lett. **B 368** (1996) 299

2. E.A. Kuraev, L.N. Lipatov and V.S. Fadin, Sov. Phys. JETP **45** (1977) 199; Ya.Ya. Balitski and L.N. Lipatov, Sov. J. Nucl. Phys. **28** (1978) 822

3. S.J. Brodsky, F. Hautmann and D.E. Soper, Phys. Rev. **D 56** (1997) 6957

4. J. Bartels, A. De Roeck and H. Lotter, Phys. Lett. **B 389** (1996) 742; J. Bartels, A. De Roeck, C. Ewerz and H. Lotter, hep-ph/9710500

5. J. Bartels, C. Ewerz, R. Staritzbichler, hep-ph/0004029

6. L3 Collab., M. Acciarri *et al.*, Phys. Lett. **B 453** (1999) 333

7. V. S. Fadin and L. N. Lipatov, Phys. Lett. **B 429** (1998) 127; G. Camici and M. Ciafaloni, Phys. Lett. **B 430** (1998) 349

8. V. S. Fadin and L. N. Lipatov, Proc. Theory Institute on Deep Inelastic Diffraction, ANL, Argonne, September 14 - 16, 1998; C. R. Schmidt, Phys. Rev. **D 60** (1999) 074003; J. R. Forshaw, D. A. Ross and A. Sabio Vera, Phys. Lett. **B 455** (1999) 273; S.J. Brodsky *et al.*, JETP Lett. **70** (1999) 15, hep-ph/99101229

9. G. Salam, JHEP **9807** (1998) 019

10. M. Ciafaloni *et al.*, Phys. Rev. **D 60** (1999) 114036; M. Ciafaloni and D. Colferai, Phys. Lett. **B 452** (1999) 372

11. R. S. Thorne, Phys. Rev. **D 60** (1999) 054031

12. G. Altarelli, R. D. Ball and S. Forte, hep-ph/0001157

13. V. T. Kim, L. N. Lipatov and G. B. Pivovarov, hep-ph/9911228 and hep-ph/9911242; V. Kim, private communication

14. N.N. Nikolaev, J. Speth and V.R. Zoller, hep-ph/0001120

15. PHOJET version 1.05c is used, R. Engel, Z. Phys. **C 66** (1995) 203; R. Engel and J. Ranft, Phys. Rev. **D 54** (1996) 4244

16. J.A.M. Vermaseren, Nucl. Phys. **B 229** (1983) 347

17. Chih-Hsun Lin and M.Wadhwa, L3 Preprint 2568, June 2000

HEAVY QUARK PRODUCTION IN γγ COLLISIONS

A. BÖHRER

Fachbereich Physik, Universität Siegen, D-57068 Siegen, Germany
E-mail: armin.boehrer@cern.ch

New results on inclusive heavy quark production in γγ collisions are presented. Charm and bottom production are investigated at LEP II energies by the experiments ALEPH, DELPHI, L3, and OPAL. The total and differential cross sections for charm quarks are measured. The contributions from the direct and single-resolved processes are separated and their fractions quantified. More detailed studies, such as the dependence of the cross section on the two-photon centre-of-mass energy and the charm structure function $F_{\gamma,c}^2$, are reported. The inclusive bottom cross section is presented. Measurements are compared to next-to-leading order calculations.

1 Introduction

The production of heavy flavour in two-photon collisions is dominated by two processes, the direct and the single-resolved process. Both contribute in equal shares to heavy flavour final states at LEP II energies. The large quark mass allows reliable perturbative calculations for the direct contribution. The single-resolved one in addition depends on the gluon density of the photon.

In this article the new measurements[1] from the LEP experiments on heavy flavour production and the comparison with next-to-leading order calculations[2] are summarized. Only the main results are given. For details the reader is referred to the original papers.

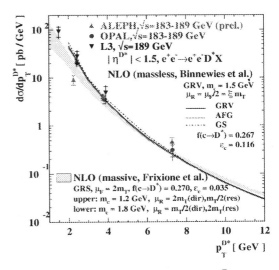

Figure 1. Differential cross section $d\sigma/dp_T^{D*}$ versus the transverse momentum p_T^{D*} of the D^{*+}.

2 Heavy Quark Identification

A clear, non-ambiguous signal for charm quark production is the presence of a D^{*+}. This gold-plated signature has been exploited by all four LEP collaborations. Both the production probability $\mathcal{P}(c \to D^{*+})$ and the branching ratio to $D^0\pi^+$ as well as the D^0 branching ratios are reasonably well-known.

The leptonic decay has been used by ALEPH ($\mu^\pm$ for charm) and L3 ($\mu^\pm$ and $e^\pm$ both for charm and bottom). Though these analyses rely on the momentum spectrum as theoretical input, the statistics is substantially increased.

3 Differential Charm Cross Section

The good detector resolution and with the low kinetic energy available, the D^{*+} method allows for a measurement of the differential cross section in transverse momentum (see Figure 1) and pseudorapidity. The three experiments, ALEPH, L3, OPAL, agree among themselves and with the theoretical expectation, that the distribution in pseudorapidity is about constant. In transverse momentum the massless approach follows the OPAL measurements, while the massive calculation is closer to the ALEPH data.

418

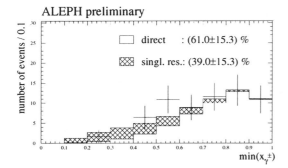

Figure 2. Differential cross section $d\sigma/dp_{\mathrm{T}}^{D*}$ versus the transverse momentum p_{T}^{D*} of the D^{*+}.

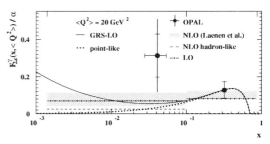

Figure 3. Charm structure function obtained with single-tagged events.

4 Charmed Particle Production

Charmed particles other than D^{*+} have only been measured by DELPHI. From the extracted signal and the efficiencies, the determined charm cross section for $D^{\pm}$, D^0, and Λ_c agree with the expectation, which include both direct and resolved processes. They are also within the errors consistent with the $(2J+1)$-relation $\sigma_{D*+} = 3 \cdot \sigma_{D^0} = 3 \cdot \sigma_{D^+}$.

5 Fraction of Direct and Single-Resolved Contributions

Direct and single-resolved contribution can be separated using the fact that in the resolved one the remnant jet carries away a part of the invariant mass available in the $\gamma\gamma$ collision. Two variables have been used: 1) $x_{\mathrm{T}} = p_{\mathrm{T}}^{D*}/W_{\mathrm{vis}}$, which is the ratio of p_{T}^{D*}, a measure for the invariant mass of the $c\bar{c}$ system, and the visible invariant mass W_{vis}, a measure for the invariant mass of the $\gamma\gamma$ system; 2) $x_\gamma^{\min}$ (see Figure 2), the minimum of $x_\gamma^{\pm} = \sum_{\mathrm{jets}}(E \pm p_z)/\sum_{\mathrm{part}}(E \pm p_z)$, a measure for the fraction of particles, which do not escape in the remnant jet.

ALEPH measures a relative contribution $r_{\mathrm{dir}} : r_{\mathrm{res}} = 62 : 38$ in their acceptance range in agreement with the NLO by Frixione (70 : 30). OPAL obtains 51 : 49.

6 Charm Cross Section as function of $W_{\gamma\gamma}$

The L3 collaboration has measured the charm cross section as function of the two-photon centre-of-mass energy. The charm-flavoured quarks are identified by their semi-leptonic decays to electrons. A parameterization of the form $\sigma_{\mathrm{tot}} = As^\varepsilon + Bs^{-\eta}$ (Pomeron + Reggeon) describes the data well. The PYTHIA Monte Carlo clearly fails, predicting only 66% of the total cross section. This may be partially attributed to next-to-leading order corrections, which are not included in PYTHIA. The Pomeron slope fitted from the data is steeper than the rise observed in $\sigma(\gamma\gamma \to q\bar{q}X)$.

7 Charm Structure Function $F_{\gamma,\mathrm{c}}^2$

When one of the scattered beam particles is detected, the event can be used to determine the charm structure function $F_{\gamma,\mathrm{c}}^2$. With 30 such single-tagged events with a D^{*+} meson, the OPAL collaboration performed a first measurement. (See the article of S. Soeldner-Rembold in this proceedings for a general overview on photon structure functions.) The HERWIG and the Vermaseren Monte Carlo, which nicely describe the data, are used for unfolding. The result is displayed in Figure 3. The comparison with the calculations shows that a point-like contribution is not sufficient to describe the data. A hadron-like part is needed. The data even exceed the models,

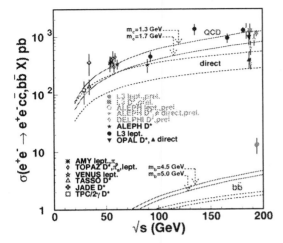

Figure 4. Charm and bottom quark production in $\gamma\gamma$ collisions as a function of the e^+e^- centre-of-mass energy. The predictions calculated in NLO are also given as solid lines. The dashed lines represent the contribution from the direct process only.

though the measurement errors are still too large to be conclusive.

8 Bottom Quark Production

Bottom production is measured by the L3 collaboration using the fact, that the momentum as well as the transverse momentum of leptons with respect to the closest jet is higher for muons and electrons from bottom than for background, which is mainly charm. Events with bottom quarks are selected with an efficiency of 1.2% and 1.0% for muons and electrons, resp. The purity is about 50%. The cross section is measured to be about three times the prediction from NLO calculations.

9 Inclusive Charm and Bottom Cross Section

The total cross section measurements are summarized in Figure 4. The results are compared to NLO calculations. Previously published results from LEP and those from lower energy colliders are also given.

10 Summary

The four LEP experiments have provided good measurements of the heavy quark production in $\gamma\gamma$ collisions. The inclusive measurement of charm is in agreement with QCD prediction with clear evidence for the gluon content in the photon. Next-to-leading order contributions seem important. The agreement among the measurements of the experiments is fair; comparing the various analysis techniques, D-meson versus lepton, the latter tends to give higher cross sections. The charm studies are at a transition to precision measurements: detailed investigations such as direct/single-resolved, $\sigma(W_{\gamma\gamma})$, $F_{\gamma,c}^2$ are performed. The bottom production cross section is predicted too low.

Acknowledgements

I thank S. Soeldner-Rembold and V. Andreev, who provided Figures 1 and 4.

References

1. Contributed papers on heavy flavour:
   \# 109, OPAL: CERN-EP/99-157;
   \# 110, OPAL: CERN-EP/99-157;
   \# 268, ALEPH Note 2000-070;
   \# 270, ALEPH Note 2000-031;
   \# 582, L3 Note 2565;
   \# 584, L3 Note 2548;
   \# 586, L3: CERN-EP/99-106;
   \# 765, DELPHI Note 2000-064.
2. M. Drees, M. Krämer, J. Zunft and P.M. Zerwas, *Phys. Lett.* B **306**, 371 (1993);
   S. Frixione, M. Krämer and E. Laenen, *hep-ph/9908483*, subm. *Nucl. Phys.* B (1999);
   J. Binnewies, B.A. Kniehl and G. Kramer, *Phys. Rev.* D **53**, 6110 (1996) and D **58**, 14014 (1998).

E155X MEASUREMENTS OF THE STRUCTURE FUNCTIONS g_2^p AND g_2^n

S. TRENTALANGE

Physics Department, UCLA, Los Angeles, CA 90095 USA
E-mail: trent@physics.ucla.edu

We report on measurements of the transverse spin structure functions for the proton and neutron (deuteron) using deep inelastic electron scattering on transversely polarized solid targets. Data cover the range 1 $GeV^2 < Q^2 < 20$ GeV^2 and $0.02 < x < 0.9$. Errors on these data are a factor of three smaller than previous results, allowing meaningful comparison of proposed models and sum rules.

1 Spin Structure Functions

Experiment E155x was designed to measure the spin structure functions $g_2^{p,n}(x, Q^2)$. This experiment has now produced the first statistically significant measurements of this quantity, testing various models and sum rules. Indeed, until it becomes possible to measure polarized Drell-Yan asymmetries, this experiment may be the only look at twist-3 contributions to the structure functions.

The polarized deep inelastic scattering (DIS) asymmetries $A_\parallel$ and $A_\perp$ for a longitudinally polarized electron beam scattering from a longitudinally or transversely polarized target are given in terms of the structure functions $g_1^{p,n}(x, Q^2)$ and $g_2^{p,n}(x, Q^2)$ by the expressions:

$$A_\parallel = \frac{\sigma^{\uparrow\downarrow} - \sigma^{\uparrow\uparrow}}{\sigma^{\uparrow\downarrow} + \sigma^{\uparrow\uparrow}}$$

$$= f_k \left[g_1[E + E'\cos(\theta)] - \frac{Q^2}{\nu}g_2 \right]$$

$$A_\perp = \frac{\sigma^{\downarrow\leftarrow} - \sigma^{\uparrow\leftarrow}}{\sigma^{\downarrow\leftarrow} + \sigma^{\uparrow\leftarrow}}$$

$$= f_k E' \sin(\theta) \left[g_1 + \frac{2E}{\nu}g_2 \right]$$

In these expressions, θ is the laboratory scattering angle of the electron, Q^2 is the 4-momentum transfer squared, ν is the energy transfer, and $x = Q^2/2M\nu$ is the Bjorken scaling variable. F_i and g_i ($i = 1, 2$) are the unpolarized and polarized structure functions of the nucleon, respectively and f_k is a function of all these quantities. From detailed consideration of the kinematic factors, it may be seen that $A_\parallel$ is mainly sensitive to g_1 while $A_\perp$ is sensitive to both g_1 and g_2. We use the measured values of $A_\perp$ from the present experiment and fits to the E155 measurements of g_1 to extract g_2.

In the naive Quark Parton Model (QPM) g_1 is related to the quark spin distribution functions:

$$g_1(x, Q^2) = \frac{1}{2} \sum_i e_i^2 (q_i^\uparrow(x, Q^2) - q_i^\downarrow(x, Q^2))$$

The index i is over the quark and antiquark flavors and e_i is the corresponding charge.

Unlike $g_1(x, Q^2)$, there is no simple interpretation for $g_2(x, Q^2)$ in the QPM. However, the functions g_2 and g_1 are related to each other at the twist-2 level by a series of sum rules due to Wandzura and Wilczek [1]

$$\int_0^1 dx\, x^{J-1} \left[\frac{J-1}{J} g_1(x) + g_2(x) \right] = 0.$$

Inverting this gives:

$$g_2^{WW}(x) = -g_1(x) + \int_x^1 \frac{g_1(y)}{y}\, dy.$$

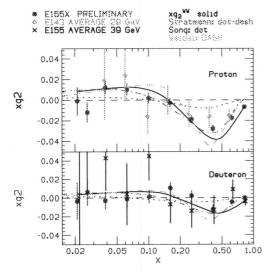

Figure 1: E155x preliminary results for the transverse spin structure functions $g_2^{p,d}(x,Q^2)$.

2 Results from E155x

Figure 1 presents preliminary findings for $g_2^{p,d}(x,Q^2)$. These results are in agreement with the predictions of the simple twist-2 model of Wandzura and Wilczek [1]. Also plotted are calculations based on bag model wavefunctions [2, 3] and chiral solitons [4].

3 Sum Rules

The transverse spin structure function $g_2(x)$ obeys a number of sum rules. Firstly, the Burkhardt-Cottingham sum rule, originally derived from the large Q^2 behavior of the virtual Compton amplitude [6], but also obtained from rotational invariance:

$$\int_0^1 g_2(x)\,dx = 0 \ .$$

The experimental results for this integral obtained by E155x are -0.004 ±0.014 (proton) and -0.006 ± 0.019 (deuteron). Both of these results are consistent with zero.

Another sum rule, due to Efremov-Teryaev-Leader [7], is

$$\int_0^1 [\,g_1^V(x) + 2g_2^V(x)\,]\,dx = 0.$$

This sum rule is exact, but pertains only to the valence quark contribution and therefore cannot be used directly. However, if we assume the sea quark contributions of the proton and neutron are similar, then we can invoke a kind of generalized Bjorken sum rule:

$$\int_0^1 [\,g_1^p(x) + 2g_2^p(x) - \\ - g_1^n(x) - 2g_2^n(x)\,]\,dx = 0 \ .$$

The E155x and E143 experimental data have been averaged and evaluated at $Q^2 = 3\ \mathrm{GeV}^2$ to give: -0.0059 ± 0.0081.

A more exact expression for g_2 was given by Cortes, Pire and Ralston [5]

$$g_2(x) = g_2^{WW}(x) - \\ - \int_x^1 \frac{dy}{y} \frac{\partial}{\partial y}\left(\frac{m_q}{M}h_T(y) + \xi(y)\right)$$

where $h_T(x)$ is a twist-2 contribution which is suppressed by the factor m_q/M and $\xi(x)$ is a twist-3 contribution containing multi-parton correlations. If we then write the (mainly) twist-3 contribution as

$$g_2(x) = g_2^{WW}(x) + \bar{g}_2(x)$$

We can define the twist-3 matrix element d_2 as

$$d_2 = 3\int_0^1 x^2 \bar{g}_2\,dx$$

This quantity is important because it is a correction to the Bjorken sum rule, the Ellis-Jaffe sum rules, and is a separate test of QCD calculations as well. Physically, it is a measure of the mean color magnetic field in the direction of the transverse spin.

In fig. 2 we give our experimental measurement of this quantity compared with a

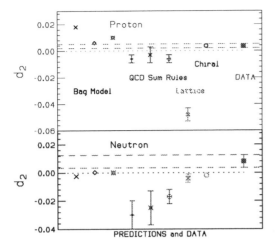

Figure 2: Twist-3 contribution d_2 experiment compared to theoretical calculations.

number of theoretical estimates of different types. The reasons for agreement and disagreement with our data are currently being discussed within the group, but arise from apparently different causes. For example, the strong disagreement of the lattice calculations can be traced to the use of perturbatively normalized operators used in the calculations. A more recent calculation (not shown) reportedly gives much better agreement for proton and neutron with experiment, albeit with larger errors.

4 Conclusions

The E155x collaboration has presented preliminary measurements of $g_2(x, Q^2)$ of the proton and neutron which are a factor of 3 more statistically significant than previous data, with systematic errors which are small compared to the statistical. These data indicate the the twist-3 matrix element d_2 is small and show good agreement with the twist-2 Wandzura-Wilczek model, the bag model of Stratmann *et al* and the chiral-soliton model of Weigel *et al*. The Burkhardt-Cottingham sum rule appears to be valid, and the data is consistent with

a generalization of the Efremov-Teryaev-Leader sum rule.

References

[1] S. Wandzura and F. Wilczek, Phys. Lett. **B72** 195 (1977).

[2] X. Song, Phys. Rev. **D54** 1955 (1996).

[3] M. Stratmann, Z. Phys. **C60** 763 (1993).

[4] H Weigel, L. Gamberg, and H. Reinhart, Phys. Rev. **D55** 6910 (1997).

[5] J.L Cortes, B. Pire and J.P. Ralston, Z. Phys. **C55** 409 (1992).

[6] H. Burkhardt and W.N. Cottingham, Ann. Phys. **56** 453 (1970).

[7] A.V. Efremov, O.V. Teryaev and E. Leader, Phys. Rev. **D55** 4307 (1997).

RECENT RESULTS ON BFKL PHYSICS

LYNNE H. ORR

Department of Physics and Astronomy, University of Rochester,
Rochester, NY 14627-0171, USA
E-mail: orr@pas.rochester.edu

W.J. STIRLING

Departments of Physics and Mathematical Sciences, University of Durham
Durham DH1 3LE, England
E-mail: W.J.Stirling@durham.ac.uk

Virtual photon scattering in e^+e^- collisions can result in events with the electron-positron pair produced at large rapidity separation in association with hadrons. The BFKL equation resums large logarithms that dominate the cross section for this process. After a brief overview of analytic BFKL resummation and its experimental status, we report on a Monte Carlo method for solving the BFKL equation that allows kinematic constraints to be taken into account. We discuss results for e^+e^- collisions using both fixed-order QCD and the BFKL approach. We conclude with some brief comments on the status of NLL calculations.

1 Introduction

Many processes in QCD can be described by a fixed order expansion in the strong coupling constant α_S. In some kinematic regimes, however, each power of α_S gets multiplied by a large logarithm (of some ratio of relevant scales), and fixed-order calculations must give way to leading-log calculations in which such terms are resummed. The BFKL equation [1] resums these large logarithms when they arise from multiple (real and virtual) gluon emissions. In the BFKL regime, the transverse momenta of the contributing gluons are comparable but they are strongly ordered in rapidity.

The BFKL equation can be solved analytically, and its solutions usually result in (parton-level) cross sections that increase as the power λ, where $\lambda = 4C_A \ln 2\, \alpha_s/\pi \approx 0.5$.[a] For example, in dijet production at large rapidity separation Δ in hadron colliders,[2] BFKL predicts for the parton-level cross section $\hat{\sigma} \, e^{\lambda\Delta}$. In virtual photon scattering, for example in e^+e^- collisions, where the

electron-positron pair at emerge with a large rapidity separation and hadronic activity in between,[3] BFKL predicts $\sigma_{\gamma^*\gamma^*} (W^2/Q^2)^\lambda$. W^2 is the invariant mass of the hadronic system (equivalently, the photon-photon center-of-mass energy) and Q^2 is the invariant mass of either photon.

2 Experimental Status and Improved Predictions

The experimental status of BFKL is ambiguous at best, with existing results being far from definitive. The data tend to lie between the predictions of fixed-order QCD and analytic solutions to the BFKL equation. This happens, for example, for the azimuthal decorrelation in dijet production at the Fermilab Tevatron[4] and for the virtual photon cross section at LEP.[5] Similar results are found in ep collissions at HERA.[b]

It is not so surprising that analytic BFKL predicts stronger effects than seen in data.

---

[a] λ is also known as $\alpha_P - 1$.

[b] One exception is the ratio of the dijet production cross sections at center of mass energies 630 GeV and 1800 GeV at the Tevatron, where the measured ratio lies above *all* predictions.[6]

Analytic BFKL solutions implicitly contain sums over arbitrary numbers of gluons with arbitrary energies, but the kinematics are leading-order only. As a result there is no kinematic cost to emit gluons, and energy and momentum are not conserved, and BFKL effects are artifically enhanced.

This situation can be remedied by a Monte Carlo implementation of solutions to the BFKL equation.[7,8] In such an implementation the BFKL equation is solved by iteration, making the sum over gluons explicit. Then kinematic constraints can be implemented directly, and conservation of energy and momentum is restored. This tends to lead to a suppression of BFKL-type effects. The Monte Carlo approach has been applied to dijet production at hadron colliders,[7,8,9] leading to better (though still not perfect) agreement with the dijet azimuthal decorrelation data.[7] Applications to forward jet production at HERA and to virtual photon scattering in e^+e^- collisions are underway; an update on the latter appears in the next section.

3 $\gamma^*\gamma^*$ Scattering: A Closer Look

BFKL effects can arise in e^+e^- collisions via the scattering of virtual photons emitted from the initial e^+ and e^-. The scattered electron and positron appear in the forward and backward regions ("double-tagged" events) with hadrons in between. With total center-of-mass energy s, photon virtuality $-Q^2$, and photon-photon invariant mass (= invariant mass of the final hadronic system) W^2, BFKL effects are expected in the kinematic regime where W^2 is large and

$$s >> Q^2 >> \Lambda_{QCD}^2.$$

At fixed order in QCD, the dominant process is four-quark production with t-channel gluon exchange (each photon couples to a quark box; the quark boxes are connected via the gluon). The corresponding BFKL contribution arises from diagrams with a gluon ladder

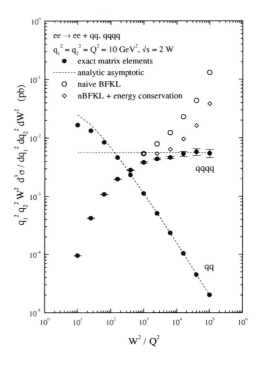

Figure 1. Exact (closed data points) and analytic asymptotic (dashed line) $e^+e^- \to e^+e^- q\bar{q}$ and $e^+e^- \to e^+e^- q\bar{q}q\bar{q}$ cross sections versus W^2/Q^2 at fixed $W^2/s = 1/4$. Also shown: analytic BFKL without (open circles) and with (open diamonds) energy conservation imposed.

attached to the t-channel gluon.

The relative contributions of fixed-order QCD and BFKL are most easily understood by looking at

$$W^2 Q_1^2 Q_2^2 \frac{d^3\sigma}{dW^2 dQ_1^2 dQ_2^2}$$

as a function of W^2/Q^2 for fixed $\sqrt{s}/W$. The asymptotic regime then corresponds to large W^2/Q^2. This quantity is shown in Figure 1 for $Q_1^2 = Q_2^2 = Q^2 = 10\ \mathrm{GeV}^2$ and $\sqrt{s} = 2W$. The solid points are the QCD calculations of two-quark ('qq') and four-quark production ('qqqq'); we see that the latter dominates for large W^2/Q^2 and approaches a constant asymptotic value. In contrast, the analytic

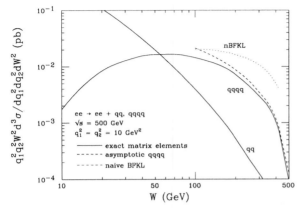

Figure 2. Exact (solid lines) and analytic asymptotic (dashed line) $e^+e^- \to e^+e^-q\bar{q}$ and $e^+e^- \to e^+e^-q\bar{q}q\bar{q}$ cross sections versus W^2/Q^2 at fixed $\sqrt{s} = 500$ GeV. Also shown: analytic BFKL (dotted line).

BFKL result, shown with open circles, rises well above that of fixed-order QCD. The diamonds show analytic BFKL with energy conservation imposed, but not exact kinematics; it can be interpreted as an upper limit for the Monte Carlo prediction, which is in progress.

It is important to note in Figure 1 that although BFKL makes a definite leading-order prediction for the behavior of the cross section as a function of W^2/Q^2, the origin in W^2/Q^2 (i.e., where BFKL meets asymptotic QCD) is *not* determined in leading order. We have chosen $W^2/Q^2 = 10^3$ GeV^2 as a reasonable value where the QCD behavior is sufficiently asymptotic for BFKL to become relevant, but another choice might be just as reasonable. Only when higher order corrections are computed can the BFKL prediction be considered unique.

From an experimental point of view, the cross section at fixed $\sqrt{s}$ is more directly relevant. Figure 2 shows $W^2Q_1^2Q_2^2 \frac{d^3\sigma}{dW^2dQ_1^2dQ_2^2}$ for a linear collider energy $\sqrt{s} = 500$ GeV. The solid lines show the exact fixed-order QCD prediction. The dashed line is the asymptotic four-quark production cross sec-

tion, and the dotted line is the analytic BFKL prediction. Now we see that all of the curves fall off at large W, but the BFKL cross section lies well above the others.

3.1 $\gamma^*\gamma^*$ Scattering at LEP

The L3 collaboration at LEP have measured the $\gamma^*\gamma^*$ cross section by dividing the double-tagged e^+e^- cross section by the $\gamma^*\gamma^*$ luminosity.[5] They present their results, for $\sqrt{s} = 183$ and 189–202 GeV, with $Q^2 = 14$ GeV^2 and $Q^2 = 15$ GeV^2, respectively, as a function of $y = \ln(W^2/Q^2)$. In this variable the asymptotic QCD four-quark cross section is flat, and the analytic QCD cross section rises, similarly to Figure 1. The data lie between the two predictions, and the higher statistics data at the higher energies show a clear rise in the data, though not as steep as predicted by analytic BFKL.

We expect that the BFKL Monte Carlo prediction (in progress) will be closer to the data. But one can also ask whether the asymptotic QCD limit for four-quark production is appropriate here. We compare the exact and asymptotic QCD curves at the LEP energy $\sqrt{s} = 183$ GeV in Figure 3 (note that this is the undivided e^+e^- cross section that includes the photon luminosity). The values of W corresponding to the LEP measurements range between about 15 and 90 GeV. Comparing four-quark predictions, we see that the exact curve is not close enough to the asymptotic in this region for the asymptotic QCD limit to be appropriate. Furthermore, the ratio of exact to asymptotic results — which is proportional to the $\gamma^*\gamma^*$ cross section — *rises* in this region. The QCD prediction is not flat at all.[c] Until the fixed-order QCD and BFKL Monte Carlo predictions are sorted out, it is not clear what we can conclude from the data.

---

[c]This does not automatically imply that fixed-order QCD describes the data, because there are unresolved normalization issues involved.

426

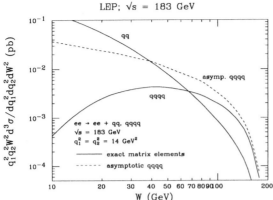

Figure 3. Exact (solid lines) and analytic asymptotic (dashed line) $e^+e^- \to e^+e^-q\bar{q}$ and $e^+e^- \to e^+e^-q\bar{q}q\bar{q}$ cross sections versus W^2/Q^2 at fixed $\sqrt{s} = 183$ GeV.

4 Status of NLL Corrections

It is apparent that, although it is not yet clear whether BFKL is necessary to describe the data in hand, leading-order analytic BFKL is not sufficient. We need BFKL at next-to-leading order (strictly speaking, next-to-leading log order). This has been accomplished after 10 years of heroic efforts by Fadin, Lipatov and many others (see [10] for a review, complete references, and more details about what follows). The bad news is that the solutions appear to be large, unstable, and capable of giving negative cross sections. The good news is that there is much progress in understanding these problems, which can mostly be traced to the fact that at NLL gluons can be close together in rapidity, leading to collinear divergences. Several methods for solving this problem are summarized in [10], and the NLL BFKL corrections appear to be coming under control.

5 Conclusions

In summary, BFKL physics is a complicated business. Tests are being performed in a variety of present experiments (Tevatron, HERA, LEP) and there is potential for the future

as well (LHC, LC). Unfortunately, comparisons between theory and experiment are not straightforward; leading order BFKL is insufficient, and subleading corrections such as kinematic constraints can be very important. A worst-case scenario which is not ruled out may be that we cannot reach sufficiently asymptotic regions in experiments to see unambiguous BFKL effects. However, reports of the demise of BFKL physics due to instability of the next-to-leading-order corrections are greatly exaggerated, and the source of the large corrections is understood and they are being brought under control. In summary, the jury on BFKL physics is still out, but there continues to be much progress.

References

1. L.N. Lipatov, Sov. J. Nucl. Phys. **23** (1976) 338; E.A. Kuraev, L.N. Lipatov and V.S. Fadin, Sov. Phys. JETP **45** (1977) 199; Ya.Ya. Balitsky and L.N. Lipatov, Sov. J. Nucl. Phys. **28** (1978) 822.
2. A.H. Mueller and H. Navelet, Nucl. Phys. **B282**, 727 (1987).
3. S.J. Brodsky, F. Hautmann and D.E. Soper, Phys. Rev. **D56**, 6957 (1997).
4. S. Abachi *et al.* [D0 Collaboration], Phys. Rev. Lett. **77**, 595 (1996).
5. M. Wadhwa for the L3 collaboration, these proceedings.
6. B. Pope for the D0 collaboration, these proceedings.
7. L.H. Orr and W.J. Stirling, Phys. Rev. **D56** (1997) 5875.
8. C.R. Schmidt, Phys. Rev. Lett. **78** (1997) 4531.
9. L.H. Orr and W.J. Stirling, Phys. Lett. **B429** (1998) 135; Phys. Lett. **B436** (1998) 372.
10. G. P. Salam, Acta Phys. Polon. **B30**, 3679 (1999); hep-ph/0005304, and references therein.

F_2^P AT LOW Q^2 AND THE TOTAL γP CROSS SECTION AT HERA

S. LEVONIAN

DESY, Hamburg, Germany
E-mail: levonian@mail.desy.de

New precise measurements of the neutral current cross section at HERA are presented in different Q^2 regimes. In the photoproduction limit the total γp cross section is found to be in agreement with the universal soft Pomeron prediction. At medium scales $1.5 \leq Q^2 \leq 150$ GeV^2 both the F_2 and the F_L structure functions are determined. These data are well described by NLO pQCD. The gluon density in the proton has been extracted from the scaling violation of F_2 at low x. The transition between the two regimes occurs at around $Q^2 \simeq 1$ GeV^2.

1 Introduction

Half a century of the extensive studies of strong interactions resulted in two different approaches to the problem: Reggeon field theory (RFT)[1] and QCD. RFT, the S-matrix theory based on the most general physical principles, has proven to be very successful in describing soft peripheral processes ($t/s \ll 1$) at high energies ($s \to \infty$). There is however no microscopic picture of underlying dynamics in it. On the other hand QCD, being *the* theory of strong interactions, has technical problems in the non-perturbative regime. pQCD is applicable to hard processes only. Lepton-proton scattering experiments at HERA have the unique possibility to contribute to the successful merging of those two approaches by performing the scan over the large available range of the photon virtuality, Q^2, and studying the interplay of short and long distance physics.

In this talk the following questions are discussed, using high statistics Neutral Current (NC) data recorded in the years 1994 to 1997 with the H1 and ZEUS apparatus in e^+p collisions at $\sqrt{s} = 300$ GeV:

1. How far up in Q^2 can one get with Regge theory starting from photoproduction?

2. How far down in Q^2 can one go with pQCD?

3. Where in Q^2 is the transition region?

2 Photoproduction limit

The total γp cross section is measured at HERA by detecting scattered positrons under very small angles $\theta < 5$ mrad with respect to the incoming e^+ beam, in a special calorimeter installed in the tunnel. This ensures $Q^2 < 0.02$ GeV^2 (with the average value of 10^{-4} GeV^2) and justifies the use of pure transverse photon flux in the equivalent photon approximation relating the ep cross section with $\sigma_{\text{tot}}^{\gamma p}$. Major systematics is approximately equally shared between the positron detector acceptance uncertainty and the precision of the hadronic final state modelling.

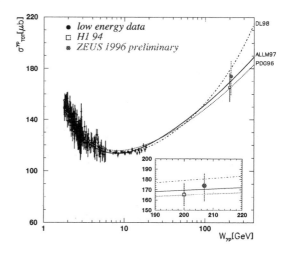

Figure 1. The total photon-proton cross section as a function of centre-of-mass energy. The insert magnifies HERA results.

New measurement of the total γp cross section at average centre-of-mass energy $W_{\gamma p} = 207$ GeV has been presented by the ZEUS collaboration [2]. Their result of $\sigma_{tot}^{\gamma p} = 172 \pm 1(stat.)_{-15}^{+13}(syst.)$ μb is shown in Fig. 1 together with the published H1 measurement [3] and the low-energy data. Also shown are three Regge motivated parameterizations [4]. In two of them the high energy behavior of $\sigma_{tot}^{\gamma p}$ is driven by the universal soft Pomeron, while the DL98 parameterization has additionally a hard Pomeron term.

3 NC cross section in DIS regime

New measurements of the deep inelastic NC cross section are available from both the H1 [5] and ZEUS [6] collaborations. High statistics data span the kinematic region of $1.5 \leq Q^2 \leq 150$ GeV^2 and Bjorken-x values $0.00003 \leq x \leq 0.2$. Using improved detection capabilities and increased HERA luminosity a high precision of typically 3% is achieved. This allowed, for the first time, to perform NLO QCD analysis of the inclusive cross section measurements using H1 data alone [5]. The gluon density at low x has been determined from the large positive scaling violation of F_2 (see Fig. 2). It was found, that NLO QCD fit describes all low x data well. Some dependence is observed however of the fit parameters on the value of Q_{min}^2 – the minimum Q^2 of the H1 data used in the fit. This is directly reflected in the steepness of the gluon density, as seen in Fig. 2,b.

In order to reduce such 'flexibility' of the QCD fit additional constraints may be imposed. Potentially powerful is the longitudinal structure function F_L, which contains independent information about gluon distribution. $F_L(x, Q^2)$ as determined using two different methods [5] is shown in Fig. 3 together with the fixed target results. The increase of $F_L(x, Q^2)$ towards low x reflects the rise of the gluon momentum distribution and is consistently described by NLO QCD fit.

4 Transition region: closing the gap

To summarize, HERA has verified that RFT works in photoproduction ($\sigma_{tot}^{\gamma p}$). New precision results also demonstrate that in DIS regime pQCD describes inclusive NC cross section (F_2, F_L). But where is the transition between the two? And how does it happen?

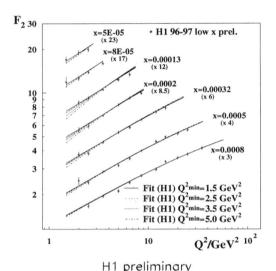

Figure 2. Effect of the Q_{min}^2 cut applied in DGLAP QCD fit to the H1 data on a) the structure function F_2 at low Q^2 and b) the gluon distribution at $Q^2 = 5$ GeV^2.

New low Q^2 ZEUS data [7] at $0.045 \leq Q^2 \leq 0.65$ GeV^2 almost completely closed the gap

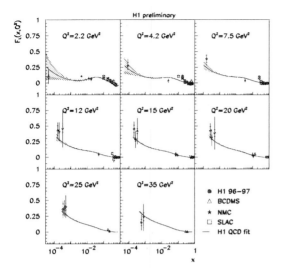

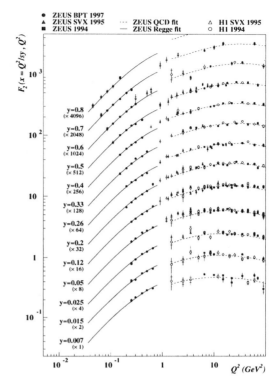

Figure 3. The longitudinal structure function $F_L(x, Q^2)$. The error bands are due to the experimental (inner) and model (outer) uncertainty of the F_L calculation using NLO QCD fit to the H1 data for $y < 0.35$ and $Q^2 > 3.5$ GeV^2.

Figure 4. Measured $F_2(Q^2)$ in bins of y.

between photoproduction and DIS. They are shown in Fig. 4 together with previous HERA measurements at higher Q^2. It is seen that low Q^2 points are described adequately by the Regge motivated fit (solid lines) while pQCD fits the data above 1.5 GeV^2 (dashed lines). The data exhibit a smooth transition at around $Q^2 \approx 1$ GeV^2 while the matching between the two theoretical fits is not perfect yet.

5 Conclusions

New precise HERA data have been used to study how the properties of strong interactions evolves with Q^2. In the photoproduction limit $\sigma_{\text{tot}}^{\gamma p}$ exhibits mild rise similar to that of hadron-hadron scattering. It is well described by the conventional Regge theory with universal soft Pomeron. Regge parameterization also describes the data in the low $Q^2 < 0.7$ GeV^2 region. In DIS regime NLO QCD is able to describe F_2 data all the way down to $Q^2 \simeq 1.5$ GeV^2. A smooth transition

from partonic to hadronic degrees of freedom occurs at around $Q^2 \simeq 1$ GeV^2. The details of the underlying dynamics is still a challenge for theory.

References

1. V.N. Gribov, *Sov. Phys. JETP* **26**, 414 (1968).
2. ZEUS Collab., Contributed Paper 411.
3. H1 Collab., S. Aid et al., *Z. Phys. C* **69**, 27 (1995).
4. R.M.Barnett et al., *Phys. Rev. D* **54**, 1 (1996); H. Abramowicz and A. Levy, *DESY* 97-251; A. Donnachie and P.V. Landshoff, *Phys.Lett.* **B437**, 408 (1998).
5. H1 Collab., Contributed Paper 415.
6. ZEUS Collab., Contributed Paper 412.
7. ZEUS Collab., Contributed Paper 294; *Phys. Lett. B* **487**, 53 (2000).

THE PROTON STRUCTURE FUNCTION F_2 AT LOW Q^2 AND LOW X

A. PELLEGRINO[‡]

Argonne National Laboratory, 9700 S.Cass Avenue, Argonne, IL 60439, USA
E-mail: antonio@mail.desy.de

The recent measurements of the proton structure function F_2 made by the H1 and ZEUS collaborations span six orders of magnitude both in the virtuality of the photon, Q^2, and in the Bjorken scaling variable, x. The data exhibit a clear transition from the (approximate) scaling behavior of F_2 to a much stronger Q^2-dependence at low Q^2. In order to further elucidate this transition, a study is made of the behavior of the derivatives, $\partial F_2/\partial \log Q^2$, evaluated for fixed x.

The recent measurements [1,2] of the proton structure function F_2 at HERA cover the range of Q^2 from 0.045 to 30000 GeV^2 and of x from 10^{-6} to 0.65. Figure 1 shows F_2 as a function of Q^2 for bins of fixed x; the value

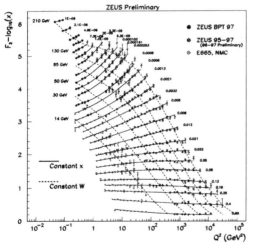

Figure 1. $F_2 - \log_{10} x$ as a function of Q^2 for bins of fixed x. The solid dots are data from the ZEUS collaboration. The open circles are results from the E665 and NMC collaborations

of F_2 in each bin of fixed x is offset by an additive factor of $(-\log_{10} x)$ to ensure that the vertical separation between the bins is proportional to x. The parameterization $A(x) + B(x)\log_{10}Q^2 + C(x)(\log_{10}Q^2)^2$ has been fit to the data; the quality of the fit is good, and the result is denoted by the solid line

[‡]On behalf of the H1 and ZEUS Collaborations.

in Fig.1. The constant $W = \sqrt{Q^2(1/x - 1)}$ points on the parameterizations are denoted by the dashed lines in the plot. In the absence of any x,Q^2 dependence in F_2, these lines would be essentially straight. Changes in the slope of the fixed-W lines, like the one observed for W above 85 GeV at $x \approx 10^{-4}$ and $Q^2 \approx 5$ GeV^2, reflect changes in the x,Q^2 dependence of F_2.

In a previous publication [3], fits to F_2 data using DGLAP equations have been shown to work well at $Q^2 > 1$ GeV^2, and a model combining Regge theory and the Vector Meson Dominance Model has been shown to work well at $Q^2 < 1$ GeV^2. The determination of the Q^2 and x at which the transition occurs is of great interest and may help in clarifying the nature of the transition. The study of the derivative of F_2, taken at fixed x, with respect to $\log Q^2$ can further elucidate this transition. The DGLAP evolution equations imply at low x the (leading order) relation:

$$\frac{\partial F_2}{\partial \log Q^2} \propto xG(x, Q^2), \qquad (1)$$

where $xG(x, Q^2)$ is the gluon momentum density. Figure 2 shows a recent determination of the logarithmic derivative of F_2 by the H1 Collaboration as a function of x at a constant W. At higher x (corresponding to higher Q^2 for fixed W), the derivatives fall with increasing x and tend to become independent of W at $x > 0.003$, in line with the expectation of Eq.(1), if $xG(x, Q^2)$ has the

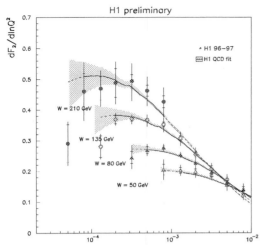

Figure 2. The derivatives of F_2 with respect to $\log Q^2$, evaluated at fixed x, shown as a function of x for fixed W.

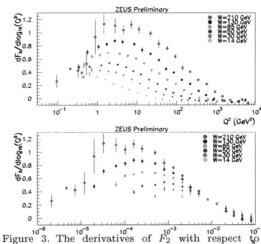

Figure 3. The derivatives of F_2 with respect to $\log Q^2$, evaluated at fixed x, shown as a function of Q^2 (upper panel) and x (lower panel) for fixed W.

form $x^{-\lambda}$ and only a slow dependence on Q^2. It should be noted that the data are successfully described by DGLAP fits down to low Q^2 values around 2-3 GeV^2.

On the other hand, at sufficiently low Q^2, F_2 must vanish as Q^2 from current conservation, which implies:

$$\frac{\partial F_2}{\partial \log Q^2} \propto Q^2 \sigma_0 . \qquad (2)$$

The plot of $\partial F_2/\partial \log Q^2$ at a constant W should show a transition from the behavior of Eq.(1) to that of Eq.(2), provided $xG(x,Q^2)$ has a weak Q^2 dependence and σ_0 has a weak x (i.e. energy) dependence. In Fig.3, the derivatives of F_2, evaluated by the ZEUS Collaboration, are shown at a constant W as a function of Q^2 and x, resp. in the upper and lower panel. While the data in the lower panel confirms the behavior expected from Eq.(1) at higher x and already observed in Fig.2, the data in the upper panel shows that at lower Q^2 the derivative falls as Q^2 decreases and tends to become independent of W at Q^2 of about 0.4 GeV^2, in line with the expectation of Eq.(2). The transition, for values of W above 85 GeV, happens at a relatively high Q^2 of 2-6 GeV^2 and correspondingly at x of $5 \cdot 10^{-4}$ to $3 \cdot 10^{-3}$.

In summary, the derivative of F_2 with respect to $\log Q^2$, $\partial F_2/\partial \log Q^2$, at fixed x, has been evaluated for HERA and fixed target data that spans six orders of magnitude in Q^2 and Bjorken x. The derivatives have been plotted at fixed values of the virtual photon-proton center-of-mass energy, W, and show a transition from the behavior expected from the DGLAP evolution equations to that expected from conservation of the electromagnetic current. While such a transition must necessarily take place, it is surprising to see that at low x, it takes place at a relatively high Q^2 between 2-6 GeV^2. Many calculations and models that go beyond the DGLAP formalism (see ref.[2] and references therein) qualitatively describe such a behavior.

References

1. H1 Collaboration, paper 294;
   ZEUS Collaboration, papers 412 and 415-416, submitted to this conference.
2. ZEUS Collaboration, J. Breitweg et. al., Phys. Lett. B487, (2000) 53.
3. ZEUS Collaboration, J. Breitweg et. al., Eur. Phys. J. C7, (1999) 609.

RECENT SPIN PHYSICS RESULTS FROM HERMES AND PLANS FOR RHIC

ANDREAS GUTE ON BEHALF OF THE HERMES COLLABORATION

Friedrich–Alexander–Universität Erlangen–Nürnberg

Physikalisches Institut, Erwin–Rommel–Str. 1, D–91058 Erlangen, Germany

(e-mail: andreas.gute@physik.uni-erlangen.de)

The HERMES experiment[1] at the DESY laboratory in Hamburg, Germany, uses a polarised internal gas target in combination with the longitudinally polarised 27.5 GeV positron beam of the HERA accelerator to study the spin structure of the nucleon via deep–inelastic lepton–nucleon scattering. Recent inclusive measurements of the spin structure function g_1 of the proton and the deuteron are presented.

Additionally HERMES focuses on semi–inclusive measurements. Results include the first observation of a single spin azimuthal asymmetry in pion production, the flavour separated quark polarisations and the first direct experimental estimation of the polarisation of gluons. Future plans from HERMES and RHIC are presented.

1 Introduction

Understanding the spin structure of the nucleon has been a challenge in hadron physics for more than twenty years. In general the nucleon spin can be expressed as a combination of three sources: the spin distribution of the quarks ($\Delta\Sigma$), the spin distribution of the gluons (ΔG) and orbital angular momenta of quarks and gluons (L_q, L_G):

$$s_z^N = \frac{1}{2} = \frac{1}{2}\Delta\Sigma + \Delta G + L_q + L_G. \quad (1)$$

NLO PQCD fits[2,3] of recent inclusive measurements show that only $\approx 30\%$ of the nucleon spin can be attributed to the quark spins and the gluon polarisation (ΔG) is only poorly constrained by the inclusive measurements.

The investigation of the nucleon spin structure can be extended through semi–inclusive measurements in which specific hadrons are observed in coincidence with the scattered lepton.

The goal of the HERMES experiment is to disentangle the different contributions to the nucleon spin using inclusive and semi–inclusive deep–inelastic scattering (DIS) measurements of polarised leptons from polarised atomic gas targets ($\vec{\mathrm{H}}$, $\vec{\mathrm{D}}$, $\overrightarrow{^3\mathrm{He}}$).

In the near future also other experiments (COMPASS, PHENIX, STAR) will start taking data to study the spin structure of the nucleon in more detail. The COMPASS experiment will also use deep–inelastic lepton–nucleon scattering whereas the RHIC SPIN PROGRAM is based on polarised proton–proton scattering.

2 Inclusive Measurements

2.1 The Spin Structure Function of the Proton

At centre–of–mass energies where weak contributions can be neglected, inclusive polarised DIS is characterised by two spin structure functions: $g_1(x, Q^2)$ and $g_2(x, Q^2)$. Here x is the Bjorken scaling variable and Q^2 is the negative squared four–momentum of the exchanged virtual photon. In leading order QCD the structure function g_1 is given by the charge weighted sum over the polarised quark (anti–quark) spin distributions Δq_f ($\Delta\bar{q}_f$). Experimentally the spin structure function g_1 is determined from a measurement of the cross section asymmetry $A_\parallel$. From this asymmetry the structure function ratio g_1^p/F_1^p can

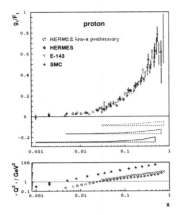

Figure 1. The structure function ratio g_1^p/F_1^p as a function of x. In the lower part the Q^2 range for the different experiments is plotted.

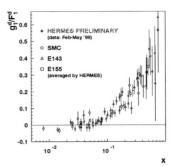

Figure 2. The structure function ratio g_1^d/F_1^d as a function of x.

be determined as

$$\frac{g_1^p}{F_1^p} = \frac{1}{1+\gamma^2} \left[\frac{A_\parallel}{D} + (\gamma - \eta)A_2 \right]. \quad (2)$$

Here D is the virtual photon depolarisation factor, γ and η are kinematical factors and A_2 is the transverse virtual photon asymmetry. The HERMES result[4] of the 1997 proton data together with results from SLAC–E143 and SMC is shown in Fig. 1. There is an excellent agreement between the three experiments. Despite the fact that the Q^2 range of SMC is about a factor of ten higher compared to HERMES and E143 the structure function ratio g_1^p/F_1^p is compatible within the statistical error. Therefore the Q^2 dependence of g_1^p/F_1^p is only weak.

2.2 The Spin Structure Function of the Deuteron

From the deuteron data HERMES accumulated in spring 1999 the structure function ratio g_1^d/F_1^d was extracted. The preliminary HERMES result shown in Fig. 2 is in nice agreement with the other experiments and the statistical errors are already comparable. But the used statistics in this plot is only a small fraction of the total HERMES $\vec{D}$ data. Therefore the statistical precision of

the HERMES result will be improved significantly after a full analysis of the 1998-2000 data.

3 Semi-Inclusive Measurements

3.1 Single–Spin Azimuthal Asymmetries

A complete description of the structure of the nucleon at leading twist requires three structure functions. Besides the already measured structure functions g_1 and F_1 the structure function h_1, called transversity, corresponds to the distribution $\delta q(x)$ of transverse quark spin in a transversely polarised nucleon. Transversity is as yet unmeasured because it belongs to the class of chiral–odd structure functions and is therefore not observable in inclusive DIS experiments. However it has been suggested that the needed sensitivity can be provided by the semi–inclusive production of pions with modest transverse momentum p_T [5], involving a chiral–odd fragmentation function.

HERMES has recently measured[6] the single–spin asymmetry for semi–inclusive pion production in DIS using an unpolarised beam and a longitudinally polarised proton target. A clear $\sin \Phi$ dependence of the single–spin asymmetry is observed for π^+. Here Φ is the azimuthal angle of the pion around the virtual photon direction, with respect to the lepton scattering plane. No such

434

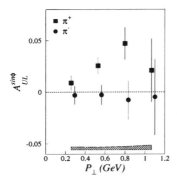

Figure 3. Target–spin analysing powers in the $\sin\Phi$ moment as a function of transverse momentum.

$\sin\Phi$ dependence is seen for π^- production.

The various contributions to the Φ dependent spin asymmetry are isolated by extracting moments of the cross section weighted by corresponding Φ dependent functions. The $\sin\Phi$ moment $A_{UL}^{\sin\Phi}$ is expected to provide sensitivity to the Collins fragmentation function in combination with different spin distribution functions. In Fig. 3 $A_{UL}^{\sin\Phi}$ averaged over x is plotted for π^+ and π^- as a function of p_T. The result is in good agreement with the theoretical expectation[5].

3.2 Quark Polarisation

Semi–inclusive measurements also allow to separate spin contributions Δq_f of quark and anti–quark flavours to the total spin of the nucleon.

The semi–inclusive asymmetry A_1^h is related to the quark polarisations $\Delta q_f/q_f$ and fragmentation functions. Measuring a set of different spin asymmetries on different targets allows to extract the quark polarisations. HERMES used a combination of six inclusive and semi–inclusive asymmetries on polarised $\overrightarrow{\text{H}}$ and ^{3}He. The extracted polarised quark distributions[7] are plotted in Fig. 4. In the same plot the predicted errors for the polarised quark distributions are shown when in addition the new deuteron data set from 1998-2000 is used.

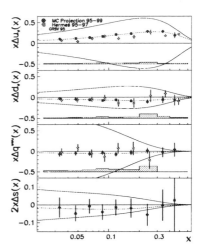

Figure 4. Measured and predicted precision of the spin distributions as a function of x.

Furthermore the possibility to identify pions, kaons and protons with the RICH over the whole kinematic range covered by the spectrometer will allow a direct measurement of sea polarisation for different flavours.

A different very precise measurement of the quark polarisations will be performed at RHIC using W production[8]. The estimated accuracy using this channel is shown in Fig. 5.

3.3 Gluon Polarisation

As described earlier, the gluon polarisation is only poorly constrained by existing data. One way to measure ΔG directly is via the photon gluon fusion process. Two useful experimental signatures of this process are the open charm production and the production of jets or rather pairs of oppositely charged hadrons with high–p_T.

HERMES has recently presented such a measurement[9]. The observed negative asymmetry shown in Fig. 6 is in contrast to the positive asymmetries typically measured in DIS from protons, where scattering from u-quarks dominates. This negative asymmetry

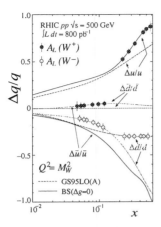

Figure 5. Estimated errors on $\Delta q/q$ after 4 months of data taking with PHENIX with a duty cycle of 40%.

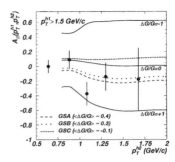

Figure 6. $A_{\parallel}$ for high–p_T hadron production measured at HERMES is compared with Monte–Carlo predictions for $\Delta G/G$ and phenomenological LO QCD fits of [10].

was interpreted using the PYTHIA Monte–Carlo Generator in leading order.

Using the assumptions and model parameters described in ref. [9] ($\Delta G/G$) was determined in LO QCD to be 0.41 ± 0.18(stat.) ± 0.03(syst.); the systematic uncertainty represents the experimental contribution only.

The accumulated statistics for open charm production is up to now not sufficient for a significant measurement of ΔG. But using the new deuteron data from 1998-2000 gives us good hope to perform an independent measurement of ΔG in the near future.

A very precise measurement of the gluon polarisation will be performed by the RHIC experiments[8].

4 Summary and Outlook

A selection of recent spin physics results of HERMES has been presented. In addition the physics potential of the high statistics deuteron data was discussed. Furthermore the projected accuracy of the measurements planned at RHIC has been shown.

After the HERA luminosity upgrade starting this September HERMES will measure with a transversely polarised target which will allow the first measurement of transversity.

Acknowledgements

This work was financially supported by the German Bundesministerium für Bildung, Wissenschaft, Forschung und Technologie (grant numbers 05 6ER 12I, 05 7ER 12P2 and 05 HH9 WE1) and the Deutsche Forschungsgemeinschaft. We gratefully acknowledge this support.

References

1. K. Ackerstaff et al., *NIM* A **417**, 230 (1998).

2. B. Adeva et al., *Phys. Rev.* D **58**, 112002 (1998).

3. K. Abe et al., *Phys. Rev.* D **58**, 112003 (1998).

4. A. Airapetian et al., *Phys. Lett.* B **442**, 484 (1998).

5. J. C. Collins, *Nucl. Phys.* B **396**, 161 (1993).

6. A. Airapetian et al., *Phys. Rev. Lett.* **84**, 4047 (2000).

7. K. Ackerstaff et al., *Phys. Lett.* B *464*, 123 (1999).

8. G. Bunce et al., *hep-ph*/0007218 (2000).

9. A. Airapetian et al., *Phys. Rev. Lett.* **84**, 2584 (2000).

10. T. Gehrmann, W. J. Stirling, *Phys. Rev.* D **53**, 6100 (1996).

UNITARITY CORRECTIONS IN PHOTO AND DIS PROCESSES

URI MAOR

HEP Department, School of Physics and Astronomy,
Raymond and Beverly Sackler Faculty of Exact Science,
Tel-Aviv University, Ramat Aviv, 69978, Israel
E-mail: maor@post.tau.ac.il

This short review aims to examine the applicability and possible need for a re-formulation of pQCD as we know it, in the limit of small Q^2 and x, when approaching the kinematic interface with the less understood npQCD dominated domain.

1 Introduction

The physics of small Q^2 and small x is associated with the search for the scale of gluon saturation implied by s-channel unitarity. One should remember that gluon saturation signals the transition from the perturbative to the non perturbative regime. We expect this transition to be preceded by SC signatures which should be experimentally visible even though the relevant scattering amplitude has not yet reached the unitarity (black disk) limit. We also know from our experience with soft Pomeron physics, that different channels have different scales at which unitarity corrections become appreciable. Specifically, the scale associated with the diffractive channels are considerably smaller than those associated with the elastic channel.

In spite of significant theoretical progress in recent years it is still not clear what the saturation scale in the present experimentally accessible kinematic region is . We recall that, while the global analysis of $F_2(x, Q^2)$ (or $\sigma_{tot}^{\gamma^* p}(W, Q^2)$) shows no conclusive deviations from DGLAP, there are dedicated HERA investigations suggesting possible deviations from the DGLAP expectations in the small Q^2 and x limits. These signatures are observed in both the fine details of $F_2(x, Q^2)$ as well as in the diffractive channels, provided Q^2 and x are sufficiently small. However, the question if these experimental signatures force upon us a reformulation of conventional pQCD is not settled.

In the following I present a detailed study of $\partial F_2 / \partial \ln Q^2$ coupled to an analysis of J/Ψ photo and DIS production [1]. Our investigation is based on the observation that these observables (in the LLA of pQCD) are proportional to $xG(x, Q^2)$ and $\left(xG(x, Q^2)\right)^2$ respectively. Being relatively well measured they may discriminate between the relevant theoretical approaches and models. We note that the calculation of the J/Ψ channels depends on a few correcting factors. However, since these corrections imply, essentially, only normalization changes, we consider this channel as a very important element in our analysis. Our study follows the observation that, presently, there is no satisfactory calculation, based on the latest p.d.f. editions of GRV98, MRS99 and CTEQ5, which provides an adequate simultaneous reproduction, at small pQCD scales, of the recent HERA data on the logarithmic slope of F_2 [2] at small Q^2 and x, as well as the abundant high energy data on J/Ψ photo and DIS production [3]. We proceed to show that when GRV98$\overline{MS}$ is corrected for SC, as defined in our previous publications [4,5], it gives a very good description of these data.

2 The small Q^2 and x behavior of $\partial F_2/\partial lnQ^2$

In the small x limit of DGLAP, we have in the LLA

$$\frac{\partial F_2(x, Q^2)}{\partial lnQ^2} = \frac{2\alpha_S}{9\pi} xG^{DGLAP}(x, Q^2). \quad (1)$$

Accordingly, a significant deviation of the data from Eq. (1), where xG^{DGLAP} is obtained from the global F_2 analysis, may serve as an experimental signature indicating the growing importance of unitarity corrections. Preliminary HERA data have recently became available [2]. As I shall show, a pQCD analysis of these data is consistent with a SC interpretation. I follow the eikonal SC formalism [4], where screening is calculated in both the quark sector, to account for the percolation of a $q\bar{q}$ through the target, and the gluon sector, to account for the screening of $xG(x, Q^2)$. The factorizable result obtained is

$$\frac{\partial F_2^{\mathrm{SC}}(x, Q^2)}{\partial lnQ^2} =$$

$$D_q(x, Q^2)D_g(x, Q^2)\frac{\partial F_2^{\mathrm{DGLAP}}(x, Q^2)}{\partial \ln Q^2}, (2)$$

where the SC damping factors have been calculated from the opacities [4]:
$\kappa_q = \frac{2\pi\alpha_S}{3Q^2}xG^{DGLAP}(x, Q^2)\Gamma(b^2)$ and $\kappa_g = \frac{4}{9}\kappa_q$. The calculation is significantly simplified if we assume a Gaussian parameterization for the two gluon non perturbative form factor, $\Gamma(b^2) = \frac{1}{R^2}e^{-b^2/R^2}$. $R^2 = 8.5 GeV^{-2}$ is determined directly from the forward slope of J/Ψ photoproduction. Our results are presented in Fig.1.

Following are some comments relating to our results:

1) The experimental data, as well as our model, are consistent with $\partial F_2/\partial lnQ^2$ (at fixed Q^2) being a monotonic increasing function of $1/x$. No deviation of this behavior has been seen even at the lowest reported values of Q^2 and x.

2) In the limit of small Q^2 and x there is a significant difference between the screened and

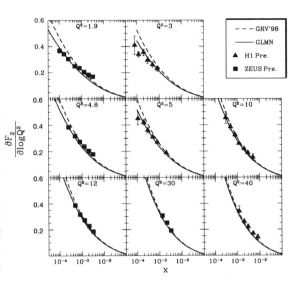

Figure 1. *x dependence of H1 and ZEUS $\partial F_2/\partial lnQ^2$ data at fixed Q^2 compared with our calculations. Experimental numbers are preliminary and were read off the plots.*

non screened values of $\partial F_2/\partial lnQ^2$. As expected the SC results are smaller and softer than the non screened input.

3) Our overall reproduction of the experimental data is very good, in particular when considering that our input is essentially parameter free. A proper χ^2 calculation requires the knowledge of the unknown theoretical errors. If we follow the standard procedure and replace the theoretical errors with the experimental ones, we obtain excellent $\chi^2/ndf = 0.75$ for 21 H1 points (with the exception of 3 points which are visibly out of line). The ZEUS data appears to have errors which are considerably smaller and as a result our χ^2/ndf is not as good, even though our reproduction of the ZEUS data is reasonable.

4) The ZEUS $Q^2 = 1.9 GeV^2$ data are somewhat softer than our predictions, which do not contain a soft non perturbative background.

5) Whereas we find strong support for the

need for SC in the small Q^2 and x limits for $\partial F_2/\partial lnQ^2$, we are unable to directly determine the gluon saturation scale from the latest HERA data. The gluon saturation scale may be theoretically estimated [4] from the contours produced at the boundary of $\kappa_g = 1$.

6) We caution against attempts to determine the gluon saturation scale from turn overs seen in the Caldwell and fixed W plots. Both have very severe kinematical correlations which produce spurious effects not connected with saturation. All parametrizations of F_2 which provide even an approximate descriptionof the data, reproduce these turn overs.

3 Photo and DIS production of J/Ψ

The t=0 differential cross section of photo and DIS production can be calculated in the dipole LLA. When using the static non relativistic approximation of the vector meson wave function, the differential cross section has a very simple form

$$\left(\frac{d\sigma(\gamma^* p \to Vp)}{dt}\right)_0 = \frac{\pi^3 \Gamma_{ee} M_V^3}{48\alpha} \times$$
$$\frac{\alpha_S^2(\bar{Q}^2)}{Q^8} \left(xG(x, \bar{Q}^2)\right)^2 \left(1 + \frac{Q^2}{M_V^2}\right), (3)$$

where in the non relativistic limit we have $\bar{Q}^2 = \frac{M_V^2 + Q^2}{4}$ and $x = \frac{4\bar{Q}^2}{W^2}$.

In this presentation I shall discuss the photo and DIS production of J/Ψ. The integrated cross section data available for this channel span a relatively wide energy range. From a theoretical point of view, its hardness (or separation) scale is comparable to the scales we have studied in our $\partial F_2/\partial lnQ^2$ analysis. To relate the integrated cross section to Eq. (3) we need to know B, the J/Ψ forward differential cross section slope. For this we may use the experimental values, which are approximately constant, $B \simeq \frac{R^2}{2}$.

SC account well for the reported moderate energy dependence [3]. Since the non screened J/Ψ photo and DIS cross sections are proportional to $\left(xG(x, Q^2)\right)^2$, the study of this channel can serve as a compatibility check supplementing our study of $\partial F_2/\partial lnQ^2$.

The main problem with the theoretical analysis of J/Ψ is the realization that the simple pQCD calculation needs to be corrected for the following reasons:

1) A correction for the contribution of the real part of the production amplitude. This correction is well understood and is given by $C_R^2 = (1 + \rho^2)$, where $\rho = ReA/ImA = tg(\frac{\pi\lambda}{2})$ and $\lambda = \partial ln(xG)/\partial ln(\frac{1}{x})$.

2) A correction for the contribution of the skewed (off diagonal) gluon distributions. This correction is calculated to be $R_g^2 = \left(\frac{2^{2\lambda+3} \Gamma(\lambda+2.5)}{\sqrt{\pi} \Gamma(\lambda+4)}\right)^2$.

3) A more controversial issue relates to the non relativistic approximation assumed for the J/Ψ Charmonium. Relativistic effects produced by the Fermi motion of the bound quarks result in a considerable reduction of the calculated pQCD cross section. We denote this correction C_F^2 and note that it is very sensitive to the value of m_c. Ref.[6] assumes that $m_c \simeq 1.5 GeV$ and obtains $C_F^2 \simeq 0.25$ with minimal energy dependence. A small change in the input value of m_c changes the above estimate significantly. We suggest, therefore, to consider C_F^2 as a free parameter. In our calculations we have used $C_F^2 = 0.68$ which corresponds to a c-quark mass of approximately $1.53 GeV$.

Our calculation of SC for J/Ψ photo and DIS production is rather similar to the Q^2 logarithmic slope calculation presented earlier. We follow our publications [5] and define transverse and longitudinal damping factors due to the screening in the quark sector i.e. the percolation of the $c\bar{c}$ through the target. The above quark damping factor *differs* from the square of the corresponding quark sector damping defined for F_2 log-

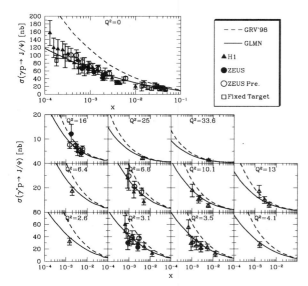

Figure 2. *Elastic photo and DIS production of J/Ψ as a function of x. H1, ZEUS and fixed target Data and our calculations. Some of the HERA data is preliminary and was read off the plots.*

arithmic slope. Our expression for D_g^2, the damping in the gluon sector, is the square of the gluon damping defined in the previous section. Our calculations compared with the experimental data are presented in Fig.2.

The theoretical reproduction of the data is excellent with a $\chi^2/n.d.f.$ which is well below 1. These excellent χ^2 values are maintained when calculating over the entire data base as well as limiting ourselves to the high energy HERA data. Note that $R^2 = 8.5 GeV^{-2}$, which is the essential parameter in the SC calculation is determined directly from the J/Ψ photoproduction forward slope. In a model such as ours, we expect a weak dependence of B on the energy which is indeed observed experimentally.

4   Conclusions

1) NLO GRV98, when screened for both the percolation of a $q\bar{q}$ pair through the target and for multigluon exchange, provides a very satisfactory simultaneous reproduction of the HERA new data on $\partial F_2/\partial lnQ^2$ and photo and DIS production of J/Ψ.

2) The question if non screened p.d.f. or other screened and non screened models [789] can reproduce the data is still open as none of these options was tested in a detailed simultaneous analysis such as the one presented in this summary.

3) In as much as I consider the need to modify conventional pQCD in the small Q^2 and x limits to be rather convincing, I wish to stress again that the present HERA data does not enable us to experimentally determine the gluon saturation scale. Such a determination may have to wait for a detailed lepton DIS study on a nuclear target where xG is considerably bigger than the proton one.

Acknowledgements:

I wish to thank my long standing collaborators (EG,EF,EL,EN) for their assistance. This research was supported by in part by the Israel Academy of Science and Humanities and by BSF grant # 98000276. This review was written while I was visiting TIFR(Mumbai). I wish to thank my hosts for their kind hospitality.

References

1. E. Gotsman,E. Ferreira, E. Levin, U. Maor and E. Naftali: paper submitted to ICHEP2000, `hep-ph/0007274`. It contains a more detailed list of references.

2. H1 and ZEUS Collaborations data on F_2 presented at ICEP2000 including compilations of earlier data.

3. H1 and ZEUS Collaborations data on J/Ψ presented at ICHEP2000 including compilations of earlier data.

4. E. Gotsman, E.M. Levin and U. Maor: *Phys. Lett.* **B425** (1998) 369; E. Gotsman, E.M. Levin, U. Maor and E. Naftali: *Nucl. Phys.* **B539** (1999) 535.

5. E. Gotsman, E.M. Levin and U. Maor: *Nucl. Phys.* **B464** (1996) 251; *Nucl. Phys.* **B493** (1997) 354; *Phys. Lett.* **B403** (1997) 120.

6. L. Frankfurt, W. Koepf and M. Strikman: *Phys. Rev.* **D54** (1996) 3194; *Phys. Rev.* **D57** (1998) 512.

7. K. Golec-Biernat and M. Wuesthoff: *Phys. Rev.* **D59** (1998) 014017.

8. A.B. Kaidalov, C. Merino and D. Pertermann: `hep-ph/0004237`; A. Capella, E.G. Ferreiro, C.A. Salgado and A.B. Kaidalov: `hep-ph/0005049`; `hep-ph/0006233`.

9. A. Donnachie and P.V. Landshoff: *Phys. Lett.* **B437** (1998) 408; *Phys. Lett.* **B470** (1999) 243.

RECENT STRUCTURE FUNCTION RESULTS FROM NEUTRINO SCATTERING AT FERMILAB

U. K. YANG,[7] T. ADAMS,[4] A. ALTON,[4] C. G. ARROYO,[2] S. AVVAKUMOV,[7]
L. DE BARBARO,[5] P. DE BARBARO,[7] A. O. BAZARKO,[2] R. H. BERNSTEIN,[3]
A. BODEK,[7] T. BOLTON,[4] J. BRAU,[6] D. BUCHHOLZ,[5] H. BUDD,[7] L. BUGEL,[3]
J. CONRAD,[2] R. B. DRUCKER,[6] B. T. FLEMING,[2] J. A. FORMAGGIO,[2] R. FREY,[6]
J. GOLDMAN,[4] M. GONCHAROV,[4] D. A. HARRIS,[7] R. A. JOHNSON,[1] J. H. KIM,[2]
B. J. KING,[2] T. KINNEL,[8] S. KOUTSOLIOTAS,[2] M. J. LAMM,[3] W. MARSH,[3]
D. MASON,[6] K. S. MCFARLAND, [7] C. MCNULTY,[2] S. R. MISHRA,[2] D. NAPLES,[4]
P. NIENABER,[3] A. ROMOSAN,[2] W. K. SAKUMOTO,[7] H. SCHELLMAN,[5]
F. J. SCIULLI,[2] W. G. SELIGMAN,[2] M. H. SHAEVITZ,[2] W. H. SMITH,[8]
P. SPENTZOURIS, [2] E. G. STERN,[2] N. SUWONJANDEE,[1] A. VAITAITIS,[2]
M. VAKILI,[1] J. YU,[3] G. P. ZELLER,[5] AND E. D. ZIMMERMAN[2]

(Presented by Un-ki Yang for the CCFR/NuTeV Collaboration)

[1] *University of Cincinnati, Cincinnati, OH 45221*
[2] *Columbia University, New York, NY 10027*
[3] *Fermi National Accelerator Laboratory, Batavia, IL 60510*
[4] *Kansas State University, Manhattan, KS 66506*
[5] *Northwestern University, Evanston, IL 60208*
[6] *University of Oregon, Eugene, OR 97403*
[7] *University of Rochester, Rochester, NY 14627*
[8] *University of Wisconsin, Madison, WI 53706*

We report on the extraction of the structure functions F_2 and $\Delta x F_3 = x F_3^\nu - x F_3^{\bar\nu}$ from CCFR ν_μ-Fe and $\bar\nu_\mu$-Fe differential cross sections. The extraction is performed in a physics model independent (PMI) way. This first measurement of $\Delta x F_3$, which is useful in testing models of heavy charm production, is higher than current theoretical predictions. The ratio of the F_2 (PMI) values measured in ν_μ and μ scattering is in agreement (within 5%) with the NLO predictions using massive charm production schemes, thus resolving the long-standing discrepancy between the two sets of data. In addition, measurements of F_L (or, equivalently, R) and $2xF_1$ are reported in the kinematic region where anomalous nuclear effects in R are observed at HERMES.

Deep inelastic lepton-nucleon scattering experiments have been used to determine the quark distributions in the nucleon. However, the quark distributions determined from μ and ν experiments[1,2] were found to be different at small values of x, because of a disagreement in the extracted structure functions. Here, we find that the neutrino-muon difference is resolved by extracting the ν_μ structure functions from CCFR neutrino data in a physics model independent (PMI) way. In addition, measurements of $\Delta x F_3$, F_L, and $2xF_1$ are presented.

The sum of ν_μ and $\bar\nu_\mu$ differential cross sections for charged current interactions on an isoscalar target is related to the structure functions as follows:

$$F(\epsilon) \equiv \left[\frac{d^2\sigma^\nu}{dxdy} + \frac{d^2\sigma^{\bar\nu}}{dxdy} \right] \frac{(1-\epsilon)\pi}{y^2 G_F^2 M E_\nu}$$
$$= 2xF_1[1 + \epsilon R] + \frac{y(1-y/2)}{1+(1-y)^2} \Delta x F_3. \quad (1)$$

Here G_F is the Fermi weak coupling constant, M is the nucleon mass, E_ν is the incident energy, the scaling variable $y = E_h/E_\nu$ is the fractional energy transferred to the hadronic vertex, E_h is the final state hadronic energy, and $\epsilon \simeq 2(1-y)/(1+(1-y)^2)$ is the polarization of the virtual W boson. The structure function $2xF_1$ is expressed in terms of

F_2 by $2xF_1(x, Q^2) = F_2(x, Q^2) \times \frac{1+4M^2x^2/Q^2}{1+R(x,Q^2)}$, where Q^2 is the square of the four-momentum transfer to the nucleon, $x = Q^2/2ME_h$ is the fractional momentum carried by the struck quark, and $R = \frac{\sigma_L}{\sigma_T}$ is the ratio of the cross-sections of longitudinally- to transversely-polarized W bosons. The ΔxF_3 term, which in leading order $\simeq 4x(s-c)$, is not present in the μ-scattering case. In addition, there is a threshold suppression originating from the production of heavy c quarks in a ν_μ charged current interaction with s quarks. For μ-scattering, there is no suppression for scattering from s quarks, but more suppression when scattering from c quarks.

In previous analyses of ν_μ data[2], structure functions were extracted by applying a slow rescaling correction to correct for the charm mass suppression in the final state. In addition, the ΔxF_3 term from a leading order charm production model was used as input in the extraction. These resulted in physics model dependent (PMD) structure functions[2]. In the new analysis reported here, slow rescaling corrections are not applied. ΔxF_3 and F_2 are extracted from two-parameter fits to the $F(\epsilon)$ distributions according to Eq. (1). However, in the $x > 0.1$ region, we extract values of F_2 with ΔxF_3 constrained to the NLO TR-VFS(MRST)[3] predictions. Since ΔxF_3 for $x > 0.1$ is small, the extracted values of F_2 are insensitive to ΔxF_3.

Fig. 1(left) shows the extracted values of ΔxF_3 as a function of x (above $Q^2 = 1$), including both statistical and systematic errors, compared to various theoretical methods for modeling heavy charm productions within a QCD framework. Fig. 1(right) shows the sensitivity to the choice of scale. With reasonable choices of scale, all the theoretical models yield similar results. However, at low Q^2, our ΔxF_3 data are higher than all theoretical models.

Our F_2 (PMI) measurements divided by the NLO TR-VFS(MRST) predictions

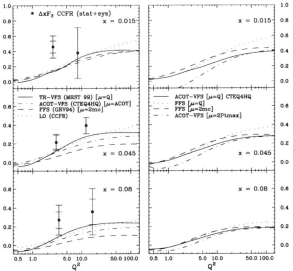

Figure 1. ΔxF_3 data as a function of x (above $Q^2 = 1$) compared with various schemes for massive charm production. (Left) TR-VFS(MRST99), ACOT-VFS(CTEQ4HQ), FFS(GRV94), and the CCFR-LO (a leading order model with a slow rescaling correction): (right) sensitivity of the theoretical calculations to the choice of scale.

are shown in Fig. 2(left). Also shown are F_2^μ and F_2^e divided by the theory predictions. Nuclear effects, target mass, and higher twist corrections are included in the calculation. As shown in Fig. 2, within 5% both the neutrino and muon structure functions are in agreement with the NLO TR-VFS(MRST) predictions, and therefore in agreement with each other, thus resolving the long-standing discrepancy between the two sets of data. A comparison using the NLO ACOT-VFS(CTEQ4HQ)[4] predictions yields similar results. Note that previously there was up to a 20% difference between the CCFR F_2 (PMD) and NMC data at $x = 0.015$, as shown in Fig. 2(right).

Recently, there has been a renewed interest in R at small x and $Q^2 < 1$, because of the large anomalous nuclear effect that has been reported by the HERMES experiment[5]. Their measurement implies a large enhancement in F_L but suppression in $2xF_1$ in heavy

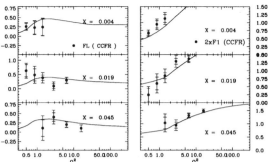

Figure 3. Preliminary measurements of F_L and $2xF_1$ as a function of Q^2 for $x < 0.1$, The curves are the predictions from a QCD inspired leading order fit to the CCFR differential cross section data with $R = R_{world}$ for neutrino scattering.

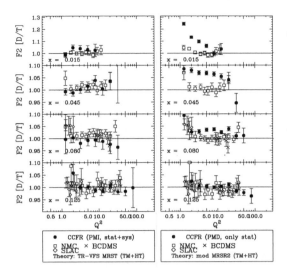

Figure 2. (Left) The ratio of the F_2^ν (PMI) data divided by the predictions of TR-VFS (MRST99) with target mass and higher twist corrections; (right) The ratio of the previous F_2^ν (PMD) data and the predictions of MRSR2. Also shown are the ratios of the F_2^μ (NMC, BCDMS) and F_2^e (SLAC) to the theoretical predictions.

nuclear targets. It is expected that any nuclear effect in R would be enhanced in the CCFR iron target with respect to the nitrogen target in HERMES, unless the origin of this effect depends on the incident probe (electron versus neutrino).

Values of F_L and $2xF_1$ are extracted from the sums of the corrected ν_μ-Fe and $\overline{\nu}_\mu$-Fe differential cross sections in different energy bins according to Eq. (1). An extraction of F_L requires knowledge of ΔxF_3. which we obtain from the NLO TR-VFS(MRST) calculation. Because of the large uncertainty in ΔxF_3 at low Q^2 region, an extrapolation of the curve which describes the measured CCFR ΔxF_3 data above $Q^2 = 1$ is used for the systematic error. Here we are interested in the relative Q^2 dependence of F_L and $2xF_1$.

Fig. 3 shows the preliminary values of F_L and $2xF_1$ as a function of Q^2 for $x < 0.1$. The inner errors include both statistical and experimental systematic errors. The outer errors represent the ΔxF_3 model errors added in quadrature. The curves are the predictions from a QCD-inspired leading order fit to the CCFR differential cross section data with $R = R_{world}$ (for neutrino scattering) which does not include the HERMES effect. Large anomalous deviations from the fit (e.g. 200 - 300%) are not seen in the CCFR data.

More details on this work can be found in reference 6 and 7.

References

1. M. Arneodo *et al.*, *Nucl. Phys.* **B483**, 3 (1997).

2. W. G. Seligman *et al.*, *Phys. Rev. Lett.* **79**, 1213 (1997).

3. R. S. Thorne and R .G. Roberts, *Phys. Lett.* **B421**, 303 (1998).

4. M. Aivazis, J. Collins, F. Olness, and W. K. Tung, *Phys. Rev.* **D50**, 3102 (1994).

5. K. Ackerstaff *et al.*, *Phys. Lett.* **B475**, 386 (1999).

6. U. K. Yang *et al.*, UR-1586, hep-ex/0009041, submitted to *Phys. Rev. Lett.*

7. U. K. Yang, Ph.D. Thesis, University of Rochester (UR-1583), 2000.

DIFFRACTIVE VECTOR MESON PRODUCTION AT HERA

BRUCE MELLADO

On behalf of the H1 and ZEUS Collaborations.

Columbia University, 538 120th Street West, New York, NY 10027, USA

E-mail: mellado@nevis1.nevis.columbia.edu

The H1 and ZEUS collaborations at HERA report new data on the diffractive production of vector mesons with a many-fold increase of luminosity compared to previous measurements. The new data include photoproduction and electroproduction of $\rho^0, \phi, J/\psi$. The available data on elastic VM production indicate that the interaction scales as $Q^2 + M_V^2$.

1 Introduction

The production of vector mesons (VM, V) at HERA is interesting for the study of non-perturbative hadronic physics, perturbative QCD (pQCD) and their interplay. Moreover, VM production is complementary to deep-inelastic scattering (DIS). DIS has shown the correctness of pQCD down to low values of $Q^2 \sim 1$ GeV^2 (where Q^2 is the virtuality of the photon exchanged), a region where a transition to non-perturbative physics is observed [1]. In the past the elastic processes were mostly treated through non-perturbative methods. These methods are successful to describe basic features of exclusive light VM production at low photon virtualities. However, in recent years a pQCD picture of the exclusive VM production is able to describe its basic features provided that $Q^2/\Lambda_{QCD}^2 \gg 1$ and $M_V/W \ll 1$, where M_V is the mass of the VM and W is the photon-proton center-of-mass energy. The photon fluctuates into a quark-antiquark pair, long before the interaction with the proton occurs. The interation occurs via gluon ladders. The steepness of the rise of the VM production with W is basically driven by the gluon density in the proton, which is probed at an effective scale $K \sim (Q^2 + M_V^2)$ at low x ($x \simeq (Q^2 + M_V^2)/W^2$).

Assuming a flavor independent production mechanism the relative production rates should scale approximately with the square of the quark charges, i.e. the relative production rates scale as $\rho^0 : \omega : \phi : J/\psi = 9 : 1 : 2 : 8$, referred here as SU(4) ratios. It is interesting to determine how the interaction changes the SU(4) ratios.

As for today the H1 and ZEUS collaborations have measured the elastic production of vector mesons $e + p \rightarrow e + V + p$, where $V = \rho^0, \omega, \phi, J/\psi$ over a wide range of W, from photoproduction ($Q^2 \simeq 0$) to $Q^2 = 100$ GeV^2 [2–10]. New results on proton-dissociative diffractive photoproduction of VM at $W \simeq 100$ GeV and $-t \leq 12$ GeV^2 [11], where t is the squared four-momentum transfer at the proton vertex, are also available. The kinematic range of DIS and VM production now overlap at HERA giving us the chance to examine more deeply fundamental issues of the physics of hadronic interactions.

2 New results

The ZEUS collaboration has presented [7] new data on the elastic electroproduction of ρ^0 using an integrated luminosity of 38 pb^{-1}. The cross section for $\gamma^* p \rightarrow \rho^0 p$ has been measured for $32 < W < 160$ GeV and $5 < Q^2 < 80$ GeV^2. The W dependence of $\sigma(\gamma^* p \rightarrow \rho^0 p)$ is measured. If one assumes the form W^δ, where δ is extracted from a fit at fixed Q^2, it is shown that δ displays a marked increase with Q^2. The Q^2 dependence of $\gamma^* p \rightarrow \rho^0 p$ is fit to a form $(Q^2 + m_\rho^2)^{-n}$. The power n is found to depend on Q^2.

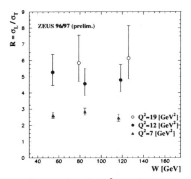

Figure 1. $R = \sigma_L/\sigma_T$ for ρ^0 production is presented for fixed Q^2 values. No marked W dependence is observed.

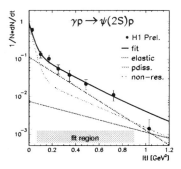

Figure 2. The corrected number of events $1/N \cdot dN/dt$ of $\psi(1S)$ candidates as a function of $|t|$. The lines correspond to different contributions to the combined fit of the t-slope.

The angular distribution of the decay products of the ρ^0 are used to measure the ratio of the production cross sections, $R = \sigma_L/\sigma_T$, where σ_L and σ_T correspond to longitudinally and transversely polarized virtual photons. The Q^2 dependence of R is parametrized as $R = \frac{1}{\xi}(Q^2/m_\rho^2)^\kappa$, where $\xi = 2.17 \pm 0.07$ and 0.75 ± 0.03. The W dependence of R is measured for three fixed values of Q^2. Fig. 1 shows that R does not depend on W, therefore the steepness of the rise of $\sigma(\gamma^* p \to \rho^0 p)$ with W appears to be independent of the photon polarization.

The ZEUS collaboration has presented [5] new results on the elastic photoproduction of J/ψ meson using two decay channels with an integrated luminosity of 38 pb^{-1} and 48 pb^{-1}. The total $\sigma(\gamma p \to J/\psi p)$ has been measured in the kinematic range $20 < W < 290$ GeV, exhibiting a steep dependence on W, which is understood within pQCD as being driven by the rise of the gluon density in the proton at low x. The differential cross-section $d\sigma/dt$ has been measured in the kinematic range $30 < W < 170$ GeV and $-t \le 1.8$ GeV^2. The $d\sigma/dt$ data is used to determine $\alpha_{I\!\!P}(t)$ in six bins of t by means of a fit of the form $d\sigma/dt \propto (W^2)^{2\alpha_{I\!\!P}(t)-2}$. The resulting values of $\alpha_{I\!\!P}(t)$ are fitted to a line, $\alpha_{I\!\!P}(t) = \alpha_{I\!\!P}(0) + \alpha'_{I\!\!P} \cdot t$, where $\alpha_{I\!\!P}(0) = 1.193^{+0.019}_{-0.015}$ and $\alpha'_{I\!\!P} = 0.105^{+0.033}_{-0.031}$. The measurement of the intercept, $\alpha_{I\!\!P}(0)$, is neither compatible

with the *soft* Pomeron nor the *hard* Pomeron alone. The slope, $\alpha'_{I\!\!P}$, is significantly smaller than that of the *soft* Pomeron and suggests the presence of small shrinkage.

The H1 collaboration has presented [6] new results on the elastic photoproduction of $\psi(2S)$ corresponding to an integrated luminosity of 38 pb^{-1} in a range $40 < W < 150$ GeV. The t-dependence of the elastic production of $\psi(2S)$ differential cross section is extracted by means of a combined fit that takes into account the contribution of the proton dissociative and non-resonant QED background (see Fig. 2). A single exponential $e^{-b|t|}$ is used to describe the elastic cross section. The fit yields $b_{\psi(2S)} = (4.5^{+1.7}_{-1.4})$ GeV^{-2}. This result is consistent with the slope parameter of the J/ψ of $b_{J/\psi} = (4.73^{+0.39}_{-0.46})$ GeV^{-2} [5]. This is consistent with the pQCD prediction despite the naive expectation that $\psi(2S)$ has more peripheral t-dependence since the two c-quarks are on average further apart from each other than in the ground state.

The ZEUS collaboration has presented [10] new results on the elastic electroproduction of J/ψ using an integrated luminosity of 75 pb^{-1} in the kinematic range $50 < W < 150$ GeV and $2 < Q^2 < 100$ GeV^2. The W dependence of the cross section is parametrized with W^δ at fixed Q^2. The obtained values of δ are consistent with those measured in photoproduction. The Q^2 dependence of the cross section is fairly well described by predictions of pQCD. The ratio of the J/ψ to ρ^0 cross sections rises rapidly with Q^2 approaching the SU(4) ratios. The ratio $R = \sigma_L/\sigma_T$ is extracted from the angular distributions of the decay products of the J/ψ and increases with Q^2. The measured R is consistent with the one obtained for ρ^0 after taking into account the suppression factor of $M_\rho/M_{J/\psi}$.

The ZEUS collaboration has produced [11] new results on the photoduction of $\rho^0, \phi, J/\psi$ where the proton dissociates into a low mass state N. The data correspond to 24 pb^{-1} at $W \simeq 100$ GeV and extended up to $-t = 12$ GeV^{-2}. The measured differential cross section $d\sigma/dt$ is parametrized as a power function $d\sigma/dt \propto (-t)^{-n}$. The fitted value of n decreases with increasing mass of the VM: $n_\rho = 3.31^{+0.12}_{-0.12}, n_\phi = 2.77^{+0.18}_{-0.18}$ and $n_{J/\psi} = 1.7^{+0.28}_{-0.28}$. A comparison of the measured differential cross sections with QCD models shows that the perturbative part of the calculation for ρ^0 and ϕ production at the t values covered in this analysis is well below the data. The pQCD prediction is in agreement with the J/ψ data, but large theoretical uncertainties remain. As illustrated in Fig. 3, the differential cross section ratio $\frac{d\sigma}{dt}(\gamma p \to \phi N)/\frac{d\sigma}{dt}(\gamma p \to \rho^0 N)$ increases with -t and approaches the SU(4) ratio at $-t \approx 3.5$ GeV^2, while in elastic electroproduction regime the ϕ/ρ^0 ratio approaches the SU(4) value at a larger scale, $Q^2 \geq 6$ GeV^2. In the case of the J/ψ this ratio is significantly smaller than the SU(4) value up to $-t$ values of 4 GeV^2. For -t$\approx$ 3.5 GeV^2 this ratio is a factor of five above the $(J/\psi)/\rho^0$ ratio for $Q^2 \approx 3.5$ GeV^2, indicating that t and Q^2 are not equivalent scales.

The decay angle analysis is performed for the ρ^0 and ϕ and demonstrates a clear deviation from S channel helicity conservation (SCHC). The measured values of the spin density matrix are in agreement with pQCD

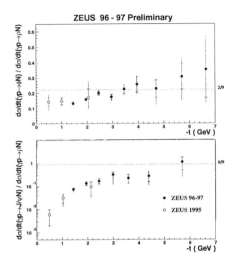

Figure 3. The ratios of the cross sections $d\sigma/dt$ for ϕ to ρ^0 and J/ψ to ρ^0 for proton-dissociative photoproduction.

predictions.

3 The scale of interaction in elastic VM production.

Recent results on the elastic VM production at HERA have shown that the W dependence of the total cross section $\sigma(\gamma^* p \to Vp)$ and the t dependence of the differential cross section $d\sigma(\gamma^* p \to Vp)/dt$ is dependent on Q^2 and M_V^2, suggesting that the observables of the interaction be functions of these two variables [12] (See other relevant publications [9,13,14]).

It has been shown that the relative production ratios reach approximately the SU(4) ratios at $Q^2 \gg M_V^2$. However, Fig. 4 shows the total cross section ratios $\sigma_\omega, \sigma_\phi, \sigma_{J/\psi}$ to σ_ρ being approximately constant with $Q^2 + M_V^2$. The ratio $\sigma_{J/\psi}/\sigma_\rho$ in Fig. 4 is systematically higher than the SU(4) ratio, possibly due to difference in the wave function of light and heavy quarks.

The universal character of the interaction is further demonstrated in Fig. 5, where $d\sigma(\gamma^* p \to Vp)/dt(t = 0) \approx b_V \cdot \sigma(\gamma^* p \to Vp) \cdot r_\rho^{SU(4)}/r_V^{SU(4)}$ is plotted as a function of x. Here b_V is extracted from the differen-

Elastic VM production at HERA

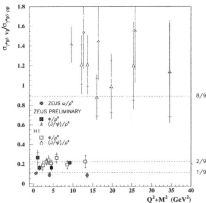

Figure 4. The total cross section ratios $\sigma_\omega, \sigma_\phi, \sigma_{J/\psi}$ to σ_{ρ^0} as a function of $Q^2 + M_V^2$ at fixed W compared to the SU(4) ratios.

tial cross section $d\sigma(\gamma^* p \to Vp)/dt$ assuming $d\sigma/dt \propto e^{-b|t|}$ and $r_V^{SU(4)}$ are the SU(4) ratios. This indicates that the effects due to the wave functions of the different VM do not play a major role. The x dependence of the VM cross sections is parametrized as $A \cdot x^{-2 \cdot \lambda_V}$, where A and λ_V are fitted in each bin of $Q^2 + M_V^2$. The behavior of the steepness of the rise of the VM production cross section as $x \to 0$ with changing $Q^2 + M_V^2$ is similar to the x dependence of the inclusive F_2^p with changing Q^2 [1]. These results indicate that the combination $Q^2 + M_V^2$ is a good choice of scale of the interaction in elastic VM production.

References

1. H1 Collab., C. Adloff et al., Nucl. Phys. B497 (1997) 3; ZEUS Collab., J. Breitweg et al. Phys.Lett. **B443** (1998) 394.
2. H1 Collab., S. Aid et al., Nucl.Phys. **B463** (1996) 3; ZEUS Collab., J. Breitweg et al., Eur.Phy.J. **C2** (1998) 247.
3. ZEUS Collab., M. Derrick et al., Z.Phys. **C73** (1996) 1, 73.
4. ZEUS Collab., M. Derrick et al., Phys.Lett. **B377** (1996) 259.
5. H1 Collab., C. Adloff et al., Phys.Lett.

Elastic VM Production at HERA

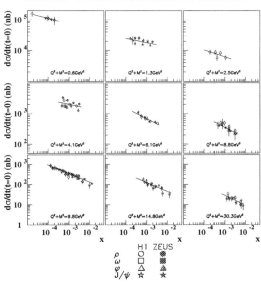

Figure 5. Elastic $d\sigma/dt(t = 0)$ for various VM as a function of x in bins of $Q^2 + M_V^2$ (See text). The solid line are fits of the form $A \cdot x^{-2 \cdot \lambda_V}$.

B483 (2000) 23; Paper 437, ICHEP200, Osaka, July 2000.

6. Abstracts 987, 985, ICHEP200, Osaka, July 2000.
7. H1 Collab., C. Adloff et al., Eur.Phys.J. **C13** (2000) 371; Paper 439, ICHEP200, Osaka, July 2000.
8. ZEUS Collab., J. Breitweg et al., accepted by Phys.Lett. B-PLB 16283.
9. H1 Collab., C. Adloff et al., Phys.Lett. **B483** (2000) 360; Abstract 793 at ICHEP98, Vancouver, July 1998.
10. H1 Collab., C. Adloff et al., Eur.Phys.J. **C10** (1999) 373; Paper 438, ICHEP200, Osaka, July 2000.
11. Paper 442, ICHEP200, Osaka, July 2000.
12. B. Mellado, talk 02c-01 at ICHEP2000, Osaka, July 2000.
13. B. Clerbaux, hep-ph/9908519.
14. B. Naroska, VM Production at HERA, XXXVth Rencontres de Moriond, March 2000, Les Arcs, France.

NEW MEASUREMENTS OF INCLUSIVE DIFFRACTION AT HERA

P. R. NEWMAN

School of Physics and Astronomy, University of Birmingham, B15 2TT, UK
E-mail: prn@hep.ph.bham.ac.uk

Two new diffractive measurements from HERA are described. ZEUS data on the diffractive structure function F_2^D at low Q^2 constrain the transition in diffraction from the perturbative high Q^2 region to the photoproduction limit. An effective pomeron intercept $\alpha_{I\!P}(0)$ is extracted from the energy dependence of the data and is compared with values from diffractive and inclusive ep interactions at lower and higher Q^2. An H1 analysis of photoproduced dijet events with low levels of hadronic activity between the jets demonstrates the presence of strongly interacting colour singlet exchanges at very high momentum transfer. The data are used to investigate the relative coupling strengths of this exchange to quarks and gluons.

1 Diffraction at HERA

Data from the HERA ep collider have led to considerable progress in the understanding of diffraction within QCD.[1] The HERA diffractive process can be viewed most generally as shown in figure 1. A photon of virtuality Q^2 (coupled to the electron) interacts with a proton at centre of mass energy W and squared four momentum transfer t to produce two distinct systems of masses M_X and M_Y at the photon and proton vertices respectively. The hermetic nature of the H1 and ZEUS detectors make it possible to control all five kinematic variables in figure 1 at some level.

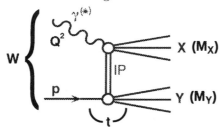

Figure 1. The generic HERA diffractive process of the type $\gamma^{(*)}p \rightarrow XY$.

Where both Q^2 and $|t|$ are small, the dynamics are similar to those of soft diffractive hadron-hadron scattering.[2] Where $|t|$ is small but Q^2 is large, perturbative QCD techniques have successfully been applied at the photon vertex. Under these circumstances, the virtual photon can be viewed as probing the structure of the diffractive exchange or 'pomeron' ($I\!P$).[3,4] The region in which $|t|$ is large is well suited to approaches that calculate the diffractive exchange itself perturbatively.[5] Each of these regions is interesting in its own right. The transitions between them are likely to be particularly revealing.

2 The Diffractive Structure Function at Very Low Q^2

Measurements of the semi-inclusive process $ep \rightarrow eXY$ at low M_Y and $|t|$ and finite Q^2 have generally been presented in the form of a diffractive structure function $F_2^D(\beta, Q^2, x_{I\!P})$, where $x_{I\!P}$ may be interpreted as the fraction of the proton longitudinal momentum transferred to the X system and β as the fraction of the exchanged momentum carried by the quark coupling to the photon. F_2^D has now been measured at HERA throughout a large part of the accessible $\beta - Q^2$ kinematic plane for $x_{I\!P} < 0.05$.[3,4,6]

Using their beam-pipe calorimeter (BPC) to detect and measure electrons scattered at very small angles, the ZEUS collaboration have recently released F_2^D data in the region $0.22 < Q^2 < 0.70$.[7] Diffractive samples are extracted both by direct tagging of leading protons in a set of Roman pot detectors inserted into the beampipe in the outgoing proton direction and by decomposition of the distribution of M_X measured in the main detector. The $x_{I\!\!P}(\equiv x/\beta)$ dependence of the data has been interpreted at fixed β and Q^2 in a Regge motivated model, whereby

$$x_{I\!\!P}F_2^D \sim A(\beta, Q^2) \ (1/x)^{2\alpha_{I\!\!P}(t)-2} \ . \quad (1)$$

The t averaged value of the effective exchange trajectory extracted by the M_X decomposition method is $\overline{\alpha_{I\!\!P}(t)} = 1.126 \pm 0.012$ (stat.) $^{+0.027}_{-0.032}$ (syst.). After correcting for the finite t values of the measurement, the value extracted for the intercept of this effective trajectory is shown in figure 2 together with other values from diffractive[2,3,4] and inclusive[8] ep measurements. The values from the inclusive data are extracted using

$$F_2 \sim B(Q^2) \ (1/x)^{\alpha_{I\!\!P}(0)-1} \ . \quad (2)$$

In the inclusive case, the energy dependence clearly becomes stronger as Q^2 increases. A similar effect is suggested by the diffractive data. In photoproduction and low Q^2 electroproduction, the effective intercepts in the diffractive and inclusive cases are compatible. However, at the largest Q^2, they differ. In fact, it has been observed[4] that the energy dependences of the diffractive and inclusive cross sections are rather similar at large Q^2, in contrast to the expectation for a simple Regge pole in equations 1 and 2.

Figure 3 shows the Q^2 dependence of diffractive ZEUS data at fixed M_X and W, presented in the form of a $\gamma^* p$ cross section. As is the case for the inclusive cross section,[8,9] a transition takes place in the region of $Q^2 \sim 1$ GeV^2 between the high Q^2 scaling region and a low Q^2 region in which the cross section saturates.

3 Rapidity Gaps Between High $p_{\scriptscriptstyle T}$ Jets in Photoproduction

Diffractive scattering where both $W^2/|t|$ and $|t|$ are large corresponds to a region where Regge asymptotics are applicable, yet the

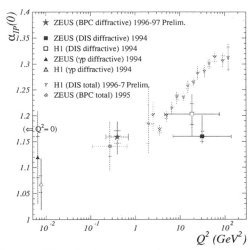

Figure 2. The effective pomeron intercept extracted from the energy dependence of selected inclusive and diffractive HERA data.

large scale set by $|t|$ encourages the use of perturbative QCD techniques.[5] The classic experimental signature is the appearance of regions devoid of hadronic activity between high $p_{\scriptscriptstyle T}$ jets. A complication arises from possible re-interactions between the beam remnants, which can destroy the rapidity gaps in a non-perturbatively calculable manner. The H1 collaboration has devised a new means of presenting the data in an attempt to decouple the underlying event from the dynamics of the hard interaction.[10] Dijet events are studied where the two highest $p_{\scriptscriptstyle T}$ jets have transverse momenta in excess of 6 GeV and 5 GeV respectively. As illustrated in figure 4, an event is classified as a 'gap' event if the sum $E\!\!\!/_t^{jets}$ of all transverse energy not allocated to the jets yet in the pseudorapidity region between the jet axes is less than a specified value, E_t^{cut}. The fraction f of dijet events for which $E\!\!\!/_t^{jets} < E_t^{cut}$ is measured as a function of E_t^{cut} and other kinematic variables. The E_t^{cut} dependence is sensitive essentially to the spectator interactions only and can be used to tune Monte Carlo simulations of the underlying event. The other kinematic variables are chosen to be sensitive to the dynamics of the hard interaction.

450

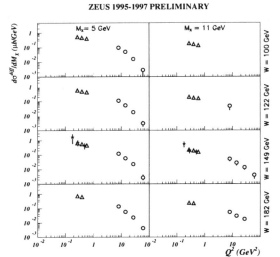

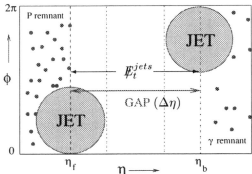

Figure 4. Illustration in $\eta-\phi$ space of the event topology and variable definitions in the rapidity gaps between jets analysis.

Figure 3. ZEUS measurements of the diffractive cross section, shown as a function of Q^2 at fixed M_X and W. The open triangles show the BPC data extracted from the M_X decomposition method. The stars show the BPC data extracted by direct leading proton tagging. The open circles are from a previous analysis.[4]

Choosing $E_t^{cut} = 1$ GeV, figure 5 shows the dependence of the gap fraction on the pseudorapidity separation of the two jets, $\Delta\eta$. The distribution is steeply falling at low values of $\Delta\eta$, where rapidity gaps arise dominantly from fluctuations in the hadronisation of standard photoproduction events. At the largest accessed values of $\Delta\eta$, the distribution becomes flatter, having a value similar to a previous ZEUS measurement[11], but significantly larger than that obtained in related measurements of $p\bar{p}$ scattering at the Tevatron.[12] The data are compared with four simulations. Those labelled 'PYTHIA' and 'HERWIG + JIMMY' are based on standard photoproduction matrix elements and contain models for multiple interactions. These simulations describe the data only at small $\Delta\eta$. The data show a clear excess over these models at large $\Delta\eta$, indicating a need for strongly interacting colour singlet exchanges at high $|t|$. Due to its highly peripheral t dependence,[13] the diffractive exchange studied in low $|t|$ diffraction at HERA[2,3,4] can-

not explain the large rapidity gaps in dijet events. Two models that contain high $|t|$ colourless exchanges are compared with the data in figure 5. The first, labelled 'HERWIG + JIMMY + BFKL' includes a calculation of elastic parton scattering $qq \to qq^a$ at large momentum transfer,[14] based on the leading order BFKL equation with α_s set to 0.17. The second, labelled 'PYTHIA + $(\gamma \times 1200)$', models the colour singlet exchange using photon exchange, tuned to the data by scaling the calculated cross section by an unphysical factor of 1200. Since the BFKL exchange couples dominantly to gluons[b] whereas photon exchange couples exclusively to quarks, the latter model serves as a control to assess the sensitivity of the data to the nature of the external couplings. Both models containing colour singlet exchanges are able to describe the $\Delta\eta$ dependence of the gap fraction.

Since the relative sizes of the gluon and quark densities in the proton vary strongly with x, a variable that is more sensitive to the nature of the external couplings is the fractional momentum x_p^{jets} of the parton entering the hard scattering from the proton. x_p^{jets} is

---

[a] Antiquark and gluon couplings of the exchange are also implied

[b] The colour factors at the external vertices result in relative couplings of 81 : 16 for gluons compared to quarks.

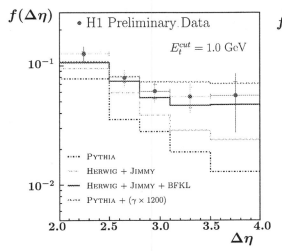

Figure 5. The gap fraction measured as a function of the pseudorapidity separation $\Delta\eta$ of the two leading jets.

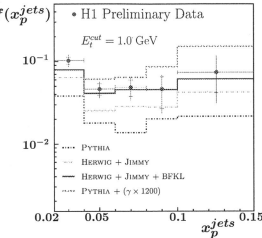

Figure 6. The gap fraction for events with $\Delta\eta > 2.5$, measured as a function of a hadron level estimator x_p^{jets} of the momentum fraction of the parton from the proton entering the hard scattering.

reconstructed from the total $E + p_z$ of the di-jet system[10]. The dependence of the gap fraction on x_p^{jets} is shown in figure 6 for the region $\Delta\eta > 2.5$, again with $E_t^{cut} = 1$ GeV. There is a signal above standard photoproduction predictions at all values of x_p^{jets}. The 'HERWIG + JIMMY' and 'PYTHIA + ($\gamma \times 1200$)' predictions differ in shape. The shape of the data distribution is more similar to the BFKL than to the photon exchange model, suggesting that the data contain a significant fraction of couplings to gluons, which at the present level of precision can be described by the BFKL calculation. Improved statistics are required before firm conclusions can be drawn.

Acknowledgments

The author thanks his colleagues from H1 and ZEUS for assistance in preparing the talk and the UK Particle Physics and Astronomy Research Council (PPARC) for funding.

References

1. P. Newman, *'Diffractive Phenomena at HERA'*, hep-ex/9901026.

H. Abramowicz, *'Diffraction and the Pomeron'*, hep-ex/0001054.

2. H1 Col., *Z. Phys.* C**74**, 221 (1997). ZEUS Col., *Z. Phys.* C**75**, 421 (1997).

3. H1 Col., *Z. Phys.* C**76**, 613 (1997).

4. ZEUS Col., *E.P.J.* C**6**, 43 (1999).

5. J. Forshaw, P. Sutton, *E.P.J.* C**1**, 285 (1998).

6. H1 Col., Paper 571, ICHEP98, Vancouver, Canada (1998).

7. ZEUS Col., Paper 875, ICHEP00, Osaka, Japan (2000).

8. H1 Col., Paper 534, ICHEP98, Vancouver, Canada (1998). ZEUS Col., *E.P.J.* C**7**, 609 (1999). H1 Col., Paper 944, ICHEP00, Osaka, Japan (2000).

9. ZEUS Col., *Phys. Let.* B**487**, 53 (2000).

10. H1 Col., Paper 962, ICHEP00, Osaka, Japan (2000).

11. ZEUS Col., *Phys. Let.* B**369**, 55 (1996).

12. D0 Col., *Phys. Let.* B**440**, 189 (1998). CDF Col., *Phys. Rev. Let.* **81**, 5278 (1998).

13. ZEUS Col., *E.P.J.* C**1**, 81 (1998).

14. A. Mueller, W. Tang, *Phys. Let.* B **284**, 123 (1992).

452

FINAL STATES IN DIFFRACTION AT HERA

M. MARTINEZ

(on behalf of the H1 and ZEUS Collaborations)
Deutsches Elektronen Synchrotron
Notkestrasse 85, 22607 Hamburg, Germany
E-mail: martinez@mail.desy.de

Hadronic final states in diffractive deep inelastic scattering have been studied by the H1 and ZEUS Collaborations. Cross sections for dijet and three-jet production have been measured and compared with different model predictions.

1 Introduction

Diffraction in deep inelastic scattering (DIS) has been interpreted in terms of the exchange of a colour-singlet object known as the Pomeron. In factorizable models, the Pomeron is assumed to have partonic structure. In this theoretical framework, the Pomeron parton densities, extracted from the HERA data, suggest a Pomeron dominated by gluons.

A number of perturbative QCD inspired models have been proposed to describe diffraction in DIS. In these models, the virtual photon dissociates into a $q\bar{q}$ or $q\bar{q}g$ state which interacts with the proton by an exchange of two gluons. The $q\bar{q}g$ final state contribution is found to dominate the diffractive cross section at high masses.

The study of the hadronic final state and, in particular, the measurements of dijet and three-jet production in diffractive DIS allow to separate the $q\bar{q}$ and $q\bar{q}g$ contributions and to test the validity of the different theoretical approaches supported by the observed agreement with inclusive measurements[1,2].

2 Energy flow in the $\gamma^* I\!\!P$ cms

The energy flow of the hadronic final state has been measured by the H1 and ZEUS Collaborations[3] as a function of the particle pseudorapidity [a], (η^*), in the $\gamma^* I\!\!P$ center-of-

---

[a] $\eta^* = -ln(tan(\frac{\theta^*}{2}))$, where θ^* is the polar angle defined with respect to the photon direction.

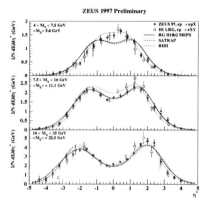

Figure 1. Measured energy flow in the $\gamma^* I\!\!P$ center-of-mass system as a function of the particle pseudorapidity, η^*, in different regions of M_X. The photon direction defines the positive pseudorapidities.

mass system . Figure 1 shows the measured energy flow in different regions of the total invariant mass, M_X. A clear dijet structure is observed for masses $M_X > 8$ GeV.

3 Dijet production

Dijet production in diffractive DIS has been studied by the H1 Collaboration[4] in the kinematic region $4 < Q^2 < 80$ GeV^2, $0.1 < y < 0.7$ and $x_{I\!\!P} < 0.025$ for jets in the region $p^*_{T,jet} > 4$ GeV and $-3 < \eta^*_{jet} < 0$, where the jets were searched for in the $\gamma^* p$ center-of-mass system using the cone algorithm with radius $R = 1$. After all selection cuts about 2500 dijet events were selected corresponding to an integrated luminosity of 18 pb^{-1}.

3.1 Dijet cross section versus $z_{I\!P}$

In a resolved gluon-dominated Pomeron model, dijet production comes from the boson-gluon-fusion hard interaction between the virtual photon and a gluon from the Pomeron. In such models, the dijet system provides direct information of the gluon content of the Pomeron, $g^D(z_{I\!P}, \mu^2)$, where $z_{I\!P}$ is the fraction of the Pomeron momentum carried by the gluon and μ is the relevant scale of the interaction.

In Figure 2, the cross section as a function of $z_{I\!P}^{\text{jets}}$ for dijet production is presented, where $z_{I\!P}^{\text{jets}}$ is reconstructed from the dijet system. The measurement is compared with the Monte Carlo (MC) predictions from a resolved Pomeron model with two different parameterizations for the gluon density in the Pomeron (also shown in the figure). The gluon distributions were extracted from a DGLAP analysis of the H1 inclusive measurements [1]. The measured cross section is well described by the model and shows large sensitivity to the gluon distribution at high $z_{I\!P}$, where a "flat" gluon solution (denoted as "H1 fit 2") is preferred. The measured $z_{I\!P}$ distribution peaks at rather low values of $z_{I\!P}$ ($z_{I\!P} \sim 0.2$). This indicates that a large fraction of the hadronic final state is not contained in the dijet system. It is interpreted in terms of the presence of a remnant with a large longitudinal momentum in the Pomeron direction.

3.2 Comparison with pQCD models

In Figure 3, the cross section as a function of $P_{T,\text{jets}}^*$, $z_{I\!P}$ and $P_{T,\text{rem}}^{(I\!P)}$ are shown, where $P_{T,\text{jets}}^*$ is the mean value of the transverse momentum of the dijets, and $P_{T,\text{rem}}^{(I\!P)}$ is defined as the total transverse momentum of the particles in the Pomeron hemisphere which are not associated to the dijet system. The measurements were performed in the region $x_{I\!P} < 0.01$, and are compared with the MC predictions from models based on colour

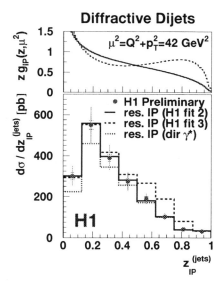

Figure 2. Measured differential cross section as a function of $z_{I\!P}$ for dijet production, compared with a resolved Pomeron model with different parameterizations for the gluon density in the Pomeron.

dipole cross sections and two-gluon exchange. As shown in the figure, the $q\bar{q}$ contribution is negligible and the cross sections are dominated by the $q\bar{q}g$ final state contribution. The Saturation model[5], taking only k_T ordered configurations into account, predicts cross sections a factor 2 too low. The model of Bartels et al.[6] (BJLW), in which the strong k_T ordering is not imposed, gives a reasonable description of the measurements when a cutoff for the gluon transverse momentum of $P_{T,g}^2 > 1.0$ GeV^2 is applied. Lower values of this cutoff are disfavored.

4 Three jet production

Exclusive three-jet production has been studied by the ZEUS Collaboration[7] in diffractive DIS in the kinematic region $5 < Q^2 < 100$ GeV^2, $200 < W < 250$ GeV, $23 < M_X < 40$ GeV and $x_{I\!P} < 0.025$, for events with $\eta_{\text{hadron}}^{\text{max}} < 3.0$. Jets were searched for, in the centre-of-mass system of the observed hadronic final state, using the exclusive k_T-

454

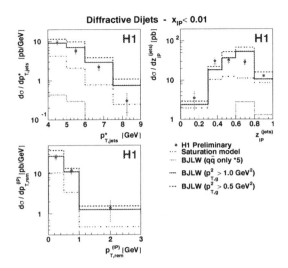

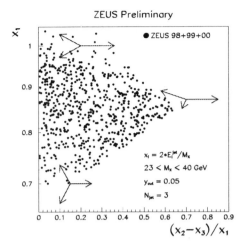

Figure 4. Distribution of the observed three-jet sample in the $(x_1, (x_2 - x_3)/x_1)$ plane.

Figure 3. Measured differential cross sections as a function of $P^*_{T,\text{jets}}$, $z_{I\!\!P}$ and $P^{(I\!\!P)}_{T,\text{rem}}$ for dijet production, compared with different pQCD-inspired models.

algorithm with $y_{\text{cut}} = 0.05$. Events with jets in the region $|\eta^{\text{jet}}_{\text{lab}}| > 2.3$ were removed from the sample, where $\eta^{\text{jet}}_{\text{lab}}$ denotes the pseudorapidity of the jet boosted back to the laboratory frame. After all requirements 678 three-jet events were selected corresponding to a luminosity of 39 pb^{-1}.

4.1 Three-jet final state topologies

The topology of the three-jet final states was described using the fractional energy variables $x_i = (2 \cdot E^{\text{jet}}_i)/M_X$, where E^{jet}_i ($i = 1 - 3$) denotes the jet energy, and the jets are sorted by their energies ($x_1 \geq x_2 \geq x_3$). Figure 4 shows the distribution of the observed three-jet sample in the $((x_2 - x_3)/x_1, x_1)$ plane. As indicated in the figure, different regions in the plane correspond to different three-jet topologies. The data show that all configuration are present, including those for which the the three jets have similar energies.

4.2 Energy flow in the event plane

The energy flow, as measured in the event plane, was studied for the three-jet sample as a function of the azimuthal angle in the plane, ϕ^*. In Figure 5, the normalized energy flow in the event plane is presented for different regions of the $((x_2 - x_3)/x_1, x_1)$ plane. By construction, the most energetic jet is at $\phi^* = 0$ while the second and third most energetic jets are located at positive and negative ϕ^* values, respectively. A clear three-jet structure is observed in all regions.

4.3 Cross section as a function of η^{jet}_i

The differential cross section as a function of the jet pseudorapidity, $d\sigma/d\eta^{\text{jet}}_i$, measured with respect to the $\gamma^* I\!\!P$ axis, was determined for three-jet production. For each event all jets were included.

The measured $d\sigma/d\eta^{\text{jet}}_i$ cross section is shown in Figure 6 together with the MC predictions based on a resolved gluon-dominated Pomeron (RAPGAP), the Saturation model (SATRAP) and pQCD calculation from Ryskin[8] (RIDI). The RAPGAP prediction for the cross section is consistent with the measurement[b]. The SATRAP and

---

[b]The measurements were not corrected for the contri-

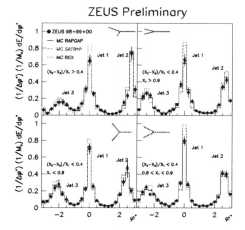

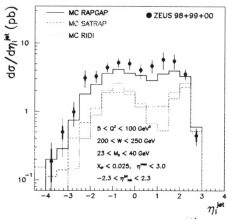

Figure 5. Measured energy distribution in the three-jet plane, normalized to the total invariant mass, as a function of the azimuthal angle, ϕ^*, in the different regions of the $((x_2 - x_3)/x_1, x_1)$ plane.

Figure 6. Differential cross section, $d\sigma/d\eta_i^{jet}$, as a function of the pseudorapidity of the three jets in three-jet production. The virtual-photon direction defines the negative hemisphere ($\eta < 0$).

RIDI absolute predictions are found to be about a factor 2 lower than the measured cross section. The shape of the measured distribution is reproduced by both RAPGAP and SATRAP, although both SATRAP and RIDI predict an asymmetric double-peak structure around $\eta^{jet} \simeq -1$ and $2 < \eta^{jet} < 2.5$, which is not observed in the data. This tendency is much more pronounced in RIDI. It reflects the presence in these models of dominant three-jet topologies where the jet in the Pomeron direction ($\eta > 0$) is emitted almost collinear to the $\gamma^* I\!\!P$ axis, and for which the second and third jet are found in the photon hemisphere ($\eta < 0$).

As mentioned in section 3.2, the cross sections for these models are dominated by the $q\bar{q}g$ contribution where the gluon is emitted in the Pomeron direction. The observed disagreement with the measured cross section suggests that larger transverse momenta of the emitted gluons with respect to the $\gamma^* I\!\!P$ axis are required in the two gluon model in order to describe the data.

bution of proton dissociative processes (estimated to be of the order of 16% of the measured cross section).

The observed good agreement between all the measured cross sections and the predictions from a resolved gluon-dominated Pomeron model supports the validity of this picture to describe diffractive interactions.

References

1. H1 Collab., C. Adloff et al., Z. Phys. C 76 (1997) 613.
2. ZEUS Collab., J. Breitweg et al., Eur. Phys. J. C 6 (1999) 43.
3. ZEUS Collab., J. Breitweg et al., ICHEP'2000 Contributed paper N-436. H1 Collab., C. Adloff et al., Phys. Lett. B 428 (1998) 206.
4. H1 Collab., C. Adloff et al., ICHEP'2000 Contributed paper N-293.
5. M. Wüsthoff and K. Golec-Biernat, Phys. Rev. D 59 (1999) 014017.
6. J. Bartels, H. Jung, M. Wüsthoff, Eur. Phys. J. C 11 (1999) 111.
7. ZEUS Collab., J. Breitweg et al., ICHEP'2000 Contributed paper N-433.
8. RIDI (version 2.0), M.G. Ryskin and A. Solano, in DESY-PROC-1999-02.

HARD SINGLE DIFFRACTION IN $\bar{P}P$ COLLISIONS AT $\sqrt{S} = 630$ AND 1800 GEV

ANDRE SZNAJDER FOR THE DØ COLLABORATION

Universidade Estadual do Rio de Janeiro(UERJ), Brazil

E-mail: sznajder@fnal.gov

Using the DØ detector, we have studied events produced in $\bar{p}p$ collisions that contain large forward regions with very little energy deposition ("rapidity gaps") and concurrent jet production at center-of-mass energies of $\sqrt{s}$ =630 and 1800 GeV. The fractions of forward and central jet events associated with such rapidity gaps are measured and compared to predictions from Monte Carlo models. For hard diffractive candidate events, we use the calorimeter to extract the fractional momentum loss of the scattered protons.

1

Please see hep-ex/9912061 for information on
the contents of this talk

DEEPLY VIRTUAL COMPTON SCATTERING AT HERA

LAURENT FAVART

On behalf of the H1 and ZEUS Collaborations
Université Libre de Bruxelles, I.I.H.E., Belgium
E-mail: lfavart@ulb.ac.be

Results on Deeply Virtual Compton Scattering at HERA measured by the H1 and ZEUS Collaborations are presented. The cross section, measured for the first time, is reported for $Q^2 > 2\,\mathrm{GeV}^2$.

1 Introduction

Along these last years, the exclusive vector meson production (as ρ and J/Ψ) has been studied extensively at HERA and has provided very interesting results, in particular, testing for the domain of applicability and the relevance of perturbative QCD in the field of diffraction (see e.g.[1,2]). Here, we report the first analyses of a similar process, the Deeply Virtual Compton Scattering, consisting in the hard diffractive scattering of a virtual photon off a proton (Fig. 1). The DVCS process offers a new and comparatively clean way to study diffraction at HERA. In comparison to vector meson production it avoids large uncertainties on the theoretical predictions due to the meson wave-function. The largest interest comes from the access it gives to the skewed parton distributions of the proton[3].

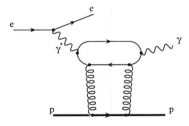

Figure 1. DVCS process.

The DVCS process contributes to the reaction $e^+p \rightarrow e^+\gamma p$, whose total cross section is dominated by the purely electromagnetic

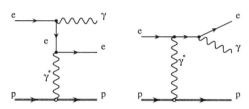

Figure 2. Bethe–Heitler process.

Bethe–Heitler process (Fig. 2).

Since the virtual photon is scattered onto mass shell in the final state it is necessary to transfer longitudinal momentum from the photon to the proton, i.e. forcing a nonforward kinematic situation[3,4]. In the picture of the two gluon exchange the fractional momentum of the gluons are therefore not equal, which implies the involvement of the skewed (or nonforward) parton distribution.

In presence of a hard scale the DVCS process can be completely calculated in perturbative QCD. In the present case, the photon virtuality Q^2, above a few GeV^2, insures the presence of a hard scale. LO QCD calculations exist, based on the two gluon exchange model[5]. The factorization theorem in perturbative QCD having been demonstrated[4,6], the DVCS process provides a unique way to extract these distributions from experimental data through the measurement of the asymmetry of the photon azimuthal angle distribution due to the interference with the Bethe–Heitler process[7].

From an experimental point of view, HERA kinematic enables the DVCS process to be studied in a large range in Q^2, W and t and offers the possibility to study this diffractive mechanism in detail.

2   Analysis strategy

Around the interaction region both experiments, H1 and ZEUS, are equipped with tracking devices which are surrounded by calorimeters. Since the proton escapes the main detector through the beam pipe only the scattered electron and photon are measured. Therefore the event selection is based on demanding two electromagnetic clusters, one in the backward and one in the central or forward part of the detector ($\theta \lesssim 140^o$ - the backward direction ($\theta = 0$) is defined as the direction of the incoming electron). If a track can be reconstructed it has to be associated to one of the clusters and determines the electron candidate. To enhance the DVCS contribution in comparison to the Bethe–Heitler process the phase space has to be restricted by demanding the photon candidate in the forward part of the detector.

The H1 analysis selects more specifically the elastic component by using, in addition, detectors which are placed close to the beam pipe and which are used to identify particles originating from proton dissociation processes.

3   Results

3.1   ZEUS

The first observation of the DVCS process was reported by the ZEUS collaboration in 1999[8]. In the analysis a photon virtuality $Q^2 > 6\,\text{GeV}^2$ is demanded. In Fig. 3 the polar angular distribution of the photon candidates is shown. A clear signal above the expectations for the Bethe–Heitler process

is observed. The LO calculation including the DVCS and the Bethe–Heitler processes achives a good description of the experimental data. A clear DVCS signal is still seen after the photon energy cut is increased. Also a shower shape analysis of the calorimetric clusters was performed that shows that the signal originates from photons and not from π^0 background.

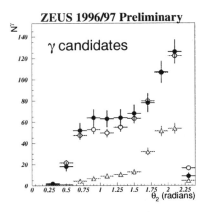

Figure 3. Distribution (uncorrected) of the polar angle of the photon candidate with an energy above 2 GeV. Data correspond to the full circles. The prediction for the Bethe–Heitler process is indicated by the open triangles. The prediction of Frankfurt et al. based on calculations including the Bethe–Heitler and the DVCS process is shown by the open circles.

3.2   H1

In the H1 analysis, the DVCS cross section is measured in the kinematic region: $2 < Q^2 < 20\,\text{GeV}^2$, $|t| < 1\,\text{GeV}^2$ and $30 < W < 120\,\text{GeV}$. The proton dissociation background has been estimated at around 10% and subtracted statistically assuming the same W and Q^2 dependence as for the elastic component. The acceptance, initial state radiation of real photons and detector effects have been estimated by MC to extract the elastic cross section.

In Fig. 4 the differential cross sections as a function of Q^2 and of W are shown. The data are compared with the Bethe–Heitler prediction alone and with the full calculation including Bethe–Heitler and DVCS. The description of the data by the calculations is good, in shape and in absolute normalization when a t slope is chosen between 7 and 10 GeV^{-2}.

It is important to notice that, at the LO, the interference term cancels when integrating over the azimuthal angle of the final state photon (as in the differential cross sections in Q^2 and of W).

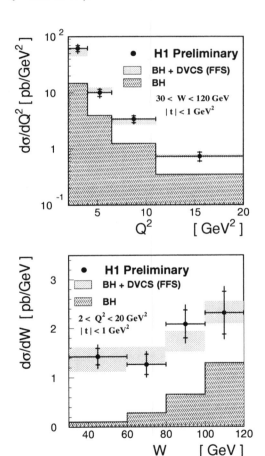

4 Conclusion

The DVCS process has been observed by the H1 and ZEUS Collaborations. Cross section measurements have been presented for the first time. The experimental results are well described by the calculations of Frankfurt et al.

References

1. P. Schleper, these proceedings.
2. P. Marage, "Hard Diffraction in Vector Meson Production at HERA", Review presented at the XXVIII International Symposium on Multiparticle Dynamics, hep-ph/9904255.
3. A. V. Radyushkin *Phys. Rev.* D **56**, 5524 (1997).
4. J. C. Collins, A. Freund, *Phys. Rev.* D **59**, 074009 (1999).
5. L. L. Frankfurt, A. Freund, M. Strikman *Phys. Rev.* D **58**, 114001 (1998) and *Phys. Rev.* D **59**, 119901E (1999).
6. X.-D. Ji and J. Osborne, *Phys. Rev.* D **58**, 094018 (1998).
7. A. Freund *Phys. Lett.* B **472**, 412 (2000).
8. P. R. B. Saull, Proceedings of the Int. Europhysics Conf. on High Energy Physics, Tampere, Finland, 15-21 July 1999, edited by K. Huitu, H. Kurki-Suonio and J. Maalampi (hep-ex/0003030).

Figure 4. The measured cross sections of the reaction $e^+p \rightarrow e^+\gamma p$ as a function of Q^2 and W are shown and compared to theoretical predictions. The uncertainty in the theoretical prediction, shown here as a shaded band is dominated by the unknown slope of the t-dependence of the DVCS part of the cross section, assuming $7 < b < 10\,\text{GeV}^{-2}$

BOSE EINSTEIN CORRELATIONS, COLOUR RECONNECTION ETC. AT THE Z^0

H. PALKA

H. Niewodniczanski Institute of Nuclear Physics, Ul. Kawiory 26A
30-055 Krakow, Poland
E-mail: henryk.palka@ifj.edu.pl

This report summarizes recent LEP papers on different aspects of particle correlations phenomena in the process $e^+e^- \to Z^0 \to hadrons$: intermittency and multiparticle correlations, oscillations of moments of the charged multiplicity distribution and Bose Einstein correlations in charged kaon and pion pairs.

1 Introduction

High statistics data samples of hadronic Z^0 decays recorded by four LEP experiments ($\sim 4 million$ events each) are valuable source of information on production mechanism of hadrons. The elementary interaction in the initial e^+e^- state is accurately known, the partonic phase of the hadronization is adequately described by perturbative QCD, thus constituting a clean environment to study the less known phase of formation of hadrons. This report reviews contributed papers on studies of charged multiplicity distribution from OPAL [1] and L3 [2], and new results on BE correlations in charged kaon pairs from OPAL [3], and in charged pion pairs from DELPHI [4], and OPAL [5].

2 Intermittency and multiparticle correlations

Intermittency [6] postulates scaling of the multiparticle spectra as a consequence of self-similarity (fractality) of the underlying production process. The intermittent behaviour of a distribution implies that at ever decreasing bin size the distribution becomes wider, which corresponds to the existence of particle correlations. The scaling hypothesis has been originally formulated in terms of the factorial moments. Currently more sophisticated tools exist like corrected modi-fied factorial moments which account for non-uniformity of single-particle distributions, or corrected modified cumulant moments used to extract the genuine multiparticle correlations. OPAL analysis [1] used these tools to study intermittency and correlations in very fine bins (up to 300 bins were used) of multidimensional space. Rapidity, azimuthal angle and transverse momentum were chosen as kinematic variables. Both the factorial moments and cumulants are measured up to the fifth order. The study of factorial moments reveals larger intermittency effect in rapidity than in azimuthal angle and no effect in transverse momentum. The saturation of factorial moments at small bin sizes is observed in rapidity and azimuthal angle. The intermittent behaviour is enhanced in multidimensional space, and no saturation effect is observed there. The values of cumulants are large and positive, indicating significant genuine multiparticle correlations up to four particles in one dimension and significant five-particle corellations in higher dimensions. Monte Carlo codes HERWIG 5.9 and JETSET 7.4 describe the data qualitatively well, however underestimate measurements for small bin sizes.

L3 analysed [2] the shape of charged multiplicity distribution in terms of the ratio $H_q = \frac{K_q}{F_q}$ of the normalized cumulant factorial moments and normalized factorial moments of order q . H_q measures the genuine q-particle

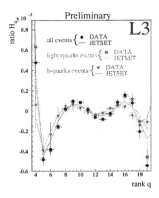

Figure 1. The H_q determined from the charged-particle multiplicity distribution ([2]).

correlation integral relative to the density integral. The H_q have been calculated for the soft gluon multiplicity in different approximations of perturbative QCD, and in the NNLLA exhibits oscillatory pattern in q. Based on the validity of the Local Parton Hadron Duality hypothesis (LPHD), such behaviour can be also expected for real charged-particle multiplicity distribution. Oscillations of the H_q have been observed by the SLD Collaboration and this have been interpreted as a confirmation of the NNLLA prediction. L3 Collaboration measurement of the H_q for all events and subsamples of light quark and b-quark tagged events (Fig. 1) confirms the oscillatory behaviour. However, since the H_q moments do not seem to be sensitive to quark mass effects and jet multiplicity analysis shows the effect only for very small y_{cut}, they conclude that the oscillations are unlikely to be related to the behaviour predicted by the NNLLA perturbative QCD calculations.

3   Bose Einstein correlations

Information on BE correlations in kaon pairs in e^+e^- data is scarce. Only one measurement exists for K^+K^- (DELPHI). For K^0K^0 three measurements were done (ALEPH, OPAL, DELPHI) but there BE correlation might be obscured by the threshold scalar resonanse $f_0(980)$. The determination of the correlation radius R_{KK} for kaons is important to clarify a possible dependence of the source dimension on mass of bosons.

OPAL [3] has presented a new measurement of $R_{K^\pm K^\pm}$ based on full statistics of hadronic Z^0 decays. Measurement was done for subsample of two-jet events selected by $Thrust > 0.95$ cut. Selected sample of identical charged kaon pairs has a purity of 48%. A reference sample was obtained by mixing. The correlation function was studied using a double ratio formed by the number of like-sign pairs normalised by a reference sample in the data, divided by the same ratio in a Monte Carlo simulation. Using a Gaussian parametrisation of the correlation function the parameters of the BE correlations were measured to be $\lambda = 0.82 \pm 0.22^{+0.17}_{-0.12}$ for the strength and $R_0 = 0.56 \pm 0.08^{+0.08}_{-0.06} fm$ for the source radius. Corrections for final-state interactions and the effect of short lived sources were not done, but they were found to be negligible. The measured kaon source radius is similar to the pion source radius when the measurement was done with the same technique (two jets,mixed reference sample,double ratio). Thus, this measurement does not support a strong mass dependence of the emitting source.

Earlier results, and particularly measurements for $\Lambda\Lambda$ pairs, suggest rather strong mass dependence, therefore a question if $R_0(\pi\pi) > R_0(KK)$ still needs to be clarified. Mass dependence of the correlation radius may be understood from the uncertainty principle [8] if the observed correlation radii correspond exactly to the actual size of the emission region. It is also obtained in a classical picture of particle production from expanding tube [9] based on the Gottfried-Bjorken hypothesis of in-out cascade (Fig. 2).

DELPHI has presented [4] new result on charged pions BE correlations as a part of the study of correlation function in hadronic

462

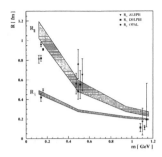

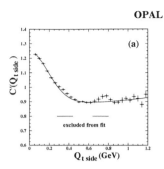

OPAL

Figure 2. Mass dependence of $R_{\parallel}$ and $R_{\perp}$ ([9]). Points represent 1-dim. results from LEP data.

Figure 3. The projection of the 3-dim. correlation function onto Q_l ([5]).

Z^0 decays and both fully hadronic and mixed hadronic-leptonic W^+W^- decays. The analysis for W-pairs is presented in another talks [7]. In the Z^0 analysis Monte Carlo like-sign pairs were used as a reference sample. The study was done for all like-sign pairs and for a subsample depleted with b quark events. The measured correlation radius is compatible for the two samples ($R_{all} = 0.573 \pm 0.016 fm$), while the strength parameter is 30% higher for non-$b\bar{b}$ subsample ($\lambda_{non-b\bar{b}} = 0.306 \pm 0.009$).

OPAL [5] has presented 3 dimensional analysis of correlations of charged pions in full Z^0 sample. Unlike pion pairs from the data are used as a reference sample, with contributions from K_s^0 and resonances subtracted and corrected for Coulomb interactions. The correlation function was analysed in three components ($Q_l, Q_{tside} and Q_{tout}$) of the four-momentum difference calculated in Longitudinally CoMoving System (LCMS). The result of the fit with 3-dimensional Goldhaber formula for two-jet events is shown in Fig. 3. The fit yields 20% elongation of the pions source, in agreement with earlier measurements of L3 and DELPHI. Such elongation is expected in the string phenomenology of $e^+e^- \to hadrons$ [10]. It is also obtained in a classical picture of the model considered in [9], where the change in the correlation radius results from the correlation between four-momentum of the particle and the space-time coordinate at which it is produced.

Acknowledgments

This study was partially supported by the KBN Grant No 2P0311116.

References

1. OPAL, CERN-EP/99-009 (April 1999), contribution 117 to this conference
2. L3, Note 2561 (July 2000), contribution 643 to this conference
3. OPAL, Note (July 2000), contribution 101 to this conference
4. DELPHI, Note DELPHI 2000-115 CONF 414 (July 2000), contribution 637 to this conference
5. OPAL, CERN-EP-2000-004 (January 2000), contribution 104 to this conference
6. A.Bialas and R.Peschanski, Nucl. Phys. B273 (1986) 703
7. N.Pukhaeva these proceedings ; A.Valassi these proceedings
8. G.Alexander et al. Phys.Lett. B452, 159 (1999)
9. A.Bialas et al., Phys.Rev. D62, 114007 (2000)
10. B.Andersson and M.Ringner, Phys.Lett. B421, 283 (1998)

CONSTITUENT QUARK AND HADRONIC STRUCTURE IN THE NEXT-TO-LEADING ORDER

FIROOZ ARASH

Institute for Studies in Theoretical Physics and Mathematics,P.O. Box 19395-5531, Tehran, Iran and Tafresh University, Tafresh, Iran E-mail: arash@vax.ipm.ac.ir

ALI N. KHORRAMIAN

Institute for Studies in Theoretical Physics and Mathematics,P.O. Box 19395-5531, Tehran, Iran and Semnan University, Semnan, Iran E-mail: khorramiana@theory.ipm.ac.ir

Structure of a constituent quark in calculated in the NLO. It is applied to generate the Structure function of Proton. It is found that $SU(2)$ symmetry breaking is due to fluctuation of binding gluon to $\bar{q}q$ pairs which contributes to the hadronic structure by as much as 3-5 percents.

We calculate the structure function of a Constituent Quark (CQ) in the Next-to-leading order in QCD. Using the convolution theorem, structure functions other hadrons can easily be obtained. We present results for Proton structure function F_2^p.

By definition a CQ is a universal building block for every hadron. Its structure is universal. At hight enough Q^2 it is the structure of a CQ which is probed in DIS experiment and at low Q^2 this structure cannot be resolved and it behaves as valence quark and hadron is viewed as the bound state of its CQs. A CQ receives it structure by dressing of a valence quark in QCD. For a U-type CQ one can write its structure function as:

$$F_2^U(z,Q^2) = \frac{4}{9}z(G_{\frac{u}{U}} + G_{\frac{\bar{u}}{U}}) +$$

$$\frac{1}{9}z(G_{\frac{d}{U}} + G_{\frac{\bar{d}}{U}} + G_{\frac{s}{U}} + G_{\frac{\bar{s}}{U}}) + ... \quad (1)$$

where all the functions on the right-hand side are the probability functions for quarks having momentum fraction z of a U-type CQ at Q^2. Defining singlet (S) and nonsinglet (NS) CQ distributions as:

$$G^S = \sum_{i=1}^{f}(G_{\frac{q_i}{CQ}} + G_{\frac{\bar{q}_i}{CQ}}) \quad (2)$$

$$G^{NS} = \sum_{i=1}^{f}(G_{\frac{q_i}{CQ}} - G_{\frac{\bar{q}_i}{CQ}}) \quad (3)$$

we evaluate the moments of these distributions in NLO. Our initial scale is $Q_0^2 = 0.283$ GeV^2 and $\Lambda = 0.22\ GeV$. The moments of the CQ structure function, $F_2^{CQ}(z,Q^2)$ are expressed completely in terms of evolution parameter $t = ln\frac{ln\frac{Q^2}{\Lambda^2}}{ln\frac{Q_0^2}{\Lambda^2}}$. The moments of valence and sea quarks in CQ are:

$$M_{\frac{valence}{CQ}} = M^{NS}(N,Q^2) \quad (4)$$

$$M_{\frac{sea}{CQ}} = \frac{1}{2f}(M^S - M^{NS}) \quad (5)$$

where $M^{S,NS}$ are singlet and nonsinglet moments. Evaluating $M_{\frac{valence}{CQ}}$ and $M_{\frac{sea}{CQ}}$ at any Q^2 or t is now straight forward. The corresponding parton distributions in a CQ are parameterized as:

$$zq_{\frac{val.}{CQ}}(z,Q^2) = az^b(1-z)^c \quad (6)$$

$$zq_{\frac{sea}{CQ}}(z,Q^2) = \alpha z^\beta(1-z)^\gamma[1+\eta z+\xi z^{0.5}] \quad (7)$$

The parameters a, b, c, α, etc. are functions of Q^2 through the evolution parameter t. This completes the structure of a CQ . The results are shown in Fig.(1).

Next we go to the structure of a hadron and we can write:

$$F_2^h(x,Q^2) = \sum_{CQ}\int_x^1 \frac{dy}{y}G_{\frac{CQ}{h}}(y)F_2^{CQ}(\frac{x}{y},Q^2) \quad (8)$$

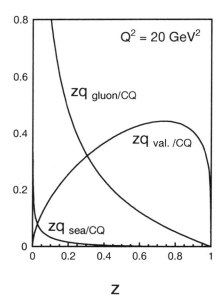

Figure 1. Various parton distributions in a Constituent Quark.

where summation runs over the number of CQ's in a particular hadron. Also notice that $G_{\frac{CQ}{h}}(y)$ is independent of the nature of the probe and its Q^2 value. In effect $G_{\frac{CQ}{h}}(y)$ describes the wave function of hadron in CQ representation containing all the complications due to confinement. From the theoretical point of view this function cannot be evaluated accurately. To facilitate phenomenological analysis, following [1] we assume a simple form for the exclusive CQ distribution in proton and pion as follows:

$$G_{UUD/p}(y_1, y_2, y_3) = l(y_1 y_2)^m y_3^n \delta(y_1+y_2+y_3-1) \quad (9)$$

After integrating out unwanted momenta, we can arrive at inclusive distribution of individual CQ:

$$G_{U/p}(y) = \frac{1}{B(m+1, n+m+2)} y^m (1-y)^n \quad (10)$$

$$G_{D/p}(y) = \frac{1}{B(n+1, 2m+2)} y^n (1-y)^{2m+1} \quad (11)$$

numerical values are: $m = 0.65$ and $n = 0.35$.

Having evaluated all ingredients needed to complet eq.(6) we calculate the proton Structure function. it turns out that the model results fall 3-5 percents bellow the experimental results. We attribute this shortfall to the presence of soft gluons in proton which binds CQs to form a physical hadron. A CQ can radiate a gluon which in turn decays into a pair of $\bar{q} - q$. After such a pair is created a $\bar{u}$ can couple to a D-type CQ to form an intermediate $\pi^- = D\bar{u}$ while the u quark combines with the other two U-type CQs to form a Δ^{++}. Similarly, a $\bar{d}d$ can fluctuate into a $\pi^+ n$ state. Since Δ^{++} state is more massive than n state, the probability of $\bar{d}d$ fluctuation will dominate and that leads to an excess of $\bar{d}d$ pairs over $\bar{u}u$. This process is responsible for the $SU(2)$ symmetry and the violation of Gottfried Sum rule [2]. The result of such a calculation is depicted in Fig. (2) and it is compared with the experimental results from Fermilab E866.

Adding this nonperturbative contribution to the constituent quark contribution will complete the evaluation of proton structure function F_2^p. As it can be seen in Fig. (3) the results of model calculation agrees rather well with HERA results in a wide range of kinematics both in x and Q^2.

References

1. R. C. Hwa and M. S. Zahir, Phys. Rev. D **23**, 2539 (1981);

2. Firroz Arash and Ali N. Khorramian, hep-ph/0003157.

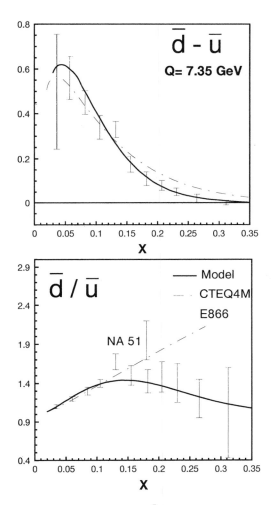

Figure 2. $\bar{d} - \bar{u}$ and $\frac{\bar{d}}{\bar{u}}$ as function of x.

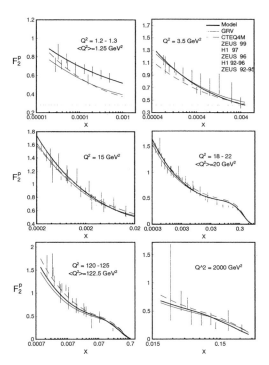

Figure 3. proton structure function, F_2^p as a function of x at various Q^2 values.

INCLUSIVE PARTICLE PRODUCTION IN TWO-PHOTON COLLISIONS AT LEP

CHRISTOPH SCHÄFER

CERN, CH-1211 Geneva 23, Switzerland
E-mail: Christoph.Schaefer@cern.ch

Two-Photon collision processes are studied with the L3 detector at the Large Electron Positron collider (LEP) at CERN. The inclusive particle production of K_s^0 and, for the first time at LEP, π^0 in the reaction $e^+e^- \to e^+e^-\gamma\gamma \to e^+e^-$ hadrons is analysed for quasi-real photons at center-of-mass energies, $\sqrt{s}$, for the incoming electron-positron pair of 189 GeV $< \sqrt{s} <$ 202 GeV. The differential cross sections of the neutral pions and kaons are measured as a function of their transverse momentum as well as their pseudo rapidity and compared to next-to-leading order perturbative QCD predictions.

1 Introduction

The LEP collider can also be seen as a photon factory, since the circulating electron and positron beams are surrounded by a cloud of photons of a wide energy spectrum. Due to the Abelian structure of QED, photons do not interact directly with themselves but can react within a "hadronic blob" (See Figure 1).

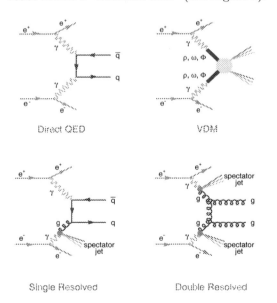

Figure 1. Some Feynman diagrams for Two-Photon collision processes.

Each photon can fluctuate into a vector meson with the same quantum numbers as the photon (vector dominance model), thus

initiating a strong interaction process with characteristics similar to hadron-hadron collisions. This process is dominant at low invariant mass of the two-photon system, i.e. where the final state hadrons have a low transverse momentum with respect to the incoming electron-positron pair. For outgoing hadrons with a high transverse momentum, direct QED processes contribute significantly as well as processes where the photon is resolved into partons (quarks and gluons). The study of inclusive single particle production of K_s^0 and π^0 in hadronic Two-Photon collisions allows to check leading order and next-to-leading order QCD predictions.

2 Event Selection

The data presented here were collected by the L3 detector in the years 1998 and 1999 at center-of-mass energies between 189 GeV and 202 GeV. The total integrated luminosity available for this analysis is 410 pb^{-1}, corresponding to an average center-of-mass energy of 194 GeV.

A detailed description of this analysis is given in [1], [2]. Events identified as hadronic Two-Photon collisions must have an energy deposit in the calorimeters of less than 40 % of the center-of-mass energy in order to reject annihilation reactions. In addition more than 5 particles have to be detected per event. In order to select only quasi-real Two-

Photon collisions, a maximum energy deposit of 70 GeV must not be exceeded in the luminosity monitors, the so-called "anti-tag" condition. The Q^2 value of these events is hence less than 8 GeV^2 with an average value of 0.015 GeV^2.

After applying these selection cuts, a background of 1 % remains, which predominantly consists of $e^+e^- \rightarrow$ hadrons and $e^+e^- \rightarrow e^+e^-\tau^+\tau^-$ events.

The π^0 is selected via its decay into a pair of photons. About $2 \cdot 10^6$ events are selected. The distribution of the reconstructed mass of the Two-Photon system shows a narrow π^0 peak with a mass of 134.6 MeV, consistent with Monte Carlo expectations.

The K_s^0 is selected via its decay into a pair of charged pions. The idea is to exploit the relatively long lifetime of the neutral kaon while observing a secondary decay vertex. The invariant mass distribution of the $\pi^+\pi^-$ system originating from this secondary vertex peaks at 497 MeV, which is in agreement with the mass of a neutral kaon. About $5 \cdot 10^5$ events are selected.

3   Data Analysis

The differential cross section as a function of the transverse momentum p_t is measured in a pseudo rapidity range of $|\eta| < 0.5$ (π^0) and $|\eta| < 1.5$ (K_s^0), respectively. (See Figure 2)

Furthermore the differential cross section as a function of $|\eta|$ is measured in a transverse momentum range of $p_t > 2$ GeV (π^0) and $p_t > 1.5$ GeV (K_s^0), respectively. (See Figure 3)

The influence of the detector acceptance, reconstruction efficiency, and the selection cuts are determined with the PHOJET Monte Carlo generator. The systematic uncertainty is derived by comparing the two Monte Carlo generators PHOJET and PYTHIA and found to be significantly smaller than the statistical error.

The cross sections shown in Figure 2 ex-

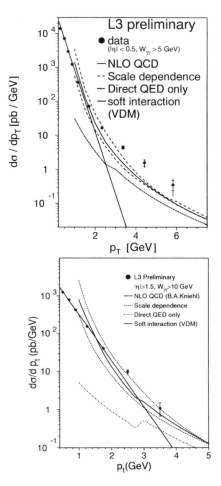

Figure 2. Differential cross section as a function of the transverse momentum, top: π^0, bottom: K_s^0

hibit an exponential behaviour at low transverse momentum, which is typical for soft interactions (VDM). At high p_t, the data points are clearly above the exponential curve, because of the existence of direct QED processes and hard QCD interaction.

The measured differential cross sections are compared to analytical next-to-leading order QCD predictions [4]. For this calculation, the flux of quasi-real photons is obtained using the Equivalent Photon Approximation. The renormalization, the factorisation as well as the fragmentation scales are taken to be equal: $\mu = M = M_F = \xi p_t$. The

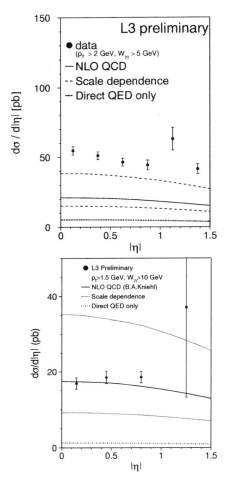

Figure 3. Differential cross section as a function of the pseudo rapidity, top: π^0, bottom: K_s^0

4 Conclusions

The inclusive production of K_s^0 and, for the first time at LEP, π^0 in Two-Photon collision processes is measured with the L3 detector. At low transverse momentum, the differential cross section $d\sigma/dp_t$ follows an exponential dependence like hadron-hadron and hadron-photon soft processes. At higher p_t, the contributions of QED and hard QCD processes are clearly visible. The available next-to-leading order QCD calculations predict cross sections smaller than observed.

Acknowledgements

I would like to thank the Two-Photon group of the L3 collaboration for giving me the results. Especially I want to thank Pablo Achard and Saverio Braccini for their help and discussions.

References

1. L3 Collaboration, Inclusive K_s^0 Meson Production in Two-Photon Collisions, Internal Note 2589, submitted to 30th ICHEP Osaka.

2. L3 Collaboration, Inclusive π^0 Production in Two-Photon Collisions at LEP, Internal Note 2594, submitted to 30th ICHEP Osaka.

3. OPAL Collaboration, Inclusive Production of Charged Hadrons and K_s^0 Mesons in Photon-Photon Collisions, Eur. Phys. J. **C 6** (1999) 253.

4. J. Binnewies, B.A. Kniehl and G. Kramer, Phys. Rev **D 53** (1996) 3573; B.A. Kniehl private communication.

uncertainty in the NLO calculations is estimated by varying the value ξ (central value: $\xi = 1$, upper dotted line: $\xi = 0.5$, lower dotted line: $\xi = 2$).

The differential cross sections for inclusive pion and kaon production as a function of the transverse momentum are compared to the OPAL measurements [3]. In case of the neutral pion production, the L3 result is corrected to be comparable with the OPAL measurements of inclusive charged hadron production (mainly charged pions). Good agreement between L3 and OPAL data is found.

COLOR TRANSPARENCY AND PION VALENCE QUARK DISTRIBUTIONS FROM DI-JET EVENTS IN FERMILAB E791

JEFFREY A. APPEL

For the Fermilab E791 Collaboration
Fermilab, PO Box 500, Batavia IL 60510, USA
E-mail: appel@fnal.gov

Diffractive, exclusive di-jet events produced by 500 GeV/c π^- scattered off nuclei were used to measure their A-dependence, and to make the first direct measurement of the valence-quark momentum distribution in pions. Data on the latter are compared to two limiting predictions for the pion light-cone wave-function. The results show that the asymptotic wave-function of perturbative QCD describes the data well for Q^2 of 10 GeV^2 and above. The measured A-dependence is consistent with observation of point-like configurations in the pion and color-transparency calculations.

1 Introduction

Color transparency (CT) is the name given to the prediction that the color fields of QCD cancel for physically-small singlet systems of quarks and gluons.[1] This color neutrality (or color screening) should lead to the suppression of initial and final state interactions of the small-sized systems in hard processes.[2] Observing color transparency requires that point-like configurations (PLC's) are formed and that the energies are high enough so that expansion of the PLC does not occur while traversing the target[3,4,5] (the "frozen" approximation). We use the A-dependence of di-jet production to test for the existence of PLC's in coherent interactions with nuclei.

Given that we find such exclusive di-jet, PLC interactions, we can use these to measure the valence quark distribution in pions. Here, we summarize two papers being submitted to Phys. Rev. Lett.[6].

2 Data Set and Event Selection

From a 10% subset of the Fermilab E791[7] 2×10^{10} recorded interactions, a selection was made to find those which had exclusively two-jets recoiling *coherently* from the carbon and platinum targets. On-line, only a single incident pion was allowed in the resolving time of the calorimeters and a loose minimum trans-verse energy requirement was made. Off-line, further selection was made by demanding that at least 90% of the incident pion momentum appear in charged particles with a total charge equal to -1. Using the JADE algorithm[8] optimized for di-jet finding, only events with two jets in the final state were accepted. The jets had to be back-to-back in the plane transverse to the beam within 20 degrees.

For each two-jet event, we calculated the transverse momentum of each jet with respect to the beam axis (k_t), the di-jet transverse momentum above the minimum for that di-jet mass (q_t), and the di-jet invariant mass (M_{di-j}). The di-jet invariant mass is related by simple kinematics to the quarks' longitudinal momentum fractions (x) in the pion infinite momentum frame: $M_{di-j}^2 = k_t^2/[x(1-x)]$. To assure clean selection of high-mass di-jet events, a minimum k_t of 1.2 GeV/c is required. The distribution of events vs x gives the square of the valence-quark wave-function, assuming that each of the two jets is a measure of the quark from which it came.

The size of a $|q\bar{q}\rangle$ system which produces di-jets with $k_t > 1.5$ GeV/c can be estimated as $1/Q \leq 0.1$ fm where $Q^2 \sim M_{di-j}^2 \sim 4k_t^2 \sim 10$ GeV^2/c^2. The distance that the $|q\bar{q}\rangle$ system travels before it expands appreciably, the coherence length, is given by $\ell_c \sim$

Table 1. Experimental results and color-transparency (CT) predictions[10] for α values in coherent pion dissociation off nuclei vs k_t.

k_t GeV/c	α	$\Delta\alpha$	α (CT)
$1.25 - 1.5$	1.64	$+0.06 \ -0.12$	1.25
$1.5 - 2.0$	1.52	± 0.12	1.45
$2.0 - 2.5$	1.55	± 0.16	1.60

$(2p_{\text{lab}})/(M_{di-j}^2 - m_\pi^2)$ [3] which is ~ 10 fm for $M_{di-j} \sim 5$ GeV/c^2. Therefore, we expect that the di-jet signal events selected in this analysis evolve from point-like configurations which will exhibit color transparency.[9,10]

3 Coherent Scattering and Color Transparency

We derive the numbers of produced di-jet events in the data for each target in three k_t bins by integrating over the coherent diffractive terms in fits of the MC-smeared distributions to the q_t^2 distributions of the di-jet events. Using the resulting yields and the known target thicknesses, we determine the ratio of cross sections for diffractive dissociation on platinum and carbon. The exponents α are then calculated using the cross section dependence $\sigma \propto A^\alpha$. The α values are listed in Table 1, as are CT theoretical predictions[10]. The α's are consistent with those predictions and above $k_t = 1.5$ GeV/c, clearly inconsistent with α values like those in $\sigma \propto A^{2/3}$ for incoherent scattering observed in other hadronic interactions.

4 Pion Light-Cone Wave-Function

The pion wave-function can be expanded in terms of Fock states:

$$\Psi = \alpha|q\bar{q}\rangle + \beta|q\bar{q}g\rangle + \gamma|q\bar{q}gg\rangle + \cdots . \quad (1)$$

For interactions in which pions transfer momentum to other particles over sufficiently

Table 2. Asymptotic (a_{as}) and CZ (a_{CZ}) wave-function contributions in fits of the data.

k_t GeV/c	a_{as}	$\Delta_{a_{as}}$	a_{CZ}	$\Delta_{a_{CZ}}$
1.25 - 1.5	0.64	+0.14 -0.12	0.36	$-\Delta_{a_{as}}$
1.5 - 2.5	1.00	+0.10 -0.14	0.00	$-\Delta_{a_{as}}$

short distances (for sufficiently high Q^2), the first component should be dominant.[11]

The pion light-cone wave-function is predicted[9,12,13] by perturbative QCD for asymptoticly large Q^2 to be

$$\Phi_{asy}(x) = \sqrt{3}x(1-x). \quad (2)$$

Using QCD sum rules, at low Q^2 Chernyak and Zhitnitsky (CZ) proposed[14]

$$\Phi_{CZ}(x) = 5\sqrt{3}(1-x)(1-2x)^2, \quad (3)$$

where x is the usual fractional momentum carried by the quark. In the measurements, we use $x = p_{jet1}/(p_{jet1} + p_{jet2})$.

For measurement of the pion wave-function, we used data from the platinum target only, since it has a sharp diffractive distribution and low background. We used events with $q_t^2 < 0.015$ GeV/c^2. In order to get a measure of the correspondence between the experimental results and the calculated light-cone wave-functions, we fit the results with a linear combination of squares of the two MC-smeared wave-functions. This assumes an incoherent combination of the two wave functions and that the evolution of the CZ function is slow (as stated in Ref. 14). The coefficients a_{as} and a_{CZ} representing the contributions of the asymptotic and CZ functions, respectively, are listed in Table 2. The results for the higher k_t window show clearly that the asymptotic wave-function describes the data very well. Thus, for $k_t > 1.5$ GeV/c, which translates to $Q^2 \sim 10$ (GeV/$c)^2$, the perturbative QCD approach that led to construction of the asymptotic wave-function is reasonable. The distribution in the lower window is consistent with a significant contribution from the CZ wave-function.

The k_t dependence of diffractive di-jets is an observable that can show how well the perturbative calculations describe the data. As shown in Ref. 10, assuming interaction via two gluon exchange and slowly-varying ϕ_{as} leads to $\frac{d\sigma}{dk_t} \sim (k_t)^n$, with n = - 6. For our data, the region above $k_t \sim 1.8$ GeV/c can be fit with $n = -6.5 \pm 2.0$ with $\chi^2/dof = 0.8$, consistent with perturbative QCD predictions. This supports the evaluation of the light-cone wave-function at large k_t.

5 Summary

We have observed pion scattering events which exhibit A^α dependence consistent with color transparency for coherent diffractive di-jet production off nuclei. These events exhibit k_t dependence transitioning from non-perturbative to the perturbative regime. In addition, using the events from the platinum target, we have made the first direct measurement of valence quark distribution in the pion. This wave-function is consistent with dominance of asymptotic form for k_t above $\sim 1.5\ GeV/c\ (Q^2 \sim 10\ GeV^2)$.

Acknowledgments

We thank S.J. Brodsky, L. Frankfurt, G.A. Miller, and M. Strikman for many fruitful discussions. We also thank the staffs of Fermilab and all the participating institutions for their assistance, and the Brazilian Conselho Nacional de Desenvolvimento Científico e Technológico, the Mexican Consejo Nacional de Ciencia y Technologica, the Israeli Academy of Sciences and Humanities, the US Department of Energy, the US-Israel Binational Science Foundation, and the US National Science Foundation for support.

References

1. F. E. Low, Phys. Rev. **D12** (1975) 163; S. Nussinov, Phys. Rev. Lett **34** (1975) 1286.
2. A.H. Mueller in Proceedings of Seventeenth Rencontre de Moriond, Les Arcs, 1982 ed. J Tran Thanh Van (Editions Frontieres, Gif-sur-Yvette, France, 1982) Vol. I, p 13; S.J. Brodsky in Proceedings of the Thirteenth Int'l Symposium on Multiparticle Dynamics, ed. W. Kittel, W. Metzger and A. Stergiou (World Scientific, Singapore 1982,) p 963.
3. G.R. Farrar, H. Liu, L.L. Frankfurt, and M.I. Strikman, Phys. Rev. Lett. **61** (1988) 686.
4. B.K. Jennings and G.A. Miller, Phys. Lett. **B236**, (1990) 209; B.K. Jennings and G.A. Miller, Phys. Rev. **D44** (1991) 692; Phys. Rev. Lett. **69** (1992) 3619; Phys. Lett. **B274** (1992) 442.
5. S.J. Brodsky and A.H. Mueller Phys. Lett. **B206**, 685 (1988).
6. E791 Collaboration, E.M. Aitala *et al.*, Fermilab-Pub-00/220-E and Fermilab-Pub-00/221-E.
7. E791 Collaboration, E.M. Aitala *et al.*, EPJdirect **C4**, 1 (1999).
8. JADE collaboration, W. Bartel *et al.*, Z. Phys. **C33**, 23 (1986).
9. G. Bertsch, S.J. Brodsky, A.S. Goldhaber, and J. Gunion, Phys. Rev. Lett. **47**, 297 (1981).
10. L.L. Frankfurt, G.A. Miller, and M. Strikman, Phys. Lett. **B304**, 1 (1993).
11. S.J. Brodsky and G.R. Farrar, Phys. Rev. Lett. **31**, 1153 (1973); G. Sterman and P. Stoler, Ann. Rev. Nuc. Part. Sci. **43**, 193 (1997).
12. S.J. Brodsky and G.P. Lepage, *Phys. Rev.* D **22**, 2157 (1980).
13. A.V. Efremov and A.V. Radyushkin, Theor. Math. Phys. **42**, 97 (1980).
14. V.L. Cernyak and A.R. Zhitnitsky, Phys. Rep. **112**, 173 (1984).

MASS, TOTAL AND TWO-PHOTON DECAY WIDTHS OF THE η_C MESON

YUICHI KUBOTA

School of Physics and Astronomy, 116 Church St. S.E., Minneapolis, MN 55455, USA
E-mail: yk@umn.edu

I will present recent measurements of the mass, total and two-photon decay widths of the η_c meson. The world average of the total decay width measurements have been substantially smaller than the theoretically expected value, but new measurements bring them into better agreement. The two-photon decay width sheds new light on the disagreement between experimental measurements and the QCD prediction for the rate of the radiative decay $J/\psi \to \gamma\eta_c$.

1 Introduction

The mass, total decay width and two-photon decay width of the η_c meson are important properties used to test our understanding of the charmonium states based on QCD. The hyperfine mass splitting between the J/ψ and η_c tells us the spin-spin coupling part of the charmonium potential energy. This in turn constrains the confining potential between charm quarks. Both the total width and two-photon partial width are predictable by QCD. In their ratio, non-perturbative contributions cancel, leading to a reliable prediction:

$$R_1 = \frac{\Gamma_{gg}}{\Gamma_{\gamma\gamma}} = \frac{9\alpha_s^2}{8\alpha^2}\frac{1 + 4.8\alpha_s/\pi}{1 - 3.4\alpha_s/\pi} = 3.4 \times 10^3.$$

The two-photon width relative to the electronic width of the J/ψ can also be predicted with minimal uncertainty arising from non-perturbative components:

$$R_2 = \frac{\Gamma_{\gamma\gamma}}{\Gamma_{J/\psi \to \ell\ell}} = \frac{4}{3}(1 + 1.96\alpha_s/\pi) = 1.57,$$

where the non-perturbative contributions are assumed to cancel each other.

The measured value[1] of $R_2 = 1.43^{+0.31}_{-0.27}$ agrees with the above theoretical expectation. The average of the R_1 measurements of $(1.8 \pm 0.6) \times 10^3$, however, does not agree with the prediction even though the cancellation of the non-perturbative contribution is thought to be better in R_1.

Theories predict the hyperfine mass splitting to be in the range 110 to 130 MeV. The

uncertainty arises mainly from the uncertainty in the form of the confining part of the charmonium potential and the value of α_s. The splitting is measured to be (117.1 ± 2.1) MeV and is consistent with the theoretical expectation, although there are 3σ- level inconsistencies among the measurements.

2 New Experimental Results

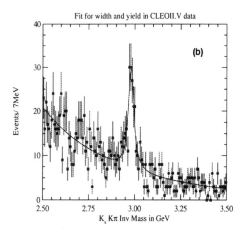

Figure 1. CLEO: $K_s K\pi$ mass distribution for the CLEO II.V data sample showing a clear η_c signal.

The CLEO and DELPHI experiments contributed papers[2] to this conference on the η_c. In both experiments, η_c's were produced in the two-photon process and observed in their decays to $K_S^0 K\pi$ mode. Figure 1 shows the η_c-candidate mass distributions for one of the two CLEO data sets. A clear η_c signal is evident. From the entire data sample, CLEO obtains the mass, total decay width and two-photon partial width. The total width is calculated from the observed width, accounting for the detector resolution of about 8 MeV. The two-photon width is obtained by comparing the rate of η_c production and the prediction of Monte Carlo simulations of the two-photon processes based on the formalism of Budnev et al.[3] Their results are

$$M_{\eta_c} = (2980.4 \pm 2.3 \pm 0.6)\text{MeV}$$
$$\Gamma_{\eta_c} = (27.0 \pm 5.8 \pm 1.4)\text{MeV, and}$$
$$\Gamma_{\eta_c}^{\gamma\gamma} = (7.6 \pm 0.8 \pm 0.4 \pm 2.3)\text{keV}.$$

The first error is the statistical error and the second is the systematic error. The third error on the two-photon width arises from the poorly measured branching fraction for $\eta_c \rightarrow K_S^0 K^\pm \pi^\mp$, which is assumed to be $(1.8 \pm 0.6)\%$.[1]

Since the mass resolution of DELPHI is much larger than the intrinsic width of the η_c, they obtain only the $\gamma\gamma$ width: $\Gamma_{\eta_c}^{\gamma\gamma} = (10.9 \pm 2.2 \pm 1.6 \pm 3.4)\text{keV}.$

I summarize the recent measurements of the mass, total width and two-photon partial width of the η_c in Table 1. The previous mass measurements from J/ψ decay experiments tended to be lower than those from $p\bar{p}$ experiments, and the two sets of measurements were inconsistent with each other $(\chi^2/\text{D.F.} > 3)$. The new CLEO measurement falls in-between the two, making the overall consistency among the mass measurements slightly better, while keeping the average value of the mass unchanged.

The ratio between the total and two-photon widths was substantially lower than what was expected from theory. Although the E760 experiment[4] agreed with the theoretical expectation, measuring a significantly larger total width than the average, other experiments, measuring narrower widths with smaller absolute errors (despite larger fractional errors), dominated the PDG world average. The new CLEO measurement as well as the preliminary E835[5] result agree with both the E760 total width and the theoretical expectation. The ratio of the two decay widths calculated from the new world averages is $(2.5 \pm 0.9) \times 10^3$ and consistent with the theoretical expectation of 3.4×10^3.

The new $\gamma\gamma$ decay width measurements are consistent with the previous world average and also consistent with the theoretical expectation.

There are, however, a few areas where QCD predictions on charmonium properties do not agree with experimental measurements. The most conspicuous one is the rate of radiative J/ψ decay to η_c. The prediction overestimates the rate by a factor of 2.5. The rate is measured by Crystal Ball[6] to be (1.3 ± 0.4) keV whereas the prediction by Shifman[7] is given by

$$\Gamma_{J/\psi \rightarrow \gamma \eta_c} = \frac{2}{9} \frac{\Gamma_{\eta_c \rightarrow \gamma\gamma}}{\Gamma_{J/\psi \rightarrow e^+ e^-}} \alpha \frac{m_{J/\psi}^4}{m_{\eta_c}^3} (1 - \frac{m_{\eta_c}^2}{m_{J/\psi}^2})^3 \tag{1}$$

$$= (3.3 \pm 0.7)\text{keV},$$

where the error is due the experimental uncertainties entering the prediction and does not include theoretical uncertainties. The ratio between the measurement and the expectation is (0.39 ± 0.15), or 4σ away from unity.

Rupak Mahapatra[14] observed that measurements of the $\gamma\gamma$ width, particularly the new CLEO measurement with a small error, can be used to test this prediction independent of the Crystal Ball measurement. When the prediction of Shifman (Eq. 1) is multiplied by the decay branching fraction $\mathcal{B}(\eta_c \rightarrow K_S^0 K^\pm \pi^\mp)$, the prediction can be tested using a different set of measurements: $\Gamma(J/\psi \rightarrow \gamma \eta_c) \times \mathcal{B}(\eta_c \rightarrow K_S^0 K^\pm \pi^\mp)$ on the left-hand side of the equation, and $\Gamma(\eta_c \rightarrow \gamma\gamma) \times \mathcal{B}(\eta_c \rightarrow K_S^0 K^\pm \pi^\mp)$ on the right-hand side. DM2 and MARK III measured the first quantity to be $(2.1 \pm 0.4) \times 10^{-2}$. The right-hand side can be calculated to be $(5.4 \pm 0.7) \times 10^{-2}$ using CLEO's measurement of

474

Table 1. Mass, Total Width and $\gamma\gamma$ Width of the η_c.

Exp.	Mass(MeV)	Total Width(MeV)	$\gamma\gamma$ Width(keV)	method
ARGUS[8]	-	-	11.3 ± 4.2	$\gamma\gamma$
CLEO90[9]	-	-	$5.9^{+2.1}_{-1.8} \pm 1.9$	$\gamma\gamma$
CBAL[6]	-	11.5 ± 4.5	-	$J/\psi, \psi'$
MARKIII[10]	2980.6 ± 1.6	-	-	J/ψ
DM2[11]	2974.4 ± 1.9	-	-	J/ψ
E760[4]	$2988.3^{+3.3}_{-3.1}$	$23.9^{+12.6}_{-7.1}$	$6.7^{+2.4}_{-1.7}$	$\bar{p}p$
PDG98[1]	2979.8 ± 2.1	13.2 ± 3.5	$7.5 \pm 1.5^*$	-
BES[12]	2976.3 ± 3.1	$11.0 \pm 8.1 \pm 4.1$	-	$J/\psi, \psi'$
E835[5]	2985.1 ± 2.1	$22.4^{+7.8}_{-6.4}$	$4.1^{+1.7}_{-1.4} \pm 1.5$	$\bar{p}p$
L3[13]	2974 ± 18	-	$6.9 \pm 1.9 \pm 2.0$	$\gamma\gamma$
DELPHI[2]	-	-	$10.9 \pm 2.7 \pm 3.4$	$\gamma\gamma$
CLEO[2]	2980.4 ± 2.4	27.0 ± 6.0	$7.6 \pm 0.9 \pm 2.3$	$\gamma\gamma$
Average**	2981.1 ± 1.7	$18.0^{+3.0}_{-2.7}$	$7.1^{+0.8}_{-0.7} \pm 2.0$	-

* Common systematic error arising from η_c decay branching ratios do not seem to be accounted for properly in the systematic error here.

** Scale factor *a la* PDG is used to account for inconsistency among measurements.

$\Gamma(\eta_c \to \gamma\gamma) \times \mathcal{B}(\eta_c \to K_S^0 K^\pm \pi^\mp)$. Their ratio is (0.39 ± 0.09), or 6σ away from unity. This discrepancy with the QCD prediction in Eq. 1 is comparable to the discrepancy between the Crystal Ball measurement and the prediction. This indicates that the QCD prediction of the radiative J/ψ decay rate deserves serious consideration.

Acknowledgments

I am grateful to my colleagues in the CLEO collaboration who contributed to the results presented here in a variety of ways.

References

1. C. Caso, *el al. Eur. Phys. Jour.* **C3**, 1 (1998).

2. #555 Study of Two-Photon Production of the η_c. #757 The η_c Production in Two-Photon Collisions at LEP Energies. Also used $2K^+2K^-$ mode.

3. V.M. Budnev, *Physics Reports* **C15**, 181 (1975).

4. T.A. Armstrong, *et al.*(E760 Collaboration), *Phys. Rev.* **D52**, 4839 (1995).

5. Michelle Stancari, Talk given at PHOTON99, May 1999, Freiburg im Breisgau, Germany. Slightly different numbers were given, for example, at BEAUTY 2000, June 2000, Valencia, Spain.

6. J. Gaiser, *et al. Phys. Rev.* **D34**, 711 (1986).

7. M. Shifman, *Z. Phys.* **C4**, 345 (1980); Erratum *ibid* **C6**, 282 (1980).

8. H. Albrecht, *et al. Phys. Lett.* **B338**, 390 (1994).

9. W.Y. Chen, *et al. Phys. Lett.* **B243**, 169 (1990).

10. R.M. Baltrusaitis, *et al. Phys. Rev.* **D33**, 629 (1986)

11. D. Bisello, *et al. Nucl. Phys.* **B350**, 1 (1991)

12. J.Z. Bai, hep-ex/0002006.

13. M. Acciarri, *et al. Phys. Lett.* **B461**, 155 (1999).

14. Rupak Mahapatra, U. of Minnesota thesis, unpublished, 2000.

LEADING BARYON PRODUCTION IN DIS AND PHOTOPRODUCTION

T. WILDSCHEK

For the H1 and ZEUS collaborations
ZEUS Wisconsin, DESY, Notkestr. 85, 22607 Hamburg, Germany
E-mail: Torsten.Wildschek@desy.de

The production of leading protons and neutrons in deep inelastic scattering (DIS) and photoproduction has been studied by the H1 and ZEUS collaborations.

1 Introduction

The H1 and ZEUS experiments have observed events with neutrons or protons carrying a large fraction of the proton beam energy. One mechanism for the production of such leading baryons is the exchange of a meson, pomeron or reggeon (figure 1). For this mechanism the hadron and the lepton vertex factorize.

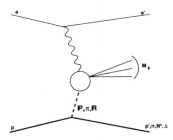

Figure 1. Diagram for the production of leading baryons via an exchange mechanism - the exchange particle can be a pomeron, pion, reggeon etc

1.1 Kinematic quantities

The two kinematic quantities specific to the study of leading protons or neutrons can be chosen as $x_L = \frac{E_{LB}}{E_p}$, the fraction of the incoming proton's momentum carried by the leading baryon, and the leading baryon's transverse momentum p_T.

1.2 Detectors

Both H1 and ZEUS use a Forward Neutron Calorimeter (FNC) with an angular accep-

tance of $\theta_n < 0.8$ mrad to detect leading neutrons. In addition, a position detector is installed in front of the ZEUS FNC. Leading protons are detected using the standard Roman pot technique.

2 Results

Only the main results are given in this paper. Please refer to the contributed papers for details.

2.1 Neutron yield

ZEUS[1] have studied the fractional neutron yield and neutron energy spectrum in different bins of photon virtuality Q^2. The fractional neutron yield increases with Q^2 and saturates at $Q^2 \approx 4$ GeV^2 (not shown in plot). This effect can be attributed to absorptive effects: as the photon's virtuality increases, its size and hence the probability of rescattering resulting in loss of the neutron decreases.

2.2 Dijet photoproduction with leading neutron

Both H1[2] and ZEUS[3] have studied photoproduction with a leading neutron and at least two jets in the final state. They have measured neutron energy spectra and the differential cross section as a function of E_T and η of the jets, and of x_γ and x_π, the fraction of the photon's and a hypothetical pion's (figure 3) momentum participating in

476

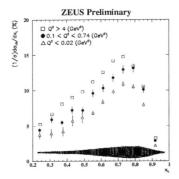

Figure 2. Fractional neutron yield as a function of x_L in photoproduction, intermediate Q^2-region and DIS. The shaded band shows the systematic uncertainty due to the acceptance uncertainty of the FNC.

the creation of the dijet system. The measurements have been compared to the predictions of the one-pion-exchange (OPE) model, as implemented in the POMPYT and RAPGAP Monte Carlo programs, and to inclusive Monte Carlo programs (PYTHIA, HERWIG).

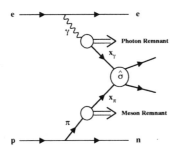

Figure 3. Diagram for resolved photoproduction. The interaction is mediated by a pion. x_γ (x_π) is the fraction of the photon's (pion's) momentum participating in the creation of the dijet system.

For $x_L > 0.5$ OPE describes both the shape and normalization of the neutron energy spectrum, while the inclusive Monte Carlo programs describe neither (figure 4). Factorization works well, as exemplified by figure 5, which shows that the shape of the neutron energy spectrum is approximately independent of x_γ.

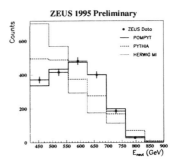

Figure 4. Energy spectrum of neutrons in photoproduction (area-normalized). One-pion-exchange Monte Carlo programs reproduce the shape of the spectrum.

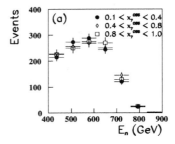

Figure 5. Neutron energy spectrum in different bins of x_γ^{OBS}. The approximate independence of the spectrum on x_γ^{OBS} is expected from the factorization hypothesis.

2.3 Leading neutron p_T distributions in DIS

ZEUS[4] have measured the p_T distributions of leading neutrons using the Forward Neutron Tracker. The shape is well described by an exponential

$$\frac{dN}{dp_T^2} \propto e^{-b(x_L)p_T^2}$$

with an x_L-dependent slope b. Figure 6 compares the measured x_L-dependence of the slope parameter b to predictions of various models.

2.4 Leading protons in photoproduction

H1[5] have studied events with a leading proton in tagged photoproduction, in the range $0.66 < x_L < 0.90$, where the diffractive

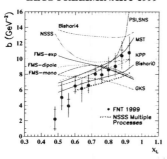

Figure 6. Measured b-slope as a function of x_L compared to various models.

process is suppressed relative to pion and reggeon exchange. They find the cross section to be approximately constant as function of W, the center-of-mass energy of the proton-photon system, and x_L, in agreement with the assumption of factorization (figure 7).

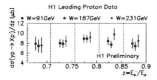

Figure 7. Differential cross section as a function of x_L (called z in this plot) in 3 bins of W.

In addition they have investigated the transition between the photoproduction and DIS regimes, comparing the process with a leading proton to the inclusive process. It is found that photoproduction appears suppressed in the proton-tagged relative to the inclusive process.

3 Summary and Conclusions

H1 and ZEUS have studied a wide range of processes involving leading protons or neutrons. In all the processes investigated, factorization works well. In leading neutron production, OPE models describe the observed neutron spectrum for $x_L > 0.5$ and the jet distributions in dijet production well. The yield of leading neutrons as a function of Q^2 rises quickly at low Q^2 and saturates above 4 GeV^2, which can be interpreted as a rescattering effect. The cross section for photoproduction with a leading proton is approximately constant as a function of W and x_L. Comparing the cross section with a leading proton and the inclusive cross section as functions of Q^2, it is found that leading proton production appears suppressed in the photoproduction regime.

Acknowledgments

I am grateful to my colleagues from the H1 and ZEUS collaborations for providing me with the material for this contribution and for helpful discussions.

References

1. ZEUS collaboration, Energetic neutron production in e^+p collisions at HERA, *contributed paper*.
2. H1 collaboration, Measurement of Dijet Cross-Sections with Leading Neutron in Photoproduction at HERA, *contributed paper*.
3. ZEUS collaboration, Measurement of dijet cross sections with a leading neutron in photoproduction at HERA, *contributed paper*.
4. ZEUS collaboration, Measurement of leading neutron p_T distributions in deep inelastic scattering at HERA, *contributed paper*.
5. H1 collaboration, Measurement of the Photoproduction Cross Section with a Leading Proton at HERA, *contributed paper*.

Parallel Session 3

Hard Interactions
(high Q^2 : DIS, jet, perturbative QCD)

Conveners: Eckhard E. Elsen (DESY) and
Alfred T. Goshaw (Duke)

JET PRODUCTION IN DIS AT HERA

JÖRG GAYLER

DESY, Hamburg, Germany
E-mail: joerg.gayler@desy.de

Data on jet production in deep inelastic e^+p scattering are presented. The results are compared with pQCD calculations. At low Q^2 no consistent description of the data over all the phase space is available yet. At high Q^2 ($\gtrsim 150$ GeV^2) the data are well described by pQCD in NLO.

1 Introduction

Inclusive deep inelastic lepton nucleon scattering, where only the scattered lepton is detected, played an important role in establishing QCD and continues to provide a well defined testing ground of perturbative QCD (pQCD). The aim of measurements of final state jets is to relate them to final state quarks and gluons and thereby to gain additional insight in the dynamics of lepton nucleon scattering.

The data presented in this talk were recorded in the years 1995 to 1997 at HERA with the H1 and ZEUS detectors where positrons of 27.5 GeV collided with protons of 820 GeV.

1.1 Kinematics

The basic Feynman diagrams describing jet production in deep inelastic scattering are shown in Fig. 1.

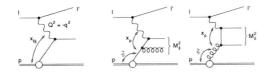

Figure 1. Processes in DIS: Born process; QCD-Compton process; and boson-gluon fusion (left to right).

Standard kinematic quantities [a] are

---

[a] Polar angles θ are measured with respect to the incident proton direction, the pseudo rapidity is given by $\eta = -\ln(\tan\theta/2)$.

$Q^2 = -q^2 = -(l - l')^2$, the virtuality of the boson exchange and the Bjorken variable $x_{Bj} = Q^2/2pq$. The momentum fraction entering the hard process of jet production with a jet-jet mass M_{jj} (see Fig. 1) is given by $\xi = x_{Bj}(1 + M_{jj}^2/Q^2)$ of which the fraction $x_p = x_{Bj}/\xi$ interacts with the exchanged boson.

In most cases the data are analysed in the Breit frame defined by the condition $2x_{Bj}\vec{p} + \vec{q} = 0$. Quark parton model like events (Fig. 1, left) exhibit no p_t in this frame apart from effects of fragmentation and decays. Jet finding is performed mostly using the inclusive k_t algorithm [1].

1.2 Multi-Jet Production in pQCD

Calculations at the parton level are available up to order α_s^2, i.e. to next to leading order (NLO) (Fig. 1 shows diagrams up to leading order (LO)). They can be compared with data after corrections for hadronisation are applied. DISENT [2] and DISASTER++ [3] have been shown [4] to agree in the kinematic range of interest here. MEPJET [5] is the only program implementing also charged current reactions and JetVip [6] allows resolved photon processes to be included.

A common ambiguity in these fixed order calculations is the choice of the renormalization scale μ_R^2. Typical quantities characterizing the process are Q^2 and E_t^2 and the agreement with the data for these hard scales and the sensitivity to scale variations is studied.

Forward jets, i.e. jets close to the proton remnant, are of special interest, because they are expected [7] to be sensitive probes of the evolution of parton densities. In particular in $\alpha_s \log(1/x)$ resummation (BFKL approach) one expects jets with larger p_t ("k_t") close to the proton remnant than in the standard $\alpha_s \log(Q^2)$ resummation (DGLAP approach), due to the strong k_t ordering in the latter case.

2 ϕ Asymmetries

A measurement of the ϕ distribution of charged particle tracks has been presented by the ZEUS [8] collaboration, for different transverse momentum cuts. Here ϕ is the azimuthal angle of the hadron production plane with respect to the positron scattering plane in the hadronic centre of mass system. Finite terms $B < 0$ and $C > 0$ were measured in the angular distribution $d\sigma/d\phi = A + B \cos(\phi) + C \cos(2\phi)$ as expected in QCD-based calculations.

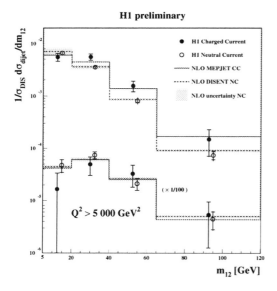

Figure 2. Distribution of jet-jet mass m_{12} (not including the proton remnant jet) for CC and NC interactions. $p_t^{lepton} > 25$ GeV, $Q^2 > 640$ GeV^2 and $Q^2 > 5000$ GeV^2 (the latter scaled by 1/100).

3 Jets in CC Interactions

Jet distributions in charged current (CC) interactions at high Q^2 are consistent with pQCD expectations (see Fig. 2). The differences to neutral current (NC) jets are mainly due to the different boson propagators [9].

4 Jets at Low and High Q^2

The E_T distribution in the Breit frame of single-inclusive jets is shown in Fig. 3 in different regions of η_{lab}. The discrepancies visible in the forward region, where the NLO corrections are huge, originate predominantly from small Q^2. The x_{Bj} dependence in the

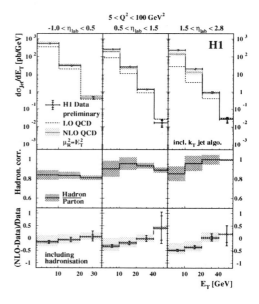

Figure 3. $d\sigma_{Jet}/dE_T$ compared to LO (dashed) and NLO (dotted) pQCD (DISENT) predictions using $\mu_R^2 = E_T^2$. The shaded band shows the sensitivity to scale variations by a factor 4. Also shown are the hadronization corrections and the relative deviations after their application.

forward and central region (Fig. 4) cannot be described with the scale $\mu_R^2 = E_T^2$. A consistent description is possible with $\mu_R^2 = Q^2$, but the susceptibility to scale variations is vastly increased (shaded band in Fig. 4) [10].

Such forward cross sections can be de-

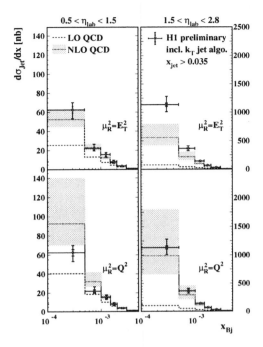

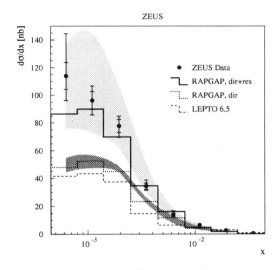

Figure 5. $d\sigma_{Jet}/dx$ for $p_z^{Breit} > 0$, $Q^2 > 10$ GeV^2, $0.5 < E_T^2/Q^2 < 2$. Comparisons with models of direct (LEPTO, RAPGAP,dir) and direct and resolved photon interactions (RAPGAP, dir+res).

Figure 4. $d\sigma_{Jet}/dx$ for $Q^2 > 5$GeV^2 in two regions of η_{lab}. NLO calculation in upper plots with $\mu_R^2 = E_T^2$, in lower with $\mu_R^2 = Q^2$.

scribed by the NLO program JetVip and by DGLAP-based QCD Monte Carlo models if the hadronic structure of the interacting virtual photon is resolved (RAPGAP [11], dir+res in Fig. 5), whereas inclusion of direct photon interactions only (RAPGAP, dir and LEPTO [12]) is insufficient [13]. In the case of resolved photons, the strong k_t ordering is effectively lost, leading to larger jet E_T close to the proton remnant. However, there are ambiguities in JetVip in the treatment of parton masses and no general solution has been found which is consistent with the H1 data in a large range of rapidities η_{lab} [10].

For detailed discussions of di-jet production at low Q^2 see the contributions [14,15].

At high Q^2 there are precise high statistics data available from H1 and ZEUS which agree with NLO calculations on the 10% level in detailed comparisons (see Fig. 6) [16]. For inclusive jets ZEUS reports [17] at $Q^2 < 250$ GeV^2 some disagreement on the 15% level for

E_T^2 and Q^2 scales (see Fig. 7), but otherwise the agreement of data and NLO calculations (DISENT) is very good [18].

5 Conclusion

The description of the available jet data is considerably improved in going from LO to NLO ($\sim \alpha_s^2$) pQCD. However some definite discrepancies remain to be resolved. They are more pronounced choosing E_T^2 as renormalization scale than for Q^2. In the latter case the effects of scale variations are large. Forward jets are better described if the hadronic structure of the virtual photon is taken into account. At high Q^2 ($\gtrsim 150$ GeV^2) the data are well described by NLO pQCD, the NLO corrections are moderate and hadronization corrections $\lesssim 10\%$. These data are well suited for quantitative QCD analyses [19].

Acknowledgments

I am grateful to T. Schörner and M. Wobisch for discussions and to E. Elsen and B. Foster for comments to the manuscript.

484

ZEUS PRELIMINARY

$d\sigma/dE_T^{Breit}$

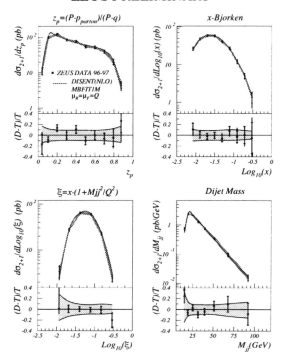

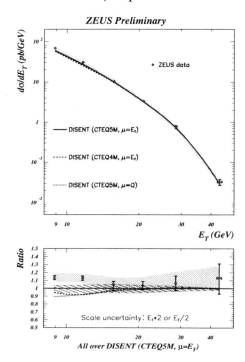

Figure 6. Di-jet distributions for $470 < Q^2 < 20000$ GeV^2, $-1 < \eta_{jet}^{lab} < 2$, $E_T^{jet1} > 8$ GeV and $E_T^{jet2} > 5$ GeV. The upper band shows the experimental energy scale uncertainty, the lower the uncertainty of the NLO calculation.

Figure 7. Inclusive jets, $Q^2 > 125$ GeV^2, $-2 < \eta_{jet}^{Breit} < 1.8$

References

1. S.D. Ellis and D.E.Soper, *Phys. Rev. D* **48**, 3160 (1993); S. Catani et al., *Nucl. Phys.* B **406**, 187 (1993).

2. S. Catani, M.H. Seymour, *Nucl. Phys.* B **485**, 291 (1997), Erratum-ibid. B **510**, 503 (1997).

3. D. Graudenz, hep-ph/9710244.

4. C. Duprel et al., Proc. on the Workshop *Monte Carlo Generators for HERA physics*, p. 142 (1999).

5. E. Mirkes and D. Zeppenfeld, *Phys. Lett.* B **380**, 205 (1996); E. Mirkes, TTP-97-39, hep-ph/9711224.

6. B. Pötter, *Comp. Phys. Commun.* **119**, 45 (1999).

7. A.H. Mueller, *Nucl. Phys. B (Proc. Suppl.)* C **18**, 125 (1990); J. Phys. G **17**, 1443 (1991).

8. ZEUS Collab., Contributed Paper 430; *Phys. Lett.* B **481**, 199 (2000).

9. H1 Collab., Contributed Paper 316.

10. H1 Collab., Contributed Paper 318.

11. H. Jung, Comp. Phys. Commun. **86**,147 (1995); http://www-h1.desy.de/~jung/rapgap.html.

12. G. Ingelman, Proc. of the Workshop *Physics at HERA*, Vol 3 (1992).

13. ZEUS Collab., Contributed Paper 429; Phys. Lett. B **474**, 223 (2000).

14. H1 Collab., Contributed Paper 317.

15. ZEUS Collab., Contributed Paper 417.

16. ZEUS Collab., Contributed Paper 420.

17. ZEUS Collab., Contributed Paper 419.

18. H1 Collab., Contributed Paper 319 and references there in.

19. E. Tassi, these proceedings.

THE STRONG COUPLING CONSTANT AND THE GLUON DENSITY FROM JET PRODUCTION IN DIS AT HERA

E. TASSI

NIKHEF, Kruislaan 409, 1098 SJ Amsterdam, The Netherlands.
E-mail: tassi@mail.desy.de

(On behalf of the H1 and ZEUS collaborations)

We present results on the determination of the strong coupling constant and the gluon density of the proton obtained in recent QCD analyses of HERA jet data. Topics include updated determinations of $\alpha_s(M_Z)$, tests of the α_s energy scale dependence, a study of the influence of HERA dijet cross sections on the extraction of the gluon density in a DGLAP fit, and a first attempt to a direct simultaneous determination of $\alpha_s(M_Z)$ and the gluon density of the proton.

1 Introduction

The study of jet production in neutral current (NC) e^+p deep inelastic scattering (DIS) at HERA provides a rich testing ground of Quantum Cromodynamics (QCD). The relevant QCD partonic processes are directly sensitive to the strong coupling constant, α_s, and the gluon density of the proton. These fundamental quantities of QCD can be determined in a QCD analysis of the measured multijet cross sections, provided that a phase-space region can be selected where the QCD predictions are little affected by the theoretical uncertainties.

In this contribution we briefly review the results on the determination of the strong coupling constant and the gluon density presented at this conference by the H1[1,2,3] and ZEUS[4] collaborations.

2 Determinations of α_s

Both the H1 and ZEUS Collaborations have presented determinations of the strong coupling constant obtained via a QCD fit of measured jet observables to the corresponding NLO QCD predictions[5] corrected for hadronisation effects. In this approach, use is made in the theoretical calculations of parton distribution functions (PDFs) extracted in re-

cent global DGLAP fits[6,7]. These PDFs depend on the $\alpha_s(M_Z)$ value used in the evolution equations, hence particular care must be placed in the QCD analysis to avoid circularity in the determination of quantities. The uncertainty of the PDFs has to be quantified including correlations. A recent DGLAP analysis[7] provides an estimate of the uncertainty in the NLO cross sections associated with the uncertainties of the PDFs. This information has been fully exploited by H1 and ZEUS in order to estimate realistically the uncertainty on α_s due to the parton densities of the proton.

The H1 experiment has determined α_s as a function of the jet transverse energy E_T using the measured inclusive double differential jet cross section, $d\sigma_{jet}/dE_T dQ^2$ (where Q^2 is the virtuality of the exchanged boson). ZEUS has made use of the dijet fraction, $R_{2+1} = \sigma_{2+1}/\sigma_{tot}$, as a function of Q^2. The results on the α_s scale dependence are presented in figure 1 and 2. The extracted values show a nice agreement with the renormalization group prediction over a wide kinematic range in the tested renormalization scale region. When extrapolated to the reference scale provided by the Z^0 boson mass the resulting values are:
H1:

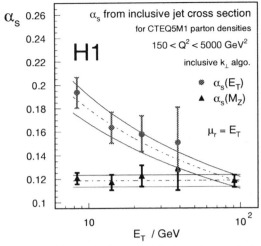

Figure 1. The scale ($\mu_r = E_T$) dependence of the strong coupling constant as determined by the H1 experiment.

$$\alpha_s(M_Z) = 0.1186(30)_{exp}\binom{+39}{-45}_{th}\binom{+33}{-23}_{pdf}$$

ZEUS (Preliminary):

$$\alpha_s(M_Z) = 0.1166\binom{+39}{-47}_{exp}\binom{+55}{-42}_{th}\binom{+12}{-11}_{pdf}$$

in good agreement with the current PDG world value. The experimental uncertainty is dominated in both experiments by the uncertainty on the jet hadronic energy scale. The theoretical uncertainty (which mostly dominates the total uncertainty) receives contributions from the residual renormalization scale dependence in the NLO QCD predictions and from the uncertainty on the estimate of the hadronisation effects. The total uncertainties are comparable to other NLO QCD determinations of α_s.

3 Determination of the proton gluon density

There exist two main sources of information on the proton gluon density at HERA: the scaling violations of the inclusive NC DIS

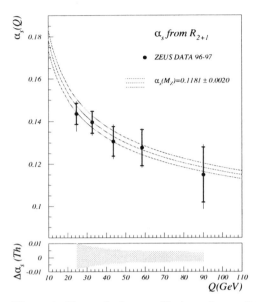

Figure 2. The scale ($\mu_r = Q$) dependence of the strong coupling constant as determined by the ZEUS experiment.

cross section and the direct information provided by the jet cross sections. The H1 experiment has used its measured inclusive and dijet cross sections in a DGLAP fit to extract the gluon density in the proton. The parton densities are evolved, from a starting scale of $Q_0^2 = 4$ GeV^2, using Mellin space evolution. The evolution equations as well as the dijet NLO QCD calculations assume $\alpha_s(M_Z) = 0.119$. The influence of the uncertainty on the strong coupling is taken into account by repeating the QCD fit for $\alpha_s(M_Z)$ values in the range 0.116 to 0.122. The QCD fit fully takes into account the statistical and systematic uncertainties. The resulting gluon density at $Q^2 = 200$ GeV^2 is presented in figure 3. The inclusion of the information provided by the dijet cross section affects the gluon density mostly at large values of x and improves the fit stability. Also shown in figure 3 are the results of the fit obtained by

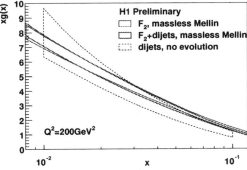

Figure 3. The gluon density and its associated uncertainty at $Q^2 = 200$ GeV^2 as obtained in a combined DGLAP fit of the inclusive and dijet cross sections (thick, full line). For comparison the gluon densities obtained using the dijet data (dashed line) or the inclusive data (thin, full line) only are shown.

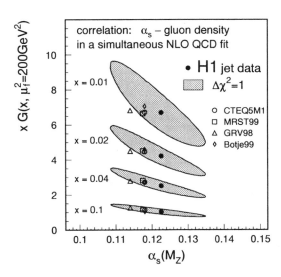

Figure 4. The correlation of the fit results for $\alpha_S(M_Z)$ and the gluon density at four different values of x, determined in a simultaneous QCD fit to the inclusive DIS cross section, the inclusive jet cross section and the dijet cross section. The error ellipses include the experimental and the theoretical uncertainties.

using only the inclusive or the dijet cross section. The precise measurements of the inclusive cross section are clearly important in reducing the gluon uncertainty.

4 Simultaneous determination of $\alpha_S(M_Z)$ and the gluon density

The H1 experiment has also presented to this conference a first attempt to a simultaneous, direct determination of the strong coupling and the gluon density. The basic idea of the analysis (see [3] for a detailed description) is to use the information provided by the NC DIS inclusive cross section and the inclusive and dijet cross sections (measured in the Breit frame) in a combined QCD fit so as to extract simultaneusly the quark and gluon densities in the proton and $\alpha_S(M_Z)$. The QCD fit is performed at a fixed factorization scale $\mu_f^2 = Q^2 = 200$ GeV^2. The results of this fit are displayed in figure 4 as a correlation plot between $\alpha_S(M_Z)$ and the gluon density evaluated at four different values of $x = 0.01, 0.02, 0.04, 0.1$. The central fit result is indicated by the full marker and the error ellipse is the contour along which the χ^2 of the fit is one unit larger than the mini-

mum (including experimental and theoretical uncertainties). The error ellipses clearly indicate that the data are very sensitive to the product $\alpha_S \cdot xg(x)$ although they do not yet allow a determination of both parameters simultaneously with useful precision.

References

1. H1 Collab., EPS99 contributed paper n. 157. Updated for ICHEP 2000.
2. H1 Collab., ICHEP 2000 contributed paper n. 315.
3. H1 Collab., ICHEP 2000 contributed paper n. 319.
4. ZEUS Collab., ICHEP 2000 contributed paper n. 420.
5. S. Catani and M. H. Seymour, *Nucl. Phys.* B **485**, 291 (1997).
6. A.D. Martin et al., *Eur. Phys. J.* C **14**, 133 (2000); H.L. Lai et al., *Eur. Phys. J.* C **12**, 375 (2000).
7. M. Botje, *Eur. Phys. J.* C**14**, 285(2000).

JET PRODUCTION AT DØ

N. VARELAS

(for the DØ Collaboration)

Department of Physics, University of Illinois at Chicago,
Chicago, IL 60607, USA
E-mail: varelas@uic.edu

We report on a new measurement of the rapidity dependence of the inclusive jet production cross section in $p\bar{p}$ collisions at $\sqrt{s} = 1.8$ TeV using 92 pb^{-1} of data collected by the DØ detector at the Tevatron collider. The differential cross sections, $\langle d^2\sigma/(dE_T d\eta)\rangle$, are presented as a function of jet transverse energy (E_T) in five pseudorapidity (η) intervals, up to $|\eta| = 3$, significantly extending previous measurements beyond $|\eta| = 0.7$. We also present recent results on the ratio of central ($|\eta| < 0.5$) inclusive cross sections from two center-of-mass energies, 0.63 TeV and 1.8 TeV, as a function of jet x_T. Experimental results are compared to next-to-leading order QCD predictions.

1 Introduction

Over the last decade, impressive progress has been made in both theoretical and experimental understanding of collimated streams of particles or "jets" resulting from inelastic hadron collisions. The Fermilab Tevatron $p\bar{p}$ Collider, operated at center-of-mass energies (CM) of 1.8 TeV and 0.63 TeV, has been a prominent arena for studying hadronic jets. Theoretically, jet production in $p\bar{p}$ collisions is understood within the framework of quantum chromodynamics (QCD) as a hard scattering of constituents of protons, the quarks and gluons (or partons) that manifest themselves as jets in the final state. The study of the inclusive jet cross sections in various kinematic regions and at two CM energies by the same experiment, constitutes a stringent test of QCD.

Perturbative QCD calculations of jet cross sections[1], using new and accurately determined parton distribution functions (PDF's)[2], add particular interest to the corresponding measurements at the Tevatron. These measurements test the short range behavior of QCD, the structure of the proton in terms of PDF's, and any possible substructure of quarks. The measurements we report are based on integrated luminosities of 92 pb^{-1} at $\sqrt{s} = 1.8$ TeV and 0.54 pb^{-1} at $\sqrt{s} = 0.63$ TeV collected by the DØ experiment during the 1994–95 Tevatron run.

At DØ, jets are reconstructed using an iterative cone algorithm with a fixed cone radius of $\mathcal{R} = 0.7$ in $\eta - \varphi$ space, where the pseudorapidity η is related to the polar angle (relative to the proton beam) θ via $\eta = \ln[\cot\theta/2]$, and φ is the azimuth. Offline data selections eliminate contamination from background caused by electrons, photons, noise, or cosmic rays. This is achieved by applying an acceptance cutoff on the z–coordinate of the interaction vertex, flagging events with large missing transverse energy, and applying jet quality criteria. Details of data selection and corrections due to noise and/or contamination are described elsewhere[3,4,5].

A correction for jet energy scale accounts for instrumental effects associated with calorimeter response, showering and noise, as well as for contributions from spectator partons, and corrects on average jets from their reconstructed to their "true" E_T. The effect of calorimeter resolution on jet cross section is removed through an unfolding procedure. In DØ, the energy scale and resolution corrections are determined mostly from data and applied in two separate steps.

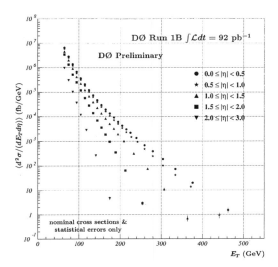

Figure 1. DØ preliminary measurement of rapidity dependence of single inclusive jet production cross section presented as a function of jet E_T in five jet $|\eta|$ intervals.

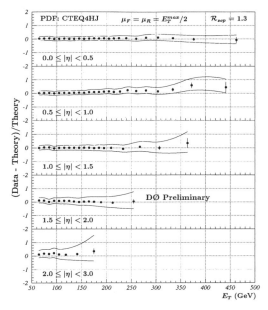

Figure 2. Comparisons of the DØ inclusive jet cross section preliminary measurements in five $|\eta|$ intervals with α_s^3 QCD predictions calculated by JETRAD with CTEQ4HJ PDF. The error bars are statistical, while the error bands indicate systematic uncertainties.

2 Inclusive Jet Cross Sections at $\sqrt{s} = 1.8$ TeV

DØ has recently completed a measurement of the rapidity dependence of the inclusive jet production cross section[4]. The differential cross section, $\langle d^2\sigma/(dE_T d\eta)\rangle$, is determined as a function of jet E_T in five intervals of $|\eta|$, up to $|\eta| = 3$, thereby significantly extending previously available measurements from CDF and DØ beyond $|\eta| = 0.7$. The cross section is calculated from the number of jets in each η-E_T bin, scaled by the integrated luminosity, selection efficiencies, and the unfolding correction. The preliminary results in each of the five $|\eta|$ regions are presented in Fig. 1. The measurement spans about seven orders of magnitude in E_T, and extends to the highest energies ever reached.

The results are compared to the α_s^3 predictions from JETRAD (Giele, *et al.* [1]), with equal renormalization and factorization scales set to $\mu = E_T^{max}/2$, and using the parton clustering parameter $\mathcal{R}_{sep} = 1.3$. Comparisons have been made using all recent PDF's of the CTEQ and MRST families.

Figure 2 shows the comparisons on a linear scale with the CTEQ4HJ PDF, which appears to best describe the data in all η intervals. The error bars are statistical, while the error bands indicate one standard deviation systematic uncertainties. Theoretical uncertainties are on the order of the systematic uncertainties. Work is currently underway to obtain a more quantitative comparison with predictions (such as a χ^2 test), taking into consideration correlations in E_T and in $|\eta|$. The extended range of the measurement promises to provide greater discrimination among different PDF's.

3 The Ratio of Inclusive Jet Cross Sections

DØ has measured the ratio of dimensionless inclusive jet cross sections ($\frac{E_T^3}{2\pi} \cdot d^2\sigma/dE_T d\eta$) at two CM energies, $\sqrt{s} = 0.63$ TeV and 1.8

490

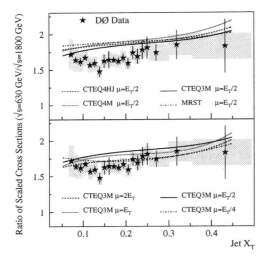

Figure 3. Ratio of dimensionless jet cross sections (numerator $\sqrt{s}$ = 630 GeV, denominator $\sqrt{s}$ = 1800 GeV) compared to NLO QCD as given by JE-TRAD. The error bars are statistical, while the error bands indicate systematic uncertainties.

Acknowledgments

We thank the staffs at Fermilab and at collaborating institutions for contributions to this work, and acknowledge support from the Department of Energy and National Science Foundation (USA), Commissariat à L'Energie Atomique and CNRS/Institut National de Physique Nucléaire et de Physique des Particules (France), Ministry for Science and Technology and Ministry for Atomic Energy (Russia), CAPES and CNPq (Brazil), Departments of Atomic Energy and Science and Education (India), Colciencias (Colombia), CONACyT (Mexico), Ministry of Education and KOSEF (Korea), CONICET and UBACyT (Argentina), A.P. Sloan Foundation, and the A. von Humboldt Foundation.

TeV, in the central region of pseudorapidity, $|\eta| < 0.5$[5,6]. The strength of this measurement is that several theoretical uncertainties (notably due to the choice of various PDF's) are reduced significantly in the ratio, as are many experimental uncertainties due to their correlated nature at the two CM energies.

Figure 3 shows the ratio of dimensionless jet cross sections as a function of jet $x_T = 2E_T/\sqrt{s}$, along with theoretical predictions from JETRAD for different choices of the input parameters (μ scales and PDF's). A covariance matrix χ^2 test for the ratio of the cross sections shows that there is no significant difference in shape between data and NLO QCD. However, the absolute values of the predictions lie systematically higher than the data throughout most of the measured x_T range, in particular between x_T of 0.1 and 0.2, where the ratio has the smallest statistical uncertainty. Choice of PDF has little effect on the prediction — only the renormalization/factorization scales change the prediction appreciably.

References

1. W.T. Giele, E.W.N. Glover, and D.A. Kosower, Phys. Rev. Lett. **73**, 2019 (1994); S.D. Ellis, Z. Kunszt, and D.E. Soper, Phys. Rev. Lett. **64**, 2121 (1990); F. Aversa et al., Phys. Rev. Lett. **65**, (1990).

2. H.L. Lai et al., (CTEQ Collaboration) Phys. Rev. **D51**, 4763 (1995); A.D. Martin et al., (MRST Collaboration) Eur. Phys. J. **C4**, 463 (1998).

3. B. Abbott et al., (DØ Collaboration), Phys. Rev. Lett. **82**, 2451, (1999).

4. L. Babukhadia, Ph.D. Thesis, University of Arizona, Tucson, Arizona, 1999 (unpublished);
http://fnalpubs.fnal.gov/archive/ 1999/thesis/t-babukhadia.ps.

5. J. Krane, Ph. D. Thesis, University of Nebraska–Lincoln, 1998 (unpublished);
http://fnalpubs.fnal.gov/techpubs/ theses.html.

6. B. Abbott et al., (DØ Collaboration), submitted to Phys. Rev. Lett, hep-ex/0008072 (2000).

DIJETS AT LARGE RAPIDITY INTERVALS

BERNARD G. POPE FOR THE D0 COLLABORATION

Department of Physics and Astronomy,
Michigan State University,
East Lansing, MI 48824-1116, USA
E-mail: pope@pa.msu.edu

Inclusive dijet production at large pseudorapidity intervals ($\Delta\eta$) between the two jets has been suggested as a regime for observing BFKL dynamics. We have measured the dijet cross section for large $\Delta\eta$ in $p\bar{p}$ collisions at $\sqrt{s} = 1800$ and 630 GeV using the DØ detector. The partonic cross section increases strongly with the size of $\Delta\eta$. The observed growth is even stronger than expected on the basis of BFKL resummation in the leading logarithmic approximation. The growth of the partonic cross section can be accommodated with an effective BFKL intercept of $\alpha_{\rm BFKL}(20\,{\rm GeV}) = 1.65 \pm 0.07$.

Inclusive dijet production at large pseudorapidity intervals in high energy $p\bar{p}$ collisions provides an excellent testing ground for Balitsky-Fadin-Kuraev-Lipatov (BFKL) dynamics [1]. We present a measurement of the dijet cross section at large $\Delta\eta$ using the DØ detector at the Fermilab Tevatron collider. We reconstruct the event kinematics using the most forward/backward jets, and measure the cross section as a function of x_1, x_2 and Q^2. We calculate the longitudinal momentum fractions of the proton and antiproton, x_1 and x_2, carried by the two interacting partons after taking $E_{T_1}(E_{T_2})$ and $\eta_1(\eta_2)$ as the transverse energy and pseudorapidity of the most forward(backward) jet, $\Delta\eta = \eta_1 - \eta_2 \geq 0$, and $\bar{\eta} = (\eta_1 + \eta_2)/2$. The momentum transfer during the hard scattering is defined as:

$$Q = \sqrt{E_{T_1} E_{T_2}} \ . \qquad (1)$$

The total dijet cross section, σ, can be factorized into the partonic cross section, $\hat{\sigma}$, and the parton distribution functions (PDF), $P(x_{1,2}, Q^2)$, in the proton and antiproton: $\sigma = x_1 P(x_1, Q^2)\, x_2 P(x_2, Q^2)\, \hat{\sigma}$. Using the BFKL prescription to sum the leading logarithmic terms $\alpha_s \ln(\hat{s}/Q^2)$ to all orders in α_s, results in an exponential rise of $\hat{\sigma}$ with $\Delta\eta$ [2]:

$$\hat{\sigma}_{\rm BFKL} \propto \frac{1}{Q^2} \cdot \frac{e^{(\alpha_{\rm BFKL}-1)\Delta\eta}}{\sqrt{\alpha_s \Delta\eta}} \ , \qquad (2)$$

where $\alpha_{\rm BFKL}$ is the BFKL intercept that governs the strength of the growth of the gluon distribution at small x.

The predicted rise of the partonic cross section with $\Delta\eta$ is difficult to observe experimentally due to the dependence of the total cross section on the PDF. To overcome this difficulty, we measure the cross section at two c.m. energies, $\sqrt{s_A} = 1800$ GeV and $\sqrt{s_B} = 630$ GeV, and take their ratio for the same values of x_1, x_2 and Q^2. This eliminates the dependence on the PDF, and reduces the ratio to that of the partonic cross sections. The latter is purely a function of the $\Delta\eta$ values:

$$R = \frac{\hat{\sigma}(\Delta\eta_A)}{\hat{\sigma}(\Delta\eta_B)} = \frac{e^{(\alpha_{\rm BFKL}-1)(\Delta\eta_A - \Delta\eta_B)}}{\sqrt{\Delta\eta_A/\Delta\eta_B}} \ . \qquad (3)$$

Thus, varying $\sqrt{s}$, while keeping x_1, x_2 and Q^2 fixed, is equivalent to varying $\Delta\eta$, which directly probes the BFKL dynamics. In addition, measurement of the ratio leads to cancellation of certain experimental uncertainties, and enables an experimental extraction of $\alpha_{\rm BFKL}$. In the DØ [3] detector, jets are identified using the uranium/liquid-argon calorimeters. These cover the range of $|\eta| \leq 4.1$, and are segmented into towers of $\Delta\eta \times \Delta\phi = 0.1 \times 0.1$ (ϕ is the azimuthal angle).

The data samples for this analysis were collected during the 1995–1996 Tevatron Collider run. Events were selected online by a three-level trigger system culminating in the software trigger requirement of a jet candidate with $E_T > 12$ GeV. The trigger was 85% efficient for jets with $E_T = 20$ GeV, and fully efficient for jets with $E_T > 30$ GeV. The integrated luminosity of the trigger was 0.7 nb^{-1} for the $\sqrt{s} = 1800$ GeV sample, and 31.8 nb^{-1} for the $\sqrt{s} = 630$ GeV sample [4].

Jets were reconstructed offline using an iterative fixed-cone algorithm with a cone radius of $\mathcal{R} = 0.7$ in (η, ϕ) space [5]. To ensure good jet reconstruction efficiency and jet energy calibration, jets were selected with $E_T > 20$ GeV and $|\eta| < 3$. A minimum pseudorapidity interval of $\Delta\eta > 2$ was required between the most forward and most backward jet. The values of x_1, x_2 and Q^2 were calculated for each event. Most of the data at $\sqrt{s} = 1800$ GeV are within $0.01 < x_{1,2} < 0.30$, and at 630 GeV, within $0.03 < x_{1,2} < 0.60$. The region of maximum overlap, $0.06 < x_{1,2} < 0.30$, was divided into six equal bins of x_1 and x_2. Due to limited statistics, only one bin in Q^2 was used: $400 < Q^2 < 1000$ GeV^2. The dijet cross section, corrected for trigger, event and jet selection inefficiencies, was computed in each (x_1, x_2, Q^2) bin.

The dijet cross section at low (x_1, x_2) is affected by the acceptance of the $E_T > 20$ GeV and $\Delta\eta > 2$ requirements. To avoid this bias, we require $x_1 \cdot x_2 > 0.01$. Similarly, the cross section at high (x_1, x_2) is biased by the $|\eta| < 3$ requirement, so that we require $x_{1,2} < 0.22$. A total of ten (x_1, x_2) bins satisfy both requirements.

The dijet cross sections for $\Delta\eta > 2$ at $\sqrt{s} = 1800$ and 630 GeV in the selected (x_1, x_2) bins are shown in Table 1. In each bin, the average values of x_1, x_2 and Q^2 are in good agreement, within the precision of our measurement, between the two c.m. energies. This ensures the cancellation of the PDF in the ratio of the cross sections. The mean pseudorapidity interval, $\langle\Delta\eta\rangle$, in the selected bins is equal to 4.6 units at 1800 GeV and 2.4 units at 630 GeV.

Table 1. The dijet cross sections for $\Delta\eta > 2$ at $\sqrt{s} = 1800$ and 630 GeV in each of the ten (x_1, x_2) bins. The minimum jet E_T is 20 GeV. The uncertainties are statistical.

x_1 range	x_2 range	σ_{1800} (nb)	σ_{630} (nb)
0.06–0.10	0.18–0.22	28.1 ± 6.9	8.4 ± 0.9
0.10–0.14	0.14–0.18	40.1 ± 9.5	8.8 ± 0.9
	0.18–0.22	$3.6 \,^{+\,4.1}_{-\,2.3}$	5.4 ± 0.6
	0.10–0.14	27.9 ± 7.3	8.4 ± 0.8
0.14–0.18	0.14–0.18	$10.4 \,^{+\,6.1}_{-\,5.0}$	5.0 ± 0.6
	0.18–0.22	$5.6 \,^{+\,4.5}_{-\,3.8}$	2.9 ± 0.5
	0.06–0.10	26.3 ± 6.6	8.6 ± 0.9
0.18–0.22	0.10–0.14	$12.5 \,^{+\,6.3}_{-\,5.4}$	6.3 ± 0.7
	0.14–0.18	$6.8 \,^{+\,5.0}_{-\,3.2}$	3.1 ± 0.4
	0.18–0.22	$2.4 \,^{+\,2.8}_{-\,1.7}$	1.7 ± 0.3

All sources of systematic uncertainty amount to 11% on the ratio of the cross sections and 3% on the BFKL intercept, yielding the final results: The mean value of the ratios of the cross sections in the ten bins is equal to $\langle R \rangle = 2.8 \pm 0.3\,(\text{stat}) \pm 0.3\,(\text{sys}) = 2.8 \pm 0.4$, $\langle \alpha_{\text{BFKL}} \rangle = 1.65 \pm 0.05\,(\text{stat}) \pm 0.05\,(\text{sys}) = 1.65 \pm 0.07$. Hence, for the same values of x_1, x_2 and Q^2, the dijet cross section at large $\Delta\eta$ increases by almost a factor of three between the two c.m. energies.

Several theoretical predictions can be compared to our measurement. Leading order QCD predicts the ratio of the cross sections to fall asymptotically toward unity with increasing $\Delta\eta$. For the $\Delta\eta$ values relevant to this analysis, the predicted ratio is $R_{\text{LO}} = 1.2$ [6].

The HERWIG MC provides an alternative prediction. It calculates the exact $2 \to 2$ subprocess, including initial and final state radiation and angular ordering of the emitted partons. Using the same (x_1, x_2, Q^2) bins as in the data yields $R_{\text{HERWIG}} = 1.6 \pm 0.1\,(\text{stat})$.

The predicted BFKL intercept for $\alpha_s(20\,\text{GeV}) = 0.17$ [6] is $\alpha_{\text{BFKL, LLA}} = 1.45$.

For $\Delta\eta_{1800} = 4.6$ and $\Delta\eta_{630} = 2.4$, Eq. (3) yields $R_{\mathrm{BFKL, LLA}} = 1.9$. It should be noted, however, that the leading log approximation may be too simplistic, and that exact quantitative predictions including the next-to-leading logarithmic [7] corrections to the BFKL kernel are not as yet available.

It is evident that the growth of the dijet cross section with $\Delta\eta$ (from $\langle\Delta\eta\rangle = 2.4$ to 4.6) is stronger in the data than in the theoretical models we considered. The measured ratio is higher by 4 standard deviations than the LO prediction, 3 deviations than the HERWIG prediction, and 2.3 deviations than the LLA BFKL prediction.

Finally, the $\Delta\eta > 2$ requirement was changed to $\Delta\eta > 1$, and the analysis was repeated. This results in a selection of fifteen unbiased (x_1, x_2) bins. The mean pseudorapidity interval in the selected bins is equal to 4.2 at 1800 GeV and 1.9 at 630 GeV. The average ratio of the 1800 and 630 GeV cross sections in the selected bins was measured to be 1.8 ± 0.1(stat)±0.1(uncorrelated sys). The results are shown in Fig. 1 as a function of the mean pseudorapidity interval at $\sqrt{s} = 630$ GeV. In the case of the $\Delta\eta > 1$ requirement, the observed ratio is once again larger than the exact LO and HERWIG predictions. It is interesting, however, that HERWIG exhibits the same qualitative behavior as the data in that the ratio of cross sections decreases as the $\Delta\eta$ requirement is relaxed, whereas the exact LO calculation predicts a very different trend. (A BFKL prediction is not shown for the case of $\Delta\eta > 1$ since the pseudorapidity interval is not sufficiently large for the formalism to be meaningful.)

The results presented in this talk have recently been published [8].

References

1. L.N. Lipatov, Sov. J. Nucl. Phys. **23**, 338 (1976);
   E.A. Kuraev, L.N. Lipatov,

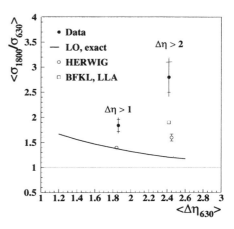

Figure 1. The ratio of the dijet cross sections at $\sqrt{s} = 1800$ and 630 GeV for $\Delta\eta > 1$ and $\Delta\eta > 2$. The minimum jet E_T is 20 GeV. The inner error bars on the data points represent statistical uncertainties; the outer bars represent statistical and uncorrelated systematic uncertainties added in quadrature. The error bars on the HERWIG predictions represent statistical uncertainties. The LO and BFKL predictions are analytical calculations.

and V.S. Fadin, Sov. Phys. JETP **44** (1976) 443; Sov. Phys. JETP **45**, 199 (1977);
Y.Y. Balitsky and L.N. Lipatov, Sov. J. Nucl. Phys. **28**, 822 (1978).

2. A.H. Mueller and H. Navelet, Nucl. Phys. **B282** 727 (1987).

3. DØ Collaboration, S. Abachi *et al.*, Nucl. Instrum. Methods Phys. Res. A **338**, 185 (1994).

4. J. Bantly *et al.*, Fermilab-TM-1995 (1997); J. Krane, J. Bantly and D. Owen, Fermilab-TM-2000 (1997).

5. B. Abbott *et al.*, Fermilab-Pub-97/242-E (1997).

6. L.H. Orr and W.J. Stirling, Phys. Lett. B **429** 135 (1998).

7. V.S. Fadin and L.N. Lipatov, Phys. Lett. B **429**, 127 (1998); G. Camici and M. Ciafaloni, Phys. Lett. B **430**, 349 (1998).

8. DØ Collaboration, B. Abbott *et al.*, Phys. Rev. Lett. **84**, 5722 (2000).

JET STRUCTURE STUDIES AT LEP AND HERA

J. WILLIAM GARY

Department of Physics, University of California, Riverside CA 92521, USA
E-mail: bill.gary@ucr.edu

A summary of some recent studies in jet physics is given. Topics include leading particle production in light flavor events in e^+e^- annihilations, an analytical treatment of gluon and quark jets at the next-to-next-to-next-to-leading order (3NLO), and various studies performed at LEP and HERA involving separated gluon and quark jets.

1 Leading particle production in separated light quark events

Separated charm (c) and bottom (b) quark events have been well studied in e^+e^- annihilations. In contrast, separated up (u), down (d) and strange (s) events have not been much studied. u, d and s quarks are copiously produced during jet development, making events in which they are produced as primary quarks more difficult to identify.

A recent study[1] by the OPAL Collaboration at LEP identifies $e^+e^- \to q\bar{q} \to hadrons$ events with q=u, d or s. The probabilities $\eta_q^i(x_p^{min.})$ are determined for the quark q=u, d, s (or charge conjugate) to appear in an identified hadron $h_i(x_p^{min.})$, where $h_i=\pi^+$, K^+, K_S^0, p=proton, or Λ (or charge conjugate) is the leading particle in the jet, i.e. it has the largest momentum, with a scaled momentum $x_p=2p/\sqrt{s}$ larger than a minimum $x_p > x_p^{min.}$, where $\sqrt{s}$=91 GeV.

The method for the measurement is presented in ref.[2]. Briefly, $e^+e^- \to hadrons$ events are divided into hemispheres using the plane perpendicular to the thrust axis. Thus an *inclusive* definition of jets is used, i.e. a jet is an event hemisphere. The single and double tag rates in the jets are measured, $N_i(x_p^{min.})/N_{had.}$ and $N_{ij}(x_p^{min.})/N_{had.}$, where $N_i(x_p^{min.})$ is the number of jets in which the highest momentum particle has $x_p > x_p^{min.}$ and is identified as π^+, K^+, K_S^0, p or Λ, and $N_{ij}(x_p^{min.})$ is the analogous quantity for events in which a

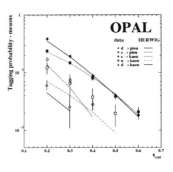

Figure 1. Tagging probabilities for pions and kaons as a function of the minimum scaled momentum $x_p^{min.}$, in comparison to Monte Carlo predictions.

hadron h_i is identified in one hemisphere and hadron a h_j is identified in the other. Supplementary information involving assumptions of isospin symmetry and the flavor independence of the strong interaction are invoked to algebraically solve for the values of η_q^i (see refs.[1,2]).

The measured probabilities η_q^i for $i=\pi$ and K are shown in figure 1 as a function of $x_p^{min.}$. From this figure, the so-called leading particle effect is clearly visible, i.e. primary quarks appear primarily as valence quarks in the highest momentum hadrons. Thus d and u quarks appear predominantly in pions, and s quarks in kaons, rather than vice versa.

Using the probabilities η_q^i, basic hadronization parameters such as the strange quark suppression factor γ_s can be determined rather directly. For example the probability factors $\eta_u^{K^\pm}$ and $\eta_s^{K^\pm}$ differ only by $s\bar{s}$ or

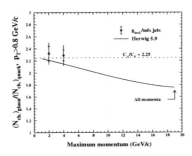

Figure 2. Ratio r of charged particle multiplicities between unbiased gluon and quark jets for particles with large transverse momenta to the jet axis defined by $p_\perp > 0.8$ GeV/c. The results are shown as a function of the softness of the the particles, defined by the maximum particle momentum used to determine r.

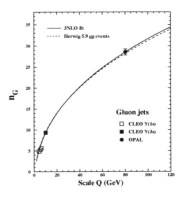

Figure 3. Measurements of unbiased gluon jet charged particle multiplicity in comparison to Monte Carlo and the 3NLO analytic predictions.

$u\bar{u}$ pair production from the sea. Their ratio thus determines γ_s. The result for γ_s is 0.422 ± 0.049 (stat.) ± 0.059 (syst.).

2 Experimental properties of gluon and quark jets from a point source

Inclusive production of hadrons in e^+e^- annihilations provides a natural source for unbiased quark jets, used – for example – in the study described in section 1. One can ask if an analogous sample of high energy unbiased gluon jets can be identified.

The answer, as discussed in ref.[3], is yes, by selecting rare events in e^+e^- annihilations in which two identified quark jets appear in the same hemisphere of an event. The opposite hemisphere, against which the two quark jets recoil, approximates an unbiased gluon jet with high accuracy[3]. Such events have been labelled $e^+e^- \to q_{tag}\bar{q}_{tag}g_{incl.}$ events. The tagged quark jets q_{tag} and $\bar{q}_{tag}$ are identified using b tagging. The recoiling hemisphere "$g_{incl.}$" is the unbiased gluon jet.

Experimental analysis of $g_{incl.}$ jets from Z^0 decays has been presented by OPAL[4]. The $g_{incl.}$ jets are compared to a sample of light (u,d and s) quark jets, also from Z^0 decays, defined using the hemisphere definition. Only one aspect of the results will be pre-

sented here, namely the ratio of soft particles at large transverse momentum $p_\perp$ between the unbiased gluon and quark jets.

Figure 2 shows the charged particle multiplicity ratio between the unbiased gluon and quark jets, r, as a function of the softness of the particles. The softness of the particles is defined by the maximum particle momentum $p_{max.}$ considered when determining r. The particles are required to have $p_\perp > 0.8$ GeV/c. Particles with $p_\perp < 0.8$ GeV/c are dominated by the effects of hadronization. The solid curve shows the prediction of the Herwig Monte Carlo.

With no explicit cut on $p_{max.}$ ("All momenta") the multiplicity ratio is predicted to be about 1.8. As softer and softer particles are selected ($p_{max.}$ is decreased), the curve approaches the QCD result $C_A/C_F=2.25$. OPAL results are shown for $p_{max.}=2$ GeV/c and 4 GeV/c. The result using $p_{max.}=4$ GeV/c is $r=2.29 \pm 0.017$ (stat. + syst.) which provides one of the most accurate current experimental determinations of C_A/C_F.

3 Analytic description of multiplicity in gluon and quark jets

An analytic description of multiplicity in unbiased gluon and quark jets has recently

496

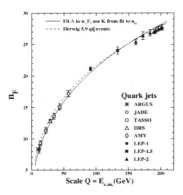

Figure 4. Measurements of unbiased quark jet charged particle multiplicity in comparison to Monte Carlo and the 3NLO analytic predictions.

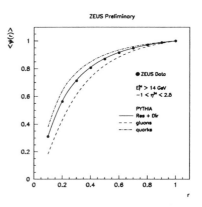

Figure 5. Jet profile of the two highest energy jets in $\gamma p \rightarrow 2\,jets + X$ events.

been performed at the next-to-next-to-next-to-leading order (3NLO) in perturbation theory[5]. Here some comparisons of the results with experiment will be discussed.

Measurements of mean multiplicity in unbiased gluon jets, n_G, are shown in figure 3. The solid curve is a fit of the 3NLO analytic prediction[5] using two free parameters: (1) an overall normalization K and (2) an effective QCD scale parameter Λ. Translating the fitted result for Λ into $\alpha_S(M_Z)$ yields $\alpha_S(M_Z)=0.14 \pm 0.01$.

Measurements of mean multiplicity in unbiased quark jets, n_F, are shown in figure 4. Making a one parameter fit of the 3NLO expression for quark jet multiplicity[5] to the data (solid curve), with Λ as the fitted parameter and with the normalization K fixed from the fit to the gluon jet data, yields $\alpha_S(M_Z)=0.135\pm0.002$, not very different from the result presented above for gluon jets. This demonstrates the consistency of the analytic approach to the growth of multiplicity with scale.

The ratio of the *slopes* of multiplicity,
$$r^{(1)} = \frac{d\langle n_G\rangle/dy}{d\langle n_F\rangle/dy}$$ where $y=\ln(Q/\Lambda)$ with Q the jet energy, has the same asymptotic limit of 2.25 as r, but is predicted[5] to have smaller pre-asymptotic corrections. Experimental results for $r^{(1)}$ from OPAL[6] and the DELPHI[7]

Collaboration at LEP are in general agreement with the 3NLO prediction $r^{(1)}\approx1.9$.

4 Substructure dependence of dijet cross sections in photoproduction at HERA

A recent study from the ZEUS Collaboration at HERA concerns the photoproduction of dijets in low Q^2 ep scattering[8].

Events with at least two jets are selected using the longitudinally invariant $k_\perp$ jet finder. Events are retained if they contain at least two jets with transverse energy $E_T>14$ GeV in the pseudo-rapidity range $-1<\eta<2.5$. The jet profiles of the two jets with highest E_T are determined. The jet profile $\Psi(r)$ is the distribution of the fraction of jet energy inside a cone of half radius r around the jet axis.

The mean profile of the selected jets is presented in figure 5. The data, shown by the points, are well represented by the Monte Carlo, indicated by the solid line. The individual contributions of gluon and quark jets are shown by the dashed and dash-dotted curves, respectively. Gluon jets are predicted to be much less collimated around the jet axis than quark jets.

Choosing a cone size $r=0.3$ radians, quark and gluon jet dominated samples

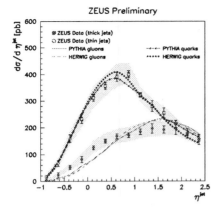

Figure 6. Rapidity distribution of "thick" and "thin" jets in comparison to the Monte Carlo predictions for gluon and quark jets in γp collisions.

are selected by requiring $\Psi(r=0.3)>0.8$ or $\Psi(r=0.3)<0.6$, respectively. The resulting samples are denoted "thin" jets and "thick" jets and have quark and gluon jet purities of about 85% and 60%, again respectively. The rapidity distributions of "thin" and "thick" jets are shown in figure 6 by the open and solid points. The curves show the Monte Carlo predictions for quark and gluon jets in γp collisions. The "thick" and "thin" jet measurements are seen to follow the predictions for gluon and quark jets quite well, demonstrating a successful separation and test of the cross sections for gluon and quark jets individually.

5 π^0, η, K^0 and charged particle multiplicities in quark and gluon jets

Last, I discuss a recent study from OPAL[6] on differences in the production rates of identified particles in gluon and quark jets.

QCD predicts that r_h – the ratio of the mean multiplicities of identified hadrons between gluon and quark jets – is the same for all hadrons h. Certain models of hadronization predict predict r_h to be larger for η mesons than for charged particles.

Three-jet events are selected using a jet finder. The jets are ordered such that jet 1 has the highest energy and jet 3 the lowest. The jets are then examined in terms of the so-called hardness scale $Q_{jet}=E_{jet}\sin(\theta_{min.}/2)$, where $\theta_{min.}$ is the smaller of the angles with respect to the two other jets.

The π^0, η, K_S^0 and charged particle rates in jets 2 and 3 are compared in their overlap region, defined by $7<Q_{jet}<30$ GeV. The multiplicity measurements are unfolded algebraically using the known quark and gluon jet content of jets 2 and 3 to obtain results corresponding to pure gluon and quark jets. The results for r_h for π^0, η and K_S^0 are then divided by the corresponding result for charged particles to obtain:

$$r_\eta/r_{ch.} = 1.09 \pm 0.12$$
$$r_{\pi^0}/r_{ch.} = 1.01 \pm 0.04$$
$$r_{K_s^0}/r_{ch.} = 0.95 \pm 0.04$$

All three results are consistent with unity, indicating no evidence for a dynamical difference in the hadronization of gluon and quark jets.

References

1. OPAL Collab., CERN-EP/99-164.
2. J. Letts and P. Mättig, *Z. Phys.* C **73**, 217 (1997).
3. J.W. Gary, *Phys. Rev.* D **49**, 4503 (1994).
4. OPAL Collab., *Phys. Lett.* B **388**, 659 (1996); *Eur. Phys. J.* C **1**, 479 (1998); *Eur. Phys. J.* C **11**, 217 (1999).
5. I.M. Dremin and J.W. Gary, *Phys. Lett.* B **B459**, 341 (1999); I.M. Dremin and J.W. Gary, hep-ph/0004215; A. Capella et al., *Phys. Rev.* D **61**, 074009 (2000).
6. OPAL Collab., CERN-EP-2000/070.
7. DELPHI Collab., *Phys. Lett.* B **449**, 383 (1999); CERN-OPEN-2000-134.
8. ZEUS Collab., Contribution #906 to ICHEP 2000.

JET FRAGMENTATION STUDIES AT TEVATRON

ANWAR AHMAD BHATTI

The Rockefeller University, 1230 York Ave, New York NY 10021, USA
E-mail: bhatti@fnal.gov

The particle multiplicity and momenta distributions within a jet, measured by CDF collaboration, are in good agreement with the Modified Leading Logarithms Approximation (MLLA) predictions. The MLLA cut-off scale is found to be $Q_{\mathrm{eff}} = 240 \pm 40$ MeV. In this framework The measured value of ratio of mulitiplicies in gluon and quark jets is $r = 1.8 \pm 0.4$. The ratio of charged hadrons multiplicity to the predicted parton multiplicity $K_{LPHD}^{charged} = 0.58 \pm 0.10$. The ratio of the charged particle multiplicity in quark and gluon jets is determined to be $1.74 \pm 0.11 \pm 0.07$ using di-jet and photon-jet data independent of MLLA formalism. The ratio of additional subjet multiplicities in quark and gluon jet $R = 1.91 \pm 0.04$ (stat) $^{+0.23}_{-0.19}$ (sys). is extracted from 1800 GeV and 630 GeV data by DØ collaboration which is consistent with Herwig Monte Carlo predictions.

1 Charge and Momenta distributions within a jet

Quantum Chromo-dynamics (QCD) is very successful in describing the hadronic process with astonishing accuracy at large momentum transfer (short distance) scales where the QCD coupling constant is small and a first few terms in perturbation theory give sufficiently accurate predictions. However, at very large distances, the coupling constant becomes large and the perturbative calculation can not be performed. The jet fragmentation can be thought as a two stage process. Immediately after hard interaction, the parton develops a shower through cascade emission of partons. This stage can be described well within perturbative QCD framework. At large time/distance scale, when partons hadronize, the process becomes non-perturbative and need phenomenological treatment. The boundary between these two stages is fuzzy and characterized by some cut off scale which may be as low as Λ_{QCD}. At large k_T, transverse momentum of particle with respect to initial parton, pQCD based shower Monte Carlo programs, *e.g.* Herwig, with phenomenological hadronization model describe the data well but in low k_T region, where most of the hadrons are produced, one need to sum the pQCD to all orders to adequately describe the data.

In MLLA, the terms in powers of leading logarithms are summed to all orders while taking care of interference between various emissions (angular ordering). The result is an infra-red stable expression for parton multiplicity and momentum distribution, valid down to a cut off parameter Q_{cutoff}, to be determined from experiment. In this framework, the ratio of hadron multiplicities in quark and gluon jets is given by ratio of their color charges $r = N_g/N_q = C_A/C_F = 9/4$. The next-to-MLLA (nMLLA) corrections[2] increase the multiplicity by a scale factor of 1.2–1.4, depending on the calculation, almost independent of initial parton energy. At nMLLA, the multiplicity ratio, $r = 1.5-1.7$ for a 100 GeV jet, with a weak energy dependence. According to local parton hadron duality hypothesis (LPHD), the partons hadronize locally and thus, the hadrons remember parton distributions. The number of hadrons is given by

$$N_{hadron} = K_{LPHD} \times N_{partons}$$

where K_{LPHD} is a parameter of order unity to be determined from the experimental data. For charged hadrons, $K_{LPHD} \sim 2/3$.

Well balanced, central di-jet events with dijet mass $80 < M_{JJ} < 630$ GeV, collected during 1993-95 by CDF collaboration at Fermilab Tevatron, are used to measure

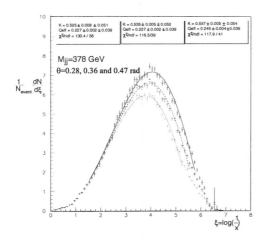

Figure 1. $\xi = \log E_{Jet}/p$ distributions in a jet for cone size θ=0.28,0.36 and 0.47 radians.

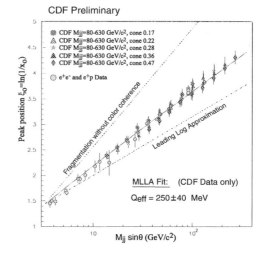

Figure 2. Peak position of the ξ distribution of charge tracks in a jet.

the charged particle multiplicity and momentum distributions. Herwig Monte Carlo predicts that $\sim 62\%$ of these jets are gluon jets at $M_{JJ} = 80$, decreasing to 22% at $M_{JJ} = 630$ GeV. The jets are reconstructed using cone algorithm with R=0.7. Charged tracks, corrected for track reconstruction inefficiency, underlying event and multiple $p\bar{p}$ interactions, in restricted cones around the jet axis are used in this analysis. Fits to $\xi = \log(E_{jet}/p)$ distribution to MLLA shape function $dN(\xi, Y)d\xi$ where $Y=E_{jet}\sin\theta/Q_{eff}$ for 27 points (9 mass ranges and 3 cones) yields $Q_{eff} = 240 \pm 40$ MeV.

The K_{LPHD} can be determined through the variation of momentum distribution with M_{JJ} through $N_h(\xi) = K(M_{jj})N^g(\xi)$ where

$$K(M_{jj}) = K_{LPHD}\left(f_g + (1 - f_g)/r\right)F_{nMLLA},$$

N^g is MLLA-predicted ξ distribution for gluon jets, f_g is the gluon fraction, determined from HERWIG, and we use $F_{nMLLA} = 1.30 \pm 0.20$ for the nMLLA scale factor.

The charged particle momenta distribution for three cone sizes (θ=0.28,0.36 and 0.47 radians) for $M_{JJ} = 378$ as a function of ξ, is shown in Fig. 1. As expected, the $K(M_{JJ}) = 0.53 \pm 0.05, 0.54 \pm 0.05, 0.56 \pm 0.05$ for three cone sizes, is almost constant. From $K(M_{JJ})$

distribution, we determine $r = 1.8 \pm 0.4$ and $K_{LPHD} = 0.58 \pm 0.05 \pm 0.8$. This ratio, r, depends on MLLA-predicted shape of ξ distribution. The variation of multiplicity with M_{JJ} yields $r = 1.7 \pm 0.3$ and $K_{LPHD} = 0.55 \pm 0.06$. The multiplicity variation with M_{JJ} is also well described by Herwig Monte Carlo after scaling by 0.89.

The peak values of the ξ distributions, $\xi_0 = Y/2 + \sqrt{cY} - c$ ($c=0.29$ for three flavors) are shown in Fig.2 along with the data from e^+e^- and e^+p colliders. The peak position is determined by fitting the distribution locally and is insensitive to the very low/high momentum particles. This is the first evidence of the predicted scaling in $E_{Jet}\sin\theta$. The extracted $Q_{eff} = 250 \pm 40$ in very good agreement with the determination from the shape of momentum distribution.

The di-jet data at low and moderate E_T is dominated by gluon-gluon scattering. The HERWIG Monte Carlo, using CTEQ4M parton distribution functions (PDF) predicts that $62 \pm 2\%$ of the jets are gluon-type at $E_T = 40$ GeV at $\sqrt{s} = 1800$ GeV, where the uncertainty is derived using different PDF. The true photon-jet sample is predicted to contain $84 \pm 2\%$ quark jets. However, $\sim 30\%$ of events tagged as γ-jet events are actually

500

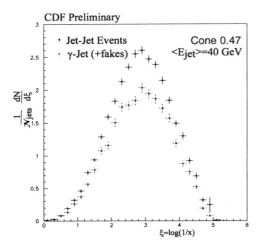

CDF Preliminary

* Jet-Jet Events   Cone 0.47
* γ-Jet (+fakes)   <Ejet>=40 GeV

$\frac{1}{N_{jets}}\frac{dN}{d\xi}$

ξ=log(1/x)

Figure 3. ξ distribution of charged tracks in a jet in di-jet and γ-jet events.

di-jet events with one of the jet fluctuating to fake a photon. The average multiplicity of di-jet sample is 5.77 ± 0.3 tracks/jet compared to 4.83 ± 0.05 tracks/jet in γ-jet sample. Using gluon fraction from Herwig Monte Carlo and measured fake rate, we determine $r = 1.74 \pm 0.11 \pm 0.07$, independent of MLLA formalism. These data show that r may depend on particle momenta, (~ 1.1 for high momentum particles), but errors are to large to be conclusive.

The CDF data are in good agreement with MLLA predictions. The measured $K_{LPHD} = 0.56 \pm 0.10$ is in good agreement with previous measurement of 0.59 ± 0.01 by TASSO. The $\alpha_s = 2\pi/[b \log(k_T/\Lambda_{QCD})]$ evaluated by substituting Λ_{QCD} with $Q_{eff} = 240 \pm 40$ MeV is in a good agreement with the world average value.

2 Sub-Jet Multiplicity

Instead of measuring the particle multiplicity which depends on the hadronization effect, one can separate the perturbative QCD effect by looking at clusters of particles, sub-jets, within a jet. The sub-jet multiplicity is both infrared and collinear safe and can be calculated to all orders in perturbative theory for

large leading and next-to-leading logarithms. The soft gluon emission or collinear parton pair preserve the jet's net flavor. The effect of $\mathcal{O}(\alpha_s)$ correction on jet flavor is small[4]. QCD predicts that gluons radiate more than quarks. Asymptotically, the ratio of sub-jets multiplicities in gluon and quark jets is expected to be equal to ratio of their color charges $C_A/C_F = 9/4$. Experimentally, the sub-jet multiplicity is less sensitive to cuts and reconstruction efficiencies than particle multiplicity.

At a $\sqrt{s}$ value, the jets are mixture of quarks and gluons. The measured sub-jet multiplicity can be written as

$$M = f_g M_g + (1 - f_g)M_q$$

where f_g the fraction of gluon jets. Using the data at two $\sqrt{s}$ values, the sub-jet multiplicity in quark and gluon jets can be determined. The gluon fraction, f, at $\sqrt{s} = 1800(630)$, determined from LO QCD Monte Carlo HERWIG with CTEQ4M PDFs, is $f_g = 0.59 \pm 0.02(0.33 \pm 0.03)$ where the uncertainty is determined by varying the PDFs.

The jets are identified using k_T algorithm with jet clustering parameter $D = 0.5$[3]. The calorimeter towers/particles are pre-clustered in pseudo-rapidity(η)-azimuthal angle(ϕ) space with minimum distance$\Delta R = \sqrt{\Delta\eta^2 + \Delta\phi^2} > 0.2$. These pre-clusters are clustered into objects by following prescription. For each object i, the distance $d_{ii} \equiv E_{T,i}^2$ is where E_T is the energy transverse to the beam. For each pairi, j, $d_{ij} \equiv min(E_{T,i}^2, E_{T,j}^2)\Delta R_{i,j}^2/D^2$. The object i and j are replaced by a new object given by their 4-vector sum provided d_{ij} in minimum of all possible combinations i, j, including d_{ii}. If $d_{ii} < d_{ij}$ for all i, j, the object i is promoted to a jet and removed from object list. This process is repeated until object list is empty.

For this study by DØ collaboration, the $p\bar{p}$ data taken at $\sqrt{s} = 630$ and 1800 GeV during 1992-1995 is used. Of the two highest transverse energy (E_T) jet in the event, those with $55 < E_T < 100$ GeV and $|\eta| < 0.5$ are

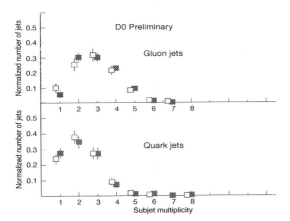

Figure 4. Fully simulated (open symbols) subjet multiplicity in quark and gluon jets compared with raw DØ data(closed symbols)

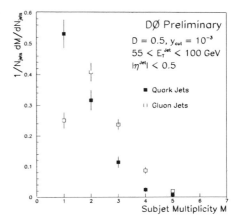

Figure 5. Fully corrected Sub-Jet multiplicity in quark and gluon jets.

used to measure subjet multiplicities.

The sub-jets multiplicity, M, within a jet is determined by running the same k_T algorithm on the pre-cluster list of a jet but with a resolution parameter $y_{cut} = 10^{-3}$. Pairs of objects with the smallest d_{ij} are merged successively until all remaining $d_{ij} > y_{cut}E_T^2$. As shown in Fig.4, the raw subject multiplicity for quark and gluon jets is well described by Herwig Monte Carlo.

The measured sub-jet multiplicity is corrected for detector effects using the corrections derived from Herwig Monte Carlo and detector simulations. The true sub-jet multiplicity (M^{true}) is determined by running the clustering procedure described above on the particle list from HERWIG Monte Carlo and the observed multiplicity (M^{obs}) is determined using the same procedure on the calorimeter towers after detector simulation. The observed jets are matched to true jets in $\eta - \phi$ space. Two dimensional correlations between M^{true} and M^{obs}, determined separately for gluon and quark jets, are used to correct the data. The procedure was tested on simulated data and works well. The corrected sub-jet multiplicities in quark and gluon jets are shown in Fig. 5.

As expected, the gluon jets have higher

subjet multiplicity than quark jets. The ratio of additional subjets in gluon to quark jets
$$R = \frac{\langle M_g \rangle - 1}{\langle M_q \rangle - 1} = 1.91 \pm 0.04(\text{stat})^{+0.23}_{-0.19}(\text{sys})$$
where the dominant source of systematic uncertainty is the gluon fraction ($^{+0.18}_{-0.12}$) uncertainty. Other sources include Jet E_T cut(± 0.12), detector smearing (± 0.08) and the unsmearing procedure (± 0.04). The measured ratio R is consistent with Herwig Monte Carlo (1.86 ± 0.04) and is slightly lower than the naive expectation of 9/4. The recent calculation of the sub-jet multiplicity [4], at next-to-leading accuracy in $\log(y_{cut})$, can not be compared because of the preclustering used in the jet finding algorithm.

References

1. Dokshitzer, Khoze, Tryon Int.J. Mod. Phy. A7(1992)1875, A. Mueller, Nucl. Phy. B213 (1983) 85.

2. Catani, Dokshitzer, Fiorani, Webber, Nucl. Phys. B377 (1992) 445. Dremin, Gary, Phys.Lett. B459 (1999), 341-346.

3. S. D. Ellis, D. E. Soper, Phys.Rev. D48 (1993) 3160-3166, hep-ph/9305266.

4. J.R. Forshaw and M.H. Seymore, JHEP 9909 (1999)009, hep-ph/9908307.

A NEXT-TO-LEADING ORDER CALCULATION OF HADRONIC THREE JET PRODUCTION

WILLIAM B. KILGORE

Physics Department, Building 510A, Brookhaven National Laboratory, Upton, NY 11973, USA
E-mail: kilgore@bnl.gov

WALTER T. GIELE

Theoretical Physics Department, Fermi National Accelerator Laboratory, Batavia, IL 60510, USA
E-mail: giele@fnal.gov

We present results of a next-to-leading order calculation of three jet production at hadron colliders. This calculation will have many applications. In addition to computing such three-jet observables as spectra, mass distributions, this calculation permits the first next-to-leading order studies at hadron colliders of jet and event shape variables.

1 Introduction

One of the difficulties in interpreting experimental results is in assessing the uncertainty to be associated with the theoretical calculation. This is particularly true in QCD where the coupling is quite strong and one expects higher order corrections to be significant.

Typically, one characterizes theoretical uncertainty by the dependence on the renormalization scale μ. Since we don't actually know how to choose μ or even a range of μ, the uncertainty associated with scale dependence is somewhat arbitrary. One motivation for performing next-to-leading order (NLO) calculations is to reduce the scale dependence associated with the calculation.

However, this is not the only benefit of an NLO calculation. There are times when the LO calculation is a bad estimator of the physical process. It may be that leading order kinematics artificially forbids the most important physical process. It could also be that the NLO corrections are simply large. Even if the overall NLO correction is relatively small, there may be regions of phase space, where NLO corrections are large. It is only in those regions of phase space where the NLO corrections are well behaved (as determined by the ratio of the NLO to LO terms) that one has confidence in the reliability of the calculation and can begin to believe the uncertainty estimated from scale dependence and it is only when one has a reliable estimate of the theoretical uncertainty that comparisons to experiment are meaningful.

2 Methods

The NLO three jet calculation consists of two parts: two to three parton processes at one-loop ($gg \to ggg$[1], $\bar{q}q \to ggg$[2], $\bar{q}q \to \bar{Q}Qg$[3], and processes related to these by crossing symmetry) and two to four parton processes ($gg \to gggg$, $\bar{q}q \to gggg$, $\bar{q}q \to \bar{Q}Qgg$, and $\bar{q}q \to \bar{Q}Q\bar{Q}'Q'$, and the crossed processes) computed at tree-level. Both of these contributions are infrared singular; only the sum of the two is infrared finite and meaningful. The Kinoshita-Lee-Nauenberg theorem[4] guarantees that the infrared singularities of the one-loop processes cancel those of the real emission processes for sufficiently inclusive observables.

In order to implement the kinematic cuts necessary to compare a calculation to experimental data one must compute the cross section numerically. Thus, we must find a numerically safe way of canceling the singularities. The method we use is the "subtraction improved" phase space slicing method[5,6,7].

3 Applications

The next-to-leading order calculation of three jet production will have a wide array of phenomenological applications.

3.1 Measurement of α_s

It should be possible to extract a purely hadronic measurement of α_s. One possibility for such a measurement would be a comparison of the three jet to two jet event rate[8]. Since both processes are sensitive to all possible initial states at tree-level, a next-to-leading order comparison should be relatively free of bias from the parton distributions. Because the measurement will be simultaneously performed over a wide range of energy scales, the running of α_s can be used to constrain the fits and enhance the precision of the combined measurement[9].

It has also been suggested that α_s can be determined from Dalitz distributions of three jet events[10].

3.2 Study Jet Clustering Algorithms

Because there are up to four partons in the final state, as many as three partons can end up in a single jet. This makes the three jet calculation sensitive to the details of jet clustering algorithms. This sort of study in pure gluon production[7] uncovered an infrared sensitivity in the commonly used iterative cone algorithms.

3.3 Study Jet Structure and Shape

Because there can be three partons clustered into a single jet, this calculation will allow truly next-to-leading order studies of the en-

ergy distribution in jets.[11] Studies of jet production in deep inelastic scattering[12] show that the next-to-leading order correction for this variable is substantial and agrees rather well with experimental measurements.

3.4 Study Event Shape Variables

There has been a long history of studying event shape variables like Thrust at e^+e^- colliders. These measurements challenge the ability of perturbative QCD to describe the data and provide a means (other than event rate) of obtaining a precise measurement of α_s. It will be interesting to see if one can make a meaningful study of such variables at hadron colliders.

3.5 Study Backgrounds to New Physics

Models of physics beyond the standard model typically involve massive states that can generate three jet signals either by associated production or by decay into three jets. To identify such signals, one must understand the pure QCD contribution to three jet production and look for deviations from the expected distributions.

4 Results

At present we have results for some of the most basic event distributions: the transverse energy spectrum and the angular distributions. The transverse energy distribution and its scale dependence has been shown in other conference proceedings[13]. We find that the next-to-leading order correction to the total rate is very small. The scale dependence of the NLO calculation, however, is a factor of two smaller than that of the LO calculation. The combination of a small NLO correction and reduced scale dependence indicates that we are obtaining a reliable calculation of three jet production.

We also have results for the angular distributions of the three jet events. Shown be-

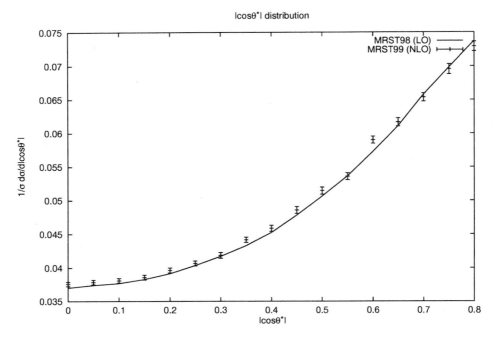

Figure 1. Distribution of events in $\cos\theta^*$, where θ^* is the angle between the leading jet and the beam axis in the three jet center of momentum. The next-to-leading order results are shown as points and the leading order results as solid lines.

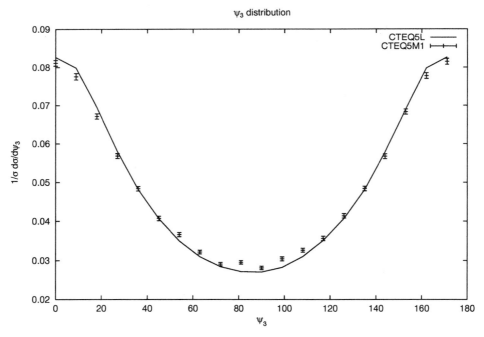

Figure 2. Distribution of events in ψ_3, where ψ_3 is the angle between the plane formed by leading jet and the beam axis and the plane formed by the second and third leading jets in the three jet center of momentum. The next-to-leading order results are shown as points and the leading order results as solid lines.

low are the distributions in $\cos\theta^*$, where θ^* is the angle between the leading jet and the beam axis in the three jet center of momentum, and ψ_3, where ψ_3 is the angle between the plane formed by leading jet and the beam axis and the plane formed by the second and third leading jets in the three jet center of momentum[14].

Again, the NLO correction is quite small. The important feature, however, is that the NLO results are more reliable than the LO results. Such distributions are particularly important for identifying (or eliminating) signals of new physics. It was the fact that the angular distributions of dijet events looked like QCD that eliminated the more exotic explanations of the famous high E_T anomaly in the one jet inclusive distribution.[15,16]

Acknowledgments

This work was supported by the US Department of Energy under grant DE-AC02-98CH10886.

References

1. Z. Bern, L. Dixon, D.A. Kosower, *Phys. Rev. Lett.* **70**, 2677 (1993) [hep-ph/9302280].

2. Z. Bern, L. Dixon and D.A. Kosower, *Nucl. Phys.* B **437**, 259 (1995) [hep-ph/9409393].

3. Z. Kunszt, A. Signer and Z. Trócsányi, *Phys. Lett.* B **336**, 529 (1994) [hep-ph/9405386].

4. T. Kinoshita, *J. Math. Phys.* **3**, 650 (1962);
   T.D. Lee and M. Nauenberg, *Phys. Rev.* **133**, 1549 (1964).

5. W.T. Giele and E.W.N Glover, *Phys. Rev.* D **46**, 1980 (1992).

6. W.T. Giele, E.W.N. Glover and D.A. Kosower, *Nucl. Phys.* B **403**, 633 (1993) [hep-ph/9302225].

7. W.B. Kilgore and W.T. Giele, *Phys. Rev.* D **55**, 7183 (1997).

8. B. Abbott et al., The D0 Collaboration, Fermilab Preprint FERMILAB-PUB-00-218-E [hep-ex/0009012].

9. W.T. Giele, E.W.N. Glover and J. Yu, *Phys. Rev.* D **53**, 120 (1996) [hep-ph/9506442].

10. A. Brandl et al., The CDF Collaboration, Proceedings of the XXXVth Rencontres de Moriond: QCD and High Energy Hadronic Interactions, Les Arcs, France, March 18–25, 2000, FERMILAB-CONF-00-170-E.

11. F. Abe et al., The CDF Collaboration, *Phys. Rev. Lett.* **70**, 713 (1993);
    S. Abachi et al., The D0 Collaboration, *Phys. Lett.* B **357**, 500 (1995);
    W.T. Giele, E.W.N. Glover and D.A. Kosower, *Phys. Rev.* D **57**, 1878 (1998) [hep-ph/9706210];
    M.H. Seymour, *Nucl. Phys.* B **513**, 269 (1998) [hep-ph/9707338].

12. N. Kauer, L Reina, J. Repond, and D. Zeppenfeld, PLB **460**, 189 (1999) [hep-ph/9904500].

13. W.B. Kilgore and W.T. Giele, Proceedings of the XXXVth Rencontres de Moriond: QCD and High Energy Hadronic Interactions, Les Arcs, France, March 18–25, 2000 [hep-ph/0009176].

14. F. Abe et al., The CDF Collaboration, *Phys. Rev.* D **54**, 4221 (1996) [hep-ex/9605004];
    S. Abachi et al., The D0 Collaboration, *Phys. Rev.* D **53**, 6000 (1996) [hep-ex/9509005].

15. F. Abe et al., The CDF Collaboration, *Phys. Rev. Lett.* **77**, 5336 (1996), PRL **78**, 4307 (1997) [hep-ex/9609011].

16. B. Abbott et al., The D0 Collaboration, *Phys. Rev. Lett.* **80**, 666 (1998) [hep-ex/9707016].

PROTON STRUCTURE AT HIGH Q^2

KUNIHIRO NAGANO

REPRESENTING THE H1 AND ZEUS COLLABORATIONS

DESY, Notkestrasse 85, 22603 Hamburg, Germany

E-mail: kunihiro.nagano@desy.de

The H1 and ZEUS experiments at HERA have measured e^+p and e^-p deep inelastic scattering cross sections at high Q^2 both for neutral and charged current interactions. Results are presented from data taken during the 1994–1997 (e^+p) and 1998–April 1999 (e^-p) running periods. Preliminary results by H1 from the e^+p data taken during the recent July 1999–2000 running period are also presented.

1 Deep inelastic scattering at HERA

Deep inelastic scattering (DIS) of leptons on nuclei has been key to our understanding of the structure of the nucleon. HERA at DESY is a unique facility which collides electrons (or positrons) with protons. During the 1994–1997 running period, HERA collided an e^+ beam of 27.5 GeV and a p beam of 820 GeV. The H1 and ZEUS experiments collected 35.6 and 47.7 pb^{-1} of luminosity, respectively. From 1998 to April 1999, HERA provided an e^- beam of 27.5 GeV with a p beam of 920 GeV. Both experiments collected about 16 pb^{-1} of luminosity. From July 1999 on, HERA has been providing an e^+ beam of 27.5 GeV and a p beam of 920 GeV. Both experiments have collected about 50 pb^{-1} of luminosity so far.

The neutral current (NC) DIS interaction, $e^\pm p \to e^\pm X$, and the charged current (CC) DIS interaction, $e^{+(-)}p \to \bar{\nu}(\nu)X$, have been studied using these data at high Q^2, typically $Q^2 > 200$ GeV^2 [1,2,3,4,5]. All four types of cross-sections in unpolarized ep DIS, i.e. NC and CC in both e^+p and e^-p, were measured.

1.1 DIS cross section formulae

NC DIS cross sections can be written as

$$\frac{d^2\sigma^{NC}(e^\pm p)}{dxdQ^2} = \frac{2\pi\alpha^2 Y_+}{xQ^4} \cdot \tilde{\sigma}_\pm^{NC},$$

$$\tilde{\sigma}_\pm^{NC} = F_2^{NC} \mp \frac{Y_-}{Y_+} xF_3^{NC} - \frac{y^2}{Y_+} F_L^{NC},$$

where α is the electromagnetic fine structure constant, x is the Bjorken scaling variable, y is the inelasticity parameter and $Y_\pm = 1 \pm (1-y)^2$. The helicity dependence of the electroweak interactions is mostly contained in $Y_\pm$. The evolution in Q^2 of the structure functions, F_2^{NC}, F_3^{NC} and F_L^{NC}, is evaluated using the next-to-leading-order (NLO) DGLAP evolution equations of QCD. F_L^{NC} is not zero in NLO QCD but is only relevant at high y, where it contributes to the cross section by up to about 10%. F_3^{NC} is dominated by a parity-violating term and contributes only at high Q^2 oppositely in e^-p and e^+p.

CC DIS cross sections can be written as

$$\frac{d^2\sigma^{CC}(e^\pm p)}{dxdQ^2} = \frac{G_F^2}{2\pi x} \left(\frac{M_W^2}{M_W^2 + Q^2}\right)^2 \cdot \tilde{\sigma}_\pm^{CC},$$

$$\tilde{\sigma}_\pm^{CC} = \frac{1}{2}\{Y_+ F_2^{CC} \mp Y_- xF_3^{CC} - y^2 F_L^{CC}\},$$

where G_F is the Fermi constant and M_W is the mass of the $W^\pm$ boson. In leading-order (LO) QCD, the reduced cross sections, $\tilde{\sigma}_\pm^{CC}$, can be written as

$$\tilde{\sigma}_+^{CC} = x[(\bar{u} + \bar{c}) + (1-y)^2(d+s)],$$
$$\tilde{\sigma}_-^{CC} = x[(u+c) + (1-y)^2(\bar{d}+\bar{s})],$$

where d is, for example, the parton density function (PDF) for the d quark. The flavor-selecting nature of the CC interaction

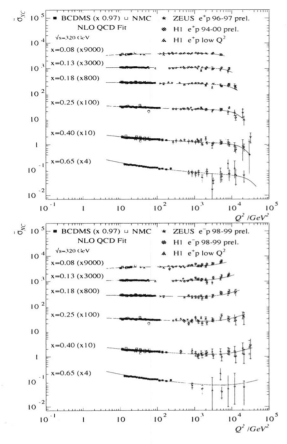

Figure 1. The reduced NC DIS cross section $\tilde{\sigma}^{NC}$ compared with the SM predictions for e^+p (upper plot) and e^-p (lower plot).

is clearly seen; only down-type quarks (anti-quarks) and up-type antiquarks (quarks) participate at LO in e^+p (e^-p) CC DIS. The kinematic suppression factor $(1-y)^2$ is applied to the contribution from quarks (anti-quarks) in e^+p (e^-p).

2 NC cross sections

2.1 Double differential cross sections

Figure 1 shows the reduced double differential cross section, $\tilde{\sigma}^{NC}$, compared to the SM prediction. A good agreement was observed between data and the SM predictions both in e^+p and e^-p. At $Q^2 = 20000$ GeV^2 and

$x = 0.4$, the new H1 e^+p data in 1999–2000 undershoot the SM expectation while the measurement in 1994–1997 exceeded the expectation [1]; consequently, the combined results show no deviation from the expectation. NLO QCD describes the x and Q^2 dependence of proton structure for $x \lesssim 0.65$ and $Q^2 \lesssim 30000$ GeV^2.

A positive (negative) interference effect between the γ and the Z^0, i.e. the effect of xF_3^{NC}, was observed in e^-p (e^+p) at high Q^2 and high x, which will be described in the following section.

2.2 Extraction of xF_3^{NC}

xF_3^{NC} can be extracted as the difference between $\tilde{\sigma}_-^{NC}$ and $\tilde{\sigma}_+^{NC}$,

$$xF_3^{NC} = \mathcal{C}_Y(\tilde{\sigma}_-^{NC} - \tilde{\sigma}_+^{NC}) + \Delta F_L,$$
$$\mathcal{C}_Y = (Y_-^{920}/Y_+^{920} + Y_-^{820}/Y_+^{820})^{-1},$$

where Y_+^{920} denotes the Y_+ factor for a proton-beam energy of 920 GeV and so on. Due to the difference in beam energies between e^+p and e^-p data, a small correction due to F_L^{NC} (denoted ΔF_L) is needed. The maximum ΔF_L is 10% at lowest Q^2 and lowest x (i.e. at highest y).

Both H1 and ZEUS extracted xF_3^{NC}; figure 2 shows xF_3^{NC} measured by H1 compared to the SM prediction. This is the first measurement of xF_3^{NC} in the high Q^2 regime. The measured xF_3^{NC} is consistent with the SM prediction. Currently, the measurement errors are limited by the number of e^-p events.

3 CC cross sections

3.1 Double differential cross sections

Figure 3 shows the reduced double differential cross section, $\tilde{\sigma}^{CC}$. A good agreement between data and the SM prediction was observed both in e^+p and e^-p. As shown in the figure, $\tilde{\sigma}^{CC}$ depends directly on PDFs in LO

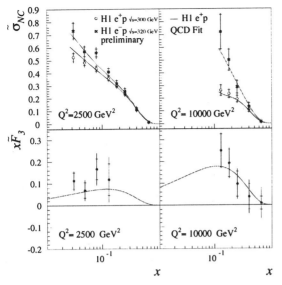

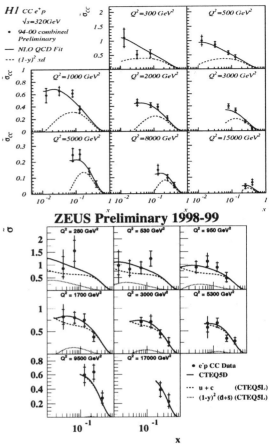

Figure 2. The reduced NC DIS cross section $\tilde{\sigma}^{NC}$ (upper plots) and the structure function xF_3^{NC} (lower plots) measured at Q^2=2500 GeV^2 (left plots) and Q^2=10000 GeV^2 (right plots) by H1.

Figure 3. The reduced CC DIS cross section $\tilde{\sigma}^{CC}$ compared with the SM predictions for e^+p (upper plot) and e^-p (lower plot).

QCD. $\tilde{\sigma}^{CC}$ is sensitive to the quark combination $d + s$ ($u + c$) at high x for fixed Q^2 in e^+p (e^-p). At low x, $\tilde{\sigma}^{CC}$ is sensitive to the antiquark combination $\bar{u}+\bar{c}$ in e^+p, while the $u + c$ combination still gives a dominant contribution in e^-p due to the suppression to antiquarks by $(1 - y)^2$. Both the quark and the antiquark combinations are required in order to obtain a good description of the data.

3.2 $d\sigma/dx$

The sensitivity of the e^+p (e^-p) CC cross section to $d(u)$-quark density at high x can be seen more clearly in the single-differential cross-section $d\sigma/dx$. Figure 4 shows the ZEUS measurement of $d\sigma/dx$ for $Q^2 > 200$ GeV^2 compared to the SM prediction with the CTEQ 4D PDF [6] both for e^+p and e^-p. For $x \gtrsim 0.3$, the e^+p data lie above the SM prediction. The d-quark density is poorly constrained from existing experimental data. The possibility of a larger d/u ratio

than currently assumed has been discussed in recent years; a modification of the CTEQ 4D PDF according to the prescription in Yang and Bodek [7] is known to give a good description of this [3] e^+p CC $d\sigma/dx$, for example.

4 Q^2 dependence of NC and CC cross sections

Q^2 sets the scale for a typical spatial resolution to probe the proton of $\lambda \sim 1/\sqrt{Q^2}$. The large center-of-mass energy of about 300 GeV at HERA allows an extension of the explorable kinematic phase space by two orders of magnitude in Q^2 compared with that

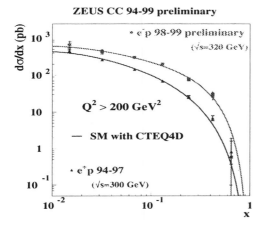

Figure 4. The CC cross section $d\sigma/dx$ measured by ZEUS compared with the SM predictions with the CTEQ 4D PDF.

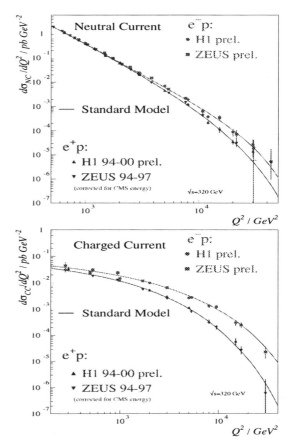

Figure 5. The single differential cross-sections $d\sigma/dQ^2$ measured at HERA for NC (upper plot) and CC DIS (lower plot).

covered by the previous fixed-target experiments. The maximum Q^2 reaches almost 10^5 GeV^2, which corresponds to a spatial resolution of about 10^{-16} cm.

Figure 5 shows the single differential NC and CC cross sections as functions of Q^2 both for e^+p and e^-p.

The measured NC cross sections agree well with the SM predictions over six orders of magnitude in $d\sigma/dQ^2$, and two orders in Q^2. NLO QCD describes the data up to very high $Q^2 \sim 40000$ GeV^2. A clear difference of cross sections in e^+p and e^-p is seen at large $Q^2 \gtrsim M_Z^2$, which is due to the xF_3^{NC} contribution.

The measured CC cross sections are consistent with the SM. The e^-p cross section is higher than e^+p for all kinematic region of $Q^2 > 200$ GeV^2; for $Q^2 \gtrsim 10000$ GeV^2, it is an order of magnitude higher. This can be explained in the SM as due to the u-quark density, which is larger than that of the d quark, and the kinematic suppression factor, which is less of an effect in e^-p.

Acknowledgments

The author is financially supported by the Japan Society for the Promotion of Science.

References

1. H1 Collab., C. Adloff *et al.*, *Eur. Phys. J.* **C13**(2000) 609.

2. ZEUS Collab., J. Breitweg *et al.*, *Eur. Phys. J.* **C11**(1999) 427.

3. ZEUS Collab., J. Breitweg *et al.*, *Eur. Phys. J.* **C12**(2000) 411.

4. H1 Collab., contributed paper #971 and #975 to ICHEP2000, Osaka.

5. ZEUS Collab., contributed paper #1049 and #1050 to ICHEP2000, Osaka.

6. H.L. Lai *et al.*, *Phys. Rev.* D **55**(1997) 1280.

7. U.K. Yang and A. Bodek, *Phys. Rev. Lett.* **82**(1999) 2467.

STRUCTURE FUNCTION AT HERA AND QCD ANALYSIS

F. ZOMER

IN2P3-CNRS and Université de Paris-Sud, Laboratoire de l'Accélérateur Linéaire, Bâtiment 200, F-91898 BP 34 Orsay Cedex, France

A preliminary determination of α_s and xg using the new inclusive structure function F_2 measurements at HERA is presented.

In this communication, preliminary results of a QCD analysis applied to the new medium Q^2 F_2 data measured at HERA are shown. In this kinematic range a statistical accuracy $\approx 1\%$ for $Q^2 \approx 10\text{GeV}^2$ and $\approx 5\%$ for $Q^2 \approx 1000\text{GeV}^2$ are presently achieved (the systematics being between 2% and 3%). Fig. 1 show the HERA F_2 data as function of Q^2, H1 and ZEUS measurements are in agreement within a relative overall uncertainty of $\approx 4\%$.

The quark and gluon density functions (pdfs) are determined from the solution of the NLO DGLAP equations and the fixed flavor scheme is used to determined the charm contribution to F_2. Since the x pdfs dependence must be parameterised at a fixed Q^2 (called Q_0^2 here), a fit is performed to the H1 and the BCDMS fixed Hydrogen target data. In order to avoid higher twist corrections the following cuts are applied to the data entering the fit: $Q^2 \geq 3.5\text{GeV}^2$ for all data sets and $y > 0.3$ for the BCDMS data (to avoid the data affected by large systematics). All correlations induced by the systematics are properly taken into account.

In fact, $\approx 10^5$ fits were performed varying simultaneously all the fit ingredients: parameterisation forms, Q_0^2, charm mass, α_s, kinematical cuts applied to the data. The choices for the fit ingredients are finally made according to a χ^2 stability criteria.

As a result of this study, a sensitivity to α_s is obtained. This is demonstrated in fig. 2 where the χ^2 variation as function of α_s for three series of fits, performed in the same conditions, are compared: H1 alone, BCDMS alone and H1 plus BCDMS. While H1 alone cannot determine precisely α_s, BCDMS alone does it but with a resulting xg not compatible with the rising of F_2 at small x measured at HERA. When both H1 and BCDMS data are fitted, the χ^2 curve gets narrower and the minimum is displaced toward higher α_s values. We found $\alpha_s = 0.1150 \pm 0.0017(exp) \pm 0.0012(model)$, where the model uncertainty is obtained by varying all fit ingredients. In addition, a renormalisation and factorisation scales uncreatainty of 0.005 is estimated. The resulting gluon density is determined with a relative accuracy of 3% at $Q^2 = 20$ GeV^2 and $x = 10^{-3}$. The gluon density obtained from the fit to the HERA data alone is in agreement with our nominal fit result, thus demonstrating the capability of the HERA data to determine xg for a fixed α_s.

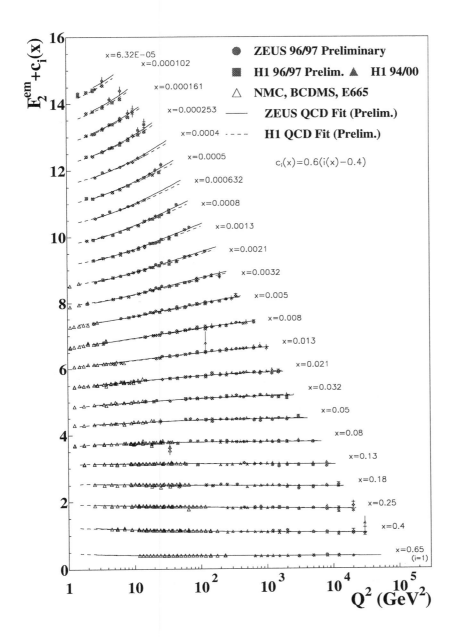

Figure 1. Preliminary F_2 measurements by H1 and ZEUS as function of x for $Q^2 = 15$ GeV^2 compared with the preliminary QCD fits.

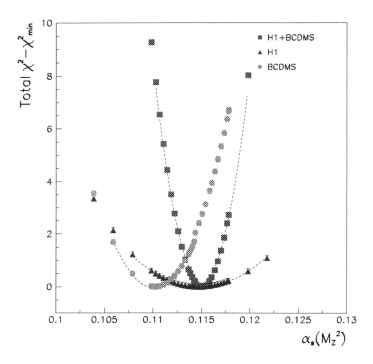

Figure 2. Determination of α_s. Comparison of different fits.

THE CHARM CONTRIBUTION TO THE STRUCTURE FUNCTION OF THE PROTON

EKATERINI TZAMARIUDAKI

for the H1 and ZEUS Collaborations
Max Planck Institute Munich, DESY Notkestrasse 85, D-22603 Hamburg, GERMANY
E-mail: katerina@mail.desy.de

Measurements of the open charm production cross section in deep inelastic scattering performed by the H1 and ZEUS experiments at HERA are presented. Final states containing open charm are identified either by full reconstruction of $D^{*\pm}$ mesons or via the semileptonic decay into electrons. The resulting differential cross sections are compared to theoretical calculations and the charm contribution to the proton structure function, $F_2^c(x, Q^2)$, is determined.

1 Introduction

Heavy quark production in electron proton scattering proceeds, in QCD, almost exclusively via the photon gluon fusion mechanism: a photon, emitted by the incoming electron, interacts with a gluon in the proton to form a quark-antiquark pair i.e. $\gamma g \to c\bar{c}$.

The kinematics of the ep interaction is described by three independent variables: the centre of mass energy $\sqrt{s}$; the squared four momentum transfer of the photon $q^2 = -Q^2$ and either one of the scaling variables $y = (q \cdot P/l \cdot P)$, the inelasticity of the ep-interaction, or Bjorken $x = Q^2/(2P \cdot l)$. Here, P and l denote the four-momenta of the proton and the electron, respectively.

The results[1,2,3] presented here are based on 34 pb^{-1} and 18.6 pb^{-1} recorded in 1996 and 1997 with the ZEUS and H1 detectors respectively.

2 Open charm production

Several schemes are used to perform the calculations. All approaches assume the scale to be hard enough to apply pQCD and to guarantee the validity of the factorisation theorem. Here, for the NLO calculations in the DGLAP scheme, the massive approach is adopted which is a fixed order (in α_s) calculation with massive quarks assuming three active flavours in the proton. The heavy quark is only produced at the perturbative level via boson-gluon fusion[4]. Based on the NLO calculations, the Monte Carlo integration program HVQDIS[5] provides the four-momenta of the outgoing partons. Charmed quarks are fragmented into charmed hadrons according to the Peterson fragmentation funcion[6].

Recently the CCFM model[7] has also been available for comparisons to the data. Here, in the parton cascade, gluons are emitted only in an angular ordered region. Due to this ordering, the unintegrated gluon distribution depends on the maximum allowed angle as well as the momentum fraction x and the transverse momentum of the propagator gluon.

3 Cross Sections

The charm content of the events is detected either by the reconstruction of $D^{*\pm}$ mesons or the identification of electrons from the semileptonic decay. $D^{*\pm}$ mesons are reconstructed in the decay $D^{*+} \to D^0 \pi^+$ with $D^0 \to K^- \pi^+$ or $D^0 \to K^- \pi^+ \pi^+ \pi^-$.

Differential cross sections for the process $ep \to D^{*+}X$ have been measured by both ZEUS[1] and H1[3]. The inclusive D^{*+} differential cross sections in the kinematic region $1 < Q^2 < 100 \ GeV^2$, $0.05 < y < 0.7$ and $p_{TD^*} > 1.5 \ GeV/c$, $|\eta_{D^*}| < 1.5$ are shown in Figure 1 for the H1 data as a function of the kinematical quantities W, x_{Bj} and Q^2 and as a function of the D^{*+} observables p_{TD^*}, η_{D^*} and $z_{D^*} = (E - p_z)_{D^*}/2yE_e$. A bin by bin correction due to QED radiation has been applied using the HECTOR program[8]. The data points are compared to the expectation of the NLO QCD calculations using the HVQDIS program[5] with the GRV98-HO[9] parton density parametrization. The uncertainty in the calculation due to the variation of the c-quark mass and due to the fragmentation is shown as light shaded area. Figure 1 also includes the prediction of the CCFM calculations using the CASCADE program, which has been made available recently, with the gluon density taken from the fit to the H1 inclusive F_2 data. For the CCFM prediction, the dark shaded band reflects the uncertainty introduced by the charm mass. The agreement in the shape of the differential cross sections of both predictions with the data is reasonable. The expectations from the CASCADE program are found to agree better in normalisation with the H1 data especially in the forward η region. The ZEUS data[1,2] agree with the expectations from the HVQDIS program using the structure function from their fit to inclusive F_2 data.

Using the semileptonic decay into electrons to tag charm events [2], a higher statistics sample is obtained, making it possible to extend the kinematic region to $1 < Q^2 < 1000$ GeV^2.

4 Charm contribution to the proton structure function $F_2^c(x, Q^2)$

Neglecting the contribution from F_L, the DIS inclusive cross section $\sigma^c(x, Q^2)$ can be expressed in terms of F_2^c:

$$\frac{d^2\sigma^c}{dx\,dQ^2} = \frac{2\pi\alpha^2}{xQ^4}\left(1 + (1-y)^2\right) \cdot F_2^c(x, Q^2)$$

where x is the Bjorken scaling variable. The visible inclusive cross sections, $\sigma_{vis}^{exp}(x, Q^2)$, in bins of x and Q^2 are converted to

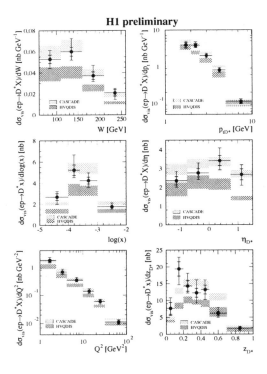

H1 preliminary

Figure 1. Differential cross section $\sigma(ep \to eD^{*+}X)$ versus W, x_{Bj}, Q^2 and p_{T,D^*}, η_{D^*}, z_{D^*}. The inner error bars account for the statistical errors and the outer error bars for the total error. The curves are described in the text.

$F_2^{c\ exp}(x, Q^2)$ by the relation:

$$F_2^{c\ exp}(x, Q^2) = \frac{\sigma_{vis}^{exp}(x, Q^2)}{\sigma_{vis}^{theo}(x, Q^2)} \cdot F_2^{c\ theo}(x, Q^2),$$

where σ_{vis}^{theo}, $F_2^{c\ theo}$ are the theoretical predictions from the model under consideration. The HVQDIS and the Riemersma $al.$[4] programs are used to calculate these quantities by both experiments. H1 has also used the recently released CASCADE program which is based on the CCFM equation to extract F_2^c and the result is in good agreement with the one extracted with the HVQDIS program within the present statistics of the data.

In Figures 2 and 3, a compilation of the different measurements of the charm contribution to the proton structure function $F_2^c(x, Q^2)$ is shown. The experimental results are compared with the predictions of F_2^c from the H1 NLO DGLAP fit to the inclusive F_2 measurements.

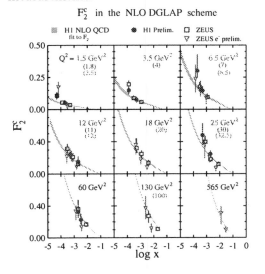

Figure 2. The charm contribution to the proton structure function F_2^c as derived from the inclusive D^{*+} meson analysis by H1 and ZEUS and from the ZEUS semileptonic decays analysis. The shaded band represents the predictions of F_2^c from the H1 NLO DGLAP fit to the inclusive F_2 measurements with m_c ranging from 1.3 to 1.5 GeV.

5 Conclusions

Results on open charm production in deep inelastic ep-scattering using the D^{*+} meson or the charm semileptonic decay into electrons for tagging have been reported. The resulting differential cross sections from ZEUS and H1 are in good agreement and they are reasonably well discribed by the NLO DGLAP calculation using the HVQDIS program. The comparison with the CCFM calculations, where made, shows good agreement with the data. The charm contribution to the proton structure function, F_2^c, has been extracted by extrapolating to the full p_T and η region. The measurements of F_2^c derived from the different experiments and using different methods are in good agreement.

References

1. ZEUS Collab., J. Breitweg *et al*, *Phys. Lett.***B407**, 402 (1997); *Eur.Phys.J.*C**12**, 1 (2000).

2. ZEUS Collab., J. Breitweg *et al*, ICHEP2000, abstract 853.

3. H1 Collab., C. Adloff *et al*, ICHEP2000, abstract 983.

4. E.Laenen *et al*, *Nucl.Phys.*B**392**, 162 (1993); E.Laenen *et al*, *Phys.Lett.*B**291**, 325 (1992); S.Riemersma, J.Smith and W.L. van Neerven, *Phys.Lett.*B**347**, 143 (1995).

5. B.W.Harris and J.Smith, *Phys.Rev.*D**57**, 2806 (1998).

6. C.Peterson *et al*, *Phys.Rev.*D**27**, 105 (1983).

7. M.Ciafaloni,*Nucl.Phys.*B **296**, 49(1998); S.Catani, F.Fiorani and G.Marchesini, *Phys. Lett.*B **234**, 339(1990); G.Marchesini, *Nucl.Phys.*B**445**, 45 (1995).

8. A. Arbuzov *et al*,*Comp.Phys.Com.***94**, 128 (1996).

9. M. Glück, E. Reya and A. Vogt, *Eur.Phys.J.*C**5**, 461 (1998).

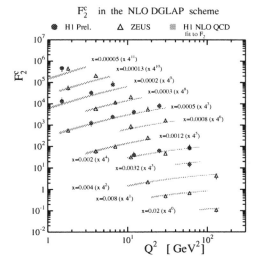

Figure 3. The charm contribution to the proton structure function F_2^c as derived from the inclusive D^{*+} meson analysis by H1 and ZEUS. The shaded band represents the predictions of F_2^c from the H1 NLO DGLAP fit to the inclusive F_2 measurements with m_c ranging from 1.3 to 1.5 GeV.

MEASUREMENT OF DIFFERENTIAL NEUTRINO-NUCLEON CROSS-SECTIONS AND STRUCTURE FUNCTIONS USING THE CHORUS LEAD CALORIMETER

R.G.C. OLDEMAN, FOR THE CHORUS COLLABORATION

NIKHEF, Amsterdam, The Netherlands

E-mail: Rolf.Oldeman@nikhef.nl

A high-statistics sample of neutrino and anti-neutrino interactions in a lead target has been recorded in the 1998 CHORUS run at CERN. Preliminary measurements of differential νN cross-sections and of the structure functions $F_2(x, Q^2)$, $xF_3(x, Q^2)$ and $R(x, Q^2)$ are presented.

1 Introduction

Neutrino-nucleon differential cross-sections and structure functions are valuable ingredients for constraining parton distribution functions and testing QCD. The comparison of $F_2^{\nu N}(x, Q^2)$ with charged lepton data provides a test of the universality of parton distributions. The structure function $xF_3^{\nu N}(x, Q^2)$ represents in leading order the valence quark densities. In QCD, its evolution is independent of the gluon density.

The only published neutrino structure functions based on samples of more than 10^5 neutrino and anti-neutrino interactions, are from CCFR[1] and CDHSW[2]. Both experiments used massive iron target calorimeters. The results on $F_2(x, Q^2)$ show a significant difference. A third independent measurement will help to clarify the disagreement. Furthermore, a comparison between lead and iron data allows to study nuclear effects. This motivated us to perform a dedicated experiment using the CHORUS detector at CERN. The analysis of the here discussed data is described in more detail elsewhere[3].

2 Data collection

Neutrinos are produced using the 450 GeV protons of the CERN-SPS accelerator, extracted every 14.4 s in two 6 ms spills on a segmented beryllium target. After focusing and charge selection, secondary pions and kaons decay in a 309 m evacuated decay tunnel. Surviving hadrons and muons stop in a 370 m shielding of iron and earth.

The CHORUS detector is described in detail elsewhere[4]. Only the detector systems that are most relevant for this analysis are described here.

The 118 ton lead-scintillator calorimeter serves both as a target and for measuring the shower energy of neutrino interactions. It consists of three sections of different size and granularity. Calorimetry is complemented using the scintillators of the muon spectrometer. The energy scale and resolution of the calorimeter system have been measured using momentum-selected pions in the energy range 8–28 GeV. For interactions inside the fiducial volume, the resolution on the shower energy is 40–60%/$\sqrt{E}$.

The muon momentum and charge are measured in a spectrometer containing six magnetized iron disks. The muon trajectory is measured at seven locations around and between the magnets. The momentum resolution, measured using test beam muons, varies from 12% at 10 GeV to 17% at 100 GeV.

The data were taken in a dedicated run in 1998 during 18 weeks. Our run selection contains about $3.6 \cdot 10^6$ recorded events in neutrino beam and $1.0 \cdot 10^6$ in anti-neutrino beam.

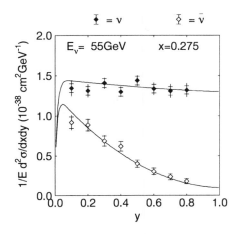

Figure 1. Example of the measured νN cross-section. The curve indicates the cross-section model used for the MC generation.

3 Differential cross-sections

Charged-current neutrino interactions are selected, containing a muon with a momentum of at least 4 GeV and a scattering angle below 300 mrad. Because of saturation of the calorimeter ADCs, events with a shower energy above 100 GeV are rejected. Events with a wrong-sign muon, corresponding to the beam contamination, are rejected. The interaction vertex is required to be reconstructed inside a central fiducial area of $180 \times 180 \, \text{cm}^2$ in the upstream part of the calorimeter. With these cuts the data are reduced to $1.1 \cdot 10^6$ neutrino interactions and $0.23 \cdot 10^6$ anti-neutrino interactions.

A Monte Carlo event sample is generated using a cross-section model based on the GRV94 LO[5] parton distributions, folding in the measured beam spectrum and using a parameterization of the detector response. A correction is applied to the cross-section model to improve the description of the data at low Q^2. The MC sample is used to correct the data for the cuts, the trigger inefficiency, the reconstruction inefficiency and the resolution smearing.

The neutrino and anti-neutrino data are binned in E_ν (10–200 GeV in 10 bins), x (0.01–0.70 in 11 bins) and y (0.05–0.95 in 9 bins). Only the 2×733 bins that are fully contained within the kinematical cuts are used for the cross-section measurement. The differential cross-sections are then determined as follows:

$$\frac{1}{E_\nu} \frac{d^2\sigma}{dx dy} = \frac{\sigma_{tot}^{\nu,\bar{\nu}}}{E_\nu} \frac{N(E_\nu, x, y)}{N(E_\nu) \Delta x \Delta y}, \quad (1)$$

where $\sigma_{tot}^{\nu,\bar{\nu}}$ is the total (anti-)neutrino-nucleon scattering cross-section[6], corrected for the non-isoscalarity of the target.

Systematic uncertainties have been evaluated by applying the following shifts: a change in the calorimeter energy scale of 5% and an offset of 150 MeV, a change of the muon momentum scale of 2.5% and an offset of 150 MeV. The 2.1% uncertainty on the total νN cross-section and the 1.4% uncertainty on the $\bar{\nu}/\nu$ cross-section have been taken into account, as well as an uncertainty of the energy dependence of the total νN cross-section of 1%/100 GeV and 0.5%/100 GeV on the $\bar{\nu}/\nu$ cross-section.

An example of the measured differential cross-sections for one value of E_ν and x is shown in Figure 1.

4 Structure function extraction

Neutrino structure functions are related to differential cross-sections as follows:

$$\frac{d\sigma^{\nu N}}{dx \, dy} = \frac{G_F^2 M_N E_\nu}{\pi (1 + Q^2/m_W^2)^2} \times \quad (2)$$

$$[\frac{y^2}{2} 2xF_1 + (1 - y - \frac{Mxy}{2E_\nu})F_2 \pm (y - \frac{y^2}{2})xF_3],$$

where the structure functions $2xF_1$, F_2 and xF_3 depend on x and Q^2 only. They are not necessarily equal for neutrino and anti-neutrino interactions.

The differential cross-section data are corrected for the non-isoscalarity of the target and for radiative effects[7].

By exploiting the y-dependence of the cross-section in each x, Q^2 bin, $2xF_1$, F_2

518

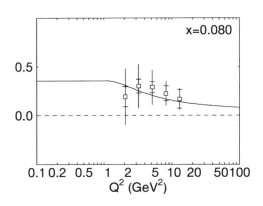

Figure 2. Example of the measurement of $R = \sigma_L/\sigma_T$ for one value of x, compared to the SLACR90 parameterization.

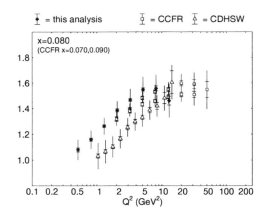

Figure 3. Example of the measured structure function $F_2(x, Q^2)$. The results are compared with those of CDHSW and CCFR.

and xF_3 can be determined independently for neutrinos and anti-neutrinos. This extraction, however, results in highly correlated values with large statistical errors.

According to the quark-parton model, the isoscalar structure functions $2xF_1$ and F_2 are equal for neutrinos and anti-neutrinos, and $\Delta xF_3 = xF_3^\nu - xF_3^{\bar\nu} = 4(s-c)$. With these assumptions, the ν–$\bar\nu$ average $2xF_1$, F_2 and xF_3 can be extracted more precisely. The strange and charm seas are taken from the GRV94 parton distributions with a systematic uncertainty of 20%.

Having extracted both $F_2(x, Q^2)$ and $2xF_1(x, Q^2)$, the ratio, R, of longitudinal to transverse cross-sections, can be determined:

$$R = \left(1 + \frac{4M^2x^2}{Q^2}\right)\frac{F_2}{2xF_1} - 1 \qquad (3)$$

The measured values of $R(x, Q^2)$ for one value of x are shown in Figure 2, and compared to the SLACR90 parameterization[8].

The most precise values of F_2 and xF_3 are obtained by constraining R to the SLACR90 parameterization. A systematic uncertainty of 20% is attributed to R. A comparison is made of the structure function $F_2(x, Q^2)$ between this measurement and

those of CCFR and CDHSW. An example for one value of x is shown in Figure 3. We observe an overall agreement with the CCFR data, while our data differ significantly from those of CDHSW.

References

1. W.G. Seligman et al., Phys. Rev. Lett. **79**, 1213 (1997).
2. J.P. Berge et al., Z. Phys. **C49**, 187 (1991).
3. R.G.C. Oldeman, PhD Thesis, University of Amsterdam, June 2000. Available at `choruswww.cern.ch/~oldeman`.
4. E. Eskut et al. Nucl. Instrum. Meth. **A401**, 7 (1997).
5. M. Gluck, E. Reya and A. Vogt, Z. Phys. **C67**, 433 (1995).
6. J.M. Conrad, M.H. Shaevitz and T. Bolton, Rev. Mod. Phys. **70**, 1341 (1998)
7. D. Y. Bardin and V. A. Dokuchaeva, JINR-E2-86-260.
8. L. W. Whitlow et al., Phys. Lett. **B250**, 193 (1990).

UNCERTAINTIES OF PARTON DISTRIBUTIONS AND THEIR PHYSICAL PREDICTIONS

WU-KI TUNG

Michigan State University, E. Lansing, MI, USA
E-mail: Tung@Pa.Msu.Edu

A concrete program to study the uncertainties of parton distributions in global QCD analysis is outlined. The emphasis is on using all available experimental constraints in the analysis. A three-step practical procedure is formulated to take into account global constraints as well as error information at the level of individual experiments. Two complementary methods to make predictions, based on the Hessian error matrix and Lagrange multiplier methods, are described.

All calculations of high energy processes with initial hadrons, whether within the Standard Model (SM) or exploring New Physics, require parton distribution functions (PDFs) as an essential input. The reliability of these calculations, which underpin both future theoretical and experimental progress, depends substantially on understanding the uncertainties of the PDFs. The assessment of uncertainties of PDFs has, therefore, become an important challenge to high energy physics in recent years.

The PDFs are derived from global analyses of experimental data from a wide range of hard processes in the framework of perturbative quantum chromodynamics (PQCD). There are sources of uncertainty from perturbation theory (e.g. higher order and power law corrections), from choices of parametrization of the non-perturbative input (i.e. initial parton distributions at some confinement scale), from uncertain nuclear corrections to experiments performed on nuclear targets, and from normal experimental statistical and systematic errors. We shall be concerned here only with these experimental errors. Even this limited problem is complicated: in a *global analysis*, the large number of data points (~ 1300 in our case) do not come from a uniform set of measurements, but consist of a collection of measurements from many experiments (~ 15) on a variety of physical processes ($\sim 5 - 6$) with diverse characteristics, precision, and error determination. This is compounded by a large number of theoretical parameters (~ 16). Recently our group at MSU[a] has developed two methods to overcome some of the difficulties associated with these complications.[2] This talk briefly summarizes preliminary results obtained with these new techniques in an ongoing project. Space limitation prevents references to other related works.[1]

1 General Approach

The complexity of the global analysis system and the imperfection of real experimental data sets (including non-textbook-like errors) precludes a "rigorous" treatment of uncertainties. We develop a three-step practical procedure to deal with this problem: (i) using an effective global χ^2 function, we obtain a global minimum (the "standard fit"), as well as well-defined sets of PDFs in the vicinity of the minimum as "alternative hypotheses" to the true PDFs; (ii) we assess the likelihood of these alternate hypotheses with respect to the individual experiments by utilizing the full statistical power of available information on the errors for each experiment; and (ii) we combine these local uncertainties into an estimated global uncertainty for either physically measurable quantities (Sec. 2), or the PDF parameters themselves (Sec. 3).

---

[a]J. Pumplin, D. Stump, and W.K. Tung

In the global analysis stage (step (i)), as in previous CTEQ analyses, we adopt the following <u>effective</u> *global* χ^2 function to represent known experimental inputs:

$$\chi_g^2 = \sum_n \chi_n^2 \quad (n \text{ labels the experiments})$$

$$\chi_n^2 = \left(\frac{1-N_n}{\sigma_n^N}\right)^2 + \sum_I w_n \left(\frac{N_n D_{nI} - T_{nI}(a)}{\sigma_{nI}^D}\right)^2$$

For the nth experiment, D_{nI}, σ_{nI}^D, and $T_{nI}(a)$ denote the data value, measurement uncertainty, and theoretical value (dependent on the theory parameters $\{a\}$) for the Ith data point; σ_n^N is the experimental normalization uncertainty; N_n is a relative normalization factor for that experiment in the global analysis; and w_n is a "prior" weighting factor based on physics considerations or on information derived from existing work. This χ_g^2 function provides an effective way of searching for global minimum solutions of the PDF parameters $\{a_i\}$, which incorporates all experimental constraints in a uniform manner while allowing some flexibility for physics input. It is particularly practical since most experiments used in the global analysis only publish an effective point-to-point (i.e. uncorrelated) systematic error along with the statistical errors.

Using the same experimental and theoretical inputs as the CTEQ5 analysis, but slightly improved parametrization of the initial distributions, we minimize the above χ_g^2 and obtain a standard set of PDF, denoted by S_0. To study the uncertainties, according to the two different methods described below, we obtain different sets of PDFs $\{S_m\}$ as alternate hypotheses. From these, we make uncertainty estimates.

2   Lagrange Multiplier Method

The idea of this method has been previously introduced.[1,2] The first step is to perform a global analysis as described above. Let X be a particular physical quantity of interest,

which depends on the PDFs. The best estimate (or prediction) of X is $X_0 = X(S_0)$. To assess the uncertainty of the predicted value for X, we first use the Lagrange multiplier method to determine how the minimum of χ_g^2 increases, *i.e.*, how the quality of the fit to the global data set decreases, as X deviates from the best estimate X_0. This is most conveniently done by introducing the Lagrange multiplier variable λ, and minimizing the function

$$\Psi(\lambda, a) = \chi_g^2(a) + \lambda(X(a) - X_0) \quad (1)$$

with respect to the PDF parameters $\{a_i\}$ for fixed values of λ. For a given value of λ, the minimum of $\Psi(\lambda, a)$ yields a set of parameters $\{a_{\min}(\lambda)\}$ that lead to the pair $\{\chi_g^2(\lambda), X(\lambda)\}$. $\chi_g^2(\lambda)$ represents the lowest achievable χ_g^2 for the global data sets if X takes the value $X(\lambda)$, taking into account all possible PDFs in the *full n dimensional PDF parameter space* represented by $\{a_j\}$. By repeating the calculation many times with different choices of λ, one obtains the dependence of χ_g^2 on X, i.e. the function $\chi_g^2(X)$.

We can obtain the maximum range of allowed values of X, say ΔX, for a given tolerance of the goodness of fit $\Delta\chi_g^2 = \chi_g^2 - \chi_0^2$ by, say, reading the numbers off a graph of χ_g^2 vs. X.[3] This is illustrated for the case of W production cross-section σ_W at the Tevatron, in Fig. 1. The points on Fig. 1 are the

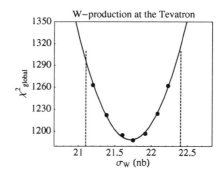

Figure 1. Minimum χ_g^2 versus $\sigma_W B$.

results of constrained fits with various values of the Lagrange multiplier λ. The solid curve is a quadratic fit to the points to provide a smooth representation of the continuous function $\chi_g^2(X)$. We see that the best fit value for σ_W is at the minimum of the curve; and the other fits follow closely an expected parabolic behavior around the minimum. This method is independent of any approximations associated with functional dependence of $\chi_g^2(a)$ and $X(a)$ on the PDF parameters $\{a_i\}$.

The effective χ_g^2 function, unlike an ideal χ^2 function, does not have *a priori* quantitative statistical significance. To assess the uncertainty on X, we go to step (ii) of the procedure outlined in Sec. 1. We select points $\{S_m\}$ on the solid curve in Fig. 1, each representing a PDF set, and regard them as alternate hypotheses to the true PDF set. We then assess the likelihood of these alternate hypotheses with respect to the individual experiments by utilizing the full statistical power of each experiment, including correlated systematic errors if available. At the individual experiment level, these error estimates are meaningful. Details of this work will be given in Ref. ??. The result on σ_W at the Tevatron is summarized in Fig. 2.

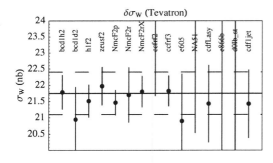

Figure 2. 68% error bars for the individual experiments. The dashed lines are estimated bounds of the global uncertainty explained in the text.

The last step is to combine the individual errors into a global measure of uncertainty of σ_W.[3] We note that the ranges shown by the error bars in Fig. 2 are not errors de-

termined independently by each experiment; rather they represent the allowed ranges of *constrained fits to the global data sets* $\{S_m\}$ imposed by individual experiments. We define the global uncertainty band, shown by the dashed lines in the figure, to be the largest range not excluded by the individual experiments by more than 2σ. In the case of σ_W for the Tevatron, the percentage error is 3%. This can be turned into an effective "tolerance" value for $\Delta\chi_g^2$ of ~ 100 by Fig. 1.

We have performed similar studies for Z production at the Tevatron. Although the alternate hypotheses PDFs $\{S_m\}$ are different from those for W production, the percentage error estimate is the same: 3%. For W, Z production at LHC, the important partonic subprocesses are different, and the uncertainty turns out to be around 6%. However, in all cases, one gets $\Delta\chi_g^2 \sim 100$, which seems to be a robust figure to gauge the current global goodness-of-fit based on the effective χ_g^2. This method can be extended to study uncertainties of any physical quantity, as well as correlated uncertainties of pairs (X, Y) and multiplets of physical variables.[3]

3 Error Matrix Method

We now turn to the more conventional error matrix approach to study uncertainties of the PDFs, as represented by the parameters $\{a_i\}$. This (in principle) straightforward method encounters some serious technical (numerical) difficulties when applied to global QCD analysis, because of complexities mentioned earlier. The main problem is that eigenvalues of the Hessian matrix vary by 5-6 orders of magnitude in most choices of PDF parameters. The highly anisotropic behavior of the χ_g^2 function, as well as non-smooth behavior of the theory values $T_I(a)$ (due to adaptive integrations used in their calculation), make conventional general purpose programs (such as MINUIT) inadequate in providing physically sensible error estimates. We have overcome this problem with a newly developed iterative method for calculating the Hessian.[2]

The basic assumption of the error matrix approach is that χ^2 can be approximated by a quadratic expansion in the fit parameters $\{a_i\}$ near the global minimum. It is true if the variation of the theory values T_I with $\{a_i\}$ is approximately linear near the minimum. Letting $\Delta\chi^2 = \chi^2 - \chi_0^2$, we have

$$\Delta\chi^2 = \tfrac{1}{2} \sum_{i,j} H_{ij}\,(a_i - a_i^0)(a_j - a_j^0). \quad (2)$$

H_{ij} has a complete set of n orthonormal eigenvectors $\{v_{ik}\}$ with eigenvalues ϵ_k : $\sum_j H_{ij} v_{jk} = \epsilon_k v_{ik}$. The eigenvectors provide a natural basis to express arbitrary variations around the minimum. In terms of a new set of parameters defined with respect to the eigenvectors, $z_i = \sqrt{\frac{\epsilon_i}{2}} \sum_j (a_j - a_j^0) v_{ji}$, one obtains $\Delta\chi^2 = \sum z_i^2$, i.e. *the surfaces of constant χ^2 are spheres in $\{z_i\}$ space, of radius $\sqrt{\Delta\chi^2}$.*

Now consider any *physical quantity X* that depends on PDFs, hence the parameters $\{a_i\}$. In the neighborhood of the global minimum, assuming the first term of the Taylor-series expansion of X gives an adequate approximation, the deviation of X from its best estimate is given by $\Delta X = X - X_0 \cong \sum_i X_i z_i$ where $X_i \equiv \partial X/\partial z_i$ are the components of the z-gradient evaluated at the global minimum, *i.e.*, at the origin in z-space. One can show that, for a given tolerance in $\Delta\chi^2$, the uncertainty of the physical quantity can be evaluated from the z-gradient vector components by the simple formula

$$(\Delta X)^2 = \Delta\chi^2 \sum_i X_i^2. \quad (3)$$

In practice, X_i is calculated by finite differences. Within the linear approximation, this equation can be reduced to an extremely compact and practical form

$$\Delta X = \sqrt{\sum_i (X(S_i^+) - X(S_i^-))^2} \quad (4)$$

where $S_i^\pm$ are PDF sets which correspond to two points in the z parameter space specified by $\{z_j^\pm = \pm\delta_{ij}\sqrt{\Delta\chi^2}/2\}$. The squared uncertainty is proportional to $\Delta\chi^2$, the tolerance on the global χ^2 function. It can be determined by a procedure similar to the one described in the previous section, now applied to each of the eigenvector directions rather than to a specific variable. The overall tolerance can be an average over the eigenvector directions. We expect it to be of the same order (~ 100) as obtained earlier.

Thus, with the reliable calculation of the Hessian and its eigenvectors, we can provide $2n + 1$ sets of PDFs, $\{S_0, S_i^\pm, i = 1, ..., n\}$ (where n is the number of PDF parameters), from which the "user" can evaluate the uncertainty associated with any physical quantity X according to Eq. 4. Preliminary results on application to the W cross-section show that they are consistent with those obtained by the Lagrange multiplier method. We are currently writing up these results,[4] and we plan to make the $2n + 1$ sets of PDFs, $\{S_0, S_i^\pm\}$, available, so they can be used by anyone to study consequences of PDF uncertainties.

We should point out that many other sources of uncertainty mentioned in the introduction are not yet included in this study. Therefore, the uncertainty estimates described here represent, at best, lower bounds.

References

1. Cf. contributions of Zomer, Botje to DIS2000 Proceedings; Alekhin, Giele & Keller to LHC Workshop [hep-ph/0005025]; and R. Brock *et al.* to RunII Workshop [hep-ph/0006148].

2. J. Pumplin, D. Stump, and W.-K. Tung, [hep-ph/0008191]

3. D. Stump *Uncertainties of Parton Distribution Functions* (to be published).

4. J. Pumplin *et al.*, *Uncertainties of Physical Observables due to PDFs* (to be published).

CONSTRAINTS ON THE PROTON'S GLUON DENSITY FROM LEPTON-PAIR PRODUCTION

EDMOND L. BERGER

High Energy Physics Division, Argonne National Laboratory, Argonne, IL 60439, USA
E-mail: berger@anl.gov

M. KLASEN

II. Institut für Theoretische Physik, Universität Hamburg, D-22761 Hamburg, Germany
E-mail: michael.klasen@desy.de

Massive lepton-pair production, the Drell-Yan process, should be a good source of independent constraints on the gluon density, free from the experimental and theoretical complications of photon isolation that beset studies of prompt photon production. We provide predictions for the spin-averaged and spin-dependent differential cross sections as a function of transverse momentum Q_T.

1 Introduction

Massive lepton-pair production, $h_1 + h_2 \rightarrow \gamma^* + X; \gamma^* \rightarrow l\bar{l}$, the Drell-Yan process, and prompt real photon production, $h_1 + h_2 \rightarrow \gamma + X$, are two of the most valuable probes of short-distance behavior in hadron reactions. They supply critical information on parton momentum densities and opportunities for tests of perturbative quantum chromodynamics (QCD). Spin-averaged parton momentum densities may be extracted from spin-averaged nucleon-nucleon reactions, and spin-dependent parton momentum densities from spin-dependent nucleon-nucleon reactions.

The Drell-Yan process has tended to be thought of primarily as a source of information on quark densities. Indeed, the mass and longitudinal momentum (or rapidity) dependences of the cross section (integrated over the transverse momentum Q_T of the pair) provide essential constraints on the *antiquark* momentum density, complementary to deep-inelastic lepton scattering from which one gains information of the sum of the quark and antiquark densities. Prompt real photon production, on the other hand, is a source of essential information on the *gluon* momentum density. At lowest order in perturbation the-

ory, the reaction is dominated at large values of the transverse momentum p_T of the produced photon by the QCD "Compton" subprocess, $q + g \rightarrow \gamma + q$. This dominance is preserved at higher orders, indicating that the experimental inclusive cross section differential in p_T may be used to determine the density of gluons in the initial hadrons.

In this contribution, we summarize recent work[1,2], in which we demonstrate that the QCD Compton subprocess, $q + g \rightarrow \gamma^* + q$ also dominates the Drell-Yan cross section in polarized and unpolarized proton-proton reactions for values of the transverse momentum Q_T of the pair that are larger than roughly half of the pair mass Q, $Q_T > Q/2$. Dominance of the qg contribution in the massive lepton-pair case is as strong if not stronger than it is in the prompt photon case. Massive lepton-pair differential cross sections are therefore an additional useful source of constraints on the the spin-averaged and spin-dependent *gluon densities*. Although the cross section is smaller than the prompt photon cross section, massive lepton pair production is cleaner theoretically since long-range fragmentation contributions are absent as are the experimental and theoretical complications associated with isolation of the real photon. As long Q_T is large,

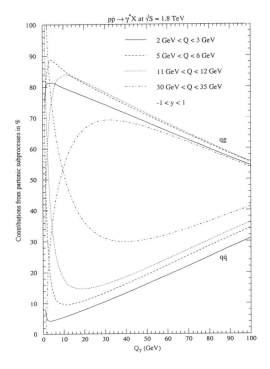

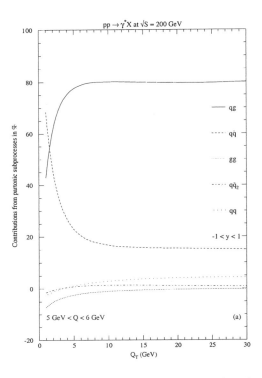

Figure 1. Contributions from the partonic subprocesses qg and $q\bar{q}$ to the invariant inclusive cross section $Ed^3\sigma/dp^3$ as a function of Q_T for $p\bar{p} \to \gamma^* X$ at $\sqrt{S} = 1.8$ TeV.

Figure 2. Contributions from the partonic subprocesses qg and $q\bar{q}$ to the invariant inclusive cross section $Ed^3\sigma/dp^3$ as a function of Q_T for $pp \to \gamma^* X$ at $\sqrt{S} = 200$ GeV.

the perturbative requirement of small $\alpha_s(Q_T)$ can be satisfied without a large value of Q. We therefore explore and advocate the potential advantages of studies of $d^2\sigma/dQdQ_T$ as a function of Q_T for modest values of Q, $Q \sim 2\,\text{GeV}$, below the range of the traditional Drell-Yan region.

2 Unpolarized Cross Sections

For $p\bar{p} \to \gamma^* + X$ at $\sqrt{S} = 1.8$ TeV and several values of the mass of the lepton-pair, we present in Fig. 1 the $q\bar{q}$ and qg perturbative contributions to the invariant inclusive cross section $Ed^3\sigma/dp^3$ as a function of Q_T. For small Q_T, the $q\bar{q}$ contribution exceeds that of qg channel. However, as Q_T grows, the qg contribution becomes increasingly important and accounts for 70 to 80 % of the rate once $Q_T \simeq Q$. (The $q\bar{q}$ contribution begins to

be felt a second time at very large Q_T owing to the valence nature of the $\bar{q}$ density in the $\bar{p}$.) Subprocesses other than those initiated by the $q\bar{q}$ and qg initial channels contribute negligibly.

Prompt photons have been observed in Fermilab Tevatron collider experiments with values of p_T extending to 100 GeV and beyond. Lepton-pair cross sections are smaller owing to the factor $\alpha_{em}/(3\pi Q^2)$ associated with the decay of the virtual photon to $\mu^+\mu^-$. It should be possible to examine massive lepton-pair cross sections in the same data sample out to Q_T of 30 GeV or more. The statistical limitation to Q_T of about 30 GeV means that the reach in x_{gluon}, the fractional light-cone momentum carried by the incident gluon, is limited presently to $2Q_T/\sqrt{S} \sim 0.033$, a factor of three less than that potentially accessible with prompt photons. It is valuable nevertheless to investigate the gluon density in the region $x_{gluon} \sim 0.033$, and less, with a process that has reduced experimental and theoretical systematic uncertainties from those of the prompt photon case.

In Fig. 10 of the first paper[1], we show a comparison with data of our calculated invariant inclusive cross section $E d^3\sigma/dp^3$ as a function of Q_T for $p + \bar{p} \to \gamma^* + X$, with $\gamma^* \to \mu^+\mu^-$, at $\sqrt{S} = 630$ GeV, with $2m_\mu < Q < 2.5$ GeV. The theoretical expectation is in good agreement with the data published by the CERN UA1 collaboration[3]. Dominance of the qg component is evident over a large interval in Q_T. It would be valuable to make a similar comparison with Tevatron data.

Results similar to those above are shown in Fig. 2 for $pp \to \gamma^* + X$ at $\sqrt{S} = 200$ GeV appropriate for the RHIC collider at Brookhaven. The fraction of the cross section attributable to qg initiated subprocesses again increases with Q_T, growing to 80 % for $Q_T \simeq Q$. Predictions of spin-averaged and spin-dependent cross sections for the energies of the RHIC collider may be found in the second paper[2]. Adopting the nominal value $E d^3\sigma/dp^3 = 10^{-3}$pb/GeV^2, we establish that the massive lepton-pair cross section may be measured to $Q_T = 14$ and 18.5 GeV in $pp \to \gamma^* + X$ at $\sqrt{S} = 200$ and 500 GeV, respectively, when $2 < Q < 3$ GeV, and to $Q_T = 11.5$ and 15 GeV when $5 < Q < 6$ GeV. In terms of reach in x_{gluon}, these values of Q_T may be converted to $x_{gluon} \simeq x_T = 2Q_T/\sqrt{S} = 0.14$ and 0.075 at $\sqrt{S} = 200$ and 500 GeV when $2 < Q < 3$ GeV, and to $x_{gluon} \simeq 0.115$ and 0.06 when $5 < Q < 6$ GeV. The smaller cross section in the case of massive lepton-pair production means that the reach in x_{gluon} is restricted to a factor of about two to three less, depending on $\sqrt{S}$ and Q, than that potentially accessible with prompt photons in the same sample of data.

In the first paper[1], we compare our spin-averaged cross sections with fixed-target data on massive lepton-pair production at large values of Q_T, and we establish that fixed-order perturbative calculations, without resummation, should be reliable for $Q_T > Q/2$.

Although the qg Compton subprocess is dominant, one might question whether uncertainties associated with the quark density compromise the possibility to determine the gluon density. In this context, it is useful to recall[4] that when the Compton subprocess is dominant, the spin-averaged cross sections for prompt photon production and for lepton-pair production may be rewritten in a form in which the quark densities do not appear explicitly, but, instead, directly in terms of the proton structure function $F_2(x, \mu_f^2)$ *measured* in spin-averaged deep-inelastic lepton-proton scattering. An analogous statement applies in the spin-dependent case where the lepton pair cross section may be expressed in terms of the $g_1(x, \mu_f^2)$ structure function measured in spin-dependent deep-inelastic lepton-proton scattering.

Acknowledgments

Work at Argonne National Laboratory is supported by the U.S. Department of Energy, Division of High Energy Physics, under Contract W-31-109-ENG-38. M.K. is supported by DFG through grant KL 1266/1-1.

References

1. E. L. Berger, L. E. Gordon, and M. Klasen, *Phys. Rev.* D **58**, 074012 (1998) and *Nucl. Phys. Proc. Suppl.* **82**, 179 (2000).
2. E. L. Berger, L. E. Gordon, and M. Klasen, *Phys. Rev.* D **62**, 014014 (2000) and hep-ph/0001190.
3. CERN UA1 Collaboration, C. Albajar *et al*, *Phys. Lett.* B**209**, 397 (1988).
4. E. L. Berger and J.-W. Qiu, *Phys. Rev.* D **40**, 778 (1989) and D **40**, 3128 (1989).

STUDIES OF THE STRONG COUPLING

ZOLTÁN TRÓCSÁNYI

University of Debrecen and
Institute of Nuclear Research of the Hungarian Academy of Sciences
H-4001 Debrecen, PO Box 51, Hungary

We summarize recent measurements of the QCD colour factors C_A and C_F, the strong coupling $\alpha_s(M_{Z^0})$ and its energy dependence.

The theory of massless QCD is $SU(3)_c$ gauge theory with one parameter, the strong coupling, α_s. The value of the α_s is not predicted by the theory. In practice, when we make a prediction, the theory is supplemented with pheonomenological models that account for the not understood low energy behaviour of the theory. As a result, there is an intrinsic uncertainty in all QCD predictions for cross sections, which are customarily translated into uncertainties of a *single* parameter, α_s. Thus, by studying the α_s, we mean (i) $\alpha_s(M_{Z^0})$ measurements for testing practical QCD; (ii) measurements of the colour factors and the energy dependence of the running coupling for testing $SU(3)$ dynamics and the particle spectrum. In this talk we present four recent results of measurements of type (i) and (ii)[1,2,3,4].

The OPAL Collaboration reported a simultaneous measurement of the QCD colour factors C_A, C_F and $\alpha_s(M_{Z^0})$ using perturbative predictions for the differential two-jet and four-jet rates at the matched next-to-leading logarithmic (NLL) and next-to-leading order (NLO) accuracy and NLO prediction for the four-jet angular correlations[1]. This was the first colour factor measurement based upon NLO, or higher perturbative accuracy. Although the measured colour factor values at a fixed energy do not depend on α_s, the systematic uncertainties do, so for a more reliable assesment of the systematics the simultaneous measurement is necessary. The results, unfolded to usual QCD parameters,

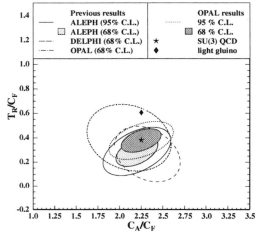

Figure 1. 67 and 95 % confidence level contours of the colour factor ratios.

$C_A = 3.02 \pm 0.25(\text{stat.}) \pm 0.44(\text{syst.})$, $C_F = 1.34 \pm 0.13(\text{stat.}) \pm 0.19(\text{syst.})$, $\alpha_s(M_{Z^0}) = 0.120 \pm 0.011(\text{stat.}) \pm 0.016(\text{syst.})$, are in agreement with $SU(3)$ predictions for the colour factors and the world average of $\alpha_s(M_{Z^0})$. The two-dimensional contour plot of the total errors for C_A/C_F and T_R/C_F in Fig. 1 shows that the main effect of including the NLO corrections is to stabilize the value and reduce the uncertainty for T_R/C_F.

The ALEPH Collaboration measured $\alpha_s(M_{Z^0})$ using four-jet final states in e^+e^- annihilation obtained using the Durham clustering algorithm. The matched NLO and resummed NLL approximations provide very good agreement with the data as seen in Fig. 2. The measured value of α_s is $\alpha_s(M_{Z^0}) = 0.1172 \pm 0.0001(\text{stat.})$

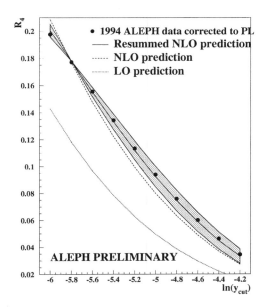

Figure 2. Comparison of measured and predicted four-jet rates with fitted value of $\alpha_s(M_{Z^0})$

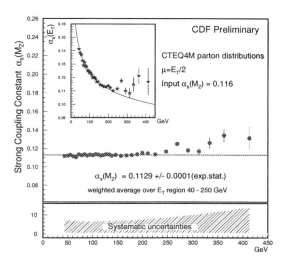

Figure 3. Energy dependence of α_s and the measured value of $\alpha_s(M_{Z^0})$

± 0.0008(had.) ± 0.0040 (theo.). The largest source of uncertainty comes form the theory and is investigated by the variation of the renormalization scale μ. The band in the plot corresponds to the variation $0.5 < \mu/E_{cm} < 2$, where E_{cm} is the center of mass energy.

Utilizing the similarity of the JADE and OPAL detectors, a new study of the energy-dependence of α_s has been performed using jet rates and jet multiplicities obtained in electron-positron annihilation at nine different energies between 35 and 189 GeV[3]. The conclusion of the very detailed work is that the running of α_s is compatible with the three-loop running of α_s with $\chi^2/\text{dof} = 1.12$. The $\alpha_s(M_{Z^0})$ value obtained from the fit of the energy dependence, $\alpha_s(M_{Z^0}) = 0.1187^{+0.0034}_{-0.0019}$, is very close to the world average and has remarkably small errors.

The CDF Collaboration also studied energy dependence of α_s over a wide energy range using the same experimental setting in a measurement based upon inclusive jet cross sections[4]. The run 1B data were binned into 33 jet transverse energy bins. The de-

fault results, shown in Fig. 3 and obtained using CTEQ4M parton distributions (PDF), show discrepancy with perturbation theory for $E_\perp > 250$ GeV. Using the CTEQ4HJ PDF set the discrepancy disappears and the $\alpha_s(M_{Z^0})$ value obtained from a fit over the whole $E_\perp$-range is close to the value obtained in the previous fit. The central values of the fit are considerably below the world average in both cases, but agree with that within the rather large systematic errors.

In summary, the new measurements of the QCD colour factors, the strong coupling and its running show a remarkable consistence of the underlying gauge theory.

References

1. OPAL Collaboration, note PN430, abstract 251 to this conference.

2. ALEPH Collaboration, note 2000-045, abstract 162 to this conference.

3. JADE and OPAL Collaborations, hep-ex/0001055, abstract 114 to this conference.

4. CDF Collaboration, abstract 1080 to this conference.

STUDIES OF $B\bar{B}$ GLUON AND $C\bar{C}$ VERTICES

TOSHINORI ABE

Stanford Linear Accelerator Center
Stanford University, Stanford, California, 94309
USA
E-mail: toshi@SLAC.Stanford.EDU

We report on several new studies of $b\bar{b}g$ and $c\bar{c}g$ vertices using 3- and 4-jet hadronic Z^0 decays by e^+e^- collision. The gluon energy spectrum is measured over the full kinematic range, providing an improved test of QCD and limits on anomalous bbg couplings. The parity violation in $Z^0 \to b\bar{b}g$ decays is consistent with electroweak theory plus QCD. New tests of T- and CP-conservation at the bbg vertex are performed. A measurement of the probability of gluon splitting into $c\bar{c}$ pairs finds $g_{c\bar{c}} = (2.45 \pm 0.29 \pm 0.53)\%$. A new measurement of the rate of gluon splitting into $b\bar{b}$ pairs yields $g_{b\bar{b}} = (2.84 \pm 0.61(stat.) \pm 0.59(syst.)) \times 10^{-3}$ (Preliminary).

1 Introduction

We present a number of precision tests of QCD in the perturbative regime using 3- and 4-jet final states from L3 and SLD collaborations.

Experimental studies of the structure of 3-jet events have typically used energy and angle distributions of energy-ordered jets. Since the gluon is expected to be the lowest-energy jet in most events, this suffices to confirm the $q\bar{q}g$ origin of such events and to determine the gluon spin. The identification of the three jets in such events would allow more complete and stringent tests of QCD. Here we present a study [1] of 3-jet final states in which two of the jets are tagged as b or $\bar{b}$ jets by the SLD collaboration. The remaining jet is tagged as the gluon jet and its energy spectrum studied over the full kinematic range. Adding a tag of the charge of the b or $\bar{b}$ jet, and exploiting the high electron beam polarization of the SLAC Linear Collider, the SLD collaboration measures [2] two angular asymmetries sensitive to parity violation in the Z^0 decay, and also construct new tests of T- and CP-conservation at the bbg vertex.

The rate of secondary heavy flavor production via gluon splitting, $g \to c\bar{c}$, $g \to b\bar{b}$ is a sensitive test of QCD, as it is suppressed strongly by the mass of the heavy quark, but is still expected to be the dominant source of secondary heavy hadrons. Here we present a measurement of the $g \to c\bar{c}$ rate [3] by the L3 collaboration and $g \to b\bar{b}$ rate [4] by the SLD collaboration. The study of events containing $b/\bar{b}$ quarks is especially useful, both as important input into measurements such as electroweak parameters [5] (R_b and A_b) in Z^0 decays and bottom production in hadron-hadron collisions, and also as a sensitive probe of any new physics that couples more strongly to heavier quarks.

2 The Gluon Energy Spectrum

Hadronic events with exactly 3 jets (JADE algorithm, $y_{cut} = 0.02$) are selected. Jet energies E_i are calculated from the inter-jet angles, and the jets are energy ordered: $E_1 > E_2 > E_3$. A cut ($n_{sig}^{(3)} \geq 2$) in exactly two of the three jets is required, and the remaining jet is tagged as the gluon jet. This yields 8196 events with an estimated purity of correctly tagged gluon jets of 91%. In 3.0% (12.9%) of these events, jet 1(2), the (second) highest energy jet, is tagged as the gluon jet, giving coverage over the full kinematic range.

The distribution of the scaled gluon energy $x_g = 2E_g/\sqrt{s}$ is corrected for non-$b\bar{b}g$ and mistag backgrounds, selection efficiency and resolution. The prediction of the JETSET parton shower simulation reproduces the data well.

The x_g spectrum is particularly sensitive to the presence of an anomalous chromomagnetic term in the strong interaction Lagrangian. A fit of the theoretical prediction including an anomalous term parametrized by a relative coupling κ, yields a value of $\kappa = -0.01 \pm 0.05$ (Preliminary), consistent with zero, and corresponding to 95% C.L. limits on such contributions to the $b\bar{b}g$ coupling of $-0.11 < \kappa < 0.08$.

3 Symmetry tests in polarized Z^0 decays to $b\bar{b}g$

Now two angles, the polar angle of the quark with respect to the electron beam direction θ_q, and the angle between the quark-gluon and quark-electron beam planes $\chi = \cos^{-1}(\hat{p}_q \times \hat{p}_g) \cdot (\hat{p}_q \times \hat{p}_e)$ are considered. The cosine x of each of these angles should be distributed as $1 + x^2 + 2A_P A_Z x$, where the Z^0 polarization $A_Z = (P_e - A_e)/(1 - P_e A_e)$ depends on that of the e^- beam P_e, and A_e and $A_P = A_{QCD} A_q$ are predicted by QCD plus electroweak theory.

Three-jet events (Durham algorithm, $y_{cut} = 0.005$) are selected and energy ordered. The 14,658 events containing a secondary vertex with mass above 1.5 GeV/c^2 in any jet are kept, having an estimated $b\bar{b}g$ purity of 84%. We calculate the momentum-weighted charge of each jet j, $Q_j = \Sigma_i q_i |\vec{p}_i \cdot \hat{p}_j|^{0.5}$, using the charge q_i and momentum $\vec{p}_i$ of each track i in the jet. We assume that the highest-energy jet is not the gluon, and tag it as the b ($\bar{b}$) if $Q = Q_1 - Q_2 - Q_3$ is negative (positive). We define the b-quark polar angle by $\cos\theta_b = -\text{sign}(Q)(\hat{p}_e \cdot \hat{p}_1)$.

The left-right-forward-backward asymmetry A^b_{LRFB} in $\cos\theta_b$ is reconstructed. A fit to the data yields an asymmetry parameter of $A_P/A_b = 0.914 \pm 0.053(stat.) \pm 0.063(syst.)$, consistent with the QCD prediction of $A_P/A_b = 0.93$.

Then tag one of the two lower energy jets is tagged as the gluon jet: if jet 2 has $n^{(3)}_{sig} = 0$ and jet 3 has $n^{(3)}_{sig} > 0$, then jet 2 is tagged as the gluon; otherwise jet 3 is tagged as the gluon. We construct the angle χ and A^χ_{LRFB}. Our measurement is consistent with the prediction, as well as with zero. A fit yields $A_\chi/A_b = -0.014 \pm 0.035 \pm 0.002$, to be compared with an expecta-

tion of -0.064.

Using these fully tagged events, observables that are formally odd under time reversal and/or CP reversal can be constructed. For example, the triple product $\cos\omega^+ = \vec{\sigma}_Z \cdot (\hat{p}_1 \times \hat{p}_2)$, formed from the directions of the Z^0 polarization $\vec{\sigma}_Z$ and the highest- and second highest-energy jets, is T_N-odd and CP-even. Since the true time reversed experiment is not performed, this quantity could have a nonzero A_{LRFB}, and a limit using events of all flavors has been previously set. A calculation including Standard Model final state interactions predicts that $A^{\omega^+}_{LRFB}$ is largest for $b\bar{b}g$ events, but is only $\sim 10^{-5}$. The fully flavor-ordered triple product $\cos\omega^- = \vec{\sigma}_Z \cdot (\hat{p}_q \times \hat{p}_{\bar{q}})$ is both T_N-odd and CP-odd.

Measured $A^{\omega^+}_{LRFB}$ and $A^{\omega^-}_{LRFB}$ are consistent with zero at all $|\cos\omega|$. Fits to the data yield 95% C.L. limits on any T_N-violating and CP-conserving or CP-violating asymmetries of $-0.045 < A^+_T < 0.016$ or $-0.082 < A^-_T < 0.012$, respectively.

4 Gluon splitting into a $c\bar{c}$ pair

L3 has reported measurements of the probability of gluon splitting into charmed quarks in hadronic Z decays using the following methods, (1) search for leptons in three jet events (2) use of neural networks and event shape variables.

Events are clustered in 3 jets (JADE algorithm, $Y_{cut} = 0.03$), with the assumption that the two charm quarks from splitting are close enough in angle to generate a single jet. Moreover, the jet from splitting is likely to be the least energetic one, due to the energy spectrum of the gluon. Signal selection is thus obtained by requesting a lepton (e or μ) in the least energetic jet in 3-jet events. The main background comes from leptons from b quarks decays. Those events are rejected using b anti-tag and a cut on the mass of the third jet. The 360 electron and 450 muon candidate events are selected. The backgrounds are estimated to be 285 ± 10 and 399 ± 11 for electron and muon sample, respectively. The estimated $g \to c\bar{c}$ efficiency is 0.40% for electron and 0.54% for muon.

For neural network analysis, events with 3 jets are initially selected, and jets are order in energy; in this way, gluon splitting generated jets should be the second or the third.

The variables chosen for the input to the neural network are: (1) $M_{jet2} + M_{jet3} - M_{jet1}$; (2) 3 different Fox-Wolfram momenta; and (3) the fraction of the energy of the second jet within a cone of 8° around its axis. The additional cut requests the presence of a minimum number of tracks displaced from the primary vertex; this is done to enrich the sample in events with gluon splitting to heavy flavors. The selection efficiency is 4.4% and the purity is 4.5%.

Combining the results from the lepton and the neural network analyses, L3 obtains $g_{c\bar{c}} = (2.45 \pm 0.29 \pm 0.53)\%$, which is consistent with the QCD prediction of 2.0007% [6].

5   Gluon splitting into a $b\bar{b}$ pair

A new measurement of the rate of gluon splitting into $b\bar{b}$ pair has been reported from the SLD collaboration. Candidate events containing a gluon splitting into a $b\bar{b}$ pair, $Z^0 \to q\bar{q}g \to q\bar{q}b\bar{b}$, where the initial $q\bar{q}$ can be any flavor, are required to have 4 jets (Durham algorithm, $y_{cut} = 0.009$). A secondary vertex is required in each of the two jets (jet1 and jet2) with the smallest opening angle in the event, yielding 547 events. This sample is dominated by background ($S/N \sim 1/5$), primarily from $Z^0 \to b\bar{b}g(g)$ events and events with a gluon splitting into a $c\bar{c}$ pair.

In order to improve the signal/background ratio, a neural network technique is used. Nine observables are chosen for inputs to the neural network. The inputs include the the P_T-corrected mass of the vertices and the angle between the vertex axes. b jets have higher P_T-corrected mass than c/uds jets, hence it is useful to separate b jets from c/uds jets. Many $Z^0 \to b\bar{b}$ background events have one b-jet which was split by the jet-finder into 2 jets so that the two found vertices are from different decay products from the same B decay. The two vertex axes tend to be

collinear. The neural network is trained using Monte Carlo samples of 1800k $Z \to q\bar{q}$ events, 1200k $Z \to b\bar{b}$ events, 780k $Z \to c\bar{c}$ events and 50k $g \to b\bar{b}$ events. A cut (neural network output > 0.6) keeps 79 events, with an estimated background of 37.8 events. Using this and the estimated efficiency for selecting $g \to b\bar{b}$ splittings of 4.99% yields a measured fraction of hadronic events containing such a splitting of $g_{b\bar{b}} = (2.84 \pm 0.61(stat.) \pm 0.59(syst.)) \times 10^{-3}$ (Preliminary). The result is consistent with the QCD prediction of 1.75×10^{-3} [6].

6   summary

Several new studies of $b\bar{b}g$ and $c\bar{c}g$ vertices using 3- and 4-jet hadronic Z^0 decays by e^+e^- collision has reported in this proceeding. The results are in good agreement with the predictions of perturbative QCD.

References

1. K. Abe et al. [SLD Collaboration], SLAC-PUB-8505
2. K. Abe et al. [SLD Collaboration], Submitted to Phys. Rev. Lett. , [hep-ex/0007051].
3. M. Acciarri et al. [L3 Collaboration], Phys. Lett. **B476**, 243 (2000) [hep-ex/9911016].
4. K. Abe et al. [SLD Collaboration], SLAC-PUB-8506.
5. The LEP Collaboration, the LEP Electroweak Working Group, The SLD Heavy Flavour and Electroweak Working Groups, CERN-EP-2000-016, January 2000.
6. D. J. Miller and M. H. Seymour, Phys. Lett. **B435**, 213 (1998) [hep-ph/9805414].

STUDIES OF B-QUARK FRAGMENTATION

V. CIULLI

Scuola Normale Superiore, P.zza dei Cavalieri 7, 56126 Pisa, Italy
E-mail: Vitaliano.Ciulli@cern.ch

I will review new studies of b-quark fragmentation performed at the Z peak by ALEPH and SLD. An improved sensitivity to distinguish between fragmentation model and more accurate measurements of the mean b-hadron scaled energy have been obtained.

1 Introduction

In e^+e^- collisions the b-quark fragmentation function is given by the normalized scaled energy distribution of b-hadrons

$$D(x) \equiv \frac{1}{\sigma}\frac{d\sigma}{dx} \qquad (1)$$

where x is the ratio of the observed b-hadron energy to the beam energy. Usually, since the b-quark mass is much larger than the QCD scale Λ, the b-quark energy prior to hadronization is calculated using perturbative QCD. Hence the b-hadron energy is related to the quark energy via model-dependent assumptions. Therefore measurement of $D(x)$ serve to constrain both perturbative QCD and model predictions. Furthermore, the uncertainty in the fragmentation function $D(x)$ must be taken into account in studies of the production and decay of heavy quarks: more accurate measurements of this function will allow increased precision test of heavy flavour physics.

At this conference new measurements have been presented by ALEPH[1] and by SLD[2].

2 ALEPH measurement

ALEPH searches for B^+ and B^0 mesons[a] in five semi-exclusive decay channels $B \rightarrow D^{(*)}\ell\nu X$. In three of them the B decays to $D^{*+}\ell\nu X$, followed by $D^{*+} \rightarrow D^0\pi^+$, and the D^0 is reconstructed in the decay channels

---

[a] Charge conjugation implied throughout

Table 1. ALEPH: results of model-dependent analysis. Errors include systematic uncertainty

Model	$\langle x_B^L \rangle$	χ^2/ndf
Peterson[3]	0.733 ± 0.006	116/94
Kartevelishvili[4]	0.746 ± 0.008	97/94
Collins[5]	0.712 ± 0.007	164/94

$D^0 \rightarrow K^-\pi^+$, $D^0 \rightarrow K^-\pi^+\pi^+\pi^-$ and $D^0 \rightarrow K^-\pi^+\pi^0$. The remaining two channels are $B \rightarrow D^0\ell\nu X$, followed by $D^0 \rightarrow K^-\pi^+$, and $B \rightarrow D^+\ell\nu X$, followed by $D^+ \rightarrow K^-\pi^+\pi^+$. Using the full data sample collected at the Z peak, about 4 million hadronic Z decays, a total of 2748 candidates have been found, with a signal purity between 63% and 90%, depending on the decay channel. The B energy is estimated from $D^{(*)}$ and lepton momentum, plus the hemisphere missing energy, due to the neutrino. The energy resolution is described by the sum of two Gaussians, with widths of 0.04 and 0.10, and 50-60% of the candidates in the core.

The mean scaled energy $\langle x_B \rangle$ is extracted from the raw x_B distribution in both a model dependent and a model-independent way. In the first case, three different fragmentation models have been used to hadronize the b-quark after the parton shower, simulated by JETSET 7.4[6]. Reconstruction efficiency and energy resolution, as well as missing pions from B^{**} and D^{**} decays, are taken into account by Monte Carlo simulation. Then, for each model, the reconstructed x_B spectrum

532

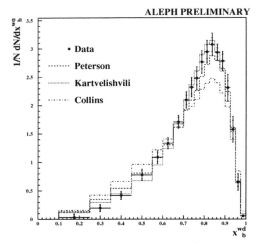

Figure 1. Scaled energy distribution of weakly-decaying B mesons as measured by ALEPH.

in Monte Carlo is compared to data and a minimization of the difference is performed by varying the model parameter. Table 1 shows the results for $\langle x_B^L \rangle$, the mean scaled energy of the leading B, the meson resulting from hadronization, prior to any decay. The fragmentation model of Kartvelishvili et al.[4] gives the better agreement with data. In the model-independent analysis, the Monte Carlo is used to calculate the efficiency, $\epsilon(x_B)$, and the resolution matrix $G(x_B, x_B^{reco})$, defined as the probability for the B meson to have a scaled energy x_B, given the measured x_B^{reco}. Hence the fragmentation function $D(x_B)$ is obtained by unfolding the measured distribution $D^{data}(x_B^{reco})$:

$$D(x_B) = \epsilon^{-1}(x_B) \cdot G(x_B, x_B^{reco}) \cdot D^{data}(x_B^{reco}). \quad (2)$$

Since G depends on the Monte Carlo fragmentation function, the procedure must be iterated, using in the Monte Carlo the above D function obtained from data, until convergence is reached. The results of this analysis are $\langle x_B^L \rangle = 0.7499 \pm 0.0065(stat) \pm 0.0069(syst)$, for the leading B meson, and $\langle x_B^{wd} \rangle = 0.7304 \pm 0.0062(stat) \pm 0.0058(syst)$, for the weakly decaying one. Figure 1 shows the resulting fragmentation function

for the weakly-decaying B meson, compared to the distributions obtained from the model-dependent analysis.

3   SLD measurement

SLD measurement is based on 350,000 Z hadronic decays collected in 97 and 98. The analysis method is the same used for an already published SLD result[7], based on a smaller data sample. A topological secondary vertex finder exploits the small and stable SLC beam spot and the CCD-based vertex detector to inclusively reconstruct b-decay vertices with high efficiency and purity. Precise vertexing allows to reconstruct accurately the b-hadron flight direction and hence the transverse momentum of tracks associated to the vertex with respect to this direction. Their invariant mass, corrected for the transverse momentum of missing particle, is used to separate b-hadrons from $udsc$ background, yielding a 98% pure b-sample with 44% efficiency.

The b-hadron energy is also measured from the invariant mass and the transverse momentum of the tracks associated to the vertex. Constraining the vertex mass to the B^0 mass, an upper limit on the mass of the missing particles is found for each reconstructed b-decay vertex, and is used to solve for the longitudinal momentum of the missing particles, and hence for the energy of the b-hadron. In order to further improve the b-sample purity and the reconstructed b-hadron energy, only vertices with low invariant mass are kept. The selection yields 4164 candidates, with an overall efficiency of 4.2% and 9.6% energy resolution in the core, which accommodate about 80% of the candidates. Moreover, both the efficiency and the energy resolution are remarkably flat in the region $x_b > 0.2$

Several fragmentation functions have been fitted to the candidates scaled energy distribution, as shown in Figure 2. A good

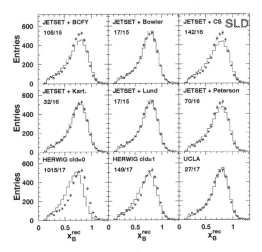

Figure 2. SLD test of fragmentation models.

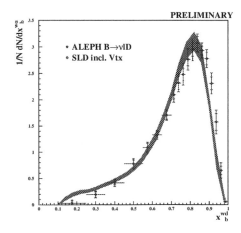

Figure 3. Scaled energy of the weakly-decaying b-hadron as measured by ALEPH and SLD.

description of the data is obtained using JETSET together with the phenomenological models of the Lund group[8], Bowler[9] and Kartvelishvili et al.[4], or using the UCLA[10] fragmentation model. Several functional forms of the true energy distribution $D(x_b)$ have been tried too, and four of them have been found consistent with data. Hence the true distribution has been obtained from equation 2, using the above eight best fitted distributions to calculate the unfolding matrix G from Monte Carlo. The resulting mean scaled energy for the weakly-decaying b-hadron is $\langle x_b^{wd} \rangle = 0.710 \pm 0.003(stat) \pm 0.005(syst) \pm 0.004(model)$.

4 Summary and conclusions

In Figure 3, the fragmentation functions measured by ALEPH and SLD are compared. A slight disagreement is observed between the two. However it must be pointed out that ALEPH selects B mesons only, while the SLD sample also includes B_s and barions, which may be responsible of the observed difference.

References

1. R Barate et al., ALEPH Collab., ALEPH 2000-68, CONF 2000-46, 2000.

2. K. Abe et al., SLD Collab., SLAC-PUB-8504, 2000.

3. C. Peterson et al., Phys. Rev. D **27**, 105 (1983).

4. V. G. Kartvelishvili et al., Phys. Lett. B **78**, 615 (1978).

5. P. D. B. Collins and T. P. Spiller, J. Phys. G **11**, 1289 (1985).

6. S. Sjöstrand, Comp. Phys. Comm. **82**, 74 (1994).

7. K. Abe et al., Phys. Rev. Lett **84**, 4300 (2000).

8. B. Andersson et al., Phys. Rep. **97**, 32 (1983).

9. M. G. Bowler, Z. Phys. C **11**, 169 (1981).

10. S. Chun and C. Buchanan, Phys. Rep. **292**, 239 (1998).

MEASUREMENTS OF THE BOTTOM QUARK MASS

G. DISSERTORI

EP Division, CERN, CH-1211 Geneva 23, Switzerland
E-mail: Guenther.Dissertori@cern.ch

I will review new measurements of the b quark mass, presented at this conference by ALEPH and DELPHI. A large set of observables has been used and detailed studies on jet algorithms have been performed. These measurements at the Z peak are consistent with the results obtained at the Υ scale when assuming the running of the b quark mass as predicted by perturbative QCD.

1 Introduction

The b quark mass is one of the fundamental parameters of the QCD Lagrangian. However, due to confinement, quarks do not appear as asymptotically free particles and therefore the definition of their mass is ambiguous. In the framework of perturbative QCD, quark masses can either be defined as the position of the pole of the quark propagator, or they can be interpreted as effective coupling constants in the Lagrangian. In the former definition the mass is called "pole mass" and does not depend on an energy scale; in the latter the mass is called "running mass", since it is a function of the renormalization scale.

The effects of the b mass become very small with increasing energy for inclusive observables such as the total cross section, since they are proportional to m_b^2/M_Z^2 ($\mathcal{O}$ (0.1%)). However, for semi-inclusive quantities such as jet rates, the effects are enhanced, up to a few percent. Such quantities are sensitive to the amount of gluon radiation, which is suppressed in the case of massive quarks.

At this conference new measurements have been presented by ALEPH[1] and by DELPHI[2]. Previously measurements had been published by DELPHI [3] and by Brandenburg *et al.* [4] who had analysed SLD data.

2 Analysis Method

The method for extracting the b quark mass is based on the measurement of the ratio $R_{b/uds} = O_b/O_{uds}$ of an infrared safe observable O computed for b and uds induced events and assuming α_s universality.

This ratio is either directly obtained by tagging b and uds induced events (DELPHI), or by measuring first the ratio $R_{b/inc} = O_b/O_{inc}$ (inc=all flavours inclusive) and then inferring from that the ratio $R_{b/uds}$, using the precise knowledge of the partial widths of the Z to b and c quarks (ALEPH). The tag of b (uds) events is mainly based on lifetime information. The ratio is then corrected for hadronization, detector and tagging biases.

The b quark mass is extracted by comparing the corrected measured ratio $R_{b/uds}^P$ with the predictions for the observable under study. In general the NLO prediction for $R_{b/uds}^P$ as a function of the b quark mass is of the form

$$R_{b/uds}^P = 1 + \frac{m_b^2}{M_Z^2}\left[b_0(m_b) + \frac{\alpha_s}{2\pi}b_1(m_b)\right] \quad (1)$$

where the coefficient functions b_0 and b_1 are obtained from the integration of the massive and massless matrix elements in terms of the pole or running mass.

Systematic uncertainties arise from imperfections of the detector modelling, from the limited knowledge of the hadronization corrections and B decays, which are particularly important for this analysis, and from theoretical ambiguities because of the renormalization scale and the quark mass scheme employed to compute b_0 and b_1.

3 Results from ALEPH

ALEPH has studied the first and second moments of a large set of event shape variables such as thrust, jet broadenings, or the transition resolution value y_3 for going from three to two jets when applying the Durham [5] algorithm. Furthermore, they have used the ratio of three-jet rates with $y_{cut} = 0.02$. Then the set of observables is reduced by requiring that the NLO contributions to the perturbative prediction be clearly smaller then the LO terms, and that the hadronization corrections do not exceed the measured mass effect in size. These requirements leave only the three-jet rate and the first moment of y_3 as observables. The latter turns out to give the smallest total uncertainties, and the result for the running mass in the $\overline{\text{MS}}$ scheme is $m_b(M_Z) = (3.27 \pm 0.22_{stat} \pm 0.22_{syst} \pm 0.38_{had} \pm 0.16_{theo})$ GeV/c^2.

4 Results from DELPHI

They have updated their previous analysis [3] by including data from 1995, improving the b-tag algorithm, and by studying also the Cambridge jet clustering algorithm [6] for the ratio of three-jet rates. It turns out that with this algorithm the perturbative expansion for $R_{b/uds}^P$ converges more rapidly in the running mass scheme than in the pole mass scheme, and the theoretical uncertainties are smaller than with the Durham algorithm. However, the measurement based on the Cambridge algorithm still suffers from rather large hadronization uncertainties. The result is $m_b(M_Z) = (2.61 \pm 0.18_{stat} \pm 0.18_{syst} \pm 0.47_{had} \pm 0.07_{theo})$GeV/$c^2$.

5 Conclusions

The new measurements of the b quark mass by ALEPH and DELPHI are in good agreement with determinations at lower scales [7], extrapolated to the Z pole by using the

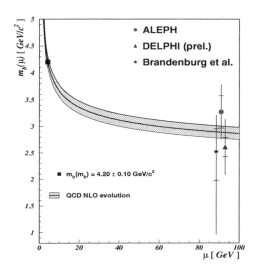

Figure 1. The running of the b quark mass

running predicted by perturbative QCD, as shown in figure 1. There are some indications for still uncontrolled biases from hadronization and/or uncomputed higher orders, since the results based on the three-jet rate tend to be systematically lower than the one obtained from the first moment of an event shape distribution.

References

1. R. Barate *et al.*, ALEPH Collab., CERN EP/2000-093, subm. to *Eur. Phys. J. C.*

2. M.J. Costa *et al.*, DELPHI Collab., DELPHI 2000-121, CONF 420, 2000.

3. P. Abreu *et al.*, DELPHI Collab., *Phys. Lett.* B **418**, 430 (1998).

4. A. Brandenburg *et al.*, *Phys. Lett.* B **468**, 168 (1999).

5. S. Catani *et al.*, *Phys. Lett.* B **269**, 179 (1991); N. Brown and J. Stirling, *Z. Phys.* C **53**, 629 (1992).

6. Yu.L. Dokshitzer *et al.*, Cavendish-HEP-97/06; JHEP 9708:001, 1997; hep-ph/9707323.

7. A. Pich, hep-ph/000118, 2000.

COMPLETE RENORMALIZATION GROUP IMPROVEMENT- AVOIDING SCALE DEPENDENCE IN QCD PREDICTIONS

C.J. MAXWELL

Centre for Particle Theory, University of Durham, Durham, DH1 3LE, UK
E-mail: C.J.Maxwell@durham.ac.uk

Using generalized dimensional analysis we show how dimensionless QCD observables can be directly related to the dimensional transmutation parameter of the theory. In this procedure there is no need to mention, let alone to arbitrarily vary, the unphysical renormalization scale μ. At the next-to-leading order in perturbation theory the result is identical to that of Grunberg's Effective Charge approach.

A difficulty with fixed-order renormalization group (RG)-improved QCD perturbation theory is that next-to-leading order (NLO) predictions depend on the renormalization scale μ which arises in removing ultraviolet divergences from the theory. Usually one chooses $\mu = xQ$, proportional to Q the physical energy scale of the process, for instance the c.m. energy in e^+e^- collisions. Typically this is then varied over some range around $\mu = Q$, $x = 1$, the so-called "physical scale", say between $x = 10$ and $x = 0.1$. The resulting fits to data of $\alpha_s(xQ)$ are then customarily [1] converted to $\alpha_s(M_Z)$ using the two-loop evolution of the coupling to yield a central value based on $x = 1$ and a "theoretical error bar" based on the variation between $x = 10$ and $x = 0.1$. Unfortunately the range of x to be considered is completely *ad hoc* , and there is no reason why the central values obtained should reflect the actual value of $\alpha_s(M_Z)$. In this talk we wish to present perturbative QCD from a rather different viewpoint in which we directly relate QCD observables to the dimensional transmutation parameter of the theory, e.g. $\Lambda_{\overline{MS}}$. In this approach Λ assumes central importance and μ and the renormalized coupling $\alpha_s(\mu)$ are manifestly irrelevant quantities which only appear in intermediate steps. Practical application of the resulting formalism to the direct extraction of $\Lambda_{\overline{MS}}$ from a wide range of e^+e^- jet observables is in progress [2].

The following discussion owes much to the treatment of [3] and closely follows that of [4]. Let us suppose that we have a generic dimensionless QCD observable $\mathcal{R}(Q)$, dependent on ths single dimensionful (energy) scale Q. Quark masses will be taken to be zero throughout our discussion. Since $\mathcal{R}(Q)$ is dimensionless, dimensional analysis clearly implies that

$$\mathcal{R}(Q) = \Phi\left(\frac{\Lambda}{Q}\right) , \qquad (1)$$

where Λ is a dimensionful scale which will turn out to be related to the dimensional transmutation parameter. An obvious proposal is to invert Eq.(1) to obtain

$$\frac{\Lambda}{Q} = \Phi^{-1}(\mathcal{R}(Q)) , \qquad (2)$$

where Φ^{-1} is the inverse function. This is indeed the basic motivation for Grunberg's Method of Effective Charges [5]. It is convenient to begin from the form for the derivative of $\mathcal{R}(Q)$ with respect to Q imposed by dimensional analysis.

$$\frac{d\mathcal{R}(Q)}{dQ} = \frac{B(\mathcal{R}(Q))}{Q} , \qquad (3)$$

where $B(\mathcal{R}(Q))$ is a dimensionless function of $\mathcal{R}$. This can be rearranged to obtain,

$$\frac{d\mathcal{R}(Q)}{d\ln Q} = B(\mathcal{R}(Q)) . \qquad (4)$$

This is a separable first-order differential equation. The boundary condition will be the crucial property of asymptotic freedom, i.e. $\mathcal{R}(\infty) = 0$. Integrating the equation then gives

$$\ln\frac{Q}{\Lambda_\mathcal{R}} = \int_0^{\mathcal{R}(Q)} \frac{dx}{B(x)} + \kappa \ . \tag{5}$$

The constant of integration has been split into $\ln\Lambda_\mathcal{R}+\kappa$, where $\Lambda_\mathcal{R}$ is a finite dimensionful scale specific to the observable $\mathcal{R}$, and κ is a universal *infinite* constant needed to implement $\mathcal{R}(\infty) = 0$. To determine κ we shall need to know the behaviour of $B(x)$ around $x = 0$. We shall assume that $\mathcal{R}$ has the perturbative expansion

$$\mathcal{R} = a\left(1 + \sum_{n>0} r_n a^n\right) , \tag{6}$$

where $a\equiv\alpha_s(\mu)/\pi$ is the RG-improved coupling. $a(\mu)$ will satisfy the beta-function equation

$$\mu\frac{\partial a}{\partial\mu} \equiv \beta(a)$$

$$= -ba^2\left(1 + ca + \sum_{n>1} c_n a^n\right) , \tag{7}$$

where $b = (33 - 2N_f)/6$, and $c = (153 - 19N_f)/12b$, are the first two coefficients of the beta-function for SU(3) QCD with N_f active (massless) flavours of quark. They are universal whereas the subsequent coefficients c_i, $(i > 1)$ are scheme-dependent, and as pointed out by Stevenson may be used to label the scheme [6]. To obtain the form of $B(x)$ around $x = 0$ we can note that there exists a scheme in which $\mathcal{R} = a$, thus $B(\mathcal{R})$ will be given by $\beta(\mathcal{R})$ from Eq.(7), with the $c_n = \rho_n$, the beta-function coefficients in this Effective Charge scheme [5]. The ρ_n are RS-invariant and Q-independent combinations of the r_i and c_i, and are determined given a complete N^{n}LO calculation. We therefore see that for κ we need

$$\kappa = -\int_0^C \frac{dx}{K(x)} , \tag{8}$$

where $K(x)$ must be such that the singularity of $1/B(x)$ at $x = 0$ in Eq.(5) is cancelled, which implies that $K(x) = -bx^2(1 +$

$cx + \Delta(x))$, where we reproduce the universal coefficients of the beta-function and $\Delta(x)$ is only constrained by the requirement that $\Delta(x)/x^2$ is finite as $x\to0$. Different choices of the upper limit of integration C, and the function $\Delta(x)$ can be absorbed into the definition of the constant of integration $\Lambda_\mathcal{R}$. Convenient choices are $C = \infty$ and $\Delta(x) = 0$. Eq.(5) can then be rewritten as

$$b\ln\frac{Q}{\Lambda_\mathcal{R}} = \int_{\mathcal{R}(Q)}^\infty \frac{dx}{x^2(1 + cx)}$$
$$+ \int_0^{\mathcal{R}(Q)} dx\left[\frac{b}{B(x)} + \frac{1}{x^2(1 + cx)}\right] \tag{9}$$

The first integral yields $F(\mathcal{R})\equiv(1/\mathcal{R}) + c\ln(c\mathcal{R}/(1 + c\mathcal{R}))$, and denoting the second integral by $G(\mathcal{R})$ and exponentiating we find the desired inverse relation of Eq.(2) to be

$$\frac{\Lambda_\mathcal{R}}{Q} = \mathcal{F}(\mathcal{R}(Q))\mathcal{G}(\mathcal{R}(Q)) , \tag{10}$$

where $\mathcal{F}(\mathcal{R})$ is the universal function $e^{-1/b\mathcal{R}}(1 + 1/c\mathcal{R})^{c/b}$, and $\mathcal{G}(\mathcal{R})\equiv e^{-G(\mathcal{R})/b}\approx1 - \rho_2\mathcal{R}/b+\ldots$. Given only a NLO calculation the NNLO RS-invariant ρ_2 will be *unknown*, and so $\mathcal{G} = 1$ is the *best* one can do. We finally need to relate the integration constant $\Lambda_\mathcal{R}$ which depends on the particular observable , to the *universal* dimensional transmutation parameter which depends only on the subtraction procedure used to remove ultraviolet divergences, $\Lambda_{\overline{MS}}$ for instance. Fortunately they can be related *exactly* given only a one-loop (NLO) perturbative calculation of the observable. To see this we note that on rearranging Eq.(10) and taking the limit as $Q\to\infty$ we obtain an operational definition of $\Lambda_\mathcal{R}$

$$\Lambda_\mathcal{R} = \lim_{Q\to\infty} Q\mathcal{F}(\mathcal{R}(Q)) , \tag{11}$$

where we have used the fact that $\mathcal{G}(0) = 1$ together with asymptotic freedom. Denoting by $a(Q)$ the $\overline{MS}$ coupling with $\mu = Q$ we see that it will satisfy the beta-function equation (7), of the same form as Eq.(4) for $\mathcal{R}(Q)$, with $\beta_{\overline{MS}}(a)$ replacing $B(\mathcal{R})$. This may be integrated following exactly the same steps as above. The constant of integration $\Lambda_{\mathcal{R}}$ will be replaced by $\tilde{\Lambda}_{\overline{MS}}$, and the coefficients ρ_i by the $\overline{MS}$ beta-function coefficients $c_i^{\overline{MS}}$. Again choosing $C = \infty$ and $\Delta(x) = 0$, we arrive at

$$\Lambda_{\overline{MS}} = \lim_{Q \to \infty} Q\mathcal{F}(a(Q)) . \tag{12}$$

Defining for convenience $r \equiv r_1^{\overline{MS}}(\mu = Q)$ and using the perturbation series of Eq.(6) it is easy to show that as $Q \to \infty$, $F(\mathcal{R}) \approx F(a) - r + \ldots$, where the ellipsis denotes terms which vanish as $Q \to \infty$. Note that r is Q-independent. Thus we find the exact relation

$$\Lambda_{\mathcal{R}} = e^{r/b}\tilde{\Lambda}_{\overline{MS}} . \tag{13}$$

The tilde reflects the convention chosen for defining κ. The standard convention [7] corresponds to translating κ by the finite shift $c\ln(b/2c)$, and so $\tilde{\Lambda}_{\overline{MS}} = (2c/b)^{-c/b}\Lambda_{\overline{MS}}$. Assembling all this we find

$$\Lambda_{\overline{MS}} = Q\mathcal{F}(\mathcal{R}(Q))\mathcal{G}(\mathcal{R}(Q))e^{-r/b}(2c/b)^{c/b} . \tag{14}$$

This equation can then be used to perform direct extractions of $\Lambda_{\overline{MS}}$. Given a NLO calculation the data for $\mathcal{R}$ can be simply fed into Eq.(14) (with $\mathcal{G} = 1$) and $\Lambda_{\overline{MS}}$ extracted bin-by-bin in a given QCD observable. To the extent that uncalculated NNLO and higher-order perturbative corrections and possible power corrections are small one should obtain the same value of $\Lambda_{\overline{MS}}$ for each observable and bin. Typically one finds a flat plateau region, with a dramatic deviation in the two-jet region where large infra-red logarithms should be important and require resummation. Such an analysis has been carried out for e^+e^- jet observables in collaboration with Stephen Burby, and the results will be published shortly [2].

In fact the result obtained at NLO is exactly the same as that using the Effective Charge scheme of Grunberg with $\mu = e^{-r/b}Q$. This corresponds to a *complete* resummation of Ultraviolet (UV) logarithms of Q, and builds the physical Q-dependence of $\mathcal{R}(Q)$ which we obtained above by integration of Eq.(4). The sense in which standard RG-improvement is *incomplete* with respect to resummation of UV logarithms has recently been discussed by us [8].

References

1. For a recent comprehensive review see S. Bethke, hep-ex/0004021.
2. S.J. Burby and C.J. Maxwell in preparation.
3. P.M. Stevenson, *Ann. Phys.* **152** 383 (1981).
4. D.T. Barclay, C.J. Maxwell and M.T. Reader, *Phys. Rev.* **D49** 3480 (1994).
5. G. Grunberg, *Phys. Rev.* **D29** 2315 (1984).
6. P.M. Stevenson, *Phys. Rev.* **D23** 2916 (1981).
7. A.J. Buras, E.G. Floratos, D.A. Ross and C.T. Sachrajda, *Nucl. Phys.* **B131** 308 (1977); W.A. Bardeen, A.J. Buras, D.W. Duke and T. Muta, *Phys. Rev* **D18** 3998 (1978).
8. C.J. Maxwell and A. Mirjalili, *Nucl. Phys.*, **B577** 209 (2000).

EVENT SHAPES AND OTHER QCD STUDIES AT LEP

SWAGATO BANERJEE

LAPP, IN2P3-CNRS, Chemin de Bellevue, BP 110, Annecy-le-Vieux, France 74941.

QCD results on jet rates, event shapes and inclusive charged particle spectra in hadronic events from e^-e^+ annihilations at LEP up to a center-of-mass energy of 208 GeV are presented.

1 Introduction

With the increase of center-of-mass energy ($\sqrt{s}$), LEP provides an ideal environment for precise tests of QCD over a wide energy range with small hadronization corrections, which scale inversely as $\sqrt{s}$. In this paper, results up to $\sqrt{s} = 208$ GeV from the LEP experiments: ALEPH [1], DELPHI [2], L3 [4,5] and OPAL [6,7] are reviewed. Radiative Z-decay events with isolated photons, provide information on the hadronic system down to $\sqrt{s}$ = 30 GeV [2,4]. From quark and gluon jet multiplicities measured at the Z peak, colour factors are obtained [3].

2 Jet Rates

Hadronic events are characterised by a multijet topology. Jets have been studied using the JADE, DURHAM and CAMBRIDGE algorithms. Figure 1 shows the different jet rates for year 2000 data using the CAMBRIDGE algorithm [5]. Satisfactory agreement with the predictions of QCD models is obtained.

NLO predictions for 4 jet rates are now available [8]. α_s measured by DELPHI [2] using CAMBRIDGE algorithm ($y_{cut} = 0.004$) are compared to $\mathcal{O}(\alpha_s^3)$ calculations in figure 2. Fit to $\frac{d\alpha_s^{-1}}{d\log E_{cm}}$ gives 1.17 ± 0.28, in agreement with expectation of 1.27 for 5 flavours.

3 Event Shapes

The energy flow in hadronic events can be conveniently studied in terms of collinear and infrared safe global event shape variables. The energy dependence comes from logarith-

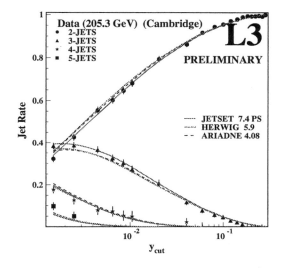

Figure 1. Jet rates using the CAMBRIDGE algorithm.

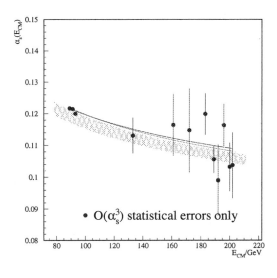

Figure 2. α_s from 4 jet rates.

mic behaviour in α_s, as well as inverse power dependence from non-perturbative effects.

For some event shape variables, like thrust (T), heavy jet mass (ρ), total and wide jet broadenings $(B_T$ and $B_W)$ and the C-parameter (C), complete theoretical calculations exist up to $\mathcal{O}(\alpha_s^2)$ and the leading and next-to-leading order terms have been resummed up to all orders. While the second order calculations work well in the multi-jet region, the resummed calculations are necessary to describe the data in the two jet region. Thus the whole distribution of the event shapes can be used, giving reduced statistical errors in the determination of α_s.

4 Determination of α_s

4.1 Power Law ansatz

Assuming the soft gluon emission is controlled by an effective coupling (α_{eff}), which differs from α_s in the infrared region, the hadronization correction to the non-perturbative component of moments as well as the differential distributions of event shapes can be parameterised using a power law ansatz. The mean values and the distributions have been simultaneously fitted to $\alpha_s(M_Z)$ and the non-perturbative parameter $\alpha_0(\mu_I)(= \frac{1}{\mu_I}\int_0^{\mu_I} dq\,\alpha_{\text{eff}}(q))$ at a scale $\mu_I = 2$ GeV, using second order calculations for the perturbative part. The results of α_s and α_0 from fits to moments and distributions are shown in the figure 3. The scatter in the measurements from different variables between the different experiments being large, universality of α_0 is verified within $\pm$ 20%.

Phenomenological study [2] shows that the relative size of the power law contribution is large for event shapes variables with large relative second order term. L3 studied [4] the second moment of the event shapes using an additional $1/s$ term in the power law parametrization. The contribution is small for ρ, negative for B_W and substantial for others.

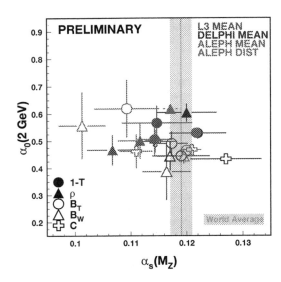

Figure 3. Summary of α_s-α_0 measurements.

4.2 Resummed LL & NLL + $\mathcal{O}\,(\alpha_s^2)$

The standard determination of α_s from event shape variables uses matching of second order predictions with resummed NLLA calculation. Quark mass effects are included in the $\mathcal{O}\,(\alpha_s^2)$ calculations: while the effect is 1% at M_Z, it scales as $1/s$ down to 0.2% at 200 GeV. Fits to the five event shape variables measured by L3 [5] at $\sqrt{s} = 206$ GeV are shown in the figure 4.

The measurements of α_s using the different event shapes are averaged for each value of $\sqrt{s}$, and evolved to M_Z for comparison in the figure 5. The average of all the measurements gives:

$$\alpha_s(M_Z) = 0.1204 \pm 0.0007(\text{exp}) \pm .0034(\text{theo})$$

From a fit to the slope of the α_s measured between 30 and 208 GeV by the L3 [5], the number of active quark flavours is obtained to be 5.1 ± 1.3 (exp) ± 1.9 (theo), in agreement with DELPHI measurements [2] of $\frac{d\alpha_s^{-1}}{d\log E_{cm}}$.

5 Charged Particle spectra

Beyond the leading logarithmic approximation, the *intrajet coherence* phenomena arising from destructive interference between the

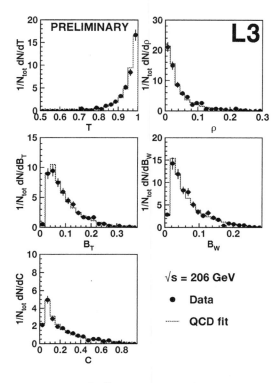

Figure 4. $\mathcal{O}\,(\alpha_s{}^2) +$ resummed calculations fit.

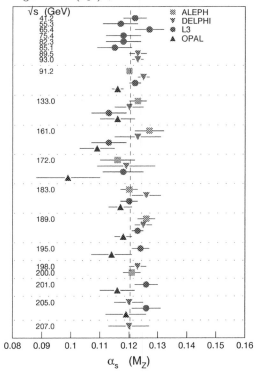

Figure 5. α_s measurements from event shapes.

soft gluon emission within the jets, reduces the phase space available for further parton emission to an angular ordered region. This dynamical suppression of the soft momenta leads to energy and emission angle ordering of successive parton radiations. It results in reduced parton multiplicities and a dip in the parton momenta in the low momentum region.

The *charged particle momentum spectrum* is studied in terms of the variable $\xi = -\ln(\frac{2|\vec{p}|}{\sqrt{s}})$ (where $\vec{p}$ is the momentum). The energy evolution of the charged particle multiplicity and the measured peak position ($\xi^\star$) (figures 6, 7) provide evidence for gluon coherence: predictions without coherence effects fail to describe the data.

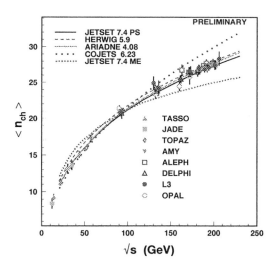

Figure 6. Energy evolution of charged particle multiplicity.

6 Quark and gluon multiplicities

The evolution of the charged particle multiplicity has been studied as a function of the opening angle (θ) between the sub-leading jets in mirror symmetric 3 jet events [3]. Correcting for the case when the gluon fragments into the leading jet, large opening angles translate into harder gluon jets, giving higher multiplicities.

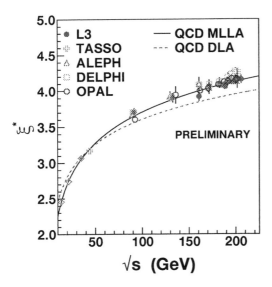

Figure 7. Energy evolution of peak position of ξ-spectra.

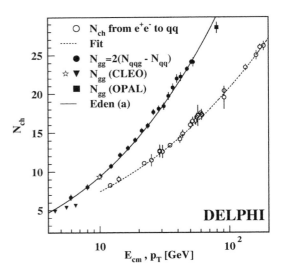

Figure 8. Charged particle multiplicity in quarks and gluon jets.

Recent theoretical calculations relate the derivatives of quark-gluon multiplicities to the QCD colour factors [9]. From a fit to the data from hadronic Z decays, DELPHI updated their analysis using the CAMBRIDGE jet algorithm, and obtained: $C_A/C_F = 2.221 \pm 0.047$ (exp) ± 0.058 (had) ± 0.075 (theo), in good agreement with the QCD prediction of 9/4.

The gluon-gluon multiplicity is extracted at scales given by $\sqrt{s}$ and p_T using the relation: $N_{gg} = 2[N_{q\bar{q}g}(\theta) - N_{q\bar{q}} - N_0]$, estimating the non-perturbative offset term (N_0) from CLEO data. Measurements of the quark and gluon jet multiplicities are shown in the figure 8.

Summary

Different aspects of hard and soft gluon radiation have been studied in hadronic events with $\sqrt{s}$ in the range 30 to 208 GeV. Evidence of gluon coherence and higher gluon colour charge have been observed. The strong coupling constant, the number of active quark flavours and QCD colour factors have been measured.

Acknowledgements

I thank all the 4 LEP experiments for sharing with me the preliminary results and thank Sunanda Banerjee, Dominique Duchesneau, John Field, Klaus Hamacher and Oliver Passon for interesting discussions.

References

1. ALEPH coll., contributed paper #164.
2. DELPHI coll., contributed paper #638.
3. DELPHI coll., contributed paper #640.
4. L3 coll., Physics letters **B489** (2000) 65.
5. L3 coll., contributed paper #630.
6. OPAL coll., Eur.Phys.J. **C16** (2000) 185.
7. OPAL coll., contributed paper #252.
8. Zoltan Trocsanyi, these proceedings.
9. P.Eden *et al.*, JHEP 09 (1998) 15. P.Eden *et al.*, Eur.Phys.J. **C11** (1999) 345.

EVENT SHAPE STUDIES AT HERA

HANS-ULRICH MARTYN

I. Physikalisches Institut der RWTH, D–52056 Aachen, Germany
(on behalf of the H1 and ZEUS collaborations)

Recent progress on the study of power corrections applied to event shape variables in deep inelastic *ep* scattering is discussed.

1 Introduction

Event shape variables in deep inelastic scattering allow to study QCD properties of the final state. Recent results based on high statistics data by the ZEUS[1] and H1[2] collaborations at HERA are discussed with emphasis on power law hadronisation effects.

A suitable frame of reference with optimal separation between the current jet and the proton remnant is the Breit system, where the spacelike γ/Z with momentum Q collides head-on with an incoming quark of momentum $Q/2$. The quark is back-scattered into the current hemisphere, while the proton fragments into the remnant hemisphere (QPM picture). Event shapes studied in the current region[1,2] are: thrust τ and τ_c, broadening B, C parameter and jet mass ρ and ρ_0. H1 also investigates two-jet rates, the transitions from $(2+1)$ to $(1+1)$ jets, in the whole phase space: y_{fJ} for a factorisable JADE and y_{kt} for the k_t jet algorithm. The analyses of both experiments are very similar. The kinematic range covers $Q = 7 - 140$ GeV, ZEUS distinguishes two x-bins at low $Q < 18$ GeV. The data are unfolded to the hadron level, where H1 takes the true masses and ZEUS assumes massless hadrons. This difference affects jet masses and two-jet rates.

2 Power corrections to mean values of event shapes

The mean value of an event shape variable F can be expressed as[3]

$$\langle F \rangle = \langle F \rangle^{\text{pert}} + a_F \mathcal{P} \, . \tag{1}$$

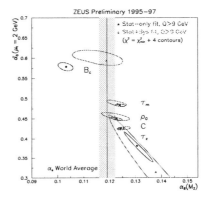

Figure 1. Correlations of $\bar{\alpha}_0$ vs $\alpha_s(M_Z)$ from power correction fits to ZEUS event shape means. Small contours include statistical errors only

The perturbative part $\langle F \rangle^{\text{pert}}$ is calculated in $\mathcal{O}(\alpha_s^2)$ using DISENT[4]. The program DISASTER++[5] yields consistent but somewhat higher mean values[2], at most a few per cent for $\langle B \rangle$ at low Q. These discrepancies have no influence on the conclusions.

Hadronisation is treated within the concept of power corrections[3] with a calculable coefficient a_F and a universal function

$$\mathcal{P} = 1.61 \frac{\mu_I}{Q} \left[\bar{\alpha}_0(\mu_I) - \alpha_s(Q) \right.$$
$$\left. -1.22 \left(\ln \frac{Q}{\mu_I} + 1.45 \right) \alpha_s^2(Q) \right] \, . \tag{2}$$

One expects a $1/Q$ behaviour (except for y_{kt}), which is multiplied by terms involving the strong coupling α_s and a non-perturbative effective coupling $\bar{\alpha}_0(\mu_I)$, defined at an infrared matching scale $\mu_I = 2$ GeV.

The Q dependences of the event shape means are well described by this ansatz. Results of fits to $\bar{\alpha}_0$ and $\alpha_s(M_Z)$ are shown as

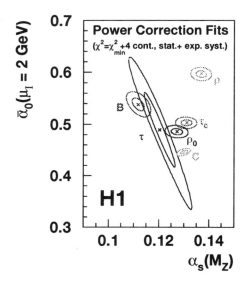

Figure 2. Correlations of $\bar{\alpha}_0$ vs $\alpha_s(M_Z)$ from power correction fits to H1 event shape means

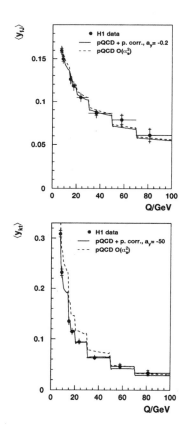

Figure 3. Mean values of two-jet rates y_{fJ} and y_{kt} vs Q of the H1 experiment

correlations in figs. 1 and 2; the renormalisation scale uncertainties (not shown) exceed the experimental errors by far. For the nonperturbative parameter one finds a universal value of $\bar{\alpha}_0 \simeq 0.5 \pm 20\%$ for most observables. ZEUS reports that fits to $\langle B \rangle$ and $\langle \tau_z \rangle$, resulting in a large spread (see fig. 1), are especially sensitive to experimental systematics and exhibit a significant x-dependence at low Q. In general, power corrections do not dependent on x; such terms may, however, arise for $\langle B \rangle$. The H1 analysis, presented in fig.2, demonstrates the strong influence of the treatment of hadrons on the jet masses ρ and ρ_0. A correction to massless hadrons, ρ_0, leads to a more consistent interpretation of power corrections. The spread of $\alpha_s(M_Z)$ is considerable for both experiments, suggesting that higher order QCD contributions are missing. This is supported by large scale uncertainties.

The energy dependence of the H1 mean two-jet rates are shown in fig. 3. They exhibit much smaller hadronisation corrections than the other variables. For the JADE algorithm $\langle y_{fJ} \rangle$ the conjectured coefficient[3] $a_{fJ} = 1$ leads to an unphysically low value of $\bar{\alpha}_0$ and is excluded by the data. Instead a small negative hadronisation contribution is preferred. In case of the k_t algorithm no firm power correction prediction exists except of a $1/Q^2$ dependence for $\langle y_{kt} \rangle$, very different from the other event shapes. Such a behaviour is supported by the H1 data, a $1/Q$ shape can be ruled out. An experimental determination of the unknown parameters, a_{kt} and $\bar{\alpha}_1$ together with α_s, suffers from large correlations[2]. In view of the existing data more theoretical work on the jet rates is needed.

3 Power corrections to spectra

Power corrections to event shape spectra lead to a shift of the pQCD prediction[6]

$$\frac{1}{\sigma_{\text{tot}}} \frac{d\sigma(F)}{dF} = \frac{1}{\sigma_{\text{tot}}} \frac{d\sigma^{\text{pert}}(F - a_F \mathcal{P})}{dF}, \quad (3)$$

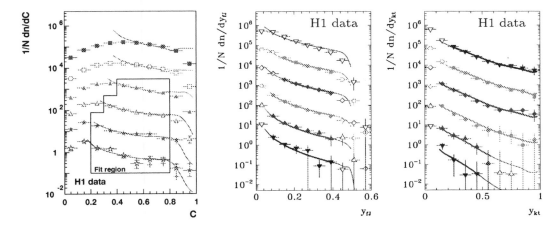

Figure 4. Fits to differential distributions of the C parameter including power corrections and the two-jet rates y_{fJ} and y_{kt} applying pQCD without hadronisation corrections. The H1 data cover a $\langle Q \rangle$ range from 15 GeV (top) to 81.3 GeV (bottom).

provided $\mu_I/Q < F < F_{max}$. The shift $a_F \mathcal{P}$ amounts to exactly the same value as for the mean values. The event shape distributions at $Q > 14$ GeV can be well described by eq. (3) within restricted regions, as shown in fig. 4. However, the fit values of $\overline{\alpha}_0$ and $\alpha_s(M_Z)$ are, in general, inconsistent (larger) to those from fits to the means. For example $(\overline{\alpha}_0, \alpha_s(M_Z)) = (0.45, 0.130)$ from $\langle C \rangle$ and $(0.62, 0.131)$ from $d\sigma/dC$. It is hoped that resummed QCD calculations[7] will improve the applicability of power corrections to DIS event shape spectra.

The analysis of mean values lead to small hadronisation corrections for the two-jet rates. In fact, at sufficiently high energies Q, the jet rate spectra can be reasonably well described by pQCD alone, *i.e.* neglecting power corrections or hadronisation contributions completely. This is shown in fig. 4 for $d\sigma/dy_{fJ}$ and $d\sigma/dy_{kt}$, using 0.116 and 0.118, respectively, for the strong coupling constant.

Summary

Event shape studies of deep inelastic scattering provide very useful information to get a better understanding of the interplay
between perturbative and non-perturbative QCD. The basic concept of approximate universal power corrections is generally supported by the HERA experiments, yielding a common parameter $\overline{\alpha}_0 \simeq 0.5 \pm 0.1$. However, there remain several open questions. The quality of the data requires further theoretical progress concerning the jet rates and x-dependence of power corrections and resummed QCD calculations.

References

1. ZEUS Coll., paper 421 to ICHEP 2000.
2. C. Adloff *et al.*, H1 Coll., Eur. Phys. J. C 14 (2000) 255 and addendum subm. to EPJC.
3. Yu.L. Dokshitzer, B.R. Webber, Phys. Lett. B 352 (1995) 451; B.R. Webber, Proc. DIS 95, Paris, France, 1995.
4. S. Catani, M.H. Seymour, Nucl. Phys. B 485 (1997) 291.
5. D. Graudenz, hep-ph/9710244.
6. Yu.L. Dokshitzer, B.R. Webber, Phys. Lett. B 404 (1997) 321.
7. V. Antonelli, M. Dasgupta, G.P. Salam, JHEP 02 (2000) 001.

MULTIPLICITY STUDIES AT HERA

N. H. BROOK

ON BEHALF OF THE ZEUS COLLAB.

H. H. Wills Physics Lab, University of Bristol, Bristol, UK
E-mail: n.brook@bristol.ac.uk

The factorial moments and the mean of the charged multiplicity distribution have been investigated in neutral-current deep inelastic scattering. The measurements are compared to data from other interactions and theoretical QCD predictions made in the double logarithmic approximation in conjunction with local parton hadron duality.

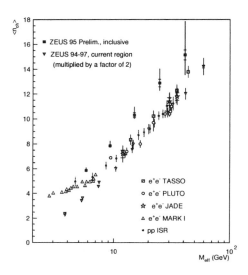

Figure 1. Mean charged particle multiplicity dependence on M_{eff} from ZEUS. Also shown are e^+e^- data ($M_{eff} = \sqrt{s_{ee}}$) and DIS data from the current region of the Breit frame ($M_{eff} = Q$) and proton-proton data from the ISR.

1 Introduction

This paper reports the results of two studies on the properties of the hadronic final state in neutral-current positron-proton deep inelastic scattering (DIS).

The event kinematics of DIS are determined by the negative square of the four-momentum of the virtual exchanged boson (γ^*), $Q^2 \equiv -q^2$, and the Bjorken scaling variable, $x = Q^2/2P \cdot q$, where P is the four-momentum of the proton. In the Quark Parton Model (QPM), the interacting quark from the proton carries four-momentum xP. The invariant mass, W, of the γ^*P system can be expressed as $W^2 = Q^2(1 - x)/x$.

2 Multiplicity Distributions

The dynamics of the hadronisation process has been investigated using the mean charged multiplicity, $\langle n_{ch} \rangle$, as a function of invariant mass, M_{eff}, in a fixed rapidity interval. The M_{eff} is essentially measuring the rapidity "along" the gluon ladder, of the parton evolution in the proton, in DIS at low-x.

DIS events were selected in the range $8 < Q^2 < 1200$ GeV^2 with $70 < W < 260$ GeV. The M_{eff} of the event was measured in a fixed pseudorapidity range of $|\eta| < 1.75$, which corresponds to a region of high acceptance of the ZEUS central tracking detector.

Figure 1 shows the mean charged multiplicity as a function of M_{eff}; the data grows linearly with $\ln(M_{eff})$. As is seen in Fig. 1, the $\langle n_{ch} \rangle$ is higher than e^+e^- data [1], low energy proton-proton data [2] and earlier ZEUS data in the current region of the Breit frame [3]. All these other processes are quark initiated, in general, while the measurements presented here are compatible with the scenario of additional gluon radiation due to a colour octet exchange for DIS events, in the low-x kinematic region under study.

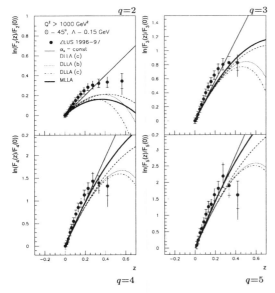

Figure 2. $F_q(z)/F_q(0)$ as a function of the scaling variable z compared to different QCD predictions: DLLA(a) [6], DLLA(b) [7], DLLA(c) [8] and MLLA [6].

3 Factorial Moments

DIS events were selected at high Q^2 ($Q^2 > 1000$ GeV^2) in order to minimise the contribution from boson gluon fusion events ($\gamma^* g \to q\bar{q}$.) This allows the current region of the Breit frame to be compared to a single hemisphere of e^+e^- annihilation data [4,5] at $\sqrt{s_{ee}} = Q$. The charged particle multiplicities, for a given observable Ω, are studied in terms of the normalised factorial moments $F_q(\Delta\Omega) = \langle n_{ch}(n_{ch} - 1)\ldots(n_{ch} - q + 1)\rangle/\langle n_{ch}\rangle^q$, inside a phase space region $\Delta\Omega$. The particle multiplicity is measured in polar rings of size Θ around the $\gamma^* P$ axis in the current region of the Breit frame. This is then expressed in terms of the phase space variable z defined as $\ln(\Theta_0/\Theta)/\ln(Q\Theta_0/2\Lambda)$, where Θ_0 is the half opening-angle of a cone around the $\gamma^* P$ axis and Λ is an effective QCD scale.

Figure 2 shows the angular moments $F_q(z)/F_q(0)$; the data rises as z increases (ie the angular window, Θ, is decreased.) Also in Fig. 2 the data is compared with var-

ious QCD predictions made in the double leading logarithmic approximation (DLLA) or modified leading logarithmic approximation (MLLA), assuming local parton hadron duality (LPHD) holds. The data is not described by the theory, particularly for low orders of the moments, though the data can be well described by leading-order event generators (not shown.) This suggests, in the kinematic region under study here, that there are large hadronisation corrections to the underlying partonic behaviour and therefore the use of LPHD is invalid. Similar observations have been made in studies in e^+e^- annihilation [9].

References

1. MARK I Collab., V. Luth et al., *Phys. Lett.* B **70**, 120 (1977); JADE Collab., W. Bartel et al., *Z. Phys.* C **20**, 187 (1983); PLUTO Collab., Ch. Berger et al., *Phys. Lett.* B **95**, 313 (1990); TASSO Collab., W. Braunschweig et al., *Z. Phys.* C **45**, 193 (1989).

2. M. Basile et al., *Nuovo Cimento* A **67**, 244 (1982); M. Basile et al., *Lett. Nuovo Cimento* **41**, 293 (1984).

3. ZEUS Collab., J. Breitweg et al., *Eur. Phys.* C **11**, 251 (1999).

4. Yu. L. Dokshitzer et al., *Sov. Phys. JETP* **68**, 1303 (1988).

5. ZEUS Collab., M. Derrick et al., *Z. Phys.* C **67**, 93 (1995).

6. Yu. L. Dokshitzer and I. M. Dremin, *Nucl. Phys.* B **402**, 139 (1993).

7. Ph. Brax, J.-L. Meunier and R. Peschanski, *Z. Phys.* C **62**, 649 (1994).

8. W. Ochs and J. Wosiek, *Phys. Lett.* B **289**, 159 (1992); *Phys. Lett.* B **304**, 144 (1993); *Z. Phys.* C **68**, 269 (1995).

9. DELPHI Collab., P. Abreu et al., *Phys. Lett.* B **457**, 368 (1999); L3 Collab.,M. Acciarri et al., *Phys. Lett.* B **428**, 186 (1998); L3 Collab., M. Acciarri et al., *Phys. Lett.* B **429**, 375 (1998)

SEARCH FOR QCD-INSTANTONS AT HERA

E.A. DE WOLF

CERN, European Organisation for Nuclear Research, CH-1211 Geneva 23, Switzerland and Universitaire Instelling Antwerpen, Universiteitsplein 1, B-2610 Antwerpen
E-mail: e.de.wolf@cern.ch

FOR THE H1 COLLABORATION

Signals of QCD instanton induced processes are searched for in deep-inelastic ep scattering a HERA in a kinematic region defined by the Bjorken scaling variables $x > 10^{-3}$, $0.1 < y < 0.6$ and polar angle of the scattered positron $\theta_{el} > 156°$. Upper limits are derived from the expected instanton-induced final state properties based on the QCDINS Monte Carlo model.

1 Instantons in DIS

After initial work by Balitsky and Braun[1], theoretical and phenomenological studies of the rôle of instantons in deep inelastic scattering (DIS) at HERA have been vigorously pursued[a] by Ringwald, Schrempp and collaborators[2].

Instanton induced processes (I-events) in DIS arise predominantly from photon gluon fusion in an instanton background (see Fig. 1) via the reaction

$$\gamma^* + g \to \sum_{n_{\text{flavours}}} (\bar{q}_R + q_R) + n_g g,$$

where q_R ($\bar{q}_R$) denotes right handed quarks[b] and g gluons. The cross section is calculated[3] to be in the range 10–100 pb for $0.1 < y < 0.9$ and $x > 10^{-3}$, which is sizable but three orders of magnitude smaller than that of "normal" DIS events.

The final states in I-events are characterized by: a current-quark jet (q'', Fig. 1), a partonic final state from I-decay consisting of $2n_f - 1$ right-handed quarks and anti-quarks. In every I-induced event, one quark anti-quark pair of all $n_f (= 3)$ flavours is simultaneously produced. In addition, on average $\langle n_g \rangle^{(I)} \sim \mathcal{O}(1/\alpha_s) \sim 3$ gluons are

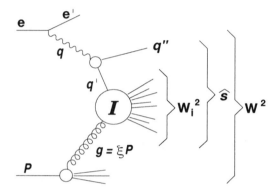

Figure 1. Dominant graph for the instanton-induced contribution to deep-inelastic ep scattering with the relevant kinematic variables as indicated: $Q'^2 = -q'^2 = -(q - q')^2$, $x' = Q'^2/2(g \cdot q')$, $W_i^2 = (q' + g)^2 = Q'^2(1 - x')/x'$.

isotropically emitted in the I-process. I-events are thus expected to show a pseudo-rapidity (η) region (with a width of about 1.1 units) densely populated with particles of high transverse momentum and uniformly distributed in azimuth. This, together with the high density of partons emitted in the I-process leads to a high particle multiplicity and large transverse energy. To simulate QCD-instanton induced scattering processes in DIS and their characteristic final states, the QCDINS Monte Carlo generator was developed.[4] It is based on instanton perturbation theory and imbedded in HERWIG.[5]

---

[a] See papers 250–252, 254, 255 submitted to this Conference.

[b] Right handed quarks are produced in instanton processes, left-handed ones in anti-instanton processes.

2 H1 results

The preliminary H1 results presented here use data taken in 1997 with the H1 detector, corresponding to an integrated luminosity of $\mathcal{L} = 15.78$ pb^{-1}. The analysis is performed in the DIS kinematic region $0.1 < y_{el} < 0.6$, $x_{el} > 10^{-3}$ and $\theta_{el} > 156°$. The total DIS sample comprises ~ 280K events.

The search strategies aim to enrich a data sample in I-induced events using cuts on selected observables while optimizing the separation power, defined as the ratio of the detection efficiencies, for I-induced and DIS-events.

The following observables[c] have been used: (1) Et_{jet}, the jet with highest E_T (cone algorithm with radius $R = 0.5$). This jet is associated with the current quark (q") in Fig. 1. (2) The virtuality of the quark entering the I-process $Q'^2 = -(q - q")^2$ where the photon (4-momentum q) is reconstructed from the scattered electron. (3) The number of charged particles n_B in the so-called "instanton band".[d] (4) The sphericity Sph calculated in the rest system of the particles outside the current jet. (5) Et_b the total transverse energy in the instanton band calculated as the scalar sum of the transverse energies and (6) Δ_b,[e] a quantity measuring the E_T weighted Φ event isotropy.

Among many studied, three scenarios are chosen based on the following criteria: (A) The highest instanton efficiency ($\epsilon_{ins} \approx 30\%$), (B) high ϵ_{ins} with reasonable background reduction and (C) highest back-

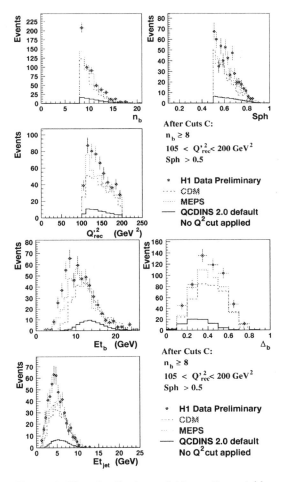

Figure 2. The distributions of kinematic variables in cut-scenario C, compared with the Color Dipole Model (CDM)[6], MEPS[7] and QCDINS.

ground reduction ($\epsilon_{DIS} \approx 0.13 - 0.16\%$) with $\epsilon_{ins} \approx 10\%$.

Table 1 summarizes the number of events in data and expected in the standard DIS Monte Carlo simulations after applying cuts (A)–(C). Distributions after cuts (C) are shown in Fig. 2. An excess of events over DIS Monte Carlos is observed. However, the size of this potential signal is comparable with the difference between CDM and MEPS. Also, the excess in transverse energy Et_b in the instanton band differs from the QCDINS expectation. Nevertheless, given the uncertainties in the I-event cross section calculation

---

[c]All observables, except sphericity, are calculated in the hadronic CMS ($\vec{q} + \vec{P} = \vec{0}$).

[d]Particles belonging to the jet q" are removed from the final state and the E_T-weighted mean pseudorapidity $\bar{\eta}$ is recalculated with the remaining ones. The instanton band is defined as $\bar{\eta} \pm 1.1$.

[e]Δ_b is defined as $\Delta_b = \frac{E_{in} - E_{out}}{E_{in}}$ where E_{in} (E_{out}) is the maximal (minimal) value of the sum of the projections on all possible axes $\vec{i}$ of all energy depositions in the band (i.e. $E_{in} = \max \sum_n |\vec{p_n} \vec{i}|$). For isotropic events (jet-like events) Δ_b is expected to be small (large).

Table 1. Measured numbers of events and expected background for 3 cut scenario's. The errors are systematics dominated.

(A) DATA: 3000		(B) DATA: 1332		(C) DATA: 549	
CDM	MEPS	CDM	MEPS	CDM	MEPS
2469^{+242}_{-238}	2572^{+237}_{-222}	1005^{+82}_{-70}	1084^{+75}_{-46}	363^{+22}_{-26}	435^{+36}_{-22}

and the modelling of the hadronic final state, an I-signal of the form predicted by QCDINS cannot be excluded at this stage of the analysis.

H1 has hence derived cross section upper limits (95% CL) by comparing the QCDINS predicted cross section, data and CDM/MEPS expectations. Results are shown in Fig. 3. Regions above the curves are excluded[f]. Instanton cross sections between 100 and 1000 pb are excluded.

Acknowledgements

I am grateful to T. Carli, S. Mikocki, A. Ringwald and F. Schrempp for help and valuable comments.

References

1. I. Balitsky, V. Braun, Phys. Lett. B438 (1993) 237.
2. S. Moch, A. Ringwald, F. Schrempp, *Nucl. Phys.* B507 (1997) 134; A. Ringwald, F. Schrempp, *Phys. Lett.* B438 (1998) 217; A. Ringwald, F. Schrempp, *Phys. Lett.* B459 (1999) 249; A. Ringwald, F. Schrempp, hep-ph/9411217, in Quarks '94, eds. D. Yu. Grigoriev et al., World Scientific, Singapore 1995; T. Carli, J. Gerigk, A. Ringwald, F. Schrempp, DESY 00-067, MPI-PhE/99-02, hep-ph/9906441, in Monte Carlo Generators for HERA Physics, eds. A. T. Doyle et al.
3. A. Ringwald, F. Schrempp, 8th International Workshop on Deep Inelastic Scattering (DIS2000), Liverpool, UK, June 2000, hep-ph/0006215.

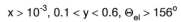

$x > 10^{-3}, 0.1 < y < 0.6, \Theta_{el} > 156°$

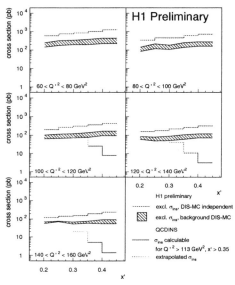

Figure 3. Upper limit on the cross section for instanton induced events as modelled by QCDINS as a function of x' in bins of Q'^2. Regions above the curves are excluded. Also shown is the instanton cross section predicted in the fiducial region $x' > 0.35$ and $Q'^2 > 113$ GeV^2.

4. A. Ringwald, F. Schrempp, *Comp. Phys. Comm.* (1999) in print; hep-ph/9911516.
5. G. Marchesini et al., *Comp. Phys. Comm.* **67** (1992) 465.
6. L. Lönnblad, *Comp. Phys. Comm.* **71** (1994) 15 .
7. H. Jung, *Comp. Phys. Comm.* **86** (1995) 147.
8. H1 Collab., *Search for QCD Instanton Induced Events in Deep-Inelastic Scattering at HERA*, paper 307, this conference.

---

[f] For a detailed description of the methods used, the reader should consult the original H1 paper. [8].

COMBINED RECOIL AND THRESHOLD RESUMMATION FOR HARD SCATTERING CROSS SECTIONS

ERIC LAENEN

NIKHEF Theory Group, Kruislaan 409
1098 SJ Amsterdam, The Netherlands

GEORGE STERMAN

C.N. Yang Institute for Theoretical Physics, SUNY Stony Brook
Stony Brook, New York 11794 - 3840, U.S.A.

WERNER VOGELSANG

RIKEN-BNL Research Center, Brookhaven National Laboratory,
Upton, NY 11973, U.S.A.

We discuss the simultaneous resummation of threshold and recoil enhancements to partonic cross sections due to soft radiation. Our method is based on a refactorization of the parton cross section near its partonic threshold. It avoids double counting, has next-to-leading logarithmic accuracy, conserves the flow of partonic energy and reproduces either threshold or recoil resummation when the other enhancements are neglected.

1 Introduction

A large class of hard-scattering cross sections in QCD may be factorized into convolutions of parton distributions and fragmentation functions with hard-scattering functions [1]. The latter are computed in perturbation theory. Intermediate infinities associated with virtual and emitted soft gluons cancel in their higher order corrections, but finite remnants, assuming the form of plus-distributions and delta-functions, may lead to large enhancements when integrated against the smooth functions in the convolutions. In physical terms, soft gluon radiation affects the hard-scattering process by reweighting the relative importance of near-threshold partonic subprocesses to the physical cross section, and by providing the final state with overall transverse momentum Q_T through its recoil. When the conservation of energy is taken into account, the sign of the cumulative effect is *a priori* unclear: enhancement effects of recoil may be offset by the suppression due to extra energy required

to produce the desired final state plus recoil radiation. A combined and consistent analysis is required [2] and in what follows we sketch our approach to this [3].

2 Electroweak annihilation

These processes are characterized at lowest order by $ab \to V$, with a, b partons and V an electroweak final state of mass Q^2 and transverse momentum $\mathbf{Q}_T$. Because of their relative simplicity these processes have been studied intensely for resummation purposes. Partonic threshold is at $z \equiv Q^2/\hat{s} = 1$ and $\mathbf{Q}_T = 0$, and the singular functions to be resummed are plus-distributions in $1 - z$ and Q_T/Q. Our method generalizes that of Ref. [4] and organizes these distributions in the cross section into field-theoretically defined matrix elements of appropiate operators,

$$\frac{d\sigma_{ab \to V}}{dQ^2 d^2\mathbf{Q}_T} = \frac{1}{S}\, \sigma^{(0)}_{ab \to V}(Q^2)\, h^{(j)}_{ab}(\alpha_s(Q^2))$$

$$\times \int dx_a d^2\mathbf{k}_a\, R_{a/a}(x_a, \mathbf{k}_a, Q)$$

$$\times \int dx_b d^2\mathbf{k}_b \, R_{b/b}(x_b, \mathbf{k}_b, Q)$$

$$\times \int dw_s d^2\mathbf{k}_s \, U_{ab}(\beta, \beta', w_s, Q, \mathbf{k}_s)$$

$$\times \delta(1 - Q^2/S - (1 - x_a) - (1 - x_b) - w_s)$$

$$\times \delta^2\left(\mathbf{Q}_T - \mathbf{k}_a - \mathbf{k}_b - \mathbf{k}_s\right) + Y_j, \qquad (1)$$

where

$$R_{f/f}(x, \mathbf{k}, 2p_0) = \frac{1}{4\sqrt{2}N_c} \int \frac{d\lambda}{2\pi} \frac{d^2\mathbf{b}}{(2\pi)^2} \times \quad (2)$$

$$e^{-i\lambda x p_0 + i\mathbf{b}\cdot\mathbf{k}} \langle f(p)|\bar{q}_f(\lambda, \mathbf{b}, 0)\gamma^+ q_f(0)|f(p)\rangle$$

is a partonic quark density at fixed energy and transverse momentum, and U_{ab} a purely eikonal function, depending on parton velocities β, and on the soft radiation's energy $w_s Q$. The x's and $\mathbf{k}$'s are defined by (2). The delta functions in the triple convolution of (1) relate the singular behavior of the various functions, and Y_j represents the matching to finite order. The all-order behavior of the above functions and the cross section can be analyzed through their eikonalized equivalents. The eikonal cross section can be expressed as an exponent E_{ab} of an integral over "webs" [5] w_{ab}

$$E_{ab} = 2 \int^Q \frac{d^{4-2\epsilon}k}{\Omega_{1-2\epsilon}}$$

$$\times w_{ab}\left(k^2, \frac{k\cdot\beta k\cdot\beta'}{\beta\cdot\beta'}, \mu^2, \alpha_s(\mu^2), \epsilon\right)$$

$$\times \left(e^{-N(k_0/Q)+i\mathbf{k}\cdot\mathbf{b}} - 1\right). \qquad (3)$$

The k^μ in the integral is the momentum (including energy and transverse momentum) contributed by the web to the final state, and it is in this way that the joint resummation for electroweak annihilation is achieved.

3 Single-particle inclusive

For definiteness we consider prompt photon production, but the results below are more general [3]. A similar refactorization as in (1) holds for single-particle inclusive cross sections. In particular, it contains the same $R_{f/f}$ functions. The arguments supporting

this refactorization are somewhat more involved than for (1), but reveal that only initial state radiation contributes to Q_T (the transverse momentum of the $2 \to 2$ parton cms frame) if no jet-definition is applied to the final state radiation, which can thus be treated in threshold resummation only. In contrast to section 2, Q_T is here an unobserved variable, and thus more akin to $1 - z$. Thus, the jointly resummed prompt photon p_T spectrum may be written as an integral over $\mathbf{Q}_T$ of a "profile function"

$$P_{ij}(N, \mathbf{Q}_T, Q) = \int d^2\mathbf{b} \, e^{-i\mathbf{b}\cdot\mathbf{Q}_T} e^{E_{ij\to\gamma k}} \quad (4)$$

where the exponential exhibits the joint resummation, as

$$\frac{p_T^3 d\sigma_{AB\to\gamma}^{(\text{resum})}}{dp_T} =$$

$$\times \sum_{ij} \frac{p_T^4}{8\pi S^2} \int_{\mathcal{C}} \frac{dN}{2\pi i} \, \tilde{\phi}_{i/A}(N)\tilde{\phi}_{j/B}(N)$$

$$\times \int_0^1 d\tilde{x}_T^2 \, (\tilde{x}_T^2)^N \, \frac{|M_{ij}(\tilde{x}_T^2)|^2}{\sqrt{1 - \tilde{x}_T^2}}$$

$$\times C_\delta^{(ij\to\gamma k)}(\alpha_s, \tilde{x}_T^2) \int \frac{d^2\mathbf{Q}_T}{(2\pi)^2} \, \Theta\left(\bar{\mu} - Q_T\right)$$

$$\times \left(\frac{S}{4\mathbf{p}_T'^2}\right)^{N+1} P_{ij}(N, \mathbf{Q}_T, Q) \quad (5)$$

with $\tilde{x}_T^2 = 4|\mathbf{p}_T - \mathbf{Q}_T/2|^2/\hat{s}$, $\tilde{\phi}_{i/A}(N)$ Mellin moments of the parton distributions, $|M_{ij}|^2$ the Born amplitudes, and $\bar{\mu}$ a cut-off restricting $\mathbf{Q}_T$ to sufficiently small values for resummation to be relevant. For notational simplicity we have suppressed all factorization and renormalization scale dependence. Cross sections computed on the basis of (5) are shown in Fig. 1 as functions of Q_T at fixed p_T. The kinematics are those of the E706 experiment [7], see [3] for details of the calculation, in particular those regarding the evaluation of the b-integral in (4). The dashed lines are $d\sigma_{\text{pN}\to\gamma X}^{(\text{resum})}/dQ_T dp_T$, with recoil neglected by fixing $\mathbf{p}_T' = \mathbf{p}_T$, thus showing how each Q_T contributes to threshold enhancement. The solid lines show the same, but now including the true recoil factor $(S/4\mathbf{p}_T'^2)^{N+1}$. The

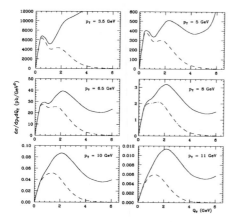

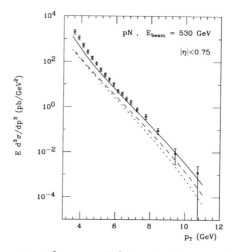

Figure 1. $d\sigma_{\mathrm{pN}\to\gamma X}/dQ_T dp_T$ at $\sqrt{s} = 31.5$ GeV, as a function of Q_T for various values of photon p_T. Dashed lines are computed without recoil ($\mathbf{p}'_T = \mathbf{p}_T$ in (5)), solid lines are with recoil.

Figure 2. $Ed^3\sigma_{\mathrm{pN}\to\gamma X}/dp^3$ for pN collisions at $\sqrt{s} = 31.5$ GeV. The dotted line represents the full NLO calculation, while the dashed and solid lines respectively incorporate pure threshold resummation [6] and the joint resummation described in this paper. Data have been taken from [7].

resulting enhancement is clearly substantial. For small p_T, the enhancement simply grows with Q_T, while for p_T above 5 GeV it has a dip at about $Q_T = 5$ GeV, which remains substantially above zero. This makes it difficult to confidently determine $\bar{\mu}$.

So far, the numerical results given in [3] are primarily to be regarded as illustrations, rather than quantitative predictions. This applies in particular to the resummed Q_T-*integrated* cross section, shown in Fig. 2 for $p_T \geq 3.5$ GeV and $\bar{\mu} = 5$ GeV. These figures demonstrate the size of the additional enhancement that recoil can produce and its potential phenomenological impact.

work.

Acknowledgements

The work of G.S. was supported in part by the National Science Foundation, grant PHY9722101. The work of E.L. is part of the research program of the Foundation for Fundamental Research of Matter (FOM) and the National Organization for Scientific Research (NWO). W.V. is grateful to RIKEN, Brookhaven National Laboratory and the U.S. Department of Energy (contract number DE-AC02-98CH10886) for providing the facilities essential for the completion of this

References

1. J.C. Collins, D.E. Soper and G. Sterman, in *Perturbative Quantum Chromodynamics*, ed. A.H. Mueller (World Scientific, Singapore, 1989).

2. H.-n. Li, *Phys. Lett.* **B454**, 328 (1999), hep-ph/9812363.

3. E. Laenen, G. Sterman and W. Vogelsang, *Phys. Rev. Lett.* **84**, 4296 (2000), hep-ph/0002078 and in preparation.

4. G. Sterman, *Nucl. Phys.* **B281**, 310 (1987); S. Catani and L. Trentadue, *Nucl. Phys.* **B327**, 323 (1989), **B353**, 183 (1991).

5. J.G.M. Gatheral, *Phys. Lett.* **B133**, 9 (1983).

6. S. Catani, M.L. Mangano, P. Nason, C. Oleari and W. Vogelsang, JHEP **9903** (1999) 025, hep-ph/9903436.

7. L. Apanasevich *et al.*, E706 Collab., *Phys. Rev. Lett.* **81**, 2642 (1998), hep-ex/9711017.

STUDY OF HADRONIC DECAYS OF THE Z BOSON AT LEP

ALESSANDRO DE ANGELIS

Dipartimento di Fisica dell'Università di Udine and INFN Trieste, Via delle Scienze 208,
I-33100 Udine, Italy – E-mail: deangelis@ud.infn.it

This report summarizes four recent papers on the characteristics of the hadronic decays of the Z by the LEP collaborations ALEPH, DELPHI and OPAL.

1 Introduction

Each of the four LEP experiments has recorded around 4 million hadronic Z decays, mostly in the period between 1990 and 1995. The analysis of the main physics topics on this subject has been essentially completed, but a few specific points are still under investigation, especially in the QCD sector.

This report summarizes four recent papers on the characteristics of the hadronic decays of the Z, submitted to this conference by the LEP collaborations OPAL[1], ALEPH[2,3] and DELPHI[4].

2 Charged multiplicities in Z decays into u, d, and s quarks

Flavour independence is a fundamental property of QCD: a breaking of the flavour symmetry should only occur due to (calculable) mass effects. An observable for testing flavour independence is the multiplicity of charged hadrons in jets originating from quarks of a specific flavour. For the light (u, d, and s) quarks the mass effects are expected to be negligible at LEP energies[5].

OPAL[1] presents a new, high statistics investigation of the flavour dependence of the strong interaction based on the mean charged multiplicity determined separately for events of primary u, d, s quarks in hadronic Z decays.

In order to identify the flavour of the primary quark, the leading particle effect is exploited. This effect is based on the correlation between the flavour of the primary quark and the type of the hadron carrying the largest momentum. Three different selections are based on K^0_S, $K^\pm$ and highly energetic charged particles. Leading K^0_S and $K^\pm$ are likely to contain a primary s quark from the Z decay. To a lesser extent, due to the requirement of $s\bar{s}$ quark pair creation during hadronization, leading K^0_S and $K^\pm$ should also emerge from primary d and u quarks, respectively. In a sample with large momentum (non identified) charged particles, events of primary u, d and s quarks are expected to be found in approximately equal proportions.

By requiring a K^0_S with momentum fraction $x_p = p/p_{beam}$ larger than 0.4, 19,359 events are selected with 58% Z$\rightarrow s\bar{s}$ and 16% Z$\rightarrow d\bar{d}$. By requiring a $K^\pm$ with $x_p > 0.5$, 18,979 events are selected with 54% Z$\rightarrow s\bar{s}$ and 22% Z$\rightarrow u\bar{u}$. By requiring a charged particle with $x_p > 0.7$, 27,909 events are selected almost equally populated in $u\bar{u}$, $d\bar{d}$ and $s\bar{s}$, but with only 4% heavy quarks.

The presence of a large momentum particle biases the multiplicity. To reduce this bias, the multiplicity measurement was performed in the hemisphere opposite to the one containing the particle used for the tagging. The final corrected multiplicities are

$$< n >_{u\bar{u}} = 17.77 \pm 0.52(stat)^{+0.86}_{-1.20}(sys)$$
$$< n >_{d\bar{d}} = 21.44 \pm 0.69(stat)^{+1.46}_{-1.17}(sys)$$
$$< n >_{d\bar{s}} = 20.02 \pm 0.14(stat)^{+0.39}_{-0.37}(sys),$$

where the $u\bar{u}$ and $d\bar{d}$ multiplicities are statistically anticorrelated (about -89%). The systematic error is dominated by uncertainties on the fragmentation. The multiplicities are consistent in 1.8σ for the ud case, in 1.5σ for us, and in 0.9σ for ds. The world average of the charge multiplicity[6] (for all flavours) is 21.07 ± 0.11.

The multiplicities were transformed into α_S ratios using a NLLA calculation[7]. Ratios of α_S for the light quark flavours were found consistent with 1 at a precision of 5 to 9%, superior to earlier investigations.

3 Results on identified hadrons

The description of the hadronization process in QCD is connected with confinement and requires nonperturbative methods which are not available. Measurements of identified hadron spectra in e^+e^- improve the understanding of hadronization, and allow tuning the free parameters in the QCD inspired Monte Carlo models. Being closer to the main event, vectors and higher order resonances are of particular interest.

More insights into the hadronization process may be obtained from the analysis of quark and gluon jets separately.

3.1 Production of $\omega(782)$

ALEPH[2] presents a measurement of the inclusive momentum distributions of the $\omega(782)$ meson. The ω cross section is extracted using the invariant mass distribution of $\pi^+\pi^-\pi^0$ triplets.

The average ω rate per event is

$$< \omega > = 0.996 \pm 0.032(stat) \pm 0.056(sys),$$

below the predictions of JETSET[8] 7.4 (1.31) and HERWIG[9] 6.1 (1.17) over the full momentum spectrum. This measurement improves substantially the world average[6] of 1.08 ± 0.09.

The muonic branching fraction of the $\omega(782)$ is measured for the first time: $BR(\omega \to \mu^+\mu^-) = (9.8 \pm 2.9 \pm 1.1) \times 10^{-5}$.

3.2 Production of π^0, η, $\eta'(958)$, K_S^0 and Λ in 2- and 3-jet events

For isoscalar mesons (η, $\eta'(958)$, $\omega(782)$, $\phi(1020)$), some theoretical models predict a special enhancement in gluon jets compared to quark jets. L3[10] found that the measured momentum spectrum in gluon jets is harder than in HERWIG and JETSET.

In a new analysis, ALEPH[3] measures the production rates of π^0, η, $\eta'(958)$, K_S^0 and Λ in hadronic events, two-jet events and each jet of three-jet events.

Jets are clustered using Durham[11] with $y_{cut} = 0.01$; this classifies 31% of the events as 3-jet. The lowest energy jet originates from a gluon with 71% probability.

The π^0 and η mesons are analyzed using the $\gamma\gamma$ decay channel, and the η' using $\eta' \to \eta\pi^+\pi^-$. K_S^0 are reconstructed from their decay into $\pi^+\pi^-$, and Λs into $p\pi^-$.

The spectra for π^0 are reasonably reproduced by JETSET and HERWIG. η and $\eta'(958)$ are well reproduced by JETSET for quark jets and gluon jets. Therefore, the JETSET description of gluon fragmentation into isoscalar mesons is in agreement with the experiment. HERWIG shows a too steep dependence on x_p for η spectra in two-jet events and each jet of three-jet events.

The spectra for K_S^0 and Λ are well reproduced by JETSET and ARIADNE[12], while HERWIG fails.

4 Rapidity-rank structure of $p\bar{p}$

The baryon sector is not well modeled by current QCD–inspired Monte Carlos; only the string fragmentation model JETSET can reproduce the octet and decuplet rates.

In JETSET hadrons results from breaks in the string stretched between the primary quarks. Baryons are formed when diquark-antidiquark pairs are created, and couple to an adjacent (anti)quark. A mechanism to attenuate the strict rapidity ordering coming from this mechanism is the so-called *pop-*

corn: a meson "pops up" inside a diquark–antidiquark pair[8].

To study the relative occurrence of popcorn, DELPHI[4] proposes a novel method based on the rapidity–rank structure of $p\bar{p}$ pairs, where rapidity is defined with respect to the thrust axis.

Proton identification is provided by Cherenkov angle measurement from the RICH and ionization energy loss in the TPC. The number of events with one p and one $\bar{p}$ in each hemisphere is 27.6 thousand. The background to this event sample can be determined from the number of events that have two p's or two $\bar{p}$'s in a given hemisphere. These events, 10.1 thousand, result mainly from misidentifications, but also from noncorrelated baryon pairs, and they are subtracted from the signal.

The charged particles in each event are ordered in rapidity. Two types of rapidity-rank configurations for $p\bar{p}$ pairs are considered: (a) when the p and $\bar{p}$ are adjacent in rapidity; (b) when the p and $\bar{p}$ have one or more mesons between them. The ratio $R = N(pM\bar{p})/(N(p\bar{p}) + N(pM\bar{p})$ is calculated, where $N(p\bar{p})$ and $N(pM\bar{p})$ represent the number of rapidity-rank configurations of each type in the data sample, and are implicitly a function of the minimum distance Δy_{min} between a baryon in the pair and a meson. The probability that a given rapidity configuration will represent the actual rank order on the string is enhanced as Δy_{min} is made larger.

The ratio $R(\Delta y_{min})$ for data is plotted in Figure 1, and compared to the predictions in the cases of no popcorn and 100% popcorn. The data are consistent with no contribution from $pM\bar{p}$ configurations. An upper limit contribution of 15% is determined at 90% confidence level.

Previous studies of the $\Lambda\bar{\Lambda}$ rapidity difference have claimed evidence for popcorn[13]. This might indicate the importance of dynamical effects not incorporated in JETSET

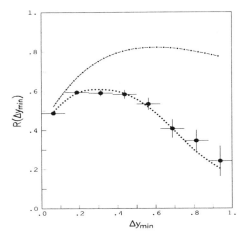

Figure 1. The relative amount, $R(\Delta y_{min})$, of the $pM\bar{p}$ configuration as a function of Δy_{min}, compared to the JETSET predictions in case of no popcorn (lower line) and full popcorn (upper line). Due to normalization, Δy_{min} is multiplied by 2/3 for $p\bar{p}$.

or simply the inadequacy of the popcorn model, although no firm conclusion can be drawn yet.

References

1. OPAL, PN396 (July 1999), contribution 105 to this conference.
2. ALEPH, note 2000-049 (June 2000), contribution 163 to this conference.
3. ALEPH, CERN-EP/99-105 (July 1999), contribution 161 to this conference.
4. DELPHI, CERN-EP/00-81 (June 2000), contribution 641 to this conference.
5. A. de Angelis, in Proc. of ICHEP 1995
6. PDG, Review of Particle Physics 2000
7. B. Webber, Phys. Lett. B143 (1984) 501
8. T. Sjöstrand, Comp. Phys. Comm. 82 (1994) 74
9. G. Marchesini et al., Comp. Phys. Comm. 67 (1992) 465
10. L3, Phys. Lett. B371 (1996) 126
11. S. Catani et al., Phys. Lett. B269 (1991) 432
12. L. Lönnblad, Comp. Phys. Comm. 71 (1992) 15
13. See the references in [4].

STUDIES OF HADRONIC DECAYS OF Z^0 BOSONS AT SLD

P.N. BURROWS

Particle Physics, Oxford University, Keble Rd., Oxford, OX1 3RH, UK
E-mail: p.burrows@physics.ox.ac.uk
(Representing the SLD Collaboration)

The latest SLD results on light-quark fragmentation and tests of hadronisation models are presented.

We have a poor understanding of hadronisation in terms of quantitative predictions from QCD. Phenomenological models provide useful insights, but no model accounts successfully for all the observed features of jet fragmentation. Heavy jet fragmentation is relatively well understood; the quark mass provides a cutoff against divergences, allowing pQCD calculations to be performed. Also, the decay signatures of B or D hadrons allow high-purity b or c jet samples to be identified with high efficiency. The study of u, d or s-jet fragmentation is less tractable both theoretically and experimentally: the quark masses are comparable with Λ_{QCD}, leading (and non-leading) light hadrons (π, K, p, ...) can be produced in jets of *any* flavour, making the isolation of u,d or s-jets problematic; a dedicated particle i.d. system is required to identify these light hadron species.

The clean $Z^0 \rightarrow$ u$\bar{\text{u}}$, d$\bar{\text{d}}$, s$\bar{\text{s}}$, c$\bar{\text{c}}$, b$\bar{\text{b}}$ events provided by SLC are ideal for quark fragmentation studies; updated results presented here are based on the complete sample of 550k Z^0 decays recorded by SLD. The unique CCD vertex detector allows high-purity b/c/uds/g jet separation with good efficiency. The unique polarised SLC electron beam allows quark/antiquark jet separation with a purity of 73% and 100% efficiency. The Cerenkov Ring Imaging Detector allows $\pi^\pm$/K$^\pm$/p$^\pm$ separation within $0.5 \leq p \leq 35$ GeV/c.

Our light-hadron fragmentation functions in inclusive-flavour jets[1] are shown in Fig. 1; the momentum coverage spans the

kinematic range in Z^0 decays. The dependence on primary jet flavour is illustrated in Fig. 2, in terms of $\xi = \ln(1/x_p)$; the peak position and height depend strongly on jet flavour, implying that light and heavy jets should be separated before performing pQCD fits to these distributions[1].

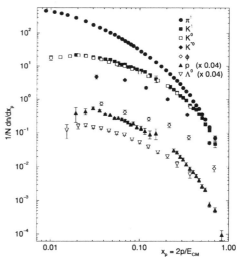

Figure 1. Light hadron fragmentation functions.

The ratio of heavy- to light-jet $\pi^\pm$, K$^\pm$ or p$^\pm$ fragmentation functions is compared with model predictions[2] in Fig. 3; HERWIG, in particular, has difficulty reproducing these ratios. Hadron (h) and antihadron ($\bar{\text{h}}$) fragmentation functions in light-*quark* jets[2] are shown separately in Fig. 4; the excess of h over $\bar{\text{h}}$ at high x_p is an indication of leading-particle production.

We have studied correlations in rapidity (y) between pairs of hadron species[3]. The excess of opposite-charge over like-charge (high-p) pairs at small Δy (Fig. 5) indicates lo-

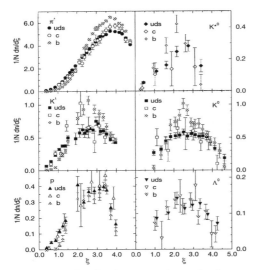

Figure 2. ξ distributions vs. primary jet flavour.

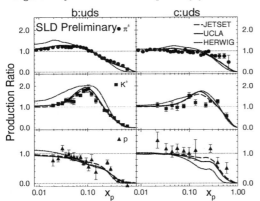

Figure 3. Flavour ratios of x_p distributions.

cal charge compensation in the fragmentation process. The excess at large Δy (most noticeable for the $K^{\pm}$) is further evidence of leading-particle production. We observe[3] a significant hadron-species and jet-flavour dependence of these long-range correlations. We have defined a new observable, the rapidity distribution for the case in which the thrust axis is signed to point in the *quark* direction. These distributions for h and $\bar{\text{h}}$ are noticeably different, indicating a preference for high-rapidity h ($\bar{\text{h}}$) to point along (against) the quark direction, again an indicator of leading particles. This effect can be enhanced by considering the distribution of the charge-ordered difference in rapidity between opposite-sign and like-sign pairs of $\pi^{\pm}$,

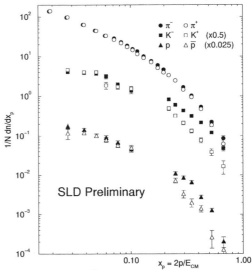

Figure 4. h vs. $\bar{\text{h}}$ production in light-quark jets.

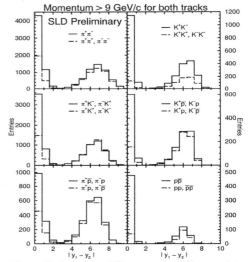

Figure 5. Rapidity differences in all-flavour jets.

$K^{\pm}$ or $p^{\pm}$; details are given in ref.[3].

We have exploited the overwhelming evidence in our data for a leading strange-particle effect to measure directly the EW coupling of the s-quark to the Z^0, via the polarised forward-backward asymmetry of strange-hadron production[4]. We find $A_s = 0.895 \pm 0.066 \pm 0.062$.

References

1. SLD, *Phys. Rev.* D**59**, 052001 (1999).
2. K. Abe *et al.*, SLAC-PUB-8507 (2000).
3. K. Abe *et al.*, SLAC-PUB-8508 (2000).
4. K. Abe *et al.*, SLAC-PUB-8408 (2000).

W PRODUCTION AND DECAY STUDIES AT THE TEVATRON

SUYONG CHOI

(for the CDF and DØ Collaborations)

Department of Physics, University of California, Riverside CA 92521, USA

E-mail: suyong@fnal.gov

We present some of the recent results from CDF and DØ on electronic branching ratio of the W, W width, and the lepton angular distribution from W decays based on data sets collected during the 1991-95 run ("Run 1b"), with integrated luminosities of 85–95 pb^{-1} per experiment.

1 Measurement of the Electronic Branching Ratio of W and W Width

The results on the W and Z production cross sections times branching ratios from CDF and DØ are shown in Fig. 1. CDF measure $\sigma_W \cdot R(W \to e\nu) = 2490 \pm 80(\text{stat} \oplus \text{syst}) \pm 90(\text{lum})$ pb and $\sigma_Z \cdot R(Z \to e^+e^-) = 249 \pm 5(\text{stat} \oplus \text{syst}) \pm 10(\text{lum})$ pb [1], where "lum" is due to uncertainty on the integrated luminosity. DØ obtain $\sigma_W \cdot R(W \to e\nu) = 2310 \pm 10(\text{stat}) \pm 50(\text{syst}) \pm 100(\text{lum})$ pb and $\sigma_Z \cdot R(Z \to e^+e^-) = 221 \pm 3(\text{stat}) \pm 4(\text{syst}) \pm 10(\text{lum})$ pb [2]. [a]

The integrated luminosity uncertainty and many other systematic uncertainties cancel in the ratio of cross sections: $R = \sigma_W \cdot B(W \to e\nu)/\sigma_Z \cdot B(Z \to e^+e^-)$. This allows a precise measurement of the electronic branching ratio of the W boson and the width of the W ($\Gamma(W)$). This follows using

$$\frac{\sigma_W \cdot B(W \to e\nu)}{\sigma_Z \cdot B(Z \to e^+e^-)} = \frac{\sigma_W}{\sigma_Z} \frac{1}{B(Z \to e^+e^-)} \frac{\Gamma(W \to e\nu)}{\Gamma(W)}$$

together with the theoretical calculation of σ_W/σ_Z [3], the measured $Z \to e^+e^-$ branching

---

[a] The CDF and DØ results must be compared with care, since the experiments use different total $p\bar{p}$ cross sections to determine their integrated luminosities. CDF use their own measurement, while DØ take the average of the CDF, E710 and E811 measurements. If the CDF normalization is used, the DØ Run 1b cross sections must be multiplied by 1.062. Note that the results in Fig. 1 have not been rescaled.

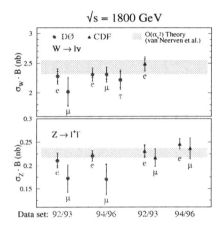

Figure 1. W and Z production cross section measurements at $\sqrt{s} = 1.8$ TeV

ratio from LEP [4], and the Standard Model value of $\Gamma(W \to e\nu)$ [5].

The measured values of R are $10.49 \pm 0.14(\text{stat}) \pm 0.21(\text{syst})$ for DØ and $10.38 \pm 0.14(\text{stat}) \pm 0.17(\text{syst})$ for CDF, using the combined electron data from Runs 1a and 1b. The main sources of systematic errors are due to uncertainties in backgrounds, efficiencies, and electron energy scale. A 1% error due to NLO electroweak radiative corrections is also included. The two R measurements are combined to yield $R = 10.42 \pm 0.18$, and the resulting branching fraction is $B(W \to e\nu) = (10.43 \pm 0.25)\%$. The W width is $\Gamma(W) = 2.171 \pm 0.052$ GeV, and it agrees with the SM prediction $\Gamma(W) = 2.0927 \pm 0.0025$ GeV[6] when the error is taken into account.

2 The Angular Distribution of Electrons in W Boson Decays at DØ

NLO perturbative QCD predicts an angular distribution of $1 \pm \alpha_1 \cos \theta^* + \alpha_2 \cos^2 \theta^*$ for the leptons in W decays, where θ^* is the polar angle in Collins-Soper frame. QCD corrections impart transverse momentum to W bosons and modify the helicity of W bosons produced. The α_1 and α_2 become functions of the transverse momentum of the W boson (P_T^W). This measurement provides a test of NLO QCD corrections which have a non-negligible contribution to the W mass measurement. DØ does not identify the charge of the electron due to lack of magnetic field in the tracking region and information about α_1 is lost.

The transformation from the lab frame to the W rest frame (Collins-Soper frame) is not calculable, since only the transverse components of the neutrino momentum are measured. Therefore, the polar angle of the electron from W decay, θ^*, is not directly measurable. However, θ^* can be inferred from correlation between the transverse mass(M_T^W) and $\cos \theta^*$ through Bayes theorem. The probability $g(M_T^W \,|\, \cos \theta^*)$ of measuring M_T^W for a given value $\cos \theta^*$ in a given P_T^W bin is inverted to give the probability of measuring $\cos \theta^*$ for a measured M_T^W, $f(\cos \theta^* | M_T^W)$, using Bayes theorem:

$$
\begin{aligned}
&f(\cos\theta^* | M_T^W) \\
&= \frac{g(M_T^W \,|\, \cos\theta^*) h(\cos\theta^*)}{\int g(M_T^W \,|\, \cos\theta^*) h(\cos\theta^*) d(\cos\theta^*)}
\end{aligned}
$$

where $h(\cos \theta^*)$ is the prior probability function, taken to be $h(\cos \theta^*) = 1 + \cos^2 \theta^*$, the expectation from $V - A$ theory without QCD modification.

A Monte Carlo simulation of the DØ detector is used to derive the probability function, $g(M_T^W \,|\, \cos \theta^*)$. The angular distribution is obtained in a given P_T^W bin by inverting g. With the unfolded angular distribu-

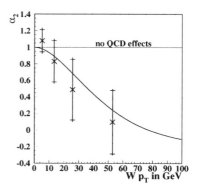

Figure 2. α_2 as a function of P_T^W compared to NLO QCD(curve) and calculation in the absence of QCD corrections (horizontal line). The vertical bars denote the total errors while the statistical errors are marked by horizontal ticks.

tions calculated, α_2 can be extracted for each P_T^W bin by comparing to samples of Monte Carlo for different values of α_2 through likelihood method. The results of this measurement along with the theoretical prediction is shown in Fig. 2. The QCD prediction is preferred by 2.3σ [7] over a $V - A$ theory without QCD corrections.

References

1. (CDF Collaboration), *Phys. Rev. Lett.* **76**, 3070 (1996); (CDF Collaboration), *Phys. Rev.* D **59**, 052002 (1999).
2. (DØ Collaboration), *Phys. Rev.* D **60**, 052003 (1999).
3. R. Hamburg, W.L. van Neerven, T. Matsuura, *Nucl. Phys.* B **359**, 343 (1991); W.L. van Neerven, E.B. Zijlstra, *Nucl. Phys.* B **382**, 11 (1992).
4. L. Montanet *et al.*, (Particle Data Group), *Phys. Rev.* D **54**, 1 (1996).
5. L. Rosner *et al.*, *Phys. Rev.* D **49**, 1363 (1994).
6. D.E. Groom *et al.*, (Particle Data Group), *Eur. Phys. J.* C **15**, 1 (2000).
7. G. Steinbrück, FERMILAB-CONF-00-094-E (2000).

STUDY OF FINAL STATE INTERACTIONS IN W^+W^- EVENTS AT LEP

NELLI PUKHAEVA*

University of Antwerp, Belgium. email: Nelli.Pukhaeva@cern.ch
* Visiting Post–doctoral Fellow of the FWO–Vlaanderen

Experimental results on colour reconnection effects and Bose-Einstein correlations in hadronic decays of WW events are reported. The data were collected by the DELPHI detector at LEP at energies between 183 and 202 GeV. An indication of correlations between like-sign particles coming from different Ws was observed.

1 Introduction

The possible presence of Bose-Einstein correlations (BEC) between decay products of different Ws and colour reconnection effects between partons of Ws in

$$e^+e^- \to W^+W^- \tag{1}$$

events has been discussed on a theoretical basis, in relation to the measurement of the W mass. These effects can induce a systematic uncertainty on the W mass measurement in the fully hadronic channel comparable with the expected accuracy of the measurement. The experimental study of colour reconnection and BEC in reaction (1) is also important for the understanding of QCD effects.

2 Colour Reconnection Effects

For the analysis of colour reconnection effects we used the charged particle multiplicity and inclusive distributions[1]. Most models predict that, in case of colour reconnection, the ratio between the multiplicity in fully hadronic WW events (4q) events and twice the multiplicity in semileptonic WW (2q) events would be smaller than 1; the difference is expected to be at the percentage level, and a deficit of the multiplicity is expected to be concentrated in the region of low momentum. Our analysis of DELPHI data shows that such deficit is not observed in the full momentum range, nor in the low-momentum intervals (Figure 1).

3 Correlations Between Particles

For the study of BEC we measured the enhanced probability for emission of two identical bosons, for which the correlation functions (R) as a function of the four–momenta difference between the particles(Q) were used. These correlation functions for like-sign particles were measured in hadronic Z decays, in semileptonic and in fully hadronic WW channels[2]. Evidence for Bose-Einstein correlations was observed in all three cases; the parameters λ and r characterizing the correlation in fully hadronic WW events agree with those in mixed hadronic-leptonic WW events, as well as in a sample of Z decays in which the contribution from $b\bar{b}$ pairs was depleted.

Averaged over all energies, the selected WW fully hadronic events contained 13% of $q\bar{q}$ events and 5% of ZZ events. The correction for these background contributions to the fully hadronic sample was done in two ways.

In the first way, the influence of the background events on $R(Q)$ in the (4q) channel was corrected by subtracting the Q-distributions for the $q\bar{q}$ and ZZ contributions from the experimental Q-distribution. The Q-distributions for the background events were estimated as follows. The Q-distributions of simulated $q\bar{q}$ and ZZ events without BEC which passed the WW (4q) selection criteria were multiplied by the Gaussian form with λ and r BEC parameters as

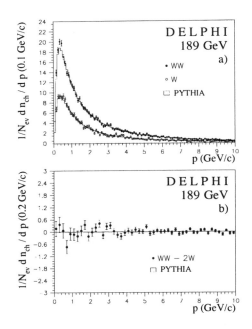

Figure 1. (a) momentum distributions for charged particles in WW semileptonic and fully hadronic events and (b) comparison of these distributions.

measured for selected high energy $q\bar{q}$ events and Z events.

For the second method to correct for the background, a sample of $q\bar{q}$ events was generated with BEC included according to the LUBOEI model with tuned parameters $\lambda=0.85$ and $r=0.5$ fm. These events were subjected to the same event and track selection criteria as the fully hadronic WW sample and the Q-distribution of the background was calculated from the events passing the selection. The obtained Q-distribution for background events was corrected for the discrepancy between the data and the simulation at Z-peak.

In the subsequent analyses the $R(Q)$ distributions for fully hadronic WW events after the background subtractions for both methods were used and the half the difference between the two methods was assigned as a systematic error due to the background correc-

tions.

Measurements were performed to extract correlations between pions from different Ws[2]. Analyses were made using comparison samples which contain only BEC for particle pairs coming from a single W boson, but not for particle pairs from different Ws. The comparison samples were analysed together with the fully hadronic WW events. Such comparison samples were constructed by the following two techniques: (a) using an event mixing method; (b) using a correlation function calculated from R_{2q} (called the Linear Scenario).

In the event mixing method the mixed hadronic-leptonic data are used to construct a comparison sample. From each selected semileptonic event, the hadronic part was boosted to the rest frame of the W candidate. The rest frames of the W candidates were determined using the energy and momenta of the Ws obtained from the kinematical fits. An event was then constructed from two W candidates by boosting the particles of the individual Ws in opposite directions. The boost vectors were determined separately for each W taking energy-momentum conservation and the fitted W candidate mass into account. It was verified using simulated events with inside Ws BEC and exactly the same selections and method, that the $R_{4q}(Q)(mixing)$, calculated using the event mixing method, was practically the same as $R_{4q}(Q)$, as expected.

In the second technique (the linear scenario) extra input from simulations is needed, in particular the fraction of pairs from different Ws as a function of Q.

Using a model independent event mixing technique, the difference between the correlation strengths of like-sign pairs for real WW ($4q$) events and for a comparison sample which contains only correlations coming from the same W boson was (see Figure 2)

$$\Delta\lambda(mixing) = 0.062 \pm 0.025 \pm 0.021 \quad (2)$$

DELPHI(preliminary)

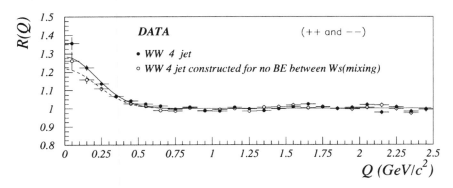

Figure 2. Measured correlation functions $R_{4q}(Q)$ (closed circles) and $R_{4q}(Q)(constructed)$ (open circles) for like-sign pairs. $R_{4q}(Q)(constructed)$ was computed from events constructed from 2-jet events using the mixing technique. The full curve shows the best fit to the Gaussian form for the data sample, the dashed curve for the constructed mixing sample.

(the first error is statistical and the second is systematical).

Another measurement of $\Delta\lambda$ obtained using a comparison of R_{4q} and R_{2q} (the linear scenario) yielded

$$\Delta\lambda(linear) = 0.077 \pm 0.026 \pm 0.020 \quad (3)$$

(the first error is statistical and the second is systematical).

4 Summary

The colour reconnection and Bose-Einstein correlation effects were studied in reaction (1) using the data collected by the DELPHI detector during the 1997, 1998 and 1999 runs with integrated luminosity of 437 pb^{-1} at centre-of-mass energies of 183, 189 and 192–202 GeV.

A deficit of charged particle multiplicity in WW fully hadronic events (which is predicted by colour reconnection models) was not observed in the data.

Measurements were performed to extract correlations between pions from different Ws

using two methods, (a) event mixing technique, (b) comparison of R_{4q} and R_{2q} (the linear scenario). Both measurements yield compatible results (see equations 2 and 3). Our data support the hypothesis of correlations between like-sign pions coming from different Ws at the level of about two standard deviations.

References

1. *Charged and Identified Particles in the Hadronic Decay of W Bosons and in $e^+e^- \to q\bar{q}$ from 130 to 200 GeV.*
   P. Abreu et al., DELPHI Collaboration, CERN-EP 2000-023, Submitted to Eur. Phys. J. C. Contributed paper No. 642.
2. *Correlations Between Particles in $e^+e^- \to W^+W^-$ Events.*
   P. Abreu et al., DELPHI Collaboration, DELPHI 2000-115 CONF 414, July, 2000. Contributed paper No. 637.

PROMPT (DIRECT) PHOTON PRODUCTION AT COLLIDERS

S. R. MAGILL

Argonne National Laboratory, 9700 S. Cass Ave., Argonne, Illinois, USA

New results from the ZEUS Collaboration on prompt (direct) photon production in ep collisions at HERA are presented along with recent results from CDF in $\bar{p}p$ collisions at the Tevatron. These results are examined with respect to the hypothesis of increased $< k_T >$ from inital-state soft gluon radiation in the proton and are both found to be consistent with previous fixed-target and collider prompt photon production results.

1 Motivation

It has been observed in many fixed-target and collider experiments that NLO QCD calculations have been unable to describe the production cross sections for prompt photons (or direct photons, since they are produced directly in the hard scattering process). At low values of photon p_T, the data is in excess of the theory, and furthermore, a comparison of experiments at fixed scaled photon transverse energy, x_T ($x_T = 2E_T/\sqrt{s}$), shows wide disagreement between data sets (Fig. 1). This would seem to indicate that

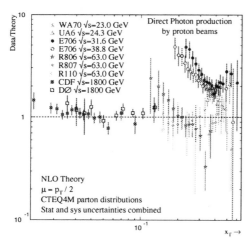

Figure 1. Prompt photon data/theory comparison vs x_T.

the difference between data and theory is not due to incomplete knowledge of proton parton distribution functions (pdfs). Scale effects are also not able to account for the mis-

match. The discrepancy has been attributed to effects of initial-state soft gluon radiation, not included in the NLO QCD calculations. These effects have been modelled in simulation programs by incorporating at the parton shower level additional $< k_T >$ from initial-state gluons in the proton. Also, resummation techniques have recently been applied to NLO QCD prompt photon production, pointing the way to better agreement of theory to data [1].

2 Production Processes and Results

At the Tevatron, a prompt photon can be produced when a quark from the $\bar{p}(p)$ interacts with a gluon from the $p(\bar{p})$. The prompt photon is a direct (hadronization-free) probe of QCD in the hard scattering process. Similarly, in resolved photoproduction at HERA, a quark from the incoming photon can interact with a gluon from the proton to produce a prompt photon. In addition, since the incoming photon also interacts in a pointlike manner (direct photoproduction), a prompt photon can be produced in a direct quark (proton) - photon interaction. At HERA, by selecting events in which the incoming photon interacts directly, $< k_T >$ effects in the proton can be studied without contributions from the incoming photon. Once these effects are understood in the proton, resolved photoproduction of prompt photons can be used to study similar effects in the photon.

2.1 Results from CDF

A recent analysis of prompt photon production by the CDF Collaboration at $\sqrt{s} = 630\ GeV$ is compared to previous UA2 results at the same energy, showing good agreement as a function of the photon p_T in Fig. 2. The CDF data at $\sqrt{s} = 630\ GeV$ is

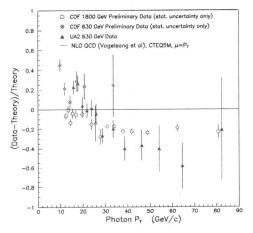

Figure 2. Comparison of data and theory for CDF(630), UA2(630), and CDF(1800) vs p_T.

also in reasonable agreement with that at $\sqrt{s} = 1800\ GeV$ as a function of p_T. Comparison with NLO QCD clearly shows the characteristic disagreement between data and theory at low p_T for all 3 data sets. By comparing the data to NLO QCD as a function of photon x_T (Fig. 3), disagreement between the 630 and 1800 data sets at constant x_T confirms that proton pdfs are not responsible for the difference between data and theory.

The CDF Collaboration also analyzed events with a prompt photon and a muon from b- or c-quark decay in the final state. The cross section for prompt photon production + c- or b-jet is reproduced adequately by the NLO QCD calculation. Any additional $< k_T >$ needed to describe prompt photon production is presumably masked by the relatively large p_T requirements for both the photon and the muon.

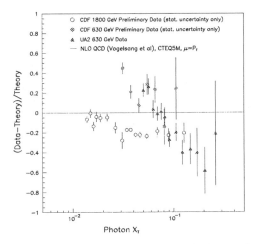

Figure 3. Comparison of data and theory for CDF(630), UA2(630), and CDF(1800) vs x_T.

2.2 Results from ZEUS

Published results [2] from ZEUS on inclusive prompt photon production in ep photoproduction as a function of photon E_T are in reasonable agreement with NLO QCD, but only when higher order corrections are included in the calculation. If the data is examined in

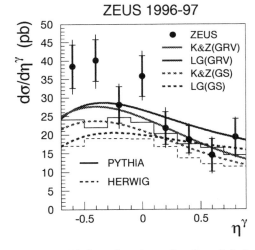

Figure 4. ZEUS data plotted as a function of photon pseudorapidity compared to NLO QCD.

more detail, the discrepancy between theory and data appears to be strongest for prompt photons detected in the negative pseudorapidity direction (incoming photon direction) as seen in Fig. 4. By varying parameters in the NLO QCD calculations, indications are that this discrepancy can be linked to insufficient high x partons in the resolved photon.

To systematically evaluate effects on prompt photon production from the proton and the photon, a final state including a prompt photon and a jet is required. The addition of a jet in the final state allows the fraction of incoming photon momentum participating in the hard scatter, x_γ, to be measured, leading to a separation of events dominated by resolved photoproduction processes (low x_γ) and direct photoproduction (in this case, $x_\gamma > 0.9$). In high x_γ events, the incoming photon is predominantly pointlike and any effects such as additional $< k_T >$ can be attributed to the proton. Shape distributions of data and simulated data are compared for the following variables: $p_\perp$ and $p_\parallel$ - the transverse and longitudinal photon-jet momentum imbalance; and $\Delta\phi$ - the azimuthal angle between the photon and jet. These variables are particularly sensitive to additional $< k_T >$ through the recoil of the prompt photon from the jet. The simulated data (PYTHIA) is generated with the p_T of initial state gluon radiation adjusted by hand (gaussian smearing) upwards of the default value of 0.44 GeV. As seen in Fig. 5, the data disfavors the default $< k_T >$ value of 0.44 GeV and also as much as 3 GeV. A $< k_T >$ value of between 1 and 2 GeV is favored by the data.

3 Summary

Recent results from CDF (prompt photon production at $\sqrt{s} = 630\ GeV$) and ZEUS (direct photoproduction of prompt photon + jet) are consistent with the need for additional $< k_T >$ of initial state soft gluon radiation from the proton. From the ZEUS inclusive prompt photon data, there are indications that current understanding of the

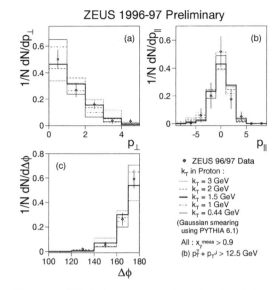

Figure 5. ZEUS data compared to simulated data with various additional $< k_T >$ values.

photon structure is also lacking. Using additional prompt photon + jet data, a systematic investigation of effects from both the proton and photon can be done by extending the analysis to low x_γ - the region dominated by resolved photoproduction.

Acknowledgments

The author wishes to thank the members of the CDF and ZEUS Collaborations for the opportunity to present these results. The support and encouragement of the management of the respective laboratories, Fermilab and DESY, is also acknowledged. Finally, the design, construction and installation of the detectors has been made possible by the efforts of many people and their respective funding agencies who are not listed as authors here.

References

1. E. Laenen in these proceedings.
2. J. Breitweg, et. al., *Phys. Lett.* B **472**, (2000).

INCLUSIVE DIRECT-PHOTON AND π^0 PRODUCTION IN PROTON–NUCLEON COLLISIONS

T. FERBEL

University of Rochester, Rochester, New York 14627

We present a study of inclusive direct-γ and π^0 production in hadronic interactions that focuses on a comparison of the ratio of γ and π^0 yields with expectations from next-to-leading order perturbative QCD (NLO pQCD).

1   Introduction and the γ/π^0 Ratio

Direct-photon production in hadronic collisions at high transverse momenta (p_T) has long been viewed as an ideal vehicle for extracting the gluon content, $G(x)$, of hadrons. The sensitivity to $G(x)$ arises from the contribution of the Compton subprocess $gq \to \gamma q$ to direct-γ production. $G(x)$ is well constrained by other data for $x < 0.25$, but not at larger x. In principle, a precise measurement of direct-photon production at fixed-target energies can constrain $G(x)$ at large x, and such data have been used in fits to global parton distribution functions (PDF).

Deviations have been noted between measured inclusive direct-photon cross sections and NLO pQCD. Ratios of data to theory for both γ and pion production indicate substantial disagreement between data and pQCD, as well as between different experiments.[1] The latter is not completely surprising, because, especially for direct-photon production, signals are often small and backgrounds large, especially at lower energies. Several experiments show better agreement with NLO pQCD than others, but the results do not provide confidence in the theory nor in the quality of all data. Although it has been suggested that deviations from theory for both γ and pion production can be ascribed to higher-order effects of initial-state soft-gluon radiation, it seems unlikely that theoretical developments alone will accommodate the observed level of scatter in data/theory. These discrepancies motivated us to consider measurements of the γ/π^0 ratio over the available range of center-of-mass energies ($\sqrt{s}$). Both experimental and theoretical uncertainties tend to cancel in such a ratio, and it is also less sensitive to the treatment of gluon radiation. A sample of the ratio of direct-γ to π^0 cross sections for both data and NLO pQCD is given for incident protons, as a function of $x_T = 2p_T/\sqrt{s}$, in Figs. 1–3. The results at $\sqrt{s} = 19.4$ GeV are displayed in Fig. 1. For all measurements, theory is high compared to data. (The NLO calculations use a single scale of $\mu = p_T/2$, CTEQ4M PDFs,[2] and BKK fragmentation functions[3] for pions.) Figure 2 shows the γ to π^0 ratio at $\sqrt{s} \approx 23 - 24$ GeV. Just as in Fig. 1, theory is high relative to data. At larger $\sqrt{s}$, the NLO value for the ratio agrees better with experiment, as seen in Fig. 3 for $\sqrt{s} = 31 - 39$ GeV. At even higher $\sqrt{s}$, theory lies slightly below the data.[1] A compilation of these results, displayed for simplicity without their uncertainties, is presented in Fig. 4. Here, the ratio of data to theory was fitted to a constant value at high-p_T, and the results plotted as a function of $\sqrt{s}$. The results suggest an energy dependence in the ratio of data to theory for γ/π^0 production, already noted in Figs. 1–3. There are also substantial differences between experiments at low $\sqrt{s}$, where the observed γ/π^0 is smallest, which makes it difficult to quantify this trend. Recognizing the presence of these differences is especially important because only the direct-photon experiments at low energy have been used in PDF fits to $G(x)$.

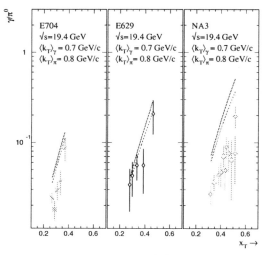

Figure 1. Comparison of γ/π^0 rates as a function of x_T for NA3, E704, and E629 at $\sqrt{s} = 19.4$ GeV. Overlayed are the results from NLO pQCD (solid) and k_T-enhanced calculations (dashed). Values of $\langle k_T \rangle$ used for the k_T-enhanced calculations are given in the legend.

2   Corrections for Soft Gluon Emission

In the absence of a rigorous theoretical treatment, a more intuitive, but often successful, phenomenological approach has been used to describe soft gluon radiation in high-p_T inclusive production,[4] and parametrized in terms of an effective k_T that provided additional transverse impulse to the outgoing partons. This provided p_T-dependent corrections to the NLO pQCD calculations. The corrections for direct-γ and π^0 production in Fermilab experiment E706 are large (and comparable) over the full range of p_T.[1] The corrections depend on the values used for $\langle k_T \rangle$, with changes of 200 MeV/c making substantial difference, and therefore making it difficult to obtain the precision needed for extracting global parton distributions. In addition, there are different ways to implement such models,[1,5] which can produce quantitative differences in the k_T-correction factors. However, it is expected that, in the ratio of γ and π^0 cross sections, the impact of k_T cor-

rections should be minimal, and this is observed in Figs. 1–3, where the dashed curves indicate the predicted ratios using previous k_T corrections.[4] Thus, it seems that the trend in Fig. 4 cannot be understood purely on the basis of corrections for k_T.

Resummed pQCD calculations for single direct-photon production are currently under development.[6,7,8,9,10,11] Two recent threshold-resummed pQCD calculations for direct photons[6,7] exhibit far less dependence on QCD scales than NLO theory, and provide an enhancement at high p_T. A method for simultaneous treatment of recoil and threshold corrections in inclusive single-γ cross sections is also being developed.[11] This approach accounts explicitly for the recoil from soft radiation in the hard-scattering, and conserves both energy and transverse momentum for the resummed radiation. The possibility of substantial enhancements from higher-order perturbative and power-law nonperturbative corrections relative to NLO are indicated at both moderate and high p_T for fixed-target energies, similar to the enhancements obtained with simple k_T-smearing.[4]

Although there are discrepancies between experiments, especially significant at low $\sqrt{s}$, there appears to be an unexplained systematic trend with energy in Fig. 4. Hopefully, this can be clarified once resummation calculations for inclusive pion production become available. Nevertheless, the recent developments in theory of direct-γ processes provide cause for optimism that the long-standing difficulties in developing an adequate description of direct-γ production can eventually be resolved, making possible a reliable extraction of $G(x)$ from such data.

This report is based on work performed with L. Apanasevich, M. Begel, C. Bromberg, G. Ginther, J. Huston, S. Kuhlmann, P. Slattery, M. Zielinski, and V. Zutshi. I also wish to thank P. Aurenche, D. Soper, and G. Sterman for helpful discussions.

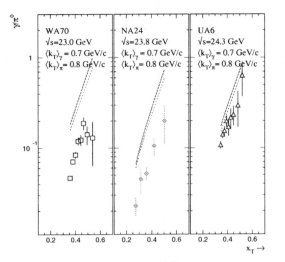

Figure 2. Comparison of γ/π^0 rates as a function of x_T for WA70, NA24, and UA6 at $\sqrt{s} \approx 23 - 24$ GeV. (See text and Fig. 1 for additional explanation.)

References

1. L. Apanasevich *et al.*, hep-ph/0007191.

2. H. L. Lai *et al.*, Phys. Rev. **D55**, 1280 (1997).

3. J. Binnewies, B. A. Kniehl, and G. Kramer, Phys. Rev. **D52**, 4947 (1995).

4. L. Apanasevich *et al.*, Phys. Rev. **D59**, 074007 (1999).

5. A. D. Martin *et al.*, Eur. Phys. J. **C4**, 463 (1998).

6. S. Catani *et al.*, JHEP **03**, 025 (1999).

7. N. Kidonakis and J. F. Owens, Phys. Rev. **D61**, 094004 (2000).

8. E. Laenen, G. Oderda, and G. Sterman, Phys. Lett. **B438**, 173 (1998).

9. H.-L. Lai and H. nan Li, Phys. Rev. **D58**, 114020 (1998).

10. H. nan Li, Phys. Lett. **B454**, 328 (1999).

11. E. Laenen, G. Sterman, and W. Vogelsang, Phys. Rev. Lett. **84**, 4296 (2000).

12. P. Aurenche *et al.*, Eur. Phys. J. **C9**, 107 (1999).

13. P. Aurenche *et al.*, Eur. Phys. J. **C13**, 347 (2000).

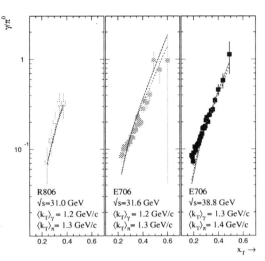

Figure 3. Comparison of γ/π^0 rates as a function of x_T for R806 at $\sqrt{s} = 31$ GeV and E706 at $\sqrt{s} = 31.6$ and 38.8 GeV. (See text and Fig. 1 for additional explanation.)

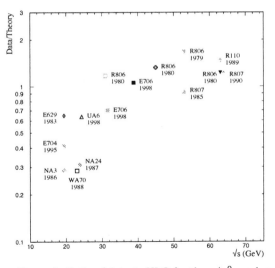

Figure 4. Ratio of data to NLO for the γ/π^0 production as a function of $\sqrt{s}$.[1] The values represent fits to the ratio of data to NLO pQCD theory, without k_T-enhancement (see text).

Parallel Session 4

High Energy Heavy Ion Collisions

Conveners: John W. Harris (Yale) and
              Johanna Stachel (Heidelberg)

E917 RESULTS ON STRANGENESS PRODUCTION IN AU+AU COLLISIONS AT AGS

W.C. CHANG, FOR THE E917 COLLABORATION

Institute of Physics, Academia Sinica, Taipei 11529, Taiwan

B.B. BACK,[1] R.R. BETTS,[1,5] J. CHANG,[3] W.C. CHANG,[3] C.Y. CHI,[4] Y.Y. CHU,[2] J.B. CUMMING,[2] J.C. DUNLOP,[7] W. ELDREDGE,[3] S.Y. FUNG,[3] R. GANZ,[5] E. GARCIA,[6] A. GILLITZER,[1] G. HEINTZELMAN,[7] W.F. HENNING,[1] D.J. HOFMAN,[1] B. HOLZMAN,[5] J.H. KANG,[9] E.J. KIM,[9] S.Y. KIM,[9] Y. KWON,[9] D. MCLEOD,[5] A.C. MIGNEREY,[6] M. MOULSON,[4] V. NANAL,[1] C.A. OGILVIE,[7] R. PAK,[8] A. RUANGMA,[6] D.E. RUSS,[6] R.K. SETO,[3] P.J. STANSKAS,[6] G.S.F. STEPHANS,[7] H.Q. WANG,[3] F.L.H. WOLFS,[8] A.H. WUOSMAA,[1] H. XIANG,[3] G.H. XU,[3] H.B. YAO,[7] C.M. ZOU[3]

[1]*Argonne National Laboratory, Argonne, IL 60439, USA*

[2]*Brookhaven National Laboratory, Upton, NY 11973, USA*

[3]*University of California, Riverside, CA 92521, USA*

[4]*Columbia University, Nevis Laboratories, Irvington, NY 10533, USA*

[5]*University of Illinois at Chicago, Chicago, IL 60607, USA*

[6]*University of Maryland, College Park, MD 20742, USA*

[7]*Massachusetts Institute of Technology, Cambridge, MA 02139, USA*

[8]*University of Rochester, Rochester, NY 14627, USA*

[9]*Yonsei University, Seoul 120-749, South Korea*

The strangeness production in Au+Au collisions at beam energies of 6, 8, and 10.8 A GeV is reported from AGS Experiment E917. We present the excitation function of the K/π ratio as well as the yield of ϕ mesons and particle ratios as a function of centrality at the highest beam energy. With reference to pp collisions, we observe a strong enhancement in the production of strangeness carrying particles in heavy ion collisions, which is demonstrated in terms of various particle production ratios.

1 Introduction

An enhanced strangeness production is one of the proposed signatures for the occurrence of a de-confined quark matter in relativistic heavy ion collisions [1]. The ϕ meson, the lightest bound state of strange quarks ($s\bar{s}$), is an interesting probe because its production is suppressed in ordinary pp interactions in accordance with the Okubo-Zweig-Iizuka (OZI) rule [2]. However, hadronic re-scattering and conversion of excitation energy of formed resonances into particles is known to be an effective mechanism for the formation of the strange hadrons in heavy ion collisions [3,4]. For example, K and ϕ mesons can be pro-

duced via the channels $\pi N \longrightarrow K \Lambda N, \phi N$ which are not available in pp collisions. In this talk, we present the measurement of kaons and ϕ mesons[a] in $Au + Au$ collisions at AGS energies. With the comparison to the non-strange π mesons, and the kaon and ϕ production in pp interaction at similar energies, we observe a strong strangeness enhancement in heavy ion collisions. This study serves as an experimental benchmark to discern any extra degree of strangeness enhancement which may occur at RHIC energies by traversing the QCD phase-transition boundary.

---

[a]It is noted that the analysis results of ϕ mesons are preliminary.

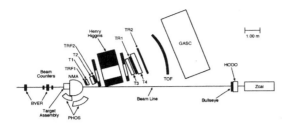

Figure 1. Detector configuration of experiment E917.

2 Experimental Setup

Experiment E917 at the AGS at Brookhaven National Laboratory, measured $Au + Au$ reactions at beam kinetic energies of 6.0, 8.0, and 10.8 A GeV in the fall of 1996. The detector configuration is schematically illustrated in Fig. 1. The experimental device consists of beamline detector arrays for global event characterization and a movable magnetic spectrometer for tracking and particle identification. The centrality of events is characterized by a zero-degree-calorimeter (ZCAL) positioned downstream at beam axis which measures the energy deposited by the beam spectator after the interaction. The typical momentum resolution ($\delta p/p$) of the spectrometer is about one percent.

Particles analyzed include strangeness-carrying particles like K^+ , K^- , ϕ , Λ and $\bar{\Lambda}$, as well as non-strange particles like π, p and $\bar{p}$ [5,6]. In order to enhance the data-taking rate of ϕ and $\bar{\Lambda}$ events, data at 10.8 A GeV was collected using a two-level triggering: a minimum-biased interaction trigger followed by an online particle identification trigger which required at least two kaons or one $\bar{p}$ to be detected in the spectrometer.

3 Excitation Function of Kaons

Kaons are the most abundant species of strangeness-carrying particles. The K/π ratio is an important indicator of strangeness enhancement in making comparison between heavy ion and pp collisions. Combining the

results from the earlier E866 series of experiments with ours, we study kaon and pion production in $Au + Au$ collisions over the full beam energy range available at the AGS [5].

In Fig. 2, the rapidity density (dN/dy) at mid-rapidity for both K^+ and K^- display a strong increase with the center-of-mass beam energy per nucleon over the pp production threshold. No sudden jump in the yield with beam energy is seen. A rising trend with beam energy exists for the K^+/π^+ ratio within the same energy range as shown in Fig. 3. This ratio seems, however, to saturate or even decrease when the beam energy is increased further as evidenced by the measurement at SPS energies. The results from pp measurements are represented by the two hashed regions in the figure. Thus there exists a strong enhancement of strangeness production in the heavy ion collisions, characterized by K^+/π^+ ratio. This is most likely the results of hadronic re-scattering and resonance formation in the nuclear fireball at AGS energies [4].

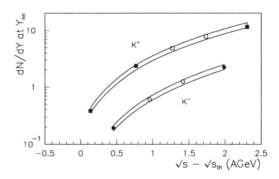

Figure 2. The yield of K^+ and K^- at mid-rapidity for central Au+Au collisions as a function of available beam energy per nucleon ($\sqrt{s} - \sqrt{s_{th}}$). Solid(open) points are measurements by E866(E917).

4 Centrality Dependence of ϕ meson Production

The ϕ mesons are reconstructed from the invariant mass spectrum of K^+K^- pairs. Fig. 4 displays this spectrum and the ϕ signal after

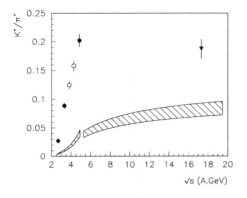

Figure 3. K^+/π^+ ratio of mid-rapidity yield in central $Au + Au$ collisions as a function of center-of-mass energy per nucleon ($\sqrt{s}$). Solid(open) circles represents E866(E917) measurement while the solid triangle is NA49 data [7]. The hashed region is the K^+/π^+ ratio from parametrized K and π yields in pp reactions [4].

event-mixed background subtraction. A relativistic Breit-Wigner parameterization, convoluted with a Gaussian experimental mass resolution of ~ 2 MeV, was used to fit the ϕ peak. The fit gives a value for the peak at $m_0 = 1018.9 \pm 0.4$ MeV/c^2. This result is in reasonable agreement with the most recent particle data book value of $m_\phi = 1019.417 \pm 0.014$ MeV/c^2.

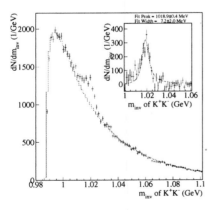

Figure 4. Invariant mass spectrum of K^+K^- pairs and the ϕ signal after background-subtraction fit with a relativistic Breit-Wigner function (inset).

The transverse mass spectra was constructed for different centrality and rapid-

ity bins and a two-parameter exponential fit was used to determine the rapidity density (dN/dy) and the inverse slope parameters (T_{inv}). The rapidity density around mid-rapidity ($y = 1.6$) shows a strong and systematic increase with centrality.

5 Centrality Dependence of Particle Ratios of ϕ meson

In order to make comparison with pp interactions, we characterized the centrality by the average number of projectile participants, $\langle N_{pp} \rangle$ which is derived from the deposited energy, E_{ZCAL}, in the zero-degree calorimeter and the total kinetic energy of beam particles (E_{beam}) using the relation $N_{pp} = 197 * (1 - E_{ZCAL}/E_{beam})$. The parameterized beam energy dependency of multiplicities of ϕ, π and K in pp reactions are taken from Ref. [8,9]. The yield of particles emitted with the rapidity range $1.2 < y < 1.6$, is used for points representing heavy ion collisions.

First, the ϕ yield per projectile participant exhibits a steady increase toward the central collisions as shown in Fig. 5. This signifies that the yield of ϕ increases nonlinearly with N_{pp}, as seen also for kaons [10]. Therefore the production of ϕ mesons in heavy-ion collisions is enhanced with respect to a simple incoherent superposition of individual nucleon-nucleon interactions.

The particle ratio of $\phi/\langle\pi\rangle$ where $\langle\pi\rangle = (\pi^+ + \pi^-)/2$, also shows a clear but less pronounced increase toward central collisions in Fig. 6. This signifies an enhancement of ϕ mesons relative to non-strange π mesons in the central collisions. The enhancement factor is 5.0 ± 1.3 comparing the most central bin of $Au + Au$ collisions with pp data. The secondary re-scattering in the central collisions could in principle contribute more significantly to the production of ϕ than that of π by overcoming the high ϕ-production threshold in pp interactions.

An increase of the K/π ratio with cen-

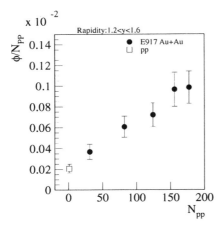

Figure 5. The normalized yield of ϕ mesons per projectile participants (N_{pp}) as a function of N_{pp}. The value for pp interactions ($N_{pp} = 1$) is obtained from the parameterization described in the text. The accepted rapidity(y) region is $1.2 < y < 1.6$.

trality was reported in $Au + Au$ at the same energy by experiment E866 [11]. It is important to compare the degree of enhancement of two strangeness-carrying particles, ϕ and K. Here we are able to use the yield of K^+ to represent overall kaons production because the K^-/K^+ ratio is found to be about 0.15 and independent of centrality in $Au + Au$ at the AGS [10]. In Fig. 7, the ϕ/K^+ ratio versus centrality is shown to be basically flat for AA collisions and a factor of 1.44 ($\pm$0.38) larger for the most central bin than the value in pp collisions. This reveals that both ϕ and K^+ (or K^-) possess a similar enhancement with centrality at this energy.

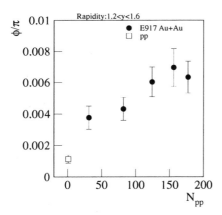

Figure 6. Same as Fig. 5 for ϕ/π ratio.

Because mesons with more strange quarks are heavier and strangeness is conserved in strong interactions, the production of multi-strange particles is in principle suppressed by increasingly higher thresholds in hadronic interactions. The different production thresholds for ϕ mesons with two strange quarks and kaons with only one strange quark, leads to an expectation of a strong dependence of the ϕ/K ratio on the centrality. However a very weak centrality dependence is indeed seen for the ratio of ϕ/K. Together with the centrality independence of the K^-/K^+ ratio, our observation gives a hint at a possible occurrence of strangeness chemical equilibrium. However, the study of the other multi-strange particles is necessary to confirm this speculation.

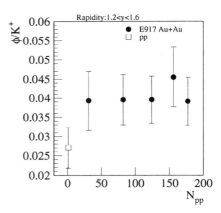

Figure 7. Same as Fig. 5 for ϕ/K^+ ratio.

One possible production mechanism for ϕ mesons is kaon coalescence. Models of density coalescence of K^+K^- into ϕ predict a decrease of $\phi/(K^+ * K^-)$ ratio scaled as the inverse of interacting volume (V_{source}) toward central collisions [12]. This is indeed seen in Fig. 8 where the $\phi/(K^+ * K^-)$ ratio is plotted as a function of N_{pp}, which is proportional to V_{source}. Under a coalescence framework of ϕ mesons constructed by A.J. Baltz and C. Dover [13], we have estimated the yield of ϕ mesons coming from K^+K^- coalescence. The freeze-out phase space distributions of kaons from the predic-

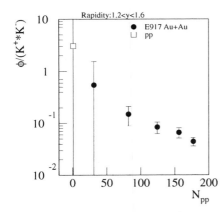

Figure 8. Same as Fig. 5 for $\phi/(K^+ * K^-)$ ratio.

tion of the cascade model RQMD V2.3 are used as the input. It is found that the rate of ϕ meson production from $K^+ K^-$ coalescence under-predicts the observed yield by about a factor of ten. This study indicates that $K^+ K^-$ coalescence in the freeze-out is not the main source for ϕ production in these collisions.

6 Summary

We study the strangeness production by the measurement of kaons and ϕ mesons in $Au + Au$ collisions at the AGS. The excitation function of K/π ratio at AGS energies shows a rise with beam energy. This demonstrates a strangeness enhancement via kaon production relative to non-strange mesons. Additionally, this enhancement of kaon production is found to be stronger in $Au + Au$ collisions compared to pp interactions. The ϕ mesons, whose production is suppressed in pp interactions, has an increasing yield per projectile participant from the peripheral $Au + Au$ collisions to the central ones. In comparison, the non-strange π mesons do not show such a strong production enhancement with centrality while kaons exhibits a similar degree of increase. Both of these observations point to the importance and effectiveness of hadronic re-scattering and resonance formation in heavy ion collisions which are able to overcome the high strangeness pro-

duction threshold and contribute in achieving strangeness enhancement.

The centrality dependence of $\phi/(K^+ * K^-)$ ratio is consistent with the scenario of ϕ coalesced from K^+ and K^-. Nevertheless, the estimated yield of ϕ from coalescence in the current framework accounts for less than 10% of the measured yield. This suggests that coalescence might not be the dominating production mechanism for ϕ mesons.

This work was supported by the U.S. Department of Energy (USA), the National Science Foundation (USA), and KOSEF (Korea).

References

1. P. Koch, B. Müller, and J. Rafelski, Z. Phys. A **324**, 453 (1986).

2. A. Shor, Phys. Rev. Lett. **54**, 1122 (1985).

3. W.S. Chung et al., Nucl. Phys. **A625**, 347 (1997).

4. J.C. Dunlop and C.A. Ogilvie, Phys. Rev. C **61**, 031901 (2000).

5. L. Ahle et al., Phys. Lett. **B476**, 1 (2000).

6. B.B. Back et al., submitted to Phys. Rev. Lett., nucl-ex/0003007.

7. J. Bächler et al., Nucl. Phys. **A661**, 45c (1999).

8. W. Cassing and E.L. Braikovskaya, Phys. Rept. **308**, 65 (1999); A.A. Sibirtsev, Nucl. Phys. **A604**, 455 (1996).

9. A.M. Rossi et al., Nucl. Phys. **B84**, 269 (1975).

10. L. Ahle et al., Phys. Rev. C **60**, 044904 (1999); L. Ahle et al., Phys. Rev. C **58**, 3523 (1998).

11. L. Ahle et al., Phys. Rev. C **59**, 2173 (1999); L. Ahle et al., Phys. Rev. C **57**, R466 (1998).

12. S. Das Gupta and A.Z. Mekjian, Phys. Rep. **72**, 131 (1981).

13. A.J. Baltz and C. Dover, Phys. Rev. C **53**, 362 (1996).

MULTI-STRANGE PARTICLE ENHANCEMENTS IN PB-PB INTERACTIONS AT 158 GEV/C PER NUCLEON

PRESENTED BY DAVID EVANS FOR THE WA97 COLLABORATION

School of Physics and Astronomy, University of Birmingham, Birmingham B15 2TT, UK,
E-mail: d.evans@bham.ac.uk

F. Antinori[e,i], H. Bakke[b], W. Beusch[e], I.J. Bloodworth[d], R. Caliandro[a], N. Carrer[i], D. Di Bari[a], S. Di Liberto[k], D. Elia[a], D. Evans[d], K. Fanebust[b], R.A. Fini[a], J. Ftáčnik[f], B. Ghidini[a], G. Grella[l], H. Helstrup[c], A.K. Holme[h], D. Huss[g], A. Jacholkowski[a], G.T. Jones[d], J.B. Kinson[d], K. Knudson[e], I. Králik[f], V. Lenti[a], R. Lietava[e], R.A. Loconsole[a], G. Løvhøiden[h], V. Manzari[a], M.A. Mazzoni[k], F. Meddi[k], A. Michalon[m], M.E. Michalon-Mentzer[m], M. Morando[i], P.I. Norman[d], B. Pastirčák[f], E. Quercigh[e], G. Romano[l], K. Šafařík[e], L. Šándor[e,f], G. Segato[i], P. Staroba[j], M. Thompson[d], T.F. Thorsteinsen[b], G.D. Torrieri[d], T.S. Tveter[h], J. Urbán[f], O. Villalobos Baillie[d], T. Virgili[l], M.F. Votruba[d] and P. Závada[j].

[a] Dipartimento I.A. di Fisica dell'Università e del Politecnico di Bari and Sezione INFN, Bari, Italy
[b] Fysisk institutt, Universitetet i Bergen, Bergen, Norway
[c] Høgskolen i Bergen, Bergen, Norway
[d] School of Physics and Astronomy, University of Birmingham, Birmingham, UK
[e] CERN, European Laboratory for Particle Physics, Geneva, Switzerland
[f] Institute of Experimental Physics, Slovak Academy of Sciences, Košice, Slovakia
[g] GRPHE, Université de Haute Alsace, Mulhouse, France
[h] Fysisk institutt, Universitetet i Oslo, Oslo, Norway
[i] Dipartimento di Fisica dell'Università and Sezione INFN, Padua, Italy
[j] Institute of Physics, Academy of Sciences of the Czech Republic, Prague, Czech Republic
[k] Dipartimento di Fisica dell'Università "La Sapienza" and Sezione INFN, Rome,Italy
[l] Dipartimento di Scienze Fisiche "E.R. Caianiello" dell'Università and INFN, Salerno, Italy
[m] Institut de Recherches Subatomiques, IN2P3/ULP, Strasbourg, France

Strange and multi-strange baryon and anti-baryon production is expected to be enhanced in heavy ion interactions if a phase transition from hadronic matter to a Quark-Gluon Plasma takes place. The production yields and transverse mass spectra of strange and multi-strange baryons and anti-baryons are presented for lead-lead interactions at 158 GeV/c per nucleon. Yields and transverse mass spectra from proton-lead and proton-beryllium interactions, where no phase transition is expected, are also presented and compared to those from lead-lead interactions.

The results show a significant enhancement of strange and multi-strange baryons in Pb-Pb interactions. The enhancement increases with strangeness content as predicted in QGP models and exceeds one order of magnitude in the case of the Ω particle. These results represent some of the best evidence, to date, for a new phase of matter being formed in these collisions.

1 Introduction

QCD predicts that, under extreme conditions of temperature and density, hadronic matter will undergo a phase transition into a deconfined plasma of quarks and gluons known as a Quark-Gluon Plamsa (QGP). It is believed that the energy densities produced in ultra-relativistic heavy ion collisions may be high enough for such a phase transition to take place. The enhancement of strange particle production in such collisions is considered to be one of the most promising signatures for QGP formation to date [1]. In particular, it is expected that the enhancement of multi-strange particles would be more pronounced than that for singly strange particles [2].

The WA97 experiment was designed to study strange and multi-strange baryon production in Pb-Pb interactions and compare the strangeness yields to those from proton induced reactions.

2 Experimental Setup

The WA97 setup, shown in figure 1, is described in detail elsewhere[3,4]. Particle tracks

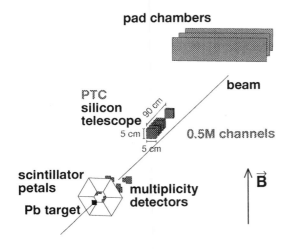

Figure 1. The WA97 experimental layout.

were reconstructed using a silicon telescope containing seven pixel planes with $75 \times 500\mu$m pixel size and ten microstrip planes with a 50 μm pitch. The telescope had a cross-section of 5×5 cm^2 and contained about 0.5×10^6 channels. The first plane of the telescope was placed 60 cm downstream of the target (90 cm for the proton bean reference runs), above the beam line and inclined at an angle (pointing to the target) in order to accept particles at central rapidity and medium transverse momentum. Pad chambers were used as a leverarm for high momentum particles.

Scintillator ("petals") detectors were used by the trigger to select the most central $\sim 40\%$ Pb-Pb interactions. Two planes of microstrip multiplicity detectors (MSD) sampled the overall charged multiplicity in the pseudorapidity region $2 \leq \eta \leq 4$ which was used in the off-line analysis. The whole setup was placed inside the 1.8 T magnetic field of the CERN Omega magnet.

In the proton reference runs, Be and Pb targets were used, both with a thickness corresponding to 8% of the interaction length. In these runs a trigger was applied demanding at least two tracks in the telescope. Data were also collected requiring at least one track in the telescope in order to study negative particle production.

K_s^0 mesons, Λ, Ξ^- and Ω^- hyperons and their anti-particles were identified by reconstructing their weak decays into final states containing only charged particles:

$$K_S^0 \rightarrow \pi^+ + \pi^-$$
$$\Lambda \rightarrow p + \pi^-$$
$$\Xi^- \rightarrow \Lambda + \pi^-$$
$$\hookrightarrow p + \pi^-$$
$$\Omega^- \rightarrow \Lambda + K^-$$
$$\hookrightarrow p + \pi^-$$

A reference sample of negatively charged particles (h^-) have also been analysed. Details of the reconstruction and weighting procedures are described in[4,5]. The acceptance windows for the particles under study are shown in reference[4].

3 Centraility Measurement

The centrality measurements are based on the charged particle multiplicity measured by the MSD. The Pb-Pb data sample is divided into four multiplicity classes. The average number of wounded nucleons in each multiplicity class is determined by a Wounded Nucleon Model[6] fit. The fit procedure has been described in detail elsewhere[7]. Figure 2 shows the results of this fit. The four multiplicity classes are indicated in the figure by dashed lines. The bin widths were chosen so as to contain similar numbers of multistrange hyperons in each bin before weighting. The data in the first multiplicity bin are corrected for the effect of the centrality trigger. The number of wounded nucleons in proton-induced reactions have been determined as an average value for inelastic collisions in the framework of the Glauber model[8].

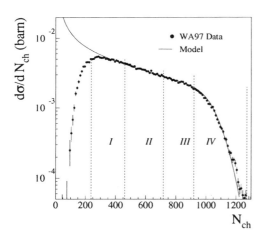

Figure 2. Wounded nucleon model fit to the multiplicity distribution.

4 Transverse Mass Spectra

The distributions of particle yields have been fitted to the expression

$$\frac{\mathrm{d}^2 N}{\mathrm{d}m_\mathrm{T}\mathrm{d}y} = f(y)m_\mathrm{T}\exp\left(-\frac{m_\mathrm{T}}{T}\right) \qquad (1)$$

where m_T is the transverse mass and y is the rapidity. The fit was performed using the method of maximum likelihood. The analysis assumed the rapidity distributions to be flat in the region $|y - y_{cm}| < 0.5$ *i.e.* $f(y) = $ constant. Preliminary data on the rapidity distributions of singly-strange particles support this assumption on the flatness of the rapidity distribution near mid-rapidity[9]. Figure 3 shows the inverse slope, T, from expression (1) for different particle species as a function of the number of participants. The squares represent data from p-Be interactions, the triangles represent data from p-Pb interactions, and the circles represent data from Pb-Pb interactions split into four centrality bins. As can be seen from figure 3, the inverse slope as a function of centrality is flat for negative

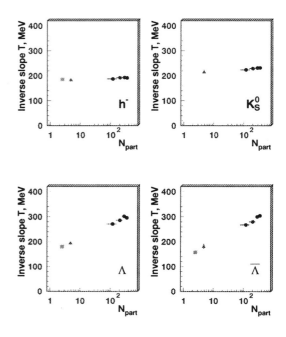

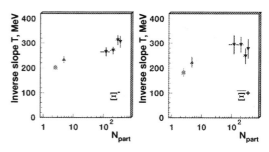

Figure 3. Inverse m_T slopes for different particles in different interactions.

particles and almost flat for K^0s while it increases with centrality for Λs and $\overline{\Lambda}$s . There is also an indication that $\langle T \rangle$ increases with centrality for Ξ^- and $\overline{\Xi}^+$ hyperons although the statistics are lower.

A compilation of inverse slope parameters measured by WA97, WA98, NA44 and NA49 (see ref.[10] and reference therein) is shown in figure 4.

A general trend is observed that the inverse slope increases linearly with particle mass. The Ω, and possible the Ξ, slopes form

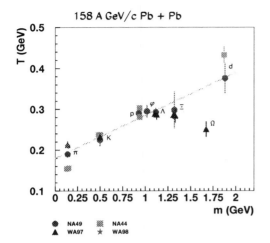

158 A GeV/c Pb + Pb

T (GeV)

d

p φ Λ
Ξ
K Ω
π

m (GeV)

NA49   NA44
WA97   WA98

Figure 4. Dependence of the m_T inverse slope patameters, T on the particle rest mass.

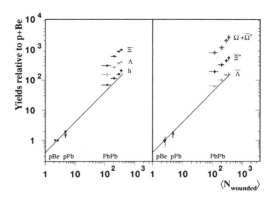

Figure 5. Particle yields relative to p-Be yields.

5 Strangeness Enhancement

Average particle yields per event have been calculated for different particles in p-Be, p-Pb and Pb-Pb interactions. Figure 5 shows the particle yield, as a function of the number of wounded nucleons, relative to that for p-Be interactions (*i.e.* the yields per participant have been rescaled so that the value for the p-Be data is set to one). All the Pb-Pb yields show a steady increase with centrality up to very central events. The particle yields in Pb-Pb interactions have been compared to a yield curve (full line) drawn through the p-Be points and proportional to the number of wounded nucleons. As can been seen from figure 5, the strange particle yields increase from p-Be to Pb-Pb faster than linearly with the number of wounded nucleons. The p-Pb points, however, lie on the line and are compatible with increasing linearly with the number of wounded nucleons.

a notable exception from this trend. This may indicate that the multi-strange particles decouple earlier than other particles.

From the average yield per wounded nucleon

$$\frac{Y}{\langle N_{\text{wound}} \rangle},$$

the strangeness enhancement, E, from p-Be to Pb-Pb interactions, is defined by

$$E = \left(\frac{Y}{\langle N_{\text{wound}} \rangle} \right)_{Pb-Pb} \bigg/ \left(\frac{Y}{\langle N_{\text{wound}} \rangle} \right)_{p-Be}.$$

The enhancement values, E, for different particles are shown in figure 6. As can be seen from figure 6, there is an enhancement of strange and multi-strange baryons going from proton induced to central Pb-Pb interactions although there is no further increase after $N_{parts} > 100$. The baryon enhancements also increase with strangeness content with the Ω being enhanced by over an order of magnitude.

6 Summary & Conclusions

In conclusion, we find that the inverse m_T slopes, T, in Pb-Pb interactions do not increase with centrality for h^- but do increase with centrality for Λs and $\overline{\Lambda}$s . The inverse m_T slopes, extracted by several experiments, generally show a linear increase with particle mass. Multi-strange hadrons, however, do not follow this trend suggesting that these

582

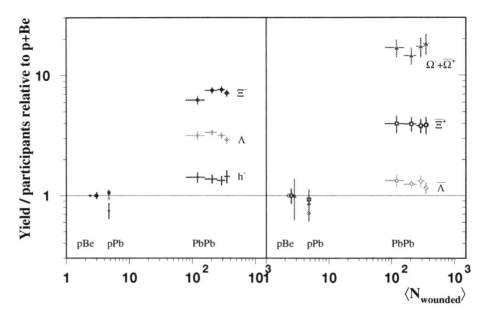

Figure 6. WA97 Particle Enhancements.

particles decouple early from the hadronic phase. There is no increase in strangeness yields per participant from p-Be to p-Pb interactions but there is a significant increase in strangeness from p-Pb to Pb-Pb interactions. The enhancement increases with strangeness content as predicted in QGP models and exceeds one order of magnitude in the case of the Ω particle.

References

1. J. Rafelski and B. Müller, Phys. Rev. Lett. **48** (1982) 1066.
   J. Rafelski and B. Müller, Phys. Rev. Lett. **56** (1986) 2334.
   P. Koch, B. Müller and J. Rafelski, Phys. Rep. **142** (1986) 167.
2. J. Rafelski, Phys. Lett. **B262** (1991) 333.
3. F. Antinori *et al.* (WA97 Collaboration), Nucl. Phys. **A590** (1995) 139.
4. E. Andersen *et al.* (WA97 Collaboration), Phys. Lett. **B433** (1998) 209.
5. E. Andersen *et al.* (WA97 Collaboration), Phys. Lett. **B449** (1999) 401.
6. A. Bialas, M. Bleszy'nski and W. Czyż, Nucl. Phys. **B111** (1976) 461.
7. N. Carrer *et al.* (WA97 & NA57 Collaborations), Nucl. Phys. **A661** (1999) 357.
8. C. Y. Wong, *Introduction to High-Engery Heavy-Ion Collisions*, World Scientific, Singapore, 1994, p. 251-264.
9. D. Evans *et al.* (WA97 Collaboration), Nucl. Phys. **A663&A664** (2000) 717.
10. J. Stachel, nucl-ex/9903007.

RESULTS OF NA49 ON PB+PB COLLISIONS AT THE CERN SPS

P. FILIP[16] - NA49 COLLABORATION

S.V. AFANASIEV[10],T. ANTICIC[21], J. BÄCHLER[6,8], D. BARNA[5], L.S. BARNBY[3],
J. BARTKE[7], R.A. BARTON[3], L. BETEV[14], H. BIAŁKOWSKA[17], A. BILLMEIER[11],
C. BLUME[8], C.O. BLYTH[3], B. BOIMSKA[17], M. BOTJE[22], J. BRACINIK[4],
F.P. BRADY[9], R. BRAMM[11], R. BRUN[6], P. BUNČIĆ[6,11], L. CARR[19], D. CEBRA[9],
G.E. COOPER[2], J.G. CRAMER[19], P. CSATÓ[5], V. ECKARDT[16], F. ECKHARDT[15],
D. FERENC[9], H.G. FISCHER[6], Z. FODOR[5], P. FOKA[11], P. FREUND[16], V. FRIESE[15],
J. FTACNIK[4], J. GÁL[5], R. GANZ[16], M. GAŹDZICKI[11], G. GEORGOPOULOS[1],
E. GŁADYSZ[7], J.W. HARRIS[20], S. HEGYI[5], V. HLINKA[4], C. HÖHNE[15], G. IGO[14],
M. IVANOV[4], P. JACOBS[2], R. JANIK[4], P.G. JONES[3], K. KADIJA[21,16],
V.I. KOLESNIKOV[10], T. KOLLEGGER[11], M. KOWALSKI[7], B. LASIUK[20],
M. VAN LEEUWEN[22], P. LÉVAI[5], A.I. MALAKHOV[10], S. MARGETIS[13],
C. MARKERT[8], B.W. MAYES[12], G.L. MELKUMOV[10], A. MISCHKE[8],J. MOLNÁR[5],
J.M. NELSON[3], G. ODYNIEC[2], M.D. OLDENBURG[11], G. PÁLLA[5],
A.D. PANAGIOTOU[1], A. PETRIDIS[1], M. PIKNA[4], L. PINSKY[12], A.M. POSKANZER[2],
D.J. PRINDLE[19], F. PÜHLHOFER[15], J.G. REID[19], R. RENFORDT[11], W. RETYK[18],
H.G. RITTER[2], D. RÖHRICH[11], C. ROLAND[8], G. ROLAND[11], A. RYBICKI[7],
T. SAMMER[16], A. SANDOVAL[8], H. SANN[8], E. SCHÄFER[16], N. SCHMITZ[16],
P. SEYBOTH[16], F. SIKLÉR[5,6], B. SITAR[4], E. SKRZYPCZAK[18], R. SNELLINGS[2],
G.T.A. SQUIER[3], R. STOCK[11], P. STRMEN[4], H. STRÖBELE[11], T. SUSA[21],
I. SZARKA[4], I. SZENTPÉTERY[5], J. SZIKLAI[5], M. TOY[2,14], T.A. TRAINOR[19],
S. TRENTALANGE[14], T. ULLRICH[20], D. VARGA[5], M. VASSILIOU[1], G.I. VERES[5],
G. VESZTERGOMBI[5], S. VOLOSHIN[2], D. VRANIĆ[6], F. WANG[2],
D.D. WEERASUNDARA[19], S. WENIG[6], A. WETZLER[11], C. WHITTEN[14], N. XU[2],
T.A. YATES[3], I.K. YOO[15], J. ZIMÁNYI[5]

[1] Department of Physics, University of Athens, Athens, Greece.
[2] Lawrence Berkeley National Laboratory, University of California, Berkeley, USA.
[3] Birmingham University, Birmingham, England.
[4] Comenius University, Bratislava, Slovakia.
[5] KFKI Research Institute for Particle and Nuclear Physics, Budapest, Hungary.
[6] CERN, Geneva, Switzerland.
[7] Institute of Nuclear Physics, Cracow, Poland.
[8] Gesellschaft für Schwerionenforschung (GSI), Darmstadt, Germany.
[9] University of California at Davis, Davis, USA.
[10] Joint Institute for Nuclear Research, Dubna, Russia.
[11] Fachbereich Physik der Universität, Frankfurt, Germany.
[12] University of Houston, Houston, TX, USA.
[13] Kent State University, Kent, OH, USA.
[14] University of California at Los Angeles, Los Angeles, USA.
[15] Fachbereich Physik der Universität, Marburg, Germany.
[16] Max-Planck-Institut für Physik, Munich, Germany.
[17] Institute for Nuclear Studies, Warsaw, Poland.
[18] Institute for Experimental Physics, University of Warsaw, Warsaw, Poland.
[19] Nuclear Physics Laboratory, University of Washington, Seattle, WA, USA.
[20] Yale University, New Haven, CT, USA.
[21] Rudjer Boskovic Institute, Zagreb, Croatia.
[22] NIKHEF, Amsterdam, Netherlands.

Recent experimental results from central Pb+Pb 160 GeV/n collisions studied by NA49 large acceptance hadron spectrometer are presented. Emphasis is put on K/π event-by-event fluctuations, the production of $\phi(1020)$ and $\Lambda^*(1520)$ and preliminary rapidity distributions of Ξ^- and $\bar{\Xi}^+/\Xi^-$ ratio.

1 Introduction

The aim of heavy ion experiments at the CERN SPS is to study properties of nuclear matter under extreme conditions created in relativistic collisions of heavy nuclei. Based on results of lattice QCD calculations[1] hadronic matter may undergo transition into a deconfined state of quarks and gluons at sufficiently high energy densities. The presence of such a deconfined phase in the space-time evolution of heavy ion collision (HIC) may exhibit itself in the final state of the collision which is characterized by momenta of detected particles. Due to hadronization and subsequent rescattering process extraction of a clear signature of the deconfined state from detected particles is a non-trivial task. The NA49 experiment[2] with its large acceptance and particle identification coverage is well suited for the analysis of hadronic signals. Results on the production of strange hadrons in Pb+Pb 160 GeV/n and 40 GeV/n collision at the CERN SPS exhibit a clear deviation from the expectations based on $p + p$ and $p + Pb$ data. Nevertheless an unambiguous signal of the existence of the deconfined phase in Pb+Pb collisions at CERN SPS from strange particle measurements has not been extracted so far. A combination of independent signals is much more suitable for such conclusions. Experimental data obtained with the NA49 detector allow the detailed study of hadronic spectra, fluctuations of single-event observables and also high precision interferometry and anisotropic flow analyses.

2 Fluctuations of K/π ratio.

Presence of a first or second order phase transition in the time evolution of HIC should exhibit itself as non-statistical fluctuations of

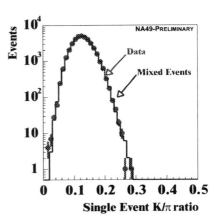

Figure 1. Distributions of single-event and mixed-event K/π ratio for central $Pb + Pb$ collisions.

single-event kaon to pion total-yield ratio[3]. NA49 has performed the study of such fluctuations in central $Pb + Pb$ 160 GeV/n events[4]. The strength of non-statistical σ_{dyn} (dynamical) fluctuations of K/π ratio defined as difference between the relative width of single-event σ_{event} and mixed-event σ_{mix} distributions (Fig.1).

$$\sigma_{dyn} = \sqrt{\sigma_{event}^2 - \sigma_{mix}^2} = 2.8\% \pm 0.5\% \quad (1)$$

has been found to be small compared to the strength of strangeness enhancement observed in $Pb + Pb$ collisions[5]. At the same time however e.g. presence of significant dynamical fluctuations of size $\sigma_{dyn} > 15\%$ cannot be ruled out if such fluctuations are present only in 5% of events selected for the analysis.

3 $\Lambda^*(1520)$ production

$\Lambda^*(1520)$ resonance has been studied in central $Pb + Pb$ collisions via its $\Lambda^* \to p + K^-$ decay channel (B.R.= 22.5%, $\Gamma = 16 MeV$). While in inelastic $p + p$ collisions analysis

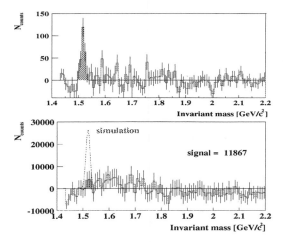

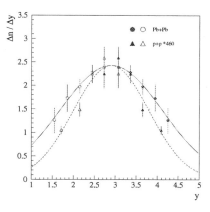

Figure 3. Rapidity distribution of Φ meson for $p + p$ inelastic and $Pb + Pb$ central collisions.

Figure 2. Invariant mass spectrum of pK^- pairs in $p + p$ (top) and $Pb + Pb$ central 160 GeV/n collisions.

reveals clear $\Lambda^*(1520)$ peak (see Fig.2) expected Λ^* signal in central $Pb + Pb$ data is surprisingly weak. Upper limit for the observed yield of Λ^* baryonic state

$$\langle \Lambda^*(1520) \rangle < 1.36 \, ; \quad 95\% \; c.l. \qquad (2)$$

in $Pb + Pb$ 160 GeV/n central collisions reveals significant suppression of the yield per participant nucleon when compared to inelastic $p + p$ collisions. A possible mechanism for this behaviour could be rescatering of Λ^* decay products ($c\tau \approx 15fm$) in interacting hadron gas created in $Pb + Pb$ collisons.

4 $\Phi(1020)$ meson production

Production of mesons with hidden flavor (e.g. $\Phi, J/\Psi$) in heavy ion collisions has attracted theoretical interest many years ago[6].

The NA49 experiment can identify Φ meson via its hadronic decay $\Phi \rightarrow K^+ + K^-$. The position of the Φ-signal in invariant mass spectrum of K^+K^- pairs has been found at 1018.7 ± 0.7 MeV in $Pb + Pb$ collisions. Both mass and width are in agreement with standard values and do not show significant shifts when compared to elementary collisions.

Distribution of $\Phi(1020)$ mesons in rapidity for central $Pb + Pb$ and inelastic $p + p$ collisions (rescaled) is shown in Figure 3. In central $Pb + Pb$ collisions rapidity distribution is wider ($\sigma_y = 1.22 \pm 0.16$) than in $p + p$ collisions ($\sigma_y = 0.89 \pm 0.06$). Inverse slope parameters of ϕ-meson transverse-mass distributions are $T = 169 \pm 17$ MeV for $p + p$ and 305 ± 15 MeV for $Pb + Pb$ collisions.

Total average ϕ-meson multiplicity in $Pb + Pb$ central collisions extracted from NA49 data is 7.6 ± 1.1 while an extrapolation of preliminary NA50 data obtained from $\phi \rightarrow \mu\mu$ channel would result in a higher value. Results of NA49 and NA50 on ϕ meson production in $Pb + Pb$ 160 GeV/n collisions are not necessarily in contradiction. NA50 identifies ϕ meson via different - dileptonic decay channel. If ϕ-meson cannot decay into K^+K^- pairs in nuclear medium due to in-medium modification of meson masses ($M_\phi < M_{K^+} + M_{K^-}$) NA49 effectively measures only ϕ mesons decaying outside the hadronic fireball. Inside the hadronic fireball ϕ meson is allowed to decay only via other hadronic channels or into dileptons. Whether this mechanism can also explain difference of NA50 and NA49 transverse mass slope parameters of ϕ-meson in $Pb + Pb$ collisions requires a more detailed quantitative study.

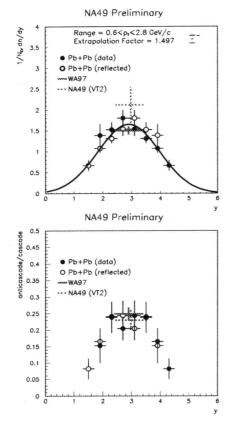

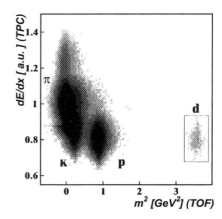

Figure 5. Combined TOF and dE/dx measurement used for identification of deuterons in $Pb + Pb$ data.

6 Deuteron production

Simultaneous TOF and dE/dx measurement allowed to study deuteron production in central $Pb + Pb$ events. By the comparison with coalescence model[10] numerical values for coalescence probability factor B_2 and "effective" size parameter R_G can be determined[9]. An interesting conclusion from the slopes of the transverse mass distribution of deuterons and nucleons is that the transverse density profile of the hadronic fireball appears to be of non-Gaussian - box shape[9].

Figure 4. Rapidity distribution of Ξ^- baryons (top) and dependence of the ratio $\bar{\Xi}^+/\Xi^-$ on rapidity (bottom) obtained from 400.000 $Pb+Pb$ central collisions.

Comparison of normalized ϕ meson yield $\langle\Phi\rangle/\langle\pi\rangle$ in $Pb + Pb$ and $p + p$ collisions gives an enhancement factor 3.0 ± 0.7 in $Pb + Pb$ data.

5 Multistrange baryons

Ξ^-, $\bar{\Xi}^+$ and Ω^-, $\bar{\Omega}^+$ multistrange baryons have been clearly identified in invariant mass distributions of their decay products. In Figure 4 rapidity distribution of Ξ^- hyperons and also the ratio $\bar{\Xi}^+/\Xi^-$ are shown. The large acceptance of NA49 allows to determine the total yields of $\langle\Xi^-\rangle = 4.4 \pm 0.3$ and $\langle\bar{\Xi}^+\rangle = 0.74 \pm 0.04$ in central $Pb + Pb$ data.

References

1. E.B.Gregory et al., *Phys.Rev.* D **62** (2000) 054508
2. S.Afanasiev et al.,*Nucl. Instr. Meth.* A **430** (1999) 1585
3. J.I.Kapusta et al., *Phys. Rev.* D **33** (1986) 1304
4. C.Roland, *PhD thesis*, Frankfurt (1999)
5. F.Sikler, Nucl.Phys. **A661** (1999) 45c.
6. A.Shor, *Phys.Rev.Lett.* **54** (1985) 1122
7. N.Willis, *Nucl.Phys.A* **661** (1999) 534c
8. R.Barton, *Proceed. of Strangeness 2000 conf.*; Berkeley, California (in print)
9. NA49 Coll. *Phys.Lett.B* **486** (2000) p.22
10. R.Scheibl et al., *Phys.Rev.C* **59** (1999) 1585

Δ^{++} PRODUCTION AND ELLIPTIC EMISSION OF K^+ AND π^+ IN 158 A·GEV PB + PB COLLISIONS

YASUO MIAKE FOR WA98 COLLABORATION

Institute of Physics, University of Tsukuba, Tsukuba, 305 Japan
E-mail: miake@tac.tsukuba.ac.jp

The Δ^{++}-resonance production and azimuthal angular correlations for K^+ and π^+ have been studied in 158 A GeV ^{208}Pb$+^{208}$Pb collisions at the CERN SPS. The Δ^{++} production was estimated from the invariant mass spectrum of pπ^+-pairs by subtracting a mixed event background. An event-by-event analysis of the azimuthal angular correlation with respect to the reaction plane shows that in semi-central collisions, K^+ mesons are preferentially emitted out of the reaction plane, while π^+ mesons are emitted in the reaction plane.

1 Hadron Production in AA Collisions

Relativistic heavy ion collisions provide a unique tool for the study of nuclear properties at high temperature and density. Nuclear matter formed in high energy heavy ion collisions is dense and energetic enough so that quarks are no longer confined to each individual nucleon but move freely in a relatively large volume. This new phase of matter is called the quark gluon plasma (QGP). According to the lattice QCD calculation, the phase transition takes place at around 200 MeV of temperature.

Enthusiasm to create the QGP in the laboratory has initiated the heavy ion programs both at BNL-AGS providing heavy ion beams at 12 - 15 AGeV and at CERN-SPS with the beams at 160 - 200 AGeV. Hadron production have been intensively studied both at the AGS and SPS. Study of the single particle spectra reveals that invariant differential cross section shows exponential in mt, as is known as "mt scaling" in pp or pA collisions. While in pp or pA the slope of the exponential are the same for all the particle species, the inverse slope of the exponential distributions in AA are found to be proportional to the mass of the particles. This striking feature of the single particle spectra in AA is observed both at AGS and SPS. The HBT two particle correlations have been studied sys-

tematically and an interesting feature is also seen in AA. For a static source of particles, the HBT correlation function is known to depend only on relative momenta of the pair; it should be independent of sum momentum of the pair. However, the measurements at AGS and SPS show a clear dependence on the sum momentum of the pair.

Both features of the data are attributed to an existence of collective flow. Using the expanding source model[1], in which local thermal equilibrium and independent transverse and longitudinal collective flow are assumed, the single particle spectra and the HBT correlations have been studied. One set of the expansion velocity β and the temperature T have been shown to explain both features of the data[2].

In the analysis of the single particle spectra, however, one problem is seen at low pt region of pion. Observed pion spectra clearly deviate from the model prediction at low pt region, which has been claimed as contributions from resonances such as Δ's, but, direct confirmation has been awaited at SPS energies.

From the success of expanding fireball model, the importance of the collective flow has been well demonstrated in central Pb+Pb collisions. From a hydrodynamical view, collective flow is governed by pressure gradients in the compressed nuclear matter during the early stage of the collision. If the phase transition from ordinary nuclear matter to QGP takes place, the equation of state should exhibit a softening due to the increased number of degrees of freedom[3]. The WA98 experiment has studied other modes of flow effect in non-central Pb+Pb collisions from an event-by-event analysis of azimuthal angular correlations.

2 WA98 Experiment

The WA98 experiment has a large acceptance photon and hadron spectrometer. Charged particles, produced in the collisions, traverse a large magnet (1.6 Tm) and are deflected into two tracking arms. For the analysis reported in this letter, the tracking arm on the positively charged particles is used. The tracking arm consists of two planes of Multi-Step Avalanche Chambers and two planes of streamer tube arrays. The particle identification is based on a measurement of momentum and time-of-flight. The tracking arm has a momentum resolution of $\Delta p/p \simeq 0.97\% + 0.16\%p + 0.023\%p^2$ (p in GeV/c) and a time-of-flight resolution of < 90 ps.

For the reference of azimuthal angular correlation analysis, the Plastic Ball detector is used[4]. It has full azimuthal coverage in the pseudo-rapidity range of $3.5 < \eta < 5.5$ and identifies pions, protons, deuterons, and tritons with kinematic energies of 50 to $250 MeV$ by $\Delta E - E$ measurement. Detailed information about the experiment can be found elsewhere[4,5,6].

3 Δ^{++} Production

The Δ^{++} production has been estimated from the invariant mass spectrum of $p\pi^+$-pairs by subtracting a mixed event background, assuming that the invariant mass spectrum of the mixed event has the same shape as the combinatorial background in the spectrum of the real event[6].

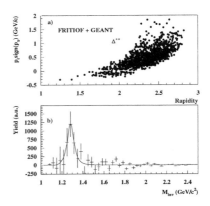

Figure 1. a) The distribution of accepted Δ^{++} obtained from Fritiof events. b) The extracted Δ^{++} resonance together with the corresponding acceptance-filtered modified Breit-Wigner.

Fig. 1a) shows the distribution of accepted Δ^{++} obtained from FRITIOF7.02 events filtered through the acceptance of the spectrometer. Fig. 1b) shows the invariant mass spectrum of the real events subtracted with the combinatorial background;

$$F(M_{inv}) - N_{ev} \cdot N_{pair} \cdot (1 - \xi)g(M_{inv}), (1)$$

where $F(M_{inv})$ is the invariant mass distribution in the real event, N_{ev} is the number of events and $g(M_{inv})$ is the unit normalized mixed event spectrum. The solid curve in the figure is the modified Breit-Wigner function with acceptance filter obtained from the best fit.

From the observed Δ^{++}/proton ratio, the $\Delta(1232)$/nucleon ratio is obtained as 0.62 ± 0.38 in central Pb+Pb collisions. Statistical error is the dominant source of the error. For the exclusive measurement of Δ's, both decayed particles need to get out of the system without further rescatterings. In order to evaluate the contribution of Δ's to the inclusive pion spectra, it is required to estimate the rescattering probability of protons from Δ. An intra-nuclear cascade model was used for this analysis. It is suggested that Δ gives the major contribution to the low pt enhancement of pions[7].

4 Elliptic Flow

The reaction plane is determined as the azimuthal direction, Φ_0, opposite to the total transverse momentum vector of fragments (p, d, and t) observed by the Plastic Ball detector[4]. The Φ_0 is determined in the laboratory as,

$$\Phi_0 = -\tan^{-1}\left(\frac{\sum_{i=1}^{N} p_{Ti}\sin(\phi_i)}{\sum_{i=1}^{N} p_{Ti}\cos(\phi_i)}\right) \quad (2)$$

where the sum runs over all fragments. Here ϕ_i and p_{Ti} are the azimuthal angle in the laboratory and the transverse momentum of the i-th fragment, respectively. The Φ_0 distribution in the laboratory has less than 2% variation. Correction of this effect has been done, but no significant difference in the results.

To test the detector bias and also to evaluate the resolution of the reaction plane determination, we divide each event randomly into two subevents and determine the azimuthal direction of the total transverse momentum in each subevent, Φ_a and Φ_b. Fig. 2 shows the $\Phi_a - \Phi_b$ correlation in (a) semi-central and (b) central collisions. Clear correlation observed in semi-central collision demonstrates that the total transverse momentum vector is well defined. In the figure, mixed-events are also shown. It clearly shows that the effect is not due to a detector bias.

Fig. 3 shows the azimuthal distributions of K^+ and π^+ mesons with respect to the reaction plane for semi-central and central Pb+Pb collisions. Since the acceptance of π^+ crosses mid-rapidity, Φ_0 was rotated by 180° for pions in the region forward of mid-rapidity. Otherwise weeker effect is seen due to the cancellation of the effect. For π^+ mesons in semi-central collisions, the azimuthal distribution indicates weak maxima at $\phi - \Phi_0 = 0^o$ and $\pm 180^o$, which indicates an enhanced emission in the reaction plane. On the other hand, the K^+ azimuthal distribu-

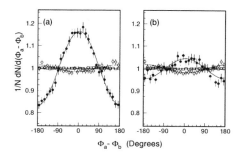

Figure 2. a) The distribution of differences between the total transverse momentum directions of two randomly divided subevents for (a) semi-central and (b) central collisions. Solid circles are for the real event. Open squares are for mixed events allowing multiple module hits, while open circles forbidding.

tion in semi-central collisions exhibits maxima at $\phi - \Phi_0 = \pm 90^o$ which shows an enhanced emission out of the reaction plane. The K^+ emission axis is found to be orthogonal to the in-plane emission axis of the π^+. In central collisions, as shown in the figure, the azimuthal distributions for both particle types are nearly flat, which is consistent with the azimuthal symmetry of the system in central collisions.

To test for detector effects, mixed-events were analyzed in the same manner as the real events. As shown as histograms in the figure, the azimuthal distributions for the mixed-events are flat, indicating that the observed anisotropies are not due to the detector bias.

The strength of the anisotropies is evaluated in terms of a Fourier expansion as,

$$\frac{1}{N}\frac{dN}{d(\phi - \Phi_0)} = 1 + 2v_1\cos(\phi - \Phi_0) + 2v_2\cos(2(\phi - \Phi_0)). \quad (3)$$

The value of the Fourier coefficients v_1 and v_2 are evaluated with the correction for the effects of the resolution of the reaction plane determination. See the reference[5] for details.

Obtained strength of the elliptic flow, v_2's are compared with other measurements. In Fig. 4, the v_2 values for protons, pions, and kaons near mid-rapidity are plotted as

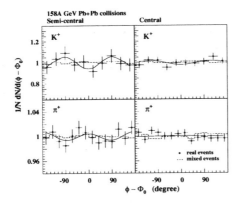

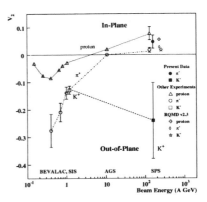

Figure 3. The azimuthal distributions of K^+ and π^+ mesons with respect to the reaction plane for semi-central and central 158 A·GeV Pb + Pb collisions. Solid circles with error bars show the real events and dashed histgrams show the mixed events.

Figure 4. Beam energy dependence of the v_2 value near mid-rapidity. Solid symbols indicate the present data. The RQMD (v2.3 cascade mode) calculations for proton, π^+ and K^+ in 158 A·GeV Pb + Pb collisions are also shown for the impact parameter range $b = 6.5 - 12$ fm with the filter of experimental acceptance.

a function of the beam energy. Striking feature is the transition from out-of-plane to in-plane emission for both pions and protons at around 5-10 A·GeV. At SPS energy, results from NA49 have shown that protons and pions exhibit in-plane emission near mid-rapidity[8]. Our π^+ data agree with the NA49 results within errors. Unlike protons and pions, the present results indicates that K^+ mesons exhibit out-of-plane emission.

As shown in the figure, the RQMD calculation agrees with the measured results for π^+ and proton, but it fails to reproduce the out-of-plane elliptic emission of K^+. These results suggest an importance of the in-medium potentials of the K^+[9]. A simple model calculation[10], in which K^+ mesons propagate through a static anisotropic distribution of nucleons with a repulsive K^+N potential, demonstrates that a final out-of-plane elliptic emission pattern can emerge from an initially isotropic azimuthal distribution of K^+. Further theoretical analysis is required.

References

1. U. Heinz, *et al.*,, Phys.Lett. B382 (1986) 181.

2. H. Appelshauser, *et al.*,, Eur. Phys. J. C2 (1998) 661.

3. D.H. Rischke, Nucl. Phys. A610(1996)88c .

4. M.M. Aggarwal,*et al.*, Nucl-ex/9807004.

5. M.M. Aggarwal, *et al.*,, Phys.Lett. B469 (1999) 30.

6. M.M. Aggarwal, *et al.*,, Phys.Lett. B477 (2000) 37.

7. Private communication with Susumu SATO.

8. H. Appelshäuser, *et al.*, Phys. Rev. Lett. 80 (1998) 4136;

9. G.Q. Li and C.M. Ko, Nucl. Phys. A594 (1995) 460; G.Q. Li, C.M. Ko and B.A. Li, Phys. Rev. Lett. 74 (1995) 235.

10. K. Enosawa, Doctor thesis at University of Tsukuba.

HADRONIZATION OF QUARK GLUON PLASMA (AND GLUON JETS) AND THE ROLE OF THE 0++ GLUEBALL AS PRIMARY HADRON PRODUCTS

PETER MINKOWSKI

Institute for Theoretical Physics, University of Bern, Sidlerstrasse 5, CH-3012 Bern, Switzerland
E-mail: mink@itp.unibe.ch

SONJA KABANA

Laboratory for High Energy Physics, University of Bern, Sidlerstrasse 5, CH-3012 Bern, Switzerland
E-mail: sonja.kabana@cern.ch

WOLFGANG OCHS

Max Planck Institut fuer Physik, Werner Heisenberg Institut, Foehringer Ring 6, D-80805, Muenchen, Germany
E-mail: wwo@mppmu.mpg.de

Signatures of dominant and central production of glueballs (binary gluonic mesons) in heavy ion and hadromic collisions are discussed. Search strategies are proposed.

1 Introduction

The three sequences of binary gluonic mesons (gb) are reviewed [1], [2], represented by the respective gb resonances with lowest mass and J^{PC} quantum numbers : 0^{++}, 0^{-+} and 2^{++}. While the 2^{++} sequence is associated through Regge analytic continuation in angular momentum of two body elastic amplitudes to the Pomeron trajectory, the triple Pomeron vertex is thought to be responsible for the multiparticle production of mainly 0^{++} glueballs. We discuss search strategies for the main 0^{++} component and also for the heavier 2^{++} state in heavy ion and hadronic inelastic scattering at high energy and high (initial) energy density under the hypothesis that they become the dominant primary systems of multiparticle production in this environment [3]. We propose in particular to impose centrality selections also in hadronic reactions, e.g. $p\bar{p}$ collisions at the Tevatron, in order to discover eventual transitory behaviour bearing

a similarity to kaon distributions in Pb Pb collisions at the SPS as analyzed in ref. [4], [5].

2 Discussion

The three J^{PC} series of binary gluonic mesons are

$$
\begin{array}{llllll}
0^{++} & 0^{++} & 2^{++} & 4^{++} & \cdots \\
2{+}{+} & & 2^{++} & 3^{++} & 4^{++} & 5^{++} \cdots \\
0^{-+} & 0^{-+} & 2^{-+} & 4^{-+} & \cdots
\end{array} \tag{1}
$$

We envisage two temporal developments of a collision :

a) quark gluon plasma formation

initial thermal equilibrium as appropriate for an expanding medium within the quark gluon plasma phase. In the further development the hadronisation pro-

592

cess from within the plasma phase coincides closely in time with chemical freezout.

b) no quark gluon plasma formation

initial thermal equilibrium occurs within the hadronic phase with subsequent chemical freezout.

It is conceivable, that for a given centre of mass collision energy both phases a) and b) above occur depending on the centrality of the collision. Then phase a) is distinguished by the independence of the thermodynamic intensive variables, mainly temperature and chemical potentials for baryon number and strangeness $T \sim T_{cr}$, μ_b and μ_s from the initial energy density, prior to thermalization ε_0 [7].

It is for the case of phase a) that we expect dominant production of gb (0^{++}) to occur upon chemical freezout. We discuss the possibility of characteristic $\pi^+\pi^-$ as well as $K\overline{K}$ invariant mass and relative momentum distributions to remain observable notwithstanding further hadronic collisions before final thermal freezout.

The following properties appear characteristic

a1) dipion invariant mass distribution :

Ordering all pions produced in rapidity and relative momentum the invariant mass spectrum corresponding to gb (0^{++}) production should show the interference pattern first observed in central production in p p collisions at $\sqrt{s} = 63 GeV$ by the AFS collaboration [6]

a2) $K\overline{K}$ invariant mass distribution :

The high mass tail of gb (0^{++}) as well as a separate peak from production of gb (2^{++}) can be observed in the invariant mass distribution of $K\overline{K}$ pairs.

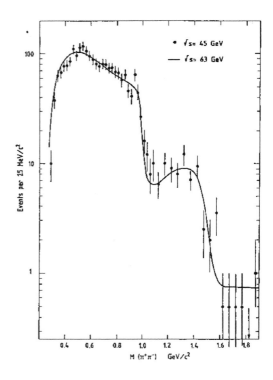

Figure 1. Invariant mass distribution of pion pairs centrally produced in pp collisions at ISR energies [6]

We further propose to look for similar invariant mass distribution in multijet dominated collisions [8].

Conclusions

We have presented search strategies for dominant production of binary gluonic mesons, mainly gb (0^{++}) and to a lesser extent gb (2^{++}) in high energy and high initial energy density heavy ion as well as hadronic collisions, whence a transition through the equilibrated quark gluon phase takes place. If these strategies prove successful, dominant glueball production may consitute a direct and clear signature of the quark gluon phase and its hadronic transition.

Acknowledgments

I thank Wolfgang Ochs and Sonja Kabana for fruitful and challenging questions and discussion.

References

1. H. Fritzsch and P. Minkowski, Nuovo Cim. 30A (1975) 393.
2. P. Minkowski and W. Ochs, European Physical Journal C9 (1999) 283, hep-ph/9811518.
3. S. Kabana and P. Minkowski, Phys.Lett. B472 (2000) 155-160, hep-ph/9907570, S. Kabana and P. Minkowski, hep-ph/9909351.
4. S. Kabana, hep-ph-0004138.
5. S. Kabana, contribution to SQM2000 and Univ. of Bern preprint BUHE-00-06.
6. AFS Coll., T. Akesson et al., Nucl. Phys. B264 (1986) 154.
7. S. Kabana and P. Minkowski, work in progress.
8. P. Minkowski and W. Ochs, Phys.Lett. B485 (2000) 139, hep-ph/0003125.

PHOTON AND LEPTON PAIR PRODUCTION IN A QUARK-GLUON PLASMA

P. AURENCHE

Laboratoire de Physique Théorique LAPTH, BP110, F-74941 Annecy-le-Vieux Cedex, France
E-mail: aurenche@lapp.in2p3.fr

We discuss the production of real or virtual photons in a quark-gluon plasma.

It has long been thought that electromagnetic probes *i.e.* real or virtual photons would provide a way to detect the formation of a quark-gluon plasma in ultra-relativistic heavy ion collisions. The energy distribution of the photons would allow to measure the temperature of the plasma provided the rate of production in the plasma exceeds that of various backgrounds. It is expected that this will occur in a small window in the GeV range for the energy of the photon. At lower values of the energy the rate is dominated by various hadron decay processes while at higher values the usual hard processes (those occurring in the very early stage of the collision before the plasma is formed), calculable by standard perturbative QCD methods, would dominate. In contrast to hadronic observables (or heavy quarkonia) which are sensitive to the late evolution of the plasma as well as to the re-hadronisation phase, the photons in the GeV range are produced soon after the plasma is formed and then they escape the plasma without further interaction.

We assume the plasma in thermal equilibrium (temperature T) with vanishing chemical potential. The rate of production, per unit time and volume, of a real photon of momentum $Q = (q_o, \boldsymbol{q})$ is

$$\frac{dN}{dt d\boldsymbol{x}} = -\frac{d\boldsymbol{q}}{(2\pi)^3 2q_o} \, 2 n_B(q_o) \operatorname{Im} \Pi^{R}{}_\mu{}^\mu(q_o, \boldsymbol{q}) \, , \tag{1}$$

while for a lepton pair of mass $\sqrt{Q^2}$ it is

$$\frac{dN}{dt d\boldsymbol{x}} = -\frac{dq_o d\boldsymbol{q}}{12\pi^4} \, \frac{\alpha}{Q^2} \, n_B(q_o) \operatorname{Im} \Pi^{R}{}_\mu{}^\mu(q_o, \boldsymbol{q}) \, , \tag{2}$$

where $\Pi^{R}{}_\mu{}^\mu(q_o, \boldsymbol{q})$ is the retarded photon polarisation tensor. The pre-factor $n_B(q_o)$ provides the expected exponential damping $\exp(-q_o/T)$ when $q_o \gg T$. This report is devoted to the study of Π^{R} which contains the strong interaction dynamics of quarks and gluons in the plasma. The theoretical framework is that of the effective theory with re-summed hard thermal loops (HTL)[1].

We briefly review the status of $\operatorname{Im} \Pi^{R}$ calculated up to the two-loop approximation. Some phenomenological consequences are mentioned. Then we turn to a discussion of higher loop corrections.

1 The two-loop approximation

Following the HTL approach[1] one distinguishes two scales: the "hard" scale, typically of order T or larger (the energy of quarks and gluons in the plasma) and the "soft" scale of order gT where g, the strong coupling, is assumed to be small. Collective effects in the plasma modify the physics at scale gT *i.e.* over long distances of $\mathcal{O}(1/gT)$. These effects lead to a modification of the propagators and vertices of the theory and one is led to introduce effective (re-summed) propagators and vertices. This is easily illustrated with the example of the fermion propagator, $S(P)$, which in the "bare" theory is simply $1/p$ (we neglect spin complications and make only a dimensional analysis). The thermal contribution to the one loop correction $\Sigma(p)$ is found to be $\Sigma(p) \sim g^2 T^2 / p$ which is of the same order as the inverse propagator when

Figure 1. One-loop contribution.

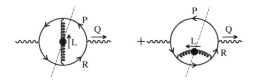

Figure 2. The dominant two-loop contributions.

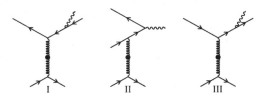

Figure 3. Physical processes included in the diagrams of Fig. 2, in the region $L^2 < 0$. I: bremsstrahlung with an antiquark. II: $q\bar{q}$ annihilation with scattering. III: bremsstrahlung with a quark.

p is of order gT. The re-summed propagator $^*S(P) = 1/(p - \Sigma(p))$ is then deeply modified for momenta of $\mathcal{O}(gT)$ whereas the thermal corrections appear essentially as higher order effects for hard momenta. Likewise, the gluon propagator and vertices are modified by hard thermal loops when the external momenta are soft[1]. One can construct an effective Lagrangian in terms of effective propagators and vertices and calculate observables in perturbation theory.

In the one-loop approximation, the photon production rate is given by the diagram shown in fig. 1 where the symbol $\bullet$ means that effective propagators and vertices are used. The result has been known for some time and can be expressed, in simplified notation, as[2,3]

$$\operatorname{Im}\Pi^R(q_o, \boldsymbol{q}) \sim e^2 g^2 T^2 \left(\ln(\frac{q_0 T}{m_q^2}) + C(\frac{Q^2}{m_q^2}) \right) \tag{3}$$

where $m_q^2 \sim g^2 T^2$ is related to the thermal mass of the quark. One notes the presence of a "large" logarithmic term $\ln(1/g)$ dominating over a "constant term" $C(Q^2/m_q^2)$.

The two-loop diagrams are displayed in fig. 2. In principle, there are more diagrams in the effective theory but only those leading to the dominant contribution are shown. All propagators and vertices should be effective but since the largest contribution arises from hard fermions it is enough, following the HTL strategy, to keep bare fermion propagators and vertices as indicated[a]. Only the gluon line needs to be effective since soft momentum L through the gluon line dominates the integrals. To evaluate these diagrams it is

---

[a] Note that for consistency of our approach, based on an expansion in terms of effective quantities, we keep the thermally generated mass in the hard limit of the effective propagator.

convenient to distinguish between the contribution arising from a time-like gluon ($L^2 > 0$) and a space like gluon ($L^2 < 0$). The first type leads to a contribution similar to eq. (3) and requires some care as counter-terms (not shown) eliminate the parts of the two-loop diagrams already contained in the one-loop diagrams[4]. We concentrate on the second case which in terms of physical processes corresponds to bremsstrahlung production of a photon or production in a quark-antiquark annihilation process where one of the quark is put off-shell by scattering in the plasma (see fig. 3). The result for hard photons is[5]

$$\operatorname{Im}\Pi^R(q_o, \boldsymbol{q})\Big|_{\mathrm{brems}} \sim e^2 \, g^2 T^2 \tag{4}$$

$$\operatorname{Im}\Pi^R(q_o, \boldsymbol{q})\Big|_{\mathrm{annil}} \sim e^2 \, g^2 T q_0 \tag{5}$$

The reason why these two-loop contributions have the same order as the one-loop one is due to the presence of strong collinear singularities. To calculate $\operatorname{Im}\Pi^R$ one has to cut the propagators as indicated by the dash-dotted lines in fig. 2. In the integration over the loop hard momentum P (with P^2, $(R + L)^2$ on shell) the denominators R^2 and $(P + L)^2$ of the un-cut fermion propagators simultane-

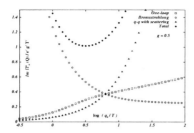

Figure 4. Comparison of various contributions to $\mathrm{Im}\,\Pi^\mu_\mu(Q)$ for a hard real photon. The comparison is made for $N = 3$ colors and $N_F = 2$ flavors.

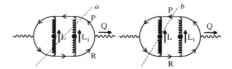

Figure 5. A ladder diagram.

ously almost vanish when $\boldsymbol{p}$ is parallel to $\boldsymbol{q}$ i.e. in the collinear configuration. This leads to an enhancement factor of type T^2/M^2_{eff} where the cut-off $M^2_{\mathrm{eff}} = m^2_q + p(p+q_0)Q^2/q^2_0$ emerges from the calculation. For the kinematic range of concern to us here, $M^2_{\mathrm{eff}} \sim g^2 T^2$ so that the two-loop diagram is enhanced by a factor $1/g^2$ which compensates the g^2 factor associated to the coupling of the gluon to the quarks. An interesting result of the calculation is the importance of process II of fig. 3 which grows with the energy of the photon and dominates over the other contributions when $q_0/T \gg 1$ as shown in fig. 4. Phenomenological applications of these results have been carried out and the two-loop processes have been included in hydrodynamic evolution codes to predict the rate of real photon production at RHIC or LHC[6]. It is found that the two-loop processes (especially the annihilation with scattering) lead to an increase by an order of magnitude compared to the one-loop processes. This may even have consequences for heavy ion collisions at SPS energies[7]. Several effects may reduce these over-optimistic predictions: lack of chemical equilibrium and more importantly higher order corrections as discussed next.

2 Higher order corrections

Since the one-loop and two-loop results are of the same order it is reasonable to worry about

the convergence of the perturbative expansion in the effective theory! The enhancement mechanism operative at two-loop could also be at work at the multi-loop level especially in ladder diagrams, an example of which is shown in fig. 5: indeed many "small" fermion denominators appear in such diagrams which can produce a pile-up of collinear singularities. A recent study of the three-loop ladder diagram shows that the extra gluon "rung" yields the results[8]

$$\mathrm{Im}\,\Pi^R\Big|_{3-\mathrm{loop}} \sim \mathrm{Im}\,\Pi^R\Big|_{2-\mathrm{loop}} \times \frac{g^2 T}{l_{min}} \quad (6)$$

where l_{min} is the largest of the cut-offs:
- $l^{(1)}_{min} = M^2_{\mathrm{eff}} q_0/p_0 r_0$, which is the collinear cut-off encountered above: it depends on the thermal quark mass and momentum ($p_0 \sim r_0 \sim T$) as well as on the external variables;
- $l^{(2)}_{min} = m_D \sim gT$, the Debye mass if the added gluon is longitudinal, or $l^{(2)}_{min} = m_{mag} \sim g^2 T$ if it is transverse.

For the kinematic configuration of interest, in the case of an extra longitudinal gluon one can check that $m_D \gg l^{(1)}_{min}$ and the Debye mass acts as a cut-off with the result that the three-loop contribution is suppressed by a factor g compared to the two-loop. On the contrary, for a transverse gluon, both regulators are of order $g^2 T$ (as long as $Q^2/q^2_0 < g^2$) and the three-loop diagram is of the same order as the two-loop one. One is therefore in a non-perturbative regime. The problem is similar to the magnetic mass problem pointed out by Linde in the perturbative calculation of the free energy[9], except that here it appears at leading order.

Another effect which can modify the collinear enhancement mechanism is related to the fermion damping rate. Indeed, including the damping rate on the fermion lines, will shift the pole of the propagators away from the real axis: this affects the enhancement mechanism based on the near-vanishing of the denominators. Ignoring the requirement of gauge invariance and concentrating only on the mathematical effect of shifting the poles to the complex plane one can do again the two-loop calculation with fermion propagators including the damping rate $\Gamma \sim g^2 T \ln(1/g)$. The result is intuitively simple as a regulator of the form[10]

$$\mathcal{M}_{\text{eff}}^2 = M_{\text{eff}}^2 + 4i\Gamma \frac{p_0(p_0 + q_0)}{q_o} \qquad (7)$$

comes out, with M_{eff}^2 defined above. This equation lends itself to a simple interpretation. It can be written as

$$\mathcal{M}_{\text{eff}}^2 = 2 \frac{p_0(p_0 + q_0)}{q_o} \left(1/\lambda_{\text{for}} + i/\lambda_{\text{mean}} \right) \quad (8)$$

where $\lambda_{\text{mean}} = 1/\Gamma$ is the mean free path of the quark in the plasma and $\lambda_{\text{for}} = 2p_0(p_0 + q_0)/M_{\text{eff}}^2 q_0$ can be shown to be the formation length of the photon. Then, if $\lambda_{\text{mean}} \gg \lambda_{\text{for}}$ the effect of the damping rate can be ignored and the corresponding higher order diagrams are suppressed. In the opposite case, re-scattering in the plasma modifies the two-loop result. This is equivalent to say that the Landau-Pomeranchuk-Migdal (LPM) effect has to be taken into account in the calculation[11]. Two interesting features emerge from the study[10]: 1) the LPM effect not only modifies the production of bremsstrahlung photon but also that of very hard photons emitted in the "annihilation with scattering" process; 2) if the virtuality $\sqrt{Q^2}$ of the hard lepton pair is large enough then one falls in the domain $\lambda_{\text{mean}} \gg \lambda_{\text{for}}$ and the perturbative calculation at two-loop is sufficient. The problems discussed above are an illustration of a more general situation concerning thermal Green's function with external momenta close to the light-cone[12].

The production mechanism of hard photons in the plasma is very complex and turns out to be non-perturbative for real or small mass virtual photons. Taking into account higher order effects to obtain a quantitative estimate remains to be done.

Acknowledgments

I thank F. Gelis, R. Kobes and H. Zaraket for a fruitful collaboration on the work discussed above. LAPTH is a CNRS laboratory (UMR 5108), associated to Université de Savoie.

References

1. E. Braaten, R.D. Pisarski, *Nucl. Phys.* B **337**, 569 (1990); B **339**, 310 (1990). J. Frenkel, J.C. Taylor, *Nucl. Phys.* B **334**, 199 (1990); B **374**, 156 (1992).
2. J.I. Kapusta, P. Lichard, D. Seibert, *Phys. Rev.* D **44**, 2774 (1991); R. Baier, H. Nakkagawa, A. Niegawa, K. Redlich, *Z. Phys.* C **53**, 433 (1992).
3. T. Altherr, P.V. Ruuskanen, *Nucl. Phys.* B **380**, 377 (1992). M.H. Thoma, C.T. Traxler, *Phys. Rev.* D **56**, 198 (1997).
4. P. Aurenche, F. Gelis, R. Kobes, H. Zaraket, *Phys. Rev.* D **60**, 076002 (1999).
5. P. Aurenche, F. Gelis, R. Kobes, H. Zaraket, *Phys. Rev.* D **58**, 085003 (1998).
6. D.K. Srivastava, *Eur. Phys. J.* C **10**, 487 (1999); M.G. Mustafa, M.H. Thoma, *Phys. Rev.* C **62**, 014902 (2000).
7. D.K. Srivastava, B. Sinha, nucl-th/0006018.
8. P. Aurenche, F. Gelis, H. Zaraket, *Phys. Rev.* D **61**, 116001 (2000).
9. A.D. Linde, *Phys. Lett.* B **96**, 289 (1980).
10. P. Aurenche, F. Gelis, H. Zaraket, hep-ph/0003326, to appear in *Phys.Rev.* D.
11. L.D. Landau, I.Ya. Pomeranchuk, *Dokl. Akad. Nauk. SSR* **92**, 735 (1953); A.B. Migdal, *Phys. Rev.* **103**, 1811 (1956).
12. F. Gelis, hep-ph/0007087.

THE PHENIX EXPERIMENT AT RHIC

SAMUEL ARONSON for the PHENIX Collaboration
Physics Department, Brookhaven National Laboratory, Upton, NY 11973, USA
E-mail: aronsons@bnl.gov

PHENIX is a large detector at the Relativistic Heavy Ion Collider (RHIC) at BNL. RHIC and PHENIX have recently operated for the first time, producing and detecting collisions of gold ions at beam energies of 30 and 65 GeV per nucleon. The current performance and future plans of PHENIX and of RHIC are presented.

1 Introduction

PHENIX is one of two large experiments at the BNL Relativistic Heavy Ion Collider (RHIC). With the start-up of heavy ion collisions at RHIC in the summer of 2000, PHENIX has commenced data taking. The physics goals, detector status and early experience with collecting data are described in this paper.

Starting in June 2000 RHIC produced collisions of gold ions at beam energies up to 65 GeV per nucleon. Luminosities of about $5 \times 10^{25}/cm^2 \cdot sec$ were achieved. The design goals of RHIC are $5 \times 10^{26}/cm^2 \cdot sec$ at 100 GeV/nucleon for heavy ions and $10^{31}/cm^2 \cdot sec$ at 250 GeV beam energy for polarized protons. Full-energy heavy ion operations and the start of polarized proton running are both expected in 2001. The first run ended in September 2000; RHIC delivered integrated luminosities of 2 to 3 μb^{-1} to each of 4 experiments - BRAHMS, PHENIX, PHOBOS and STAR.

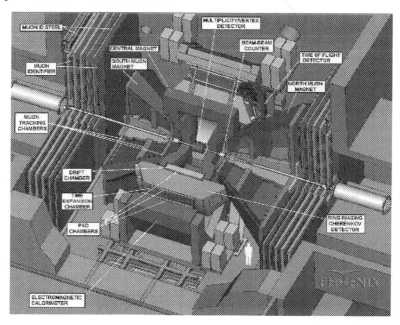

Figure 1.Cutaway view of the PHENIX detector.

2 The PHENIX Detector

The PHENIX detector[1], comprising 4 large aperture spectrometers plus detectors for global event characterization, is depicted in Figure 1. Two central spectrometers view the collision around $\eta = 0$ with high precision tracking, calorimetry and particle identification. Two spectrometers covering the pseudorapidity ranges $1.1 < |\eta| < 2.4$ detect muons. Silicon strip and pad detectors, beam-beam counter (BBC) arrays and zero-degree calorimeters (ZDC) provide collision triggers and charged multiplicity distributions. In all, PHENIX has 12 separate subsystems and about 400,000 channels of electronic readout.

3 Physics Goals

The PHENIX detector design allows for broad research programs in both heavy ion and polarized proton collisions. In relativistic heavy ion collisions, the program is aimed at a study of strongly interacting matter under extreme conditions. In particular, PHENIX will search for and study the so-called Quark-Gluon Plasma (QGP) and the phase transitions that separate the QGP from ordinary hadronic matter. PHENIX is unique among the present generation of RHIC detectors in being able to detect leptons and photons among the particles created in heavy ion collisions. These probes carry information from the early stages of the collision, while hadronic secondaries primarily report on the later stages, after the QGP expands, cools and rehadronizes. Table 1 indicates the probes to which PHENIX is sensitive, and the time-ordered stages of the collision to which they pertain.

In polarized proton collisions, the main focus of the PHENIX program is to study

Table 1. PHENIX physics goals ordered by the timescale in the collision from early to late.

TIMESCALE	PROBE
Initial collision	**Hard scattering** - single "jet" via leading particle; γ+ "jet"
Deconfinement	**High-mass vector mesons** - J/ψ, ψ' screening; Y (non) screening
Chiral restoration	**Low-mass vector mesons** - ρ, ω, ϕ mass, width; ϕ branching ratios
QGP thermalization	**Photons** - π^0, η, η'; continuum direct γ **Dileptons** - non-resonant: 1-3 GeV; soft continuum: < 1GeV **Heavy quark production** - open charm; open charm via single lepton
Hadronization	**Hadrons** - HBT interferometry, π/K; - strangeness production; spectra of identified hadrons
Hydrodynamics	**Global variables** - E_T, dN/dy

the spin of the proton. RHIC provides a unique opportunity to study polarized structure functions, including that of the gluons. Very high energy polarized proton collisions also allow PHENIX to use parity violating asymmetries (e.g., in W production) as sensitive probes of new physics. Figure 2 shows the PHENIX statistical power in the measurement of ΔG, the polarized gluon structure function. The curves indicate different possible values of ΔG.

Figure 2. π^0 asymmetry measurement in polarized proton running versus different estimates of ΔG

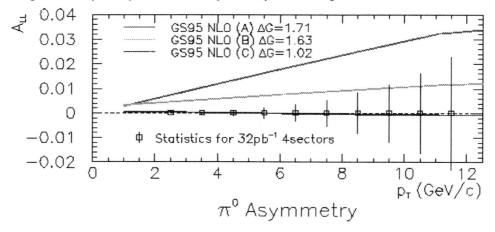

π^0 Asymmetry

4 The 2000 Run

In the year 2000 run PHENIX operated its two mid-rapidity spectrometers, with about half their apertures being read out. From mid-June to mid-September RHIC delivered to PHENIX about 3 μb^{-1} of integrated luminosity. From this exposure PHENIX recorded about 3 million events.

Initial analysis of these data indicates very good detector performance and very good correspondence between the subsystems of the mid-rapidity spectrometers. Figure 3 shows a $\gamma\gamma$ mass distribution from the electromagnetic calorimeter, typical of the performance of PHENIX subsystems in the 2000 run.

With the data collected and reconstructed, PHENIX expects to produce several physics results in the next few months, including:

- $dN_{ch}/d\eta$ in the pseudorapidity interval $|\eta| \leq 0.35$ versus the number of participants;
- $dE_T/d\eta$ versus the number of participants in the same η interval;
- Spectra of identified charged and neutral hadrons out to $p_T > 5$ GeV/c;
- Collective effects such as elliptic flow.

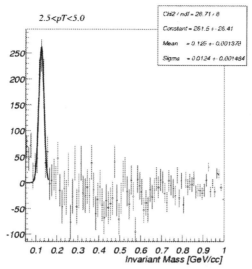

Figure 4. γγ invariant mass reconstructed in the calorimeter. Combinatorial background has been subtracted using mixed events.

5 Future Plans, Outlook

Work is now under way to install the first of two muon spectrometers and to complete the readout of the mid-rapidity spectrometers. These enhancements will be in place for the next run, expected to start in March of 2001. Increased aperture in the central arms and the added muon capability will allow PHENIX to start collecting dilepton events, in particular at the J/ψ and ψ′. This will give PHENIX access to one of the most interesting putative signatures of the QGP, suppression of J/ψ and ψ′ production[2] via color screening.

Beyond that, PHENIX will complete its second muon spectrometer and begin on a program of upgrades. These upgrades will primarily target improved particle identification, triggering and background rejection in the lepton signatures. Such detector enhancements, plus increased bandwidth for data acquisition, will allow PHENIX to pursue rare phenomena at the highest luminosities RHIC will eventually deliver.

Acknowledgments

This work is supported in part by grant DE-AC02-98CH10886 from the U. S. Department of Energy

References

1. D. P. Morrison in *Proceedings of the Yamada Conference XLVIII on Quark Matter '97*, eds. T. Hatsuda, Y. Miake, S. Nagamiya and K. Yagi (Yamada Science Foundation and Elsevier Science Publishers, Tsukuba, 1998).

2. T. Matsui and H. Satz, *Phys. Lett.* **B178**, 416 (1986).

BARYON STOPPING: A LINK BETWEEN ELEMENTARY P+P INTERACTIONS AND CONTROLLED-CENTRALITY P+A AND A+A COLLISIONS

FERENC SIKLÉR FOR THE NA49 COLLABORATION

CERN, CH-1211 Geneva 23, Switzerland

E-mail: Ferenc.Sikler@cern.ch

New data on p+p interactions from the NA49 detector are used to study particle yields and distributions as function of final state proton momentum. The observed correlation functions are then folded with net proton momentum distributions from controlled-centrality p+A and A+A collisions measured with the same detector. Some resulting predictions for pion and kaon yields in p+A and A+A reactions are compared to data.

1 Motivation

We study nuclear reactions in order to elucidate the strong interaction process and study its development in time and space with the hope to establish experimental links between elementary and complex interactions.

2 The experiment, centrality

The NA49 experiment is a large acceptance detector for charged hadrons[1]. The tracking of particles is based on four large volume time projection chambers, the particle identification is possible via the measurement of their specific energy loss (dE/dx).

One of the key parameters in the systematics of hadron production is the centrality of the collision: the number of collisions per participant nucleon.

In case of p+A reactions the centrality is known to be correlated with the number of "grey" particles – slow protons and deuterons – which are measured by a centrality detector surrounding the target. Together with those grey particles identified in the acceptance of the tracking system, an overall acceptance of 50% has been achieved. Their number can be correlated with the number of participating nucleons, using simulation.

In Pb+Pb reactions the deposited energy in the zero degree calorimeter was used to select events with given centrality. It was then

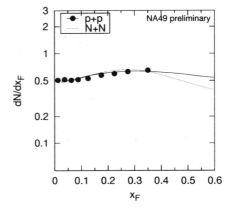

Figure 1. Feynman-x distribution of net protons (p – p̄) in the p+p reaction. The lines through the measured points are to guide the eye. An isospin corrected curve is also given (N+N), serving as a base of comparison to A+A.

correlated with the average number of collisions a nucleon undergoes, using simulation.

The data presented here are preliminary results from p+p, p+Al, p+Pb and Pb+Pb reactions at 158 GeV·A energy.

Charged particle yields measured in the forward hemisphere will be shown. This choice is only partially due to the good acceptance and particle identification capabilities: when comparing p+A collisions to symmetric reactions, in this hemisphere it is possible to study the fragmentation of the projectile that went through multiple collisions.

Two main topics are discussed: stopping

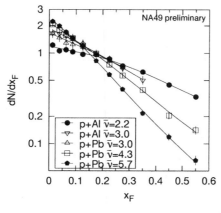 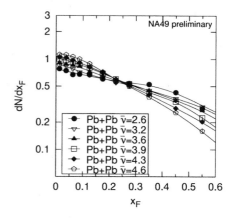

Figure 2. Feynman-x distribution of net protons ($p - \bar{p}$) with different average number of collisions ($\bar{\nu}$) in a) p+Al and p+Pb collisions, b) Pb+Pb collisions. Lines are to guide the eye.

and baryon transfer at first, then correlations and predictions.

3 Stopping and baryon transfer

The longitudinal momentum distribution of net protons from non-single-diffractive p+p interactions has been measured (Fig. 1) and plotted in the rest frame of the collision (Feynman-x distribution, x_F).

This can be compared to the measurements of the net proton distribution in centrality selected p+A reactions (Fig. 2.a). With increasing centrality (increasing average number of collisions) the final state protons appear to be more and more stopped, having the biggest effect in the most central sample. The evolution is smooth, connecting p+p via p+Al to p+Pb.[a]

In Pb+Pb collisions a similar gradual evolution of stopping is seen (Fig. 2.b), although not reaching the high level observed in most central p+Pb.

4 Correlations

A very characteristic systematics of the net proton distribution as function of $\bar{\nu}$ is ob-

served in p+A and A+A reactions.

What about p+p collisions? Although here $\nu = 1$ by definition, the degree of inelasticity of the interaction can be characterized by the x_F of the final state proton. This is exemplified in the following by inspecting the correlations of hadronic variables with x_F^p.

4.1 Pion density

In a minimum bias p+p collision there are three positively and three negatively charged pions in the forward hemisphere on average. The average number of charged pions $\langle \pi \rangle$ strongly correlates with the x_F of the final state proton (Fig. 3.a): a fast proton will be accompanied by a few, a slow one by many pions. If the proton is almost at rest in the center of mass system, five pions are produced on average.[b]

Using the $\langle \pi \rangle - x_F^p$ correlation measured in p+p, predictions can be made for other reactions by folding their – above discussed – proton distribution with this correlation curve. For a reaction that has slower protons

---

[a] Note the similarity of p+Al and p+Pb in the overlap region at $\bar{\nu} \approx 3$.

[b] The slight decrease at low x_F values is due to the fact that here the proton may not be the fastest baryon that is likely to be a neutron. The exclusion of fast neutrons would strengthen the correlations and predictions.

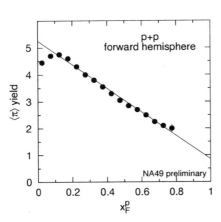
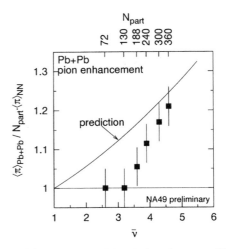

Figure 3. *a)* Average number of charged pions ($\langle\pi\rangle = (\pi^+ + \pi^-)/2$) in the forward hemisphere in p+p collision if the fastest proton has longitudinal momentum x_F^p. *b)* Pion enhancement in Pb+Pb collisions relative to minimum bias p+p as function of $\bar{\nu}$. Measured points and predicted line using p+p correlations are shown.

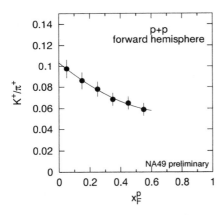
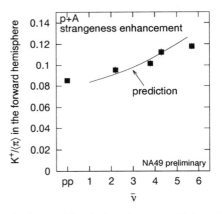

Figure 4. *a)* Ratio of number of positive kaons and pions in the forward hemisphere in p+p collisions if the fastest proton has longitudinal momentum x_F^p. *b)* Strangeness enhancement in p+A collisions in the forward hemisphere as function of $\bar{\nu}$. Measured points and predicted line using p+p correlations are shown.

one would extrapolate bigger pion density. This is what happens in more and more central Pb+Pb collisions: the observed increase of pion density with centrality is close to the prediction from p+p interactions (Fig. 3.b).

4.2 Strangeness

A similar study can be performed concerning the strangeness content of the collision. Here the K/π ratio in the forward hemisphere is presented. The minimum bias p+p collisions show a correlation of this quantity with the x_F of the final state proton: the slower the final state proton the higher is the corresponding strangeness content (Fig. 4.a).

Folding this with the stopping curve of p+A and Pb+Pb reactions a prediction can be made on K$^+$/$\langle\pi\rangle$ as function of number of collisions. The p+A data are quite close to the line of the prediction (Fig. 4.b), also in the case of Pb+Pb collisions a strangeness enhancement is predicted, though the data

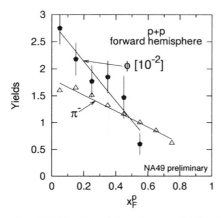 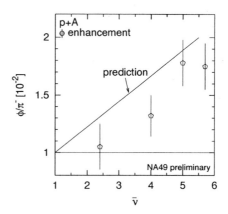

Figure 6. *a)* Yields of negative pions and ϕ in the forward hemisphere in p+p collision if the fastest proton has longitudinal momentum x_F^p. *b)* ϕ enhancement in p+A collisions in the forward hemisphere as function of $\bar{\nu}$. Measured points and predicted line using p+p correlations are shown.

indicate higher values (Fig. 5).[c]

Another example is the $\phi(1020)$ meson that carries hidden strangeness. The ratio ϕ/π^- in the forward hemisphere increases if the event has a slower proton (Fig. 6.a).

---

[c]This could be understood partially: in p+p an asymmetric selection has been performed by fixing the fastest proton in the projectile hemisphere only. In Pb+Pb reactions the proton distribution is event-by-event symmetric thus requiring similar stopping for protons on both sides.

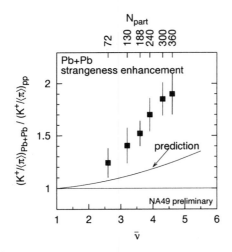

Figure 5. Strangeness enhancement in Pb+Pb collisions as function of $\bar{\nu}$. Measured points and predicted line using p+p correlations are shown.

This leads to a prediction of ϕ enhancement in p+A that is in agreement with the data (Fig. 6.b).

5 Conclusions and outlook

The NA49 experiment identifies particles in the forward hemisphere with controlled centrality. Hadronic reactions with different targets, projectiles and energies are studied.

It was shown that the leading proton can characterize the interaction: its final momentum correlates with measured quantities, like multiplicity and particle ratios. Using this internal structure of the p+p interaction, predictions can be made on p+A and A+A collisions. This is to be confronted with the usual method comparing with minimum bias p+p collision.

Recently the particle identification capabilities of the NA49 detector have been extended to neutral hadrons and photons[2]. This should improve the selectivity of the correlation studies discussed above.

References

1. S. Afanasiev *et al.* , Nucl. Instrum. Meth. **A430** (1999) 210.

2. *Status and Future...*, Addendum-6, CERN/SPSC 2000-033, August 2000.

HADRONIC CENTRALITY DEPENDENCE IN NUCLEAR COLLISIONS

SONJA KABANA

Laboratory for High Energy Physics, University of Bern, Sidlerstrasse 5, 3012 Bern, Switzerland, E-mail: sonja.kabana@cern.ch

The kaon number density in nucleus+nucleus and p+p reactions is investigated for the first time as a function of the initial energy density ϵ and is found to exhibit a discontinuity around $\epsilon=1.3$ GeV/fm^3. This suggests a higher degree of chemical equilibrium for $\epsilon > 1.3$ GeV/fm^3. It can also be interpreted as reflection of the same discontinuity, appearing in the chemical freeze out temperature (T) as a function of ϵ. The $N^{\alpha \sim 1}$ dependence of (u,d,s) hadrons, with N the number of participating nucleons, also indicates a high degree of chemical equilibrium and T saturation, reached at $\epsilon > 1.3$ GeV/fm^3. Assuming that the intermediate mass region (IMR) dimuon enhancement seen by NA50 is due to open charm ($D\overline{D}$), the following observation can be made: a) Charm is not equilibrated. b) $J/\Psi/D\overline{D}$ suppression -unlike $J/\Psi/DY$- appears also in S+A collisions, above $\epsilon \sim 1$ GeV/fm^3. c) Both charm and strangeness show a discontinuity near the same ϵ. d) J/Ψ could be formed mainly through $c\bar{c}$ coalescence. e) The enhancement factors of hadrons with u,d,s,c quarks may be connected in a simple way to the mass gain of these particles if they are produced out of a quark gluon plasma (QGP). We discuss these results as possible evidence for the QCD phase transition occuring near $\epsilon \sim 1.3$ GeV/fm^3.

1 Introduction

The quark-gluon plasma phase transition predicted by QCD [3] may occur and manifest itself in ultrarelativistic nuclear collisions through discontinuities in the initial energy density (ϵ_i) dependence of relevant observables. A major example of a discontinuity is seen in the $J/\Psi/DY$ [4] discussed e.g. in [5,6]. We investigate for the first time the dependence of strangeness production, in particular of kaons, on the initial energy density ϵ_i [11,10]. The degree of equilibrium achieved in nuclear collisions has been intensively studied comparing hadron ratios and densities to models (see e.g. [6,7,8,9]). We investigate here if chemical equilibrium is achieved, examining an other aspect of equilibrium states, namely the volume (V) independence of hadron densities (ρ).[a]

2 Results and discussion

The kaon density (ρ_K=(K per collision)/V) at the thermal freeze out in nuclear reactions, investigated as a function of the initial energy

density ϵ_i (figure 1) (see [10] for calculation details), exhibits a dramatic changeover around $\epsilon=1.3$ GeV/fm^3, saturating for higher ϵ values, while it is falling below. The syst. error on ϵ_i is estimated to be $\sim 30\%$. It is assumed that the number of nucleons participating in the collision (N) is proportional to the volume of the particle source at the thermal freeze out [10]. The new results from Si+Au at 14.6 A GeV and p+p at 158 A GeV shown in figure 1, which are not included in [10], have been estimated using data from [12] and methods described in [10]. Furthermore, ρ_K rises with N respectively with V below $\epsilon=1.3$ GeV/fm^3 while it does not depend on N respectively on V above $\epsilon=1.3$ GeV/fm^3. To illustrate this, two values of V are noted on figure 1. The changes of $K^{\pm}$ and $\pi^{\pm}$ with N within the Pb+Pb system, have been first realized in [13]. A similar behaviour as the one seen in figure 1, can be inferred for pions as well as for the K/π ratio (S.K. work in progress).

The N^{α} exponent of hadrons with (u,d,s) quarks above $\epsilon=1.3$ GeV/fm^3, do not depend on the particle mass (figure 3). At $\epsilon > 1.3$ GeV/fm^3 α is near to one, as expected in case of a chemically equilibrated state, assuming

---

[a] Results of the NA52 experiment shown in this talk can be found in [1,2].

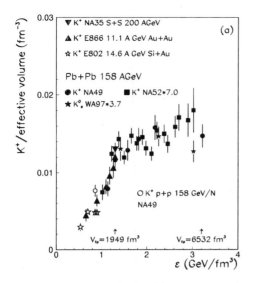

Figure 1. Initial energy density (ϵ) dependence of the K^+ multiplicity over the effective volume of the particle source at thermal freeze out [10].

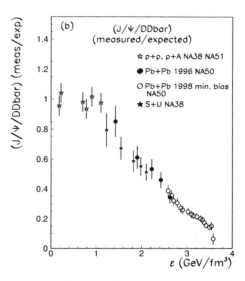

Figure 2. Initial energy density (ϵ) dependence of the $J/\Psi/D\overline{D}$ (measured/'expected') ratio [10].

$N \sim V$. The deviations seen in ϕ, π^0 and $\overline{p}$ may be due to the transverse momentum acceptance. Therefore, figure 3 supports the assumption of a high degree of chemical equilibrium reached above $\epsilon = 1.3$ GeV/fm^3, among hadrons with u,d,s quarks. The N^α exponent of kaons is found to depend strongly on $\sqrt{s}$ for kaons (figure 4). Therefore, below $\epsilon = 1.3$ GeV/fm^3, ρ_k (figure 1 and figure 4), ρ_π and the K/π ratio, show an increase with increasing N respectively with V.

Figures 1, 3 and 4 can be interpreted in two ways. Firstly, kaons may achieve a higher degree of chemical equilibrium only for $\epsilon > 1.3$ GeV/fm^3, and may not be fully equilibrated below [10]. The equilibration of strangeness is expected in a QGP and its observation at $\epsilon \sim 1.3$ GeV/fm^3 could therefore be a sign of a transition to QGP. In this case, it is a transition from a non equilibrated hadron gas to an equilibrated QGP. It is therefore not a well defined phase transition in the thermodynamic sense.

Secondly, kaons can be in fact chemically equilibrated also below $\epsilon = 1.3$ GeV/fm^3, and

the change respectively the constancy of ρ_K with V_{fo} and ϵ_i observed in figure 1, can be a result of the increase of the freeze out temperature with ϵ_i below $\epsilon = 1.3$ GeV/fm^3, respectively of the stability of T_{fo} above 1.3 GeV/fm^3. This dependence of T_{fo} on ϵ_i, namely rising until it reaches a critical T_c value and saturating above for all reactions, would strongly support the QCD phase transition appearing at $\epsilon \sim 1.3$ GeV/fm^3. This interpretation fully agrees with thermal models which suggest that particle ratios at freeze out are compatible with thermalization even in A+A collisions at 1 A GeV [7]. However the first interpretation is not in gross disagreement with [7], because there the thermal model description is modified (introducing e.g. $\rho_k \sim V$) in order to describe the data at 1 A GeV.

Furthermore, the correct interpretation can be corroborated by further investigations discussed in the following. The nonzero baryochemical potential (μ_B), which in the reactions shown in figure 1, happens to change with ϵ_i, makes the intepretation of figure 1 difficult. Therefore, it appears that the dependence of the temperature at chemical

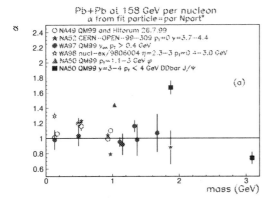

Figure 3. The parameter α, resulting from the N^α fit to hadron yields shown as a function of the mass of the particles in the region $\epsilon > 1.3$GeV/fm^3 at SPS. N is the number of participating nucleons.

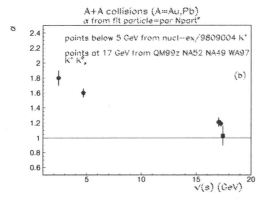

Figure 4. The parameter α, resulting from the N^α fit to hadron yields shown as a function of $\sqrt{s}$ for kaons. N is the number of participating nucleons.

freeze out extrapolated to $\mu_b = 0$, on ϵ_i, would help to identify and prove the QCD phase transition. A rising and then a for ever saturating freeze out temperature above ϵ 1.3 GeV/fm^3 is a strong argument that the QCD phase transition occurs at this ϵ, and figure 1 is a direct consequence of it.

The question if the QCD phase transition appears at the critical ϵ_i in any volume, or if there is additionally a critical initial volume of the particle source above which the transition takes place, can be answered comparing QGP signatures in systems with different volumes but the same ϵ_i. For example comparing $p+p$, e^+e^- etc collisions to heavy ion col-

lisions e.g. at the same ϵ. This is not yet done for the signature of the J/Ψ suppression and it has to be clarified e.g. using Tevatron data [10]. For the signature of strangeness enhancement it is suggested by figure 1 in [8] that there is indeed a critical initial volume, only above which strangeness is enhanced over $p + \bar{p}$ at the same ϵ_i. This conclusion follows, if we assume that Tevatron reaches at least ϵ_i values similar to SPS A+A collisions [14] and if figure 1 in [8] is not biased by the model calculation [8].

If strangeness is indeed not equilibrated at $\epsilon < 1.3$ GeV/fm^3, this may explain the decrease of the double ratio $(K/\pi)(A+A/p+p)$ with increasing $\sqrt{s}$. In particular, a larger strangeness annihilation is enforced by equilibrium at SPS reducing the strange particle yield. However the assumption of non equilibrium of $s\bar{s}$ at low ϵ is not nessecary here, since the above observation can be possibly traced back to e.g. the variation of $\mu_B(A + A)/\mu_B(p + p)$ with $\sqrt{s}$ in A+A collisions. Furthermore, in the context of QGP formation, it seems irrelevant to discuss e.g. $s\bar{s}$ enhancement in A+B over p+p collisions in a nonequilibrium situation. It is the very establishment of equilibrium in the (u,d,s) sector, which can reveal informations on QGP.

The kaon number densities in p+p and A+B collisions in figure 1 are similar, when compared at the same ϵ_i. See also [15] for a discussion of universality of pion phase space densities.

Our prediction for the N dependence of hadrons at RHIC and LHC is the N^1 thermal limit, as long as hadron yields are dominated by low transverse momentum particles. Furthermore, if the changeover of ρ_k at $\epsilon = 1.3$ GeV/fm^3 shown in figure 1 is due to the QCD phase transition, we predict for RHIC and LHC the same total strangeness (or kaon) number density and the same freeze out temperature, -after correction for the

μ_B dependence-, as for $\epsilon =$ 1.3-3.0 GeV/fm^3. If this change is however due to the onset of equilibrium in a hadronic gas, and the QCD phase transition takes place at higher ϵ, it may manifest itself through a second changeover of hadron number densities, ratios and freeze out temperatures -after correction for the different μ_B- e.g. in RHIC above $\epsilon \sim 3$ GeV/fm^3.

Assuming that the IMR dimuon enhancement seen by NA50 is due to open charm, the following observations can be made: a) open charm appears not to be equilibrated ($\alpha = 1.7$) (figure 3) [10]. b) The $J/\Psi/D\overline{D}$ ratio deviates from p+p and p+A data also in S+U collisions (figure 2), above $\epsilon \sim 1$ GeV/fm^3. c) It therefore appears that both charm and strangeness show a discontinuity near the same $\epsilon \sim 1$ GeV/fm^3 [10], similar to the critical $\epsilon_c \sim$ 1-2 GeV/fm^3 predicted by QCD [3,5]. d) The N dependence of the $J/\Psi/D\overline{D}$ ratio can be interpreted as the J/Ψ being formed through $c,\overline{c}$ coalescence [10].

e) Finally, the enhancement factors of hadrons with u,d,s,c quarks may be connected in a simple way to the mass gain of these particles in the quark gluon plasma (table below) [16]. T_q are the enhancement factors of the lightest mesons with u,d,s,c quarks (π, K, D), if they are produced out of a quark gluon plasma (e.g. $g+g \to s+\overline{s}$ (1)), as compared to their direct production from hadron interactions away from the transition point (e.g. $p + p \to K^+ + \Lambda + p$ (2)). The gain is taken proportional to $m_{particle} - m_{quarks}$, as this expresses the different thresholds of reactions (1) and (2). In the table below the predicted enhancement factors (T_q) of hadrons with u,d,s,c quarks from a QGP are compared to the experimentally measured ones (E_q), and are found to be similar. (Definitions: $th_q = m_0 - m_q$, $m_{u,d}$=7 MeV, m_s=175 MeV, m_c=1.25 GeV, $m_0 = m(\pi, K, D)$, $T_q = \sqrt{th_q/th_{u,d}}$, $E = \frac{(A+B)}{(N+N)}$, $E_q = E/E_{u,d}$).

Acknowledgments I would like to thank

q	th_q	T_q	E		E_q
u,d	133	1	π/N	~ 1.12	1
s	320	1.55	K/N	~ 2	1.79
c	615	2.15	$D\overline{D}$	~ 3	2.68

Prof. K Pretzl and Prof. P Minkowski for stimulating discussions and the Schweizerischer Nationalfonds for their support.

References

1. S. Kabana et al (NA52 coll.), submitted to ICHEP2000 and preprint BUHE-00-07.
2. S. Kabana et al (NA52 coll.), proceedings of the conference SQM2000 and Univ. of Bern preprint BUHE-00-05.
3. E. Laermann, Nucl. Phys. A610 (96) 1c.
4. M.Abreu et al. (NA50 coll.),CERN-EP-2000-013.
5. H. Satz, hep-ph/0007069.
6. P. Braun Munzinger, J. Stachel, nucl-th/0007059.
7. J. Cleymans, H. Oeschler, K. Redlich, Phys. Lett. B485 (00) 27.
8. F. Becattini et al, hep-ph/0002267.
9. J. Rafelski, hep-ph/0006085.
10. S. Kabana, hep-ph-0004138.
11. S. Kabana, proceedings of the conference SQM2000 and Univ. of Bern preprint BUHE-00-06.
12. E802 nucl-ex/9903009, Nucl. Phys. A661 (99) pages: 45, 439, 448, Phys. Lett. B332 (94) 258, M. Tannenbaum priv. communication.
13. G. Ambrosini et al. (NA52 Coll.), New J. of Phys. (1999) 1, 22 and New J. of Phys. (1999) 1, 23, S. Kabana et al, (NA52 coll.) Nucl. Phys. A661 (99) 370c.
14. J. Bartke et al, (NA35 coll.) Z. Phys. C 48 (1990) 191.
15. D. Ferenc et al, Phys. Lett. B457 (99) 347.
16. S. Kabana, P. Minkowski, to be published.

NEW RESULTS IN RELATIVISTIC NUCLEAR PHYSICS AT JINR (DUBNA)

A. I. MALAKHOV

Joint Institute for Nuclear Research, 141980, Dubna, Moscow reg., Russia
E-mail: malakhov@lhe.jinr.ru

New results of investigations in relativistic nuclear physics at Laboratory of High Energies, Joint Institute for Nuclear Research are presented. The correlation data on production of the cumulative proton pairs produced in the interactions of nuclear beams of the novel accelerator Nuclotron with nuclear targets are presented. The transverse interaction radius for studying process was determined in this experiment. We have obtained the new data in Synchrophasotron polarized deuteron beam at our Laboratory. The measurement result of the fragmentation of tensor polarized deuterons into cumulative pions are discussed. The predictions made on basis of the generalization of our results for the production cross section of secondary particles in ultrarelativistic heavy ion collisions are presented too. The results of some other experiments carried out at the Laboratory are also discussed.

1 Introduction

The scientific program of the Laboratory of High Energies (LHE), Joint Institute for Nuclear Research (JINR) is presently concentrated on investigations of interactions of relativistic nuclei in the energy region from a few hundred MeV to a few GeV per nucleon with the aim of searching for manifestations of quark and gluon degrees of freedom in nuclei, asymptotic laws for nuclear matter at high energy collisions as well as on the study of the spin structure of lightest nuclei. Experiments along these lines are being carried out using beams of the Synchrophasotron – Nuclotron accelerator complex.

2 Accelerator Complex

The accelerator complex at LHE is the basic facility of JINR for generation of proton, polarized deuteron (also neuteron/proton) and multicharged ion (nuclear) beams in energy range up to 6 GeV/u. General view of the facility is shown in figure 1.

The Nuclotron is based on the unique technology of superconducting magnetic system, which was proposed and investigated at Laboratory[1]. Before last year for physics experiments we have used only internal beam

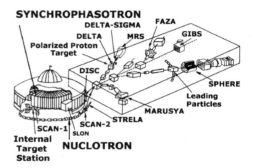

Figure 1. The accelerator complex of the Laboratory of High Energies, Joint Institute for Nuclear Research.

of the Nuclotron. In the end 1999 was obtained the external beam of the Nuclotron with help of the beam slow extraction system constructed on the base of the superconducting elements. Now we have begun the experiments at the Nuclotron external beams. Some parameters of the Synchrophasotron and Nuclotron beeams are presented in table.

3 Measurements with Relativistic Nuclei

The inclusive spectra were measured for deuteron fragmentation into cumulative π^- mesons on a nuclear target, and the relevant cross section was investigated as function of the atomic number A of a target nucleus at SPHERE setup with Synchrophasotron beams. The deuteron beam momentum was varied between 7.3 and 8.9 GeV/c, and data were collected for the π^- meson momenta of $p_\pi = 3.0$ and $4.9\ GeV/c$. The shapes of the pion spectra are found to be similar for all investigated target. The cross section shows a characteristic peripheral dependence on the target atomic number A for $A > 12$ ($d\sigma_\pi \sim A^{0.4}$) and a steeper decrease toward a hydrogen target (figure 2).

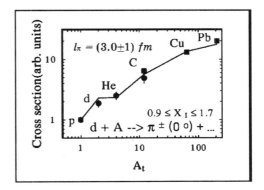

Figure 2. Cross section for production of cumulative pions as function of atomic number A of the nucleus.

Table. Some beam parameters of the Synchrophasotron and Nuclotron.

Beam	Intensity (particles per cycle)		
	Synchro-phasotron	Nuclotron	
		available	have to be
p	$4 \cdot 10^{12}$	$2 \cdot 10^{10}$	10^{13}
n	10^{10}		10^{13}
n↑	10^6		10^{11}
d	10^{12}	$5 \cdot 10^{13}$	10^{13}
d↑	$2 \cdot 10^9$	$3 \cdot 10^8$	$5 \cdot 10^{10}$
t	10^9	$4 \cdot 10^5$	10^{10}
^{3}He	$2 \cdot 10^{10}$		$5 \cdot 10^{11}$
^{4}He	$5 \cdot 10^{10}$	$8 \cdot 10^8$	$2 \cdot 10^{12}$
^{7}Li	$2 \cdot 10^9$		$5 \cdot 10^{12}$
^{12}C	10^9	10^8	$2 \cdot 10^{12}$
^{16}O	$5 \cdot 10^7$		10^{10}
^{20}Ne	10^4		$5 \cdot 10^9$
^{24}Mg	$5 \cdot 10^6$		$5 \cdot 10^{11}$
^{28}Si	$3 \cdot 10^4$		10^{10}
^{32}S	10^3		10^{10}
^{40}Ar			10^{10}
^{56}Fe			10^{11}
^{84}Kr		10^3	$5 \cdot 10^8$
^{131}Xe			$2 \cdot 10^8$
^{181}Ta			10^8
^{209}Bi			10^8
^{238}U			10^8
Energy (AGeV)	4.5	5.2	6.0

The proposed theoretical interpretation of the observed A dependence takes into account the interaction of the deuteron and of the emitted pion with target nucleons. The best description of the data[2] is obtained under the assumption that the pion-formation length is $l_\pi = (3 \pm 1) fm$.

M.I.Podgoretsky suggested a method to measure the velocity (and size) of the source by the interference method. This method allows one to get direct experimental evidence for the nonstationary state of the pion generation volume. This evidence was first obtained with the setup GIBS at LHE for central Mg-Mg interactions[3] at 4.4 GeV/c. The pictures were taken in streamer chamber with a magnesium target inside, exposed to the extracted magnesium beam from the Synchrophasotron.

The figure 3 displays rapidities ($Y = 0.5 \ln[(1 + \beta)/(1 - \beta)]$) of sources Y_{source} obtained by fitting the correlation function for subset of pions from different region of the kinematic spectrum.

Average rapidities of these subsets Y_{subset} are plotted on the horizontal axis. The data are given for subsets moving both along the reaction axis and across it. It is seen that pions from different regions of the kinematic spectrum are emitted by different "elements" of source, moving with respect to

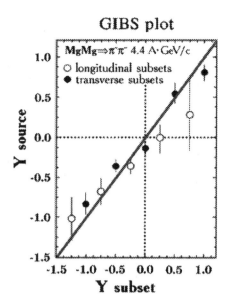

Figure 3. Rapidities Y_{source} of the pion production volume elements corresponding to different pion subset moving with different rapidities Y_{subset} along and across the reaction axis in the Mg-Mg rest frame.

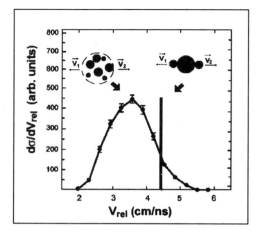

Figure 4. Distribution of relative velocities for the coincident fragments from $\alpha + Au$ collisions measured at correlation angles 150° - 180°. The vertical line shows the expected maximum position for fragments evaporation from the nucleus surface. The experimental distribution is shifted to lower velocities, corresponding to the volume distribution of the expanded system.

one another with almost highest possible velocities. For a stationary source, all points must be on the horizontal axis.

A number of experiments have been conducted with the proton and alpha beams of Synchrophasotron on 4π-device FAZA. The following main result is obtained: the hot target spectator expands before the fragment emission. The break-up density is found to be $\sim 1/3$ of normal one (figure 4)[4]. It is possible to interpreted this effect as observation of the gas - liquid faze transition in nuclear matter.

In the processes of investigation is a target fragmentation into two cumulative protons with help of the SKAN setup at the internal beam of the Nuclotron. The goal of experiment is measuring of transfer dimension of nucleus-nucleus interaction region. Method is the measurement correlation of cumulative protons, emitted at small relative momentum. Correlations of protons, emitted in the angle interval between $106\text{-}112^o$ in laboratory system, are studied in the reactions

$d + C \rightarrow p + p + ...$ and $d + Cu \rightarrow p + p + ...$ ($E_d = 2 \ A \cdot GeV$). As result were obtained approximately the same transfer dimensions for dC and dCu interactions[5]: $r_{dC} = 3.0^{+0.5}_{-0.4}$ fm and $r_{dCu} = 2.6^{+0.8}_{-0.7}$ fm.

4 Investigations with polarized beams

The tensor analyzing power A_{yy} for the cumulative pion production $d \uparrow +^{12}C \rightarrow \pi^{\pm}(0, 135, 178 mrad) + ...$ has been measured with tensor polarized deuteron beam of the Synchrophasotron at the SPHERE setup. This experiment is focused on "cumulatively produced pions", which are produced beyond the kinematically nucleon-nucleon collisions (figure 5).

The measured values of A_{yy} are in disagreement with result of our impulse approximation calculation that is based on a single $NN \rightarrow \pi NN$ interaction and takes into account the internal motion of nucleons in the deuteron[6].

New results for the np spin-dependent

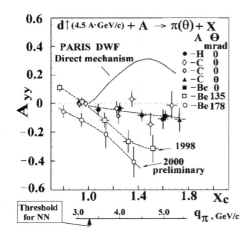

Figure 5. A_{yy} vs x_c and q_π for the reaction $d + A \to \pi^-(0, 135, 178mrad) + ...$

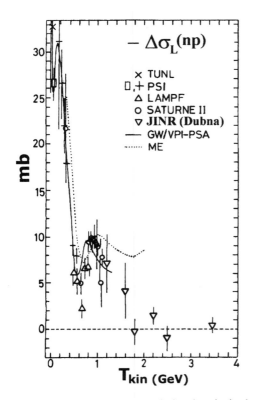

Figure 6. Energy dependence of the $\Delta\sigma_L(np)$ obtained with neutron polarized beam.

total cross section difference $\Delta\sigma_L(np)$ at neutron beam kinetic energies 1.59, 1.79, and 2.20 GeV were obtained (figure 6). A quasi-monochromatic neutron beam was produced by break-up of accelerated and extracted vector polarized deuterons from Synchrophasotron. The neutrons were transmitted through a large proton polarized target. The values of $\Delta\sigma_L(np)$ were measured as difference between the np total cross sections for parallel and antiparallel beam and target polarization, both oriented along the beam momentum. A fast decrease of $\Delta\sigma_L(np)$ with increasing energy above 1.1 GeV was confirmed[7].

5 Asymptotics in Relativistic Nuclear Physics

The principles of symmetry and self-similarity have been used to obtain an explicit analytical expression for inclusive cross section of production particles, nuclear fragments and antinuclei in relativistic nuclear collisions in the central rapidity region ($y = 0$). The results are in agreement with available experimental data. It is shown that the effective number of nucleons participating in nuclear collisions decreases with increasing energy and the cross section tends to a constant value equal both for particles and antiparticles. The analysis of the obtained results makes it possible to predict asymptotic behaviour of production cross section of particles, nuclear fragments and antinuclei.

One example of such prediction is presented in figure 7 from Dubna up to LHC energies[8,9].

6 Conclusions

We have the interesting research program with relativistic beams of nuclei, polarized deuterons and neutrons in JINR (Dubna) and the new perspectives of investigations with extracted nuclear beams of the Nuclotron.

614

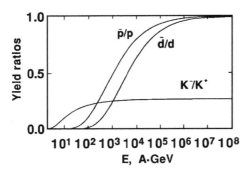

Figure 7. Predictions of the production cross section ratios for antiparticles to particles versus laboratory collision energy.

Acknowledgments

I express my sincere gratitude to leaders of all projects at the Syncrophasotron-Nuclotron accelerator complex for assignment of information for this report and I.I.Migulina for help in preparation of manuscript. The work is supported by grants of the Russian Fund of Fundamental Research 00-02-26109 and 00-02-16580.

References

1. A.D.Kovalenko. Proceedings of the Intern. Symposium "The 50th anniversary of the discovery of phase stability principle" 12-15 July, 1994, Editors: A.M.Baldin et al., *Dubna* (1996) 44-55.

2. Yu.S.Anisimov et al. *Phys.of Atomic Nuclei*. Vol.**60**, No.6 (1997) 957-963.

3. M.Kh.Anikina et al. *Physics Letters* B **397** (1997) 30-36.

4. V.A.Karnaukhov. *Preprint of JINR*, Dubna, **P1-99-193** (1999).

5. Yu.S.Anisimov et al. *JINR Rapid Communications* No.**5(91)-98**, Dubna (1998) 25-32.

6. S.Afanasiev et al. *Physics Letters* B **445** (1998) 14-19.

7. V.I.Sharov et al. *Eur. Phys. J.* C **1**, (2000) 255-265.

8. A.M.Baldin and A.I.Malakhov. *JINR Rapid Communications* No.**1**(87)-98, Dubna (1998) 5-12.

9. A.I.Malakhov. Proceeding of the 29th International Conference on High Energy Physics, Vancouver, Canada, 23-29 July 1998, Editors: Alan Astbury et al., *World Scientific*, v.**II**, 1497-1500.

ANISOTROPIC COLOR SUPERCONDUCTOR

J. HOŠEK

Dept. Theoretical Physics, Nuclear Physics Institute,25068 Řež (Prague), Czech Republic
E-mail: hosek@ujf.cas.cz

We argue that the QCD matter not far above a critical confinement-deconfinement baryon density and low temperatures can develop spontaneously the condensates of spin-one quark Cooper pairs. Depending upon their color these condensates characterize two distinct anisotropic color-superconducting phases. For them we derive the generic form of the quasiquark dispersion laws and the gap equation. We also visualize the soft Nambu-Goldstone modes of spontaneously broken global symmetries, and demonstrate an unusual form of the Meissner effect.

1 Basic picture

With QCD as the microscopic theory of strong interactions it became mandatory to pursue, if possible, both experimentally and theoretically, all corners of its phase diagram[1]. Here we restrict our attention to that of high baryon densities and very low temperatures: Not far above a critical confinement-deconfinement baryon density $n_c \sim 5n_{nucl.matter} \sim 5 \times 0.72/fm^3$ and at low T the deconfined QCD matter should be a rather strongly interacting quantum many-colored-quark system. Its detailed actual behavior in the considered region depends solely upon the details of the effective interactions relevant there.

For definiteness (and because we think it is both natural and simplifying) we assume that the strong (but nonconfining) gluon interactions dress the tiny quark masses m_u, m_d (we restrict our discussion to the case of two light flavors) into a common larger effective mass m_*, and become weak. Residual interaction between the massive (quasi)quark excitations $\psi^a_{\alpha A}$ (a - color, α - Dirac, A - flavor $SU(2)$ indices) can then be described by appropriate short-range (approximately contact) four-fermion interactions $\mathcal{L}_{int}$. Some pieces of $\mathcal{L}_{int}$ even have a solid theoretical justification: the instanton-mediated interaction of t'Hooft[2], and the Debye-screened chromoelectric one-gluon exchange. For our case of fragile ordered phases to be discussed

below one should keep in mind also yet unknown effective local four-fermion interactions due to the exchanges of heavy collective excitations eventually existing in more robust phases. The resulting effective Lagrangian

$$\mathcal{L}_{eff} = \overline{\psi}(i\gamma^\mu D_\mu - m_* + \mu\gamma_0)\psi$$
$$- \frac{1}{4}F_{a\mu\nu}F^{a\mu\nu} + \mathcal{L}_{int} \qquad (1)$$

in which the gluon interactions are treated perturbatively (and neglected in the lowest approximation) defines a relativistic version of the Landau Fermi-liquid concept. The quark chemical potential of interest is of the order of $\mu \sim 500 MeV$, whereas m_* is to be determined experimentally.

By assumption, $\mathcal{L}_{eff}$ is exactly $SU(3)_c \times SU(2)_I \times U(1)_V \times O(3)$ invariant. There is no approximate $SU(2)$ chiral symmetry of (1) which could be broken spontaneously. Hence, there should be no Nambu-Goldstone(NG) pions in dense and cold deconfined phase(s) of QCD. It would be also misleading to think of ψ and m_* as of the constituent quark and of the constituent mass. There is nothing they might constitute.

At present, there are no experimental data, either real or the lattice ones which would check our assumption. For our considerations it is not, however, essential. An alternative picture, and in fact the more commonly discussed one is that in the cold deconfined QCD matter the u, d quarks stay approximately massless at the Lagrangian level as they were in the confined phase. Discussion of the superfluid[a] phases presented below applies also to this case. On top of that it is, however, obligatory to ask (and to answer) how the (approximate) chiral symmetry is realized in this case.

The cold and dense deconfined QCD matter should exist in the interiors of the neutron stars, and optimistically also in the early stages of the relativistic heavy-ion collisions studied experimentally with much effort at present. Consequently, its theoretical studies are more than an intellectual challenge.

2 Isotropic superconductors

In principle, the behavior of a cold and dense deconfined QCD matter governed by (1) should be similar to that of any non-relativistic low-T Landau Fermi-liquid (say of electrons in metals or of the atoms of liquid ^3He). The differences are rather technical: (1) Characteristic energies given by the chemical potential $\mu \sim O(500 MeV)$ require the relativistic description. (2) The quarks carry, besides spin, also the flavor and color. (3) The gauge fields are both Abelian (photon) and non-Abelian (gluons). (4) The origin of the effective interactions is different.

In fact, the behavior of any (non-relativistic and relativistic) quantum many-fermion liquid of the Landau type is uniquely dictated by "theorems": (1) When scaling the fermion momenta towards the Fermi surface all interactions but one become irrelevant[3,4]. If this happens, such systems should behave thermodynamically as a corresponding non-interacting system of fermions. For example, the specific heat should grow linearly with T. Such a behavior is indeed observed in the low-T electron systems in metals, and in the liq-

uid ^3He. We are not aware of any experimental data in the relativistic systems. (2) The four-fermion interaction attracting fermions with opposite momenta at the Fermi surface is the only exception: Even if arbitrarily small, it causes the (Cooper) instability of the filled Fermi sea with respect to spontaneous condensation of the fermion Cooper pairs with opposite momenta into a more energetically favorable ground state. The new ground state, being by construction and by definition translationally invariant, has in the simplest nonrelativistic case of the ordinary local BCS-type four- fermion interaction the property

$$\langle \psi_\alpha^+(x)(\sigma_2)_{\alpha\beta}\psi_\beta^+(x)\rangle = \Delta \neq 0 \qquad (2)$$

It clearly exhibits spontaneous breakdown of the $U(1)$ phase symmetry generated by the operator of the particle number. Given a four-fermion interaction, the BCS theory provides for the microscopic quantitative understanding of the system.

Potential relevance of the physics of superconductors for the cold and dense QCD matter was recognized long ago[5,6,7]. Recent abrupt increase of interest in this idea was driven by two influential papers[8,9]: By explicit calculations they found that the realistic four-fermion interaction of the t'Hooft type [2] gives rise to the color-superconducting isotropic (spin-zero) phase I characterized by the condensate

$$\langle \overline{\psi}_{\alpha a A}(x)\epsilon^{ab3}(\tau_2)_{AB}(\gamma_5 C)_{\alpha\beta}\overline{\psi}_{\beta b B}(x)\rangle = \Delta \qquad (3)$$

which is phenomenologically quite appealing: $\Delta \sim 100 MeV$. The condensate (3) breaks spontaneously the $SU(3)_c \times U(1)$ symmetry down to $SU(2)_c$. Consequently, when the gauge interactions are switched off, there are 5+1 NG gapless collective excitations in the spectrum. Clearly, they can be only exited by the quark bilinears. When the gauge interactions are switched on 5 gluons acquire masses by the underlying Higgs mechanism[10]. Recently, properties of this phase (and of its

---

[a] We do not sharply distinguish between superconductivity and superfluidity in many-fermion systems. The superconductor is a superfluid having the perturbative long-range gauge interactions switched on. Superfluidity in many-boson systems (for example in ^4He) is, however, an entirely different story.

three-flavor relative) were studied in great detail.

Pauli principle allows for yet another isotropic, color-superconducting phase II. It is characterized by the condensate (we introduce $T_3 = \tau_3\tau_2$)

$$\langle \overline{\psi}_{\alpha a A}(x)(T_3)_{AB}(\gamma_5 C)_{\alpha\beta}\overline{\psi}_{\beta b B}(x)\rangle = \Delta_a\delta_{ab} \tag{4}$$

While Δ in (3) corresponds to the vacuum expectation value (vev) of the spin-0, color triplet, isospin-0 Higgs field, the condensate (4) corresponds to the vev of the spin-0, color sextet, isospin-1 Higgs field. The phase II is theoretically interesting mainly by spontaneous breakdown of truly global isospin symmetry. Since there is nobody who might "eat" them, the two corresponding NG gapless excitations remain in the physical spectrum, and become thermodynamically important. Generic form of the excitations in this phase is the same as in the phase I. Different is merely their counting.

3   Anisotropic superconductors

The Pauli principle itself allows for yet another two condensates, both having spin one[11]:   (1)The anisotropic color-superconducting phase III is characterized by

$$\langle \overline{\psi}_{\alpha a A}(x)(\tau_2)_{AB}(\gamma_0\gamma_3 C)_{\alpha\beta}\overline{\psi}_{\beta b B}(x)\rangle = \Delta_a\delta_{ab} \tag{5}$$

which corresponds to the vev of isospin-0, color sextet antisymmetric tensor Higgs field $\Phi_{ab\alpha\beta}(x)$ describing spin-1. (2)The phase IV is characterized by the condensate

$$\langle \overline{\psi}_{\alpha a A}(x)(T_3)_{AB}\epsilon^{ab3}(\gamma_0\gamma_3 C)_{\alpha\beta}\overline{\psi}_{\beta b B}(x)\rangle = \Delta \tag{6}$$

which corresponds to the vev of isospin-1, color triplet, antisymmetric tensor Higgs field $\Phi^c_{I\alpha\beta}(x)$.

Clearly, only the detailed behavior of $\mathcal{L}_{int}$ can select which one out of the four physically distinct ordered phases is the most energetically favorable one.

The anisotropic phases are interesting for they break down spontaneously the rotational symmetry like ferromagnets. In relativistic systems this is certainly not a very frequent phenomenon. It is possible only at finite quark density which itself breaks down explicitly the Lorentz invariance. The translational invariance of the ground state must of course remain inviolable.

In order to find out the generic features of the excitations of anisotropic superconductors we have analyzed a model having all necessary properties but less indices. It is defined by its Lagrangian

$$\mathcal{L}_{eff} = \overline{\psi}(i\gamma^\mu D_\mu - m_* + \mu\gamma_0)\psi - \frac{1}{4}F_{\mu\nu}F^{\mu\nu} + \mathcal{L}_{int} \tag{7}$$

where

$$\mathcal{L}_{int} = G[(\overline{\psi}\gamma_0\psi)^2 - (\overline{\psi}\gamma_0\vec{\tau}\psi)^2] \tag{8}$$

The Lagrangian (7) is $U(1)$ gauge invariant, and $SU(2)_V \times O(3)$ globally invariant. The interaction (8) is chosen in such a way that the isotropic (spin-0) condensate

$$\langle \overline{\psi}_{\alpha A}(x)(T_3)_{AB}(\gamma_5 C)_{\alpha\beta}\overline{\psi}_{\beta B}(x)\rangle \tag{9}$$

identically vanishes. Thus, either (7) can be treated perturbatively, or the interaction (8) gives rise to the anisotropic condensate

$$\langle \overline{\psi}_{\alpha A}(x)(\tau_2)_{AB}(\gamma_0\gamma_3 C)_{\alpha\beta}\overline{\psi}_{\beta B}(x)\rangle = \Delta \tag{10}$$

Introducing the field ($\psi^C = C\overline{\psi}$)

$$q_{\alpha A} = \frac{1}{\sqrt{2}}\begin{pmatrix} \psi_{\alpha A} \\ (\tau_2)_{AB}\psi^C_{\alpha B} \end{pmatrix} \tag{11}$$

we define a new self-consistent perturbation theory by the bilinear Lagrangian

$$\mathcal{L}'_0 = \overline{q}S^{-1}(p)q = \overline{q}S_0^{-1}(p)q - \mathcal{L}_\Delta \tag{12}$$

and the new interaction $\mathcal{L}'_{int} = \mathcal{L}_{int} + \mathcal{L}_\Delta$ where

$$\mathcal{L}_\Delta = \overline{q}\begin{pmatrix} 0 & \gamma_0\gamma_3\Delta \\ \gamma_0(\gamma_0\gamma_3\Delta)^+\gamma_0 & 0 \end{pmatrix}q \tag{13}$$

Finding $S(p)$ amounts to finding the form of the quasiquark dispersion laws. They have the form

$$E_{(1)}(\vec{p}) = \left(\epsilon_p^2 + |\Delta|^2 + \mu^2 + D^2(\vec{p})\right)^{1/2} \quad (14)$$

$$E_{(2)}(\vec{p}) = \left(\epsilon_p^2 + |\Delta|^2 + \mu^2 - D^2(\vec{p})\right)^{1/2} \quad (15)$$

where $\epsilon_p = \sqrt{\vec{p}^2 + m_*^2}$, and

$$D^2(\vec{p}) = 2(\epsilon_p^2\mu^2 + (p_1^2 + p_2^2 + m_*^2)|\Delta|^2)^{1/2}. \quad (16)$$

The equations (14) and (15) explicitly demonstrate spontaneous breakdown of the rotational symmetry of (7).

Requirement that $\mathcal{L}_0'$ gives zero contribution to $S^{-1}(p)$ in the lowest self-consistent approximation results in the equation for the gap Δ:

$$\Delta + 2\Delta G \int \frac{d^3p}{(2\pi)^3} \left(\frac{1}{E_{(1)}(\vec{p})} + \frac{1}{E_{(2)}(\vec{p})} \right)$$

$$\left(1 - \frac{4(p_1^2 + p_2^2)}{E_{(1)}(\vec{p})E_{(2)}(\vec{p})} \right) = 0 \quad (17)$$

Numerical analysis of Eq.(17) yet remains to be done. Here we simply assume that for the properly regularized integral[8] in (17)$\Delta \neq 0$ does exist. It is interesting to note that $E_{(2)}$ vanishes at $p_1^2 + p_2^2 = |\Delta|^2 + \mu^2 - m_*^2, p_3 = 0$. Fortunately for the gap equation the circle of vanishing $E_{(2)}$ does not lie on the Fermi surface where the interaction is relevant.

For the gauge interaction switched off the condensate (10) breaks spontaneously the global $U(1) \times SU(2) \times O(3)$ symmetry down to $SU(2) \times O(2)$. Consequently, there should exist $1 + 2$ NG quark-composite excitations. Their quantum numbers are found by analyzing the Goldstone commutator. In terms of the field q and of the Pauli matrices Γ_i which operate in the space of q the NG composites have the form $\bar{q}\gamma_0\gamma_3(\Gamma_1 - i\Gamma_2)q, i\bar{q}\gamma_0\gamma_2(\Gamma_1 + i\Gamma_2)q, i\bar{q}\gamma_0\gamma_1(\Gamma_1 + i\Gamma_2)q$.

Finally, it would be desirable to know how the anisotropic condensate (10) influences the behavior of the gauge field when perturbatively switched on. We plan to address this question within the microscopic description (7) in a future work. Here we present a straightforward analysis within the Higgs approach taking for granted that the condensate (10) is a vev of the antisymmetric order parameter $\Phi^{\mu\nu}$ describing the spin one. Requirement that only its time-space components propagate fixes the form of its kinetic term which has to be gauged. The self-interaction of the field $\Phi^{\mu\nu}$ is chosen in such a way that $\langle\Phi_{03}\rangle = \Delta$. The resulting effective Lagrangian has the form

$$\mathcal{L}_H = -(D^\lambda\Phi_{\lambda\mu})^+ D_\nu\Phi^{\nu\mu} - V(\Phi) \quad (18)$$

in which we have ignored for simplicity the fact that (18) should be only $O(3)$ invariant. The mass term of the gauge field following from (18) has the form

$$\mathcal{L}_{mass} = e^2|\Delta|^2(A_0^2 - A_3^2) = e^2|\Delta|^2 A_\mu A^\mu$$
$$+ e^2|\Delta|^2(A_1^2 + A_2^2) \quad (19)$$

We think that the anisotropic Meissner-Higgs effect (19) is peculiar but not unexpected. The first term $e^2|\Delta|^2 A_\mu A^\mu$ is the effect independent of the spin of the order parameter, whereas the second term represents an anisotropic "anti-Meissner-Higgs" effect due to the spin.

4 Conclusion

It is gratifying to observe that QCD has a corner in its phase diagram which is accessible experimentally, and which should be full of new phenomena associated with macroscopic quantum ordered phases of the superfluid type.

Although the analogy between superconductivity and the low-T deconfined QCD matter seems rather strong, it falters in one important respect: For studying the color-superconducting phases we can safely forget about using external chromomagnetic and external chromoelectric fields. We believe that due to the macroscopic quantum nature of these phases it is nevertheless justified to speculate that their experimental signatures might be brighter than those of the 'ordinary' quark-gluon plasma above a superconducting T_c.

Acknowledgments

Part of the work was done during the program "QCD at Finite Baryon Density" organized by the INT in Seattle. I am grateful to the Institute for the financial support, and to Mike Alford, Michael Buballa, Krishna Rajagopal, Edward Shuryak, Jac Verbaarschot, and Uwe Wiese for many discussions. The work was also supported by the Committee for the CERN-CR Cooperation, and by the grant GACR 202/0506.

References

1. F. Wilczek, What QCD Tells Us About Nature - and Why We Should Listen, hep-ph/9907340.

2. G. t'Hooft, *Phys. Rev.* D **14**, 3432 (1976).

3. J. Polchinski, Effective Field Theory and the Fermi Surface, hep-th/9210046 v2.

4. N. Evans, S. D. H. Hsu and M. Schwetz, An Effective Field Theory Approach to Color Superconductivity, hep-ph/9808444.

5. J. C. Collins and M. J. Perry, *Phys. Rev. Lett.* **34**, 1353 (1975).

6. S. C. Frautschi, Asymptotic freedom and color superconductivity in dense quark matter, in *Proceedings of the Workshop on Hadronic Matter at Extreme Energy Density*, ed. N. Cabibbo (Erice, Italy, 1978).

7. D. Bailin and A. Love, *Phys. Rep.* **107**, 325 (1984).

8. M. Alford, K. Rajagopal and F. Wilczek, *Phys. Lett.* B **422**, 247 (1998).

9. R. Rapp, T. Schaefer, E. Shuryak and M. Velkovsky, *Phys. Rev. Lett.* **81**, 53 (1998).

10. D. H. Rischke, Debye screening and Meissner effect in a two-flavor color superconductor, nucl-th/0000040.

11. J. Hosek, Macroscopic quantum phases of deconfined QCD matter at finite density, hep-ph/9812516, and to be published.

Parallel Session 5

Tests of the Electroweak Gauge Theory

Conveners: Robert B. Clare (MIT) and
Wolfgang Hollik (Karlsruhe)

THE Z^0 LINESHAPE AND LEPTON ASYMMETRIES AT LEP

GÜNTER DUCKECK

LMU München, Sektion Physik, Am Coulombwall 1, 85748 Garching, Germany
E-mail: guenter.duckeck@physik.uni-muenchen.de

The final results of the Z^0 lineshape measurements by the four LEP experiments ALEPH, DELPHI, L3 and OPAL are presented. The combination procedure and the common uncertainties are briefly described and the combined results and their interpretation are discussed.

1 Introduction

One of the main goals of the LEP e^+e^- collider at CERN was to make precise measurements of the properties of the Z^0 gauge boson. From 1989 to 1995 LEP was operated at centre-of-mass energies close to the mass of the Z^0. The four experiments, ALEPH, DELPHI, L3 and OPAL, performed precision measurements of the total cross-sections for the inclusive hadronic (quark-pair) final states and the cross-sections and forward-backward asymmetries of the three charged leptons, (e, μ, τ) at these energies [1]. From these measurements the properties of the Z^0 boson, such as its mass, total and partial decay widths, and the couplings to fermions can be determined. These observables provide stringent tests of the Standard Model (SM) of electroweak interactions.

2 Data analysis

2.1 Cross-section and asymmetry measurements

Each experiment recorded about 4×10^6 Z^0 decays from 1989 – 1995. About 2/3 of the luminosity was devoted to high statistics runs at the peak of the Z^0 resonance and 1/3 was taken at several off-peak energy points within $m_Z \pm 3$ GeV. Three ingredients are needed for the measurement of the total cross-sections:
1) Event selections count the four distinguished final states, e^+e^-, $\mu^+\mu^-$, $\tau^+\tau^-$ and $q\bar{q}$. Due to the clear signatures of the events the selections have high efficiencies and low

backgrounds. Systematic effects can in general be controlled at a level similar to the statistical uncertainty, which is about 0.05 % for $q\bar{q}$ and 0.2 % for the three charged leptons.
2) The luminosity is determined using small-angle Bhabha-scattering as a reference process. This reaction is dominated by t-channel photon exchange and precisely calculable. Important milestones have been the upgrades of the detectors with new luminometers based on calorimeters with Si-layers, which was undertaken in 1992/93 by all four experiments, and the reduction of theoretical uncertainty [2] to about 0.06 %.
3) The LEP energy is precisely calibrated using the method of resonant depolarization [3]. This technique is based on the precise knowledge of $g - 2$ for electrons and has an intrinsic precision of $\mathcal{O}(0.1\,\mathrm{MeV})$. However, since this calibration cannot be performed during physics data-taking, several corrections must be applied, which account for tidal deformations, leakage currents and interaction-point specific effects and degrade the precision.

2.2 Parameterization of the Z^0 resonance

The energy dependence of the total cross-section is parametrized as a Breit-Wigner resonance

$$\sigma^Z_{f\bar{f}} = \sigma^0_{f\bar{f}} \frac{s\Gamma^2_Z}{(s - m^2_Z)^2 + s^2\Gamma^2_Z/m^2_Z} , \quad (1)$$

which describes the Z^0 boson with mass m_Z, total width Γ_Z and final-state dependent peak cross-section $\sigma^0_{f\bar{f}}$. The peak cross-section is related to the partial widths by $\sigma^0_{f\bar{f}} \propto \Gamma_{ee}\Gamma_{ff}/\Gamma^2_Z$.

For each LEP experiment the full data set consists of about 200 individual cross-section and asymmetry measurements. In a χ^2 fit the experiments determine nine parameters: the three basic Breit-Wigner terms m_Z, Γ_Z and σ^0_h, the three ratios of the Z^0 partial widths, $R_\ell \equiv \Gamma_{had}/\Gamma_{\ell\ell}$ and the three lepton pole-asymmetries $A^{0,\ell}_{FB}$ for $\ell = e$, μ, τ. These parameters are termed pseudo-observables, since they rely on the programs TOPAZ0 [4] and ZFITTER [5] to calculate the effect of s-channel γ exchange, QED radiative corrections, and small non-factorizable contributions from higher-order SM corrections and imaginary parts of the couplings. Additional t-channel corrections for e^+e^- final states are calculated in the SM using the ALIBABA program [6]. Correlated uncertainties between the cross-sections or asymmetries arise from the event selections, luminosity, LEP energy and t-channel corrections and are accounted for in the fit.

3 Combination procedure

The results of the four experiments are combined at the level of the nine pseudo-observables accounting for correlations. Four sources give rise to common errors between the experiments:

1) The luminosity theory uncertainty affects σ^0_h. In case of ALEPH, DELPHI and L3 it amounts to 0.061%. OPAL includes an additional light pair correction [7] which reduces the error to 0.054%.

2) Uncertainties in the t-channel corrections have been evaluated [6] and translate into common errors of 0.11% for R_e and 0.0014 for $A^{0,e}_{FB}$.

3) Other theoretical uncertainties are small. Photon radiation and fermion-pair radiation are treated in $\mathcal{O}(\alpha^3)$ [8]; the residual uncertainties are less than 20%.

4) The LEP energy errors are specified in

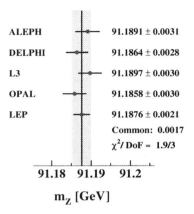

Figure 1. LEP measurements of m_Z.

terms of a covariance matrix [3] relating the cross-sections and asymmetries at the various energy points. They are translated into effective uncertainties on the pseudo-observables by comparing full fits to cross-sections and asymmetries with different scale factors applied to the LEP energy errors. This results in errors of 1.7 MeV for m_Z, 1.2 MeV for Γ_Z and 0.011 nb for σ^0_h.

Overall, most parameters are still statistics limited; only for m_Z and σ^0_h systematic uncertainties dominate.

4 Results

Combining the four LEP experiments, accounting for common errors and correlations, results in

$$
\begin{aligned}
m_Z &= 91.1875 \pm 0.0021\,\text{GeV}, \\
\Gamma_Z &= 2.4952 \pm 0.0023\,\text{GeV}, \\
\sigma_{had} &= 41.540 \pm 0.037\,\text{nb}, \\
R_\ell &= 20.767 \pm 0.025, \\
A^0_{FB} &= 0.01714 \pm 0.00095,
\end{aligned}
\tag{2}
$$

when lepton universality is assumed. Correlations between the five parameters are in general small ($\leq 10\%$), except for Γ_Z and σ^0_h with -30% and R_ℓ and σ^0_h with 18%. The average has a χ^2 per degree of freedom of 36.5/31, which corresponds to a probability of 23%. The consistency of the experiments

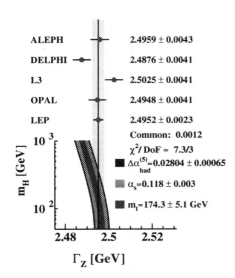

Figure 2. LEP measurements of Γ_Z.

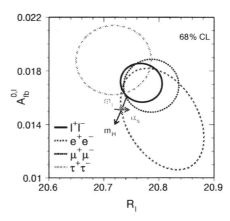

Figure 3. Contours in the $R_\ell - A_{FB}^{0,\ell}$ plane.

is demonstrated in Figs. 1 and 2 for m_Z and Γ_Z. All parameters agree well with the SM predictions, the results of R_ℓ and $A_{FB}^{0,\ell}$ for the three charged lepton species are consistent as shown in Fig. 3.

The ratio $R_\ell \equiv \Gamma_{had}/\Gamma_{\ell\ell}$ is particularly sensitive to the strong coupling constant $\alpha_s(m_Z^2)$, since effects from the top quark and the Higgs boson largely cancel. The result corresponds to $\alpha_s(m_Z^2) = 0.123 \pm 0.004^{+0.003}_{-0.001}$, where the second reflects the residual effect of varying $m_H = 150^{+850}_{-60}$ GeV.

The parameters in (2) can be transformed to determine the partial Z^0 decay widths. Of particular interest is the invisible width $\Gamma_{inv} = 499.0 \pm 1.5$ MeV. Assuming the decay width for a single neutrino generation from the SM one can determine the number of light neutrino generations

$$N_\nu = \frac{\Gamma_{inv}/\Gamma_{\ell\ell}}{(\Gamma_{\nu\nu}/\Gamma_{\ell\ell})_{SM}} = 2.984 \pm 0.008 . \quad (3)$$

Alternatively, assuming three neutrino species, one can derive an upper limit for additional invisible decay modes of the Z^0: $\Gamma_{inv}^{new} < 2.0$ MeV @ 95% CL.

Acknowledgements

I wish to thank my colleagues from the LEP experiments and the electroweak working group as well as the ZFITTER and TOPAZ0 teams for the fruitful collaboration.

References

1. ALEPH Collab., *Eur. Phys. J.* C **14** 1 (2000);
   DELPHI Collab., CERN-EP-2000-037;
   L3 Collab., CERN-EP-2000-022;
   OPAL Collab., PN442;
   LEP-EWWG, contr. paper PL06-279.
2. B.F.L. Ward *et al.*, *Phys. Lett.* B **450** 262 (1999).
3. R. Billen *et al.*, *Eur. Phys. J.* C **6**, 187 (1999).
4. G. Montagna *et al.*, *Nucl. Phys.* B **401** 3 (1993); hep-ph/9804211.
5. D. Bardin *et al.*, *Phys. Lett.* B **255** 290 (1991); hep-ph/9908433.
6. W. Beenakker *et al.*, *Nucl. Phys.* B **349** 323 (1991); W. Beenakker *et al.*, *Phys. Lett.* B **425** 199-207 (1998).
7. G. Montagna *et al.*, *Nucl. Phys.* B **547** 39 (1999).
8. G. Montagna *et al.*, *Phys. Lett.* B **406** 243 (1997);
   A.B. Arbuzov, hep-ph/9907500.

TAU POLARIZATION AT LEP

J.M. RONEY

Dept. of Physics and Astronomy, University of Victoria,
Victoria, B.C. V3P 3WP, Canada
E-mail: mroney@uvic.ca

The tau polarization measurements performed at LEP are described and a summary of the current status is given. The results can be interpreted as one of the most precise determinations of the weak mixing angle: $\sin^2\theta_{\text{eff}}^{\text{lept}} = 0.23160 \pm 0.00040$.

Parity violation in the weak neutral current results in a longitudinal polarization of final state fermion−anti-fermion pairs produced in Z decay, with the τ lepton being the only fundamental fermion whose polarization is experimentally accessible at LEP. This is achieved by measuring the momentum spectra of the visible τ decay products. The τ polarization, $\mathcal{P}_\tau$, is given by

$$\mathcal{P}_\tau \equiv (\sigma_+ - \sigma_-)/(\sigma_+ + \sigma_-)$$

where $\sigma_{+(-)}$ represents the cross section for producing positive (negative)-helicity τ^- leptons which, at these high energies are very good approximations for the right-handed and left-handed chiral states. Moreover, the inequality of the Z coupling to left-handed and right-handed initial state electrons results in a polarization of the Z itself, which can be determined by measuring the angular dependence of $\mathcal{P}_\tau$. For the unpolarized e^+e^- beams at LEP the dependence of $\mathcal{P}_\tau$ on the angle θ_{τ^-} between the e^- beam and the final state τ^-, assuming vector and axial-vector couplings, can be expressed to lowest order as:

$$\mathcal{P}_\tau(\cos\theta_{\tau^-}) = \frac{\langle\mathcal{P}_\tau\rangle\,(1+\cos^2\theta_{\tau^-}) + +\frac{8}{3}A_{\text{FB}}^{\text{pol}}\cos\theta_{\tau^-}}{(1+\cos^2\theta_{\tau^-}) + \frac{8}{3}A_{\text{FB}}\cos\theta_{\tau^-}}, \quad (1)$$

where $\langle\mathcal{P}_\tau\rangle$ is the τ polarization averaged over all production angles, $A_{\text{FB}}^{\text{pol}}$ is the forward-backward polarization asymmetry, which gives the average polarization of

the Z, and A_{FB} is the forward-backward asymmetry of the τ-pairs [1].

The standard model can be used to make predictions for $\langle\mathcal{P}_\tau\rangle$ and $A_{\text{FB}}^{\text{pol}}$ in terms of $\sqrt{s}$, the mass and width of the Z, and the vector and axial-vector couplings of the Z to the electron and the τ lepton. The measurement of $\mathcal{P}_\tau$ is directly related to $g_{V\tau}/g_{A\tau}$ and that of $A_{\text{FB}}^{\text{pol}}$ to g_{Ve}/g_{Ae} [1]. In the improved Born approximation[2], which accounts for the most significant weak radiative corrections, and neglecting the contributions of the photon propagator, γ-Z interference and other photonic radiative corrections, the standard model predicts for $\sqrt{s}=m_Z$:

$$\langle\mathcal{P}_\tau\rangle = -\mathcal{A}_\tau \quad \text{and} \quad A_{\text{FB}}^{\text{pol}} = -\frac{3}{4}\mathcal{A}_e. \quad (2)$$

The four LEP experiments measure the $\cos\theta_{\tau^-}$ dependence of $\mathcal{P}_\tau$ and express their τ polarization results in terms of $\mathcal{A}_\tau$ and $\mathcal{A}_e$ using ZFITTER[3] to correct for the contributions of the photon propagator, γ-Z interference and electromagnetic radiative corrections for initial state and final state radiation, as well as for the $\sqrt{s}$ dependence of the τ polarization measurements. This latter feature is important since the off-peak data are included in the event samples for all experiments.

As a function of $\cos\theta_{\tau^-}$, $\mathcal{P}_\tau(\cos\theta_{\tau^-})$ provides nearly independent determinations of both $\mathcal{A}_\tau$ and $\mathcal{A}_e$. Consequently, the τ polarization measurements provide not only a determination of $\sin^2\theta_{\text{eff}}^{\text{lept}}$ but also test the hypothesis of the universality of the couplings of the Z to the electron and τ lepton.

Each LEP experiment measures $\mathcal{P}_\tau$ using the five τ decay modes $e\nu\bar{\nu}$, $\mu\nu\bar{\nu}$, $\pi\nu$, $\rho\nu$ and $a_1\nu$[4,5,6,7]. The five decay modes do not all have the same sensitivity to the τ polarization. The $\tau \to \pi\nu$ mode has the highest sensitivity because it is a two body decay involving a spinless particle, whereas the $\tau \to e\nu\bar{\nu}$ and $\tau \to \mu\nu\bar{\nu}$ modes have substantially lower sensitivities because the τ decays to three fermions, two of which are undetected neutrinos. The $\tau \to \rho\nu$ and $\tau \to a_1\nu$ decays have reduced sensitivity because they involve spin-1 particles. Much of this sensitivity reduction can be regained by using those kinematic properties of the ρ and a_1 decays which are related to the parent's spin orientation. The $\tau \to \rho\nu$ (high branching ratio of 25%) and $\tau \to \pi\nu$ (high sensitivity) channels dominate the combined polarization measurement.

In all analyses, a value of $\mathcal{P}_\tau$ is extracted from the data by fitting linear combinations of positive and negative helicity distributions in kinematic variables appropriate to each τ decay channel to the data, where the two distributions are obtained from Monte Carlo simulation. For example, in the $\tau \to \mu\nu\bar{\nu}$, $\tau \to e\nu\bar{\nu}$ and $\tau \to \pi\nu$ channels, the energy of the charged particle decay divided by the beam energy is the appropriate kinematic variable. For the $\tau \to \rho\nu$ and $\tau \to a_1\nu$ channels, an 'optimal variable'[8], ω, is employed by all experiments. Using Monte Carlo distributions in the fitting procedure allows for simple inclusion of detector effects and their correlations, efficiencies and backgrounds. Any polarization dependence in the backgrounds from other τ decays are automatically incorporated into these analyses. The systematic errors associated with the detector then amount to uncertainties in how well the detector response is modelled by the Monte Carlo simulation of the detector, whereas the errors associated with uncertainties in the underlying physics content in the distributions

arise from uncertainties in the Monte Carlo generators of the signal and backgrounds. Because the systematics uncertainties associated with detector modelling (such as those arising from the uncertainty in momentum and energy scales and calorimeter response) dominate the systematic errors, the four LEP experiments provide measurements with very little correlation between them.

In addition to a five-channel analysis, some of the LEP experiments augment their measurements with a analyses of: the $\tau^- \to \pi^- 2\pi^0\nu$ channel (ALEPH and DELPHI); an inclusive hadronic channel (DELPHI and L3); and event acollinearity (ALEPH and L3). ALEPH also employs the tau flight direction information[8] to gain sensitivity in some hadronic channels. Unique among the LEP measurements is OPAL's global event analysis which incorporates the intrinsic correlation between the helicities of the τ^+ and τ^- produced in the same Z decay, an effect which is accounted for by the other experiments by applying a correction to the statistical errors of the fit results. A major appeal of this global analysis technique is that the evaluation of the systematic errors fully incorporates all correlations between the systematic uncertainties in the different channels automatically. For the channel-by-channel analyses of ALEPHI, L3 and DELPHI, the correlation in the systematic errors between channels are taken into account in the combination. DELPHI also exploits the advantages of a combined analysis in a separate neural network analysis of its 1993-1995 one-prong data set.

The LEP combination is made using the overall results from each experiment based on complete analyses of the entire LEP 1 data set. This ensures that correlated errors between channels are properly taken into account since they are treated correctly by the individual experiments. The DELPHI[5] and L3[6] measurements are final and have been published. The ALEPH results[4] are prelimi-

A_{lepton} (LEP)=0.1464±0.0032

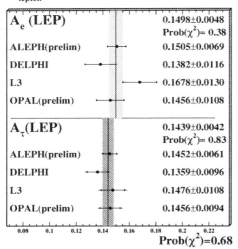

A_e (LEP)		0.1498±0.0048
		Prob(χ^2)= 0.38
ALEPH(prelim)		0.1505±0.0069
DELPHI		0.1382±0.0116
L3		0.1678±0.0130
OPAL(prelim)		0.1456±0.0108
A_τ(LEP)		0.1439±0.0042
		Prob(χ^2)= 0.83
ALEPH(prelim)		0.1452±0.0061
DELPHI		0.1359±0.0096
L3		0.1476±0.0108
OPAL(prelim)		0.1456±0.0094

Prob(χ^2)=0.68

Figure 1. Summary of LEP τ polarization results. Statistical and systematic errors have been added in quadrature.

nary and are the same as those presented at ICHEP 1998. The OPAL results[7] are preliminary and, for the first time, include the data over the entire solid angle. They are presented for the first time at this conference. The ALEPH and OPAL results are expected to be published later this year.

Figure 1 summarizes the most recent results for A_τ and A_e obtained by the four LEP collaborations[4,5,6,7] and their combination. The correlated systematic effects have been studied and found to be small and are neglected in the combined A_τ and A_e values quoted. The statistical correlation between the extracted values of A_τ and A_e is small ($\leq 5\%$), and is neglected.

The average values for A_τ and A_e:

$$A_\tau = 0.1439 \pm 0.0042 \qquad (3)$$
$$A_e = 0.1498 \pm 0.0048, \qquad (4)$$

are compatible, in agreement with lepton universality. Assuming $e - \tau$ universality, the values for A_τ and A_e can be combined. This combination is performed neglecting any pos-

sible common systematic error between A_τ and A_e within a given experiment, as these errors are also estimated to be negligibly small. The combined result of A_τ and A_e is:

$$A_\ell = 0.1464 \pm 0.0032. \qquad (5)$$

Within the standard model this can be expressed as:

$$\sin^2\theta_{\text{eff}}^{\text{lept}} = 0.23160 \pm 0.00040, \qquad (6)$$

which represents one of the most precise determinations of the weak mixing angle available. It is in agreement with all other high precision measurements of $\sin^2\theta_{\text{eff}}^{\text{lept}}$ [9,10,11].

Acknowledgments

I express my sincerest thanks to fellow members of the LEP tau polarization working group (J.-C. Brient, P. Garcia-Abia, W. Lohmann, F. Mattoras and D. Reid) for their contributions to this paper.

References

1. S. Jadach and Z. Wąs in *Z Physics at LEP1*, CERN 89-08, edited by G. Altarelli *et al.*, Vol. 1 (1989) 235.

2. M. Consoli, W. Hollik and F. Jegerlehner in *Z Physics at LEP1*, CERN 89-08, edited by G. Altarelli *et al.*, Vol. 1 (1989) 7.

3. D. Bardin, *et al.*, CERN-TH. 6443/92 (1992).

4. ALEPH Note 98-067 (ICHEP'98 Abs.939, ICHEP'00 Abs.441)

5. DELPHI Collab.,P. Abreu *et al.* Eur. Phys. J. **C14** (2000) 585.

6. L3 Collab., M. Acciarri et al., Phys. Lett. **B429** (1998) 387.

7. OPAL Note PN438 (ICHEP'00 Abs.205)

8. M. Davier *et al.*, Phys. Lett. **B306** (1993) 411.

9. T. Abe, Proceed. this conference.

10. E. Migliore, Proceed. this conference.

11. A. Gurtu, Proceed. this conference.

b AND c QUARK ASYMMETRIES AT LEP I

ERNESTO MIGLIORE

CERN, CH-1211, Geneva 23, Switzerland
E-mail: Ernesto.Migliore@cern.ch

Measurements of the forward-backward production asymmetries of b and c quarks in electron-positron annihilation at LEP I are reviewed together with their impact on the determination of the effective electroweak mixing angle $\sin^2 \theta^\ell_{W,eff}$.

1 Introduction

In the process $e^+e^- \rightarrow f\bar{f}$ at the Z peak, the angular distribution of the fermion can be expressed as

$$\frac{d\sigma}{\cos\theta} \propto 1 + \cos^2\theta + \frac{8}{3}\mathrm{A}^f_{FB}\cos\theta,$$

where $\cos\theta$ is the production angle of the fermion with respect to the incoming electron beam. The term A^f_{FB} accounts for the production asymmetry due to the parity violating couplings of the Z boson to the fermions.

Usually A^f_{FB} is factorized as the product $\mathrm{A}^f_{FB} = 3/4\,\mathcal{A}_e\mathcal{A}_f$ with $\mathcal{A}_f = \frac{2\,x_f}{1+(x_f)^2}$, $x_f = \frac{v_f}{a_f} = 1 - 4|Q_f|\sin^2\theta_W$ where θ_W is the electroweak mixing angle. Most of the electroweak radiative corrections can be accounted for by the very same relations after introducing an effective mixing angle for each fermion type, $\theta^f_{W,eff}$. The dependence of $\mathcal{A}_f$ on $\sin^2\theta^f_{W,eff}$ is shown in Figure 1. $\mathcal{A}_f$ is larger for quarks thus giving a better statistical sensitivity to the measurement of the asymmetry in hadronic Z decays (A^q_{FB}). On the other hand, since the measured asymmetry is the product of a quantity rapidly varying with $\sin^2\theta_W$ ($\mathcal{A}_l$) and one almost constant ($\mathcal{A}_q$), the measurement of A^q_{FB} is mostly sensitive to the effective leptonic mixing angle $\sin^2\theta^\ell_{W,eff}$.

2 Experimental techniques

Among all the final state decays of the Z boson, those into heavy quark pairs, $c\bar{c}$ and

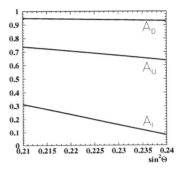

Figure 1. Dependence of the factor $\mathcal{A}_f$ on $\sin^2\theta_W$ for up-type ($\mathcal{A}_U$), down-type ($\mathcal{A}_D$) quarks and leptons ($\mathcal{A}_l$).

$b\bar{b}$, are the most easily tagged by the experiments. Therefore the experimental effort in the measurement of the quark asymmetry has been concentrated in the measurement of A^c_{FB} and A^b_{FB}. In the following the three most relevant techniques are reviewed: jet charge, lepton tag and D meson tag. The three methods differ in the way of tagging the flavour of the decay of the Z boson and, as a consequence, in the way to separate the quark and anti-quark hemispheres. In common is the use of the thrust axis of the event to estimate the direction of the primary quark.

2.1 A^b_{FB} using the jet charge

Decays of the Z boson into $b\bar{b}$ pairs are selected by exploiting the long lifetime of b-hadrons ($\gamma\beta c\tau \sim 2$ mm) and the capability of silicon strip vertex detectors to reconstruct the tracks originating from their decay. The determination of the charge of the parent quark is done using the charge flow $Q_{FB} = Q_F - Q_B$ where:

$$Q_{hem} = \frac{\Sigma q_i |\vec{p}_i \cdot \vec{T}|^\kappa}{\Sigma |\vec{p}_i \cdot \vec{T}|^\kappa} \quad (hem = F, B).$$

Q_{hem} is the momentum weighted average charge in the forward ($\vec{p}_i \cdot \vec{T} > 0$) or backward ($\vec{p}_i \cdot \vec{T} < 0$) hemisphere as defined by the thrust axis $\vec{T}$, and the exponent k is typically in the range $0.5 \div 2$.

In a pure $b\bar{b}$ sample, the average charge flow is related to the asymmetry via the relation

$$\langle Q_{FB} \rangle \propto \delta_b A_{FB}^b.$$

Both the charge separation δ_b, which would be $2q_b$ if neither hadronization nor reinteractions take place, and the fraction of $b\bar{b}$ events, which appears in the realistic case of a not 100 % $b\bar{b}$-pure sample, are determined directly from the data. The main sources of systematic uncertainties are related to the knowledge of the sample composition, the determination of the charge correlation between the hemispheres, which affects the determination of δ_b, and the charge separation for the residual $c\bar{c}$ events. Each term is in the range $\Delta A_{FB}^b = \pm 0.001 \div 0.002$, which should be compared to a statistical uncertainty $\Delta A_{FB}^b = \pm 0.004 \div 0.005$.

2.2 A_{FB}^c and A_{FB}^b using the lepton tag

Leptons with high momentum (p) and transverse momentum with respect to the axis of the closest jet (p_T), are characteristic of prompt semileptonic decays of heavy flavour hadrons. Their charge is moreover correlated with that of the parent quark. Cascade decays ($b \to c \to \ell$) and $B\bar{B}$ mixing dilute the charge correlation. The observed asymmetry is

$$A_{FB}^{obs} = \frac{N(\ell^+) - N(\ell^-)}{N(\ell^+) + N(\ell^-)},$$

i.e. the difference between positive and negative lepton tagged events in $\cos\theta_T$ bins. Since

the prompt component cannot be completely isolated by a selection based just on p and p_T cuts, the expression for the asymmetry includes the contributions of right and wrong sign components (f_b, f_{bc}), of the $B\bar{B}$ mixing, of charm events (f_c) and background (f_{bgd}):

$$A_{FB}^{obs} = \left[(1 - 2\overline{\chi})(f_b - f_{bc})A_{FB}^b + f_c A_{FB}^c \right] + f_{bgd} A_{FB}^{bgd}.$$

The separation of the different contributions is achieved on a statistical basis using the predictions from decay models and kinematic spectra of the heavy flavour hadrons.

In the past two years the largest experimental effort has been concentrated on the lepton tag analysis[1]. The b-tagging techniques have been applied to isolate samples enriched in $b\bar{b}$ events and additional variables, such as the impact parameter significance with respect to the primary vertex[a] of the lepton track or the jet charge in the hemisphere opposite to the lepton, were used to increase the separation between prompt and cascade leptons. The improvement in the statistical error thus obtained by the experiments amounts typically to 10 %. The main systematics arise from the uncertainty in the decay models and in the branching ratios of the heavy flavour hadrons, used to predict the lepton spectra, and from the experimental control of the residual background asymmetry of misidentified leptons. In the case of the b quark asymmetry all the systematics are below $\Delta A_{FB}^b = \pm 0.002$, which should be compared to a statistical error $\Delta A_{FB}^b = \pm 0.004 \div 0.006$. In the case of the c quark asymmetry the main systematics is related to the knowledge of the semileptonic branching ratios and fragmentation parameters, $\Delta A_{FB}^c = \pm 0.003$, while the statistical error is in the range $\Delta A_{FB}^c = \pm 0.006 \div 0.009$.

---

[a]The impact parameter L is defined as the distance of closest approach of a track to the vertex, its significance is the ratio of L to its error (L/σ_L).

Figure 2. Measurements of the pole asymmetries corrected for pure Z boson exchange ($A_{FB}^{c,0}$) obtained in the different channels. The average value obtained from the full electroweak fit is also shown for comparison.

Figure 3. Measurements of the pole asymmetries corrected for pure Z boson exchange ($A_{FB}^{b,0}$) obtained in the different channels. The average value obtained from the full electroweak fit is also shown for comparison.

2.3 A_{FB}^c using the D meson tag

D mesons are produced only in decays of the Z boson into $c\bar{c}$ and $b\bar{b}$ pairs. The charge state of the fully reconstructed meson can be related with that of the parent quark. On the other hand the statistics is limited by the number of final states that can be reconstructed. The most relevant channel is $D^{*+} \to D^0\pi_{sl}^+$, tagged by the presence of a low momentum pion. The main difficulties in the analysis are the determination of the relative fraction of D mesons produced in Z decays to $c\bar{c}$ and $b\bar{b}$ pairs, and the knowledge of the asymmetry of the combinatorial background. These effects typically amount to $\Delta A_{FB}^c = \pm0.001 \div 0.002$, compared to a statistical error $\Delta A_{FB}^c = \pm0.009$.

3 Impact on the electroweak fit

The measured asymmetries are corrected for QED and QCD effects, and the term corresponding to a pure Z boson exchange, $A_{FB}^{q,0}$, is extracted. The average values are shown in Figures 2 and 3 together with the contributions of the individual channels. The different determinations are consistent with each other and are dominated by the the statistical error. Figure 4 shows their impact on the determination of $\sin^2\theta_{W,eff}^\ell$. It can be noted

that $A_{FB}^{b,0}$ represents the most precise determination of $\sin^2\theta_{W,eff}^\ell$ performed at LEP.

References

1. ALEPH Collaboration, ALEPH 99-076, CONF 99-048;
   DELPHI Collaboration, DELPHI 2000-101, CONF 400.

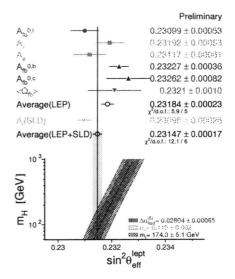

Figure 4. Different determinations of the effective electroweak mixing angle for lepton, $\sin^2\theta_{W,eff}^\ell$, from LEP and SLD measurements.

R_b, R_c MEASUREMENTS AT SLD AND LEP-I

D. SU

Stanford Linear Accelerator Center, P.O.Box 4349, Stanford, CA 93409, USA
E-mail: sudong@slac.stanford.edu

This report summarizes the measurements of R_b, R_c at SLD and LEP-I. These measurements are sensitive probes of the Z^0 couplings to heavy quarks, which provide precision tests of the Standard Model of electroweak interactions at $\sim 0.3\%$ and $\sim 2\%$ level respectively.

1 Introduction

The abundant production of Z^0's at the $e^+e^- \rightarrow Z^0$ resonance peak provides an ideal opportunity for tests of the SM through precision electroweak measurements, where possible new physics could manifest themselves through radiative corrections. In particular, the heavy quark production fractions in Z^0 hadronic decays, $R_b = \frac{\Gamma(Z^0 \rightarrow b\bar{b})}{\Gamma(Z^0 \rightarrow Hadrons)}$ and $R_c = \frac{\Gamma(Z^0 \rightarrow c\bar{c})}{\Gamma(Z^0 \rightarrow Hadrons)}$ are observables with clean theoretical interpretations. The relatively democratic production of all quark flavors in the Z^0 decays combined with our ability to tag b and charm hadron decays, offer the possibility to test the Z^0 coupling to the individual quark flavors at high precision.

The measurements reviewed in this report are from the ALEPH,DELPHI, L3 and OPAL experiments at LEP, and the SLD experiment at SLC. The LEP experiments each accumulated 4.4M Z^0's from the LEP-I runs during 1989-1995. SLD has accumulated a total of 450K Z^0's from 1992 to 1998. These measurements rely on good capabili-

ties in heavy quark identification using silicon vertex detectors. The characteristics of the LEP/SLD vertex detectors and the the event primary vertex (PV) resolution are summarized in Table.1.

2 R_b Measurements

2.1 Measurement methods

The modern R_b measurements generally adopt the double tag technique to reach the interesting precision of $< \sim 1\%$. Events are divided into two hemispheres and a b-tag algorithm is applied to each hemisphere. The measured hemisphere tag rate and event double tag rate allow the extraction of both R_b and the b-tag efficiency ϵ_b from the data. Only the small background tagging efficiencies for uds and charm hemispheres, $\epsilon_{uds}, \epsilon_c$, and the b-tag hemisphere correlation need to be estimated using Monte Carlo (MC). R_c is also taken as the SM value.

It is essential to develop a high efficiency and high purity b-tag to enable the double tag scheme to achieve the necessary statistical and systematic precision. One commonly used tag is the hemisphere impact parameter probability tag first established by ALEPH [1] which is still used in the current ALEPH [2], DELPHI [3] and L3 [4] measurements. A major improvement in b-tag purities for most of the current analyses is to incorporate vertex mass information to further suppress the charm background, as first introduced by SLD [6]. The SLD implementation also used the precise vertex flight di-

Table 1. The Vertex Detector parameters and primary vertex resolution at LEP/SLD.

Experiment	LEP	SLD
Detector type	Double−sided Silicon−strips	CCDs
Inner radius	6.3cm	2.7cm
Spatial resolution	$8\,\mu m$	$4\,\mu m$
High-P $\sigma_{impact}(r\phi)$	$15\,\mu m$	$8\,\mu m$
High-P $\sigma_{impact}(rz)$	$25\,\mu m$	$10\,\mu m$
Mult.Scat. σ_{impact}	$\frac{60-110\,\mu m}{P}$	$\frac{30\,\mu m}{P}$
<Beam Spot>	$\sigma_y = 10\,\mu m$	$\sigma_{x,y} = 4\,\mu m$
Event by event PV	$\sigma_{x,z} = 50\,\mu m$	$\sigma_z = 12\,\mu m$

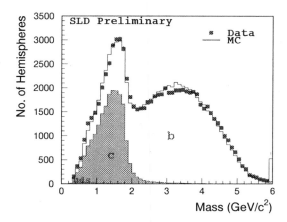

Figure 1. The P_t-corrected vertex mass distributions compared between data and MC for the current SLD R_b analysis.

rection to estimate a P_T correction from the missing neutrals to achieve a 'P_T corrected' vertex mass tag (see Fig.1) with boosted efficiency as well as the desired ~98% b-purity.

Another modern development in the R_b analyses is to combine various discrimination variables to enhance the b-tag performance. One such approach is the multivariant analysis introduced by DELPHI [8], and are used by both DELPHI[3] and ALEPH[2] in their current analyses. Both implementations use 5 different tags constructed from a set of observables using lifetime, decay kinematics, event shape and lepton information. A more compact form of multi-variable tag is to combine the information using Fisher discriminant (DELPHI[3]) or neural network (NN) (OPAL[5],SLD[7]) to construct a single discriminating variable and still follow the simple double tag analysis. The OPAL[5] and L3[4] analyses have included high P_T lepton tags as a separate tag and used a simple 'OR' combination with the lifetime tags.

2.2 New Preliminary Results

At ICHEP-2000, all LEP R_b results are final for publications and essentially unchanged for the last two years. The only new preliminary result is from SLD [7], which is updated to include the last 150K Z^0's not previously used. An improved b-tag is applied to data and MC from a recent new reconstruction. The new b-tag uses a neural network to optimize the track to vertex association, and a second neural network to construct a $c - b$ separation variable, incorporating vertex decay length, multiplicity and momentum information in addition to the P_T corrected vertex mass. This tag variable provides b-tag at one extreme and also a high performance c-tag at the other extreme. The b-tagging efficiency and purity achieved for the bulk 350K Z^0 data from the 97-98 run are 61.8% and 98.3% respectively.

2.3 Systematics and Result Summary

With the R_b measurements now reaching an impressive precision of ~0.5% for even individual measurements, it is crucial to ensure that the systematic evaluation is convincing. The breakdown of the R_b measurement errors are listed in Table 2.

For most of the correlated systematic errors, there is generally a consistent evaluation among the measurements following the standard procedure [9]. The uncertainties due to physics modeling for $\epsilon_c, \epsilon_{uds}$ are extensively checked and yield good consistency between measurements. The rather complex hemisphere correlation systematic evaluation has been subject to intense scrutiny, especially in the areas of primary vertex and QCD gluon radiation effects. The discussions in the OPAL publication [5] is particularly detailed. The general finding is that the MC typically describes the correlation fairly well, although it is difficult to be convinced that all possible significant sources are investigated.

A major common systematic source is the uncertainty in the $g \to c\bar{c}$, $g \to b\bar{b}$ rates which can cause a primary light quark event to be tagged. There are quite few recent mea-

Table 2. The R_b measurement b-tag performance and measurement uncertainty comparison. Tagging performance numbers refer to the main vertex/lifetime tags only.

Experiment	ALEPH	DELPHI	L3	OPAL	SLD
b-tag method	multivar.	multivar.	impact+ℓ	vtx-NN+ℓ	vtx-mass NN
b-tag efficiency	19.6%	29.5%	23.7%	20.9%	61.8%
b-tag purity	98.5%	98.5%	84.0%	97.9%	98.3%
$\delta R_b \times 10^{-5}$					
statistics	87	67	150	112	94
$\epsilon_c, \epsilon_{uds}$ physics	39	25	218	74	44
Hemisphere correlation	36	28	116	71	23
$g \to b\bar{b}$	38	27	11	25	22
$g \to c\bar{c}$	22	8	13	17	18
Detector effects	46	13	43	25	42
Event selection	7	9		33	70
Internal (MC stat. etc.)	47	33	81	59	14
$\delta R_c \pm 0.005$	10	12	108	35	17

surements on both $g \to c\bar{c}$ [10,11] and $g \to b\bar{b}$ [12]. These results are tabulated in Table 3. The measurements are well within the range of uncertainty assumed by the standard procedure [9]. The $g \to QQ$ rates would at first sight imply a more significant problem for R_b. The generally low tagging efficiency for the heavy hadrons from gluon splitting, which are mostly at low momentum, reduces the R_b systematic sensitivity to this effect.

Table 3. The $g \to c\bar{c}$, $g \to b\bar{b}$ measurements compared to the current standard recommendations from the LEP HF group.

Expt.	$g \to c\bar{c}$ ($\times 10^{-2}$)	$g \to b\bar{b}$ ($\times 10^{-3}$)
ALEPH	$3.23 \pm 0.48 \pm 0.53$	$2.77 \pm 0.42 \pm 0.57$
DELPHI		$2.1 \pm 1.1 \pm 0.9$ $3.3 \pm 1.0 \pm 0.8$
L3	$2.45 \pm 0.29 \pm 0.53$	
OPAL	$3.20 \pm 0.21 \pm 0.38$	$3.07 \pm 0.53 \pm 0.97$
SLD		$2.84 \pm 0.61 \pm 0.59$
LEP Std	3.19 ± 0.46	2.51 ± 0.63

The detector systematics mainly cover tracking resolution and efficiency effects. The MC track impact parameters are typically 'smeared' to accommodate unsimulated vertex detector misalignment effects. Although this random smearing tend to bring MC to agree with data overall, the underlying effects are typically local and can affect tracks coherently going through the same detector region in a jetty environment. The treatment for the resolution effect among the experiments unfortunately vary significantly, reflecting very different levels of subjective judgments. Given the fact that underestimated tracking resolution systematic was the main culprit driving the previously high R_b result in 1995-1996, this is perhaps an area which should deserve more attention for better confidence in the systematics.

A possible surprise in systematics is in the event selection flavor bias which has received relatively little attention. The theoretical calculations [13] indicate that the use of the running quark mass compared to the pole mass or massless quark for the 3 or 4 jet rate calculations can have significant effects for b events. This can affect R_b measurements through event selection bias or hemisphere correlation. This specific source of systematics is so far only evaluated by SLD explicitly, which currently assigned an event selection bias systematic for this effect as large as all other systematics combined.

The current R_b measurement results are listed in Fig.2. All measurements are in good agreement with the SM and the combined world average of 0.21651 ± 0.00069 is consistent with the SM expectation of 0.2158 for the currently known top quark mass.

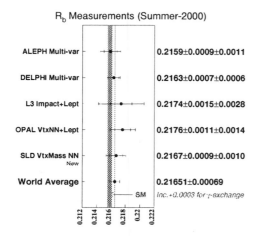

R_b Measurements (Summer-2000)

ALEPH Multi-var   0.2159±0.0009±0.0011

DELPHI Multi-var   0.2163±0.0007±0.0006

L3 Impact+Lept   0.2174±0.0015±0.0028

OPAL VtxNN+Lept   0.2176±0.0011±0.0014

SLD VtxMass NN   0.2167±0.0009±0.0010
New

World Average   0.21651±0.00069

SM   inc.+0.0003 for γ-exchange

Figure 2. R_b^0 measurement results. The inner and outer error bars represent statistical and total errors respectively.

3 R_c Measurements

The measurement of R_c also requires an efficient and pure charm tag to achieve good precision. This has turned out to be a more challenging task than tagging b's. Exclusive charm reconstruction is a clean tag, but the usable branching fractions are somewhat limited and the reconstruction random combinatorial background is also not negligible. The charm hadrons produced in b decays adds further complication. The charm hadrons have shorter lifetimes and smaller decay charged multiplicities compared to the B hadrons, which make the inclusive charm tagging difficult. Without a clean and high efficiency inclusive charm tag at LEP, the R_c measurement techniques are therefore rather diverse:

Lepton Spectrum Analysis: A specialized analysis by ALEPH [14] fits the lepton P, P_T spectra after subtracting $b\bar{b}$ contributions using lepton spectra in hemispheres opposite a b-tag. The main systematics sources are the MC simulation of the $c \to \ell$ branching ratio and spectrum.

Charm Counting: All primary charm hadron decay chains will eventually involve a

weak decay via D^0, D^+, D_s or Λ_c (and a very small fraction of Ξ_c, Ω_c). R_c can be measured [16,15,11] from the sum of production cross sections of the these four charm hadrons using fully reconstructed decays of well-known modes. The systematic limitation come from charm fragmentation simulation and charm decay branching ratio uncertainties.

Exclusive/Inclusive $D^{(*)}$ Cross Tags: Another widely used technique at LEP is the exclusive/inclusive D^* cross tag [14,15,17] which partially calibrates the tagging efficiencies from data. The analysis uses a fully reconstructed $D^{*+} \to D^0\pi^+$ charm tag or a b-tag in one hemisphere, to calibrate a more inclusive charm tag in the other hemisphere by identifying the transition pions π_* from the $D^{*+} \to D^0\pi_*^+$ decays without exclusive D^0 reconstruction. The production fractions $f(c \to D^{*+}X)$ and $f(b \to D^{*+}X)$ are extracted from data together with R_c. The main systematic sources are background subtraction, c/b separation, and $\mathrm{Br}(D^0 \to K^-\pi^+)$ error.

Inclusive Vertex Double Tag: The SLD measurement of R_c [7] is the only true high efficiency inclusive double tag measurement. The inclusive vertex charm tag is a product of the same neural network $c - b$ separation tag as described in section 2.2. The essence of $c - b$ separation power can be seen from Fig.1 where the low mass region is already clearly dominated by charm. The operating point of the hemisphere charm tag gives a $b : c : uds$ composition of 15.6:83.8:0.6 and a charm-tag efficiency of 17.4%. The key successes is the strong suppression of uds background. This measurement is not only statistically most precise, it has also achieved the lowest systematic as the charm region b, c tag efficiencies are all measured from data similar to an R_b analysis.

The various R_c measurements are compared in Fig.3. The world average of $R_c = 0.1709 \pm 0.0034$ is in good agreement with

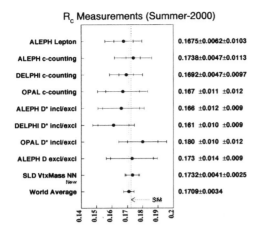

R_c Measurements (Summer-2000)

ALEPH Lepton	0.1675±0.0062±0.0103
ALEPH c-counting	0.1738±0.0047±0.0113
DELPHI c-counting	0.1692±0.0047±0.0097
OPAL c-counting	0.167 ±0.011 ±0.012
ALEPH D* incl/excl	0.166 ±0.012 ±0.009
DELPHI D* incl/excl	0.161 ±0.010 ±0.009
OPAL D* incl/excl	0.180 ±0.010 ±0.012
ALEPH D excl/excl	0.173 ±0.014 ±0.009
SLD VtxMass NN New	0.1732±0.0041±0.0025
World Average	0.1709±0.0034

Figure 3. R_c measurement results. The inner and outer error bars represent statistical and total errors respectively.

the SM prediction of 0.1723 at the present precision. The main reason for the shift of the LEP R_c measurements toward the confirmation of the SM since the 'R_b, R_c crisis' in 1995-1996 is that the D^{*+} production fraction $f(c \to D^{*+})$ is now measured at LEP and noticeably lower than the previously assumed values based on low energy measurements by CLEO and ARGUS.

4 Conclusions

The continuous effort of the last 9 years at LEP and SLD have yielded precision tests of the standard model for the $Zb\bar{b}, Zc\bar{c}$ couplings through R_b and R_c measurements, reaching impressive levels of precision at ±0.34% for R_b and ±2% for R_c. The results are in good agreement with the SM predictions. The heavy flavor tagging techniques pushed by these measurements have far reaching effects for many other measurements at present and in the future.

On behalf of the LEP and SLD collaborations, I would like to thank the CERN and SLAC accelerator departments for their dedicated effort on LEP and SLC to make these measurements possible.

References

1. ALEPH Collab.: D. Buskulic *et al.*, *Phys. Lett.* B **313**, 535 (1993).
2. ALEPH Collab.: R. Barate *et al.*, *Phys. Lett.* B **B401**, 150,163 (1997).
3. DELPHI Collab., P. Abreu *et al.*, *Euro. Phys. Journ.* **C9**, 367 (1999).
4. L3 Collab.: M. Acciarri *et al.*, *Euro. Phys. Journ.* **C13**, 47 (2000).
5. OPAL Collab.: G. Abbiendi *et al.*, *Euro. Phys. Journ.* **C8**, 217 (1999).
6. SLD Collab.: K. Abe *et al.*, *Phys. Rev. Lett.* **80**, 660 (1998).
7. SLD Collab., SLAC-PUB-8667, contribution #739 to ICHEP-2000.
8. P. Billoir *et al.*, *Nucl. Instrum. Methods* A **360**, 532 (1995).
9. LEP Heavy Flavor Working Group, LEP HF/99-01;98-01;96-01.
10. OPAL Collab.: K. Akers *et al.*, *Phys. Lett.* B **353**, 595 (1995); OPAL Collab.: G. Abbiendi *et al.*, *Euro. Phys. Journ.* **C13**, 1 (2000); L3 Collab.: M. Acciarri *et al.*, *Phys. Lett.* B **476**, 243 (2000).
11. ALEPH Collab., CERN-EP/99-094, contribution #178 to ICHEP-2000.
12. DELPHI Collab.: P. Abreu *et al.*, *Phys. Lett.* B **405**, 202 (1997); ALEPH Collab.: R. Barate *et al.*, *Phys. Lett.* B **434**, 437 (1998); DELPHI Collab.: P. Abreu *et al.*, *Phys. Lett.* B **462**, 425 (1999); OPAL Collab.: G. Abbiendi *et al.*, CERN-EP/2000-123; SLD Collab.: K. Abe *et al.*, SLAC-PUB-8506, contribution #691 to ICHEP-2000.
13. W. Bernreuther *et al.*, *Phys. Rev. Lett.* **79**, 189 (1997); G. Rodrigo *et al.*, *Phys. Rev. Lett.* **79**, 193 (1997).
14. ALEPH Collab., R. Barate *et al.*, *Euro. Phys. Journ.* **C4**, 557 (1998).
15. DELPHI Collab.: P. Abreu *et al.*, *Euro. Phys. Journ.* **C12**, 225 (2000).
16. OPAL Collab.: G. Alexander *et al.*, *Z. Phys.* C **72**, 1 (1996).
17. OPAL Collab.: G. Ackerstaff *et al.*, *Euro. Phys. Journ.* **C1**, 439 (1998).

THE FINAL SLD RESULTS FOR A_{LR} AND A_{LEPTON}

TOSHINORI ABE

Stanford Linear Accelerator Center
Stanford University, Stanford, California, 94309
USA
E-mail: toshi@SLAC.Stanford.EDU

Representing the SLD Collaboration

We present the final measurements of the left-right cross-section asymmetry A_{LR} for Z boson production by e^+e^- collisions and Z boson-lepton coupling asymmetry parameters A_e, A_μ, and A_τ in leptonic Z decays with the SLD detector at the SLAC Linear Collider. Using the complete sample of polarized Z bosons collected at SLD, we get $A_{LR} = 0.15056 \pm 0.00239$, $A_e = 0.1544 \pm 0.0060$, $A_\mu = 0.142 \pm 0.015$, and $A_\tau = 0.136 \pm 0.015$. The $A_{LR}(\equiv A_e)$ and A_e results are combined and we find $A_e = 0.1516 \pm 0.0021$. Assuming lepton universality, we obtain a combined effective weak mixing angle of $\sin^2 \theta_W^{eff} = 0.23098 \pm 0.00026$. Within the context of the SM, our result prefers a light Higgs mass.

1 Introduction

The SLD collaboration has reported a series of A_{LR} measurement [1] and A_e, A_μ, and A_τ measurements [2] in the production and decay of Z bosons by e^+e^- collisions. A_{LR} is the single best measurement of the effective weak mixing angle ($\sin^2 \theta_W^{eff}$) and has remarkably small systematic error. The measurements of A_e, A_μ, and A_τ improve precision for the effective weak mixing angle measurement. These measurements also provide a test of lepton universality and SLD makes the only direct measurement of A_μ. In the context of the Standard Model (SM), these measurements provide the best sensitivity to the Higgs mass and favor a light Higgs. In this letter, we will present the final results of these measurements at SLD.

2 Asymmetry measurements at SLD

Polarization-dependent differential cross section for $e^-_{L,R} + e^+ \to Z^0 \to f\bar{f}$ is expressed as follows

$$\frac{d\sigma}{dx} \propto (1 - P_e A_e)(1 + x^2) + 2A_f(A_e - P_e)x$$

where $x = \cos\theta$ is the direction of the outgoing fermion with respect to the electron beam direction. The signed longitudinal polarization of the electron beam is shown as P_e with the convention that left-handed bunches have negative sign. The asymmetry parameter is defined as

$$A_f = 2v_f a_f / (v_f^2 + a_f^2)$$

where v_f and a_f are the effective vector and axial-vector couplings of the Z boson to the fermion (flavor "f") current, respectively. The SM assumes lepton universality and lepton asymmetry parameters are directly related to the effective weak mixing angle

$$A_l \equiv \frac{2\left[1 - 4\sin^2\theta_W^{eff}\right]}{1 + \left[1 - 4\sin^2\theta_W^{eff}\right]^2}.$$

The polarized electron beam at SLD allows for measurements of the lepton asymmetry parameters with two different techniques. One is a left-right asymmetry. The left-right asymmetry is the cross-section asymmetry for the production of Z-bosons from left-handed and right-handed electron beams. This is sensitive to the initial state coupling ($e^+e^- \to Z$). The other is a polarized forward-backward asymmetry. This asymmetry is a double asymmetry which is formed by taking the difference in the number of forward and backward events for left-handed and right-handed beam polarization. This is sensitive to the final state coupling ($Z \to e^+e^-, \mu^+\mu^-, \tau^+\tau^-$).

From hadronic final states, we measure A_e using the left-right asymmetry, which is known as A_{LR} measurement. Leptonic final states provide A_e from the left-right asymmetry and A_e, A_μ, and A_τ from the polarized forward-backward asymmetry. We compare these asymmetry measurements to test lepton universality. Assuming universality, we combine A_{LR} with A_e, A_μ, and A_τ to derive our grand average effective weak mixing angle.

SLD has collected polarized Z data from 1992 to 1998. We collect about 530K polarized Z events with about 75% electron beam polarization.

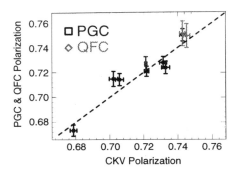

Figure 1. Comparison of PGC and QFC polarizations to the one by CKV.

3 A_{LR} measurement

The event selection for the A_{LR} measurement requires a hadronic signature and discriminates against beam background, two photon, and e^+e^- final states events. The selection efficiency is about 91% for hadronic final states. There is a small amount of $\tau^+\tau^-$ final state (0.3%) which is not background. The background fraction is only 0.04%. Using selected events, we extract the measured cross-section asymmetry A_m as follows

$$A_m = \frac{N_L - N_R}{N_L + N_R}$$

where N_L (N_R) is a number of selected events for left-handed (right-handed) electron beam. We need two more steps to evaluate the result. First, we correct background and small machine/beam related asymmetries, which are small, and divide by the measured polarization as follows,

$$A_{LR} = \frac{A_m}{P_e} + \frac{1}{P_e}\delta A_m = \frac{A_m}{P_e} + O(10^{-4}).$$

Next, A_{LR} is converted to the Z-pole result by applying γZ interference and initial state radiation corrections

$$A_{LR}^0 = A_{LR} + \delta A_{EW}.$$

This calculation requires knowledge of our luminosity weighted mean center-of-mass energy. The relative size of the correction is about 2%.

The electron polarization plays an important role in the measurement. There are three detectors to measure the electron polarization.

Our primary polarimeter is the Cherenov detector (CKV), which detects Compton-scattered electrons. Polarized Gamma Counter (PGC) and Quartz Fiber Calorimeter (QFC) are used to assist in the calibration of CKV and detect Compton-scattered photons. Fig. 1 shows a comparison of measured polarizations with these detectors. The measurements of electron polarization by these detectors are consistent and the systematic error from the calibration uncertainty is 0.40%. The obtained total systematic uncertainty from polarization measurement is 0.50%, which is the biggest systematic error source for the A_{LR} measurement.

We performed two more additional systematic error checks in the 1997-98 run, energy scale and positron polarization. For precise understanding of average center-of-mass energy, we did a Z-pole scan. We measured two off-peak data points and obtained the average center-of-mass energy of $\sqrt{s} = 91.237 \pm 0.029$ GeV. The uncertainty of the center-of-mass energy leads to systematic error of 0.39% on A_{LR}^0 due to energy dependence of γZ interference and initial state radiation correction. This is the second biggest systematic error source of the A_{LR} measurement. In the past, we had assumed positron polarization is zero. We now have measured directly the positron polarization with a Møller polarimeter in End Station A and obtained a positron polarization of $(-0.02 \pm 0.07)\%$ which is consistent with zero.

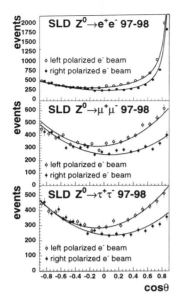

Figure 2. Polar-angle distributions for Z decays to e, μ, and τ pairs.

The final result of the A_{LR} measurement is

$$A_{LR}^0 = 0.15138 \pm 0.00216,$$
$$\sin^2 \theta_W^{eff} = 0.23097 \pm 0.00027.$$

The systematic error in the effective weak mixing angle measurement is about 0.0001. Our error is still dominated by the statistical error.

4 A_e, A_μ, and A_τ measurements

Leptonic Z decay candidates are required to have low charged multiplicity and two back-to-back leptons (or in the case of the tau-pair events, the tau decay products). The selection efficiencies for each lepton species are $70 \sim 75\%$ and their purities are about 99% except tau (about 95%). Fig. 2 shows angular distributions of selected leptonic final states for left- and right-handed electrons (Selection efficiency is corrected in the figure).

Using these events, A_e and A_μ or A_τ are simultaneously determined from an unbinned maximum liklihood functions include Z, γZ, and γ cross-section terms, and initial state radiation effects. We find the results for A_e, A_μ, and A_τ from

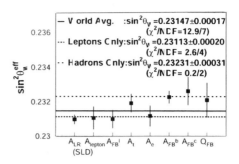

Figure 3. The world effective weak mixing angle measurements.

leptonic Z decay events are

$$A_e = 0.1544 \pm 0.0060 ,$$
$$A_\mu = 0.142 \pm 0.015 , \text{ and}$$
$$A_\tau = 0.136 \pm 0.015.$$

5 The SLD grand average result

The A_{LR} measurement measures the initial state coupling ($A_{LR}^0 \equiv A_e$). Hence we combine the A_{LR}^0 result and A_e from purely leptonic final states taking care of small effects due to correlations in systematic uncertainties, and obtain

$$A_{LR}^0 + A_e = 0.1516 \pm 0.0021.$$

Our results are consistent with lepton universality. Therefore we can assume universality and the obtained our grand average result of the lepton asymmetry parameter and the effective weak mixing angle are

$$A_l = 0.15130 \pm 0.00207,$$
$$\sin^2 \theta_W^{eff} = 0.23098 \pm 0.00026.$$

6 World effective weak mixing angle measurements

Now we compare our result with other measurements [3]. Fig. 3 shows the world effective weak mixing angle measurements. The world average value is $\sin^2 \theta_W^{eff} = 0.23147 \pm 0.00017$. Results with leptonic asymmetry are consistent each other ($\chi^2/NDF = 2.6/4$) and the average value is $\sin^2 \theta_W^{eff} = 0.23113 \pm 0.00020$. Results with hadronic technique ($\sin^2 \theta_W^{eff} = 0.23231 \pm$

640

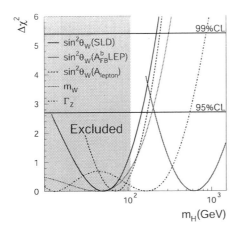

Figure 4. Higgs mass plot by technique.

0.00031) are self-consistent ($\chi^2/NDF = 0.2/2$). However there is 3σ difference between leptons only and hadrons only results.

Since the effective weak mixing angle is very sensitive to Higgs mass, it is interesting to derive the allowed Higgs mass region. We use the measured Z boson [3] and top quark [4] masses, a determination of $\alpha(M_Z^2)$ [5], and the ZFITTER 6.23 program [6] to obtain the results. Fig. 4 shows the allowed Higgs mass regions given by the individual measurements. The SLD result prefers a light Higgs mass [7]. On the other hand, the result by the hadron technique expects a heavy Higgs mass. There are several measurements sensitive to the Higgs mass, W mass (m_W) and Z width (Γ_Z) measurements. The allowed Higgs mass regions by the measurements are also shown in Fig. 4. The most sensitive measurement to Higgs mass is given by the effective weak mixing angle by the lepton technique. The curve given by W mass measurement and the result is in good agreement with the one by the lepton technique. These results prefer a light Higgs mass.

7 Conclusions

The SLD collaboration has finalized its very precise measurement of the weak mixing angle and the result is $\sin^2\theta_W^{eff} = 0.23098 \pm 0.00026$. The SLD error is equivalent to an W mass measure-

ment uncertainty of 38MeV assuming the SM, which is equal to the error of global direct m_W measurements from FERMILAB and LEP II. The SLD/LEP lepton asymmetry measurements are self-consistent. In the context of the SM, our data prefers a light Higgs mass.

References

1. K. Abe *et al.* [SLD Collaboration], Phys. Rev. Lett. **84**, 5945 (2000);
   K. Abe *et al.* [SLD Collaboration], Phys. Rev. Lett. **78**, 2075 (1997);
   K. Abe *et al.* [SLD Collaboration], Phys. Rev. Lett. **73**, 25 (1994);
   K. Abe *et al.* [SLD Collaboration], Phys. Rev. Lett. **70**, 2515 (1993).
2. K. Abe *el al.* [SLD Collaboration], SLAC-PUB-8618, submitted to Phys. Rev. Lett. ;
   K. Abe *et al.* [SLD Collaboration], Phys. Rev. Lett. **79**, 804 (1997).
3. The LEP Collaboration, the LEP Electroweak Working Group, The SLD Heavy Flavour and Electroweak Working Groups, CERN-EP/2000-016, January 2000.
4. T. Affolder *et al.* [CDF Collaboration], Submitted to Phys. Rev. D [hep-ex/0006028].
   B. Abbott *et al.* [D0 Collaboration], Phys. Rev. **D58**, 052001 (1998).
5. A. D. Martin, J. Outhwaite and M. G. Ryskin, [hep-ph/0008078].
6. D. Bardin *et al.* DESY 99-070 (1999) [hep-ph/9908433].
7. There is a nice summary about the effective weak mixing angle measurement and Higgs mass issue. See http://www-sldnt.slac.stanford.edu/alr/ .

MEASUREMENTS OF PARITY-VIOLATION PARAMETERS AT SLD

MASAKO IWASAKI

Department of Physics, University of Oregon, Eugene, OR 97403
E-mail: masako@slac.stanford.edu

Representing the SLD Collaboration

We present direct measurements of the parity-violation parameters A_b, A_c, and A_s at the Z^0 resonance with the SLD detector. The measurements are based on approximately 530k hadronic Z^0 events collected in 1993-98. Obtained results are $A_b = 0.914 \pm 0.024$ (SLD combined: preliminary), $A_c = 0.635 \pm 0.027$ (SLD combined: preliminary), and $A_s = 0.895 \pm 0.066(stat.) \pm 0.062(sys.)$.

1 Introduction

In the Standard Model, the Z^0 coupling to fermions has both vector (v_f) and axial-vector (a_f) components. Measurements of fermion asymmetries at the Z^0 resonance probe a combination of these components given by $A_f = 2v_f a_f/(v_f^2 + a_f^2)$. The parameter A_f expresses the extent of parity violation at the $Zf\bar{f}$ vertex and its measurement provides a sensitive test of the Standard Model.

At the Stanford Linear Collider (SLC), the ability to manipulate the longitudinal polarization of the electron beam allows the isolation of A_f through formation of the left-right forward-backward asymmetry:

$$\tilde{A}_{FB}^f(z)$$
$$= \frac{[\sigma_L^f(z) - \sigma_L^f(-z)] - [\sigma_R^f(z) - \sigma_R^f(-z)]}{[\sigma_L^f(z) + \sigma_L^f(-z)] + [\sigma_R^f(z) + \sigma_R^f(-z)]}$$
$$= |P_e| A_f \frac{2z}{1 + z^2},$$

where P_e is the longitudinal polarization of the electron beam, and $z = \cos\theta$ is the direction of the outgoing fermion relative to the incident electron.

The measurements described here are based on 530k Z^0-decay events taken in 1993-98 with the SLC Large Detector (SLD)[1]. The average electron polarization is $|P_e| = 73\pm0.5\%$[2]. Polarized electron beams, a small and stable SLC interaction region, the high resolution CCD vertex detector[3], and the excellent particle identification with Čerenkov Ring imaging Detector (CRID)[4] provide precision electroweak measurements.

2 A_b measurements

In order to tag the b-quark, topologically reconstructed mass of the secondary vertex[5] is used. The secondary vertex is reconstructed with charged tracks, and its invariant mass is calculated. To account for neutral particles and missing tracks, the vertex mass is corrected: we calculate the P_T-corrected mass M_{P_T} by estimating a missing P_T from the acolinearity between the momentum sum of the vertex and the direction of the vertex flight path. Applying the cut of $M_{P_T} > 2\ GeV/c^2$, we identify the b-quark with 98% purity and 50% efficiency.

To determine the b-quark charge, we uses 4 different methods: 1) vertex charge, 2) jet charge, 3) cascade kaon and 4) lepton.

The vertex-charge analysis uses the track charge sum of the secondary vertex to identify the charge of the primary quark[6]. We introduce the Neural Network technique to reject background and to associate the tracks to the secondary vertices. It improves the b-tagging efficiency to 57%. In this analysis, we reconstruct the tracks which has hits in the vertex detector only. By adding such tracks, we enhance the charge separation performance to 83%. The b-tagging purity and correct charge probability are estimated us-

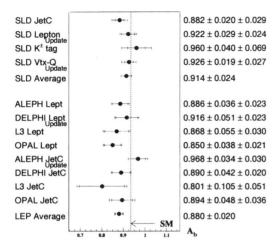

SLD JetC		$0.882 \pm 0.020 \pm 0.029$
SLD Lepton Update		$0.922 \pm 0.029 \pm 0.024$
SLD $K^{\pm}$ tag		$0.960 \pm 0.040 \pm 0.069$
SLD Vtx-Q Update		$0.926 \pm 0.019 \pm 0.027$
SLD Average		0.914 ± 0.024
ALEPH Lept		$0.886 \pm 0.036 \pm 0.023$
DELPHI Lept Update		$0.916 \pm 0.051 \pm 0.023$
L3 Lept		$0.868 \pm 0.055 \pm 0.030$
OPAL Lept		$0.850 \pm 0.038 \pm 0.021$
ALEPH JetC Update		$0.968 \pm 0.034 \pm 0.030$
DELPHI JetC		$0.890 \pm 0.042 \pm 0.020$
L3 JetC		$0.801 \pm 0.105 \pm 0.051$
OPAL JetC		$0.894 \pm 0.048 \pm 0.036$
LEP Average		0.880 ± 0.020

Figure 1. The world A_b measurements (Summer 2000). LEP measurements are derived from $A_b = 4A_{FB}^{0,b}/(3A_e)$ using $A_e = 0.1500 \pm 0.0016$ (the combined SLD A_{LR} and LEP A_{lepton}).

ing opposite hemisphere information.

In the jet-charge analysis, we use the net momentum-weighted jet-charge[7]. The track charge sum and difference between the two hemispheres are used to extract the analyzing power from data, thereby reducing MC dependencies and lowering systematic effects.

In the kaon analysis, we use the charged kaon in the decay $\bar{B} \to D \to K^-$, to determine the b charge[8]. CRID is used to identify $K^{\pm}$ with high-impact parameter tracks. The charges of the kaon candidates are summed in each hemisphere and the difference between the two hemisphere charges is used to determine the polarity of the thrust axis for the b-quark direction.

Electrons and muons are used to identify the charge and direction of the primary b quark[9]. Geometrical information is used to separate cascade and prompt leptons. In the electron analysis, we also use the Neural Network for source classification.

Fig.1 shows the preliminary results from the SLD and LEP measurements, where the LEP measurements are derived from $A_b = 4A_{FB}^{0,b}/(3A_e)$ using $A_e = 0.1500 \pm 0.0016$ (the combined SLD A_{LR} and LEP A_{lepton}). The

combined preliminary SLD result for A_b is obtained as $A_b = 0.914 \pm 0.024$.

3 A_c measurements

At the SLD, four different techniques are used to measure the A_c: 1) inclusive charm-asymmetry measurement with kaon charge and vertex charge, 2) lepton, 3) exclusively reconstructed D^* and D-mesons, and 4) using P_T spectrum of soft-pion from D^*.

In the inclusive charm analysis, c-quarks are tagged using intermediate P_T-corrected-mass vertices[10]. It provides 82% purity and 29% efficiency for $Z^0 \to c\bar{c}$ events. A b veto is applied to reject any event with high vertex mass in either hemisphere. For the hemispheres with a secondary vertex, a secondary track identified as $K^{\pm}$ from the CRID, or a non-zero vertex charge, is used to sign the charm quark direction. The background is mostly b events and its fraction is constrained by the double-tag calibration. This analysis has significantly high statistical power and the systematic errors are still very much under control.

We also measure the charm asymmetry with traditional technique using electrons and muons which not only tag the c events but also determine the c-quark direction from the lepton[9].

The exclusive reconstruction of charmed mesons provide the cleanest technique for the charm-asymmetry measurements[11]. We use four decay modes to identify D^{*+}: the decay $D^{*+} \to \pi_s^+ D^0$ followed by $D^0 \to K^-\pi^+$, $D^0 \to K^-\pi^+\pi^0$ (Satellite resonance), $D^0 \to K^-\pi^+\pi^-\pi^+$, or $D^0 \to K^-l^+\nu_l$ (l =e or μ). We also identify D^+ and D^0 mesons via the decay of $D^+ \to K^-\pi^+\pi^+$ and $D^0 \to K^-\pi^+$ (not from D^{*+}). In this analysis, we reject $Z^0 \to b\bar{b}$ events using P_T-corrected mass of the reconstructed vertices. The random-combinatoric background can be estimated from the mass sidebands.

The soft-pions from the decay $D^{*+} \to$

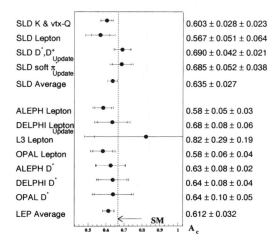

SLD K & vtx-Q	$0.603 \pm 0.028 \pm 0.023$
SLD Lepton	$0.567 \pm 0.051 \pm 0.064$
SLD D^{*}, D^{+} Update	$0.690 \pm 0.042 \pm 0.021$
SLD soft π Update	$0.685 \pm 0.052 \pm 0.038$
SLD Average	0.635 ± 0.027
ALEPH Lepton	$0.58 \pm 0.05 \pm 0.03$
DELPHI Lepton Update	$0.68 \pm 0.08 \pm 0.06$
L3 Lepton	$0.82 \pm 0.29 \pm 0.19$
OPAL Lepton	$0.58 \pm 0.06 \pm 0.04$
ALEPH D^{*}	$0.63 \pm 0.08 \pm 0.02$
DELPHI D^{*}	$0.64 \pm 0.08 \pm 0.04$
OPAL D^{*}	$0.64 \pm 0.10 \pm 0.05$
LEP Average	0.612 ± 0.032

Figure 2. The world A_c measurements (Summer 2000). LEP measurements are derived from $A_c = 4A_{FB}^{0,c}/(3A_e)$ using $A_e = 0.1500 \pm 0.0016$ (the combined SLD A_{LR} and LEP A_{lepton}).

$D^0\pi_s^+$ are also used to tag c-quarks[11]. To determine the D^* direction, charged tracks and neutral clusters are clustered into jets. We also reject the $b\bar{b}$ background using P_T-corrected-mass of reconstructed vertices. Using the momenta transverse to the jet axis (P_T) for tracks, we select the soft-pion candidates which have small P_T value. The largest systematic uncertainty is the choice of the background P_T shape.

Fig.2 shows the preliminary results from the SLD and LEP measurements. The combined preliminary SLD result for A_c is obtained as $A_c = 0.635 \pm 0.027$.

4 Measurement of A_s

In this analysis, we use high-momentum strange particles[12]. We require both event hemispheres have $K^{\pm}$ with $p > 9$ GeV/c, or K_s^0 with $p > 5$ GeV/c. CRID is used to identify $K^{\pm}$. To determine the s-quark charge, we require at least one hemisphere have $K^{\pm}$. The heavy-quark background are rejected by identifying B and D decay vertices. We obtain 66% purity for $Z^0 \to s\bar{s}$ events.

From the 1993-98 SLD data, we get

the result of $A_s = 0.895 \pm 0.066(stat.) \pm 0.062(sys.)$. As a test of d-type quark universality, we compare it with the SLD combined A_b measurement: $A_b/A_s = 1.02 \pm 0.10$. These are consistent within the error.

5 Conclusion

SLD produces world class measurements of parity-violation parameters. The SLD measurements of A_c and A_s are now the most precise single measurements in the world. The measured A_b, A_c and A_s results are consistent with the Standard Model.

References

1. SLD Collab., K. Abe *et al.*, Phys. Rev. **D53** (1996) 1023.
2. SLD Collab., K. Abe *et al.*, Phys. Rev. Lett. **84** (2000) 5945.
3. K. Abe *et al.*, Nucl. Inst. Meth. **A400** (1997) 287.
4. K. Abe *et al.*, Nucl. Inst. Meth. **A343** (1994) 74.
5. D. Jackson, Nucl. Inst. Meth. **A388** (1997) 247.
6. SLD Collab., K. Abe *et al.*, SLAC-PUB-8542, contribution to ICHEP 2000, Osaka, Japan.
7. SLD Collab., K. Abe *et al.*, SLAC-PUB-7886, contribution to ICHEP 98, Vancouver, Canada.
8. SLD Collab., K. Abe *et al.*, SLAC-PUB-8200, contribution to EPS 99, Tampere, Finland.
9. SLD Collab., K. Abe *et al.*, [hep-ex/0009064], SLAC-PUB-8516, contribution to ICHEP 2000, Osaka, Japan.
10. SLD Collab., K. Abe *et al.*, [hep-ex/9907065], SLAC-PUB-8199, contribution to EPS 99, Tampere, Finland.
11. SLD Collab., K. Abe *et al.*, [hep-ex/0009035], submitted to Phys. Rev. D.
12. SLD Collab., K. Abe *et al.*, [hep-ex/0006019], submitted to Phys. Rev. Lett.

NEW R VALUES IN 2-5 GEV FROM THE BESII AT BEPC

Z. G. ZHAO REPRESENTING THE BES COLLABORATION

Institute of High Energy Physics, 19 Yuquan Road, Beijing 100039, China
E-mail: zhaozg@pony1.ihep.ac.cn

We report the preliminary R values for all the 85 energy points scanned in the energy region of 2-5 GeV with the upgraded Beijing Spectrometer (BESII) at Beijing Electron Positron Collider (BEPC). On average, the uncertainties of the R values we measured are $\sim 7\%$. The new R values has a significant impact on the prodicted mass of the Higgs (m_H) from the global fit to the electroweak data, and will also contribute to the interpretation of the E821 $g - 2$ experient.

1 Introduction

The QED running coupling constant evaluated at the Z pole, $\alpha(M_Z^2)$, and the anomalous magnetic moment of the muon, $a_\mu = (g - 2)/2$, are two fundamental quantities to test the Standard Model(SM) [1,2]. $\alpha(M_Z^2)$, as one of the three input parameters in the global fit to the electroweak dadat, is sensitive to the predicted mass of the Higgs. Theoretically, a_μ is sensitive to large energy scales and very high order radiative corrections [3]. Any deviation between the SM predicted value of anomalous magnetic moment of the muon, a_μ^{SM}, and that from the experimentally measured one, a_μ^{Exp}, may hint new physics. However, the uncertainties in both $\alpha(M_Z^2)$ and a_μ^{SM} are dominanted by the hadronic vacuum polarization, which cannot be reliably calculated but are related to R values through dispersion relations [2]. Here R is the lowest order cross section for $e^+e^- \to \gamma^* \to$ hadrons, which is defined as $R = \sigma(e^+e^- \to \text{hadrons})/\sigma(e^+e^- \to \mu^+\mu^-)$, where the denominator is the lowest-order QED cross section, $\sigma(e^+e^- \to \mu^+\mu^-) = \sigma_{\mu\mu}^0 = 4\pi\alpha^2/3s$.

Since the uncertainties in $\alpha(M_Z^2)$ and a_μ^{SM} are dominated by the errors of the values of R in the cm energy range below 5 GeV [2], it is crucial to significantly reduce the uncertainties in the R values measured about 20 years ago with a precision of about 15-20% in the energy region of 2-5 GeV [1,3].

2 R scan with BESII at BEPC

Following the first R scan with 6 energy points in 2.6-5 GeV range done in 1998 [4], the BES collaboration did a finer R scan with 85 energy points in the energy region of 2-4.8 GeV. To understand the beam associated background, separated beam running was done at 24 energy points and single beam running for both e^- and e^+ was done at 7 energy points distributed over the whole scanned energy region. Special runs were taken at the J/ψ to determine the trigger efficiency.

The scan was done with BESII, a conventional collider detector which has been described in detail in ref. 5.

3 Data Analsis

The R values from the BESII scan data are measured by observing the final hadronic events inclusively, i.e. the value of R is determined from the number of observed hadronic events (N_{had}^{obs}) by the relation

$$R = \frac{N_{had}^{obs} - N_{bg} - \sum_l N_{ll} - N_{\gamma\gamma}}{\sigma_{\mu\mu}^0 \cdot L \cdot \epsilon_{had} \cdot (1 + \delta)}, \quad (1)$$

where N_{bg} is the number of beam associated background events; $\sum_l N_{ll}$, $(l = e, \mu, \tau)$ and $N_{\gamma\gamma}$ are the numbers of misidentified lepton-pairs from one-photon and two-photon processes events respectively; L is the integrated luminosity; δ is the radiative correction; ϵ_{had} is the detection efficiency for hadronic events.

The hadronic event selection identifies one photon multi-hadron production from all other possible contamination mechanisms. Cosmic rays, lepton pair production, two-photon process and beam associated processes are the backgrounds involved in our measurement. Clear Bhabha events are first rejected. Then the hadronic events are selected based on charged track information. Cuts on fiducial, vertex, track fit quality, maximum and minimum energy deposition, momentum and time-of-flight are applied to select hadronic events. Special attention is paid to two-prong events. Additional cuts are utilized to further reject cosmic ray, Bhabha and beam associated backgrounds [4].

The cosmic rays and part of the lepton pair production events are directly removed by the event selection. The remaining background from lepton pair production and two-photon processes is then subtracted out statistically according to a Monte Carlo simulation.

Most of the beam associated background events are rejected by a vertex cut. The same hadronic event selection criteria are applied to the separated-beam data, and the number of separated-beam events, N_{sep}, surviving these criteria are obtained. The number of the beam associated background events, N_{bg}, in the corresponding hadronic event sample is given by $N_{bg} = f \times N_{sep}$, where f is the ratio of the product of the pressure at the collision region times the integrated beam currents for colliding beam runs and that for the separated beam runs.

The beam associated background can also be subtracted by fitting the event vertex along the beam direction with a Gaussian for real hadronic events and a polynomial of degree two for the background. The difference between R values obtained using these two methods to subtract the beam associated background is about $(0.3 \sim 2.3)\%$, depending on the energy. This difference was taken into account in the systematic uncertainty.

The integrated luminosity is determined by large-angle Bhabha events using only the energies deposited in BSC.

A cross check using only dE/dx information from the MDC to identify electrons was generally consistent with the BSC measurement; the difference was taken into account in the overall systematic error of 1.5-2.6%.

A great effort has been made by the Lund group and BES collaboration to develop the formalism using the basic Lund Model Area Law directly for a Monte Carlo simulation, which removes the extreme high energy approximation in string fragmenation in JET-SET [6]. The final state simulation in LU-ARLW is exclusive as opposed to JETSET's inclusive method, and LUARLW has only one free parameter in the fragmentaion function. Above 3.77 GeV, the production of charmed mesons are included in the generator based on Eichten Model [7,8].

The parameters in LUARLW are tuned with R scan data to reproduce distributions of kinematic variables such as multiplicity, sphericity, angular and momentum distributions, etc.

We find that the same set of parameters can be applied to the energy region below open charm, and another set of parameters can be used for the energies above it. Parameters are also tuned point by point for the continuum and we find that the detection efficiencies determined are consistent within 2%, which is taken into account in the systematic errors.

Radiative corrections determined using four different schemes [4] agreed with each other to within 1% below charm threshold. Above charm threshold, where resonances are important, the agreement is within 1-3%. For the measurements reported here, we use the formalism of Ref. 9 and include the differences with the other schemes in the systematic error of less than 4%.

The errors from differnt sources are listed in table 1.

646

Table 1. Error Sources for E_{cm}=3.0 GeV. Adding the systematic and statistic errors in quadrature gives a total error of 5.8%.

Source	N_{had}	L	ϵ_{had}	$1+\delta$	Stat.
Err.(%)	3.3	2.3	3.0	1.3	2.5

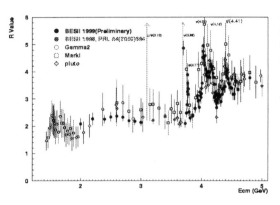

Figure 1. R values below 5 GeV.

To further improve the measurement of R values at BEPC, one needs better performance from the detector and a better handle on the uncertainty arising from the hadronic event generator, as well as higher machine luminosity, particularly for the energies below 3.0 GeV.

The preliminary R values obtained at all 85 energy are graphically displayed in Fig. 1, together with the 6 energy points measured in the first scan and those measured by MarkI, $\gamma\gamma2$ and Pluto about twenty years ago. The preliminary R values from BESII have an average uncertainty of about 7% and are slightly lower than that from the previously measurements. The two to three factor improvement in precision of the R values in 2-5 GeV has a significant impact on the global fit to the electroweak data for the determination of m_H. The preliminary fit results show that the predicted m_H is significant increased with the preferred value lying just above the LEP2 excluded region, and the new χ^2 pro-

file of the fit accommodates the LEP2 bound on the mass more comfortably [10,11]. On the other hand, BESII R values can also greatly contribute to the interpretation of the E821 $g-2$ measurement [3].

We would like to thank the staff of the BEPC accelerator and IHEP Computing Center for their efforts. We also wish to acknowledge useful discussions with B. Andersson, H. Burkhardt, M. Davier, B. Pietrzyk, T. Sjöstrand, A. D. Martin, M. L. Swartz. We especially thank M. Tigner for major contributions not only to BES but also to the operation of the BEPC.

References

1. A. Blondel, Proc. of the 28th Int. Conf. on High Energy Physics, Warsaw, Poland, July 1996.
2. Z. G. Zhao, Proc. of LP99, SLAC, USA, July 1999.
3. Robert Carey, "New Results from g-2 Experiment", talk given at ICHEP2000, Osaka, Japan, July 2000.
4. J. Z. Bai, et. al. the BES collaboration, *Phys. Rev. Lett.* **84**, 594 (2000).
5. J.Z. Bai *et al.*, (BES Collab.), *Nucl. Instrum. Methods* A **344**, 319 (1994); J.Z. Bai *et al.*, (BES Collab.), "The Upgraded Beijing Spectrometer", accepted by NIM.
6. B. Andersson and Haiming Hu, "Few-body States in Lund String Fragmentation Model" hep-ph/9910285.
7. E. Eichten, et.,al., *Phys. Rev.* D **21**, 203 (1980).
8. J.C. Chen, et, al., *Phys. Rev.* D **62**, 034003 (2000).
9. G. Bonneau and F. Martin, *Nucl. Phys.* B **27**, 387 (1971).
10. B. Pietrzyk, "The Global Fit to the Electroweak Data", talk given at ICHEP2000, Osaka, Japan, July 2000.
11. A. Martin, et. al., hep-ph/0008078.

RECENT THEORETICAL DEVELOPMENTS IN LEP 2 PHYSICS

GIAMPIERO PASSARINO

Dipartimento di Fisica Teorica, Università di Torino, Italy
INFN, Sezione di Torino, Italy
E-mail: giampiero@to.infn.it

Recent theoretical developments in e^+e^--annihilation into fermion pairs are summarized. In particular, two-fermion production, DPA for $W - W$ signal, single-W production and $Z - Z$ signal

After an illustrious career LEP stops running rather soon so it is unlikely there will be any more data in this energy region and we all must try to do the best we can to get the most accurate measurements and the most precise predictions we can.

From the point of view of theory there is of course no deep reason why the theory uncertainty should be reduced below that of the experimental precision, but it is surely a useful target as the theory error has to be added in quadrature in looking for deviations from the standard model.

In this talk the most recent theoretical developments connected with LEP 2 physics will be shortly reviewed. As the LEP 2 community has written a report that has just come out I refer the interested reader to that report [1] where one of the goals was to summarize and review critically the progress made in theoretical calculations and their implementation in computer programs since the 1995 workshop on *Physics at LEP2*.

- $e^+e^- \to \overline{f}f(\gamma, \text{pairs})$

On the basis of comparisons of various calculations, theoretical uncertainties have been estimated and compared with those for the final LEP 2 data analysis. In the following list we summarize the present status of theoretical and experimental accuracy as given in the report of the 2f Working Group of the LEP 2/MC Workshop [2] to which we refer for more details:

1. $e + e^- \to \overline{q}q(\gamma)$ 0.3% / 0.1%-0.2%

2. $e + e^- \to \mu^+\mu^-(\gamma)$ 0.4% / 0.4%-0.5%

3. $e + e^- \to \tau^+\tau^-(\gamma)$ 0.4% / 0.4%-0.6%

4. $e + e^- \to e^+e^-(\gamma)$ (endcap) 0.5% / 0.1%

5. $e+e^- \to e^+e^-(\gamma)$ (barrel) 2.0% / 0.2%

6. $e + e^- \to e^+e^-(\gamma)$ 3.0% / 1.5%

7. $e + e^- \to l^+l^-$ 1.0% / 0.5%

8. $e + e^- \to \overline{\nu}\nu(\gamma)$ 4.0% / 0.5%

First entry is the present theoretical uncertainty, second one is the experimental precision tag. The total hadronic and leptonic cross-sections are now predicted to the total precision tag of 0.2%, (excluding pairs) by ZFITTER [3] and KKMC [4].

- News for Pairs in e^+e^- annihilation

Shortly before and during this workshop a lot of new codes for pair corrections at LEP 2 were developed. Before 1999, only the diagram-based pair correction with $s' = M_{\text{prop}}^2$ could be calculated by ZFITTER and TOPAZ0 [5].

Common exponentiation of IS-γ and ISNS$_\gamma$ pairs for energies away from the Z-peak as well as optional ISS$_\gamma$ pairs were implemented in both codes in 1999 (see [2] for their definition). Now ZFITTER has been upgraded to include explicit FS$_\gamma$ with the possibility of mass cuts. Furthermore, the new GENTLE/4fan [6] offers even more options with mass cuts on all pairs and inclusion of pairs from virtual Z and swapped FS diagrams and a new combination of KKMC and KORALW is being developed.

The main achievements in this area can be summarized as follows: a proposal for a signal definition which can be, to better than 0.1% accuracy defined either based on cuts or on diagrams. The determination of efficiency corrections using full event generators has been checked for GRC4f [17] to a precision of 0.1%, from a comparison of real pair cross-sections with GENTLE. However, problems of pairing ambiguities for four identical fermions become increasingly important with the larger ZZ cross-sections at high energies. From varying pairing algorithms, a worst-case difference of 0.8 per mill was found for inclusive hadrons at 206 GeV. Furthermore, differences for pair corrections between s' definitions via the propagator or primary pair mass in the diagram-based approach have been determined and GENTLE – ZFITTER both find them to be about 0.3(1.1) per mill for high s' hadrons (muons).

Maximum differences for the diagram-based pair correction of 1.7(1.5) per mill for inclusive hadrons (muons) and 0.2(0.4) per mill for high s' hadrons (muons) between any two of the programs GENTLE, ZFITTER and TOPAZ0 have been found. Compared to the LEP-combined statistical precision of the measurements all these differences are small. Even the 1.7 per mill difference is only about half of the expected LEP-combined statistical error.

Finally, a first complete calculation of pair corrections for Bhabha scattering has been done by LABSMC [7].

The conclusion for the inclusion of pair effects in the two-fermion cross-section are as follows: with the exception of the 1.7 per mill (tag of 1.1 per mill) difference for inclusive hadrons, all theoretical uncertainties are well below the experimental precision tags. Especially for the case of Bhabha scattering it would be highly desirable to have more than one code predicting the effects of secondary pairs. Note that improvements are still expected in GENTLE, TOPAZ0 and KKMC + KORALW.

In the following we will discuss items that are related to four-fermion production.

- *WW* signal: the CC03 class

While the CC03 cross-section is not an observable, it is nevertheless a useful quantity at LEP 2 energies where it can be classified as a pseudo-observable. It contains the interesting physics, such as the non-abelian couplings and the sensitivity of the total cross-section to M_W near the W-pair threshold. The goal of this common definition is to be able to combine the different final state measurements from different experiments so that the new theoretical calculations can be checked with data at a level better than 1%.

It is worth summarizing the status of the WW cross-section prior to the 2000 Winter Conferences. Nominally, any calculation for $e^+e^- \to WW \to 4f$ was a tree level calculation including as much as possible of the universal corrections in some sort of Improved Born Approximation (IBA). A CC03 cross-section, typically in the G_F-scheme, with universal ISR QED and non-universal ISR/FSR QED corrections produces a curve that been used for the definition of the standard model prediction with a $\pm 2\%$ systematic error assigned to it. However, we have clear indications that non-universal electroweak corrections for WW(CC03) cross-section are not small and even larger than the experimental LEP accuracy. Furthermore, one should stress the importance of photon reconstruction at LEP 2 accuracy.

Recently [8], a new electroweak $\mathcal{O}(\alpha)$ CC03 cross-section has become available, in the framework of double-pole approximation (DPA), showing a result that is $2.5 \div 3\%$ smaller than the old CC03 cross-section. This is a big effect since the combined experimental accuracy of LEP experiments is even smaller.

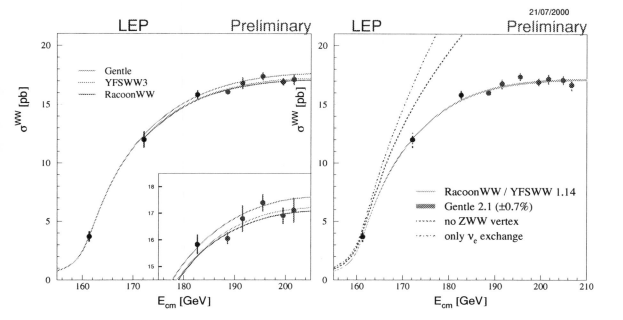

DPA emerges from the CC03 diagrams upon projecting the W-boson momenta in the matrix element to their on-shell values. This means that the DPA is based on the residue of the double resonance, which is a gauge-invariant quantity. In contrast to the CC03 cross-section, the DPA is theoretically well-defined. DPA provides a convenient framework for the inclusion of radiative corrections, but should not be applied for Born-level calculations. Summarizing we may say that the at present only workable approach for evaluating the radiative corrections to resonance-pair-production processes, involves the so-called leading-pole approximation. This approximation restricts the complete pole-scheme expansion to the term with the highest degree of resonance.

Conclusions for CC03 are as follows: the data are in good agreement with the predictions of RacoonWW [9] and YFSWW3 [10] (see also BBC [11]). At the time of Winter 2000 predictions of YFSWW3 were about 0.5 % − 0.7 % higher, somewhat larger than intrinsic DPA uncertainty. The main source of this

discrepancy is found, RacoonWW and YF-SWW3 differ only by about 0.3 % at LEP 2 energies for total cross-sections and within 1 % in angular and invariant-mass distributions. There is a general satisfaction with the progress induced by new DPA calculations. Nevertheless, the theoretical uncertainty could probably be improved somewhat in the future.

- single-W production

A fairly large amount of work has been done in the last years on the topic of single-W production. The experimental community agreed on some setup to define the single-W production and now this has been formalized in one of the LEP EWWG meetings; there, it was decided to have a signal definition as follows:

1. $ee\nu\nu$, t-channel only, $E(e^+) > 20$ GeV, $|\cos\theta(e^+)| < 0.95$, $|\cos\theta(e-)| > 0.95$;

2. $e\nu\mu\nu$, t-channel only, $E(\mu^+) > 20$ GeV;

3. $e\nu\tau\nu$, t-channel only, $E(\tau^+) > 20$ GeV;

4. $e\nu ud$, t-channel only, $M(ud) > 45 GeV$;

5. $e\nu cs$, t-channel only, $M(cs) > 45 GeV$.

The main problems in dealing with single-W production are the correct choice of the energy scale in couplings and the proper treatment of QED radiation in processes that are not dominated by annihilation diagrams..

For the energy scale in couplings we have now an exact calculation [12] based on the massive formulation of the Fermion-Loop scheme (FL) which, at the Born-level (no QED)is known to be at the 1% level of accuracy (see WTO [13]). No program includes $\mathcal{O}(\alpha)$ electroweak radiative corrections. Note that the FL-scheme developed in [14] and refined in [15] makes the approximation of neglecting all masses for the incoming and outgoing fermions in the processes $e^+e^- \to n$ fermions. The recent development, however, goes beyond this approximation.

A description of single-W processes by means of the FL-scheme is mandatory because FL is the only known QFT consistent scheme that preserves gauge invariance and, moreover, single-W production is a process that depends on several scales: the single-resonant s-channel exchange of W-bosons, the exchange of W-bosons in t-channel, the small scattering angle peak of outgoing electrons.

A correct treatment of the multi-scale problem can only be achieved via FL-scheme and a naive rescaling cannot reproduce the full answer for all situations, all kinematical cuts.

The effect of QED on the total cross-sections are between 7% and 10% at LEP 2 energies. Furthermore, grc4f and SWAP [8] have estimated that if one uses the wrong energy scale s in the IS structure functions, the ISR effect is overestimated of about 4%. SWAP [16] estimates that the effects due to non-s-scales predict a lowering of the Born cross-section of about 8%. SWAP results show a good agreement with those of grc4f [17].

Conclusions for single-W are as follows: although we register substantial improvements upon the standard treatment of QED ISR, the problem is not yet fully solved for processes where the non-annihilation component is relevant. A solution of it should rely on the complete calculation of the $\mathcal{O}(\alpha)$ correction. At the moment, a total upper bound of $\pm 5\%$ th. uncertainty should be assigned to single-W.

We could say that QED in single-W is understood at a level better than 4% but we are presently unable to quantify this assertion.

- ZZ signal

NC02 is $e^+e^- \to ZZ$, (t and u channel), with all Z decay modes allowed. Since the interferences between the crossings are not double-resonant, it is customary to consider them as background and to define the ZZ signal from the absolute squares of the double-resonant diagrams only. The choice is based on the observation that $R_{uucc/uuuu} = 2.06$, $R_{ddss/dddd} = 2.08$.

Compared to the experimental uncertainty [18] on the NC02 ZZ cross-section a difference of about 1% between theoretical predictions is acceptable. The global estimate of theoretical uncertainty is 2%, again acceptable. However, it would be nice to improve upon the existing calculations.

Conclusions for NC02 are as follows: for the NC02 cross-section we have a 1% variation, obtained by changing the input parameter set in GENTLE and in ZZTO [20] and by varying from the standard GENTLE approach for ISR to the complete lowest order corrections. We estimate the real uncertainty to be 2%. Furthermore, ZZTO which is a FL calculation (with universal ISR, FSR_{QED} FSR_{QCD} and running masses) agrees rather well with YFSZZ [19], roughly below the typical DPA accuracy of 0.5%, and the latter features leading pole approximation, on $\mathcal{O}(\alpha^2)$ leading-logarithms YFS exponentia-

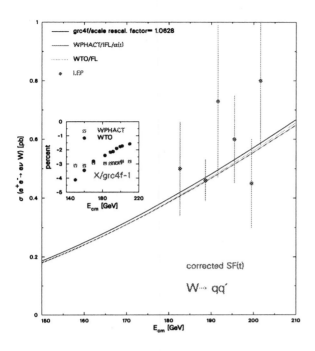

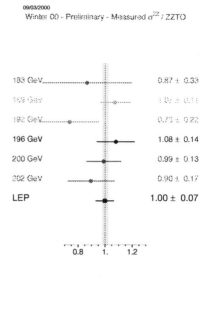

tion (EEX).

The implementation of a DPA calculation, in more than one code, in the NC02 Z-pair cross-section will bring the corresponding accuracy at the level of 0.5%, similar to the CC03 case.

• Conclusions

To gauge the priorities of this rather short summary one should remember that the experimental situation [21] is rather different for WW when compared to other processes. For W-pairs, LEP (ADLO) is able to test the theory to below 1%, i.e. , below the old uncertainty of $\pm 2\%$ established in 1995. Thus the CC03-DPA, including non-leading electroweak corrections, constitutes a very important theoretical development. However, ADLO cannot test single-W or ZZ-signal to an equivalent level, since their total cross-section is of the order of $1\,pb$ or less, 20 times smaller than that of W-pair production.

Acknowledgments

I would like to acknowledge the precious help and collaboration from all participants in the LEP 2/MC Workshop and, in particular, from Martin Grünewald. I thank W. Hollik for inviting me to this lively session and the LEP EWWG for providing some of the figures.

References

1. LEP 2/MC Workshop, hep-ph/0007180, hep-ph/0005309, hep-ph/0006259, to appear as a whole in a Cern Yellow report.
2. Two Fermion Working Group (Michael Kobel et al.). Jul 2000, hep-ph/0007180.
3. D. Bardin, P. Christova, M. Jack, L. Kalinovskaya, A. Olchevski, S. Riemann, T. Riemann, DESY-99-070, Aug 1999, hep-ph/9908433.
4. S. Jadach, B.F.L. Ward, Z. Was. Comput. Phys. Commun.130:260-325,2000, hep-ph/9912214; S. Jadach,

B.F.L. Ward, Z. Was, Nucl. Phys. Proc. Suppl. 89:106-111,2000, hep-ph/0006359.

5. G. Montagna et al., Comput. Phys. Commun. 76(1993)328; G. Montagna et al., Nucl. Phys. B401(1993)3.

6. D. Bardin, J. Biebel. D. Lehner, Leike, A. Olchevski and T. Riemann Comput. Phys. Commun. 104(1997)161.

7. A.Arbuzov, hep-ph/9907298.

8. Four Fermion Working Group (M.W. Grünewald and G. Passarino,). Jul 2000, hep-ph/0005309.

9. A. Denner et al., hep-ph/0007245; A. Denner et al., Nucl.Phys.Proc.Suppl. 89(2000)100, hep-ph/0006309; A. Denner et al., hep-ph/0006307; A. Denner et al., hep-ph/9912447; A. Denner et al., J. Phys. G26(2000)593, hep-ph/9912290.

10. M. Skrzypek et al., Comput. Phys. Commun. 94(1996)216; M. Skrzypek et al., Phys. Lett. B372(1996)289; S. Jadach et al., Comput. Phys. Commun. 119(1999)272; M. Skrzypek et al., M. Skrzypek and Z. Wąs, Comput. Phys. Commun. 125(2000)8.

11. W. Beenakker, F.A. Berends and A.P. Chapovsky, Nucl. Phys. B548 (1999) 3.

12. G. Passarino, Nucl. Phys. B574 (2000)451; G. Passarino, Nucl. Phys B578 (2000)3;
G. Passarino, hep-ph/9810416; E. Accomando, A. Ballestrero, E. Maina, Phys. Lett. B479(2000)209.

13. G. Passarino, Comput. Phys. Commun. 97(1996)261.

14. E.N. Argyres et al., Phys. Lett. B358(1995)339.

15. W. Beenakker et al., Nucl. Phys. B500(1997)255.

16. G. Montagna et al., hep-ph/0005121.

17. T. Ishikawa et al., GRACE manual, KEK report 92-19, 1993.

18. S. Mele, Talk 05c-03.

19. S. Jadach, W. Płaczek and B.F.L. Ward, Phys. Rev. D56(1997)6939.

20. http://www.to.infn.it/giampier/zzto.

21. A. Gurtu, plenary talk 05.

WW CROSS SECTIONS AND W BRANCHING RATIOS

A. EALET

CNRS/IN2P3, Centre de Physique des Particules, Marseille, FRANCE
E-mail:ealet@cppm.in2p3.fr

Data at collision energies up to 208 GeV collected by LEP experiments have been analysed to extract the W pair production cross section. Combining all LEP2 centre-of-mass energies has allowed the determination of the W decay branching ratios into leptons and hadrons and to derive the value of the CKM matrix element $|V_{cs}|$. An update of direct $|V_{cs}|$ measurements is also presented.

1 Introduction

Since 1997, LEP has increased its centre-of-mass energy from 161 to 208 GeV in 2000. An integrated luminosity of 450 pb^{-1} has been recorded at different energies as summarised in Table 1. In 2000, LEP has been running at centre-of-mass energies between 200 GeV and 208 GeV. Data have been averaged in 2 sets with mean centre-of-mass energies of 204.9 and 206.7 GeV. Results presented here use an integrated luminosity for each set respectively of 60 and 30 pb^{-1}.

In this paper, the determination of the WW production cross section is reviewed and the W branching ratio into leptons and hadrons is extracted.

Most of the results quoted in this paper are the combined values of the four experiments. They are preliminary unless specified otherwise. These results are compared with new theoretical predictions recently available through the work of the LEP2 Monte-Carlo workshop[1].

2 Event Selections

At LEP, Ws are pair-produced and can decay leptonically or hadronically, giving three possible final states. The selections are based on the ones developed by the four experiments at 189 GeV[2], scaled by the centre-of-mass energy.

The fully leptonic channels WW $\rightarrow l\nu l\nu$ (11 % of the decays) are characterized by 2 acoplanar leptons and missing momentum. The selections are in general cut based analyses and give typical efficiencies of 60-80 % for a low background level of $\approx$ 150 fb.

The semileptonic channels account for 44 % of final states and are characterised by 2 hadronic jets and an isolated lepton. Selections are based on a well identified and isolated lepton (e, μ or τ jet) and missing momentum. To improve upon cut based selections, the experiments use multivariate techniques based on probability or likelihood. Very high purity ($\approx$ 90 %) are achieved in those analyses.

The fully hadronic channel has four hadronic jets, no missing energy with a decay fraction of 46 %. However, analyses have also to fight against a huge QCD background ($e^+e^- \rightarrow q\bar{q}\gamma(gg)$). To enhance background rejection, all experiments use sophisticated analyses based on neural network techniques or likelihood analyses. The efficiencies are around 85 % with a typical background of 1-2 pb.

3 Results and interpretation

3.1 Method

The WW cross section in each channel is extracted from a maximum likelihood fit using the number of observed events, a cross-efficiency matrix between channels, the luminosity and the expected number of background events. The signal WW events are produced only via the so-called CC03

Table 1. LEP recorded energies and luminosity

YEAR	1996		1997	1998	1999				2000	
$\sqrt{s}$ (GeV)	161	172	183	189	192	196	200	202	205	207
lumi.(pb^{-1})	10	10	55	180	30	80	80	40	60	30

diagrams[2]. The interference with other 4 fermions final states is corrected in the cross section measurement. Combining all channels in one fit allows the extraction of the total WW production cross section and of the leptonic and hadronic W branching fractions.

3.2 WW cross sections

The 4 experiments have given preliminary results at all energies[3,4]. The combined results of the LEP experiments for the total WW production cross section at the various centre-of-mass energies are shown in figure 1, included two new points at 205 and 207 GeV. The experimental points are compared with new calculations using the Double Pole Approximation (DPA)[5], RACOONWW[6] and YFSWW 1.14[7]. These new codes have been shown to agree at a level better than 0.4 %[1]. The gray band of figure 1 represents the theoretical uncertainty, which decreases with the centre-of-mass energy from 0.7 % at 170 GeV to 0.4 % at 210 GeV. Since the DPA calculation can only be used away from the production threshold, the GENTLE model [8], tuned to reproduce DPA predictions, is used below 170 GeV for a complete comparison with the experimental results. The measured cross sections are in good agreement with the actual predictions.

3.3 W Branching ratios

Using the data on the full range of centre-of-mass energies, a fit to extract the W branching ratios can be performed with and without the assumption of lepton universality.

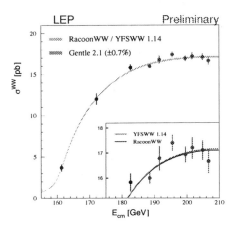

Figure 1. WW cross section results of the LEP combined data compared to the predictions of RacoonWW, YFSWW and GENTLE.

ALEPH and OPAL have used all data from 161 to 207 GeV whereas DELPHI and L3 have only used data up to 202 GeV. The results from each experiment are given in Table 2 together with the LEP combined value. Assuming lepton universality, the leptonic and hadronic branching ratios are:

$$B(W \rightarrow l\nu) = 10.74 \pm 0.10\%$$
$$B(W \rightarrow qq) = 67.78 \pm 0.32\%$$

Results agree well within the SM expectations. The quoted errors include systematic uncertainty derived from the error calculated at 189 GeV. The correlated errors between the different channels of each experiment are taken into account in the averaging procedure together with the QCD component which is assumed to be fully correlated among experiments.

Table 2. W branching ratios results in % using all data from 161 GeV to 208 GeV for ALEPH and OPAL and up to 202 GeV for DELPHI and L3.

Exp	e ν	$\mu\nu$	$\tau\nu$	qq
A	11.19	11.05	10.53	67.22
	± 0.34	± 0.32	± 0.42	± 0.53
D	10.33	10.68	11.28	67.81
	± 0.45	± 0.34	± 0.56	± 0.61
L	10.22	9.87	11.64	68.47
	± 0.36	± 0.38	± 0.51	± 0.59
O	10.52	10.56	11.15	67.84
	± 0.37	± 0.35	± 0.49	± 0.62
LEP	10.62	10.60	11.07	67.78
	± 0.20	± 0.18	± 0.25	± 0.32

3.4 $|V_{cs}|$ from $B(W \to qq)$

Within the Standard Model, the hadronic branching ratio can be related to the CKM matrix element $|V_{cs}|$ using

$$\frac{B(W \to qq)}{1 - B(W \to qq)} = (\sum_{i,j} |V_{ij}|^2)(1 + \alpha_s(m_W^2)/\pi).$$

without need for a CKM unitarity constraint. Using the world average value of α_s and the experimental values of CKM matrix elements involving light quarks [9], the above result is then interpreted as

$$|V_{cs}| = 0.989 \pm 0.016$$

which is the best existing measurement.

3.5 $|V_{cs}|$ from R_c

A direct measurement of $|V_{cs}|$ can be made by measuring the charm content of hadronic W decays.

$$R_c^W = \frac{\Gamma(W \to cX)}{\Gamma W \to qq)} = \frac{|V_{cd}|^2 + |V_{cs}|^2 + |V_{cb}|^2}{\sum_{i,j=u,d,s} |V_{cij}|^2}$$

OPAL has used data up to 189 GeV[10] to provide a new measurement of:
$R_c^W = 0.48 \pm 0.042 \pm 0.032$ leading to the direct estimation:

$$|V_{cs}| = 0.93 \pm 0.08 \pm 0.06$$

Combining this result to previous published values from ALEPH, DELPHI and L3 at lower energies gives an average direct LEP mesurement of

$$|V_{cs}| = 0.95 \pm 0.08$$

Acknowledgments

Many thanks to the 4 LEP experiments and to the LEP 4-f working group, expecially to Marco Verzocchi, for preparing the preliminary results presented here.

References

1. M.W Grunewald, G. Passarino et al., 1999/2000; hep-ph/0005309.
2. ALEPH coll., *Phys. Lett.* B **484**, 205 (2000); DELPHI coll., *Phys. Lett.* B **479**,89 (2000); L3 coll., CERN-EP/2000-104; OPAL coll. CERN-EP/2000-101.
3. ALEPH coll., ALEPH 2000-005 CONF 2000-002; DELPHI coll., DELPHI 2000-140 CONF 439; L3 coll., L3 Note 2514; OPAL coll., physics Note PN437 and PN420.
4. ALEPH coll., ALEPH 2000-071 CONF 2000-049; DELPHI coll., DELPHI 2000-142 CONF 441; L3 coll., L3 Note 2599; OPAL coll., physics Note PN437.
5. W.Beenakker et al., *Nucl. Phys.* B**548**, 3 (1999); see also[1] for discussions.
6. A. Denner et al., *Phys. Lett.* B **475**, 127 (2000); BI-TP 2000/06, hep-ph/0006307.
7. S. Jadach et al., *Phy. Rev.* D **61** 113010 (2000); UTHEP 00-0101.hep-ph/0007012.
8. D.Bardin et al., *Comp. Phy. Comm.* **104**, 161 (1997). version 2.1 described in [1].
9. Particle Data group, D.E. Groom et al. *Eur. Phys. J.* C **15**, 1 (2000).
10. OPAL coll., CERN-EP 2000/100.

CHARGED BOSON TRIPLE GAUGE COUPLINGS AT LEP
$WW\gamma$ WWZ AND W POLARISATION

S. JÉZÉQUEL

LAPP-IN2P3 Chemin de Bellevue 74941 Annecy-Le-Vieux Cedex, France
E-mail: jezequel@lapp.in2p3.fr

The study of WWZ and WWγ vertices is presented based on LEP2 data collected up to 202 GeV (475 pb^{-1} per experiment). WW pair and single W productions are used to measure possible deviations (quantified by the terms of a general Lagrangian) from the charged boson Triple Gauge Couplings (TGC) predicted by the Standard Model. In addition, the measurement of the W polarisation in W pair production gives access to deviations without using any model.

1 Introduction

The most general Lagrangian being Lorentz invariant for the WWV (V=Z or γ) interaction is described in these papers[1,2,3,4]. It contains seven terms: (g_1^V, κ_V and λ_V) conserve C and P, g_5^V violates C and P but conserves CP and (g_4^V, $\tilde{\kappa}_V$ and $\tilde{\lambda}_V$) which violates CP. Taking into account LEP1 constraints, the base line analysis[1] at LEP2 measures 3 parameters (Δg_1^Z, $\Delta\kappa_\gamma$, λ_γ) which are the most likely to deviate from the Standard Model (C and P conservation, $U(1)_{em}$ gauge symmetry). They are equal to zero at the Standard Model. The $SU(2)_L \times U(1)_Y$ gauge symmetry implies the following constraint:

$$\Delta\kappa_Z = \Delta g_1^Z + \Delta\kappa_\gamma \tan^2\theta_w; \lambda_Z = \lambda_\gamma$$

This analysis supposes that only one, two or three terms are different from the Standard Model. The study is extended to the measurement of each of the 14 terms individually.

2 WW analysis

The selection of WW events is the same as for the cross section[5] but restricted to well reconstructed events[6]. S-channel for W pair production is sensitive to Δg_1^Z, $\Delta\kappa_\gamma$ and λ_γ contrary to the t-channel. Anomalous couplings affect quadratically the differential cross section. This effect is looked for through the total WW cross section and the angular distributions of the 5 angles of the event. Cross section are extracted from adjustement of the expected cross-section which is a function of couplings to the observed one. To have the most precise measurement of one/many couplings out of the angles, an unbinned maximum likelihood is used by ALEPH (analytical Particle Density Function convoluted with a detector resolution function) and L3 (PDF mapped from simulated events). A second solution is the Optimal Observable[7] method where the five kinematic variables are projected onto 1 or 2 parameters per TGC coupling. DELPHI does a binned maximum likelihood fit to $\mathcal{O}_i^1$ and $\mathcal{O}_{ij}^2$ while OPAL (resp. ALEPH) fits the $\mathcal{O}_i^1$ and $\mathcal{O}_{ij}^2$ (resp. $\mathcal{O}_{ii}^2$) average.

For the TGC measurement, there are four main systematics: the theoretical uncertainty on the WW cross section ($\pm$ 2 %), the fragmentation model of jets (comparison JETSET/HERWIG), the approximate models on final state interaction (Bose-Einstein and Color Reconnection effect).

3 Single W analysis

The single W channel is made of many s- and t-channel diagrams. The measured cross section up to 202 GeV for this definition is shown in figure 1. The theoretical uncertainty on the cross section (as defined during the LEP

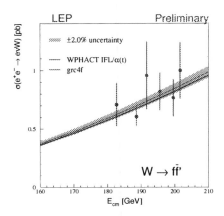

Figure 1. Measured and theoretical single W cross section at LEP2

Table 1. Systematics

Systematics	Δg_1^Z	$\Delta\kappa_\gamma$	λ_γ
σ (WW)	0.012	0.055	0.014
Fragmentation	0.013	0.051	0.014
Color Reconnec.	0.003	0.012	0.005
Bose-Einstein	0.006	0.020	0.006
σ (Weν)	-	0.049	0.067

Table 2. Results for one free parameter with errors

Δg_1^Z	$-0.002^{+0.070}_{-0.067}$
$\Delta\kappa_\gamma$	$-0.024^{+0.028}_{-0.027}$
λ_γ	$-0.002^{+0.030}_{-0.030}$

the 3 parameter fit result is shown.

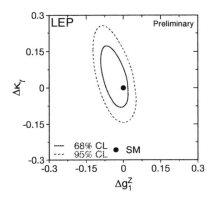

Figure 2. Exclusion contour plot for Δg_1^Z-$\Delta\kappa_\gamma$

Workshop[9]) was reduced to 5% but remains the main systematic for this channel.

This cross section measurement is translated into TGC[8] one ($\Delta\kappa_\gamma$ and λ_γ). ALEPH and OPAL improve the limits by including kinematic information based on visible energy and angle of the W.

4 Global results

The presented results[10] include all LEP data (except single W from OPAL for 99 data) up to 202 GeV (475 pb^{-1} per experiment). The likelihood of each measurement including statistical and uncorrelated errors were added. The table 1 shows the amplitude of the main systematics which are correlated between experiments and energies and no more negligible for the combination of LEP results. Because of non gaussian likelihood, the correlated errors were included with an approximate method: the likelihood curve is described as a gaussian with a width which is a function of the TGC value.

For the baseline analysis, supposing that only one parameter deviates from Standard Model, the results including all systematics are presented in table 2. The figure 2 presents one example out of three of exclusion contour for 2 free parameters analysis. In the note[10],

Measurements are done on the other 14 parameters. ALEPH scanned individually the C, P or CP violating parameters. L3 looked at the CP conserving terms ($\Delta\kappa_Z$, λ_Z and g_5^Z). The result on $g_5^Z = 0.05\pm0.17\pm0.08$ is the ALEPH+L3 combination. OPAL measures the CP violating terms[11] with the Spin Density Method applying $SU(2)_L \times U(1)_Y$ constraint.

5 W Polarisation in WW pair production

W polarisation in WW pair production is measured independantly of any Lagrangian definition. L3[12] extracts the W helicity fractions from the $\cos\theta^*$ distributions of $W \to l\nu$ decays splitted in $\cos\theta_W$ bins (183-202 GeV data). The measured fraction of longitudinal W is $\sigma_L/\sigma_{tot} = 25.9 \pm 3.5$ % (24.8 % at the Standard Model). OPAL[11] applies the Spin Density Method on all W decays (183-189 GeV data) obtaining $\sigma_L/\sigma_{tot} = 21.0 \pm 3.3(stat.) \pm 1.5(syst.)$ % (25.7 % at the Standard Model).

6 Conclusion

The four LEP experiments have measured with the WW and single W events, the 3 parameters (Δg_1^Z, $\Delta\kappa_\gamma$ and λ_γ) independantly. The 95 % exclusion limits are : $-0.077 < \Delta g_1^Z < 0.030$, $-0.130 < \Delta\kappa_\gamma < 0.130$ and $-0.094 < \lambda_\gamma < 0.024$. They have been measured simultaneously too. These measurements become limited by systematics. Most of the other eleven parameters have been measured. None of these measurement has shown any deviation from the Standard Model. In W pairs, the distributions of W polarisation are consistent with the Standard Model expectation.

To improve these limits, the year 2000 LEP data will be included and systematics need to be reduced.

Acknowledgments

I want to thank my ALEPH, DELPHI, L3 and OPAL colleagues for making available their preliminary results, and the TGC LEP Working group for the combination results.

References

1. Physics at LEP2, edited by G. Altarelli, T. Sjostrand and F. Zwirner, CERN 96-01 Vol.1, 525.

2. K. Hagiwara, R.D. Peccei, D. Zeppenfeld and H. Hikasa, *Nucl Phys.* B **282** (1987) 253.

3. M. Bilenky, J.L. Kneur, F.M. Renard and D. Schldknecht, *Nucl. Phys.* B **409** (1993) 22; *Nucl. Phys.* B **419** (1994) 240.

4. K. Gaemers and G. Gounaris, *Z. Phys.* C **1** (1979) 259.

5. A. Ealet "WW cross sections and W decay branching ratios", these proceedings.

6. ALEPH Collaboration, ALEPH 2000-015 CONF 2000-012.
   ALEPH Collaboration, ALEPH 2000-055 CONF 2000-037.
   DELPHI COllaboration DELPHI 2000-146 CONF 445.
   L3 Collaboration L3 Note 2567.
   OPAL Collaboration CERN-EP/98-167, *Eur. Phys. J.* C **8** (1999) 191-215.
   OPAL Collaboration PN441.

7. M. Diehl and O. Nachtmann, *Z. Phys.* C **62** (1994) 397.

8. ALEPH Collaboration, ALEPH 2000-054 CONF 2000-036.
   DELPHI Collaboration DELPHI 2000-143 CONF 442.
   L3 Collaboration L3 Note 2518.
   OPAL Collaboration PN427.

9. G. Passarino "Theoretical developments in LEP 2 physics", these proceedings.

10. The LEP Collaborations and the LEP TGC Working Group LEPEWWG/TGC/2000-02.

11. OPAL Collaboration The OPAL Collaboration, G. Abbiendi et al.CERN-EP-2000-113, Submitted to Eur. Phys. J. C.

12. L3 Collaboration L3 Note 2574.

MEASUREMENT OF ANOMALOUS COUPLINGS $ZZ\gamma$, $Z\gamma\gamma$ AND ZZZ

C.MATTEUZZI

Dipartimento di Fisica, Università di Milano and INFN sez.Milano, Italy
E-mail: clara.matteuzzi@cern.ch

The anomalous couplings $Z\gamma\gamma$, $ZZ\gamma$ and ZZZ have been studied at LEP 2 with the reactions $e^+e^- \to q\bar{q}\gamma$, $e^+e^- \to \nu\bar{\nu}\gamma$ and $e^+e^- \to ZZ$ at $\sqrt{s}$ from 189 to 202 GeV. Limits are derived on three gauge neutral bosons couplings combining the results of the 3 experiments L3, OPAL and DELPHI.

1 Introduction

In the Standard Model of Electroweak Interactions, at the lowest order in perturbation theory, couplings between neutral gauge bosons Z and γ are not allowed and the quantum corrections are too small to be observed[1]. Therefore an experimental evidence of their presence would be a signature of new physics beyond the Standard Model (SM).

The $ZZ\gamma$ and $Z\gamma\gamma$ couplings can be studied through the process $e^+e^- \to Z\gamma$, when $Z \to \bar{q}q$ or $Z \to \bar{\nu}\nu$. The ZZZ coupling can be studied through ZZ production with subsequent Z decay into leptons and quarks.

2 $Z\gamma\gamma$ and $ZZ\gamma$ Couplings

The theoretical description of the anomalous $ZZ\gamma$ and $Z\gamma\gamma$ interactions is done in a model independent way. The most general vertex function requiring the Lorentz and the $U(1)_{em}$ gauge invariance is given in[2]. The vertex function contains 8 indipendent anomalous couplings: h_i^Z and h_i^γ ($i = 1, \cdots, 4$). The CP violating amplitudes (h_1^V and h_2^V) do not interfere with the SM ones. In order to avoid that the cross section for the $Z\gamma$ production in terms of anomalous couplings violates unitarity, the couplings can be parametrized by a form factor that depend on $\sqrt{s}$. A common choice of the form factor[3] is

$$h_i^V = \frac{h_{i0}^V}{(1 + s/\Lambda^2)^n}$$

where h_{i0}^V are constants, Λ is the scale of new physics and $n > 3/2(5/2)$ for $h_{1,3}^V(h_{2,4}^V)$. In general the form factors are complex numbers[1]. At LEP 2, a scale of new physics $\Lambda = \infty$ is assumed, justified by the fact that at LEP $\sqrt{s}$ is fixed and known.

The most sensitive kinematical region is where a very energetic photon is produced at large polar angle. The main effect of the anomalous couplings is to modify the SM cross section, generally increasing it, except for h_3^γ and h_4^γ because of the presence of a strong interference term. For these couplings, one can expect strong limits if the statistics is sufficiently large. The anomalous couplings modify also the shape of some angular variables, as the decay angle of the Z in its rest frame ($\alpha^\star$). This variable is directly related to the average Z polarization, and the anomalous couplings enhances the longitudinal Z's production, as can be seen in figure 1 where the ratio of the longitudinal to the total cross section is shown as a function of four of the eight different couplings. The distribution of $\cos \alpha^\star$ is shown in figure 2 where the SM behaviour is compared to the one for $h_3^\gamma = 0.3$.

The weaker coupling of the Z with the initial fermions, gives a stronger sensitivity to h_i^γ with respect to h_i^Z.

All the three experiments, DELPHI L3 and OPAL use the channels with two jets and an energetic γ ($q\bar{q}\gamma$ events), and the single γ final state ($\nu\bar{\nu}\gamma$ events). In the $q\bar{q}\gamma$ events, the background is mainly due to $q\bar{q}$ where the photon is a π^0, and to a very small contam-

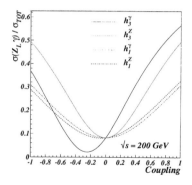

Figure 1. Cross section of the $q\bar{q}\gamma$ channel as a function of the anomalous couplings obtained with the calculation described in[3].

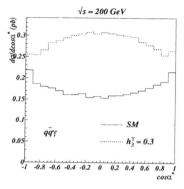

Figure 2. Angle of the q with the Z direction in the Z rest frame in the $q\bar{q}\gamma$ channel.

ination from the 4 fermions processes. The background is negligible in the single γ final state if a large polar angle region is selected. All the observed distributions in both channels are in good agreement with the SM expectations. The rate observed in each experiment is given in table 1, compared to the SM predictions.

The process $e^+e^- \rightarrow f\bar{f}\gamma$ in presence of anomalous couplings was calculated with different generators by the three experiments, as described in [4],[5],[6]. The higher order radiative corrections have been taken into account.

In order to extract the anomalous couplings the three experiments use total and differential cross sections as described in details in

	DELPHI	L3	OPAL
$\nu\bar{\nu}\gamma$	198	267	370
expected	196.5	294	412
$q\bar{q}\gamma$	1074	956	1525
expected	1086	978	1577
Luminosity(pb^{-1})	380	230	176
$\sqrt{s}$ GeV	189-202	192-202	189

Table 1. Observed and expected rates in the three LEP experiments.

Table 2. 95% CL intervals for 1 parameter fits.

h_1^γ	h_2^γ	h_3^γ	h_4^γ
-0.10,+0.03	-0.04,+0.06	-0.075,+0.004	0.005,+0.056
h_1^Z	h_2^Z	h_3^Z	h_4^Z
-0.13,+0.04	-0.04,+0.08	-0.16,+0.07	0.04,+0.10

[4],[5],[6].

The results are then combined summing the likelihoods[7] from each experiment, determined fitting one or two parameters at the time (while the others are fixed to their SM values). The 95% confidence interval and the central values fitted are reported in tables 2 and 3. The combined likelihood curves are shown in figure 3 and 4. Only the (h_3^γ,h_4^γ) results are shown as an example. Very good limits are obtained for the couplings h_3^γ and h_4^γ. This is a direct consequence of the strong interference between the anomalous amplitudes and the SM.

Several sources of systematics effects have been studied by the three experiments. The systematic uncertainties are included in the combination. The statistical error is at present the dominant uncertainty.

h_1^γ vs h_1^γ	h_3^γ vs h_4^γ
-0.21,+0.10	-0.20,+0.13
-0.11,+0.10	-0.11,+0.12

Table 3. 95% CL intervals for the 2 parameter fits.

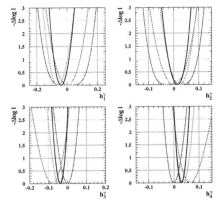

Figure 3. Combination of the likelihoods curves for the 1 parameter fits.

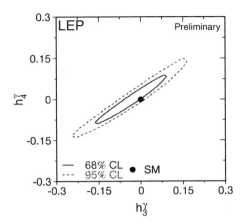

Figure 4. Contours for the 2 parameter fits.

Table 4. 95% CL intervals for 1 parameter fits.

f_4^γ	f_4^Z
-0.41,+0.39	-0.66,+0.68

f_5^γ	f_5^Z
-0.89,+0.84	-1.06,+0.51

$f_4^\gamma vs f_4^Z$	$f_5^\gamma vs f_5^Z$
-0.40,+0.38	-0.89,+0.86
0.66,+0.68	-1.06,+0.69

Table 5. 95% CL intervals for 2 parameter fits.

3 ZZZ Couplings

In order to describe the possible new physics, the vertex VZZ ($V=Z,\gamma$) is parametrized as suggested in[2]. There are 4 couplings: two f_4^V and two f_5^V. The main effect of the presence of anomalous couplings is an increase in the cross section and a change in the shape of the Z production angle.

The reaction considered by the three experiments is $e^+e^- \to \bar{f}f\bar{f}f$, with $f = \ell, q, \nu$.

The three experiments use different generators to describe the effect of anomalous couplings [8,9,10].

In tables 4 and 5 the combined 95% limits for 1 and 2-parameter fits are given, and in figure 5 the contours for the (f_4^γ, f_4^Z) are shown.

4 Conclusions

The couplings h_i^V (i =1,4 $V=\gamma/Z$) and $f_{4,5}^V$ between neutral gauge bosons have been measured by the three LEP experiments DELPHI, L3 and OPAL with data taken at $\sqrt{s}$ up to 202 GeV and using the reactions $e^+e^- \to q\bar{q}\gamma$, $e^+e^- \to \nu\bar{\nu}\gamma$, $e^+e^- \to ZZ \to \bar{f}ff\bar{f}$. No deviation from the Standard Model expectations was found and limits are given on the anomalous coupling parameters.

References

1. G.J. Gounaris, J. Layssac and F.M. Renard, hep-ph/0003143 (2000).

2. K.Hagiwara, *etal.*, Nucl. Phys. **B282** 253 (1987)

3. U. Baur and E. Berger, Phys. Rev. **D47** (1993) 4889.

4. DELPHI coll., Contribution to ICHEP2000,

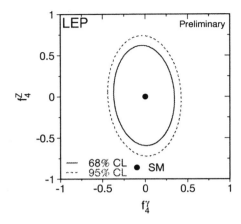

Figure 5. Contours for the 2 parameter fits to f_4^V.

DELPHI 2000-138 CONF 437.

5. L3 coll., Contribution to ICHEP2000, L3 Note 2552 July 2000.

6. OPAL coll., CERN-EP-2000-067.

7. LEPEWWG/TGC/2000-02.

8. DELPHI coll., Contribution to ICHEP2000, DELPHI 2000-145 CONF 444 July 2000.

9. L3 coll., Contribution to ICHEP2000, L3 Note 2579 July 2000.

10. OPAL coll., Contribution to ICHEP2000,PN423 (3/2000).

MEASUREMENTS OF QUARTIC GAUGE BOSON COUPLINGS AT LEP

S. SPAGNOLO

Rutherford Appleton Laboratory, Chilton, Didcot, Oxon, OX11 0QX, UK
E-mail: Stefania.Spagnolo@cern.ch

At the LEP 2 center-of-mass energies quartic gauge boson couplings (QGCs) involving at least one photon can be studied in final states with two W bosons and one photon, two acoplanar photons and two photons and a Z boson. Preliminary determinations of event rates and of distributions of kinematic variables, which are in good agreement with the Standard Model predictions, have been used to constrain the strength of anomalous contributions to quartic couplings of the gauge bosons.

1 Introduction

The study of QGCs has been initiated very recently at LEP 2[1]. Four quartic gauge boson vertices are predicted in the Standard Model (SM) with fixed couplings, $W^+W^-W^+W^-$, $W^+W^-Z^0Z^0$, $W^+W^-Z^0\gamma$ and $W^+W^-\gamma\gamma$, but their contributions to processes studied at LEP 2 is negligible. On the other hand, anomalous contributions to effective QGCs arising from physics beyond the SM could lead to measurable effects.

This report will describe the constraints on anomalous neutral and charged QGCs, obtained from the study of the production of the final states $W^+W^-\gamma$, $Z^0\gamma\gamma$ and two photons with missing energy (E_T). The diagrams in figure 1 show the processes involving quartic gauge vertices which are regarded as signal in the following discussion. Table 1 summarises the data sets used in QGC analyses.

The formalism of anomalous QGCs involving at least one photon[2,3] has recently received new contributions[4,5]. After neglecting operators leading to triple gauge couplings[a], two Lorentz structures, corresponding to C and P conserving operators of dimension 6 are required to describe the anomalous vertex $WW\gamma\gamma$, which contributes to the $WW\gamma$ and $\gamma\gamma+E_T$ final states. Therefore, two parameters a_0 and a_c, normalised to the square of the energy scale Λ of the underlying new physics, are introduced. Similarly, in the neutral sec-

		$\sqrt{s}$ /GeV	$\mathcal{L}$ /pb^{-1}
$WW\gamma$	L3	189	180
	OPAL	189	180
$\gamma\gamma+E_T$	ALEPH	189-202	430
	L3	183-202	450
	OPAL	189	180
$Z\gamma\gamma$	L3	130-202	500
	OPAL	130-208	580

Table 1. Data samples used for the QGCs analyses.

tor two couplings a_0' and a_c' parametrise the anomalous $ZZ\gamma\gamma$ vertex which contributes to $q\bar{q}\gamma\gamma$ production through the third diagram in fig. 1. Originally explicit custodial SU(2) was imposed to the effective Lagrangian with the relations $a_0' = a_0$ and $a_c' = a_c$, although the four parameters are, in general, independent.

2 Measurements at LEP 2

$WW\gamma$ production at LEP 2 proceeds mainly via initial state radiation (ISR) in W pair events. Further contributions come from final state radiation (FSR) and radiation of the photon off a W boson (WR). The signature of anomalous QGCs is an excess of high energy photons in W pair events. Therefore, the event selection is based on a search for an isolated photon in four-fermion events consistent with two hadronic W decays or one hadronic and one leptonic W decay. While the background from FSR and WR can be

---

[a] Already strongly constrained by TGC studies.

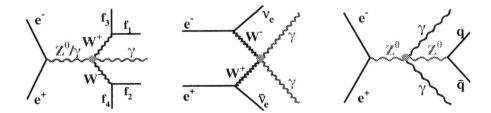

Figure 1. The processes involving quartic gauge boson couplings at LEP2.

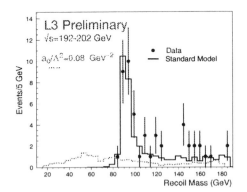

Figure 2. Distribution of the mass recoiling against the two acoplanar photons observed by L3 in the analysis of the $\gamma\gamma + E\!\!\!/_T$ channel.

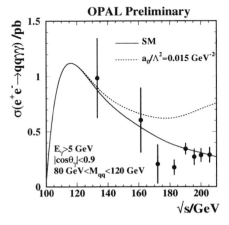

Figure 3. Cross section for $q\bar{q}\gamma\gamma$ production as a function of the center-of-mass energy.

reduced by requiring the invariant mass of the two pairs of fermions to match M_W, ISR is an irreducible background to $WW\gamma$ production via a quartic gauge vertex. L3 and OPAL have measured the cross section within similar signal definitions based on kinematic cuts on the photon energy, E_γ, and polar angle, θ_γ, on the separation between the photon and the fermions from the W decays and, in the case of OPAL, on the invariant mass of fermion pairs. The measurements, based on 42 and 17 events selected, respectively, by L3 and OPAL, are in agreement with the SM predictions, within the statistical uncertainty. The constraints on a_0 and a_c[1,6], listed in table 2, have been derived from the comparison of the event rate and of the shape of the distributions of E_γ (L3) and of E_γ and $\cos\theta_\gamma$ (OPAL) with the predictions of the

EEWWG[4] Monte Carlo (MC)[b].

In general, the theoretical predictions for the processes relevant for QGC measurements at LEP 2 are affected by uncertainties related to the consistent treatment of all the contributing diagrams and of the interference between them, which are taken into account, as systematic uncertainties, in the experimental results on anomalous QGCs.

The W fusion diagram, shown in fig. 1, competes with the large SM background from doubly-radiative return to the Z^0 in the final

---

[b]$WW\gamma$ production can be achieved also via the quartic vertex $W^+W^-Z^0\gamma$. Recently a set of Lorentz structures, C and P conserving, associated to it has been introduced[5] but previously only a CP violating coupling a_n was considered[3]. The combined L3 and OPAL constraint on this parameter is -0.45 GeV$^{-2} < a_n/\Lambda^2 < 0.41$ GeV^{-2} at 95% C.L..

state with two acoplanar photons. The signal from anomalous QGCs in the WW$\gamma\gamma$ vertex can be isolated from the dominant background by restricting the signal acceptance to values of the invariant mass, recoiling against the two-photon system, lower than, typically, 70-80 GeV where the SM predicts a negligible cross-section (fig. 2). Since, no events are observed in the signal region[1,6,7] by ALEPH, L3 and OPAL, the bounds on a_0 and a_c reported in table 2 are derived using the predictions from the ENUNUGANO calculation[4].

Finally, the anomalous ZZ$\gamma\gamma$ vertex might enhance the rate of events with two jets and two photons, which arise in the SM mainly from doubly-radiative return to the Z^0. The cross section for $q\bar{q}\gamma\gamma$ has been measured at several center of mass energies (fig. 3) by L3 and OPAL within signal definitions based on cuts on the photon energies and polar angles and on the di-jet invariant mass, which is required to be close to M_Z. In addition to the event rate, the shape of the distributions of variables like the energy of the second most energetic photon, E_{γ_2}, and the polar angle, θ_f, of the most forward photon are sensitive to QGCs. The predictions for the contributions of anomalous QGCs, obtained with the EEZGG[4] MC, are fit to the measured cross sections and distributions (E_{γ_2} is used by L3 and OPAL and $\cos\theta_f$ by OPAL only) and the bounds on the value of a_0' and a_c' summarised in table 2 are derived[8,9].

3 Conclusions

Preliminary results on anomalous QGCs from LEP 2 data have been presented. Since the effect of anomalous QGCs increases with the center-of-mass energy, the present constraints are weak compared to the foreseen sensitivity of future colliders, but the full potential of the LEP 2 data is not yet fully exploited and future improvements of the combined LEP results can be expected.

95% C.L. bounds on anomalous QGCs		
Charged vertex W$^+$W$^-\gamma\gamma$		
	a_0/Λ^2	a_c/Λ^2
WWγ L	[-0.045, 0.045]	[-0.08,0.13]
O	[-0.070, 0.070]	[-0.13,0.19]
$\gamma\gamma$ A	[-0.045,0.042]	[-0.115,0.115]
+$E\!\!\!/_T$ L	[-0.041,0.040]	[-0.12,0.12]
O	[-0.086,0.085]	[-0.23,0.23]
All	[-0.037,0.036]	[-0.077,0.095]
Neutral vertex Z^{0}Z$^0\gamma\gamma$		
	$(a_0'/\Lambda^2)\cdot 10^2$	$(a_c'/\Lambda^2)\cdot 10^2$
Z$\gamma\gamma$ L	[-0.6,0.6]	[-0.6,1.0]
O	[-0.6,0.8]	[-0.8,1.2]
All	[-0.48,0.56]	[-0.52,0.99]

Table 2. Preliminary constraints on anomalous QGCs from Aleph (A), L3 (L) and Opal (O). The preliminary combined results (All) are supplied by the LEP TGC working group. Units are GeV^{-2}.

References

1. OPAL Collaboration, G. Abbiendi *et al.*, Phys. Lett. **B471** (1999) 293; M.Thomson, proceedings of EPS-HEP '99, Tampere, July 1999.

2. G.Bélanger, F.Boudjema, Phys. Lett. **B288** (1992) 201; Phys. Lett. **B288** (1992) 210.

3. O.J.P.Éboli *et al.*, Nucl. Phys. **B411** (1994) 381; G.Abu Leil and W.J.Stirling, J. Phys. **G21** (1995) 517.

4. W.J.Stirling, A.Werthenbach, Phys. Lett. **B466** (1999) 369; Eur. Phys. J. **C14** (2000) 103 and references therein.

5. G.Bélanger *et al.*, Eur. Phys. J. **C13** (2000) 283.

6. The L3 Collaboration, ICHEP abs. 520, (2000), CERN-EP/2000-097.

7. The ALEPH Collaboration, ICHEP abs. 286, (2000), ALEPH 2000-050 CONF 2000-032.

8. The OPAL Collaboration, ICHEP abs. 572, (2000), OPAL PN452.

9. The L3 Collaboration, ICHEP abs. 505, (2000), CERN-EP/2000-006.

FERMION PAIR CROSS-SECTIONS AND ASYMMETRIES AND LIMITS ON NEW PHYSICS

G. DELLA RICCA

INFN - AREA di Ricerca, Padriciano 99, I-34012 Trieste, ITALY
E-mail: Giuseppe.Della-Ricca@ts.infn.it

The measurements of hadron and lepton pair production cross-sections and lepton pair forward-backward asymmetries performed by the ALEPH, DELPHI, L3 and OPAL collaborations are summarized. The results obtained with data taken at centre-of-mass energies between 130 and 207 GeV, including the events recorded in the early summer 2000 at 205 and 207 GeV, show no significant deviations from the Standard Model predictions. This allows to constrain new physics phenomena like contact interactions, additional heavy gauge bosons Z′, models of gravity in extra dimensions, exchange of sneutrinos or squarks in R-parity violating models and leptoquarks.

1 Fermion pair production at LEP2

Since 1995 LEP has been operated at centre-of-mass energies, $\sqrt{s}$, well above the Z resonance, allowing to study fermion pair production at 130 and 136 GeV (1995), 161 and 172 GeV (1996), 183 GeV (1997), 189 GeV (1998), 192, 196, 200 and 202 GeV (1999) and between 205 and 207 GeV in 2000 (Tab. 1).

Year	$\sqrt{s}$ [GeV]	$\mathcal{L}$/exp. [pb^{-1}]
1995-97	130-136	6/ 6
1996	161/172	10/10
1997	183	55
1998	189	175
1999	192/196	30/80
	200/202	80/40
2000	205/207	60/30

Table 1. Average integrated luminosity per experiment for the different centre-of-mass energies. The 2000 values at the conference time are shown.

2 Standard Model interpretations

The LEP averages of the ALEPH,[1] DELPHI,[2] L3[3] and OPAL[4] hadron and lepton pair cross-section and lepton pair asymmetry measurements from 130 to 207 GeV are compared to the Standard Model (SM) predictions in Fig. 1, for both inclusive and non-radiative events, as a function of $\sqrt{s}$. The measured cross-sections and asymmetries are consistent with the SM expectations.

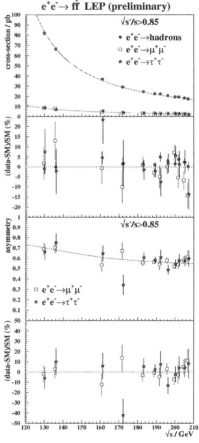

Figure 1. LEP averaged cross-sections (top half) and asymmetries (bottom half) for the $e^+e^- \to q\bar{q}(\gamma)$, $\mu^+\mu^-(\gamma)$, $\tau^+\tau^-(\gamma)$ processes measured at energies from 130 up to 207 GeV. The curves are the SM expectations. The lower plots show the differences between the measurements and the SM predictions.

2.1 Running of α_{em}

The non-radiative cross-section and asymmetry measurements have been used by OPAL[4] to measure the electro-magnetic coupling constant α_{em} at LEP2 energies, including in the fits the data collected up to 207 GeV. The measured values of α_{em} are shown in Fig. 2 (open dots). The result for $1/\alpha_{em}$ obtained

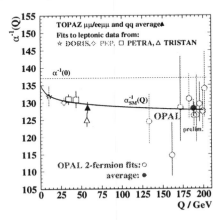

Figure 2. Fitted values of $1/\alpha_{em}$ as a function of $Q=\sqrt{s}$. The OPAL[4] results at 130-207 GeV are shown with the open circles. The solid line shows the SM expectation.

from a combined fit of all the data at the luminosity-weighted centre-of-mass energy of 190.6 GeV (full dot) is $128.4^{+2.5}_{-2.3}$, in good agreement with the SM expectation of 127.9.

2.2 S-matrix fits: γ-Z interference term

The high energy data from 130 to 202 GeV were also included by L3[3] in a fit carried out in the framework of the S-matrix approach.[5] At the lowest order, the total cross-section $\sigma^0_{f,\text{tot}}$ is given by Eq. (1), where the parameters r_f, j_f and g_f determine the Z-exchange, γ-Z interference and γ-exchange contributions, respectively.

$$\sigma^0_{f,a}(s) = \frac{4}{3}\pi\alpha^2 \left[\frac{g^a_f}{s} + \frac{sr^a_f + (s - \overline{m}^2_Z)j^a_f}{(s - \overline{m}^2_Z)^2 + \overline{m}^2_Z\overline{\Gamma}^2_Z} \right]$$

$$a = \text{tot, fb} \quad \text{and} \quad f = \text{had, e, } \mu, \tau \quad (1)$$

The correlation between the parameters m_Z and $j^{\text{tot}}_{\text{had}}$ is shown in Fig. 3. A significant

improvement on the precision on $j^{\text{tot}}_{\text{had}}$ is obtained when the high energy data are included in the fit: compared to the Z^0 data alone, the error on $j^{\text{tot}}_{\text{had}}$ is reduced by almost a factor five. The result of the fit for $j^{\text{tot}}_{\text{had}}$ is then 0.31 ± 0.13, in good agreement with the SM prediction.

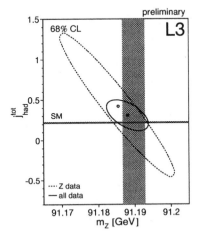

Figure 3. Probability contour plots in the $m_Z - j^{\text{tot}}_{\text{had}}$ plane, as measured by L3.[3] The horizontal band shows the SM prediction for $j^{\text{tot}}_{\text{had}}$, the vertical one the value of m_Z resulting from a model dependent fit assuming the SM prediction for $j^{\text{tot}}_{\text{had}}$.

3 Limits upon new phenomena

Cross-sections and asymmetries can be used to improve the constraints on several extensions of the SM: here the very general contact term formalism, the models including an additional gauge boson Z', gravity in extra dimensions, R-parity violating sneutrino exchange and leptoquark exchange will be reviewed.

3.1 Contact interactions

A fit for contact interactions between leptons, assuming lepton universality in the couplings was made by DELPHI[2] using data at all energies from 130 to 202 GeV for the e^+e^-, $\mu^+\mu^-$ and $\tau^+\tau^-$ final states. Lower limits at 95% CL on the scale of the interactions in the effective Lagrangian of the four-fermion interactions are given in Tab. 2. Similar results

model	Λ^+ [TeV]	Λ^- [TeV]
LL	9.4	7.3
RR	9.0	7.4
VV	17.8	13.6
AA	10.2	12.8
RL	8.8	6.4
LR	8.8	6.4

Table 2. 95% CL lower limits on the scale Λ of contact interactions, for $e^+e^- \to l^+l^-$ final states, assuming lepton universality, obtained by DELPHI.[2]

have been obtained by OPAL,[4] where also the non-radiative hadronic final states have been studied.

3.2 Additional gauge boson Z'

Additional heavy neutral vector bosons are predicted in many theories, like the E(6) and L-R Symmetric GUT models, and the Sequential SM extension. Fits were made to the mass of a Z', $M_{Z'}$, to the mass of the Z^0, M_Z, and to the mixing angle between the two bosonic fields, $\Theta_{ZZ'}$, using cross-section and asymmetry measurements from LEP1 and LEP2. The fitted value of M_Z was

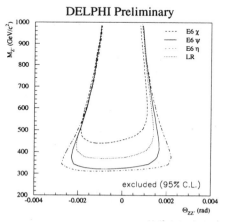

DELPHI Preliminary

excluded (95% C.L.)

--- E6 χ
— E6 ψ
··· E6 η
···· LR

Figure 4. The 95% C.L. allowed regions in the $M_{Z'} - \Theta_{ZZ'}$ plane for the χ, ψ, η and LR models, as measured by DELPHI.[2]

found to be in agreement with the value obtained from fits to the data with no additional

gauge boson. No evidence was found for the existence of a Z' boson in any of the models. The 95% CL limit contours in the plane $M_{Z'} - \Theta_{ZZ'}$ obtained by DELPHI[2] using data from 130 to 207 GeV are shown in Fig. 4. The best limits at 95% CL from the four LEP

Z' model	$m_{Z'}$ [GeV]
$E_6(\chi)$	753 (O^4)
$E_6(\psi)$	410 (A^1)
$E_6(\eta)$	486 (O^4)
$E_6(I)$	510 (A^1)
LRS	635 (O^4)
SSM	1000 (L^3)

Table 3. One-dimensional lower limits at 95% CL on the Z' mass, for several models.

experiments, using data up to 202 GeV, are summarized in Tab. 3.

3.3 Gravity in extra dimensions

Limits on the energy scale M_s of models of gravity in extra dimensions have been obtained for the two cases, $\lambda = \pm 1$, of either constructive or destructive interference between the gravitational and the SM processes. Fits to the e^+e^-, $\mu^+\mu^-$ and $\tau^+\tau^-$ final states differential cross-sections, and to the $q\bar{q}$ inclusive cross-section, R_b ratio, and b, c jet-charge forward-backward asymmetries were performed using data collected from 130 to 202 GeV by ALEPH.[1] The fits gave results compatible with the SM. The 95% CL lower limits on M_s are 0.94(1.17) TeV for $\lambda = -1(+1)$, respectively. The deviations in the $\mu^+\mu^-$ differential cross-sections for the two values of λ are compared to the data collected by DELPHI[2] at $\sqrt{s} \approx 192\text{-}202$ GeV in Fig. 5.

3.4 R-parity violating sneutrino exchange

SUSY theories with R-parity violation have in addition terms in the Lagrangian of the form $\lambda_{ijk} L_i L_j \overline{E}_k$. The parameters of interest are the dimensionless Yukawa couplings, λ_{ijk},

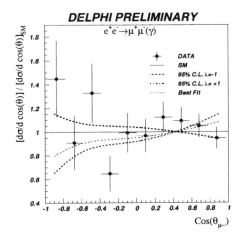

Figure 5. The weighted averages of the deviations from the SM angular distributions for the DELPHI $\mu^+\mu^-$ differential cross-section measurements.[2]

between the superfields of different generations (L denotes a lepton doublet, $\overline{E}$ a lepton singlet superfield), i, j and k ($i,j,k=1,2,3$), together with the mass of the sneutrino exchanged, $m_{\widetilde{\nu}}$. Figure 6 shows the limits obtained by ALEPH[1] using the measured e^+e^- (λ_{121} or λ_{131}), $\mu^+\mu^-$ ($\lambda_{131}=\lambda_{232}$) and $\tau^+\tau^-$ ($\lambda_{121}=\lambda_{233}$) differential cross-sections. If only one λ is non zero, DELPHI[2] has also obtained 95% CL upper limits on λ, assuming a sneutrino mass of 100(200) GeV/c^2, of 0.21(0.28) for the $\mu^+\mu^-$, and 0.48(0.66) for the $\tau^+\tau^-$ channel, respectively.

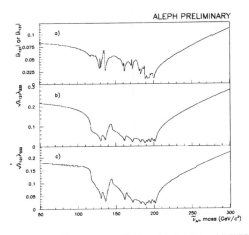

Figure 6. 95% CL upper limits obtained by ALEPH[1] on $|\lambda_{ijk}|$ versus the assumed $\widetilde{\nu}$ mass, for $\widetilde{\nu}$-couplings to a) e^+e^-, b) $\mu^+\mu^-$ and c) $\tau^+\tau^-$ pairs.

3.5 Leptoquark exchange

Limits on leptoquarks were derived from all $q\overline{q}$ cross-sections and from the jet-charge asymmetries measurements up to 202 GeV by ALEPH.[1] The 95% CL limits for first or

scalar LQ mass [GeV/c^2]		vector LQ mass [GeV/c^2]	
$S_0(L)$	380	$V_0(L)$	618
$S_0(R)$	56	$V_0(R)$	137
$\widetilde{S}_0(R)$	128	$\widetilde{V}_0(R)$	331
$S_{1/2}(L)$	120	$V_{1/2}(L)$	144
$S_{1/2}(R)$	99	$V_{1/2}(R)$	169
$\widetilde{S}_{1/2}(L)$	-	$\widetilde{V}_{1/2}(L)$	105
$S_1(L)$	319	$V_1(L)$	515

Table 4. 95% CL lower limits on the leptoquark mass for first or second generation, as obtained by ALEPH.[1]

second generation leptoquark mass are summarized in Tab. 4. In SUSY theories with an R-parity violating term of the form $\lambda'_{1jk}L_1Q_j\overline{D}_k$ ($j,k=1,2,3$), the $S_0(L)$ and the $\widetilde{S}_{1/2}(L)$ leptoquarks are equivalent to a $\tilde{d}$ and a $\tilde{u}$ squark, respectively, and the limits in terms of the leptoquark couplings are then equivalent to limits in terms of λ'_{1jk}.

Acknowledgments

I would like to thank the LEP Collaborations for providing me with their latest preliminary results, and the LEP ElectroWeak Working Group for combining them.

References

1. ALEPH Collaboration, contributions **188**, **654** to ICHEP 2000.
2. DELPHI Collaboration, contributions **647**, **652**, **653** to ICHEP 2000.
3. L3 Collaboration, contributions **411**, **526**, **529** to ICHEP 2000.
4. OPAL Collaboration, contributions **149**, **150**, **168**, **184** to ICHEP 2000.
5. A. Leike *et al.*, *Phys. Lett.* B**273**-91 513. S. Riemann, *Phys. Lett.* B**293**-92 451.

HEAVY FLAVOUR PRODUCTION AT LEP 2

S. ARCELLI

EP Division, CERN, CH-1211 Geneva 23. E-mail: Silvia.Arcelli@cern.ch

The ALEPH, DELPHI, L3 and OPAL collaborations have measured the cross-sections and the forward-backward asymmetries for heavy-flavour quark production in $e^+e^- \rightarrow Z^\circ/\gamma \rightarrow q\bar{q}$ events at LEP 2, at centre-of-mass energies well above the Z° resonance. The measurements are in good agreement with the Standard Model predictions and are used to constraint the parameters describing the $\gamma - Z^\circ$ interference, and to place limits on contact interactions.

1 Introduction

At LEP 2 energies, $b\bar{b}$ and $c\bar{c}$ production proceeds through an s-channel exchange of a photon or a Z° boson and their interference. Measurements of heavy flavour production have been performed by the four LEP collaborations at several centre-of-mass energies between 130 and 209 GeV. The observables which are measured are the cross section ratios $R_b = \sigma(b\bar{b})/\sigma(had)$ and $R_c = \sigma(c\bar{c})/\sigma(had)$, where $\sigma(had)$ is the total $q\bar{q}$ cross-section, and the forward backward asymmetries $A_{FB}^{b\bar{b}}$ and $A_{FB}^{c\bar{c}}$. Both published and preliminary results [1] are available.

At energies above the Z° resonance, a large fraction of two-fermion events is characterised by hard initial-state radiation. In order to select "non-radiative" events, the effective centre of mass energy after radiation $\sqrt{s'}$ is measured event-by-event from the angles of the jets and from isolated photons observed in the detector. Typically $\sqrt{s'/s}$ is required to be larger than 0.85, which keeps a residual contamination of radiative events of 5-10% depending on the experiment and on the centre of mass energy. Another important background comes form W-pair production, which is rejected on the basis of topological information. The residual contamination ranges between 5-15%.

2 Cross-section Ratios

Due to the low available statistics, double-tagging techniques cannot be applied as at LEP 1, and event-tagging techniques are used to measure R_b and R_c. Furthermore, the priority is given to maximising the efficiency, at the expense of a lower purity compared to the measurements at the Z° resonance. After flavour identification and subtraction of background from radiative $q\bar{q}$ and W-pair events, R_b or R_c are then derived from the tagged fraction N_{tag}/N_{had} according:

$$\frac{N_{tag}}{N_{had}} = \epsilon_b R_b + \epsilon_c R_c + \epsilon_{uds}(1 - R_b - R_c).$$

As flavour identification is based on single-tag techniques, all the tagging efficiencies ϵ_i have to be estimated from MonteCarlo.

2.1 R_b Measurements

The identification of $b\bar{b}$ events is based on "lifetime" tags, which rely on the presence of secondary vertices significantly displaced from the e^+e^- interaction point. ALEPH and L3 both use the measured impact parameters in the event to form a probability that the tracks originate from the primary vertex; this probability is then taken as the tagging variable. The OPAL b-tagging is based on the decay length significance L/σ_L of the secondary vertices; a "folded-tag" is employed, reducing the dependence on the light-quark efficiencies and the modelling of the detector resolution. DELPHI uses a very efficient (72%) and pure (90%) technique, which combines displaced vertex quantities such as track impact parameters, the mass of the secondary vertex and the rapidity of tracks

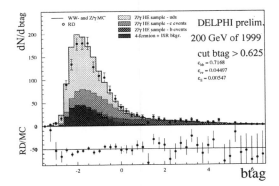

Figure 1. The distribution of the DELPHI b-tagging variable at $\sqrt{s} = 200$ GeV.

within the jets. The distribution of this tagging variable is shown in figure 1.

2.2 R_c Measurements

Measurements of R_c have been performed by ALEPH at $\sqrt{s} = 183 - 189$ GeV. In a first stage, $b\bar{b}$ events are efficiently rejected using their b-tagging tool, which allows to reduce the the b-contribution to the sample to $\approx 4\%$. Then c-tagging is performed using a neural network based on lifetime information, track rapidities, and the information on leptons, slow pions and D mesons in the jet. By cutting on the sum of the neural network outputs in the two jets a charm purity of 41% is obtained, with a charm efficiency of 59%.

3 Asymmetries

In addition to flavour tagging, the identification of the primary quark direction signed by its charge is needed to measure the forward-backward asymmetries, $A_{FB}^{b\bar{b}}$ and $A_{FB}^{c\bar{c}}$. Due to the back-to-back topology of the events, the event thrust axis provides a very good approximation of the primary quark direction. Two main techniques, which result in largely uncorrelated measurements and of similar precision, are then used to determine the primary quark charge:

- A "Jet Charge" technique in conjunction with a lifetime tag; this method has a high efficiency, but a considerable ($\approx 30\%$) probability of charge misidentification.

- The charge of high p, p_t leptons originating from heavy-quark semileptonic decays; in this case the essentially unambiguous charge identification compensates the lower efficiency of this technique.

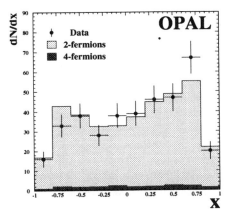

Figure 3. The $x = -\text{sign}(Q_{FB}) \cdot \cos\theta_T$ distribution for OPAL data collected at $\sqrt{s} = 189$ GeV.

3.1 Jet Charge Measurements

ALEPH, DELPHI and OPAL have measured $A_{FB}^{b\bar{b}}$ using jet charge techniques. Lifetime-tagged events are divided in two hemispheres, and the charge flow $Q_{FB} = Q_F - Q_B$ is calculated from the jet charges in the forward and backward thrust hemispheres, with a probability of correct primary quark charge identification of $\approx 65\text{-}75\%$. By fitting the $x = -\text{sign}(Q_{FB}) \cdot \cos\theta_T$ distribution, shown in figure 3 for OPAL data, the observed asymmetry is determined and then corrected for charm and light quark contamination and for charge dilution in order to extract the b-quark asymmetry $A_{FB}^{b\bar{b}}$.

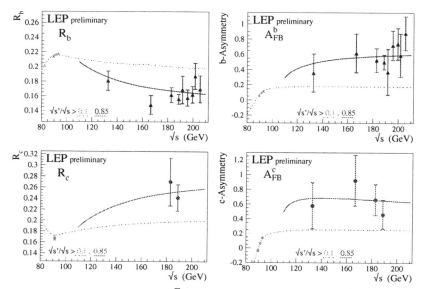

Figure 2. Combined LEP results for R_b R_c $A_{FB}^{b\bar{b}}$ and $A_{FB}^{c\bar{c}}$. The lines show the Standard Model predictions for "non radiative" events ($\sqrt{s/s'} > 0.85$, solid lines) and for inclusive events ($\sqrt{s/s'} > 0.1$, dotted lines). The precise LEP 1 measurements [2] are also shown.

ALEPH measures $A_{FB}^{c\bar{c}}$ from the charm enriched sample using for primary quark charge identification a neural network which combines the information from jet charge, the charge of leptons, kaons and slow pions and the charge of leading particles in the jets. The asymmetry is then determined by counting the forward (x > 0) and backward (x < 0) hemispheres and correcting for background and charge misassignment.

3.2 Lepton measurements

Measurements using leptons are performed by OPAL and L3. L3 uses high p, p_t muons and electrons and measures the b-asymmetry from a likelihood fit to the $-Q_l \cdot \cos\theta_T$ distribution, accounting for cascade decays and background from other flavours. OPAL measures simultaneously $A_{FB}^{b\bar{b}}$ and $A_{FB}^{c\bar{c}}$ using neural networks both to identify prompt electrons and muons, and to distinguish "direct-b" and "direct-c" leptonic decays from other sources. To increase the sensitivity to the

charm asymmetry, also slow pions are used in the measurement, which are selected by a dedicated neural network. The neural network outputs are used, along with the lepton and slow pion charge and the thrust direction, to fit for $A_{FB}^{b\bar{b}}$ and $A_{FB}^{c\bar{c}}$.

4 Combined results

The results from the four experiments have been combined taking into account the correlated systematic uncertainties and converting, when necessary, the measurements to a common signal definition (i.e. full angular acceptance). The dependencies of the individual measurements on the other parameters have been explicitly accounted for in the combination. The combined results are shown in figure 2. Good agreement is observed with the Standard Model predictions as derived from ZFITTER[3]. The measurements are all statistically limited. The dominant systematic errors arise from the understanding of the detector resolution and from uncertainties on b and c-decay physics modelling.

5 Interpretations

The combined results presented here have been used both to constraint the $\gamma - Z^\circ$ interference for heavy quarks in the framework of the S-matrix formalism, and to derive limits on new physics in the context of contact interactions. For these interpretations, the R_b and R_c measurements have been translated into absolute cross-sections using the total hadronic cross-section at each centre-of-mass energy.

5.1 S-Matrix Results

Within the S-matrix approach [4], the total cross section and the forward-backward cross section can be expressed as a function of the Z° boson mass and width, M_{Z° and Γ_{Z°, and of three parameters which are related to the strength of the pure photon exchange ($g^{tot,fb}$), the pure Z° exchange ($r^{tot,fb}$) and the $\gamma - Z^\circ$ interference ($j^{tot,fb}$):

$$\sigma^{tot,fb}(s) \propto \frac{g^{tot,fb}}{s} + \frac{j^{tot,fb}(s - M_Z^2) + r^{tot,fb}s}{(s - M_Z^2)^2 + M_Z^2\Gamma_Z^2}$$

$$A_{FB}(s) \propto \frac{\sigma^{fb}}{\sigma^{tot}}$$

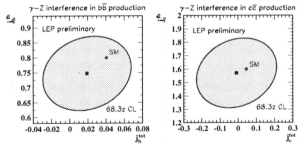

Figure 4. S-matrix fit results of the $\gamma - Z^\circ$ interference parameters $j^{tot,fb}$, for $b\bar{b}$ (left) and $c\bar{c}$ (right) production.

The fit uses the combined LEP 2 heavy flavour results presented here and the corresponding LEP 1 averages. While the parameters related to the Z° exchange and $M_{Z^\circ}, \Gamma_{Z^\circ}$

are mostly determined by LEP 1 data, heavy flavour results at LEP 2 allow to constraint the terms describing the $\gamma - Z^\circ$ interference. The photon terms are fixed in the fit. The results are shown in figure 4. Good agreement with the Standard Model prediction is observed.

5.2 Contact Interactions

Contact interactions provide a convenient framework to parametrise virtual effects from new physics characterised by a large mass scale $\Lambda >> \sqrt{s}$ and by a coupling strength g, which by convention is set to $g^2 = 4\pi$. Limits at 95 % C.L. on the characteristic interaction scale Λ have been set using both the $b\bar{b}$ and $c\bar{c}$ cross-sections and the asymmetries. They range between 2 TeV and 14 TeV, depending on the helicity structure which is assumed for the contact interaction. These results specific to heavy flavours are particularly interesting since they are inaccessible to $p\bar{p}$ or ep colliders.

References

1. ALEPH Collaboration, EPS-HEP99 **6**, 694; (a)nd references therein; ALEPH internal note ALEPH 2000-046 CONF 2000-029.
   DELPHI Collaboration, ICHEP abstract 648 (2000) and references therein;
   L3 Collaboration, ICHEP abstract 666 (2000) and references therein;
   OPAL Collaboration, ICHEP abstract 152 (2000).

2. The LEP collaborations *et al.*, CERN-EP/2000-016.

3. ZFITTER V6.23, D. Bardin *et al.*, Preprint hep-ph/9908433.

4. A. Leike *et al.*, Phys. Lett. **B 273** (1991) 513.

ZZ CROSS SECTION MEASUREMENTS

SALVATORE MELE

EP Division, CERN, CH-1211, Genève 23, Switzerland*
E-mail: Salvatore.Mele@cern.ch

Results on the cross section measurement of Z boson pair–production at LEP[1,2] are presented. The more general case of neutral–current four–fermion production and the particular case of ZZ events enriched in b quarks are also discussed. All the results agree with the Standard Model predictions.

1 Introduction

LEP was successfully operated in the years from 1997 through 2000 at centre–of–mass ($\sqrt{s}$) energies from 183 GeV up to 208 GeV. This allowed each of its four experiments to collect more than 540 pb^{-1} of data above the Z boson pair–production threshold, as summarised in Table 1.

This process tests the Standard Model of electroweak interactions in the neutral–current sector and is sensitive to New Physics scenarios such as couplings between neutral gauge bosons[4] or extra space dimensions[5].

The results presented here refer to a particular subset of all the possible diagrams for neutral–current four–fermion production, denoted as NC02 and depicted in Figure 1. The Figure also shows the ZZ production cross section as calculated with the YFSZZ and ZZTO programs[3]. These diagrams define the Z pair production signal, also defined by some experiments with a wider part of the full four–fermion phase space compatible with Z pair–production.

The related topics of general neutral–current four–fermion production and b quark content in Z–pair events are also investigated, and discussed in the following.

Detailed accounts of the data sets, analysis techniques and results of each experiment can be found elsewhere[1,2]. All the results at $\sqrt{s} = 183$ and 189 GeV are published while the others are preliminary.

---

*On leave of absence from INFN Sezione di Napoli, I-80126, Napoli, Italy.

2 Data analysis

The four experiments devised analysis strategies that rely on the identification of the signatures of the pair–production of two particles with equal masses, compatible with Z bosons.

Multivariate techniques are used for the largest statistic fully hadronic final state (49% of the Z pair decays), where a large background is expected from the QCD and W pair–production processes. Event shape variables are used to reject the first, focusing on the signal four–jet topology, also common to hadronic decays of W pairs. This background can be discriminated thanks to the different boson mass that reflects into the topology of the decay products. Moreover, W decays lack b quarks, present in 39% of the fully hadronic decays of the Z pairs. The use of the b tag techniques developed for the Higgs search hence increases the signal purity.

The final state with a Z decaying into hadrons and the other into neutrinos is the second most populated (28%) and is investigated with hadronic events with large missing energy. The hermeticity of the detectors allows to reconstruct the four–momentum of the Z decaying into neutrinos. The main backgrounds are the production of a Z in association with an undetected high energy and low polar angle initial state radiation photon, semileptonic decay of W pairs with the charged lepton escaping detection or the more general case of a W decaying into

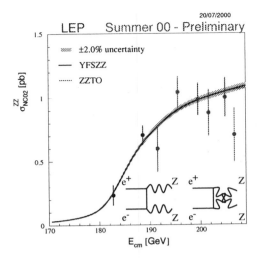

Figure 1. Measured and predicted cross sections for the Z boson pair–production process via the two diagrams shown in the lower right corner.

Table 1. LEP $\sqrt{s}$ and sum of the integrated luminosities ($\mathcal{L}$) collected by the four experiments together with the measured ($\sigma_{NC02}^{ZZ\ Exp}$) and expected ($\sigma_{NC02}^{ZZ\ Th}$) ZZ cross sections. Data collected in the year 2000 are grouped into the two last energy bins. The first uncertainty is statistical, the second systematic.

$\sqrt{s}$ GeV	$\mathcal{L}$ pb^{-1}	$\sigma_{NC02}^{ZZ\ Exp}$ pb	$\sigma_{NC02}^{ZZ\ Th}$ pb
182.7	221	$0.23 \pm 0.08 \pm 0.02$	0.25
188.7	686	$0.70 \pm 0.07 \pm 0.03$	0.65
191.5	114	$0.60 \pm 0.18 \pm 0.04$	0.77
195.6	310	$1.04 \pm 0.12 \pm 0.04$	0.90
199.6	326	$0.98 \pm 0.12 \pm 0.04$	0.98
201.7	152	$0.88 \pm 0.18 \pm 0.04$	1.01
205.0	241	$1.00 \pm 0.15 \pm 0.05$	1.04
206.8	123	$0.70 \pm 0.21 \pm 0.05$	1.06

hadrons produced together with a non resonant system of a low polar angle electron and a neutrino. Sequential cuts or multivariate techniques that enforce the signal topology of an undetected Z suppress these backgrounds.

High signal purity is achieved in the lower statistic (14%) final state with hadrons and charged leptons, which benefits of the high resolution measurements of the lepton momenta. Kinematic fits in the hypothesis of an equal mass of the lepton–lepton and hadron systems are performed, requiring then this mass to be compatible with the Z mass. Topological variables such as the angles of the leptons and the hadronic jets are also used.

The final states with two charged leptons and two neutrinos (4%) and four charged leptons (1%) are penalised by low statistics, even though identified with good purity thanks to the high lepton resolutions. The background from lepton pair–production and four–fermion processes is mainly rejected by requiring the lepton invariant and recoil masses to be compatible with the Z mass.

The undetectable final state with four neutrinos accounts only for 4% of the Z–pair decays.

3 Results

The four experiments measured the Z pair–production cross section at all the $\sqrt{s}$ above threshold[1,2]. Combined results are presented in Table 1 and Figure 1. This average takes into account sources of common and correlated systematic uncertainties, mainly due to uncertainties on the background cross sections and modelling, as well as uncorrelated ones, dominated by Monte Carlo statistics and detector related effects, in particular for the b tag procedure.

All the measurements agree with the Standard Model predictions. The ratio between the measured and predicted cross sections is formed at each $\sqrt{s}$, and its average over all the $\sqrt{s}$ yields 0.99 ± 0.06, what reveals an overall agreement within the combined accuracy of 6%, mainly statistical.

4 Four–fermion production

The DELPHI Collaboration extends its Z pair–production analysis to the case of a Z boson produced in association with a pair of fermions from a virtual photon. Final states with two quarks and either a muon– or a neutrino–pair are analysed. In the latter case the mass of the hadronic system must be below 60 GeV. The cross sections over the full data sample are respectively expected to be $0.19 - 0.25$ pb and $0.13 - 0.16$ pb, decreasing with $\sqrt{s}$. The measured cross sections read:

$$\sigma^{183\,\mathrm{GeV}-208\,\mathrm{GeV}}_{\mathrm{e^+e^-}\to Z\gamma^*\to\mu^+\mu^- q\bar{q}} = 0.22 \pm 0.05 \pm 0.02\,\mathrm{pb}$$

$$\sigma^{183\,\mathrm{GeV}-208\,\mathrm{GeV}}_{\mathrm{e^+e^-}\to Z\gamma^*\to\nu\bar{\nu}q\bar{q}} = 0.19 \pm 0.06 \pm 0.02\,\mathrm{pb}\,,$$

where the first uncertainty is statistical and the second systematic.

5 B quark content

The experimental investigation of Z–pair final states containing b quarks validates the capability of the LEP experiments to detect the production and decay of the Higgs boson.

These two process have a similar topology, as the Higgs boson production would preferentially manifests as a pair of b quarks from the decay of a heavy object that recoils against a Z. The expected cross sections are also similar, as illustrated in Figure 2. The Figure also shows the results of the measurement by the L3 Collaboration, in agreement with the Standard Model predictions. In this measurement Z decays into hadrons, neutrinos and charged leptons are considered in association with the b quark pair.

The OPAL Collaboration measures the branching ratio of the Z into b quarks in the selected Z–pair events, als finding agreement with the value measured at LEP at the Z resonance.

In conclusion, if such rare processes with these topologies can be observed, a Higgs boson light enough to be produced at LEP will not escape detection.

Acknowledgements

I wish to thank my L3 and LEP colleagues working on these subjects for all the construc-

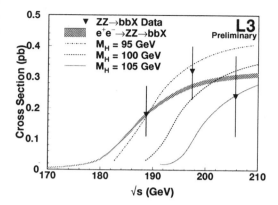

Figure 2. Measured and predicted cross sections for neutral–current four–fermion events compatible with Z pairs with at least a b quark pair. Data are presented for the average $\sqrt{s}$ of the years 1998, 1999 and 2000. The computed cross section for the production of the Standard Model Higgs boson is also presented for different mass hypotheses.

tive discussions we had throughout the last years and for sharing with me their preliminary results.

References

1. ALEPH Collab., R. Barate *et al.*, *Phys. Lett.* B **469**, 287 (1999), Contr. #289[a]; Contr. #287; Contr. #188; DELPHI Collab., P. Abreu *et al.*, Preprint CERN-EP/2000-89, Contr #659. Contr. #459; Contr. #460; L3 Collab., Contr. #502; Contr. #411; OPAL Collab., G. Abbiendi *et al.*, *Phys. Lett.* B **476**, 256 (2000), Contr. #153; Contr #154; Contr. #168.
2. L3 Collab., M. Acciarri *et al. Phys. Lett.* B **450**, 281 (1999); *Phys. Lett.* B **465**, 363 (1999).
3. G. Passarino, these proceedings
4. C. Matteuzzi, these proceedings.
5. G. Landsberg, these proceedings.

---

[a]Notation for contributed papers to this conference

COHERENT EXCLUSIVE EXPONENTIATION FOR PRECISION MONTE CARLO CALCULATIONS OF FERMION PAIR PRODUCTION / PRECISION PREDICTIONS FOR (UN)STABLE W^+W^- PAIRS

B.F.L. WARD[A,B,C], S. JADACH[A,B,D,E], W. PŁACZEK[B,F], M. SKRZYPEK[B,E], AND Z. WĄS[B,E]

[A] *Department of Physics and Astronomy, University of Tennessee, Knoxville, TN 37996-1200, USA*

[B] *TH Div., CERN, CH-1211 Geneva 23, Switzerland*

[C] *SLAC, Stanford UNiversity, Stanford, CA 94309, USA*

[D] *Theory Div., DESY, D-22603 Hamburg, Germany*

[E] *Institute of Nuclear Physics, ul. Kawiory 26a, PL-30-055 Krakow, Poland*

[F] *Institute of Computer Science, Jagellonian University, ul. Nawojki 11, 30-072 Krakow, Poland*

UTHEP-00-0901, SEPT., 2000

Presented by B.F.L. Ward at ICHEP2000, Osaka, Japan, July 27, 2000

We present the new Coherent Exclusive Exponentiation (CEEX), in comparison to the older Exclusive Exponentiation (EEX) and the semi-analytical Inclusive Exponentiation (IEX), for the process $e^+e^- \to f\bar{f}+n\gamma$, $f = \mu, \tau, d, u, s, c, b$, with validity for centre of mass energies from τ lepton threshold to 1 TeV. We analyse $2f$ numerical results at the Z-peak, 189 GeV and 500 GeV. We also present precision calculations of the signal processes $e^+e^- \to 4f$ in which the double resonant W^+W^- intermediate state occurs using our YFSWW3-1.14 MC. Sample $4f$ Monte Carlo data are explicitly illustrated in comparison to the literature at LEP2 energies. These comparisons show that a TU for the signal process cross section of 0.4% is valid for the LEP2 200 GeV energy. LC energy results are also shown.

1 Introduction

At the end of the LEP2 operation, the total cross section for the process $e^-e^+ \to f\bar{f}+n\gamma$ will have to be calculated with the precision 0.2% - 1%, depending on the event selection. In addition, the awarding of the 1999 Nobel Prize to G. 't Hooft and M. Veltman emphasises the importance of the on-going precision studies of the Standard Model processes $e^+e^- \to W^+W^- + n(\gamma) \to 4f + n(\gamma)$ at LEP2 energies, as well as the importance of the planned future higher energy studies of such processes in LC physics programs.

In what follows, we present precision predictions for both sets of processes, using our new coherent exponentiation (CEEX) [1] theory ($\mathcal{KK}$ MC) for the former set and our older and firmly established exclusive exponentia-

tion (EEX) [2] theory (YFSWW3-1.14 MC [4]) for the latter set. Both CEEX and and EEX are based on the YFS exclusive exponentiation theory of Yennie, Frautschi and Suura [3]. A detailed description [1,4,2] of our two approaches to the precision exponentiation theory may be found in Refs. [1,4,2]. As we indicate below, we have compared our $\mathcal{KK}$ MC calculations with with EEX, its semianalytical partner IEX, and ZFITTER 6.21 [5] and we have compared our YFSWW3-1.14 MC calculations with RacoonWW [6] and with the Beenakker *et al.*[7] semi-analytical approach.

The paper proceeds as follows. In Sec. 2 we discuss the implementation of CEEX in our $\mathcal{KK}$ MC in relation to EEX. In Sec. 3 we present some of its new results for $2f + n(\gamma)$ processes at high energies. In Sec. 4 we present the EEX theory realization in our

YFSWW3-1.14 MC. In Sec. 5 we present some of its new results on $WW + n(\gamma) \rightarrow 4f + m(\gamma)$ processes at high energies. Sec. 6 contains our summary remarks.

2 $\mathcal{KK}$ MC

The main differences between CEEX and EEX are best illustrated by focusing on the process of interest, which is

$$e^-(p_1, \lambda_1) + e^+(p_2, \lambda_2) \rightarrow f(q_1, \lambda_1') + \bar{f}(q_2, \lambda_2')$$
$$+ \gamma(k_1, \sigma_1) + ... + \gamma(k_n, \sigma_n). \tag{1}$$

The respective EEX total cross section

$$\sigma = \sum_{n=0}^{\infty} \int_{m_\gamma} d\Phi_{n+2} \, e^{Y(m_\gamma)} D_n(q_1, q_2, k_1, ..., k_n) \tag{2}$$

corresponds to the attendant $\mathcal{O}(\alpha^1)$ distributions D_n as given in Ref. [2] by formulas such as, for $n = 0, 1$, $D_0 = \bar{\beta}_0$ and $D_1(k_1) = \bar{\beta}_0 \tilde{S}(k_1) + \bar{\beta}_1(k_1)$, where the real soft factors $\tilde{S}(k)$ are defined as usual [2]. The important point is that the IR-finite building blocks $\bar{\beta}_n$, for example, $\bar{\beta}_0 = \sum_\lambda |\mathcal{M}_\lambda^{\text{Born}}|^2$, in the multi-photon distributions are all in terms of $\sum_{spin} |...|^2$! Here, $\lambda =$ fermion helicities and $\sigma =$ photon helicity. In contrast, in the analogous $\mathcal{O}(\alpha^1)$ case of CEEX

$$\sigma = \sum_{n=0}^{\infty} \int_{m_\gamma} d\Phi_{n+2}$$
$$\sum_{\lambda, \sigma_1, ..., \sigma_n} |e^{B(m_\gamma)} \mathcal{M}_{n, \sigma_1, ..., \sigma_n}^\lambda (k_1, ..., k_n)|^2 \tag{3}$$

the differential distributions for $n = 0, 1$ photons are, for example, $\mathcal{M}_0^\lambda = \hat{\beta}_0^\lambda$, $\lambda =$ fermion helicities and $\mathcal{M}_{1, \sigma_1}^\lambda(k_1) = \hat{\beta}_0^\lambda \mathfrak{s}_{\sigma_1}(k_1) + \hat{\beta}_{1, \sigma_1}^\lambda(k_1)$, with the IR-finite building blocks $\hat{\beta}_0^\lambda = (e^{-B} \mathcal{M}_\lambda^{\text{Born+Virt.}})|_{\mathcal{O}(\alpha^1)}$ and $\hat{\beta}_{1,\sigma}^\lambda(k) = \mathcal{M}_{1,\sigma}^\lambda(k) - \hat{\beta}_0^\lambda \mathfrak{s}_\sigma(k)$. Explicitly, this time everything is in terms of $\mathcal{M}$-spin-amplitudes! This is the basic difference

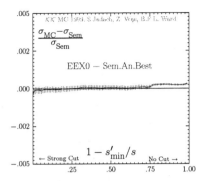

Figure 1. Results for 189 GeV in the $\mu\bar{\mu}$ channel, for $v < 0.999$. We plot the difference between the $\mathcal{KK}$ MC result and semi-analytical (IEX) result divided by the latter.

between EEX/YFS AND CEEX. Complete expressions for spin amplitudes with CEEX exponentiation, n_γ arbitrary, are given in Phys. Lett. **B449**, 97 (1999) for the $\mathcal{O}(\alpha^1)$ case and in CERN-TH/2000-087,UTHEP-99-09-01, for the $\mathcal{O}(\alpha^2)$ case, all are based on GPS spinor conventions as given in CERN-TH-98-235, hep-ph/9905452.

3 Results: CEEX

In Figs. 1, 2 and 3 we show the baseline technical precision test with the $\bar{\beta}_0$ level matrix element and physical precision tests of σ_{tot}, A_{FB}, and the IFI at LEP2 energies as effected in the LEP2 MC Workshop [8]. With these and related tests we achieve the technical precision tag of 0.02% at LEP2 energies, the physical tags of 0.2%(0.2 − 0.4%) for the σ_{tot}(the A_{FB}), and firm control on the IFI [1]:

we see that the IFI $\cong$ 1.5% for energy cut 0.3, that a $|cos\theta| < 0.9$ cut reduces the IFI by 25%, and that the IFI is very small at the Z return, for example.

4 YFSWW3-1.14 MC

Starting from the underlying process of interest, Eq.(1), its cross section, Eq.(2), **and the attendant W^+W^- produc-**

Physical Precision Tests
(a)

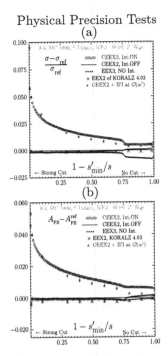

Figure 2. Absolute predictions for σ_{tot}, A_{FB}: $\mu\bar{\mu}$, 189 GeV.

Physical Precision Tests
(a)

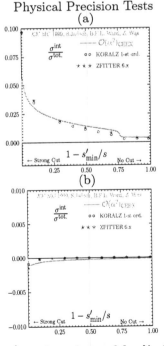

Figure 3. s'-cut dependence of $\delta\sigma$, No θ-cut: (a), 189 GeV; (b), M_Z.

tion and decay, $e^-(p_1) + e^+(p_2) \rightarrow W^-(q_1) + W^+(q_2)$, $W^-(q_1) \rightarrow f_1(r_1) + \bar{f}_2(r_2)$, $W^+(q_2) \rightarrow f_1'(r_1') + \bar{f}_2'(r_2')$, we may isolate the Leading Pole Approximation ($LPA_{a,b}$) as follows:

$$
\mathcal{M}_{4f}^{(n)}(p_1, p_2, r_1, r_2, r_1', r_2', k_1, ..., k_n) \overset{LPA}{=>}
$$

$$
\mathcal{M}_{LPA}^{(n)}(p_1, p_2, r_1, r_2, r_1', r_2', k_1, ..., k_n)
$$

$$
= \sum_{\gamma \; Part'ns} \mathcal{M}_{Prod}^{(n),\lambda_1\lambda_2}(p_1, p_2, q_1, q_2, k_1, ..., k_a)
$$

$$
\times \frac{1}{D(q_1)} \mathcal{M}_{Dec_1,\lambda_1}^{(n)}(q_1, r_1, r_2, k_{a+1}, ..., k_b)
$$

$$
\times \frac{1}{D(q_2)} \mathcal{M}_{Dec_2,\lambda_2}^{(n)}(q_2, r_1', r_2', k_{b+1}, ..., k_n),
$$

$$(4)$$

in an obvious notation [4] for the $W^\pm$ propagator denominators $D(q_i)$, etc. Here, we can identify two different realizations, $LPA_{a,b}$, of the leading pole residues in Eq. (4) by following the prescriptions of Eden et al. [9] and Stuart [10]: in $\mathcal{M} = \sum_j \ell_j A_j(\{q_k q_l\})$, the complete set of spinor covariants $\{\ell_j\}$ may (b) or may not (a) be evaluated at the pole positions for the respective Lorentz scalar functions $\{A_j(\{q_k q_l\})\}$, as these latter already realize the analyticity properties of the S-matrix by themselves. We do both.

The standard YFS methods [2] (EEX-Type) give us the corresponding analog of Eq.(2). In realizing the exact $\mathcal{O}(\alpha)$ corrections in the latter equation in the LPA, we have chosen, for our renormalization scheme, the G_μ-Scheme of Fleischer et al. [11] in version 1.13 and the schemes A and B in version 1.14, where in A only the hard EW correction has α_{G_μ} whereas in B the entire $\mathcal{O}(\alpha)$ correction has $\alpha(0)$. The analysis in Ref. [12] tells us that the schemes A and B are improvements over the G_μ scheme in version 1.13, as we have verified in the context of the LEP2 MC Workshop comparisons with Denner et al. As a consequence, we have a

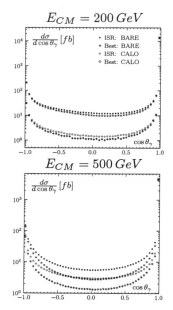

$E_{CM} = 200\,GeV$

$E_{CM} = 500\,GeV$

Figure 4. Distribution of $cos\theta_\gamma$ with respect to the e^+ beam

no cuts		σ_{tot} [fb]	
final state	program	Born	best
$\nu_\mu\mu^+\tau^-\bar{\nu}_\tau$	YFSWW3	219.770(23)	199.995(62)
	RacoonWW	219.836(40)	199.551(46)
	(Y−R)/Y	−0.03(2)%	0.22(4)%
$u\bar{d}\mu^-\bar{\nu}_\mu$	YFSWW3	659.64(07)	622.71(19)
	RacoonWW	659.51(12)	621.06(14)
	(Y−R)/Y	0.02(2)%	0.27(4)%
$ud\bar{s}\bar{c}$	YFSWW3	1978.18(21)	1937.40(61)
	RacoonWW	1978.53(36)	1932.20(44)
	(Y−R)/Y	−0.02(2)%	0.27(4)%

Table 1. Total cross sections, CC03 from RacoonWW, YFSWW3, $\sqrt{s} = 200$ GeV without cuts. Statistical errors – last digits in ( ), etc.⇒ 0.4% TU.

−0.3 ÷ −0.4% shift of the **NORMALISATION** of version 1.14 relative to version 1.13. See G. Passarino [13] for more details and references.

5   Results: YFSWW3-1.14

In Fig. 4, we show the hardest photon angular distribution, both at 200 GeV and at 500 GeV. We see that the NL EW correction is relevant both for the **BARE** and **CALO** event selections as defined Ref. [14] away from the beam direction. Similar effects are discussed in Refs. [4], where we find that the EW NL correction at LEP2 energies is large, $\sim -2\%$, and is in general a non-trivial function of the kinematical variables. The authors in Ref. [6] have reached the analogous conclusion.

Indeed, in Table 1 we show a comparison between the results from RacoonWW and YFSWW3-1.14, where we have chosen the case of without cuts, as carried out in the context of the LEP2 MC Workshop. From these results and others similar to them we arrive at the theoretical precision tag of 0.4% at 200 GeV for the WW signal cross section at LEP2. See G. Passarino [13] for more details and references.

6   Conclusions

Our conclusion for the CEEX $\mathcal{KK}$ MC discussion is that the CEEX is a clear upgrade path for the EEX in a spin amplitude level MC. We have shown that, for LEP2, the total TU is 0.2%(0.2-0.4%) for $\sigma_{tot}(A_{FB})$, for typical cuts – for the LC at 0.5 TeV, these are a factor of 2 worse, and for $\gamma\gamma^*$ the TU is 0.3% for LEP2 (there is no firm result for LC). The IFI (ISR⊗FSR) is included and under firm control. Our conclusions for YFSWW3-1.14 are that the EW NL correction [11] in $\mathcal{O}(\alpha)$, which is also realized in RacoonWW, is important both for the normalisation and for the differential distributions. The TU at 200 GeV, based on comparisons with RacoonWW, is 0.4%.

Acknowledgements

The authors all thank the CERN TH Div. and the CERN ALEPH, DELPHI, OPAL and L3 Collaborations for

In the figure: ISR: BARE, Best: BARE, ISR: CALO, Best: CALO

$\frac{d\sigma}{d\cos\theta_\gamma}\,[fb]$

$\cos\theta_\gamma$

their support and kind hospitality during the course of this work. S. J. thanks the DESY TH. Div. for its support and kind hospitality while a part of this work was done. This work was supported in part by the Polish Government grants KBN 2P03B08414 and 2P03B14715, the Maria Skłodowska-Curie Joint Fund II PAA/DOE-97-316, and by the US Department of Energy Contracts DE-FG05-91ER40627 and DE-AC03-76ER00515.

References

1. S. Jadach, B. F. L. Ward and Z. Was, DESY-99-106, CERN-TH-2000-087, UTHEP-99-0901, and references therein.

2. S. Jadach and B. F. L. Ward, Phys. Lett. B274, 470(1992); Comp. Phys. Commun. 56, 351 (1990); S. Jadach, B. F. L. Ward and Z. Was, *ibid.* 79, 531 (1994); S. Jadach *et al.*, *ibid.* 102, 229 (1997), and references therein.

3. D. R. Yennie, S. C. Frautschi and H. Suura, Ann. Phys. (NY) 13, 379 (1961).

4. S. Jadach *et al.*, Phys. Lett. B417, 326 (1998); Phys. Rev. D61, 113010 (2000); UTHEP-00-0101.

5. D. Bardin *et al.*, hep-ph/9908433, and references therein.

6. A. Denner *et al.*, hep-ph/9912261, 9912290, 9912447; Phys. Lett. B475, 127 (2000); BI-TP 2000/06; hep-ph/0006307.

7. W. Beenakker *et al.*, hep-ph/99023 33, 9811481, and references therein.

8. M. Kobel *et al.*, hep-ph/0007180.

9. R. J. Eden *et al.*, *The Analytic S-Matrix*, (Cambridge University Press, Cambridge, 1966).

10. R. G. Stuart, Nucl. Phys. B498, 28 (1997); Eur. Phys. J. C4, 259 (1998); hep-ph/9706431, 9706550.

11. J. Fleischer *et al.*, Zeit. Phys. C42, 409 (1989), and references therein.

12. B. F. L. Ward, Phys. Rev. D36, 939 (1987).

13. G. Passarino, in these *Proceedings*.

14. M. W. Gruenwald *et al.*, hep-ph/0005309.

MEASUREMENT OF e − γ INTERACTIONS AT LEP

C. PALOMARES

CIEMAT, Avda. Complutense 22, 28040 Madrid, Spain
E-mail: Carmen.Palomares@cern.ch

This report shows the studies of different eγ interaction processes at LEP. The cross-section of the quasi-real Compton scattering has been measured at centre-of-mass energies between 20 GeV and 185 GeV, using the L3 detector at LEP. The production of single neutral intermediate vector bosons in Compton scattering is analysed by the DELPHI and OPAL experiments. The production of single excited electrons in a eγ interaction has been consider as well.

1 Quasi-real Compton Scattering

The quasi-real Compton scattering is a specific configuration of the reaction $e^+e^- \rightarrow e^+e^-\gamma$, shown in Figure 1, like Bremsstrahlung and radiative Bhabha scattering at finite angle. When p^2 or q^2 are not close to zero, particles will be deflected at a finite angle. Quasi-real Compton scattering corresponds to the situation in which $p^2 \rightarrow 0$ with $|p^2| << |q^2|$. An electron and a photon are detected in the central region while the other electron is scattered at almost zero degree (dashed line in Figure 1).

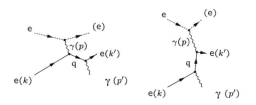

Figure 1. Diagrams contributing to the quasi-real Compton scattering.

One can take advantage of the overconstrained kinematics in the process. The energies of the observed electron and the photon are defined through the measurement of their scattered angle in the Laboratory frame. Since the events considered are dominated by quasi-real photons, the total transverse momentum is close to zero.

2. e − γ interactions at LEP

The quasi-real Compton scattering, $e^\pm\gamma \rightarrow e^\pm\gamma$, has been studied at LEP at effective centre-of-mass energies, $\sqrt{s'}$, larger than 20 GeV.

On the other hand, the process $e^\pm\gamma \rightarrow e^\pm Z/\gamma^\star$ has been measured for the first time in e^+e^- collisions at LEP. This process is the same as ordinary Compton scattering with the outgoing real photon replaced by a virtual photon $\gamma^\star$ or a Z. In the analyses of this process made by OPAL and DELPHI Collaborations [2] [3] the $Z/\gamma^\star$ is detected via its decay into hadrons.

The signature of the eγ interactions produced at LEP will in both cases consist of an electron escaping along the beam pipe and a second electron observed in the detector, together with a photon or two fermions coming from $Z/\gamma^\star$ decay.

2.1 $e^\pm\gamma \rightarrow e^\pm\gamma$

The L3 experiment has studied the quasi-real Compton scattering at centre-of-mass energies of the eγ collision between 20 and 185 GeV [1]. The analysis is based on data corresponding to an integrated luminosity of 634.6 pb^{-1}. The prediction of Compton scattering is based on Monte Carlo events generated with the program TEEGG, taking into account corrections of order $\mathcal{O}(\alpha^4)$. The main background contributions are expected from the reaction $e^+e^- \rightarrow \gamma\gamma(\gamma)$ (where one photon is converted) or from Bhabha scattering (where one track is not reconstructed).

The identification of electrons and photons in the final state requires clusters in the electromagnetic calorimeter with an energy larger than 2 GeV. Both electron and photon are required to be observed within $|\cos\theta| < 0.94$. The scattering angle in the centre-of-mass system of the $e\gamma$ pair has to satisfy $|\cos\theta^\star| < 0.8$ to reduce the contribution from low angle Bhabha scattering. To ensure the selection of quasi-real photons, the transverse momentum of the final state system has to be lower than 15% of the beam energy.

In total 7335 candidates are selected. Figure 2 shows the measurements of the total cross-section compared with the QED prediction for the Compton scattering. The results are in good agreement with the theoretical prediction.

2.2 $e^\pm\gamma \to e^\pm Z/\gamma^\star \to e^\pm q\bar{q}$

The LEP reaction involving the so-called electroweak Compton scattering, $e^+e^- \to e^+e^-q\bar{q}$, gets contributions from many intermediate states, like: $e^+e^- \to ZZ \to e^+e^-q\bar{q}$ and $e^+e^- \to \gamma\gamma \to e^+e^-q\bar{q}$. The signal definition made by OPAL is based on Lorentz invariant variables: The Mandelstam variable t, the squared four-momentum transferred to the quasi-real photon, p^2, and the invariant mass of the two fermions in the final state, $M_{f\bar{f}}$. The kinematic limits are: $|t| > 400$ GeV^2, $|p^2| < 10$ GeV^2, $M_{f\bar{f}} > 5$ GeV. This definition of the kinematic region of the signal is an effective cut on the centre-of-mass energy of the $e\gamma$ system at $\sqrt{s'} \geq 20.6$ GeV.

DELPHI defines the signal region topologically, requiring at least one electron to be above the acceptance of the DELPHI electromagnetic calorimeter, i.e., $|\cos\theta_e| < 0.985$, and having an energy $E_e > 4$ GeV. Only the phase space region where the invariant mass of the two fermions is larger than 15

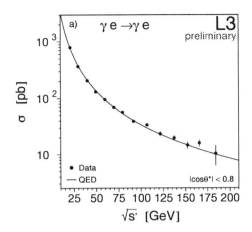

Figure 2. Measured total cross-section of Compton scattering inside the angular range $|\cos\theta^\star| < 0.8$ as a function of the effective centre-of-mass energy $\sqrt{s'}$. The solid line shows the QED prediction.

GeV is considered. In both cases, the cross-section is compared with the one obtained from GRC4F and PYTHIA programs.

The electron is identified through its signal in the electromagnetic calorimeter. Due to the hadronic activity of the event, a minimum number of tracks and calorimeter clusters is required. Two jets are reconstructed making use of the Durham algorithm. According to the cut on the signal definition, the invariant mass of the hadronic system is required to be larger than 5 GeV. The analysis made by DELPHI takes a smaller signal region cuting in $M_{jj} > 15$ GeV. A kinematic fit of the event is then performed assuming a topology of signal events with two jets and two electrons, one of them going unobserved along the beam pipe. Fits with a small χ^2 probability are rejected. To ensure that the selected events are consistent with the signal definition, both experiments apply cuts according to each acceptance region. OPAL requires $\sqrt{s'} = M_{\gamma e} \geq 25$ GeV and $|t| > 500$ GeV^2, and DELPHI fixes the geometrical range, $|\cos\theta_e| < 0.985$.

The main contribution to the background

Table 1. Preliminary results on the cross-section of the electroweak Compton scattering measured by OPAL experiment at $\sqrt{s} = 189$ GeV

$\sqrt{s} = 189$ GeV (Preliminary)		
σ (pb)	$\gamma^\star$ee	Zee
Measured	$4.6 \pm 0.9 \pm 0.6$	$1.5 \pm 0.3 \pm 0.3$
SM	3.06 ± 0.04	1.19 ± 0.02

comes from the process $e^+e^- \to e^+e^-\gamma\gamma \to e^+e^- +$ hadrons. Other sizeable sources of background are the hadronic events $e^+e^- \to q\bar{q}$, processes involving two fermions in the final state and four-fermion processes like the W-pair production. In order to reject these events several cuts on the missing momentum and on angular variables are applied.

The cross-section of the process $e\gamma \to eZ/\gamma^\star \to eq\bar{q}$ has been calculated separately for $\gamma^\star$ee-like and Zee-like events, with a cut at a hadronic mass of 60 GeV.

The last results presented by the OPAL Collaboration correspond to the measurement of the cross-section at a centre-of-mass energy of 189 GeV. The numbers are shown in Table 1. DELPHI has analysed the collected data at energies between 183 and 202 GeV. The results are shown in Figure 3.

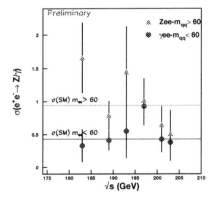

Figure 3. Measured cross-section of the process $e\gamma \to eZ/\gamma^\star \to eq\bar{q}$ for 183 GeV$< \sqrt{s} < 202$ GeV. The red points correspond to the low mass region and the triangles to the Zee-like events. The lines show the SM prediction in both cases.

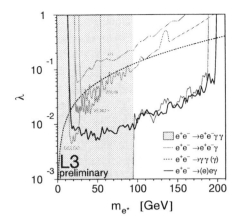

Figure 4. The upper limit on the coupling λ at 95% confidence level as a function of $m_{e^\star}$, derived from the measurement of Compton scattering in L3 (solid line). The dashed and dotted lines show the analogous results from the L3 analyses of the processes $e^+e^- \to \gamma\gamma(\gamma)$ and $e^+e^- \to e^+e^-(\gamma)$. The results from other collider experiments are included as indicated.

2.3 Production of single excited electrons

The Compton scattering process is sensitive to the production of an excited electron, $e^\star$, decaying into an electron-photon pair: $e^\pm\gamma \to e^\star \to e^\pm\gamma$. Since the number of Compton scattering events agrees with the QED prediction, an upper limit for the number of expected $e^\star$-events can be determined. Taking into account that $\sigma(e\gamma \to e^\star \to e\gamma) \propto \frac{\lambda^2}{m_{e^\star}^2}$, an upper limit to the $e^\star e\gamma$ coupling, λ, as a function of $m_{e^\star}$ is obtained, Figure 4.

References

1. L3 Collab., *Measurement of quasi-real Compton scattering at LEP*, L3 note 2562 (2000).

2. DELPHI Collab., *Single Intermediate Vector Boson production at $\sqrt{s} = 183$ GeV and 189 GeV*, DELPHI 2000-143 CONF 442/1 (2000).

3. OPAL Collab., *Measurement of $e\gamma \to eZ/\gamma^\star$*, OPAL Physics Note PN392.

MULTIPHOTON PRODUCTION AND TESTS OF QED AT LEP-II

M. WINTER

IReS, IN2P3/ULP, 23 rue du loess, BP 28, F-67037 Strasbourg, France
E-mail: marc.winter@cern.ch

Data collected by the 4 LEP collaborations from 1995 to 2000 at collision energies ranging from 130 to 208 GeV were used to measure the cross-section of the process $e^+e^- \rightarrow \gamma\gamma(\gamma)$. QED predictions for this reaction were tested with a few per-cent accuracy and manifestations of physics beyond the Standard Model (SM) were investigated. Preliminary lower bounds on the cut-off parameter $\Lambda_\pm$, the mass of an excited electron, the string mass scale underlying low-scale Quantum Gravity and on energy scales expressing various contact interactions were derived.

1 Introduction

Pairs of collinear photons produced in e^+e^- interactions have a clear topology, easy to isolate from all backgrounds. The reaction cross-section is sizeable (of O(10 pb) at LEP-II) and its electroweak corrections (occuring only at the one-loop level) are small. The process provides therefore one of the cleanest ways of testing QED predictions as well as manifestations of physics beyond the SM.

The production of collinear photons in e^+e^- collisions is described within QED by the t-channel exchange of an electron. The differential cross-section w.r.t. the production angle of one photon reads

$$\frac{d\sigma}{d\Omega} = \left(\frac{\alpha^2}{s} \cdot \frac{1+\cos^2\theta^*}{1 + \frac{2m_e^2}{s} - \cos^2\theta^*} \right) (1 + \delta_{RC})$$

where the first term stands for the Born-level cross-section [1] (m_e being the electron mass and s the square of the collision energy). δ_{RC} expresses the dominant radiative corrections, amounting to $\sim$ 3 - 6 %.

$\theta^\star$ is defined in the $\gamma\gamma$ rest-frame[a] in order to reduce deviations from the Born-level expression, and is derived from the production angles of the two most energetic photons in the laboratory-frame, i.e.

$$\cos\theta^* = \left| \frac{\sin\frac{\theta_1 - \theta_2}{2}}{\sin\frac{\theta_1 + \theta_2}{2}} \right|.$$

---

[a] assuming that initial state radiation is due to the emission of a single photon along the beam axis.

Table 1. Data sets used for the results presented at the conference. The range of collision energies covered by the data analysed in each collaboration is shown with the corresponding integrated luminosity.

expt.	years	$\sqrt{s}$(GeV)	L(pb^{-1})
ALEPH	1995-99	130 - 202	496
DELPHI	1995-00	130 - 208	454
L3	1995-98	130 - 189	256
OPAL	1995-00	130 - 208	479

2 Data analysis

LEP experiments have measured the differential cross-sections as well as their values integrated over the detector acceptance, at collisions energies ranging from about 91 GeV (at LEP-I) up to more than 208 GeV (achieved at LEP-II in 2000).

The final states of interest were selected by requiring at least two photons produced nearly back-to-back and carrying close to the beam energy. In order to increase the size of the selected sample, events where one of the two energetic photons was converted in the detector material, were also retained. The mean residual background (mainly due to Bhabha and Compton-like scattering) contaminating the selected data sample was estimated from Monte-Carlo simulations to be less than 1%. The results presented at the conference [2] rely on the data sets summarised in Table 1.

Table 2. 95% C.L. lower bounds (in TeV) on the parameters expressing physics beyond the SM extracted from the cross-sections measured in each experiment ($M_{e^\star}$, Λ_7) or from the LEP average ($\Lambda_\pm$, M_s).

	O	L	D	A
Λ_+	LEP: 0.44			
Λ_-	LEP: 0.36			
$M_e^\star/\sqrt{\kappa}$	0.34	0.28	0.32	0.34
M_s	LEP: 0.96 ($\lambda = +1$)			
M_s	LEP: 0.78 ($\lambda = -1$)			
Λ_7	0.74	0.70	–	0.71

3 Results

The cross-sections observed at each collision energy were corrected for radiative corrections in order to extract the Born-level values. The latter compared well with those predicted by QED. This is illustrated by the $\chi^2/d.o.f.$ between the measured and expected total cross-sections of the 4 experiments for all data sets collected since 1995, which amounts to 23.8/23.

The agreement of the measurements with the SM values was used to set limits on parameters expressing physics beyond it and including it as an effective low-energy theory. Several phenomenological models were considered, and their free parameters fitted to the measured cross-sections.

The 95% C.L. lower bounds derived from the fit results are summarised in Table 2. The parameter $\Lambda_\pm$ stands for the energy scale at which QED could break down [3]. $M_{e^\star}$ stands for the mass of an excited electron exchanged in the t-channel, the parameter κ expressing the ratio between the coupling $e^\star e\gamma$ and the standard $ee\gamma$ coupling [4]. M_s represents the string mass scale related to low-scale Quantum Gravity [5], whose predictions depend on an additional parameter λ of order unity. Λ_7 is the energy scale associated to contact interactions expressed by a dimension-7 operator added to the QED Lagrangian [6].

4 Summary - Conclusion

The cross-section of the process $e^+e^- \to \gamma\gamma(\gamma)$ measured at LEP-II energies ranging up to 208 GeV agrees well with QED. Lower bounds on parameters expressing physics beyond the SM were derived, which improve the limits found at LEP-I by a factor 2.5 -3.

Acknowledgments

The representatives of the LEP collaborations are gratefuly anknowledged for their help in collecting the material for the conference.

References

1. I.Harris and L.M.Brown, *Phys. Rev.* **105**, 1656 (1957); F.A.Berends and R.Gastmans, *Nucl. Phys.* B **61**, 414 (1973);

2. OPAL coll., *OPAL physics note PN 433*, contributed paper Nr.215, and *OPAL physics note PN 420*, contributed paper Nr.184.
   M.Acciari et al. (L3 coll.), *Phys. Lett.* B **475**, 198 (2000).
   S.Andriga et al. (DELPHI coll.), *DELPHI physics note 2000-131 CONF 430*, contributed paper Nr.651, and P.Abreu et al. (DELPHI coll.), *CERN-EP/2000-094* (submitted to *Phys. Lett.* B), contributed paper Nr.269.
   ALEPH coll., *ALEPH physics note 2000-008 CONF 2000-005*, contributed paper Nr.390.

3. S.D.Drell, *Ann. Phys.* **4**, 75 (1958).

4. F.E.Low, *Phys. Rev. Lett.* **14**, 238 (1965), A.Blondel et al., *CERN-EP/87-050* and *CERN 87-08* (1987).

5. G.F.Giudice et al., *Nucl. Phys.* B **544**, 3 (1999); K.Agashe and N.G.Deshpande, *Phys. Lett.* B **456**, 60 (1999); J.Hewett, *Phys. Rev. Lett.* **82**, 4765 (1999).

6. O.J.P.Eboli et al., *Phys. Lett.* B **271**, 274 (1991).

MEASUREMENT OF THE BEAM ENERGY AT LEP2

P.B. RENTON

(ON BEHALF OF THE LEP ENERGY WORKING GROUP)

Nuclear Physics Lab., University of Oxford, Oxford OX13RH, U.K.,
E-mail: p.renton1@physics.ox.ac.uk

The status of the determination of the beam energy at LEP2 is discussed.

1 Extrapolation method

At LEP1 the uncertainties due to the beam energy on the Z mass and width are 1.7 and 1.2 MeV respectively, to be compared to the combined experimental errors from all other sources of 1.2 and 2.0 MeV. The main tools used for this excellent precision were *resonant depolarisation*, which gives the instantaneous beam energy to a fraction of an MeV, and extensive logging and monitoring for the many environmental effects which were found to influence the beam energy[1].

At LEP2 the determination of the mass of the W boson sets the requirements The experimental uncertainty from the combined LEP data is expected to be $\Delta m_W \simeq 30$ to 40 MeV, depending on progress with the understanding of final state interaction effects. This sets the target of $\Delta m_W \simeq 10$-15 MeV for the desired uncertainty from the LEP beam energy. Unfortunately there is insufficient transverse polarisation above $E_b \simeq 60$ GeV to use this method directly at LEP2.

The method currently used to determine the beam energy is based on the readings of 16 NMR probes, which are located in dipole magnets around the LEP ring. These are read out almost continuously and are calibrated assuming a linear relationship between the NMR field and the beam energy from resonant depolarisation, in the region E_b 40 to 60 GeV. There are 20 fills with 2 or more energy points and the residuals of a linear fit to all the data is shown in fig 1. It can be seen that the residuals are less than

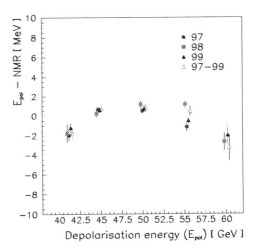

Figure 1. Residuals of NMR energy versus E_{pol}.

2 MeV and reproducible from year to year. The additional corrections applied to determine E_b for physics fills are described in [2]. The main disadvantages with this method is that it only samples 16 out of 3300 dipole magnets and the uncertainty in how any non-linearities will extrapolate.

The ring dipole magnets contain single turn flux loops which can be used to make a direct measurement of the field integral seen by the beam (in fact 96.5%), using dedicated machine current cycles [2]. The flux-loop integral is determined for each LEP octant and the mean then found. The beam energy is proportional to the total field integral but the flux-loop is not accurate enough to make a precise independent measurement of E_b. In-

Figure 2. Differences in MeV between the measured and extrapolated flux-loop values for the 1999 data.

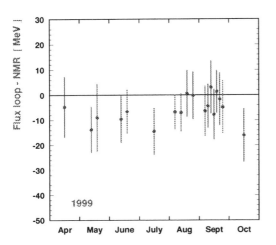

Table 1. Error breakdown for 1999 data.

source	error (MeV)
NMRs plus extrap.	19
E_{pol}, e^+e^- diff, optics diff	4
other(tide, correctors..)	3
IP specific (RF,dispersion)	5
TOTAL	21

Figure 3. Layout of the spectrometer.

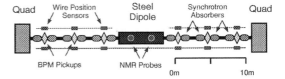

stead it is calibrated against NMR measurements in the region 40-60 GeV. At high energies ($\simeq$ 100 GeV) the extrapolated values can be compared with the directly measured flux-loop values. The results for the data of 1999 are shown in fig 2.

It can be seen that the difference flux-loop - NMR is stable with time and, on average, is slightly negative, but there is no indication of any large non-linearity. It is also stable year-to-year, both averaged over all octants, and octant by octant. Based on these studies an extrapolation error of 15 MeV at LEP2 physics energies is assigned. The flux-loop is not used directly in assigning the central energy values, but is used as a cross-check and to assign the extrapolation error.

Central values and errors on E_b have been determined up to 1999, and a breakdown of the errors for 1999 is given in Table 1. In previous years the total error on E_b was 25 MeV in 1997 and 20 MeV in 1998, with a correlation $\simeq$ 70-90 % between years[2]. These uncertainties translate into an uncertainty on the W mass of $\Delta m_W \simeq \pm 17$ MeV.

2   Magnetic Spectrometer

In order to try and improve the accuracy the LEP Spectrometer project was conceived in 1997. The idea is to measure the deflection of the beam in the LEP ring by measuring the trajectory before and after a special steel dipole magnet. The beam energy is proportional to the field integral of the magnet divided by the angular deflection. A schematic of the apparatus is shown in fig 3. The magnet used was accurately mapped outside of the LEP tunnel and later in situ to ensure the field remained stable. Good accuracy and stability is achieved and an error on $E_b \simeq 3$ MeV from this source is estimated.

The most demanding requirement is that on the Beam Position Monitors (BPMs), which must achieve an accuracy of $\simeq 1\mu$m. Special electronics were needed and the BPMs are shielded against synchrotron radiation and must work in a very hostile environment. To achieve the required 10^{-4} accuracy the device is calibrated against resonant depolarisation. From measurements made in 1999 a preliminary attempt has been made to determine E_b and its error. The uncertainties which are related to the NMRs and

Figure 4. Comparison of Spectrometer and NMRs at physics energies.

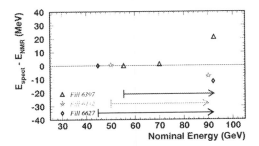

the normalisation are $\simeq 3$ MeV. The uncertainties in computing E_b from the spectrometer readings (RF model etc) amount to 5.5 MeV and the estimated uncertainty from the BPMs is $\simeq 8$ MeV; giving a total estimated systematic uncertainty at physics energies of $\simeq 11$ MeV. The uncertainty in the region of 40-60 GeV is about 6 MeV, and this is compatible with the 8 MeV rms observed between the spectrometer and resonant depolarisation measurements in the energy range.

Spectrometer measurements of E_b have been obtained from three fills in 1999, in which they were normalised to resonant depolarisation at low energy and then a ramp to physics energies made. The spectrometer measurements are compared to the usual NMR model in fig 4. The mean difference E_{spect}-$E_{NMR} = 0.5$ MeV, with an rms of 15 MeV. This is the first direct measurement at the LEP2 energy scale and supports the extrapolation method.

For 2000 more measurements are being made to gain statistics and understand systematics. However the early data obtained seem to show some internal BPM inconsistencies, the cause of which is yet to be resolved.

3 Beam energy from Q_s v V_{RF}

A second direct method has also been employed. In this, E_b is extracted from the energy loss $U_0 \propto E_b^4/\rho$, by measuring the synchrotron tune Q_s versus V_{RF}:

$$Q_s^2 = \left(\frac{\alpha_c h}{2\pi E_b}\right)\sqrt{(e^2 V_{RF}^2 - U_0^2)}$$

In practice a more detailed model is used. Three measurements have been made in 1998/9 above 80 GeV and these give E_{QS}-$E_{NMR} = -9$ MeV, with an rms of 11 MeV; again confirming the NMR method. More measurements are being performed in 2000 to gain statistics and measure the effective LEP bending radius ρ.

4 Beam energy from event data

Since the Z boson mass is well known a study of $e^+e^- \to f\bar{f}$ events, in which there is an ISR photon such that the $f\bar{f}$ system 'returns' to the Z, can be used to estimate E_b. A new study by DELPHI using $\mu^+\mu^-$ events [3] gives E_{data}-$E_{NMR} = 80 \pm 101$(stat) ± 58(syst) MeV. An earlier study using $q\bar{q}$ events by ALEPH [4] gives E_{data}-$E_{NMR} = -15 \pm 95$(stat) ± 40(syst). The average is E_{data}-$E_{NMR} = 27 \pm 77$ MeV; again in good agreement with the NMR method.

5 Summary and Outlook

The contribution of the beam energy to the uncertainty on m_W is $\Delta m_W \simeq 17$ MeV. The final step is to complete and fully analyse the direct measurements from the Spectrometer and Q_s v V_{RF}.

References

1. R. Assmann *et al*, LEP Energy Working Group, *Eur.Phys. J.* **C 6**, 187 (1999).
2. A. Blondel *et al*, LEP Energy Working Group, *Eur.Phys. J.* **C 11**, 573 (1999).
3. DELPHI Collaboration; paper 759 submitted to this conference.
4. R. Barate *et al*, The ALEPH Collaboration; *Phys. Lett.B* **464** 339 (1999).

BOSE-EINSTEIN CORRELATIONS IN W-PAIR EVENTS AT LEP

A. VALASSI

CERN, EP Division, 1211 Genève 23, Switzerland
E-mail: Andrea.Valassi@cern.ch

An overview of measurements of Bose-Einstein correlations in W-pair events at LEP is given. The results presented are based on data collected at centre-of-mass energies between 172 and 202 GeV. The review concentrates on the search for Bose-Einstein correlations between pions from different W's in $e^+e^- \rightarrow W^+W^- \rightarrow q\bar{q}q\bar{q}$ events. No agreement is reached in the results of the four experiments.

1 Introduction

The existence of Bose-Einstein correlations (BEC) in reactions leading to hadronic final states is well known. BEC are observed as an enhancement of the production of multiple identical bosons close in momentum space, first reported for pairs of charged pions in p$\bar{\text{p}}$ collisions[1]. They have also been studied for two or more identical bosons in hadronic Z decays at LEP[2], including $\pi^\pm$, K^0_s and $K^\pm$.

This talk reviews recent studies of BEC[3,4,5,6] for charged pion pairs produced in $e^+e^- \rightarrow W^+W^-$ events at LEP. "Intra-W" BEC (or BEI, BEC Inside a W) between pions from the same $W \rightarrow q\bar{q}$ decay are observed unambiguously. As expected, these closely resemble BEC in $Z \rightarrow q\bar{q}$ decays, when only udsc quark flavours are considered.

Since the average separation between the two W decays at LEP2 is $\lesssim 0.1$ fm, smaller than the hadronisation scale of $\sim$ 1 fm, "inter-W" BEC (or BEB, BEC Between W's) could also be expected between pions from different W's in WW $\rightarrow q\bar{q}q\bar{q}$ (4q) events. The theoretical framework is, however, still unclear. In the absence of exact nonperturbative QCD calculations for the symmetrized production amplitude of multiple mesons from hadronic W decays, BEC are described only by phenomenological models. Many models exist, with contradictory predictions[7] about BEB. This note will therefore concentrate on the experimental measurements of BEB in W-pair events at LEP.

The question whether BEB exist is particularly important as the cross-talk between the two W decays could bias the W mass (m_W) measurement in 4q events. Different models of BEC predict different values for the shift[7]. Using MC events with full detector simulation, the LEP experiments obtain[8] m_W biases from 20 to 67 MeV, conservatively taken as systematic errors on the individual measurements. In the LEP combination, a common BEC systematic error of 25 MeV in the 4q channel is assumed[9]. The impact of this error is more than halved when combining 4q events to WW $\rightarrow \ell\nu q\bar{q}$ (2q) events. While presently the BEC error is not the limiting factor to the LEP measurement of m_W, a better understanding of BEC in W decays is important in view of the future reduction of the other, currently larger, systematics.

2 Overview of analysis methods

BEC in identical boson pairs are often studied by the two-particle correlation function

$$R(Q) = \frac{\rho(Q)}{\rho_0(Q)} \quad . \qquad (1)$$

Here, ρ is the two-boson density in the presence of BEC, while the reference ρ_0 should ideally describe pair-production if there were no BEC. Since BEC are largest at small four-momentum difference $Q = \sqrt{-(p_1 - p_2)^2}$, in a simplified approach the effect can be studied in terms of this variable alone[10].

In the studies reviewed in this note, Bose-Einstein (BE) enhancements of production rates are parametrised using functions like

$$R(Q) \sim (1 + \lambda e^{-r^2 Q^2}) \quad , \qquad (2)$$

which describe source distributions of radius r and BE "strength" λ. Fit results for r and λ will not be compared here, since the various analyses differ in detector acceptance, choice of fit function and definition of R.

The choice of the reference ρ_0 and the definition of R are distinctive features of each BEC analysis. The simplest is to derive ρ_0 from MC events with no BEC ("standard MC"). Data can also be used: for instance, the density for unlike-sign pion pairs, ρ^{+-}, is often taken as a reference to study BEC in like-sign pairs, $\rho^{++,--}$. Since unlike-sign pion pairs are not free from correlations other than BEC, such as those due to resonance decays, the correlation function is frequently computed as a double ratio of the form

$$R^{\mathrm{d.r.}}(Q) = \frac{(\rho^{++,--}/\rho^{+-})^{\,\mathrm{data}}}{(\rho^{++,--}/\rho^{+-})^{\,\mathrm{standard\ MC}}}, \quad (3)$$

where resonance correlations in the MC cancel those in data. Additional corrections may be introduced to describe final-state Coulomb interactions[11], different in like- and unlike-sign pairs and usually absent in the MC.

In the study of BEB, instead of comparing 4q data to a reference with no BEC at all, one can finally define[12] a function like

$$R^{\mathrm{mix}}(Q) = \frac{(\rho^{++,--})^{\,4q}}{(\rho^{++,--})^{\,\text{"mixed" 2q}}}, \quad (4)$$

where a reference with BEI and no BEB is obtained by "mixing" hadronically decaying W's from pairs of different 2q data events. Different mixing procedures exist; in general, the two W's, chosen of opposite charges, are boosted to be approximately back-to-back in the lab frame, as in 4q events. A double-ratio version of Eq. (4) can also be used.

3 Experimental results at LEP2

The analyses presented use data collected by the four experiments at 172–202 GeV (Tab. 1). Signal efficiency and background contamination are 70–90% and 20% (mainly $q\bar{q}$) for 4q, 50–75% and 6% for 2q events. Up to ~7000 selected WW events per experiment are used, including both channels.

Exper.	$\sqrt{s}$ [GeV]	$\mathcal{L}$ [pb^{-1}]	Status
ALEPH[3]	172–202	479	Prel.
DELPHI[4]	183–202	437	Prel.
L3[5]	189	177	Publ.
OPAL[6]	172–189	250	Prel.

Table 1. Data samples used in the analyses presented. The L3 analysis is submitted for publication, those from the other experiments are preliminary.

All analyses are performed on pairs of like-sign charged pions, with generally loose pion selection criteria. In each event, every pion enters several pairs, which introduces bin-to-bin correlations in some distributions. Where necessary, these are corrected for by all experiments, using various techniques.

The four experiments simulate BEC using different MC models, differently tuned. WW MC events with BEC are needed for model-dependent analyses or cross-checks of model-independent assumptions. BEC also exist in the $q\bar{q}$ background, although they may differ in the four experiments depending on the anti-b tagging criteria used in the selection. The simulation of BEC in $q\bar{q}$ events selected as 4q is especially important, as[3] they look more similar to BEC for 4q events in BEI+BEB MC than in BEI-only MC.

In ALEPH[3], a double ratio R^* of like- and unlike-sign pion pairs for data over standard MC is used, similar to that of Eq. (3). The analysis of BEB is based on the comparison of R^* for data to that expected from MC with BEI-only or BEI+BEB, simulated according to a specific model of BEC tuned on Z data. Background is added to the WW signal in computing R^* for the MC. Fits to the $R^*(Q)$ distributions are performed using a function similar to Eq. (2). The results are compared in terms of the integral of the BE signal, $I = \int_0^\infty \lambda e^{-r^2 Q^2} dQ = \frac{\sqrt{\pi}}{2}\frac{\lambda}{r}$. The value for I in the BEI-only MC is compatible with that in data, while I in data and in the MC with BEI+BEB are inconsistent at the level of 2.2σ. This includes systematics, dominated by the tuning of BEC MC parameters using Z data. It is concluded that data support the BEI model considered, while the BEB model is disfavoured. An alternative analysis, using a double ratio R^m of 4q and mixed 2q events in data over standard MC, yields qualitatively similar results (Fig. 1).

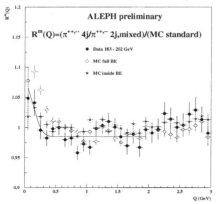

Figure 1. ALEPH[3] distributions of R^m for data, BEI-only MC ("inside") and BEI+BEB MC ("full").

In the DELPHI analysis[4], also presented in another talk[13], correlation functions for like-sign pion pairs R_{4q} and R_{4q}^{mix} are built for both 4q events and mixed 2q events, as single ratios over standard MC (Fig. 2a). Unlike-sign pairs are also used for qualitative comparisons (Fig. 2b). Expected backgrounds are subtracted from data in computing R_{4q} and R_{4q}^{mix}. By construction, $R_{4q}(Q)$ and $R_{4q}^{mix}(Q)$ are thus expected to be equal (different) in the absence (presence) of BEB. Signal MC events with BEC (BEI-only, or BEI+BEB) are used to verify these hypotheses and for systematic studies. A simultaneous fit to $R_{4q}(Q)$ and $R_{4q}^{mix}(Q)$ determines a common BE source radius and two different BE strengths, which are expected to be equal in the absence of BEB. The fit to their difference $\Delta\lambda^{mix} = \lambda_{4q} - \lambda_{4q}^{mix}$ yields

$$\Delta\lambda^{mix} = 0.062 \pm 0.025 \pm 0.021 \qquad (5)$$

It is concluded that data support BEB at the level of $\sim 2\sigma$. The largest systematic errors come from the check that $\Delta\lambda^{mix} = 0$ in BEI-only MC, and from bin-to-bin correlations.

In L3[5], the correlation functions for like-sign charged pions are built taking mixed 2q

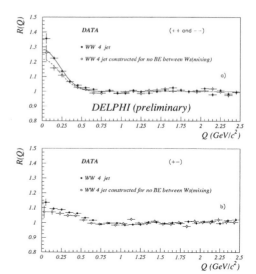

Figure 2. DELPHI[4] distributions of R_{4q} (black circles) and R_{4q}^{mix} (hollow circles) for a) like-sign and b) unlike-sign pion pairs, as a function of Q.

events as a reference. A single ratio D is built as in Eq. (4). The main analysis uses a double ratio D', where D for data is divided by that for BEI-only MC events. Backgrounds are subtracted from data in the calculation of D'. By construction, it is thus expected that $D'(Q) = 1$ in the absence of BEB. Signal MC with BEI-only is used to verify this hypothesis (Fig. 3a). Unlike-sign pairs are also used for qualitative comparison (Fig. 3b). The $D'(Q)$ distribution is fitted for the BEB strength Λ, expecting $\Lambda = 0$ in the absence of BEB. The fit yields

$$\Lambda = 0.001 \pm 0.026 \pm 0.015 \quad , \qquad (6)$$

indicating that data are compatible with the absence of BEB. Systematics are dominated by the selection of charged tracks and the inclusion of the low-purity $\tau\nu q\bar{q}$ channel. The fit is repeated for MC events with a model of BEB, yielding $\Lambda = 0.127 \pm 0.007$, where only statistical errors are given. The BEB model considered is disfavoured by more than 4σ.

In OPAL[6], WW $\rightarrow \ell\nu q\bar{q}$, WW $\rightarrow q\bar{q}q\bar{q}$ and non-radiative high-energy $(Z/\gamma)^* \rightarrow q\bar{q}$ decays are analysed. Correlation functions for these samples, built from double ratios as

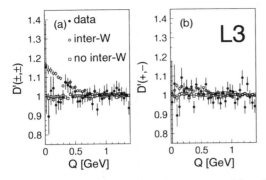

Figure 3. L3[5] distributions of D' for like-sign (a) and unlike-sign (b) pion pairs. Data are compared to BEI-only ("intra-W") and BEI+BEB ("inter-W") MC.

in Eq. (3) and adjusted for resonance multiplicities observed in data, are unfolded as sums of contributions from three categories: pion pairs coming from the same W, from different W's, or from hadronic Z decays. BEC parameters for each category are determined in a simultaneous fit, correcting for the differences in BEC between $q\bar{q}$ events selected as $(Z/\gamma)^*$ and those selected as WW. The BEC strength for the second category, λ^{diff}, directly measures BEB and is expected to be zero in their absence. MC events with BEC are only used for qualitative comparison to the data (Fig. 4). Fits are performed with different assumptions about BEC source sizes in the three categories. The main fit yields

$$\lambda^{\text{diff}} = 0.05 \pm 0.67 \pm 0.35 \quad , \qquad (7)$$

where systematics mainly come from tracking for low-Q like-sign pairs and the HERWIG simulation of resonances. While the result is compatible with the absence of BEB, OPAL conclude that, at the current precision, they are unable to determine whether BEB exist.

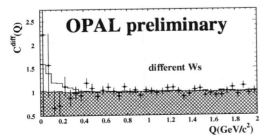

Figure 4. Unfolded correlation for pions from different W's in OPAL[6]. Data (crosses) are compared to MC with BEI-only (hatched), or BEI+BEB (open).

4 Conclusions

Bose-Einstein correlations between like-sign charged pion pairs produced in W-pair events at LEP are studied by the four experiments using different techniques. The existence of BEC for pions coming from the same W is firmly established, while no agreement is reached in the results of the four experiments about the possible existence of BEC between pions coming from different W's. The existence of inter-W BEC is supported by DELPHI at the 2σ level, whereas models of inter-W BEC are disfavoured by L3 and ALEPH at the 4σ and 2.2σ level, respectively. No conclusion is reached by OPAL at the present level of statistical accuracy.

References

1. G.Goldhaber et al., *Phys. Rev.* **120** (1960) 300.

2. ALEPH, *Phys. Rep.* **294** (1998) 1; DELPHI, *Phys. Lett.* B **471** (2000) 460; L3, *Phys. Lett.* B **458** (1999) 517; OPAL, CERN-EP/2000-004; and ref. therein.

3. ALEPH 2000-039 CONF 2000-059, and ref.[1,2] therein.

4. DELPHI 2000-115 CONF 414, and ref.[16] therein.

5. L3, CERN-EP/2000-084.

6. OPAL PN393 (1999), and ref.[9] therein.

7. L.Lönnblad, T.Sjöstrand, *Eur. Phys. J.* C **2** (1998) 165, and ref. therein.

8. ALEPH 2000-018 CONF 2000-015; DELPHI 2000-149 CONF 446; L3 Note 2575; OPAL PN422; and ref. therein.

9. The LEP WW Working Group, Note LEPWWG/MW/00-01 in preparation. *http://lepewwg.web.cern.ch/LEPEWWG*.

10. M.G.Bowler, *Z. Phys.* C **29** (1985) 617, and ref. therein.

11. M.Gyulassy et al., *Phys. Rev.* C **20** (1979) 2267.

12. S.V.Chekanov, E.A. De Wolf and W. Kittel, *Eur. Phys. J.* C **6** (1999) 403.

13. N.Pukhaeva, these proceedings.

COLOUR RECONNECTION IN W DECAYS

PAUL DE JONG

NIKHEF, P.O.Box 41882, 1009 DB Amsterdam, the Netherlands
E-mail: Paul.de.Jong@cern.ch

The studies of colour reconnection in $e^+e^- \rightarrow W^+W^- \rightarrow q\bar{q}'q\bar{q}'$ events at LEP are reviewed. It is shown that the analysis of the particle- and energy flow between jets is sensitive to realistic model predictions. The effects on the W mass measurement are discussed. Most results are preliminary.

1 Colour Reconnection

With some 450 pb^{-1} per experiment already recorded at $\sqrt{s} = 183 - 202$ GeV, and more to come at higher energies, each of the four LEP experiments have selected up to now some 3200 WW $\rightarrow qqqq$ and some 2500 WW $\rightarrow qq\ell\nu$ candidate events. The mass and width of the W boson are measured from the kinematics of W decay products. Any energy-momentum exchange between W decay products not well simulated in Monte Carlo will affect the W mass and width measurement. Conventional MC's treat the two $q\bar{q}'$ systems in a WW $\rightarrow q\bar{q}'q\bar{q}'$ event as independent. However, QCD interconnections, or colour reconnection (CR) can be expected [1]. Perturbative CR effects are estimated to be small [2]; CR is a non-perturbative hadronization uncertainty that can only be studied in the context of various models.

CR models being used in these studies are those implemented in PYTHIA, ARIADNE and HERWIG. The models in PYTHIA, SK I, SK II and SK II', are based on reconfiguration of overlapping or crossing strings [2]. In the SK I model, the probability of reconnection is calculated as $P_{\text{reco}} = 1 - e^{(-k_i O)}$, where O is the overlap of two finite strings, and k_i is a free parameter. In the SK II and II' models, the string has no lateral dimension, and strings are reconnected when they cross. The ARIADNE models are based on rearrangement of colour dipoles if this reduces the string length [3]. It should be noted that these models also affect LEP 1 data and

are disfavoured from an OPAL study of the properties of quark- and gluon jets [4].

A reconfiguration of the colour flow is expected to change the (charged) particle multiplicity (typically decreasing it by 0.2 to 0.9 units), especially at low momentum, and more specifically between jets associated to the same W.

2 Multiplicities

2.1 Inclusive Charged Multiplicity

The charged multiplicity in WW events is measured by all four LEP experiments from charged tracks in the tracking system [5,6,7,8]. The track multiplicity distribution is corrected to a charged particle multiplicity distribution by a matrix unfolding procedure. Alternatively, the multiplicity as a function of momentum (fragmentation function) or p_T is determined and corrected bin-by-bin. The results are shown in Table 1.

The difference $\Delta < N_{ch} >=< N_{ch}^{4q} > -2 < N_{ch}^{2q} >$ is also given in Table 1. DELPHI prefers to quote the ratio $R =< N_{ch}^{4q} > /2 < N_{ch}^{2q} >= 0.990 \pm 0.015 \pm 0.011$. Combining the results, it can be concluded that $\Delta < N_{ch} >$ is consistent with 0 within an error of 0.3-0.4. A proper average is difficult due to differences in the definition, and the correlated systematics; the size of these correlated systematics (0.2-0.3), which is of the same size as the CR effects, limits the sensitivity of this method.

Table 1. Average charged multiplicity in $qqqq$ events, $< N_{ch}^{4q} >$, in $qq\ell\nu$ events, $< N_{ch}^{2q} >$, and the difference $\Delta < N_{ch} >=< N_{ch}^{4q} > -2 < N_{ch}^{2q} >$, as measured by the four LEP experiments. The ALEPH results are quoted within detector acceptance and not corrected to full phase space; DELPHI prefers to quote the ratio $R =< N_{ch}^{4q} > /2 < N_{ch}^{2q} >$, see text.

	$< N_{ch}^{4q} >$	$< N_{ch}^{2q} >$	$\Delta < N_{ch} >$
OPAL 183 GeV	$39.4 \pm 0.5 \pm 0.6$	$19.3 \pm 0.3 \pm 0.3$	$+0.7 \pm 0.8 \pm 0.6$
OPAL 189 GeV	$38.31 \pm 0.24 \pm 0.37$	$19.23 \pm 0.19 \pm 0.19$	$-0.15 \pm 0.44 \pm 0.38$
L3 183-202 GeV	$37.90 \pm 0.14 \pm 0.41$	$19.09 \pm 0.11 \pm 0.21$	$-0.29 \pm 0.26 \pm 0.30$
DELPHI 183 GeV	$38.11 \pm 0.57 \pm 0.44$	$19.78 \pm 0.49 \pm 0.43$	
DELPHI 189 GeV	$39.12 \pm 0.33 \pm 0.36$	$19.49 \pm 0.31 \pm 0.27$	
ALEPH 183-202 GeV	$35.75 \pm 0.13 \pm 0.52$	$17.41 \pm 0.12 \pm 0.15$	$+0.93 \pm 0.27 \pm 0.29$

2.2 Fragmentation Function

By studying the particle multiplicity as a function of $x_p = 2p/\sqrt{s}$, or $\xi = -\log(x_p)$, one can study the low momentum region $p < 1$ GeV where CR effects predominantly reside, but at the cost of reduced statistics. No significant effects at low x_p are observed by any of the experiments.

2.3 Heavy Hadrons

Massive particles, like $K^\pm$ or (anti)protons, are more sensitive to CR effects than pions, by a factor 2 to 3. However, this is counterbalanced by the decreased statistics. DELPHI[6] and OPAL[9] have studied the production of heavy hadrons in $qqqq$ and $qq\ell\nu$ events and observe no significant differences.

3 Particle Flow

A more promising technique to study CR appears to be the study of the particle- or energy flow between jets from the same W and between different W's, in analogy to studies of the string effect[10].

The construction of the particle flow is explained in Figure 1. A jet-finder is used to construct four jets in $qqqq$ events. Each pair of jets defines a plane onto which all reconstructed particles in the event are projected; for the energy flow weighted with the particle energy. In the preliminary studies submitted

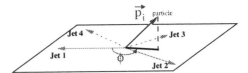

Figure 1. Construction of the particle flow.

to this conference, L3[7] and ALEPH[11] use strong cuts on the angles between jets to select topologies with well separated jets and planar-like events, and obtain a selection efficiency of $\sim 15\%$; OPAL[12] uses less restrictive cuts and a jet-pairing likelihood that gives a higher efficiency of $\sim 42\%$, but selects also topologies with less clear separation between CR models. The flow is symmetrized with respect to the choice of jet-pairs, and particle angles between jets are rescaled between 0 and 1.

CR models indeed show a depletion of the particle flow between jets from the same W, and an increase between jets from different W's, as naively expected. It is convenient to average the flow in the two regions between jets from the same W (regions j1-j2 and j3-j4), and to do the same for the flow in the two regions between jets from different W's (regions j2-j3 and j4-j1). Subsequently, the ratio of these within-W/between-W flows is taken as a function of the rescaled particle angle.

L3 has studied 176 pb^{-1} of data taken

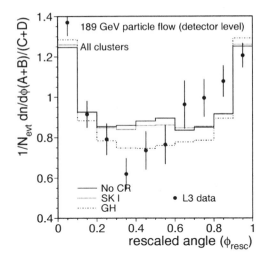

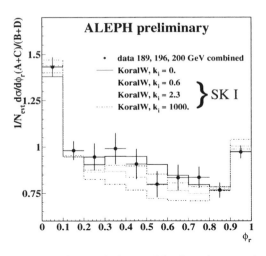

Figure 2. Ratio of the particle flows between jets from the same W and between jets from different W's, as a function of the rescaled angle, for L3 data at $\sqrt{s} = 189$ GeV and Monte Carlo.

Figure 3. Ratio of the particle flows between jets from the same W and between jets from different W's, as a function of the rescaled angle, for ALEPH data at $\sqrt{s} = 189 - 200$ GeV and Monte Carlo.

at $\sqrt{s} = 189$ GeV. The ratio of the particle flow between jets from the same W and between jets from different W's is shown in Figure 2, for data, PYTHIA without CR, and PYTHIA with SK I and GH, as a function of the rescaled angle ϕ_{resc}. With this data sample only, a sensitivity of 3.5 σ to SK I ($k_i = 1000$) and 1.0 σ to SK I ($k_i = 0.6$) is reached. Varying the fraction of reconnected events, and calculating the χ^2 for the data-MC comparison, a fraction of $\sim$40% of reconnected events in the SK I model is favoured, and the No-CR scenario is disfavoured at 1.7 σ.

ALEPH has analyzed 347 pb^{-1} of data taken at $\sqrt{s} = 189 - 200$ GeV. Their particle flow ratio is shown in Figure 3, for data, KORALW without CR, and KORALW with the SK I model for various values of k_i. Varying k_i, ALEPH finds the best data-MC agreement for $k_i \approx 0.25$, and puts a 1 σ upper limit on k_i of 1.4 which corresponds to 45% of reconnected events at $\sqrt{s} = 189$ GeV.

OPAL has studied 183 pb^{-1} of data taken at $\sqrt{s} = 189$ GeV, and find sensitivities of 4.0 σ for SK I ($k_i = 100$), 1.1 σ for

SK I ($k_i = 0.9$), 0.4 σ for SK II and II', and 0.5-1.8 σ for AR2 and AR3. As a cross-check, OPAL uses the strong cuts like L3 and ALEPH, and observes slightly smaller sensitivities. The actual data is ambiguous, and prefers some reconnection in the default analysis, but no CR in the cross-check analysis. This, and in particular the role of the background, will be further studied.

DELPHI is also working on a similar analysis, but was not yet able to submit results to this conference.

With the full LEP 2 data sample, and combining all experiments, a further gain in sensitivity by a factor ~ 3.5 can be expected.

4 Effect on M_{W}^{qqqq}

The estimates for $\Delta M_{\text{W}}^{qqqq}$ from the individual experiments calculated with their own Monte Carlo samples are summarized in Table 2 [13]. A difference in reconnection probability in the SK II model can be expected from differences in the parton shower cutoff scale Q_0 [2], which ranges between 1.0 GeV (L3) and 1.9 GeV (OPAL).

Common samples of KORALW + JET-

Table 2. Experimental estimates of ΔM_W^{qqqq}, in MeV, from the four LEP experiments, as calculated with their own implementations of various CR models at $\sqrt{s} = 189$ GeV. The fraction of reconnected events in each sample is given between brackets.

	OPAL	L3	DELPHI	ALEPH
SK I	$+66 \pm 8$ (35.1%)	$+29 \pm 15$ (32.1%)	$+46 \pm 2$ (35.9%)	$+30 \pm 10$ (29.2%)
	($k_i = 0.9$)	($k_i = 0.6$)	($k_i = 0.65$)	($k_i = 0.65$)
SK II	$+3 \pm 8$ (19.8%)	-5 ± 15 (32.4%)	-2 ± 5	$+6 \pm 8$ (29.2%)
SK II'	$+10 \pm 8$ (17.6%)	-33 ± 15 (28.8%)		$+4 \pm 8$ (26.7%)
AR 2	$+85 \pm 8$ (50.3%)	$+106 \pm 26$	$+28 \pm 6$	$+21 \pm 19$
AR 3	$+140 \pm 10$ (62.3%)	$+170 \pm 26$	$+55 \pm 6$	$+34 \pm 34$
HERWIG				$+20 \pm 10$

SET Monte Carlo events, with (SK I) and without colour reconnection, have been generated for the four experiments; each experiment then passed these samples though detector simulation, event selection and analysis procedures. The resulting mass shifts found by the experiments were equal within errors, and a correlation between experiments of close to 100% was found. In view of this, further LEP collaboration will be needed to fully understand the differences in Table 2, especially in the ARIADNE estimates. For the LEP M_W combination, a CR systematic error of 50 MeV was used, fully correlated between experiments.

Estimates of the CR effect on the W width in the $qqqq$ channel range between $+40$ and $+70$ MeV in the SK models [13].

The studies of the particle flow between jets have proven to be sensitive to realistic CR model predictions. These studies will thus directly measure the amount of CR in the data. In models with a free parameter, such as SK I, this parameter can be measured from data; a calibration curve of mass shift versus k_i can be used to estimate the CR systematic error. Already ALEPH, with a 1σ upper limit on k_i of 1.4, puts a 1σ upper limit on ΔM_W^{qqqq} in the SK I model of 40 MeV [11]. With the full LEP 2 data sample and combining all experiments, the CR systematic error on M_W is likely to be below that, and, most important, actually measured from data.

References

1. G. Gustafson, U. Pettersson and P. Zerwas, *Phys. Lett.* B **209**, 90 (1988).
2. T. Sjöstrand and V. Khoze, *Z. Phys.* C **62**, 281 (1994); *Eur. Phys. J.* C **6**, 271 (1999).
3. G. Gustafson and J. Häkkinen, *Z. Phys.* C **64**, 659 (1994); L. Lönnblad, *Z. Phys.* C **70**, 107 (1996).
4. OPAL Collaboration, G. Abbiendi *et al.*, *Eur. Phys. J.* C **11**, 217 (1999).
5. ALEPH Collaboration, ALEPH Note 2000-058 CONF 2000-038.
6. DELPHI Collaboration, P. Abreu *et al.*, CERN-EP/2000-023.
7. L3 Collaboration, L3 Note 2560 (2000).
8. OPAL Collaboration, G. Abbiendi *et al.*, *Phys. Lett.* B **453**, 153 (1999); OPAL Collaboration, OPAL Physics Note 417.
9. OPAL Collaboration, OPAL Physics Note 412 (1999).
10. D. Duchesneau, preprint LAPP-EXP 2000-02 (2000).
11. ALEPH Collaboration, paper #242 submitted to this conference.
12. OPAL Collaboration, OPAL Physics Note 448 (2000).
13. ALEPH Collaboration, R. Barate *et al.*, CERN-EP/2000-045; OPAL Collaboration, G. Abbiendi *et al.*, CERN-EP/2000-099; N. Kjaer, M. Mulders, A. Straessner, private communication.

W MASS MEASUREMENTS USING
FULLY HADRONIC EVENTS AT LEP2

SASCHA SCHMIDT-KÄRST

III. Physikalisches Institut, RWTH Aachen
Physikzentrum, 52056 Aachen, Germany
E-mail: S.Schmidt-Kaerst@cern.ch

Since 1996 each LEP experiment recorded data corresponding to an integrated luminosity around $460 - 500 \, \text{pb}^{-1}$ at centre-of-mass energies between 161 GeV and 202 GeV. The experimental methods to measure the mass of the W boson in the fully hadronic channel are discussed and the preliminary combined value of the four LEP experiments is presented: $M_W(qqqq) = 80.432 \pm 0.073$ GeV. The LEP combined result for the mass of the W boson, obtained from all channels and all data recorded at centre-of-mass energies above the W-pair threshold, is $M_W = 80.427 \pm 0.046$ GeV. The width of the W boson is measured to be $\Gamma_W = 2.12 \pm 0.11$ GeV.

1 Introduction

The study of the properties of the W boson is one of the main physics goals of the LEP2 program. The mass of the W boson is indirectly measured exploiting its relation with the Z boson mass and couplings within the Standard Model, precisely measured at LEP1 and SLC. The direct measurement of the W boson mass by the LEP and TEVATRON[1] experiments and the comparison with its indirect determination from fits to electroweak data is therefore an important consistency check of the Standard Model.

The mass of the W boson was first directly measured at LEP at the energy threshold for W-pair production[2], exploiting the sensitivity of the W^+W^- cross-section on M_W at this centre-of-mass energy of 161 GeV. At higher energies from 172 GeV up to 202 GeV, M_W is measured by direct reconstruction of the W decay products. In 46% of the time, both W bosons decay hadronically. The experimental methods of the direct reconstruction of M_W in the fully hadronic channel, $W^+W^- \to qqqq$, are discussed in this paper. The analysis exploiting the semileptonic and fully leptonic channel are presented elsewhere[4]. The combination of the preliminary results for the fully hadronic channel is presented, including data collected in 1999 at centre-of-mass energies between 192 GeV and 202 GeV [3]. Combining all data and mass information from all decay channels, preliminary results for the mass, M_W, and width, Γ_W, of the W boson are reported.

2 Experimental methods

2.1 Event Selection

Fully hadronic $W^+W^- \to qqqq$ decays are selected as high multiplicity hadronic events with no missing energy. In order to separate the signal from the dominating QCD background, $e^+e^- \to Z/\gamma \to 4 \, \text{jets}$, the LEP experiments apply multivariate analyses. The signal efficiencies ranges between 83% and 89%, and comparing with the remaining background, the signal is selected with purities up to 85%. Details of the selection are described elsewhere in these proceedings[5].

2.2 Invariant masses and kinematic fits

Using jet cluster algorithms, the selected events are forced into four jets. The raw invariant masses of the W bosons, calculated from the observed decay products, are improved imposing kinematic constraints as energy and momentum conservation, $\sum(E, \vec{p}) = (\sqrt{s}, 0)$ (4C), and the requirement of equal masses of pairs of jets (1C). Since the jet angles are experimentally better determined than their energies, the kinematic fit mainly improves the en-

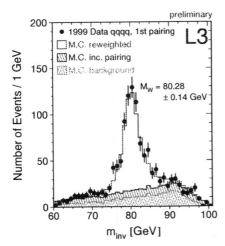

Figure 1. Reconstructed invariant mass distribution of the first pairing of selected fully hadronic W^+W^- events after applying a 5C kinematic fit.

ergy resolution. The L3 and OPAL collaborations apply a 5C fit yielding an average mass $\langle M \rangle$ for each event, whereas the ALEPH and DELPHI collaborations impose only the four energy and momentum constraints, taking into account afterwards both masses M_1 and M_2. The ALEPH collaboration rescales the masses with respect to the ratio of the LEP beam energy to the sum of measured jet energies: $M_i' = M_i \cdot E_{beam}/(E_1 + E_2)$.

2.3 Jet-Pairing

Forcing the hadronic event into four jets, three combinations of associating the jets to the two W bosons have to be taken into account. Only the correct pairing contains information on the mass of the W boson, whereas the other two incorrect pairings result in a combinatorial background. The L3 and OPAL collaborations arrange the jet pairings in descending order with respect to the 5C fit probabilities, $P_1 > P_2 > P_3$. Since the second highest probability fit still contains significant mass information, the first and second pairing satisfying cuts on P_1 and P_2 are used to determine the W boson mass. Figure 1 shows the mass information contained in the first pairing of $W^+W^- \rightarrow qqqq$ events obtained by L3 after applying a 5C kinematic fit. The ALEPH collaboration ap-

plies a dedicated algorithm to decide on the best pairing: Ordering the jet pairings with respect to the corresponding matrix element $|\mathcal{M}|^2$ for W^+W^- production, the first or second pairing satisfying cuts as regards a window around the W boson mass and the sum of di-jet angles is chosen. The OPAL and DELPHI collaborations also include 5-jet configurations if they are more likely than the 4-jet hypothesis. OPAL selects the preferred permutation among ten possible pairing combinations, whereas the DELPHI analysis to determine M_W takes into account all possible jet pairings. In general, the correct combination is assigned in typically up to 90% of cases by the different approaches.

2.4 Methods to determine M_W

Various methods are applied and cross-checked by the LEP experiments to determine M_W. The ALEPH, L3 and OPAL collaborations apply likelihood fits comparing the measured and expected mass spectra. The Monte Carlo events are reweighted calculating the ratio of the matrix elements for various values of M_W with respect to the generated mass: $\omega_i = |\mathcal{M}_i(M_W^{fit})|^2/|\mathcal{M}_i(M_W^{gen})|^2$. Biases in the reconstructed mass spectra, e.g. due to detector resolution and event selection, are included in the Monte Carlo distributions. ALEPH determines the mass from a 2-dimensional fit to the two rescaled masses from the 4C kinematic fit, whereas L3 and OPAL perform a 1-dimensional fit of the average mass distribution from the 5C fit. Additionally, OPAL fits an asymmetric Breit-Wigner-Function to the average mass spectrum to determine M_W. This method needs to be calibrated using several Monte Carlo samples and ,unlike in the reweighting method, the fitted mass has to be corrected for biases. DELPHI uses a convolution method to determine the mass, taking into account all possible pairings of the 4-jet and 5-jet configurations. The fitted masses M_1 and M_2 and their uncertainties are interpreted as 2-dimensional probability density functions. They are convoluted

700

with a 2-dimensional Breit-Wigner-Function $BW(M_1, M_2, M_W)$, yielding an event likelihood $\mathcal{L}(M_W)$. Experimental biases and effects due to initial state radiation have to be corrected afterwards. A comparable approach is also applied by the OPAL collaboration.

3 Results

The preliminary results for the W boson mass using the fully hadronic channel (Fig. 2) include all data used for direct reconstruction ($\sqrt{s} = 172$ GeV $- 202$ GeV). While the statistical uncertainties of the individual experiments are comparable, the systematic uncertainties show large differences, reflected by the differences of the total errors given in Figure 2. For final state interactions (FSI) such as Bose-Einstein Correlations[6] and Colour Reconnection[7], which affect only the fully hadronic W^+W^- decay channel, the experiments quote different estimates for their effect on the measured value M_W. Common tests using Monte Carlo events with and without Colour Reconnection yield that all experiments show equal sensitivity to this FSI. Assuming also an equal sensitivity to Bose-Einstein Correlations, a common uncertainty for both FSI is used in the combination of the four LEP results. A typical breakdown of the contributions to the systematic uncertainty of M_W using only fully hadronic events is given in table 1. The LEP combination is a simultaneous fit to all individual results from the experiments for each channel and year, accounting for a correct treatment of correlated errors. As an experimental test of a possible mass shift due to FSI, the difference between the values for M_W from the hadronic and semileptonic decay channels, $\Delta M_W = M_W(qqqq) - M_W(qq\ell\nu)$, is determined. Setting the systematic uncertainty due to FSI to zero, the LEP combined value is found to be $\Delta M_W = 5 \pm 51$ MeV. Thus no experimental indication of a mass shift due to Colour Reconnection and Bose-Einstein Correlations is found. Some of the systematic uncertainties like the LEP beam energy uncertainty, the

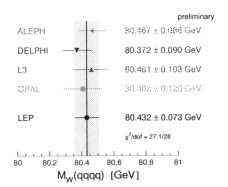

Figure 2. Preliminary results for M_W from direct mass reconstruction using fully hadronic events.

QED radiation effects (ISR/FSR) and particularly the mass shift evaluated using different hadronization models are correlated between the fully hadronic and semileptonic final state. Consequently, they contribute significantly to the total systematic uncertainty of the combined mass value from all channels. Combining all measurements of M_W from the direct reconstruction using all W^+W^- decay channels with the threshold measurement at 161 GeV yields the preliminary LEP result $M_W = 80.427 \pm 0.046$ GeV. Figure 3 compares the results from the individual LEP experiments and their combined value, respecting correlations between the different channels and experiments. The large uncertainty due to FSI reduces the weight of the fully hadronic channel to 27% when combined with the semileptonic and fully leptonic channel. It should be noted that if there were no systematic uncertainties, the statistical precision of the measurement would be

Table 1. Typical breakdown of the contributions to the systematic uncertainty on M_W using only fully hadronic events.

Source	ΔM_W[MeV]
ISR/FSR	10
Hadronization	23
Detector systematics	7
LEP beam energy	17
Colour Reconnection	50
Bose-Einstein Correlations	25
Other	5
Total Systematic	64
Statistical	34

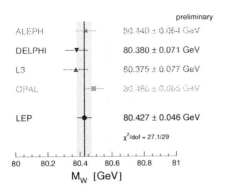

Figure 3. Preliminary results for M_W, combining all W^+W^- decay channels and the measurements using direct reconstruction as well as the threshold measurement at $\sqrt{s} = 161$ GeV.

25 MeV, to be compared to the actual value of 30 MeV. The systematic uncertainty of the combined M_W value is dominated by the uncertainty describing the hadronization, correlated between the fully hadronic and semileptonic final state. The method of direct reconstruction is also used to determine the width of the W boson, Γ_W. The LEP combination is based on a similar procedure as described for the mass value. Taking into account the correlated systematic uncertainties, a value of $\Gamma_W = 2.12 \pm 0.11$ GeV is obtained. This result is in good agreement with the direct determination of the W boson width $\Gamma_W = 2.05 \pm 0.13$ GeV performed by the CDF collaboration[8].

4 Conclusions

Preliminary measurements for the mass and width of the W boson, including data recorded by the LEP experiments in 1999, are reported. Using the direct reconstruction method in the fully hadronic channel, the mass is found to be $M_W(qqqq) = 80.432 \pm 0.073$ GeV. No experimental indication of a mass shift due to FSI is found comparing with the result for M_W from the semileptonic channel only. The LEP combined measurement of the mass and width of the W boson are $M_W = 80.427 \pm 0.046$ GeV and $\Gamma_W = 2.12 \pm 0.11$ GeV. An additional luminosity of $200 \, \text{pb}^{-1}$ per experiment is expected

for the year 2000 run. Thus the uncertainty on the W boson mass could be reduced to $\Delta M_W = 30 - 40$ MeV.

Acknowledgements

It is a pleasure to thank all collaborators of the LEP electroweak and W working group who provided much of the information presented here as well as Arno Straessner and Martin Grünewald for valuable discussions. This work has been supported by the German Bundesministerium für Bildung und Forschung.

References

1. CDF collaboration, hep-ex/0007044, submitted to Phys. Rev. D; DØ collaboration, hep-ex/9907028.

2. ALEPH Collaboration, Phys. Lett. **B 401** (1997) 347; DELPHI Collaboration, Phys. Lett. **B 397** (1997) 158; L3 Collaboration, Phys. Lett. **B 398** (1997) 223; OPAL Collaboration, Phys. Lett. **B 389** (1996) 416.

3. ALEPH Collaboration, ALEPH 2000-018 CONF 2000-015; ALEPH Collaboration, CERN-EP/2000-045, submitted to Eur. Phys. J.; DELPHI Collaboration, DELPHI 2000-147 CONF 446; DELPHI Collaboration, DELPHI 2000-144 CONF 446; L3 Collaboration, L3 Note 2575; OPAL Collaboration, OPAL PN422; OPAL Collaboration, CERN-EP/2000-099, submitted to Phys. Lett. B.

4. R. Ströhmer, *"W mass measurements using semileptonic and fully leptonic events at LEPII"*, these proceedings.

5. A. Ealet, *"WW cross-sections and W decay branching fractions"*, these proceedings.

6. A. Valassi, *"Bose-Einstein correlations in W decays"*, these proceedings.

7. P. de Jong, *"Colo(u)r reconnection in W decays"*, these proceedings.

8. CDF collaboration, hep-ex/0004017, submitted to Phys. Rev. Lett.

W MASS MEASUREMENTS USING SEMILEPTONIC AND FULLY LEPTONIC EVENTS AT LEPII

R. STRÖHMER

Lehrstuhl Schaile, Ludwig-Maximilians-Universität München, Am Coulombwall 1, D-85748 Garching, Germany

E-mail: Raimund.Stroehmer@Physik.Uni-Muenchen.DE

All 4 LEP experiments have preliminary results for the W-mass which include the 1999 data [1,2,3]. The preliminary combined result from semileptonic and fully leptonic events is $M_W = 80.427 \pm 0.051$ GeV. Combining this results with the measurement from fully hadronic events and the threshold measurement at $\sqrt{s} = 161$ GeV one obtains a preliminary result of $M_W = 80.427 \pm 0.046$ GeV.

1 Introduction

At the end of the 1999 data taking period each LEP experiment had collected more then 7000 W-pairs at center-of-mass energies from 161 to 202 GeV. (The integrated luminosity per experiment above the W-pair threshold was in the range of 460 - 500 pb^{-1}). 44 % of the W pairs decay semileptonically and 10 % decay fully leptonically. Both channels are free from the problems which arises from the fact that both W's decay into hadrons in the fully hadronic channel. Therefore there are no systematic uncertainties due to final state interactions between the decay products of the two W. There are also no problems to associate which jets and which particles originate from which W except for the hadronic τ decays when the τ decays close to one of the jets from the hadronic W decay.

2 M_W from semileptonic W decays

2.1 Event selection

The event selection is based on the event topology of two jets and one lepton. In the case of electrons and muons the lepton is identified by a loose lepton identification and isolation cuts. Taus are identified as isolated low multiplicity jets. The selections have both a high purity and a high efficiency. The cross contamination from e and μ in $q\bar{q}\tau\nu$ events does not pose a problem since in this channel the mass information is extracted from the hadronic system which is not influenced by the misidentification of an e or μ as a τ (see also figure 3). On the contrary, it is better to use $q\bar{q}e\nu$ and $q\bar{q}\mu\nu$ which fail the lepton identification as $q\bar{q}\tau\nu$ events than not to use them at all.

2.2 Kinematic fit

For each event the momentum of the partons can be determined from the measured jet and lepton momenta and their errors in a kinematic fit. The 4 constraints from energy and momentum conservation are always used in these fits. A possible 5th constraint is the assumption that both W bosons have the same mass which is equivalent to the assumption that the energies of the $q\bar{q}$ and of the $l\nu$ pair are equal to the beam energy (beam energy constraint). Due to the finite width of the W-mass this constraint does not exactly reflect the momentum of partons; however if one is interested in the average W-mass in an event it improves the resolution.

For the electron and muon channel there are 3 unknowns (the neutrino momentum). This allows both, a fit with two constraints using the beam constraint, from which on gets one mass M^{BC} and it's error σ^{BC}, and a fit with one remaining constraint, which gives both the mass and error of the hadronically decaying W ($M^{q\bar{q}}$, $\sigma^{q\bar{q}}$) and of the leptoni-

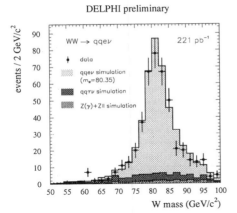

Figure 1. M^{BC} from $q\bar{q}e\nu$ events

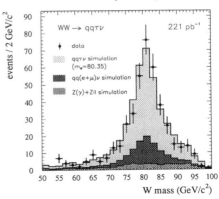

Figure 3. M^{BC} from $q\bar{q}\tau\nu$ events

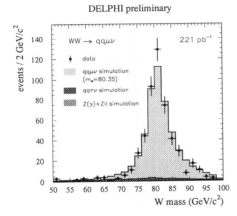

Figure 2. M^{BC} from $q\bar{q}\mu\nu$ events

cally decaying W ($M^{l\nu}$, $\sigma^{l\nu}$).

In the tau channel only the τ direction can be determined. Due to the neutrino(s), in the τ decays the energy of the τ cannot be determined. Because of this extra unknown one can only determine the mass and its error of the hadronic decaying W using the beam constraint (M^{BC}, σ^{BC}). Figures 1 to 3 show as an example M^{BC} the mass from the beam constraint fits for electrons, muons and taus from DELPHI. One can see that in the $q\bar{q}\tau\nu$ channel the contamination from $q\bar{q}e\nu$ and $q\bar{q}\mu\nu$ events has the same shape (mass information) as the signal.

2.3 Mass extraction

There are different approaches to extract the value of M_W from the event masses and errors. One approach is to compare the distribution in data with that predicted from Monte Carlo for different values of M_W and Γ_W. Since it is not feasible to produce Monte Carlo events for every considered value of M_W a reweighting technique is used to produce the Monte Carlo predictions for arbitrary values of M_W and Γ_W. Each event (with W boson masses M_W^1 and M_W^2) is weighted with the ratio of the probabilty that such a mass combination of M_W^1 and M_W^2 is produced for the desired values of M_W and Γ_W by the probability that M_W^1 and M_W^2 are produced for the values of M_W and Γ_W used to generate the original Monte Carlo sample. The W-mass is then determined by finding the value of M_W for which the Monte Carlo and Data distributions agree best. The value of Γ_W is fixed by the standard model prediction as a function of M_W. ALEPH used for electrons and muons a 3-dimensional distribution in M^{BC}, σ^{BC} and $M^{q\bar{q}}$. For taus they used a 2-dimensional distribution in M^{BC}, σ^{BC}. L3 and OPAL used 1-dimensional distributions in M^{BC}. In OPAL the analysis was done separately in 4 bins of σ^{BC}.

A 1-dimensional convolution fit was used

by DELPHI and OPAL to extract the W-mass. For each event, M^{BC} and σ^{BC} can be used to estimate the likelihood as a function of M_W, that such an event is produced, by the convolution of a physics function (e.g. a Breit Wigner function with M_W and Γ_W) and a resolution function (e.g. a gaussian with width σ^{BC}). The W-mass is then determined by maximizing the overall likelihood, the product of the event likelihoods. The main effect from initial state radiation can be included in the proper definition of the physics function. But potential biases due to details of the fragmentation, detector resolution, backgrounds and event selection can only be taken into account by a calibration curve which is obtained by performing the full analysis with Monte Carlo samples produced at different values of M_W.

2.4 Systematic uncertainties

As we have seen in Sec. 2.3 the mass extraction relies on the comparison of data with Monte Carlo prediction, either direct as in the case of the reweighting approach or indirect via the calibration curve in the convolution fit. Thus the primary systematic uncertainties are due to the uncertainties in the Monte Carlo simulation.

- Detector simulation: Each year also data at the Z^0 resonance are taken. This is used to measure the energy scale and the energy and angular resolution for jets and leptons.

- Hadronisation: The systematic error is determined from the comparison of different Monte Carlo generators (PYTHIA, HERWIG, ARIADNE) and different Monte Carlo tunes. This is cross-checked by analyzing Z^0 calibration data emulating a W in W^+W^- after a suitable Lorentz boost (DELPHI) and by propagating the difference between data and Monte Carlo in fragmentation related observables as weights through the mass analysis and

Table 1. Typical values for systematic errors

Source	Error
Detector Systematic	11 MeV
Hadronisation	26 MeV
LEP Beam Energy	17 MeV
ISR/FSR	8 MeV
Other	5 MeV
Total Systematic	35 MeV
Statistical	38 MeV
Total	51 MeV

thus estimating the effect of the difference in M_W (ALEPH).

- Beam Energy: Detailed studies of the uncertainty in the beam energy are made by the LEP energy working group.

- Initial and final state radiation: The difference between the ISR treatment in KORALW and EXCALIBUR or between different orders in KORALW are used to estimate the ISR uncertainties.
  The full $\mathcal{O}(\alpha)$ effects are not included yet. Calculations using double-pole approximations indicate possible mass shifts of about 10 MeV, but full simulations were not available at the time of the analysis.

Table 1 shows typical errors for the systematic errors discussed above. In the combination of the 4 LEP experiments the correlation between the different experiments and the different center-of-mass energies have been properly taken into account.

3 Fully leptonic W deacys

In 10 % of all W pairs both W's decay leptonically. Because of the two undetected neutrinos in these events, one has 6 unknowns and only 5 possible constraints. Thus one cannot reconstruct the event completely in a kinematic fit. Instead one can look at quantities which are sensitive to the W-mass. One such quantity is the lepton energy. It's spectrum is given by:

$$E_\ell = \frac{\sqrt{s}}{4} + \cos\theta_\ell^* \sqrt{\frac{s}{16} - \frac{M_W^2}{4}}$$

Where θ_ℓ^* is the angle between the W and its decay products in the W restframe and s is the center-of-mass energy. Another quantity can be calculated when one reconstructs the event under the assumption that the neutrinos are in the same plane as the charged leptons. If the assumption is nearly true, one reconstructs the right mass, otherwise one gets a larger mass. This procedure yields a distribution (called pseudo-mass) with an edge which depends on M_W. The following results were obtained:

ALEPH (E_ℓ, 189 GeV data):

$$M_W^{l\nu l\nu} = 81.81 \pm 0.67(\text{stat.}) \pm 0.20(\text{syst.}) \text{ GeV.}$$

OPAL preliminary
(E_ℓ, pseudo-mass, 183-202 GeV):

$$M_W^{l\nu l\nu} = 80.27 \, {}^{+0.51}_{-1.62}(\text{stat.}) \pm 0.14(\text{syst.}) \text{ GeV.}$$

The curvature of the Likelihood curve at $M_W = 80.5$ GeV corresponds to a parabolic error of ~ 0.5 GeV.

4 Results

Figure 4 shows the results of the four LEP experiments in the semileptonic channel. Figure 5 compares the direct and indirect measurements of M_W. The LEP2 result is the combination of all decay modes and of the threshold measurement at 161 GeV.

5 Conclusions

The direct measurements of M_W have reached the same precision as the indirect prediction from electroweak fits. The measurement starts to be dominated by systematic uncertainties. We still expect an improvement of the LEP2 M_W measurement with this year's (2000) data and with refined analysis.

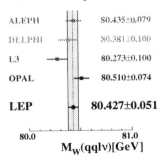

Figure 4. Preliminary results from semileptonic W^+W^- decays, for ALEPH and OPAL the fully leptonic channel is included.

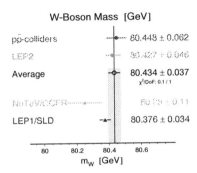

Figure 5. Comparison of different M_W determinations.

Acknowledgments

I would like to thank the LEP Electroweak WG for providing much of the information presented here.

References

1. ALEPH Collaboration, CERN EP 2000-45 submited to Eur. Phys. J. C.
2. OPAL Collaboration, CERN EP 2000-99 submitted to Phys. Lett. B
3. Preliminary M_W results: ALEPH Collab.: ALEPH 2000-018 CONF 2000-015, DELPHI Collab.: DELPHI 2000-144 CONF 443, DELPHI 2000-149 CONF 446, L3 Collab.: L3 Note 2514 and OPAL Collab.: PN422, PN447.

RECENT ELECTROWEAK RESULTS FROM ν − N SCATTERING AT NUTEV

JAEHOON YU

MS357, FNAL, P.O.Box 500, Batavia, IL 60510, USA

NuTeV is a neutrino-nucleon deep inelastic scattering experiment which took its data in the 1996-97 Fermilab TeVatron 800 GeV fixed target run. We present the results of two recent electroweak analyses in this paper, using this data set. We present a precision measurement of $\sin^2 \theta_W$, using the Paschos-Wolfenstein relationship, which minimizes large systematic uncertainties caused by charged current production of charm quarks in the final state. The result of the measurement is $\sin^2 \theta_W = 0.2253 \pm 0.0019 \pm 0.0010$. The resulting mass of the W boson is $M_W = 80.26 \pm 0.11 \mathrm{GeV/c^2}$ and is comparable to direct measurements from TeVatron and LEP-II. In addition, we present a measurement of the cross section for the inverse muon decay process. The measured cross section is $\sigma_{\mathrm{IMD}}/E_\nu = (14.2 \pm 2.9) \times 10^{-42} \mathrm{cm^2/GeV}$ which is consistent with its Standard Model prediction.

1 Introduction

Neutrino-nucleon deep inelastic scattering is an excellent testing field of the Standard Model (SM), especially the electroweak sector, because neutrinos interact only weakly. Neutrino deep inelastic processes provide a precise measurement of the weak mixing angle, $\sin^2 \theta_W$, and, in turn, provide the mass of W boson, M_W, within the context of the Standard Model, at precision comparable to direct measurements.

In addition, in the momentum transfer range of these processes, the $\sin^2 \theta_W$ measurement is sensitive to light quark couplings and hence to any anomalies in the quark couplings caused by extra bosons or other non-SM couplings. In other words, this measurement is sensitive to new physics, such as compositeness of quarks or neutrino oscillations. Other electroweak measurements in ν − N scattering, such as the inverse muon decay process, also provide additional tools to probe the Standard Model and to look for new physics.

In this paper, we present the results of the $\sin^2 \theta_W$ measurement, as well as the recent cross section measurement for the inverse muon decay (IMD) process.

2 Precision Measurement of $\sin^2 \theta_W$

The coupling between the light quarks in charged current (CC) interactions of neutrino and nucleon is, to first order, proportional to the weak isospin while the neutral current (NC) interactions are proportional to weak isospin as well as the mixing angle. The mixing angle needs to be introduced to provide probability distributions of the NC events mediated by neutral heavy vector boson and a electromagnetic gauge boson. Therefore, the ratio between CC and NC cross sections is proportional to $\sin^2 \theta_W$.

The most important element in this measurement is separating NC from CC events. This separation is obtained statistically by using the event length variable. The event length is defined for each event to be the number of consecutive calorimeter [1] scintillation counters with energy deposition above $1/4$ of that of a single muon. NC events have short length due to the absence of muons in the event, while CC events are long. Thus, the experimentally measured ratio $\mathcal{R}_{meas} = N_{short}/N_{long}$, represents the NC to CC cross section ratio, $\mathcal{R}^{\nu(\bar{\nu})}$.

The Monte Carlo (MC) incorporates a leading order (LO) corrected Quark-Parton-Model (QPM), using LO parton distribution functions from the CCFR structure function measurements, together with a detailed detector simulation. The detailed MC also takes into account radiative corrections, target isovector and higher order QCD effects, and heavy quark production.

There are two major sources of systematic uncertainty in the CCFR measurements [2,3]. The first is the theoretical uncertainty due to mass threshold effects in the CC production of charm quarks which results in $\delta\sin^2\theta_W = 0.0027$. The second source is the lack of precise knowledge of the ν_e flux in the beam from neutral kaon decays. The size of this uncertainty is 2.9% uncertainty in N_{ν_e} which results in $\delta\sin^2\theta_W = 0.0015$.

Minimizing the two largest systematic uncertainties requires a technique insensitive to the sea quark distributions and a beamline which minimizes the number of electron neutrinos, especially those resulting from K_L decays. In order to minimize the uncertainty due to the charm quark production, NuTeV uses the Paschos-Wolfenstein parameter [4]:

$$\mathcal{R}^- = \frac{\sigma_{NC}^\nu - \sigma_{NC}^{\bar\nu}}{\sigma_{CC}^\nu - \sigma_{CC}^{\bar\nu}} = \frac{R^\nu - rR^{\bar\nu}}{1 - r} =$$
$$(g_L^2 - g_R^2) = \rho\left(\frac{1}{2} - \sin^2\theta_W\right). \quad (1)$$

Since $\sigma^{\nu q} = \sigma^{\bar\nu\bar q}$ and $\sigma^{\bar\nu q} = \sigma^{\nu\bar q}$, the effect of scattering off the sea quarks cancels by taking the differences in the neutrino and antineutrino cross sections. However, the measurement of this quantity is complicated due to the fact that the NC final states look identical for ν and $\bar\nu$. NuTeV built a Sign Selected Quadrupole Train (SSQT) to select either ν or $\bar\nu$ beam at a given running period to accomplish discrimination of NC events in the two beams. The uncertainty caused by the ν_e flux is minimized by reducing neutral secondary particle acceptances.

The extraction of $\sin^2\theta_W$ in NuTeV is based on a data sample of 1.3 million neutrino and 0.3 million antineutrino events passing a set of fiducial and energy cuts. The resulting ratios of NC candidates (short length)

to CC candidates (long length), $\mathcal{R}_{meas}^{\nu(\bar\nu)}$, are 0.4198 ± 0.008(stat) and 0.4215 ± 0.0017(stat) for neutrino and antineutrino, respectively. The variable $\mathcal{R}_{meas}$ is then related to $\sin^2\theta_W$, using a detailed MC.

To extract $\sin^2\theta_W$ from these measured ratios, we form a linear combination of $\mathcal{R}_{meas}^\nu$ and $\mathcal{R}_{meas}^{\bar\nu}$:

$$\mathcal{R}_{meas}^- \equiv \mathcal{R}_{meas}^\nu - \alpha\mathcal{R}_{meas}^{\bar\nu}, \quad (2)$$

where α is determined using the MC such that $\mathcal{R}_{meas}^-$ is insensitive to small changes in the CC cross sections due to the charm mass threshold effect. For this measurement, the value of α is found to be 0.514. This technique essentially employs the third expression in Eq. 1 instead of the second which requires separate background estimates and flux normalizations for neutrino and antineutrino cross sections. This technique cancels out a large number of systematics by taking ratios separately in neutrino and antineutrino modes while at the same time largely canceling the uncertainties related to charm quark production from the sea quark scattering. The remaining small uncertainty due to heavy quark production results from the scattering off the d-valence quark which is Cabbibo suppressed. The dominant uncertainty in this measurement is the data statistics.

The preliminary result from the NuTeV $\sin^2\theta_W$ measurement in the on-shell renormalization scheme is :

$$\sin^2\theta_W^{(on-shell)} =$$
$$0.2253 \pm 0.0019\text{(stat)} \pm 0.0010\text{(syst)}. \quad (3)$$

The small residual dependence of this result on M_{top} and M_H comes from the leading terms in the electroweak radiative corrections [5]. Within the on-shell renormalization scheme, together with the Standard Model prediction of ρ, the $\sin^2\theta_W$ is related to M_W and M_Z by $\sin^2\theta_W \equiv 1 - \frac{M_W^2}{M_Z^2}$, implying:

$$M_W = 80.26 \pm 0.10\text{(stat)} \pm 0.05\text{(syst)}$$

$$+0.073 \times \left(\frac{M_{top}^2 - (175 GeV)^2}{(100 GeV)^2} \right)$$

$$-0.025 \times log_e \left(\frac{M_H}{150 GeV} \right). \quad (4)$$

This result is in good agreement with and is in comparable precision to other M_W measurements.

3 Inverse Muon Decay Cross Section Measurement

The inverse muon decay (IMD) process provides a clean test of the weak sector because it is a purely leptonic reaction: $\nu_\mu + e^- \rightarrow \mu^- + \nu_e$. Since the process involves only one muon with little deflection, its signature is very distinct. Furthermore, this reaction only occurs in the ν_μ beam, because there are no intrinsic positrons in atoms. Thus any indication of the equivalent reaction in the $\bar{\nu}_\mu$ beam would be due to a lepton number violating (LNV) reaction with $\Delta \mathcal{L} = 2$: $\bar{\nu}_\mu + e^- \rightarrow \mu^- + \bar{\nu}_e$. The IMD process also provides the tools for estimating the background contributions to LNV ($\Delta \mathcal{L} = 2$) processes in anti-neutrino mode. The Standard Model prediction of the cross section for this process in the limit of $s \gg m_\mu^2$:

$$\sigma(\nu_\mu + e^- \rightarrow \mu^- + \nu_e) = G_F^2 s/\pi$$
$$= 17.2 \times 10^{-42} E_\nu cm^2. \quad (5)$$

The candidate event are required to satisfy the following cuts: 1) the event must be within the fiducial volume, 2) the μ^- must be well measured and fully contained in neutrino running mode, 3) the hadronic energy must be less than 3 GeV, and 4) the transverse momentum of muons, p_T^μ, must be lower than P_T^{Max} where $P_T^{Max} = 0.059 + E_\mu/671.1$ GeV/c, which is designed to retain 95% of IMD events. The overall cut efficiency is found to be approximately 81%.

The background taken into account for this measurement are quasi-elastic scattering ($\nu_\mu + n \rightarrow \mu^- + p$), resonance production ($\nu_\mu + N \rightarrow \mu^- + R$), and coherent meson production off the nucleus. All of these processes are expected to be produced equally by high energy neutrinos and antineutrinos. Therefore, the background from these processes can be estimated from the antineutrino data.

We employ two different methods for the background subtraction. The first is a direct subtraction method which uses events with low hadronic energy in antineutrino mode. This method is limited by statistics in antineutrino mode. We have also developed a Monte Carlo based method which is, at the moment, dominated by various modeling systematics. However, the results from these two background subtraction methods are in good agreement.

One of the most important factors in this measurement is the modeling of low hadronic energy events. The MC used in the background method includes a detailed detector model together with low hadronic cross section models for the background processes. In addition, the MC modeling includes Fermi motion [6] as well as Pauli suppression [7].

Using the MC based background subtraction method, we obtain $N_{IMD} = 1238 \pm 116(stat) \pm 300(syst)$ which results in an energy normalized IMD cross section:

$$\sigma_{IMD}/E_\nu = (16.5 \pm 1.5(stat) \pm 4.0(syst))$$
$$\times 10^{-42} cm^2/GeV. (6)$$

Figure 1 shows the comparison between data (solid circles) after background subtraction and the Standard Model MC predictions of IMD signal (histograms). The data is well described by the MC.

Using the direct background subtraction method using the data in antineutrino running mode, we obtain $N_{signal} = 1066 \pm 188(stat) \pm 107(syst)$ which results in a cross section value of

$$\sigma_{IMD}/E_\nu = (14.2 \pm 2.5(stat) \pm 1.4(syst))$$
$$\times 10^{-42} cm^2/GeV. (7)$$

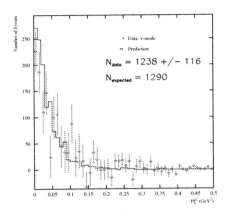

Figure 1. Comparisons between data (solid circle) and MC (histogram) for IMD events after MC based background subtraction.

As can be seen, this method provides a more precise cross section measurement and is in good agreement with the Standard Model prediction: $\sigma_{IMD}/E_\nu = 17.2 \times 10^{-42} \text{cm}^2/\text{GeV}$. This result is also in good agreement with other precise measurements of IMD cross sections [8]. The most important significance of this measurement, however, is that we now have developed a method to precisely estimate the background for Lepton Number Violation measurements.

4   Conclusions

NuTeV has measured $\sin^2 \theta_W = 0.2253 \pm 0.0019(\text{stat}) \pm 0.0010(\text{syst})$ which is equivalent to $M_W^{(\text{on-shell})} = 80.26 \pm 0.11 \text{GeV}/c^2$. This measurement is the first of its kind using the Paschos-Wolfenstein relationship and provides a measurement of the W boson mass in comparable precision to that from direct measurements.

NuTeV also presents the IMD cross section using a direct subtraction method: $\sigma_{\text{IMD}}/E_\nu = (14.2 \pm 2.5(\text{stat}) \pm 1.4(\text{syst})) \times 10^{-42} \text{cm}^2/\text{GeV}$. This result is consistent with other measurements and Standard Model prediction. In addition, it provides a tool to handle backgrounds for Lepton Number Violating (LNV) processes. The analysis for the LNV process is in progress, and we expect to present the results in the near future.

References

1. D. A. Harris and J. Yu et al., Nucl. Inst. Meth. **A447**, 377(2000)
2. C. Arroyo, B. J. King et al., Phys. Rev. Lett. **72**, 3452 (1994)
3. K. S. McFarland et al., CCFR, Eur. Phys. Jour. **C1**, 509 (1998)
4. E. A. Paschos and L. Wolfenstein, Phys. Rev. **D7**, 91 (1973)
5. D. Yu. Bardin, V. A. Dokuchaeva, JINR-E2-86-260 (1986)
6. A. Bodek and J. L. Ritchie, Phys. Rev. **D23**, 1070 (1981)
7. C. H. Llewellyn Smith and J. S. Bell, Nucl. Phys. **B28**, 317 (1971)
8. G. Fiorillo for Charm II Collaboration, Proceedings for HEP95, Brussels, Belgium, High Energy Physics *518 (1995)

THE GLOBAL FIT TO ELECTROWEAK DATA

B. PIETRZYK

LAPP, IN2P3, ch. de Bellevue, 74941 Annecy-le-Vieux Cedex, France.
E-mail: pietrzyk@lapp.in2p3.fr

The global fit to electroweak data is given. The Standard model is in a good shape and is consistent. The new BES measurements of R_{had} are used in the determination of the hadronic contribution to the running of α. Using these measurements, data still prefer low m_H, but less low.

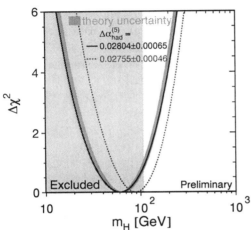

Figure 1. $\Delta\chi^2$ distribution of the Standard Model fit results as a function of the Higgs mass m_H. Official fit results are shown by the solid line, the dotted line represents fits results with the $\Delta\alpha_{had}^{(5)}(s)$ obtained here.

Osaka 2000

	Measurement	Pull	Pull -3 -2 -1 0 1 2 3
m_Z [GeV]	91.1875 ± 0.0021	.05	
Γ_Z [GeV]	2.4952 ± 0.0023	-.42	
σ_{hadr}^0 [nb]	41.540 ± 0.037	1.62	
R_l	20.767 ± 0.025	1.07	
$A_{fb}^{0,l}$	0.01714 ± 0.00095	.75	
A_e	0.1498 ± 0.0048	.38	
A_τ	0.1439 ± 0.0042	-.97	
$\sin^2\theta_{eff}^{lept}$	0.2321 ± 0.0010	.70	
m_W [GeV]	80.427 ± 0.046	.55	
R_b	0.21653 ± 0.00069	1.09	
R_c	0.1709 ± 0.0034	-.40	
$A_{fb}^{0,b}$	0.0990 ± 0.0020	-2.38	
$A_{fb}^{0,c}$	0.0689 ± 0.0035	-1.51	
A_b	0.922 ± 0.023	-.55	
A_c	0.631 ± 0.026	-1.43	
$\sin^2\theta_{eff}^{lept}$	0.23098 ± 0.00026	-1.61	
$\sin^2\theta_W$	0.2255 ± 0.0021	1.20	
m_W [GeV]	80.452 ± 0.062	.81	
m_t [GeV]	174.3 ± 5.1	-.01	
$\Delta\alpha_{had}^{(5)}(m_Z)$	0.02804 ± 0.00065	-.29	

Figure 2. Input values to the fit and the pulls.

Beautiful experimental results have been presented in this session. These results are used in the global fit to electroweak data in which the validity of the Standard Model (SM) is assumed. In this fit the consistency of the Standard Model is tested. The value of the Higgs mass is extracted from the measurements of radiative corrections.

Figure 1 shows the $\Delta\chi^2$ distribution of the fit results. The central value of the Higgs mass is

$$m_H = 60^{+52}_{-29} \text{ GeV}$$

The one-sided 95%CL (90% two-sided) upper limit on m_H is 165 GeV.

This value was $66.5\pm^{+60}_{-33}$ in Moriond in spring 2000, so the change is small. The largest contribution to the change comes from the change in the value of m_W. In this analysis the most advanced fitting programs ZFITTER and TOPAZ0 are used. The band around the solid line represents theoretical uncertainties obtained by comparison of results of two fitting programs, and also by varying the different implementations of higher order corrections in these programs in order to estimate the size of the remaining even higher order terms. The shadowed region is excluded by direct searches.

Input values to the fit and the pulls are shown in figure 2. The χ^2/d.o.f. is 21/15, giving a reasonable probability of 12%. The only pull above 2 comes from the $A_{fb}^{0,b}$ measurement. The pull of the A_{LR} measurement

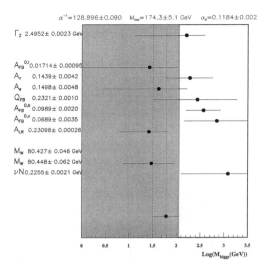

Figure 3. The results of the Higgs mass fit to different measurements.

is 1.6.

The results of the Higgs mass fit to different measurements is shown in figure 3. A reasonable distribution of the results of these fits around the central value is observed. Three measurements $A_{FB}^{0,l}$, A_{LR} and m_W give central value of m_H of about 30 GeV.

The numerical values of fit results are given in reference[1].

Very good consistency of the Standard Model is observed in figures 4, 5, 6 by comparing fit results with the direct measurements.

Figure 7 shows the fit results and the Standard Model predictions in the M_W, Γ_l plane. The Standard Model prediction is also shown when radiative corrections are restricted to the running of α alone. The genuinely electroweak radiative corrections are clearly seen. These corrections, from measurements of which the Higgs mass value is obtained, depend also strongly on the top contribution.

Thus, in order to determine the central value and the error of the Higgs mass one has to know precisely the central values and the uncertainties on α and m_t. The strong

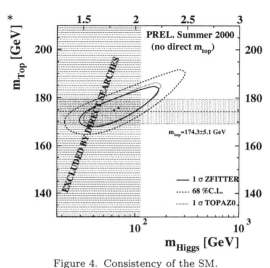

Figure 4. Consistency of the SM.

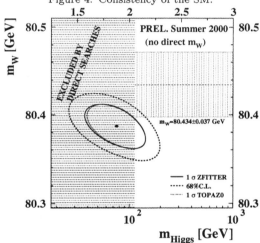

Figure 5. Consistency of the SM.

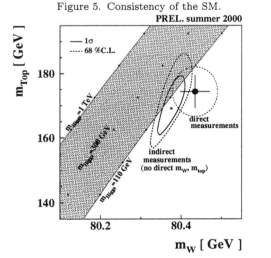

Figure 6. Consistency of the SM.

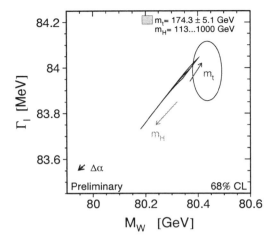

Figure 7. The fit results and the Standard Model predictions in the M_W, Γ_l plane.

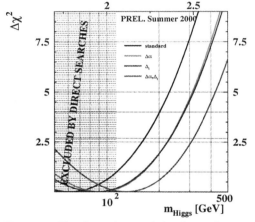

Figure 8. The dependence of the fit results on the variation of central values of α and m_t.

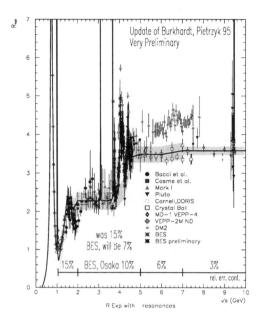

Figure 9. The results of R_{had} measurements with 85 preliminary and 6 published points of BES.

dependence of the fit results on the variation of central values of α and m_t is shown in figure 8. The central value of α^{-1} is changed by 0.09 (1σ), and the central value of m_t by 5.1 GeV (1σ) in the direction which increases the value of m_H in the fit output.

α is running :

$$\alpha(s) = \frac{\alpha(0)}{1 - \Delta\alpha_l(s) - \Delta\alpha_{had}^{(5)}(s) - \Delta\alpha_{top}(s)}$$

The contribution of the leptonic loops $\Delta\alpha_l(s)$ is precisely known[2]. The contribution of quark loops is obtained by integrating the R_{had} distribution measured in e^+e^- annihi-

lation.

$$\mathrm{Re}\,\Pi_{\gamma\gamma}(s) = \frac{\alpha\,s}{3\pi}\,P\int_{4m_\pi^2}^{\infty}\frac{R_{had}(s')}{s'(s'-s)}ds'$$

The BES collaboration reported[3] at this conference the new preliminary measurements of R_{had} in the c.m.s region 2-5 GeV. The 85 preliminary points together with the 6 published ones (figure 9) are used in a very preliminary update of the results of calculations in reference[4]. The value of $\Delta\alpha_{had}^{(5)}(s) = 0.02755 \pm 0.00046$ is obtained corresponding to $\alpha^{-1} = 128.945 \pm 0.060$. The central value is changed by 0.7σ in comparison with the published[4] value of 0.0280 ± 0.0007.

The values of R_{had} measured by BES are generally lower than the earlier data used in the previous analysis both in the low energy region (figure 10) and in the charmonium region. The solid line in figure 10, used in the $\Delta\alpha_{had}^{(5)}(s)$ analysis, is now closer to the perturbative QCD prediction[5] (figure 10).

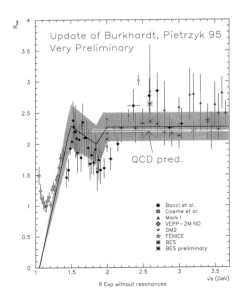

Figure 10. The results of R_{had} measurements in the low e^+e^- c.m.s. energy region.

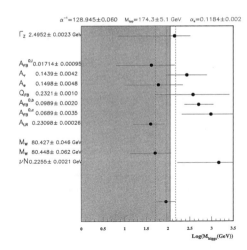

Figure 11. The results of the Higgs mass fit to different measurements using the value of $\Delta\alpha_{had}^{(5)}(s)$ obtained here.

Fits of the Higgs mass to different channels are repeated using the new value of α (figure 11). The central values of different measurements are shifted to the higher values and those of $A_{FB}^{0,l}$, A_{LR} and m_W give the $\Delta\chi^2$ minima at 40-50 GeV.

The change in the $\Delta\chi^2$ distribution using the new value of $\Delta\alpha_{had}^{(5)}(s)$ is shown by the dotted line in figure 1. The $\Delta\chi^2$ minimum moves to about 88 GeV and the one-sided 95%CL upper limit moves to 210 GeV.

In conclusion :

- the Standard Model is still in good shape
- the Standard Model is consistent
- the most significant changes are :
    - BES R measurements, not yet included in the official fit results
    - m_W from LEP
    - Z lineshape is now final, but preliminary numbers were already very good
- the data prefer a low Higgs mass
  official fit results $\rightarrow m_H = 60^{+52}_{-29}$ GeV,
  $$m_H < 165 \text{ GeV}$$
- using BES R measurements, the data still prefer low m_H, but less low

thanks to :

Gunter Quast with his ZF package, Zhengguo Zhao and his BES team, LEPP EWWG with Martin Gruenewald and Bob Clare, LEP Collaborations and Helmut Burkhardt.

References

1. http://alephwww.cern.ch/~pietrzyk/fit-osaka00.ps; the small difference between numbers presented in Osaka and given here are explained there.
2. M. Steinhauser,
   Phys. Lett. **B429**(1998)158.
3. Z. Zhao, *Measurements of R at 2-5 GeV*, this proceedings.
4. H. Burkhardt and B. Pietrzyk,
   Phys. Lett. **B356**(1995)398.
5. M. Davier and A. Höcker,
   Phys. Lett **B419**(1998)419.

PROSPECTS FOR W BOSON PHYSICS IN RUN II
AT THE FERMILAB TEVATRON

A. V. KOTWAL

FOR THE CDF AND DØ COLLABORATIONS

Physics Department, Duke University, Durham, NC 27708-0305, USA
E-mail: kotwal@phy.duke.edu

Run I (1992-96) at the Fermilab Tevatron collider produced impressive results in W boson physics, which we review here. The Tevatron accelerator and the CDF and DØ collider experiments are undergoing major upgrades in preparation for Run II, which will deliver at least a factor of 20 more data. Prospects for W boson physics from Run II are discussed.

1 Introduction

The Tevatron upgrade will increase substantially the luminosity, along with an energy increase to approximately 980 GeV. Table 1 shows the typical parameters for Run II (starting March 2001).

Table 1. Comparison of Run 1 and Run 2.

	Run 1	Run 2A	Run 2B
$\mathcal{L}$ (/cm^2/s)	5×10^{30}	10^{32}	5×10^{32}
$\int \mathcal{L}dt$/exp	120/pb	2/fb	15/fb
crossings	3.5 μs	396 ns	132 ns

Table 2. Expected W event yields for CDF II with 2 fb^{-1} of data[1]: events when Run Ib detector configuration is assumed (II column), improvement factors when increase in cross-section (III column) and increased acceptance (last column) are taken into account. For $W\gamma$, the cuts are $E_T^\gamma > 10$ GeV and $|\eta^\gamma| < 1$.

Channel	Events	$\frac{\sigma(2)}{\sigma(1.8)}$	$\frac{A(|\eta|<2)}{A(|\eta|<1)}$
$W \to e\nu$	1.4×10^6	1.12	2.0
$W \to e\nu$	6.5×10^5	1.12	2.6
$W\gamma$	1.5×10^3	1.13	1.5
$WW \to l\nu l\nu$	77	1.17	2.1
$WZ \to l\nu ll$	10	1.22	4.4

The CDF Run II detector will have a redesigned tracking system, a new scintillator-based time-of-flight detector, a new plug calorimeter extending to $|\eta| < 3.6$ and an enhanced central and forward muon system. The DØ detector will have a new magnetic spectrometer covering the region $|\eta| < 3$, scintillator-based central and forward preshower detectors and a new forward muon system with pixel readout. Table 2 demonstrates the enhanced electroweak physics potential with Run 2.

2 W mass and width

The CDF/DØ preliminary combined measurement [2,3] is $M_W = 80.454 \pm 0.062$ GeV. In Run 2, uncertainties from the W production and decay model, radiative corrections and PDFs should be substantially reduced by the combination of new theoretical calculations and experimental results like the boson p_T measurements and the W charge asymmetry and W and Z rapidity measurements. The latter will benefit from the improved rapidity coverage and will be limited by statistics. DØ has demonstrated that the inclusion of large-rapidity leptons reduces the PDF sensitivity ($\delta M_W \sim 7$ MeV). A Run 2A measurement with 25 MeV precision should be within reach. Together with a top mass uncertainty of 2 GeV, m_H should be constrained within 50% (see Fig. 1).

The direct measurement of the W width from the high transverse mass events [4] is $\Gamma_W = 2.05 \pm 0.13$ GeV. The indirect determination of the total W width (Γ_W (CDF) = 2.064 ± 0.084 GeV [5], Γ_W (DØ) = 2.152 ± 0.066 GeV [6]) from the ratio of the partial leptonic cross sections is systematics-dominated (un-

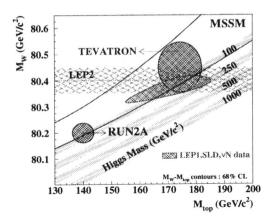

Figure 1. Measurements of the W and top mass (68% C.L.) compared with the indirect constraints, theoretical predictions and the expected Run 2A measurement uncertainties.

certainties due to σ_W/σ_Z and QED radiative corrections amount to 35 MeV). The direct method is likely to improve more rapidly with higher statistics. Its projected Run 2 precision, assuming 1(10) fb^{-1} of data is 48(20) MeV from the $W \to e\nu$ channel alone [1]. A combined overall precision of 10 MeV should be within reach.

3 High mass searches

Searches for heavy $W' \to l\nu$ decays yielded the limits of $M_{W'} > 720$ GeV [7] (DØ) and $M_{W'} > 660$ GeV [8], 652 GeV [9] (CDF) at 95% C.L. The limit is projected [1] to improve to 990 GeV with 2 fb^{-1}. Similar increases in sensitivity are expected for the Z', compositeness, large extra dimensions and technicolor. Current limits are $M_{Z'} > 690$ GeV [10] (CDF), $M_{Z'} > 670$ GeV [11] (DØ preliminary), and compositeness $\Lambda > 3.1 - 6.3$ TeV [12] (CDF), $\Lambda > 3.3 - 6.1$ TeV [13] (DØ).

4 Anomalous gauge couplings

A DØ simultaneous fit [14] of the most sensitive distributions to anomalous WWV cou-

plings has resulted in the 68% C.L. limits of $\lambda_\gamma = 0.00^{+0.10}_{-0.09}$ and $\Delta\kappa_\gamma = -0.08^{+0.34}_{-0.34}$ under the HISZ assumptions for $\Lambda = 2$ TeV. For Run 2, the substantial increase in statistics should provide an improvement of order 3(5) over the present coupling limits for integrated luminosities of 1(10) fb^{-1}. The improved rapidity coverage is expected to provide sensitivity to observe the amplitude zero effect in $p\bar{p} \to W\gamma$ production with 1 fb^{-1} of data [1].

References

1. D. Amidei et. al., 1996, FNAL-Pub-96/082, *TeV2000 Study Group Report*.
2. B. Abbott et. al. (DØ Collaboration), *Phys. Rev. Lett.* **84**, 222 (2000); hep-ex/9908057, submitted to *Phys. Rev. D*.
3. T. Affolder et. al. (CDF Collaboration), hep-ex/0007044, submitted to *Phys. Rev. D*.
4. T. Affolder et. al. (CDF Collaboration), hep-ex/0004017, submitted to *Phys. Rev. Lett.*
5. F. Abe et. al. (CDF Collaboration), *Phys. Rev. D* **52**, 2624 (1995).
6. B. Abbott et. al. (DØ Collaboration), *Phys. Rev. D* **61**, 72001 (2000).
7. S. Abachi et. al. (DØ Collaboration), *Phys. Rev. Lett.* **76**, 3271 (1996).
8. F. Abe et. al. (CDF Collaboration), *Phys. Rev. Lett.* **84**, 5716 (2000).
9. F. Abe et. al. (CDF Collaboration), *Phys. Rev. Lett.* **74**, 2900 (1995).
10. F. Abe et. al. (CDF Collaboration), *Phys. Rev. Lett.* **79**, 2192 (1997).
11. DØ Collaboration, *Search for heavy particles decaying into e^+e^- pairs in $p\bar{p}$ collisions*, submitted to this conference.
12. F. Abe et. al. (CDF Collaboration), *Phys. Rev. Lett.* **79**, 2198 (1997).
13. B. Abbott et. al. (DØ Collaboration), *Phys. Rev. Lett.* **82**, 4769 (1999).
14. B. Abbott et. al. (DØ Collaboration), *Phys. Rev. D* **60**, 72002 (1999).

PROSPECTS ON ELECTROWEAK PHYSICS FROM THE LHC

PRATIBHA VIKAS

University of Minnesota, EP Division CERN, CH-1211 Geneva 23, SWITZERLAND
E-mail: Pratibha.Vikas@cern.ch

The abundant production of gauge bosons, gauge boson pairs and top quarks at the LHC will offer the opportunity for comprehensive and challenging tests of theoretical predictions in the electroweak sector. Some issues which influence these measurements followed by prospects on some possible measurements by the ATLAS and CMS experiments at the Large Hadron Collider (LHC), at CERN are discussed.

1 Introduction

The LHC luminosity in the first three years of operation is expected to be $10^{33} cm^{-2} s^{-1}$ ("low luminosity") giving $10 fb^{-1}$/year/experiment. The luminosity is then expected to rise to its design value of $10^{34} cm^{-2} s^{-1}$ ("high luminosity") giving $100 fb^{-1}$/year/experiment. Table 1 lists the expected cross sections and event yields for some processes during the "low luminosity" phase. As can be inferred from Table 1, the LHC can be considered a W and Z boson and t and b quark factory enabling ATLAS [1] and CMS [2] to perform precision tests of theory in the electroweak sector. In most cases, these measurements are expected to significantly improve the results obtained at previous machines. Thanks to the high statistics, the statistical errors and systematic errors which scale down with statistics will be negligible for these measurements. The uncertainties will be dominated by systematic errors arising from the knowledge of the detector and the physics.

2 Main Sources of Uncertainties in Electroweak Measurements

Due to the large statistics at the LHC, harder cuts can be imposed in event selections and large control samples (e.g. $Z \to ll$) will be available to study the detector response. This will enable tight controls on systematic effects. The three major sources of systematic errors expected at the LHC are:

1. Lepton Energy and Momentum Scale: The knowledge of the lepton energy and momentum scales is related to the calibration of the tracker, electromagnetic calorimeter and the muon spectrometer. This is the dominant source of uncertainty on the m_W measurement at the Tevatron where it is known with an accuracy of 0.1%. The LHC goal is to ascertain these to 0.02% utilising the high statistics $Z \to ll$ sample in addition to the use of the E/p of isolated electrons. Since the mass of the Z boson is close to that of the W, the extrapolation error from the measurement to the calibration region will be small. The Tevatron experiments do not have sufficient Z's and have to rely on J/ψ and π^0's. Results from preliminary studies on ATLAS with 500 000 $Z \to ee$ events with full simulation indicate that the desired accuracy is difficult but not impossible to achieve [5].

2. Jet Energy Scale: The knowledge of the jet energy scale contributes to the systematic error in the measurement of m_t. This depends not only on the understanding of the detector but also the knowledge of physics processes like fragmentation, gluon radiation, etc. It is known to a precision of 3% on the Tevatron using mainly events with a γ or a Z decaying into leptons and balanced by a high p_T jet. The LHC goal is to determine this with a 1% accuracy. This can be achieved with $W \to jj$ events from $t \to bW$ for light quark jet calibration in addition to the Tevatron method. $t\bar{t}$ final states with $t \to bl\nu$ and

Table 1. Some expected cross-sections and event yields at the LHC during the "low luminosity" operation.

Process	σ (pb)	Events/second/experiment	Events/year/experiment
$W \to e\nu$	1.5×10^4	15	10^8
$Z \to e^+e^-$	1.5×10^3	1.5	10^7
$t\bar{t}$	800	0.8	10^7
$b\bar{b}$	5×10^8	5×10^5	10^{12}
$H(m_H = 700 \text{ GeV})$	1	10^{-3}	10^4
Inclusive jets $(p_T > 200 \text{ GeV})$	10^5	10^2	10^9

$t \to bjj$ events are relatively clean and have a high rate at the LHC ($\sim 0.3Hz$).

3. Absolute Luminosity: The knowledge of the absolute luminosity contributes to the uncertainty on all cross section measurements. Several methods (very forward two-photon e^+e^- pairs, central two photon production of $\mu^+\mu^-$ pairs) are foreseen to determine the absolute luminosity on ATLAS and CMS. Presently, studies indicate that the precision achievable is 10%, which will make this the dominant source of systematic error on all cross section measurements. If the theoretical understanding of Z and W production improves, a 1% precision can be obtained using their production rates [6].

3 Measurements in the top-quark Sector

The exploration of the properties of the t-quark has only just begun at the Tevatron. The expected event yield of top events on ATLAS and CMS will be 1000 times that at the Tevatron after the first three years of "low luminosity" operation, thanks to the higher cross-section and luminosity. The large statistics will allow precision measurements of the mass, production cross-section, branching ratios and searches for exotic decays.

At LHC startup in 2005, the top mass is expected to be known with a precision of ~ 3 GeV from the Tevatron experiments [9]. At the LHC [3], the statistical error is expected

to be < 100 MeV and the precision will be limited by systematic effects. The precision of 1% on the jet energy scale translates to a ~ 2 GeV error on m_t. A promising channel which is very clean and can be used at the highest luminosity is $t\bar{t} \to bW$, followed by $b \to J/\psi X$ and $W \to l\nu$ [7]. With the 1000 events/year at high luminosity, CMS expects a systematic uncertainty on m_t of ≤ 1 GeV.

4 W Mass Measurement

At LHC startup, the precision on the W mass is expected to be ~ 30 MeV from LEP2 [8] and the Tevatron [9]. The radiative corrections in the SM prediction of m_W are proportional to m_t^2 and $\log m_H$, where m_H is the mass of the Higgs boson. Hence, measurements of m_t and m_W constrain the mass of the SM Higgs and MSSM h bosons. In order to make meaningful comparisons with theory, the errors on m_t and m_W ought to be comparable in any χ^2 test - a 2 GeV accuracy on m_t implies that m_W should be known within ~ 15 MeV. This will impose a $\sim 30\%$ constraint on m_H and if and when the Higgs boson is found, will be an important consistency check of the theory.

At low luminosity, around 300 million W events/year/experiment are expected at the LHC. Studies indicate that using W decays to both leptonic channels and combining measurements from ATLAS and CMS, the LHC uncertainty on m_W is reducable to the goal of ~ 15 MeV [3].

5 Determination of TGC's

The study of Triple-Gauge Couplings (TGC's) ($WW\gamma$ or WWZ) may yield hints of "New Physics" since, many of these processes are expected to give anamolous contibutions to TGC's. In the SM, these couplings are described by five parameters g_1^Z, λ_γ, λ_Z (0 in the SM) and κ_γ and κ_Z (1 in the SM). At LHC startup, the expected precision on these is $\sim 10\%$ from LEP2 [8] and the Tevatron [9]. The LHC experiments will benefit from the large statistics and high centre-of-mass energies, since the sensitivity to anomalous couplings also scales up with energy. Studies on ATLAS and CMS [3] indicate an improvement in the sensitivity of up to two(four) orders of magnitude for anamolous $WW\gamma/WWZ(ZZ\gamma)$ couplings with respect to the current limits.

6 Drell-Yan Production of Lepton Pairs

In the SM, the production of lepton pairs in hadron-hadron collisions is described by the s-channel exchange of γ or Z bosons. At the Z peak, the Z dominates the exchange but at higher energies both the γ and Z contribute. Fermion pair production above the Z-pole is a rich search field for new phenomena at present and future high energy colliders. In the presence of "New Physics", the differential cross-section is given by

$$\frac{d\hat\sigma}{d\Omega} \sim |\gamma_s + Z_s + \text{New Physics ?!}|^2, \quad (1)$$

where many proposed types of "New Physics" processes can lead to observable effects by adding new amplitudes or through their interference with neutral currents of the SM. The LHC experiments [3] will be able to make precision measurements with events near the Z-pole and high mass pairs (110-400 GeV). For the first time, a sizeable sample of very high mass pairs (400-4000 GeV) will also be available for tests of the SM and searches of new phenomena at the TeV scale.

7 Conclusions

The ATLAS and CMS experiments will be able to perform several precision electroweak measurements. A non-exhaustive selection has been presented here. Thanks to the large event samples and their design, the precision on most of these measurements will outperform those from current experiments with just one or two years of operation. The precision on many of these measurements will be limited by physics knowledge thus warranting improved theoretical calculations.

Acknowledgements

The author would like to thank the members of the Electroweak and Top working groups in the 1999 LHC SM workshop.

References

1. ATLAS collaboration, *Technical Proposal*, CERN/LHC/94-93 (1994).
2. CMS collaboration, *Technical Proposal*, CERN/LHC/94-38 (19994).
3. G. Altarelli and M.L. Mangano (ed.), *Electroweak Physics* in CERN 2000-004 (May, 2000) and all the references therein.
4. G. Altarelli and M.L. Mangano (ed.), *Top Physics* in CERN 2000-004 (May, 2000) and all the references therein.
5. ATLAS collaboration, *Detector and Physics Performance Technical Design Report Vol I and II*, CERN/LHC/99-15 (May, 1999).
6. M.Dittmar, F. Pauss, D. Zurcher, *Phys. Rev.* D **56**, 7284 (1997).
7. A. Kharchilava, *Phys. Rev.* B **476**, 73 (2000).
8. *Physics at LEP2*, CERN 96-01.
9. D. Amidei and R. Brock (ed.), FERMILAB-Pub-96/082.

Parallel Session 6

Light Flavor Physics
(Kaon, Muon)

Conveners: Anthony R. Barker (Colorado) and
Stefano Bertolini (INFN & SISSA)

NEW RESULT ON DIRECT CP VIOLATION FROM NA48

B. GORINI

CERN, EP Division, 1211 Geneva 23, Switzerland

NA48 experiment aims at measuring the direct CP Violation parameter $\mathrm{Re}(\epsilon'/\epsilon)$ in the decays of neutral kaons with an accuracy of 2×10^{-4}. A result based on 1997 data sample has recently been published, confirming the occurrence of direct CP Violation, with an error on $\mathrm{Re}(\epsilon'/\epsilon)$ of 7×10^{-4}. A new preliminary result based on the much larger 1998 data sample is presented. The principle of the measurement, the data analysis and the main systematics are discussed.

1 Physics Motivations

Indirect CP violation occurring in the mixing of K^0 and $\bar{K}^0$ eigenstates (parametrised by ε) is known since 1964 when the decay of the long-lived neutral kaons (K_L) into two pions has been first observed[1].

CP violation can also arise from the direct decay of the K_L CP-odd component K_2 into two pions. This effect is parametrised by ϵ'. The Standard Model of weak interactions can account for such a "direct" CP violation as an effect of a complex phase in the CKM matrix.

As ε and ϵ' have nearly the same phase, one can investigate direct CP violation by measuring $\mathrm{Re}(\epsilon'/\epsilon)$ which is given by:

$$\mathrm{Re}(\epsilon'/\epsilon) \simeq \frac{1}{6}\left\{1 - \frac{\frac{\Gamma(K_L \to \pi^0 \pi^0)}{\Gamma(K_S \to \pi^0 \pi^0)}}{\frac{\Gamma(K_L \to \pi^+ \pi^-)}{\Gamma(K_S \to \pi^+ \pi^-)}}\right\}$$

$$= \frac{1}{6}(1 - \mathrm{R}) \qquad (1)$$

where R is the so-called double ratio. Previous measurements of $\mathrm{Re}(\epsilon'/\epsilon)$ can be found in [2,3,4,5].

2 The NA48 Experiment

In order to minimise the systematic uncertainties NA48 uses almost collinear K_L and K_S beams to measure concurrently, with the same detector and in the same fiducial region, the four relevant decay modes. In such an approach the acceptance, efficiency and flux factors of the four widths cancel out (at first

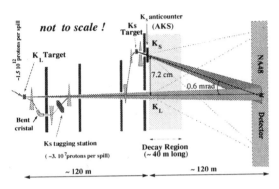

Figure 1. The combined K_S/K_L beam.

order) in the double ratio R, keeping Montecarlo corrections small.

2.1 The Combined K_S/K_L Beam

A long baseline, high intensity beam and a short baseline, low intensity beam of neutral kaons are produced simultaneously (see figure 1). The second beam is produced with an angle of 0.6 mrad relative to the first one, in order for the two to be spatially superimposed at the level of the NA48 detector.

The first beam originates from a primary proton beam with 450 GeV/c of momentum impinging on a Beryllium target. Given its long baseline it is essentially a K_L beam.

A $\sim 10^5$ less intense proton beam is selected from the non interacting primary protons, exploiting channelling through a bent crystal, and is directed to impinge on a second beryllium target, located 110 m downstream of the K_L target. Given the fact that the decay rate $\Gamma(K_S \to 2\pi)$ is $\sim 10^6$ larger than $\Gamma(K_L \to 2\pi)$, the number of detected $K_L \to 2\pi$ decays coming from this second beam is to all effects negligible. We can thus consider it as a pure K_S beam.

2.2 The Detectors

Neutral decays $K \to 2\pi^0 \to 4\gamma$ are reconstructed with a quasi-homogeneous liquid krypton calorimeter, which provides the measurement of energy, time of arrival, and impact position of the 4 photons.

The detector for charged decays is based on a magnetic spectrometer with an invariant mass resolution of 2.5 MeV/c^2 at the kaon mass and on a hodoscope, consisting of two planes of scintillator counters, which provides a measurement of the pions arrival time with a resolution better than 200 ps.

A muon veto system is installed at the end of the apparatus to identify $k_{\mu3}$ decays.

2.3 The Tagging Procedure

A system of scintillation counters, called the tagging system, is installed on the secondary proton beam upstream of the K_S target to distinguish K_S from K_L decays. If a proton signal is measured in the tagging system in coincidence (in a time window of ± 2 ns) with the measured arrival time in the detector of the kaon decay products, the event is assumed to originate from a K_S. Otherwise the event is assumed to be a K_L decay.

The tagging inefficiency (i.e. the probability of missing the coincidence for a K_S decay) and the mistagging probability (i.e. the fraction of K_L events with an accidental coincidence), are measured with $K \to \pi^+\pi^-$ decays (which can be tagged very precisely from the reconstructed vertical position of the decay vertex) to be respectively $(1.95 \pm 0.05) \times 10^{-4}$ and $(11.05 \pm 0.01)\%$.

Indirect methods are used to check the possible differences of inefficiency and mistagging probability between neutral and charged decays. Given the cancellations induced by the double ratio technique, the corrections on R depend only on those differences and turn out to be $(0 \pm 3) \times 10^{-4}$ and $(+1 \pm 8) \times 10^{-4}$ respectively.

2.4 1998 Data Analysis

Kaon decays between 70 and 170 GeV are selected for the final analysis to minimise the differences in K_L and K_S production spectra. The fiducial region for decay vertices starts at the position of the last anticounter on the KS beam line (AKS) and is 3.5 K_S lifetimes long. The upstream edge of the fiducial volume for K_S decays is defined requiring no signal in the AKS counter: no software cut on the reconstructed vertex position is applied to avoid effects due to the difference in resolution tails between charged and neutral apparatus. The absolute energy/position scale of the neutral apparatus is fixed by imposing that the reconstructed position of the AKS match the measured one.

To ensure the cancellation of the acceptance terms in R, besides the differences between K_S and K_L energy spectra, the final analysis is performed in 20 energy bins. Moreover, to ensure the same cancellation despite of the difference between K_S and K_L vertex distributions, K_L events are weighted, according to their reconstructed vertex position, with a function given by the ratio of K_S to K_L decay distributions. This procedure reduces the acceptance correction on R to $(31 \pm 9) \times 10^{-4}$, at the price of an increase of 35 % in the statistical error.

A correction of $(-1 \pm 11) \times 10^{-4}$ on R is needed to account for possible differences between K_S and K_L in the charged trigger efficiency which is measured to be $\sim 99.7\%$. Given the large neutral trigger efficiency ($\sim 99.9\%$) no correction is needed for the neutral events. The dead time induced on charged events by the drift chambers readout ($\sim 25\%$) and by the charged trigger ($\sim 5\%$) is

Table 1. Corrections to R $[10^{-4}]$

Charged Trigger	-1	$\pm\,11$
Accidental tagging	+1	$\pm\,8$
Tagging inefficiency	-	$\pm\,3$
Energy scale	-	$\pm\,10$
Charged vertex	+2	$\pm\,2$
Acceptance	+31	$\pm\,9$
Neutral background	-7	$\pm\,2$
Charged background	+19	$\pm\,3$
Beam scattering	-10	$\pm\,3$
Accidental activity	+2	$\pm\,12$
Total	+37	$\pm\,24$

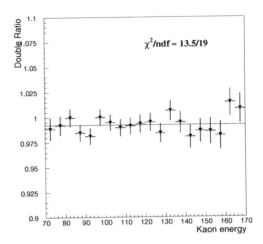

Figure 2. R versus kaon energy.

applied to the neutral events as well to symmetrise possible systematic effects.

The background in charged events is mainly due to $K_{\mu3}$ and K_{e3} decays. The residual level of background after the final analysis cuts is measured to be $(19 \pm 3) \times 10^{-4}$. The background in neutral decays is due to $K_L \rightarrow 3\pi^0$ decays with two non-reconstructed photons. The residual level is measured to be $(7 \pm 2) \times 10^{-4}$.

Effects arising from possible asymmetric losses induced by accidental activity in K_L and K_S events are estimated by overlaying good events with events collected with a pseudo-random trigger proportional to the beam intensity. The relative correction on R is $(2 \pm 6) \times 10^{-4}$. Possible additional effects on R due to the different geometrical characteristics of the two beams are estimated to be $< 10 \times 10^{-4}$.

Table 1 summaries the correction to R.

3 Result

The 1998 data sample is composed of 7.46 millions of accepted $K_S \rightarrow \pi^+\pi^-$, 4.87 millions of $K_L \rightarrow \pi^+\pi^-$, 1.80 millions of $K_S \rightarrow \pi^0\pi^0$ and 1.14 millions of $K_L \rightarrow \pi^0\pi^0$.

The preliminary measurement based on this data sample is:

$$\mathrm{Re}(\epsilon'/\epsilon) = (12.2 \pm 2.9_{(\mathrm{stat})} \pm 4.0_{(\mathrm{syst})}) \times 10^{-4} \qquad (2)$$

Figure 2 shows R as a function of the kaon energy.

Combining this preliminary result with the previous NA48 published result based on data collected during 1997 [5] one gets:

$$\mathrm{Re}(\epsilon'/\epsilon) = (14.0 \pm 4.3) \times 10^{-4} \qquad (3)$$

This result confirms a non-zero, positive value for $\mathrm{Re}(\epsilon'/\epsilon)$.

References

1. J.H.Christenson, J.W.Cronin, V.L.Fitch and R.Turlay, *Phys. Rev. Lett.* B **13**, 138 (1964).
2. Gibbons et al., *Phys. Rev. Lett.* **70**, 1203 (1993).
3. G.Barr et al., *Phys. Lett.* B **371**, 233 (1993).
4. A. Alavi-Harati et al., *Phys. Rev. Lett.* **83**, 22 (1999)
5. V. Fanti et al, *Phys. Lett.* B **465**, 335 (1999).

ϵ'/ϵ RESULTS FROM KTEV

E. BLUCHER

The Enrico Fermi Institute, The University of Chicago,
5640 South Ellis Avenue, Chicago, IL 60611, USA
E-mail: blucher@hep.uchicago.edu

The current status of the measurement of the direct CP violation parameter ϵ'/ϵ from the KTeV experiment at Fermilab is described.

1 Introduction

Since the 1964 discovery of CP violation in the decay $K_L \rightarrow \pi^+\pi^-$,[1] there has been a significant effort to distinguish between direct CP violation in the decay amplitude (parameterized by ϵ') and indirect CP violation arising from an asymmetry between $K^0 \rightarrow \overline{K}^0$ and $\overline{K}^0 \rightarrow K^0$ mixing (parameterized by ϵ). The ratio ϵ'/ϵ can be determined from the double ratio of the 2-pion decay rates of K_L and K_S:

$$\mathrm{Re}(\epsilon'/\epsilon) \simeq \frac{1}{6} \left[\frac{\dfrac{\Gamma(K_L \rightarrow \pi^+\pi^-)}{\Gamma(K_S \rightarrow \pi^+\pi^-)}}{\dfrac{\Gamma(K_L \rightarrow \pi^0\pi^0)}{\Gamma(K_S \rightarrow \pi^0\pi^0)}} - 1 \right]. \quad (1)$$

$\epsilon'/\epsilon \neq 0$ is an unambiguous indication of direct CP violation.

Recent measurements from KTeV[2] and NA48[3] have firmly established the existence of direct CP violation. The KTeV and NA48 experiments, as well as the KLOE experiment, are now working to measure ϵ'/ϵ at the $(1-2) \times 10^{-4}$ level. The current status of the ϵ'/ϵ analysis of the KTeV experiment[4] is the subject of this paper.

2 The KTeV ϵ'/ϵ Analysis

The KTeV experiment (Fig. 1) collected data in 1996, 1997 and 1999. The first ϵ'/ϵ result[2] was based on $K \rightarrow \pi^0\pi^0$ data collected during 1996 and $K \rightarrow \pi^+\pi^-$ data collected at the beginning of the 1997 run. The analysis described below uses the remaining 1997 data

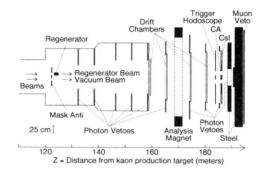

Figure 1. Diagram of the KTeV detector.

sample, which is approximately three times larger than that used for the first result.

The $K \rightarrow \pi\pi$ event reconstruction and background subtraction for the full 1997 sample are similar to that described in Ref. 2. The background levels are $\sim 0.1\%$ in the $\pi^+\pi^-$ mode and $\sim 1\%$ in the $\pi^0\pi^0$ decay mode. After subtracting background, we use a Monte Carlo simulation to correct for the acceptance difference resulting from the very different decay vertex distributions for the K_L and K_S decays. Much of the analysis effort is focused on evaluating the systematic uncertainty in this acceptance correction.

The most critical global check on the acceptance correction is the comparison of data and Monte Carlo vertex distributions in the beam (z) direction. Figure 2 shows this comparison for the $K_L \rightarrow \pi\pi$ signal modes, as well as for the much larger $\pi e\nu$ and $3\pi^0$ samples. There is good agreement between data

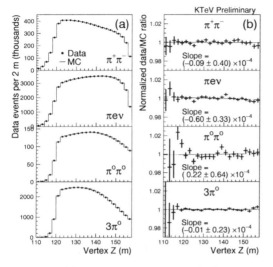

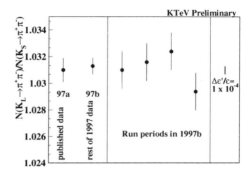

Figure 2. (a) 1997 Data – Monte Carlo comparisons of vacuum-beam z distributions for $\pi^+\pi^-$, $\pi e\nu$, $\pi^0\pi^0$, and $3\pi^0$ decays (data used in the previous analysis[2] are excluded); (b) straight-line fits to the data/MC ratios.

Figure 3. Stability of acceptance-corrected vacuum beam (K_L) / regenerator beam ("K_S") ratio for $\pi^+\pi^-$. Only statistical errors are shown. The shift in this ratio corresponding to 1×10^{-4} in ϵ'/ϵ is shown.

and Monte Carlo.[a] Including other sources of uncertainty, the total systematic error for the $K \to \pi^+\pi^-$ decay mode is about 2/3 of the corresponding error in KTeV's previously published ϵ'/ϵ analysis. Systematic studies for the $K \to \pi^0\pi^0$ decay mode are still in progress.

As in KTeV's first ϵ'/ϵ analysis, the analysis of the 1997 data has been done "blind", keeping the value of ϵ'/ϵ hidden with an unknown offset. Since the $K \to \pi^+\pi^-$ analysis is essentially complete, we have removed the offset from the acceptance-corrected $K_L \to \pi^+\pi^-/K_S \to \pi^+\pi^-$ ratio. Figure 3 shows the consistency of this ratio between the previously published data (97a) and the current data sample (97b); the stability of the charged ratio throughout the 1997b data sample also is shown. The $K_L \to \pi^0\pi^0/K_S \to \pi^0\pi^0$ ratio and ϵ'/ϵ for the 1997 data sample are still hidden.

---

[a] In KTeV's previous analysis,[2] a slope of $(-1.60 \pm 0.63) \times 10^{-4}$ m^{-1} in the ratio of data/MC z distributions for $K \to \pi^+\pi^-$ resulted in the single largest systematic uncertainty in ϵ'/ϵ.

3 Conclusions

Based on about 1/4 of the data collected during the Fermilab 1996-1997 fixed-target run, KTeV measured $\mathrm{Re}(\epsilon'/\epsilon) = (28.0 \pm 3.0 \text{ (stat)} \pm 2.8 \text{ (syst)}) \times 10^{-4}$. The analysis of the remaining 3/4 of KTeV's 1996-1997 data sample is almost complete. The statistical error on $\mathrm{Re}(\epsilon'/\epsilon)$ from this independent data sample will be 1.7×10^{-4}. The full KTeV data sample (1996-1999) will reduce the statistical uncertainty on $\mathrm{Re}(\epsilon'/\epsilon)$ to 1×10^{-4}. Significant work will be required to reduce the systematic error to a similar level.

References

1. J.H. Christenson, J.W. Cronin, V.L. Fitch, and R. Turlay, Phys. Rev. Lett. **13**, 138 (1964).

2. A. Alavi-Harati et al. [KTeV Collaboration], Phys. Rev. Lett. **83**, 22 (1999).

3. V. Fanti et al. [NA48 Collaboration], Phys. Lett. **B465**, 335 (1999); B. Gorini, these proceedings.

4. The KTeV collaboration includes Arizona, UCLA, UCSD, Chicago, Colorado, Elmhurst, Fermilab, Osaka, Rice, Rutgers, Virginia, and Wisconsin.

$\varepsilon'_K/\varepsilon_K$ AT NEXT-TO-LEADING IN $1/N_C$ AND TO LOWEST ORDER CHPT

JOHAN BIJNENS

Department of Theoretical Physics 2, Lund University
Sölvegatan 14A, S 22362 Lund, Sweden

JOAQUIM PRADES

Departamento de Física Teórica y del Cosmos, Universidad de Granada
Campus de Fuente Nueva, E-18002 Granada, Spain

We report on a calculation of $\varepsilon'_K/\varepsilon_K$ at next-to-leading order in the $1/N_c$ expansion and to lowest order in Chiral Perturbation Theory. We also discuss the chiral corrections to our results and give the result of including the two known chiral corrections.

1 Introduction

Recently, direct CP violation in the Kaon system has been unambiguously established by KTeV at Fermilab and by NA48 at CERN [1]. The present world average is

$$\mathrm{Re}\,(\varepsilon'_K/\varepsilon_K) = (19.3 \pm 2.4) \cdot 10^{-4}\,. \quad (1)$$

Recent reviews and predictions for this quantity in the Standard Model and earlier references are in [2]. Here, we report on a calculation [3] of this quantity in the chiral limit and next-to-leading (NLO) order in $1/N_c$. We also discuss the changes when the known chiral corrections -final state interactions (FSI) and $\pi_0 - \eta$ mixing- are included.

Direct CP-violation in the $K \to \pi\pi$ decay amplitudes is parameterized by

$$\frac{\varepsilon'_K}{\varepsilon_K} = \frac{1}{\sqrt{2}} \left[\frac{A[K_L \to (\pi\pi)_{I=2}]}{A[K_L \to (\pi\pi)_{I=0}]} \right.$$
$$\left. - \frac{A[K_S \to (\pi\pi)_{I=2}]}{A[K_S \to (\pi\pi)_{I=0}]} \right]\,. \quad (2)$$

$K \to \pi\pi$ amplitudes can be decomposed into definite isospin amplitudes as

$$i\,A[K^0 \to \pi^0\pi^0] \equiv \frac{a_0}{\sqrt{3}}\,e^{i\delta_0} - \sqrt{\frac{2}{3}}\,a_2\,e^{i\delta_2}\,,$$

$$i\,A[K^0 \to \pi^+\pi^-] \equiv \frac{a_0}{\sqrt{3}}\,e^{i\delta_0} + \frac{a_2}{\sqrt{6}}\,e^{i\delta_2}\,. \quad (3)$$

with δ_0 and δ_2 the FSI phases.

We want to predict a_0 and a_2 to NLO order in $1/N_c$ and to lowest order in CHPT.

2 Short-Distance Scheme and Scale Dependence

The procedure to obtain the Standard Model effective action $\Gamma_{\Delta S=1}$ below the W-boson mass has become standard and explicit calculations have been performed to two-loops [4]. The full process implies choices of short-distance scheme, regulators, and operator basis. Of course, physical matrix elements cannot depend on these choices.

The Standard Model $\Gamma_{\Delta S=1}$ effective action at scales ν somewhat below the charm quark mass, takes the form [5]

$$\Gamma_{\Delta S=1} \sim \sum_{i=1}^{10} C_i(\nu) \int \mathrm{d}^4 x\, Q_i(x) + \mathrm{h.c.} \quad (4)$$

where $Q_i(x)$ are four-quark operators and $C_i = z_i + \tau y_i$ are Wilson coefficients. In the presence of CP-violation, $\tau \equiv -V_{td}V_{ts}^*/V_{ud}V_{us}^*$ gets an imaginary part.

At low energies, it is more convenient to use an effective action $\Gamma_{\Delta S=1}^{LD}$ which uses different degrees of freedom. Different regulators and/or operator basis can be more practical too. The effective action $\Gamma_{\Delta S=1}^{LD}$ depends on all these choices and in particular on the scale μ_c introduced to regulate the divergences, analogous to ν in (4) and on effective couplings g_i, which are the equivalent of the Wilson coefficients in (4). Matching conditions between the effective field theories of (4) and $\Gamma_{\Delta S=1}^{LD}$ are obtained by requiring that S-matrix elements of asymptotic states are the same at some perturbative scale.

$$\langle 2|\Gamma_{\Delta S=1}^{LD}|1\rangle = \langle 2|\Gamma_{\Delta S=1}|1\rangle . \qquad (5)$$

The matching conditions fix *analytically* the short-distance behavior of the couplings g_i

$$g_i(\mu_c, \cdots) = \mathcal{F}(C_i(\nu), \alpha_s(\nu), \cdots) . \qquad (6)$$

This was done explicitly in [6] for $\Delta S = 2$ transitions and used in [3] for $\Delta S = 1$ transitions.

2.1 The Heavy X-Boson Method

For energies below the charm quark mass, we use an effective field theory of heavy color-singlet X-bosons coupled to QCD currents and densities[3,6,7]. For instance, the effective action reproducing

$$Q_1(x) = [\bar{s}\gamma^\mu(1-\gamma_5)d]\,[\bar{u}\gamma_\mu(1-\gamma_5)u]\,(x)$$

is

$$\Gamma_X \equiv g_1(\mu_c, \cdots) \int \mathrm{d}^4 y\, X_1^\mu \left\{[\bar{s}\gamma_\mu(1-\gamma_5)d]\,(x)\right.$$
$$\left. + [\bar{u}\gamma_\mu(1-\gamma_5)u]\,(x)\right\} . \qquad (7)$$

Here the degrees of freedom of quarks and gluons above the scale μ_c have been integrated out. The advantage of this method is that two-quark currents are unambiguously identified and that QCD densities are much easier to match than four-quark operators.

We use a 4-dimensional Euclidean cut-off μ_c to regulate UV divergences. We can now calculate $\Delta S = 1$ Green's functions with the X-boson effective theory consistently.

3 Long-Distance–Short-Distance Matching

Let's study the $\Delta S = 1$ two-point function

$$\Pi(q^2) \equiv i\int \mathrm{d}^4 x\, e^{iq\cdot x} \langle 0|T(P_i^\dagger(0) P_j(x) e^{i\Gamma x}|0\rangle .$$

The P_i are pseudoscalar sources with quantum numbers describing $K \to \pi$ amplitudes.

Taylor expanding the off-shell amplitudes $K \to \pi$ obtained from these Green's

functions, in external momentum and π, and K masses, one can obtain the couplings of the CHPT Lagrangian. These predict $K \to \pi\pi$ at a given order. This is unambiguous.

At leading order in the $1/N_c$ expansion the contribution to the Green function $\Pi(q^2)$ is factorizable. This only involves strong two-point functions and is model independent.

The non-factorizable contribution, is NLO in the $1/N_c$ expansion. It involves the integration [8] of strong four-point functions $\Pi_{P_i P_j J_a J_b}$ over the momentum Euclidean r_E that flows through the currents/densities J_a and J_b from 0 to ∞, schematically written as

$$\Pi(q^2) \sim \int \frac{\mathrm{d}^4 r_E}{(2\pi)^4} \Pi_{P_i P_j J_a J_b}(q_E, r_E). \qquad (8)$$

We separate long- from short-distance physics with a cut-off μ in r_E. The short distance part can be treated within OPE QCD.

Recently, it was emphasized that dimension eight operators may be numerically important for low values of the cut-off scale[9]. This issue can be studied straightforwardly in our approach.

There is *no* model dependence in our evaluation of $K \to \pi$ amplitudes at NLO in $1/N_c$ within QCD up to now. The long distance part from 0 up to μ remains For very small values of μ one can use CHPT but it starts to be insufficient already at relatively small values of μ. Too small to match with the short-distance part. The first step to enlarge the CHPT domain is to use a good hadronic model for intermediate energies. We used the ENJL model[10]. It has several good features -it includes CHPT to order p^4, for instance- and also some drawbacks as explained in [6]. Work is in progress to implement the large N_c constraints on three- and four-point functions along the lines of [11].

4 ε'_K in the Chiral Limit

To a very good approximation,

$$|\varepsilon'_K| \simeq \frac{1}{\sqrt{2}} \frac{\mathrm{Re}\, a_2}{\mathrm{Re}\, a_0} \left\{ -\frac{\mathrm{Im}\, a_0}{\mathrm{Re}\, a_0} + \frac{\mathrm{Im}\, a_2}{\mathrm{Re}\, a_2} \right\} . (9)$$

728

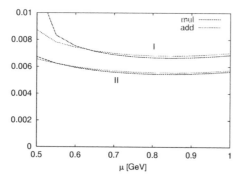

Figure 1. Plot showing the matching between short- and long-distances for $\varepsilon'_K/\varepsilon_K$. Notice the wide plateau. Curves I and II correspond to different choices in the running of α_s to two-loops.

Acknowledgments

This work is partially supported by the European Union TMR Network EURODAPHNE (Contract No. ERBFMX-CT98-0169), the Swedish Science Foundation (NFR), CICYT, Spain (Grant No. AEN-96/1672), and Junta de Andalucía (Grant No. FQM-101).

The lowest order CHPT values for $\mathrm{Re}\,a_0$ and $\mathrm{Re}\,a_2$ are obtained from a fit[12] to $K \to \pi\pi$ and $K \to \pi\pi\pi$ amplitudes to order p^4. Our results[7] reproduce the $\Delta I = 1/2$ enhancement within 40 %. We use the *experimental* lowest order CHPT values[12] for $\mathrm{Re}\,a_I$ to predict ε'_K as shown in Figure 1.

For the two dominant operators and $\varepsilon'_K/\varepsilon_K$ we obtain at NLO in $1/N_c$ and in the chiral limit[3,7]

$$B_{6\chi}^{(1/2)NDR}(2\,\mathrm{GeV}) = 2.5 \pm 0.4$$

$$B_{8\chi}^{(3/2)NDR}(2\,\mathrm{GeV}) = 1.35 \pm 0.20$$

$$\left|\frac{\varepsilon'_K}{\varepsilon_K}\right|_\chi = (60 \pm 30) \cdot 10^{-4}. \quad (10)$$

5 Higher Order CHPT Corrections

The rôle of FSI in the standard[2] predictions of $\varepsilon'_K/\varepsilon_K$ has been recently studied[13]. We took a different strategy. The ratio $\mathrm{Im}\,a_I/\mathrm{Re}\,a_I$ has no FSI to all orders, thus FSI only affects the ratio $\mathrm{Re}\,a_2/\mathrm{Re}\,a_0$.

Among the isospin breaking effects only π^0-η mixing is under control and is known [14] to order p^4. Other real p^4 and higher electromagnetic corrections are mostly unknown.

Including π^0-η mixing and FSI our result (10) becomes

$$\left|\frac{\varepsilon'_K}{\varepsilon_K}\right| = (34 \pm 18) \cdot 10^{-4}. \quad (11)$$

References

1. A. Alavi-Harati et al., *Phys. Rev. Lett.* **83**, 22 (1999); V. Fanti et al., *Phys. Lett.* B **465**, 335 (1999); A. Ceccucci, CERN Particle Physics Seminar (29 February 2000) hhtp://www.cern.ch/NA48

2. A.J. Buras et al., hep-ph/0007313; M. Ciuchini and G. Martinelli, hep-ph/0006056; S. Bertolini et al, hep-ph/0002234; T. Hambye et al., *Nucl. Phys.* B **564**, 391 (2000); E. Pallante et al, València preprint IFIC/00-31

3. J. Bijnens and J. Prades, *J. High Energy Phys.* **06**, 035 (2000)

4. A.J. Buras et al, *Nucl. Phys.* B **400**, 37, 75 (1993); M. Ciuchini et al, *Nucl. Phys.* B **415**, 403 (1994)

5. A.J. Buras, hep-ph/9806471.

6. J. Bijnens and J. Prades, *J. High Energy Phys.* **01**, 002 (2000)

7. J. Bijnens and J. Prades, *J. High Energy Phys.* **01**, 023 (1999)

8. J. Bijnens and J. Prades, *Phys. Lett.* B **342**, 331 (1995); *Nucl. Phys.* B **444** 523 (1995)

9. V. Cirigliano et al, hep-ph/0007196

10. J. Bijnens, *Phys. Rep.* **265**, 369 (1996) and references therein

11. S. Peris, M. Perrottet, and E. de Rafael, *J. High Energy Phys.* **05**, 011 (1998)

12. J. Kambor, J. Missimer, and D. Wyler, *Phys. Lett.* B **261**, 496 (1991)

13. E. Pallante and A. Pich, *Phys. Rev. Lett.* **84**, 2568 (2000); hep-ph/0007208

14. G. Ecker et al, *Phys. Lett.* B **477**, 88 (2000)

THE STANDARD MODEL PREDICTION FOR ε'/ε

E. PALLANTE

Facultat de Física, Universitat de Barcelona, Diagonal 647, E-08028 Barcelona, Spain,
E-mail: pallante@ecm.ub.es

A. PICH AND I. SCIMEMI

Departament de Física Teòrica, IFIC, Apt. Correus 2085, E-46071, València, Spain E-mail:
pich@chiral.ific.uv.es, scimemi@hal.ific.uv.es

We review fundamental aspects of a new Standard Model analysis of ε'/ε which takes into account the strong enhancement induced by final state interactions.

1 Introduction

The study of non–leptonic $K \to \pi\pi$ decays is of great importance in the understanding of CP violation mechanisms within the Standard Model and beyond. In particular, a crucial quantity is the parameter ε'/ε which measures the magnitude of the direct CP violation in the Kaon system. The experimental situation has been greatly improved recently, after the measurement by NA48 at CERN and KTeV at Fermilab. The new quoted experimental world average[1] is $\mathrm{Re}\,(\varepsilon'/\varepsilon) = (19.3 \pm 2.4) \cdot 10^{-4}$, providing a clear evidence of the existence of direct CP violation with a non–zero value of ε'/ε. The theoretical prediction of ε'/ε still suffers from many uncertainties which mainly affect the determinaton of the long–distance contributions to $K \to \pi\pi$ matrix elements and the matching with the short–distance part. Recently, it has been observed[2,3] that soft final state interactions (FSI) of the two pions in the final state play an important role in the determination of ε'/ε. From the measured $\pi - \pi$ phase shifts one can easily infer that FSI generate a strong enhancement of the predicted value of ε'/ε by roughly a factor of two[2,3], providing a good agreement with the experimental value. Here, we discuss a few basic aspects of a new Standard Model evaluation of ε'/ε which has been proposed in Refs.[2,3,4] and includes FSI effects. In addition to the large infrared logarithms generated by FSI effects there are the well known large ultraviolet logarithms that govern the short–distance evolution of the Wilson coefficients in the OPE. Both these logarithms need to be resummed and included in the evaluation of ε'/ε. In order to include all these effects in a consistent way one has to find a unique framework with a well defined power counting. The framework chosen in Ref.[4] is the large–N_C expansion[6,7]. In Sec. 2 we review the calculation of long–distance $K \to \pi\pi$ matrix elements with the inclusion of FSI effects, while Sec. 3 is devoted to the evaluation of ε'/ε.

2 $K \to \pi\pi$ matrix elements

The long–distance realization of matrix elements among light pseudoscalar mesons such as $K \to \pi\pi$ can be realized with ChPT, as an expansion in powers of momenta of the external particles and light quark masses[3]. The $K \to \pi\pi$ amplitudes with $I = 0, 2$ generated by the lowest–order ChPT lagrangian are

$$A_0 = -\frac{G_F}{\sqrt{2}} V_{ud} V_{us}^* \sqrt{2} f$$
$$\left\{ \left(g_8 + \frac{1}{9} g_{27} \right) (M_K - M_\pi^2) - \frac{2}{3} f^2 e^2 g_{EM} \right\},$$

$$A_2 = -\frac{G_F}{\sqrt{2}} V_{ud} V_{us}^* \frac{2}{9} f \left\{ 5 g_{27} (M_K - M_\pi^2) \right.$$
$$\left. -3 f^2 e^2 g_{EM} \right\}, \tag{1}$$

where g_8, g_{27} and g_{EM} are the chiral couplings and the isospin decomposition of Ref.[3] has been used. FSI effects start at next–to–leading order in the chiral expansion. To resum those effects the Omnès approach for $K \to \pi\pi$ decays has been proposed in Ref.[2] and discussed in detail in Ref.[3]. For CP–conserving amplitudes, where $e^2 g_{EM}$ corrections can be safely neglected, the most general Omnès solution for the on–shell amplitude can be written as follows

$$\mathcal{A}_I = \left(M_K^2 - M_\pi^2\right) \Omega_I(M_K^2, s_0) \, a_I(s_0) \quad (2)$$
$$= \left(M_K^2 - M_\pi^2\right) \Re_I(M_K^2, s_0) \, a_I(s_0) \, e^{i\delta_0^I(M_K^2)}.$$

The Omnès factor $\Omega_I(M_K^2, s_0)$ provides an evolution of the amplitude from low energy values (the subtraction point s_0), where the ChPT momentum expansion can be trusted, to higher energy values, through the exponentiation of the infrared effects due to FSI. It can be split into the dispersive contribution $\Re_I(M_K^2, s_0)$ and the phase shift exponential [a]. Taking a low subtraction point $s_0 = 0$, we have shown[3] that one can just multiply the tree–level formulae (1) with the experimentally determined Omnès exponentials [3]. The two dispersive correction factors thus obtained [3] are $\Re_0(M_K^2, 0) = 1.55 \pm 0.10$ and $\Re_2(M_K^2, 0) = 0.92 \pm 0.03$. One more subtlety has to be solved when applying the Omnès procedure to the CP–violating $K \to \pi\pi$ amplitudes relevant for the estimate of the direct CP violation parameter ε'/ε. The complete derivation of the Omnès solution for $K \to \pi\pi$ amplitudes makes use of Time–Reversal invariance, so that it can be strictly applied only to CP–conserving amplitudes. However, working at the first order in the Fermi coupling, the CP–odd phase is fully contained in the ratio of CKM matrix elements $\tau = V_{td} V_{ts}^* / V_{ud} V_{us}^*$ which multiplies

---
[a] For the electroweak penguin operator Q_8, the lowest order chiral contribution is a constant proportional to $e^2 g_{EM}$, where the explicit $SU(3)$ breaking is induced by the quark charge matrix, instead of the term $M_K^2 - M_\pi^2$.

the short–distance Wilson coefficients. Decomposing the isospin amplitude as $\mathcal{A}_I = \mathcal{A}_I^{CP} + \tau \, \mathcal{A}_I^{C\cancel{P}}$, the Omnès solution can be derived for the two amplitudes $\mathcal{A}_I^{CP}$ and $\mathcal{A}_I^{C\cancel{P}}$ which respect Time–Reversal invariance. In a more standard notation, $\text{Re}A_I \approx A_I^{CP}$ and $\text{Im}A_I = \text{Im}(\tau) A_I^{C\cancel{P}}$, where the absorptive phases have been already factored out through $A_I = A_I \, e^{i\delta_0^I}$.

3 The parameter ε'/ε

The direct CP violation parameter ε'/ε can be written in terms of the definite isospin $K \to \pi\pi$ amplitudes as follows

$$\frac{\varepsilon'}{\varepsilon} = e^{i\Phi} \frac{G_F \omega}{2|\epsilon|} \left[\frac{\text{Im}A_0}{\text{Re}A_0} - \frac{\text{Im}A_2}{\text{Re}A_2} \right], \quad (3)$$

where the phase $\Phi = \Phi_{\varepsilon'} - \Phi_\varepsilon \simeq 0$ and $\omega = \text{Re}A_2/\text{Re}A_0$. Since the hadronic matrix elements are quite uncertain theoretically, the CP–conserving amplitudes $\text{Re}A_I$, and thus the factor ω, are usually set to their experimentally determined values; this automatically includes the FSI effect. All the rest in the numerator has been *theoretically* predicted mostly via short–distance calculations, which therefore do not include FSI corrections. This produces a mismatch which can be easily corrected by introducing in the numerator the appropriate dispersive factors $\Re_I$ for FSI effects. The evaluation of ε'/ε proposed in Ref.[4] proceeds through the following steps:

- all the short–distance Wilson coefficients are evolved at next-to-leading logarithmic order[10,11] down to the charm quark mass scale $\mu = m_c$. All gluonic corrections of $\mathcal{O}(\alpha_s^n t^n)$ and $\mathcal{O}(\alpha_s^{n+1} t^n)$ are already known. Moreover, the full m_t/M_W dependence (at lowest order in α_s) has been taken into account. This provides the resummation of the large ultraviolet logarithms $t \equiv \ln(M/m)$, where M and m refer to any scales appearing in the evolution from M_W down to m_c.

• At the scale $\mu \sim 1$ GeV the $1/N_C$ expansion can be safely implemented. At this scale the logarithms which govern the evolution of the Wilson coefficients remain small $\sim \ln(m_c/\mu)$ so that the $1/N_C$ expansion has a clear meaning within the usual perturbative expansion in powers of α_s. In the large-N_C limit both the Wilson coefficients $C_i(\mu)$ and the long–distance matrix elements $\langle Q_i(\mu)\rangle_I$ can be computed and the matching at the scale $\mu \leq m_c$ can be done *exactly*.

• The Omnès procedure can be applied to the individual matrix elements $\langle Q_i\rangle_I$. Since the FSI effect is next–to–leading in the $1/N_C$ expansion one can include it via the realization $\langle Q_i(\mu)\rangle_I \sim \langle Q_i(\mu)\rangle_I^{N_C\to\infty} \times \Re_I$, while avoiding a double counting.

The large-N_C realization of the matrix elements $\langle Q_i(\mu)\rangle_I$, with $i \neq 6,8$ is always a product of the matrix elements of colour–singlet vector or axial currents. Each of them being an observable, the corresponding matrix element is renormalization scale and scheme independent. The same is true for the corresponding Wilson coefficients in the large–N_C limit, so that the matching is exact. The large–N_C realization of $\langle Q_{6(8)}(\mu)\rangle_I$ scales like the inverse of the squared fermion mass, being the product of colour–singlet scalar and pseudoscalar currents. Conversely, the Wilson coefficients of the operators Q_6 and Q_8 scale proportionally to the square of a quark mass in the large–N_c limit, so that again the matching is exact. The connection between the tree level ChPT amplitudes (1) and the large–N_C realization of the operators Q_i can be clarified as follows. At the lowest non trivial order in the chiral expansion, the large–N_C realization of an operator Q_i gives the contribution of Q_i itself to the chiral couplings g_8, or g_{27} or g_{EM} (according to its transformation properties) in the large–N_C limit. In this sense, the Omnès solution formulated in Sec. 2 can be applied directly to the large–N_C matrix elements of the operators Q_i with the dispersive factors

$\Re_I(M_K^2, 0)$ already estimated. A preliminary Standard Model analysis of ε'/ε gives $\varepsilon'/\varepsilon = (17\pm6)\cdot 10^{-4}$, where the error is dominated by the $1/N_C$ approximation. Further refinement and details of the analysis will be given elsewhere[4].

References

1. NA48 collaboration (V. Fanti et al.), hep-ex/9909022; http://www.cern.ch/NA48/Welcome.html; KTeV collaboration (A. Alavi–Harati et al.) *Phys. Rev. Lett.* **83**, 22 (1999).
2. E. Pallante and A. Pich, *Phys. Rev. Lett.* **84**, 2568 (2000).
3. E. Pallante and A. Pich, hep-ph/0007208.
4. E. Pallante, A. Pich and I. Scimemi, in preparation.
5. R. Omnès, *Nuovo Cimento* 8, 316 (1958); N.I. Muskhelishvili, *Singular Integral Equations*, Noordhoof, Groningen, 1953; F. Guerrero and A. Pich, *Phys. Lett.* B **412**, 382 (1997).
6. G. 't Hooft, *Nucl. Phys.* B **72**, 461 (1974); *Nucl. Phys.* B **75**, 461 (1974).
7. E. Witten, *Nucl. Phys.* B **149**, 285 (1979); *Nucl. Phys.* B **160**, 57 (1979); Ann. Phys. 128, 363 (1980).
8. F.J. Gilman and M.B. Wise, *Phys. Rev.* D **20**, 2392 (1979); *Phys. Rev.* D **21**, 3150 (1980)
9. A.J. Buras, *Weak Hamiltonian, CP Violation and Rare Decays*, Proc. 1997 Les Houches Summer School, Vol. I, p. 281 [hep-ph/9806471].
10. A.J. Buras, M. Jamin and M.E. Lautenbacher, *Nucl. Phys.* B **408**, 209 (1993); *Phys. Lett.* B **389**, 749 (1996).
11. M. Ciuchini et al., *Phys. Lett.* B **301**, 263 (1993); Z. Phys. C **68**, 239 (1995).

FIRST RESULTS FROM $\phi \to K_L K_S$ DECAYS WITH THE KLOE DETECTOR

THE KLOE COLLABORATION [1]

presented by A.PASSERI

INFN Sezione Roma III and Dipartimento di Fisica Università "Roma Tre",
via della Vasca Navale 84, 00146 roma, Italy
E-mail: passeri@roma3.infn.it

The KLOE experiment has collected 2.4 pb^{-1} of integrated luminosity during the commissioning of the DAΦNE ϕ-factory in 1999. The performance of the detector has been studied using the $\phi \to K_L K_S$ decays collected during this period, yielding also first measurements of relevant K parameters such as masses and lifetimes. A clean $K_S \to \pi^+\pi^-$ sample is used to select $K_L \to \pi^+\pi^-$ CP-violating decays and $K_L \to K_S$ regeneration events in the detector material. Results on the regeneration probability in a beryllium-alluminum alloy and carbon-fiber plus aluminum composite are presented.

1 Experimental setup

The DAΦNE ϕ factory [2] has come into operation in july 1999, delivering 2.4 pb^{-1} of integrated luminosity to the KLOE experiment during its commissioning period. Such data, taken at the energy of the ϕ resonance, allowed us to perform a careful check of the KLOE detector operation and performances [3]. Typical values of the beam parameters were 20+20 bunches, a current of 300 mA per bunch and a lifetime of 30 minutes. The average luminosity was $10^{30}\,cm^{-2}s^{-1}$. The beams cross with a period multiple of 2.7 ns and with an angle of 25 mrad in the horizontal plane, which implies a 13 MeV center of mass boost relative to the experiment.

The KLOE detector was described elsewhere[3]. The trigger requirement during most of the data taking was at least two calorimeter energy deposits, with an average efficiency of 90% for most ϕ decays and of ∼84% for $\phi \to K_L K_S$ events with kaons decaying in charged particles. The trigger rate was typically 1.5 kHz mostly due to cosmic rays. The luminosity was measured with an accuracy of ∼5% using Bhabha scattering events in the angular region $22^o < \theta < 158^o$.

About 7 million ϕ decays were collected during data taking. Data were filtered against cosmics and machine background, and classified using several tagging algorithms. This analysis used only events containing a candidate K_S charged decay, defined as: one vertex with two opposite curvature tracks, $\rho = (x^2 + y^2)^{1/2} < 4$ cm, $|z| < 8$ cm, two pion invariant mass $400 < m < 600$ MeV, and momentum $50 < p < 120$ MeV.

2 Measurement of the K_S lifetime

The position of the secondary vertex (SV) of $\pi^+\pi^-$ pairs from K_S decays, together with the knowledge of the average position of e^+e^- primary vertex (PV) allows the measurement of the K_S decay length in the ϕ frame, λ_S.

The PV position was reconstructed run-by-run from Bhabha scattering events with a typical resolution of $60 \div 70\,\mu m$ in x and y coordinates, and of ∼ $150\,\mu m$ in z. To avoid the systematic distortion of the λ_S distribution due to the SV resolution, λ_S is extracted from a fit to the distribution of its projection along the K_S momentum direction $\vec{p_S}$, and finally Lorentz-transformed to the ϕ system using the run-average ϕ boost as measured from Bhabha events.

This analysis was performed using only a subsample of events for which the position of the beam spot had been computed. The selection required the simultaneous presence of the K_S candidate vertex and of a second two-track vertex outside a sphere of 11 cm radius. If a third vertex was present the event was rejected. The mass and momentum cuts on

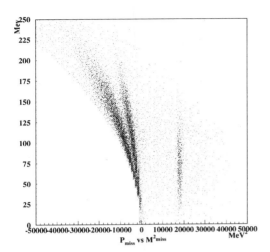

Figure 1. Correlation between missing momentum and squared missing mass for K_L charged decays

K_S candidates were tightened respectively in the interval $(495,499)$ MeV/c^2 and $(105,114)$ MeV/c, while the K_S polar angle had to l in the region above $45°$. The final selected sample is 6866 decays. Their λ distribution was fit to four different functions: an exponential smeared with two or three gaussians, with or without an extra non-smearing gaussian modeling a zero-lifetime component. The fit statistical error on λ_S was taken as the largest fit error, while the fit systematic error was taken as half the spread of the four λ_S values. The preliminary result is then:

$$\lambda_S = 5.78 \pm 0.08\,(stat) \pm 0.10\,(syst)\,mm$$

3 Measurement of the $K_L \to K_S$ regeneration cross section

Due to the different absorbtion cross section of K^o and $\bar{K}^o$, a pure beam of K_L mesons will regenerate[4] K_S mesons when traversing the detector. Thus $K_L \to K_S \to \pi\pi$ is a potential source of background for the rare $K_L \to \pi\pi$ CP-violating decays. It has been shown [5] that for a thin regenerator, $t < \lambda_S$, the coherent regeneration is negligible. In-elastic regeneration can also be neglected in 110 MeV/c K_L-nuclei interactions.

$K_S \to \pi^+\pi^-$ candidates were selected as described in sect. 1 with the additional requirement of the presence of at least one more vertex, with two unlike sign tracks, and tighter cuts on K_S candidates mass and momentum. The ϕ vertex (the K_L origin) was defined as the distance of closest approach between the K_S direction and the beam line (z axis). A K_L associated charged decay was then defined to be any second vertex reconstructed with two unlike sign tracks found in a cone opposite to the K_S direction in the ϕ reference system. This cut selected 134997 events. Fig.1 shows the correlation of the missing momentum p_{mis} and the missing mass M_{mis}^2 computed assuming the pion mass for all charged particles: $K_L \to \pi^+\pi^-\pi^o$ decays populate the region of $M_{mis}^2 = m_{\pi^o}^2$; $K_L \to \pi\ell\nu$ decays are clearly separated in two bands with $M_{mis}^2 < 0$; $K_L \to \pi^+\pi^-$ decays are peaked around $p_{mis} = 0$, $M_{mis}^2 = 0$; $K_L \to K_S \to \pi^+\pi^-$ "elastic" events are expected to populate the band with $p_{mis} = (-M_{mis}^2)^{1/2}$.

To select $K_L \to \pi^+\pi^-$ decays and $K_L \to K_S \to \pi^+\pi^-$ elastic events we first required the K_L vertex invariant mass to lie in the region $m = 497.7 \pm 4$ MeV, the measured K_L mass resolution being $\sigma_m = 1.1$ MeV. A further background rejection was obtained by requiring $-6 < \Delta|\vec{p}| < +12$ MeV, where $\Delta|\vec{p}| = |\vec{p}_S| - |\vec{p}_L|$. A sample of 930 events was obtained. Their radial distribution is shown in fig. 2, together with a control sample obtained from the sidebands of the K_L mass peak. The peaks around $r = 10$ cm and $r = 28$ cm are due to $K_L \to K_S$ regeneration in the spherical beam pipe (bp) and in the cylindrical drift chamber inner wall (cw). Two regions of interest were defined of width $(-2,+4)$ cm around the regenerator positions. The number of regenerated events (background subtracted) is then: $N_{reg}^{bp} = 123 \pm 13$, $N_{reg}^{cw} = 122 \pm 12$.

Finally $K_L \to \pi^+\pi^-$ were separated from the semileptonic decays and regenerations by requiring the angle ω between the K_S and

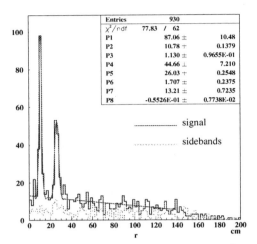

Entries	930
χ^2/ndf	77.83 / 62
P1	87.06 ± 10.48
P2	10.78 ∓ 0.1379
P3	1.130 ± 0.9655E-01
P4	44.66 ⊥ 7.210
P5	26.03 ∓ 0.2548
P6	1.707 ± 0.2375
P7	13.21 ± 0.7235
P8	-0.5526E-01 ± 0.7738E-02

.... signal

.... sidebands

Figure 2. Radial distribution of for $K_L \to \pi\pi$ decays and regenerated events, and of the side bands sample

K_L momentum to be less than $75\,mrad$. A sample of 279 events was obtained.

3.1 Results

The number of $K_L \to K_S \to \pi^+\pi^-$ regenerated events is related to the regeneration cross section on nuclei, σ_{reg}, by

$$N_{reg} = N_L \; \varepsilon \; B_{S\pi\pi} \; \sigma_{reg} \; n \; t \qquad (1)$$

with $N_L = N_{SL} \; e^{-r/\lambda}$ the number of K_L mesons reaching the regenerator, ε the detection efficiency, $B_{S\pi\pi}$ the $K_S \to \pi^+\pi^-$ branching fraction, n the number of nuclei per unit volume, t the regenerator thickness. The spherical bp, of radius $10\,cm$, $0.50\,mm$ thick, is made of AlBeMet, an alloy of 61% Beryllium and 39% Aluminum: $(nt)^{bp} = 4.93 \; 10^{21} \; cm^{-2}$. The cylindrical cw is made of Carbon 0.75 mm thick, 60%-fiber and 40%-epoxy and has a 0.20 mm thick Aluminum shield: $(nt)^{cw} = 6.97 \; 10^{21} \; cm^{-2}$.

The detection efficiency for charged regenerated events is the same as for the selected $K_L \to \pi^+\pi^-$ sample, the only difference being the final ω cut. Such efficiency, expressed in terms of $dN_{L\pi\pi}/dr$, of the decay length in the laboratory λ_L, and of the number of $K_S K_L$ events, can be substituted in eq.1, giving:

$$\sigma_{reg} = \frac{B_{L\pi\pi}}{B_{S\pi\pi}} \; \frac{1}{\lambda \; dN_{L\pi\pi}/dr} \; \frac{N_{reg}}{\langle nt \rangle} \qquad (2)$$

Using the measured $dN_{L\pi\pi}/dr$ we got:

$$\sigma^{Be-Al} = 75.7 \pm 9.6_{stat} \pm 10.6_{syst} \; mb$$

$$\sigma^{C-Al} = 51.9 \pm 6.2_{stat} \pm 5.3_{syst} \; mb$$

Three main categories of systematic error sources have been considered: a) event counting method: regions of interest definition and background parameterization; b) efficiency evaluation: the assumption that the efficiency for regenerated K_S charged decays is the same as for $K_L \to \pi^+\pi^-$ decays (tested with monte carlo), and a possible contamination of regenerated K_S decays in the $K_L \to \pi^+\pi^-$ sample (checked in the data); c) knowledge of the regenerator thicknesses, which turned out to be the dominant contribution.

3.2 Discussion

With the present data it is not possible to evaluate separately the regeneration cross section on Beryllium, Carbon and Aluminun nuclei unless we make additional hypotheses on its dependence upon the atomic mass. However a reasonable agreement is observed with a previous measurement on Beryllium [7] made with the CDM-2 detector at VEPP-2M. On the other hand theoretical calculations [6] based on the eikonal approximation can hardly be accomodated.

References

1. For the list of KLOE authors refer to hep-ex/0006035, 29 Jun 2000

2. S.Guiducci et al., DAΦNE Operating Experience, Proc. of PAC99, N.Y. 1999.

3. S.Bertolucci, A Status Report of KLOE at DAΦNE, hep-ex/0002030

4. A.Pais and O.Piccioni, Phys.Rev. 100 (1955) 1487

5. K.Kleinknecht,Fort.fürPhys.21(1973)57

6. R.Baldini and A.Michetti, LNF-96-008 (IR), 16.2.1996.

7. E.P.Solodov, Proc. of the 1999 Workshop on Kaon Physics, p.311.

THE $\Delta I = 1/2$ RULE AND ε'/ε

HAI-YANG CHENG

Institute of Physics, Academia Sinica, Taipei, Taiwan 115, R.O.C.
E-mail: phcheng@ccvax.sinica.edu.tw

The $\Delta I = 1/2$ rule and direct CP violation ε'/ε in kaon decays are studied within the framework of the effective Hamiltonian approach in conjunction with generalized factorization for hadronic matrix elements.

1 Difficulties with the Chiral Approach for $K \to \pi\pi$

Conventionally the $K \to \pi\pi$ matrix elements are evaluated under the factorization assumption so that $\langle O(\mu) \rangle$ is factorized into the product of two matrix elements of single currents, governed by decay constants and form factors. However, the information of the scale and γ_5-scheme dependence of $\langle O(\mu) \rangle$ is lost in the factorization approximation. To implement the scale dependence, it has been advocated that a physical cutoff Λ_c, which is introduced to regularize the quadratic (and logarithmic) divergence of the long-distance chiral loop corrections to $K \to \pi\pi$ amplitudes, can be identified with the renormalization scale μ of the Wilson coefficients [1]. However, this chiral approach faces several difficulties: (i) The long-distance evolution of meson loop contributions can only be extended to the scale of order 600 MeV, whereas the perturbative evaluation of Wilson coefficients cannot be reliably evolved down to the scale below 1 GeV. The conventional practice of matching chiral loop corrections to hadronic matrix elements with Wilson coefficient functions at the scale $\mu = (0.6-1.0)$ GeV requires chiral perturbation theory and/or perturbative QCD be pushed into the regions beyond their applicability. (ii) It is quite unnatural to match the quadratic scale dependence of chiral corrections with logarithmic μ dependence of Wilson coefficients. This means that it is necessary to apply the same renormalization scheme to regularize short-distance $c(\mu)$

and long-distance chiral corrections. (iii) It is not clear how to address the issue of γ_5-scheme dependence in the chiral approach. (iv) While the inclusion of chiral loops will make a large enhancement for A_0, the predicted A_2 is still too large compared to experiment. This implies that nonfactorized effects other than chiral loops are needed to explain A_2. Therefore, *not all the long-distance nonfactorized contributions to hadronic matrix elements are fully accounted for by chiral loops.* (v) Finally, this approach based on chiral perturbation theory is not applicable to heavy meson decays. Therefore, it is strongly desirable to describe the nonleptonic decays of kaons and heavy mesons within the same framework.

2 Generalized Factorization

The scale and scheme problems with naive factorization will not occur in the full amplitude since $\langle Q(\mu) \rangle$ involves vertex-type and penguin-type corrections to the hadronic matrix elements of the 4-quark operator renormalized at the scale μ. Schematically, weak decay amplitude = naive factorization + vertex-type corrections + penguin-type corrections+spectator contributions+$\cdots$,. where the spectator contributions take into account the gluonic interactions between the spectator quark of the kaon and the outgoing light meson. In general, the scheme- and μ-scale-independent effective Wilson coefficients have the form [2,3]:

$$c_i^{\text{eff}}(\mu_f) = c_i(\mu) + \frac{\alpha_s}{4\pi}\left(\gamma_V^T \ln\frac{\mu_f}{\mu} + \hat{r}_V^T\right)_{ij} \quad (1)$$

$$\times\, c_j(\mu) + \text{penguin}-\text{type corrections.}$$

For kaon decays under consideration, there is no any heavy quark mass scale between m_c and m_K. Hence, the logarithmic term emerged in the vertex corrections to 4-quark operators is of the form $\ln(\mu_f/\mu)$. We will set $\mu_f = 1$ GeV in order to have a reliable estimate of perturbative effects on effective Wilson coefficients. Writing

$$\sum_i c_i(\mu)\langle Q_i(\mu)\rangle = \sum_i a_i\langle Q_i\rangle_{\text{VIA}}, \quad (2)$$

the effective parameters can be rewritten as

$$a_{2i}^{\text{eff}} = z_{2i}^{\text{eff}} + \left(\frac{1}{N_c} + \chi_{2i}\right) z_{2i-1}^{\text{eff}}, \quad (3)$$

$$a_{2i-1}^{\text{eff}} = z_{2i-1}^{\text{eff}} + \left(\frac{1}{N_c} + \chi_{2i-1}\right) z_{2i}^{\text{eff}},$$

with χ_i being nonfactorized terms. Contrary to charmless B decays where χ_i are short-distance dominated and hence calculable in $m_b \to \infty$ limit, the nonfactorized effects in kaon decays arise mainly from the soft gluon exchange, implying large nonfactoized corrections to naive factorization.

Since nonfactorized effects in $K \to \pi\pi$ decays are not calculable by perturbative QCD, it is necessary to make some assumptions. We assume that

$$\chi_{LL} \equiv \chi_1 = \chi_2 = \chi_3 = \chi_4 = \chi_9 = \chi_{10},$$
$$\chi_{LR} \equiv \chi_5 = \chi_6 = \chi_7 = \chi_8, \quad (4)$$

and $\chi_{LR} \neq \chi_{LL}$. As shown in [4], the nonfactorized term χ_{LL}, assuming to be real, can be extracted from $K^+ \to \pi^+\pi^0$ decay to be $\chi_{LL} = -0.73$ at $\mu_f = 1$ GeV. A large negative χ_{LL} necessary for suppressing A_2 will enhance A_0 by a factor of 2. Although no constraints on χ_{LR} can be extracted from $K^0 \to \pi\pi$, we find that the ratio A_0/A_2 and direct CP violation ε'/ε are not particularly sensitive to the value of χ_{LR} [5].

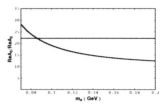

Figure 1. The ratio of $\text{Re}A_0/\text{Re}A_2$ versus m_s (in units of GeV) at the renormalization scale $\mu = 1$ GeV, where the solid (dotted) curve is calculated in the NDR (HV) scheme and use of $\chi_{LR} = -0.1$ has been made. The solid thick line is the experimental value for $\text{Re}A_0/\text{Re}A_2$.

3 $K \to \pi\pi$ isospin amplitudes

we plot in Fig. 1 the ratio A_0/A_2 as a function of m_s at the renormalization scale $\mu = 1$ GeV. Specifically, we obtain

$$\frac{\text{Re}A_0}{\text{Re}A_2} = \begin{cases} 15.2 & \text{at } m_s\,(1\,\text{GeV}) = 125\,\text{MeV}, \\ 13.6 & \text{at } m_s\,(1\,\text{GeV}) = 150\,\text{MeV}, \\ 12.7 & \text{at } m_s\,(1\,\text{GeV}) = 175\,\text{MeV}. \end{cases} \quad (5)$$

It is clear that the strange quark mass is favored to be smaller and that the prediction is renormalization scheme independent, as it should be.

It is instructive to see the anatomy of the $\Delta I = 1/2$ rule. In the absence of QCD corrections, we have $\text{Re}A_0/\text{Re}A_2 = 5/\sqrt{2} = 0.9$. With the inclusion of lowest-order short-distance QCD corrections to the Wilson coefficients z_1 and z_2 evaluated at $\mu = 1$ GeV, A_0/A_2 is enhanced from the value of 0.9 to 2.0, and it becomes 2.3 if $m_s(1\,\text{GeV}) = 150$ MeV and QCD penguin as well as electroweak penguin effects are included. This ratio is suppressed to 1.7 with the inclusion of the isospin-breaking effect, but it is increased again to the value of 2.0 in the presence of final-state interactions with $\delta_0 = 34.2°$ and $\delta_2 = -6.9°$. At this point, we have $\text{Re}A_0 = 7.7 \times 10^{-8}$ GeV and $\text{Re}A_2 = 3.8\times10^{-8}$ GeV. Comparing with the experimental values $\text{Re}\,A_0 = 3.323 \times 10^{-7}$ GeV, $\text{Re}\,A_2 = 1.497 \times 10^{-8}$ GeV, we see that the conventional calculation based on the effective Hamiltonian and naive factoriza-

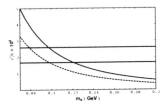

Figure 2. Direct CP violation ε'/ε versus m_s (in units of GeV) at the renormalization scale $\mu = 1$ GeV, where the solid (dotted) curve is calculated in the NDR (HV) scheme and use of $\text{Im}(V_{td}V_{ts}^*) = 1.29 \times 10^{-4}$ and $\chi_{LR} = -0.1$ has been made. The solid thick lines are the world average value for ε'/ε with one sigma errors.

tion predicts a too small $\Delta I = 1/2$ amplitude by a factor of 4.3 and a too large $\Delta I = 3/2$ amplitude by a factor of 2.5. In short, it is a long way to go to achieve the $\Delta I = 1/2$ rule within the conventional approach. Our analysis indicates that there are two principal sources responsible for the enhancement of $\text{Re}A_0/\text{Re}A_2$: the vertex-type as well as penguin-type corrections to the matrix elements of four-quark operators and nonfactorized effect due to soft-gluon exchange, which is needed to suppress the $\Delta I = 3/2$ $K \to \pi\pi$ amplitude.

4 Direct CP violation ε'/ε

For direct CP violation, we find for $\text{Im}(V_{td}V_{ts}^*) = (1.29 \pm 0.30) \times 10^{-4}$ (see Fig. 2)

$$\frac{\varepsilon'}{\varepsilon} = \begin{cases} 1.56 \pm 0.39 \ (1.02 \pm 0.26) \times 10^{-3} \\ 1.07 \pm 0.27 \ (0.70 \pm 0.18) \times 10^{-3} \\ 0.78 \pm 0.20 \ (0.51 \pm 0.13) \times 10^{-3} \end{cases} \tag{6}$$

at $m_s(1 \text{ GeV})$=125 MeV, 150 MeV, 175 MeV, respectively, in the NDR scheme, where the calculations in the HV scheme are shown in parentheses. The theoretical uncertainties of direct CP violation come from the uncertainties of $\text{Im}(V_{td}V_{ts}^*)$, Ω_{IB}, and strong phase shifts. Experimentally, the world average including NA31 [6], E731 [7], KTeV [8] and NA48 [9] results is

$$\text{Re}(\varepsilon'/\varepsilon) = (1.93 \pm 0.24) \times 10^{-3}. \tag{7}$$

From Fig. 2 we observe that, contrary to the case of A_0/A_2, the prediction of ε'/ε shows some scale dependence; roughly speaking, $(\varepsilon'/\varepsilon)_{\text{NDR}} \approx 1.5 \, (\varepsilon'/\varepsilon)_{\text{HV}}$. Direct CP violation involves a large cancellation between the dominant y_6^{eff} and y_8^{eff} terms. The scale dependence of the predicted ε'/ε is traced back to the scale dependence of the effective Wilson coefficient y_6^{eff}. It seems to us that the difference between $y_6^{\text{eff}}(\text{NDR})$ and $y_6^{\text{eff}}(\text{HV})$ comes from the effects of order α_s^2, which is further amplified by the strong cancellation between QCD penguin and electroweak penguin contributions, making it difficult to predict ε'/ε accurately. It appears to us that the different results of ε'/ε in NDR and HV schemes can be regarded as the range of theoretical uncertainties.

References

1. W.A. Bardeen, A.J. Buras, and J.-M. Gérard, *Phys. Lett.* **B192**, 138 (1987).
2. A. Ali and C. Greub, *Phys. Rev.* **D57**, 2996 (1998); A. Ali, G. Kramer, and C.D. Lü, *Phys. Rev.* **D58**, 094009 (1998).
3. H.Y. Cheng and B. Tseng, *Phys. Rev.* **D58**, 094005 (1998).
4. H.Y. Cheng, *Chin. J. Phys.* **38**, 783 (2000).
5. H.Y. Cheng, hep-ph/9910291, to appear in Chin. J. Phys. (2000).
6. NA31 Collaboration, D. Barr *et al.*, *Phys. Lett.* **B317**, 233 (1993).
7. E731 Collaboration, L.K. Gibbons *et al.*, *Phys. Rev. Lett.* **70**, 1203 (1993).
8. KTeV Collaboration, A. Alavi-Harati *et al. Phys. Rev. Lett.* **83**, 22 (1999); *Phys. Rev.* **D55**, 22 (1999).
9. NA48 Collaboration, V. Fanti *et al.*, *Phys. Lett.* **B465**, 335 (1999); B. Gorini in these proceedings.

ε'/ε IN THE STANDARD MODEL WITH HADRONIC MATRIX ELEMENTS FROM THE CHIRAL QUARK MODEL

M. FABBRICHESI

INFN and SISSA, via Beirut 4, I-34014 Trieste, Italy
E-mail: marco@he.sissa.it

I discuss the estimate of the CP-violating parameter ε'/ε based on hadronic matrix elements computed in the chiral quark model. This estimate suggested, before the current experimental results, that the favored value of ε'/ε in the standard model is of the order of 10^{-3}. I briefly review the physical effects on which this result is based and summarize current estimates.

If we imagine to be back in 1997—looking at the experimental results for the ratio ε'/ε and its theoretical estimates—we will find ourselves in a rather confusing situation in which the theoretical estimates favor values of the order of 10^{-4} and the experiments disagree by more than 3σ of their errors, and, moreover, do not rule out the super-weak scenario in which ε'/ε vanishes (for a review, see, [1]).

That was the situation when we decided to assess our theoretical understanding and possibly provide a new estimate. The crucial point was, and still is, that, if there is no sizable cancellation between some of the relevant effective operators, the order of magnitude of ε'/ε is bound to be of the order of 10^{-3}. A simple argument for this is presented in [2]. The problem is that any cancellation, or the lack thereof, among the operators heavily depends on the size of the hadronic matrix elements and, in 1997, there was no estimate of them that was free of hard-to-control assumptions.

Was it possible to improve on this situation? We wanted to estimate the hadronic matrix elements in a systematic manner without having first to solve QCD (not even by lattice simulation). To do this we needed a model that would be simple enough to understand its dynamics and, at the same time, not too simple so as to still include what we thought was the relevant physics. We chose the chiral quark model [3] in which all

coefficients of the relevant chiral lagrangian are parameterized in terms of just three parameters: the quark and gluon condensates, and the quark constituent mass. The model makes possible a complete estimate of all matrix elements, it includes non-factorizable effects, chiral corrections and final-state interaction, all of which we thought to be relevant. In order to determine the three free

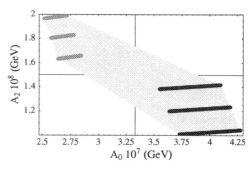

Figure 1. Fitting the $\Delta I = 1/2$ selection rule. The bars represent values according to the given ranges of the model-dependent parameters and other inputs.

parameters of the model, the experimental CP-conserving, isospin $I = 0$ and 2 components of the $K \to \pi\pi$ amplitudes, respectively A_0 and A_2, are fitted to obtain the values reported in [4] for the parameters. The systematic uncertainty of this approach is included by varing the fit by 30% around the experimental values of the amplitudes. Notice that the parameter values turn out to be rather close to those found by independent estimates, even though *a priori* they could have been any number. Moreover, the $\Delta I = 1/2$ rule is reproduced in a natural manner (see [5] for a discussion). This rule is such a fundamental feature of kaon physics that no estimate of ε'/ε can be said reliable unless it also reproduces this selection rule.

These results are stable under changes of the renormalization scale and γ_5-scheme (see [4] for details).

Having fixed the model-dependent parameters, we can proceed and compute the ratio ε'/ε. As it can been seen from fig. 2, the gluon penguin operator Q_6 dominates all other operators so that the final value of CP-violating ratio turns out to be of the expected order of 10^{-3}, and the standard model does not mimic the super-weak scenario. This is the main result of our analysis; its publication in 1997 [7] correctly predicted the current experimental results. The present estimate is an update of the short-distance inputs which also contains an improved treatment of the uncertainties. To estimate the uncertainty of

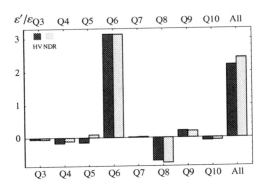

Figure 2. Contribution of the various operators of the $\Delta S = 1$ four-quark effective Hamiltonian to ε'/ε.

our result we can vary, according to a Gaussian distribution, all the short-distance inputs and by a flat distribution the model-dependent parameters to obtain the distribution of values shown in fig. 3. Such a distribution gives the value

$$\varepsilon'/\varepsilon = (2.2 \pm 0.8) \times 10^{-3}, \qquad (1)$$

in good agreement with the current experi-

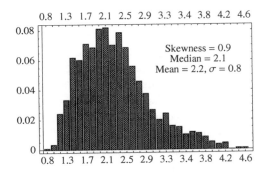

Figure 3. Distribution of values of ε'/ε at the varying of the input parameters.

mental average [6]

$$\varepsilon/\varepsilon' = (1.9 \pm 0.46) \times 10^{-3}, \qquad (2)$$

where the error has been inflated according to the Particle Data Group procedure to be used when averaging over experimental data with substantially different central values. In a more conservative approach all inputs are varied with uniform probability over their whole ranges to obtain

$$0.9 \times 10^{-3} < \varepsilon'/\varepsilon < 4.8 \times 10^{-3}. \qquad (3)$$

Given the intrinsec difficulty of the computation, I do not expect in the near future smaller uncertainties.

It is easy to go back into the computation and understand the final result. Chiral loops and final-state interactions both tends to enhance the A_0 amplitudes by making the gluon penguin contribution larger. Larger gluon penguins dominate the contribution of the electro-weak sector in ε'/ε and no effective cancellation between the two occurs. Non-factorizable (soft) gluon corrections make A_2 smaller. They play an important role in the $\Delta I = 1/2$ rule and in the determination of the model-dependent parameters although not directly in ε'/ε where only penguin operators enter. Most of these effects can be summarized by saying that the bag factor B_6 of the the gluon operator Q_6 is much larger (at a given scale) than its vacuum-saturation value of 1.

Many of the points suggested by the chiral quark model analysis have been taken up by other groups after the current experiments favored a value of ε'/ε of the order of 10^{-3}. In particular, chiral corrections [8,9], non-factorizable effects [10], final-state interactions [11] and effective-model estimates [13] have been discussed recently.

In Fig. 4 current estimates [4,8,9,12,13] are summarized; the same figure shows that, nowadays, contrarily to what is still too often repeated in papers and seminars, most standard model estimates agree with the experiments and with the prediction of the chiral quark model.

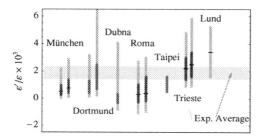

Figure 4. Experiments vs. theoretical prediction in the year 2000. References are quoted in the text.

Because of its simplicity, the chiral quark model is clearly not the final word and it can now been abandoned—as a ladder used to climb a wall after we are on the other side—as we work for better estimates, in particular, those from the lattice simulations.

Acknowledgments

It is a pleasure to thank my collaborators S. Bertolini and J. O. Eeg and my former students V. Antonelli and E. I. Lashin for the work done together.

References

1. S. Bertolini, J. Eeg and M. Fabbrichesi, *Rev. Mod. Phys.* **72**, 65 (2000).
2. M. Fabbrichesi, *Nucl. Phys. B* (Proc. Suppl.) **86**, 322 (2000).
3. K. Nishijima, *N. Cim.* **11**, 698 (1959); F. Gursey, *N. Cim.* 16 (1960) 23 and Ann. Phys. **12**, 91 (1961); J. A. Cronin, *Phys. Rev.* **161**, 1483 (1967); S. Weinberg, *Physica* **96A**, 327 (1979); A. Manohar and H. Georgi, *Nucl. Phys* **B234**, 189 (1984); D. Espriu et al., *Nucl. Phys* **B345**, 22 (1990).
4. S. Bertolini et al., `hep-ph/0002234`; M. Fabbrichesi, `hep-ph/0002235`.
5. S. Bertolini et al., *Nucl. Phys.* **B514**, 63 (1998).
6. KTeV Collaboration (A. Alavi-Harati et al.), *Phys. Rev. Lett.* **83**, 22 (1999); NA48 Collaboration (A. Ceccucci), `http://www.cern.ch/NA48/`.
7. S. Bertolini et al., *Nucl. Phys.* **B514**, 93 (1998).
8. T. Hambye et al., *Nucl. Phys.* **B564**, 391 (2000); `hep-ph/0001088`
9. A. A. Bel'kov et al., `hep-ph/9907335`.
10. H.-Y. Cheng, *Mod. Phys. Lett.* **A14**, 2453 (1999).
11. E. Pallante and A. Pich, *Phys. Rev. Lett.* **84**, 2568 (2000); E. A. Paschos, `hep-ph/9912230`; A. J. Buras et al., *Phys. Lett.* **B480**, 80 (2000).
12. S. Bosch et al., *Nucl. Phys.* **B565**, 3 (2000); M. Ciuchini et al., *Nucl. Phys.* **B573**, 201(2000).
13. J. Bijnens and J. Prades, `hep-ph/0005189`

THE WEAK OPE AND DIMENSION-EIGHT OPERATORS

E. GOLOWICH

Physics Department, University of Massachusetts, Amherst MA 01003, USA
E-mail: golowich@physics.umass.edu

We discuss recent work which identifies a potential flaw in standard treatments of weak decay amplitudes, including that of ϵ'/ϵ. The point is that (contrary to conventional wisdom) dimension-eight operators contribute to weak amplitudes at order $G_F \alpha_s$ and without $1/M_W^2$ suppression. The effect of dimension-eight operators is estimated to be at the 100% level in a sum rule determination of the operator $\mathcal{Q}_7^{(6)}$ for $\mu = 1.5$ GeV, suggesting that presently available values of μ are too low to justify the neglect of these effects.

1 Motivation

1.1 Calculating Kaon Weak Amplitudes

The modern approach to calculating a kaon weak nonleptonic amplitude $\mathcal{M}$ involves use of the operator product expansion,

$$\mathcal{M} = \sum_d \sum_i \mathcal{C}_i^{(d)}(\mu) \, \langle \mathcal{Q}_i^{(d)} \rangle_\mu \ , \quad (1)$$

in which the nonleptonic weak hamiltonian $\mathcal{H}_W$ is expressed as a linear combination of local operators $\mathcal{Q}_i^{(d)}$. There is a sum over the dimensions (starting here at $d = 6$) of the local operators and a sum over all operators of a common dimension. In practice, the following hybrid methodology is employed:

1. The Wilson coefficients $\mathcal{C}_i^{(d)}(\mu)$ are calculated in $\overline{MS}$ renormalization.

2. The operator matrix elements $\langle \mathcal{Q}_i^{(d)} \rangle_\mu$ are calculated in cutoff renormalization at the energy scale μ. The term 'cutoff' means specifically that μ serves as a 'separation scale' which distinguishes between short-distance and long-distance physics. Three different approaches falling into this category are quark models, $1/N_c$ expansion methods, and lattice-QCD evaluations.[a]

The reason for this hybrid approach is that it is not practical to carry out the (low energy) kaon matrix element evaluations with

---
[a] A list of references is given elsewhere.[1]

$\overline{MS}$ renormalization. Typical choices for the scale μ fall in the range $0.5 \leq \mu(\text{GeV}) \leq 3$, the lower part used in quark-model and $1/N_c$ evaluations and the upper part in lattice simulations.

The purpose of this talk is to describe some recent results:[1]

1. In a pure cutoff scheme, dimension-eight operators occur in the weak hamiltonian at order $G_F \alpha_s/\mu^2$, μ being the separation scale. This can be explicitly demonstrated (see Sect. 2) in a calculation involving a LR weak hamiltonian.

2. In dimensional regularization (DR), the $d = 8$ operators do *not* appear explicitly in the hamiltonian at order $G_F \alpha_s$. However, the use of a cutoff scheme for the calculation of the matrix elements of dimension-six operators requires a careful matching onto DR for which dimension-eight operators *do* play an important role.

These findings mean that hybrid evaluations, in the sense described above, of kaon matrix elements at low μ will contain (unwanted) contributions from dimension-eight operators. At the very least, this will introduce an uncertainty of unknown magnitude into the evaluation.

2 Cutoff Renormalization

2.1 ϵ'/ϵ in the Chiral Limit

The determination of ϵ'/ϵ can be shown to depend upon the matrix elements $\langle(\pi\pi)_0|\mathcal{Q}_6^{(6)}|K\rangle$ and $\langle(\pi\pi)_2|\mathcal{Q}_8^{(6)}|K\rangle$.[2] In the chiral limit of vanishing light-quark mass, the latter matrix element (as well as that of operator $\mathcal{Q}_7^{(6)}$) can be inferred from certain vacuum expectation values, $\langle 0|\mathcal{O}_{1,8}^{(6)}|0\rangle \equiv \langle\mathcal{O}_{1,8}^{(6)}\rangle$, where $\mathcal{O}_{1,8}^{(6)}$ are dimension-six four-quark operators.[3] The use of soft-meson techniques to relate physical amplitudes to those in the world of zero light-quark mass is a well-known procedure of chiral dynamics.

2.2 Sum Rules for $\langle\mathcal{O}_{1,8}^{(6)}\rangle$

Numerical values for $\langle\mathcal{O}_{1,8}^{(6)}\rangle$ in cutoff renormalization can be obtained from the following sum rules,[3]

$$\frac{16\pi^2}{3}\langle\mathcal{O}_1^{(6)}\rangle_\mu^{(\text{c.o.})} = \int_0^\infty ds\, s^2 \ln\frac{s+\mu^2}{s}\Delta\rho$$

$$2\pi\langle\alpha_s\mathcal{O}_8^{(6)}\rangle_\mu^{(\text{c.o.})} = \int_0^\infty ds\, s^2\frac{\mu^2}{s+\mu^2}\Delta\rho ,$$

$$(2)$$

where $\Delta\rho(s)$ is the difference of vector and axialvector spectral functions, and $\Delta\Pi(Q^2)$ is the corresponding difference of isospin polarization functions ($\mathcal{I}m\,\Delta\Pi = \pi\Delta\rho$).

2.3 Physics of a LR Operator

One can probe the influence of $d = 8$ operators by considering the K-to-π matrix element $\mathcal{M}(p)$,

$$\mathcal{M}(p) = \langle\pi^-(p)|\mathcal{H}_{\text{LR}}|K^-(p)\rangle , \quad (3)$$

where $\mathcal{H}_{\text{LR}}$ is a LR hamiltonian obtained by flipping the chirality of one of the quark pairs in the usual LL hamiltonian $\mathcal{H}_{\text{W}}$. The reason for defining such a LR operator is that, in leading chiral order, its K-to-π matrix element is nonzero and yields information on $\langle\mathcal{O}_1^{(6)}\rangle$ and $\langle\mathcal{O}_8^{(6)}\rangle$.

To demonstrate this, we proceed to the chiral limit to find

$$\mathcal{M} \equiv \mathcal{M}(0) = \lim_{p=0}\mathcal{M}(p)$$
$$= \frac{3G_F M_W^2}{32\sqrt{2}\pi^2 F_\pi^2}\int_0^\infty dQ^2\,\frac{Q^4}{Q^2+M_W^2}\,\Delta\Pi .$$

$$(4)$$

This result is *exact* — it is not a consequence of any model. Information about $\langle\mathcal{O}_1^{(6)}\rangle$ and $\langle\mathcal{O}_8^{(6)}\rangle$ is obtained by performing an operator product expansion on $\Delta\Pi(Q^2)$. Working to first order in α_s we have

$$\mathcal{M} = \frac{G_F}{2\sqrt{2}F_\pi^2}\left[\langle\mathcal{O}_1^{(6)}\rangle_\mu^{(\text{c.o.})}\right.$$
$$\left. +\frac{3}{8\pi}\ln\frac{M_W^2}{\mu^2}\langle\alpha_s\mathcal{O}_8^{(6)}\rangle_\mu + \frac{3}{16\pi^2}\frac{\mathcal{E}_\mu^{(8)}}{\mu^2} + \cdots\right]$$

$$(5)$$

The three additive terms in Eq. (5) are proportional respectively to the quantities $\langle\mathcal{O}_1^{(6)}\rangle$, $\langle\mathcal{O}_8^{(6)}\rangle$ and $\mathcal{E}^{(8)}$. The last of these ($\mathcal{E}^{(8)}$) contains the effect of the $d = 8$ contributions.[b] For dimensional reasons, $\mathcal{E}^{(8)}$ must be accompanied by an inverse squared energy. This turns out to be the factor μ^{-2}.

In Table 1 we display the numerical values (in units of 10^{-7} GeV^2) of the three terms of Eq. (5) for various choices of μ. Observe for the lowest values that the dimension-eight term dominates the contribution from $\langle\mathcal{O}_1^{(6)}\rangle$. Only when one proceeds to a sufficiently large value like $\mu = 4$ GeV is the $d = 8$ influence suppressed.

3 Dimensional Regularization

Suppose one wishes to express the entire analysis in terms of $\overline{MS}$ quantities. To do so requires converting matrix elements in cutoff renormalization to those in $\overline{MS}$ renormalization. Recall, in dimensional regularization

---

[b]Although the $d = 8$ LL operators arising from $\mathcal{Q}_2^{(6)}$ have been determined[1], to our knowledge the individual $d = 8$ LR operators comprising $\mathcal{E}^{(8)}$ have not.

Table 1. Eq. (5) in units of 10^{-7} GeV^2.

μ (GeV)	Term 1	Term 2	Term 3
1.0	-0.12	-3.84	0.64
1.5	-0.28	-3.49	0.30
2.0	-0.44	-3.24	0.17
4.0	-0.89	-2.63	0.04

one calculates in d dimensions and for dimensional consistency introduces a scale $\mu_{\text{d.r.}}$.

The dimensionally regularized matrix element for $\langle \mathcal{O}_1^{(6)} \rangle$ is found from the d-dimensional integral,[3]

$$\langle \mathcal{O}_1^{(6)} \rangle_\mu^{(\text{d.r.})} = \langle \mathcal{O}_1^{(6)} \rangle_\mu^{(\text{c.o.})}$$
$$+ \frac{d-1}{(4\pi)^{d/2}} \frac{\mu_{\text{d.r.}}^{4-d}}{\Gamma(d/2)} \int_{\mu^2}^\infty dQ^2 \, Q^d \, \Delta\Pi(Q^2) \, . \quad (6)$$

The term in Eq. (6) containing the integral is proof that the dimensionally regularized matrix element $\langle \mathcal{O}_1^{(6)} \rangle_\mu^{(\text{d.r.})}$ will contain *short-distance* contributions. As written, this term becomes divergent for four dimensions and also is scheme-dependent. In the $\overline{\text{MS}}$ approach, the divergent factor $2/\epsilon - \gamma + \ln(4\pi)$ is removed. The NDR scheme involves a certain procedure for treating chirality in d-dimensions. The final result is a relation (given here to $\mathcal{O}(\alpha_s)$) between the cutoff and $\overline{\text{MS}}$-NDR matrix elements,

$$\langle \mathcal{O}_1^{(6)} \rangle_\mu^{(\overline{\text{MS}}-\text{NDR})} = \langle \mathcal{O}_1^{(6)} \rangle_\mu^{(\text{c.o.})}$$
$$+ \frac{3}{8\pi} \left[\ln \frac{\mu_{\text{d.r.}}^2}{\mu^2} - \frac{1}{6} \right] \langle \alpha_s \mathcal{O}_8^{(6)} \rangle_\mu$$
$$+ \frac{3}{16\pi^2} \cdot \frac{\mathcal{E}_\mu^{(8)}}{\mu^2} + \dots \quad (7)$$

The effect of the $d = 8$ contribution to the weak OPE now appears in the $d = 6$ $\overline{\text{MS}}$-NDR operator matrix element. Note also that the parameter $\mu_{\text{d.r.}}$ is distinct from the separation scale μ.

4   Evaluation of $B_{7,8}^{(3/2)}$

To suppress the effect of dimension-eight operators on the determinations of Eq. (2), one should evaluate the two sum rules for $\langle \mathcal{O}_{1,8} \rangle_\mu^{(\text{c.o.})}$ at a large value of μ (e.g. $\mu \geq$ 4 GeV) and then use renormalization group equations to run the matrix elements down to lower values of μ (e.g. $\mu = 2$ GeV).[4] Alternative approaches might involve the finite energy sum rule framework[5] or QCD-lattice simulations at sufficiently large μ.

5   Concluding Remarks

This talk has dealt with an important aspect of calculating kaon weak matrix elements, the role of dimension-eight operators. In this regard, Eq. (7) is of special interest. It reveals that the relation between $\overline{\text{MS}}$-NDR and cutoff matrix elements will involve not only mixing between operators of a given dimension but also mixing between operators of differing dimensions. The net result of our work is that existing work on ϵ'/ϵ will be affected, especially for methods which take $\mu \leq 2$ GeV.

Acknowledgments

This work was supported in part by the National Science Foundation.

References

1. V. Cirigliano, J.F. Donoghue and E. Golowich, *Dimension-eight operators in the weak OPE*, hep-ph/0007196.
2. For a review see A. J. Buras, *Weak hamiltonian, CP violation and rare decays*, hep-ph/9806471.
3. J.F. Donoghue and E. Golowich, *Phys. Lett.* B **478**, 172 (2000).
4. V. Cirigliano, J.F. Donoghue and E. Golowich, work in progress.
5. E. Golowich and K. Maltman, work in progress.

ADDITIONAL ISOSPIN-BREAKING EFFECTS IN ϵ'/ϵ

G. VALENCIA

Department of Physics, Iowa State University, Ames, IA 50011, USA
E-mail: valencia@iastate.edu

In this talk I discuss the role of isospin violation in the calculation of ϵ'/ϵ. I point out that there is considerable uncertainty in the quantity Ω_{IB}.

The analysis of the decay modes $K^0 \to \pi\pi$ in the isospin limit indicates that there are two weak amplitudes corresponding to $\Delta I = 1/2$ and $\Delta I = 3/2$. They are labeled A_0 and A_2 because they correspond to a two pion state with isospin 0 and 2 respectively. It is known empirically that $\omega \equiv |A_2/A_0| \sim 1/22$ is quite small. In terms of these amplitudes one finds the expression,

$$\left|\frac{\epsilon'}{\epsilon}\right| = \frac{\omega}{\sqrt{2}\epsilon \mathrm{Re}A_0}\left[\mathrm{Im}A_0 - \frac{1}{\omega}\mathrm{Im}A_2\right]. \quad (1)$$

The qualitative importance of isospin violation is that it introduces two different ways to get an effective $\Delta I = 3/2$ transition amplitude (which we continue to call A_2). The weak amplitude can be $\Delta I = 3/2$, but it can also be $\Delta I = 1/2$ if accompanied by a $\Delta I = 1$ isospin violating factor arising from the quark mass difference $m_d - m_u$. Numerically, these two sources of $\Delta I = 3/2$ transitions could be similar. It is conventional to define,

$$\Omega_{IB} \equiv \frac{1}{\omega}\frac{\mathrm{Im}A_2^{IB}}{\mathrm{Im}A_0}. \quad (2)$$

The calculation of Ω_{IB} is equivalent to the calculation of the difference between the $K^0 \to \pi^+\pi^-$ and $K^0 \to \pi^0\pi^0$ amplitudes that is proportional to $m_d - m_u$. This problem is well suited to the formalism of chiral perturbation theory.

The leading order (p^2) weak, $\Delta I = 1/2$, chiral Lagrangian generates identical vertices for $K^0 \to \pi^+\pi^-$ and $K^0 \to \pi^0\pi^0$ because at lowest order the weak chiral Lagrangian does not accommodate quark mass factors. However, it also generates a non-zero $K^0 \to \pi^0\eta$

Figure 1. $K^0 \to \pi^0\pi^0$ via $\pi^0 - \eta$ mixing.

vertex. The lowest order strong chiral Lagrangian, on the other hand, contains a factor of the quark masses, and induces an isospin violating $\pi^0 - \eta$ mixing term [1]. Combining these elements as in the diagram of Figure 1, one obtains the leading order prediction [2,3]

$$\Omega_\eta = \frac{1}{3\sqrt{2}\omega}\left(\frac{m_d - m_u}{m_s - \hat{m}}\right) \sim 0.13 \quad (3)$$

where $\hat{m} = (m_d + m_u)/2$.

The question is then to study what happens beyond leading order. It should be obvious to anyone familiar with chiral perturbation theory that it is impossible to find a precise number for the next-to-leading order corrections to Eq. 3 because the result contains unknown low energy constants. For this reason I prefer to think of our calculation [4] as an estimate for the uncertainty in the leading order result, Eq. 3.

Referring to Figure 1 again, we can see that there are several possibilities to generate an amplitude at order p^4. They are:

- Including p^4 corrections to the $\pi^0 - \eta$ mixing vertex.

- Replacing the tree level exchange of an η with a two pion loop and an embedded $\pi^0 - \eta$ mixing.

- Including (isospin conserving) p^4 corrections to the weak $K^0 \to \pi^0\eta$ vertex.

• Including direct, isospin violating, $K \to \pi\pi$ vertices that occur at p^4 since at this order the weak (octet) chiral Lagrangian admits quark mass factors. In specific models this is equivalent to replacing the η in Figure 1 with a different (heavy) meson.

At this stage I want to emphasize two points. First, the first two types of p^4 corrections can be calculated unambiguously, without unknown low-energy constants. This has been done explicitly by Ecker *et. al.* [5] recently. This calculation, however, is completely irrelevant for an estimate of ϵ' as it is incomplete and precisely ignores the two types of corrections that give rise to the large theoretical uncertainty. Second, it is the last two types of p^4 corrections to Eq. 3 that introduce unknown low energy constants, which translate into uncertainty in ϵ'. What I will do next is to discuss a specific model that gives rise to a large direct vertex correction to Eq. 3, and that illustrates that the theoretical uncertainty in Ω_{IB} is large.

It is convenient to use the weak chiral Lagrangian at order p^4 of Kambor *et. al.*[6]. In terms of that notation one finds that the contributions to Ω_{IB} of the form of a direct weak vertex are given by [4]

$$\Omega_{IB} = 0.12 \text{GeV}^2 \text{Im}\left(2\frac{E_1}{c_2} - \frac{E_{10}}{c_2} + \cdots\right) \quad (4)$$

The E_i in Eq. 4 are the unknown low energy constants and in this talk I will consider only two of them. The c_2 is the coefficient of the leading order, octet, weak chiral Lagrangian so that the expectation of dimensional analysis is that

$$\frac{E_i}{c_2} \sim 1 \text{GeV}^{-2}. \quad (5)$$

That is, the natural size of the p^4 corrections to Ω is 0.12. At this very simple level we already see that there is a large uncertainty in Ω as the p^4 corrections are expected to be just as large as the leading order result, Eq. 3.

Of course, chiral perturbation theory does not tell us what the value of the E_i's is, so we need to turn to models for more detailed estimates. The early calculations of Donoghue *et. al.*[2] and of Buras and Gerard [3] can be interpreted as one such model. In addition to the leading order result of Eq. 3, these early papers considered the effect of $\pi^0 - \eta'$ mixing. The exchange of the heavy η' constituting a resonance model for the E_i's involved. With a certain prescription to include the effect of $\eta - \eta'$ mixing, they found that [2,3]

$$\Omega_{\eta+\eta'} \sim 0.28 \quad (6)$$

in agreement with our earlier statement of large p^4 corrections to the leading order result. From the point of view of this model, it is hard to estimate an uncertainty as one doesn't know what exactly has been left out. Our new model can also be interpreted as an answer to this question: the octet of scalar resonances has been left out. And as we will see, their contribution can be just as large as that of the η'.

The new low energy feature that we consider is the scalar octet containing the $a_0(980)$ and the $K_0^\star(1430)$. The model that we use follows from considering isospin breaking effects in the hadronization of the penguin operator. To achieve this one starts from the short distance expression for the penguin operator, and interpret it as the product of two scalar densities. The next step is to extract these densities from the strong chiral Lagrangian in the usual form. One immediately finds that to obtain a weak chiral Lagrangian of order p^4 with this ansatz, one needs to consider the strong chiral Lagrangian at order p^6. Unfortunately the coefficients of $\mathcal{L}_S^{(6)}$ are not known. It is here where the scalar octet comes in, as a resonance saturation model for some of these coefficients.

The specific model we consider [4], introduces three types of couplings between the scalar octet and the pions (kaons). Two of them couple one scalar to arbitrary numbers of pions derivatively (c_d) and via quark masses (c_m) [7]. The third one couples two scalars to arbitrary numbers of pions via a quark mass (d_m) in such a way that in the soft pion limit it constitutes a quark-mass induced $SU(3)$ breaking in the masses of the scalar octet [8].

The two coefficients c_d and c_m are determined by imposing the condition that the scalar octet saturate [7] the coefficients L_5 and L_8 of Gasser and Leutwyler [1]. The third constant, d_m, is determined from [8]

$$d_m = \frac{M_{K_0^\star}^2 - M_{a_0}^2}{2(M_\pi^2 - M_K^2)}. \tag{7}$$

It can easily be argued that $SU(3)$ breaking in the masses of the scalar octet is much more complicated than implied by this simple model. For this reason we prefer to determine d_m by using quark model calculated [9] masses for the $K_0^\star$ and the a_0 instead of the physical masses.

It is important to point out that the sign of the contribution to Ω_{IB} of the scalar octet is opposite to that of the η', so that they tend to cancel out. This feature follows directly from the sign of d_m, which depends only on the fact that the mass splitting of the $K_0^\star$ and the a_0 is such that the $K_0^\star$ is heavier. Numerically we find that

$$\frac{E_1}{c_2} = -1.6\text{GeV}^{-2} \ , \quad \frac{E_{10}}{c_2} = -0.8\text{GeV}^{-2}. \tag{8}$$

Notice how these numbers are of the size expected from dimensional analysis. With these numbers one obtains a contribution to Ω_{IB} from the scalar octet equal to -0.27, which can be added to the contributions from the η and the η' to obtain in this model

$$\Omega_{IB} = 0.13 + 0.15 - 0.27 = 0.01. \tag{9}$$

To conclude, isospin violation can significantly affect the value of ϵ'/ϵ. Referring to the talk by M. Fabbrichesi [10], in the scenario in which ϵ' is the result of delicate cancella-

tions between different operators, isospin violation can affect the result by factors of 2 or 3. In the scenario in which ϵ' is dominated by one operator, isospin violation amounts to a more modest correction of order 30%.

What should we use for Ω_{IB} in numerical estimates of ϵ'? In my opinion the best we can do at this time is to use the leading order (in chiral perturbation theory) calculation as a central value, and treat the various model calculations of next-to-leading order effects as an estimate for the uncertainty, I would then propose

$$\Omega_{IB} = 0.13 \pm 0.12. \tag{10}$$

Acknowledgments This work was done in collaboration with Susan Gardner and supported in part by DOE under contract number DE-FG02-92ER40730.

References

1. J. Gasser and H. Leutwyler, *Nucl. Phys.* **B250** (1985) 465.
2. J. F. Donoghue *et. al.*, *Phys. Lett.* **179B** (1986) 361.
3. A. J. Buras and J. M. Gerard, *Phys. Lett.* **192B** (1987) 156.
4. S. Gardner and G. Valencia, *Phys. Lett.* **B466** (1999) 355.
5. G. Ecker *et. al.*, *Phys. Lett.* **B477** (2000) 87.
6. J. Kambor, J. Missimer and D. Wyler, *Nucl. Phys.* **B346** (1990) 17.
7. G. Ecker *et al.*, *Nucl. Phys.* **B321** (1989) 311.
8. G. Amorós, J. Bijnens and P. Talavera, hep-ph/9907264.
9. S. Godfrey and N. Isgur, *Phys. Rev.* **D32** (1985) 189.
10. M. Fabbrichesi, these proceedings.

NEW RESULTS OF KAON ELECTROMAGNETIC DECAYS FROM KTEV

YAU W. WAH

The Enrico Fermi Institute and Department of Physics
5640 South Ellis Ave, Chicago, IL 60637, USA
E-mail: wah@hep.uchicago.edu

New results on branching fraction and form factors of three kaon Dalitz decays are presented. Due to the higher flux, higher energy, and high acceptance of the Fermilab kaon beam and the KTeV detector, all modes are at least one to two orders of magnitude better in statistics and superior in resolutions than before. We make the first experimental measurement of the parameter α from the D'Ambrosio, Isidori, and Portoles form factor, finding α = -1.54 ± 0.09. This measurement of α limits the CKM parameter ρ > - 0.2.

1 Kaon Dalitz decays and tennis matches

The nomenclature for different kaon Dalitz decays are the same as that are used in tennis matches, e.g. singles are $K_L \to e^+e^-\gamma$ and $K_L \to \mu^+\mu^-\gamma$ doubles are $K_L \to 4e$ and $K_L \to 4\mu$, and mixed double is $K_L \to e^+e^-\mu^+\mu^-$. We presented here new results on $K_L \to \mu^+\mu^-\gamma$, $K_L \to 4e$, and $K_L \to e^+e^-\mu^+\mu^-$.

2 $K_L \to \mu^+\mu^-\gamma$

The $K_L \to \mu^+\mu^-\gamma$ decay is a probe into the long distance physics associated with the intermediate $K_L \to \gamma^*\gamma$ vertex and could be described by pseudoscalar meson pole models [1,2].

The interest in $K_L \to \mu^+\mu^-\gamma$ centers on the the $K_L \to \mu^+\mu^-$ decay mode. The $K_L \to \mu^+\mu^-$ rate receives small contributions from short distance processes that are sensitive to the CKM parameter ρ [3,4,5]. However, the rate is dominated by long distance contributions involving two-photon intermediate states. The total rate can be decomposed according to $B(K_L \to \mu^+\mu^-)$ = $|ReA|^2 + |ImA|^2$. The experimental value $B(K_L \to \mu^+\mu^-)$ = $(7.18 \pm 0.17) \times 10^{-9}$ is almost completely saturated by the absorptive component calculated to be $|ImA|^2$ = $(7.07 \pm 0.18) \times 10^{-9}$ [6,7]. $|ImA|^2$ is the uni-

tarity bound and corresponds to the case of two real intermediate photons. Short distance and $K_L \to \gamma^*\gamma^*$ contributions make up the dispersive component, $ReA = ReA_{SD} + ReA_{LD}$, which is limited to $|ReA_{exp}|^2 < 3.7 \times 10^{-10}$ (90% C.L.). ReA_{LD} can be calculated using the form factor measured from $K_L \to \mu^+\mu^-\gamma$ and ρ is limited with the expression [8]:

$$\rho > 1.2 - \max[\frac{|ReA_{exp}| + |ReA_{LD}|}{3 \times 10^{-5}}$$
$$\times (\frac{\overline{m}_t(m_t)}{170 \text{GeV}})^{-1.55} \times (\frac{|V_{cb}|}{0.040})^{-2}]. \quad (1)$$

Also presented are measurements of the form factor according to the models of Bergström, Massó, and Singer (BMS) [1]

$$f(x) = \frac{1}{1 - 0.418x} + \frac{C\alpha_{K^*}}{1 - 0.308x}(\frac{4}{3}$$
$$- \frac{1}{1 - 0.418x} - \frac{1}{9}\frac{1}{1 - 0.405x}$$
$$- \frac{2}{9}\frac{1}{1 - 0.238x}) \quad (2)$$

and of D'Ambrosio, Isidori, and Portolés (DIP) [2]

$$f(x) = 1 + \alpha \left(\frac{x}{x - 2.40}\right), \quad (3)$$

where $x = (m_{\mu\mu}/m_{K_L})^2$ and C is a dimensionless constant [9]. Each model of the $\gamma^*\gamma$ form factor includes a single free parameter: α_{K^*} for BMS and α for DIP. Both parameters can be determined from the

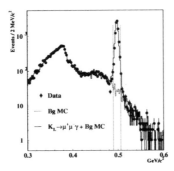

Figure 1. The reconstructed $m_{\mu^+\mu^-\gamma}$ distribution after final cuts. The large peak centered at 380 MeV/c² is $K_L \to \pi^+\pi^-\pi^0$. $K_{\mu 3}$ dominates the background from ~400 MeV/c² out to 600 MeV/c² with a slight enhancement at 450 MeV/c² from $K_{\mu 3\gamma}$.

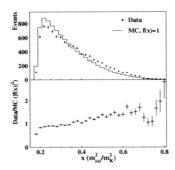

Figure 2. The dimuon mass distributions for data and for Monte Carlo with no form factor (top). The data/Monte Carlo ratio is a direct measurement of the form factor (bottom). The Monte Carlo is normalized to the total number of data events.

differential and integrated decay rates (i.e. from the dimuon mass distribution and the branching ratio). The final reconstructed $m_{\mu^+\mu^-\gamma}$ distribution is shown in Fig. 1.

There are 9327 events in the mass window 490 MeV/c² < $m_{\mu^+\mu^-\gamma}$ < 506 MeV/c². The normalized Monte Carlo reproduces the mass distribution extremely well over a wide range and predicts 221.9 ± 14.9 background events under the signal mass peak. The final $K_L \to \pi^+\pi^-\pi^0$ sample is background free and contains 210660 events in the mass window 486 MeV/c² < $m_{\mu^+\mu^-\gamma}$ < 510 MeV/c². Monte Carlo acceptances for the signal and normalization modes were (7.895 ± 0.018)% and

(3.765±0.003)% respectively, where both errors are statistical. These acceptances together with the $K_L \to \pi^+\pi^-\pi^0$ branching ratio [10] and the numbers reported above give $B(K_L \to \mu^+\mu^-\gamma) = (3.66 \pm 0.04_{stat}) \times 10^{-7}$.

In Fig. 2 is shown a model-independent measurement of the form factor as a function of x (bin-by-bin values for $|f(x)|^2$ are given in [15]). The model parameters α_{K*} and α were measured from the shape of the x distribution by making a log-likelihood comparison with Monte Carlo generated with various parameter values. In this way, the parameters were measured to be $\alpha_{K*}{}^{Shape} = -0.193^{+0.035}_{-0.049}$ and $\alpha^{Shape} = -1.73^{+0.14}_{-0.18}$, where systematic errors dominated by the placement of the cut on track momentum are included. Integrating the differential decay rate using Eq.2 or 3, $B(K_L \to \mu^+\mu^-\gamma)$ was found as a function of α_{K*} or α, yielding $\alpha_{K*}{}^{BR} = -0.117^{+0.038}_{-0.037}$ and $\alpha^{BR} = -1.38 \pm 0.12$. The shape and branching ratio measurements were combined to give $\alpha_{K*} = -0.157^{+0.025}_{-0.027}$ and $\alpha = -1.53 \pm 0.09$.

Using the measured form factor parameters with the DIP model, the limit is $|ReA_{LD}|_{DIP} < 2.07 \times 10^{-5}$. In forming the $|ReA_{exp}| + |ReA_{LD}|_{DIP}$ limit, the measured values of $|ReA_{exp}|^2$ and $|ReA_{LD}|_{DIP}$ are converted into Gaussian distributions, the product of which forms a two-dimensional probability distribution. The contour of $|ReA_{exp}| + |ReA_{LD}|_{DIP}$ under which 90% of the probability in the positive (physical) quadrant lies is the limit. This procedure yields the result $\rho > -0.2$ which is close to the combined limit from $|V_{ub}|$, B mixing, and ϵ.

3 $K_L \to 4e$

We observed 436 $K_L \to 4e$ events with a background of 3.5 event. We measured $B(K_L \to 4e)$ to be $(3.73 \pm 0.18 \text{ (stat)} \pm 0.24 \text{ (sys)}) \times 10^{-8}$. Figure 3 shows the reconstructed invariant mass of the two pairs of electron-positron.

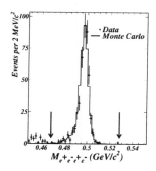

Figure 3. The 4e invariant mass distribution after all cuts except the invariant mass cut. The dots represent the data and the histogram represents the Monte Carlo simulation. There are 436 events in the signal region between the arrows. The decay $K_L \to \pi^{\pm} e^{\pm} \nu e^+ e^-$ is seen in the lower mass region.

4 $K_L \to e^+ e^- \mu^+ \mu^-$

We have measured the branching ratio for the decay $K_L \to e^+ e^- \mu^+ \mu^-$ to be $(2.50 \pm 0.41 \,(\text{stat}) \pm 0.15 \,(\text{sys})\,) \times 10^{-9}$. Figure 4 shows the fit of the data to the scaled background simulation. We estimate 0.03 events background from $K_L \to \pi^+ \pi^- \pi_D^0$ and $K_L \to \pi^+ \pi^- e^+ e^-$ decays in the signal region. Note that the VMD model of Quigg and Jackson [16] which predicts 2.37×10^{-9} is in better agreement with our result than more recent predictions. We have placed an upper limit on the CP violating term of Uy [17], of $(\frac{g_2}{h_2})^2 \leq 2.77 \times 10^{-4}$ at 90% C.L, using the branching ratio. We have also searched for lepton flavor violation in like-sign lepton decays, and placed a limit $\mathcal{B}(K_L \to e^{\pm} e^{\pm} \mu^{\mp} \mu^{\mp}) \leq 1.36 \times 10^{-10}$ at 90% confidence level.

References

1. L. Bergstrom, E. Masso and P. Singer, Phys. Lett. **131B**, 229 (1983).

2. G. D'Ambrosio, G. Isidori and J. Portoles, Phys. Lett. B **423**, 385 (1998).

3. G. Belanger and C. Q. Geng, Phys. Rev. D **43**, 140 (1991);

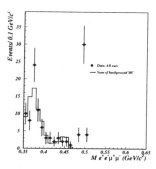

Figure 4. $M_{ee\mu\mu}$ for data (dots) and the scaled sum of $K_L \to \pi^+ \pi^- \pi_D^0$ and $K_L \to \pi^+ \pi^- e^+ e^-$ background simualtions (line), with $P_t^2 \leq 250$ (MeV/c)^2.

4. J. L. Ritchie and S. G. Wojcicki, Rev. Mod. Phys. **65**, 1149 (1993).

5. L. Wolfenstein, Phys. Rev. Lett. **51**, 1945 (1983).

6. D. Ambrose et al., Phys. Rev. Lett. **84**, 1389 (2000).

7. L. M. Sehgal, Phys. Rev. **183**, 1511 (1969).

8. A. J. Buras and R. Fleischer, in *Heavy Flavours II*, edited by A. J. Buras and M. Lindner, (World Scientific, 1997), p. 65.

9. Ref [11,12,13,14] use the value $C = 2.5$ calculated with numbers from the 1988 Particle Data Group. Here we use $C = 2.3$ based on information found in [10].

10. C. Caso et al., European Phys. Journal C **3** 458, (1998).

11. M. B. Spencer et al., Phys. Rev. Lett. **74**, 3323 (1995); V. Fanti et al., Z. Phys. **C76**, 653 (1997).

12. G. D. Barr et al., Phys. Lett. B **240**, 283 (1990)

13. K. E. Ohl et al.,Phys. Rev. Lett.**65**, 1407 (1990);

14. V. Fanti et al., Phys. Lett. B **458**, 553 (1999).

15. B. Quinn, Ph.D. Thesis, The University of Chicago (June, 2000).

16. Quigg, Jackson, UCRL-18487 (1968)

17. Z. Uy, Phys. Rev. **D43**, 802 (1991).

RARE K^0 DECAYS
(PRELIMINARY)

V. KEKELIDZE
(FOR THE NA48 COLLABORATION)
Joint Institute for Nuclear Research, Dubna, 141980, Russia
E-mail: kekel@sunse.jinr.ru

The preliminary results are obtained in the NA48 experiment at CERN SPS: the first evidence for the decay of $K_S \to \pi^+\pi^-e^+e^-$, an upper limit estimation for $Br(K_S \to \pi^0 e^+e^-)$, and the measurement of the branching ratios of K_S decays into two γ's, and K_L decays into $e^+e^-e^+e^-$, $\pi^0\gamma\gamma$ and $e^+e^-\gamma\gamma$.

1 Experimental Technique

The NA48 detector is designed for the precision measurement of the direct CP violation in the neutral kaon decays. The two beams, K_S and K_L, are produced on two different targets located consequently 6m and 126m upstream of the beginning of the decay region contained in 90m long vacuum tank. The both beams are collimated almost parallel along the OZ axis. The accepted kaon decays corresponds to the beam momentum range from 60 GeV/c^2 to 170 GeV/c^2. The main detector consists of a magnetic spectrometer, a liquid krypton electro-magnetic calorimeter (LKr), an iron-scintillator hadron calorimeter and a muon identification system.

The magnetic spectrometer includes four drift chambers, dipole magnet with momentum kick of 267 MeV/c, and two plastic scintillator hodoscope planes, and provides the mean momentum resolution of $\delta p/p \simeq 0.5 \oplus 0.009(GeV/c)^{-1} \cdot p$.

The LKr has a structure of 13212 square towers of $2 \times 2 cm^2$ cross section and 27 radiation length each, formed by copper-beryllium ribbons and contained into the liquid krypton. The energy resolution is $\sigma(E)/E \simeq 0.100/E \oplus 0.032/\sqrt{E} \oplus 0.005$, where E is in GeV. A decay vertex position (Z coordinate) of a neutral particle with mass M^0 is reconstructed according to an expression $Z_V = Z_{LKr} - \sqrt{\sum_{i,j} \cdot E_i \cdot E_j \cdot r_{ij}^2 / M^0}$, where Z_{LKr} is Z-coordinate of the LKr front plane,

$E_{i,j}$ is a reconstructed energy of i-,j- clusters, and r_{ij} is a distance between those clusters.

A more complete description of the apparatus can be found elsewhere [1].

Presented results are based on the data recorded in 1997 - 1999 at different trigger and beam intensity conditions.

2 Search for $K_S^0 \to \pi^0 e^+e^-$

A study of $K_S^0 \to \pi^0 e^+e^-$ decay is important for the estimation of indirect CP-violation contribution to the decay $K_L^0 \to \pi^0 e^+e^-$ and test of some models. The predictions of χPT models for this branching ratio are at the level of 10^{-9}, while an experimental limit [2] is $1.1 \cdot 10^{-7}$.

The data recorded during a special high intensity K_S run in 1999 have been used for this measurement. 83960 The decay $K_S \to \pi^0 \pi_D^0 \to \pi^0 e^+e^-\gamma$ (83960 selected events) was used as a normalisation channel. A comparison of two electron subsystem invariant mass (M_{ee}) spectra obtained for the normalisation channel, with the one predicted [3] for $K_S^0 \to \pi^0 e^+e^-$, is shown in fig. 1. No events have been found in the region of $M_{ee} > 165 MeV/c^2$, where less than 0.3 background events of the normalisation channel are expected. Other possible background is negligible. That allows to calculate a model independent upper limit on $Br(K_S^0 \to \pi^0 e^+e^-, M_{ee} > 165 MeV/c^2)$ as

$$8.3 \cdot 10^{-8}, \quad at \ 90\% \quad C.L.$$

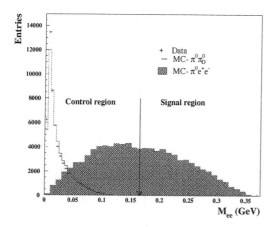

Figure 1. M_{ee} spectra for the signal and background.

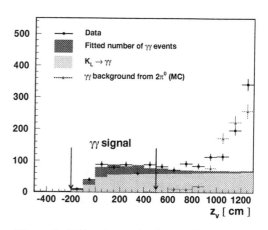

Figure 2. Different contribution to Z_V spectrum.

A limit obtained in the whole kinematic region using M_{ee} spectra of model [3], is:

$$< 1.6 \cdot 10^{-7}.$$

Systematic errors are caused mainly by the difference between the simulated (MC) and experimental spectra (fig. 1), radiative corrections and uncertainties of the LKr cluster energy reconstruction.

3 Study of $K_S \to \pi^+\pi^- e^+ e^-$ Decays

Study of this decay is important for the estimation of CP-conserving contribution to the $K_L \to \pi^+\pi^- e^+ e^-$ decay. A measurement of an asymmetry $A = cos\phi \cdot sin\phi$ (here ϕ is an angle between two planes, one defined by 2 pions and another - by two electrons, both at the kaon CM system) could check the CP-conservation in this decay.

The data for the present measurement has been recorded in 1998. An acceptance was calculated according to the model [4]. There have been reconstructed 52 events of $K_S \to \pi^+\pi^- e^+ e^-$ decay and 89 events of the decay $K_L \to \pi^+\pi^-\pi_D^0 \to \pi^+\pi^- e^+ e^-(\gamma)$ which is used as a normalisation channel. The major background is expected from this channel, and is less than one. The radiative

corrections, LKr cluster reconstruction, and MC deviation from the experimental spectra have been considered as the major sources of the systematic. The asymmetry A has been found to be consistent with zero. The obtained $Br(K_S \to \pi^+\pi^- e^+ e^-)$ is:

$$(5.1 \pm 0.9_{tat} \pm 0.3_{syst}) \cdot 10^{-5}.$$

4 Measurement of $Br(K_S \to \gamma\gamma)$

This measurement could be a critical test of χPT models, because no short distance effects are contributed to the relevant calculations.

The data recorded during the high intensity run ($6 \cdot 10^9$ ppp on K_S target) in 1999 has been used. The events have been selected having only 2 LKr clusters with an energy $3 < E_{cl} < 100 GeV$. To extract $K_S \to \gamma\gamma$ decays a Z_V distribution for the selected events (fig. 2) has been analysed. The major contribution to this spectrum came from $K_L \to 2\gamma$ (which number has been estimated as 294 events) and $K_S \to 2\pi^0 \to 2\gamma(2\gamma)$ decays. The latter one is used as a normalisation channel with the Z_V spectrum obtained by the MC. A signal region was chosen in the restricted Z_V interval which is close to the end of the K_S final collimator. The result of

752

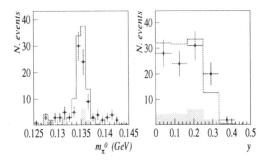

Figure 3. Two γ invariant mass and y spectra.

148 $K_S \to \gamma\gamma$ decays has been obtained using the maximum likelihood method (binned over energy and Z_V) to fit to the Z_V distribution, which leads to:

$$Br(K_S \to 2\gamma) = (2.6 \pm 0.4_{stat} \pm 0.2_{syst}) \cdot 10^{-6}.$$

5   Study [a] of $K_L \to e^+e^-e^+e^-$

The obtained branching ratio is based on 132 reconstructed events:

$$(3.67 \pm 0.32_{stat} \pm 0.23_{syst} \pm 0.08_{norm}) \cdot 10^{-8}.$$

It has been observed 19 events of the $K_L \to \mu^+\mu^-e^+e^-$ decay.

6   Study of $K_L \to \pi^0 2\gamma$ Decays

The measurement of $Br(K_L \to \pi^0\gamma\gamma)$ allows to test χPT calculations of the order of p^6 taking into account the vector meson exchange mechanism, and estimate a CP-conserving intermediate state in $K_L \to \pi^0 e^+e^-$ decay. The p^4 order diagrams can not describe neither a tail in the two gamma invariant mass ($M_{\gamma\gamma}$) distribution at the low masses ($< 240 MeV/c^2$), nor the spectrum of photon energy ($y = |E_1^* - E_2^*|/M_K$) in the kaon CM system.

---

[a] More detail analysis is presented in the contributed paper No. 229.

A part of the data recorded in 1998 and data of 1999 runs have been used for this analysis. In total 1397 $K_L \to \pi^0\gamma\gamma$ decays have been selected, 30 of which were estimated as the background ones. An effective vector coupling constant $a_v = -0.45$ has been used in the MC efficiency calculations. That has led to the reasonable agreement between the experiment and MC in $M(\pi^0 \to 2\gamma)$ and y spectra at $M_{\gamma\gamma} < 240 Me4V/c^2$ (fig. 3). The corresponding branching ratio (at $a_v = -0.45$) has been evaluated as:

$$(1.51 \pm 0.05_{stat} \pm 0.20_{syst}) \cdot 10^{-6}.$$

7   Study of $K_L \to e^+e^-2\gamma$

This study is important for the estimation of the background in searching for a CP-violating $K_L \to \pi^0 e^+e^-$ decay.

It have been selected 492 events of $K_L \to e^+e^-\gamma\gamma$ decay, and 29879 events of $K_L \to e^+e^-\gamma$ decay used as a normalisation channel, among the data recorded in 1997 and 1998 runs. The major background contribution was estimated to be from $K_s \to \pi^0\pi_D^0 \to \pi^0 e^+e^-(\gamma)$ (3.8 ev.), $K_L \to e^+e^-\gamma + (\gamma)_{bremss.}$ (9.9 ev.) and $K_L \to all~others$ (11.4 ev.). Systematic sources related to an uncertainty in α_{K^*}, the external bremsstrahlung and applied cut variations, have been taken into account. The obtained branching ratio at $E_\gamma^* > 5 MeV$, is:

$$(6.32 \pm 0.31_{stat} \pm 0.20_{sys} \pm 0.29_{norm}) \cdot 10^{-7}.$$

References

1. V.Fanti et al., *Phys.Lett.* **B465**, 335(1999).
2. Rev.Part.Phys., *Eur.Phys.J.*C15(2000).
3. G.D'Ambrosio, G.Ecker, G.Isidori and J.Portoles *JHEP*08, 004(1998).
4. P.Heiliger and L.M.Sehgal, *Phys.Rev.*D48, 4146 (1983); *Phys.Rev.*D60,79902(1999).

RECENT RESULTS FROM THE BNL E787 EXPERIMENT

TAKESHI K. KOMATSUBARA

for the E787 Collaboration
KEK-IPNS, Oho 1-1, Tsukuba, Ibaraki 305-0801, Japan
E-mail: takeshi.komatsubara@kek.jp

Recent results from a rare kaon decay experiment E787 at the BNL-AGS on $K^+ \to \pi^+ \nu \bar{\nu}$, $K^+ \to \mu^+ \nu \gamma$, and $K^+ \to \pi^+ \pi^0 \gamma$ decays are reported.

1 $K^+ \to \pi^+ \nu \bar{\nu}$

1.1 Theoretical Motivation

The $K^+ \to \pi^+ \nu \bar{\nu}$ decay is a flavor changing neutral current process induced in the Standard Model (SM) by loop effects in the form of penguin and box diagrams. The decay is sensitive to top-quark effects and provides an excellent route to determine the absolute value of V_{td} in the Cabibbo-Kobayashi-Maskawa matrix. With the constraints from other K and B decay experiments the SM prediction of the branching ratio is $(0.82 \pm 0.32) \times 10^{-10}$, and using only the results on $B_d - \bar{B}_d$ and $B_s - \bar{B}_s$ mixing a branching ratio limit $< 1.67 \times 10^{-10}$ can be extracted[1]. New physics beyond the SM could affect the branching ratio[2].

1.2 E787 Detector

E787[a] measures the charged track emanating from stopped K^+ decays. The E787 detector (Figure1) is a solenoidal spectrometer with a 1.0 Tesla field directed along the LESB3 beam line[3]. Slowed by a BeO degrader, kaons stop in the scintillating-fiber target at the center of the detector. A delayed coincidence requirement (> 2nsec) between the stopping kaon and the outgoing pion times helps to reject backgrounds of pions scattered into the detector or kaons decaying in flight. Charged decay products pass through the drift cham-

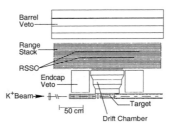

Figure 1. Side view of the upper half of the E787 detector.

ber, lose energy by ionization loss and stop in the Range Stack made of plastic scintillators and straw chambers. Momentum, kinetic energy and range are measured to reject the backgrounds by kinematic means. For further rejection of μ^+ tracks, the output pulse-shapes of the Range Stack counters are recorded and analyzed so that the decay chain $\pi^+ \to \mu^+ \to e^+$ can be identified in the stopping scintillator. $K^+ \to \pi^+ \pi^0$ and other decay modes with extra particles (γ, e, ...) are vetoed by the coincident signals in the hermetic shower counters.

1.3 Result and Prospect

E787 took data on $K^+ \to \pi^+ \nu \bar{\nu}$ from 1995 to 1998. In the 1995 data set, one event was observed[4] in the signal region. The new result[5] from the 1995-1997 data set is shown in Figure2. The same one event was observed and no new events were found in the signal region. The new value for the branching ratio is $(1.5^{+3.4}_{-1.2}) \times 10^{-10}$. Compared to the result from 1995, 2.1 times more kaon ex-

---

[a]E787 is a collaboration of BNL, Fukui, KEK, Osaka, Princeton, TRIUMF, Alberta and British Columbia.

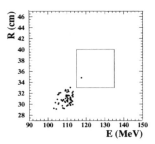

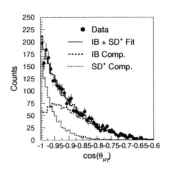

Figure 2. Range vs kinetic energy plot of the final $K^+ \to \pi^+\nu\bar{\nu}$ sample from the E787 1995-1997 data set. The box indicates the signal acceptance region.

Figure 3. Counts vs. $\cos(\theta_{\mu\gamma})$ of the $K^+ \to \mu^+\nu\gamma$ events, and various fits as described in the text.

posure (3.2×10^{12}) and 30% better acceptance are achieved. The background level, 0.08 ± 0.02 events, corresponding to a branching ratio of 1.2×10^{-11}, is improved by a factor of 2.5. The new result provides a constraint $1.07 \times 10^{-4} < |V_{ts}^*V_{td}| < 1.39 \times 10^{-3}$ without reference to the B system.

The kaon exposure in the E787 1998 data set is comparable to 1995-1997, and the analysis is ongoing. The sensitivity for the entire E787 data is expected to reach 0.7×10^{-10}, which is comparable to the SM prediction.

E949[6] continues the experimental study of $K^+ \to \pi^+\nu\bar{\nu}$ at the AGS based on the experience of E787 and is expected to reach a sensitivity of 10^{-11} or less in two to three years of operation. An engineering run of E949 is scheduled for 2001.

2 Photon Detection in E787

The Barrel Veto counter (BL) in Figure1 detects photons in the study of radiative kaon decays. The counter consists of 48 azimuthal sectors and 4 radial layers, made of lead/scintillator 14 radiation lengths in depth, and covers a solid angle of about 3π sr. The position of BL hits along the beam line is measured with ADC and TDC information from phototubes on both ends of each 2-m long module. The energy and direction of the photons from stopped kaon decays are

determined from the offline clustering in the BL and the decay vertex position in the target.

3 $K^+ \to \mu^+\nu\gamma$

The decay $K^+ \to \mu^+\nu\gamma$ can proceed via internal bremsstrahlung (IB) and structure dependent decay (SD). The latter is sensitive to the electroweak structure of the kaon because the photon is emitted from intermediate states. E787 has made the first measurement[7] of the SD$^+$ component in $K^+ \to \mu^+\nu\gamma$ decay, which is proportional to the square of the absolute value of the sum of the Vector and Axial form factors ($|F_V + F_A|$).

The SD$^+$ component peaks at high muon and photon energy. With a total kaon exposure of 9.2×10^9 and 1.5×10^6 triggers for $K^+ \to \mu^+\nu\gamma$, 2693 events are observed in the signal region where the μ^+ kinetic energy is > 137MeV and the γ energy is > 90MeV. The distribution of the opening angle between μ^+ and γ ($\cos(\theta_{\mu\gamma})$) for background-subtracted data, shown in Figure3 with the Monte Carlo distributions for IB and SD$^+$ components superimposed, clearly indicates that the SD$^+$ component is present. Detailed fits yield $|F_V + F_A| = 0.165 \pm 0.007 \pm 0.011$, which corresponds to an SD$^+$ branching ratio of $(1.33\pm0.12\pm0.18) \times 10^{-5}$, and a 90% confidence level limit $-0.04 < F_V - F_A < 0.24$.

4 $K^+ \to \pi^+\pi^0\gamma$

The decay $K^+ \to \pi^+\pi^0\gamma$ in which the photon is directly emitted (DE) is sensitive to the low energy hadronic interactions of mesons. E787 has performed a new measurement[8] of $K^+ \to \pi^+\pi^0\gamma$ decay, with significantly higher statistics than before and improved kinematic constraints using the K^+ decays at rest. The DE component is isolated kinematically with the variable W^2, which is reconstructed from the opening angle between π^+ and γ $(\cos(\theta_{\pi^+\gamma}))$, π^+ energy and momentum (E_{π^+}, P_{π^+}), and γ energy (E_γ) as $W^2 = E_\gamma^2 \times (E_{\pi^+} - P_{\pi^+} \times \cos(\theta_{\pi^+\gamma}))/(m_{K^+} \times m_{\pi^+}^2)$ in the stopped K^+ decays and is directly related to the observables in the E787 detector.

With a total kaon exposure of 1.8×10^{11} and 1.1×10^7 "3gamma" triggers to detect the charged track and three γ clusters in the BL, 19836 events survived all selection cuts including requiring that the kinematically fitted π^+ momentum be between 140 and 180 MeV/c. Figure4 shows the W spectrum of the signal events. The DE component is measured to be $(1.8 \pm 0.3)\%$ of the IB component, yielding a branching ratio for DE of $(4.7 \pm 0.8 \pm 0.3) \times 10^{-6}$ in the π^+ kinetic energy range 55-90 MeV[b]. This result can be understood by purely magnetic contributions in the framework of Chiral Perturbation Theory.

Acknowledgments

This research was supported in part by the U.S. Department of Energy under Contracts No. DE-AC02-98CH10886, W-7405-ENG-36, and grant DE-FG02-91ER40671, by the Ministry of Education, Science, Sports and Culture of Japan through the Japan-U.S. Cooperative Research Program in High Energy Physics and under the Grant-in-Aids for Scientific Research, for Encouragement of Young Scientists and for JSPS Fellows, and by the Natural Sciences and Engineering Research Council and the National Research Council of Canada.

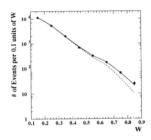

Figure 4. W spectrum of the observed $K^+ \to \pi^+\pi^0\gamma$ events and best fits to IB+DE (solid curve) and IB alone.

References

1. G. Buchalla and A.J. Buras, *Nucl. Phys.* B **548**, 309 (1999).
2. Y. Nir and M.P. Worah, *Phys. Lett.* B **423**, 319 (1998) and references therein; A.J. Buras *et al.*, *Nucl. Phys.* B **566**, 3 (2000).
3. J. Doornbos *et al.*, *Nucl. Instr. Meth.* A **444**, 546 (2000).
4. S. Adler *et al.*, *Phys. Rev. Lett.* **79**, 2204 (1997).
5. S. Adler *et al.*, *Phys. Rev. Lett.* **84**, 3768 (2000).
6. The information on E949 is available from http://www.phy.bnl.gov/e949/.
7. S. Adler *et al.*, *Phys. Rev. Lett.* **85**, 2256 (2000).
8. S. Adler *et al.*, hep-ex/0007021.
9. Review of Particle Physics, Particle Data Group, D.E. Groom *et al.*, *Eur. Phys. J.* C **15**, 1 (2000).

---

[b]Previous experiments used decay-in-flight techniques. The current Particle Data Group average[9] is $(1.8 \pm 0.4) \times 10^{-5}$.

SEARCH FOR T-VIOLATING TRANSVERSE MUON POLARIZATION IN $K^+ \to \pi^0 \mu^+ \nu$ DECAY

M. ABE[1], M. AOKI[2], I. ARAI[3], Y. ASANO[1], T. BAKER[2], M. BLECHER[4],
M.D. CHAPMAN[2], P. DEPOMMIER[6], M.P. GRIGORJEV[5], M. HASINOFF[8], K. HORIE[9],
H.C. HUANG[10], Y. IGARASHI[2], T. IKEDA[3], J. IMAZATO[2], A.P. IVASHKIN[5],
J.H. KANG[11], M.M. KHABIBULLIN[5], A.N. KHOTJANTSEV[5], Y.G. KUDENKO[5],
Y. KUNO[2], J.-M. LEE[11], A.S. LEVCHENKO[5], J.A. MACDONALD[7], D.R. MARLOW[12],
C.R. MINDAS[12], O.V. MINEEV[5], C. RANGACHARYULU[13], H.M. SHIMIZU[2],
S. SHIMIZU[9], Y.-H. SHIN[11], Y.-M. SHIN[13], K.S. SIM[14], A. SUZUKI[3], A. WATANABE[3],
AND T. YOKOI[2]
(KEK-E246 COLLABORATION)

[1] *Institute of Applied Physics, University of Tsukuba, Ibaraki 305-0006, Japan*
[2] *IPNS, High Energy Accelerator Research Organization (KEK), Ibaraki 305-0801, Japan*
[3] *Institute of Physics, University of Tsukuba, Ibaraki 305-0006, Japan*
[4] *Department of Physics, Virginia Polytechnic Institute and State University, VA 24061-0435, U.S.A.*
[5] *Institute for Nuclear Research, Russian Academy of Sciences, Moscow 117312, Russia*
[6] *Laboratoire de Physique Nucléaire, Université de Montréal, Montréal, Québec H3C 3J7, Canada*
[7] *TRIUMF, Vancouver, British Columbia V6T 2A3 , Canada*
[8] *Department of Physics and Astronomy, University of British Columbia, Vancouver V6T 1Z1, Canada*
[9] *Department of Physics, Osaka University, Osaka 560-0043, Japan*
[10] *Department of Physics, National Taiwan University, Taipei 106, Taiwan*
[11] *Department of Physics, Yonsei University, Seoul 120-749, Korea*
[12] *Department of Physics, Princeton University, NJ 08544, U.S.A.*
[13] *Department of Physics, University of Saskatchewan, Saskatoon S7N 5E2, Canada*
[14] *Department of Physics, Korea University, Seoul 136-701, Korea*

presented by M. Aoki
E-mail: masaharu.aoki@kek.jp

An experiment searching for a T-violating transverse muon polarization (P_T) in $K^+ \to \pi^0 \mu^+ \nu$ decay is underway at the 12-GeV proton synchrotron at KEK (KEK-PS). This search is sensitive to new mechanisms of T or CP violation beyond the Standard Model. The expected sensitivity from 1996-98 data is presented based on the published result from 1996-97 data and the new result from 1998 data.

1 Introduction

The transverse muon polarization (P_T) in the $K^+ \to \pi^0 \mu^+ \nu$ ($K_{\mu 3}^+$) decay is the polarization component normal to the decay plane, and is a T-odd observable defined by

$$P_T = \frac{\boldsymbol{\sigma}_\mu \cdot (\boldsymbol{p}_{\pi^0} \times \boldsymbol{p}_{\mu^+})}{|\boldsymbol{p}_{\pi^0} \times \boldsymbol{p}_{\mu^+}|}, \qquad (1)$$

where $\boldsymbol{\sigma}_\mu$ is the μ^+ spin vector and $\boldsymbol{p}_{\pi^0}$ and $\boldsymbol{p}_{\mu^+}$ are the momentum vectors of the π^0 and μ^+, respectively. Since any spurious effect from a final state interaction is known to be small[1], a non-zero value of P_T would be a signal of a violation of time reversal invariance (T). In addition, because the contribution[2] to P_T from the standard model (SM) is only

10^{-7}, a non-zero value of P_T would be evidence for new physics beyond the SM. In particular, it would also provide important clues to a CP-violation mechanism from non-SM origins[3] assuming CPT invariance.

2 KEK-E246 Detector

A separated K^+ beam $(\pi^+/K^+ \simeq 6)$ of 660 MeV/c is produced by a 12 GeV proton beam from the KEK-PS. The typical intensity of the K^+ beam was 3.0×10^5 per 0.6 s spill duration with 3 s repetition.

Schematic side and end views of the whole E246 detector as well as one sector of the polarimeter are shown in Fig. 1. Kaons are distinguished from the predominant pions by means of a Fitch type Čerenkov counter, slowed down in the degrader and then stopped in a target array of 256 scintillating fibers located at the center of the detector. $K^+_{\mu3}$ decays are observed by detecting μ^+'s and π^0's with a super-conducting toroidal magnetic spectrometer and a CsI(Tl) photon detector, respectively.

Photons from the π^0 decays are detected with an array of 768 CsI(Tl) crystals with PIN photodiode readout[4]. The solid angle coverage is about 75% of 4π. In order to increase the total number of usable events, we identify π^0's not only as two photons with the correct invariant mass (2γ events) but also as one of the photons with an energy greater than 70 MeV (1γ events) which sufficiently preserves the directional information of the parent π^0.

A muon emerging from the target passes through one of 12 identical holes of the CsI(Tl) photon detector, and enters one of 12 identical magnet gaps. The momentum and flight direction are measured using multiwire proportional chambers at the entrance (C2) and exit (C3 and C4) of the magnet gap along with the active target fibers and an array of 32 thin ring-shaped plastic scintillators. The momentum resolution ($\sigma_p = 2.6$ MeV/c)

at $P_\mu = 205$ MeV/c is adequate to remove the predominant background of $K^+ \to \pi^+\pi^0$ ($K_{\pi2}$) decay. K_{e3} events are eliminated by means of time-of-flight.

The μ^+ is bent by 90° in the magnet gap and enters one of the 12 identical polarimeters. In the polarimeter, the muon is slowed down by a wedge-shaped Cu block and comes to rest in a stack of pure Al plates. The transverse component of the muon polarization is preserved by the azimuthal magnetic field shaped by a pair of shim plates with very high symmetry relative to the median plane. The average strength of the magnetic field is 130 G. The e^+ from $\mu^+ \to e^+\nu\bar{\nu}$ decay is detected by one of 12 identical e^+ counters. The T-violating asymmetry A_T was extracted using a double ratio defined as

$$A_T = \frac{1}{4}\left[\frac{(N_{\text{cw}}/N_{\text{ccw}})_{\text{fwd}}}{(N_{\text{cw}}/N_{\text{ccw}})_{\text{bwd}}} - 1\right], \quad (2)$$

where N_{cw} and N_{ccw} are the e^+ counting rates at the clockwise and counterclockwise e^+ counters, respectively, and subscripts fwd/bwd indicate the direction of π^0. Then P_T was computed as $P_T = A_T/(\alpha\langle\cos\theta_T\rangle)$ by using the experimentally determined analyzing power, α, and the kinematical angular attenuation factor, $\langle\cos\theta_T\rangle$.

3 Analysis

The results from the 1996-97 data have already been reported elsewhere[5]. Two independent analyses were performed and the consistencies between them were investigated in detail. The systematic error for the 1996-97 data was also analyzed and reported to be $\delta P_T = 0.0009$, which is more than a factor of five smaller than the statistical error.

The analysis of the 1998 data was performed following the same methodology as the previous analysis: 1) two independent analyses, 2) blind analysis and 3) combination of the two analysis results. The consistency check between the two analyses provides an estimation of the systematic errors

758

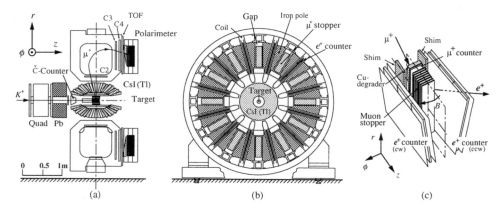

Figure 1. Experimental setup; (a) cross section side view, (b) end view and (c) schematic view of one sector of the polarimeter.

associated with the analysis and greatly improves the reliability of the result. The policy of the blind analysis was maintained to remove any possible human biases.

In both analyses, the angular attenuation factors, $\langle \cos \theta_T \rangle$, which measure the overall quality of the events are comparable to those in the 1996-97 data. A null asymmetry check by means of $A_0 = [(N_{cw}/N_{ccw})_{fwd+bwd} - 1]/2$ was also performed in both analyses and showed consistency with a null result. The results from the 1998 data were then obtained for both analyses (A1 and A2):

$$P_T^{1\gamma+2\gamma}(\text{A1}) = -0.0010 \pm 0.0059$$
$$P_T^{1\gamma+2\gamma}(\text{A2}) = 0.0002 \pm 0.0056 .$$

These are consistent with zero transverse polarization. By assuming the fraction of overlapped events between A1 and A2 in the 1998 data to be the same as in the 1996-97 data, the statistical error from a combined 1996-98 result is expected to be $\Delta \text{Im} \xi = 0.011$ for the T-violation physics parameter $\text{Im} \xi$[5]. The systematic error is still smaller than the statistical error by a factor of 4.

4 Summary

KEK-PS E246 has now almost completed the analysis of the 1998 data. The expected sta-

tistical sensitivity to $\text{Im} \xi$ from the combined 1996-98 data should be about 0.011. The physics run will continue until the end of next year. Together with the KEK-PS intensity upgrade, the final sensitivity is expected to be $\Delta \text{Im} \xi \simeq 0.007$.

References

1. A.R. Zhitnitskii, Yad. Fiz. **31**, 1024 (1980) [Sov. J. Nucl. Phys. **31**, 529 (1980)].

2. I.I. Bigi and A.I. Sanda, *CP Violation*, (Cambridge University Press, Cambridge, England, 2000).

3. R. Garisto and G. Kane, Phys. Rev. D **44**, 2038 (1991); G. Bélanger and C.Q. Geng, Phys. Rev. D **44**, 2789 (1991); M. Kobayashi, T.-T. Lin and Y. Okada, Prog. Theor. Phys. **95**, 361 (1995); C.Q. Geng and S.K. Lee, Phys. Rev. D **51**, 99 (1995); M. Fabbrichesi and F. Vissani, Phys. Rev. D **55**, 5334 (1997); G.-H. Wu and J.N. Ng, Phys. Lett. B **392**, 93 (1997).

4. D.V. Dementyev, *et al.*, Nucl. Instrum. Methods Phys. Res., Sect. A **440**, 151 (2000).

5. M. Abe *et al.*, Phys. Rev. Lett. **83**, 4253 (1999).